DIE GRUNDLEHREN DER

MATHEMATISCHEN WISSENSCHAFTEN

IN EINZELDARSTELLUNGEN MIT BESONDERER
BERÜCKSICHTIGUNG DER ANWENDUNGSGEBIETE

HERAUSGEGEBEN VON

J. L. DOOB · E. HEINZ · F. HIRZEBRUCH
E. HOPF · H. HOPF · W. MAAK · S. MAC LANE
W. MAGNUS · F. K. SCHMIDT · K. STEIN

GESCHÄFTSFÜHRENDE HERAUSGEBER

B. ECKMANN UND B. L. VAN DER WAERDEN
ZÜRICH

BAND 60

SPRINGER-VERLAG
BERLIN · HEIDELBERG · NEW YORK
1966

THE NUMERICAL TREATMENT
OF DIFFERENTIAL EQUATIONS

BY

DR. LOTHAR COLLATZ
O. PROFESSOR IN THE UNIVERSITY OF HAMBURG

2ND PRINTING OF THE 3RD EDITION

TRANSLATED FROM A SUPPLEMENTED VERSION OF THE
SECOND GERMAN EDITION
BY P.G. WILLIAMS, B. SC.
MATHEMATICS DIVISION, NATIONAL PHYSICAL LABORATORY,
TEDDINGTON, ENGLAND

WITH 118 DIAGRAMS
AND 1 PORTRAIT

SPRINGER-VERLAG
BERLIN · HEIDELBERG · NEW YORK
1966

Geschäftsführende Herausgeber:

Prof. Dr. B. Eckmann
Eidgenössische Technische Hochschule Zürich

Prof. Dr. B. L. van der Waerden
Mathematisches Institut der Universität Zürich

ISBN 3-540-03519-2 Springer-Verlag Berlin Heidelberg New York
ISBN 0-387-03519-2 Springer-Verlag New York Heidelberg Berlin

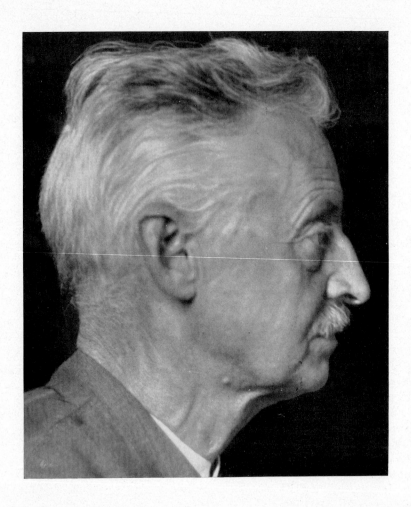

C. Runge

Collatz, Numerical treatment
3rd edition, 2nd printing

Springer-Verlag Berlin · Heidelberg · New York

From the preface to the first edition

This book constitutes an attempt to present in a connected fashion some of the most important numerical methods for the solution of ordinary and partial differential equations. The field to be covered is extremely wide, and it is clear that the present treatment cannot be remotely exhaustive; in particular, for partial differential equations it has only been possible to present the basic ideas, and many of the methods developed extensively by workers in applied fields — hydrodynamics, aerodynamics, etc. —, most of which have been developed for specific problems, have had to be dismissed with little more than a reference to the literature.

However, the aim of the book is not so much to reproduce these special methods, their corresponding computing schemes, etc., as to acquaint a wide circle of engineers, physicists and mathematicians with the general methods, and to show with the aid of numerous worked examples that an idea of the quantitative behaviour of the solution of a differential equation problem can be obtained by numerical means with nothing like the trouble and labour that widespread prejudice would suggest. This prejudice may be partly due to the kind of mathematical instruction given in technical colleges and universities, in which, although the theory of differential equations is dealt with in detail, numerical methods are gone into only briefly. I have always observed that graduate mathematicians and physicists are very well acquainted with theoretical results, but have no knowledge of the simplest approximate methods. If approximate methods were more well known, perhaps many problems would be solved with their aid which hitherto have simply not been tackled, despite the fact that interest in their solution has existed throughout. Especially with partial differential equations it has been the practice in many applied fields to restrict attention to the simplest cases — sometimes even to the cases for which the solution can be obtained in closed form —, while advancing technology demands the treatment of ever more complex problems. Further, considerable effort has often been put into the linearization of problems, because of a diffidence in tackling non-linear problems directly; many approximate methods are, however, immediately applicable also to non-linear problems, though clearly heavier computation is only to be expected; nevertheless, it is my belief that there will be a great increase in the importance of non-linear problems in the future.

As yet, the numerical treatment of differential equations has been investigated far too little, both in theoretical and practical respects, and approximate methods need to be tried out to a far greater extent than hitherto; this is especially true of partial differential equations and non-linear problems. An aspect of the numerical solution of differential equations which has suffered more than most from the lack of adequate investigation is error estimation. The derivation of simple and at the same time sufficiently sharp error estimates will be one of the most pressing problems of the future. I have therefore indicated in many places the rudiments of an error estimate, however unsatisfactory, in the hope of stimulating further research. Indeed, in this respect the book can only be regarded as an introduction.

Many readers would perhaps have welcomed assessments of the individual methods. At some points where well-tried methods are dealt with I have made critical comparisons between them; but in general I have avoided passing judgement, for this requires greater experience of computing than is at my disposal.

Hannover, December 1950

LOTHAR COLLATZ

From the preface to the second edition

In this new edition I have incorporated, so far as they have been accessible to me, the advances which have been made since the publication of the first edition. With the intense active interest which is now being taken in the numerical solution of differential equations the world over, new results are being obtained in a gratifyingly large number of topics. I always welcome especially the derivation of new error estimates, and in the present edition I have in fact been able to include error estimates with a large number of examples for which not even the rudiments of an error estimate were given in the first edition. May such further progress be made that in the future an error estimate will be included as a matter of course in the numerical treatment of any differential equation of a reasonable degree of complexity.

In spite of the fact that the book has been allowed to expand, considerable care has been necessary in choosing what extra material should be taken in; with such a large field to cover it is hardly possible to achieve completeness. It is certain that many readers will notice the omission of something which in their opinion ought to have been included. In such cases I am always grateful for criticism and interesting suggestions.

My very especial thanks are due to Dr. JOHANN SCHRÖDER, Dr. JULIUS ALBRECHT and Dr. HELMUT BARTSCH, who have inspected the proof-sheets with great care and in doing so have made numerous valuable suggestions for improvement.

Hamburg, Summer 1954

LOTHAR COLLATZ

Preface to the third edition

This English edition was translated from the second German edition by Mr. P. G. WILLIAMS, B. Sc., Mathematics Division, National Physical Laboratory, Teddington, England. It differs in detail from the second edition in that throughout the book a large number of minor improvements, alterations and additions have been made and numerous further references to the literature included; also new worked examples have been incorporated. Mr. WILLIAMS has made a series of suggestions for improving the presentation, which I gratefully acknowledge. My especial thanks are due to him, to his wife, Mrs. MARION WILLIAMS, and to my assistant, Dr. PETER KOCH, for the proof-reading, and also to Springer-Verlag for their continued ready compliance with all my wishes.

Hamburg, Summer 1959

LOTHAR COLLATZ

Contents

Chapter I

Mathematical preliminaries and some general principles

Chapter II

Initial-value problems in ordinary differential equations

Contents

Chapter III

Boundary-value problems in ordinary differential equations

Chapter IV

Initial- and initial-/boundary-value problems in partial differential equations

Chapter V

Boundary-value problems in partial differential equations

Chapter VI

Integral and functional equations

Appendix

Chapter I

Mathematical preliminaries
and some general principles

In this chapter we collect together some mathematical results which will be needed later and state some general approximation principles which are applicable in all the following chapters.

Some notes on the numerical examples

1. The numerical examples will be used solely for illustrating the methods. Consequently it is sufficient in many cases to exhibit only the early stages of the computation and it is permissible to simplify the calculations by using a rather large finite-difference interval or by retaining only a few terms in a Ritz approximation, etc. The results obtained with such simplifications are often quite crude, but it should be borne in mind that their accuracy can always be improved by using a smaller interval or taking in more terms as the case may be (to save space this will not be stated explicitly in each individual case). The explanatory treatment of the examples in the text should enable the reader to effect such improvement in accuracy without difficulty, though this may not be necessary in many cases; in technical applications, for instance, quite a low accuracy, permitting an error of several per cent maybe, is often quite sufficient.

2. To study accuracy I have computed the results of many of the examples to more significant figures than are given in the tables and elsewhere. These are rounded values and hence anyone who works through an example with the number of significant figures given is liable to arrive at a slightly different result.

3. *One should be wary of drawing general conclusions as to the merit or demerit of a method on the basis of individual examples, for a great deal of experience is needed before a sound assessment can be made.* Furthermore, the efficacy of a method is strongly dependent on the computing technique of the individual, the degree to which he is accustomed to the method, the resources at his disposal, and many other factors.

4. Several important checks are mentioned in the text. However, the numerous checks which were, in fact, always applied during the calculation of the examples are generally not reproduced in the interests

of economy of space. *In carrying out a computation, whether for a new problem or as a check on results already calculated, one should apply as many current checks as possible;* the beginner usually regards checks as superfluous until he makes a deep-rooted mistake whose location and correction takes longer than a proper computation carried out with current checks. One should beware the hasty calculation and heed the well-known proverb: "More haste, less speed."

Of course, a study of possible checking techniques is a necessary preliminary. For any calculation it is important to consider how one can control effectively which sources of error shall be present and how suitable checks can be kept on the errors arising from these sources. Only when sufficient checks have been satisfied to inspire confidence in the accuracy of the calculation up to the current point should one proceed further. Even though checks have been satisfied, it is still possible that the calculation may have gone astray; the experienced computer knows that he can never be too suspicious of the calculation in hand. Whenever possible the results should be confirmed by a second calculation based on a different method or by a recalculation performed by another person.

5. Elementary calculations such as the solution of algebraic equations or systems of linear equations and the evaluation of elementary integrals have been omitted. Hence the actual work involved in a calculation is often considerably greater than a glance at the printed reproduction would suggest and the reader is warned not to be misled by this.

§ 1. Introduction to problems involving differential equations

1.1. Initial-value and boundary-value problems in ordinary differential equations

The general solution of an n-th order differential equation

$$F\big(x, y(x), y'(x), y''(x), \ldots, y^{(n)}(x)\big) = 0 \tag{1.1}$$

for a real function $y(x)$ normally depends on n parameters $c_1, \ldots, c_n$. In an initial-value problem these parameters are determined by prescribing the values

$$y_0^{(\nu)} = y^{(\nu)}(x_0) \qquad (\nu = 0, 1, 2, \ldots, n-1) \tag{1.2}$$

at a fixed point $x = x_0$. If the conditions are based on more than one point x, then the problem is called a boundary-value problem. The "boundary conditions" can have the form

$$\left. V_\nu\big(y_{x_1}, y'_{x_1}, \ldots, y^{(n-1)}_{x_1}, y_{x_2}, y'_{x_2}, \ldots, y^{(n-1)}_{x_2}, \ldots, y_{x_k}, y'_{x_k}, \ldots, y^{(n-1)}_{x_k}\big) = 0 \right\} \tag{1.3}$$
$$(\nu = 0, \ldots, n-1),$$

where the x_s ($x_1 < x_2 < \cdots < x_k$, say) are prescribed points which may include $\pm \infty$ and the $y_{x_s}^{(\varrho)}$ denote the values of the ϱ-th derivative of $y(x)$ at the points $x = x_s$:

$$y_{x_s}^{(\varrho)} = \left(\frac{d^\varrho y}{d x^\varrho} \right)_{x=x_s}.$$

F and V_ν are given functions which, in general, will be non-linear. To solve the "boundary-value problem" (1.1), (1.3) is to find that function $y(x)$ which satisfies equations (1.1) and (1.3). The boundary conditions need not be restricted to the form (1.3); for instance, we can impose a condition such as

$$\int_{x_1}^{x_2} V\left(x, y(x), \ldots, y^{(n-1)}(x)\right) dx = 0, \tag{1.4}$$

where V is a prescribed integrable function.

Depending on the functions F and V_ν, such a boundary-value problem may have no solutions, one solution, several solutions or even infinitely many; for example, the problem

$$y'' + y = 0; \quad y(0) = y(\pi) = 1$$

has no solution and the problem

$$y'' + y = 0; \quad y(0) = 1; \quad y(\pi) = -1$$

has infinitely many solutions.

Further, as an example in which the length of the range of integration is an additional unknown, we mention the gunnery problem in which, for a fixed muzzle velocity v_0, the angle of elevation ϑ_0 is to be determined so as to hit a given target. If the projectile has unit mass and for simplicity the problem is assumed to be two-dimensional, then the co-ordinates of the projectile after a time t satisfy the fourth-order system of differential equations

$$\ddot{x} = -w \cos \vartheta, \quad \ddot{y} = -w \sin \vartheta - g(y).$$

Here, dots denote differentiation with respect to time, the acceleration g due to gravity is a given function of the height y, the air resistance w is a given function of height and projectile velocity, and the angle between the trajectory and the horizontal is denoted by $\vartheta = \tan^{-1}(dy/dx)$. With the firing point as the origin of the co-ordinates and (x_1, y_1) as the co-ordinates of the target we have (corresponding to the four constants of integration and the unknown time of flight t_1) the five boundary conditions

$$t = 0: \quad x = 0, \quad y = 0, \quad \dot{x}^2 + \dot{y}^2 = v_0^2;$$
$$t = t_1: \quad x = x_1, \quad y = y_1.$$

Note that in the condition $\dot{x}^2 + \dot{y}^2 = v_0^2$ we have a non-linear boundary condition.

1.2. Linear boundary-value problems

We shall deal in greater detail with the special class of boundary-value problems *for which the differential equation and the boundary conditions are linear (these are called linear boundary-value problems)* and for which the boundary conditions are based on only two points, say $x_1 = a$ and $x_2 = b$, where $a < b$. The differential equation may be written in the form

$$L[y] = r(x), \qquad (1.5)$$

where

$$L[y] = \sum_{r=0}^{n} f_r(x) y^{(r)} = f_0(x) y + f_1(x) y' + f_2(x) y'' + \cdots + f_n(x) y^{(n)}, \qquad (1.6)$$

and similarly the boundary conditions (assumed to be linearly independent) may be written

$$U_\nu[y] = \gamma_\nu \qquad (\nu = 1, 2, \ldots, n), \qquad (1.7)$$

where

$$U_\nu[y] = \sum_{k=0}^{n-1} \left(\alpha_{\nu, k} \, y^{(k)}(a) + \beta_{\nu, k} \, y^{(k)}(b) \right). \qquad (1.8)$$

The $f_\nu(x)$ and $r(x)$ are given functions which we normally assume to be continuous and the γ_ν, $\alpha_{\nu,k}$, $\beta_{\nu,k}$ are given constants.

The differential equation is called homogeneous when $r(x) \equiv 0$ in the interval $a \leq x \leq b$ and otherwise inhomogeneous; similarly a boundary condition is called homogeneous when the γ_ν associated with it are zero and otherwise inhomogeneous. The boundary-value problem is called homogeneous when the differential equation and all the boundary conditions are homogeneous.

When the differential equation is of even order $n = 2m$, the boundary conditions can be divided into "essential" and "suppressible"[1,2] (the senses in which they are essential or suppressible, respectively, will be explained in Ch. III §§ 5, 6 when we deal with Ritz's *method). As many as possible linearly independent linear combinations of the $2m$ given boundary conditions (1.7) are formed such that derivatives of the m-th and higher orders are removed. By this process, k (say) linearly independent boundary conditions are obtained which contain only derivatives of up to the $(m-1)$-th order. These are called "essential" boundary conditions. A further $2m - k$*

---

[1] Following E. Kamke: Math. Z. **48**, 67—100 (1942), who called the two types of boundary condition "wesentlich" and "restlich". Biezeno-Grammel: Technische Dynamik, Vol. I, 2nd ed., p. 136, Berlin 1953, uses a different terminology; from mechanical ideas he derives the terms "geometrische" and "dynamische" for boundary conditions in second- and fourth-order problems corresponding to our "essential" and "suppressible", respectively.

[2] For special problems the following distinctions are also found in the literature: For differential equations of the second order $(m = 1)$ one calls the conditions

$$y(a) = \alpha, \qquad y(b) = \beta \qquad \text{boundary conditions of the first kind,}$$
$$y'(a) = \alpha, \qquad y'(b) = \beta \qquad \text{boundary conditions of the second kind,}$$
$$c_1 y(a) + c_2 y'(a) = \alpha, \ d_1 y(b) + d_2 y'(b) = \beta \qquad \text{boundary conditions of the third kind or}$$
$$\text{(where } |c_1| + |c_2| > 0 \text{ and } |d_1| + |d_2| > 0) \qquad \text{sometimes Sturm's boundary conditions.}$$

boundary conditions, which may be termed "suppressible", can be specified
such that no more "essential" boundary conditions can be obtained from
them by linear combination and such that the combined "essential" and
"suppressible" boundary conditions are equivalent to the original system
of 2m boundary conditions.

For example, if the boundary conditions $y(0) + y'(1) = 1$, $y(1) + 2y'(1) = 3$ are given with a second-order differential equation $(m = 1)$, they can be combined linearly to give the new boundary condition $2y(0) - y(1) = -1$ in which the first derivative no longer appears. No more linear combinations of these boundary conditions can be found which contain no derivatives and hence we have one "essential" boundary condition $2y(0) - y(1) = -1$ and one "suppressible" boundary condition, say $y(0) + y'(1) = 1$.

A homogeneous boundary-value problem (sometimes called completely homogeneous) will usually possess only the trivial solution $y(x) \equiv 0$. One therefore considers those problems in which a parameter λ occurs, either in the differential equation or in the boundary conditions, and investigates the values of λ, the so-called "eigenvalues", for which the boundary-value problem has a "non-trivial" solution, i.e. a solution which does not vanish identically. Such a solution is called an eigenfunction and the whole problem is called an "eigenvalue problem" (see Example III in § 1.2 of Ch. III and also § 8 of that chapter).

1.3. Problems in partial differential equations

Examples of initial-value and boundary-value problems in partial differential equations which will acquaint the reader with the types of problems that arise will be found in Ch. IV, § 1 and Ch. V, § 1. In this section we go directly to the general formulation and pose the problem of determining a function $u(x_1, x_2, \ldots, x_n)$ of n independent variables which satisfies a partial differential equation

$$F(x_1, \ldots, x_n, u, u_1, \ldots, u_n, u_{11}, \ldots, u_{nn}, \ldots) = 0 \quad \text{in } B \quad (1.9)$$

and certain boundary conditions

$$V_\mu(x_1, \ldots, x_n, u, u_1, \ldots, u_n, u_{11}, \ldots, u_{nn}, \ldots) = 0 \quad \text{on } \Gamma_\mu. \quad (1.10)$$

The subscripts on u denote partial differentiation with respect to the x_i; for example,

$$u_j = \frac{\partial u}{\partial x_j}, \quad u_{jk} = \frac{\partial^2 u}{\partial x_j \partial x_k}. \quad (1.11)$$

B is a given region of the $(x_1, \ldots, x_n)$ space, the Γ_μ are $(n-1)$-dimensional "hyper-surfaces" of this space and F and the V_μ are given functions which will be assumed to be continuous.

Here also the linear problems play a special rôle. A partial differential equation is said to be linear when it is linear in u and the derivatives of u, i.e. when it has the form

$$L[u] \equiv \sum_{\alpha_1 + \cdots + \alpha_n \leq m} A_{\alpha_1, \alpha_2, \ldots, \alpha_n} \frac{\partial^{\alpha_1 + \cdots + \alpha_n} u}{\partial x_1^{\alpha_1} \ldots \partial x_n^{\alpha_n}} = r(x_1, \ldots, x_n), \quad (1.12)$$

where the $A_{\alpha_1, \alpha_2, \ldots, \alpha_n}$ and r are given functions of $x_1, \ldots, x_n$ and m is the order of the differential equation (in so far as a derivative of the m-th order actually appears). As with ordinary differential equations the terms homogeneous and inhomogeneous are applied to the equations with $r \equiv 0$ and $r \not\equiv 0$, respectively.

A partial differential equation is called quasi-linear when it is linear in the highest order derivatives occurring (say of the m-th order), i.e. when it has the form

$$\sum_{\alpha_1 + \cdots + \alpha_n = m} A_{\alpha_1, \ldots, \alpha_n} \frac{\partial^{\alpha_1 + \cdots + \alpha_n} u}{\partial x_1^{\alpha_1} \ldots \partial x_n^{\alpha_n}} = r,$$

where r and the $A_{\alpha_1, \ldots, \alpha_n}$ are now given functions of $x_1, \ldots, x_n$, u and partial derivatives of u of up to and including the $(m-1)$-th order.

If the boundary conditions are linear, i.e. linear in u and its partial derivatives, then we write them in the form

$$\left. \begin{aligned} U_\mu(x_1, \ldots, x_n, \, u, u_1, \ldots, u_n, \, u_{11}, \ldots, u_{nn}, \ldots) = \gamma_\mu \\ \text{on } \Gamma_\mu \quad (\mu = 1, \ldots, k), \end{aligned} \right\} \quad (1.13)$$

where the γ_μ are given functions of position on the Γ_μ and U_μ is linear and homogeneous in u and its derivatives. Again we say the boundary conditions are homogeneous when $\gamma_\mu \equiv 0$ and inhomogeneous when $\gamma_\mu \not\equiv 0$.

A boundary-value problem is termed linear when the differential equation and all the boundary conditions are linear. Any linear boundary-value problem can be reduced to one in which either the differential equation or the boundary conditions (which should not be self-contradictory, of course) are homogeneous. This is achieved by introducing a new function $u^* = u - u_0$, where u_0 is a function satisfying either the inhomogeneous differential equation or the inhomogeneous boundary conditions, respectively.

§2. Finite differences and interpolation formulae

We assume here that the reader is familiar with the foundations of the calculus of finite differences so that we may be brief.

2.1. Difference operators and interpolation formulae

Suppose that we know the values $f_\nu = f(x_\nu)$ of a function $f(x)$ at the $(N+1)$ equidistant points $x_\nu = x_0 + \nu h$, where $\nu = 0, 1, 2, \ldots, N$ (sometimes ν may be non-integral). h is variously called the interval of tabulation, pivotal interval, step size, or just the interval and is always taken to be positive. For any function $F(x)$ we can define the difference operators Δ, ∇, δ for the interval h as follows:

$$\Delta F(x) = F(x+h) - F(x),$$
$$\nabla F(x) = F(x) - F(x-h),$$
$$\delta F(x) = F(x + \tfrac{1}{2}h) - F(x - \tfrac{1}{2}h);$$

we call

$$\Delta f_k = f_{k+1} - f_k \qquad \text{forward differences,}$$
$$\nabla f_k = f_k - f_{k-1} \qquad \text{backward differences,}$$

and

$$\delta f_k = f_{k+\frac{1}{2}} - f_{k-\frac{1}{2}} \quad \text{central differences.}$$

Each of these may be extended to higher differences; thus, for example,

$$\Delta^2 f_k = \Delta(\Delta f_k) = \Delta(f_{k+1} - f_k) = f_{k+2} - 2f_{k+1} + f_k,$$

and in general

$$\Delta^p f_k = \Delta(\Delta^{p-1} f_k), \quad \nabla^p f_k = \nabla(\nabla^{p-1} f_k), \quad \delta^p f_k = \delta(\delta^{p-1} f_k) \quad (p=1,2,3,\ldots).$$

To extend these definitions to $p=0$ we write

$$\Delta^0 f_k = \nabla^0 f_k = \delta^0 f_k = f_k.$$

For the general non-central differences we have

$$\Delta^p f_k = \sum_{\varrho=0}^{p} (-1)^\varrho \binom{p}{\varrho} f_{k+p-\varrho}, \qquad \nabla^p f_k = \sum_{\varrho=0}^{p} (-1)^\varrho \binom{p}{\varrho} f_{k-\varrho}.$$

The "first" $(p=1)$ and "higher" $(p>1)$ differences are often conveniently written down in the "difference table" of the function $f(x)$ in which the difference of two entries appears opposite the gap between them as in Table I/1.

Table I/1. The "sloping" difference table of a function $f(x)$

x	f	First ∇f	Second $\nabla^2 f$	Third $\nabla^3 f$	differences $\ldots$
x_{-2}	f_{-2}		$\nabla^2 f_{-1}$		$\ldots$
		∇f_{-1}		$\nabla^3 f_0$	
x_{-1}	f_{-1}		$\nabla^2 f_0$		$\ldots$
		∇f_0		$\nabla^3 f_1$	
x_0	f_0		$\nabla^2 f_1$		$\ldots$
		∇f_1		$\nabla^3 f_2$	
x_1	f_1		$\nabla^2 f_2$		$\ldots$
		∇f_2		$\nabla^3 f_3$	
x_2	f_2		$\nabla^2 f_3$		$\ldots$

Occasionally, however, it is more convenient to write the differences $V^p f_k$ ($p = 0, 1, 2, 3, \ldots$) with the same k all on the same line as in Table I/2.

Table I/2. *The "horizontal" difference table of* $f(x)$

x	f	First ∇f	Second $\nabla^2 f$	differences $\ldots$
x_{-1}	f_{-1}	∇f_{-1}	$\nabla^2 f_{-1}$	$\ldots$
x_0	f_0	∇f_0	$\nabla^2 f_0$	$\ldots$
x_1	f_1	∇f_1	$\nabla^2 f_1$	$\ldots$

In the theory of interpolation, polynomials are derived having the property that they coincide with the function $f(x)$ at certain points x_ν; we reproduce here only those polynomials which will be used later. With the abbreviation $u = \dfrac{x - x_0}{h}$ "NEWTON's interpolation polynomial with backward differences" can be written

$$N_p(x) = f_0 + \frac{u}{1!} \nabla f_0 + \frac{u(u+1)}{2!} \nabla^2 f_0 + \\ + \frac{u(u+1)(u+2)}{3!} \nabla^3 f_0 + \cdots + \frac{u(u+1)\ldots(u+p-1)}{p!} \nabla^p f_0; \tag{2.1}$$

it takes the values of $f(x)$ at the points $x_0, x_{-1}, \ldots, x_{-p}$. There is a point $x = \xi$ inside the smallest interval of the x axis containing the points $x, x_0, x_{-1}, \ldots, x_{-p}$ such that the remainder term at non-tabular points is given by[1]

$$R_{p+1}(x) = f(x) - N_p(x) = \frac{u(u+1)\ldots(u+p)}{(p+1)!} h^{p+1} f^{(p+1)}(\xi). \tag{2.2}$$

The corresponding Newtonian interpolation formula with forward differences is

$$N_p^*(x) = f_0 + \frac{u}{1!} \Delta f_0 + \frac{u(u-1)}{2!} \Delta^2 f_0 + \cdots + \binom{u}{p} \Delta^p f_0 + \\ + \binom{u}{p+1} h^{p+1} f^{(p+1)}(\xi). \tag{2.3}$$

STIRLING's interpolation polynomial, which we write down only for even p, is

$$St_p(x) = f_0 + u \frac{\nabla f_0 + \nabla f_1}{2} + \frac{u^2}{2!} \nabla^2 f_1 + \frac{u(u^2-1)}{3!} \frac{\nabla^3 f_1 + \nabla^3 f_2}{2} + \\ + \frac{u^2(u^2-1)}{4!} \nabla^4 f_2 + \cdots + \frac{u^2(u^2-1)\ldots\left(u^2 - \left(\frac{p}{2}-1\right)^2\right)}{p!} \nabla^p f_{p/2}; \tag{2.4}$$

[1] For the whole of this section see, for example, G. SCHULZ: Formelsammlung zur praktischen Mathematik. Sammlung Göschen, Vol. 1110. Leipzig and Berlin 1945; and FR. A. WILLERS: Methoden der praktischen Analysis, 2nd ed. Berlin 1950, particularly p. 76 et seq.

it coincides with $f(x)$ at the points $x_{-p/2}, \ldots, x_{-1}, x_0, x_1, \ldots, x_{p/2}$ and the remainder term at other points is given by

$$R_{p+1}(x) = f(x) - St_p(x) = \binom{u + \dfrac{p}{2}}{p+1} h^{p+1} f^{(p+1)}(\xi).$$

From STIRLING's polynomial we can obtain a symmetric formula[1] whose remainder term is of higher order:

$$\left. \begin{aligned} \frac{f(x_0 + u\,h) + f(x_0 - u\,h)}{2} &= f_0 + \frac{u^2}{2!} \nabla^2 f_1 + \frac{u^2(u^2 - 1)}{4!} \nabla^4 f_2 + \cdots + \\ &+ \frac{u^2(u^2 - 1) \ldots \left(u^2 - \left(\dfrac{p}{2} - 1\right)^2\right)}{p!} \nabla^p f_{p/2} + R_p^{**}, \end{aligned} \right\} \quad (2.5)$$

where

$$R_p^{**} = \frac{u^2(u^2 - 1) \ldots \left(u^2 - \left(\dfrac{p}{2}\right)^2\right)}{(p + 2)!} h^{p+2} f^{(p+2)}(\xi).$$

2.2. Some integration formulae which will be needed later

From the polynomials of the preceding section we can obtain formulae for the approximate integration of $f(x)$ over specified intervals by replacing $f(x)$ by one of these polynomials in the required interval. Thus by integrating NEWTON's interpolation formula (2.1), (2.2) with respect to x over the interval x_0 to $x_0 + h$ we obtain

$$\left. \begin{aligned} \int_{x_0}^{x_0+h} f(x)\,dx &= h\left[f_0 + \frac{1}{2}\nabla f_0 + \frac{5}{12}\nabla^2 f_0 + \frac{3}{8}\nabla^3 f_0 + \cdots\right] + \\ &+ \int_{x_0}^{x_0+h} R_{p+1}(x)\,dx = h\sum_{\varrho=0}^{p} \beta_\varrho \nabla^\varrho f_0 + S_{p+1}, \end{aligned} \right\} \quad (2.6)$$

where

$$\beta_\varrho = \frac{1}{\varrho!} \int_0^1 u(u + 1) \ldots (u + \varrho - 1)\,du \qquad (\varrho = 1, 2, \ldots). \quad (2.7)$$

The first few β_ϱ are

$$\beta_0 = 1, \ \beta_1 = \frac{1}{2}, \ \beta_2 = \frac{5}{12}, \ \beta_3 = \frac{3}{8}, \ \beta_4 = \frac{251}{720}, \ \beta_5 = \frac{95}{288}, \ \beta_6 = \frac{19087}{60480}.$$

For S_{p+1} we have the estimate

$$\left. \begin{aligned} |S_{p+1}| &= \left| \int_{x_0}^{x_0+h} \frac{u(u+1) \ldots (u+p)}{(p+1)!} h^{p+1} f^{(p+1)}(\xi)\,dx \right| \\ &\leq h^{p+2} \beta_{p+1} |f^{(p+1)}|_{\max}. \end{aligned} \right\} \quad (2.8)$$

---

[1] STEFFENSEN, J. F.: Interpolation, p. 29. Baltimore 1927.

If we replace $f(x)$ by the Newtonian polynomial based on the points $x_1, x_0, x_{-1}, \ldots, x_{-p+1}$, i.e. the polynomial

$$f_1 + \frac{u-1}{1!} \nabla f_1 + \frac{(u-1)\,u}{2!} \nabla^2 f_1 + \cdots + \frac{(u-1)\,u \ldots (u+p-2)}{p!} \nabla^p f_1,$$

then integration over the interval x_0 to $x_0 + h$ gives

$$\int_{x_0}^{x_0+h} f(x)\,dx = h \left[f_1 - \frac{1}{2} \nabla f_1 - \frac{1}{12} \nabla^2 f_1 - \frac{1}{24} \nabla^3 f_1 - \cdots \right] + S_{p+1}^*$$

$$= h \sum_{\varrho=0}^{p} \beta_\varrho^* \nabla^\varrho f_1 + S_{p+1}^*, \qquad (2.9)$$

where

$$\beta_\varrho^* = \frac{1}{\varrho!} \int_0^1 (u-1)\,u\,(u+1) \ldots (u+\varrho-2)\,du.$$

The first few β_ϱ^* are

$$\beta_0^* = 1, \quad \beta_1^* = -\frac{1}{2}, \quad \beta_2^* = -\frac{1}{12}, \quad \beta_3^* = -\frac{1}{24}, \quad \beta_4^* = -\frac{19}{720},$$

$$\beta_5^* = -\frac{3}{160}, \quad \beta_6^* = -\frac{863}{60480}.$$

An estimate for S_{p+1}^* is given by

$$|S_{p+1}^*| = \left| \int_{x_0}^{x_0+h} \frac{(u-1)\,u\,(u+1) \ldots (u+p-1)}{(p+1)!}\, h^{p+1} f^{(p+1)}(\xi)\,dx \right| \qquad (2.10)$$

$$\leq h^{p+2} |\beta_{p+1}^*|\, |f^{(p+1)}|_{\max}.$$

If we apply (2.9) to the function $\tilde{f}(x) = f(2x_0 + h - x)$, we obtain

$$\int_{x_0}^{x_0+h} \tilde{f}(x)\,dx = \int_{x_0}^{x_0+h} f(2x_0 + h - x)\,dx = h \sum_{\varrho=0}^{p} \beta_\varrho^* (-1)^\varrho \nabla^\varrho f_0 + \tilde{S}_{p+1} \qquad (2.11)$$

$$= h \left(f_0 + \frac{1}{2} \nabla f_1 - \frac{1}{12} \nabla^2 f_2 + - \cdots \right) + \tilde{S}_{p+1}$$

with a corresponding form for the remainder term $\tilde{S}_{p+1}$.

The β_ϱ and β_ϱ^* are connected by the relation

$$\beta_\varrho + \beta_{\varrho+1}^* = \beta_{\varrho+1} \qquad (\varrho = 0, 1, 2, \ldots), \qquad (2.12)$$

for

$$\beta_\varrho + \beta_{\varrho+1}^* = \int_0^1 \left[\binom{u+\varrho-1}{\varrho} + \binom{u+\varrho-1}{\varrho+1} \right] du = \int_0^1 \binom{u+\varrho}{\varrho+1} du = \beta_{\varrho+1}.$$

By adding equations (2.12) for $\varrho = 0, 1, 2, \ldots, p-1$, together with $\beta_0 = \beta_0^*$ $(=1)$, we obtain the relation

$$\sum_{\varrho=0}^{p} \beta_\varrho^* = \beta_p. \tag{2.13}$$

Integration of STIRLING's formula (2.5) over the interval $x_0 - h$ to $x_0 + h$ gives

$$\left.\begin{aligned}
\int_{x_0-h}^{x_0+h} f(x)\,dx &= h\left[2f_0 + \frac{1}{3}\nabla^2 f_1 - \frac{1}{90}\nabla^4 f_2 + \frac{1}{756}\nabla^6 f_3 - \cdots\right] + S_p^{**}\\
&= h\sum_{\varrho=0}^{p/2} \beta_\varrho^{**}\nabla^{2\varrho} f_\varrho + S_p^{**},
\end{aligned}\right\} \tag{2.14}$$

where

$$\beta_\varrho^{**} = \frac{2}{(2\varrho)!}\int_0^1 u^2(u^2-1)(u^2-4)\ldots\left(u^2-(\varrho-1)^2\right)du. \tag{2.15}$$

An upper bound for the magnitude of the remainder term is

$$|S_p^{**}| \leqq h^{p+3}\left|\beta_{\frac{p}{2}+1}^{**}\right|\,|f^{(p+2)}|_{\max}. \tag{2.16}$$

For $p=2$ (2.14) reduces to SIMPSON's rule

$$\int_{x_0-h}^{x_0+h} f(x)\,dx = h\left[2f_0 + \frac{1}{3}\nabla^2 f_1\right] + S_2^{**}, \tag{2.17}$$

where

$$|S_2^{**}| \leqq \frac{h^5}{90}\,|f^{(IV)}|_{\max}. \tag{2.18}$$

2.3. Repeated integration

If we integrate NEWTON's interpolation formula (2.1), (2.2) over the interval x_0 to x $(x = x_0 + uh)$, we obtain the indefinite integral

$$\left.\begin{aligned}
\int_{x_0}^{x} f(x)\,dx &= h\int_0^u N_p(x(u))\,du + \int_{x_0}^{x} R_{p+1}\,dx = {}'f(x) - {}'f(x_0)\\
&= h\left[u f_0 + \frac{u^2}{2}\nabla f_0 + \frac{3u^2+2u^3}{12}\nabla^2 f_0 + \cdots\right] + R_{1,\,p+1},
\end{aligned}\right\} \tag{2.19}$$

and a second integration gives

$$\left.\begin{aligned}
\int_{x_0}^{x}\int_{x_0}^{x} f(x)\,(dx)^2 &= {}''f(x) - {}''f(x_0) - {}'f(x_0)(x-x_0)\\
&= h^2\left[\frac{u^2}{2}f_0 + \frac{u^3}{6}\nabla f_0 + \cdots\right] + R_{2,\,p+1},
\end{aligned}\right\} \tag{2.20}$$

where $'f$ is an indefinite integral of $f(x)$ $(={}^{(0)}f)$, $''f$ is a indefinite second integral of $f(x)$ and in general $^{(\nu)}f$ denotes a ν-th repeated indefinite integral of $f(x)$ or, shortly, a ν-th integral of $f(x)$, so that

$$^{(\nu)}f = \int^x {}^{(\nu-1)}f\, dx \qquad (\nu = 1, 2, 3, \ldots). \tag{2.21}$$

Similarly we define (with $R_{0,\,p+1} = R_{p+1}$)

$$R_{\nu,\,p+1} = \int_{x_0}^x R_{\nu-1,\,p+1}\, dx \qquad (\nu = 1, 2, 3, \ldots). \tag{2.22}$$

By further integrations we obtain the general n-th integral

$$\left.\begin{aligned}
\int_{x_0}^x \cdots \int_{x_0}^x f\,(d\,x)^n &= {}^{(n)}f(x) - \sum_{\nu=0}^{n-1} {}^{(n-\nu)}f(x_0)\,\frac{(x-x_0)^\nu}{\nu!} \\
&= h^n \sum_{\varrho=0}^{p} P_{n,\,\varrho}(u)\, V^\varrho f_0 + R_{n,\,p+1}(x),
\end{aligned}\right\} \tag{2.23}$$

where the $P_{n,\,\varrho}(u)$ are defined by the n-th integrals

$$\left.\begin{aligned}
P_{n,\varrho}(u) &= \int_0^u \cdots \int_0^u \frac{u\,(u+1)\ldots(u+\varrho-1)}{\varrho!}\,(d\,u)^n \quad \text{for} \quad \varrho \geq 1 \\
P_{n,0}(u) &= \frac{1}{n!}\,u^n \qquad\qquad\qquad\qquad (n = 1, 2, \ldots).
\end{aligned}\right\} \tag{2.24}$$

For small n and ϱ the $P_{n,\varrho}$ have already been exhibited as polynomials in u in the formulae (2.19), (2.20) above; the polynomials for several higher values of n and ϱ are given in Table I/3.

Table I/3. *The polynomials $P_{n,\,\varrho}(u)$*

	$\varrho=0$	$\varrho=1$	$\varrho=2$	$\varrho=3$	$\varrho=4$
$n=1$	u	$\frac{1}{2}\,u^2$	$\frac{1}{12}(3u^2+2u^3)$	$\frac{1}{24}(4u^2+4u^3+u^4)$	$\frac{1}{720}(90u^2+110u^3+45u^4+6u^5)$
$n=2$	$\frac{1}{2}\,u^2$	$\frac{1}{6}\,u^3$	$\frac{1}{24}(2u^3+u^4)$	$\frac{1}{360}(20u^3+15u^4+3u^5)$	$\frac{1}{1440}(60u^3+55u^4+18u^5+2u^6)$
$n=3$	$\frac{1}{6}\,u^3$	$\frac{1}{24}\,u^4$	$\frac{1}{240}(5u^4+2u^5)$	$\frac{1}{720}(10u^4+6u^5+u^6)$	$\frac{1}{10\,080}(105u^4+77u^5+21u^6+2u^7)$
$n=4$	$\frac{1}{24}\,u^4$	$\frac{1}{120}\,u^5$	$\frac{1}{720}(3u^5+u^6)$	$\frac{1}{5040}(14u^5+7u^6+u^7)$	

The value of the n-th indefinite integral (2.23) when $x=x_1$ is

$$\left.\begin{aligned}
\int_{x_0}^{x_1}\int_{x_0}^x \cdots \int_{x_0}^x f\,(d\,x)^n &= {}^{(n)}f(x_1) - \sum_{\nu=0}^{n-1} {}^{(n-\nu)}f(x_0)\,\frac{h^\nu}{\nu!} \\
&= h^n \sum_{\varrho=0}^{p} \beta_{n,\varrho}\, V^\varrho f_0 + R_{n,\,p+1}(x_1),
\end{aligned}\right\} \tag{2.25}$$

where the coefficients $\beta_{n,\varrho}$ denote the numbers

$$\beta_{n,\varrho} = \int\limits_0^1 \int\limits_0^u \cdots \int\limits_0^u \frac{u(u+1)(u+2)\cdots(u+\varrho-1)}{\varrho!} (du)^n = P_{n,\varrho}(1). \quad (2.26)$$

These coefficients $\beta_{n,\varrho}$ include the numbers β_ϱ of (2.7) as the special case $n=1$, i.e. $\beta_\varrho = \beta_{1,\varrho}$. The first few $\beta_{n,\varrho}$ are given in Table I/4. From (2.2) and (2.26) we see that an estimate for the remainder term in (2.25) is given by

$$|R_{n,\,p+1}(x_1)| \le \beta_{n,\,p+1} h^{p+n+1} |f^{(p+1)}|_{\max}. \quad (2.27)$$

Table I/4. *The numbers* $\beta_{n,\varrho}$

	$\varrho=0$	$\varrho=1$	$\varrho=2$	$\varrho=3$	$\varrho=4$	$\varrho=5$
$n=1$	1	$\frac{1}{2}$	$\frac{5}{12}$	$\frac{3}{8}$	$\frac{251}{720}$	$\frac{95}{288}$
$n=2$	$\frac{1}{2}$	$\frac{1}{6}$	$\frac{1}{8}$	$\frac{19}{180}$	$\frac{3}{32}$	$\frac{863}{10080}$
$n=3$	$\frac{1}{6}$	$\frac{1}{24}$	$\frac{7}{240}$	$\frac{17}{720}$	$\frac{41}{2016}$	$\frac{731}{40320}$
$n=4$	$\frac{1}{24}$	$\frac{1}{120}$	$\frac{1}{180}$	$\frac{11}{2520}$	$\frac{89}{24192}$	$\frac{5849}{181440}$

If we take the upper limit in (2.25) to be x_{-1} instead of x_1, we obtain

$$\left.\begin{aligned}
\int\limits_{x_0}^{x_{-1}}\cdots\int\limits_{x_0}^x f\,(dx)^n &= {}^{(n)}f(x_{-1}) - \sum_{v=0}^{n-1} {}^{(n-v)}f(x_0)\frac{(-h)^v}{v!} \\
&= (-1)^{n+\varrho} h^n \sum_{\varrho=0}^p \gamma_{n,\varrho} \nabla^\varrho f_0 + R_{n,\,p+1}(x_{-1}),
\end{aligned}\right\} \quad (2.28)$$

where

$$\left.\begin{aligned}
\gamma_{n,\varrho} &= (-1)^{n+\varrho} \int\limits_0^{-1}\int\limits_0^u \cdots \int\limits_0^u \binom{u+\varrho-1}{\varrho} (du)^n \\
&= \int\limits_0^1\int\limits_0^u \cdots \int\limits_0^u \binom{u}{\varrho} (du)^n = (-1)^{n+\varrho} P_{n,\varrho}(-1).
\end{aligned}\right\} \quad (2.29)$$

The first few numbers $\gamma_{n,\varrho}$ are given in Table I/5.

Table I/5. *The numbers* $\gamma_{n,\varrho}$

	$\varrho=0$	$\varrho=1$	$\varrho=2$	$\varrho=3$	$\varrho=4$
$n=1$	1	$\frac{1}{2}$	$-\frac{1}{12}$	$\frac{1}{24}$	$-\frac{19}{720}$
$n=2$	$\frac{1}{2}$	$\frac{1}{6}$	$-\frac{1}{24}$	$\frac{1}{45}$	$-\frac{7}{480}$
$n=3$	$\frac{1}{6}$	$\frac{1}{24}$	$-\frac{1}{80}$	$\frac{1}{144}$	$-\frac{47}{10080}$
$n=4$	$\frac{1}{24}$	$\frac{1}{120}$	$-\frac{1}{360}$	$\frac{1}{630}$	

Clearly we can also obtain these numbers by integrating NEWTON's formula (2.3) with forward differences:

$$\left.\begin{array}{l} \int\limits_{x_0}^{x_1}\int\limits_{x_0}^{x}\cdots\int\limits_{x_0}^{x} f(d\,x)^n = {}^{(n)}f(x_1) - \sum_{\nu=0}^{n-1} {}^{(n-\nu)}f(x_0)\,\frac{h^\nu}{\nu!} \\[3mm] \qquad\qquad = h^n \sum_{\varrho=0}^{p} \gamma_{n,\varrho}\,\Delta^\varrho f_0 + S. \end{array}\right\} \tag{2.30}$$

An estimate for the remainder term S can easily be derived by integration of the remainder term in (2.3):

$$|S| \leq h^{p+n+1}|\gamma_{n,\,p+1}|\,|f^{(p+1)}|_{\max}. \tag{2.31}$$

In addition to the coefficients introduced already we shall need the quantities $\beta^*_{n,\varrho}$ which are obtained in the same way as the $\beta_{n,\varrho}$ except that the lower limit x_0 in (2.19) to (2.26) is replaced by x_{-1}. Performing the individual integrations afresh, we obtain

$$\left.\begin{array}{l} \int\limits_{x_{-1}}^{x} f(x)\,d\,x = h\int\limits_{-1}^{u} N_p(x(u))\,d\,u + \int\limits_{x_{-1}}^{x} R_{p+1}d\,x = {}'f(x) - {}'f(x_{-1}) \\[3mm] = h\left[(u+1)\,f_0 + \frac{u^2-1}{2}\,\nabla f_0 + \frac{2u^3+3u^2-1}{12}\,\nabla^2 f_0 + \cdots\right] + R^*_{1,\,p+1}, \end{array}\right\} \tag{2.32}$$

$$\left.\begin{array}{l} \int\limits_{x_{-1}}^{x}\int\limits_{x_{-1}}^{x} f(d\,x)^2 = {}''f(x) - {}''f(x_{-1}) - {}'f(x_{-1})\,(x-x_{-1}) \\[3mm] = h^2\left[\frac{(u+1)^2}{2}\,f_0 + \frac{u^3-3u-2}{6}\,\nabla f_0 + \cdots\right] + R^*_{2,\,p+1}, \end{array}\right\} \tag{2.33}$$

and generally, for n integrations,

$$\left.\begin{array}{l} \int\limits_{x_{-1}}^{x}\cdots\int\limits_{x_{-1}}^{x} f(d\,x)^n = {}^{(n)}f(x) - \sum_{\nu=0}^{n-1} {}^{(n-\nu)}f(x_{-1})\,\frac{(x-x_{-1})^\nu}{\nu!} \\[3mm] \qquad = h^n \sum_{\varrho=0}^{p} P^*_{n,\varrho}(u)\,\nabla^\varrho f_0 + R^*_{n,\,p+1}(x), \end{array}\right\} \tag{2.34}$$

where the polynomials $P^*_{n,\varrho}$ are given by the n-th indefinite integrals

$$\left.\begin{array}{l} P^*_{n,\varrho}(u) = \int\limits_{-1}^{u}\cdots\int\limits_{-1}^{u}\binom{u+\varrho-1}{\varrho}(d\,u)^n \quad \text{for } \varrho \geq 1 \\[3mm] P^*_{n,0}(u) = \frac{1}{n!}\,(u+1)^n \quad (n=1,2,\ldots). \end{array}\right\} \tag{2.35}$$

Table I/6 gives the polynomials $P^*_{n,\varrho}$ for a few values of n and ϱ. For $x=x_0$ (2.34) becomes

$$\left.\begin{array}{l} \int\limits_{x_{-1}}^{x_0}\cdots\int\limits_{x_{-1}}^{x} f(d\,x)^n = {}^{(n)}f(x_0) = \sum_{\nu=0}^{n-1} {}^{(n-\nu)}f(x_{-1})\,\frac{h^\nu}{\nu!} \\[3mm] \qquad = h^n \sum_{\varrho=0}^{p} \beta^*_{n,\varrho}\,\nabla^\varrho f_0 + R^*_{n,\,p+1}(x_0). \end{array}\right\} \tag{2.36}$$

Table I/6. *The polynomials* $P^*_{n,\varrho}(u)$

	$\varrho=0$	$\varrho=1$	$\varrho=2$
$n=1$	$u+1$	$\frac{1}{2}\,(u^2-1)$	$\frac{1}{12}\,(2u^3+3u^2-1)$
$n=2$	$\frac{1}{2}\,(u+1)^2$	$\frac{1}{6}\,(u^3-3u-2)$	$\frac{1}{24}\,(u^4+2u^3-2u-1)$
$n=3$	$\frac{1}{6}\,(u+1)^3$	$\frac{1}{24}\,(u^4-6u^2-8u-3)$	$\frac{1}{240}\,(2u^5+5u^4-10u^2-10u-3)$
$n=4$	$\frac{1}{24}\,(u+1)^4$	$\frac{1}{120}\,(u^5-10u^3-20u^2-15u-4)$	$\frac{1}{720}\,(u^6+3u^5-10u^3-15u^2-9u-2)$

	$\varrho=3$	$\varrho=4$
$n=1$	$\frac{1}{24}\,(u^4+4u^3+4u^2-1)$	$\frac{1}{720}\,(6u^5+45u^4+110u^3+90u^2-19)$
$n=2$	$\frac{1}{360}\,(3u^5+15u^4+20u^3-15u-7)$	$\frac{1}{1440}\,(2u^6+18u^5+55u^4+60u^3-$
$n=3$	$\frac{1}{720}\,(u^6+6u^5+10u^4-15u^2-14u-4)$	$-38u-17)$
$n=4$	$\frac{1}{5040}\,(u^7+7u^6+14u^5-35u^3-49u^2-28u-6)$	

Corresponding to (2.22) and (2.26) we have

$$R^*_{\nu,\,p+1}=\int\limits_{x_{-1}}^{x} R^*_{\nu-1,\,p+1}\,dx \qquad (\nu=1,2,3,\ldots) \tag{2.37}$$

(with $R^*_{0,\,p+1}=R^*_{p+1}$) and

$$\left.\begin{aligned}\beta^*_{n,\,\varrho}&=\int\limits_{-1}^{0}\int\limits_{-1}^{u}\cdots\int\limits_{-1}^{u}\binom{u+\varrho-1}{\varrho}\,(du)^n \quad\text{for } \varrho\geqq 1,\\[4pt]\beta^*_{n,\,0}&=\frac{1}{n!}\,.\end{aligned}\right\} \tag{2.38}$$

Here also the coefficients $\beta^*_{n,\,\varrho}$ contain the numbers β^*_ϱ of (2.9) as the special case $n=1$ since $\beta^*_\varrho=\beta^*_{1,\,\varrho}$. The first few $\beta^*_{n,\,\varrho}$ are given in Table I/7. Exactly as in (2.12) it can be shown that

$$\beta_{n,\,\varrho}+\beta^*_{n,\,\varrho+1}=\beta_{n,\,\varrho+1} \qquad (\varrho=0,1,2,\ldots)\ (n=1,2,3,\ldots). \tag{2.39}$$

By adding equations (2.39) for $\varrho=0,1,2,\ldots,p-1$, together with $\beta_{n,\,0}=\beta^*_{n,\,0}\left(=\frac{1}{n!}\right)$, we obtain a generalized form of (2.13), namely

$$\sum_{\varrho=0}^{p}\beta^*_{n,\,\varrho}=\beta_{n,\,p}\,. \tag{2.40}$$

Table I/7. *The numbers* $\beta^*_{n,\,\varrho}$

	$\varrho=0$	$\varrho=1$	$\varrho=2$	$\varrho=3$	$\varrho=4$
$n=1$	1	$-\frac{1}{2}$	$-\frac{1}{12}$	$-\frac{1}{24}$	$-\frac{19}{720}$
$n=2$	$\frac{1}{2}$	$-\frac{1}{3}$	$-\frac{1}{24}$	$-\frac{7}{360}$	$-\frac{17}{1440}$
$n=3$	$\frac{1}{6}$	$-\frac{1}{8}$	$-\frac{1}{80}$	$-\frac{1}{180}$	$-\frac{11}{3360}$
$n=4$	$\frac{1}{24}$	$-\frac{1}{30}$	$-\frac{1}{360}$	$-\frac{1}{840}$	$-\frac{83}{120960}$

From (2.2), (2.37), (2.38) we see that limits for the remainder term $R^*_{n,\,p+1}$ are given by

$$|R^*_{n,\,p+1}(x_0)| \leq h^{p+n+1}|\beta^*_{n,\,p+1}|\,|f^{(p+1)}|_{\max}. \tag{2.41}$$

Finally we derive some repeated integration formulae from STIRLING's interpolation formula (2.5). With

$$\Phi(x) = \tfrac{1}{2}\,[f(x_0 + u\,h) + f(x_0 - u\,h)] = \tfrac{1}{2}\,[f(x) + f(2\,x_0 - x)] \tag{2.42}$$

we have

$$
\left.
\begin{aligned}
\int_{x_0}^{x} \Phi\,dx &= \frac{1}{2}\,['f(x_0 + u\,h) - 'f(x_0 - u\,h)] \\
&= h\left[u f_0 + \frac{u^3}{6}\,\nabla^2 f_1 + \frac{-5u^3 + 3u^5}{360}\,\nabla^4 f_2 + \right. \\
&\quad \left. + \frac{28u^3 - 21u^5 + 3u^7}{15\,120}\,\nabla^6 f_3 + \cdots\right] + R^{**}_{1,\,p},
\end{aligned}
\right\} \tag{2.43}
$$

$$
\left.
\begin{aligned}
\int_{x_0}^{x}\!\int_{x_0}^{x} \Phi(d\,x)^2 &= \frac{1}{2}\,[''f(x_0 + u\,h) + ''f(x_0 - u\,h)] - ''f(x_0) \\
&= h^2\left[\frac{u^2}{2}\,f_0 + \frac{u^4}{24}\,\nabla^2 f_1 + \frac{-5u^4 + 2u^6}{1440}\,\nabla^4 f_2 + \cdots\right] + R^{**}_{2,\,p}, \\[6pt]
\int_{x_0}^{x}\!\!\int\!\!\int \Phi(d\,x)^3 &= \frac{1}{2}\,['''f(x_0 + u\,h) - '''f(x_0 - u\,h)] - ''f(x_0)\,(x - x_0) \\
&= h^3\left[\frac{u^3}{6}\,f_0 + \frac{u^5}{120}\,\nabla^2 f_1 + \frac{-7u^5 + 2u^7}{10080}\,\nabla^4 f_2 + \cdots\right] + R^{**}_{3,\,p}, \\[6pt]
\int_{x_0}^{x}\!\!\int\!\!\int\!\!\int \Phi(d\,x)^4 &= \frac{1}{2}\,[''''f(x_0 + u\,h) + ''''f(x_0 - u\,h)] - \\
&\quad - ''''f(x_0) - ''f(x_0)\,\frac{(x - x_0)^2}{2} \\
&= h^4\left[\frac{u^4}{24}\,f_0 + \frac{u^6}{720}\,\nabla^2 f_1 + \frac{-14u^6 + 3u^8}{120960}\,\nabla^4 f_2 + \cdots\right] + R^{**}_{4,\,p}.
\end{aligned}
\right\} \tag{2.44}
$$

We write down the general n-th repeated integral of $\Phi(x)$ only for $x = x_1$:

$$
\int_{x_0}^{x_1}\!\int_{x_0}^{x}\!\cdots\!\int_{x_0}^{x} \Phi(d\,x)^n = h^n \sum_{\varrho=0}^{p/2} \beta^{**}_{n,\,\varrho}\,\nabla^{2\varrho} f_\varrho + R^{**}_{n,\,p}(x_1)
$$

$$
\left.
= \begin{cases}
\dfrac{1}{2}\,[^{(n)}f(x_0 + h) + {}^{(n)}f(x_0 - h)] - \displaystyle\sum_{\varrho=0}^{(n/2)-1} \dfrac{h^{2\varrho}}{(2\varrho)!}\,{}^{(n-2\varrho)}f(x_0) \\
\hfill \text{for } n \text{ even,} \\[10pt]
\dfrac{1}{2}\,[^{(n)}f(x_0 + h) - {}^{(n)}f(x_0 - h)] - \displaystyle\sum_{\varrho=0}^{(n-3)/2} \dfrac{h^{2\varrho+1}}{(2\varrho+1)!}\,{}^{(n-1-2\varrho)}f(x_0) \\
\hfill \text{for } n \text{ odd,}
\end{cases}
\right\} \tag{2.45}
$$

where

$$R_{\nu,p}^{**} = \int_{x_0}^{x} R_{\nu-1,p}^{**} \, dx \tag{2.46}$$

and

$$R_{0,p}^{**} = R_p^{**} \quad \text{from (2.5)}.$$

The first few numbers $\beta_{n,\varrho}^{**}$ are given in Table I/8.

Table I/8. *Some values of the numbers* $\beta_{n,\varrho}^{**}$

	$\varrho=0$	$\varrho=1$	$\varrho=2$	$\varrho=3$
$n=1$	1	$\frac{1}{6}$	$-\frac{1}{180}$	$\frac{1}{1512}$
$n=2$	$\frac{1}{2}$	$\frac{1}{24}$	$-\frac{1}{480}$	$\frac{31}{120\,960}$
$n=3$	$\frac{1}{6}$	$\frac{1}{120}$	$-\frac{1}{2016}$	
$n=4$	$\frac{1}{24}$	$\frac{1}{720}$	$-\frac{11}{120\,960}$	

Table I/9. *The numbers* $r_{n,p}^{**}$

	$p=2$	$p=4$
$n=1$	$\frac{2}{15}$	$\frac{10}{21}$
$n=2$	$\frac{1}{20}$	$\frac{31}{168}$
$n=3$	$\frac{1}{84}$	
$n=4$	$\frac{11}{5040}$	

From (2.5) an estimate for the remainder term for even p is

$$\left| R_{n,p}^{**} \right| \leqq r_{n,p}^{**} \frac{h^{p+n+2}}{(p+2)!} \left| f^{(p+2)} \right|_{\max}, \tag{2.47}$$

where

$$r_{n,p}^{**} = \int_0^1 \int_0^u \cdots \int_0^u \left| u^2 (u^2-1)(u^2-4) \cdots \left(u^2 - \left(\frac{p}{2}\right)^2\right) \right| (du)^n ; \tag{2.48}$$

the first few values of these numbers are given in Table I/9.

2.4. Calculation of higher derivatives

From a differential equation

$$y^{(n)} = f(x, y, y', \ldots, y^{(n-1)}) \tag{2.49}$$

we can calculate the higher derivatives $y^{(n+1)}, y^{(n+2)}, \ldots$ by repeated differentiation (assuming that f is differentiable a sufficient number of times with respect to each of its arguments). Thus for the first derivative we have

$$y^{(n+1)} = \frac{\partial f}{\partial x} + \frac{\partial f}{\partial y} y' + \frac{\partial f}{\partial y'} y'' + \cdots + \frac{\partial f}{\partial y^{(n-1)}} f.$$

We now introduce the notation

$$\left. \begin{array}{l} y = u_0, \quad y' = u_1, \quad y'' = u_2, \ldots, \quad y^{(n)} = u_n = f, \\[2mm] f_x = \dfrac{\partial f}{\partial x}, \quad f_\nu = \dfrac{\partial f}{\partial u_\nu}, \quad f_{\mu\nu} = \dfrac{\partial^2 f}{\partial u_\mu \, \partial u_\nu}, \quad f_{x\nu} = \dfrac{\partial^2 f}{\partial x \, \partial u_\nu}, \ldots \end{array} \right\} \tag{2.50}$$

and also the symbolic operator D whose operation on any function $\varphi(x, u_0, u_1, \ldots, u_{n-1})$ is defined by

$$D\varphi = \frac{\partial \varphi}{\partial x} + u_1 \frac{\partial \varphi}{\partial u_0} + u_2 \frac{\partial \varphi}{\partial u_1} + \cdots + u_{n-1} \frac{\partial \varphi}{\partial u_{n-2}} + f \frac{\partial \varphi}{\partial u_{n-1}}, \quad (2.51)$$

so that

$$y^{(n+1)} = Df. \quad (2.52)$$

For the higher derivatives it is convenient to introduce further operators E and F. By the operator $D^{(r)}$ we understand the operator which results from formal expansion of the power

$$\left(\frac{\partial}{\partial x} + u_1 \frac{\partial}{\partial u_0} + u_2 \frac{\partial}{\partial u_1} + \cdots + u_{n-1} \frac{\partial}{\partial u_{n-2}} + f \frac{\partial}{\partial u_{n-1}} \right)^r,$$

treating the u_ν and f as constant factors so that they can be collected together as coefficients in front of each differential operator; thus, for example,

$$D^{(2)} = \frac{\partial^2}{\partial x^2} + u_1^2 \frac{\partial^2}{\partial u_0^2} + \cdots + u_{n-1}^2 \frac{\partial^2}{\partial u_{n-2}^2} + f^2 \frac{\partial^2}{\partial u_{n-1}^2} + 2u_1 \frac{\partial^2}{\partial x \partial u_0} +$$

$$+ 2u_2 \frac{\partial^2}{\partial x \partial u_1} + \cdots + 2u_1 u_2 \frac{\partial^2}{\partial u_0 \partial u_1} + \cdots + 2f u_{n-1} \frac{\partial^2}{\partial u_{n-2} \partial u_{n-1}}.$$

If we now differentiate Df with respect to x, the differentiation of the derivatives $\frac{\partial f}{\partial x}, \frac{\partial f}{\partial u_\nu}$ occurring in the individual products gives precisely $D^{(2)}f$; in addition there are the terms arising from the differentiation of the factors u_ν and f. Thus

$$\left. \begin{aligned} y^{(n+2)} &= \frac{d}{dx} Df \\ &= D^{(2)}f + u_2 f_0 + u_3 f_1 + \cdots + u_{n-1} f_{n-3} + f f_{n-2} + f_{n-1} Df = Ef, \end{aligned} \right\} (2.53)$$

say, where E is the operator defined by the preceding expression.

For a further differentiation it is expedient to derive first the formula

$$\frac{d}{dx} (D^{(2)}f) = D^{(3)}f +$$
$$+ 2[u_2 Df_0 + u_3 Df_1 + u_4 Df_2 + \cdots + u_{n-1} Df_{n-3} + f Df_{n-2} + Df \cdot Df_{n-1}]$$

by writing down $D^{(2)}f$ in detail and applying the operator D. Then if we define another operator F by

$$\left. \begin{aligned} Ff &= D^{(3)}f + \\ &+ 3[u_2 Df_0 + u_3 Df_1 + \cdots + u_{n-1} Df_{n-3} + f Df_{n-2} + Df \cdot Df_{n-1}] + \\ &\quad + [u_3 f_0 + u_4 f_1 + \cdots + u_{n-1} f_{n-4} + f f_{n-3}], \end{aligned} \right\} (2.54)$$

we can write

$$y^{(n+3)} = Ff + f_{n-1} Ef + f_{n-2} Df. \quad (2.55)$$

2.5. Hermite's generalization of Taylor's formula

Equations (2.57), (2.58) below constitute a generalization[1] of Taylor's formula; they give a relation between the derivatives of a function $f(x)$ at two points a and b, where $f(x)$ has continuous derivatives of the $(k+m+1)$-th order in the interval $\langle a, b\rangle$ (k, m are non-negative integers).

If $g(x) = (x-a)^k (x-b)^m$, then, by integration by parts $k+m$ times, we have

$$\int_a^b f'(x) \, g^{(k+m)}(x) \, dx = [f'(x) \, g^{(k+m-1)}(x) - f''(x) \, g^{(k+m-2)}(x) + \\ + \cdots + (-1)^{k+m-1} f^{(k+m)}(x) \, g(x)]_a^b + \\ + (-1)^{(k+m)} \int_a^b f^{(k+m+1)}(x) \, g(x) \, dx. \qquad (2.56)$$

Since $g^{(k+m)}(x) = (k+m)!$, the value of the integral on the left-hand side is

$$(k+m)! \, (f(b) - f(a)).$$

It remains to calculate the values of the derivatives of g at the boundary points a, b. By Leibniz's rule for the differentiation of a product we have

$$g^{(r)}(x) = \sum_{\varrho=0}^{r} \binom{r}{\varrho} \binom{k}{\varrho} \varrho! \, (x-a)^{k-\varrho} \binom{m}{r-\varrho} (r-\varrho)! \, (x-b)^{m-r+\varrho}.$$

Bearing in mind that the binomial coefficients $\binom{k}{\varrho}$ and $\binom{m}{r-\varrho}$ are zero for $\varrho > k$ and $\varrho < r-m$, respectively, we see that $\sum\limits_{\varrho=0}^{r}$ can be replaced by $\sum\limits_{\varrho=r-m}^{k}$ when $r \le k+m$ and that the sum is zero when $r > k+m$. If we now put $x=a$, only the term with $\varrho=k$ remains and we have

$$g^{(r)}(a) = \binom{r}{k} k! \binom{m}{r-k} (r-k)! \, (a-b)^{m-r+k} = r! \binom{m}{r-k} (a-b)^{m+k-r};$$

similarly

$$g^{(r)}(b) = r! \binom{k}{r-m} (b-a)^{m+k-r}.$$

---

[1] Different proofs are given by S. Hermite: Œuvres 3, 438 (1912). — Kowalewski, G.: Dtsch. Math. 6, 349—351 (1942). — Obreschkoff, N.: Abh. Preuss. Akad. Wiss., Math.-naturw. Kl. 1940, Nr. 4, 1—20. — Pflanz, E.: Z. Angew. Math. Mech. 28, 167—172 (1948). — Beck, E.: Z. Angew. Math. Mech. 30, 84—93 (1950).

With these values for the derivatives at the end-points and with h denoting the interval length $b-a$, (2.56) becomes

$$(k+m)!\,(f(b)-f(a))+\left[\sum_{\nu=1}^{k+m}(-1)^{\nu}f^{(\nu)}\,g^{(k+m-\nu)}\right]_{a}^{b}$$

$$=\sum_{\nu=0}^{k+m}(k+m-\nu)!\,h^{\nu}\left\{(-1)^{\nu}\binom{k}{k-\nu}f^{(\nu)}\,(b)-\binom{m}{m-\nu}f^{(\nu)}\,(a)\right\}$$

$$=(-1)^{k+m}\int_{a}^{b}f^{(k+m+1)}\,g\,dx.$$

Still bearing in mind that $\binom{k}{k-\nu}=0$ when $\nu>k$ and $\binom{m}{m-\nu}=0$ when $\nu>m$, we see that, after division by $(k+m)!$,

$$\sum_{\nu=0}^{k}(-1)^{\nu}f^{(\nu)}(b)\,\frac{h^{\nu}}{\nu!}\,\frac{\binom{k}{\nu}}{\binom{k+m}{\nu}}=\sum_{\nu=0}^{m}f^{(\nu)}(a)\,\frac{h^{\nu}}{\nu!}\,\frac{\binom{m}{\nu}}{\binom{k+m}{\nu}}+R_{k,m}\qquad(2.57)$$

with the remainder term given by

$$R_{k,m}=\frac{(-1)^{k+m}}{(k+m)!}\int_{a}^{b}(x-a)^{k}\,(x-b)^{m}\,f^{(k+m+1)}\,(x)\,dx.\qquad(2.58)$$

This result includes TAYLOR'S theorem as the special case $k=0$.

When $k=1$ and $m=2$, we have the formula

$$f(b)-\frac{h}{3}\,f'(b)=f(a)+\frac{2h}{3}\,f'(a)+\frac{h^{2}}{6}\,f''(a)+R_{1,2},\qquad(2.59)$$

which will be used later in Ch. II, § 1.5.

Formula (2.57), like TAYLOR'S formula, can be extended to more independent variables; in fact the extension can be derived using precisely the same device as is used for TAYLOR'S formula. For simplicity we consider the extension only for a function $u(x, y)$ of two independent variables and for two points (x_0, y_0), (x_1, y_1). All we have to do is to apply formula (2.57) to the function

$$f(t)=u\left(x_0+t(x_1-x_0),\ y_0+t(y_1-y_0)\right)$$

in the interval $\langle a, b\rangle=\langle 0, 1\rangle$; then, assuming that all derivatives occurring are continuous, we have

$$\left.\begin{aligned}\sum_{\nu=0}^{k}(-1)^{\nu}\left(h\frac{\partial}{\partial x}+l\frac{\partial}{\partial y}\right)^{\nu}\frac{u(x_1, y_1)}{\nu!}\,\frac{\binom{k}{\nu}}{\binom{k+m}{\nu}}\\=\sum_{\nu=0}^{m}\left(h\frac{\partial}{\partial x}+l\frac{\partial}{\partial y}\right)^{\nu}\frac{u(x_0, y_0)}{\nu!}\,\frac{\binom{m}{\nu}}{\binom{k+m}{\nu}}+R,\end{aligned}\right\}\qquad(2.60)$$

where $h = x_1 - x_0$, $l = y_1 - y_0$ and the symbolic notation $\left(h \dfrac{\partial}{\partial x} + l \dfrac{\partial}{\partial y} \right)^{\nu}$ is that generally used in TAYLOR's formula[1]. The form of the remainder term R may be found from (2.58).

§3. Further useful formulae from analysis

3.1. GAUSS's and GREEN's formulae for two independent variables

GAUSS's integral theorem in two dimensions: Let B be a closed, finite region of the (x, y) plane bounded by a piecewise-smooth curve Γ without double points. We denote the inward normal to this boundary curve by ν and the arc length measured anti-clockwise from a fixed point on the curve by s (see Fig. I/1). If $f(x, y)$ and $g(x, y)$ are continuous functions with continuous first partial derivatives in B, then[2]

$$\left. \begin{aligned} & \iint_B \frac{\partial f(x, y)}{\partial x}\, dx\, dy \\ & \qquad = - \int_\Gamma f(x, y) \cos(\nu, x)\, ds, \end{aligned} \right\} \quad (3.1)$$

$$\left. \begin{aligned} & \iint_B \frac{\partial g(x, y)}{\partial y}\, dx\, dy \\ & \qquad = - \int_\Gamma g(x, y) \cos(\nu, y)\, ds. \end{aligned} \right\} \quad (3.2)$$

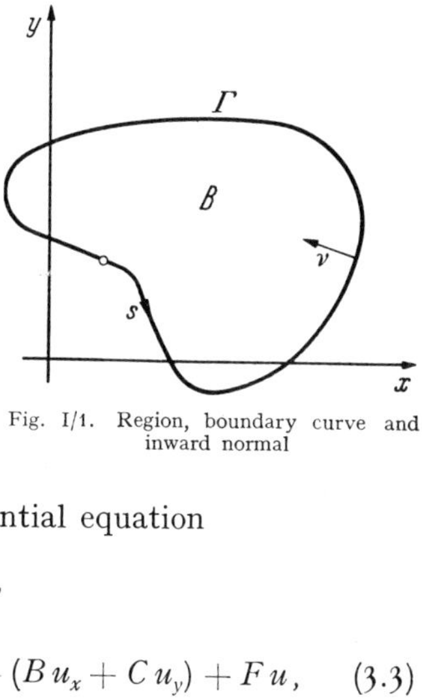

Fig. I/1. Region, boundary curve and inward normal

To apply these formulae to the differential equation

$$L[u] = r(x, y),$$

where

$$L[u] = - \frac{\partial}{\partial x}(A u_x + B u_y) - \frac{\partial}{\partial y}(B u_x + C u_y) + F u, \quad (3.3)$$

in which A, B, C, F are continuous functions with A, B, C also possessing continuous derivatives of at least the first order, we put

$$f = (A \psi_x + B \psi_y)\varphi$$

in (3.1) and

$$g = (B \psi_x + C \psi_y)\varphi$$

in (3.2) and add the two resulting equations (here subscripts denote partial differentiation, e.g. $u_x = \partial u / \partial x$). We then see that any two

[1] Cf., for example, H. v. MANGOLDT and K. KNOPP: Einführung in die Höhere Mathematik, Vol. II, 10th ed., p. 355. Stuttgart 1947.

[2] See, for instance, H. v. MANGOLDT and K. KNOPP: Einführung in die Höhere Mathematik, Vol. III, 10th ed., p. 346. Stuttgart 1948.

functions $\varphi(x, y)$ and $\psi(x, y)$ with continuous partial derivatives of up to the second order satisfy GREEN's formula

$$
\left.
\begin{aligned}
&\iint_B [A\,\varphi_x\psi_x + B\,(\varphi_x\psi_y + \varphi_y\psi_x) + C\,\varphi_y\psi_y + F\,\varphi\psi]\,dx\,dy - \\
&\quad - \iint_B \varphi\,L\,[\psi]\,dx\,dy = -\int_\Gamma \varphi\,L^*[\psi]\,ds,
\end{aligned}
\right\} \quad (3.4)
$$

where L^* is defined by

$$
L^*[\psi] = (A\,\psi_x + B\,\psi_y)\cos(\nu, x) + (B\,\psi_x + C\,\psi_y)\cos(\nu, y). \quad (3.5)
$$

If φ and ψ are interchanged in (3.4) and the resulting equation is subtracted from (3.4), the first integrals cancel out and we obtain the equation

$$
\iint_B (\varphi\,L\,[\psi] - \psi\,L\,[\varphi])\,dx\,dy = \int_\Gamma (\varphi\,L^*[\psi] - \psi\,L^*[\varphi])\,ds. \quad (3.6)
$$

This is often known as GREEN's formula also. In the special case $A \equiv C \equiv 1$, $B \equiv F \equiv 0$, equations (3.4) and (3.6) become

$$
\iint_B (\varphi_x\psi_x + \varphi_y\psi_y)\,dx\,dy + \iint_B \varphi\,V^2\psi\,dx\,dy = -\int_\Gamma \varphi\,\frac{\partial\psi}{\partial\nu}\,ds \quad (3.7)
$$

and

$$
\iint_B (\varphi\,V^2\psi - \psi\,V^2\varphi)\,dx\,dy = -\int_\Gamma \left(\varphi\,\frac{\partial\psi}{\partial\nu} - \psi\,\frac{\partial\varphi}{\partial\nu}\right)ds, \quad (3.8)
$$

where V^2 is the usual symbol for LAPLACE's differential operator

$$
V^2\varphi = \frac{\partial^2\varphi}{\partial x^2} + \frac{\partial^2\varphi}{\partial y^2}
$$

and $\partial/\partial\nu$ denotes differentiation along the inward normal:

$$
\frac{\partial\psi}{\partial\nu} = \frac{\partial\psi}{\partial x}\cos(\nu, x) + \frac{\partial\psi}{\partial y}\cos(\nu, y). \quad (3.9)
$$

3.2. Corresponding formulae for more than two independent variables

We also need GAUSS's integral theorem in space, say the space S_m of m dimensions. Let B be a closed, finite region in the $(x_1, x_2, \ldots, x_m)$ space bounded by a surface Γ [an $(m-1)$-dimensional hypersurface] which may consist of a finite number of "faces", each having continuous tangential hyperplanes. Let v be a given vector field in B with components $a_1, a_2, \ldots, a_m$, each being a continuous function of $x_1, x_2, \ldots, x_m$ with continuous partial derivatives of the first order. If $d\tau = dx_1\,dx_2 \ldots dx_m$ is the volume element in B, dS the surface element on Γ, ν the inward normal to Γ and $v_\nu = (v, \nu)$ the component of v along the inward normal (i.e. the scalar product of v with the unit vector ν directed inwards from and perpendicular to Γ), then GAUSS's integral theorem states that

$$\int_B \operatorname{div} v \, d\tau = -\int_\Gamma v_\nu \, dS \qquad (3.10)$$

i.e.

$$\int_B \sum_{i=1}^m \frac{\partial a_i}{\partial x_i} \, d\tau = -\int_\Gamma \sum_{i=1}^m a_i \cos(\nu, x_i) \, dS. \qquad (3.11)$$

From this we now derive a formula which will be useful when we deal with the differential equation

$$L[u] = -\sum_{i,\,k=1}^m \frac{\partial}{\partial x_i}\left(A_{ik} \frac{\partial u}{\partial x_k}\right) + q u = r, \qquad (3.12)$$

where the coefficients q and r are continuous and the A_{ik} have continuous first derivatives. If, in (3.11), we put

$$a_i = \varphi \sum_{k=1}^m A_{ik} \frac{\partial \psi}{\partial x_k},$$

then, for two functions φ and ψ with continuous second derivatives, we have [in the notation of (3.12)]

$$\left.\begin{aligned}
\int_B \varphi L[\psi] \, d\tau &= \int_B \left(-\varphi \sum_{i,\,k=1}^m \frac{\partial}{\partial x_i}\left(A_{ik}\frac{\partial \psi}{\partial x_k}\right) + q\varphi\psi\right) d\tau \\
&= \int_B \left\{-\sum_{i=1}^m \frac{\partial}{\partial x_i}\left(\varphi \sum_{k=1}^m A_{ik}\frac{\partial \psi}{\partial x_k}\right) + \sum_{i,\,k=1}^m A_{ik}\frac{\partial \varphi}{\partial x_i}\frac{\partial \psi}{\partial x_k} + q\varphi\psi\right\} d\tau \\
&= J[\varphi,\psi] + \int_\Gamma \varphi L^*[\psi] \, dS,
\end{aligned}\right\} \quad (3.13)$$

where

$$J[\varphi,\psi] = \int_B \left(\sum_{i,\,k=1}^m A_{ik}\frac{\partial \varphi}{\partial x_i}\frac{\partial \psi}{\partial x_k} + q\varphi\psi\right) d\tau, \qquad (3.14)$$

$$L^*[\psi] = \sum_{i,\,k=1}^m A_{ik}\frac{\partial \psi}{\partial x_k}\cos(\nu, x_i). \qquad (3.15)$$

3.3. Co-normals and boundary-value problems in elliptic differential equations

If we introduce a positive number A and a direction σ so that

$$\sum_{i=1}^m A_{ik}\cos(\nu, x_i) = A_k^* = A\cos(\sigma, x_k),$$

then (3.15) can be written

$$L^*[\psi] = A\frac{\partial \psi}{\partial \sigma}.$$

The half-line running from Γ in the direction σ is called the "co-normal"[1].

---

[1] See, for example, A. G. WEBSTER and G. SZEGÖ: Partielle Differential-gleichungen der mathematischen Physik, p. 311. Leipzig and Berlin 1930.

If the matrix A_{ik} is symmetric $(A_{ik}=A_{ki})$, then, by interchanging φ and ψ in (3.13) and subtracting the resulting equation from (3.13), we obtain GREEN's formula

$$\left.\begin{aligned}
\int_B (\varphi L[\psi] - \psi L[\varphi])\, d\tau &= \int_\Gamma (\varphi L^*[\psi] - \psi L^*[\varphi])\, dS \\
&= \int_\Gamma A\left(\varphi \frac{\partial \psi}{\partial \sigma} - \psi \frac{\partial \varphi}{\partial \sigma}\right) dS.
\end{aligned}\right\} \quad (3.16)$$

In the special case

$$A_{ik} = \delta_{ik} = \begin{cases} 0 & \text{for} \quad i \neq k \\ 1 & \text{for} \quad i = k, \end{cases}$$

we have $A = 1$ and the co-normal σ coincides with the inward normal ν. If, in addition, $q = 0$, then with

$$\nabla^2 = \sum_{i=1}^m \frac{\partial^2}{\partial x_i^2}$$

the formulae (3.13), (3.16) become

$$-\int_B \varphi \nabla^2 \psi\, d\tau = \int_B (\operatorname{grad} \varphi, \operatorname{grad} \psi)\, d\tau + \int_\Gamma \varphi \frac{\partial \psi}{\partial \nu}\, dS, \qquad (3.17)$$

$$\int_B (\varphi \nabla^2 \psi - \psi \nabla^2 \varphi)\, d\tau = -\int_\Gamma \left(\varphi \frac{\partial \psi}{\partial \nu} - \psi \frac{\partial \varphi}{\partial \nu}\right) dS, \qquad (3.18)$$

where $\partial/\partial \nu$ denotes, in the usual way, differentiation in the direction of the inward normal:

$$\frac{\partial \varphi}{\partial \nu} = \sum_{i=1}^m \frac{\partial \varphi}{\partial x_i} \cos(\nu, x_i). \qquad (3.19)$$

If the differential equation is of elliptic type, i.e. the matrix A_{ik} is positive definite, so that for arbitrary real numbers $p_1, p_2, \ldots, p_m$, not all zero,

$$\sum_{i,k=1}^m A_{ik} p_i p_k > 0, \qquad (3.20)$$

the co-normal points into the interior of the region B since the scalar product of the co-normal vector with the inward normal is then positive:

$$\sum_{k=1}^m A_k^* \cos(\nu, x_k) = \sum_{i,k=1}^m A_{ik} \cos(\nu, x_i) \cos(\nu, x_k) > 0. \qquad (3.21)$$

Boundary conditions of the form

$$A_1 u + A_2 L^*[u] = A_1 u + A_2 A \frac{\partial u}{\partial \sigma} = A_3, \qquad (3.22)$$

where the A_1, A_2, A_3 are given functions on the boundary Γ, are often associated with differential equations of the type (3.12). In the case

$A_2 = 0$, $A_1 \neq 0$, the problem of determining u is called the first boundary-value problem; when $A_1 = 0$, $A_2 \neq 0$, it is called the second boundary-value problem and when neither A_1 nor A_2 are zero it is called the third boundary-value problem. In the case $A_2 \neq 0$ it is usually assumed, as will be done in this book, that the boundary $\varGamma$ is piecewise smooth, i.e. consists of a finite number of sections $\varGamma_\varrho$ each of which is an $(m-1)$-dimensional closed hypersurface with an $(m-2)$-dimensional boundary $\varGamma_\varrho^*$ such that the inward normal ν to $\varGamma_\varrho$ is continuous and tends to a definite limiting direction as any specific boundary point P on $\varGamma_\varrho^*$ is approached from a point inside $\varGamma_\varrho$.

3.4. GREEN'S functions

GREEN's functions are often of great value for theoretical investigations but they can also be used effectively for establishing formulae which are useful in numerical work. For ordinary differential equations one occasionally uses the GREEN's function directly, but for partial differential equations direct application is usually avoided because the associated GREEN's functions are usually either too complicated or even not specifiable explicitly at all.

Let us consider the boundary-value problem (1.12), (1.13). There are classes of such boundary-value problems for which it may be shown that there exists a GREEN's function $G(x_1, \ldots, x_n, \xi_1, \ldots, \xi_n)$, or shortly $G(x_i, \xi_i)$, with the following property:

For any continuous function $r(x_1, \ldots, x_n)$ the boundary-value problem (1.12), (1.13) with homogeneous boundary conditions, i.e.

$$L[u] = r, \qquad U_\mu[u] = 0, \tag{3.23}$$

is equivalent to

$$u(x_j) = \int_B G(x_j, \xi_j)\, r(\xi_j)\, d\xi_j. \tag{3.24}$$

Thus the boundary-value problem (3.23) may be solved with the aid of the GREEN's function by means of (3.24), or in other words, a function calculated from (3.24) with given r satisfies the boundary-value problem (3.23). If a GREEN's function exists, then the boundary-value problem (1.12), (1.13) with inhomogeneous boundary conditions is soluble, for it can be reduced to the case with homogeneous boundary conditions by the introduction of a new function u^* as in § 1.3. The detailed theory of GREEN's functions is covered in text books on differential equations[1].

We note here only two simple examples of GREEN's functions which will be used later[2].

---

[1] See, for example, R. COURANT and D. HILBERT: Methoden der math. Physik, Vol. I, 2nd ed., p. 302 et seq. Berlin 1931.

[2] A short collection of GREEN's functions can be found in L. COLLATZ: Eigenwertaufgaben mit technischen Anwendungen, p. 425/426. Leipzig 1949.

1. The GREEN's function for the boundary-value problem

$$L[u] = -u'' = r(x), \qquad u(0) = u(l) = 0 \qquad (3.25)$$

is

$$G(x, \xi) = \begin{cases} \dfrac{x}{l}(l - \xi) & \text{for } x \leq \xi, \\[2mm] \dfrac{\xi}{l}(l - x) & \text{for } x \geq \xi. \end{cases} \qquad (3.26)$$

2. If $\dfrac{nl}{2\pi}$ is not an integer, then

$$L[u] = -u'' - n^2 u = r(x), \qquad u(0) - u(l) = u'(0) - u'(l) = 0 \qquad (3.27)$$

has the GREEN's function

$$G(x, \xi) = -\frac{1}{2n \sin \dfrac{nl}{2}} \times \begin{cases} \cos n\left(\dfrac{l}{2} - \xi + x\right) & \text{for } x \leq \xi, \\[3mm] \cos n\left(\dfrac{l}{2} - x + \xi\right) & \text{for } x \geq \xi. \end{cases} \qquad (3.28)$$

3.5. Auxiliary formulae for the biharmonic operator

A formula which is used repeatedly in biharmonic problems, namely (3.38) below, can be derived easily from (3.8). Let u, v, w be three functions of x, y (all functions mentioned here should possess derivatives of orders as high as occur) such that

$$\nabla^4 u = \nabla^4 v \quad \text{in } B \qquad (3.29)$$

and

$$u = w, \qquad \frac{\partial u}{\partial v} = \frac{\partial w}{\partial v} \qquad \text{on } \Gamma, \qquad (3.30)$$

where the region B, boundary Γ and inward normal v are as used in § 3.2 and ∇^4 is the biharmonic operator defined by

$$\nabla^4 u \equiv \frac{\partial^4 u}{\partial x^4} + 2\frac{\partial^4 u}{\partial x^2 \partial y^2} + \frac{\partial^4 u}{\partial y^4}. \qquad (3.31)$$

If we demand further that u shall satisfy the boundary-value problem

$$\nabla^4 u = p(x, y) \quad \text{in } B \qquad (3.32)$$

with

$$u = f(s), \qquad u_v = g(s) \quad \text{on } \Gamma \quad \left(u_v = \frac{\partial u}{\partial v}\right), \qquad (3.33)$$

then v and w will be functions satisfying the differential equation and boundary conditions, respectively.

Let us define

$$D[\varphi, \psi] = \iint_B \nabla^2 \varphi \, \nabla^2 \psi \, dx \, dy, \qquad D[\varphi] = D[\varphi, \varphi], \qquad (3.34)$$

for any two functions φ and ψ. Then

$$D[\varphi + \psi] = D[\varphi] + 2D[\varphi, \psi] + D[\psi]. \tag{3.35}$$

Further from Schwarz's inequality we have

$$(D[\varphi, \psi])^2 \leqq D[\varphi] D[\psi]. \tag{3.36}$$

If we replace φ by $V^2\varphi$ in (3.8), we obtain

$$\left.\begin{array}{l} D[\varphi, \psi] = \iint\limits_B V^2\varphi\, V^2\psi\, dx\, dy \\[2mm] \qquad = \iint\limits_B \psi\, V^4\varphi\, dx\, dy + \int\limits_\Gamma \left(\psi(V^2\varphi)_\nu - \psi_\nu\, V^2\varphi\right) ds. \end{array}\right\} \tag{3.37}$$

We now calculate

$$D[v - w] = D[v - u + u - w] = D[v - u] + D[u - w] + \\ + 2D[v - u,\, u - w].$$

If $\varphi = v - u$, $\psi = u - w$, then $V^4\varphi = 0$ from (3.29) and $\psi = \psi_\nu = 0$ on Γ from (3.30); putting these values in (3.37), we have

$$D[v - u,\, u - w] = 0.$$

Hence, if u, v, w satisfy (3.29) and (3.30),

$$D[v - w] = D[v - u] + D[u - w]. \tag{3.38}$$

All three terms in this equation are non-negative so that

$$D[v - u] \leqq D[v - w] \quad \text{and} \quad D[u - w] \leqq D[v - w]. \tag{3.39}$$

If the boundary-value problem (3.32), (3.33) possesses a solution $u(x, y)$, then from (3.37) we can obtain a formula for the value $u(x_0, y_0)$ at the point (x_0, y_0) by using the "fundamental" solution

$$\varrho(x, y, x_0, y_0) = r^2 \ln r, \quad \text{where} \quad r = + \sqrt{(x - x_0)^2 + (y - y_0)^2}. \tag{3.40}$$

We have

$$\varrho_r = \frac{\partial \varrho}{\partial r} = r(2 \ln r + 1), \quad V^2\varrho = \varrho_{rr} + \frac{1}{r} \varrho_r = 4(\ln r + 1), \quad (V^2\varrho)_r = \frac{4}{r}, \quad V^4\varrho = 0.$$

We now form $D[\varrho, u] - D[u, \varrho]$ using (3.37), integrate over the region which consists of that part of B outside of a small circle C with radius δ and centre (x_0, y_0) and then let δ tend to zero. Since the contribution from $\int\limits_C u(V^2\varrho)_r\, ds$ becomes $8\pi u(x_0, y_0)$ and all other contributions from integrals around C tend to zero as $\delta \to 0$, we have

$$8\pi u(x_0, y_0) = \iint\limits_B \varrho\, V^4 u\, dx\, dy + \int\limits_\Gamma [u_\nu\, V^2\varrho - u(V^2\varrho)_\nu + \varrho(V^2 u)_\nu - \varrho_\nu\, V^2 u]\, ds. \tag{3.41}$$

§4. Some error distribution principles

We describe here some approximate methods which use various principles for distributing the error as uniformly as possible throughout the domain of the solution. These are generally applicable in all the following chapters, including Ch. VI, but for convenience they are described here with reference to partial differential equations.

4.1. General approximation. "Boundary" and "interior" methods

Problems in differential equations are often attacked by assuming as an approximation to the solution $u(x_1, \ldots, x_n)$ or $y(x)$ of an initial-value or boundary-value problem (1.9), (1.10) an expression of the form

$$u \approx w(x_1, \ldots, x_n, a_1, \ldots, a_p) \tag{4.1}$$

which depends on a number of parameters $a_1, \ldots, a_p$ and is such that, for arbitrary values of the a_ϱ,

(1) the differential equation is already satisfied exactly ("boundary" method),

or (2) the boundary conditions are already satisfied exactly ("interior" method),

or (3) w satisfies neither the differential equation nor the boundary conditions, in which case we speak of a "mixed" method.

One then tries to determine the parameters a_ϱ so that w satisfies

in case (1) (boundary method) the boundary conditions,
in case (2) (interior method) the differential equation,
in case (3) (mixed method) the boundary conditions and the differential equations

as accurately as possible in some sense yet to be defined (this is done in § 4.2).

For ordinary differential equations interior methods are used mostly, for if we did know the general solution of the differential equation, the fitting of the parameters to the boundary conditions would still require the solution of a set of simultaneous (possibly non-linear) equations. For partial differential equations, on the other hand, both boundary and interior methods are used, but in general boundary methods are to be preferred since their use, in so far as integration is involved, requires the evaluation of integrals over the boundary rather than throughout the region. This applies also when collocation is used: the two types of method offer the alternatives of boundary collocation and collocation throughout the region, the former being the more acceptable and less prone to error (see Ia of the following section). Mixed methods (case 3) are used when the differential equations and boundary conditions are rather complicated.

If, in case (1), we insert the approximation w in a boundary condition $V_\mu = 0$, we are left with an error function

$$\left. \begin{aligned} \varepsilon_\mu(x_1, \ldots, x_n, \ a_1, \ldots, a_p) \\ = V_\mu(x_1, \ldots, x_n, \ w, w_1, \ldots, w_n, \ w_{11}, \ldots, w_{n1}, \ldots) \end{aligned} \right\} \qquad (4.2)$$

defined on Γ_μ (the subscripts to w denote precisely the same partial derivatives of w as were defined for u in (1.11), e.g. $w_{12} = \dfrac{\partial^2 w}{\partial x_1 \partial x_2}$). Similarly, for case (2), we have the error function

$$\varepsilon(x_j, a_\varrho) = F(x_j, w, w_1, \ldots, w_{11}, \ldots) \qquad (4.3)$$

defined in the region B and for case (3) we have two error functions ε_μ and ε. The parameters a_ϱ must then be determined so that these error functions ε_μ and ε approximate the zero function as nearly as possible on Γ_μ and in B, respectively. Various principles can be formulated for doing this, which we now describe; for brevity we will usually refer only to the error function ε for case (2), but what is said is naturally applicable also to ε_μ for case (1) and to both ε and ε_μ for case (3).

4.2. Collocation, least-squares, orthogonality method, partition method, relaxation

I a. Pure collocation. The error ε is made to vanish at p points $P_1, \ldots, P_p$ — the "collocation" points. One tries to distribute these p points fairly uniformly over the region B or boundary surfaces Γ_μ. If the co-ordinates of the P_ϱ are $x_{1\varrho}, \ldots, x_{n\varrho}$, then the equations for determining the a_σ read

$$\varepsilon(x_{1\varrho}, x_{2\varrho}, \ldots, x_{n\varrho}, \ a_1, \ldots, a_p) = 0 \qquad (\varrho = 1, \ldots, p). \qquad (4.4)$$

In general, collocation should be used as a boundary method wherever possible, for a reasonable uniformity in the distribution of the collocation points can be achieved more easily on the boundaries Γ_μ than in the region B. (It is, for example, easier to arrange p points uniformly around the circumference of a circle than throughout its interior.) Few investigations have been made into the suitable choice of collocation points.

I b. Collocation with derivatives. The calculation is sometimes simplified if ε is not equated to zero at p points but at q $(<p)$ points, the number of the equations for the a_σ then being made up to p by equating to zero derivatives of ε at several points which may or may not coincide with any of the first q points.

II a. Least-squares method. Here we require that the mean square error shall be as small as possible:

$$J = \int\limits_B \varepsilon^2 \, d\tau + \sum_{\mu=1}^{k} \int\limits_{\Gamma_\mu} \varepsilon_\mu^2 \, dS = \text{minimum.} \qquad (4.5)$$

For a boundary method (case 1) the first integral, over the region B, would not appear and for an interior method (case 2) the second term, involving integrals over the boundary surfaces, would be absent. The requirement that J should be a minimum leads to the well-known necessary conditions

$$\frac{1}{2}\frac{\partial J}{\partial a_\varrho} = \int_B \varepsilon \frac{\partial \varepsilon}{\partial a_\varrho} d\tau + \sum_{\mu=1}^{k} \int_{\Gamma_\mu} \varepsilon_\mu \frac{\partial \varepsilon_\mu}{\partial a_\varrho} dS = 0 \qquad (\varrho = 1, \dots, p), \qquad (4.6)$$

which constitute p equations, non-linear in general, for the determination of the a_σ.

IIb. Least-squares with weighting functions[1]. Let $P(x_1, \dots, x_n)$ and $P_\mu(x_1, \dots, x_n)$ be chosen as positive weighting functions in B and on Γ_μ, respectively. Then instead of (4.5) we require that

$$J = \int_B P \varepsilon^2 d\tau + \sum_{\mu=1}^{k} \int_{\Gamma_\mu} P_\mu \varepsilon_\mu^2 dS = \text{minimum.} \qquad (4.7)$$

III. Orthogonality method. For an interior method (and similarly for a boundary method) we choose p linearly independent functions $g_\varrho(x_1, \dots, x_n)$ and require that ε shall be orthogonal to these functions in the region B, i.e.

$$\int_B \varepsilon g_\varrho d\tau = 0 \qquad (\varrho = 1, \dots, p); \qquad (4.8)$$

the g_ϱ are often chosen to be the first p functions of a complete system of functions in B.

IV. Partition method. The region B (and similarly each bounding surface Γ_μ for a boundary method) is partitioned into p sub-regions $B_1, \dots, B_p$ and we require that the integral of ε over each sub-region be zero, i.e.

$$\int_{B_\varrho} \varepsilon d\tau = 0 \qquad (\varrho = 1, \dots, p). \qquad (4.9)$$

V. Relaxation. We choose a set of values of the a_ϱ and calculate the corresponding values of ε (or ε_μ) at a large number of points in B (or on Γ_μ); these values are called the "residuals" in relaxational parlance. We then note what changes are produced in these residuals by altering, or "relaxing", each a_ϱ in turn by an amount δa_ϱ say. It is frequently quite easy to see from this how much we must alter the individual a_ϱ by in order to make the magnitudes of the residuals as small as possible. This method allows one considerable latitude in the actual calculation; with practice it often leads to the required result more quickly than any other method.

---

[1] Picone, M.: Analisi quantitativa ed esistenziale nei problemi di propagazione. Atti del 1° Congresso dell'Unione Mathematica Italiana 1937.

This list of principles can easily be extended but those already mentioned are, in fact, the ones which have been most used hitherto. While on the subject of error distribution, we should also mention the principle of the smallest maximum error, which is of importance for boundary-value problems with elliptic differential equations (see Ch. V).

Combinations and variations of these principles may also be used; for example, one can choose $q \ (> p)$ points $P_1, \ldots, P_q$ in B and require that

$$ J = \sum_{\sigma=1}^{q} [\varepsilon (P_\sigma, a_1, \ldots, a_p)]^2 = \text{minimum}, $$

so that the a_ϱ are determined by the equations $\dfrac{\partial J}{\partial a_\varrho} = 0 \ (\varrho = 1, \ldots, p)$.
Many further variants can be devised; see, for example, Note 3 in Ch. III, § 5.3.

4.3. The special case of linear boundary conditions

We now assume that the boundary conditions are linear and of the form (1.13), although the differential equation may still be non-linear. We may therefore take the approximate solution w to be a linear expression

$$ w = v_0(x_1, \ldots, x_n) + \sum_{\varrho=1}^{p} a_\varrho v_\varrho (x_1, \ldots, x_n) \tag{4.10} $$

in the parameters a_ϱ with v_0 satisfying the inhomogeneous boundary conditions and the v_ϱ the corresponding homogeneous conditions, i.e.

$$ U_\mu(v_0) = \gamma_\mu, \qquad \begin{pmatrix} \mu = 1, \ldots, k \\ \varrho = 1, \ldots, p \end{pmatrix}. \tag{4.11} $$
$$ U_\mu(v_\varrho) = 0 \qquad\qquad\qquad\qquad \tag{4.12} $$

This ensures that the w given by (4.10) satisfies the prescribed boundary conditions (1.13). The v_ϱ are naturally assumed to be linearly independent.

All the principles described in § 4.2 can, of course, be applied to problems of this particular type, but we single out for mention here a noteworthy special case of the orthogonality method (principle III of § 4.2).

III a. GALERKIN'S method. This is the special case of the orthogonality method in which the functions v_ϱ are used for the g_ϱ; thus the equations (4.8), with ε replaced by its explicit formula (4.3), here read

$$ \int_B F(x_1, \ldots, x_n, w, w_1, \ldots, w_n, \ldots) \, v_\varrho (x_1, \ldots, x_n) \, d\tau = 0 $$
$$ (\varrho = 1, \ldots, p). \tag{4.13} $$

These are GALERKIN'S equations and are easily remembered (cf. § 5.3, Ch. III).

If the differential equation is linear as well as the boundary conditions, then the equations determining the a_ϱ are also linear for any of the principles listed in § 4.2. Further, the least-squares method can be regarded as the special case of the orthogonality method in which $g_\varrho = \partial \varepsilon / \partial a_\varrho$.

As in (1.12), let the linear differential equation be written

$$L[u] = r, \tag{4.14}$$

where $L[u]$ is a linear expression in u and its partial derivatives and $r(x_1, \ldots, x_n)$ is a position function in B. The error function ε is now linear in the a_ϱ:

$$\varepsilon = L[w] - r = \sum_{\varrho=1}^{p} a_\varrho L[v_\varrho] + L[v_0] - r = \sum_{\varrho=1}^{p} a_\varrho V_\varrho + V_0 - r, \tag{4.15}$$

where we have put $V_\varrho(x_1, \ldots, x_n) = L[v_\varrho]$.

I. If collocation is used, p collocation points P_ϱ with co-ordinates $x_{1\varrho}, \ldots, x_{n\varrho}$ are chosen and the a_ϱ found from the p equations

$$\sum_{\nu=1}^{p} v_{\varrho\nu} a_\nu = T_\varrho, \tag{4.16}$$

where $\quad v_{\varrho\nu} = V_\nu(x_{1\varrho}, x_{2\varrho}, \ldots, x_{n\varrho})$

$$\left. \vphantom{\sum} \right\} \quad (\varrho = 1, \ldots, p).$$

and $\quad T_\varrho = r(x_{1\varrho}, \ldots, x_{n\varrho}) - V_0(x_{1\varrho}, \ldots, x_{n\varrho}) \tag{4.17}$

III. Similarly, if the orthogonality method is used, p functions $g_\varrho(x_1, \ldots, x_n)$ are chosen and the a_ϱ determined from the p equations

$$\sum_{\nu=1}^{p} w_{\varrho\nu} a_\nu = t_\varrho, \tag{4.18}$$

where $\quad w_{\varrho\nu} = \int_B g_\varrho V_\nu \, d\tau$

$$\left. \vphantom{\int} \right\} \quad (\varrho = 1, \ldots, p).$$

and $\quad t_\varrho = \int_B g_\varrho (r - V_0) \, d\tau \tag{4.19}$

As mentioned above, these include as special cases GALERKIN'S[1] equations when $g_\varrho = v_\varrho$ (principle III a) and the equations for the least-squares method when $g_\varrho = L[v_\varrho] = V_\varrho$ (principle II).

---

[1] GALERKIN, B. G.: Reihenentwicklungen für einige Fälle des Gleichgewichts von Platten und Balken. Wjestnik Ingenerow Petrograd 1915, H. 10 [Russian]. — HENCKY, H.: Eine wichtige Vereinfachung der Methode von RITZ zur angenäherten Behandlung von Variationsaufgaben. Z. Angew. Math. Mech. 7, 80—81 (1927). — DUNCAN, W. J.: The principles of the Galerkin method. Rep. and Mem. No. 1848 (3694), Aero. Res. Comm. 1938, 24 pp.

IV. If the region B is divided into sub-regions B_ϱ, then for the determination of the a_ϱ by the partition method we have the p equations

$$\sum_{\nu=1}^{p} z_{\varrho\nu} a_\nu = \zeta_\varrho, \qquad (4.20)$$

where
$$z_{\varrho\nu} = \int_{B_\varrho} V_\nu \, d\tau \qquad (\varrho = 1, \dots, p).$$

and
$$\zeta_\varrho = \int_{B_\varrho} (r - V_0) \, d\tau \qquad (4.21)$$

4.4. Combination of iteration and error distribution

Consider the problem (1.9), (1.10) with the differential equation in the form

$$M[u] = P[u],$$

where M and P are functions of the x_j, u and its derivatives. We define an iterative scheme in which a sequence of approximations u_n is generated from an arbitrary function u_0 by the equations

$$\begin{aligned} M[u_{n+1}] &= P[u_n], \\ U_\mu[u_{n+1}] &= \gamma_\mu \quad (\mu = 1, \dots, k) \end{aligned} \quad (n = 0, 1, \dots).$$

u_0 can still be chosen to depend on p parameters $a_1, \dots, a_p$, in which case u_1 will also be a function of these a_ϱ; we can then demand that u_0 and u_1 shall be as close as possible, i.e. that the difference

$$\zeta = \zeta(a_\varrho, x_j) = u_1 - u_0$$

shall be as small as possible. The error distribution principles[1] de-scribed in § 4.2 are at our disposal for satisfying this requirement.

If $P[u]$ is linear in u and contains no derivatives of u, i.e. the differential equation has the form

$$M[u] = p(x_j) + q(x_j) u,$$

then
$$q(x_j)\zeta = p(x_j) + q(x_j) u_1 - M[u_1]$$

and, apart from the factor $q(x_j)$ multiplying ζ, the method coincides with the ordinary error distribution methods of § 4.2 only now they are applied to the first iterate u_1 obtained from u_0. With a suitable choice of M and P we may expect better results from this method than

---

[1] NovoŽILOV, V. V.: On an approximate method of solution of boundary problems for ordinary differential equations. Akad. Nauk SSSR. Prikl. Mat. Meh. **16**, 305—318 (1952) [Russian]. [Reviewed in Math. Rev. **13**, 993 (1952); also in Zbl. Math. **46**, 343.] NovoZILOV gives particular prominence to least-squares and collocation.

if we had applied the error distribution methods directly to the error function formed with u_0. An example using a combination of iteration and collocation will be found in Ch. III, § 4.8, I.

§ 5. Some useful results from functional analysis

In this section we first prove a fundamental general theorem on iteration which can be used effectively in several places in this book. In order to avoid placing needless restrictions on its range of application from the start, it is necessary to employ a general formulation. The representation of functional analysis can be used with advantage for this purpose, for, although it seems rather abstract at first, it proves to be very fruitful. Certain manifolds of functions (or other "elements") are considered and each "element" is regarded as a "point" of a "space". The introduction in § 5.1 is purposely kept broad.

5.1. Some basic concepts of functional analysis with examples

Let S be an abstract space containing elements denoted by $f, f_1, f_2, \ldots$ In this book these elements will be continuous functions but this fact is not used in the general theorem of § 5.2. The symbol $\in$ used in $f \in S$ signifies that f "is contained in" S. For any two elements f_1, f_2 of the space (or, shortly, for $f_1, f_2 \in S$) we define a "distance" as a real number $||f_1 - f_2||$ with the following properties:

1. Symmetry: $||f_1 - f_2|| = ||f_2 - f_1||$.

2. Positive definiteness: $||f_1 - f_2|| \geq 0$ for any $f_1, f_2 \in S$ and $||f_1 - f_2|| = 0$ if and only if $f_1 = f_2$. $\qquad$ (5.1)

3. Triangular inequality: $||f_1 - f_2|| \leq ||f_1 - f_3|| + ||f_3 - f_2||$ for $f_1, f_2, f_3 \in S$. $\qquad$ (5.2)

For the space S^0 of single-valued continuous functions $f(x)$ of the real variable x in the closed interval $\langle a, b \rangle$ we might define the "distance" between f_1 and f_2 as the maximum absolute value of the difference $f_1 - f_2$:

$$||f_1 - f_2|| = \max_{a \leq x \leq b} |f_1(x) - f_2(x)|; \qquad (5.3)$$

it can be seen immediately that this definition possesses the three properties listed above. For the same space we could define a more general distance by

$$||f_1 - f_2|| = \max_{a \leq x \leq b} \frac{|f_1(x) - f_2(x)|}{W(x)}, \qquad (5.4)$$

where $W(x)$ is a fixed, positive, continuous function in $\langle a, b \rangle$, e.g. e^x.

For the applications in this book there is usually a commutative addition of elements, and with respect to this addition S forms an additive group. S contains a zero element Θ such that, for $f_1 \in S$, $\Theta + f_1 = f_1$ and to each f_1 there corresponds an "inverse element" $-f_1$ with $f_1 + (-f_1) = \Theta$. (In the above example $S = S^0$, the function $f(x) \equiv 0$ is the zero element.) For such function spaces one can work with the "norm" $\|f\|$ of an element f instead of the "distance"; the norm is the distance from the zero element:

$$\|f\| = \|f - \Theta\|. \tag{5.5}$$

Further, we need the idea of the completeness of a space S or sub-space F of S. A sub-space F of S (which may be improper, i.e. coincident with S) is termed complete when to every sequence $f_1, f_2, \ldots$ of elements in F such that

$$\lim_{m, n \to \infty} \|f_m - f_n\| = 0 \tag{5.6}$$

there is a limit element f such that

$$\lim_{n \to \infty} \|f - f_n\| = 0 \tag{5.7}$$

which also belongs to F.

It is an important fact that the space S^0 of continuous functions in $\langle a, b \rangle$, considered as an example above, is complete[1] under the distance definition (5.4), as also is the sub-space of these functions for which $u_1(x) \leq f(x) \leq u_2(x)$, where $u_1(x)$ and $u_2(x)$ are two fixed, continuous functions in $\langle a, b \rangle$ with $u_1 \leq u_2$. This can be seen from (5.4) and (5.6), which imply that

$$|f_m - f_n| \leq \varrho \|f_m - f_n\|,$$

where $\varrho = \max\limits_{a \leq x \leq b} W(x)$, and hence the sequence $f_n(x)$ is uniformly convergent; as is well known, the limit function of a uniformly convergent sequence of continuous functions in a closed interval is also continuous and hence belongs to the space considered, which is therefore complete.

We now define an "operator" (or "transformation") T which associates a unique element Tf of a sub-space F^* of S with each element f of a sub-space F (the inverse mapping of F^* onto F need not be unique, nor need F^* be different from F).

In the space S^0 we can construct a considerable variety of operators, for example,

$$Tf = \int_a^x f(\xi)\, d\xi, \quad Tf = f(x)\, W(x), \quad Tf = \sin(f(x));$$

on the other hand, $Tf = df/dx$ need not be an admissible operator, for, if the sub-space F is chosen appropriately, differentiation can lead to functions not belonging to S^0.

---

[1] BANACH, ST.: Théorie des opérations linéaires. Warsaw 1932. Recent impression New York 1949, p. 11.

An operator T is said to be Lipschitz-bounded in F with reference to the chosen distance definition if there is a "Lipschitz constant" K such that

$$\| T f_1 - T f_2 \| \leq K \| f_1 - f_2 \| \tag{5.8}$$

for all $f_1, f_2 \in F$.

To give an example, let us consider again the space S^0 of continuous functions in $\langle a, b \rangle$ with the distance definition (5.4) and define an operator T by

$$T f(x) = \int_a^x G(x, \xi) f(\xi) \, d\xi,$$

where $G(x, \xi)$ is a given, continuous, bounded function with $|G(x, \xi)| \leq C$ for $a \leq x, \xi \leq b$. This operator is Lipschitz-bounded under the distance definition (5.4) since

$$\| T f_1 - T f_2 \| = \max_{a \leq x \leq b} \frac{\left| \int_a^x G(x, \xi) [f_1(\xi) - f_2(\xi)] \, d\xi \right|}{W(x)}$$

$$\leq \| f_1 - f_2 \| \max_{a \leq x \leq b} \frac{\int_a^x |G(x, \xi)| W(\xi) \, d\xi}{W(x)}.$$

If we put $\gamma = \max\limits_{a \leq x \leq b} \dfrac{\int_a^x W(\xi) \, d\xi}{W(x)}$, then γC can be used as the Lipschitz constant K.

5.2. The general theorem on iterative processes

Suppose that we require the solutions of the equation

$$f = T f \tag{5.9}$$

(or the "fixed points" or fixed-point elements of the operator T). Let us set up the iterative process[1]

$$u_{n+1} = T u_n \quad (n = 0, 1, 2, \ldots) \tag{5.10}$$

whereby we proceed from a function $u_0 \in F$ and form $u_1 = T u_0$, $u_2 = T u_1$ and so on up to u_{n+1} as long as u_n lies in F. Then, if F is complete and T Lipschitz-bounded in F, we have for $0 < n < m$

$$\left. \begin{aligned} \| u_m - u_n \| = \| T u_{m-1} - T u_{n-1} \| \leq K \| u_{m-1} - u_{n-1} \| \leq \cdots \\ \leq K^r \| u_{m-r} - u_{n-r} \| \quad (\text{for } 0 \leq r \leq m, n), \end{aligned} \right\} \tag{5.11}$$

---

[1] KANTOROVICH, L.: The method of successive approximations for functional equations. Acta Math. **71**, 63—97 (1939). — WEISSINGER, J.: Zur Theorie und Anwendung des Iterationsverfahrens. Math. Nachr. **8**, 193—212 (1952).

and from (5.2), (5.11), it follows that

$$\|u_m - u_n\| \leq \sum_{s=n}^{m-1} \|u_{s+1} - u_s\|$$

$$\leq \sum_{t=1}^{m-n} K^t \|u_n - u_{n-1}\| \leq K^{n-1} \sum_{t=1}^{m-n} K^t \|u_1 - u_0\|.$$

We now make the crucial assumption that $K < 1$, so that

$$\sum_{t=1}^{m-n} K^t \leq \frac{K}{1-K} \quad \text{and} \quad \|u_m - u_n\| \leq \frac{K^n}{1-K} \|u_1 - u_0\|. \qquad (5.12)$$

We observe now that an important sub-space of S is the "sphere" Σ containing all the elements h in S with

$$\|h - u_1\| \leq \frac{K}{1-K} \|u_1 - u_0\|, \qquad (5.13)$$

for it has the property that, if it is contained in the sub-space F, then the iteration process can be carried on indefinitely, i.e. no u_n falls outside of F; this follows from (5.12) with $n = 1$:

$$\|u_m - u_1\| \leq \frac{K}{1-K} \|u_1 - u_0\|, \qquad (5.14)$$

for this implies that u_m lies in the sphere Σ (and hence also in F) for $m = 1, 2, 3, \ldots$. It follows further from (5.12) that

$$\lim_{m,\, n \to \infty} \|u_m - u_n\| = 0$$

since $K < 1$, and hence, on account of the completeness of F, there exists a limit element u in F such that

$$\lim_{n \to \infty} \|u - u_n\| = 0. \qquad (5.15)$$

Since u lies in F, we can form Tu; then from (5.2), (5.8) we have

$$\|Tu - u\| \leq \|Tu - Tu_n\| + \|Tu_n - u\| \leq K\|u - u_n\| + \|u_{n+1} - u\|$$

for all n. Now according to (5.15) the right-hand side tends to zero as $n \to \infty$; hence

$$\|Tu - u\| = 0,$$

so that, from the "distance" property (5.1), $Tu = u$.

Thus the limit element u is a solution of (5.9), and the existence of a solution is demonstrated.

Suppose now that there exists another solution v of (5.9), so that $v = Tv$. Then from (5.8) we have

$$\|u - v\| = \|Tu - Tv\| \leq K\|u - v\|,$$

and since $K < 1$, we must have $\|u - v\| = 0$, i.e. $u = v$. Thus (5.9) can have only one solution in F, and the uniqueness of u in F is also demonstrated.

Finally, it follows from (5.2), (5.14) that

$$\|u - u_1\| \leq \|u - u_m\| + \|u_m - u_1\| \leq \|u - u_m\| + \frac{K}{1 - K} \|u_1 - u_0\|,$$

and since, according to (5.15), the term $\|u - u_m\| \to 0$ as $m \to \infty$, we have

$$\|u - u_1\| \leq \frac{K}{1 - K} \|u_1 - u_0\|, \tag{5.16}$$

i.e. u itself lies in the sphere Σ. This formula provides an error estimate.

Summary: *Let there be an equation $f = Tf$ (5.9) for an element f of a space S. If we can define a distance satisfying the three conditions of § 5.1 and can choose a sub-space F so that*

1. *Tf is defined uniquely for all f in F,*

2. *F is complete under the chosen distance definition,*

3. *T satisfies a Lipschitz-condition in F with $K < 1$,*

4. *besides the chosen u_0, F contains the whole sphere Σ defined by (5.13),*

then the iteration (5.10) may be continued indefinitely and the sequence u_n converges in the sense of (5.15) to an element u in Σ which is the unique solution of the given equation (5.9) in the sub-space F. At the same time, (5.16) gives an estimate for the error in the approximation u_1[1].

In many cases it is obvious that all u_n lie in F and the condition 4 can be omitted. Closer error estimates can be obtained in several cases by using pseudo-metric spaces[2]; the distance $\|f - g\|$ between two elements f, g of such a space is an element of a semi-ordered space and thus is not restricted to being a real number, cf. Ch. III, § 4.8.

5.3. The operator T applied to boundary-value problems

Consider the boundary-value problem with the differential equation

$$L[u] = \varphi(x_1, \ldots, x_n, u) \quad \text{in } B \tag{5.17}$$

---

[1] Another general method based on functional analysis has been investigated by P. C. ROSENBLOOM: The method of steepest descent. Proc. Symp. Appl. Math. VI, New York-Toronto-London 1956, pp. 127—176.

[2] KUREPA, L.: Tableaux ramifiés d'ensembles, espaces pseudo-distanciés. C. R. **198**, 1563—1565 (1934). — SCHRÖDER, J.: Nichtlineare Majoranten beim Verfahren der schrittweisen Näherung. Arch. Math. **7**, 471—484 (1956). — Neue Fehlerabschätzungen für verschiedene Iterationsverfahren. Z. Angew. Math. Mech. **36**, 168—181 (1956). — Über das Newtonsche Verfahren. Arch. Rational Mechanics **1**, 154—180 (1957). — Das Iterationsverfahren bei allgemeinerem Abstandsbegriff. Math. Z. **66**, 111—116 (1956).

and the linear boundary conditions as in § 1.3

$$U_\mu[u] = \gamma_\mu \quad \text{on } \Gamma_\mu \qquad (\mu = 1, \ldots, k), \tag{5.18}$$

where $L[u]$ and $U_\mu[u]$ are linear homogeneous differential expressions in a function $u(x_1, \ldots, x_n)$ and φ is a given function with a continuous derivative with respect to u and which we will assume to be continuous with respect to its other arguments. We can proceed from an arbitrarily chosen function u_0 according to the iterative formulae

$$\left. \begin{array}{l} L[u_{q+1}] = \varphi(x_j, u_q) \quad \text{in } B \\ U_\mu[u_{q+1}] = \gamma_\mu \quad \text{on } \Gamma_\mu \end{array} \right\} \qquad (q = 0, 1, 2, \ldots). \tag{5.19}$$

Thus u_{q+1} is determined from u_q by solution of a linear boundary-value problem.

Now let us assume that for any continuous functions r, γ_μ the problem

$$L[\psi] = r \quad \text{in } B, \qquad U_\mu[\psi] = \gamma_\mu \quad \text{on } \Gamma_\mu \tag{5.20}$$

always possesses a unique solution ψ; this will be the case, for example, when a GREEN's function $G(x_j, \xi_j)$ exists (see § 3.4). Then the iteration cycle can be repeated indefinitely provided that $\varphi(x_j, u_q)$ is defined for all $u = u_q$.

With the aid of a function $\tilde{u}$ satisfying the inhomogeneous boundary conditions, and also possessing continuous derivatives of as high an order as is necessary for the formation of $L[\tilde{u}]$, we can transform the problem into one with homogeneous boundary conditions (this transformation is used only in establishing an error estimate and is not needed in carrying out the computation). Corresponding to a function f we can define a function g by

$$L[g] = \varphi(x_j, \tilde{u} + f) - L[\tilde{u}] \quad \text{in } B, \qquad U_\mu[g] = 0 \quad \text{on } \Gamma_\mu; \tag{5.21}$$

we write this correspondence as $g = Tf$, thus defining an operator T. If now we put $u = \tilde{u} + v$ and $u_q = \tilde{u} + v_q$, then

$$v_{q+1} = Tv_q \quad (q = 0, 1, \ldots) \quad \text{and} \quad v = Tv, \tag{5.22}$$

and since

$$u_q - u = v_q - v, \tag{5.23}$$

the error at the q-th stage is unaltered by the transformation from u to v.

To determine the Lipschitz constant K of the transformation T, we form

$$\left. \begin{array}{l} L[g_1 - g_2] = \varphi(x_j, \tilde{u} + f_1) - \varphi(x_j, \tilde{u} + f_2), \\ U_\mu[g_1 - g_2] = 0, \end{array} \right\} \tag{5.24}$$

where $g_1 = Tf_1$ and $g_2 = Tf_2$. We then choose a domain D of the $(x_1, \ldots, x_n, u)$ space which is convex with respect to u and contains a solution u and the iterates u_q to it and assume that φ satisfies the Lipschitz condition

$$|\varphi(x_j, z) - \varphi(x_j, z^*)| \leq N(x_j)|z - z^*| \tag{5.25}$$

in D. Since φ possesses a continuous derivative with respect to u, we could put $N = \max\limits_{u} \left|\dfrac{\partial \varphi}{\partial u}\right|$. From (5.24) and (5.25) we have

$$|L[H]| \leq N|h| \quad \text{in } B, \quad U_\mu[H] = 0 \quad \text{on } \Gamma_\mu, \tag{5.26}$$

where $h = f_1 - f_2$ and $H = Tf_1 - Tf_2$. From here we can proceed in two ways.

1. Using the GREEN's function. Let $W(x_i)$ be a positive (or possibly non-negative) function in B and, as in (5.4), let the norm of f be defined by

$$\|f\| = \underset{\text{in } B}{\text{upper lim}} \left|\frac{f}{W}\right|. \tag{5.27}$$

Now the inequality in (5.26) can be replaced by the equation

$$L[H] = \vartheta N|h|, \quad \text{where} \quad |\vartheta| \leq 1,$$

which can be solved for H by means of the GREEN's function, as in (3.24):

$$H = \int\limits_{B} \vartheta(\xi_j)\, G(x_j, \xi_j)\, N(\xi_j)\, |h(\xi_j)|\, d\xi_j;$$

hence

$$|H| \leq \int\limits_{B} |G(x_j, \xi_j)\, N(\xi_j)\, h(\xi_j)|\, d\xi_j \leq \|h\| \int\limits_{B} |G(x_j, \xi_j)|\, N(\xi_j)\, W(\xi_j)\, d\xi_j, \tag{5.28}$$

and we can use

$$K = \underset{\text{in } B}{\text{upper lim}} \frac{\int\limits_{B} |G(x_j, \xi_j)|\, N(\xi_j)\, W(\xi_j)\, d\xi_j}{W(x_j)} \tag{5.29}$$

as Lipschitz constant.

If the boundary-value problem is linear, so that the differential equation (5.17) can be put in the form

$$L[u] = p(x_j)\, u + q(x_j), \tag{5.30}$$

then (5.25) holds with $N(x_j) = |p(x_j)|$.

A simple, if sometimes crude, error estimate can be given when a non-negative GREEN's function exists in B and the eigenvalue problem

$$L[z] = \lambda z \quad \text{in } B, \quad U_\mu[z] = 0 \quad \text{on } \Gamma_\mu \tag{5.31}$$

possesses a non-negative eigenfunction $z(x_j)$ in B corresponding to the eigenvalue $\lambda = \lambda_z$. We can then choose $W(x_j) = z(x_j)$, so that, since

$$\lambda_z \int_B G(x_j, \xi_j)\, z(\xi_j)\, d\xi_j = z(x_j) \tag{5.32}$$

from (3.24), (5.29) simplifies to

$$K = \frac{\max\limits_{\text{in } B} N(x_j)}{\lambda_z}. \tag{5.33}$$

2. Using a monotonic property. Here we make use of the concept of a boundary-value problem of monotonic type, thus limiting ourselves to real values (see §§ 5.4, 5.6). For our present purposes we modify the definition slightly and say that the problem (5.17), (5.18) has a monotonic character when

$$L[v] \le L[w] \quad \text{in } B, \quad U_\mu[v] = U_\mu[w] = 0 \quad \text{on } \Gamma_\mu \tag{5.34}$$

implies that $v \le w$, v and w being two functions with continuous partial derivatives of as high an order as is required to form L and U_μ.

Now let $z(x_j)$ be a function such that

$$L[z] = A(x_j) \ge \alpha > 0 \quad \text{in } B, \quad U_\mu[z] = 0 \quad \text{on } \Gamma_\mu, \tag{5.35}$$

where α is constant. Then under our assumptions it follows from

$$|L[\psi]| \le D = \text{const} \quad \text{in } B, \quad U_\mu[\psi] = 0 \quad \text{on } \Gamma_\mu \tag{5.36}$$

that $|\psi|$ is majorized by zD/α:

$$|\psi| \le \frac{zD}{\alpha}; \tag{5.37}$$

for we have

$$\left.\begin{aligned}
L\left[\frac{-zD}{\alpha}\right] &\le -D \le L[\psi] \le D \le L\left[\frac{zD}{\alpha}\right], \\
U_\mu\left[\pm\frac{zD}{\alpha}\right] &= U_\mu[\psi] = 0,
\end{aligned}\right\} \tag{5.38}$$

and hence, from the monotonic property,

$$-\frac{zD}{\alpha} \le \psi \le \frac{zD}{\alpha}. \tag{5.39}$$

We now apply (5.36), (5.37) to (5.26); this gives

$$|H| \le \frac{|Nh|_{\max} z}{\alpha}. \tag{5.40}$$

Using the same norm as defined in (5.27) with a similar function $W(x_j)$, we obtain

$$\|H\| \le \frac{|Nz|_{\max}}{\alpha} \|h\|; \tag{5.41}$$

we can therefore use

$$K = \frac{|Nz|_{\max}}{\alpha} \qquad (5.42)$$

as Lipschitz constant.

Detailed numerical examples will be found in Ch. III, § 4.8. The methods of § 5.2 are also applied to a non-linear integral equation in Example III of Ch. VI, § 1.5.

5.4. Problems of monotonic type[1]

Let R be a real space of elements $u, v, \ldots, f, g, \ldots$. These elements may be real numbers, vectors with real components or real-valued functions of real variables and the signs $\leq, <, >, \geq$ shall have their usual significance; applied to vectors, for example, the sign $\leq$ signifies that the inequality holds for all components. (In the terminology of the theory of abstract spaces, R is a semi-ordered space.)

Let there be given an operator T which associates with each element of a proper or improper sub-space M of R a unique element of an image space M^* which is also a sub-space of R. The operator T need not be linear. Now let f be a given element of R; then we ask for the solution u (an element of M) of the equation $Tu = f$. We say this problem is "of monotonic type" (or T is a monotonic operator) when $Tv \leq Tw$ implies that $v \leq w$ for arbitrary elements v, w in M. More accurately, we should say that the problem is "written as a problem of monotonic type", for it can certainly happen that a particular problem of analysis can be formulated in different ways such that one formulation is of monotonic type while another is not.

If we assume the existence of a solution u, there is the possibility of "bracketing" it when the problem is of monotonic type: if v_1 and v_2 are two approximate solutions in M such that

$$T v_1 \leq f \leq T v_2, \qquad (5.43)$$

then

$$v_1 \leq u \leq v_2. \qquad (5.44)$$

There is also the possibility of deducing an error estimate for results obtained by the relaxation method. With the relaxation technique we define the "defect" or, in the customary terminology of relaxation, the "residual", $d = d(v) = Tv - f$ of an approximation v and try to make it approximate the zero element as closely as possible by making alterations in v, usually in the nature of small corrections. If we find that $d \geq 0$, we know that $v \geq u$ and similarly $v \leq u$ follows from $d \leq 0$.

---

[1] COLLATZ, L.: Aufgaben monotoner Art. Arch. Math. **4**, 366—376 (1952).

The monotonic method requires that we know that a solution u exists; this knowledge can be acquired in several ways:

1. It is often possible to make sure that a solution exists by starting the iteration procedure of § 5.2, possibly with quite a crude first iterate, and appealing to the general theorem of § 5.2 (cf. Ch. III, § 4.8, Ex. II).

2. With the aid of topology and functional analysis, generalizations of BROUWER'S fixed-point theorem can be used to establish results concerning the existence of a fixed point for a wide range of problems; for example, non-linear boundary-value problems for elliptic differential equations have been considered by SCHAUDER and LERAY[1] and non-linear integral equations by ROTHE[2], among others.

3. For certain classes of linear and non-linear boundary-value problems the existence of a solution is assured by special theorems.

5.5. Application to systems of linear equations of monotonic type

Consider a system of real linear equations, which may be written in matrix form

$$A x = r, \tag{5.45}$$

where $A = (a_{jk})$ is the matrix of coefficients, $x = (x_j)$ the column matrix (or vector) of the unknowns and $r = (r_j)$ the column matrix of the right-hand sides. A corresponds to the operator T. The problem of determining x is of monotonic type when $A z \geq 0$ implies that $z \geq 0$ for any real vector z. Such a matrix A is called "monotonic" or "of monotonic type", as also is the system of equations $A x = r$. By reversing the sign of z we see also that $A z \leq 0$ implies $z \leq 0$ if A is monotonic; therefore $A z = 0$ implies that $z = 0$, and the determinant of A cannot be zero. For a monotonic matrix A, $A y \leq r \leq A w$ implies that $y \leq x \leq w$, for this is equivalent to: $A(y - x) \leq 0 \leq A(w - x)$ implies that $y - x \leq 0 \leq w - x$; in particular, we can conclude from $|A x| \leq A y$ that $|x| \leq y$ (corresponding to the definition in § 5.4, $|x| \leq y$ signifies that $|x_j| \leq y_j$ for $j = 1, \ldots, n$).

A necessary and sufficient condition for a matrix A to be monotonic is that all elements of the inverse matrix A^{-1} be non-negative[3]; for practical work, however, we need simpler criteria which do not involve

---

[1] SCHAUDER, J.: Zur Theorie stetiger Abbildungen in Funktionalräumen. Math. Z. **26**, 47—65 (1927) and notes thereto 417—431. — LERAY, J., and J. SCHAUDER: Topologie et équations fonctionelles. Ann. Sci. École norm. sup. **51**, 45—78 (1934).

[2] ROTHE, E.: Zur Theorie der topologischen Ordnung und der Vektorfelder in Banachschen Räumen. Comp. Math. **5**, 117—197 (1938).

[3] Another criterion which includes our Theorem 1 is given by A. OSTROWSKI: Über die Determinanten mit überwiegender Hauptdiagonale. Comm. Math. Helv. **10**, 69—96 (1937).

determinants and can be applied easily, even if they are only sufficient conditions. Such a criterion[1,2] is furnished by:

Theorem 1. *If an $n \times n$ real matrix A is such that*

1. *$a_{jk} \leq 0$ for $j \neq k$,*
2. *A does not "decompose",*
3. *there exist non-zero vectors y and r such that $y > 0$, $r \geq 0$ and $Ay = r$,*

then A is a monotonic matrix.

We say that a matrix $A = (a_{jk})$ "decomposes" when for some integer m in $1 \leq m \leq n - 1$ we can separate the integers $1, \ldots, n$ into two groups $\varrho_1, \ldots, \varrho_m$; $\sigma_1, \ldots, \sigma_{n-m}$ in such a way that $a_{\varrho_\nu \sigma_\mu} = 0$ for $\nu = 1, \ldots, m$; $\mu = 1, \ldots, n - m$.

Proof. For $x \geq 0$ to follow from $Ax \geq 0$ we need only show that the two assumptions

(a) z is a vector with at least one negative component, say $z_q < 0$,
(b) $Az \geq 0$

are contradictory.

From assumption (b) it follows that $A[(1 - \lambda) y + \lambda z] \geq 0$ if the real parameter λ lies in the interval $0 \leq \lambda \leq 1$, y being the vector of condition 3. Now under assumption (a) we can choose $\lambda = \varLambda$ in $0 < \lambda < 1$ so that the vector $w = (1 - \varLambda) y + \varLambda z$ has no negative components, but has at least one zero component and is not the zero vector. For since $y_q > 0$ and $z_q < 0$, there is a $\lambda = \lambda_q$ in $\langle 0, 1 \rangle$ such that $(1 - \lambda_q) y_q + \lambda_q z_q = 0$, and similarly for any other negative component of z, say z_r, there is a λ_r in $\langle 0, 1 \rangle$ such that $(1 - \lambda_r) y_r + \lambda_r z_r = 0$; we then let $\varLambda$ be the smallest of these values λ_r (there are at most n of them); w is not the zero vector since with $0 < \varLambda < 1$, $Ay \geq 0$ and $\dfrac{1 - \varLambda}{\varLambda} y = -z$ imply that $Az \leq 0$ and with assumption (b) this implies that $Az = Ay = 0$, in contradiction to condition 3 that r is a non-zero vector.

We now use conditions 1 and 2 to contradict the deduction from (b) that $Aw \geq 0$. Let $w_{\varrho_1}, \ldots, w_{\varrho_m}$ and $w_{\sigma_1}, \ldots, w_{\sigma_{n-m}}$ be the zero and non-zero (and therefore positive) components of w, respectively. Now

---

[1] A similar criterion is demonstrated geometrically by G. SCHULZ: Interpolationsverfahren zur numerischen Quadratur gewöhnlicher Differentialgleichungen. Z. Angew. Math. Mech. **12**, 53 (1932). The conditions which SCHULZ obtains are: $\det A \neq 0$, $a_{jj} > 0$, $a_{ij} < 0$ for $j \neq k$, and our condition 3. with $r > 0$.

[2] That the criterion is not necessary, even in the sense that at least one rearrangement (by row or column interchanges) of a monotonic matrix must satisfy it, is easily shown by simple examples such as the monotonic matrix

$$\begin{pmatrix} 1 & -2 & 8 \\ -2 & 1 & -2 \\ 8 & -2 & 1 \end{pmatrix}.$$

$1 \leq m \leq n - 1$, so that, since A does not decompose (condition 2), there is at least one element $a_{\varrho_\nu \sigma_\mu}$ which is non-zero and therefore negative (condition 1). Then calculating the ϱ_ν-th component of $A w$, we obtain

$$(A w)_{\varrho_\nu} = \sum_{k=1}^{n} a_{\varrho_\nu k} w_k = \sum_{\mu=1}^{n-m} a_{\varrho_\nu \sigma_\mu} w_{\sigma_\mu} < 0, \tag{5.46}$$

so that at least one component of $A w$ is negative.

If the vector y of condition 3 satisfies the stronger condition in which $r > 0$, condition 2 may be relaxed. For in this case $A\big((1 - \lambda) y + \lambda z\big) > 0$ if $0 \leq \lambda < 1$, so that $A w > 0$; ϱ_ν is therefore chosen arbitrarily this time and we cannot say that there is at least one negative $a_{\varrho_\nu \sigma_\mu}$; however, since $a_{\varrho_\nu \sigma_\mu} \leq 0$, (5.46) holds with $\leq$ instead of $<$ and we still have a contradiction, this time with $A w > 0$.

As the special case $y_j = 1$, Theorem 1 includes a criterion which is used very frequently in applications: condition 3 becomes the weak "row-sum" criterion[1]

3*.

$$\sum_{k=1}^{n} a_{jk} \begin{cases} \geq 0 & \text{for all } j \\ > 0 & \text{for at least one } j = j_0; \end{cases}$$

consequently conditions 1, 2, 3* are sufficient for A to be monotonic. The case where $r > 0$ in condition 3 corresponds to the ordinary row-sum criterion as defined below. We may therefore state the less general

Theorem 2. *If the coefficients a_{jk} of an $n \times n$ matrix A satisfy the conditions*

1. *sign distribution*[2]: $a_{jj} > 0$, $a_{jk} \leq 0$ for $j \neq k$,

2a. *the "weak row-sum criterion"*:

$$\sum_{k=1}^{n} a_{jk} \begin{cases} \geq 0 & \text{for } j = 1, 2, \ldots, n, \\ > 0 & \text{for at least one } j = j_0, \end{cases} \tag{5.47}$$

and 2b. *the non-decomposition of A,*

or, instead of 2a. and 2b., the stronger condition

2c. *the "ordinary row-sum criterion"*:

$$\sum_{k=1}^{n} a_{jk} > 0 \quad \text{for } j = 1, \ldots, n, \tag{5.48}$$

then A is monotonic and in particular $\det A \neq 0$.

---

[1] Collatz, L.: Aufgaben monotoner Art. Arch. Math. **3**, 373 (1952).

[2] It is convenient to include $a_{jj} > 0$ as a condition of the theorem although it can be deduced from the remaining conditions and could therefore have been omitted.

The solution x of the system of real equations $Ax = r$ can then be bracketed by $v_1 \leq x \leq v_2$ provided that two approximations v_1 and v_2 can be found such that $A v_1 \leq r \leq A v_2$.

Systems of equations of monotonic type occur frequently in the solution of boundary-value problems by finite differences. As these systems of equations are often solved by iteration we conclude this section with mention of a theorem which will be used later in this context.

A series of iteration procedures can be defined by writing the given matrix A as the sum of two matrices B and C:

$$A = B + C;$$

then starting from an arbitrarily chosen vector x_0 we determine a sequence of vectors x_k from the iterative formula

$$B x_{k+1} + C x_k = r \qquad (k = 0, 1, \ldots). \tag{5.49}$$

We assume that $\det A \neq 0$ and $\det B \neq 0$. If $B = (b_{jk})$ is chosen as the diagonal matrix with

$$b_{jk} = \begin{cases} a_{jk} & \text{for } j = k \\ 0 & \text{for } j \neq k, \end{cases} \qquad c_{jk} = \begin{cases} 0 & \text{for } j = k \\ a_{jk} & \text{for } j \neq k, \end{cases} \tag{5.50}$$

the components of the current iterate are calculated directly from the preceding iterate; this iterative scheme is usually called the "total-step process". If B is chosen to be triangular with

$$b_{jk} = \begin{cases} a_{jk} & \text{for } j \geq k \\ 0 & \text{for } j < k, \end{cases} \qquad c_{jk} = \begin{cases} 0 & \text{for } j \leq k \\ a_{jk} & \text{for } j < k, \end{cases} \tag{5.51}$$

then the components of the current iterate are calculated successively; this is usually called the "single-step process".

The following theorem[1] gives sufficient conditions for the applicability of these processes.

Theorem 3. *If the matrix A of the system of equations (5.45) satisfies the ordinary row-sum criterion*

$$\sum_{\substack{k=1 \\ k \neq j}}^{n} |a_{jk}| < a_{jj} \qquad (j = 1, \ldots, n),$$

or if it does not decompose and satisfies the weak row-sum criterion

$$\sum_{\substack{k=1 \\ k \neq j}}^{n} |a_{jk}| \begin{cases} \leq a_{jj} & \text{for } j = 1, \ldots, n \\ < a_{jj} & \text{for at least one } j = j_0, \end{cases}$$

then $\det A \neq 0$ and both the total-step and single-step iterative processes will converge when applied to the equations (5.45).

---

[1] Cf., for example, Math. Z. **53**, 149—161 (1950).

Several matrices occur in later chapters for which we show that the conditions of Theorem 2 are satisfied. We therefore observe here that those conditions are included in the conditions of Theorem 3 as the special case for which the sign distribution of condition 1. of Theorem 2 holds and that the above-mentioned iterative processes therefore converge for these matrices.

5.6. Non-linear boundary-value problems

As in (5.17), (5.18), consider the linear or non-linear differential equation

$$L[u] + F(x_j, u) = 0 \quad \text{in } B \tag{5.52}$$

for a function $u(x_1, \ldots, x_n)$ of n real variables $x_1, \ldots, x_n$ subject to the linear boundary conditions

$$U_\mu[u] = \gamma_\mu \quad \text{on } \Gamma_\mu \qquad (\mu = 1, \ldots, k). \tag{5.53}$$

Here $F(x_1, \ldots, x_n, u)$ is a given function in B possessing a continuous partial derivative with respect to u. Let us define an operator T (not the same operator as in § 5.3) by

$$T v = L[v] + F(x_j, v) \tag{5.54}$$

with a domain of definition D restricted to the domain of functions v which satisfy the boundary conditions $U_\mu[v] = \gamma_\mu$ and possess partial derivatives of order sufficiently high for the formation of the differential expressions $U_\mu[v]$ and $L[v]$.

Then, for any two functions v and w in D, we have

$$T v - T w = L[\varepsilon] + \varepsilon A(x_j), \tag{5.55}$$

where $\varepsilon = v - w$ and $A(x_j)$ is obtained from TAYLOR's theorem

$$F(x_j, v) - F(x_j, w) = \varepsilon \left(\frac{\partial F}{\partial u} \right)_{(x_j, u = \widetilde{w})} = \varepsilon A(x_j), \tag{5.56}$$

$\widetilde{w}(x_j)$ being a point intermediate between v and w; for fixed v and w, $A(x_j)$ is a function only of position. Now let H be a domain of the $(x_1, \ldots, x_n, u)$ space which is convex with respect to u and contains v and w (and hence also $\widetilde{w}$); the assumption that A is non negative in H can be fruitful in several applications. One will frequently choose for H a domain which is known to contain a solution u of (5.52), (5.53) and also an approximation to it, say v; then, with $w = u$, ε is the error $\zeta = v - u$ in the approximation v. If, on the basis of further special properties of the problem (5.52), (5.53), we can show that in H

$$L[\varepsilon] + \varepsilon A(x_j) \geqq 0 \quad \text{in } B, \tag{5.57}$$

$$U_\mu[\varepsilon] = 0 \quad \text{on } \Gamma_\mu, \tag{5.58}$$

with non-negative $A(x_j)$, implies that $\varepsilon \geqq 0$, then the problem (5.52), (5.53) is of monotonic type in H. The non-linear problem is accordingly reduced to the discussion of a linear problem with homogeneous boundary conditions, cf. Ch. III § 4.9 and Ch. V § 4.1.

Chapter II
Initial-value problems in ordinary differential equations

§ 1. Introduction

First of all, in §§ 1.1—1.3, we discuss some quite general points mainly concerning accuracy. In §§ 1.4—1.7 some crude methods are described which would be used only if a rough idea of the solution over a fairly short range were wanted quickly.

1.1. The necessity for numerical methods

Even with quite simple differential equations it can happen that their solutions are not expressible in closed form and that a numerical approach is the most convenient way of dealing with the problem. For example, the differential equation

$$\frac{dy}{dx} = x + x^2 + y^2$$

does not possess a closed solution in terms of elementary functions, although its solution can be expressed in a complicated way in terms of little-tabulated Bessel functions of fractional order. Of course, when the coefficients appearing in the differential equation are empirical functions (such as the air resistance in external ballistics problems), some kind of approximate method, whether graphical, numerical or involving the use of an analogue machine, is the only way of obtaining a solution at all.

General literature for this chapter
(in chronological order)

RUNGE, C., and H. KÖNIG: Numerisches Rechnen. Berlin 1924.

LINDOW, M.: Numerische Infinitesimalrechnung. Berlin and Bonn 1928.

SCARBOROUGH, JAMES B.: Numerical mathematical analysis. 416 pp., in particular Ch.s XI—XIII. Baltimore and London 1930.

LEVY, H., and E. A. BAGGOT: Numerical studies in differential equations, Vol. 1. London 1934.

KAMKE, E.: Differentialgleichungen, Lösungsmethode und Lösungen, Vol. 1, Ch. A § 8: Numerische, graphische und instrumentelle Integrationsverfahren, 3rd ed. 666 pp. Leipzig 1944.

SANDEN, H. V.: Praxis der Differentialgleichungen, 3rd ed. Berlin 1945.

SCHULZ, GÜNTHER: Formelsammlung zur praktischen Mathematik. Sammlung Göschen, Vol. 1110. Berlin and Leipzig 1945.

WILLERS, FR. A.: Methoden der praktischen Analysis, 2nd ed. 410 pp. Berlin 1950.

HARTREE, D. R.: Numerical analysis. 287 pp. Oxford 1952.

MINEUR, H.: Techniques de Calcul numérique. 605 pp. Paris et Liège 1952.

MILNE, W. E.: Numerical solution of differential equations. 275 pp. New York-London 1953.

KOPAL, ZDENĚK: Numerical analysis. 556 pp. London 1955.

1.2. Accuracy in the numerical solution of initial-value problems

In an initial-value problem we have to determine approximately in some interval $x_0 \leq x \leq \xi$ that solution $y(x)$ of an n-th order differential equation

$$y^{(n)} = f(x, y, y', \ldots, y^{(n-1)}) \tag{1.1}$$

which has prescribed "initial" values

$$y^{(\nu)}(x_0) = y_0^{(\nu)} \qquad (\nu = 0, 1, 2, \ldots, n-1) \tag{1.2}$$

at the "initial" point $x = x_0$. The existence and uniqueness of such a solution $y(x)$ in this interval will be assumed. Most approximate methods in current use yield approximations $y_1, \ldots, y_k, \ldots$ to the values $y(x_1), \ldots, y(x_k), \ldots$ of the exact solution at a number of discrete points $x_1, \ldots, x_k, \ldots$.

The choice of method from among the numerous approximate methods available and the whole arrangement of the calculation is governed decisively by the number of steps, i.e. the number of points x_k- and the accuracy desired. In initial-value problems conditions particularly unfavourable to accuracy are met with; not only is a lengthy calculation involved, in which inaccuracies at the beginning of the calculation influence all subsequent results (such is also the case, for example, when a large set of linear equations is solved by elimination), but inaccuracies in the individual $y_1, y_2, \ldots$ cause additional increases in the error; it can happen that solutions of the differential equation which lie close together at $x = x_1$ diverge considerably from one another at a subsequent point $x = x_n$. This last remark, sounding as it does rather trite, is nevertheless of such consequence as to be worthy of amplification by an example. Consider the initial-value problem

$$\left.\begin{array}{l} y'' = 10y' + 11y \\ y = 1, \quad y' = -1 \quad \text{for} \quad x = 0, \end{array}\right\} \tag{1.3}$$

whose solution $y(x)$ is to be calculated approximately in the interval $0 \leq x \leq 3$. The exact solution is $y(x) = e^{-x}$; this has the value $y(3) \approx 0.0498$ at $x = 3$, but the usual approximate methods will give completely false values for it. The general solution of the given differential equation is $y = c_1 e^{-x} + c_2 e^{11x}$, where c_1, c_2 are constants of integration. Now if we calculate successively the values y_1, y_2, ... at the points x_1, x_2, ... by any approximate method, the unavoidable rounding errors alone are bound to generate a component $c_2 e^{11x}$; even if we had an ideal method with no inherent error, this component, whose value at some point $x = \zeta$ is ε, say, would grow to $\varepsilon \times e^{22} \approx 3 \cdot 6 \times 10^9 \varepsilon$ by the time the point $x = \zeta + 2$ was reached, so that, if we had been computing to seven decimals, for example, the exact solution e^{-x} would already be completely swamped by it. Fortunately the conditions are usually not so unfavourable as in this "viscious" example, but it should always be borne in mind that carrying out an approximate step-by-step integration of an initial-value problem without some idea of the behaviour of the solution is like skating on ice of unknown thickness.

Accordingly we distinguish two cases:

Case 1. The solution is wanted only for a short interval and with moderate accuracy. Here we can use any of the crude methods of § 1.5; also analogue or graphical methods[1] could be used. In favourable cases with not many steps the calculation could even be performed with a slide-rule if need be.

Case 2. The solution is wanted for a long interval (or for a short interval with higher accuracy). Here, even if quite low accuracy is sufficient in the solution, the calculation must be very accurate and should certainly be based on one of the more accurate methods such as the finite-difference or Runge-Kutta methods. Moreover the accuracy to which we work must be substantially higher than that to which the results are required (for example, two or three additional decimals, the so-called "guarding figures", must be carried); consequently a slide-rule is no longer adequate and the calculation has to be carried out with the aid of a calculating machine. Also it is essential to exercise the utmost care at the beginning of the calculation to guard against the introduction of avoidable errors.

1.3. Some general observations on error estimation for initial-value problems

If approximate values y_ν of the solution $y(x)$ of the initial-value problem (1.1), (1.2) have been calculated at the points x_ν by some approximate method, then the question of the magnitude of the error

$$\varepsilon_\nu = y_\nu - y(x_\nu) \qquad (\nu = 1, 2, \ldots) \tag{1.4}$$

---

[1] WILLERS, FR. A.: Graphische Integration, pp. 45—104. Sammlung Göschen. Berlin and Leipzig 1920. — Methoden der praktischen Analysis, 2nd ed., p. 327 et seq. Berlin 1950. — SANDEN, H. v.: Praxis der Differentialgleichungen, 3rd ed., pp. 6—18. Berlin 1945. — KAMKE, E.: Differentialgleichungen, Lösungsmethoden und Lösungen, 3rd ed., Vol. I. Ch.A § 8. Leipzig 1944.

is of great importance. It might at least be possible to estimate the order of magnitude of this error ε_ν and find out how many decimals can be safely guaranteed in the results.

While we possess simple and useful error estimates for many problems in other spheres of practical analysis, we are still far from possessing such estimates for initial-value problems, at any rate as far as integration over anything but a very short interval (with only a few steps) is concerned. For many methods — the finite-difference method, for example — rigorous limits have been established for the error, but for a large number of steps these limits are usually so wide that they far exceed the actual error in magnitude and hence their practical value is illusory.

This difficulty is fundamental. It can even happen that approximate methods fail completely so rapidly do the errors grow, as the example (1.3) shows, and since a strict upper bound for the error must cover all cases, including the most unfavourable, we cannot expect it to give an accurate estimate of the magnitude of the error in a favourable case where the error remains small. Probably better limits will be established only if, instead of considering general differential equations such as $y'=f(x,y)$ or $y''=f(x,y,y')$, we treat each case on its merits so that we can use special properties of the function f. Nevertheless general error limits are derived in §§ 1, 4 and 5 in order to exhibit the underlying principles on which other error limits can be based.

Thus for integration over long intervals, we must usually be satisfied at present with "indications" of the actual error. Such indications are:

I. Comparison of two approximations with different lengths of step. If we perform the calculation with a step h, obtaining approximate values $y_1, y_2, \ldots$ at the points $x_p = x_0 + ph$, and then repeat the calculation with a step $\tilde{h}$, say $\tilde{h} = 2h$, obtaining values $\tilde{y}_1, \tilde{y}_2, \ldots, \tilde{y}_p$ corresponding to the previous values $y_2, y_4, \ldots, y_{2p}$, differences $\delta_p = y_{2p} - \tilde{y}_p$ will be revealed.

We now define a certain order k of an approximate method. Assuming that the values $y_1, y_2, \ldots, y_p$ are exact, i.e. $y_\nu = y(x_\nu)$ for $\nu = 0, 1, \ldots, p$, we calculate y_{p+1} by the method in question. If we now expand $y(x_{p+1})$ and y_{p+1} into Taylor series based on the point $x = x_p$ thus

$$y(x_{p+1}) = y(x_p) + \frac{h}{1!} y'(x_p) + \frac{h^2}{2!} y''(x) + \cdots,$$

$$y_{p+1} = y(x_p) + \frac{h}{1!} a_1 + \frac{h^2}{2!} a_2 + \cdots,$$

there will be a last term whose coefficient is the same in both series. The corresponding exponent of h is called the order k of the method; thus

$$a_1 = y'(x_p), \ldots, a_k = y^{(k)}(x_p), \quad \text{but} \quad a_{k+1} \neq y^{(k+1)}(x_p). \quad (1.5)$$

It is customary to reason (by no means rigorously) as follows: in the first calculation, with a step length h, an error proportional to h^{k+1} is introduced at each step, so that at the point $\xi = x_0 + 2nh$ the error is $A \times 2n h^{k+1} = A(\xi - x_0) h^k$, where A is a constant of proportionality; and in the second calculation, with a step length $2h$, an error proportional to $(2h)^{k+1}$ is introduced at each of the n steps up to the same point ξ, so that the error is $A \times n(2h)^{k+1} = A(\xi - x_0) 2^k h^k$, where A is the same constant of proportionality; thus we assume that approximately

$$y(\xi) \approx y_{2n} - A(\xi - x_0) h^k \approx \tilde{y}_n - A(\xi - x_0) 2^k h^k. \qquad (1.6)$$

Solving for an approximate value of A, we find the error in the calculation with the small interval to be one $(2^k - 1)$-th part of the difference between the results of the two calculations:

$$A(\xi - x_0) h^k \approx \frac{\tilde{y}_n - y_{2n}}{2^k - 1} \quad \text{and} \quad y(\xi) \approx y_{2n} - \frac{\tilde{y}_n - y_{2n}}{2^k - 1}. \qquad (1.7)$$

II. The terminal check. Here we calculate the values

$$f_k = f(x_k, y_k, y_k', \ldots, y_k^{(n-1)}) \qquad (1.8)$$

from the approximate solution and use them in the approximate evaluation of the repeated integral in the formula

$$\left.\begin{aligned} y(x) &= y_0 + y_0'(x - x_0) + y_0'' \frac{(x - x_0)^2}{2!} + \cdots + y_0^{(n-1)} \frac{(x - x_0)^{n-1}}{(n-1)!} + \\ &+ \int_{x_0}^{x} \cdots \int_{x_0}^{\xi_3} \int_{x_0}^{\xi_2} f\left(\xi_1, y(\xi_1), y'(\xi_1), \ldots, y^{(n-1)}(\xi_1)\right) d\xi_1 d\xi_2 \ldots d\xi_n, \end{aligned}\right\} \quad (1.9)$$

which is derived by repeated integration from (1.1), (1.2) (it may be added that most of the values f_k would have been found during the original calculation, anyway). Provided that we use a sufficiently accurate quadrature formula, the value for $y(x)$ obtained from (1.9) will be more accurate than that obtained in the approximate solution; it is then usually assumed that these values agree with the exact values to the same accuracy as that to which they agree with each other, an assumption which is not proved of course. For practical application of this type of terminal check, see § 2.5.

1.4. Differential equations of the first order. Preliminaries

Consider the differential equation

$$y' = \frac{dy}{dx} = f(x, y). \qquad (1.10)$$

Let $f(x, y)$ be a given continuous function in a domain D of the real (x, y) plane; for certain considerations we will assume that it is bounded in D, i.e.

$$|f(x, y)| \leq M \quad \text{in } D, \qquad (1.11)$$

and satisfies a Lipschitz condition there, i.e. there is a constant K such that for all pairs of points (x, y_1), (x, y_2) in the domain D we have

$$|f(x, y_1) - f(x, y_2)| \leq K|y_1 - y_2|. \qquad (1.12)$$

We require that solution $y(x)$ of the differential equation (1.10) in the domain D which passes through a given initial point $x = x_0$, $y = y_0$ (Fig. II/1); in actual fact we require an approximation to this solution, i.e. we have to calculate approximations y_n to the values $y(x_n)$ of the exact solution at the points x_n $(n = 1, 2, \ldots)$. The difference $\Delta x_n = x_{n+1} - x_n$ is called the step length or simply the step; it is usually denoted by h and is always taken to be positive. It often suffices to keep h constant (i.e. independent of n).

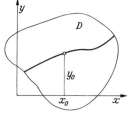

Many methods are based on the equation

$$y(x_{n+1}) = y(x_n) + \int_{x_n}^{x_{n+1}} f(x, y(x)) \, dx, \qquad (1.13)$$

Fig. II/1. Solution of a differential equation of the first order

which is obtained from (1.10) by integration, the integral on the right hand side being replaced by some approximate expression [1].

1.5. Some methods of integration

We discuss briefly a rather crude approximate method — a summation method — which would be used only when no great accuracy is required and only a small number of integration steps are involved. In this method, which we may call the "polygon" method, formula (1.13) is used in the crudest possible way: f is assumed to be constant at its initial value f_n over the step interval $\langle x_n, x_{n+1} \rangle$. Thus we calculate the next y_{n+1} from y_n by the formula

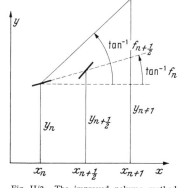

where
$$y_{n+1} = y_n + h f_n, \qquad (1.14)$$
$$f_n = f(x_n, y_n). \qquad (1.15)$$

Somewhat more accurate is the "improved polygon method" in which we use the slope at an intermediate point (see Fig. II/2): we find an

Fig. II/2. The improved polygon method

---

[1] The basic idea was used by LEONHARD EULER: Institutiones calculi integralis. Petersburg 1768—1770, published in Leonardi Euleri Opera omnia, series prima, Vol. 11, Leipzig and Berlin 1913, pp. 424—427; Vol. 12 (1914) pp. 271—274 (in Latin).

approximate intermediate point

$$x_{n+\frac{1}{2}} = x_n + \frac{h}{2}, \qquad y_{n+\frac{1}{2}} = y_n + \frac{1}{2} h f_n \qquad (1.16)$$

by using (1.14), (1.15) and then replace the f_n in (1.14) by

$$f_{n+\frac{1}{2}} = f(x_{n+\frac{1}{2}}, y_{n+\frac{1}{2}}), \qquad (1.17)$$

thus obtaining the improved approximation

$$y_{n+1} = y_n + h f_{n+\frac{1}{2}}. \qquad (1.18)$$

In the "improved Euler-Cauchy method" we first of all define rough values at x_{n+1} by

$$\tilde{y}_{n+1} = y_n + h f_n; \qquad \tilde{f}_{n+1} = f(x_{n+1}, \tilde{y}_{n+1}) \qquad (1.19)$$

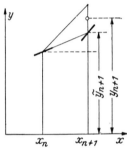

Fig. II/3. The improved Euler-Cauchy method

and then use the arithmetic mean of f_n and $\tilde{f}_{n+1}$ as an intermediate slope (see Fig. II/3). In this way we obtain the approximation

$$y_{n+1} = y_n + \frac{h}{2} \left(f_n + \tilde{f}_{n+1} \right). \qquad (1.20)$$

DUFFING's method[1] is based on formula (2.59) of Ch. I. The interval $\langle a, b \rangle$ now becomes $\langle x_n, x_{n+1} \rangle$, the function $f(x)$ becomes $y(x)$, and $f'(x)$ becomes $y'(x) = f[x, y(x)]$. If we neglect the remainder term, we obtain the approximate equation

$$y_{n+1} = \frac{h}{3} f(x_{n+1}, y_{n+1}) + y_n + \frac{2h}{3} f(x_n, y_n) + \frac{h^2}{6} \left[\frac{d}{dx} f(x, y) \right]_{(x_n, y_n)}.$$

In general, y_n being already calculated, this is a non-linear equation for y_{n+1} and is usually solved iteratively: an estimated value of y_{n+1} is first inserted in the right-hand side, which then yields a new value of y_{n+1} (see also Ex. 3 in § 5.10).

LOTKIN[2] goes one step further in this direction and proposes as an accurate method the use of the Euler-Maclaurin quadrature formula in

[1] DUFFING, G.: Zur numerischen Integration gewöhnlicher Differentialgleichungen 1. and 2. Ordnung. Forsch.-Arb. Ing.-Wes. 224, 29—50 (1920). The derivations and improvements of DUFFING's method given here are to be found in E. PFLANZ: Bemerkungen über die Methode von G. DUFFING zur Integration von Differentialgleichungen. Z. Angew. Math. Mech. 28, 167—172 (1948). The method is extended by E. BECK: Zwei Anwendungen der Obreschkoffschen Formel. Z. Angew. Math. Mech. 30, 84—93 (1950).

[2] LOTKIN, M.: A new integrating procedure of high accuracy. J. Math. Phys. 31, 29—34 (1952).

the form (with $b - a = h$)

$$\int_a^b \varphi(t)\, dt = \frac{h}{2} \left(\varphi(a) + \varphi(b) \right) + \frac{h^2}{10} \left(\varphi'(a) - \varphi'(b) \right) +$$
$$+ \frac{h^3}{120} \left(\varphi''(a) + \varphi''(b) \right) + R;$$

here the remainder term R is of the seventh order. Starting from an approximate value

$$y_{r+1}^{[0]} = y_r + h f_r + \frac{h^2}{2} f_r' + \frac{h^3}{6} f_r'',$$

which determines the starting values $f_{r+1}^{[0]}, f_{r+1}'^{[0]}, f_{r+1}''^{[0]}$, we calculate successive iterates $y_{r+1}^{[\sigma]}$ from the formula

$$y_{r+1}^{[\sigma+1]} = y_r + \frac{h}{2} (f_r + f_{r+1}^{[\sigma]}) + \frac{h^2}{10} (f_r' - f_{r+1}'^{[\sigma]}) + \frac{h^3}{120} (f_r'' + f_{r+1}''^{[\sigma]}).$$

Implicit in the presentation of this method is the assumption that $f' = d f / d x = f_x + f_y f$ and f'' may be calculated sufficiently easily.

Another method, which is sometimes effective also for differential equations of higher order, is to replace the differential equation by a simpler one solvable in closed form; for example, we can often obtain an approximate solution by replacing variable coefficients by piecewise-constant coefficients[1].

A simple example

The initial-value problem

$$y' = y - \frac{2x}{y}, \qquad y(0) = 1 \tag{1.21}$$

possesses an exact solution, namely

$$y = \sqrt{2x + 1}, \tag{1.22}$$

so that we can immediately assess the accuracy of the approximate methods for this particular case. Application of the approximate formulae (1.14) to (1.20) is so straightforward that the *modus operandi* is self-evident from the tables reproduced here.

Table II/1 gives the working for the improved polygon method with $h = 0.2$. As can be recommended for other calculations, we have indicated at the head of each column, numbered ①, ②, ③, ..., how its contents are obtained from previously calculated numbers in other columns; once this has been done, the calcula-

---

[1] In the case $y'' = f(y)$, for example, $f(y)$ can be replaced by linear functions over suitable intervals in y; then an approximation for y can be obtained by piecing together the solutions obtained in these intervals. Cf. K. KLOTTER: Technische Schwingungslehre, 2nd ed., Vol. I, p. 154. Berlin-Göttingen-Heidelberg 1951.

tion can proceed in an entirely systematic fashion without further reference to the formulae. This facilitates the actual computation considerably and renders it suitable for an assistant.

Table II/1. *Numerical integration of a first-order differential equation by the improved polygon method*

x_n	y_n	$-\dfrac{2x_n}{y_n}$	$\dfrac{h}{2}f_n$	$y_{n+\frac12}$	$x_{n+\frac12}$	$-4\dfrac{x_{n+\frac12}}{y_{n+\frac12}}$	$hf_{n+\frac12}$	y_{n+1}
①	②	③	④	⑤	⑥	⑦	⑧	⑨
(from ⑨)		$=-\dfrac{2\times①}{②}$	$=\dfrac{②+③}{10}$	$=②+④$	$=①+0.1$	$=-\dfrac{4\times⑥}{⑤}$	$=\dfrac{2\times⑤+⑦}{10}$	$=②+⑧$
0	1	0	0·1	1·1	0·1	−0·36364	0·18364	1·18364
0·2	1·18364	−0·33794	0·084570	1·26821	0·3	−0·94622	0·15902	1·34266
0·4	1·34266	−0·59583	0·074683	1·41734	0·5	−1·41109	0·14236	1·48502
0·6	1·48502	−0·80807	0·067695	1·55272	0·7	−1·80329	0·13021	1·61523
0·8	1·61523	−0·99057	0·062466	1·67770	0·9	−2·14580	0·12096	1·73619
1	1·73619	−1·15195	0·058424	1·79461	1·1	−2·45178	0·11374	1.84993
1·2	1·84993	−1·29735	0·055258	1·90519	1·3	−2·72939	0·10810	1·95803
1·4	1·9580	−1·4300	0·05280	2·0108	1·5	−2·9838	0·1038	2·0618
1·6	2·0618	−1·5520	0·05098	2·1128	1·7	−3·2185	0·1007	2·1625
1·8	2·1625	−1·6647	0·04978	2·2123	1·9	−3·4353	0·0989	2·2615
2	2·2615							

In order to compare the accuracy obtained by the various methods, we have carried more decimals than would otherwise be significant.

Table II/2 shows the corresponding calculations for the improved Euler-Cauchy method with the same step length $h = 0·2$; here the integration is taken only as far as the point $x = 1$.

Table II/2.
Numerical integration of a first-order equation by the improved Euler-Cauchy method

x_n	y_n	$-\dfrac{2x_n}{y_n}$	$\dfrac{h}{2}f_n$	$\tilde{y}_{n+1}$	x_{n+1}	$-\dfrac{2x_{n+1}}{\tilde{y}_{n+1}}$	$\tilde{f}_{n+1}$	$\dfrac{h}{2}(f_n+\tilde{f}_{n+1})$	y_{n+1}
①	②	③	④	⑤	⑥	⑦	⑧	⑨	⑩
(from ⑩)		$=\dfrac{-2\times①}{②}$	$=\dfrac{②+③}{10}$	$=②+2\times④$		$=-\dfrac{2\times⑥}{⑤}$	$=⑤+⑦$	$=④+\dfrac{⑧}{10}$	$=②+⑨$
0	1	0	0·1	1·2	0·2	−0·33333	0·8667	0·18667	1·18667
0·2	1·18667	−0·33708	0·084959	1·35658	0·4	−0·58972	0·7669	0·16165	1·34832
0·4	1·34832	−0·59333	0·075499	1·49932	0·6	−0·80036	0·6990	0·14540	1·49372
0·6	1·49372	−0·80336	0·069036	1·63179	0·8	−0·98052	0·6513	0·13416	1·62788
0·8	1·6279	−0·9829	0·06450	1·7569	1	−1·1384	0·6185	0·12635	1·7543

In Table II/3 comparison is made with the exact solution (1.22) and also with results obtained by using the more accurate methods discussed in §§ 2 and 3; below each result the error is written in brackets (the error in an approximation ξ to a true value x is defined as $\xi - x$).

Table II/3. *Comparison of results obtained by various methods*

x	Exact solution $\sqrt{2x+1}$	Improved Euler-Cauchy $\mid$ polygon method		Runge-Kutta method		Central-difference method
		$h=0\cdot2$	$h=0\cdot2$	$h=0\cdot4$	$h=0\cdot2$	$h=0\cdot1$
0·2	1·183 216 0	1·186 67 $(+0\cdot003\,45)$	1·183 64 $(+0\cdot000\,42)$		1·183 229 3 $(+0\cdot000\,013\,3)$	1·183 221 5 $(+0\cdot000\,005\,5)$
0·4	1·341 640 8	1·348 32 $(0\cdot006\,68)$	1·342 66 $(0\cdot001\,02)$	1·342 066 $(+0\cdot000\,425)$	1·341 666 9 $(0\cdot000\,026\,2)$	1·341 649 1 $(0\cdot000\,008\,4)$
0·6	1·483 239 7	1·493 72 $(0\cdot010\,5)$	1·485 02 $(0\cdot001\,78)$		1·483 281 5 $(0\cdot000\,041\,8)$	1·483 250 8 $(0\cdot000\,011\,1)$
0·8	1·612 451 5	1·627 88 $(0\cdot015\,4)$	1·615 23 $(0\cdot002\,78)$	1·613 449 $(0\cdot000\,997)$	1·612 514 0 $(0\cdot000\,062\,4)$	1·612 466 5 $(0\cdot000\,014\,9)$
1	1·732 050 8	1·754 3 $(0\cdot022\,33)$	1·736 19 $(0\cdot004\,14)$		1·732 141 9 $(0\cdot000\,091\,1)$	1·732 071 3 $(0\cdot000\,020\,5)$
1·2	1·843 908 9		1·849 93 $(0\cdot006\,02)$	1·846 00 $(0\cdot002\,09)$	1·844 040 1 $(0\cdot000\,131)$	1·843 937 $(0\cdot000\,028)$
1·4	1·949 358 9		1·958 03 $(0\cdot008\,67)$		1·949 547 $(0\cdot000\,188)$	1·949 398 $(0\cdot000\,039)$
1·6	2·049 390 2		2·061 81 $(0\cdot012\,4)$	2·053 67 $(0\cdot004\,28)$	2·049 660 $(0\cdot000\,270)$	2·049 445 $(0\cdot000\,055)$
1·8	2·144 761 1		2·162 52 $(0\cdot017\,8)$		2·145 148 $(0\cdot000\,387)$	2·144 839 $(0\cdot000\,078)$
2	2·236 068 0		2·261 45 $(0\cdot025\,4)$	2·244 86 $(0\cdot007\,89)$	2·236 624 $(0\cdot000\,556)$	2·236 179 $(0\cdot000\,111)$

The Runge-Kutta method involves about twice the amount of work per step as the central-difference method so that it is reasonable to compare the Runge-Kutta results with $h = 0\cdot2$ with those of the central-difference method with $h = 0\cdot1$.

1.6. Error estimation

We assume here that the function $f(x, y)$ in the differential equation (1.10) is bounded as in (1.11) and satisfies the Lipschitz condition (1.12) in the domain D as in § 1.4 and also that it possesses bounded partial derivatives in so far as it satisfies the conditions

$$\left| \frac{df(x, y)}{dx} \right| = \left| \frac{\partial f}{\partial x} + f \frac{\partial f}{\partial y} \right| \leqq N_1; \qquad \left| \frac{d^p f}{dx^p} \right| \leqq N_p \qquad (p = 1, 2, \ldots). \qquad (1.23)$$

For simplicity we will ignore rounding errors in obtaining the following error estimates, i.e. we estimate only the "inherent" or "truncation" errors. Rounding errors will, of course, be propagated throughout the calculation, but we take the view that, knowing an upper bound for the inherent errors from one of the error estimates, we can calculate with a number of decimals sufficient to ensure that the rounding errors remain within this bound. In principle it would not be difficult to derive similar estimates for the rounding errors.

We define the "error" in the approximate value y_n to be the quantity

$$\varepsilon_n = y_n - y(x_n);\qquad\qquad(1.24)$$

from (1.13) the error increment $\Delta\varepsilon_n$ over one step is therefore

$$\Delta\varepsilon_n = \varepsilon_{n+1} - \varepsilon_n = \underbrace{y_{n+1} - y_n}_{\Delta y_n} - \int_{x_n}^{x_{n+1}} f(x, y(x))\, dx.\qquad(1.25)$$

The expression for the difference Δy_n is characteristic of each individual method, and further analysis in which we insert this expression into (1.25) must be referred to specific methods.

I. Polygon method. From (1.14) the expression for Δy_n is

$$\Delta y_n = h f_n;$$

therefore, with $F(x) = f[x, y(x)]$, (1.25) becomes

$$\Delta\varepsilon_n = h\big(f_n - F(x_n)\big) + J_n,\qquad\qquad(1.26)$$

where

$$J_n = h F(x_n) - \int_{x_n}^{x_{n+1}} F(x)\, dx.\qquad\qquad(1.27)$$

Since we have assumed that $f(x, y)$ satisfies the Lipschitz condition (1.12), limits for the first term on the right of (1.26) can be obtained from

$$|f_n - F(x_n)| = |f(x_n, y_n) - f(x_n, y(x_n))| \le K |y_n - y(x_n)| = K |\varepsilon_n|.\qquad(1.28)$$

For the second term J_n we first transform the integral by integrating by parts:

$$\int_{x_n}^{x_{n+1}} 1 \cdot F(x)\, dx = [(x - x_{n+1}) F]_{x_n}^{x_{n+1}} - \int_{x_n}^{x_{n+1}} (x - x_{n+1})\frac{dF}{dx}\, dx;$$

it then follows that

$$\left| h F(x_n) - \int_{x_n}^{x_{n+1}} F(x)\, dx \right| = \left| \int_{x_n}^{x_{n+1}} (x - x_{n+1})\frac{dF}{dx}\, dx \right|$$

$$\le N_1 \int_{x_n}^{x_{n+1}} (x_{n+1} - x)\, dx = \frac{1}{2} N_1 h^2,\qquad\qquad(1.29)$$

where N_1 is as defined in (1.23).

Taking absolute values in (1.26) and using (1.28), (1.29), we obtain

$$|\varDelta \varepsilon_n| = |\varepsilon_{n+1} - \varepsilon_n| \leq hK|\varepsilon_n| + \tfrac{1}{2} N_1 h^2,$$

and since

$$|\varepsilon_{n+1} - \varepsilon_n| \geq |\varepsilon_{n+1}| - |\varepsilon_n|,$$

we arrive at the "recursive error estimate"

$$|\varepsilon_{n+1}| \leq (1 + hK)|\varepsilon_n| + \tfrac{1}{2} N_1 h^2. \tag{1.30}$$

Thus if an estimate for ε_n is known, ε_{n+1} can also be estimated.

We can now easily derive an "independent error estimate", i.e. one in which limits for $|\varepsilon_{n+1}|$ are determined without knowledge of limits for the preceding errors. The inequality (1.30) has the form

$$|\varepsilon_{n+1}| \leq a|\varepsilon_n| + b, \text{ where } a \geq 0,\ b \geq 0,\ \varepsilon_0 = 0, \quad (n = 0, 1, 2, \dots). \tag{1.31}$$

Consequently

$$|\varepsilon_1| \leq b, \quad |\varepsilon_2| \leq a|\varepsilon_1| + b \leq b(1 + a),$$

and in general, as can be proved immediately by induction,

$$|\varepsilon_n| \leq b(1 + a + a^2 + \cdots + a^{n-1}) = \frac{b(a^n - 1)}{a - 1}. \tag{1.32}$$

We have assumed here that $a \neq 1$.

If we now insert the values $a = 1 + hK$, $b = \tfrac{1}{2} h^2 N_1$, corresponding to (1.30), we obtain, for $K > 0$, the independent error estimate

$$|\varepsilon_n| \leq \frac{h N_1}{2K} [(1 + hK)^n - 1]. \tag{1.33}$$

Since $1 + u < e^u$ for $u > 0$, we can also use the estimate

$$|\varepsilon_n| \leq \frac{h N_1}{2K} (e^{K(x_n - x_0)} - 1), \tag{1.34}$$

where nh has been replaced by $x_n - x_0$. From this it follows in particular that at a fixed point x_n the error limits tend to zero as $h \to 0$, i.e. the method is convergent.

1.7. Corresponding error estimates for the improved methods

II. Improved polygon method. We follow the same reasoning here so we may be brief. In this case the expression for $\varDelta y_n$ is, from (1.16) to (1.18),

$$\varDelta y_n = y_{n+1} - y_n = hf(x_{n+\frac{1}{2}}, y_n + \tfrac{1}{2} h f_n),$$

and hence (1.25) becomes

$$\Delta\varepsilon_n = h\,\Phi_n + J_n^*,\qquad(1.35)$$

where

$$\Phi_n = f\!\left(x_{n+\frac{1}{2}},\,y_n + \tfrac{1}{2}h f(x_n,y_n)\right) - f\!\left(x_{n+\frac{1}{2}},\,y(x_{n+\frac{1}{2}})\right)\qquad(1.36)$$

and

$$J_n^* = h F(x_{n+\frac{1}{2}}) - \int_{x_n}^{x_{n+1}} F(x)\,dx.\qquad(1.37)$$

For the bounds of Φ_n we employ the Lipschitz condition (1.12) again:

$$|\Phi_n| \le K\,|y_n + \tfrac{1}{2}h f(x_n,y_n) - y(x_{n+\frac{1}{2}})|$$

$$= K\,|y_n - y(x_n) + \tfrac{1}{2}h\,[f(x_n,y_n) - F(x_n)] - [y(x_{n+\frac{1}{2}}) - y(x_n) - \tfrac{1}{2}h F(x_n)]|;$$

formula (1.29) applied over an interval of length $\tfrac{1}{2}h$ yields

$$\left|\tfrac{1}{2}h F(x_n) - [y(x_{n+\frac{1}{2}}) - y(x_n)]\right| \le \tfrac{1}{2}N_1\left(\tfrac{h}{2}\right)^2;$$

hence for Φ_n we have the estimate

$$|\Phi_n| \le K\left\{|\varepsilon_n| + \tfrac{1}{2}h K\,|\varepsilon_n| + \tfrac{1}{8}h^2 N_1\right\}.\qquad(1.38)$$

For the quadrature error J_n^* there are the known[1] bounds

$$|J_n^*| \le \frac{h^3}{24}N_2,\qquad(1.39)$$

where N_2 is as defined in (1.23). Consequently from (1.35), (1.38), (1.39) it follows that for $K>0$

$$\left.\begin{aligned}
&|\varepsilon_{n+1}| - |\varepsilon_n| \le |\Delta\varepsilon_n| \le h K\left\{(1 + \tfrac{1}{2}h K)\,|\varepsilon_n| + \tfrac{1}{8}h^2 N_1\right\} + \tfrac{1}{24}h^3 N_2 \\
&\text{i.e.}\quad |\varepsilon_{n+1}| \le |\varepsilon_n|\left\{1 + h K + \tfrac{1}{2}h^2 K^2\right\} + \tfrac{1}{8}h^3(K N_1 + \tfrac{1}{3}N_2).
\end{aligned}\right\}\qquad(1.40)$$

This recursive error estimate also has the form (1.31), and using (1.32) we immediately obtain the independent estimate

$$|\varepsilon_n| \le \frac{1}{8}h^2\left(N_1 + \frac{N_2}{3K}\right)\frac{(1 + h K + \tfrac{1}{2}h^2 K^2)^n - 1}{1 + \tfrac{1}{2}h K}.\qquad(1.41)$$

As with the polygon method the convergence of the method follows directly from this independent error estimate; here, however, the error at any fixed point x tends to zero like h^2 as $h \to 0$.

III. Improved Euler-Cauchy method. Proceeding exactly as in the previous case, we have from (1.19), (1.20)

$$\Delta\varepsilon_n = h\,\Psi_n + J_n^{**},$$

---

[1] See, for example, R. Courant: Differential and Integral Calculus, 2nd ed., Vol. I, p. 347. Glasgow 1937.

where

$$\Psi_n = \tfrac{1}{2}\left[f(x_n, y_n) + f(x_{n+1}, y_n + h f_n) - F(x_n) - F(x_{n+1})\right]$$

and

$$J_n^{**} = \tfrac{1}{2} h\left[F(x_n) + F(x_{n+1})\right] - \int_{x_n}^{x_{n+1}} F(x)\,dx.$$

From the Lipschitz condition it follows that

$$|\Psi_n| \leq \tfrac{1}{2} K(|\varepsilon_n| + |y(x_{n+1}) - y_n - h f(x_n, y_n)|).$$

Now[1] by TAYLOR's theorem there is a point $x = \xi_n$ in $\langle x_n, x_{n+1}\rangle$ such that

$$y(x_{n+1}) - y_n - h f(x_n, y_n) = y(x_n) + h y'(x_n) + \frac{h^2}{2} y''(\xi_n) -$$

$$- y_n - h f(x_n, y_n) = -\varepsilon_n + h[f(x_n, y(x_n)) - f(x_n, y_n)] + \frac{h^2}{2} y''(\xi_n),$$

and hence, since $|y''(\xi_n)| \leq N_1$ from (1.23), we have

$$|y(x_{n+1}) - y_n - h f(x_n, y_n)| \leq |\varepsilon_n| + h K |\varepsilon_n| + \frac{h^2}{2} N_1.$$

For the quadrature error J_n^{**} in the trapezium approximation used in this method there is the well-known estimate

$$|J_n^{**}| \leq \frac{1}{12} h^3 N_2.$$

Altogether, we have

$$|\varepsilon_{n+1}| - |\varepsilon_n| \leq |\Delta \varepsilon_n| \leq h |\Psi_n| + |J_n^{**}| \leq \frac{h K}{2}\left(|\varepsilon_n|(2 + h K) + \frac{h^2}{2} N_1\right) + \frac{h^3}{12} N_2,$$

i.e.

$$|\varepsilon_{n+1}| \leq |\varepsilon_n|\left(1 + h K + \frac{h^2 K^2}{2}\right) + \frac{1}{4} h^3\left(K N_1 + \frac{1}{3} N_2\right).$$

Again the recursive error estimate arrived at has the form of the inequality (1.31) so that the independent estimate can be immediately written down from (1.32):

$$|\varepsilon_n| \leq \frac{h^2}{4}\left(N_1 + \frac{N_2}{3 K}\right) \frac{(1 + h K + \tfrac{1}{2} h^2 K^2)^n - 1}{1 + \tfrac{1}{2} h K}. \tag{1.42}$$

These limits are precisely twice as wide as the limits (1.41) for the improved polygon method.

Once more we can infer the convergence of the method. In this case also, the error limits at a fixed point x tend to zero quadratically as $h \to 0$.

§ 2. The Runge-Kutta method for differential equations of the n-th order

The Runge-Kutta method is a much-used method for accurate integration. The general formulae derived in §§ 2.1—2.3 by some rather laborious calculations need

----

[1] The estimate given here constitutes a sharpening of the estimate given in the first German edition; I am indebted for it to a written communication from Dr. WILLY RICHTER, Neuchâtel.

not concern the reader whose interest lies chiefly in the numerical integration of specific problems; he will find the computational formulae for equations of up to the fourth order collected together in § 2.4.

2.1. A general formulation

Suppose that the n-th order differential equation

$$y^{(n)} = f(x, y, y', y'', \ldots, y^{(n-1)}) \tag{2.1}$$

is to be integrated subject to prescribed initial values

$$y_0^{(\nu)} = y^{(\nu)}(x_0) \qquad (\nu = 0, 1, 2, \ldots, n-1) \tag{2.2}$$

at the point $x = x_0$. Approximate values will be derived for y and its derivatives $y^{(\nu)}$ at the point $x_1 = x_0 + h$ one step ahead.

Rough approximations can be obtained by using the prescribed initial values in Taylor series truncated at the terms in $y_0^{(n-1)}$: for $y_1^{(\nu)}$ we have the value

$$y_0^{(\nu)} + \frac{h}{1!} y_0^{(\nu+1)} + \frac{h^2}{2!} y_0^{(\nu+2)} + \cdots + \frac{h^{n-\nu-1}}{(n-\nu-1)!} y_0^{(n-1)} = T_\nu(1),$$

the notation T_ν being defined, somewhat more generally, by

$$\left. \begin{aligned} T_\nu(\alpha) &= y_0^{(\nu)} + \frac{\alpha h}{1!} y_0^{(\nu+1)} + \frac{(\alpha h)^2}{2!} y_0^{(\nu+2)} + \cdots + \\ &\quad + \frac{(\alpha h)^{n-\nu-1}}{(n-\nu-1)!} y_0^{(n-1)} \qquad (\nu = 0, 1, 2, \ldots, n-1). \end{aligned} \right\} \tag{2.3}$$

In order to obtain better approximations we add corrections to these truncated Taylor series; we derive quantities $k, k', k'', \ldots, k^{(n-1)}$ such that the approximation

$$y_1^{(\nu)} = T_\nu(1) + \frac{\nu!}{h^\nu} k^{(\nu)} \qquad (\nu = 0, 1, \ldots, n-1) \tag{2.4}$$

to the ν-th derivative $y^{(\nu)}(x_1)$ is as accurate as possible (the factor $\nu! h^{-\nu}$ will be found convenient later).

First of all we describe a general formulation which includes many well-known numerical methods of integration as special cases.

The "corrections" $k^{(\nu)}$ are expressed as linear combinations

$$k^{(\nu)} = \sum_{\varrho=1}^{r} \gamma_{\nu\varrho} k_\varrho \qquad (\nu = 0, 1, \ldots, n-1) \tag{2.5}$$

of certain auxiliary quantities $k_1, k_2, \ldots, k_r$; these are defined in terms of the values of the function f at intermediate points in such a way that they can be calculated successively:

$$
\left.
\begin{aligned}
k_1 &= \frac{h^n}{n!} f\left(x_0, T_0(0), T_1(0), \ldots, T_{n-1}(0)\right), \\
k_2 &= \frac{h^n}{n!} f\left(x_0 + \alpha_1 h, T_0(\alpha_1) + a_{10} k_1, T_1(\alpha_1) + a_{11} k_1, \ldots, \right. \\
&\qquad\qquad\qquad\qquad \left. T_{n-1}(\alpha_1) + a_{1, n-1} k_1\right), \\
k_3 &= \frac{h^n}{n!} f\left(x_0 + \alpha_2 h, T_0(\alpha_2) + a_{20} k_1 + b_{20} k_2, \ldots, \right. \\
&\qquad\qquad\qquad\qquad \left. T_{n-1}(\alpha_2) + a_{2, n-1} k_1 + b_{2, n-1} k_2\right).
\end{aligned}
\right\} \quad (2.6)
$$

From the definition of the $T_\nu(\alpha_\varrho)$ we know that the r "points"

$$
x = x_0 + \alpha_\varrho h, \quad y = T_0(\alpha_\varrho), \quad y' = T_1(\alpha_\varrho), \ldots, y^{(n-1)} = T_{n-1}(\alpha_\varrho)
$$

$(\varrho = 1, \ldots, r)$ of the $(x, y, y', \ldots, y^{(n-1)})$ space lie near the true solution. The constants $\alpha, a_{\nu\varrho}, b_{\nu\varrho}, \ldots, \gamma_{\nu\varrho}$ are at our disposal for making the Taylor series for the $y_1^{(\nu)}$ coincide with the Taylor series for the $y^{(\nu)}(x_1)$ to as high an order as possible, i.e. for making the power of h in the last identical terms as high as possible. With the notation of (2.3) the Taylor series for $y^{(\nu)}(x_1)$ can be written

$$
\left.
\begin{aligned}
y^{(\nu)}(x_1) &= T_\nu(1) + \frac{h^{n-\nu}}{(n-\nu)!} y_0^{(n)} + \frac{h^{n-\nu+1}}{(n-\nu+1)!} y_0^{(n+1)} + \cdots \\
&= T_\nu(1) + \frac{\nu!}{h^\nu}\left[\binom{n}{\nu}\frac{h^n y_0^{(n)}}{n!} + \binom{n+1}{\nu}\frac{h^{n+1} y_0^{(n+1)}}{(n+1)!} + \cdots\right],
\end{aligned}
\right\} \quad (2.7)
$$

which becomes

$$
\left.
\begin{aligned}
y^{(\nu)}(x_1) &= T_\nu(1) + \frac{\nu!}{h^\nu}\left[\binom{n}{\nu}\frac{h^n}{n!} f + \binom{n+1}{\nu}\frac{h^{n+1}}{(n+1)!} Df + \right. \\
&\quad \left. + \binom{n+2}{\nu}\frac{h^{n+2}}{(n+2)!} Ef + \binom{n+3}{\nu}\frac{h^{n+3}}{(n+3)!} (Ff + f_{n-1} Ef + f_{n-2} Df) + \cdots\right]
\end{aligned}
\right\} \quad (2.8)
$$

when the expressions (2.52), (2.53), (2.55) of Ch. I are inserted (when no particular argument values are specified for f, the values at the initial point $x = x_0$ are to be taken). We determine the $k^{(\nu)}$ as accurately as possible by choosing them so that when their Taylor expansions are inserted in the expressions (2.4), the latter coincide with (2.8) to as high an order as possible:

$$
\left.
\begin{aligned}
k^{(\nu)} &= \frac{h^n}{n!}\left[\binom{n}{\nu} f + \binom{n+1}{\nu}\frac{h}{n+1} Df + \binom{n+2}{\nu}\frac{h^2}{(n+1)(n+2)} Ef + \right. \\
&\quad \left. + \binom{n+3}{\nu}\frac{h^3}{(n+1)(n+2)(n+3)} (Ff + f_{n-1} Ef + f_{n-2} Df) + \cdots\right].
\end{aligned}
\right\} \quad (2.9)
$$

2.2. The special Runge-Kutta formulation

Here we discuss the special case of the general formulation of § 2.1 which gives rise to the Runge[1]-Kutta[2] formulae. These have found favour on account of their simple coefficients and symmetrical form; they are exhibited for differential equations of up to the fourth order in the tables in § 2.4. Some other formulae of Runge-Kutta type are collected together in Table I of the appendix.

The basic idea due to C. RUNGE[3] was applied to first-order equations in more accurate form by HEUN[4] and KUTTA[5] and extended to second-

---

[1] CARL DAVID TOLMÉ RUNGE, one of the most notable of German applied mathematicians, also known as a physicist through his investigations on spectral series, was one of the pioneers in the application of mathematical methods to the numerical treatment of technical problems. E. TREFFTZ writes [Z. Angew. Math. Mech. **6**, 423—424 (1926)] of him (translated from the German): "If RUNGE succeeded in bridging the gap between mathematics and technology, his success was due to two characteristics, which mark a true applied mathematician: his profound mathematical knowledge ..., and his unflagging energy in perfecting his methods with particular regard to their practical usefulness."

RUNGE was born in Bremen on the 30th August, 1856 and spent his early childhood in Havanna, where his father was in charge of the administration of the Danish consulate. From 1876 to 1880 he studied first in Munich and then in Berlin, where he took his doctor's degree in 1880; he became an unsalaried lecturer at the University of Berlin in 1883 and was later, 1886, appointed as professor of mathematics in the Technische Hochschule, Hannover. From 1904 to 1924 he was professor of applied mathematics in Göttingen, where he died on the 3rd January, 1927.

His interests outside of scientific spheres were many and varied. He undertook several very lengthy journeys; in particular he accompanied SCHWARZSCHILD on an expedition to Algeria in connection with the solar eclipse of 1906 He also travelled to New York in 1909, where he was exchange professor in the Columbia University for the winter semester. Right into his old age he continued to engage in sports, gymnastics, swimming and rowing. Recalling some facets of his character L. PRANDTL [Naturwiss. **15**, 227—229 (1927)] writes (translated from the German): "He was of a kindly disposition, yet strongly independent in his opinions, even to severe condemnations of that which appeared to him unfair or narrow-minded. With regard to himself he was extremely modest."

Details can be found in the biography by his daughter IRIS RUNGE: Carl Runge und sein wissenschaftliches Werk. Göttingen 1949.

[2] MARTIN WILHELM KUTTA, born on the 3rd November, 1867 in Pitschen (Upper Silesia), studied in Breslau from 1885 to 1890, then went to Munich, where he took his doctor's degree in 1901 and became an unsalaried lecturer in 1902. He spent 1898—1899 in Cambridge. In 1910 he was appointed to Aachen and in 1911 to Stuttgart as ordinary professor of mathematics (emeritus 1935). He died on the 25th December, 1944 in Fürstenfeldbruck (near Munich), where he was staying with his brother.

[3] RUNGE, C.: Math. Ann. **46**, 167—178 (1895) and Nachr. Ges. Wiss. Göttingen, Math.-phys. Kl. **1905**, 252—257.

[4] HEUN, K.: Z. Math. Phys. **45**, 23—38 (1900).

[5] KUTTA, W.: Z. Math. Phys. **46**, 435—453 (1901).

order equations by NYSTRÖM[1]. ZURMÜHL continued the extension, first to third-order equations[2] and then to equations of the n-th order[3].

In this particular formulation[4] we use just four auxiliary quantities k_1, k_2, k_3, k_4. We choose $\alpha_1=\alpha_2=\frac{1}{2}$, $\alpha_3=1$, so that k_2 and k_3 are based on the mid-point of the step interval and k_4 on the last [k_1 is based on the first, being defined always as in (2.6)]. The $a_{\nu\varrho}, b_{\nu\varrho}, \ldots$ are chosen in an obvious way for k_2: the additional terms in the arguments of f merely extend the truncated Taylor series. For k_4 these constants are chosen so that the additional terms are calculated in a similar fashion but from k_3 instead of k_1. The definition of k_3 differs from that of k_2 only in the last argument of f, where k_2 is used instead of k_1, so that $k_2=k_3$ when f does not depend explicitly on $y^{(n-1)}$. Thus

$$
\begin{aligned}
k_2 = &\frac{h^n}{n!} f\left[x_0 + \frac{h}{2}, \quad T_0\left(\frac{1}{2}\right) + \frac{k_1}{2^n}, \quad T_1\left(\frac{1}{2}\right) + \binom{n}{1}\frac{1!\,k_1}{2^{n-1}h},\right. \\
&\left. T_2\left(\frac{1}{2}\right) + \binom{n}{2}\frac{2!\,k_1}{2^{n-2}h^2}, \ldots,\right. \\
&\left. T_{n-2}\left(\frac{1}{2}\right) + \binom{n}{n-2}\frac{(n-2)!\,k_1}{4\,h^{n-2}}, \quad T_{n-1}\left(\frac{1}{2}\right) + \frac{n!\,k_1}{2\,h^{n-1}}\right], \\
k_3 = &\frac{h^n}{n!} f\left[x_0 + \frac{h}{2}, \quad T_0\left(\frac{1}{2}\right) + \frac{k_1}{2^n}, \quad T_1\left(\frac{1}{2}\right) + \binom{n}{1}\frac{1!\,k_1}{2^{n-1}h},\right. \\
&\left. T_2\left(\frac{1}{2}\right) + \binom{n}{2}\frac{2!\,k_1}{2^{n-2}h^2}, \ldots,\right. \\
&\left. T_{n-2}\left(\frac{1}{2}\right) + \binom{n}{n-2}\frac{(n-2)!\,k_1}{4\,h^{n-2}}, \quad T_{n-1}\left(\frac{1}{2}\right) + \frac{n!\,k_2}{2\,h^{n-1}}\right], \\
k_4 = &\frac{h^n}{n!} f\left[x_0 + h, \quad T_0(1) + k_3, \quad T_1(1) + \binom{n}{1}\frac{1!\,k_3}{h},\right. \\
&\left. T_2(1) + \binom{n}{2}\frac{2!\,k_3}{h^2}, \ldots,\right. \\
&\left. T_{n-2}(1) + \binom{n}{n-2}\frac{(n-2)!\,k_3}{h^{n-2}}, \quad T_{n-1}(1) + \frac{n!\,k_3}{h^{n-1}}\right].
\end{aligned} \tag{2.10}
$$

(For the practical evaluation of these arguments of f see § 2.4.)

---

[1] NYSTRÖM, E. J.: Über die numerische Integration von Differentialgleichungen. Acta Soc. Sci. Fenn. **50**, Nr. 13, 1—55 (1925).

[2] ZURMÜHL, R.: Zur numerischen Integration gewöhnlicher Differentialgleichungen zweiter und höherer Ordnung. Z. Angew. Math. Mech. **20**, 104—116 (1940).

[3] ZURMÜHL, R.: Runge-Kutta-Verfahren zur numerischen Integration von Differentialgleichungen n-ter Ordnung. Z. Angew. Math. Mech. **28**, 173—182 (1948).

[4] The general formulation of § 2.1 is pursued further by J. ALBRECHT: Beiträge zum Runge-Kutta-Verfahren. Z. Angew. Math. Mech. **35**, 100—110 (1955).

2.3. Derivation of the Runge-Kutta formulae

We now expand k_1, k_2, k_3, k_4 into Taylor series based on the point $(x_0, y_0, y_0', \ldots, y_0^{(n-1)})$. This is trivial for k_1 since it is already in the required form, but for the other k_ν the expansion presents a rather laborious calculation, for, in general, all the $n+1$ arguments of f are different from their values at x_0, as for instance, in

$$\frac{n!}{h^n} k_2 = f\left(x_0 + \frac{h}{2}, \quad y_0 + y_0'\frac{h}{2} + y_0''\frac{h^2}{4 \cdot 2!} + y_0'''\frac{h^3}{8 \cdot 3!} + \cdots + f\frac{h^n}{2^n \cdot n!}, \right.$$
$$\left. y_0' + y_0''\frac{h}{2} + \cdots, \cdots\right).$$

We shall continue the expansions of the required values of f only as far as the terms in h^3. It may help the reader if we first work through a simple case, say with $n=2$: with $y'=u$ and subscripts denoting derivatives of f we have

$$\frac{2}{h^2} k_2 = f\left(x_0 + \frac{h}{2}, \quad y_0 + \frac{h}{2}u + \frac{h^2}{8}f, \quad u + \frac{h}{2}f\right)$$

$$= f + \frac{h}{2}f_x + \left(\frac{hu}{2} + \frac{h^2}{8}f\right)f_y + \frac{h}{2}ff_u + \frac{h^2}{8}f_{xx} + \left(\frac{h^2 u}{4} + \frac{h^3 f}{16}\right)f_{xy} +$$

$$+ \frac{h^2}{4}ff_{xu} + \left(\frac{h^2 u^2}{8} + \frac{h^3 uf}{16}\right)f_{yy} + \left(\frac{h^2 uf}{4} + \frac{h^3 f^2}{16}\right)f_{yu} + \frac{h^2 f^2}{8}f_{uu} + \cdots.$$

(Again all function values are evaluated at $x = x_0$ unless otherwise specified.) In general, using the notation of (2.51), (2.53), (2.55) in Ch. I and neglecting powers of h greater than the third, we obtain the expressions

$$k_1 = \frac{h^n}{n!}f,$$

$$k_2 = \frac{h^n}{n!}\left[f + \frac{h}{2}Df + \frac{h^2}{8}(Ef - f_{n-1}Df) + \frac{h^3}{48}(Ff - 3DfDf_{n-1})\right],$$

$$k_3 = \frac{h^n}{n!}\left[f + \frac{h}{2}Df + \frac{h^2}{8}(Ef + f_{n-1}Df) + \right.$$
$$\left. + \frac{h^3}{48}(Ff + 3DfDf_{n-1} + 3f_{n-1}(Ef - f_{n-1}Df))\right],$$

$$k_4 = \frac{h^n}{n!}\left[f + hDf + \frac{h^2}{2}Ef + \right.$$
$$\left. + \frac{h^3}{6}\left(Ff + \frac{3}{4}f_{n-1}(Ef + f_{n-1}Df) + \frac{3}{2}f_{n-2}Df\right)\right].$$

$$(2.11)$$

The corrections $k^{(0)}, k^{(1)}, \ldots, k^{(n-1)}$ from which the new values of y and its derivatives $y^{(\nu)}$ at $x = x_0 + h$ will be calculated are now formed by linearly combining these four expressions for the k_ϱ as in (2.5). The coefficients of the combination are then determined by equating factors of terms involving powers of h up to the $(n+2)$-th with the corresponding factors in (2.9); we find that

from $\dfrac{h^n}{n!} f$ 　　　$\gamma_{\nu 1} + \gamma_{\nu 2} + \gamma_{\nu 3} + \gamma_{\nu 4} = \dbinom{n}{\nu}$,

from $\dfrac{h^{n+1}}{n!} Df$ 　　$\dfrac{1}{2} \gamma_{\nu 2} + \dfrac{1}{2} \gamma_{\nu 3} + \gamma_{\nu 4} = \dbinom{n+1}{\nu} \dfrac{1}{n+1}$,

from $\dfrac{h^{n+2}}{n!} Ef$ 　　$\dfrac{1}{8} \gamma_{\nu 2} + \dfrac{1}{8} \gamma_{\nu 3} + \dfrac{1}{2} \gamma_{\nu 4} = \dbinom{n+2}{\nu} \dfrac{1}{(n+1)(n+2)}$,

from $\dfrac{h^{n+2}}{n!} f_{n-1} Df$ 　$- \dfrac{1}{8} \gamma_{\nu 2} + \dfrac{1}{8} \gamma_{\nu 3} = 0.$

These four equations for the $\gamma_{\nu 1}, \dots, \gamma_{\nu 4}$ have the solution

$$\gamma_{\nu 1} = \binom{n+2}{\nu} \frac{(n-\nu)^2}{(n+1)(n+2)},$$

$$\gamma_{\nu 2} = \gamma_{\nu 3} = \binom{n+2}{\nu} \frac{2(n-\nu)}{(n+1)(n+2)}, \qquad (2.12)$$

$$\gamma_{\nu 4} = \binom{n+2}{\nu} \frac{2-(n-\nu)}{(n+1)(n+2)}.$$

Substituting back into (2.5), we obtain the final expressions for the $k^{(\nu)}$:

$$k^{(\nu)} = \binom{n+2}{\nu} \frac{1}{(n+1)(n+2)} \times$$
$$\times [(n-\nu)^2 k_1 + 2(n-\nu)(k_2 + k_3) + (2 - (n-\nu)) k_4]. \qquad (2.13)$$

With these expressions for the $k^{(\nu)}$ and with the k_1, k_2, k_3, k_4 given by (2.11), let us now compare the next terms in the Taylor series for $\dfrac{h^\nu}{\nu!} y_1^{(\nu)}$ and $\dfrac{h^\nu}{\nu!} y^{(\nu)}(x_1)$, i.e. the terms in h^{n+3}. The term from $\dfrac{h^\nu}{\nu!} y_1^{(\nu)}$ is the term in (2.13) after substitution from (2.11), i.e.

$$\frac{h^{n+3}}{n!} \binom{n+2}{\nu} \frac{1}{(n+1)(n+2)} \times$$
$$\times \left[\frac{4-q}{12} Ff + \frac{1}{4} f_{n-1} Ef + \frac{1-q}{4} f_{n-1}^2 Df + \frac{2-q}{4} f_{n-2} Df \right],$$

where $q = n - \nu$; the corresponding term from $\dfrac{h^\nu}{\nu!} y^{(\nu)}(x_1)$ is the last term in (2.9), which may be written

$$\frac{h^{n+3}}{n!} \binom{n+2}{\nu} \frac{1}{(n+1)(n+2)} \frac{1}{q+3} [Ff + f_{n-1} Ef + f_{n-2} Df].$$

These expressions are identical if, and only if, $q = 1$, i.e. $\nu = n - 1$. We have therefore obtained identity of the two Taylor series as far as the terms in h^{n+2} for $y, hy', \dots, \dfrac{h^{n-2}}{(n-2)!} y^{(n-2)}$ and as far as the terms in h^{n+3} for $\dfrac{h^{n-1}}{(n-1)!} y^{(n-1)}$.

Thus, as a method for determining the ordinate y (cf. the remarks at the end of the next subsection), the Runge-Kutta method is of the $(n+2)$-th order (in the sense of § 1.3) for $n>1$ and of the fourth order for $n=1$. Actually this assertion about the case $n=1$ does not follow immediately from our formulae, for they show only that the method is of at least the fourth order when $n=1$, but it can in fact be shown (which we do not do here) that the terms in h^5 in the Taylor series for y_1 and $y(x_1)$ are not identical.

2.4. Hints for using the Runge-Kutta method

For practical calculation it is usually more convenient to work with the quantities

$$v_\nu = \frac{h^\nu}{\nu!} y^{(\nu)} \qquad (\nu = 0, 1, \ldots, n-1) \tag{2.14}$$

than with the derivatives $y^{(\nu)}$ direct and to transform the function f and the truncated Taylor series T_ν correspondingly. We denote the initial values of the v_ν by

$$v_{\nu,0} = \frac{h^\nu}{\nu!} y_0^{(\nu)} \tag{2.15}$$

and introduce the notation

$$f\left(x, y, \frac{1!}{h} v_1, \frac{2!}{h^2} v_2, \ldots, \frac{(n-1)!}{h^{n-1}} v_{n-1}\right) = \Phi(x, v_0, v_1, v_2, \ldots, v_{n-1}), \tag{2.16}$$

$$\left.\begin{aligned}
t_\nu(\alpha) &= \frac{h^\nu}{\nu!} T_\nu(\alpha) = v_{\nu,0} + \alpha \binom{\nu+1}{\nu} v_{\nu+1,0} + \alpha^2 \binom{\nu+2}{\nu} v_{\nu+2,0} + \cdots + \\
&\quad + \alpha^{n-\nu-1} \binom{n-1}{\nu} v_{n-1,0} \qquad (\nu = 0, 1, \ldots, n-1).
\end{aligned}\right\} \tag{2.17}$$

With this new notation, (2.10) becomes

$$\left.\begin{aligned}
k_1 &= \frac{h^n}{n!} \Phi(x_0, v_{0,0}, v_{1,0}, v_{2,0}, \ldots, v_{n-1,0}), \\
k_2 &= \frac{h^n}{n!} \Phi\left(x_0 + \frac{h}{2}, t_0\left(\frac{1}{2}\right) + \frac{k_1}{2^n}, t_1\left(\frac{1}{2}\right) + \frac{k_1}{2^{n-1}}\binom{n}{1},\right. \\
&\quad t_2\left(\frac{1}{2}\right) + \frac{k_1}{2^{n-2}}\binom{n}{2}, \ldots, t_{n-2}\left(\frac{1}{2}\right) + \frac{k_1}{4}\binom{n}{n-2}, t_{n-1}\left(\frac{1}{2}\right) + \frac{k_1}{2}\binom{n}{n-1}\right), \\
k_3 &= \frac{h^n}{n!} \Phi\left(x_0 + \frac{h}{2}, t_0\left(\frac{1}{2}\right) + \frac{k_1}{2^n}, t_1\left(\frac{1}{2}\right) + \frac{k_1}{2^{n-1}}\binom{n}{1},\right. \\
&\quad t_2\left(\frac{1}{2}\right) + \frac{k_1}{2^{n-2}}\binom{n}{2}, \ldots, t_{n-2}\left(\frac{1}{2}\right) + \frac{k_1}{4}\binom{n}{n-2}, t_{n-1}\left(\frac{1}{2}\right) + \frac{k_2}{2}\binom{n}{n-1}\right), \\
k_4 &= \frac{h^n}{n!} \Phi\left(x_0 + h, t_0(1) + k_3, t_1(1) + k_3\binom{n}{1}, t_2(1) + k_3\binom{n}{2}, \ldots,\right. \\
&\quad t_{n-2}(1) + k_3\binom{n}{n-2}, t_{n-1}(1) + k_3\binom{n}{n-1}\right),
\end{aligned}\right\} \tag{2.18}$$

while the formula (2.13) for the corrections $k^{(\nu)}$ remains unaltered. The new values

$$v_{\nu,1} = \frac{h^\nu}{\nu!}\, y_1^{(\nu)}$$

are then found from the transformed version of (2.4), namely

$$v_{\nu,1} = t_\nu(1) + k^{(\nu)}. \tag{2.19}$$

We discern the following

Computing procedure: First of all, the right-hand side of the given differential equation (2.1) and the initial values (2.2) are transformed by the introduction of the auxiliary variables v_ν of (2.14) into a function Φ and initial values $v_{\nu,0}$, respectively, as in (2.16), (2.15). The values of Φ needed in (2.18) for determing k_1, k_2, k_3, k_4 are calculated with the aid of the truncated Taylor series $t_\nu(\frac{1}{2})$, $t_\nu(1)$ defined in (2.17); then the corrections $k^{(\nu)}$ are found from the k_ϱ using (2.13); finally (2.19) gives the new values $v_{\nu,1}$. These serve as starting values for the next step in the continuation of the solution.

Tables II/4—II/7 exhibit the appropriate formulae for first-, second-, third- and fourth-order equations (coefficients for corresponding computing schemes[1] for n-th order differential equations where $5 \leq n \leq 10$

Table II/4. *Runge-Kutta scheme for differential equations of the first order.*

$$y' = f(x, y)$$

x	y	$k_\nu = h \cdot f(x, y)$	Correction
x_0	y_0	k_1	$k = \frac{1}{6}(k_1 + 2k_2 + 2k_3 + k_4)$
$x_0 + \frac{1}{2}h$	$y_0 + \frac{1}{2}k_1$	k_2	
$x_0 + \frac{1}{2}h$	$y_0 + \frac{1}{2}k_2$	k_3	
$x_0 + h$	$y_0 + k_3$	k_4	
$x_1 = x_0 + h$	$y_1 = y_0 + k$		

Table II/5. *Runge-Kutta scheme for differential equations of the second order.*

$$y'' = f(x, y, y')$$

x	y	$h\,y' = v_1$	$k_\nu = \frac{h^2}{2} f\left(x, y, \frac{v_1}{h}\right)$	Correction
x_0	y_0	v_{10}	k_1	$k = \frac{1}{3}(k_1 + k_2 + k_3)$
$x_0 + \frac{1}{2}h$	$y_0 + \frac{1}{2}v_{10} + \frac{1}{4}k_1$	$v_{10} + k_1$	k_2	
$x_0 + \frac{1}{2}h$	$y_0 + \frac{1}{2}v_{10} + \frac{1}{4}k_1$	$v_{10} + k_2$	k_3	$k' = \frac{1}{3}(k_1 + 2k_2 + 2k_3 + k_4)$
$x_0 + h$	$y_0 + v_{10} + k_3$	$v_{10} + 2k_3$	k_4	
$x_1 = x_0 + h$	$y_1 = y_0 + v_{10} + k$	$v_{11} = v_{10} + k'$		

---

[1] ZURMÜHL, R.: Runge-Kutta-Verfahren zur numerischen Integration von Differentialgleichungen n-ter Ordnung. Z. Angew. Math. Mech. **28**, 173—182 (1948).

have been calculated by ZURMÜHL[1]). The actual calculations are best set out in tables of similar form.

Table II/6. *Runge-Kutta scheme for differential equations of the third order.*

$$y''' = f(x, y, y', y'')$$

x	y	$hy' = v_1$	$\dfrac{h^2}{2}y'' = v_2$	$k_\nu = \dfrac{h^3}{6}f\left(x, y, \dfrac{v_1}{h}, \dfrac{2v_2}{h^2}\right)$	
x_0	y_0	v_{10}	v_{20}	k_1	k
$x_0 + \frac{1}{2}h$	$y_0 + \frac{1}{2}v_{10} + \frac{1}{4}v_{20} + \frac{1}{8}k_1$	$v_{10} + v_{20} + \frac{3}{4}k_1$	$v_{20} + \frac{3}{2}k_1$	k_2	k'
$x_0 + \frac{1}{2}h$	$y_0 + \frac{1}{2}v_{10} + \frac{1}{4}v_{20} + \frac{1}{8}k_1$	$v_{10} + v_{20} + \frac{3}{4}k_1$	$v_{20} + \frac{3}{2}k_2$	k_3	k''
$x_0 + h$	$y_0 + v_{10} + v_{20} + k_3$	$v_{10} + 2v_{20} + 3k_3$	$v_{20} + 3k_3$	k_4	
$x_1 = x_0 + h$	$y_1 = y_0 + v_{10} + v_{20} + k$	$v_{11} = v_{10} + 2v_{20} + k'$	$v_{21} = v_{20} + k''$		

where $k = \frac{1}{20}(9k_1 + 6k_2 + 6k_3 - k_4)$,

$k' = k_1 + k_2 + k_3$,

$k'' = \frac{1}{2}(k_1 + 2k_2 + 2k_3 + k_4)$.

Table II/7. *Runge-Kutta scheme for differential equations of the fourth order.*

$$y^{IV} = f(x, y, y', y'', y''')$$

x	y	$hy' = v_1$
x_0	y_0	v_{10}
$x_0 + \frac{1}{2}h$	$y_0 + \frac{1}{2}v_{10} + \frac{1}{4}v_{20} + \frac{1}{8}v_{30} + \frac{1}{16}k_1$	$v_{10} + v_{20} + \frac{3}{4}v_{30} + \frac{1}{2}k_1$
$x_0 + \frac{1}{2}h$	$y_0 + \frac{1}{2}v_{10} + \frac{1}{4}v_{20} + \frac{1}{8}v_{30} + \frac{1}{16}k_1$	$v_{10} + v_{20} + \frac{3}{4}v_{30} + \frac{1}{2}k_1$
$x_0 + h$	$y_0 + v_{10} + v_{20} + v_{30} + k_3$	$v_{10} + 2v_{20} + 3v_{30} + 4k_3$
$x_1 = x_0 + h$	$y_1 = y_0 + v_{10} + v_{20} + v_{30} + k$	$v_{11} = v_{10} + 2v_{20} + 3v_{30} + k'$

$\dfrac{h^2}{2}y'' = v_2$	$\dfrac{h^3}{6}y''' = v_3$	$k_\nu = \dfrac{h^4}{24}f\left(x, y, \dfrac{v_1}{h}, \dfrac{2v_2}{h^2}, \dfrac{6v_3}{h^3}\right)$	
v_{20}	v_{30}	k_1	k
$v_{20} + \frac{3}{2}v_{30} + \frac{3}{2}k_1$	$v_{30} + 2k_1$	k_2	k'
$v_{20} + \frac{3}{2}v_{30} + \frac{3}{2}k_1$	$v_{30} + 2k_2$	k_3	k''
$v_{20} + 3v_{30} + 6k_3$	$v_{30} + 4k_3$	k_4	k'''
$v_{21} = v_{20} + 3v_{30} + k''$	$v_{31} = v_{30} + k'''$		

where $k = \frac{1}{15}(8k_1 + 4k_2 + 4k_3 - k_4)$,

$k' = \frac{1}{5}(9k_1 + 6k_2 + 6k_3 - k_4)$,

$k'' = 2(k_1 + k_2 + k_3)$,

$k''' = \frac{2}{3}(k_1 + 2k_2 + 2k_3 + k_4)$.

---

[1] Similar schemes can be found in E. BUKOVICS: Eine Verbesserung und Verallgemeinerung des Verfahrens von BLAESS zur numerischen Integration gewöhnlicher Differentialgleichungen. Öst. Ing.-Arch. **4**, 338—349 (1950).

Simplifications in particular cases. If the function $f(x, y, y', \ldots, y^{(n-1)})$ in the differential equation (2.1) does not depend explicitly on $y^{(n-1)}$, then, as noted in § 2.2, $k_2 = k_3$. In such cases there is consequently one less row to compute in the Runge-Kutta scheme. If f does not depend on some derivative, $y^{(r)}$ say, then only the end value v_{r1} in the column corresponding to $y^{(r)}$ need be calculated since the three auxiliary values between v_{r0} and v_{r1} are superfluous.

Rule of thumb for the size of step length h. A rough guide which is often used for finding a reasonable length of step is that h should be such that k_2 and k_3 are approximately equal; to be more specific, the difference between k_2 and k_3 should not exceed in magnitude a few per cent of the difference between k_1 and k_2, otherwise a smaller step should be taken. The ratio $\dfrac{k_2 - k_3}{k_1 - k_2}$ is a measure of "sensitiveness" (cf. the "step index" S defined in § 3.4, which can also serve as a guide to the length of step for the Runge-Kutta method with $n = 1$. S is approximately twice as big as the above measure of sensitiveness).

Indications of the magnitude of the error. A satisfactory estimate for the error $|y_1^{(v)} - y^{(v)}(x_1)|$ does not yet exist for the general case (cf. the remarks in § 1.3) but a rough guide which is often employed is provided by a comparison between solutions calculated with steps of length h and $2h$, respectively, as discussed in § 1.3. For the case $n = 1$ the Taylor series for the approximation to y coincides with the Taylor series for the exact solution as far as the term in h^4, so that, from (1.7), the error in the calculation with steps of length h should be roughly $\frac{1}{15}$ of the difference between the results of this calculation and those of the calculation with steps of length $2h$. For the cases with $n > 1$ not all the quantities $y, y', \ldots, y^{(n-1)}$ are determined to the same order of accuracy at each step, but over a large number of steps it may be expected that the influence of the least accurately calculable derivative, namely $y^{(n-1)}$, will determine the overall accuracy; since the method is always of the fourth order (§ 1.3, I) for this derivative, we may still appropriately use the factor $\frac{1}{15}$ as long as no better error estimates exist.

For the first-order equation $y' = f(x, y)$ BIEBERBACH[1] has used a Taylor series expansion to establish the error estimate

$$|y_1 - y(x_1)| < \frac{6MN|x_1 - x_0|^5 |N^5 - 1|}{|N - 1|},$$

[1] BIEBERBACH, L.: Theorie der Differentialgleichungen, 3rd ed., p. 54. Berlin 1930. — On the remainder of the Runge-Kutta formula in the theory of ordinary differential equations. Z. Angew. Math. Phys. **2**, 233–248 (1951); this paper also contains error estimates for n-th order equations (2.1) and for systems of first-order equations. Other error estimates have been given by E. BUKOVICS: Beiträge zur numerischen Integration, II. Mh. Math. **57**, H. 4 (1953), and J. ALBRECHT: Beiträge zum Runge-Kutta-Verfahren. Z. Angew. Math. Mech. **35**, 100–110 (1955).

in which M and N are numbers such that

$$|f(x,y)| < M, \qquad \left|\frac{\partial^{j+k} f}{\partial x^j \partial y^k}\right| < \frac{N}{M^{k-1}} \quad \text{for} \quad j+k \leq 3, \ |x-x_0| N < 1$$

in a domain $|x-x_0| < a$, $|y-y_0| < b$ with $a \geq h$ and $b > Ma$.

2.5. Terminal checks and iteration methods

The incorporation of current checks[1] in the calculation is to be strongly recommended. A very good check is provided by a second calculation with a double-length step; it is best used as a current check (with the two calculations carried out concurrently) so that any errors which arise are noticed as soon as possible. In addition we can apply the following terminal check when the calculation has been taken as far as is required. First we construct the difference table of a convenient multiple of f, say $h^n f$ (this in itself may reveal some errors by a lack of smoothness in the higher differences); the differences can then be used in a finite-difference quadrature formula to evaluate the repeated integral occurring in (1.9). In general, differences of a sufficiently high order do not exist near the ends of the table, so that one cannot use the same formula throughout the table.

In the case of a differential equation of the first order, for example, one would use the following formulae:

at the beginning of the table

$$\int_{x_r}^{x_{r+1}} f \, dx \approx h \sum_{\varrho=0}^{p} (-1)^\varrho \beta_\varrho^* \, \nabla^\varrho f_{r+\varrho} = h \left[f_r + \frac{1}{2} \nabla f_{r+1} - \frac{1}{12} \nabla^2 f_{r+2} + \right.$$
$$\left. + \frac{1}{24} \nabla^3 f_{r+3} - \frac{19}{720} \nabla^4 f_{r+4} + \frac{3}{160} \nabla^5 f_{r+5} - \cdots \right], \qquad (2.20)$$

in the middle of the table

$$\int_{x_{r-1}}^{x_{r+1}} f \, dx \approx h \left[2 f_r + \frac{1}{3} \nabla^2 f_{r+1} - \frac{1}{90} \nabla^4 f_{r+2} + \frac{1}{756} \nabla^6 f_{r+3} - \cdots \right], \qquad (2.21)$$

at the end of the table

$$\int_{x_{r-1}}^{x_r} f \, dx \approx h \left[f_r - \frac{1}{2} \nabla f_r - \frac{1}{12} \nabla^2 f_r - \frac{1}{24} \nabla^3 f_r - \cdots \right]. \qquad (2.22)$$

These are respectively formulae (2.11), (2.14) and (2.9) of Ch. I.

---

[1] Cf. E. LINDELÖF: Remarques sur l'integration .... Acta Soc. Sci. Fenn. A2 1938, Nr. 13, 11. — SANDEN, H. v.: Praxis der Differentialgleichungen, 3rd ed., p. 29. Berlin 1945.

For differential equations of higher order, terminal checks can be applied in a variety of ways. For instance, instead of using formula (1.9), we could check all derivatives individually by using

$$y^{(v)}(x_{r+1}) = y^{(v)}(x_r) + \int_{x_r}^{x_{r+1}} y^{(v+1)}(x)\,dx \qquad (2.23)$$

with $y^{(n)} = f$ and evaluating the integrals by the formulae (2.20) to (2.22) just quoted.

Another way is to evaluate the integral in formula (1.9) by the repeated-integration formulae (2.30), (2.45) of Ch. I; thus at the beginning of the table we use

$$y_{r+1} \approx y_r + h y_r' + \frac{h^2}{2} y_r'' + \cdots + \frac{h^{n-1}}{(n-1)!} y_r^{(n-1)} + h^n \sum_{\varrho=0}^{p} \gamma_{n,\varrho} \Delta^\varrho f_r \qquad (2.24)$$

and in the middle

$$y_{r+1} \approx \begin{cases} \left\{ \begin{aligned} & y_{r-1} + 2 \sum_{\varrho=0}^{\frac{n-3}{2}} \frac{h^{2\varrho+1}}{(2\varrho+1)!} y_r^{(2\varrho+1)} \\ & 2y_r - y_{r-1} + 2 \sum_{\varrho=1}^{\frac{n}{2}-1} \frac{h^{2\varrho}}{(2\varrho)!} y_r^{(2\varrho)} \end{aligned} \right\} + \\ + 2h^n \sum_{\varrho=0}^{p/2} \beta_{n,\varrho}^{**} \nabla^{2\varrho} f_{r+\varrho} \begin{cases} \text{for } n \text{ odd,} \\ \text{for } n \text{ even.} \end{cases} \end{cases} \qquad (2.25)$$

If corresponding formulae are written down for each derivative occurring in the differential equation, they can be used as the basis of an iterative method of solution: if we have an approximate solution, say the v-th approximation, denoted by

$$y_r^{[v]}, y_r'^{[v]}, \dots, y_r^{(n-1)[v]}$$

we can, in general, obtain a better approximation, the $(v+1)$-th, by calculating the corresponding f values $f_r^{[v]}$ and using these in the right-hand sides of (2.24), (2.25) and the corresponding formulae for the derivatives which occur.

2.6. Examples

I. A differential equation of the first order. Consider again the example of § 1.5, i.e. the initial-value problem (1.21)

$$y' = y - \frac{2x}{y}; \qquad y(0) = 1.$$

We calculate a solution first with a step length $h = 0\cdot2$. The calculation for the first few steps is reproduced in Table II/8, in which an extra column is kept for recording the values of the auxiliary quantity $2x/y$

needed in the calculation of $hf(x, y)$. As far as the point $x = 0.6$ the actual results of the calculation are distinguished by bold-face type; for steps beyond this point only the results are reproduced.

Table II/8. *A Runge-Kutta calculation for a first-order equation*

x	y	$\dfrac{2x}{y}$	$\dfrac{h}{2}f = 0 \cdot 1\left(y - \dfrac{2x}{y}\right)$	$3k$ and k
0	**1**	0	0·1	
0·1	**1·1**	0·181818	0·0918182	0·5496877
0·1	**1·091818**	0·183181	0·0908637	0·1832292
0·2	**1·181727**	0·338488	0·0843239	
0·2	**1·1832292**	0·338058	0·0845171	
0·3	**1·267746**	0·473281	0·0794465	0·4753128
0·3	**1·262676**	0·475181	0·0787495	0·1584376
0·4	**1·340728**	0·596691	0·0744037	
0·4	**1·3416668**	0·596274	0·0745393	
0·5	**1·416206**	0·706112	0·0710094	0·4248538
0·5	**1·412676**	0·707831	0·0704845	0·1416179
0·6	**1·482636**	0·809369	0·0673267	
0·6	**1·4832815**	0·809016	0·0674269	
0·8	1·6125140		0·0620282	
1	1·7321419		0·0577512	
1·2	1·8440401		0·0542565	
1·4	1·949547		0·051334	
1·6	2·049660		0·048846	
1·8	2·145148		0·046698	
2	2·236624		0·044828	

For checking, and also for assessing the accuracy, the calculation is repeated with a double-length step $h = 0.4$; the results are given in Table II/9.

Table II/9. *Corresponding calculation with a double-length step $h = 0.4$*

x	y	$\dfrac{2x}{y}$	$\dfrac{h}{2}f = 0 \cdot 2\left(y - \dfrac{2x}{y}\right)$	$3k$ and k
0	**1**	0	0·2	
0·2	**1·2**	0·333333	0·1733333	1·0261974
0·2	**1·173333**	0·340909	0·1664848	0·3420658
0·4	**1·332970**	0·600164	0·1465612	
0·4	**1·3420658**	0·596096	0·1491940	
0·6	**1·491260**	0·804689	0·1373142	0·8141486
0·6	**1·479380**	0·811151	0·1336458	0·2713829
0·8	**1·609358**	0·994185	0·1230346	
0·8	**1·6134487**			
1·2	1·84600			
1·6	2·05367			
2	2·24486			

We now collect together (in Table II/10) the results for the points $x = 0.4 \times k$ $(k = 1, 2, \ldots, 5)$ and calculate corrections as in (1.7). Comparison with the exact solution $y = \sqrt{2x + 1}$ yields the actual errors in the values $\tilde{y}$ obtained with $h = 0.2$ and in the "improved" values $[y] = \tilde{y} - \delta$; the latter are seen to be considerably smaller than the former.

Table II/10. *Comparison of the accuracies of the Runge-Kutta solution and the improved solution*

x	Results of Runge-Kutta calculation with		$\delta = \frac{1}{15}(\tilde{\tilde{y}} - \tilde{y})$	$[y] = \tilde{y} - \delta$	Error in $\tilde{y}$	Error in $[y]$
	$h = 0.2$ $\tilde{y}$	$h = 0.4$ $\tilde{\tilde{y}}$				
0.4	1.341 6669	1.342 0658	0.000 0266	1.341 6403	+0.000 0262	−0.000 0004
0.8	1.612 5140	1.613 4487	0.000 0623	1.612 4517	0.000 0624	+0.000 0001
1.2	1.844 040	1.845 99	0.000 130	1.843 910	0.000 131	+0.000 001 5
1.6	2.049 660	2.053 64	0.000 266	2.049 395	0.000 270	0.000 004 4
2	2.236 624	2.244 80	0.000 545	2.236 079	0.000 556	0.000 010 9

II. A System of first-order differential equations. Here we consider the Euler equations for the motion of an unsymmetrical top (principle moments of inertia A, B, C) subject to a frictional resistance proportional to the instantaneous angular velocity vector u_1, u_2, u_3. With differentiation in time denoted by a dot the equations run [1]

$$A \dot{u}_1 = (C - B) u_2 u_3 - \varepsilon u_1,$$
$$B \dot{u}_2 = (A - C) u_3 u_1 - \varepsilon u_2,$$
$$C \dot{u}_3 = (B - A) u_1 u_2 - \varepsilon u_3.$$

If $B = 2A$, $C = 3A$, $\varepsilon = 0.6$ A/sec., we obtain the system

$$\dot{u}_1 = u_2 u_3 - 0.6 u_1, \qquad \dot{u}_2 = - u_1 u_3 - 0.3 u_2, \qquad \dot{u}_3 = \tfrac{1}{3} u_1 u_2 - 0.2 u_3.$$

Let the initial angular velocity be given by $u_1 = 1$, $u_2 = 1$, $u_3 = 0$ at $t = 0$.

Application of the Runge-Kutta method occasions little difficulty. We have only to set up three schemes, one for each unknown, and it suffices to exhibit the first two steps with the step $h = 0.2$ and, for comparison, one step with the double length interval $h = 0.4$. This is done in Table II/11.

III. A differential equation of higher order. In the calculation of the laminar boundary layer on a flat plate parallel to a stream, the following boundary-value problem arises:

$$\frac{d^3 \zeta}{d \xi^3} = - \zeta \frac{d^2 \zeta}{d \xi^2}; \qquad \zeta(0) = \zeta'(0) = 0, \qquad \zeta' \to 2 \quad \text{as} \quad \xi \to \infty.$$

---

[1] See, for example, Handbuch der Physik, Vol. V, p. 405, Berlin 1927, or A. Föppl: Vorlesungen über technische Mechanik, Vol. 4 (Dynamik), 10th ed., p. 209. München und Berlin 1944, or J. Wittenburg: Dynamics of Systems of Rigid Bodies, Stuttgart 1977

Table II/11. Runge-Kutta method applied to a system of three first-order differential equations

t	u_1	u_2	u_3	$u_2 u_3$	$\frac{h}{2}\dot u_1$	$3k$ and k	$-u_1 u_3$	$\frac{h}{2}\dot u_2$	$3k$ and k	$\frac{1}{3}u_1 u_2$	$\frac{h}{2}\dot u_3$	$3k$ and k
0	1	1	0	0	−0·06		0	−0·03		0·33333	0·033333	
0·1	0·94	0·97	0·033333	0·032333	−0·053167	−0·322194	−0·031133	−0·032233	−0·191523	0·30393	0·029727	0·178546
0·1	0·946833	0·967767	0·029727	0·028877	−0·053933	−0·107398	−0·02815	−0·031848	−0·063841	0·30544	0·029549	0·059515
0·2	0·892134	0·936304	0·059098	0·055534	−0·047994		−0·05272	−0·033361		0·27843	0·026661	
0·2	**0·892602**	**0·936159**	**0·059515**	0·055572	−0·047985		−0·05312	−0·033397		0·27854	0·026664	
0·3	0·844617	0·902762	0·086179	0·07780	−0·042897	−0·259764	−0·07279	−0·034362	−0·205048	0·25416	0·023692	0·142967
0·3	0·849705	0·901797	0·083207	0·07504	−0·043478	−0·086588	−0·07070	−0·034124	−0·068349	0·25542	0·023878	0·047656
0·4	0·805646	0·867911	0·107271	0·09310	−0·039029		−0·08642	−0·034679		0·23308	0·021163	
0·4	**0·806014**	**0·867810**	**0·107171**	$2u_2 u_3$			$-2u_1 u_3$			$\frac{2}{3}u_1 u_2$		
0·2	1	1	0	0	−0·12		0	−0·06		0·66667	0·066667	
0·2	0·88	0·94	0·066667	0·12533	−0·093067	−0·581606	−0·11733	−0·068133	−0·396675	0·55147	0·052480	0·322239
0·4	0·906933	0·931867	0·052480	0·09781	−0·099051	−0·193869	−0·09519	−0·065431	−0·132225	0·56343	0·054244	0·107413
0·4	0·801898	0·869138	0·108488	0·18858	−0·077370		−0·17399	−0·069547		0·46464	0·042124	
0·4	**0·806131**	**0·867755**	**0·107479**									

This may be reduced to the initial-value problem[1]

$$y''' = -y\,y'', \qquad y(0) = y'(0) = 0, \qquad y''(0) = 1.$$

The Runge-Kutta calculations for $h = 0.5$ are set out in Table II/12; for the steps from $x = 1$ to $x = 3$ only the results are given.

Table II/12. *Solution of a third-order differential equation by* RUNGE-KUTTA

x	y	$v_1 = \frac{1}{2}y'$	$v_2 = \frac{1}{8}y''$	$yv_2 = -6k$	k	$\dfrac{k_2+k_3}{2k''}$ $20k$	$\dfrac{k'}{k''}$ k
0	0	0	0·125	0	0		
0·25	0·03125		0·125	0·0039063	−0·0006510	−0·0012969	−0·0012969
0·25	0·03125		0·1240234	0·0038756	−0·0006459	−0·0051444	−0·0025722
0·5	0·1243541		0·1230622	0·0153033	−0·0025506	−0·0052308	−0·0002615
0·5	**0·1247385**	**0·2487031**	**0·1224278**	0·0152715	−0·0025453		
0·75	0·2793798		0·1186099	0·0331372	−0·0055229	−0·0108378	−0·0133831
0·75	0·2793798		0·1141435	0·0318894	−0·0053149	−0·0329269	−0·0164635
1	0·4905545		0·1064831	0·0522358	−0·0087060	−0·0792285	−0·0039614
1	**0·4919080**	**0·4801756**	**0·1059643**	0·0521247	−0·0086874		
1·5	1·0679173	0·6605306	0·0722886		−0·0128664		
2	1·787924	0·767123	0·035606		−0·010610		
2·5	2·581205	0·811538	0·012286		−0·005285		
3	3·400614	0·824154	0·003316		−0·001880		

Table II/13. *Corresponding calculation with double-length step*

x	y	$v_1 = y'$	$v_2 = \frac{1}{2}y''$	k
0	0	0	0·5	0
0·5	0·125		0·5	−0·020833
0·5	0·125		0·468750	−0·019531
1	0·480469		0·441406	−0·070694
1	**0·491425**	**0·959636**	0·424289	−0·069502
1·5	1·068627		0·320036	−0·114000
1·5	1·068627		0·253289	−0·090224
2	1·785126		0·153618	−0·091409
2	**1·787377**	**1·534488**	0·139609	−0·083178
2·5	2·579126		0·014842	−0·012760
2·5	2·579126		0·120469	−0·103568
3	3·357906		−0·171096	+0·191508
3	**3·379570**	**1·614200**	0·077446	

[1] MOHR, E.: Dtsch. Math. **4**, 485 (1939). — SCHLICHTING, H.: Grenzschicht-Theorie. 483 pp. Karlsruhe 1951. — Modern Developments in Fluid Dynamics (ed. S. GOLDSTEIN), Vol. I, p. 135. London: Oxford University Press 1938.

For comparison the calculation was repeated with the double-length step $h = 1$; this is exhibited in Table II/13, from which several auxiliary columns have been omitted.

For the last step, from $x = 2$ to $x = 3$, an alternation in sign appears in the v_2 and k columns, indicating that the interval is much too large and that we can no longer have any confidence in the values obtained; these values for the point $x = 3$ do, in fact, show a considerable deviation from those obtained by the calculation with the smaller interval.

§3. Finite-difference methods for differential equations of the first order

Among the approximate methods which exist at present, the finite-difference methods are the most accurate in general. Among these, the interpolation methods are superior to the extrapolation methods in that they give a considerable improvement in accuracy with only a moderate increase in the amount of labour involved. Consequently, interpolation methods are in far more extensive use nowadays than extrapolation methods. Of the interpolation methods, the central-difference method has several advantages over the Adams interpolation method: the computations are simpler, the convergence is more rapid and smaller error limits can be derived. Naturally only very general advice can be given to guide the computer in his choice of a suitable method. *Finite-difference methods are very suitable when the functions being dealt with are smooth and the differences decrease rapidly with increasing order; calculations with these methods are best carried out with a fairly small length of step. On the other hand, if the functions are not smooth, perhaps given by experimental results, or if we want to use a larger step, then the Runge-Kutta method is to be preferred; it is also advantageous to use this method when we have to change the length of step frequently, particularly when this change is a decrease. Clearly we should not choose too large a step even for the Runge-Kutta method.* (Cf.[1] the remarks in § 2.4.)

3.1. Introduction

Suppose that the differential equation (1.10)

$$y' = f(x, y)$$

is to be integrated numerically with the initial condition $y = y_0$ at $x = x_0$. As in § 1.4, let the range of integration be covered by the equally spaced points $x_0, x_1, \ldots, x_n$ with the constant difference $h = \Delta x_\nu = x_{\nu+1} - x_\nu$

---

[1] MILNE, W. E.: Note on the Runge-Kutta method. Research Paper RP 2101, J. Res. Nat. Bur. Stand. **44**, 549—550 (1950), gives examples for which the central-difference method yields substantially better results than the Runge-Kutta method.

(the step length) and let y_ν be an approximation to the value $y(x_\nu)$ of the exact solution at the point x_ν. The finite-difference methods are based on the integrated form (1.13)

$$y(x_{r+1}) = y(x_r) + \int_{x_r}^{x_{r+1}} f(x, y(x)) \, dx \qquad (3.1)$$

of equation (1.10), as were the crude methods of § 1.5, but here the integral is approximated more accurately. Suppose that the integration has already been carried as far as the point $x = x_r$ so that approximations $y_1, \ldots, y_{r-2}, y_{r-1}, y_r$, and hence also approximate values $f_r = f(x_r, y_r)$, are known. We now have to calculate y_{r+1}. In finite-difference methods, formulae for doing this are derived by replacing the integrand in (3.1) by a polynomial $P(x)$ which takes the values f_ν at a certain number of points x_ν and then integrating this polynomial over the interval x_r to x_{r+1}. This basic idea can be used in a variety of ways (cf. § 3.3)[1] but we always need to have a sequence of approximations f_ν before we can start the step-by-step procedures defined by the finite-difference formulae (see § 3.3); consequently the finite-difference methods have two distinct stages: the first is the calculation of "starting values", in which the first few approximations $y_1, y_2, \ldots$, the "starting values" (we reserve the term "initial values" for values at the initial point $x = x_0$), sufficiently many to calculate the values f_ν required for the first application of the finite-difference formula, are obtained by some other means; then follows the main calculation, in which the finite-difference formula is used to continue the solution step by step as far as required. The starting values should be calculated as accurately as possible (cf. § 1.2).

3.2. Calculation of starting values

The starting values needed for the main calculation can be obtained in a variety of ways. As has already been stressed, particular care must be exercised in the calculation of these starting values, for the whole calculation can be rendered useless by inaccuracies in them.

We now mention several possible ways of obtaining starting values; anyone with little experience might restrict himself to the first two only to begin with.

I. Using some other method of integration. Provided that it is sufficiently accurate, any method of integration which does not require starting values (as distinct from initial values) can be used. Bearing in mind the high accuracy desired, one would normally choose the

---

[1] The interpolation idea is used by W. Quade: Numerische Integration von gewöhnlichen Differentialgleichungen nach Hermite. Z. Angew. Math. Mech. **37**, 161—169 (1957).

Runge-Kutta method; further, one would work preferably with a step of half the length to be used in the main calculation and with a greater number of decimals.

II. Using the Taylor series for $y(x)$. If the function $f(x, y)$ is of simple analytical form, we can determine the derivatives $y'(x_0)$, $y''(x_0)$, $y'''(x_0)$, ... by differentiation of the differential equation (1.10); starting values can then be calculated from the Taylor series

$$y(x_\nu) = y(x_0) + \frac{\nu h}{1!} y'(x_0) + \frac{(\nu h)^2}{2!} y''(x_0) + \cdots, \tag{3.2}$$

of which as many terms are taken as are necessary for the truncation not to affect the last decimal carried (always assuming that the series converges).

Several of the finite-difference methods need three starting values, and for these it suffices to use (3.2) for $\nu = \pm 1$; this usually possesses advantages over using (3.2) for $\nu = 1, 2$, particularly as regards convergence.

Example. For the example (1.21)

$$y' = y - \frac{2x}{y}, \qquad y(0) = 1$$

of § 1.5 we have

$$y'y - y^2 + 2x = 0,$$
$$y''y + y'^2 - 2y'y + 2 = 0,$$
$$y'''y + 3y''y' - 2y''y - 2y'^2 = 0,$$
$$\cdots \cdots \cdots \cdots \cdots \cdots$$

Putting $x = 0$ and $y = 1$ we obtain successively the initial values of the derivatives:

$$y'(0) = 1; \quad y''(0) = -1; \quad y'''(0) = 3; \quad y^{IV}(0) = -15; \quad y^V(0) = 105;$$

[generally, $y^{(n)}(0) = (-1)^{n-1} \times 1.3.5. \ldots (2n - 3)$ for $n = 2, 3, \ldots$]. With $h = 0.1$, for example, we have the starting values

$$y(\pm 0.1) = 1 \pm 0.1 - \frac{1}{2!}(0.1)^2 \pm \frac{3}{3!}(0.1)^3 - \frac{15}{4!}(0.1)^4 \pm \frac{105}{5!}(0.1)^5 - \left.\begin{array}{r} \\ - \frac{945}{6!}(0.1)^6 \pm \frac{10395}{7!}(0.1)^7 - \cdots, \end{array}\right\} \tag{3.3}$$

which to seven decimals are

$$y(0.1) = 1.0954451,$$
$$y(-0.1) = 0.8944272.$$

II a. MILNE'S starting procedure. W. E. MILNE[1] has given formulae which bring in the derivatives at the point x_1. They are obtained by eliminating $y^{IV}(x_0)$, $y^V(x_0)$, $y^{VI}(x_0)$ between the Taylor series for $y(x_1)$,

---

[1] MILNE, W. E.: A note on the numerical integration of differential equations. Research paper RP 2046, J. Res. Nat. Bur. Stand. **43**, 537—542 (1949).

$y'(x_1)$, $y''(x_1)$, $y'''(x_1)$ truncated after the term in $y^{VI}(x_0)$. The result is the formula

$$y(x_1) - y(x_0) = \frac{h}{2}\left[y'(x_1) + y'(x_0)\right] - \frac{h^2}{10}\left[y''(x_1) - y''(x_0)\right] +$$
$$+ \frac{h^3}{120}\left[y'''(x_1) + y'''(x_0)\right] + R_6,$$

where

$$R_6 = -\frac{h^7}{100\,800}\, y^{VII}(\xi).$$

The intermediate point ξ in this remainder term lies in the interval $x_0 \leqq \xi \leqq x_1$.

This formula is applicable when the higher derivatives may be expressed in simple analytical form in terms of x and y. For the same degree of accuracy, fewer derivatives need be calculated than with method II but since y_1 appears on the right-hand side, the formula is an equation, rather than an explicit expression, for y_1 and for non-linear differential equations y_1 will usually have to be determined by iteration (cf. § 3.3 method II). The formula can also be used for the main calculation. If we also expand $y(x_2)$ into a Taylor series and truncate after the term in $y^{VI}(x_0)$, we can derive in like manner the formula

$$y(x_2) - 2y(x_1) + y(x_0) = 7h\left[y'(x_1) - y'(x_0)\right] - 3h^2\left[y''(x_1) + y''(x_0)\right] +$$
$$+ \frac{h^3}{12}\left[11y'''(x_1) - 5y'''(x_0)\right] + \frac{1}{480}h^7 y^{VII}(\xi).$$

III. Using quadrature formulae. Using formulae (2.11) and (2.14) of Ch. I, we build up rough values of y_ν, hf_ν, hVf_ν, ... for the first few values of ν and then improve them by an iterative process. Various schemes can be arranged for doing this; we give here a procedure[1] which is suitable for the construction of two (y_1, y_2) or three (y_1, y_2, y_3) starting values. The procedure is completely described by the following formulae (for two starting values, only the formulae framed in dots are needed and the B equations are to be used also for $\nu = 0$):

A. *Rough values.*

$$
\begin{aligned}
&1.\ \tilde{y}_1 = y_0 + hf_0, \quad \text{thence} \quad \tilde{f}_1 = f(x_1, \tilde{y}_1); \quad V\tilde{f}_1 = \tilde{f}_1 - f_0,\\
&2.\ y_1^{[0]} = y_0 + h\left(f_0 + \tfrac{1}{2}V\tilde{f}_1\right), \quad \text{thence} \quad f_1^{[0]},\ Vf_1^{[0]},\\
&3.\ y_2^{[0]} = y_0 + 2hf_1^{[0]}, \quad \text{thence} \quad f_2^{[0]},\ Vf_2^{[0]},\ V^2f_2^{[0]},\\
&4.\ y_1^{[1]} = y_0 + h\left(f_0 + \tfrac{1}{2}Vf_1^{[0]} - \tfrac{1}{12}V^2f_2^{[0]}\right),\\
&\quad y_2^{[1]} = y_0 + h\left(2f_1^{[0]} + \tfrac{1}{3}V^2f_2^{[0]}\right), \text{thence } f_1^{[1]}, f_2^{[1]}, Vf_1^{[1]}, Vf_2^{[1]}, V^2f_2^{[1]},\\
&5.\ y_3^{[1]} = y_1^{[1]} + h\left(2f_2^{[1]} + \tfrac{1}{3}V^2f_3^{[1]}\right), \text{thence } f_3^{[1]}, Vf_3^{[1]}, V^2f_3^{[1]}, V^3f_3^{[1]}.
\end{aligned}
$$

$$(3.4)$$

[1] Other iterative schemes can be found in G. SCHULZ: Interpolationsverfahren zur numerischen Integration gewöhnlicher Differentialgleichungen. Z. Angew. Math. Mech. **12**, 44—59 (1932). — TOLLMIEN, W.: Über die Fehlerabschätzung beim Adamsschen Verfahren. Z. Angew. Math. Mech. **18**, 83—90 (1938), in particular p. 87.

B. *Iterative improvement for* $\nu=1, 2, \ldots$ (or $\nu=0, 1, \ldots$)

$$\left.\begin{aligned} y_1^{[\nu+1]} &= y_0 + h\left(f_0 + \tfrac{1}{2}\nabla f_1^{[\nu]} - \tfrac{1}{12}\nabla^2 f_2^{[\nu]} + \tfrac{1}{24}\nabla^3 f_3^{[\nu]}\right), \\ y_2^{[\nu+1]} &= y_0 + h\left(2 f_1^{[\nu]} + \tfrac{1}{3}\nabla^2 f_2^{[\nu]}\right), \\ y_3^{[\nu+1]} &= y_1^{[\nu+1]} + h\left(2 f_2^{[\nu]} + \tfrac{1}{3}\nabla^2 f_3^{[\nu]}\right). \end{aligned}\right\} \qquad (3.5)$$

Thus we alternately improve the three y values and revise the function values $f_j^{[\nu]}=f(x_j, y_j^{[\nu]})$ and their differences. This starting process should be carried out with a sufficiently small step length (see § 3.4).

Table II/14. *Iterative calculation of starting values*

From		x	y	$hf=0{\cdot}1(x+y)$	$h\nabla f$	$h\nabla^2 f$	$h\nabla^3 f$
(3.4)	1.	0·1	0	0·01	0·01		
	2.	0·1	0·005	0·0105	0·0105		
	3.	0·2	0·021	0·0221	0·0116	0·0011	
(3.4)	4.	0·1	0·005158	0·0105158	0·0105158		
		0·2	0·021367	0·0221367	0·0116209	0·0011051	
	5.	0·3	0·049800	0·0349800	128433	12224	0·0001173
Iteration (3.5)	$\nu=2$	0·1	0·005170	0·0105170	0·0105170		
		0·2	0·021400	221400	116230	0·0011060	
		0·3	0·049851	349851	128451	12221	0·0001161
	$\nu=3$	0·1	0·005171	105171	105171		
		0·2	0·021403	221403	116232	11061	
		0·3	0·049858	349858	128455	12223	1162
	$\nu=4$	0·3	0·049859	349859	128456	12224	1163

Example. With $h=0{\cdot}1$ application of this starting procedure to the initial-value problem

$$y' = x + y, \qquad y(0) = 0$$

yields the numbers in Table II/14. The value of h is, in fact, rather large (cf. the remarks on the length of step interval in § 3.4); with $h=0{\cdot}05$ so many iterations would not have been necessary.

3.3. Formulae for the main calculation

We now describe how the next approximate value y_{r+1} can be obtained once the values $y_1, y_2, \ldots, y_r$ at the points $x_1, x_2, \ldots, x_r$ have been computed.

I. The Adams extrapolation method[1,2]. In the "extrapolation methods", which we consider first, the function $f(x, y(x))$ under the integral in equation (3.1) is replaced by the interpolation polynomial[3] $P(x)$ which takes the values $f_{r-p}, \ldots, f_{r-1}, f_r$ at the points $x_{r-p}, \ldots, x_{r-1}, x_r$ [where $f_\varrho = f(x_\varrho, y_\varrho)$]. In effect we evaluate the integral by means of the quadrature formula (2.6) of Ch. I; thus, with y_{r+1} and y_r replacing $y(x_{r+1})$ and $y(x_r)$, (3.1) becomes

$$\left. \begin{aligned} y_{r+1} &= y_r + h \sum_{\varrho=0}^{p} \beta_\varrho \nabla^\varrho f_r \\ &= y_r + h\left(f_r + \frac{1}{2}\nabla f_r + \frac{5}{12}\nabla^2 f_r + \frac{3}{8}\nabla^3 f_r + \frac{251}{720}\nabla^4 f_r + \cdots\right), \end{aligned} \right\} \quad (3.6)$$

in which the β_ϱ are given generally by formula (2.7) in Ch. I. The exact solution $y(x)$ satisfies the corresponding exact form

$$y(x_{r+1}) = y(x_r) + h\sum_{\varrho=0}^{p} \beta_\varrho \nabla^\varrho f(x_r, y(x_r)) + S_{p+1}. \quad (3.7)$$

(2.8) of Ch. I gives an estimate for the remainder term S_{p+1}.

There are occasions when the extrapolation formula (3.6) is used in a somewhat different form (sometimes called the "Lagrangian" form) in which the differences $\nabla^\varrho f_r$ are expressed in terms of the function values f_s. If the terms involving each individual function value are collected together we obtain new coefficients $\alpha_{p\varrho}$ which depend on the number p of differences retained in (3.6):

$$y_{r+1} = y_r + h\sum_{\varrho=0}^{p} \alpha_{p\varrho} f_{r-\varrho}. \quad (3.8)$$

[1] JOHN COUCH ADAMS, the English astronomer, born on the 5th June 1819 in Laneast, became a Fellow of St. Johns College, Cambridge, and tutor in mathematics there; then in 1849 he was appointed Director of the Observatory and in 1858 Lowndean Professor of Astronomy and Geometry in the University. He died on the 22nd January 1892 in London. He was one of the discoverers of the planet Neptune. As early as 1841 he tried to explain the perturbations in the motion of Uranus by the influence of an unknown planet and to calculate the orbit of this planet by first assuming it to be circular and then modifying it to elliptical form by deriving corrections from the perturbations of Uranus. In 1844 he communicated his results to Prof. CHALLIS and asked him if he would look out for the planet in the calculated position.

[2] BASHFORTH, F., and J. C. ADAMS: An attempt to test the theories of capillary action, p. 18. Cambridge 1883.

[3] Trigonometric interpolation polynomials are recommended by W. QUADE: Numerische Integration von Differentialgleichungen bei Approximation durch trigonometrische Ausdrücke. Z. Angew. Math. Mech. **31**, 237−238 (1951); exponential sums are used by P. BROCK and F. J. MURRAY: The use of exponential sums in step-by-step integration. M. T. A. C. **6**, 63−78 (1952).

The values of the first few $\alpha_{p\varrho}$ are given in Table II/15 (a check on the calculation of the $\alpha_{p\varrho}$ is given by the relation $\sum\limits_{\varrho=0}^{p}\alpha_{p\varrho}=1$).

Table II/15. *The numbers* $\alpha_{p\varrho}$

	$\varrho=$			
	0	1	2	3
$p=1$	$\dfrac{3}{2}$	$-\dfrac{1}{2}$		
$p=2$	$\dfrac{23}{12}$	$-\dfrac{16}{12}$	$\dfrac{5}{12}$	
$p=3$	$\dfrac{55}{24}$	$-\dfrac{59}{24}$	$\dfrac{37}{24}$	$-\dfrac{9}{24}$

In order to calculate y_{r+1} from (3.6) we need the values which are "boxed" in Table II/16. The values of $f, \nabla f, \ldots$ associated with the function f are conveniently tabulated with the factor h; the coefficients which we multiply them by in (3.6) are noted at the heads of the corresponding columns. When y_{r+1} is being calculated, all the numbers above the dotted line are known, and each further step in the calculation yields in turn another line of entries "parallel" to the dotted line. For convenience, the differences are often arranged as in Table I/2 so that this line of new entries is horizontal (cf. the example in § 3.5).

Table II/16. *The Adams extrapolation method*

x	$\times 1$ y	$\times 1$ hf	$\times\dfrac{1}{2}$ $h\nabla f$	$\times\dfrac{5}{12}$ $h\nabla^2 f$	$\times\dfrac{3}{8}$ $h\nabla^3 f$
x_{r-1}	y_{r-1}	hf_{r-1}		$\boxed{h\nabla^2 f_r}$	$\boxed{h\nabla^3 f_r}$
x_r	$\boxed{y_r}$	$\boxed{hf_r}$	$\boxed{h\nabla f_r}$		
x_{r+1}	$\boxed{y_{r+1}}$				

If we use the equation

$$y(x_{r+1}) = y(x_{r-1}) + \int\limits_{x_{r-1}}^{x_{r+1}} f(x, y(x))\,dx \tag{3.9}$$

instead of (3.1) and, as above, replace the integrand by the polynomial $P(x)$ which takes the values f_ϱ at $x=x_\varrho$ $(\varrho=r-p, \ldots, r)$, we obtain NYSTRÖM's extrapolation formula

$$\begin{aligned}
y_{r+1} &= y_{r-1}+h\left[2f_r+\frac{1}{3}\nabla^2 f_r+\frac{1}{3}\nabla^3 f_r+\frac{29}{90}\nabla^4 f_r+\frac{14}{45}\nabla^5 f_r+\cdots\right]\\
&= y_{r-1}+h\left[2f_r+\frac{1}{3}(\nabla^2 f_r+\nabla^3 f_r+\nabla^4 f_r+\nabla^5 f_r)-\right.\\
&\qquad\left. -\frac{1}{90}(\nabla^4 f_r+2\nabla^5 f_r)+\cdots\right],
\end{aligned} \tag{3.10}$$

which follows immediately from (2.32) of Ch. I with $u = 1$ (we have only to replace x_{-1}, x_0, x by x_{r-1}, x_r, x_{r+1}, respectively). If truncated after the term in $V^3 f_r$, this formula has simpler coefficients than the Adams formula (3.6). The corresponding difference scheme is shown in Table II/17 with the values needed for the calculation of y_{r+1} boxed as before.

Table II/17. NYSTRÖM's *extrapolation method*

x	y	$\times 1$ hf	hVf	$\times \frac{1}{3}$ hV^2f	$\times \frac{1}{3}$ hV^3f
x_{r-1}	$\boxed{y_{r-1}}$	hf_{r-1}	hVf_{r-1}	$\boxed{hV^2f_r}$	$\boxed{hV^3f_r}$
x_r	y_r	$\boxed{hf_r}$	hVf_r		
x_{r+1}	$\boxed{y_{r+1}}$	hf_{r+1}			

II. The Adams interpolation method. Here the integrand $f(x, y(x))$ in the equation (3.1) is replaced by the polynomial $P^*(x)$ which takes the values $f_{r-p+1}, \ldots, f_{r-1}, f_r, f_{r+1}$ at the points $x_{r-p+1}, \ldots, x_{r-1}, x_r, x_{r+1}$. Then from the quadrature formula (2.9) of Ch. I it follows that

$$\left. \begin{aligned} y_{r+1} &= y_r + h \sum_{\varrho=0}^{p} \beta_\varrho^* V^\varrho f_{r+1} \\ &= y_r + h \left(f_{r+1} - \frac{1}{2} V f_{r+1} - \frac{1}{12} V^2 f_{r+1} - \frac{1}{24} V^3 f_{r+1} - \cdots \right), \end{aligned} \right\} \quad (3.11)$$

where the β_ϱ^* are given generally by formula (2.9) of Ch. I.

For the exact solution $y(x)$ we have the corresponding formula

$$y(x_{r+1}) = y(x_r) + h \sum_{\varrho=0}^{p} \beta_\varrho^* V^\varrho f(x_{r+1}, y(x_{r+1})) + S_{p+1}^*$$

with remainder term S_{p+1}^*, for which an estimate is given by (2.10) of Ch. I.

Formula (3.11) is also used occasionally in Lagrangian form corresponding to (3.8):

$$y_{r+1} = y_r + h \sum_{\varrho=0}^{p} \alpha_{p\varrho}^* f_{r+1-\varrho}. \qquad (3.12)$$

The first few values of the $\alpha_{p\varrho}^*$ are given in Table II/18.

As with the $\alpha_{p\varrho}$, here also we have the check $\sum_{\varrho=0}^{p} \alpha_{p\varrho}^* = 1$. From (2.7), (2.13) of Ch. I it follows that

$$\alpha_{p0}^* = \beta_p. \qquad (3.13)$$

The quantities which appear in (3.11) are indicated in Table II/19, where the finite-difference scheme is set out in the same way as for the extrapolation method.

Table II/18. *The numbers $\alpha^*_{p\varrho}$*

	$\varrho =$				
	0	1	2	3	4
$p = 1$	$\dfrac{1}{2}$	$\dfrac{1}{2}$			
$p = 2$	$\dfrac{5}{12}$	$\dfrac{8}{12}$	$-\dfrac{1}{12}$		
$p = 3$	$\dfrac{9}{24}$	$\dfrac{19}{24}$	$-\dfrac{5}{24}$	$\dfrac{1}{24}$	
$p = 4$	$\dfrac{251}{720}$	$\dfrac{646}{720}$	$-\dfrac{264}{720}$	$\dfrac{106}{720}$	$-\dfrac{19}{720}$

In the application of (3.11) the difficulty arises that the quantities depending on $f_{r+1} = f(x_{r+1}, y_{r+1})$ which appear on the right-hand side are not yet known. Consequently the unknown y_{r+1} appears on both

Table II/19. *The Adams interpolation method*

			$\times 1$	$\times 1$	$\times\left(-\dfrac{1}{2}\right)$	$\times\left(-\dfrac{1}{12}\right)$	$\times\left(-\dfrac{1}{24}\right)$
x	y	hf	$h\nabla f$	$h\nabla^2 f$	$h\nabla^3 f$		
x_{r-1}	y_{r-1}	hf_{r-1}		$h\nabla^2 f_r$		
			$h\nabla f_r$		$h\nabla^3 f_{r+1}$	
x_r	y_r	hf_r		$h\nabla^2 f_{r+1}$		
			$h\nabla f_{r+1}$			
x_{r+1}	y_{r+1}	hf_{r+1}				

sides of the equation and only in special cases is it possible to solve this equation exactly for y_{r+1}. Equation (3.11) is, however, very suitable for the iterative calculation of y_{r+1} provided that h is sufficiently small. We put an approximate value $y^{[\sigma]}_{r+1}$ in the right-hand side, forming $f^{[\sigma]}_{r+1} = f(x_{r+1}, y^{[\sigma]}_{r+1})$ and the differences $\nabla^\varrho f^{[\sigma]}_{r+1} = \nabla^{\varrho-1} f^{[\sigma]}_{r+1} - \nabla^{\varrho-1} f_r$, and then calculate

$$y^{[\sigma+1]}_{r+1} = y_r + h \sum_{\varrho=0}^{p} \beta^*_\varrho \nabla^\varrho f^{[\sigma]}_{r+1} \qquad (\sigma = 0, 1, 2, \ldots) \qquad (3.14)$$

as the $(\sigma + 1)$-th iterate.

III. Central-difference interpolation method. If we integrate both sides of the differential equation (1.10) over the interval $x_r - h$ to $x_r + h$ using STIRLING's interpolation formula, as in Ch. I (2.14), (2.15), we obtain (with p even)

$$y(x_{r+1}) - y(x_{r-1}) = h \sum_{\varrho=0}^{p/2} \beta_{\varrho}^{**}\, V^{2\varrho} f\big(x_{r+\varrho}, y(x_{r+\varrho})\big) + S_p^{**}.$$

If the remainder term is neglected, we have an equation relating the approximations y_{ϱ}, namely

$$\left.\begin{aligned}
y_{r+1} - y_{r-1} &= h \sum_{\varrho=0}^{p/2} \beta_{\varrho}^{**}\, V^{2\varrho} f_{r+\varrho} \\
&= h\left(2 f_r + \frac{1}{3}\, V^2 f_{r+1} - \frac{1}{90}\, V^4 f_{r+2} + \frac{1}{756}\, V^6 f_{r+3} - \cdots\right).
\end{aligned}\right\} \quad (3.15)$$

Usually this formula is truncated after the term in V^2, which gives SIMPSON's rule:

$$y_{r+1} = y_{r-1} + h\left(2 f_r + \frac{1}{3}\, V^2 f_{r+1}\right) = y_{r-1} + \frac{h}{3}\left(f_{r-1} + 4 f_r + f_{r+1}\right). \quad (3.16)$$

An estimate for the remainder term S_2^{**} in the corresponding formula

$$y(x_{r+1}) = y(x_{r-1}) + h\left[2 f(x_r, y(x_r)) + \tfrac{1}{3} V^2 f(x_{r+1}, y(x_{r+1}))\right] + S_2^{**} \quad (3.17)$$

for the exact solution is given by (2.18) in Ch. I.

As with formula (3.11), (3.16) also includes the unknown y_{r+1} on both sides of the equation, so that here also one determines y_{r+1} iteratively in general. Thus the next approximation $y_{r+1}^{[\sigma+1]}$ is obtained from the current value $y_{r+1}^{[\sigma]}$ according to the formula

$$y_{r+1}^{[\sigma+1]} = y_{r-1} + h\big(2 f_r + \tfrac{1}{3} V^2 f_{r+1}^{[\sigma]}\big). \quad (3.18)$$

The form of the tabular scheme for this method can be seen in Table II/20.

Table II/20. *The central-difference method*

x	$\times 1$ y	$\times 2$ $h f$	$h V f$	$\times \dfrac{1}{3}$ $h V^2 f$
x_{r-1}	$\boxed{y_{r-1}}$	$h f_{r-1}$		$h V^2 f_r$
			$h V f_r$	
x_r	y_r	$\boxed{h f_r}$		$\boxed{h V^2 f_{r+1}}$
			$h V f_{r+1}$	
x_{r+1}	$\boxed{y_{r+1}}$	$h f_{r+1}$		

LINDELÖF[1] has suggested a method based on formula (3.15) in which the term in $V \cdot f_{r+2}$ is taken into account as well. He rewrites the equation

---

[1] LINDELÖF, E.: Remarques sur l'intégration numérique des équations différentielles ordinaires. Acta Soc. Sci. Fenn. A 2 **1938**, Nr. 13 (21 pp.). He also gives a further refinement of the method.

in the form

$$y_{r+1} = y_{r-1} + h\left[2f_r + \tfrac{1}{3}\left(V^2 f_r + V^3 f_r\right)\right] + \delta$$

with

$$\delta = \frac{h}{3}\left(V^4 f_{r+1} - \frac{1}{30} V^4 f_{r+2}\right),$$

and then, assuming that the solution has already been computed up to the point $x = x_r$, uses

$$y^{[0]}_{r+j+1} = y_{r+j-1} + h\left[2 f_{r+j} + \tfrac{1}{3}\left(V^2 f_{r+j} + V^3 f_{r+j}\right)\right]$$

for $j = 0$ and $j = 1$ to obtain tentative values for the next two points. These are then used to build up the difference table temporarily so that approximate values of $V^4 f_{r+1}$ and $V^4 f_{r+2}$ are available for the correction δ to y_{r+1}. The new value of y_{r+1} extends the final difference table up to the point $x = x_{r+1}$.

IV. Mixed extrapolation and interpolation methods. With the methods II and III, the f_{r+1} on the right-hand sides of equations (3.11) and (3.16), respectively, appears as an unknown as well as y_{r+1} and must be either estimated or calculated from an extrapolation formula. Milne[1] recommends the latter procedure. A rough value y^*_{r+1} is calculated from an extrapolation formula, then $f^*_{r+1} = f(x_{r+1}, y^*_{r+1})$ is formed and the difference table completed temporarily so that sufficient differences are available to determine y_{r+1} from an interpolation formula. If need be, this value of y_{r+1} can be still further improved by iteration using the interpolation formula[2]. In particular, Milne gives the formulae

$$y^*_{r+1} = y_{r-3} + 4h f_{r-1} + \frac{8h}{3} V^2 f_r,$$

$$y_{r+1} = y_{r-1} + 2h f_r + \frac{h}{3} V^2 f^*_{r+1},$$

the second of which is the formula (3.16) of the central-difference method.

3.4. Hints for the practical application of the finite-difference methods

I. Estimation of the highest difference occurring in an interpolation method. If the requisite starting values have been obtained by one of the starting procedures described in §3.2, then, in order to begin the iterations (3.14) or (3.18), we must estimate f_{r+1}; equivalently, we can estimate the highest difference occurring $V^p f_{r+1}$ ($p = 2$ for the central-difference method), which is much easier. Once the calculation is

---

[1] Milne, W. E.: Numerical solution of differential equations, p. 65. New York and London 1953.

[2] A further variant is mentioned by P. O. Löwdin: On the numerical integration of ordinary differential equations of the first order. Quart. Appl. Math. **10**, 97—111 (1952).

properly under way there exists a series of values of $V^p f$ from which we can easily extrapolate for a good estimate of the next value by noting the trend of either the $V^p f$ or $V^{p+1} f$ values. The better the estimate, the less work there is involved in the iteration.

II. Length of step h. The step length h should be kept small enough for

(a) the iteration [(3.14) or (3.18)] to converge sufficiently rapidly — more explicitly, to settle to the required accuracy after one or two cycles — and for

(b) the first term neglected in the formula being used [(3.6), (3.11) or (3.15)] to have a negligible effect to the accuracy required.

In §§ 4.1, 4.3 a significant factor in predicting convergence and estimating the error is found to be the quantity (the "step index")

$$S = k h, \quad \text{where} \quad k = \left| \frac{\partial f}{\partial y} \right|.$$

For moderate accuracy S should be of the order 0·05 to 0·1, the smaller value being preferable for the starting calculations. In the example of § 3.2, III, the step was chosen so that $S = 0·1$ and consequently rather too many iterations were needed. If k varies considerably, it is advisable to adjust the length of step so that S remains approximately constant. We call $h = \text{const}/k$ the "natural step length", where the constant is chosen according to the accuracy required, say in the range 0·05 to 0·1 as mentioned above[1]. If the step used in the calculation is considerably longer than the natural step, then the differences do not decrease sufficiently rapidly with increasing order, the iteration converges too slowly (so that three cycles, or even more in certain circumstances, must be computed before the numbers settle) and the highest differences carried show such large fluctuations that confidence in their accuracy is no longer justified. On the other hand, if too small a step is used, the calculation runs extremely smoothly without any difficulty but, on account of the large number of steps required to cover the same range, more work is, in fact, involved.

III. Change of step length[2]. Doubling the step ($\bar{h} = 2h$) merely requires the values $y_r, y_{r-2}, y_{r-4}, \ldots$ to be tabulated afresh and a new difference table of the corresponding values of $\bar{h} f$ to be constructed.

Halving the step, on the other hand, is more laborious and requires a new starting calculation. If the calculation has proceeded up to the point x_r, so that $f_r, V f_r, V^2 f_r, V^3 f_r$ are known, then interpolation by formula (2.1) of Ch. I gives the intermediate values

$$\left. \begin{aligned} h f_{r-\frac{1}{2}} &= h \left(f_r - \frac{1}{2} V f_r - \frac{1}{8} V^2 f_r - \frac{1}{16} V^3 f_r \right), \\ h f_{r+\frac{1}{2}} &= h \left(f_r + \frac{1}{2} V f_r + \frac{3}{8} V^2 f_r + \frac{5}{16} V^3 f_r \right). \end{aligned} \right\} \tag{3.19}$$

---

[1] COLLATZ, L.: Natürliche Schrittweite … Z. Angew. Math. Mech. **22**, 216—225 (1942).

[2] Formulae for an arbitrary change in step length and for calculation with arbitrary non-equidistant abscissae (also for differential equations of higher order) are given by P. W. ZETTLER-SEIDEL: Improved Adams method of numerical integration of differential equations. Lecture at Internat. Math. Congr., Cambridge (U.S.A.), 1950.

Then the values $\tilde{h}f_j$ $(j = r - 1, r - \frac{1}{2}, r, r + \frac{1}{2},$ or $j = 0, 1, 2, 3,$ if a new numbering is adopted) which correspond to the points at the smaller interval $\tilde{h} = \frac{1}{2}h$, together with their differences, serve as starting data for the refining iteration of (3.5).

IV. Simplification of the iterative procedure in the interpolation methods. The computation involved in the iterations (3.14) and (3.18) is simplified if only the changes in y_{r+1} through each cycle of the iteration are calculated[1]. These can be determined quite simply as follows. Consider, for example, the iteration (3.14); we have

$$y_{r+1}^{[\sigma+1]} - y_{r+1}^{[\sigma]} = h \sum_{\varrho=0}^{p} \beta_\varrho^* [\nabla^\varrho f_{r+1}^{[\sigma]} - \nabla^\varrho f_{r+1}^{[\sigma-1]}].$$

Now since the changes

$$\delta = \nabla^\varrho f_{r+1}^{[\sigma]} - \nabla^\varrho f_{r+1}^{[\sigma-1]} \qquad (3.20)$$

are independent of ϱ, they can be taken outside of the summation; hence, using the result (2.13) of Ch. I, we obtain

$$y_{r+1}^{[\sigma+1]} = y_{r+1}^{[\sigma]} + h\beta_p \delta. \qquad (3.21)$$

For the central-difference method the iteration (3.18) may be replaced by a similarly modified iteration:

$$y_{r+1}^{[\sigma+1]} = y_{r+1}^{[\sigma]} + \frac{1}{3} h \delta^*, \qquad (3.22)$$

where

$$h\delta^* = h\nabla^2 f_{r+1}^{[\sigma]} - h\nabla^2 f_{r+1}^{[\sigma-1]}. \qquad (3.23)$$

V. Development of unevenness[2] with the central-difference method. Formula (3.16) provides a direct relation between a value y_{r-1} and the next but one value y_{r+1}, but between consecutive values of y there exists only an indirect link through the differential equation. In the course of a calculation over a large number of steps this can cause (in consequence of irregularities in empirically defined functions, for example) the two approximate solutions represented by the interlaced sequences $y_{r-4}, y_{r-2}, y_r, \ldots$ and $y_{r-3}, y_{r-1}, y_{r+1}, \ldots$ separately to diverge slightly from one another. Thus an unevenness develops in the y values and since it also affects the f values through the differential equation, it is accentuated in the differences of f — in fact, the building up of these irregularities is first noticed as fluctuations in the values of $\nabla^2 f$. When these irregularities reach significant proportions, they can be removed by a smoothing process[3], coupled, possibly, with a new starting iteration, after which the calculation is continued as before.

---

[1] STOHLER, K.: Eine Vereinfachung bei der numerischen Integration gewöhnlicher Differentialgleichungen. Z. Angew. Math. Mech. **23**, 120—122 (1943).

[2] Compare with the theory in § 4.7.

[3] Cf., for example, L. COLLATZ and R. ZURMÜHL: Beiträge zu den Interpolationsverfahren der numerischen Integration von Differentialgleichungen erster und zweiter Ordnung. Z. Angew. Math. Mech. **22**, 42—55 (1942). They give another smoothing procedure which is more systematic.

The smoothing can often be accomplished very simply as follows. We make small corrections $\pm \varepsilon$ in the values of $h\nabla^3 f$, say, so that the values $h\nabla^3 f_{r-j} + (-1)^j \varepsilon$, i.e. $h\nabla^3 f_{r-3} - \varepsilon$, $h\nabla^3 f_{r-2} + \varepsilon$, $h\nabla^3 f_{r-1} - \varepsilon$, $h\nabla^3 f_r + \varepsilon$, run smoothly. For consistency we must make further corrections $\pm \frac{1}{2}\varepsilon$, $\pm \frac{1}{4}\varepsilon$, $\pm \frac{1}{8}\varepsilon$ in the second, first, and zero-th differences; thus we replace

$$h\nabla^k f_{r-j} \quad \text{by} \quad h\nabla^k f_{r-j} + (-1)^j \frac{\varepsilon}{2^{3-k}}. \tag{3.24}$$

Finally we have to modify the y values so as to obtain the required corrections $\pm \frac{1}{8}\varepsilon$ in the f values. This can be done by varying a y value, say y_r, by a small amount δ and noting what change ζ this produces in f_r; then the correction to be added to y_{r-j} is $(-1)^j \frac{\delta}{\zeta} \cdot \frac{\varepsilon}{8}$ (cf. the example in § 3.5).

3.5. Examples

I. Extrapolation method. If the magnetic characteristic of a coil wound on an iron core is assumed to be of cubic form, the sudden application of a periodic voltage

$$e = e_0 \sin \omega t$$

across the coil gives rise to the initial-value problem[1]

$$e = iR + \frac{d\psi}{dt}, \quad \text{where} \quad i = a\psi + b\psi^3,$$

with $\psi(0) = 0$ (notation: voltage e, current i, resistance R, magnetic flux ψ).

With reduced variables x, y defined by

$$e_0 y = aR\psi, \quad x = aRt$$

the initial-value problem becomes

$$y' = -y - 2y^3 + \sin 2x; \quad y(0) = 0$$

for the cases with $be_0^2 = 2a^3R^2$, $2aR = \omega$.

This problem will be treated by the Adams extrapolation method. We obtain the necessary starting values from a power series solution found by the method of undetermined coefficients. This solution is

$$y = x^2 - \frac{1}{3}x^3 - \frac{1}{4}x^4 + \frac{1}{20}x^5 + \frac{13}{360}x^6 - \frac{733}{2520}x^7 + \frac{1903}{6720}x^8 + \cdots,$$

in which sufficient terms have been given to calculate $y(x)$ at $x = \pm 0.1$, ± 0.2 to six decimals. These starting values, and also the corresponding values of hf with the differences required to proceed with the main calculation, are shown in Table II/21. Further values of y_{r+1} are calculated from (3.6) truncated after the term in $\nabla^4 f_r$.

---

[1] See, for example, K. Küpfmüller: Einführung in die theoretische Elektrotechnik, 4th ed., p. 401. Berlin-Göttingen-Heidelberg 1952.

Table II/21. *A non-linear problem treated by the Adams extrapolation method*

x	y	$2y^3$	$\sin 2x$	hf	$h\nabla f$	$h\nabla^2 f$	$h\nabla^3 f$	$h\nabla^4 f$
−0·2	0·042253	0·000 15	−0·38942	−0·043 182				
−0·1	0·010 308	0·000 00	−0·19867	−0·020 898	22 284			
0	0	0	0	0	20 898	−1386		
0·1	0·009 643	0·000 00	0·198 67	0·018 903	18 903	−1995	−609	
0·2	0·036 951	0·000 10	0·38942	0·035 237	16 334	−2569	−574	+35
0·3	0·079 082	0·000 99	0·564 64	0·048 457	13 220	−3114	−545	+29
0·4	0·132 657	0·004 67	0·717 36	0·058 003	9 546	−3674	−560	−15
0·5	0·193 687	0·014 53	0·841 47	0·063 325	5 322	−4224	−550	+10
0·6	0·257 710	0·034 23	0·932 04	0·064 010				

II. Interpolation method.

II. Interpolation method. We consider again the problem (1.21)

$$y' = y - \frac{2x}{y}, \qquad y(0) = 1$$

and use it now to illustrate the central-difference method. Sufficient starting values for this method have already (§ 3.2) been calculated for $h = 0·1$ by the Taylor series method. Thus y_{-1}, y_0, y_1 are known and the corresponding function values f_{-1}, f_0, f_1, and their differences $\nabla f_0, \nabla f_1, \nabla^2 f_1$ can be calculated; this completes the first three rows of Table II/22 and we can now start the main calculation.

The iteration (3.18) for the first step of the main calculation is

$$y_2^{[\sigma+1]} = y_0 + h\left(2f_1 + \tfrac{1}{3}\nabla^2 f_2^{[\sigma]}\right); \tag{3.25}$$

to start it we must first estimate the new second difference $\nabla^2 f_2$. If we have nothing to go on, the previous value $\nabla^2 f_1$ may be taken as a first approximation; but in the present case it is better to make use of the third difference $\nabla^3 f_1$, which can be estimated quite easily from the derivatives $y^{IV}(0) = -15$, $y^{V}(0) = 105$ already calculated for the starting values. Since $f = y'$, we have $h\nabla^3 f_1 = h^4 y^{IV}(0) + \cdots$, so that we may expect that $h\nabla^3 f_1 \approx (0·1)^4 \times y^{IV}(0) = -0·0015$; the next difference is of opposite sign [being approximately proportional to $y^{V}(0)$, which is positive], so $|h\nabla^3 f|$ will begin to decrease as the calculation progresses.

If we try $h\nabla^3 f_1 \approx -0·001$, our first estimated second difference (a separate column is provided in the table for these estimates) is

$$h\nabla^2 f_2^{[0]} = 0·003 - 0·001 = 0·002,$$

and from (3.25) the first iterate is

$$y_2^{[1]} = 1 + 2\times 0·091 287 09 + \tfrac{1}{3}\times 0·002 = 1·183 240 8;$$

the row is completed by forming the corresponding $hf_2^{[1]}$, $h\nabla f_2^{[1]}$, $h\nabla^2 f_2^{[1]}$.

Table II/22. Application of the central-difference method to a non-linear equation of the first order

x	y	$+\dfrac{2x}{y}$	$hf=0.1\left(y-\dfrac{2x}{y}\right)$	$h\nabla f$	$h\nabla^2 f$	$h\nabla^3 f$	Estimated value of $h\nabla^2 f$	$h\delta^*$
−0·1	0·8944272	−0·2236068	0·11180340	−0·01180340				−0·00005556
	2223							0237
0	2215	0	0·1	−0·00871291	0·00309049			
0·1	1·0954451	0·1825742	0·09128709		0·00309049			−0·00017913
0·2	1·1832408	0·3380546	0·08451862	−0·00676847	0·00194444		0·002	821
	2223	0598	51625	77095	94196	−0·00114853		37
	2215	0601	51614					
0·3	1·2649774	0·4743168	0·07906606	−0·00545008	0·00132087		0·0015	
	9177	3392	5785	4587	31228	−0·00062968		+0·00003617
	9150	3402	5748					174
	9149	3402	5747					9
0·4	1·3416364	0·5962867	0·07453497	−0·00452250	0·00093617		0·0009	
	6485	814	3671	52067	93800	−0·00037428		2321
	6491	811	3680		69679		0·00072	116
0·5	1·4142285	0·7070993	0·07071292	−0·00383288	69557			6
	208	1032	1176	82510	53351	−0·00024243		2951
	204	1034	1170		53510		0·000504	151
0·6	1·4832405	0·8090394	0·06742011	−0·00329159	41968			
	503	341	2162	29000	66	16047	0·0042	32
	508	338	2170	−0·00287032	33828			
0·7	1·5492038	0·9036900	0·06455138	34	33876	11544	0·0033	828
	37	01	36	−0·00253206	0·00027574			45
0·8	1·6124635	0·9922705	0·06201930	53158	72	8090		
	63	688	75	−0·00225584	0·0002304	−0·00006304	0·000276	−0·00000026
	65	687	78	86	1924	−0·000453		
0·9	1·6733353	1·0756959	0·05976394	−0·0020235	1652	380	0·000231	−0·000006
	52	60	92	−0·0018331	1404	272	0·000192	4
1	1·7320713	1·154687	0·0577384	−0·0016679	5	253	165	2
1·1	1·7888763	1·229823	0·0559053	−0·0015275	0·0001231	−0·0000174	138	4
1·2	1·843937	1·301563	0·0542374	74	1063	168		
1·3	1·897397	1·370298	0·0527099	−0·0014043	950	113	0·000123	+0·000001
	398	298	7100	12980	827	123	105	13
1·4	1·949398	1·436341	0·0531057	12030	750	77	0·000096	10
1·5	2·000044	1·499967	500077	11203	660	90	84	13
1·6	2·049445	1·561398	488047	10453	0·0000608	−0·0000052	75	0
1·7	2·097681	1·620837	476844	9793			66	−0·0000001
1·8	2·144839	1·678448	466391	−0·0009185			0·0000609	
1·9	2·190981	1·734383	456598					
2	2·236179	1·788766	0·0474413					
1·8	2·144849	1·678440	0·0466409	−0·0009188	0·0000601	−0·0000065	0·0000546	−0·000002
1·9	2·190982	1·788766	0·0447411	8644	544	57	495	0
2·1	2·236178	1·841716	437767	8149	495	49	453	0
2·2	2·280483	1·893330	430618	7696	453	42	417	1
2·3	2·323948	1·943699	422922	7280	416	37		+0·0000001
2·4	2·366621	1·992904	415642	−0·0006895	0·0000385	−0·0000033	0·0000384	
2·5	2·408546	2·041015	0·0408747					
	2·449762							

Another auxiliary column is provided for recording the differences of the successive iterates for hV^2f as defined in (3.23). In the present case hV^2f_2 has changed by

$$h\delta^* = 0.001\,944\,44 - 0.002 = -0.000\,055\,56;$$

from (3.22) a third of this difference, i.e. -0.0000185, added to $y_2^{[1]}$ yields the new value $y_2^{[2]} = 1.183\,2223$.

Now we have only to work out the corresponding function values $f_2^{[2]}$ — the new differences need not be worked out since the change in the second difference required to form $h\delta^*$ is the same as the change in $f^{[\sigma]}$ itself [as in (3.20)] —, then $y_2^{[3]}$ can be calculated from (3.22). On forming $f_2^{[3]}$, we find that the change $h\delta^*$ is now only $-0.000\,000\,11$, so that one third of it is smaller in magnitude than 0.5×10^{-7} and no longer affects the y values; $y_2^{[3]}$ is therefore taken as the final value $y_2 = y_2^{[3]}$ and the corresponding row of differences filled in. Further steps can now be dealt with in a similar manner.

At first the estimation of the new values of hV^2f occasions a little difficulty, but after a number of steps a good idea of the run of the third differences can be formed and the calculation then proceeds quite happily; progress is extremely rapid and in fact the steps from $x = 1$ to $x = 2$ (with the exception of the point $x = 1.3$) are each accomplished in one row of computation. Gradually, however, the irregularities described in § 3.4 creep in, showing themselves in the third differences, which are alternately too large and too small; with the last decimal as unit the third differences for the points $x = 1.7$ to $x = 2.0$ run (with a factor h)

$$-123, \quad -77, \quad -90, \quad -52.$$

We try altering them alternately by $\pm\varepsilon$, say $-\varepsilon, \varepsilon, -\varepsilon, \varepsilon$, and find that $\varepsilon = -13 \times 10^{-7}$ gives the considerably smoother sequence

$$-110, \quad -90, \quad -77, \quad -65.$$

The remaining differences at $x = 2.0$ are now corrected correspondingly in accordance with (3.24); for example, the first difference hVf is altered by $\frac{1}{4}\varepsilon \approx -0.000\,000\,3$ to $-0.000\,918\,8$.

To find the corrections to the y values which will give the correct changes $\pm \varepsilon/8$ in the hf values, we alter y at $x = 1.8$ by $\delta = 0.000010$, say, and note that the new y value $2.144\,849$ produces a change $\zeta = 0.000\,001\,8$ in hf. To alter hf by $\varepsilon/8$ instead of ζ, we must therefore change the values y_{r-j} at $x = 2.0 - jh$ by $\delta' = \dfrac{\delta}{\zeta} \cdot \dfrac{\varepsilon}{8} \approx 0.000001$. When these changes are made at $x = 1.9$ and $x = 2.0$ the calculation proceeds smoothly again until the irregularities begin to creep in once more.

For comparison, the solution obtained here is tabulated in Table II/3 alongside the results obtained by other methods. Of particular interest is the comparison with the Runge-Kutta method with step interval $h = 0.2$, which, as is mentioned in § 1.5, corresponds roughly to the central-difference method with $h = 0.1$ in that it involves approximately the same amount of computation. In this example the central-difference method shows up to advantage but one should bear in mind the warning given on page 1 of the danger of making hasty general assessments of methods.

3.6. Differential equations in the complex plane

Let a function $w = w(z)$ of a complex variable z be defined by the differential equation

$$w' = F(z, w)$$

and the initial condition

$$w(z_0) = w_0.$$

Here F will be assumed to be an analytic function of z and w. We wish to calculate $w(z)$ numerically over some desired region of the z plane. There are various ways open to us. For instance, all quantities involved can be split into real and imaginary parts

$$z = x + iy, \quad w = u + iv, \quad F = U + iV$$

and the integration performed parallel to the real or imaginary axis; thus, integrating parallel to the real axis, we have

$$\frac{\partial u}{\partial x} = U, \qquad \frac{\partial v}{\partial x} = V, \tag{3.26}$$

a pair of simultaneous first-order equations for the functions $u(x, y_0)$ and $v(x, y_0)$, which can be treated by the methods of § 2 and § 3.

Another way is to introduce a lattice of points in the z plane defined by $z = z_0 + jh + ikl$, where $j, k = 0, \pm 1, \pm 2, \ldots$ and h, l are the mesh widths, and derive formulae which use the values of w at a group of mesh points to give approximate values of w at neighbouring mesh points.

H. E. SALZER[1] gives such formulae for differential equations of the first and second order. He uses a square mesh ($h = l$) and obtains, for example, approximations for the values at the points $Q = (0, 2), (1, 2), (2, 1)$ and $(2, 0)$ from the values at the four points $P(j, k) = (0, 0), (0, 1), (1, 0)$ and $(1, 1)$.

Firstly some ordinary extrapolation formulae of such form which are exact for polynomials of the third (and less) degree are derived:

$$\left.\begin{array}{l} F_2 = (2 - i) F_0 + (2 + 4i) F_1 - 2i F_i + (-3 - i) F_{1+i}, \\ F_{2+i} = 2i F_0 + (-3 + i) F_1 + (2 + i) F_i + (2 - 4i) F_{1+i}, \\ F_{1+2i} = -2i F_0 + (2 - i) F_1 + (-3 - i) F_i + (2 + 4i) F_{1+i}, \\ F_{2i} = (2 + i) F_0 + 2i F_1 + (2 - 4i) F_i + (-3 + i) F_{1+i}. \end{array}\right\} \tag{3.27}$$

For compactness, argument values are here denoted by subscripts.

Further, formulae for approximating

$$\Phi(z) = C + \int_0^z F(\zeta)\, d\zeta$$

---

[1] SALZER, H. E.: Formulas for numerical integration of first and second order differential equations in the complex plane. J. Math. Phys. **29**, 207–216 (1950).

are given:

$$24\,\Phi_0/h \;\; = 0 + C,$$
$$24\,\Phi_1/h \;\; = (9 + 5i)\,F_0 + (9 - 5i)\,F_1 + (3 + i)\,F_i + (3 - i)\,F_{1+i} + C,$$
$$24\,\Phi_i/h \;\; = (5 + 9i)\,F_0 + (1 + 3i)\,F_1 + (-5 + 9i)\,F_i + (-1 + 3i)\,F_{1+i} + C,$$
$$24\,\Phi_{1+i}/h = (8 + 8i)\,F_0 + (4 + 4i)\,F_1 + (4 + 4i)\,F_i + (8 + 8i)\,F_{1+i} + C.$$

$$(3.28)$$

These likewise are exact for polynomials of up to the third degree.

If we know approximate values of w, and hence of F, at the four points P (to start the calculation we must first calculate w at three points by some starting technique, say a series solution), we can obtain approximations to F at the points Q from the extrapolation formulae (3.27). Then, using the approximation to the differential equation obtained from (3.28), we calculate w at the points Q and, if necessary, improve these values of w by repeating the process with revised values of F [1].

3.7. Implicit differential equations of the first order

Occasionally it happens that when a differential equation is presented in the form

$$F(x, y, y') = 0 \tag{3.29}$$

it cannot be solved for y' in closed form. There are various ways of dealing with this situation:

1. In general, (3.29) can be transformed as follows into a pair of simultaneous explicit differential equations of the first order [2], which can then be integrated numerically by one of the methods of § 1, § 2 or § 3. We assume that the required solution is such that $y''(x) \neq 0$; then to $t = y'(x)$ there is an inverse function $x = x(t)$. Differentiation of (3.29) with respect to t yields

$$F_x \frac{dx}{dt} + F_y t \frac{dx}{dt} + F_t = 0,$$

---

[1] In addition to the "four-point formulae" quoted above SALZER (see previous footnote) gives formulae for 3 to 9 points and also formulae for repeated integration which can be used for second-order differential equations; thus for

$$\Psi(z) = A\,z + B + \int_0^z \int_0^\zeta F(s)\,ds\,d\zeta$$

he obtains the formulae

$$120\,\Psi_0/h^2 \;\; = 0 + A\,z_0 + B,$$
$$120\,\Psi_1/h^2 \;\; = (33 + 13i)\,F_0 + (12 - 12i)\,F_1 + (8 + 2i)\,F_i + (7 - 3i)\,F_{1+i} + A\,z_1 + B,$$
$$120\,\Psi_i/h^2 \;\; = (-33 + 13i)\,F_0 + (-8 + 2i)\,F_1 + (-12 - 12i)\,F_i +$$
$$+ (-7 - 3i)\,F_{1+i} + A\,z_i + B,$$
$$120\,\Psi_{1+i}/h^2 = 56i\,F_0 + (4 + 20i)\,F_1 + (-4 + 20i)\,F_i + 24i\,F_{1+i} + A\,z_{1+i} + B.$$

[2] KAMKE, E.: Differentialgleichungen reeller Funktionen, 2nd ed. (436 pp.), p. 112. Leipzig 1944.

so that, if we assume further that $F_x + F_y t \neq 0$, we obtain the system

$$
\left.
\begin{aligned}
\frac{dx}{dt} &= -\frac{F_t}{F_x + F_y t} = f(x, y, t), \\
\frac{dy}{dt} &= \frac{dy}{dx}\frac{dx}{dt} = -\frac{t F_t}{F_x + F_y t} = t f(x, y, t).
\end{aligned}
\right\}
\tag{3.30}
$$

The initial conditions are $x = x_0$, $y = y_0$ at $t = y_0'$, where y_0' is determined from $F(x_0, y_0, y_0') = 0$. Solution of a set of two first-order equations by a finite-difference method normally requires two difference tables, but since the differences of tf can be expressed in terms of the differences of f [e.g. $\nabla^\varrho (t_k f_k) = t_k \nabla^\varrho f_k + \varrho h \nabla^{\varrho-1} f_{k-1}$, where h is the step length], the computation for the equations (3.30) can be arranged[1] so as to use only the difference table of f.

2. Differentiation of (3.29) with respect to x yields the second-order equation

$$
y'' = -\frac{F_x + y' F_y}{F_{y'}},
\tag{3.31}
$$

provided we assume that $F_{y'} \neq 0$. This is of explicit form and can be treated by the usual methods for second-order differential equations.

3. If we define

$$
g(x, y, y') = y' - \frac{F(x, y, y')}{F_{y'}(x_0, y_0, y_0')},
\tag{3.32}
$$

we can set up the iterative process

$$
y_\nu(x) = y_0 + \int_{x_0}^{x} y_\nu'(x)\, dx; \quad y_{\nu+1}'(x) = g(x, y_\nu(x), y_\nu'(x)) \quad (\nu = 0, 1, 2, \ldots).
\tag{3.33}
$$

Weissinger[1] gives a convergence proof and error estimates for this method.

§4. Theory of the finite-difference methods

We start here by investigating the convergence of the iterations required in the interpolation methods of the preceding section, then describe how error estimates can be derived for all finite-difference methods. Formula (4.43) deserves particular mention as being a fundamental error formula for the central-difference method. § 4.7 deals with the danger of instability in finite-difference methods.

4.1. Convergence of the iterations in the main calculation

Here, and in the rest of § 4, it will be assumed that the function $f(x, y)$ appearing in the differential equation (1.10) satisfies a Lipschitz condition (1.12) with constant K. In practice K is usually taken to be the largest absolute value of the derivative $\partial f/\partial y$ within the domain under consideration, i.e.

$$
K = k_{\max} \quad \text{where} \quad k = \left| \frac{\partial f}{\partial y} \right|.
\tag{4.1}
$$

---

[1] Weissinger, J.: Numerische Integration impliziter Differentialgleichungen. Z. Angew. Math. Mech. 33, 63—65 (1953).

With this assumption, quite a simple analysis suffices to examine the convergence of the iterations required in the interpolation methods. First of all we investigate the iterations (3.14) and (3.18) which occur in the main calculation.

For this we consider the equations in their Lagrangian form, in which the differences are expressed in terms of the function values, as in (3.12) for the interpolation formula (3.11). Using (3.13) we can write the iteration equations for the Adams interpolation method (3.14) in the form

$$y_{r+1}^{[\sigma+1]} = y_r + h\left[\beta_p f_{r+1}^{[\sigma]} + \sum_{\varrho=1}^{p} \alpha_{p\varrho}^* f_{r+1-\varrho}\right] \qquad (\sigma = 0, 1, 2, \ldots). \qquad (4.2)$$

If the corresponding expression for $y_{r+1}^{[\sigma]}$ is subtracted from this equation, we see that the change

is given by
$$\left.\begin{aligned}
\delta^{[\sigma]} &= y_{r+1}^{[\sigma+1]} - y_{r+1}^{[\sigma]} \\
\delta^{[\sigma]} &= h\beta_p\left[f(x_{r+1}, y_{r+1}^{[\sigma]}) - f(x_{r+1}, y_{r+1}^{[\sigma-1]})\right].
\end{aligned}\right\} \qquad (4.3)$$

The Lipschitz condition (1.12) provides limits for the right-hand side:

and it follows that
$$\left.\begin{aligned}
|\delta^{[\sigma]}| &\leq hK\beta_p |\delta^{[\sigma-1]}|, \\
|\delta^{[\sigma]}| &\leq (hK\beta_p)^\sigma |\delta^{[0]}| \qquad (\sigma = 0, 1, 2, \ldots).
\end{aligned}\right\} \qquad (4.4)$$

We now suppose that h is so small that

$$Kh < \frac{1}{\beta_p}; \qquad (4.5)$$

then the geometric series $|\delta^{[0]}| \sum_{\sigma=0}^{\infty} (hK\beta_p)^\sigma$, which majorizes the series $\sum_{\sigma=0}^{\infty} \delta^{[\sigma]}$, has a ratio $h\beta_p K$ which is less than one and therefore converges. Consequently the series

$$\lim_{\sigma\to\infty} y_{r+1}^{[\sigma]} = y_{r+1}^{[0]} + \delta^{[0]} + \delta^{[1]} + \delta^{[2]} + \cdots$$

converges (absolutely, in fact) to a value y_{r+1}, which provides a solution of (3.11). Thus the inequality (4.5) represents a sufficient condition for the convergence of the iteration. For the values of p normally used we have

$$\left.\begin{aligned}
Kh &< \frac{12}{5} = 2 \cdot 4 \qquad \text{for} \quad p = 2, \\
Kh &< \frac{8}{3} \approx 2 \cdot 67 \qquad \text{for} \quad p = 3, \\
Kh &< \frac{720}{251} \approx 2 \cdot 87 \qquad \text{for} \quad p = 4.
\end{aligned}\right\} \qquad (4.6)$$

For the central-difference method (3.15) the rearranged form of the iteration formula (3.18) is

$$y_{r+1}^{[\sigma+1]} = y_{r-1} + \frac{h}{3}\left(f_{r-1} + 4f_r + f_{r+1}^{[\sigma]}\right). \qquad (4.7)$$

The same considerations as for the Adams interpolation method now yield

$$\delta^{[\sigma]} = \frac{h}{3}\{f(x_{r+1}, y_{r+1}^{[\sigma]}) - f(x_{r+1}, y_{r+1}^{[\sigma-1]})\}$$

for the change $\delta^{[\sigma]}$ of (4.3) and hence a sufficient condition for the convergence of the iteration in the central-difference method is

$$Kh < 3. \tag{4.8}$$

4.2. Convergence of the starting iteration

For investigating the convergence of the starting iteration (3.5) we make use of a theorem in matrix theory. This is concerned with a sequence of sets of quantities

$$x_1^{[\nu]}, x_2^{[\nu]}, \ldots, x_p^{[\nu]} \qquad (\nu = 0, 1, 2, \ldots),$$

which are bounded successively thus

$$|x_\varrho^{[\nu]}| \leq \sum_{\sigma=1}^{p} |A_{\varrho\sigma}| |x_\sigma^{[\nu-1]}| \qquad (\nu = 1, 2, \ldots; \varrho = 1, 2, \ldots, p).$$

The theorem states[1]: For the convergence of the p series

$$\sum_{\nu=0}^{\infty} |x_\varrho^{[\nu]}| \qquad (\varrho = 1, 2, \ldots, p)$$

it is sufficient that the absolute values of all the characteristic roots $\varkappa$ of the matrix

$$A = \begin{pmatrix} |A_{11}| & |A_{12}| & \ldots & |A_{1p}| \\ \cdot & \cdot & \cdot & \cdot & \cdot & \cdot & \cdot & \cdot \\ |A_{p1}| & |A_{p2}| & \ldots & |A_{pp}| \end{pmatrix} \tag{4.9}$$

be less than unity, i.e. $|\varkappa_\sigma| < 1$ for $\sigma = 1, \ldots, p$.

The characteristic roots $\varkappa_B$ of a matrix B with elements b_{jk} are defined as the roots $\varkappa$ of the "characteristic equation" of B:

$$\begin{vmatrix} b_{11} - \varkappa & b_{12} & \ldots & b_{1p} \\ b_{21} & b_{22} - \varkappa & \ldots & b_{2p} \\ \cdot & \cdot & \cdot & \cdot & \cdot & \cdot & \cdot \\ b_{p1} & b_{p2} & \ldots & b_{pp} - \varkappa \end{vmatrix} = 0. \tag{4.10}$$

We now turn to the starting iteration. The equation (3.5) which describes it becomes

$$y_\varrho^{[\nu+1]} = y_0 + h\left[a_{\varrho 0}f_0 + \sum_{\sigma=1}^{p} a_{\varrho\sigma}f_\sigma^{[\nu]}\right] \qquad (\varrho = 1, 2, \ldots, p; \nu = 1, 2, \ldots) \tag{4.11}$$

[1] See, for example, L. COLLATZ: Eigenwertaufgaben, p. 311. Leipzig 1949.

when written in terms of the function values. The values of the $a_{\varrho\sigma}$ for $p=2$, i.e. using only the formulae of (3.5) within the dotted frame, are given in Table II/23 and for $p=3$, i.e. using the complete set of formulae (3.5), in Table II/24.

Table II/23. The $a_{\varrho\sigma}$ for $p=2$

	$\sigma=$		
	0	1	2
$\varrho=1$	$\dfrac{5}{12}$	$\dfrac{8}{12}$	$-\dfrac{1}{12}$
$\varrho=2$	$\dfrac{1}{3}$	$\dfrac{4}{3}$	$\dfrac{1}{3}$

Table II/24. The $a_{\varrho\sigma}$ for $p=3$

	$\sigma=$			
	0	1	2	3
$\varrho=1$	$\dfrac{9}{24}$	$\dfrac{19}{24}$	$-\dfrac{5}{24}$	$\dfrac{1}{24}$
$\varrho=2$	$\dfrac{1}{3}$	$\dfrac{4}{3}$	$\dfrac{1}{3}$	0
$\varrho=3$	$\dfrac{3}{8}$	$\dfrac{9}{8}$	$\dfrac{9}{8}$	$\dfrac{3}{8}$

From (4.11) we find that the changes

$$\delta_\varrho^{[\nu]} = y_\varrho^{[\nu+1]} - y_\varrho^{[\nu]}$$

are given by

$$\delta_\varrho^{[\nu]} = h \sum_{\sigma=1}^{p} a_{\varrho\sigma} \{f(x_\sigma, y_\sigma^{[\nu]}) - f(x_\sigma, y_\sigma^{[\nu-1]})\} \tag{4.12}$$

$$(\varrho = 1, 2, \dots, p; \ \nu = 1, 2, \dots);$$

using the Lipschitz condition (1.12) we obtain the inequalities

$$|\delta_\varrho^{[\nu]}| \leq K h \sum_{\sigma=1}^{p} |a_{\varrho\sigma}| |\delta_\sigma^{[\nu-1]}|. \tag{4.13}$$

The convergence of the series

$$\lim_{\nu\to\infty} y_\varrho^{[\nu]} = y_\varrho^{[0]} + \delta_\varrho^{[0]} + \delta_\varrho^{[1]} + \delta_\varrho^{[2]} + \cdots$$

is assured by the above-mentioned matrix theorem provided that the absolute values of all characteristic roots μ of the matrix A with elements $Kh|a_{\varrho\sigma}|$ are less than unity, i.e. provided that the absolute values of all characteristic numbers μ^* of the matrix

$$A^* = \begin{pmatrix} |a_{11}| \dots |a_{1p}| \\ \cdot \quad \cdot \quad \cdot \quad \cdot \quad \cdot \quad \cdot \\ |a_{p1}| \dots |a_{pp}| \end{pmatrix}$$

are less than $(Kh)^{-1}$.

Given the matrix A^*, which we are, this condition provides an upper bound for Kh. Thus for $p=2$

$$A^* = \begin{pmatrix} \dfrac{8}{12} & \dfrac{1}{12} \\ \dfrac{4}{3} & \dfrac{1}{3} \end{pmatrix}$$

and the μ^* are the roots of

$$\begin{vmatrix} \dfrac{8}{12} - \mu^* & \dfrac{1}{12} \\[2mm] \dfrac{4}{3} & \dfrac{1}{3} - \mu^* \end{vmatrix} = 0, \quad \text{namely} \quad \mu^* = \frac{1}{2} \pm \frac{1}{6}\sqrt{5} \approx \begin{cases} 0\cdot873 \\ 0\cdot127. \end{cases}$$

Consequently convergence is ensured by choosing h such that

$$Kh < \frac{1}{0\cdot873}, \quad \text{i.e.} \quad Kh < 1\cdot14. \tag{4.14}$$

For $p = 3$ the corresponding equation is

$$\begin{vmatrix} 19 - \tau & 5 & 1 \\ 32 & 8 - \tau & 0 \\ 27 & 27 & 9 - \tau \end{vmatrix} = 0,$$

where $\tau = 24\,\mu^*$, whose largest root is $\mu^* \approx 1\cdot24$, and the iteration certainly converges if

$$Kh < \frac{1}{1\cdot24}, \quad \text{i.e.} \quad Kh < 0\cdot8. \tag{4.15}$$

The limits (4.14), (4.15) on Kh for the starting iteration are more restrictive than the limits (4.5), (4.8) for the iterations in the main calculation. This fact accords with the particular sensitivity of the starting iteration.

4.3. Recursive error estimates

In this and the following section we describe[1] how error estimates for the finite-difference methods can be derived; we may note that

---

[1] For literature on error estimation see the following list: MISES, R. v.: Zur numerischen Integration von Differentialgleichungen. Z. Angew. Math. Mech. **10**, 81—92 (1930). — SCHULZ, G.: Interpolationsverfahren zur numerischen Integration gewöhnlicher Differentialgleichungen. Z. Angew. Math. Mech. **12**, 44—59 (1932). — Fehlerabschätzung für das Störmersche Integrationsverfahren. Z. Angew. Math. Mech. **14**, 224—234 (1934). — COLLATZ, L.: Natürliche Schrittweite bei numerischer Integration von Differentialgleichungssystemen. Z. Angew. Math. Mech. **22**, 216—225 (1942). — Differenzenschemaverfahren zur numerischen Integration von gewöhnlichen Differentialgleichungen n-ter Ordnung. Z. Angew. Math. Mech. **29**, 199—209 (1949). — HAMEL, G.: Zur Fehlerabschätzung bei gewöhnlichen Differentialgleichungen erster Ordnung. Z. Angew. Math. Mech. **29**, 337—341 (1949). — WEISSINGER, J.: Eine verschärfte Fehlerabschätzung zum Extrapolationsverfahren von ADAMS. Z. Angew. Math. Mech. **30**, 356—363 (1950). — Eine Fehlerabschätzung für die Verfahren von ADAMS und STÖRMER. Z. Angew. Math. Mech. **32**, 62—67 (1952). — TOLLMIEN, W.: Bemerkung zur Fehlerabschätzung beim Adamsschen Interpolationsverfahren. Z. Angew. Math. Mech. **33**, 151—155 (1953). — UHLMANN, W.: Fehlerabschätzungen bei Anfangswertaufgaben gewöhnlicher Differentialgleichungssysteme 1. Ordnung. Z. Angew. Math. Mech. **37**, 88—99 (1957). — Fehlerabschätzungen bei Anfangswertaufgaben mit einer gewöhnlichen Differentialgleichung höherer Ordnung. Z. Angew. Math. Mech. **37**, 99—111 (1957). — VIETORIS, L.: Der Richtungsfehler einer durch das Adamssche Interpolationsverfahren gewonnenen Näherungslösung einer Gleichung $y' = f(x, y)$. Öst. Akad. Wiss., Math.-naturw. Kl., S.-Ber. IIa **162**, 157—167, 293—299 (1953).

the general remarks on error estimation made in § 1.3 apply here. The first rigorous error bounds were obtained by R. v. MISES[1].

To simplify matters we make the following assumptions:

1. *The number of decimals carried in the calculation is sufficient for rounding errors to be neglected.*

2. *The iterations* (3.5), (3.14), (3.18) *are always repeated until the numbers settle to the number of decimals carried so that it can be assumed that the values* y_r *are exact solutions of the corresponding equations without the bracketed superscripts.*

As will appear later, the following will be needed in the derivation of the error estimates:

1. A Lipschitz constant K [as in (4.1)].

---

[1] RICHARD EDLER VON MISES was one of the most eminent of applied mathematicians. The modern broad conception of applied mathematics is due to him, and his exceedingly numerous and diverse contributions to this comprehensive applied mathematics have had no little influence on its present-day importance; he published fundamental and pioneering work in almost every constituent subject: in practical analysis, geometry, probability theory, mathematical statistics, and in various branches of mechanics, from strength of materials and theory of machines to the mechanics of plastic media and hydro- and aerodynamics. Over and above these specific accomplishments he strove for philosophical understanding; he set out his unified "positivistic" view of the world, which was influenced by ERNST MACH, in a book "KLEINES LEHRBUCH DER POSITIVMUS", which embraced even religion, art and poetry. He had a great love of German literature and was particularly interested in the works of the Austrian poet RAINER MARIA RILKE; v. MISES was one of the greatest connoisseurs of RILKE and possessed one of the most notable collections of his works.

v. MISES was born in Lemberg, Austria, on the 19th April 1883. He obtained his doctorate at the Technische Hochschule in Vienna in 1908 and in the same year qualified as lecturer in Brünn. In 1909 at the age of 26 he took up a post as extraordinary professor at Strassburg. After the first world war, in which he saw flying service and also designed a large aeroplane of some 600 h.p. bearing his name, he went as professor first (1919) to Dresden and then (1920) to the University of Berlin, where he set up and directed the Institut für Angewandte Mathematik, which was to become famous later. Also in 1920 he founded the journal ZEITSCHRIFT FÜR ANGEWANDTE MATHEMATIK UND MECHANIK, the forerunner of many similar journals founded later; he was editor until 1934. During the years 1933—1939 he worked at the University of Istanbul, where, at the request of the Turkish government, he founded an institute for pure and applied mathematics. In 1939 he accepted an invitation to Harvard University, Cambridge, Mass., where he remained, becoming GORDON McKAY Professor of Aerodynamics and Applied Mathematics in 1943.

To the last years of his life he was exceedingly productive in the scientific field and retained his great versatility and nimbleness of mind. Highly honoured with the honorary degrees of numerous colleges, loved and respected by a vast number of friends and pupils from all parts of the world (the author of this book was one of his pupils), he died in Boston on the 14th July 1953, leaving a widow, HILDA GEIRINGER, his colleague for many decades.

2. Error limits Y_ϱ for the starting values y_ϱ. If, for example, the necessary starting values are calculated by method II of § 3.2 (Taylor series method), the error can usually be estimated very easily; the maximum rounding error, i.e. $\frac{1}{2}\times10^{-d}$ for a d decimal number, will often provide a suitable upper bound. If the iteration method (method III of § 3.2) is used to obtain the starting values, the error can be estimated by a special method[1] (cf. § 4.5).

3. An upper bound for the absolute value of a certain derivative $f^{(q)} = \dfrac{d^q f[x, y(x)]}{dx^q}$. When $f(x, y)$ has a simple analytic form, the estimation of $f^{(q)}$ from the explicit expression obtained by differentiating f is usually quite straightforward. Mostly, however, this method is very complicated and in fact sometimes may not be possible at all (for example, when empirical laws are involved). In such cases we must be content with an approximate value for $|f^{(q)}|_{\max}$ inferred from the difference table, using the fact that

$$f^{(q)}\left(x_k, y(x_k)\right) \approx \frac{1}{h^q}\, \nabla^q f_{k+\frac{q}{2}}\,;$$

if the q-th differences run smoothly, we can get a fair idea of the maximum absolute value of $f^{(q)}$. Of course, a rigorous error estimate cannot be obtained by this method.

We investigate first the Adams interpolation method, which is based on the formula (3.11). In Lagrangian form this formula reads [as in (3.12)]

$$y_{r+1} = y_r + h \sum_{\varrho=0}^{p} \alpha_{p\varrho}^{*} f_{r+1-\varrho}. \tag{4.16}$$

A similar relation, but with a remainder term S_{p+1}^{*}, holds for the exact solution:

$$y(x_{r+1}) = y(x_r) + h \sum_{\varrho=0}^{p} \alpha_{p\varrho}^{*} f\left(x_{r+1-\varrho}, y(x_{r+1-\varrho})\right) + S_{p+1}^{*}. \tag{4.17}$$

For this remainder term, or "truncation error", there exists the estimate (2.11) of Ch. I:

$$|S_{p+1}^{*}| \leq C^{*}, \quad \text{where} \quad C^{*} = h^{p+2} |\beta_{p+1}^{*}|\, |f^{(p+1)}|_{\max}. \tag{4.18}$$

For the error

$$\varepsilon_r = y_r - y(x_r)$$

subtraction of (4.17) from (4.16) yields the relation

$$\varepsilon_{r+1} = \varepsilon_r + h \sum_{\varrho=0}^{p} \alpha_{p\varrho}^{*} \{f_{r+1-\varrho} - f\left(x_{r+1-\varrho}, y(x_{r+1-\varrho})\right)\} - S_{p+1}^{*}.$$

---

[1] See also p. 57−58 of G. Schulz: Interpolationsverfahren …. Z. Angew. Math. Mech. **12**, 44−59 (1932), which has already been mentioned.

The differences of the f values which occur in the summation can be estimated[1] by means of the Lipschitz condition (1.12):

$$|\varepsilon_{r+1}| \leq |\varepsilon_r| + hK \sum_{\varrho=0}^{p} |\alpha_{p\varrho}^*| |\varepsilon_{r+1-\varrho}| + C^*. \tag{4.19}$$

If the equality sign is written in place of "$\leq$", an equation results which determines recursively a sequence Y_r of upper limits for the errors ε_r once upper limits Y_s for the first p errors ε_s ($s=0, 1, \ldots, p-1$) are known. In principle, therefore, an error estimate can be obtained from the equations

$$\left. \begin{aligned} (1 - hK\beta_p)\, Y_{r+1} &= \left(1 + hK|\alpha_{p1}^*|\right) Y_r + hK \sum_{\varrho=2}^{p} |\alpha_{p\varrho}^*|\, Y_{r+1-\varrho} + C^* \\ (r &= p-1, p, p+1, \ldots), \end{aligned} \right\} \tag{4.20}$$

where β_p has been substituted for α_{p0}^* in accordance with (3.13).

Here h is to be chosen so small that $Kh\beta_p < 1$. This is the same condition as that which ensures a convergent iteration in the main calculation, namely (4.5).

---

[1] WEISSINGER, J.: Z. Angew. Math. Mech. **30**, 356—363 (1950); **32**, 62—67 (1952), has succeeded in refining this estimate somewhat by performing additional manipulations before taking absolute values. He writes down the equation for $r, r-1, \ldots, r-l$, where $l \geq p$, and adds, obtaining (with the different truncation errors distinguished by the notation $s_{\sigma, p+1}^*$)

$$\varepsilon_{r+1} = \varepsilon_{r-l} + h \sum_{\sigma=0}^{p+l} b_\sigma \{ f_{r+1-\sigma} - f(x_{r+1-\sigma}, y(x_{r+1-\sigma})) \} - \sum_{\sigma=0}^{l} s_{\sigma, p+1}^*,$$

where

$$b_\sigma = \begin{cases} \displaystyle\sum_{\tau=0}^{\sigma} \alpha_{p\tau}^* & \text{for} \quad 0 \leq \sigma \leq p \\[2ex] \displaystyle\sum_{\tau=0}^{p} \alpha_{p\tau}^* = 1 & \text{for} \quad p \leq \sigma \leq l \\[2ex] \displaystyle\sum_{\tau=\sigma-l}^{p} \alpha_{p\tau}^* & \text{for} \quad l \leq \sigma \leq p+l. \end{cases}$$

Taking absolute values now yields

$$|\varepsilon_{r+1}| \leq |\varepsilon_{r-l}| + hK \sum_{\sigma=0}^{p+l} |b_\sigma| |\varepsilon_{r+1-\sigma}| + (l+1)\, C^*.$$

This is the equation which corresponds to (4.19) and the further considerations leading to an independent error estimate which are applied to (4.19) in § 4.4 apply here quite analogously: z is to be determined now from

$$z^{l+p} - z^{p-1} - Kh \sum_{\sigma=0}^{l+p} |b_\sigma|\, z^{l+p-\sigma} = 0$$

instead of from (4.28).

Precisely similar considerations applied to the Adams extrapolation method (3.6) lead to the equation

$$Y_{r+1} = Y_r + h K \sum_{\varrho=0}^{p} |\alpha_{p\varrho}| \, Y_{r-\varrho} + C \qquad (r = p, \, p+1, \ldots), \quad (4.21)$$

where

$$C = h^{p+2} \beta_{p+1} |f^{(p+1)}|_{\max}; \qquad (4.22)$$

applied to the central-difference method they give

$$(1 - \tfrac{1}{3} K h) \, Y_{r+1} = \tfrac{4}{3} K h \, Y_r + (1 + \tfrac{1}{3} K h) \, Y_{r-1} + C^{**} \qquad (r = 1, 2, \ldots), \quad (4.23)$$

where

$$C^{**} = \tfrac{1}{90} h^5 |f^{IV}|_{\max}. \qquad (4.24)$$

Here again the condition on h, namely $Kh < 3$, is identical with the convergence condition (4.8) for the iteration in the main calculation.

4.4. Independent error estimates

Practical application of these recursive estimates is laborious so we now establish an upper bound for Y_s which does not entail the calculation of the previous Y_ϱ. We forfeit something thereby, as might be expected, for the error limits so obtained generally prove to be less precise than those calculated recursively.

Again we deal with the Adams interpolation method first. The error limits Y_r for this method satisfy the linear inhomogeneous difference equation (4.20) with constant positive coefficients. The solution of this equation which is determined by the starting values $Y_0, Y_1, \ldots, Y_{p-1}$ is majorized by any particular solution W_r such that

$$W_\varrho \geqq Y_\varrho \geqq 0 \quad \text{for} \quad \varrho = 0, 1, \ldots, p-1, \qquad (4.25)$$

for these inequalities remain valid for all positive ϱ on account of the positive coefficients in (4.20).

Now the general solution of a linear inhomogeneous difference equation with constant coefficients can be given in closed form. In precisely the same way as with the corresponding type of differential equation the solution can be written as the sum of a particular solution $W^{(1)}$ of the inhomogeneous equation and the general solution $W^{(2)}$ of the homogeneous equation. In the present case $W^{(1)}$ can be taken as a constant W^* and by substitution in (4.20) we find that

$$W^* = - \frac{C^*}{K h \sum_{\varrho=0}^{p} |\alpha_{p\varrho}^*|}. \qquad (4.26)$$

The well-known method for solving the homogeneous equation is to assume a solution of the form

$$W_r = z^r; \qquad (4.27)$$

from (4.20) we must have

$$(1 - K h \beta_p) z^{r+1} = \left(1 + K h |\alpha_{p1}^*|\right) z^r + K h \sum_{\varrho=2}^{p} |\alpha_{p\varrho}^*| z^{r+1-\varrho},$$

which yields the "characteristic equation"

$$z^p - z^{p-1} = K h \sum_{\varrho=0}^{p} |\alpha_{p\varrho}^*| z^{p-\varrho} \qquad (4.28)$$

for z.

This equation always has a positive root z greater than unity, for the left-hand side is zero, and therefore smaller than the positive right-hand side, when $z = 1$, and, since the coefficient $K h \beta_p$ of the z^p on the right-hand side is less than unity [assumption (4.5)], the left-hand side must be greater than the right-hand side for sufficiently large values of z. In the following we take z to be the smallest of the roots of (4.28) which are greater than unity. We can then determine a constant A so that the

$$W_\varrho = W^* + A z^\varrho \qquad (4.29)$$

satisfy the inequalities (4.25) for $\varrho = 0, 1, \ldots, p-1$ and hence also for all positive ϱ. If all $|\varepsilon_\varrho|$ with $\varrho = 0, 1, \ldots, p-1$ are less than ε, we can put $A = \varepsilon - W^*$ (W^* is negative), thus obtaining the error limit

$$|\varepsilon_r| \le \varepsilon z^r - W^*(z^r - 1); \qquad (4.30)$$

with the values substituted from (4.18), (4.26) this becomes

$$|\varepsilon_r| \le \varepsilon z^r + \gamma_p^* h^{p+1} \frac{1}{K} |f^{(p+1)}|_{\max} (z^r - 1) \qquad (r = 1, 2, \ldots), \quad (4.31)$$

where the γ_p^* are defined by

$$\gamma_p^* = \frac{|\beta_{p+1}^*|}{\sum\limits_{\varrho=0}^{p} |\alpha_{p\varrho}^*|} \qquad (4.32)$$

and are given numerically for the first few values of p by

$$\left.\begin{aligned}
\gamma_1^* &= \frac{1}{12} \approx 0{\cdot}0833, \\[4pt]
\gamma_2^* &= \frac{1}{28} \approx 0{\cdot}0357, \\[4pt]
\gamma_3^* &= \frac{19}{1020} \approx 0{\cdot}0186, \\[4pt]
\gamma_4^* &= \frac{27}{2572} \approx 0{\cdot}0105.
\end{aligned}\right\} \qquad (4.33)$$

This method of finding a solution W_r of the difference equation which majorizes the error limits Y_r can also be applied to the Adams extrapolation formula. The appropriate difference equation (4.21) now has the particular solution

$$W = -\frac{C}{Kh\sum_{\varrho=0}^{p}|\alpha_{p\varrho}|},$$ (4.34)

and for the solution (4.27) of the corresponding homogeneous equation we determine z as the smallest positive root greater than unity of the equation

$$z^{p+1} - z^{p} = Kh\sum_{\varrho=0}^{p}|\alpha_{p\varrho}|z^{p-\varrho}.$$ (4.35)

If ε has the same significance as for the interpolation method above, we obtain here the limits

$$|\varepsilon_r| \leqq \varepsilon z^r - W(z^r - 1).$$ (4.36)

With the values (4.22), (4.34) this reads

$$|\varepsilon_r| \leqq \varepsilon z^r + \gamma_p h^{p+1}\frac{1}{K}|f^{(p+1)}|_{\max}(z^r - 1) \qquad (r = 1, 2, \ldots),$$ (4.37)

where

$$\gamma_p = \frac{\beta_{p+1}}{\sum_{\varrho=0}^{p}|\alpha_{p\varrho}|}.$$ (4.38)

Numerical values for the first few γ_p are

$$\left.\begin{array}{l}\gamma_1 = \dfrac{5}{24} \approx 0{\cdot}2033,\\[2mm]\gamma_2 = \dfrac{9}{88} \approx 0{\cdot}1023,\\[2mm]\gamma_3 = \dfrac{251}{4800} \approx 0{\cdot}0523.\end{array}\right\}$$ (4.39)

Finally we derive in a similar fashion an estimate for the error in the central-difference method. A particular solution of the pertinent difference equation (4.23) is

$$W^{**} = -\frac{C^{**}}{2Kh},$$ (4.40)

and for the solution of the corresponding homogeneous difference equation we again use the smallest positive root z greater than unity of its characteristic equation, which here reads

$$3(z^2 - 1) = Kh(z^2 + 4z + 1).$$ (4.41)

For the determination of A in the particular solution

$$W_\varrho = W^{**} + A z^\varrho$$

we ought strictly to distinguish between the cases in which the limits $|\varepsilon_\varrho| \leqq \varepsilon$ are known for $\varrho = 0, 1, 2$ and $\varrho = 0, \pm 1$, respectively, but since the second case can be reduced to the first by re-numbering the ϱ values (displacing them by 1), we need only write the error estimate in the one form

$$|\varepsilon_r| \leqq \varepsilon z^r - W^{**}(z^r - 1) \tag{4.42}$$

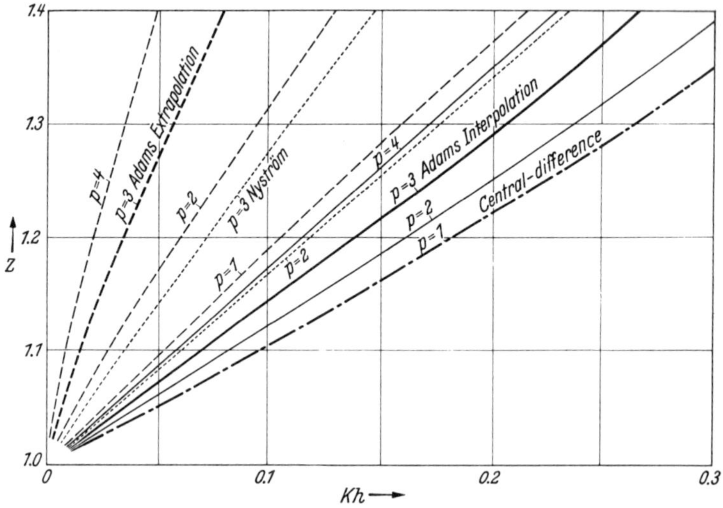

Fig. II/4. Curves of the quantity z, which determines the growth of the error limits, plotted against hK for various methods and orders of approximation. — — — ADAMS' extrapolation method (3.6), ······· NYSTRÖM's extrapolation method (3.10), ——— ADAMS' interpolation method (3.11), —·—·— Central-difference method (3.16)

corresponding to the first case; substituting from (4.24) and (4.40), we have

$$|\varepsilon_r| \leqq \varepsilon z^r + \frac{h^4}{180} \frac{1}{K} |f^{(4)}|_{\max} (z^r - 1). \tag{4.43}$$

For large values of r the growth of the error limits (4.31), (4.37), (4.43) is determined by the power z^r. Consequently the usefulness of the error estimates for the various methods may be compared by comparing the corresponding values of z for given Kh. This is done in Fig. II/4, where curves of z against Kh are drawn for small Kh for the three methods which we have been considering and also for NYSTRÖM'S extrapolation method; except for the central-difference method, for which $p = 2$, the curves for several values of p are shown for each method[1].

---

[1] Although not exactly, the curve for the Adams interpolation method with $p = 1$ coincides to within the accuracy of Fig. II/4 with the curve for the central-difference method.

For ease of comparison the curves for $p=3$ are printed more heavily; we choose $p=3$ so that the truncation error is of the same order (h^4) as the central-difference method for all the methods compared. We see that the z values are smallest, and therefore the error limits most effective, for the central-difference method. It is unfortunate that the z values increase with p so that, although in general the calculation will be more accurate for larger p (so long as the differences do not start increasing or fluctuating), the error limits for large r become less precise as p increases.

4.5. Error estimates for the starting iteration (3.5)

We describe briefly how one can estimate the error in the starting iteration (3.5). For the exact solution we have

$$\left.\begin{aligned}
y(x_1) &= y_0 + h(F_0 + \tfrac{1}{2}VF_1 - \tfrac{1}{12}V^2 F_2 + \tfrac{1}{24}V^3 F_3) + \tilde{S}_1, \\
y(x_2) &= y_0 + h(2F_1 \qquad\quad + \tfrac{1}{3}V^2 F_2) \qquad\qquad + \tilde{S}_2, \\
y(x_3) &= y(x_1) + h(2F_2 \qquad + \tfrac{1}{3}V^2 F_3) \qquad\qquad + \tilde{\tilde{S}}_3,
\end{aligned}\right\} \quad (4.44)$$

in which we have introduced the notation $F_\nu = f(x_\nu, y(x_\nu))$ [the notation $\tilde{S}_p$ for the remainder terms is not the same as in Ch. I (2.11)].

If $N_p = \left|\dfrac{d^p f}{dx^p}\right|_{\max}$ [cf. (1.23)], we have from Ch. I (2.10)

$$|\tilde{S}_1| \leq h^5 |\beta_4^*| N_4 = \frac{19}{720} h^5 N_4 \qquad (4.45)$$

and from Ch. I (2.18)

$$|\tilde{S}_2| \leq \frac{h^5}{90} N_4, \qquad |\tilde{\tilde{S}}_3| \leq \frac{h^5}{90} N_4. \qquad (4.46)$$

Now we imagine the differences in (4.44) expressed in terms of the function values so that we can subtract it from (4.11), the corresponding form of (3.5); the coefficients $a_{\varrho\sigma}$ are given in Tables II/23, II/24. Using the Lipschitz condition to estimate the function differences which arise on subtraction, we find that the absolute values of the errors $|\varepsilon_j| = |y_j - y(x_j)|$ satisfy the inequalities

$$|\varepsilon_\varrho| - hK \sum_{\sigma=1}^{3} |a_{\varrho\sigma}| |\varepsilon_\sigma| \leq |\tilde{S}_\varrho| \qquad (\varrho = 1, 2, 3), \qquad (4.47)$$

where $\tilde{S}_3 = \tilde{S}_1 + \tilde{\tilde{S}}_3$.

We can replace the right-hand sides by the upper bounds given by (4.45), (4.46), and we then consider the corresponding set of equations; these determine quantities which we will denote by Y_ϱ. In matrix form we have

$$Y - hKBY = S, \qquad (4.48)$$

where

$$Y = \begin{pmatrix} Y_1 \\ Y_2 \\ Y_3 \end{pmatrix}, \quad B = \begin{pmatrix} |a_{11}| & |a_{12}| & |a_{13}| \\ |a_{21}| & |a_{22}| & |a_{23}| \\ |a_{31}| & |a_{32}| & |a_{33}| \end{pmatrix}, \quad S = \begin{pmatrix} S_1 \\ S_2 \\ S_3 \end{pmatrix}$$

and

$$|\tilde{S}_1| \leqq S_1 = \frac{19}{720} h^5 N_4, \quad |\tilde{S}_2| \leqq S_2 = \frac{8}{720} h^5 N_4, \quad |\tilde{S}_3| \leqq S_3 = \frac{27}{720} h^5 N_4.$$

If I denotes the unit matrix, the solution of the matrix equation (4.48) is given by

$$Y = (I + hKB + h^2 K^2 B^2 + \cdots) S \tag{4.49}$$

provided that the matrix series converges. Now the condition ensuring the convergence of the initial iteration (§ 4.2), i.e. that the absolute values of all the characteristic roots of the matrix hKB shall be less than unity, also ensures the convergence of this matrix series[1], so (4.49) is the solution and the limit matrix is the inverse of the original matrix $I - hKB$. Clearly this limit matrix can have only non-negative coefficients so the system must be monotonic (see Ch. I, § 5.5) and it follows that the Y_ϱ are upper limits for the $|\varepsilon_\varrho|$.

These limits can be found by solving the equations (4.48) but it is quicker, though less precise, of course, to derive upper limits for the Y_ϱ from the solution (4.49). If b is the largest element of B, then any element of B^2 is at most $3b^2$ and in general the elements of B^q cannot exceed $\frac{1}{3}(3b)^q$; thus the matrix series in (4.49) is majorized by the series $I + \sum_{s=0}^{\infty} \frac{1}{3}(3bhK)^s J$, where $J = (j_{\varrho\sigma})$ is the matrix with $j_{\varrho\sigma} = 1$. If $3bhK < 1$, we can sum the geometric series to obtain the error limits

$$|\varepsilon_\varrho| \leqq Y_\varrho \leqq S_\varrho + \frac{hKb}{1 - 3hKb}(S_1 + S_2 + S_3) \quad (\varrho = 1, 2, 3). \tag{4.50}$$

4.6. Systems of differential equations

W. Richter[2] extends the above results to the system

$$y'_\nu(x) = f_\nu(x, y_1(x), \ldots, y_n(x)) \quad (\nu = 1, \ldots, n)$$

with prescribed initial values $y_\nu(0)$.

Let the f_ν satisfy Lipschitz conditions of the form

$$|f_\nu(x, y_1, \ldots, y_n) - f_\nu(x, y_1^*, \ldots, y_n^*)| \leqq K \sum_{\mu=1}^{n} |y_\mu - y_\mu^*| \quad (\nu = 1, \ldots, n)$$

---

[1] See, for example, L. Collatz: Eigenwertaufgaben mit technischen Anwendungen, p. 311. Leipzig 1949.

[2] Richter, Willy: Examination de l'erreur commise dans la méthode de M. W. E. Milne ... (43 pages). Diss. Neuchâtel 1952.

in a certain domain D of the $(x, y_1, \ldots, y_n)$ space and let

$$R = \max_{v, D} \frac{h^5}{180} \left| \frac{d^4 f_v(x)}{d x^4} \right| ;$$

further, put $q = \dfrac{h \, K \, n}{3}$ and $Q = \dfrac{2 (1 + 28 q) \, R}{6 q + 20 q^2}$. For the approximations Y_{vr} to $y_v(r h)$ obtained for $r = -1, +1, +2$ from a starting iteration (convergent for $q < 1$) and for $r > 2$ from MILNE'S formulae (see § 3.3, IV) it is deduced that there exist error limits ω_r ($|Y_{vr} - y_v(r h)| \leqq \omega_r$) with the upper bounds [corresponding to (4.30)]

$$\omega_r \leqq \omega z_1^{r-1} + Q(z_1^{r-1} - 1),$$

where $\omega = \max(\omega_{-1}, \omega_1, \omega_2)$ and z_1 is the positive root of

$$z^4 - 4q(1 + 2q) z^3 - (1 + q + 4q^2) z^2 - 8q^2 z - q = 0.$$

4.7. Instability in finite-difference methods

It can happen that an approximate solution calculated by a finite-difference method is unstable[1] even though the differential equation is inherently stable. This is particularly so when the difference equation used is of higher order than the differential equation, for it then has more independent solutions than the differential equation and among them there may be increasing solutions (which, on account of the ever-present rounding errors, finally determine the behaviour of the approximate solution) even when the differential equation possesses only decreasing solutions.

The following theory, developed by RUTISHAUSER[2], offers an explanation of the phenomenon. Several simplifying assumptions are made.

We can survey the situation in a rough way by assuming that f_y for the differential equation (1.10) can be treated as piecewise constant. We then consider the calculation of the y_r, say by the central-difference method (3.16), over an interval J of constant f_y. Let $y_r + \eta_r$ be another solution of the equation (3.16) for which the η_r are small, i.e. in the nature of perturbations from the solution y_r, in fact so small that quadratic terms in the η_r are negligible in comparison with the linear terms. Then these perturbations η_r satisfy the linearized "variation equation"

$$\eta_{r+1} = \eta_{r-1} + \frac{h}{3} f_y (\eta_{r-1} + 4\eta_r + \eta_{r+1}).$$

---

[1] The development of unevenness with the central-difference method which was mentioned in § 3.4, V [for which reference was made to L. COLLATZ and R. ZURMÜHL: Z. Angew. Math. Mech. **22**, 46 (1942)] is also due to such a condition of instability. For further examples of unstable behaviour see J. TODD: Solution of differential equations by recurrence relations. Math. Tables and Other Aids to Computation **4**, 39—44 (1950).

[2] RUTISHAUSER, H.: Über die Instabilität von Methoden zur Integration gewöhnlicher Differentialgleichungen. Z. Angew. Math. Phys. **3**, 65—74 (1952). — See also W. LINIGER: Stabilität der Differenzenschemaverfahren. Diss. ETH Zürich 1956.

Here we have already used the assumption that f_y is constant in J. This linear homogeneous difference equation can be solved in the usual way by assuming a solution of the form $\eta_r = \lambda^r$; we find that λ must satisfy the quadratic equation

$$\lambda^2 \left(1 - \frac{H}{3}\right) - \frac{4H}{3}\lambda - \left(1 + \frac{H}{3}\right) = 0, \qquad (4.51)$$

where $H = h f_y$, whose roots are

$$\lambda_1 = 1 + H + \frac{H^2}{2} + \frac{H^3}{6} + \frac{H^4}{24} + \cdots \approx e^H$$

$$\lambda_2 = -1 + \frac{H}{3} - \frac{H^2}{18} + \cdots \approx -e^{-\frac{1}{3}H}.$$

We now treat similarly a perturbation of the exact solution $y(x)$ of the differential equation $y' = f(x, y)$. We select another solution $y(x) + \eta(x)$ and, on the assumption that $\eta(x)$ is small in the same sense as for η_r above, derive the differential equation

$$\eta'(x) = f_y \eta(x)$$

for $\eta(x)$, which must therefore be proportional to $e^{f_y x}$. Such a function changes by a factor e^H over an interval of length h. Now if $f_y > 0$, the component λ_2^r in the central-difference solution dies away exponentially and since $\lambda_1 \approx e^H$, a perturbation η_r grows at approximately the same rate as the same perturbation in the solution of the differential equation. If, on the other hand, $f_y < 0$, then the differential equation is stable, i.e. small perturbations die away, but, in general, small perturbations η_r in the solution of the difference equation increase exponentially; a component proportional to λ_2^r is always introduced by the inevitable rounding errors. Consequently the method will be described as unstable for the case $f_y < 0$. This does not necessarily mean that the method is unusable for this case but it is advisable to estimate the error which may arise through instability of the method as being roughly of the order $e^{-\frac{1}{3}h f_y r} \times 10^{-q}$ at the r-th step of the integration, where q is the number of decimals carried.

In the work referred to, RUTISHAUSER also shows that for the Runge-Kutta method and the Adams extrapolation method no instability need be feared provided that h is chosen sufficiently small[1].

---

[1] In a note RUTISHAUSER shows that, with increasing h, instability first sets in much later for the interpolation method than for the extrapolation method. A similar result was found by A. R. MITCHELL and J. W. CRAGGS: Stability of difference relations in the solution of ordinary differential equations. Math. Tables and Other Aids to Computation 7, 127—129 (1953).

4.8. Improvement of error estimates by use of a weaker Lipschitz condition

The estimates which we have derived for the errors in finite-difference solutions of $y' = f(x, y)$ can lead to better results if the Lipschitz condition (1.12) on $f(x, y)$, which involves absolute values, is replaced by the weaker condition[1]

$$\frac{f(x, y_1) - f(x, y_2)}{y_1 - y_2} \leqq L.$$

This affords a more realistic and accurate discription of the actual situation than the condition which also specifies a lower limit for the quotient, particularly when a negative value can be chosen for L.

Consider the more general system

$$y'_j = f_j(x, y_1, y_2, \ldots, y_n) \qquad (j = 1, 2, \ldots, n) \qquad (4.52)$$

with the initial conditions $y_j(x_0) = y_{j0}$, where the f_j are given continuous (real) functions in a domain D of the $(x, y_1, y_2, \ldots, y_n)$ space which contains the initial point $(x_0, y_{10}, \ldots, y_{n0})$. For any two points $(x, \bar{y}_1, \ldots, \bar{y}_n)$, $(x, \underline{y}_1, \ldots, \underline{y}_n)$ in D with the same x let the function

$$
\left.
\begin{aligned}
&L^*(x, \bar{y}_1, \ldots, \bar{y}_n, \underline{y}_1, \ldots, \underline{y}_n) \\
&= \frac{\sum\limits_{j=1}^{n} [f_j(x, \bar{y}_1, \ldots, \bar{y}_n) - f_j(x, \underline{y}_1, \ldots, \underline{y}_n)] (\bar{y}_j - \underline{y}_j)}{\sum\limits_{j=1}^{n} (\bar{y}_j - \underline{y}_j)^2}
\end{aligned}
\right\}
\qquad (4.53)
$$

satisfy the condition

$$L^*(x, \bar{y}_1, \ldots, \bar{y}_n, \underline{y}_1, \ldots, \underline{y}_n) \leqq L. \qquad (4.54)$$

If we take two sets of functions $\bar{y}_j(x)$ and $\underline{y}_j(x)$ which, for $x_0 \leqq x \leqq x_0 + a$, are differentiable and lie within D and insert them into the differential equations, we will be left with the "error functions"

$$
\left.
\begin{aligned}
\bar{\varepsilon}_j(x) &= \bar{y}'_j(x) - f_j(x, \bar{y}_1, \ldots, \bar{y}_n) \\
\underline{\varepsilon}_j(x) &= \underline{y}'_j(x) - f_j(x, \underline{y}_1, \ldots, \underline{y}_n).
\end{aligned}
\right\}
\qquad (4.55)
$$

Now suppose that $\bar{y}_j(x) \geqq \underline{y}_j(x)$ for $j = 1, 2, \ldots, n$ and with

$$\bar{y}_j(x) - \underline{y}_j(x) = z_j(x) \geqq 0, \qquad \bar{\varepsilon}_j(x) - \underline{\varepsilon}_j(x) = \varepsilon_j(x)$$

define

$$+\sqrt{\sum_{j=1}^{n} z_j^2(x)} = z(x), \qquad +\sqrt{\sum_{j=1}^{n} \varepsilon_j^2(x)} = \varepsilon(x), \qquad \frac{\sum\limits_{j=1}^{n} z_j(x)\, \varepsilon_j(x)}{z(x)} = \tilde{\varepsilon}(x).$$

---

[1] See H. ELTERMANN: Fehlerabschätzung bei näherungsweiser Lösung von Systemen von Differentialgleichungen erster Ordnung. Math. Z. **62**, 469–501 (1955). Here we shall only sketch the basic idea of the error estimate. The assumptions actually made by ELTERMANN are slightly weaker still.

According to SCHWARZ's inequality, $\varepsilon(x) \geq |\tilde{\varepsilon}(x)|$; we will assume that $z(x) > 0$ for $x_0 > x \geq x_0 + a$. From (4.55) it follows that

$$\varepsilon_j(x) = z_j'(x) - f_j(x, \bar{y}_1(x), \ldots, \bar{y}_n(x)) + f_j(x, \underline{y}_1(x), \ldots, \underline{y}_n(x));$$

then multiplication by $\dfrac{z_j(x)}{z(x)}$ and summation over j from 1 to n yields

$$\tilde{\varepsilon}(x) = z'(x) - L^*(x) z(x), \tag{4.56}$$

where $L^*(x)$ is written for the function $L^*\big(x, \bar{y}_1(x), \ldots, \bar{y}_n(x), \underline{y}_1(x), \ldots, \underline{y}_n(x)\big)$. This is a linear differential equation for $z(x)$ with the solution

$$z(x) = z(x_0) \exp\left(\int_{x_0}^{x} L^*(t)\, dt \right) + \int_{x_0}^{x} \tilde{\varepsilon}(s) \exp\left(\int_{s}^{x} L^*(t)\, dt \right) ds.$$

Finally, using (4.54) and the fact that $|\tilde{\varepsilon}(x)| \leq \varepsilon(x)$, we arrive at the estimate

$$z(x) \leq z(x_0)\, e^{L(x - x_0)} + \int_{x_0}^{x} \varepsilon(s)\, e^{L(x-s)}\, ds. \tag{4.57}$$

If $\bar{y}_j(x)$ and $\underline{y}_j(x)$ take the prescribed initial values, then $z(x_0) = 0$ and the first term on the right-hand side disappears. [The resulting upper limit for $z(x)$ implies the uniqueness of the solution of the initial-value problem, for if $\bar{y}_j(x)$ and $\underline{y}_j(x)$ are both solutions of the problem, $\varepsilon(s) \equiv 0$ and hence also $z(x) \equiv 0$.] If $\bar{y}_j$ and $\underline{y}_j$ are respectively approximate and exact values of the solution, then (4.57) provides an error estimate for $\bar{y}_j$.

4.9. Error estimation by means of the general theorem on iteration

Error estimates which depend only on knowledge of a Lipschitz function $L(x)$ defined by

$$|f(x, y) - f(x, y^*)| \leq L(x) |y - y^*| \tag{4.58}$$

and not on bounds for the higher derivatives $f^{(q)}$ [cf. (1.23)] may be established by means of the general theory of iterative processes discussed in Ch. I, § 5.2. We introduce an operator T such that

$$T\varphi(x) = y_0 + \int_{x_0}^{x} f(\xi, \varphi(\xi))\, d\xi \tag{4.59}$$

and define a norm in an interval $\langle x_0, z \rangle$ by

$$\|\varphi\| = \max_{\langle x_0, z \rangle} \frac{|\varphi(x)|}{W(x)}, \tag{4.60}$$

where $W(x)$ is a fixed positive function at our disposal. If f satisfies the condition (4.58), we can find a Lipschitz constant K for the operator

T as follows:

$$|T\varphi_1 - T\varphi_2| = \left| \int_{x_0}^{x} [f(\xi, \varphi_1(\xi)) - f(\xi, \varphi_2(\xi))]\, d\xi \right|$$

$$\leq \int_{x_0}^{x} L(\xi)\, |\varphi_1(\xi) - \varphi_2(\xi)|\, d\xi$$

$$\leq \int_{x_0}^{x} L(\xi)\, \|\varphi_1 - \varphi_2\|\, W(\xi)\, d\xi;$$

therefore

$$\|T\varphi_1 - T\varphi_2\| \leq K\|\varphi_1 - \varphi_2\|,$$

where

$$K = \max_{\langle x_0, z\rangle} \frac{\int_{x_0}^{x} L(\xi)\, W(\xi)\, d\xi}{W(x)}. \qquad (4.61)$$

Provided that $K < 1$, we can now apply the estimate (5.16) of Ch. I — at least, in principle. By choosing $W(x) = e^{\lambda x}$ with a suitable value of λ we can, in fact, ensure that the condition $K < 1$ is satisfied; for numerical purposes functions other than $e^{\lambda x}$ may be more suitable.

If $v(x)$ is any approximation to $y(x)$ with $v(x_0) = y_0$, we define the corresponding "defect" $d(x)$ as the error in satisfying the differential equation:

$$d(x) = v'(x) - f(x, v(x)).$$

We can express the integral $D(x)$ of the defect as

$$D(x) = \int_{x_0}^{x} d(\xi)\, d\xi = v(x) - v(x_0) - \int_{x_0}^{x} f(\xi, v(\xi))\, d\xi = v - Tv = u_0 - u_1,$$

where

$$v(x) = u_0(x), \qquad u_1 = Tv.$$

Then (5.16) of Ch. I gives an error estimate for the function $u_1(x) = v(x) - D(x)$.

Example. For the Example I of § 2.6, namely

$$y' = f(x, y) = y - \frac{2x}{y}, \qquad y(0) = 1,$$

the approximation

$$u_0 = \frac{10x + 7}{3x + 7}$$

yields

$$u_1 = 1 + \frac{353}{150}\, x - \frac{3}{10}\, x^2 - \frac{49}{9}\, \ln\left(1 + \frac{3x}{7}\right) + \frac{343}{500}\, \ln\left(1 + \frac{10}{7}\, x\right).$$

Let the sub-space F (domain of definition of the operator T as used in §§ 5.1, 5.2) consist of the continuous functions $u(x)$ in $0 \leq x \leq 0.4$ which satisfy

$$|u - u_0| \leq 0.001.$$

Then in F, $L \leq 1.447$ $\left(\text{this is the maximum value of } \left|\dfrac{\partial f}{\partial y}\right| = 1 + \dfrac{2x}{y^2} \text{ in } F\right)$
and with $W(x) = e^{2x}$

$$\frac{\int\limits_0^x W(\xi)\, d\xi}{W(x)} = \frac{1 - e^{-2x}}{2} \leq 0.2753 = \varrho;$$

this choice of $W(x)$ also gives $\|u_1 - u_0\| = 0.0006$ and with $K = L\varrho = 0.398$, so that $\dfrac{K}{1 - K} = 0.662$, (5.16) of Ch. I yields

$$\|y - u_1\| \leq 0.662 \times 0.0006$$
$$= 0.0004.$$

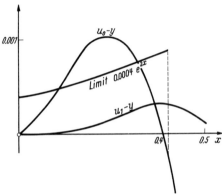

Since the functions w with $\|w - u_1\| \leq 0.0004$ lie in F, all the assumptions are satisfied and the required error estimate in $0 \leq x \leq 0.4$ is

$$|y - u_1| \leq 0.0004\, e^{2x}. \quad (4.62)$$

Fig. II/5. Error estimation by the general theorem on iteration

The actual errors $u_0 - y$ and $u_1 - y$ are compared with this limit in Fig. II/5.

§5. Finite-difference methods for differential equations of higher order

5.1. Introduction

Firstly we remark that the general observations made at the beginning of §3 on the relative merits of the various approximate methods still hold when the methods are applied to differential equations of orders higher than the first.

Secondly we note that for the most part the extension to higher orders is achieved by straightforward modifications of the considerations for the first-order case (§§ 3, 4). We can therefore be brief.

An obvious way of dealing with an n-th-order differential equation (2.1) is to convert it into a system of n first-order equations

$$y_1' = y_2, \quad y_2' = y_3, \quad y_3' = y_4, \ldots, \quad y_{n-1}' = y_n, \quad y_n' = f(x, y_1, y_2, y_3, \ldots, y_n)$$

and treat these by one of the numerical methods already described for first-order equations. However, if we do this, not only is the amount of labour involved in the computation greatly increased (thus with the finite-difference methods, for example, n difference tables will be needed), but also the accuracy suffers; in the case of the finite-difference methods this loss in accuracy is due to the fact that n functions are replaced by polynomials instead of their full Taylor expansions in powers of h and consequently n truncation errors are commited[1]. Thus the treatment of an n-th-order differential equation as it stands offers fundamental advantages over the use of the equivalent system of first-order equations. For the same reasons a given system of differential equations of low order can be transformed with advantage into a system of fewer equations of correspondingly higher order, provided that no analytical difficulties attend the transformation and that the functions appearing in the transformed equations are not too complicated.

As in § 2, consider the n-th-order differential equation (2.1)

$$y^{(n)} = f(x, y, y', \ldots, y^{(n-1)})$$

with the initial values (2.2). We assume that the function f satisfies a Lipschitz condition of the form

$$\left| f(x, y, y', \ldots, y^{(n-1)}) - f(x, y^*, y^{*\prime}, \ldots, y^{*(n-1)}) \right| \leq \sum_{\nu=0}^{n-1} K_\nu \left| y^{(\nu)} - y^{*(\nu)} \right| \quad (5.1)$$

for all pairs of points $(x, y, y', \ldots, y^{(n-1)})$, $(x, y^*, y^{*\prime}, \ldots, y^{*(n-1)})$ in a convex domain D of the $(x, y, y', \ldots, y^{(n-1)})$ space in which the solution and all approximations lie. In practice we often put

$$K_\nu = \left| \frac{\partial f}{\partial y^{(\nu)}} \right|_{\text{max in } D} \qquad (\nu = 0, 1, \ldots, n - 1). \qquad (5.2)$$

Sometimes it suffices to assume that f satisfies the simpler but cruder condition

$$\left| f(x, y, y', \ldots, y^{(n-1)}) - f(x, y^*, y^{*\prime}, \ldots, y^{*(n-1)}) \right| \leq K \sum_{\nu=0}^{n-1} \left| y^{(\nu)} - y^{*(\nu)} \right|, \quad (5.3)$$

where

$$K = \max_\nu K_\nu.$$

As with first-order equations, the calculation consists of two quite separate steps:

1. Approximations $y_\nu, y'_\nu, \ldots, y_\nu^{(n-1)}$, from which values of f_ν can be calculated, are obtained by some other means for the first few points (calculation of "starting values").

2. The solution is continued step by step by the finite-difference formulae; these give the values of y and its derivatives at the point x_{r+1} once the values at $x_r, x_{r-1}, \ldots$ are known ("main calculation"). Again we give several such step-by-step procedures (cf. § 5.4 et seq.).

---

[1] Collatz, L., and R. Zurmühl: Zur Genauigkeit verschiedener Integrationsverfahren bei gewöhnlichen Differentialgleichungen. Ing.-Arch. **13**, 34—36 (1942).

5.2. Calculation of starting values

Substantially the same means are at our disposal here as for differential equations of the first order (§ 3.2):

I. Using a "self-starting" method of integration. As for first-order equations, the Runge-Kutta method is foremost in this category and, because of the particular importance of accuracy in the starting values, is best used with a step of half the length of the main step.

II. Using the Taylor series for $y(x)$ and its derivatives. (As in II and IIa of § 3.2.)

III. Using quadrature formulae. A set of starting values can also be determined by an iterative procedure based on the formulae mentioned in § 2.5 for use as a terminal check. For the first point $x = x_1$ we use formulae corresponding to (2.24); thus for $y_1, y_1', \ldots, y_1^{(n-1)}$ we have

$$\left. \begin{aligned} y_1^{(m)} = y_0^{(m)} + h\, y_0^{(m+1)} + \frac{h^2}{2}\, y_0^{(m+2)} + \cdots + \frac{h^{n-m-1}}{(n-m-1)!}\, y_0^{(n-1)} + \\ + h^{n-m} \sum_{\varrho=0}^{p} \gamma_{n-m,\varrho}\, \varDelta^\varrho f_0. \end{aligned} \right\} \quad (5.4)$$

The main calculation will normally use second and perhaps also third differences, so we need further starting values at x_2 and perhaps also at x_3. For these we use the formulae corresponding to (2.25):

$$y_{r+1}^{(m)} = \begin{cases} y_{r-1}^{(m)} + 2 \displaystyle\sum_{\varrho=0}^{\frac{n-m-3}{2}} \frac{h^{2\varrho+1}}{(2\varrho+1)!}\, y_r^{(m+2\varrho+1)} \\ 2 y_r^{(m)} - y_{r-1}^{(m)} + 2 \displaystyle\sum_{\varrho=0}^{\frac{n-m-2}{2}} \frac{h^{2\varrho}}{(2\varrho)!}\, y_r^{(m+2\varrho)} \end{cases} \left. \begin{aligned} + 2h^{n-m}(\beta_{n,0}^{**} f_r + \\ + \beta_{n,1}^{**} \nabla^2 f_{r+1}) \\ (r = 1, 2), \end{aligned} \right\} \quad (5.5)$$

in which the upper alternative applies when $n - m$ is odd and the lower when $n - m$ is even. These formulae are used iteratively in the usual manner: first, rough values of $y_1^{(m)}, y_2^{(m)}, y_3^{(m)}$ are obtained, say from their Taylor expansions truncated after the term in $y_0^{(n)}$, and the corresponding values f_1, f_2, f_3 and their differences evaluated; these values are then inserted in the right-hand sides of (5.4), (5.5) to give the next approximations, on which the process can be repeated. We now examine this method in detail for the special case of a second-order equation.

5.3. Iterative calculation of starting values for the second-order equation $y'' = f(x, y, y')$

As in § 3.2, we suggest a preliminary scheme suitable for constructing rough values at two or three points with which to start the iteration. If starting values are required at only two points x_1, x_2, the calculation represented by the equations framed in dots suffices.

A. *Rough values*

1. $\tilde{y}_1 = y_0 + h y_0' + \frac{1}{2} h^2 f_0$, $h \tilde{y}_1' = h y_0' + h^2 f_0$;

   thence $\tilde{f}_1 = f(x_1, \tilde{y}_1, \tilde{y}_1')$, $\nabla \tilde{f}_1 = \tilde{f}_1 - f_0$;

2. $y_1^{[0]} = y_0 + h y_0' + h^2 (\frac{1}{2} f_0 + \frac{1}{6} \nabla \tilde{f}_1)$,

   $h y_1'^{[0]} = h y_0' + h^2 (f_0 + \frac{1}{2} \nabla \tilde{f}_1)$,   thence $f_1^{[0]}$, $\nabla f_1^{[0]}$;

3. $y_2^{[0]} = 2 y_1^{[0]} - y_0 + h^2 f_1^{[0]}$,

   $h y_2'^{[0]} = h y_0' + 2 h^2 f_1^{[0]}$,   thence $f_2^{[0]}$, $\nabla f_2^{[0]}$, $\nabla^2 f_2^{[0]}$.

$$(5.6)$$

4. $y_3^{[0]} = 2 y_2^{[0]} - y_1^{[0]} + h^2 (f_2^{[0]} + \frac{1}{12} \nabla^2 f_2^{[0]})$,

   $h y_3'^{[0]} = h y_1'^{[0]} + h^2 (2 f_2^{[0]} + \frac{1}{3} \nabla^2 f_2^{[0]})$.

$$\left.\right\} \quad (5.7)$$

B. *Iterative improvement for* $\nu = 1, 2, \ldots$

$y_1^{[\nu+1]} = y_0 + h y_0' + h^2 (\frac{1}{2} f_0 + \frac{1}{6} \nabla f_1^{[\nu]} - \frac{1}{24} \nabla^2 f_2^{[\nu]} \quad + \frac{1}{45} \nabla^3 f_3^{[\nu]})$

$h y_1'^{[\nu+1]} = \qquad h y_0' + h^2 \quad (f_0 + \frac{1}{2} \nabla f_1^{[\nu]} - \frac{1}{12} \nabla^2 f_2^{[\nu]} \quad + \frac{1}{24} \nabla^3 f_3^{[\nu]})$

$y_2^{[\nu+1]} = 2 y_1^{[\nu+1]} \quad - y_0 \quad + h^2 (f_1^{[\nu]} + \frac{1}{12} \nabla^2 f_2^{[\nu]})$

$h y_2'^{[\nu+1]} = \qquad h y_0' + \qquad h^2 (2 f_1^{[\nu]} + \frac{1}{3} \nabla^2 f_2^{[\nu]})$

$$(5.8)$$

$y_3^{[\nu+1]} = 2 y_2^{[\nu+1]} - y_1^{[\nu+1]} + h^2 (f_2^{[\nu]} \quad + \frac{1}{12} \nabla^2 f_3^{[\nu]})$

$h y_3'^{[\nu+1]} = \qquad h y_1'^{[\nu+1]} + h^2 (2 f_2^{[\nu]} + \frac{1}{3} \nabla^2 f_3^{[\nu]})$.

$$\left.\right\} \quad (5.9)$$

The improved values obtained at each stage are used to re-calculate the function values $f_j^{[\nu]} = f(x_j, y_j^{[\nu]}, y_j'^{[\nu]})$; the differences of these values are then formed for use in the next cycle. Very few cycles will be necessary if the step interval h is chosen sufficiently small; it should not be too large in any case, as has been stressed before (cf. § 3.4).

The scheme for calculating rough values which was given in § 3.2 incorporated an iteration on the first two points; this gave better values with which to calculate the rough value of y for the third point. Better values for starting the iteration B can be obtained here in a similar way: the rough values for the first two points are improved by an iteration using the formulae framed in B before step 4 is carried out; then the improved values are used in step 4 to give better rough values for the third point.

It may be noticed that the form of the equations in A and B is such that if y' does not occur explicitly in the differential equation, i.e.

it has the form $y''=f(x, y)$, then the calculations to find rough and improved values of the y_j' are not needed at all.

We now investigate the convergence of this starting iteration, proceeding along lines quite analogous to those of § 4.2. We first express all the differences occurring in (5.8), (5.9) in terms of function values. The iterative scheme then reads

$$\left.\begin{aligned} y_j^{[\nu+1]} &= y_0 + j\,h\,y_0' + h^2 \sum_{k=0}^{3} \sigma_{jk}\,f_k^{[\nu]} \\ y_j'^{[\nu+1]} &= y_0' \qquad\quad + h \sum_{k=0}^{3} \tau_{jk}\,f_k^{[\nu]} \end{aligned}\right\} \quad (j=1,2,3), \qquad (5.10)$$

in which the σ_{jk} and τ_{jk} have the values given in Tables II/25 and II/26. $\left(\text{Checks on the calculation of these numbers are } \sum\limits_{k=0}^{3} \sigma_{jk}=\tfrac{1}{2}j^2,\ \sum\limits_{k=0}^{3} \tau_{jk}=j.\right)$

Table II/25. *The coefficients σ_{jk}*

σ_{jk}	$k=$ 0	1	2	3
$j=1$	$\frac{97}{360}$	$\frac{19}{60}$	$-\frac{13}{120}$	$\frac{1}{45}$
2	$\frac{28}{45}$	$\frac{22}{15}$	$-\frac{2}{15}$	$\frac{2}{45}$
3	$\frac{39}{40}$	$\frac{27}{10}$	$\frac{27}{40}$	$\frac{3}{20}$

Table II/26. *The coefficients τ_{jk}*

τ_{jk}	$k=$ 0	1	2	3
$j=1$	$\frac{3}{8}$	$\frac{19}{24}$	$-\frac{5}{24}$	$\frac{1}{24}$
2	$\frac{1}{3}$	$\frac{4}{3}$	$\frac{1}{3}$	0
3	$\frac{3}{8}$	$\frac{9}{8}$	$\frac{9}{8}$	$\frac{3}{8}$

If we subtract from equations (5.10) the corresponding equations with ν replaced by $\nu-1$ and define $f_0^{[\nu]}=f_0$, we obtain

$$\delta_j^{[\nu]} = y_j^{[\nu+1]} - y_j^{[\nu]} = h^2 \sum_{k=1}^{3} \sigma_{jk}(f_k^{[\nu]} - f_k^{[\nu-1]}),$$

$$\delta_j'^{[\nu]} = y_j'^{[\nu+1]} - y_j'^{[\nu]} = h \sum_{k=1}^{3} \tau_{jk}(f_k^{[\nu]} - f_k^{[\nu-1]}).$$

Bounds for the function differences involved here are provided by the Lipschitz condition (5.1) and taking absolute values we have

$$\left.\begin{aligned} |\delta_j^{[\nu]}| &\leq h^2 \sum_{k=1}^{3} |\sigma_{jk}|\,(K_0|\delta_k^{[\nu-1]}| + K_1|\delta_k'^{[\nu-1]}|) \\ |\delta_j'^{[\nu]}| &\leq h \sum_{k=1}^{3} |\tau_{jk}|\,(K_0|\delta_k^{[\nu-1]}| + K_1|\delta_k'^{[\nu-1]}|). \end{aligned}\right\} \qquad (5.11)$$

Hence the quantities $v_j^{[\nu]}$ defined by

$$v_j^{[\nu]} = K_0|\delta_j^{[\nu]}| + K_1|\delta_j'^{[\nu]}|$$

satisfy the set of inequalities

$$v_j^{[\nu]} \leq \sum_{k=1}^{3} (K_0 h^2|\sigma_{jk}| + K_1 h|\tau_{jk}|)\,v_k^{[\nu-1]} \qquad (j=1,2,3). \qquad (5.12)$$

If we regard the (positive) numbers

$$K_0 h^2 |\sigma_{jk}| + K_1 h |\tau_{jk}| = \alpha_{jk} \qquad (5.13)$$

as the elements of a matrix A, we can apply the theorem quoted in § 4.2 exactly as we did there for the first-order case. This guarantees the convergence of the series

$$\sum_{\nu=0}^{\infty} v_j^{[\nu]}$$

and hence also the absolute convergence of the iterations for y_j and y_j', provided that all characteristic roots of the matrix A are less than 1 in absolute value.

For a numerical investigation of these characteristic roots we introduce the positive parameter p defined by

$$K_1 h = p K_0 h^2.$$

Then the characteristic roots of a second matrix

$$B = \frac{1}{K_0 h^2} A = (|\sigma_{jk}| + p |\tau_{jk}|)$$

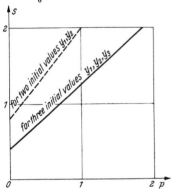

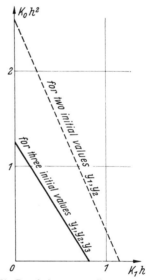

Fig. II/6. Plots of the auxiliary function $s(p)$

Fig. II/7. Boundaries corresponding to a sufficient condition for the starting iteration (5.8), (5.9)

will be functions of p only; let the greatest in absolute value be denoted by s (since B is a non-zero matrix with non-negative elements, s will in fact be positive[1]). The function $s(p)$ can be calculated point by point; the results are shown graphically in Fig. II/6 for the cases of two and three starting values.

Now the condition that all characteristic roots of A should be less than 1 in absolute value is equivalent to the condition

$$s K_0 h^2 < 1.$$

---

[1] See G. FROBENIUS: S.-B. Preuss. Akad. Wiss., Math.-phys. Kl. 1. HBd. 1912, pp. 456—477, in particular p. 457.

Hence the region of the $(K_1 h, K_0 h^2)$ plane for which convergence of the starting iteration is assured (Fig. II/7) is bounded by the curve with the parametric representation (parameter p)

$$K_0 h^2 = \frac{1}{s(p)}, \qquad K_1 h = \frac{p}{s(p)}. \tag{5.14}$$

Assuming that h is chosen so small that this convergence condition is satisfied, we can estimate the errors in the starting values

$$\varepsilon_j = y_j - y(x_j), \qquad \varepsilon_j' = y_j' - y'(x_j) \qquad (j = 1, 2, 3).$$

Again we proceed along the same lines as for the first-order case (§ 4.5) and can be brief.

For the exact solution we have

$$\left. \begin{aligned}
y(x_1) &= y(x_0) + h\,y'(x_0) + h^2(\tfrac{1}{2} F_0 + \tfrac{1}{6} V F_1 - \tfrac{1}{24} V^2 F_2 + \tfrac{1}{45} V^3 F_3) + \widetilde{T}_1, \\
y(x_2) &= 2y(x_1) - y(x_0) + h^2(F_1 + \tfrac{1}{12} V^2 F_2) + \widetilde{\widetilde{T}}_2, \\
y(x_3) &= 2y(x_2) - y(x_1) + h^2(F_2 + \tfrac{1}{12} V^2 F_3) + \widetilde{\widetilde{T}}_3, \\
y'(x_1) &= y'(x_0) + h(F_0 + \tfrac{1}{2} V F_1 - \tfrac{1}{12} V^2 F_2 + \tfrac{1}{24} V^3 F_3) + \widetilde{S}_1, \\
y'(x_2) &= y'(x_0) + h(2F_1 + \tfrac{1}{3} V^2 F_2) + \widetilde{S}_2, \\
y'(x_3) &= y'(x_1) + h(2F_2 + \tfrac{1}{3} V^2 F_3) + \widetilde{S}_3,
\end{aligned} \right\} \tag{5.15}$$

where $F_\nu = f(x_\nu, y(x_\nu), y'(x_\nu))$. With the notation

$$\left| \frac{d^p f}{d x^p} \right|_{\max} = N_p$$

the estimates of Ch. I (2.31), (2.46), (2.47) for the remainder terms occurring in (5.15) read

$$|\widetilde{T}_1| \leq \frac{7}{480} h^6 N_4 = s_1, \qquad |\widetilde{\widetilde{T}}_2| \leq \frac{1}{240} h^6 N_4 = \tilde{s}_2, \qquad |\widetilde{\widetilde{T}}_3| \leq \frac{1}{240} h^6 N_4 = \tilde{s}_3,$$

$$|\widetilde{S}_1| \leq \frac{19}{720} h^5 N_4 = s_1', \qquad |\widetilde{S}_2| \leq \frac{1}{90} h^5 N_4 = s_2', \qquad |\widetilde{S}_3| \leq \frac{1}{90} h^5 N_4 = \tilde{s}_3'.$$

With all differences expressed in terms of function values we subtract the two sets of equations (5.15) and (5.8), (5.9) for the exact and approximate values, respectively, to obtain the error equations

$$\left. \begin{aligned}
\varepsilon_j &= y_j - y(x_j) = h^2 \sum_{k=1}^{3} \sigma_{jk}(f_k - F_k) + \vartheta_j s_j \\
\varepsilon_j' &= y_j' - y'(x_j) = h \sum_{k=1}^{3} \tau_{jk}(f_k - F_k) + \vartheta_j' s_j'
\end{aligned} \right\} \qquad (j = 1, 2, 3),$$

where

$$s_2 = \tilde{s}_2 + 2s_1, \quad s_3 = \tilde{s}_3 + 2\tilde{s}_2 + 5s_1, \quad s_3' = \tilde{s}_3' + s_1', \quad |\vartheta_j| \leq 1, \quad |\vartheta_j'| \leq 1.$$

Taking absolute values and using the Lipschitz condition (5.1) for the function differences $f_k - F_k$, we find that

$$
\left.
\begin{aligned}
|\varepsilon_j| &\leq h^2 \sum_{k=1}^{3} |\sigma_{jk}| \left(K_0 |\varepsilon_k| + K_1 |\varepsilon_k'| \right) + s_j \\
|\varepsilon_j'| &\leq h \sum_{k=1}^{3} |\tau_{jk}| \left(K_0 |\varepsilon_k| + K_1 |\varepsilon_k'| \right) + s_j'
\end{aligned}
\right\}
\quad (j = 1, 2, 3). \quad (5.16)
$$

Exactly as with the first-order case, the matrix of this system of inequalities is monotonic and the quantities Y_j, Y_j' determined from the equations

$$
\left.
\begin{aligned}
Y_j &= h^2 \sum_{k=1}^{3} |\sigma_{jk}| \left(K_0 Y_k + K_1 Y_k' \right) + s_j \\
Y_j' &= h \sum_{k=1}^{3} |\tau_{jk}| \left(K_0 Y_k + K_1 Y_k' \right) + s_j'
\end{aligned}
\right\}
\quad (j = 1, 2, 3) \quad (5.17)
$$

provide upper limits for the absolute values of the errors ε_j, ε_j'.

We can either solve these linear equations for the Y_j, Y_j' or derive upper bounds for the Y_j, Y_j' by the method described at the end of § 4.5 for the first-order case. The matrices

$$
Y = \begin{pmatrix} Y_1 \\ Y_2 \\ Y_3 \\ Y_1' \\ Y_2' \\ Y_3' \end{pmatrix}, \quad
A = \begin{pmatrix}
h K_0 |\sigma_{11}| & h K_0 |\sigma_{12}| & h K_0 |\sigma_{13}| & h K_1 |\sigma_{11}| & h K_1 |\sigma_{12}| & h K_1 |\sigma_{13}| \\
\cdots & \cdots & \cdots & \cdots & \cdots & \cdots \\
h K_0 |\sigma_{31}| & h K_0 |\sigma_{32}| & \cdots & h K_1 |\sigma_{31}| & \cdots & \cdots \\
K_0 |\tau_{11}| & K_0 |\tau_{12}| & K_0 |\tau_{13}| & K_1 |\tau_{11}| & K_1 |\tau_{12}| & K_1 |\tau_{13}| \\
\cdots & \cdots & \cdots & \cdots & \cdots & \cdots \\
K_0 |\tau_{31}| & K_0 |\tau_{32}| & \cdots & K_1 |\tau_{31}| & \cdots & \cdots
\end{pmatrix}, \quad
S = \begin{pmatrix} s_1 \\ s_2 \\ s_3 \\ s_1' \\ s_2' \\ s_3' \end{pmatrix}
\quad \left.\right\} (5.18)
$$

correspond here to the matrices Y, KB, S of § 4.5, and if a is the maximum value of the elements of A, the method of § 4.5 yields

$$
\left.
\begin{aligned}
|\varepsilon_j| &\leq Y_j \leq s_j + \frac{h a}{1 - 6 h a} \sum_{k=1}^{3} (s_k + s_k') \\
|\varepsilon_j'| &\leq Y_j' \leq s_j' + \frac{h a}{1 - 6 h a} \sum_{k=1}^{3} (s_k + s_k')
\end{aligned}
\right\}
\quad (j = 1, 2, 3). \quad (5.19)
$$

We now describe several methods for the main calculation.

5.4. Extrapolation methods

Suppose that approximate values y_ϱ, y_ϱ', ..., $y_\varrho^{(n-1)}$ for the exact solution $y(x_\varrho)$, $y'(x_\varrho)$, ..., $y^{(n-1)}(x_\varrho)$ at the points x_ϱ (for $\varrho = 1, ..., r$) are known and that it is required to calculate approximate values y_{r+1}, y_{r+1}', ..., $y_{r+1}^{(n-1)}$ at the next point.

If in (2.25) of Ch. I we replace n by $n - m$ and x_0 by x_r and use the fact that $y^{(n)} = f$, we obtain a formula for the m-th derivative of the exact solution, namely

$$
\left.\begin{aligned}
y^{(m)}(x_{r+1}) = \sum_{\nu=0}^{n-m-1} \frac{h^\nu}{\nu!}\, y^{(m+\nu)}(x_r) + \\
+ h^{n-m} \sum_{\varrho=0}^{p} \beta_{n-m,\varrho} \nabla^\varrho F(x_r) + R_{n-m,p+1},
\end{aligned}\right\} \tag{5.20}
$$

in which we have used the notation

$$
f\big(x, y(x), y'(x), \ldots, y^{(n-1)}(x)\big) = F(x). \tag{5.21}
$$

An estimate for the remainder term $R_{n-m,p+1}$ is given in Ch. I (2.27). By neglecting this remainder term we obtain FALKNER'S extrapolation formula[1] for the approximate value $y_{r+1}^{(m)}$:

$$
y_{r+1}^{(m)} = \sum_{\nu=0}^{n-m-1} \frac{h^\nu}{\nu!}\, y_r^{(m+\nu)} + h^{n-m} \sum_{\varrho=0}^{p} \beta_{n-m,\varrho} \nabla^\varrho f_r, \tag{5.22}
$$

where $f_r = f(x_r, y_r, y_r', \ldots, y_r^{(n-1)})$. For the special case $n=1$, $m=0$, this reduces to the Adams extrapolation formula (3.6) for differential equations of the first order. For second-order equations ($n=2$, $m=0,1$) we have

$$
\left.\begin{aligned}
y_{r+1} &= y_r + h\,y_r' + h^2\left[\frac{1}{2}f_r + \frac{1}{6}\nabla f_r + \frac{1}{8}\nabla^2 f_r + \frac{19}{180}\nabla^3 f_r + \cdots\right] \\
y_{r+1}' &= y_r' \qquad + h\left[f_r + \frac{1}{2}\nabla f_r + \frac{5}{12}\nabla^2 f_r + \frac{3}{8}\nabla^3 f_r + \cdots\right].
\end{aligned}\right\} \tag{5.23}
$$

A general formula for the coefficients $\beta_{n,\varrho}$ is given by (2.26) of Ch. I.

There are, however, many other ways in which we can set up extrapolation formulae. For example, by linearly combining certain formulae based on (2.23) of Ch. I we can construct a formula which involves only y and f values, i.e. all intermediate derivatives can be eliminated. Such a formula is particularly suitable when the differential equation has the form $y^{(n)} = f(x, y)$.

To derive this formula we take the set of $n-1$ equations obtained from (2.23) of Ch. I by putting $x = x_{-1}, x_{-2}, \ldots, x_{-n+1}$ in turn and multiplying by $(-1)^{q+1}\binom{n}{q+1}$ for $x = x_{-q}$, and add them. With $^{(n)}f = y$ the resulting equation reads

$$
\left.\begin{aligned}
y(x_1) - \binom{n}{1}y(x_0) + \binom{n}{2}y(x_{-1}) - \cdots + (-1)^n y(x_{-n+1}) = \nabla^n y(x_1) \\
= h^n \sum_{\varrho=0}^{p} \zeta_{n,\varrho} \nabla^\varrho f_0 + \tilde{R},
\end{aligned}\right\} \tag{5.24}
$$

where

$$
\zeta_{n,\varrho} = P_{n,\varrho}(1) + \sum_{q=1}^{n-1} (-1)^{q+1}\binom{n}{q+1} P_{n,\varrho}(-q) \tag{5.25}
$$

---

[1] FALKNER, V. M.: A method of numerical solution of differential equations. Phil. Mag. (7) **21**, 621—640 (1936).

and

$$\tilde{R} = R_{n,p+1}(x_1) + \sum_{q=1}^{n-1} (-1)^{q+1} \binom{n}{q+1} R_{n,p+1}(x_{-q}). \qquad (5.26)$$

The first few values of the $\zeta_{n,\varrho}$ are given in Table II/27.

Table II/27. *The numbers* $\zeta_{n,\varrho}$

	$\varrho=0$	$\varrho=1$	$\varrho=2$	$\varrho=3$	$\varrho=4$	$\varrho=5$
$n=1$	1	$\frac{1}{2}$	$\frac{5}{12}$	$\frac{3}{8}$	$\frac{251}{720}$	$\frac{95}{288}$
$n=2$	1	0	$\frac{1}{12}$	$\frac{1}{12}$	$\frac{19}{240}$	$\frac{3}{40}$
$n=3$	1	$-\frac{1}{2}$	0	0	$\frac{1}{240}$	
$n=4$	1	-1	$\frac{1}{6}$	0	$-\frac{1}{720}$	

$$(5.27)$$

With the remainder term omitted (5.24) provides an extrapolation formula for n-th-order equations. It includes the Adams extrapolation formula (3.6) and STÖRMER's[1] extrapolation formula

$$
\begin{aligned}
y_{r+1} &= 2y_r - y_{r-1} + \\
&+ h^2 \left(f_r + \frac{1}{12} \nabla^2 f_r + \frac{1}{12} \nabla^3 f_r + \frac{19}{240} \nabla^4 f_r + \frac{3}{40} \nabla^5 f_r + \cdots \right) \\
&= 2y_r - y_{r-1} + h^2 \left[f_r + \frac{1}{12} (\nabla^2 f_r + \nabla^3 f_r + \nabla^4 f_r + \nabla^5 f_r) - \right. \\
&\left. - \frac{1}{240} (\nabla^4 f_r + 2\nabla^5 f_r) + \cdots \right]
\end{aligned}
\right\} \quad (5.28)
$$

as the particular cases $n=1$ and $n=2$, respectively; the latter formula has been used a great deal on account of its convenient coefficients (particularly when fourth and higher differences are neglected).

For third-order equations the coefficients are even more convenient; in fact, since $\zeta_{32}=\zeta_{33}=0$, there are no terms in $\nabla^2 f$ and $\nabla^3 f$:

$$y_{r+1} = 3y_r - 3y_{r-1} + y_{r-2} + h^3 \left(f_r - \frac{1}{2} \nabla f_r + \frac{1}{240} \nabla^4 f_r + \cdots \right). \qquad (5.29)$$

---

[1] CARL STÖRMER, the Norwegian mathematician, was born on the 3rd September, 1874 in Skien (Norway). He studied in Christiania 1892—1898, then went to Paris 1898—1900, where he became an unsalaried lecturer in 1899, and later (1902) to Göttingen. In 1903 he was appointed professor of pure mathematics in the University of Oslo.

In order to confirm his theory of the aurora borealis, STÖRMER and his colleagues spent several years calculating numerous orbits of electrons in the earth's magnetic field [STÖRMER, C.: Z. Astrophysik 1, 237—274 (1930)]. The computed orbits were reproduced very closely by E. BRÜCHE in his experimental work.

STÖRMER remarked of the calculations: "One must have ample time and patience, for the calculations are extremely long." 4500 working hours were needed for 120 orbits.

If fourth differences are neglected, we have a very simple yet accurate formula which involves only the first differences; provided we do not need higher differences for the formulae for y' and y'', which may appear explicitly in the differential equation, the difference table need not be extended beyond the first difference.

For fourth-order equations also we obtain a very simple formula when fourth differences are neglected; the formula (5.24) with $n=4$ reads

$$\left.\begin{aligned} y_{r+1} = 4y_r - 6y_{r-1} + 4y_{r-2} - y_{r-3} + \\ + h^4 \left(f_r - \nabla f_r + \frac{1}{6} \nabla^2 f_r - \frac{1}{720} \nabla^4 f_r + \cdots \right). \end{aligned}\right\} \quad (5.30)$$

5.5. Interpolation methods

Again suppose that approximate values $y_\varrho^{(m)}$ for the exact solution $y^{(m)}(x_\varrho)$ $(m=0, 1, \ldots, n-1)$ at the points x_ϱ $(\varrho=1, \ldots, r)$ are known and that it is required to calculate approximate values for $y^{(m)}(x_{r+1})$. For the exact solution, (2.36) of Ch. I with $n-m$ in place of n, x_{r+1} in place of x_0, and $y^{(n)}=f$, gives

$$\left.\begin{aligned} y^{(m)}(x_{r+1}) = \sum_{v=0}^{n-m-1} \frac{h^v}{v!} y^{(m+v)}(x_r) + \\ + h^{n-m} \sum_{\varrho=0}^{p} \beta^*_{n-m, \varrho} \nabla^\varrho F(x_{r+1}) + R^*_{n-m, p+1}, \end{aligned}\right\} \quad (5.31)$$

where F is defined in (5.21), the $\beta^*_{n-m, \varrho}$ are given by (2.38) of Ch. I, and the remainder term is bounded by the limits (2.41) of Ch. I. Omitting the remainder term, we obtain the Adams interpolation formula for $y_{r+1}^{(m)}$:

$$y_{r+1}^{(m)} = \sum_{\sigma=0}^{n-m-1} \frac{h^\sigma}{\sigma!} y_r^{(m+\sigma)} + h^{n-m} \sum_{\varrho=0}^{p} \beta^*_{n-m, \varrho} \nabla^\varrho f_{r+1}. \quad (5.32)$$

The case $n=1$, $m=0$ has already been dealt with under the Adams interpolation method (3.11) for differential equations of the first order (§ 3.3, II). For $n=2$ we have

$$\left.\begin{aligned} y_{r+1} = y_r + h y_r' + \\ + h^2 \left[\frac{1}{2} f_{r+1} - \frac{1}{3} \nabla f_{r+1} - \frac{1}{24} \nabla^2 f_{r+1} - \frac{7}{360} \nabla^3 f_{r+1} - \cdots \right] \\ y_{r+1}' = y_r' + h \left[f_{r+1} - \frac{1}{2} \nabla f_{r+1} - \frac{1}{12} \nabla^2 f_{r+1} - \frac{1}{24} \nabla^3 f_{r+1} - \cdots \right]. \end{aligned}\right\} \quad (5.33)$$

In (5.32) the unknown $y_{r+1}^{(m)}$ appears on both sides (as an argument of f_{r+1} on the right-hand side) and therefore, as in (3.14), we use the equation in an iterative form by putting the values for the v-th iterate into the right-hand side to evaluate the $(\nu+1)$-th iterate in the usual way. Thus the iterative form of (5.32) is

$$y_{r+1}^{(m)\,[\nu+1]} = \sum_{\sigma=0}^{n-m-1} \frac{h^\sigma}{\sigma!}\, y_r^{(m+\sigma)} + h^{n-m} \sum_{\varrho=0}^{p} \beta_{n-m,\varrho}^{*}\, \nabla^\varrho f_{r+1}^{[\nu]}, \qquad (5.34)$$

in which the iteration superscripts are again enclosed in square brackets to distinguish them from the derivative superscripts.

Again there are occasions when we need the Lagrangian form of the equations in which the differences $\nabla^\varrho f$ are expressed in terms of the function values; corresponding to the form (3.12) of (3.11) we have the Lagrangian form

$$y_{r+1}^{(m)} = \sum_{\varrho=0}^{n-m-1} \frac{h^\varrho}{\varrho!}\, y_r^{(m+\varrho)} + h^{n-m} \sum_{\sigma=0}^{p} \beta_{n-m,\sigma,p}^{*}\, f_{r+1-\sigma} \qquad (5.35)$$

of (5.32), in which the new coefficients are given by

$$\beta_{q,\sigma,p}^{*} = (-1)^\sigma \sum_{\varrho=\sigma}^{p} \beta_{q,\varrho}^{*} \binom{\varrho}{\sigma}; \qquad (5.36)$$

the corresponding iterative form is

$$\left. \begin{aligned} y_{r+1}^{(m)\,[\nu+1]} = \sum_{\varrho=0}^{n-m-1} \frac{h^\varrho}{\varrho!}\, y_r^{(m+\varrho)} + \\ + h^{n-m} \Big(\beta_{n-m,0,p}^{*}\, f_{r+1}^{[\nu]} + \sum_{\sigma=1}^{p} \beta_{n-m,\sigma,p}^{*}\, f_{r+1-\sigma} \Big). \end{aligned} \right\} \quad (5.37)$$

Just as in § 3.3 for differential equations of the first order, other formulae of the interpolation type can be derived for equations of higher order. Thus, for instance, from (2.45) of Ch. I we have

$$\left. \begin{aligned} y^{(m)}(x_{r+1}) &= -\,y^{(m)}(x_{r-1}) + 2 \sum_{\varrho=0}^{\frac{n-m}{2}-1} \frac{h^{2\varrho}}{(2\varrho)!}\, y^{(m+2\varrho)}(x_r) + \\ &\quad + 2h^{n-m} \sum_{\varrho=0}^{\frac{p}{2}} \beta_{n-m,\varrho}^{**}\, \nabla^{2\varrho} F(x_{r+\varrho}) + 2R_{n-m,p}^{**} \end{aligned} \right\} \begin{aligned} &\text{for} \\ &\text{even} \\ &n-m \end{aligned}$$

$$(5.38)$$

$$\left. \begin{aligned} y^{(m)}(x_{r+1}) &= y^{(m)}(x_{r-1}) + 2 \sum_{\varrho=0}^{\frac{n-m-3}{2}} \frac{h^{2\varrho+1}}{(2\varrho+1)!}\, y^{(m+1+2\varrho)}(x_r) + \\ &\quad + 2h^{n-m} \sum_{\varrho=0}^{\frac{p}{2}} \beta_{n-m,\varrho}^{**}\, \nabla^{2\varrho} F(x_{r+\varrho}) + 2R_{n-m,p}^{**} \end{aligned} \right\} \begin{aligned} &\text{for} \\ &\text{odd} \\ &n-m. \end{aligned}$$

The remainder term lies within the limits given in Ch. I (2.47); if it is omitted, we obtain the corresponding equations for the approximation

$y_{r+1}^{(m)}$, namely

$$
y_{r+1}^{(m)} = -\,y_{r-1}^{(m)} + 2\sum_{\varrho=0}^{\tfrac{n-m-2}{2}} \frac{h^{2\varrho}}{(2\varrho)!}\,y_r^{(m+2\varrho)} +
$$
$$
+\,2h^{n-m}\sum_{\varrho=0}^{\tfrac{p}{2}} \beta_{n-m,\varrho}^{**}\,\nabla^{2\varrho} f_{r+\varrho}
$$
$$\left.\begin{array}{c}\text{for}\\ \text{even}\\ n-m\end{array}\right\}$$

$$
y_{r+1}^{(m)} = y_{r-1}^{(m)} + 2\sum_{\varrho=0}^{\tfrac{n-m-3}{2}} \frac{h^{2\varrho+1}}{(2\varrho+1)!}\,y_r^{(m+1+2\varrho)} +
$$
$$
+\,2h^{n-m}\sum_{\varrho=0}^{\tfrac{p}{2}} \beta_{n-m,\varrho}^{**}\,\nabla^{2\varrho} f_{r+\varrho}
$$
$$\left.\begin{array}{c}\text{for}\\ \text{odd}\\ n-m,\end{array}\right\}$$

(5.39)

which describe the interpolation method known as the central-difference method. Normally these equations are used with the second sums truncated after the term with $\varrho=1$; if the next term with $\varrho=2$ were included, the unknown y_{r+2} would appear in addition to the unknown y_{r+1} and application of the method would be very involved[1]. Thus for $n=2$, for example, we use the formulae

$$
\begin{aligned}
y_{r+1} &= 2y_r - y_{r-1} + h^2\!\left(f_r + \frac{1}{12}\nabla^2 f_{r+1}\right) \\
y_{r+1}' &= y_{r-1}' \qquad\quad + h\!\left(2f_r + \frac{1}{3}\nabla^2 f_{r+1}\right).
\end{aligned}
\right\}
$$

(5.40)

The quantities appearing in these equations for the calculation of the values at x_{r+1} are framed in Table II/28, which also gives the required factors at the heads of the appropriate columns. The quantities in

---

[1] At the end of § 3.3, III we mentioned a method for first-order equations due to LINDELÖF which took into account the fourth difference $\nabla^4 f_{r+2}$; he applies the same idea to second-order equations, rewriting the formula in a similar way:

$$
y_{r+1} = 2y_r - y_{r-1} + h^2\left(f_r + \frac{1}{12}\left(\nabla^2 f_r + \nabla^3 f_r\right)\right) + \delta,
$$

where

$$
\delta = \frac{h^2}{12}\left(\nabla^4 f_{r+1} - \frac{1}{20}\nabla^4 f_{r+2}\right).
$$

If the calculation has proceeded as far as the point x_r, temporary values for the next two points are found from the formula

$$
y_{r+j+1}^{[0]} = 2y_{r+j} - y_{r+j-1} + h^2\left(f_{r+j} + \frac{1}{12}\left(\nabla^2 f_{r+j} + \nabla^3 f_{r+j}\right)\right) \quad \text{for } j=0 \text{ and } 1.
$$

These are used to build up the difference table temporarily so that a value for δ can be calculated; this value of δ determines an improved value of y_{r+1} and the difference table can be filled in permanently up to the point x_{r+1}. LINDELÖF gives (loc. cit.) a further refinement of this procedure.

broken frames are those which are not needed when the differential equation does not depend explicitly on y', i.e. when it has the form $y'' = f(x, y)$.

Table II/28. *Central-difference method for differential equations of the second order.*
$$y'' = f(x, y, y')$$

x			Factors for y	1		$\frac{1}{12}$
			Factors for $h y'$	2		$\frac{1}{3}$
	y	$h y'$	$h^2 y'' = h^2 f$	$h^2 \nabla f$	$h^2 \nabla^2 f$

	y	$h y'$	$h^2 y'' = h^2 f$	$h^2 \nabla f$	$h^2 \nabla^2 f$
x_{r-2}	y_{r-2}	$h y'_{r-2}$	$h^2 f_{r-2}$		
x_{r-1}	y_{r-1}	$h y'_{r-1}$	$h^2 f_{r-1}$	$h^2 \nabla f_{r-1}$	$h^2 \nabla^2 f_r$
x_r	y_r	$h y'_r$	$h^2 f_r$	$h^2 \nabla f_r$	$h^2 \nabla^2 f_{r+1}$
x_{r+1}	y_{r+1}	$h y'_{r+1}$	$h^2 f_{r+1}$	$h^2 \nabla f_{r+1}$	

To use the equations (5.40) for the calculation of the currently next values y_{r+1} and y'_{r+1}, we can estimate the as yet unknown value $\nabla^2 f_{r+1}$ by extrapolation from the sequence of $\nabla^2 f$ values already calculated (as in § 3.4, I). If we take $\nabla^2 f_r + \nabla^3 f_r$ as the initial value $\nabla^2 f_{r+1}^{[0]}$, i.e. we assume the third differences to be constant and use $\nabla^3 f_r$ in place of $\nabla^3 f_{r+1}$, then we obtain for $y_{r+1}^{[1]}$ the value given by STÖRMER's extrapolation formula (5.28) truncated after the term in $\nabla^3 f$. Proceeding from this estimated value $\nabla^2 f_{r+1}^{[0]}$, we improve on it successively by means of the iterative scheme

$$
\left.
\begin{aligned}
y_{r+1}^{[\nu+1]} &= 2 y_r - y_{r-1} + h^2\left(f_r + \frac{1}{12} \nabla^2 f_{r+1}^{[\nu]}\right) \\
h y'^{[\nu+1]}_{r+1} &= h y'_{r-1} + h^2\left(2 f_r + \frac{1}{3} \nabla^2 f_{r+1}^{[\nu]}\right)
\end{aligned}
\right\}
\quad (\nu = 0, 1, 2, \ldots). \quad (5.41)
$$

The step interval h is again to be chosen so small that

(a) the iterations converge sufficiently rapidly, say in one or two cycles, if possible,

(b) the terms following $\frac{h^2}{12} \nabla^2 f_{r+1}$ and $\frac{h}{3} \nabla^2 f_{r+1}$ in (5.40), namely $-\frac{h^2}{240} \nabla^4 f_{r+2}$ and $-\frac{h}{90} \nabla^4 f_{r+2}$, respectively, affect the approximations $y_{,+1}$ and y'_{r+1} as little as possible.

For reference we write out the formulae (5.39) in detail for differential equations of up to the fourth order (for an equation of order m

use the first m formulae with $n = m$):

$$y_{r+1}^{(n-1)} = y_{r-1}^{(n-1)} + h\left(2f_r + \frac{1}{3}\nabla^2 f_{r+1}\right)$$

$$y_{r+1}^{(n-2)} = 2y_r^{(n-2)} - y_{r-1}^{(n-2)} + h^2\left(f_r + \frac{1}{12}\nabla^2 f_{r+1}\right)$$

$$y_{r+1}^{(n-3)} = y_{r-1}^{(n-3)} + 2h\,y_r^{(n-2)} + h^3\left(\frac{1}{3}f_r + \frac{1}{60}\nabla^2 f_{r+1}\right)$$

$$y_{r+1}^{(n-4)} = 2y_r^{(n-4)} - y_{r-1}^{(n-4)} + h^2 y_r^{(n-2)} + h^4\left(\frac{1}{12}f_r + \frac{1}{360}\nabla^2 f_{r+1}\right).$$

(5.42)

Example. For the initial-value problem

$$y''' = -y\,y'', \qquad y(0) = y'(0) = 0, \qquad y''(0) = 1,$$

which arises in boundary layer theory (Example III, § 2.6), we put

$$y'h = u, \qquad y''\frac{h^2}{2} = v, \qquad y'''\frac{h^3}{6} = w.$$

Translated into these quantities the equations (5.42) read

$$v_{r+1} = v_{r-1} + \qquad\quad (6w_r + \nabla^2 w_{r+1}),$$

$$u_{r+1} = 2u_r - u_{r-1} + \left(6w_r + \frac{1}{2}\nabla^2 w_{r+1}\right),$$

$$y_{r+1} = y_{r-1} + 2u_r + \left(2w_r + \frac{1}{10}\nabla^2 w_{r+1}\right).$$

The values of y, y', y'' at the points $x = \pm0\cdot1$, $\pm0\cdot2$ are calculated from the power series[1]

$$y = \frac{1}{2!}x^2 - \frac{1}{5!}x^5 + \frac{11}{8!}x^8 - \frac{375}{11!}x^{11} + \cdots.$$

This enables us to fill in the computing scheme in Table II/29 down to the dotted line. The values of $\nabla^3 w$ are recorded so that we can more easily estimate the

Table II/29. *A non-linear third-order equation treated by the central-difference method*

x	y	$u = hy'$	$v = \dfrac{h^2}{2}y''$	$w = \dfrac{h^3}{6}y'''$ $= -\dfrac{h}{3}yv$	∇w	$\nabla^2 w$	$\nabla^3 w$	Estima of the new value $\nabla^2 w$
$-0\cdot2$	$0\cdot02000267$	$-0\cdot02000667$	$0\cdot00500667$	$-0\cdot000003338$				
$-0\cdot1$	$0\cdot00500008$	$-0\cdot01000042$	$0\cdot00500083$	$-\quad\quad 833$	$+2505$			
0	0	0	$0\cdot005$	0	$+\ 833$	-1671		
$0\cdot1$	$0\cdot00499992$	$0\cdot00999958$	$0\cdot00499917$	$-0\cdot000000833$	$-\ 833$	-1667	$+\ 4$	
$0\cdot2$	$0\cdot01999733$	$0\cdot01999333$	$0\cdot00499334$	$-\quad\quad 3328$	-2495	-1662	$+\ 5$	
$0\cdot3$	$0\cdot04497975$	$0\cdot02996628$	$0\cdot00497755$	$-\quad\quad 7463$	-4134	-1639	$+23$	-164
$0\cdot4$	$0\cdot07991495$	$0\cdot03989366$	$0\cdot00494698$	$-0\cdot000013178$	-5715	-1581	$+58$	-158
$0\cdot5$	$0\cdot12474056$							

[1] The convergence of this series has been investigated by A. OstrowskI: Sur le rayon de convergence de la série de Blasius. C. R. Acad. Sci., Paris **227**, 580—582 (1948).

next value of $\nabla^2 w$. The procedure is completely analogous to that for a first-order equation as in Example II, § 3.5, so fewer lines of working are given.

Astronomers[1] employ a method in which columns of the sums Σf, $\Sigma^2 f$, ... are used in addition to the columns of the differences δf, $\delta^2 f$, ... (these sum columns are formed in such a way that the $\Sigma^{n-1} f$ column contains the differences of the $\Sigma^n f$ column). In the case of a second-order differential equation $y'' = f(x, y)$, for instance, we build up the columns

x	y	$h^2 \Sigma^2 f$	$h^2 \Sigma f$	$h^2 f$	$h^2 \delta f$	$h^2 \delta^2 f$	$h^2 \delta^3 f$

This we do by using the formula

$$y_r = h^2 \left(\Sigma^2 f_r + \frac{1}{12} f_r - \frac{1}{240} \delta^2 f_r \right), \tag{5.43}$$

which follows by two successive summations from the first formula of (5.40) with the next term included, i.e.

$$\delta^2 y_r = h^2 \left(f_r + \frac{1}{12} \delta^2 f_r - \frac{1}{240} \delta^4 f_r \right)$$

in central-difference notation. Two arbitrary initial constants are introduced thereby and these may be determined as follows: if we have found y_r for $r = -1, 0, 1, 2$ by some starting procedure (cf. § 5.2), then we know also f_r for $r = -1, 0, 1, 2$ and hence $\delta^2 f_r$ for $r = 0, 1$; $\Sigma^2 f_r$ can then be calculated for $r = 0, 1$ from the equation (5.43) and this provides initial values for the two columns of Σf and $\Sigma^2 f$. If the main calculation has progressed as far as the values y_r, f_r, $\delta^2 f_{r-1}$, we first estimate values $\delta^2 f_r^{[0]}$ and $\delta^2 f_{r+1}^{[0]}$, then build $\delta^2 f_r^{[0]}$ up to $f_{r+1}^{[0]}$ and calculate $y_{r+1}^{[0]}$ from (5.43). This gives an improved value $f_{r+1}^{[1]} = f(x_{r+1}, y_{r+1}^{[0]})$ and hence also an improved value $\delta^2 f_r^{[1]}$, which can be used to make a better estimate $\delta^2 f_{r+1}^{[1]}$. These new values are then substituted in (5.43) to give an improved value $y_{r+1}^{[1]}$, and so on.

5.6. Convergence of the iteration in the main calculation

As in § 4.1 we investigate the convergence of the iterations involved in the interpolative integration formulae by first putting them in Lagrangian form with all differences expressed in terms of function values. Thus for the Adams interpolation method (5.32) we use the

---

[1] v. OPPOLZER, T. R.: Lehrbuch zur Bahnbestimmung der Kometen und Planeten, 2nd ed. Leipzig 1880. — HERRICK, S.: Step-by-step integration of $\ddot{x} = f(x, y, z, t)$ without a "corrector". Math. Tables and Other Aids to Computation 5, 61—67 (1951).

expression (5.37) for the $(\nu+1)$-th iterate. If we subtract the corresponding expression for the ν-th iterate, we obtain an expression for the iterative change

$$\delta^{(m)[\nu]} = y_{r+1}^{(m)[\nu+1]} - y_{r+1}^{(m)[\nu]},$$

namely

$$\delta^{(m)[\nu]} = h^{n-m}\beta_{n-m,0,p}^* (f_{r+1}^{[\nu]} - f_{r+1}^{[\nu-1]}).$$

From (5.36) and Ch. I (2.40) we see that the factors $\beta_{q,0,p}^*$ which occur here are given by

$$\beta_{q,0,p}^* = \sum_{\varrho=0}^{p} \beta_{q,\varrho}^* = \beta_{q,p}.$$

Using the Lipschitz condition (5.1), we have

$$|\delta^{(m)[\nu]}| \leq h^{n-m}\beta_{n-m,p}\sum_{\varrho=0}^{n-1} K_\varrho |\delta^{(\varrho)[\nu-1]}| \qquad (m=0,1,\ldots,n-1), \qquad (5.44)$$

and if these n inequalities are multiplied by $K_0, K_1, \ldots, K_{n-1}$, respectively, and summed, we find that the quantity

$$v^{[\nu]} = \sum_{\varrho=0}^{n-1} K_\varrho |\delta^{(\varrho)[\nu]}| \qquad (5.45)$$

satisfies the inequality

$$v^{[\nu]} \leq C v^{[\nu-1]}, \qquad (5.46)$$

where

$$C = h^n\beta_{n,p}K_0 + h^{n-1}\beta_{n-1,p}K_1 + \cdots + h\beta_{1,p}K_{n-1}. \qquad (5.47)$$

This last inequality has the same form as (4.4) in § 4.1 and by the same reasoning as used there it can be shown that the condition $C<1$ is sufficient for the convergence of $\sum_{\nu=0}^{\infty} v^{[\nu]}$. Since we can assume the K_ϱ to be positive, this condition is also sufficient for the absolute convergence of each of the series $\sum_{\nu=0}^{\infty} \delta^{(m)[\nu]}$ of the iteration process.

For $p=1$ and 2 this sufficient condition for convergence reads

(for $p=1$) $\dfrac{1}{2} hK_{n-1} + \dfrac{1}{6} h^2 K_{n-2} + \dfrac{1}{24} h^3 K_{n-3} + \cdots + h^n\beta_{n,1}K_0 < 1,$

(for $p=2$) $\dfrac{5}{12} hK_{n-1} + \dfrac{1}{8} h^2 K_{n-2} + \dfrac{7}{240} h^3 K_{n-3} + \cdots + h^n\beta_{n,2}K_0 < 1.$

The $\beta_{n,p}$ decrease as p increases and the restriction on h becomes correspondingly less strict.

The central-difference method (5.39) can be treated in precisely the same way and if fourth and higher differences are neglected, the corresponding sufficient condition for convergence is

$$2 \sum_{\nu=0}^{n} h^{n-\nu} \left| \beta^{**}_{n-\nu,1} \right| K_\nu < 1,$$

or, written out more fully,

$$\frac{1}{3} h K_{n-1} + \frac{1}{12} h^2 K_{n-2} + \frac{1}{60} h^3 K_{n-3} + \cdots + 2 h^n \left| \beta^{**}_{n,1} \right| K_0 < 1. \tag{5.48}$$

5.7. Principle of an error estimate for the main calculation

Since error estimates for the various types of extrapolation and interpolation methods can all be derived in similar fashion, we restrict ourselves to the consideration of one particular method, the interpolation method characterized by formula (5.32). By subtracting (5.31) from (5.32) and expressing the differences in terms of function values as in (5.35), we find that the error

$$\varepsilon_r^{(m)} = y_r^{(m)} - y^{(m)}(x_r)$$

is given by

$$\varepsilon_{r+1}^{(m)} = \sum_{\nu=0}^{n-m-1} \frac{h^\nu}{\nu!} \varepsilon_r^{(m+\nu)} + h^{n-m} \sum_{\sigma=0}^{p} \beta^*_{n-m,\sigma,p} \left(f_{r+1-\sigma} - F(x_{r+1-\sigma}) \right) - R^*_{n-m,p+1},$$

in which the notation (5.21) is used. Since the function f satisfies the Lipschitz condition (5.1), we have

$$\left| \varepsilon_{r+1}^{(m)} \right| \leq \sum_{\nu=0}^{n-m+1} \frac{h^\nu}{\nu!} \left| \varepsilon_r^{(m+\nu)} \right| + h^{n-m} \sum_{\sigma=0}^{p} \sum_{\nu=0}^{n-1} \left| \beta^*_{n-m,\sigma,p} \right| K_\nu \left| \varepsilon_{r+1-\sigma}^{(\nu)} \right| + \left| R^*_{n-m,p+1} \right|,$$

i.e.

$$\left| \varepsilon_{r+1}^{(m)} \right| - h^{n-m} \sum_{\nu=0}^{n-1} K_\nu \beta_{n-m,p} \left| \varepsilon_{r+1}^{(\nu)} \right| \leq e_m, \tag{5.49}$$

where

$$e_m = \sum_{\nu=0}^{n-m-1} \frac{h^\nu}{\nu!} \left| \varepsilon_r^{(m+\nu)} \right| + h^{n-m} \sum_{\sigma=1}^{p} \sum_{\nu=0}^{n-1} K_\nu \left| \beta^*_{n-m,\sigma,p} \right| \left| \varepsilon_{r+1-\sigma}^{(\nu)} \right| + \left| R^*_{n-m,p+1} \right|.$$

In (5.49) we have a system of n inequalities satisfied by the n absolute errors $\left| \varepsilon_{r+1}^{(m)} \right|$ with right-hand sides e_m for which we know upper bounds provided that we know upper bounds for the earlier absolute errors $\left| \varepsilon_r^{(m)} \right|, \left| \varepsilon_{r-1}^{(m)} \right|, \ldots$. The theorem of Ch. I, § 5.5 which gives sufficient conditions for a monotonic matrix is applicable here; in (5.49), $1 - h^{n-m} K_m \beta_{n-m,p} > 0$ [since h will be chosen to satisfy the convergence condition $C < 1$ in (5.47)] and all other coefficients on the left-hand side are negative.

Upper limits $Y_r^{(m)}$ for the absolute errors $(Y_r^{(m)} \geq \left| \varepsilon_r^{(m)} \right|)$ can therefore be determined from the system of equations

$$Y_{r+1}^{(m)} - h^{n-m} \sum_{\nu=0}^{n-1} K_\nu \beta_{n-m,p} Y_{r+1}^{(\nu)} = E_m, \tag{5.50}$$

where

$$E_m = \sum_{\nu=0}^{n-m-1} \frac{h^\nu}{\nu!} Y_r^{(m+\nu)} + h^{n-m} \sum_{\sigma=1}^{p} \sum_{\nu=0}^{n-1} K_\nu \left| \beta^*_{n-m,\sigma,p} \right| Y_{r+1-\sigma}^{(\nu)} + \left| R^*_{n-m,p+1} \right|.$$

Thus it is possible to carry out a recursive error estimate.

The step-by-step calculation of these error limits is tedious and does not provide a quick general guide to the growth of the error. However, we can also derive an independent error estimate as we did for first-order equations in § 4.4 and although its strict application is perhaps as tedious as the recursive estimate, we can use it crudely to get a rough idea of the growth of the error.

The derivation of the independent error estimate depends on the fact that systems of linear difference equations, of which (5.50) is an example, can be solved in the same way as a single linear difference equation (cf. § 4.4). We first obtain a particular solution of the inhomogeneous equations (5.50), namely $Y_r^{(m)} = $ constant $= \gamma_m$, then add to this solutions of the corresponding homogeneous equations. For these we assume the forms

$$Y_r^{(m)} = c_m z^r;$$

the constants c_m are then determined from the system of linear equations

$$\left. c_m z^p - h^{n-m} \left(\sum_{\nu=0}^{n-1} K_\nu c_\nu \right) \left(\sum_{\sigma=0}^{p} |\beta^*_{n-m,\sigma,p}| z^{p-\sigma} \right) - z^{p-1} \sum_{\nu=0}^{n-m-1} \frac{h^\nu}{\nu!} c_{m+\nu} = 0 \right\} \quad (5.51)$$
$$(m = 0, 1, \ldots, n-1),$$

which are obtained by substituting the above form in (5.50) and dividing through by z^{r+1-p}. These homogeneous equations will have a non-trivial solution only if their determinant

$$D = \begin{vmatrix} z^p - z^{p-1} - h^n K_0 \Phi_0(z) & -z^{p-1} \dfrac{h}{1!} - h^n K_1 \Phi_0(z) & -z^{p-1} \dfrac{h^2}{2!} - h^n K_2 \Phi_0(z) \cdots \\ \cdots - z^{p-1} \dfrac{h^{n-1}}{(n-1)!} - h^n K_{n-1} \Phi_0(z) & & \\ -h^{n-1} K_0 \Phi_1(z) & z^p - z^{p-1} - h^{n-1} K_1 \Phi_1(z) & -z^{p-1}\dfrac{h}{1!} - h^{n-1} K_2 \Phi_1(z) \cdots \\ & \cdots - z^{p-1}\dfrac{h^{n-2}}{(n-2)!} - h^{n-1} K_{n-1} \Phi_1(z) & \\ -h^{n-2} K_0 \Phi_2(z) & -h^{n-2} K_1 \Phi_2(z) & z^p - z^{p-1} - h^{n-2} K_2 \Phi_2(z) \cdots \\ & \cdots - z^{p-1}\dfrac{h^{n-3}}{(n-3)!} - h^{n-2} K_{n-1} \Phi_2(z) & \\ \cdots & \cdots & \cdots \\ -h K_0 \Phi_{n-1}(z) & -h K_1 \Phi_{n-1}(z) & -h K_2 \Phi_{n-1}(z) \cdots \\ & \cdots z^p - z^{p-1} - h K_{n-1} \Phi_{n-1}(z) & \end{vmatrix} \quad (5.5\ldots)$$

vanishes (printing limitations have made it necessary to spread each row of this determinant onto two lines). The $\Phi_m(z)$ introduced in the determinant to shorten the notation are polynomials in z defined by

$$\sum_{\sigma=0}^{p} |\beta^*_{n-m,\sigma,p}| z^{p-\sigma} = \Phi_m(z)$$

and depend upon n and p as well as upon m and z.

$D = 0$ is an algebraic equation for z, in general of the (np)-th degree, and has in general np roots, say

$$z_1, z_2, \ldots, z_{np}.$$

When these are all distinct, the general solution of the difference equation (5.50) is

$$Y_r^{(m)} = \gamma_m + \sum_{\nu=1}^{np} c_{m,\nu} z_\nu^r \quad (m = 0, 1, \ldots, n-1; \; r = 0, 1, 2, \ldots). \tag{5.53}$$

In the case of multiple roots the coefficients become polynomials in r; thus if $z_1 = z_2 = \cdots = z_q$, say, then we must use $(c_{m,1} + c_{m,2} r + \cdots + c_{m,q} r^{q-1}) z_1^r$ in place of

$$\sum_{\nu=1}^{q} c_{m,\nu} z_\nu^r.$$

Eventually, as r increases, the growth of the error limits is determined entirely by the root z with greatest absolute value; we can therefore use this root as a rough general guide to the rate at which the errors may be expected to grow. Even this is quite complicated to deal with in general so we now restrict ourselves to the important case of second-order differential equations. For these, equation (5.52) reads

$$z^{1-p} D = z^{p+1} - 2z^p + z^{p-1} - z \left(h^2 K_0 \, \Phi_0(z) + h K_1 \, \Phi_1(z) \right) + h^2 K_0 \, \Phi_0(z) + \left. \right\} \tag{5.54}$$
$$+ (h K_1 - h^2 K_0) \, \Phi_1(z) = 0.$$

For given p we can draw the curves $z = $ constant on a plane with $h^2 K_0$ and $h K$ as co-ordinates; these curves will in fact be straight lines since equation (5.54) is linear in $h^2 K_0$ and $h K_1$ [1].

5.8. Instability of finite-difference methods

The rough approximations of § 4.7 can obviously be applied in similar fashion to differential equations of higher order so we will be brief; in fact we limit ourselves to a brief description of the calculations for a particular case, namely the central-difference method (5.40) for second-order equations [2].

With the same notation (perturbations η_r in y_r, η_r' in y_r') and assumptions (in particular that f_y and $f_{y'}$ are constant) as in § 4.7 we obtain here the variation equations

$$\eta_{r+1} = 2\eta_r - \eta_{r-1} + \frac{h^2}{12} f_y (\eta_{r+1} + 10\eta_r + \eta_{r-1}) +$$
$$+ \frac{h^2}{12} f_{y'} (\eta'_{r+1} + 10\eta'_r + \eta'_{r-1}),$$
$$\eta'_{r+1} = \eta'_{r-1} \qquad + \frac{h}{3} f_y (\eta_{r+1} + 4\eta_r + \eta_{r-1}) +$$
$$+ \frac{h}{3} f_{y'} (\eta'_{r+1} + 4\eta'_r + \eta'_{r-1}).$$

Assuming the forms $\eta_r = p \lambda^r$, $\eta_r' = q \lambda^r$, we obtain two linear homogeneous equations for p and q; for a unique solution their determinant must

---

[1] Cf. Figure 6 in L. COLLATZ and R. ZURMÜHL: Z. Angew. Math. Mech. 22, 55 (1942).

[2] RUTISHAUSER, H.: Z. Angew. Math. Phys. 3, 65—74 (1952). — COLLATZ, L.: Z. Angew. Math. Phys. 4, 153—154 (1953).

vanish, and hence λ must satisfy the quartic

$$(\lambda - 1)\,[12\,(\lambda^2 - 1)\,(\lambda - 1) - A\,(\lambda + 1)\,(\lambda^2 + 10\lambda + 1)$$
$$- 4B\,(\lambda - 1)\,(\lambda^2 + 4\lambda + 1)] = 0,$$

where $A = h^2 f_y$, $B = h f'_y$.

Under the assumptions made in § 4.7, a perturbation $\eta(x)$ from $y(x)$ satisfies the differential equation

$$\eta'' = f_y\,\eta + f_{y'}\,\eta'.$$

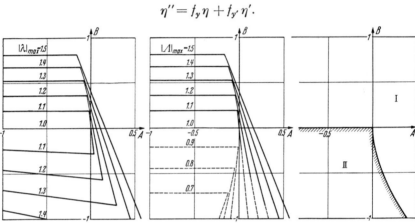

Fig. II/8. Regions of stability for the central-difference method (5.40)

This has solutions of the form $\eta(x) = e^{\mu x}$, where $\mu h = \varrho$ is a solution of the quadratic

$$\varrho^2 - B\varrho - A = 0.$$

Consequently the perturbation $\eta(x)$ increases (or decreases when $\mu < 0$) by a factor $\varLambda = e^\varrho$ over an interval of length h. If we now compare the absolute values of the roots λ with the values $\varLambda$ for given A and B (Fig. II/8), we can distinguish two regions I and II in the (A, B) plane. In region I, which contains in particular the region $f_y \geq 0$, the maximum values of $|\lambda|$ and $|\varLambda|$ coincide to a high accuracy (stable region); on the other hand, in region II there exists the danger of instability.

5.9. Reduction of initial-value problems to boundary-value problems

By raising the order of the differential equation and introducing further boundary conditions (which may be chosen with considerable arbitrariness), we can transform an initial-value problem into a boundary-value problem, which can then be treated by one of the numerous methods at our disposal[1] (cf. Ch. III). For example, by

---

[1] DE G. ALLEN, D. N., and R. T. SEVERN: The application of relaxation methods to the solution of non-elliptic partial differential equations   Quart. J. Mech. Appl. Math. 4, 209—222 (1951).

putting $y = u'$ we can transform

$$y' = f(x, y), \qquad y(x_0) = y_0$$

into the second-order boundary-value problem

$$u'' = f(x, u'), \qquad u'(x_0) = y_0, \qquad u(x_1) = u_1,$$

where x_1 and u_1 are arbitrary.

The transformation can usually be accomplished in several ways; for instance, the initial-value problem

$$y'' = f(x), \qquad y(a) = y_a, \qquad y'(a) = y_a'$$

is equivalent to both of the boundary-value problems

$$u^{IV} = f(x); \quad u''(a) = y_a, \quad u'''(a) = y_a', \quad u(b) = 0, \quad u'(b) = 0$$

and

$$y^{IV} = f''(x); \quad y(a) = y_a, \quad y'(a) = y_a', \quad y''(b) = f(b), \quad y'''(b) = f'(b).$$

As yet, insufficient evidence exists for one to be able to recommend particularly any definite type of transformation.

5.10. Miscellaneous exercises on Chapter II

1. Expand the function $y(x)$ defined by

$$y' = x^2 + y^2, \qquad y(0) = -1$$

into a power series and hence calculate its values for $x = \pm 0.1, \ \pm 0.2$.

2. With the values obtained in Exercise 1 as starting values (thus with $h = 0.1$) use the Adams interpolation formula (3.11) truncated after the third difference to calculate approximations y_r at several further points.

3. Determine the "order" (in the sense of § 1.3) of DUFFING's method in § 1.5.

4. Apply the Runge-Kutta method to the following second-order problem, for which the function $f = -y^3$ does not depend on y' and the Runge-Kutta scheme therefore reduces to three rows:

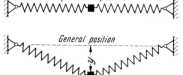

Fig. II/9. Non-linear oscillations of a mass between two springs

$$y'' = -y^3; \qquad y(0) = 0.2, \qquad y'(0) = 0. \tag{5.55}$$

(This equation is satisfied by the small oscillations of a mass which is attached to two springs in the manner of Fig. II/9 and then disturbed from the position of rest[1].)

---

[1] The solution of this differential equation can be approximated in other ways, for example, by reduction to a quadrature, and can also be given in closed form in terms of elliptic functions. See, for instance, K. KLOTTER: Einführung in die Technische Schwingungslehre, 2nd ed., Vol. I, p. 138. Berlin-Göttingen-Heidelberg, Springer 1951.

The solution represents a periodic oscillation and if the period is T, it satisfies the symmetry conditions

$$y(x + T) = y(x) = y(-x) = -y\left(\frac{T}{2} - x\right).$$

Thus we need only calculate $y(x)$ for $0 \leq x \leq \dfrac{T}{4}$, i.e. as far as its first zero, for $y(x)$ is then known for all x. (Here $\frac{1}{4}T$ is approximately 10; if the quarter period is sub-divided into 5 or 10 intervals, the step interval will be $h = 2$ or $h = 1$.)

5. Calculate an approximate solution of the initial-value problem

$$y''' = -xy; \qquad y(0) = 0, \qquad y'(0) = 1, \qquad y''(0) = 0$$

for $0 \leq x \leq 1$ using the simple extrapolation formula obtained from (5.29) by neglecting fourth and higher differences. Compare the end value with the exact value $y(1)$.

6. Use the Runge-Kutta method with $h = 0.2$ to compute the solution for the interval $0 \leq x \leq 2$ of the problem from boundary layer theory, Example III, § 2.6:

$$y''' = -yy''; \qquad y(0) = y'(0) = 0, \qquad y''(0) = 1.$$

Compare the end value with the exact value given by the power series solution.

7. Apply the least-squares method of Ch. I, § 4.2 to the problem

$$y' - y = 0; \qquad y(0) = 1$$

for the interval $0 \leq x \leq 1$ assuming an approximate solution of the form $w_n = \sum\limits_{\nu=0}^{n} a_\nu x^\nu$ with $a_0 = 1$. The method requires that

$$J[w_n] = \int\limits_0^1 (w_n' - w_n)^2 \, dx = \text{minimum}$$

and the a_ν are to be determined from the equations

$$\frac{\partial J}{\partial a_\varrho} = 0 \qquad (\varrho = 1, 2, \ldots, n).$$

Compare the approximate solution for $n = 2$ and $n = 3$ with the exact solution.

5.11. Solutions of the exercises in § 5.10

1. Insertion of $y = -1 + a_1 x + a_2 x^2 + \cdots$ into the differential equation yields

$$a_1 + 2a_2 x + 3a_3 x^2 + \cdots = x^2 + 1 - 2a_1 x + x^2(a_1^2 - 2a_2) + \cdots,$$

and equating coefficients, we find that $a_1 = 1$, $a_2 = -a_1 = -1$, ... To the eighth power in x the solution is

$$y = -1 + x - x^2 + \frac{4}{3}x^3 - \frac{7}{6}x^4 + \frac{6}{5}x^5 -$$

$$- \frac{37}{30}x^6 + \frac{404}{315}x^7 - \frac{3321}{2520}x^8 + \cdots,$$

which converges well for small $|x|$. The required values are given in Table II/30.

Table II/30

x	y
-0.2	$-1.253 016 9$
-0.1	$-1.111 513 3$
0.1	$-0.908 822 5$
0.2	$-0.830 881 3$

2. The results are reproduced in Table II/31, in which the values above the dotted lines are "starting values".

3. A discrepancy appears first with the term in h^4. The method is therefore of the third order[1].

---

[1] See, for instance, F. A. WILLERS: Methoden der praktischen Analysis, p. 307. Berlin and Leipzig 1928, or in the translation: Practical Analysis, p. 377. New York, Dover Publications, Inc. 1948.

Table II/31.

Adams interpolation formula (3.11) *applied to a non-linear first-order equation*

x	y	$hf(x,y)$	$h\nabla f$	$h\nabla^2 f$	$h\nabla^3 f$	$h\nabla^4 f$
−0·2	−1·2530169	0·16100514				
−0·1	−1·1115133	0·12454618	−0·03645896	0·01191278		
0	−1	0·1	−0·02454618	0·00814201	−0·00377077	
0·1	−0·9088225	0·08359583	−0·01640417	0·00584471	−0·00229730	0·00147347
0·2	−0·8308813	0·07303637	−0·01055946			
				0·0044666	−0·0013781	0·0009192
0·3	−0·7612061	0·06694347	−0·0060929	0·0035615	−0·0009051	0·0004730
0·4	−0·6957879	0·06441208	−0·0025314	0·0029827	−0·0005788	0·0003263
0·5	−0·6313749	0·06486343	+0·0004513			

4. Two calculations are carried out with $h=1$ and $h=2$, respectively; they are continued until y changes sign. The results are given in Tables II/32, II/33; all three lines of the Runge-Kutta calculation are given for each step up to $x=2$ but for the remaining steps only the first line is reproduced.

Table II/32. *A non-linear oscillation computed by the Runge-Kutta method with* $h=1$

x	y	y'	$k_j=-\tfrac{1}{2}y^3$	$\begin{array}{c}k\\k'\end{array}$
0	**0·2**	**0**	−0·004	
0·5	0·199		−0·0039403	−0·0039602
1	0·1960597		− 37683	−0·0078432
1	**0·1960398**	**−0·0078432**	−0·0037671	
1·5	0·1911764		− 34936	−0·0035848
2	0·1847030		− 31506	−0·0069640
2	**0·1846118**	**−0·0148072**	−0·0031458	
3	0·1669257	−0·0202935	− 23256	
4	0·1445867	−0·0241146	− 15113	
5	0·1191990	−0·0264402	− 8468	
6	0·0920821	−0·0276418	− 3904	
7	0·0641510	−0·0281344	− 1320	
8	0·0359308	−0·0282697	− 232	
9	0·0076499	−0·0282844	− 2	
10	−0·0206345	−0·0282829	+	

Table II/33. *The same computation with* $h=2$

x	y	$v=2y'$	$k_j=-2y^3$	$\begin{array}{c}k\\k'\end{array}$
0	**0·2**	**0**	−0·016	
1	0·196		−0·0150591	−0·0153727
2	0·1849409		− 126511	−0·0296292
2	**0·1846273**	**−0·0296292**	− 125868	
4	0·1446297	−0·0482341	− 60507	
6	0·0921318	−0·0552814	− 15641	
8	0·0359779	−0·0565370	− 931	
10	−0·0205907	−0·0565634		

5. The requisite starting values, separated from the following values by a dotted line in Table II/34, can be calculated by means of the power series

$$y = x - \frac{2x^5}{5!} + \frac{2 \cdot 6 x^9}{9!} - \frac{2 \cdot 6 \cdot 10 x^{13}}{13!} + - \cdots;$$

Table II/34. *Application of the extrapolation formula* (5.29)

x	y	$f = -xy$
−0·2	−0·19999467	−0·039998934
0	0	0
0·2	0·19999467	−0·039998934
0·4	0·39982934	−0·159931736
0·6	0·59870429	−0·359225728
0·8	0·79454288	−0·635634304
1	0·98336567	

this series is also used to calculate the exact values in Table II/35. The main calculation is carried out with the formula

$$y_{r+1} = 3y_r - 3y_{r-1} + y_{r-2} + \frac{h^3}{2}(f_r + f_{r-1})$$

with $h = 0·2$.

Table II/35. *Exact values*

x	y
0·4	0·399829339
1	0·983366383

6. Table II/36 gives the results of the calculation. The power series is used to calculate

$$y(0·8) = 0·3173143, \qquad y(1) = 0·4919304, \qquad y(2) = 1·78809.$$

Table II/36. *An accurate Runge-Kutta solution of the Blasius equation*

x	y	$y'h$	$y''\frac{h^2}{2}$	x	y	$y'h$	$y''\frac{h^2}{2}$
0	0	0	0·02	1·2	0·7003698	0·2241645	0·0150553
0·2	0·0199973	0·0399867	0·0199733	1·4	0·9388576	0·2520533	0·0127849
0·4	0·0799148	0·0797873	0·0197879	1·6	1·2028830	0·2751759	0·0103241
0·6	0·1793564	0·1189319	0·0192940	1·8	1·4875610	0·2933679	0·0078912
0·8	0·3173139	0·1566757	0·0183708	2	1·7880597	0·3068941	0·0056885
1	0·4919296	0·1920829	0·0169527				

7. For $n = 3$ we obtain the equations

$$-\frac{1}{2} + \frac{1}{3}a_1 + \frac{1}{4}a_2 + \frac{1}{5}a_3 = 0,$$

$$-\frac{2}{3} + \frac{1}{4}a_1 + \frac{8}{15}a_2 + \frac{2}{3}a_3 = 0,$$

$$-\frac{3}{4} + \frac{1}{5}a_1 + \frac{2}{3}a_2 + \frac{33}{35}a_3 = 0$$

with the solution

$$a_1 = \frac{9000}{8884}, \qquad a_2 = \frac{3780}{8884}, \qquad a_3 = \frac{2485}{8884}.$$

Table II/37 gives w_2 and w_3 for several values of x, together with their errors (by comparison with the exact solution $y = e^x$). For comparison, w_1 is also included.

Table II/37. *An application of the least-squares method*

x	$y = e^x$	$w_1(x)$	$w_2(x)$	Error $w_2 - y$	$w_3(x)$	Error $w_3 - y$
0·2	1·221 403	1·3	1·207 23	−0·014 17	1·221 869	+0·000 466
0·4	1·491 825	1·6	1·481 93	−0·009 89	1·491 202	−0·000 623
0·6	1·822 119	1·9	1·824 10	+0·001 98	1·821 427	−0·000 692
0·8	2·225 541	2·2	2·233 74	+0·008 20	2·225 970	+0·000 429
1	2·718 282	2·5	2·710 85	−0·007 43	2·718 258	−0·000 024

Chapter III

Boundary-value problems in ordinary differential equations

§ 1. The ordinary finite-difference method

The basis of the finite-difference method is the replacement of all derivatives occuring by the corresponding difference quotients; this is applicable to any problem in differential equations.

1.1. Description of the finite-difference method

We divide the interval $\langle a, b \rangle$ in which the solution $y(x)$ of the boundary-value problem is sought into n equal parts of length

$$h = \frac{b - a}{n}.$$

The points $x_i = a + ih$ are called the "pivotal points" and h is called the "pivotal interval" or merely the "interval".

We characterize the value of a function at the point $x_i = a + ih$ by a subscript i (i not necessarily integer); thus y_i denotes the value of the required function $y(x)$ at the point x_i. Further we denote an approximation to y_i by Y_i (see Fig. III/1).

The first stage of the finite-difference method is to set up a system of "finite equations" from which the Y_i can be determined. A finite equation is obtained by writing down the differential equation for a pivotal point x_i and replacing all derivatives which occur by difference quotients in some specified way; each derivative is thereby represented by a certain linear expression in the Y_i.

Thus (cf. Ch. I, § 2.1) we can replace the derivative $y'(x_i)$ by

$$\frac{Y_{i+1} - Y_{i-1}}{2h}, \tag{1.1}$$

the second derivative $y''(x_i)$ by

$$\frac{Y_{i+1} - 2Y_i + Y_{i-1}}{h^2},\tag{1.2}$$

a derivative $y^{(k)}(x_i)$ of even order k by the k-th difference quotient

$$\frac{1}{h^k}\,\Delta^k Y_{i-\frac{k}{2}},\tag{1.3}$$

and a derivative $y^{(p)}(x_i)$ of odd order p by the arithmetic mean of two p-th difference quotients:

$$\frac{1}{2}\,\frac{1}{h^p}\left(\Delta^p Y_{i-\frac{p+1}{2}} + \Delta^p Y_{i-\frac{p-1}{2}}\right).\tag{1.4}$$

The boundary conditions also are put into the form of finite equations; these must also be included in the system of equations for the Y_i.

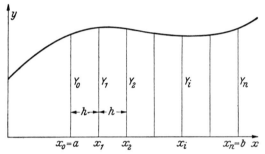

Fig. III/1. Notation used in the finite-difference method

For example, if we have a boundary-value problem of even order k with k boundary conditions at the points $x=a$ and $x=b$, we can set up a finite equation for each of the pivotal points x_i (for $i=0, 1, \ldots, n$) and likewise for each boundary condition. In general, besides the pivotal values Y_i inside the interval $\langle a, b\rangle$, values of Y at the exterior points $x_{k/2}, x_{k/2+1}, \ldots, x_{-1}, x_{n+1}, \ldots, x_{n+k/2}$ will be involved; thus we have $n+k+1$ equations for the same number of unknowns

$$Y_{-\frac{k}{2}},\quad Y_{-\frac{k}{2}+1},\ldots,\quad Y_{n+\frac{k}{2}}.$$

This system of finite equations for the Y_i is linear if and only if the boundary-value problem is linear. Nothing can be asserted about the solubility of the system without being more specific about the boundary-value problem.

Note. We can often specify finite-difference representations other than those expressed in (1.1) to (1.4) and thereby arrive at different finite equations. For example, the term $(f y')'$ can be dealt with either by differentiating the product and replacing y' and y'' by their finite-difference representations (1.1) and (1.2):

$$f_i'\,\frac{Y_{i+1} - Y_{i-1}}{2h} + f_i\,\frac{Y_{i+1} - 2Y_i + Y_{i-1}}{h^2},$$

or by replacing the two differentiations in $(fy')'$ by the finite-difference representation (1.1):

$$\frac{f_{i+1}\dfrac{Y_{i+2}-Y_i}{2h}-f_{i-1}\dfrac{Y_i-Y_{i-2}}{2h}}{2h}.$$

We can halve the interval in this last expression, which then reads

$$\frac{f_{i+\frac12}(Y_{i+1}-Y_i)-f_{i-\frac12}(Y_i-Y_{i-1})}{h^2}=\frac{1}{h^2}\,\varDelta\!\left(f_{i-\frac12}\varDelta Y_{i-1}\right). \qquad (1.5)$$

The equations which can be set up in these various ways will in general yield slightly different results.

1.2. Examples of boundary-value problems of the second order

The following worked examples are intended to familiarize the reader with the practical application of the finite-difference method and to introduce several little artifices and expedients which can often be employed with advantage. A further example will be found in Exercise 1 of § 8.11.

I. A linear boundary-value problem of the second order. Consider the bending of a strut (Fig. III/2) with flexural rigidity $EJ(\xi)$ and axial compressive

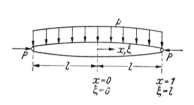

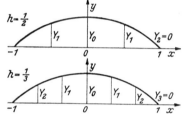

Fig. III/2. The bending of a strut

Fig. III/3. Finite-difference notation for the bending moment distribution

load P by a distributed transverse load $p(\xi)$, ξ being the co-ordinate along the axis of the strut. The bending moment distribution $M(\xi)$ satisfies the differential equation

$$\frac{d^2M}{d\xi^2}+\frac{P}{EJ(\xi)}M=-p(\xi).$$

We will assume here that the transverse load is a constant p but that the flexural rigidity is a variable

$$EJ(\xi)=\frac{EJ_0}{1+\left(\dfrac{\xi}{l}\right)^2}.$$

If we take $P=\dfrac{EJ_0}{l^2}$ and introduce dimensionless variables $x=\dfrac{\xi}{l}$, $y=\dfrac{M}{l^2p}$, the differential equation becomes

$$y''+(1+x^2)\,y=-1, \qquad (1.6)$$

where dashes denote differentiation with respect to x. We take $M=0$ at each end, corresponding to smoothly-hinged end supports, so that the boundary conditions are

$$y(-1)=y(1)=0. \qquad (1.7)$$

The interval $\langle-1,1\rangle$ is first subdivided rather coarsely into four, so that $h=\frac12$ (Fig. III/3); then, making use of the boundary conditions

$(Y_{\pm 2}=0)$ and the symmetry $(Y_1=Y_{-1})$, we have

$$(\text{for } x = 0) \qquad \frac{Y_1 - 2Y_0 + Y_1}{h^2} + Y_0 = -1,$$

$$(\text{for } x = \tfrac{1}{2}) \qquad \frac{0 - 2Y_1 + Y_0}{h^2} + \frac{5}{4} Y_1 = -1.$$

With $h^2 = \tfrac{1}{4}$ the solution of these two simultaneous equations is

$$Y_0 = \frac{59}{61} = 0.967, \qquad Y_1 = \frac{44}{61} = 0.721.$$

Thus, even though it may be rather rough, we have already obtained some idea of the bending moment distribution and with a trivial amount of computation.

To obtain more accurate values we must choose a smaller pivotal interval; thus if we take $h = \tfrac{1}{3}$ (pivotal values Y_0, Y_1, Y_2 as in Fig. III/3), which gives the system of equations

$$9(2Y_1 - 2Y_0) \quad + \quad Y_0 = -1,$$

$$9(Y_2 - 2Y_1 + Y_0) + \frac{10}{9} Y_1 = -1,$$

$$9(Y_1 - 2Y_2) \quad + \frac{13}{9} Y_2 = -1.$$

we obtain the values

$$Y_0 = \frac{53347}{56237} = 0.9486, \qquad Y_1 = \frac{47259}{56237} = 0.8404, \qquad Y_2 = \frac{29088}{56237} = 0.5172.$$

When a large number of subdivisions is used, we do not treat the difference equations as a set of simultaneous equations but take advantage of the linearity of the problem and treat them as an initial-value problem (cf. § 4.3). We start from a value $Y_0^{(1)} = 1$ and calculate successively from the difference equations the values $Y_1^{(1)}$, $Y_2^{(1)}$, ..., $Y_n^{(1)}$ (for the case $h = 1/n$), then repeat the procedure, starting from a different initial value $Y_0^{(2)} = 0$, say, and so obtain a second set of values $Y_1^{(2)}$, $Y_2^{(2)}$, ..., $Y_n^{(2)}$. Then the linear combination

$$Y_i = \frac{Y_i^{(1)} Y_n^{(2)} - Y_i^{(2)} Y_n^{(1)}}{Y_n^{(2)} - Y_n^{(1)}}$$

of these two sets of values is the required solution; for it satisfies the difference equations on account of their linearity and also satisfies both the boundary conditions. In the example under consideration we obtain for $h = 0.1$ the values exhibited in Table III/1.

Improvement by h^2-extrapolation. If the finite-difference method has been applied to a particular problem with several different values of h, the results so obtained can often be improved in the following way.

We assume that at a fixed point x the error in the finite-difference method tends to zero quadratically (for the present case) with h (cf.

Table III/1. *The linear combination of two independent solutions of the difference equations*

i	$Y_i^{(1)}$	$Y_i^{(2)}$	Y_i	i	$Y_i^{(1)}$	$Y_i^{(2)}$	Y_i
0	1	0	0·933 591	6	0·641 440	−0·174 096	0·587 281
1	0·99	−0·005	0·923 923	7	0·513 688	−0·233 539	0·464 065
2	0·960 001	−0·019 950	0·894 924	8	0·368 282	−0·299 502	0·323 935
3	0·910 018	−0·044 692	0·846 617	9	0·206 836	−0·370 553	0·168 492
4	0·840 116	−0·078 946	0·779 082	10	0·031 647	−0·444 898	0
5	0·750 468	−0·122 286	0·692 510				

§ 3.3). We can then write approximately

$$y - Y \approx Ch^2, \qquad y - Y^* \approx Ch^{*2},$$

where Y and Y^* are the values calculated with the intervals h and h^*, respectively. Eliminating the constant C, we derive a new approximate value

$$\overline{Y} = \frac{Y h^{*2} - Y^* h^2}{h^{*2} - h^2} = Y + \frac{h^2}{h^{*2} - h^2}(Y - Y^*), \qquad (1.8)$$

which will in general be a better value. Table III/2 shows the result of applying this h^2-extrapolation procedure in the present example to the values at $x = 0$. One would therefore take $y(0) \approx 0.9321$ as the new approximate value.

Table III/2. h^2-extrapolation

h	Y	
$\frac{1}{2}$	0·967 2	$\overline{Y} = 0.9337$
$\frac{1}{3}$	0·948 61	
$\frac{1}{10}$	0·933 591	$\overline{Y} = 0.932\,106$

II. A non-linear boundary-value problem of the second order.

The boundary-value problem

$$y'' = \frac{3}{2} y^2; \quad y(0) = 4, \quad y(1) = 1 \qquad (1.9)$$

possesses two solutions, one of which, namely

$$y = \frac{4}{(1 + x)^2}, \qquad (1.10)$$

is expressible in elementary terms while the other involves elliptic functions.

Here also we start by using a coarse subdivision of the interval to get a rough idea of the solution. With $h = \frac{1}{2}$ we have the single non-linear equation

$$\frac{4 - 2Y_1 + 1}{h^2} = \frac{3}{2} Y_1^2,$$

which gives for $y\left(\frac{1}{2}\right)$ the approximation

$$Y_1 = \frac{1}{3}\left(-8 \pm \sqrt{184}\right) = \begin{cases} 1\cdot8549 \text{ (error } +4\cdot3\%) \\ -7\cdot188. \end{cases}$$

With $h = \frac{1}{3}$ we have two non-linear equations

$$9(4 - 2Y_1 + Y_2) = \frac{3}{2} Y_1^2,$$

$$9(Y_1 - 2Y_2 + 1) = \frac{3}{2} Y_2^2$$

for the unknowns Y_1, Y_2 (see Fig. III/4). These equations represent two parabolas in the (Y_1, Y_2) plane (see Fig. III/5) and their points of intersection give the required approximate values:

$$Y_1 = 2\cdot2950 \text{ (error } + 2\cdot0\%)$$
$$Y_2 = 1\cdot4680 \text{ (error } + 1\cdot9\%)$$

and

$$Y_1 = -4\cdot70$$
$$Y_2 = -9\cdot72.$$

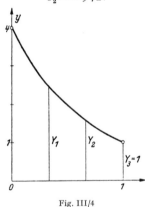

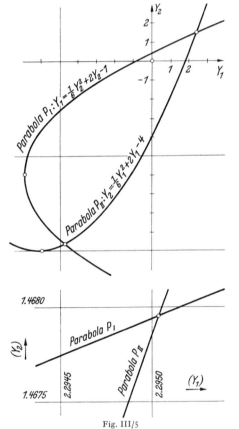

Fig. III/4

Fig. III/5

Fig. III/4. Notation for the finite-difference method applied to the non-linear problem of Example II

Fig. III/5. Solution of the algebraic equations

For finer subdivisions we treat the problem as an initial-value problem, as in Example I, by starting from $Y_0 = 4$ and a guessed value of Y_1 and calculating the remaining Y_i from the difference equation. This is repeated with several different values of Y_1 so that we can interpolate between them for a value which will give a value of Y_n satisfying the prescribed boundary condition $Y_n = 1$. The calculations with a coarse subdivision can be used to find a rough value of Y_1 for a finer subdivision.

With $h = \frac{1}{5}$ we try $Y_1 = 3$ and $Y_1 = 2 \cdot 8$ and calculate the first two Y_i columns of Table III/3 from the difference equations

$$Y_{i+1} = 0 \cdot 06 \, Y_i^2 + 2 Y_i - Y_{i-1} \quad (i = 1, 2, 3, 4). \tag{1.11}$$

Table III/3. *Solution of a non-linear problem by interpolation*

i			Y_i		
(−1)				(6·165 36)	
0	4	4	4	4	4
1	3	2·8	2·795 3	2·794 64	− 2·513 8
2	2·540	2·070 4	2·059 4	2·057 87	− 8·648 4
3	2·467	1·598 0	1·578 0	1·575 19	−10·295 3
4	2·759	1·278 8	1.246 0	1·241 38	− 5·582 6
5	3·509	1·057 7	1·007 1	1·000 03	1

From the end values 3·509 and 1·0577 we extrapolate linearly to find the better value 2·7953 for Y_1, which we use to build up the better approximations in the third Y_i column; further interpolation yields $Y_1 = 2 \cdot 794\,64$ (from this we build up the fourth column; the extra bracketed value at x_{-1} is calculated for an error estimate in § 3.4). The last column contains the results of applying the same method to the other solution.

III. An eigenvalue problem. Consider the longitudinal vibrations of a cantilever (Fig. III/6). Let its length be l, density ϱ, modulus of elasticity E and area of cross-section $F(x)$, where the co-ordinate x is chosen along its axis with the origin at the free end. The displacement y satisfies the differential equation

$$- E(F') \, y' = \omega^2 \varrho F \, y$$

and the boundary conditions

$$y'(0) = 0, \quad y(l) = 0.$$

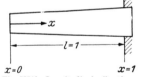

Fig. III/6. Longitudinal vibrations of a cantilever

It is required to find the natural frequencies ω. Many other physical problems, for example, the torsional vibrations of shafts, give rise to eigenvalue problems of similar form.

Let the cross-sectional area increase linearly with x from F_0 at the tip to $2 F_0$ at the base:

$$F(x) = F_0 \left(1 + \frac{x}{l}\right);$$

then with $\lambda = \omega^2 \dfrac{\varrho}{E}$ and $l = 1$ we have the fully homogeneous problem

$$- (1 + x) \, y'' - y' = - [(1 + x) \, y']' = \lambda (1 + x) \, y; \quad y'(0) = 0, \quad y(1) = 0.$$

The finite-difference method can be applied in various ways. To begin with, there are several different forms of finite equation which can be used (cf. the note in § 1.1); we choose the form

$$(1 + j h) \frac{Y_{j+1} - 2 Y_j + Y_{j-1}}{h^2} + \frac{Y_{j+1} - Y_{j-1}}{2h} + \varLambda (1 + j h) Y_j = 0 \atop (j = 0, 1, \ldots, n-1), \Bigg\} \quad (1.12)$$

where $h = 1/n$ and $\varLambda$ is an approximate value for λ. Corresponding to the boundary conditions we put

$$\frac{Y_1 - Y_{-1}}{2h} = 0, \quad \text{i.e.} \quad Y_{-1} = Y_1, \quad \text{and} \quad Y_n = 0.$$

Thus we have n linear homogeneous equations for the n unknowns $Y_0, Y_1, \ldots, Y_{n-1}$. For a non-trivial solution the determinant of the coefficients must vanish. This requirement yields an algebraic equation of the n-th degree for $\varLambda$, whose n roots $\varLambda_1, \varLambda_2, \ldots, \varLambda_n$, arranged in increasing order of magnitude, are regarded as approximations to the first n eigenvalues $\lambda_1, \lambda_2, \ldots, \lambda_n$.

With $h = \tfrac{1}{2}$, for example, we obtain the homogeneous equations

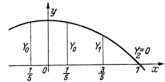

$$(-8 + \varLambda) Y_0 + 8 Y_1 = 0,$$

$$5 Y_0 + \left(-12 + \frac{3}{2} \varLambda \right) Y_1 = 0.$$

Fig. III/7. A subdivision which yields a more accurate finite-difference representation of the boundary condition $y'(0) = 0$

The condition for a non-trivial solution yields

$$0 = \begin{vmatrix} -8 + \varLambda & 8 \\ 5 & -12 + \dfrac{3}{2} \varLambda \end{vmatrix}$$

$$= \frac{3}{2} \varLambda^2 - 24 \varLambda + 56,$$

from which we obtain

$$\varLambda = 8 \pm \frac{4}{3} \sqrt{15}, \quad \text{i.e.} \quad \begin{cases} \varLambda_1^{(2)} = 2 \cdot 836 \ (\text{error} -12 \%) \\ \varLambda_2^{(2)} = 13 \cdot 164 \ (\text{error} -43 \%). \end{cases}$$

Here $\varLambda_k^{(m)}$ denotes the approximation to the k-th eigenvalue given by the calculation with interval $h = 1/m$.

A variation is to use a subdivision of the x axis which does not include $x = 0$ as one of its points; such a subdivision is obtained, for example, if we mark off points from $x = 1$ with $h = \tfrac{2}{5}$, as in Fig. III/7. The corresponding finite equations read

$$-7 Y_0 + 7 Y_1 + \varLambda \frac{24}{25} Y_0 = 0, \qquad 7 Y_0 - 16 Y_1 + \varLambda \frac{32}{25} Y_1 = 0$$

and yield the approximate values

$$\varLambda = \frac{25}{48} \left(19 \pm \sqrt{172} \right) = \begin{cases} 3 \cdot 0651 \ (\text{error} \ -4 \cdot 8 \%) \\ 16 \cdot 727 \ (\text{error} \ -27 \ \%). \end{cases}$$

In this way we represent the boundary condition $y' = 0$ at $x = 0$ more accurately than when $x = 0$ is a pivotal point; this is shown by the fact that the above value approximates the first eigenvalue more closely than the value obtained from a

Table III/4. *Solution of an eigenvalue problem (longitudinal vibrations of a cantilever) by interpolation*

ϱ	Y_3	Y_2	Y_1	Y_0	Y_{-1}	$Y_{-1}-Y_1$	By interpolation	Error in $\Lambda^{(5)}$
1·88	1·990 59	2·858 46	3·490 45	3·780 41	3·630 76	0·140 31		
1·875	1·985 294	2·837 255	3·438 349	3·679 867	3·463 963	0·025 614	$\varrho_1 = 1·87387$; $\Lambda_1^{(5)} = 3·1533$	−2·0%
1·874	1·984 235	2·833 020	3·427 968	3·659 899	3·430 984	0·003 016		
1·2	1·270 59	0·493 02	−0·828 93	−1·667 81	−1·210 60	−0·381 67		
1·16	1·228 235	0·386 403	−0·934 488	−1·639 210	−0·970 608	−0·036 120	$\varrho_2 = 1·155 79$; $\Lambda_2^{(5)} = 21·105$	− 8%
1·1558	1·223 788	0·375 418	−0·944 777	−1·634 919	−0·944 872	−0·000 095		
0	0	−1·133 333	0	1·339 394	0	0	$\varrho_3 = 0$; $\Lambda_3^{(5)} = 50$	−20%
−1·1558	−1·223 788	0·375 418	0·944 777	−1·634 919	0·944 872	0·000 095	$\varrho_4 = −1·15579$; $\Lambda_4^{(5)} = 78·9$	−35%
−1·874	−1·984 235	2·833 020	−3·427 968	3·659 899	−3·430 984	−0·003 016	$\varrho_5 = −1·87387$; $\Lambda_5^{(5)} = 96·8$	−52%

calculation with $h = \frac{1}{3}$ in which $x = 0$ is a pivotal point, even though this interval

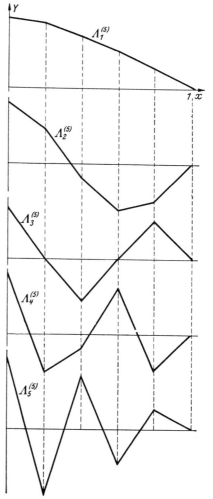

Fig. III/8. Approximations to the normal modes of vibration (longitudinal) of a cantilever

is smaller. The results for $h = \frac{1}{3}$ are

$$\Lambda_1^{(3)} = 3·0413 \ (\text{error } -5·5\%),$$
$$\Lambda_2^{(3)} = 18 \ (-22\%),$$
$$\Lambda_3^{(3)} = 32·96 \ (-48\%).$$

Here also it is convenient to reduce the boundary-value problem to an initial-value problem when a large number of pivotal points are used. We estimate the value of Λ from the calculations with a coarse subdivision and then, starting from $Y_n = 0$ and $Y_{n-1} = 1$ (a factor is disposable here), use (1.12) to calculate $Y_{n-2}, Y_{n-3}, \ldots$. If we repeat the calculation with slightly different values of Λ, it is usually quite feasible to interpolate among them to locate the value which gives $Y_1 = Y_{-1}$.

If we put $2 - \Lambda h^2 = \varrho$, the equations (1.12) for $h = \frac{1}{5}$ read

$$
\begin{aligned}
1 \cdot 7\, Y_3 &= Y_4 \varrho \times 1 \cdot 8 \\
1 \cdot 5\, Y_2 &= Y_3 \varrho \times 1 \cdot 6 - 1 \cdot 7\, Y_4 \\
1 \cdot 3\, Y_1 &= Y_2 \varrho \times 1 \cdot 4 - 1 \cdot 5\, Y_3 \\
1 \cdot 1\, Y_0 &= Y_1 \varrho \times 1 \cdot 2 - 1 \cdot 3\, Y_2 \\
0 \cdot 9\, Y_{-1} &= Y_0 \varrho \qquad\quad - 1 \cdot 1\, Y_1 .
\end{aligned}
$$

The calculations for the first two eigenfunctions using these equations in the above manner are given in Table III/4. From the form of the equations it can be seen that changing the sign of ϱ merely changes the signs of Y_3, Y_1, Y_{-1} and leaves Y_2, Y_0 unaltered. These sign changes therefore provide us with two more eigenfunctions without any further calculation; and since $\varrho = 0$ gives $Y_1 = Y_{-1} = 0$, this value yields a fifth eigenfunction (cf. the table). These approximations to the first five exact eigenfunctions are depicted in Fig. III/8.

IV. Infinite interval. As an example with a boundary condition at infinity we consider the boundary-value problem

$$
y'' = \frac{1+x}{2+x}\, y; \qquad y(0) = 1, \qquad y(\infty) = 0 .
$$

Here $y(x)$ may be interpreted as the temperature difference between an infinitely long rod and its surroundings, one end $x = 0$ of the rod being kept at unit temperature and the surroundings at zero temperature. The heat loss to the surroundings is assumed to be proportional to $\varphi(x)\, y$, where $\varphi = \dfrac{1+x}{2+x}$.

By means of the difference equations

$$
Y_{i+1} - 2\, Y_i + Y_{i-1} - h^2\, \frac{1+ih}{2+ih}\, Y_i = 0 \qquad (i = 1, 2, \ldots)
$$

we can express successively Y_2, Y_3, ... as functions (linear when the differential equation is linear, as here) of Y_1; this first unknown pivotal value is then to be determined by the boundary condition $y(\infty) = 0$. Such a boundary condition can be translated into a finite condition in various ways:

1. If we replace the condition $y(\infty) = 0$ by $Y_n = 0$, we obtain an approximate value for Y_1, and hence also for Y_2, Y_3, ..., Y_{n-1}, depending on n. The values of Y_1 for a series of values of n have differences which approximate closely to a geometric sequence, as is shown in Tables III/5 and III/6 for $h = 1$ and $h = \frac{1}{2}$. This facilitates the extrapolation to $n = \infty$ and indicates that the accuracy of this extrapolation, which yields the results $y(1) \approx 0 \cdot 447$ and $0 \cdot 444$ for $h = 1$ and $\frac{1}{2}$, respectively,

Table III/5. *Solution by extrapolation from finite boundary conditions.* $h = 1$

Y_n	$Y_n = 0$ yields	Differences	Extrapolation
$Y_2 = \dfrac{8}{3}\,Y_1 - 1$	$Y_1 = 0\cdot375$		
		$0\cdot060$	
$Y_3 = \dfrac{19}{3}\,Y_1 - \dfrac{11}{4}$	$Y_1 = 0\cdot435$		
		$0\cdot010$	$Y_1 = 0\cdot4470$
$Y_4 = \dfrac{226}{15}\,Y_1 - \dfrac{67}{10}$	$Y_1 = 0\cdot4447$		
		$0\cdot0018$	
$Y_5 = \dfrac{1636}{45}\,Y_1 - \dfrac{487}{30}$	$Y_1 = 0\cdot44652$		

Table III/6. *Solution by extrapolation from finite boundary conditions.* $h = \tfrac{1}{2}$

Y_n	$Y_n = 0$ yields	Differences	Extrapolation
$Y_2 = 2\cdot15 Y_1 - 1$	$Y_1 = 0\cdot465$		
$Y_3 = 3\cdot6583\ Y_1 - 2\cdot1667$	$Y_1 = 0\cdot593$		
$Y_4 = 5\cdot8200\ Y_1 - 3\cdot7202$	$Y_1 = 0\cdot6392$	$0\cdot046$	$Y_1 = 0\cdot6718$
$Y_5 = 9\cdot0728\ Y_1 - 5\cdot9714$	$Y_1 = 0\cdot6582$	$0\cdot0190$	and hence
$Y_6 = 14\cdot090 Y_1 - 9\cdot3836$	$Y_1 = 0\cdot6660$	$0\cdot0078$	$Y_2 = 0\cdot4444$
$Y_7 = 21\cdot925 Y_1 - 14\cdot672$	$Y_1 = 0\cdot6692$	$0\cdot0032$	

is at least as good as the finite-difference method used. To acquire greater accuracy and confidence in the results one would repeat the calculations with smaller h. Fig. III/9 shows the approximate solution with $h = \tfrac{1}{2}$.

2. For large values of x the solution behaves like constant $\times e^{-x}$. (In complicated cases we put $1/x = u$ and study the behaviour of the solutions of the differential equation for small u.) Thus for large n we have

$$Y_n \approx A\,e^{-nh},$$
$$Y_{n+1} \approx A\,e^{-(n+1)h},$$

which suggests that we use

$$Y_{n+1} - e^{-h}\,Y_n = 0$$

Table III/7. *Method 2*

n	Y_1
2	$0\cdot665$
3	$0\cdot6681$
4	$0\cdot6702$
5	$0\cdot6710$
6	$0\cdot6713$

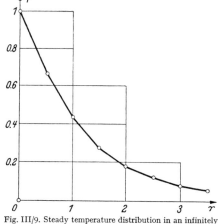

Fig. III/9. Steady temperature distribution in an infinitely long rod

as a finite boundary condition. Using the expressions for Y_n in terms of Y_1 calculated above in 1. for $h = \tfrac{1}{2}$, we obtain the Y_1 values of Table III/7; the extrapolated value again yields $y(1) \approx Y_2 = 0\cdot444$.

1.3. A linear boundary-value problem of the fourth order

Under the usual assumptions the transverse displacement $\eta(x)$ of an elastically embedded rail subject to a distributed transverse load (Fig. III/10) satisfies the differential equation

$$\frac{d^2}{d\xi^2}\left(E\,J(\xi)\,\frac{d^2\eta}{d\xi^2}\right) + K\eta = q(\xi).$$

Here the ξ co-ordinate is taken along the axis of the undeformed rail, $EJ(\xi)$ denotes the flexural rigidity, K is the elastic constant of the bed material and $q(\xi)$ is the

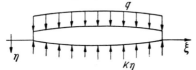

Fig. III/10. Bending of a transversely loaded, elastically embedded rail

load density. For a rail with both ends freely supported, the bending moment M and the shear force Q vanish at the end points $\xi = a$ and $\xi = b$; thus the boundary conditions are

$$\frac{d^2\eta}{d\xi^2} = \frac{d^3\eta}{d\xi^3} = 0 \quad \text{for} \quad \xi = a \quad \text{and} \quad \xi = b.$$

For the present numerical treatment of this fourth-order boundary-value problem we shall assume the following parabolic distributions of flexural rigidity and load density:

$$E\,J(\xi) = E\,J_0\,(2 - x^2); \qquad q(\xi) = q_0\,(2 - x^2),$$

where $x = \xi/l$, $2l$ being the length of the rail and $x = 0$ the mid-point. Non-dimensionalizing η and K as well, we have the dimensionless set of quantities

$$x = \frac{\xi}{l}, \qquad y = \frac{E\,J_0}{q_0\,l^4}\,\eta, \qquad k = \frac{l^4}{E\,J_0}\,K;$$

when $k = 40$ the boundary-value problem for y reads

$$\left.\begin{array}{l}[(2 - x^2)\,y'']'' + 40\,y = 2 - x^2, \\ y''(\pm 1) = y'''(\pm 1) = 0 \end{array}\right\} \tag{1.13}$$

(dashes denote differentiation with respect to x).

We could now follow the procedure of § 1.1 and without further ado take $h = 1/n$ and write down the difference equations

$$\frac{1}{h^4}\,\varDelta^2\,[(2 - (i - 1)^2\,h^2)\,\varDelta^2\,Y_{i-2}] + 40\,Y_i = 2 - i^2\,h^2 \qquad (i = 0, 1, 2, \ldots, n),$$

in which we put $Y_{-i} = Y_i$ because of the symmetry of the problem, and boundary conditions

$$Y_{n+1} - 2\,Y_n + Y_{n-1} = 0,$$
$$Y_{n+2} - 2\,Y_{n+1} + 2\,Y_{n-1} - Y_{n-2} = 0.$$

We would then have $n + 3$ equations for the same number of unknowns. However, it is rather more convenient to work with the equivalent system of second-order differential equations obtained by introducing the auxiliary quantity

$$v = (2 - x^2)\,y'';$$

in any case this quantity will be of interest in the calculation of the bending stresses, for it is effectively the negative of the bending moment.

The second-order system for v and y derived from (1.13) is

$$v'' + 40y - 2 + x^2 = 0, \left.\vphantom{\begin{matrix}a\\b\end{matrix}}\right\}$$
$$(2 - x^2)\, y'' - v = 0; \tag{1.14}$$

and if the symmetry of the problem is used to halve the range of x, the boundary conditions become

$$v(1) = v'(1) = y'(0) = v'(0) = 0. \tag{1.15}$$

With Y_i, V_i denoting the approximate values of y_i, v_i the finite-difference representations of these boundary conditions in the case $h = \frac{1}{2}$ read

$$V_2 = 0, \quad V_3 = V_1, \quad Y_{-1} = Y_1, \quad V_{-1} = V_1,$$

and with these relations taken into account the difference equations corresponding to the equations (1.14) reduce to

$$4(2V_1 - 2V_0) \qquad + 40Y_0 - 2 \quad = 0,$$
$$4(-2V_1 + V_0) \qquad + 40Y_1 - \frac{7}{4} = 0,$$
$$4 \times 2V_1 \qquad\qquad + 40Y_2 - 1 \quad = 0,$$
$$2 \times 4(2Y_1 - 2Y_0) - V_0 \qquad\qquad = 0,$$
$$\frac{7}{4} \times 4(Y_2 - 2Y_1 + Y_0) - V_1 \quad = 0$$

Solution of this system of equations yields

$$Y_0 = \frac{3726}{81440} = 0 \cdot 0457515$$
$$Y_1 = \frac{3421}{81440} = 0 \cdot 0420064$$
$$Y_2 = \frac{2666}{81440} = 0 \cdot 0327358$$

$$V_0 = -\frac{488}{8144} = -0 \cdot 0599214$$
$$V_1 = -\frac{315}{8144} = -0 \cdot 0386788.$$

Again for smaller values of h it is better, as in Examples I and II, to calculate $Y_1, V_1, Y_2, V_2, \ldots$ recursively in terms of Y_0 and V_0 and derive two linear equations for Y_0 and V_0 from the conditions $V_n = 0$ and $V_{n+1} = V_{n-1}$. For $h = \frac{1}{5}$ we obtain successively

$$Y_1 = \qquad\qquad Y_0 + 0 \cdot 01 \quad V_0$$
$$V_1 = -\ 0 \cdot 8 \quad Y_0 + \qquad\quad V_0 + 0 \cdot 04$$
$$Y_2 = \quad 0 \cdot 983674\, Y_0 + 0 \cdot 0404082\, V_0 + 0 \cdot 00081633$$
$$V_2 = -\ 3 \cdot 2 \quad Y_0 + 0 \cdot 984 \quad V_0 + 0 \cdot 1584$$
$$Y_3 = \quad 0 \cdot 897782\, Y_0 + 0 \cdot 0922076\, V_0 + 0 \cdot 00507613$$
$$V_3 = -\ 7 \cdot 173878\, Y_0 + 0 \cdot 903347 \quad V_0 + 0 \cdot 349094$$
$$Y_4 = \quad 0 \cdot 636918\, Y_0 + 0 \cdot 166040 \quad V_0 + 0 \cdot 0178504$$
$$V_4 = -\ 12 \cdot 54821 \quad Y_0 + 0 \cdot 675162 \quad V_0 + 0 \cdot 597266$$
$$Y_5 = \quad 0 \cdot 005930\, Y_0 + 0 \cdot 259730 \quad V_0 + 0 \cdot 0481913$$
$$V_5 = -\ 19 \cdot 01360 \quad Y_0 + 0 \cdot 181313 \quad V_0 + 0 \cdot 871278$$
$$V_6 = -\ 25 \cdot 45249 \quad Y_0 - 0 \cdot 728104 \quad V_0 + 1 \cdot 108183.$$

The equations $V_5 = 0$ and $V_6 = V_4$, i.e.

$$- 19 \cdot 013\,60 \, Y_0 + 0 \cdot 181\,313 \, V_0 + 0 \cdot 871\,278 = 0,$$
$$12 \cdot 868\,28 \, Y_0 + 1 \cdot 403\,266 \, V_0 - 0 \cdot 510\,917 = 0,$$

then yield

$$Y_0 = 0 \cdot 045\,331\,7 \quad \text{and} \quad V_0 = - 0 \cdot 051\,611\,5.$$

With these values of Y_0 and V_0 the remaining Y_i, V_i become

$$
\begin{aligned}
Y_1 &= 0 \cdot 044\,8156, & V_1 &= - 0 \cdot 047\,8769, \\
Y_2 &= 0 \cdot 043\,3224, & V_2 &= - 0 \cdot 037\,4473, \\
Y_3 &= 0 \cdot 041\,0152, & V_3 &= - 0 \cdot 022\,7336, \\
Y_4 &= 0 \cdot 038\,1534, & V_4 &= - 0 \cdot 008\,0441, \\
Y_5 &= 0 \cdot 035\,0551, & V_5 &= \quad 0.
\end{aligned}
$$

The correction (1.8) can also be applied here if we assume that the inherent errors tend to zero quadratically with the interval h. With $h = \frac{1}{5}$, $h^* = \frac{1}{2}$ it gives the improved value $Y_0 = 0 \cdot 045\,251\,7$.

1.4. Relaxation

The finite-difference method always entails the solution of a large number of simultaneous equations when the pivotal interval is small. A numerical solution of such a system of equations can be effected by the method[1] to be described now. It is applicable very generally; it can be used for linear and non-linear problems, for ordinary and partial differential equations, for the ordinary finite-difference method and also for the improved finite-difference methods (see § 2 and Ch. V, § 2).

The procedure may be outlined as follows: 1. An initial approximation is found either by estimation or by rough calculation, say by interpolation from values obtained with a coarse subdivision; 2. this initial approximation is inserted in the differential equation to discover where, i.e. at which pivotal points, the difference equation is already satisfied reasonably well and where there are still outstanding residual errors or "residuals" as they are usually called; 3. corrections are then made to the initial approximation at the points where the large residuals occur in such a way as to reduce their magnitude ("relaxation" of the initial

---

[1] This method of solving systems of linear equations had already been used by GAUSS when it was taken up by PH. L. SEIDEL: Münch. Akad. Abh. 1874, 81 − 108; its convergence was investigated later by R. v. MISES and H. POLLACZEK-GEIRINGER: Praktische Verfahren der Gleichungsauflösung. Z. Angew. Math. Mech. 9, 62 − 77 (1929), and in recent years it has been applied to a very wide range of problems and also expounded in two books by R. V. SOUTHWELL: Relaxation methods in engineering science, a treatise on approximate computation. London: Oxford University Press 1943; Relaxation methods in theoretical physics, a continuation of the treatise. London: Oxford University Press 1946, 248 pp.; further D. N. DE G. ALLEN: Relaxation methods. New York 1954, 257 pp. See also our Ch. V, § 1.6.

approximation). This relaxation usually produces new residuals at neighbouring pivotal points, but by continually decreasing the magnitude of the currently largest residual we try to approach the solution of the difference equations.

Experience is essential in making economical use of the method, for one has to appreciate the overall effect of making many varied sequences and combinations ("group relaxation") of corrections to the approximate solution to be able to "feel ones way" quickly towards the exact solution. In acquiring such experience by trying various corrections the beginner usually gets the impression that the method is difficult, for he often has to carry out a long calculation before the residuals become sufficiently small; but it should be borne in mind that the method permits the experienced "relaxer" to produce the solution very quickly to an accuracy sufficient for most technical applications.

A warning will not be out of place here, namely of the fallaciousness of the assumption that if the residuals are reduced to zero to a certain number of decimals, then the current relaxed values give the solution to the same number of decimals; particularly with a large number of equations the exact solution can differ widely from the values which ostensibly satisfy the equations to the required accuracy. In this connection it is important to note that in regions where several neighbouring residuals have the same sign an overall correction ("block relaxation") of considerable magnitude is usually needed to "liquidate" these residuals (cf. Ch. V, § 1.6).

Every computer will no doubt devise his own way of arranging the work when he has familiarized himself with the method. Consequently the layout in Tables III/8, III/9 is only offered as a suggestion.

For several classes of boundary-value problem one can determine rigorous limits for the error in a relaxation solution, cf. Ch. V, § 1.6.

Example I. A linear boundary-value problem. For the problem

$$y'' = -1 - (1 + x^2)\, y, \qquad y(\pm 1) = 0 \qquad (1.16)$$

of Example I, § 1.2 we choose $h = \frac{1}{5}$ and find rough starting values for the relaxation procedure by interpolating graphically (say) between the values given by the ordinary finite-difference method with $h = \frac{1}{2}$. Thus from the results of Example I, § 1.2 we obtain the initial approximation (with $x_r = 0.2 r$)

$$y_0 = 0.97, \; y_1 = 0.94, \; y_2 = 0.80, \; y_3 = 0.60, \; y_4 = 0.34.$$

The difference equations using (1.2) with $h = 1/m$ read

$$y_{k+1} - 2 y_k + y_{k-1} = -h^2 \left(1 + (1 + k^2 h^2)\, y_k\right),$$
$$y_{-1} = y_1, \qquad y_m = 0 \qquad (k = 0, 1, \ldots, m-1);$$

Table III/8. *Examples of various relaxation procedures applied to the finite-difference method for the bent strut problem*

	y_k values and corrections					z_k values					Changes $z_k - y_k$				
	y_0	y_1	y_2	y_3	y_4	z_0	z_1	z_2	z_3	z_4	z_0-y_0	z_1-y_1	z_2-y_2	z_3-y_3	z_4-y_4
Method 1. Starting values	0·97	0·94	0·80	0·60	0·34	0·979	0·925	0·809	0·606	0·331	9	−15	9	6	−9 $\}\times 10^{-3}$
Correction		−0·01				−0·01		−0·005			−10	+10	−5		
New values	0·97	0·93	0·80	0·60	0·34						−1	−5	4	6	−9
Method 2. Starting values	0·97	0·94	0·80	0·60	0·34	0·9794	0·9246	0·8086	0·6063	0·3311	94	−154	86	63	−89 $\}\times 10^{-4}$
Corrections ×10⁴	+94 −214 −42 +84 −94	−214 −42 +84 +94	−42 +84 −94	+84 −94	−94						−94	+47 +107	−107 +21	−21 −42	+42 +47
New values	0·9528	0·9134	0·7948	0·5990	0·3306	0·95246	0·91280	0·79464	0·59900	0·33034	−34	−60	−16	0	−16 $\}\times 10^{-5}$
Method 3. Starting values	0·93	0·89	0·77	0·58	0·32	0·9286	0·8885	0·7729	0·5809	0·3205	−14	−15	+29	+9	+5 $\}\times 10^{-4}$
Corrections ×10⁴	−14 −44 +14 +32 +42	−44 +14 +32 +42	+14 +32 +42	+32 +42	+42						+14	−7 +22	+22 +7	+7 −16	+16 −21
New values	0·9330	0·8944	0·7788	0·5874	0·3242	0·93306	0·89450	0·77897	0·58748	0·32434	6	10	17	8	14 $\}\times 10^{-5}$
Interpolation	0·93798	0·89922	0·78306	0·59056	0·32596	0·93798	0·89922	0·78306	0·59057	0·32595	0	0	0	+1	−1 $\}\times 10^{-5}$

they may be written in the form

$$y_k = z_k,$$

where

$$z_k = \frac{y_{k-1} + y_{k+1}}{2} +$$
$$+ \frac{h^2}{2}\left(1 + (1 + \right.$$
$$\left. + k^2 h^2)\, y_k\right).$$

(1.17)

The quantity $z_k - y_k$ represents the residual error and would normally be called the "residual" and denoted by R_k. Here it will be called the "change"[1]. This terminology arises from the close connection (cf. Ch. V, § 1.6) with the associated iteration procedure for solving such equations, i.e. the iteration defined by $y_k^{[\sigma+1]} = z_k^{[\sigma]}$; thus $z_k^{[\sigma]} - y_k^{[\sigma]}$ is the "change" $y_k^{[\sigma+1]} - y_k^{[\sigma]}$ produced by the $(\sigma+1)$-th cycle of the iteration. Calculation of residuals is equivalent to performing one cycle of the iteration procedure and noting the changes produced. If the y_k values satisfied the

---

[1] *Translator's note.* This unusual terminology is a direct translation from the German; it is used here, in spite of its seeming awkwardness and the existence of a familiar alternative, in order that the emphasis in the German edition on the relationship with the iteration procedure may be preserved.

Table III/9. *A relaxation solution of the system of non-linear equations arising in the finite-difference treatment of a non-linear problem*

	y_k values and corrections				z_k values				Changes $z_k - y_k$			
	y_1	y_2	y_3	y_4	z_1	z_2	z_3	z_4	$z_1 - y_1$	$z_2 - y_2$	$z_3 - y_3$	$z_4 - y_4$
												$\times 10^{-3}$
Correction	0·50 −0·02	0·85	1·05	1·1	0·478	0·842 −0·01	1·052	1·105	−22	−8	+2	+5
	0·48	0·85	1·05	1·1	0·477	0·832	1·052	1·105	−3	−18	+2	+5
Corrections	−0·02 0·46	−0·02 0·83	1·05	1·1	0·467	0·821	1·042	1·105	+7	+9	−8	+5
Corrections	0·46	−0·01 0·82	−0·01 1·04	1·1	0·462	0·815	1·037	1·100	+2	−5	−3	0
Corrections	−0·006 0·454	−0·013 0·807	−0·012 1·028	−0·005 1·095	0·4550	0·8058	1·0269	1·0938	+10	−12	−11	−12
Corr. $\times 10^4$	14 0·4526	50 0·8020	60 1·0220	44 1·0906	0·45245	0·80190	1·02187	1·09052	−15	−10	−13	−8
												$\times 10^{-4}$
Corr. $\times 10^5$	60 0·45200	90 0·80110	100 1·02100	66 1·08994	0·45198	0·80106	1·02104	1·08998	+2	−4	+4	+4
												$\times 10^{-5}$

difference equations to the number of decimals carried, the changes would be zero.

In Table III/8 the y_k values are recorded in the columns on the left, the corresponding z_k values, calculated from them by (1.17), in the central columns and the changes $z_k - y_k$ in the columns on the right. We now describe several relaxation procedures for reducing the magnitude of these changes.

1. Point relaxation. Here we simply make such corrections in the y_k as seem appropriate, dealing with one point at a time. Thus (cf. Table III/8) we select the change $z_1 - y_1 = -0.015$ as being particularly large and try making a correction of -0.01 in $y_1 (= y_{-1})$. The effect of this correction is to alter z_0 by -0.01 (since $y_{-1} = y_1$) and z_2 by -0.005 but to leave its own z value z_1 unaltered since z_k is not influenced by the term in h^2 to the number of decimals carried currently; in turn the changes are altered as follows: $z_0 - y_0$ by -0.01, $z_1 - y_1$ by 0.01, $z_2 - y_2$ by -0.005, and we can write down the new changes immediately without having to recalculate the z_k values. Continuing this procedure of making suitable corrections to the y_k, we try to decrease the magnitude of the changes as much as possible.

2. Special block relaxations. Again we first calculate the z_k values and the changes $z_k - y_k$ corresponding to the starting values for the y_k, but this time to one more decimal place. Working in fourth decimal units, we begin the relaxation by adding the correction $+94$ to y_0, which reduces the first change $z_0 - y_0$ to zero (apart from the error due to the neglect of the term in h^2, cf. method 1. above); the second change $z_1 - y_1$ is therefore altered to -107. We now reduce the remaining changes to zero successively by means of special block relaxations, i.e. relaxations with simultaneous identical corrections at neighbouring points (cf. Ch. V, § 1.6), whose effects depend on the following special property of the approximate equation $z_k - y_k = \frac{1}{2}(y_{k-1} + y_{k+1}) - y_k$: if identical corrections, say of $+2\alpha$, are made to the first r y_k values, only the r-th and $(r+1)$-th changes are altered, $z_r - y_r$ by $-\alpha$ and $z_{r+1} - y_{r+1}$ by α. Thus we remove the new change $z_1 - y_1 = -107$ by adding -214 to y_0 and y_1. At the same time $z_2 - y_2$ is altered by -107 to the new value $+21$ (see Table III/8), which can likewise be removed by adding $+42$ to y_0, y_1, y_2; and so on. By adding all the corrections we obtain new y_k values, from which we calculate new z_k values and then new changes; these are seen to have been reduced considerably, but now they all have the same sign and consequently the new y_k values still differ markedly from the required values. One would therefore repeat the process on the new y_k values with further repetitions if necessary. The method depends on the special form of the difference equation for this example but similar methods can be devised for many other types of difference equation.

3. *Bracketing and interpolation.* Here we aim to find two approximate solutions, one with only non-negative changes and the other with only non-positive changes. An approximation satisfying the later requirement has already been found by method 2. To find the other approximation we try new starting values $y_0 = 0.93$, $y_1 = \cdots$ (as in Table III/8) which are likely to be too small; the changes still have varying signs but one cycle of the procedure of method 2. yields new y_k values ($y_0 = 0.933$, $y_1 = \cdots$) for which all changes turn out positive. We obtain a new approximation (last row of Table III/8: $y_0 = 0.93798$, $y_1 = \cdots$) with very small changes simply by interpolating between the approximations with all positive and all negative changes, respectively. The bracketing nature of this method allays the fear that the approximate values still differ considerably from the exact values, a fear which must otherwise — with the first two methods, for example — always exist.

Example II. A non-linear boundary-value problem. Suppose that we are required to find the steady temperature distribution in a homogeneous rod of length l in which, as a consequence of a chemical reaction say or some such heat-producing process, heat is generated at a rate $f(y)$ per unit time per unit length, $f(y)$ being a given function of the excess temperature y of the rod over the temperature of the surroundings. If the ends of the rod, $x = 0$ and $x = l$, are kept at given temperatures, we are to solve the first boundary-value problem

$$y'' = -c\, f(y); \qquad y(0) = y_0, \qquad y(l) = y_l;$$

dashes denote differentiation with respect to x, which is measured along the axis of the rod, and c is a given constant.

For this example we choose an exponential law $c f(y) = 1 + e^y$ for the heat generation and $y(0) = 0$, $y(1) = 1$ as boundary conditions.

The difference equations here are very similar in form to the corresponding equations (1.17) for Example I: with $h = 1/m$ they read

$$y_k = z_k, \quad \text{where} \quad z_k = \frac{y_{k+1} + y_{k-1}}{2} + \frac{h^2}{2}(1 + e^{y_k}), \quad (k = 1, 2, \ldots, m-1),$$

$$y_0 = 0, \qquad y_m = 1.$$

We first get a rough idea of the solution with $h = \tfrac{1}{2}$, deriving from the single equation

$$e^{y_1} = 8 y_1 - 5$$

the two real solutions $y_1 = \begin{cases} 0.9474 \\ 2.903 \,. \end{cases}$

We shall restrict attention to the stable temperature distribution. This corresponds here to the solution with the smaller y value, from which we obtain by graphical interpolation the starting values $y_1 = 0.5$, $y_2 = \cdots$, $y_4 = 1.1$ (cf. Table III/9) for a relaxation solution with $h = \frac{1}{5}$.

The relaxation of these y_k values proceeds along essentially the same lines as for the linear boundary-value problem, the only marked difference being that on account of the non-linearity the new y_k values must be written down and from them the new z_k values calculated each time a group of corrections is made. Table III/9 needs little explanation as it is set out in the same way as Table III/8; after the first few corrections, the individual steps are omitted and sequences of corrections combined and written effectively as group corrections. By continuing the relaxation we could obtain a still more accurate solution of the difference equations but this would be rather pointless in view of the limitations of the ordinary finite-difference approximation.

§2. Refinements of the ordinary finite-difference method

In solving problems numerically by the "ordinary finite-difference method" described in § 1, we have seen that a very good idea of the behaviour of the solution can generally be obtained by using a large pivotal interval h with the attendant advantages of relatively short calculations and fewer unknowns. However, we have also seen that refining the subdivision $(h \to 0)$ is a slowly convergent process so that to obtain accurate values one would have to use a very small pivotal interval. The amount of labour is greatly increased thereby, and consequently for accurate work the ordinary finite-difference method does not compare favourably with other methods. We therefore describe now various ways of improving the finite-difference method.

2.1. Improvement by using finite expressions which involve more pivotal values

We can speed up the convergence of the finite-difference method by replacing the derivatives by "finite expressions" which are more general than the difference quotients of § 1.1 and represent the derivatives more accurately. For example, the departure of the expression

$$\frac{1}{12h} \left(-y_{i+2} + 8y_{i+1} - 8y_{i-1} + y_{i-2} \right) \tag{2.1}$$

from $y'(x_i)$ amounts at most to $C_1 h^4$, where $C_1 = \frac{1}{30} |y^{(5)}|_{\max \text{ in } \langle x_{i-2}, x_{i+2} \rangle}$, whereas the departure of the difference quotient

$$\frac{1}{2h} \left(y_{i+1} - y_{i-1} \right)$$

from $y'(x_i)$ can be as large as $C_2 h^2$, where $C_2 = \frac{1}{6} |y'''|_{\max \text{ in } \langle x_{i-1}, x_{i+1} \rangle}$.

Such finite expressions are easily derived by means of TAYLOR's theorem. For example, if we put

$$y'_i \approx \sum_{\varrho = -2}^{+2} c_\varrho \, y_{i+\varrho}, \tag{2.2}$$

where the c_ϱ are constants to be determined, and expand each term of the sum into a Taylor series:

$$y_{i+\varrho} = y_i + \varrho h y_i' + \frac{\varrho^2 h^2}{2!} y_i'' + \cdots,$$

we obtain for the sum the expansion

$$y_i' \approx y_i \sum_{\varrho=-2}^{2} c_\varrho + h y_i' \sum_{\varrho=-2}^{2} \varrho c_\varrho + \frac{h^2}{2!} y_i'' \sum_{\varrho=-2}^{2} \varrho^2 c_\varrho + \frac{h^3}{3!} y_i''' \sum_{\varrho=-2}^{2} \varrho^3 c_\varrho + \cdots,$$

and by solving the five linear equations

$$\sum_{\varrho=-2}^{2} c_\varrho = \sum_{\varrho=-2}^{2} \varrho^2 c_\varrho = \sum_{\varrho=-2}^{2} \varrho^3 c_\varrho = \sum_{\varrho=-2}^{2} \varrho^4 c_\varrho = 0, \qquad \sum_{\varrho=-2}^{2} \varrho c_\varrho = \frac{1}{h}$$

for the five unknown coefficients c_ϱ we can make the expansion reduce to y_i' apart from terms in h^5 and higher powers of h. The values of c_ϱ which we obtain yield the finite expression (2.1).

2.2. Derivation of finite expressions

We now generalize the method described in § 2.1. First of all we introduce the following

Definition: *Consider the n-th order homogeneous linear differential expression*

$$L[y] = \sum_{\nu=0}^{n} f_\nu(x) y^{(\nu)} \tag{2.3}$$

in y with given continuous functions $f_\nu(x)$ as coefficients. Then the linear combination

$$A = \sum_{k=1}^{p} C_k y(x_i + \alpha_k h), \tag{2.4}$$

where C_k, α_k are constants, is called a finite expression of the r-th approximation for the differential expression $L[y]$ at the point $x = x_i$ if the factors multiplying the quantities $y^{(\nu)}(x_i)$ in the Taylor expansion of the expression A are $f_\nu(x_i)$ for $0 \leq \nu \leq n$ and zero for $n+1 \leq \nu \leq n+r$.

We then state the following

Theorem: *A finite expression of the r-th approximation (in fact, infinitely many) can be derived for any given linear differential expression $L[y]$ (2.3) at any point $x = \xi$ and for any non-negative integer r. For an arbitrary choice of $q = r+n+1$ distinct points $\xi + \alpha_k h$ $(k = 1, 2, \ldots, q)$, q quantities C_k can be determined by solving a system of linear equations so that, for every function $u(x)$ with a continuous q-th derivative,*

$$A = \sum_{k=1}^{q} C_k u(\xi + \alpha_k h) = L[u(\xi)] + D \vartheta \frac{h^{r+1} |u^{(q)}|_{\max}}{q!}, \tag{2.5}$$

where $|\vartheta| \leq 1$, D is a polynomial in h of at most the n-th degree and depends on the choice of the α_i but not on u, and $|u^{(q)}|_{\max}$ is the absolute maximum of the q-th derivative of u in an interval containing all the points $\xi + \alpha_k h$.

To prove this we need only apply TAYLOR's theorem to the expression on the left-hand side of (2.5). This gives

$$
\begin{aligned}
A = \quad & u(\xi) \left[\quad C_1 + \cdots + \quad C_q \right] + \\
& + h\, u'(\xi) \left[\alpha_1 \quad C_1 + \cdots + \alpha_q \quad C_q \right] + \\
& + \frac{h^2}{2!}\, u''(\xi) \left[\alpha_1^2 \quad C_1 + \cdots + \alpha_q^2 \quad C_q \right] + \cdots + \\
& + \frac{h^{n+r}}{(n+r)!}\, u^{(n+r)}(\xi) \left[\alpha_1^{n+r} C_1 + \cdots + \alpha_q^{n+r} C_q \right] + R,
\end{aligned}
$$

where the remainder term R may be written in the form

$$
R = \frac{h^{n+r+1}}{(n+r+1)!}\, \vartheta \cdot \left| u^{(n+r+1)} \right|_{\max} \sum_{\varrho=1}^{q} \left| \alpha_\varrho^{n+r+1} C_\varrho \right|
$$

with $|\vartheta| \leq 1$. Comparison of this expansion with (2.3) yields $(n+r+1)$ linear equations for the C_ϱ, namely

$$
\sum_{\varrho=1}^{q} \alpha_\varrho^k C_\varrho = \begin{cases} \dfrac{k!}{h^k}\, f_k(\xi) & \text{for} \quad 0 \leq k \leq n, \\ 0 & \text{for} \quad n+1 \leq k \leq n+r. \end{cases} \tag{2.6}
$$

These equations always possess a solution, for the determinant of their coefficients

$$
\begin{vmatrix} 1 & 1 & \ldots 1 \\ \alpha_1 & \alpha_2 & \ldots \alpha_q \\ \alpha_1^2 & \alpha_2^2 & \ldots \alpha_q^2 \\ \cdots\cdots\cdots\cdots\cdots \\ \alpha_1^{r+n} & \alpha_2^{r+n} & \ldots \alpha_q^{r+n} \end{vmatrix} \tag{2.7}
$$

is a Vandermonde determinant, which, being the product of the differences of the distinct numbers $\alpha_1, \ldots, \alpha_q$, is never zero[1].

Some simple finite expressions for the lower order derivatives using equidistant points $\xi + kh$ are listed in Table III of the appendix. From them finite expressions for any linear differential expression of up to the fourth order can be obtained by superposition[2].

---

[1] See, for instance, PERRON, O.: Algebra, Vol. I, p. 92. 3rd ed. Berlin and Leipzig 1951, or AITKEN, A. C.: Determinants and matrices, 6th ed. p. 41. London: Oliver & Boyd 1949.

[2] Closed form solutions of the system of equations (2.6) for $n = 1$ and $n = 2$ can be found in L. COLLATZ: Das Differenzenverfahren mit höherer Approximation für lineare Differentialgleichungen. Schr. Math. Sem. u. Inst. Angew. Math. Univ. Berlin 3, 1—34 (1935), and for general n in E. PFLANZ: Über die Bildung finiter Ausdrücke für die Lösung linearer Differentialgleichungen. Z. Angew. Math. Mech. 17, 296—300 (1937).

Expressions for non-equidistant pivotal points are given by E. PFLANZ: Allgemeine Differenzenausdrücke für die Ableitungen einer Funktion $y(x)$. Z. Angew. Math. Mech. 29, 379—381 (1949).

2.3. The finite-difference method of a higher approximation

This method follows precisely the same lines as for the ordinary finite-difference method described in § 1 (which can be regarded as a first approximation) only now the derivatives are replaced by the finite expressions of a higher approximation given in Table III of the appendix. However, on account of the extra unknowns appearing in each equation based on the higher approximation, there are always more unknowns than such equations and further equations of a lower order of approximation have to be brought in at the boundary points; these can be equations corresponding to the differential equation, for example. This device is in fact used in the following example.

Example I. For the strut problem of Example I, § 1.2, i.e.

$$y'' + (1 + x^2)\, y + 1 = 0, \qquad y(\pm 1) = 0,$$

the finite equations of the third approximation (see Table III of the appendix) for the points $x = 0$ and $x = \tfrac{1}{2}$ of the coarse subdivision with $h = \tfrac{1}{2}$ read

$$\frac{-0 + 16 Y_1 - 30 Y_0 + 16 Y_1 - 0}{12 h^2} + 1 \cdot Y_0 + 1 = 0,$$

$$\frac{-Y_3 + 16 \cdot 0 - 30 Y_1 + 16 Y_0 - Y_1}{12 h^2} + \frac{5}{4} Y_1 + 1 = 0;$$

here we have already used the symmetry condition $Y_i = Y_{-i}$ and the boundary condition $Y_2 = 0$. To eliminate Y_3 we bring in another equation, namely the ordinary difference equation for the point $x = 1$:

$$\frac{Y_3 - 2 \cdot 0 + Y_1}{h^2} + 2 \cdot 0 + 1 = 0.$$

Solution of these equations yields

$$Y_0 = \frac{731}{787} = 0 \cdot 928\,844,$$

$$Y_1 = \frac{543}{787} = 0 \cdot 689\,962.$$

The more accurate values obtained by taking $h = \tfrac{1}{3}$ are

$$x = 0, \qquad Y_0 = \frac{2\,723\,933}{2\,924\,440} = 0 \cdot 931\,437,$$

$$x = \frac{1}{3}, \qquad Y_1 = \frac{2\,410\,848}{2\,924\,440} = 0 \cdot 824\,379,$$

$$x = \frac{2}{3}, \qquad Y_2 = \frac{296\,031}{584\,888} = 0 \cdot 506\,133.$$

Example II. An eigenvalue problem. For the eigenvalue problem

$$(1 + x)\, y'' + y' + \lambda (1 + x)\, y = 0, \qquad y'(0) = y(1) = 0,$$

treated in Example III of § 1.2, the equation for the point $x = \frac{3}{5}$ (with $h = \frac{2}{5}$) reads

$$\frac{5}{6}(-Y_3 - 30Y_1 + 16Y_0 - Y_0) + \frac{5}{24}(-Y_3 - 7Y_0) + \frac{8}{5}\Lambda Y_1 = 0$$

with the notation of Fig. III/7 (Y_3 is the approximate value of $y(\frac{7}{8})$).

The unknowns Y_0 and Y_3 are eliminated by using the ordinary difference equations at the points $x = \frac{1}{5}$ and $x = 1$:

$$\frac{15}{2}(Y_1 - Y_0) + \frac{5}{4}(Y_1 - Y_0) + \frac{6}{5}\Lambda Y_0 = 0,$$

$$\frac{25}{2}(Y_3 + Y_1) + \frac{5}{4}(Y_3 - Y_1) = 0, \quad \text{i.e.} \quad Y_3 = -\frac{9}{11}Y_1.$$

For $\mu = \frac{24}{25}\Lambda$ we obtain the equation

$$\begin{vmatrix} -7 + \mu & 7 \\ 583 & -1275 + 88\mu \end{vmatrix} = 0 \quad \text{and from it} \quad \Lambda = \begin{cases} 3\cdot0968 & (\text{error} \quad - \quad 3\cdot8\%) \\ 19\cdot29 & (\text{error} \quad -16 \quad \%). \end{cases}$$

2.4. Basic formulae for Hermitian methods[1]

Here we consider another method of setting up finite-difference equations of greater accuracy, i.e. with truncation errors of a higher order than the ordinary finite-difference equations. The equations we obtain are often particularly convenient for differential equations of simple form, but sometimes the practical application of the method is rather complicated. *The gain in accuracy over the ordinary method is obtained, not by including more pivotal values, as in the method of a higher approximation, but by basing the derivation of each individual difference equation on the fact that the differential equation is satisfied at several points, rather than just one as in the other methods.* This gives rise to formulae which involve the values of derivatives at more than one point in addition to the values of the function, and by analogy with HERMITE's interpolation formula we may call them Hermitian formulae and methods based upon them Hermitian methods (see footnote); as an additional justification for this terminology it may be noted that

---

[1] *Translator's note.* The German name "Mehrstellenverfahren" does not translate conveniently, for "more-point methods" would be rather misleading. The name preferred here is suggested by the use of the term "Lagrangian methods" for methods involving formulae expressed in terms of function values — by analogy with LAGRANGE's interpolation formula; here the analogy is with HERMITE's interpolation formula, which, in addition to the values of the function, also uses the values of the derivative at several points:

$$f(x) = \sum f(x_i)\,h_i(x) + \sum f'(x_i)\,H_i(x) + R;$$

cf. HOUSEHOLDER, A. S.: Principles of numerical analysis, p. 194. New York: McGraw-Hill 1953. We do not imply that the method described here for solving differential equations is directly due to HERMITE.

HERMITE's generalization of TAYLOR's theorem (cf. Ch. I, § 2.5) provides another example of such a formula. To be more specific, the method is based on certain expressions of the form

$$P = \sum_{\nu=-p}^{p} (a_\nu y_{i+\nu} + A_\nu y_{i+\nu}^{(k)}); \tag{2.8}$$

these are derived once and for all (cf. Table III of the appendix).

We form linear combinations of the values of y and its k-th derivative at neighbouring pivotal points $x_{i+\nu}$ and determine the weighting factors a_ν, A_ν so that the coefficients of powers of h in the Taylor series for the expression P centered on the point x_i all vanish to as high a power of h as possible. Consider, for example, the case $k=1$ for the first derivative. The ordinary difference quotient (1.1) for this case yields the relation

$$y_{i+1} - y_{i-1} - 2h y_i' \approx 0,$$

and by substituting the Taylor series for y_{i+1} and y_{i-1} centered on x_i we find the error term to be proportional to $h^3 y'''$. To obtain a formula with an error term of higher order without introducing more pivotal points, it is natural to try bringing in the extra "information" provided by the values y_{i+1}' and y_{i-1}'. We take $i=0$ for simplicity and try the general form

$$P = a_{-1} y_{-1} + a_0 y_0 + a_1 y_1 + A_{-1} y_{-1}' + A_0 y_0' + A_1 y_1';$$

then with each term expanded into a Taylor series centered on x_0 we have

$$P = \begin{cases} y_0 \ (a_{-1} + a_0 + a_1) \\ + \ h y_0' \left(-a_{-1} \quad + a_1 + \frac{1}{h} [\quad A_{-1} + A_0 + A_1] \right) \\ + \frac{1}{2!} h^2 y_0'' \left(\quad a_{-1} \quad + a_1 + \frac{2}{h} [-A_{-1} \quad + A_1] \right) \\ + \frac{1}{3!} h^3 y_0''' \left(-a_{-1} \quad + a_1 + \frac{3}{h} [\quad A_{-1} \quad + A_1] \right) \\ + \frac{1}{4!} h^4 y_0^{IV} \left(\quad a_{-1} \quad + a_1 + \frac{4}{h} [-A_{-1} \quad + A_1] \right) + \cdots. \end{cases}$$

By putting the quantities inside the round brackets equal to zero we obtain a set of linear equations for the unknowns a_ν, A_ν. Since the equations are homogeneous, a factor remains undetermined and we therefore solve in terms of one of the unknowns, say a_1:

$$a_{-1} = -a_1, \quad a_0 = 0, \quad A_{-1} = A_1 = -\frac{h}{3} a_1 = \frac{1}{4} A_0.$$

Thus we obtain the formula

$$\frac{P}{a_1} = y_1 - y_{-1} - \frac{h}{3}(y_1' + 4y_0' + y_{-1}') = 0 + 0(h^4). \qquad (2.9)$$

We treat the general case of a derivative of arbitrary order k in a similar fashion. Thus by substituting in (2.8) the Taylor series centered on the point x_i we have

$$\left.\begin{array}{l} P = \sum_{\nu=-p}^{p}\left\{a_\nu\left(y_i + \nu h\, y_i' + \nu^2 \dfrac{h^2}{2!}\, y_i'' + \cdots\right) + \\[2mm] \qquad\qquad + A_\nu\left(y_i^{(k)} + \nu h\, y_i^{(k+1)} + \nu^2 \dfrac{h^2}{2!}\, y_i^{(k+2)} + \cdots\right)\right\}, \end{array}\right\} \qquad (2.10)$$

and if we define

$$b_\nu = \frac{k!}{h^k}\, A_\nu, \qquad (2.11)$$

the requirement that the factors multiplying y_i, y_i', y_i'', ... up to as high an order as possible, say the $(k+r)$-th, shall be zero leads to the equations

$$\left.\begin{array}{ll} \sum\limits_{\nu=-p}^{p} a_\nu \nu^\varkappa = 0 & (\varkappa = 0, 1, 2, \ldots, k-1) \\[4mm] \sum\limits_{\nu=-p}^{p}\left\{a_\nu \nu^\varkappa + b_\nu \dbinom{\varkappa}{k}\nu^{\varkappa-k}\right\} = 0 & (\varkappa = k, k+1, \ldots, k+r) \end{array}\right\} \qquad (2.12)$$

for the a_ν and b_ν.

Clearly we do not want an expression P for which all the b_ν vanish. This can be avoided by demanding that b_0, say, i.e. the coefficient corresponding to the point x_i, does not vanish. Since a factor is indeterminate in P, we can put $b_0 = 1$; then (2.12) becomes an inhomogeneous system of $(k+r+1)$ linear equations for the unknowns a_ν, b_ν. If the number of these unknowns is sufficiently large, which can be arranged merely by choosing p sufficiently large, arbitrarily many solutions will exist. [The existence of infinitely many solutions follows from the fact that to any finite expression derived as in § 2.2 there corresponds a solution of (2.12).] Hermitian expressions of the form (2.8) for the lower order derivatives (up to the fourth order) are given in Table III of the appendix.

2.5. The Hermitian method in the general case

Consider the differential equation[1,2,3]

$$y^{(n)} = f(x, y, y', y'', \ldots, y^{(n-1)}). \qquad (2.13)$$

---

[1] By using the associated GREEN's function E. J. NYSTRÖM: Zur numerischen Lösung von Randwertaufgaben bei gewöhnlichen Differentialgleichungen. Acta Math. (Stockh.) **76**, 157—184 (1944), has developed a special Hermitian method

Here we use formula (2.8) with $k = n$ and write down the equations

$$\sum_{\nu=-p}^{p} (a_\nu Y_{i+\nu} + A_\nu Y_{i+\nu}^{(n)}) = 0 \qquad (2.14)$$

for all pivotal points inside the interval, say $x_1, x_2, \ldots, x_{q-1}$.

We now substitute throughout for the $Y_\nu^{(n)}$ in terms of the lower derivatives $Y_\nu^{(n-1)}, Y_\nu^{(n-2)}, \ldots, Y_\nu', Y_\nu$ by using the differential equation. The $Y_\nu^{(s)}$ $(1 \leq s \leq n-1)$ appear now as further unknowns; we therefore write down for each interior pivotal point the formula corresponding

for the problem $y'' = f(x, y)$, $y(x_a) = y_a$, $y(x_b) = y_b$; he gives formulae for one to four interior pivotal points (not necessarily equidistant), in particular, the following formulae for three equidistant interior points $x_i = x_a + ih$ $(i = 1, 2, 3)$, where $h = \frac{1}{4}(x_b - x_a)$:

$$y_1 = \frac{3y_0 + y_4}{4} - \frac{h^2}{480} [27 y_0'' + 332 y_1'' + 222 y_2'' + 132 y_3'' + 7 y_4''] + R_1,$$

$$y_2 = \frac{y_0 + y_4}{2} - \frac{h^2}{30} [y_0'' + 16 y_1'' + 26 y_2'' + 16 y_3'' + y_4''] \qquad + R_2,$$

$$y_3 = \frac{y_0 + 3y_4}{4} - \frac{h^2}{480} [7 y_0'' + 132 y_1'' + 222 y_2'' + 332 y_3'' + 27 y_4''] + R_3,$$

where $y_i = y(x_i)$ and R_1, R_2, R_3 are remainder terms.

[2] For second-order equations F. STÜSSI: Numerische Lösung von Randwert-problemen mit Hilfe der Seilpolygongleichung. Z. Angew. Math. Phys. **1**, 53−70 (1950), has derived the Hermitian formula

$$y_{i-1} - 2 y_i + y_{i+1} \approx \frac{h^2}{12} (y_{i-1}'' + 10 y_i'' + y_{i+1}'')$$

purely from mechanical considerations based on the funicular polygon of graphical statics; he had already used the formula in 1935: Die Stabilität des auf Biegung beanspruchten Trägers. Abh. Internat. Vereinig. f. Brückenbau u. Hochbau **3**, Zürich 1935, pp. 401−420, in particular p. 413. In numerous other works he has also derived special Hermitian formulae for ordinary differential equations of the fourth order and the biharmonic equation, always using an approach based purely on ideas from statics; see also F. STÜSSI: Baustatik, Vol. I. Basel 1946. The Hermitian formula for the second derivative quoted here can also be found in SCH. E. MIKELADZE: Über die Lösung von Randwertproblemen mit der Differenzen-methode. C. R. (Doklady) Acad. Sci. URSS. **28**, 400−402 (1940) (Russian), where formulae of the form

$$y(a + \alpha h) \pm y(a - \alpha h) = \sum_{\lambda=0}^{n-1} \frac{1 \pm (-1)^\lambda}{\lambda!} \alpha^\lambda h^\lambda y^{(\lambda)}(a) +$$

$$+ \frac{h^n}{(n-1)!} \sum_{i=0}^{r} A_i [y^{(n)}(a + t_i h) + y^{(n)}(a - t_i h)] + \text{remainder term}$$

are used; for $n > 2$ these are of a somewhat different form from those given in the present book.

[3] A variant of the Hermitian method originally due to NUMEROV has been applied to the special equation $y'' = q(x) y$ (cf. G. STRACKE: Bahnbestimmung der Planeten und Kometen, § 77. Berlin 1929). In this method a system of equations

to (2.14) for all derivatives which occur in the right-hand side of the differential equation. With the corresponding constants a_ν, A_ν distinguished by dashes these equations read

$$
\left.
\begin{aligned}
\sum_\nu [a'_\nu Y_{i+\nu} + A'_\nu Y^{(n-1)}_{i+\nu}] &= 0, \\
\sum_\nu [a''_\nu Y_{i+\nu} + A''_\nu Y^{(n-2)}_{i+\nu}] &= 0, \\
\cdots \cdots \cdots \cdots \cdots \cdots \\
\sum_\nu [a^{(n-1)}_\nu Y_{i+\nu} + A^{(n-1)}_\nu Y'_{i+\nu}] &= 0.
\end{aligned}
\right\}
\qquad (2.15)
$$

The boundary points require special attention. Here it may be necessary to use unsymmetric formulae in order to eliminate surplus unknowns. This is illustrated in Example II below in § 2.6 (see also § 2.7).

2.6. Examples of the Hermitian method

I. Inhomogeneous problem of the second order. Let us consider again the strut problem of Example I, § 1.2, i.e.

$$
y'' + (1 + x^2)\, y + 1 = 0, \qquad y(\pm 1) = 0,
$$

and use first the interval $h = \tfrac{1}{2}$ (Fig. III/3). From Table III of the appendix we obtain the equations

$$
Y_{i+1} - 2 Y_i + Y_{i-1} - \frac{h^2}{12}(Y''_{i+1} + 10 Y''_i + Y''_{i-1}) = 0 \qquad (i = 0, 1) \qquad (2.16)
$$

and from the differential equation

$$
Y''_i = -(1 + i^2 h^2)\, Y_i - 1.
$$

is set up for approximate pivotal values R_j of the function

$$
r(x) = y(x) - \frac{h^2}{12} y''(x) = y\left(1 - \frac{h^2}{12} q\right)
$$

instead of for approximate pivotal values Y_j of the solution $y(x)$. By Taylor's theorem we can show that the second central-difference of $r(x)$ has the expansion

$$
\delta^2 r(x) = r(x+h) - 2r(x) + r(x-h) = h^2 y''(x) - \frac{1}{240} h^6 y^{VI}(x) +
$$
$$
+ \text{ higher order terms.}
$$

Now $y'' = q y = q r \left(1 - \dfrac{h^2}{12} q\right)^{-1}$, so that if we retain only the first term on the right-hand side of the above formula, we obtain the difference equation

$$
\delta^2 R_j = R_{j+1} - 2 R_j + R_{j-1} = -\frac{h^2 q}{1 - \dfrac{h^2}{12} q}\, R_j
$$

to be satisfied by the approximate pivotal values of $r(x)$. STRACKE uses this formula for initial-value problems, but obviously it can also be used for boundary-value problems.

In addition we have the boundary condition $Y_2 = 0$ and symmetry condition $Y_{-i} = Y_i$, so that finally, after substitution, we have the two equations

$$2Y_1 - 2Y_0 + \frac{1}{48}\left(12 + 10Y_0 + \frac{5}{2}Y_1\right) = 0,$$

$$-2Y_1 + Y_0 + \frac{1}{48}\left(12 + Y_0 + \frac{25}{2}Y_1\right) = 0$$

for Y_0 and Y_1. Their solution is

$$Y_0 = \frac{4368}{4709} \approx 0.927585; \qquad Y_1 = \frac{3240}{4709} \approx 0.688044.$$

As in § 1.1, we treat the problem as an initial-value problem when the chosen pivotal interval is small; thus with $h = \frac{1}{5}$ we use the equations

$$300\,(2Y_1 - 2Y_0) \qquad\quad + 12 + 2.08\,Y_1 + 10 \quad Y_0 \qquad\qquad = 0,$$
$$300\,(Y_2 - 2Y_1 + Y_0) + 12 + 1.16\,Y_2 + 10.4\,Y_1 + \qquad Y_0 = 0,$$
$$300\,(Y_3 - 2Y_2 + Y_1) + 12 + 1.36\,Y_3 + 11.6\,Y_2 + 1.04\,Y_1 = 0,$$
$$300\,(Y_4 - 2Y_3 + Y_2) + 12 + 1.64\,Y_4 + 13.6\,Y_3 + 1.16\,Y_2 = 0,$$
$$300\,(Y_5 - 2Y_4 + Y_3) + 12 + 2 \quad Y_5 + 16.4\,Y_4 + 1.36\,Y_3 = 0$$

to express $Y_1, Y_2, \dots$ successively in terms of Y_0:

$$Y_1 = -\quad 0.0199309 + \quad 0.9799362\,Y_0,$$
$$Y_2 = -\quad 0.0788659 + \quad 0.9190145\,Y_0,$$
$$Y_3 = -\quad 0.1738941 + \quad 0.8154636\,Y_0,$$
$$Y_4 = -\quad 0.2990990 + \quad 0.6677412\,Y_0,$$
$$302\,Y_5 = -134.1494 \quad + 143.9457 \quad Y_0.$$

The boundary condition $Y_5 = 0$ then yields the value

$$Y_0 = 0.9319450,$$

from which we calculate the remaining Y_i:

$$Y_1 = 0.8933157,$$
$$Y_2 = 0.7776050,$$
$$Y_3 = 0.5860732,$$
$$Y_4 = 0.3231992.$$

As before (§ 1.2) we can use the values Y, Y^* calculated with two different intervals, h and h^* say, to obtain improved values $\bar{Y}$ by a formula similar to (1.8). Here, however, we assume that the error is of order h^4 (cf. § 3.3), so that the corresponding formula is

$$\overline{Y} = Y + \frac{h^4}{h^{*4} - h^4} (Y - Y^*); \tag{2.17}$$

with $h = \frac{1}{5}$, $h^* = \frac{1}{2}$ we obtain for $x = 0$

$$\overline{Y} = 0{\cdot}931\,945 + \frac{(\frac{1}{5})^4}{(\frac{1}{2})^4 - (\frac{1}{5})^4} (0{\cdot}931\,945 - 0{\cdot}9276) = 0{\cdot}932\,060.$$

II. An eigenvalue problem. With $h = \frac{2}{5}$ (notation as in Fig. III/7) we calculate two eigenvalues of the longitudinal vibration problem

$$- (1 + x) y'' - y' = \lambda (1 + x) y; \qquad y'(0) = y(1) = 0$$

of Example III, § 1.2. First of all we write down the Hermitian equation for the second derivative at the interior points $x = \frac{1}{5}$, $x = \frac{3}{5}$:

$$Y_1 - Y_0 - \frac{1}{75} (Y_1'' + 10 Y_0'' + Y_{-1}'') = 0,$$

$$- 2 Y_1 + Y_0 - \frac{1}{75} (Y_2'' + 10 Y_1'' + Y_0'') = 0.$$

Here we have used the lower order approximation $Y_0 - Y_{-1} = 0$ for the boundary condition $y'(0) = 0$; this condition also implies that $Y_0' = - Y_{-1}'$ to the same order of approximation, but no similar relation holds for the second derivatives. These relations are used below when we write down the differential equation at the point $x = -\frac{1}{5}$. The second derivatives occuring in the Hermitian equations above are expressed in terms of the lower derivatives by writing down the differential equation for each of the four pivotal points:

$$Y_2'' + \frac{1}{2} Y_2' = 0, \qquad\qquad Y_1'' + \frac{5}{8} Y_1' + \Lambda Y_1 = 0,$$

$$Y_0'' + \frac{5}{6} Y_0' + \Lambda Y_0 = 0, \qquad Y_{-1}'' - \frac{5}{4} Y_0' + \Lambda Y_0 = 0;$$

Λ is an approximate value for λ.

We still have to eliminate the three remaining first derivatives Y_0', Y_1', Y_2' and for this we use Hermitian formulae for the first derivative — symmetric formulae for the points $x = \frac{1}{5}, \frac{3}{5}$ and a lower order unsymmetric formula (see Table III of the appendix) for the boundary point $x = 1$:

$$Y_1 - Y_0 - \frac{2}{15} (Y_1' + 3 Y_0') = 0,$$

$$- Y_0 - \frac{2}{15} (Y_2' + 4 Y_1' + Y_0') = 0,$$

$$- \frac{2}{3} Y_1 - \frac{2}{15} (Y_2' + Y_1') = 0.$$

The eliminations yield finally two linear homogeneous equations for Y_0 and Y_1, namely

$$(- 6870 + 11\,\mu)\,Y_0 \qquad + (6995 + \mu)\,Y_1 = 0,$$
$$(4812 + \mu)\,Y_0 + (- 11\,158 + 10\,\mu)\,Y_1 = 0,$$

where $384\Lambda = 5\mu$, and from the usual determinant condition for a non-trivial solution we obtain

$$\Lambda = \begin{cases} 3{\cdot}1677 & (\text{error} \; - 1{\cdot}6\%), \\ 21{\cdot}111 & (\text{error} \; - 6 \; \%). \end{cases}$$

2.7. A Hermitian method for linear boundary-value problems

For a "first" boundary-value problem whose differential equation does not involve y' explicitly, as in Example I of § 2.6, for instance, substitution from the differential equation into (2.14) immediately yields a set of equations for the Y_ν; on the other hand, if y' does appear in the differential equation, the set of equations derived from (2.14) and (2.15) contain the unknowns Y'_ν in addition to the Y_ν. We now describe how a set of equations for the Y_ν alone can be obtained directly for a linear boundary-value problem of the n-th order, even when the derivatives $y', \ldots, y^{(n-1)}$ all appear in the differential equation[1].

If the functions $f_k(x)$ in the linear differential equation

$$L[y] \equiv \sum_{k=0}^{n} f_k(x)\, y^{(k)} = r(x) \tag{2.18}$$

possess continuous derivatives of the k-th order, the equation may be written in the form

$$L[y] \equiv \sum_{k=0}^{n} [g_k(x)\, y]^{(k)} = r(x). \tag{2.19}$$

This may be shown very easily by construction, for if we define sets of functions $f_k^{[r]}$ successively by forming the sequence of differential expressions

$$L_1 = L - (f_n\, y)^{(n)} = \sum_{k=0}^{n-1} f_k^{[1]}\, y^{(k)},$$

$$L_2 = L_1 - (f_{n-1}^{[1]}\, y)^{(n-1)} = \sum_{k=0}^{n-2} f_k^{[2]}\, y^{(k)},$$

$$\cdots\cdots\cdots\cdots\cdots\cdots\cdots\cdots\cdots$$

$$L_n = L_{n-1} - (f_1^{[n-1]}\, y)' = f_0^{[n]}\, y = (f_0^{[n]}\, y)^{(0)},$$

which decrease in order from $(n-1)$ down to zero, then the result follows immediately by addition, and $g_k = f_k^{[n-k]}$ with $f_n^{[0]} = f_n$. For $n = 2$ we have

$$L[y] \equiv (f_2\, y)'' + [(f_1 - 2f'_2)\, y]' + (f_0 - f'_1 + f''_2)\, y = r(x). \tag{2.20}$$

Instead of the formula (2.15) for the lower derivatives, we use here Hermitian formulae which, though very similar, differ in that the coefficients of the derivative values (corresponding to $A'_\nu, A''_\nu, \ldots, A_\nu^{(n-1)}$) are the same for each derivative and equal to those in (2.14) for the

---

[1] Following a somewhat differently derived method due to H. SASSENFELD: Ein Summenverfahren für Rand- und Eigenwertaufgaben linearer Differentialgleichungen. Z. Angew. Math. Mech. **31**, 240—241 (1951); presented in detail and called "Quadraturverfahren" by R. ZURMÜHL: Praktische Mathematik für Ingenieure und Physiker, 2nd ed. p. 447 et seq. Berlin-Göttingen-Heidelberg 1957.

highest derivatives; thus we derive Hermitian equations of the form

$$\sum_{\nu=-p}^{p} (c_\nu^{(k)} Y_{i+\nu} + A_\nu Y_{i+\nu}^{(k)}) = 0 \qquad (k = 1, \ldots, n-1) \qquad (2.21)$$

(we may include the equation with $k = n$ if we put $c_\nu^{(n)} = a_\nu$). The coefficients $c_\nu^{(k)}$ are determined in the usual way by making the Taylor expansions of the left-hand sides vanish to as high an order as possible; for $n = 2$, for example, we establish the formulae

$$\left.\begin{aligned}
\frac{12}{h^2} (y_1 - 2y_0 + y_{-1}) - (y_1'' + 10 y_0'' + y_{-1}'') &= 0(h^4), \\
\frac{1}{2h} (-y_2 + 14 y_1 - 14 y_{-1} + y_{-2}) - (y_1' + 10 y_0' + y_{-1}') &= 0(h^4), \\
(y_1 + 10 y_0 + y_{-1}) - (y_1 + 10 y_0 + y_{-1}) &= 0.
\end{aligned}\right\} \qquad (2.22)$$

[The last equation is included for convenience in writing down equation (2.23) below for specific examples.]

We now form the expression

$$\sum_{\nu=-p}^{p} A_\nu (L[y])_{x=x_{i+\nu}}$$

with $L[y]$ in the form (2.19); by virtue of (2.21) we can replace

$$\sum_{\nu=-p}^{p} A_\nu [g_k y]_{x=x_{i+\nu}}^{(k)} \quad \text{by} \quad -\sum_{\nu=-p}^{p} c_\nu^{(k)} (g_k y)_{x=x_{i+\nu}};$$

hence, using the differential equation at the points $x_{i-p}, \ldots, x_{i+p}$, we obtain

$$\sum_{\nu=-p}^{p} \sum_{k=0}^{n} c_\nu^{(k)} (g_k Y)_{x=x_{i+\nu}} = -\sum_{\nu=-p}^{p} A_\nu r(x_{i+\nu}), \qquad (2.23)$$

which provides the required linear relation between the Y_ν.

The number of equations obtained by writing this equation down for successive pivotal points x_i together with equations representing the boundary conditions must always be less than the number of unknown pivotal values and consequently further equations are needed. These can be obtained by using unsymmetric expressions for points near the boundaries; for example, in the case $n = 2$ we can use the set of formulae[1]

$$\frac{12}{h^2} (y_2 - y_1 - y_0 + y_{-1}) - (y_2'' + 11 y_1'' + 11 y_0'' + y_{-1}'') = 0(h^4),$$

$$\frac{4}{h} (y_2 + 3 y_1 - 3 y_0 - y_{-1}) - (y_2' + 11 y_1' + 11 y_0' + y_{-1}') = 0(h^4),$$

$$(y_2 + 11 y_1 + 11 y_0 + y_{-1}) - (y_2 + 11 y_1 + 11 y_0 + y_{-1}) = 0.$$

---

[1] These, together with formulae for differential equations of the fourth order, are to be found in R. ZURMÜHL: Praktische Mathematik für Ingenieure und Physiker, 2nd ed. Berlin-Göttingen-Heidelberg 1957.

The boundary conditions themselves likewise require special formulae. If, for example, a boundary condition for a second-order problem specifies the value of y' or a linear combination of y and y' at a boundary, a representative finite equation can be derived from a formula such as

$$- 85\, y_0 + 108\, y_1 - 27\, y_2 + 4\, y_3 - 66\, h\, y_0' - 18\, h^2\, y_0'' = 0\, (h^5);$$

we substitute for y_0'' in terms of y_0' and y_0 by means of the differential equation and then for y_0' in terms of y_0 by means of the boundary condition (in conformity with the usual notation the y_i are to be replaced by the approximations Y_i when the remainder term is omitted).

§3. Some theoretical aspects of the finite-difference methods

3.1. Solubility of the finite-difference equations and convergence of iterative solutions

In §§ 1 and 2 we described various ways of setting up a system of linear equations for any given linear boundary-value problem. Here we consider the solubility of such a system but naturally have to restrict the generality — after all, not every linear boundary-value problem is soluble. If we focus attention on certain classes of boundary-value problems, a few additional assumptions suffice to enable us to prove the existence and uniqueness of the solution of the finite-difference equations.

We start with a simple example. For the boundary-value problem

$$\left.\begin{aligned} - (f\, y')' + g\, y &= r\,(x); \\ y\,(a) = y_a, \qquad y\,(b) &= y_b \end{aligned}\right\} \tag{3.1}$$

the ordinary finite-difference method using (1.5) yields the equations

$$\left.\begin{aligned} \frac{1}{h^2}\, [- f_{i+\frac12}\, Y_{i+1} + (f_{i+\frac12} + f_{i-\frac12})\, Y_i - f_{i-\frac12}\, Y_{i-1}] + g_i\, Y_i - r_i &= 0 \\ (i = 1, 2, \ldots, n-1) \\ Y_0 = y_a, \qquad Y_n = y_b. \end{aligned}\right\} \tag{3.2}$$

If we assume that $f(x) > 0$ and $g(x) \geq 0$, this system of equations satisfies the conditions of Theorem 2 in Ch. I, § 5.5 (in addition to the sign distribution, the weak row-sum criterion is satisfied and the matrix of coefficients does not decompose); hence the system possesses a uniquely determined solution for arbitrary values of y_a, y_b and the r_i. Further, by Theorem 3 in Ch. I, § 5.5 this solution may be computed iteratively in single or total steps, i.e. the single-step and total-step iterations converge.

We now describe another, quite different, way of showing that a unique solution of the system (3.2) exists. With this method the convergence of the single-step iteration follows at the same time. The method is also applicable to certain differential equations of higher order, for which, in general, the weak row-sum criterion is no longer satisfied. We use the fact that the system of $n-1$ equations in (3.2) (with Y_0, Y_n not included among the unknowns) is symmetric and identify it with the system of equations

$$\frac{\partial Q}{\partial Y_i} = 0 \qquad (i = 1, 2, \ldots, n-1)$$

which constitute the necessary conditions for a minimum of the quadratic function

$$Q = \frac{1}{2} \sum_{\nu=0}^{n-1} f_{\nu+\frac{1}{2}} \left(\frac{Y_{\nu+1} - Y_\nu}{h} \right)^2 + \sum_{\nu=1}^{n-1} \left(\frac{1}{2} g_\nu Y_\nu^2 - r_\nu Y_\nu \right) \tag{3.3}$$

of the $n-1$ variables $Y_1, Y_2, \ldots, Y_{n-1}$ (Y_0, Y_n being fixed at the values y_a, y_b).

Now under our assumptions that $f(x) > 0$ and $g(x) \geq 0$, the corresponding "homogeneous" quadratic form

$$Q^* = \frac{1}{2} \sum_{\nu=0}^{n-1} f_{\nu+\frac{1}{2}} \left(\frac{Y_{\nu+1} - Y_\nu}{h} \right)^2 + \frac{1}{2} \sum_{\nu=1}^{n-1} g_\nu Y_\nu^2 \quad \text{with} \quad Y_0 = Y_n = 0$$

obtained from Q by putting r_ν, Y_0, Y_n equal to zero is positive definite: obviously it cannot take negative values if the Y_ν are real and it is zero only when all the Y_ν are zero; for $Q^* = 0$ implies that

$$Y_{\nu+1} - Y_\nu = 0, \quad \text{so that} \quad Y_\nu = \text{constant},$$

and since Y_0, Y_n are zero here, the constant must be zero. The determinant of the quadratic form Q^* must therefore be positive. But this determinant is also the determinant of the system of equations (3.2) and hence that system has a unique solution for any values of r_ν, y_a, y_b. According to a well-known theorem[1], the fact that Q^* is positive definite implies also that the solution can be computed by single-step iteration.

If, instead of (3.1), we have more complicated boundary conditions, say

$$- A y(a) + y'(a) = R \tag{3.4}$$

at $x = a$, then the corresponding finite-difference equation reads

$$- A Y_0 + \frac{Y_1 - Y_{-1}}{2h} = R.$$

[1] Mises, R. v., and H. Pollaczek-Geiringer: Praktische Verfahren der Gleichungsauflösung. Z. Angew. Math. Mech. 9, 58–77 (1929).

If we also write down the difference equation of (3.2) with $i = 0$, i.e.

$$-f_{\frac{1}{2}} Y_1 + (f_{\frac{1}{2}} + f_{-\frac{1}{2}} + g_0 h^2) Y_0 - f_{-\frac{1}{2}} Y_{-1} = r_0 h^2,$$

we can eliminate Y_{-1} between it and the previous equation. The resulting relation

$$(f_{\frac{1}{2}} + \alpha) Y_0 - f_{\frac{1}{2}} Y_1 = \beta,$$

where

$$\alpha = \frac{(g_0 h^2 + 2h A f_{-\frac{1}{2}}) f_{\frac{1}{2}}}{f_{\frac{1}{2}} + f_{-\frac{1}{2}}} \quad \text{and} \quad \beta = \frac{(r_0 h^2 - 2h f_{-\frac{1}{2}} R) f_{\frac{1}{2}}}{f_{\frac{1}{2}} + f_{-\frac{1}{2}}},$$

between Y_0 and Y_1 can be added to the set of symmetric equations of (3.2) without disturbing the symmetry. Thus we can still identify the system with the minimum equations $\partial Q/\partial Y_i = 0$ $(i = 0, 1, \ldots, n-1)$ for a quadratic function Q. The appropriate quadratic function now has the form

$$Q = \frac{1}{2} \sum_{\nu=0}^{n-1} f_{\nu+\frac{1}{2}} \left(\frac{Y_{\nu+1} - Y_{\nu}}{h} \right)^2 + \frac{1}{2} \alpha \left(\frac{Y_0}{h} \right)^2 + \sum_{\nu=1}^{n-1} \left(\frac{1}{2} g_\nu Y_\nu^2 - r_\nu Y_\nu \right) - \frac{\beta}{h^2} Y_0.$$

The quadratic form Q^* which results when we put $Y_n = 0$, $r_\nu = 0$, $R = 0$ in Q will be positive definite if, in addition to the previous assumptions $f > 0$ and $g \geq 0$, we assume further that $A \geq 0$; for then $\alpha \geq 0$. This ensures also that the weak row-sum criterion is still satisfied; it will not be satisfied, however, if $g \equiv 0$ and y' is prescribed $(A = 0)$ at both endpoints (the "second" boundary-value problem).

In general the system of equations arising in the application of the Hermitian method of § 2.4 is not symmetric and the quadratic form approach is no longer available. In many cases, however, it can be shown that the weak row-sum criterion and the other conditions of Theorem 2 in Ch. I, § 5.5 are satisfied. As an example we choose the special boundary-value problem

$$-y'' + f(x) y = r(x); \quad y(a) = y_a, \quad y(b) = y_b.$$

With $h = \frac{b-a}{n}$, $x_i = a + ih$ the associated finite equations (cf. Ex. I, § 2.6) read

$$\left. \begin{array}{l} Y_{i+1} - 2Y_i + Y_{i-1} - \dfrac{h^2}{12} (Y_{i+1}'' + 10 Y_i'' + Y_{i-1}'') = 0 \\[2mm] Y_i'' = f_i Y_i - r_i; \quad Y_0 = y_a, \quad Y_n = y_b \end{array} \right\} \quad (i = 1, 2, \ldots, n-1)$$

and yield the system

$$Y_{i+1} \left(1 - \frac{h^2}{12} f_{i+1} \right) - Y_i \left(2 + \frac{h^2 \cdot 10}{12} f_i \right) + Y_{i-1} \left(1 - \frac{h^2}{12} f_{i-1} \right)$$

$$= -\frac{h^2}{12} (r_{i+1} + 10 r_i + r_{i-1}).$$

For $f(x) > 0$ and sufficiently small h, for example, $h < (12/f)^{\frac{1}{2}}$, the row-sum criterion is satisfied.

3.2. A general principle for error estimation with the finite-difference methods in the case of linear boundary-value problems

In principle we can always calculate error bounds for the finite-difference methods, for the refined methods of § 2 as well as for the ordinary method of § 1. *However, these bounds may be expected to be reasonably close to the actual error and to predict its order of magnitude correctly only when the solution is known fairly accurately; in particular, we need to have an approximate quantitive idea of the behaviour of the higher derivatives, which are critical for the error,* or if we are using maximum values of these derivatives, we must choose upper limits with as little over-estimation as possible. Although in general no error estimates exist which are both sufficiently accurate and sufficiently simple, we can at least describe a general procedure by which an error estimate is possible.

As far as we are concerned here the ordinary finite-difference method and the method of a higher approximation can be regarded as special cases of the Hermitian method of § 2.5, and we therefore consider a system of equations of Hermitian form. Thus all equations set up by the finite-difference methods for the N unknowns Y_i, Y_i', ..., $Y_i^{(n-1)}$ $(i = 1, ..., p$, say) constitute linear relations between these unknowns and the methods therefore lead to systems of equations of the form

$$A^{(\varrho)} = \sum_{\nu, k} a_{\nu, k}^{(\varrho)} Y_\nu^{(k)} = r^{(\varrho)} \qquad (\varrho = 1, 2, 3, ..., N). \qquad (3.5)$$

We naturally assume that these equations have been set up so that the determinant of the coefficients $a_{\nu, k}^{(\varrho)}$ does not vanish, for otherwise we could not have determined an approximate solution.

We now form the same expressions $A^{(\varrho)}$ for the exact solution $y(x)$ of the boundary-value problem, whose existence and uniqueness we also assume. We then follow the usual procedure of expanding each expression into a Taylor series in the values of y and its derivatives at a point $x = x_\varrho$ and use the differential equation and boundary conditions for raising the order of the remainder terms. The resulting equations are of the form

$$\sum_{\nu, k} a_{\nu, k}^{(\varrho)} y^{(k)} (x_\nu) = r^{(\varrho)} + \vartheta_\varrho h^{n_\varrho} D_\varrho y^{(n_\varrho)} (\xi_\varrho) \qquad (\varrho = 1, 2, ..., N), \qquad (3.6)$$

where $|\vartheta_\varrho| \leq 1$, D_ϱ and n_ϱ are determinable quantities, and ξ_ϱ is an unknown point within the range of pivotal points involved in the ϱ-th equation.

We now derive a system of equations for the errors

$$\varphi_\nu^{(k)} = Y_\nu^{(k)} - y^{(k)} (x_\nu) \qquad (3.7)$$

by subtracting equations (3.6) from (3.5):

$$\sum_{\nu,\,k} a_{\nu,\,k}^{(\varrho)}\, \varphi_\nu^{(k)} = -\,\vartheta_\varrho\, h^{n_\varrho}\, D_\varrho\, y^{(n_\varrho)}(\xi_\varrho). \tag{3.8}$$

Since the determinant was assumed above to be non-zero, these equations can be solved. Thus, in so far as the finite-difference method is applicable at all, the possibility of an error estimate depends on the fact that the systems (3.5) and (3.6) have the same determinant.

For the practical application of such an estimate we often put

$$\vartheta_\varrho\, y^{(n_\varrho)}(\xi_\varrho) = \vartheta_\varrho^*\, |y^{(n_\varrho)}|_{\max}, \tag{3.9}$$

then solve the equations (3.8) for the $\varphi_\nu^{(k)}$ and from the results derive upper bounds for the $\varphi_\nu^{(k)}$ using the fact that $|\vartheta^*|\leq 1$. Values for the derivatives $y^{(n_\varrho)}$ are obtained either by expressing these derivatives in terms of lower derivatives, which are known more accurately, by differentiating the differential equation (the order of the derivatives can be reduced still further by repeated integration of the differential equation) or, less accurately, by using the differences of the Y_ϱ in a finite-difference derivative formula.

This is as much as can be said in the general case. That more can be proved[1] if special assumptions are made is shown by the example in the next section.

3.3. An error estimate for a class of linear boundary-value problems of the second order

In illustration of the general procedure just described in § 3.2 we derive a general error estimate for the ordinary finite-difference method applied to the special class of boundary-value problems defined by

$$L[y] = y'' + f_1 y' + f_0 y = r(x); \quad y(a) = y_0, \quad y(b) = y_n, \tag{3.10}$$

where y_0, y_n are given boundary values, f_0, f_1, r are functions with continuous second derivatives and $f_0 \leq 0$ (for positive f_0 the boundary-value problem may not possess a solution). Let the interval h be chosen so small that

$$\left|\frac{h}{2} f_1\right| < 1 \quad \text{for} \quad a \leq x \leq b. \tag{3.11}$$

---

[1] For many boundary-value problems of monotonic type error estimates can be obtained in a quite different way which does not require knowledge of bounds for the higher derivatives; cf. L. COLLATZ: Aufgaben monotoner Art. Arch. Math. **3**, 375 (1952), in which a numerical example with error limits is also given.

The corresponding boundary-value problem for the difference equations (see § 1.1) reads

$$A[Y_i] = Y_{i+1} - 2Y_i + Y_{i-1} + f_{1,i}\frac{h}{2}(Y_{i+1} - Y_{i-1}) + f_{0,i}h^2 Y_i = h^2 r_i \left.\begin{matrix} \\ (i = 1, 2, \ldots, n-1), \\ \\ \end{matrix}\right\} \quad (3.12)$$
$$Y_0 = y_0, \quad Y_n = y_n.$$

Taylor expansion of $A[y_i]$ yields (cf. Table III of the appendix)

$$A[y_i] = h^2 L[y]_{x=x_i} + R_i, \quad (3.13)$$

where

$$|R_i| \leq R^* = \frac{h^4}{12}\left(2|f_1 y'''|_{\max} + |y^{(4)}|_{\max}\right). \quad (3.14)$$

From (3.12) and (3.13) the error $\varphi_i = Y_i - y_i$ satisfies the difference boundary-value problem

$$A[\varphi_i] = -R_i, \quad \varphi_0 = \varphi_n = 0. \quad (3.15)$$

This is a set of $n+1$ linear equations for the $n+1$ quantities φ_i with right-hand sides for which there exist upper limits as given by (3.14). From these equations we shall derive upper limits for the quantities $|\varphi_i|$. The matrix of coefficients satisfies the conditions of Theorem 2 in Ch. I, § 5.5 (sign distribution, weak row-sum criterion, non-decomposition), so that

$$|Y_i - y_i| = |\varphi_i| \leq w_i, \quad (3.16)$$

where the w_i are quantities determined from equations $-A[w_i] = S_i$ with $S_i \geq |R_i|$. Such quantities w_i can be obtained by solving the equations $-A[w_i] = 1$; for then we need only put $w_i = R^* w_i^*$, where R^* is defined in (3.14). In many cases another way is possible which avoids the necessity of solving a second system of equations. This depends on being able to find quantities w_i^* for which the numbers $-A[w_i^*] = \sigma_i$ all turn out positive (see the following numerical example); we can then write

$$|\varphi_i| \leq w_i = \frac{w_i^* R^*}{(\sigma_i)_{\min}}. \quad (3.17)$$

Numerical example. Let us consider a simple example for which we know the exact solution so that the error limits (3.17) can be compared with the actual errors. Such an example is provided by the boundary-value problem

$$y'' - \frac{2}{x^2}y + \frac{1}{x} = 0, \quad y(2) = y(3) = 0,$$

which has the exact solution

$$y = \frac{1}{38}\left(19x - 5x^2 - \frac{36}{x}\right).$$

The pivotal values Y_1, Y_2 corresponding to a coarse subdivision with $h = \frac{1}{3}$ are calculated from the ordinary finite-difference equations

$$\frac{-2Y_1 + Y_2}{h^2} - \frac{2 \times 9}{49} Y_1 + \frac{3}{7} = 0, \qquad \frac{Y_1 - 2Y_2}{h^2} - \frac{2 \times 9}{64} Y_2 + \frac{3}{8} = 0$$

as

$$Y_1 = \frac{217}{4932} = 0 \cdot 043\,998 \quad (\text{error} - 0 \cdot 000\,279, \text{ i.e. } - 0 \cdot 6\%),$$

$$Y_2 = \frac{208}{4932} = 0 \cdot 042\,174 \quad (\text{error} - 0 \cdot 000\,225, \text{ i.e. } - 0 \cdot 5\%).$$

For the error estimate we need first a value for $|y^{(4)}|_{\max}$; the maximum of the fourth difference quotient is often used for this, but in the present simple case it is better to estimate y' approximately and use the fact that

$$y^{(4)} = \frac{4}{x^4} (4y - 2xy' - x),$$

which follows from the differential equation; we find that

$$|y^{(4)}|_{\max} \approx 0 \cdot 71.$$

Hence from (3.14)

$$R^* \approx \left(\frac{1}{3}\right)^4 \times \frac{0 \cdot 71}{12} \approx 0 \cdot 000\,73.$$

We need further a solution of the equations $-\Lambda[w_i^*] = \sigma_i$ with positive w_i^* and σ_i. Here these equations read

$$\frac{100}{49} w_1^* - w_2^* = \sigma_1, \qquad -w_1^* + \frac{65}{32} w_2^* = \sigma_2,$$

and we need look no further than the simple values $w_1^* = w_2^* = 1$; they yield $\sigma_1 = \frac{51}{49}$, $\sigma_2 = \frac{33}{32}$.

Then (3.17) yields the error estimate

$$|Y_i - y_i| \leq \frac{32}{33} \times 0 \cdot 000\,73 = 0 \cdot 000\,71.$$

At x_1, for example, these limits are about two and a half times greater than the actual error.

3.4. An error estimate for a non-linear boundary-value problem

In principle, the method described in § 3.2 may also be applied to non-linear boundary-value problems provided only that the solution y and its derivatives $y^{(v)}$ are known, or can be estimated, sufficiently accurately. Usually, of course, we will not be in a position to make the necessary estimates of y and $y^{(v)}$ beforehand and will have to infer their quantitive behaviour approximately from the values calculated by the finite-difference method. Consequently a rigorous error estimate is out of the question and we must be satisfied with approximate error limits. In practice, the calculation of such approximate limits is often so laborious and the results so crude that it is preferable to dispense with them altogether and rely on the indication of the order of magnitude of the error provided by the results of two or more calculations with different intervals. Thus the procedure does not warrant more explanation than is afforded by the following simple example. The problem we consider is that of Example II of § 2, i.e.

$$y'' = \tfrac{3}{2} y^2; \qquad y(0) = 4, \qquad y(1) = 1. \tag{3.18}$$

With $h = \tfrac{1}{5}$ the approximate pivotal values Y_i satisfy the equations

$$Y_{i+1} - 2Y_i + Y_{i-1} - \tfrac{3}{2} h^2 Y_i^2 = 0 \qquad (i = 1, 2, 3, 4), \tag{3.19}$$

while the exact values y_i satisfy the inhomogeneous equations

$$y_{i+1} - 2y_i + y_{i-1} - \tfrac{3}{2} h^2 y_i^2 = R_i \qquad (i = 1, 2, 3, 4), \tag{3.20}$$

where

$$|R_i| \leq \frac{h^4}{12} |y^{(4)}|_{\max \text{ in } \langle x_{i-1}, x_{i+1} \rangle}. \tag{3.21}$$

The equations for the error $\varphi_i = Y_i - y_i$ obtained by subtracting (3.19) from (3.20) can be written in the form

$$a_i \varphi_i - \varphi_{i-1} - \varphi_{i+1} = R_i \qquad (i = 1, 2, 3, 4), \tag{3.22}$$

where

$$a_i = 2 + \tfrac{3}{2} h^2 (Y_i + y_i), \qquad \varphi_0 = \varphi_5 = 0. \tag{3.23}$$

With $\alpha = a_1 a_2 - 1$, $\beta = a_3 a_4 - 1$, $D = \alpha\beta - a_1 a_4$ the explicit solution of equations (3.22) for the four unknowns φ_i reads

$$\left. \begin{aligned}
D\varphi_1 &= R_1 (a_2\beta - a_4) + R_2\beta &&+ R_3 a_4 &&+ R_4, \\
D\varphi_2 &= R_1\beta &&+ R_2 a_1\beta &&+ R_3 a_1 a_4 + R_4 a_1, \\
D\varphi_3 &= R_1 a_4 &&+ R_2 a_1 a_4 + R_3 a_4 \alpha &&+ R_4 \alpha, \\
D\varphi_4 &= R_1 &&+ R_2 a_1 &&+ R_3 \alpha &&+ R_4 (a_3 \alpha - a_1).
\end{aligned} \right\} \tag{3.24}$$

We complete the estimate only for the solution without zeros and further restrict attention to the point $i = 2$ (i.e. $x = 0\cdot4$). As mentioned earlier, we have to make some approximations. Thus we replace the unknown exact solution $y(x)$ and its derivatives everywhere by the approximate values given by the finite-difference method: the $Y_i + y_i$ in the quantities a_i of (3.23) are replaced by $2Y_i$, and pivotal values of the fourth derivative

$$y^{\mathrm{IV}} = 3 y'^2 + \tfrac{9}{2} y^3$$

are calculated approximately by replacing the derivatives y_i' by the difference quotients $\dfrac{Y_{i+1} - Y_{i-1}}{2h}$. The approximate values of the y_i' so obtained are exhibited in Table III/10 together with the "local" maxima

Table III/10. *Quantities required in the error estimate* (3.24)

i	x_{i-1}	y_{i-1}'	$y^{IV}_{\max}$ in $\langle x_{i-1}, x_{i+1}\rangle$	Y_i	a_i
1	0	8·43	501	2·79464	2·335
2	h	4·86	169	2·05787	2·246
3	$2h$	3·05	68	1·57519	2·189
4	$3h$	2·04	30	1·24138	2·149

of y^{IV} derived from them. The latter are used in the bounds (3.21) for the R_i, and from (3.24) with $\alpha = 4\cdot25$, $\beta = 3\cdot71$, $D = 10\cdot73$, we obtain the estimate

$$|\varphi_2| \leq 0\cdot0926.$$

This limit is about five and a half times greater than the actual error $\varphi_2 = 0\cdot01705$.

§4. Some general methods

A series of general methods, especially methods based on error distribution principles, were discussed in Ch. I, § 4.2; it will therefore suffice here to refer to these methods and give examples. More general boundary-value problems for ordinary differential equations have already been formulated in Ch. I, § 1.

For our approximation to the required solution $y(x)$ we choose here a function $w(x, a_1, \ldots, a_p)$ which depends on several parameters $a_1, \ldots, a_p$ and which already satisfies the boundary conditions independently of the choice of parameter values. By inserting w into the differential equation we obtain a function $\varepsilon(x, a_1, \ldots, a_p)$ which represents the residual error. We then determine the a_p in accordance with one of the principles mentioned in Ch. I, § 4.2 so that this error function approximates the zero function as closely as possible.

4.1. Examples of collocation

The collocation method has a very wide range of applicability, is simple to use and requires no special previous knowledge. In many cases the method yields much more than a rough quantitative idea of the solution, in fact the accuracy of the results can sometimes be quite remarkable in view of the primitive nature of the method and the slight amount of computation involved.

Example I. For the problem of Example I, § 1.2:

$$L[y] = y'' + (1 + x^2)\, y + 1 = 0, \qquad y(\pm 1) = 0$$

we use the functions $x^{2n}(1 - x^2)$ as satisfying the boundary conditions and the symmetry about $x = 0$; thus our approximating function w is of the form

$$w = a_1(1 - x^2) + a_2(x^2 - x^4) + a_3(x^4 - x^6) + \cdots,$$

and substituted in L it yields

$$L[w] = 1 + a_1\left(-2 + (1 - x^4)\right) + a_2\left(2 - 12x^2 + x^2(1 - x^4)\right) +$$
$$+ a_3\left(12x^2 - 30x^4 + x^4(1 - x^4)\right) + \cdots.$$

If, for example, we use a two-term expression, i.e. with only two parameters a_1 and a_2, we can make $L[w] = 0$ at four symmetrically placed points, say $x = \pm\frac{1}{4}$ and $x = \pm\frac{3}{4}$; this leads to the linear equations

$$\left(\text{for } x = \frac{1}{4}\right) \quad 1 - a_1\frac{257}{256} + a_2\frac{5375}{4096} = 0,$$

$$\left(\text{for } x = \frac{3}{4}\right) \quad 1 - a_1\frac{337}{256} - a_2\frac{17881}{4096} = 0$$

with the solution $a_1 = 0.929254$, $a_2 = -0.051146$. In Table III/11 these results are compared with those obtained by using one-term and three-term expressions, respectively, with suitable collocation points.

Table III/11. *Comparison of collocation results as the number of parameters increases*

p-term expression	Collocation points x	Results: values of the a_ν		
$p = 1$	$\frac{1}{2}$	$a_1 = 0.94118$		
$p = 2$	$\frac{1}{4}$ and $\frac{3}{4}$	$a_1 = 0.92925,$	$a_2 = -0.05115$	
$p = 3$	$\frac{1}{6}, \frac{3}{6}$ and $\frac{5}{6}$	$a_1 = 0.932088,$	$a_2 = -0.034108,$	$a_3 = -0.030221$

Example II. Infinite interval. Another problem considered in § 1.2 was that of determining the steady temperature distribution in an infinitely long rod (Example IV); this leads to the boundary-value problem

$$L[y] = (2 + x)\, y'' - (1 + x)\, y = 0; \qquad y(0) = 1, \qquad y(\infty) = 0.$$

In order to satisfy the conditions at infinity, we choose an approximation of the form

$$y \approx w = \sum_{\nu=1}^{p} a_\nu e^{-\nu x}.$$

However, this function does not satisfy the boundary condition at $x=0$ for arbitrary values of the a_ν and we must therefore include the equation

$$\sum_{\nu=1}^{p} a_\nu = 1$$

with the collocation equations.

For $p=3$ the error function is

$$\varepsilon = L[w] = a_1 e^{-x} + a_2 (7 + 3x) e^{-2x} + a_3 (17 + 8x) e^{-3x}.$$

If, to start with, we drop the term in a_3, we can choose just one collocation point. Choosing $x=1$, we have

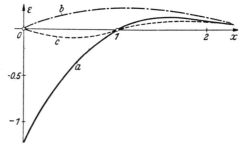

$$a_1 e^{-1} + a_2 10 e^{-2} = 0,$$

$$a_1 + a_2 = 1,$$

and hence

$$a_1 = \frac{10}{10 - e},$$

$$a_2 = -\frac{e}{10 - e};$$

Fig. III/11. Curves of the residual error obtained by various collocations

this yields in particular

$$y(1) \approx w(1) = \frac{9}{e(10 - e)} = 0.455.$$

To get some idea of how closely the differential equation is satisfied, we can draw the corresponding ε curve (Fig. III/11, curve a). If $x=0$ is chosen as collocation point instead of $x=1$, we obtain with

$$a_1 = \tfrac{7}{6}, \qquad a_2 = -\tfrac{1}{6}$$

an ε curve (curve b in Fig. III/11) which appears to be an improvement on curve a. In spite of this, the y values are worse; for example,

$$w(1) = \frac{7e - 1}{6e^2} = 0.405.$$

When a_3 is included and $x=0$ and $x=1$, say, are used as collocation points, we have

$$a_1 = 1.3605, \qquad a_2 = -0.4768, \qquad a_3 = 0.1163;$$

the corresponding ε curve is shown in Fig. III/11 as curve c. At $x=1$ we have the good approximation

$$y(1) \approx w(1) = \frac{153 e^2 - 168 e + 15}{10 e^2 (15 - e) (e - 1)} = 0.4418.$$

4.2. An example of the least-squares method

An example of a linear boundary-value problem is treated by the least-squares method in § 6.4, so here we confine ourselves to a non-linear example, namely the problem (1.9):

$$y'' = \tfrac{3}{2} y^2; \quad y(0) = 4, \quad y(1) = 1.$$

For an approximate solution we use the expression

$$\varphi = 4 - 3x + a(x - x^2),$$

which satisfies both boundary conditions for arbitrary a. As the residual error we define

$$\varepsilon = -2\varphi'' + 3\varphi^2 = 3(16 - 24x + 9x^2) + 2a(2 + 12x - 21x^2 + 9x^3) + $$
$$+ 3a^2(x^2 - 2x^3 + x^4).$$

For the least-squares method we now form

$$J[\varphi] = \int\limits_0^1 \varepsilon^2\,dx = \frac{1}{70}\,(42966 + 19740a + 3348a^2 + 101a^3 + a^4);$$

since a check on this calculation is desirable, we repeat it with $a = 3$:

$$\varepsilon = 3(20 - 24x^2 + 9x^4), \qquad \int\limits_0^1 \varepsilon^2\,dx = \frac{135126}{70}.$$

The least-squares equation

$$70\,\frac{\partial J[\varphi]}{\partial a} = 4a^3 + 303a^2 + 6696a + 19740 = 0$$

then yields

$$a = \begin{cases} -3{\cdot}4671 \\ -36{\cdot}1 \pm 10{\cdot}8\,i, \end{cases}$$

of which only the real value is significant. With this value $a = -3{\cdot}4671$ the least-squares approximation to the solution at $x = \tfrac{1}{2}$ is

$$\varphi\left(\tfrac{1}{2}\right) = 1{\cdot}7332.$$

4.3. Reduction to initial-value problems

As a general method for the numerical solution of boundary-value problems which is very useful on occasions we may mention the technique of reducing the problem to two or more initial-value problems and treating these by one of the methods of Ch. II. (See also the examples of § 1.2.)

In the simplest case of a second-order linear differential equation

$$L[y] = r(x) \tag{4.1}$$

with boundary conditions

$$y(a) = y_a, \tag{4.2}$$

$$y(b) = y_b \tag{4.3}$$

at the points $x = a$ and $x = b$, we compute the solutions y_1 and y_2 of the two initial-value problems

$$\left. \begin{array}{lll} L[y_1] = r(x); & y_1(a) = y_a, & y_1'(a) = 0, \\ L[y_2] = 0; & y_2(a) = 0, & y_2'(a) = 1. \end{array} \right\} \tag{4.4}$$

This yields two values $y_1(b)$ and $y_2(b)$ at $x = b$. On account of the linearity of the problem, the solution of the differential equation with the initial conditions $y(a) = y_a$, $y'(a) = y_a'$ is given by

$$y(x) = y_1(x) + y_a' y_2(x).$$

Then if y_a' is calculated from

$$y_1(b) + y_a' y_2(b) = y_b,$$

the boundary condition at $x = b$ will also be satisfied and we have the required solution of the boundary-value problem.

For a fourth-order linear boundary-value problem with equation $L[y] = r(x)$ and boundary conditions

$$y(a) = y_a, \quad y'(a) = y_a', \quad y(b) = y_b, \quad y'(b) = y_b',$$

one calculates numerical solutions y_1, y_2, y_3 of the following three initial-value problems:

$$\left. \begin{array}{llll} y_1 \text{ from } & L[y_1] = r; \; y_1(a) = y_a, \; y_1'(a) = y_a', \; y_1''(a) = y_1'''(a) = 0, \\ y_2 \text{ from } & L[y_2] = 0; \; y_2(a) = y_2'(a) = 0, \; y_2''(a) = 1, \; y_2'''(a) = 0, \\ y_3 \text{ from } & L[y_3] = 0; \; y_3(a) = y_3'(a) = y_3''(a) = 0, \; y_3'''(a) = 1. \end{array} \right\} \tag{4.5}$$

The solution of the original boundary-value problem is then given by

$$y(x) = y_1(x) + y''(a) y_2(x) + y'''(a) y_3(x)$$

if $y''(a)$ and $y'''(a)$ are calculated from the two simultaneous linear equations

$$y_1(b) + y''(a) y_2(b) + y'''(a) y_3(b) = y_b,$$
$$y_1'(b) + y''(a) y_2'(b) + y'''(a) y_3'(b) = y_b'.$$

Generally, the solution $y(x)$ of an n-th-order linear boundary-value problem can always be obtained as a linear combination of the solutions $y_1, y_2, \ldots$ of at most $n+1$ initial-value problems, the constants c_ν of the combination being calculated from a system of linear equations in at most n unknowns. If the boundary conditions do not permit a convenient special choice of initial conditions such as in (4.4), (4.5), the conditions

$$L[y_k] = 0, \quad y_k^{(\nu)}(a) = \left\{ \begin{array}{ll} 0 \text{ for } \nu \ne k - 1 \\ 1 \text{ for } \nu = k - 1 \end{array} \right\} \quad \begin{array}{l} (\nu = 0, 1, \ldots, n - 1) \\ (k = 1, 2, \ldots, n) \end{array}$$

will always suffice; the functions $y_1, y_2, \ldots, y_n$ which they define form a "fundamental system" for the homogeneous differential equation, and with the additional function y_{n+1} determined from

$$L[y_{n+1}] = r(x), \quad y_{n+1}^{(\nu)}(a) = 0 \quad (\nu = 0, 1, \ldots, n - 1)$$

the required solution can be obtained in the form

$$y(x) = y_{n+1}(x) + \sum_{\nu=1}^{n} c_\nu y_\nu(x). \tag{4.6}$$

When we come to non-linear boundary-value problems, linear combination is no longer applicable and we resort to interpolation. Thus several different solutions $y_1, y_2, \ldots$ are computed satisfying all the boundary conditions at the point $x = a$ and we interpolate among them to find one which satisfies the boundary conditions at $x = b$.

4.4. Perturbation methods

Perturbation methods can sometimes be employed with advantage when a boundary-value problem with a known, or easily derivable solution can be found in the "neighbourhood" of the given boundary-value problem[1], i.e. such that the values of the coefficients in its differential equation differ only slightly from the corresponding values in the given equation. For simplicity we consider here only those "neighbouring" problems which have the same boundary conditions as the given problem, but perturbation methods can also be used when the boundary conditions are "disturbed" as well as the differential equation. Attention is further restricted to linear boundary conditions in the form given in Ch. I (1.7).

---

[1] Applications of perturbation methods to eigenvalue problems can be found in the following papers: MEYER ZUR CAPELLEN, W.: Methode zur angenäherten Lösung von Eigenwertproblemen mit Anwendungen auf Schwingungsprobleme. Ann. Phys. (5) **8**, 297—352 (1931). — Genäherte Berechnung von Eigenwerten. Ing.-Arch. **10**, 167—174 (1939). — RELLICH, F.: Störungstheorie der Spektralzerlegung. Math. Ann. **113**, 600—619 (1936); **114**, 677—685 (1937); **116**, 555—570 (1939); **117**, 356—382 (1940); **118**, 462—484 (1942). — NAGY, B. v. Sz.: Perturbations des transformations autoadjointes dans l'espace de Hilbert. Comm. Math. Helv. **19**, 347—366 (1946). — Perturbations des transformations linéaires fermées. Acta Sci. Math. Szeged **14**, 125—137 (1951). — SCHRÖDER, J.: Fehlerabschätzungen zur Störungsrechnung bei linearen Eigenwertproblemen mit Operatoren eines Hilbertschen Raumes. Math. Nachr. **10**, 113—128 (1953). — Fehlerabschätzungen zur Störungsrechnung für lineare Eigenwertprobleme bei gewöhnlichen Differentialgleichungen. Z. Angew. Math. Mech. **34**, 140—149 (1954) (with a summary of results in a directly applicable form). — SCHÄFKE, FR. W.: Über Eigenwertprobleme mit 2 Parametern. Math. Nachr. **6**, 109—124 (1951). — Verbesserte Konvergenz- und Fehlerabschätzungen für die Störungsrechnung. Z. Angew. Math. Mech **33**, 255—259 (1953).

We introduce a perturbation parameter ε in such a way that the differential equation

$$G(\varepsilon, x, y, y', y'', \ldots, y^{(n)}) = 0 \qquad (4.7)$$

has a known solution when $\varepsilon = 0$ and reproduces the given differential equation when $\varepsilon = 1$:

$$G(1, x, y, y', \ldots, y^{(n)}) = F(x, y, y', \ldots, y^{(n)}) = 0. \qquad (4.8)$$

The differential equation with $\varepsilon = 0$ we call the "undisturbed" equation and that with $\varepsilon = 1$ the "disturbed" equation.

We now assume that the solution $\varphi = y(x, \varepsilon)$ of the boundary-value problem

$$G(\varepsilon, x, y, \ldots, y^{(n)}) = 0, \quad U_\nu[y] = \gamma_\nu \qquad (\nu = 1, 2, \ldots, n) \qquad (4.9)$$

may be expanded in powers of ε:

$$y(x, \varepsilon) = y_0(x) + \varepsilon y_1(x) + \varepsilon^2 y_2(x) + \cdots \qquad (4.10)$$

(that this is permissible can probably not be proved with the generality considered so far). The first term $y_0(x)$ is the known solution of the undisturbed equation satisfying the boundary conditions (1.7) of Ch. I. If the remaining coefficients y_1, y_2, ... satisfy the corresponding homogeneous boundary conditions, so that

$$U_\nu[y_0] = \gamma_\nu, \quad U_\nu[y_j] = 0 \qquad (\nu = 1, 2, \ldots, n; \; j = 1, 2, \ldots), \qquad (4.11)$$

then the power series (4.10) will satisfy the inhomogeneous conditions.

We now replace y in the differential equation (4.9) by the series (4.10):

$$G\left(\varepsilon, x, \sum_{r=0}^{\infty} \varepsilon^r y_r, \sum_{r=0}^{\infty} \varepsilon^r y_r', \ldots, \sum_{r=0}^{\infty} \varepsilon^r y_r^{(n)}\right) = 0. \qquad (4.12)$$

Expansion of the left-hand side in powers of ε using TAYLOR's theorem (assumed to be a valid procedure) then yields

$$G\left(\varepsilon, x, \sum_{r=0}^{\infty} \varepsilon^r y_r, \ldots, \sum_{r=0}^{\infty} \varepsilon^r y_r^{(n)}\right) = \sum_{s=0}^{\infty} \varepsilon^s G_s = 0, \qquad (4.13)$$

where the coefficients G_s are differential expressions involving the functions y_0, y_1, ..., y_s and their derivatives; for example,

$$G_0 = G(0, x, y_0, y_0', \ldots, y_0^{(n)}) = 0$$

and

$$G_1 = \left\{ \frac{\partial G}{\partial \varepsilon} + y_1 \frac{\partial G}{\partial y} + y_1' \frac{\partial G}{\partial y'} + \cdots + y_1^{(n)} \frac{\partial G}{\partial y^{(n)}} \right\}_{\varepsilon=0}. \qquad (4.14)$$

When we have calculated y_1, y_2, ... up to y_{s-1}, the equation $G_s = 0$ constitutes, in general, an n-th-order differential equation for y_s, from which y_s can be determined with the boundary conditions (4.11). Thus $y_1, y_2, y_3, \ldots$ can be calculated successively starting with y_1, which is calculated from y_0 by means of the linear boundary-value problem with the differential equation $G_1 = 0$ from (4.14) and the homogeneous boundary conditions from (4.11). Such a perturbation method is to be recommended only when the boundary-value problems for the y_s are of simple form.

Example. We consider again the example of the transversely loaded strut (I of § 1.2), and write the differential equation in the form

$$y'' + (1 + \varepsilon x^2)\, y + 1 = 0, \qquad y(\pm 1) = 0.$$

Then for $\varepsilon = 1$ we have the equation to be solved and for $\varepsilon = 0$ we have a simple differential equation with elementary solution:

$$y_0'' + y_0 + 1 = 0, \qquad y_0(\pm 1) = 0$$

with the solution

$$y_0 = \frac{\cos x}{\cos 1} - 1.$$

Substituting the power series (4.10) for y in the differential equation, we obtain

$$\sum_{n=0}^{\infty} \varepsilon^n (y_n'' + y_n) + \sum_{n=1}^{\infty} \varepsilon^n x^2 y_{n-1} + 1 = 0,$$

and hence

$$y_n'' + y_n + x^2 y_{n-1} = 0 \qquad (n = 1, 2, \ldots)$$

since the coefficients of the powers of ε must all be zero. With the boundary conditions $y_n(\pm 1) = 0$ we have a sequence of boundary-value problems from which the functions $y_1, y_2, \ldots$ can be determined successively.

The first function y_1 satisfies

$$y_1'' + y_1 = x^2 - x^2 \frac{\cos x}{\cos 1}, \qquad y_1(\pm 1) = 0,$$

and is given by

$$y_1 = x^2 - 2 + \frac{1}{12 \cos 1} \left[(3x - 2x^3) \sin x - 3x^2 \cos x \right] + A \cos x,$$

where

$$A = \frac{1}{12 \cos^2 1} (15 \cos 1 - \sin 1).$$

The determination of further y_j is more laborious.

For the value at $x = 0$ the first approximation gives

$$y_0(0) = \frac{1 - \cos 1}{\cos 1} = 0 \cdot 8508 \quad (\text{error} -9\%)$$

and the second

$$y_0(0) + y_1(0) = \frac{1}{\cos 1} \left(\frac{9}{4} - 3 \cos 1 - \frac{1}{12} \tan 1 \right) = 0 \cdot 9241 \quad (\text{error} -0 \cdot 8\%).$$

4.5. The iteration method or the method of successive approximations

In this method the differential equation (1.1) of Ch. I for a general boundary-value problem is put into the form

$$F_1(x, y, y', \ldots, y^{(n)}) = F_2(x, y, y', \ldots, y^{(m)}), \qquad (4.15)$$

where $m < n$ and F_1 is a convenient, simple function such that the boundary-value problem

$$F_1(x, z, z', \ldots, z^{(n)}) = r(x) \qquad (4.16)$$

with the boundary conditions (1.3) of Ch. I, which we write shortly as

$$V_\nu(z) = 0 \qquad (\nu = 1, 2, \ldots, n), \qquad (4.17)$$

can be solved readily for any right-hand side $r(x)$. If, as is usually the case, the given differential equation can be solved for the highest derivative which occurs in terms of the lower derivatives, i.e. it can be put into the form

$$y^{(n)} = \varphi(x, y, y', \ldots, y^{(n-1)}), \qquad (4.18)$$

then one obvious rearrangement of the form (4.15) has $F_1 = y^{(n)}$ and $F_2 = \varphi$.

We can now define an iterative procedure which determines a sequence of functions $y_0(x)$, $y_1(x)$, $y_2(x)$, ... in the following manner: $y_0(x)$ is chosen arbitrarily, then $y_1(x)$, $y_2(x)$, ... calculated successively as the solutions of the boundary-value problems

$$\left.\begin{array}{l} F_1\big(x, y_{k+1}(x), y'_{k+1}(x), \ldots, y^{(n)}_{k+1}(x)\big) \\ = F_2\big(x, y_k(x), y'_k(x), \ldots, y^{(m)}_k(x)\big), \\ V_\nu(y_{k+1}) = 0 \end{array}\right\} \quad (k = 0, 1, 2, \ldots). \qquad (4.19)$$

With this generality nothing can be asserted about the convergence of the sequence $y_k(x)$ to the solution $y(x)$ of the boundary-value problem; it can happen that the sequence does not converge at all. When it does converge, the effectiveness of the method is often influenced considerably by the form of the rearrangement (4.15) of the given differential equation and by the choice of starting function $y_0(x)$; the method is generally more effective the closer $y_0(x)$ is to $y(x)$.

The method is often used graphically, particularly in the simple form mentioned for equations of the form (4.18). If, for example, we have a second-order equation

$$y'' = \varphi(x, y, y'), \qquad (4.20)$$

we first find a starting function $y_0(x)$, perhaps by means of the finite-difference method, then use a funicular polygon method[1] to perform

---

[1] See, for instance, E. KAMKE: Differentialgleichungen, Lösungsmethoden und Lösungen, Vol. I, 3rd ed., p. 164 et seq. Leipzig 1944.

the successive integrations which generate the sequence of functions $y_1(x)$, $y_2(x)$, ...

If F_1, F_2, V_ν are all linear in y and its derivatives, and if certain further assumptions are made, we can specify conditions for the convergence of the iteration.

In this case the n-th-order boundary-value problem may be written in the form.

$$\left.\begin{array}{l} M[y] - a\,N[y] = r(x), \\ U_\mu[y] = \gamma_\mu \end{array}\right\} \quad (\mu = 1, 2, ..., n), \qquad (4.21)$$

where M and N are linear differential expressions and the $U_\mu[y] = \gamma_\mu$ are the linear boundary conditions (1.7) of Ch. I. The corresponding iterative scheme is defined by

$$M[y_{k+1}] = a\,N[y_k] + r(x), \quad U_\mu[y_{k+1}] = \gamma_\mu \quad (k = 0, 1, 2, ...), \qquad (4.22)$$

and to examine its convergence we introduce the sequence of functions $z_k(x)$ defined by

$$z_k(x) = y_k(x) - y(x);$$

these satisfy the corresponding homogeneous iterative scheme

$$M[z_{k+1}] = a\,N[z_k], \quad U_\mu[z_{k+1}] = 0.$$

We now make the assumption that $z_1(x)$ possesses an eigenfunction expansion

$$z_1(x) = \sum_{j=1}^{\infty} c_j \eta_j(x)$$

in terms of the eigenfunctions $\eta_j(x)$ of the corresponding homogeneous eigenvalue problem

$$M[y] - \lambda N[y] = 0, \quad U_\mu[y] = 0, \qquad (4.23)$$

and also that this expansion may be differentiated term by term n times. Then

$$z_2(x) = \sum_{j=1}^{\infty} \frac{a}{\lambda_j} c_j \eta_j(x)$$

is a solution of the boundary-value problem

$$M[z_2] = a\,N[z_1], \quad U_\mu[z_2] = 0,$$

and in general

$$z_{k+1}(x) = \sum_{j=1}^{\infty} \left(\frac{a}{\lambda_j}\right)^k c_j \eta_j(x). \qquad (4.24)$$

Hence the iteration converges for any c_i provided that $|a| < |\lambda_1|$, where λ_1 is the eigenvalue of (4.23) with smallest absolute value; when $|a| \geq |\lambda_1|$, the iteration diverges for $c_1 \neq 0$, i.e. in general.

4.6. Error estimation by means of the general iteration theorem

In many cases error estimates can be obtained by means of the general theorem on iterative processes of Ch. I, § 5; for linear boundary-value problems and also for many types of non-linear problem this theorem can be used precisely as described in Ch. I, § 5.3. This will be amplified to some extent here, although in practice the method is effective only for problems which are not far removed from linear; for otherwise the Lipschitz constant K required for the estimate can easily turn out to be greater than one. In any case, great care should be exercised in the determination of K because any increase in K will have a considerable adverse effect on the estimate on account of the factor $\frac{K}{1-K}$.

Consider the k-th-order ordinary differential equation

$$L[u] = F(x, u, u', \ldots, u^{(s)}) \tag{4.25}$$

with linear boundary conditions

$$U_\mu[u] = \gamma_\mu \qquad (\mu = 1, \ldots, k) \tag{4.26}$$

at two points a and b $(b > a)$. Here $L[u]$ is a linear homogeneous differential expression in u of order $k > s$ and F is a given function which we shall assume satisfies a Lipschitz condition of the form

$$\left. \begin{array}{c} \left| F(x, u_1, u_1', \ldots, u_1^{(s)}) - F(x, u_2, u_2', \ldots, u_2^{(s)}) \right| \\[2mm] \leq \sum_{\sigma=0}^{s} A_\sigma(x) \left| u_1^{(\sigma)} - u_2^{(\sigma)} \right| \end{array} \right\} \tag{4.27}$$

with $A_\sigma(x) \geq 0$. Further we shall assume that the boundary-value problem

$$L[u] = r(x), \qquad U_\mu[u] = 0 \qquad (\mu = 1, \ldots, k) \tag{4.28}$$

can be solved for any $r(x)$ by means of a GREEN's function $G(x, \xi)$ (cf. Ch. I, § 3.4):

$$u(x) = \int_a^b G(x, \xi)\, r(\xi)\, d\xi. \tag{4.29}$$

Finally we assume that this formula for the solution of (4.28) may be differentiated s times with respect to x, so that

$$u^{(\sigma)}(x) = \int_a^b G^{(\sigma)}(x, \xi)\, r(\xi)\, d\xi \qquad (\sigma = 0, 1, \ldots, s), \tag{4.30}$$

where

$$G^{(\sigma)}(x, \xi) = \frac{\partial^\sigma G(x, \xi)}{\partial x^\sigma}.$$

We now define a transformation T (cf. Ch. I, § 5.3) by $T f = g$ where

$$\left.\begin{array}{l} L[g] = F(x, f, f', \ldots, f^{(s)}) \\ U_\mu[g] = \gamma_\mu \quad (\mu = 1, \ldots, k). \end{array}\right\} \tag{4.31}$$

Then the difference $H = g_1 - g_2 = T f_1 - T f_2$ between two functions g_1, g_2 derived by this transformation from two functions f_1, f_2 satisfies the boundary-value problem

$$L[H] = F(x, f_1, f_1', \ldots, f_1^{(s)}) - F(x, f_2, f_2', \ldots, f_2^{(s)}),$$
$$U_\mu[H] = 0 \quad (\mu = 1, \ldots, k),$$

and hence by (4.27)

$$\left.\begin{array}{l} |H^{(\tau)}(x)| = \left| \int_a^b G^{(\tau)}(x, \xi) \, [F(\xi, f_1, \ldots, f_1^{(s)}) - F(\xi, f_2, \ldots, f_2^{(s)})] \, d\xi \right| \\[2mm] \leq \int_a^b |G^{(\tau)}(x, \xi)| \sum_{\sigma=0}^s A_\sigma(\xi) \, |h^{(\sigma)}(\xi)| \, d\xi, \end{array}\right\} \tag{4.32}$$

where $h = f_1 - f_2$. If

$$\|f\| = \underset{\text{in } a \leq x \leq b}{\text{upper limit of }} \frac{1}{W(x)} \sum_{\sigma=0}^s A_\sigma(x) \, |f^{(\sigma)}(x)|, \tag{4.33}$$

where $W(x)$ is a positive, or possibly non-negative, function in $a \leq x \leq b$, is used as the definition of the norm of a function possessing s continuous derivatives, then

$$\sum_{\tau=0}^s A_\tau(x) \, |H^{(\tau)}(x)| \leq \|h\| \sum_{\tau=0}^s A_\tau(x) \int_a^b |G^{(\tau)}(x, \xi)| \, W(\xi) \, d\xi. \tag{4.34}$$

A Lipschitz constant K for the transformation T is therefore given by

$$K = \underset{\text{in } a \leq x \leq b}{\text{upper limit of }} \frac{1}{W(x)} \sum_{\sigma=0}^s A_\sigma(x) \int_a^b |G^{(\sigma)}(x, \xi)| \, W(\xi) \, d\xi. \tag{4.35}$$

4.7. Special case of a non-linear differential equation of the second order

As an example we take a second-order differential equation in the form

$$L[u] = -u'' = F(x, u, u') \tag{4.36}$$

with the boundary conditions

$$u(a) = u_a, \quad u(b) = u_b.$$

Without loss of generality we put $a = 0$, and from Ch. I (3.26) the GREEN's function is then given by

$$G(x, \xi) = \begin{cases} \dfrac{x}{b}(b - \xi) & \text{for} \quad 0 \leq x \leq \xi, \\[2mm] \dfrac{\xi}{b}(b - x) & \text{for} \quad \xi \leq x \leq b. \end{cases} \tag{4.37}$$

For simplicity we will use constant A_0, A_1 in the Lipschitz condition (4.27). If u' does not appear explicitly in the differential equation (4.36), then $A_1 = 0$ and the eigenfunction $\sin \dfrac{\pi x}{b}$ may be chosen for $W(x)$. Since

$$\frac{\displaystyle\int_a^b G(x, \xi) \sin \dfrac{\pi \xi}{b} \, d\xi}{\sin \dfrac{\pi x}{b}} = \frac{b^2}{\pi^2}, \tag{4.38}$$

(4.35) yields the Lipschitz constant

$$K = \frac{A_0 b^2}{\pi^2}. \tag{4.39}$$

If, however, $A_1 \neq 0$, then this choice for $W(x)$ does not yield a finite upper limit in (4.35) and we must choose some other function. Let us try $W(x) = 1 - \alpha z$, where $z = \dfrac{x}{b}\left(1 - \dfrac{x}{b}\right)$; then $W > 0$ in $0 \leq x \leq b$ for $\alpha < 4$.

With

$$\left. \begin{aligned} k_0 &= \max_{0 \leq x \leq b} \frac{\displaystyle\int_0^b G(x, \xi) W(\xi) \, d\xi}{W(x)} = \max_{0 \leq z \leq \frac{1}{4}} \frac{b^2 z \left(1 - \dfrac{\alpha}{6} - \dfrac{\alpha}{6} z\right)}{2(1 - \alpha z)} \\[3mm] k_1 &= \max_{0 \leq x \leq b} \frac{\displaystyle\int_0^b G'(x, \xi) W(\xi) \, d\xi}{W(x)} = \max_{0 \leq z \leq \frac{1}{4}} b \frac{\dfrac{6 - \alpha}{12} - z + \dfrac{\alpha}{2} z^2}{1 - \alpha z} \end{aligned} \right\} \tag{4.40}$$

the Lipschitz constant (4.35) for this case is

$$K = A_0 k_0 + A_1 k_1. \tag{4.41}$$

The calculation is simplified considerably by choosing $\alpha = 0$; then $W = 1$ and we have

$$K = \frac{A_0 b^2}{8} + \frac{A_1 b}{2}. \tag{4.42}$$

As α increases, k_0 increases but k_1 decreases, so that it is often possible to obtain a sharper estimate for non-zero α; for instance, with

$$\alpha = \frac{1}{4}\left(15 - \sqrt{33}\right) \approx 2 \cdot 314$$

we have[1]

$$K = \beta \left(\frac{A_0 b^2}{2} + A_1 b \right), \tag{4.43}$$

where

$$\beta = \frac{9 + \sqrt{33}}{48} \approx 0 \cdot 3072.$$

For the solution of the differential equation (4.36) with the boundary conditions

$$u(0) = u_0, \quad u'(b) = u'_b \tag{4.44}$$

we need the GREEN's function

$$G(x, \xi) = \begin{cases} x & \text{for} \quad 0 \leq x \leq \xi, \\ \xi & \text{for} \quad \xi \leq x \leq b \end{cases}$$

instead of (4.37). In this case we choose

$$W(x) = \sin \frac{\pi x}{2b}$$

when u' does not appear in (4.36), i.e. when $A_1 = 0$; this gives the Lipschitz constant

$$K = \frac{4 A_0 b^2}{\pi^2}. \tag{4.45}$$

Again we must use some other function for $W(x)$ when $A_1 \neq 0$; with $W(x) = 1$ we obtain

$$K = A_0 \frac{b^2}{2} + A_1 b \tag{4.46}$$

and with[2] $W(x) = 3 - 6x + 4x^2$

$$K = 0 \cdot 636 A_0 b^2 + \tfrac{4}{9} A_1 b. \tag{4.47}$$

---

[1] Cf. F. LETTENMEYER: Über die von einem Punkt ausgehenden Integralkurven einer Differentialgleichung 2. Ordnung. Dtsch. Math. **7**, 56—74 (1944). LETTENMEYER considers the sequence of functions generated iteratively from [as starting funktion $u_0(x)$] the linear function which satisfies the boundary conditions; he proves convergence to the uniquely determined solution of the boundary-value for the case

$$A_0 \frac{b^2}{\pi^2} + A_1 \frac{4}{\pi^2} b < 1.$$

Comparison shows that the optimum constant b^2/π^2 appears also in our (4.39) for the case $A_1 = 0$, but that for the case $A_1 \neq 0$ our factor multiplying A_0 in (4.43) is less favourable than LETTENMEYER's while our factor multiplying A_1 is more favourable than his.

[2] For the boundary conditions (4.44) LETTENMEYER (loc. cit.) obtains the condition $\frac{4}{\pi^2} A_0 b^2 + \frac{2}{\pi} A_1 b < 1$. The optimum constant $\frac{4}{\pi^2}$ appears also in our (4.45) for the case $A_1 = 0$, as before, and for the case $A_1 \neq 0$ our factors multiplying A_0 and A_1 in (4.47) are again, as in (4.43), less and more favourable, respectively, than LETTENMEYER's.

4.8. Examples of the iteration method with error estimates

I. A linear problem. The bending of a strut (Example I of § 1.2) is again used as an example; the boundary-value problem reads

$$y'' = -1 - (1 + x^2)\, y, \qquad y(\pm 1) = 0,$$

and several iteration schemes suggest themselves.

A. The simplest is defined by

$$y_{k+1}'' = -1 - (1 + x^2)\, y_k, \qquad y_{k+1}(\pm 1) = 0 \qquad (k = 0, 1, 2, \ldots).$$

We start with the function

$$y_0(x) = A(1 - x^2),$$

which satisfies the boundary conditions, and determine the constant A so that the first iterate

$$y_1(x) = \frac{15 + 14A}{30} - \frac{1 + A}{2}\, x^2 + \frac{A}{30}\, x^6$$

is of the same order of magnitude as $y_0(x)$. The condition

$$y_0(0) = y_1(0)$$

yields $A = \frac{15}{16}$ and $y_0(0) = y_1(0) = A = 0\cdot9375$ (error $0\cdot0054$). With this value of A we have $32\,y_1 = 30 - 31\,x^2 + x^6$.

A second iteration yields

$$32\,y_2 = \frac{75379}{2520} - 31\,x^2 + \frac{1}{12}\,x^4 + \frac{31}{30}\,x^6 - \frac{1}{56}\,x^8 - \frac{1}{90}\,x^{10},$$

whose value at $x = 0$ is $y_2(0) = \frac{75379}{80640} \approx 0\cdot93476$ (error $0\cdot0027$).

To facilitate the evaluation of the next iterate, we simplify y_2 by neglecting the two highest powers of x and approximating the remaining coefficients by simpler values:

$$y_2^* = \frac{1}{379}\,(354 - 367\,x^2 + x^4 + 12\,x^6), \quad \text{for which} \quad y_2^*(0) = \frac{354}{379} = 0\cdot93404.$$

A further iteration then yields

$$379\,y_3^* = \frac{297009}{840} - \frac{733}{2}\,x^2 + \frac{13}{12}\,x^4 + \frac{61}{5}\,x^6 - \frac{13}{56}\,x^8 - \frac{2}{15}\,x^{10},$$

whose initial value is

$$y_3^*(0) = \frac{297009}{318360} \approx 0\cdot932934 \quad (\text{error } 0\cdot00088).$$

B. We obtain rather better results if the term $-y$ is taken over to the left-hand side, i.e. if we use the iteration formula

$$y_{k+1}'' + y_{k+1} = -1 - x^2\, y_k, \qquad y_{k+1}(\pm 1) = 0 \qquad (k = 0, 1, \ldots).$$

With the same starting procedure as in **A**, i.e. putting

$$y_0(x) = A(1 - x^2), \qquad y_0(0) = y_1(0), \qquad (4.48)$$

we find that

$$y_1(x) = -1 + A(26 - 13 x^2 + x^4) + \frac{1 - 14A}{\cos 1} \cos x;$$

here

$$y_1(0) = A = \frac{1 - \cos 1}{14 - 25 \cos 1} = 0.933\,505 \quad (\text{error } 0.001\,45).$$

C. The iteration can be improved still further by making a constant-coefficient approximation for the term $x^2 y$, say $\frac{1}{4} y$, and taking this over to the left-hand side as well. We obtain the slightly modified iteration

$$y_{k+1}'' + \frac{5}{4} y_{k+1} = -1 + \left(\frac{1}{4} - x^2\right) y_k, \qquad y_{k+1}(\pm 1) = 0 \qquad (k = 0, 1, \ldots),$$

which, with the same starting procedure as before, yields

$$y_1(x) = -\frac{4}{5} + A \frac{1761 - 1085 x^2 + 100 x^4}{125} + \frac{100 - 776A}{125 \cos \varrho} \cos \varrho x,$$

where $\varrho = \frac{1}{2} \sqrt{5}$. Compared with iterations **A** and **B** we have

$$y_1(0) = A = \frac{100(1 - \cos \varrho)}{776 - 1636 \cos \varrho} = 0.932\,456 \quad (\text{error } 0.000\,40);$$

the error in the first iterate here is less than the error in the third iterate of method **A**, so it clearly pays to give careful consideration to the choice of iteration scheme.

Error estimates. The method of error estimation is illustrated for the iteration in **A**, since the evaluation of the limits is by far the simplest in this case. We adopt the method 1. of § 5.3 in Ch. I, which is directly applicable; in particular, we introduce the norm defined in (5.27) of Ch. I, i.e.

$$\|f\| = \underset{\text{in } -1 \leq x \leq 1}{\text{upper limit of}} \left|\frac{f}{W}\right|,$$

where $W(x)$ is non-negative for $-1 \leq x \leq 1$. We apply the method to the pair of the successive iterates $y_1(x)$, $y_2(x)$, and show how great the effect of a suitable choice for the norm is.

The error estimate derived by using a non-negative eigenfunction z for W as defined in Ch. I (5.31) is very easily found with a negligible amount of calculation since the GREEN's function is not involved explicitly; the limits so obtained are, however, rather wide. For

$$z'' + \lambda z = 0, \qquad z(\pm 1) = 0$$

we have the non-negative eigenfunction $z = \cos \dfrac{\pi}{2} x$ and the correspond-ing eigenvalue is $\lambda_z = \dfrac{\pi^2}{4}$. With $p = 1 + x^2$, $|p|_{max} = 2$, the formula (5.33) of Ch. I yields the value $K = \dfrac{8}{\pi^2}$, which is rather close to 1; this gives a correspondingly large value for

$$\frac{K}{1-K} = \frac{8}{\pi^2 - 8} \approx 4\cdot 28.$$

We now calculate the "distance" separating y_1 and y_2:

$$\|y_1 - y_2\| = \underset{\text{in } -1 \leq x \leq 1}{\text{upper limit of}} \left| \frac{y_1 - y_2}{\cos \dfrac{\pi}{2} x} \right| = 0\cdot005\,55\,;$$

(5.16) of Ch. I then yields

$$\|y_2 - y\| \leq 4\cdot28 \times 0\cdot005\,55 = 0\cdot0237,$$

from which we infer the error estimate

$$|y_2 - y| \leq 0\cdot0237 \cos \frac{\pi}{2} x \qquad \text{for} \qquad |x| \leq 1.$$

To obtain a smaller Lipschitz constant and hence a better error estimate, we use the GREEN's function method of Ch. I (5.29). The appropriate GREEN's function for this case was given in Ch. I (3.26).

$$G(x, \xi) = \begin{cases} \dfrac{(x+1)(1-\xi)}{2} & \text{for} \quad -1 \leq x \leq \xi \\[2mm] \dfrac{(\xi+1)(1-x)}{2} & \text{for} \quad \xi \leq x \leq 1. \end{cases} \qquad (4.49)$$

Choosing $W(x) = 1$, i.e. using the norm $\|f\| = \underset{-1 \leq x \leq 1}{\max} |f|$, and with $N(x) = 1 + x^2$, we have

$$K = \underset{|x| \leq 1}{\max} \int_{-1}^{1} G(x, \xi)(1 + \xi^2)\, d\xi = \underset{|x| \leq 1}{\max} \frac{1}{12}(7 - 6x^2 - x^4) = \frac{7}{12}.$$

This value of K is much better than the value $8/\pi^2$ obtained above; it gives $\dfrac{K}{1-K} = \dfrac{7}{5} = 1\cdot4$ and in fact yields narrower error limits for y_1 at the point $x = 0$ than those obtained above for y_2. With

$$\|y_1 - y_2\| = \underset{|x| \leq 1}{\max} |y_1 - y_2| = \frac{221}{80\,640} \approx 0\cdot002\,74$$

(5.16) of Ch. I now gives the limits for y_2 as

$$|y_2 - y| \leq \frac{221}{57\,600} \approx 0\cdot003\,84.$$

The estimate using the GREEN's function is better still if a non-constant function is chosen for W, in this case a polynomial, say, for convenience in integration; a suitable choice here is $W = 1 - x^2$. We then have to evaluate the integral:

$$\int_{-1}^{1} G(x, \xi)(1 + \xi^2)(1 - \xi^2)\, d\xi = \frac{1}{30}(14 - 15 x^2 + x^6)$$

$$= \frac{(1 - x^2)(14 - x^2 - x^4)}{30}.$$

Substituting in (5.29) of Ch. I, we obtain

$$K = \max_{|x| \leq 1} \frac{14 - x^2 - x^4}{30} = \frac{7}{15},$$

and hence the favourable value

$$\frac{K}{1 - K} = \frac{7}{8}.$$

Finally, with $\|y_1 - y_2\| = \dfrac{11}{2520}$, we have the error estimate

$$|y_2 - y| \leq \frac{77}{20\,160}(1 - x^2) \approx 0.003\,82\,(1 - x^2).$$

A similar calculation with $W = 1 - \frac{1}{2}x^2$ (which gives $K = \frac{21}{40}$) leads to the estimate

$$|y_2 - y| \leq 0.003\,25\,(1 - \tfrac{1}{2}x^2).$$

At $x = 0$ this estimate exceeds the actual error $0.002\,71$ by less than 20%.

By using a more general definition of distance (cf. the paper by J. SCHRÖDER in the references given in Ch. I, §5.2) we can obtain limits as close as $|y_2 - y| \leq 0.003\,04\,(1 - x^2)$ without going beyond polynomials of the second degree for W.

II. Non-linear oscillations. It is required to calculate a periodic solution of

$$\left. \begin{array}{l} -y'' - 6y - y^2 = \frac{3}{2}\cos x, \\[4pt] y(0) - y(2\pi) = y'(0) - y'(2\pi) = 0 \end{array} \right\} \tag{4.50}$$

(forced oscillations of a system with a non-linear restoring force). The problem lies sufficiently near to the linear problem obtained by neglecting the y^2 term for the iteration defined by

$$-y''_{k+1} - 6y_{k+1} = y_k^2 + \tfrac{3}{2}\cos x, \qquad y_{k+1} \text{ with period } 2\pi \qquad (k = 0, 1, \ldots)$$

to converge satisfactorily and allow an error estimate to be made.

The solution

$$y_0(x) = -0.3 \cos x$$

of the linear problem is an obvious choice for the starting function. A short calculation then yields

$$y_1(x) = -\frac{3}{400}(1 + 40\cos x + 3\cos 2x),$$

$$y_2(x) = \frac{-3}{320\,000}(805{\cdot}5 + 32\,240\cos x +$$
$$+ 2418\cos 2x - 240\cos 3x - 2{\cdot}7\cos 4x),$$

$$y_1(x) - y_2(x) = \frac{3}{320\,000}(5{\cdot}5 + 240\cos x +$$
$$+ 18\cos 2x - 240\cos 3x - 2{\cdot}7\cos 4x),$$

$$|y_1(x) - y_2(x)| \leqq \frac{3}{320\,000}\,380 = 0{\cdot}003\,56.$$

For the error estimate we use the basic formula (5.29) of Ch. I with $W = 1$; the appropriate GREEN's function is given by (3.28) of Ch. I. With $n = \sqrt{6}$ we have

$$2n\sin n\pi\int_0^{2\pi}|G(x,\xi)|\,d\xi = 2\int_0^{\pi}|\cos nx|\,dx = 2\,\frac{\sin(\sqrt{6}\pi) + 4}{\sqrt{6}} = 2\times 2{\cdot}0361;$$

then from (5.25), (5.29) of Ch. I

$$K = \frac{2{\cdot}0361}{\sqrt{6}\sin(\sqrt{6}\pi)}\,N = 0{\cdot}8418\,N.$$

We now need a value for N, the Lipschitz constant for the function $y^2 + \frac{3}{2}\cos x$. Clearly an upper bound for $|2y|$ will suffice, and if we assume that $|y| \leqq 0{\cdot}4$, i.e. we try to find a solution in the subspace F of all continuous functions $y(x)$ of period 2π whose members satisfy $|y(x)| \leqq 0{\cdot}4$, then we can use the value

$$N = |2y|_{\max} = 0{\cdot}8;$$

this gives

$$K = 0{\cdot}8418\times 0{\cdot}8 = 0{\cdot}673, \qquad \frac{K}{1-K} = 2{\cdot}06.$$

The "sphere" Σ defined by

$$|y - y_2| \leqq 0{\cdot}003\,56\times 2{\cdot}06 = 0{\cdot}007\,35$$

lies entirely in F, and hence there is a solution in this subspace, namely the function towards which the sequence $y_0, y_1, y_2, \ldots$ is converging. Incidently this shows that there are no other periodic solutions of (4.50) with $|y| \leqq 0{\cdot}4$.

The fact that y lies in the sphere Σ gives us a smaller upper bound for $|y|_{\max}$:

$$|y| \leqq |y_1| + |y_1 - y_2| + 0{\cdot}007\,35 \leqq 0{\cdot}3409.$$

We can now repeat the above calculation with more refined values:

$$N = 2|y|_{\max} = 0\cdot6818, \qquad K = 0\cdot8418 \times 0\cdot6818 = 0\cdot585,$$

$$\frac{K}{1-K} = 1\cdot352.$$

Thus we obtain the better error estimate

$$|y - y_2| \leq 0\cdot003\,56 \times 1\cdot352 = 0\cdot004\,82.$$

This in turn gives a still smaller upper bound for $|y|_{\max}$, and the process can be repeated; however, little is gained thereby, for with $K = 0\cdot570$ we obtain $|y - y_2| \leq 0\cdot004\,72$.

4.9. Monotonic boundary-value problems for second-order differential equations

Consider the boundary-value problems with the differential equation

$$L[y] = - [p(x)\, y'(x)]' = - F(x, y),$$

where $p(x)$ is positive and possesses a continuous derivative in the interval $a \leq x \leq b$, and with the boundary conditions

(i) $$\qquad\qquad U_1[y] = y(a) = y_a, \qquad U_2[y] = y(b) = y_b,$$

or

(ii)
$$U_1[y] = y'(a) - c\,y(a) = \gamma_a,$$
$$U_2[y] = y'(b) + d\,y(b) = \gamma_b$$

with $c \geq 0$, $d \geq 0$. It will be shown that these problems are monotonic when $F(x, y)$ satisfies certain conditions.

Such problems were considered generally in Ch. I, § 5.6, where it was shown that they were of monotonic type if

$$L[\varepsilon] + \varepsilon\, A(x) \geq 0 \quad \text{with} \quad A(x) \geq 0 \quad \text{and} \quad U_\mu[\varepsilon] = 0 \quad (\mu = 1, 2) \qquad (4.51)$$

implied that $\varepsilon \geq 0$. We now show by indirect proof that this is true of the above problems, excluding only the case with $A(x) \equiv 0$, $c = d = 0$. Suppose that the continuous function ε has a negative minimum value which it assumes at the point $x = \xi$.

1. If ξ is an interior point of the open interval (a, b), then $p(\xi)\,\varepsilon'(\xi) = 0$. Since ε is continuous, there is an interval (α, β) with $a \leq \alpha < \xi < \beta \leq b$ in which $\varepsilon < 0$ and hence $(p\varepsilon')' \leq 0$ from (4.51). This implies that $\varepsilon' \geq 0$ for $\alpha < x \leq \xi$ and $\varepsilon' \leq 0$ for $\xi \leq x < \beta$; hence ε is constant in (α, β), for otherwise, since $x = \xi$ is a minimum, there would exist ϱ with $\varepsilon(\varrho) > \varepsilon(\xi)$, $\xi < \varrho < \beta$ and hence also $\tilde{\varrho}$ with $\varepsilon'(\tilde{\varrho}) > 0$, $\xi < \tilde{\varrho} \leq \varrho$. It follows that ε must be constant in (a, b) also. In case (i) we therefore have $\varepsilon \equiv 0$, which contradicts the assumption of a negative minimum value. In case

(ii) we have $c\varepsilon(a) = d\varepsilon(b) = 0$ since $\varepsilon'(a) = \varepsilon'(b) = 0$; if c and d are not both zero, then again $\varepsilon \equiv 0$; if $c = d = 0$, then $A \not\equiv 0$ (since we are excluding the case $c = d = 0$, $A(x) \equiv 0$), so that there exists σ in $a < \sigma < b$ with $A(\sigma) > 0$, and since $\varepsilon' \equiv 0$ and $\varepsilon < 0$, (4.51) is contradicted at $x = \sigma$. Consequently ξ cannot be a point of the open interval (a, b).

2. The alternative, that ξ is a boundary point, say $\xi = a$, can be ruled out immediately in case (i), for it requires that $\varepsilon(a) < 0$. In case (ii) this implies that $\varepsilon'(a) \leq 0$; but, since $\xi = a$ is a minimum, we must also have $\varepsilon'(a) \geq 0$; hence $\varepsilon'(a) = 0$, and we can use the same arguments to obtain a contradiction as in 1. but with $x = a$ taking over the role played there by $x = \xi$. Therefore ε cannot be negative in $\langle a, b \rangle$.

This result may be stated as follows:

The boundary-value problem

$$Tu \equiv - [p(x) u'(x)]' + F(x, u) = 0 \tag{4.52}$$

with the boundary conditions

(i)
$$u(a) = u_a, \quad u(b) = u_b \tag{4.53}$$

or

(ii)
$$\begin{cases} u'(a) - c u(a) = \gamma_a \\ u'(b) + d u(b) = \gamma_b, \end{cases} \tag{4.54}$$

where $p(x)$ is positive and possesses a continuous derivative in (a, b), $u_a, u_b, \gamma_a, \gamma_b$ are given constants and $c \geq 0$, $d \geq 0$, is of monotonic type in a domain H of the (x, u) space [defined by $a \leq x \leq b$, $u_0(x) \leq u \leq u_1(x)$, say] when $\dfrac{\partial F}{\partial u}$ exists in H and in case (i) $\dfrac{\partial F}{\partial u} \geq 0$ and in case (ii) either $\dfrac{\partial F}{\partial u} > 0$ or $\dfrac{\partial F}{\partial u} \geq 0$ and $c^2 + d^2 > 0$.

Under these circumstances, if the boundary-value problem possesses a solution $y(x)$ in H and we have two functions y_1 and y_2 which satisfy the boundary conditions and are such that $T y_1 \leq 0 \leq T y_2$, then we know that $y_1(x) \leq y(x) \leq y_2(x)$.

Example. The problem

$$y'' = 6x y^2, \quad y(0) = y(1) = 1$$

is of monotonic type in the domain H defined by $0 \leq x \leq 1$, $0 \leq y \leq 1$. Suppose that the existence of a solution in H has been established, say by considering the initial-value problem

$$Ty = - y'' + 6x y^2 = 0; \quad y(0) = 1, \quad y'(0) = \sigma$$

with variable σ. For y we can make the approximation

$$w = 1 + \sum_{\nu=3}^{5} a_\nu(x - x^\nu),$$

which satisfies the boundary conditions for arbitrary a_ν; the term with $\nu = 2$ is omitted since $y''(0) = 0$; more terms can be added as desired, of course. The functions

$$w_1 = 1 - (x - x^3)$$
$$w_2 = 1 - 0\cdot43\,(x - x^4)$$

give $Tw_1 \leqq 0$, $Tw_2 \geqq 0$, respectively, and hence $w_1 \leqq y \leqq w_2$; these limits may easily be improved by using more accurate expressions for w.

§5. RITZ's method for second-order boundary-value problems

For many boundary-value problems of even order it is possible to specify an integral expression $J[\varphi]$ which can be formed for all functions φ of a certain class and which has a minimum value for just that function y which solves the boundary-value problem. Consequently solution of the boundary-value problem is equivalent to minimizing $J[\varphi]$. In RITZ's method[1] the solution of this variational problem is approximated by a linear combination of suitably chosen functions. This method shares with the finite-difference method a very favoured position among the methods for the approximate solution of boundary-value problems.

5.1. EULER's differential equation in the calculus of variations

The means of formulating the required expression $J[\varphi]$ are furnished by the calculus of variations. Consider, for example, the simple variational problem of determining in the domain of all functions $\varphi(x)$ with continuous derivatives in $a \leqq x \leqq b$ and prescribed values $\varphi(a) = A$, $\varphi(b) = B$ at the end points (the domain of "admissible" functions) that function which minimizes the value of the integral

$$J[\varphi] = \int_a^b F(x, \varphi, \varphi')\,dx, \tag{5.1}$$

in which F is a given function possessing continuous derivatives with respect to each of its arguments. It may be shown[2] that if a solution

---

[1] WALTER RITZ, born 22 February 1878 in Sion (Switzerland) in the Rhone valley, son of the artist Raphael Ritz, studied in Zürich, then in Göttingen, where in 1902 he obtained his doctor's degree under Voigt; he then worked in Leyden under H. A. Lorentz, in Paris under A. Cotton, in Tübingen under F. Paschen and in 1908 went back to Göttingen, where he died on the 7 July 1909. After a poorly healed pleurisy in 1900 his zeal for his scientific work was continually in conflict with consideration for his state of health, until eventually his health was sacrificed. (Obituary by PIERRE WEISZ in W. RITZ: Gesammelte Werke. Paris 1911.) — Habilitationsschrift von RITZ über eine neue Methode zur Lösung gewisser Variationsprobleme der mathematischen Physik. J. Reine Angew. Math. **135**, H. 1 (1908). — Ann. Phys., Lpz. **28**, 737 (1909).

[2] Cf., for example, G. GRÜSZ: Variationsrechnung, p. 11. Leipzig 1938, or R. COURANT: Differential and Integral Calculus Vol. II, p. 495 et seq. London: Blackie 1936.

$\varphi = y(x)$ which gives the integral a value not greater than that given by any other admissible function $\varphi(x)$ exists at all, then $y(x)$ must necessarily satisfy the second-order differential equation

$$- \frac{d}{dx}\left(\frac{\partial F}{\partial \varphi'}\right) + \frac{\partial F}{\partial \varphi} = 0 \tag{5.2}$$

[and also, of course, the boundary conditions $y(a) = A$, $y(b) = B$]. This is EULER's equation for the functional (5.1).

5.2. Derivation of EULER's conditions

For the application of RITZ's method to the solution of boundary-value problems we must reverse the procedure of § 5.1 and try to write the differential equation of the given boundary-value problem as the Euler equation of some variational problem; this will yield the appropriate expression $J[\varphi]$. For example, for the second-order linear boundary-value problem

$$L[y] = -\frac{d}{dx}(p\,y') + q(x)\,y = r(x) \tag{5.3}$$

with the boundary conditions

$$\left.\begin{array}{l} U_1[y] = \alpha_0\,y(a) + \alpha_1\,y'(a) = \gamma_1, \\ U_2[y] = \beta_0\,y(b) + \beta_1\,y'(b) = \gamma_2 \end{array}\right\} \tag{5.4}$$

it can be seen immediately that with

$$F = p\,\varphi'^2 + q\,\varphi^2 - 2r\,\varphi \tag{5.5}$$

the Euler equation (5.2) is identical with the given differential equation (5.3). The boundary conditions here are rather more general than in § 5.1 and $J[\varphi]$ must be modified slightly (by the addition of certain "boundary" terms), otherwise a solution of the variational problem with the new domain of admissible functions satisfying the new boundary conditions (assuming not both of α, β are zero) will not exist in general; this modification does not affect the Euler equation. We demonstrate by giving a short derivation of the Euler conditions for the more general expression

$$J[\varphi] = \int_a^b (p\,\varphi'^2 + q\,\varphi^2 - 2r\,\varphi)\,dx + A_0\,\varphi_a^2 + B_0\,\varphi_b^2 + 2A_1\,\varphi_a + 2B_1\,\varphi_b, \tag{5.6}$$

in which the more compact notation φ_a has been used for $\varphi(a)$, etc.

At first we restrict the domain of the admissible functions to those functions φ with continuous second derivatives which satisfy the boundary conditions (5.4); we shall see later that this domain can sometimes be enlarged to include functions satisfying fewer boundary conditions, or even none at all.

Since we do not here go into the question of the existence of a minimum of $J[\varphi]$ with respect to the given domain of admissible functions, we will assume that there is a smallest value of J and that there is a function Y among the admissible functions φ for which J attains this value, i.e.

$$J[Y] \leq J[\varphi]. \tag{5.7}$$

As yet, this function Y has nothing to do with the boundary-value problem (5.3), (5.4).

If η is a function with continuous second derivatives which satisfies the homogeneous boundary conditions

$$\left.\begin{array}{l} \alpha_0 \eta_a + \alpha_1 \eta_a' = 0, \\ \beta_0 \eta_b + \beta_1 \eta_b' = 0 \end{array}\right\} \tag{5.8}$$

corresponding to the inhomogeneous conditions (5.4), then the function $\varphi = Y + \varepsilon \eta$, where ε is an arbitrary constant, satisfies the inhomogeneous conditions (5.4) and is an admissible function; it therefore satisfies the relation (5.7). If we allow ε to run through a range of positive and negative values, $J[\varphi]$ defines a function $\Phi(\varepsilon)$ of the parameter ε which possesses a continuous derivative with respect to ε and which attains a minimum value, namely the smallest value of $J[\varphi]$, at $\varepsilon = 0$. Hence the derivative with respect to ε must vanish at this point:

$$\Phi'(0) = \left[\frac{dJ[\varphi(\varepsilon)]}{d\varepsilon}\right]_{\varepsilon=0} = \left[\frac{dJ[Y+\varepsilon\eta]}{d\varepsilon}\right]_{\varepsilon=0} = 0. \tag{5.9}$$

$[\delta J$ is often written for the expression $\varepsilon \Phi'(0)$ and is called the "first variation" of J.]

Substituting $\varphi = Y + \varepsilon \eta$ in (5.6), we obtain

$$\Phi(\varepsilon) = J[Y + \varepsilon \eta]$$
$$= \int_a^b \{p(Y' + \varepsilon\eta')^2 + q(Y + \varepsilon\eta)^2 - 2r(Y + \varepsilon\eta)\} dx + A_0(Y_a + \varepsilon\eta_a)^2 + \cdots,$$

and hence

$$\tfrac{1}{2}\Phi'(0) = \int_a^b (p\,Y'\eta' + q\,Y\eta - r\eta)\,dx + A_0 Y_a \eta_a + B_0 Y_b \eta_b + A_1 \eta_a + B_1 \eta_b.$$

The integral of the first term in the integrand can be transformed by integration by parts:

$$\int_a^b p\,Y'\eta'\,dx = p_b Y_b' \eta_b - p_a Y_a' \eta_a - \int_a^b (p\,Y')'\eta\,dx. \tag{5.10}$$

The equation $\Phi'(0) = 0$ then reads

$$\int_a^b \eta\left[-(p\,Y')' + q\,Y - r\right]dx + \eta_a W_a + \eta_b W_b = 0, \tag{5.11}$$

where
$$W_a = A_0 Y_a + A_1 - p_a Y_a', \qquad W_b = B_0 Y_b + B_1 + p_b Y_b'. \qquad (5.12)$$

Now equation (5.11) is to hold for any function $\eta(x)$ which possesses a continuous second derivative and satisfies the boundary conditions (5.8). This is possible only if the factor multiplying η in the integrand vanishes identically, for otherwise we can always construct a function $\eta = \eta^{**}$ satisfying the above conditions but for which (5.11) does not hold: let $\eta = \eta^*$ be a function satisfying the above conditions and for which (5.11) does hold, and let x_0 be a point in (a, b) where the factor is positive, say; by continuity it must be positive also for $|x - x_0| \leq \delta$ with $\delta > 0$, and we put

$$\eta^{**} = \begin{cases} \eta^* + (x - x_0 - \delta)^3 (x - x_0 + \delta)^3 & \text{for} \quad |x - x_0| \leq \delta \\ \eta^* & \text{otherwise}. \end{cases} \qquad (5.13)$$

Consequently we must have

$$(-p Y')' + q Y - r = 0. \qquad (5.14)$$

This differential equation, known as the Euler equation of the variational problem corresponding to (5.4), (5.6), therefore furnishes a necessary condition to be satisfied by the solution $Y(x)$ of that problem.

From (5.11) and (5.14) it follows that another necessary condition is

$$\eta_a W_a + \eta_b W_b = 0. \qquad (5.15)$$

For further discussion we distinguish three cases:

1. Case $\alpha_1 \neq 0, \beta_1 \neq 0$. Here, given any values η_a, η_b we can calculate η_a', η_b' from (5.8). From all functions η with continuous second derivatives and any given boundary values η_a, η_b we can select one with the boundary derivative values η_a', η_b' calculated from (5.8); this function will then satisfy the homogeneous boundary conditions. Thus for any given η_a, η_b there is a function η with the boundary values η_a, η_b and satisfying all the conditions of admission. Now (5.15) can be valid for all η_a, η_b only if

$$W_a = W_b = 0.$$

These equations represent two boundary conditions [see (5.12)] which Y must necessarily satisfy.

In order that the solution of the variational problem of minimizing (5.6) shall also solve the boundary-value problem (5.3), (5.4), we have only to put

$$A_0 = -\frac{\alpha_0}{\alpha_1} p_a, \quad A_1 = \frac{\gamma_1}{\alpha_1} p_a, \quad B_0 = \frac{\beta_0}{\beta_1} p_b, \quad B_1 = -\frac{\gamma_2}{\beta_1} p_b, \qquad (5.16)$$

for this ensures that the boundary conditions on Y are identical with the conditions (5.4) [we have already arranged that the differential equations (5.3) and (5.14) are identical]; p_a, p_b are here assumed to be non-zero, for otherwise the points a, b, respectively, would be singular points of the differential equation.

Since we have made $W_a = W_b = 0$ by using the values determined from (5.16), we have $\eta_a W_a + \eta_b W_b = 0$ for any function η with a continuous second derivative; thus (and this is an important new aspect) the boundary conditions can be suppressed and the solution in the so widened domain of admissible functions will satisfy the boundary conditions automatically. In this case ($\alpha_1 \neq 0$, $\beta_1 \neq 0$) the boundary conditions (5.4) are "suppressible" according to the definition in § 1.2 of Ch. I; the sense in which this type of boundary condition is suppressible is now explained.

2. Case $\alpha_1 = \beta_1 = 0$, $\alpha_0 \neq 0$, $\beta_0 \neq 0$. Here our domain of admissible functions is restricted to those with the fixed values

$$\varphi(a) = \frac{\gamma_1}{\alpha_0}, \qquad \varphi(b) = \frac{\gamma_2}{\beta_0} \tag{5.17}$$

at the end points, and from (5.8) we now have $\eta_a = \eta_b = 0$ for all the functions η. Consequently the expression $\eta_a W_a + \eta_b W_b$ vanishes however W_a and W_b are constituted and we may therefore choose A_0, B_0, A_1, B_1 as we please. Thus no "boundary terms" are needed in this case, for we may take all these constants to be zero and use just the integral term in $J[\varphi]$. We cannot, however, extend the domain of admissible functions as in the previous case, and hence the boundary conditions are essential. These boundary conditions are "essential" according to the definition in § 1.2 of Ch. I; and that terminology also is now explained.

3. Case in which just one of α_1, β_1 is zero. We can combine what has been said for the two previous cases; for example, if $\alpha_1 = 0$, $\beta_1 \neq 0$, we obtain the necessary conditions

$$\eta_a = 0, \quad W_b = 0.$$

We summarize these results in the following

Theorem. *The second-order linear boundary-value problem*

$$\left. \begin{array}{c} -(p(x) y')' + q(x) y = r(x), \quad \text{where} \quad p(a) \neq 0, \quad p(b) \neq 0, \\ \alpha_0 y(a) + \alpha_1 y'(a) = \gamma_1, \\ \beta_0 y(b) + \beta_1 y'(b) = \gamma_2, \end{array} \right\} \tag{5.18}$$

including the boundary conditions, may be written as the necessary conditions to be satisfied by the solution of a variational problem $J[\varphi] = $ minimum. In the case $\alpha_1 \neq 0$, $\beta_1 \neq 0$,

$$J[\varphi] = \int_a^b (p\,\varphi'^2 + q\,\varphi^2 - 2r\,\varphi)\,dx + R_a + R_b, \tag{5.19}$$

where

$$R_a = \frac{p(a)}{\alpha_1} \left(-\alpha_0 \varphi^2(a) + 2\gamma_1 \varphi(a) \right), \quad R_b = \frac{p(b)}{\beta_1} \left(\beta_0 \varphi^2(b) - 2\gamma_2 \varphi(b) \right), \quad (5.20)$$

and no boundary conditions need be imposed on the admissible functions φ. In the case $\alpha_1 = 0$ we may put $R_a = 0$, but we must restrict the admissible functions by demanding that they satisfy the "essential" boundary condition $\alpha_0 \varphi(a) = \gamma_1$. Similarly, when $\beta_1 = 0$, we may omit the term R_b but must insist on the boundary condition $\beta_0 \varphi(b) = \gamma_2$. This theorem does not, however, answer the question whether the function $y(x)$ is actually a solution of the variational problem so formulated or indeed whether $J[\varphi]$ possesses a minimum value at all [1].

5.3. The Ritz approximation [2]

Having formulated the variational problem (5.19), (5.20) corresponding to the given boundary-value problem (5.18), we now use RITZ's method for its approximate solution. This consists in the reduction of the problem to an ordinary minimum problem by substituting a linear combination of suitable functions for $\varphi(x)$, say

$$\varphi(x) = v_0(x) + \sum_{\nu=1}^{p} a_\nu v_\nu(x), \qquad (5.21)$$

so that $J[\varphi]$ becomes an ordinary function of the unknown coefficients of the combination. If we had an "essential" boundary condition at $x = a$, say, i.e. $\alpha_0 y(a) = \gamma_1$, $\alpha_1 = 0$, then we would choose $v_0(x)$ to satisfy this boundary condition and the $v_\nu(x)$ to satisfy the corresponding homogeneous condition $v_\nu(a) = 0$; this ensures that $\alpha_0 \varphi(a) = \gamma_1$ for all values of the a_ν. On the other hand, if all the boundary conditions were "suppressible", we could omit v_0 and choose the v_ν regardless of any boundary conditions, but we would have to take into account the boundary terms R_a and R_b as defined in (5.20).

On account of the linearity of the expression assumed for φ, the equations for determining the a_ν so as to minimize $J[\varphi]$ are also linear. We have

$$\left. \begin{aligned} \frac{1}{2} \frac{\partial J[\varphi]}{\partial a_\nu} &= \int_a^b (p\varphi' v_\nu' + q\varphi v_\nu - r v_\nu)\, dx + \\ &+ \frac{p(a) v_\nu(a)}{\alpha_1} \left(-\alpha_0 \varphi(a) + \gamma_1 \right) + \frac{p(b) v_\nu(b)}{\beta_1} \left(\beta_0 \varphi(b) - \gamma_2 \right) = 0. \end{aligned} \right\} \quad (5.22)$$

---

[1] See Exercise 6 of § 8.11.

[2] Error estimates for RITZ's method can be found in NICOLAS KRYLOFF: Les méthodes de solution approchée des problèmes de la physique mathématique. Mémorial. Sci. Math., Paris **49** (1931).

Of the numerous shorter papers by KRYLOFF on error estimates of this kind we mention without giving their titles just a few published in the C. R. Acad. Sci., Paris **180**, 1316−1318 (1925); **181**, 86−88 (1925); **183**, 476−479 (1926); **186**, 298−300, 422−425 (1928).

If the v_ν possess continuous second derivatives, integration by parts yields

$$\left.\begin{aligned}\int_a^b v_\nu\{L[\varphi] - r\}\,dx - \frac{p(a)\,v_\nu(a)}{\alpha_1}\,\{U_1[\varphi] - \gamma_1\} + \\ + \frac{p(b)\,v_\nu(b)}{\beta_1}\,\{U_2[\varphi] - \gamma_2\} = 0 \quad (\nu = 1, 2, \ldots, p),\end{aligned}\right\} \quad (5.23)$$

in which the notation of (5.3), (5.4) has been used. Written in full with the expression (5.21) inserted for φ the equations for the a_ν read

$$\left.\begin{aligned}\sum_{k=1}^p a_k\left\{\int_a^b v_\nu L[v_k]\,dx - \frac{p(a)\,v_\nu(a)}{\alpha_1}\,U_1[v_k] + \frac{p(b)\,v_\nu(b)}{\beta_1}\,U_2[v_k]\right\} \\ = -\int_a^b v_\nu\{L[v_0] - r\}\,dx + \frac{p(a)\,v_\nu(a)}{\alpha_1}\,\{U_1[v_0] - \gamma_1\} - \\ - \frac{p(b)\,v_\nu(b)}{\beta_1}\,\{U_2[v_0] - \gamma_2\} \quad (\nu = 1, 2, \ldots, p).\end{aligned}\right\} \quad (5.24)$$

When we have only "essential" boundary conditions, the boundary terms in (5.23) disappear and we are left with equations identical with GALERKIN's equations (see Ch. I, § 4.3).

Note 1. The differential equation (5.3), which is of self-adjoint form, is a special case of the general second-order linear equation

$$L[y] = f_2 y'' + f_1 y' + f_0 y = r(x). \quad (5.25)$$

However, when $f_2(x) \neq 0$ [a zero of $f_2(x)$ is a singular point of the differential equation], the general equation (5.25) may be transformed into the special form (5.3); thus if we multiply by

$$\varrho(x) = \exp \int_{x_0}^x \frac{f_1(\xi) - f_2'(\xi)}{f_2(\xi)}\,d\xi$$

we obtain

$$(f_2 \varrho y')' + f_0 \varrho y = r \varrho. \quad (5.26)$$

The considerations of this section are therefore also applicable to every non-singular linear differential equation of the second order.

Note 2. In principle, any second-order differential equation (including non-linear equations) may be written as the Euler equation of some variational problem[1]. In practice the derivation of the appropriate expression $J[\varphi]$ often occasions considerable difficulty; in general it requires the solution of a partial differential equation.

Note 3. The Ritz method can also be combined with other approximate methods; for example, one can demand that the approximation function (5.21) satisfies the differential equation exactly at $q(<p)$ points (combination with the collocation method); this immediately eliminates q parameters and the Ritz calculation with $p - q$ parameters is correspondingly simpler, while the accuracy

---

[1] BOLZA, O.: Vorlesungen über Variationsrechnung. Leipzig 1909. Reprinted 1949, 705 + 10 pp.

of the approximation so obtained, as shown by examples, can be just as good as that of the pure Ritz calculation with p parameters provided that the additional q conditions are suitably chosen[1].

5.4. Examples of the application of Ritz's method to boundary-value problems of the second order

Example I. A linear inhomogeneous boundary-value problem. From (5.19) the variational problem corresponding to the strut problem of § 1.2 (Example I)

reads
$$y'' + (1 + x^2)\, y + 1 = 0, \qquad y(\pm 1) = 0$$
$$J[\varphi] = \int_{-1}^{1} \left(\varphi'^{\,2} - (1 + x^2)\, \varphi^2 - 2\varphi\right) dx = \text{min.},$$

where φ must satisfy the "essential" boundary conditions $\varphi(\pm 1) = 0$. The simplest form we can assume for φ is a polynomial expression satisfying the boundary conditions and exploiting the symmetry of the solution; thus we write

$$\varphi = \sum_{\nu=1}^{p} a_\nu v_\nu(x) = \sum_{\nu=1}^{p} a_\nu (1 - x^{2\nu}).$$

Such an expression with two terms $(p = 2)$:

$$\varphi = a(1 - x^2) + b(1 - x^4) \qquad \text{with} \qquad \varphi' = -2ax - 4bx^3$$

(in which we have used a, b instead of a_1, a_2) yields

$$\frac{1}{8}\, J[\varphi] = \frac{19}{105}\, a^2 + \frac{10}{45} \cdot 2ab + \frac{1244}{3465}\, b^2 - \frac{1}{3}\, a - \frac{2}{5}\, b. \qquad (5.27)$$

A simple yet significant check can be made on this calculation by putting $a = b = A$, so that $\varphi = A(2 - x^2 - x^4)$, and calculating J again for this function. We should get the same result, namely

$$\frac{1}{8}\, J = \frac{379}{385}\, A^2 - \frac{11}{15}\, A,$$

by putting $a = b = A$ in (5.27). We now calculate a and b from

$$\frac{1}{8}\, \frac{\partial J[\varphi]}{\partial a} = \frac{38}{105}\, a + \frac{20}{45}\, b - \frac{1}{3} = 0,$$

$$\frac{1}{8}\, \frac{\partial J[\varphi]}{\partial b} = \frac{20}{45}\, a + \frac{2488}{3465}\, b - \frac{2}{5} = 0,$$

obtaining

$$a = \frac{4200}{4252} = 0.987\,77, \qquad b = -\frac{231}{4252} = -0.054\,33.$$

---

[1] See, for example, Th. Pöschl: Über eine mögliche Verbesserung der Ritzschen Methode. Ing.-Arch. **23**, 365—372 (1955).

The second approximation by this method is therefore

$$y_2 = \frac{1}{4252}(3969 - 4200\,x^2 + 231\,x^4);$$

in particular we have

$$y_2(0) = \frac{3969}{4252} = 0\cdot933\,443.$$

If more accurate values are required, we can use a three-term expression:

$$\varphi = a(1 - x^2) + b(1 - x^4) + c(1 - x^6).$$

This yields

$$\frac{1}{8}J[\varphi] = \frac{19}{105}a^2 + \frac{10}{45}2ab + \frac{1244}{11 \times 315}b^2 + \frac{93}{385}2ac + \frac{19846}{143 \times 315}2bc +$$
$$+ \frac{25943}{143 \times 315}c^2 - \frac{1}{3}a - \frac{2}{5}b - \frac{3}{7}c,$$

which we can check similarly by putting $a = b = c = A$; with $\varphi = A(3 - x^2 - x^4 - x^6)$ we obtain

$$\frac{1}{8}J = \frac{3764}{1287}A^2 - \frac{122}{105}A.$$

The linear equations

$$\frac{1}{16}\frac{\partial J[\varphi]}{\partial a} = \frac{19}{105}a + \frac{10}{45}b + \frac{93}{385}c - \frac{1}{6} = 0,$$

$$\frac{1}{16}\frac{\partial J[\varphi]}{\partial b} = \frac{10}{45}a + \frac{1244}{11 \times 315}b + \frac{19846}{143 \times 315}c - \frac{1}{5} = 0,$$

$$\frac{1}{16}\frac{\partial J[\varphi]}{\partial c} = \frac{93}{385}a + \frac{19846}{143 \times 315}b + \frac{25943}{143 \times 315}c - \frac{3}{14} = 0,$$

i.e.

$$8151\,a + 10010\,b + 10881\,c - 7507\cdot5 = 0,$$
$$10010\,a + 16172\,b + 19846\,c - 9009 = 0,$$
$$10881\,a + 19846\,b + 25943\,c - 9652\cdot5 = 0,$$

have the solution

$$a = 0\cdot9664778,$$
$$b = -0\cdot00473781,$$
$$c = -0\cdot02966958.$$

These values define the third approximation $\varphi = y_3$, which yields in particular

$$y_3(0) = a + b + c = 0\cdot9320704.$$

II. An eigenvalue problem.

Corresponding to the eigenvalue problem

$$-[(1 + x)\,y']' = \lambda(1 + x)\,y, \quad y'(0) = y(1) = 0,$$

associated with the longitudinal vibrations of a rod (Example III, §1.2), we have from (5.19) the formal variational problem

$$J[\varphi] = \int_0^1 [(1 + x)\,\varphi'^2 - \lambda(1 + x)\,\varphi^2]\,dx = \text{extremum}, \quad \varphi(1) = 0.$$

Here λ is formally treated as a constant parameter (justification of this procedure is considered in §§ 8.7 and 8.8). The admissible functions φ need satisfy only the "essential" boundary condition $\varphi(1) = 0$, but to obtain better results we choose an expression for φ which also satisfies the "suppressible" boundary condition $\varphi'(0) = 0$:

$$\varphi = a(1 - x^2) + b(1 - x^3).$$

The boundary terms which should be added to the integral in $J[\varphi]$ are zero in this case. With Λ as the corresponding approximate value for λ we have

$$J[\varphi] = \frac{7}{3} a^2 + 2 \cdot \frac{27}{10} ab + \frac{33}{10} b^2 - \Lambda\left(\frac{7}{10} a^2 + 2 \cdot \frac{163}{210} ab + \frac{243}{280} b^2\right).$$

Repeating the calculation with $-a = b = 1$, i.e. $\varphi = x^2 - x^3$, as a check, we obtain

$$J = \frac{7}{30} - \frac{13}{840} \Lambda,$$

which is also the value obtained by putting $-a = b = 1$ in $J[\varphi]$.

The equations

$$\frac{1}{2} \frac{\partial J}{\partial a} = a\left(\frac{7}{3} - \Lambda\frac{7}{10}\right) + b\left(\frac{27}{10} - \Lambda\frac{163}{210}\right) = 0,$$

$$\frac{1}{2} \frac{\partial J}{\partial b} = a\left(\frac{27}{10} - \Lambda\frac{163}{210}\right) + b\left(\frac{33}{10} - \Lambda\frac{243}{280}\right) = 0$$

have a non-trivial solution if and only if the determinant of their coefficients vanishes:

$$\begin{vmatrix} \dfrac{7}{3} - \Lambda\dfrac{7}{10} & \dfrac{27}{10} - \Lambda\dfrac{163}{210} \\[2ex] \dfrac{27}{10} - \Lambda\dfrac{163}{210} & \dfrac{33}{10} - \Lambda\dfrac{243}{280} \end{vmatrix} = 0.$$

This yields the approximations

$$\Lambda = \begin{cases} 3 \cdot 218\,524, \\ 25 \cdot 334 \end{cases}$$

for the first two eigenvalues.

By suppressing the second row and column of the above determinant we obtain for comparison the "first approximation" to the smallest eigenvalue; thus the single-term expression $\varphi = a(1 - x^2)$ yields

$$\frac{7}{3} - \Lambda\frac{7}{10} = 0, \quad \text{so that} \quad \Lambda = \frac{10}{3} \approx 3 \cdot 333.$$

Approximations for the corresponding eigenfunctions can be found by solving the homogeneous equations with the appropriate value of Λ

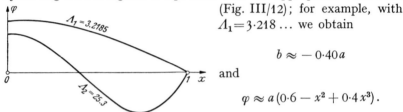

(Fig. III/12); for example, with $\Lambda_1 = 3 \cdot 218 \ldots$ we obtain

$$b \approx -0 \cdot 40\, a$$

and

$$\varphi \approx a\,(0 \cdot 6 - x^2 + 0 \cdot 4\,x^3).$$

Fig. III/12. Approximate eigenfunctions

III. A non-linear boundary-value problem. For the Example II of § 1.2

$$y'' = \frac{3}{2}\,y^2; \qquad y\,(0) = 4, \qquad y\,(1) = 1$$

the corresponding variational problem according to (5.1), (5.2) is

$$J[\varphi] = \int_0^1 (\varphi'^2 + \varphi^3)\,dx = \text{extremum}, \qquad \varphi\,(0) = 4, \qquad \varphi\,(1) = 1.$$

Fig. III/13. Solution of the non-linear equations

As in (5.21), we assume a form

$$\varphi = 4 - 3\,x + a_1\,(x - x^2) + a_2\,(x - x^3)$$

which satisfies the boundary conditions for arbitrary a_1, a_2. Then from

$$\frac{\partial J}{\partial a_1} = \int_0^1 [2\varphi'(1 - 2x) + 3\varphi^2\,(x - x^2)]\,dx = 0,$$

$$\frac{\partial J}{\partial a_2} = \int_0^1 [2\varphi'(1 - 3x^2) + 3\varphi^2\,(x - x^3)]\,dx = 0$$

we obtain the two non-linear equations

$$1407 + 490\,a_1 + 726\,a_2 + 9\,a_1^2 + 27\,a_1\,a_2 + \frac{41}{2}\,a_2^2 = 0,$$

$$1302 + 484\,a_1 + 750\,a_2 + 9\,a_1^2 + \frac{82}{3}\,a_1\,a_2 + 21\,a_2^2 = 0.$$

By linear combination of these equations two new equations can be produced which do not contain terms in a_1^2 and a_2^2, respectively, and which can therefore be easily solved for a_1 in terms of a_2 and for a_2 in terms of a_1, respectively:

$$a_1 = \frac{-105 + 24\,a_2 + \frac{1}{2}a_2^2}{6 - \frac{1}{3}a_2}, \qquad a_2 = \frac{2856 + 368\,a_1 + \frac{9}{2}a_1^2}{129 - \frac{20}{3}a_1};$$

this facilitates the determination of a_1 and a_2, see Fig. III/13. Table III/12 gives the values obtained for a_1, a_2 and compares specimen values of the solution with those obtained by means of a single-term expression (without the term in a_2).

Table III/12. *Approximate solution of a non-linear problem by* RITZ's *method*

x	Single-term approximation		Two-term approximation	
	The everywhere positive	The other	The everywhere positive	The other
	solution		solution	
	$9\,a_1 = -245 \pm \sqrt{47\,362}$		$a_1 = -7\cdot07004$	$a_1 = -32\cdot20$
	$a_1 = -3\cdot0413$	$a_1 = -51\cdot40$	$a_2 = \ \ 2\cdot72044$	$a_2 = -12\cdot60$
$\frac{1}{4}$	$2\cdot6798$ (error $+4\cdot7\%$)	$-\ 6\cdot38$	$2\cdot56197$ (error $+0\cdot08\%$)	$-\ 5\cdot75$
$\frac{1}{2}$	$1\cdot7397$ (error $-2\cdot1\%$)	$-10\cdot35$	$1\cdot7627$ (error $-0\cdot86\%$)	$-10\cdot28$
$\frac{3}{4}$	$1\cdot1798$ (error $-9\cdot7\%$)	$-\ 7\cdot88$	$1\cdot31701$ (error $+0\cdot84\%$)	$-\ 8\cdot43$

§6. RITZ's method for boundary-value problems of higher order

We now turn to problems in which higher order derivatives appear; we derive the Euler equations generally, including non-linear cases, but we can now be brief.

6.1. Derivation of higher order Euler equations

Consider the variational problem

$$J = J[\varphi] = \int_a^b F(x, \varphi, \varphi', \dots, \varphi^{(m)})\, dx + A\big(\varphi(a), \varphi'(a), \dots, \varphi^{(m-1)}(a)\big) + \\ + B\big(\varphi(b), \varphi'(b), \dots, \varphi^{(m-1)}(b)\big) = \text{minimum} \tag{6.1}$$

with respect to the domain of admissible functions $\varphi(x)$ which possess continuous derivatives of order $2m$ and satisfy certain prescribed linear (and linearly independent) boundary conditions

$$U_\mu[\varphi] = \gamma_\mu \qquad (\mu = 1, 2, \dots, k) \tag{6.2}$$

at the points a and b; these boundary conditions may not contain derivatives of order higher than $m - 1$. F is a given continuous function of $x, \varphi, \varphi', \dots, \varphi^{(m)}$ with continuous partial derivatives of up to the $(m + 1)$-th order, and A, B are given continuous functions of the boundary values of $\varphi, \varphi', \varphi'', \dots, \varphi^{(m-1)}$ with continuous first partial derivatives.

We assume that the minimum value of J exists and that there is a function $y(x)$ among the admissible functions $\varphi(x)$ for which J takes this minimum value:

$$J[y] \leq J[\varphi]. \tag{6.3}$$

Then the one-parameter family of admissible functions defined by $\varphi(x) = y(x) + \varepsilon\eta(x)$, where $\eta(x)$ is any function possessing a continuous derivative of order $2m$ and satisfying the homogeneous boundary conditions

$$U_\mu[\eta] = 0 \qquad (\mu = 1, 2, \ldots, k), \tag{6.4}$$

must also satisfy the inequality (6.3). For this family of functions $J[\varphi]$ is a function $\Phi(\varepsilon)$ of the parameter ε; it can have a minimum at $\varepsilon = 0$ only if the derivative $\Phi'(0)$ vanishes.

We now evaluate $\Phi'(0)$:

$$\begin{aligned}
\Phi'(0) = \frac{d}{d\varepsilon}\Bigg\{ & \int_a^b F(x, y + \varepsilon\eta, y' + \varepsilon\eta', \ldots, y^{(m)} + \varepsilon\eta^{(m)})\, dx + \\
& + A(x + \varepsilon\eta, \ldots, y^{(m-1)} + \varepsilon\eta^{(m-1)})_{x=a} + \\
& + B(y + \varepsilon\eta, \ldots, y^{(m-1)} + \varepsilon\eta^{(m-1)})_{x=b}\Bigg\}_{\varepsilon=0} \\
= \int_a^b & (F_\varphi\eta + F_{\varphi'}\eta' + \cdots + F_{\varphi^{(m)}}\eta^{(m)})\, dx + A_\varphi\eta(a) + \cdots + \\
& + A_{\varphi^{(m-1)}}\eta^{(m-1)}(a) + B_\varphi\eta(b) + \cdots + B_{\varphi^{(m-1)}}\eta^{(m-1)}(b),
\end{aligned} \tag{6.5}$$

where the subscripts denote partial derivatives, for example,

$$F_{\varphi^{(i)}} = \frac{\partial F}{\partial \varphi^{(i)}}.$$

The integrals of the individual terms of the integrand can be repeatedly transformed by integration by parts until the factor η appears in place of $\eta^{(i)}$; thus we have

$$\begin{aligned}
\int_a^b F_{\varphi'}\eta'\, dx &= [\eta F_{\varphi'}]_a^b - \int_a^b \eta \frac{d}{dx} F_{\varphi'}\, dx, \\
\int_a^b F_{\varphi''}\eta''\, dx &= \left[\eta' F_{\varphi''} - \eta \frac{d}{dx} F_{\varphi''}\right]_a^b + \int_a^b \eta \frac{d^2}{dx^2} F_{\varphi''}\, dx
\end{aligned} \tag{6.6}$$

and so on. With all the terms transformed the condition $\Phi'(0) = 0$ becomes

$$\int_a^b \eta \left[F_\varphi - \frac{d}{dx} F_{\varphi'} + \frac{d^2}{dx^2} F_{\varphi''} - \cdots + (-1)^m \frac{d^m}{dx^m} F_{\varphi^{(m)}} \right] dx + S = 0, \tag{6.7}$$

where S is the boundary expression defined by

$$
\left.
\begin{aligned}
S = &\left[\eta F_{\varphi'} + \left(\eta' F_{\varphi''} - \eta \frac{d}{dx} F_{\varphi''} \right) + \cdots + \right. \\
&+ \left(\eta^{(m-1)} F_{\varphi^{(m)}} - \eta^{(m-2)} \frac{d}{dx} F_{\varphi^{(m)}} + \eta^{(m-3)} \frac{d^2}{dx^2} F_{\varphi^{(m)}} + \cdots + \right. \\
&\left. + (-1)^{m-1} \eta \frac{d^{m-1}}{dx^{m-1}} F_{\varphi^{(m)}} \right) \Bigg|_a^b + A_\varphi \eta(a) + \cdots + \\
&+ A_{\varphi^{(m-1)}} \eta^{(m-1)}(a) + B_\varphi \eta(b) + \cdots + B_{\varphi^{(m-1)}} \eta^{(m-1)}(b).
\end{aligned}
\right\} \quad (6.8)
$$

Just as in § 5.2, we infer from the arbitrariness of η that the bracketed factor of the integrand must vanish [we have only to put

$$
(x - x_0 - \delta)^{2m+1} (x - x_0 + \delta)^{2m+1}
$$

in place of the third power used in (5.13)]; this yields the Euler equation

$$
F_\varphi - \frac{d}{dx} F_{\varphi'} + \frac{d^2}{dx^2} F_{\varphi''} - \cdots + (-1)^m \frac{d^m}{dx^m} F_{\varphi^{(m)}} = 0, \quad (6.9)
$$

which again represents a necessary condition to be satisfied by the solution of the variational problem. From (6.7) the boundary expression S must vanish also.

We now assume that the given functions A, B appearing in (6.1) are quadratic functions of φ and its derivatives; then the expression S is linear in the boundary values of y and its derivatives, and is linear and homogeneous in the boundary values of η and its derivatives. The k equations provided by the homogeneous boundary conditions (6.4) can be solved for k of the $2m$ boundary values $\eta(a), \eta'(a), \ldots, \eta^{(n-1)}(a)$, $\eta(b), \eta'(b), \ldots, \eta^{(m-1)}(b)$ in terms of the remaining ones. These $2m - k$ remaining boundary values of derivatives of η may be called "free boundary values"[1] and will be denoted in any order by $\eta_1, \eta_2, \ldots, \eta_{2m-k}$. If we express all boundary values of η (and its derivatives) appearing in S in terms of the free boundary values, then S can be written in the form

$$
S = \sum_{\nu=1}^{2m-k} \eta_\nu W_\nu[y] = 0, \quad (6.10)
$$

where $W_\nu[y]$ is a function of the boundary values of y, y', ... Since the η_ν may be chosen arbitrarily, and since there exists a corresponding admissible function, it follows from (6.10) that

$$
W_\nu[y] = 0 \quad (\nu = 1, 2, \ldots, 2m - k). \quad (6.11)
$$

These equations, which must be satisfied by the solution y of the variational problem (6.1), (6.2), constitute $2m - k$ additional boundary

---

[1] KAMKE, E.: Math. Z. **48**, 70 (1942).

conditions to be associated with the $2m$-th-order differential equation (6.9), thus making up the number of boundary conditions to the required number $2m$. We assert nothing about the existence of a solution of this boundary-value problem in the present generality; we also defer discussion of the different types of boundary condition (6.2) and (6.11) till the next section.

6.2. Linear boundary-value problems of the fourth order

For illustration, and also because of their important applications, we consider in detail the fourth-order boundary-value problems with the self-adjoint differential equation

$$L[y] = (g_2 y'')'' - (g_1 y')' + g_0 y = r(x) \tag{6.12}$$

and two linearly independent non-contradictory boundary conditions at each of the points $x = a$, $x = b$. These boundary conditions are divided into "essential" and "suppressible" according to the definitions in Ch. I, § 1.2. For fourth-order problems the "essential" boundary conditions are those which contain no second or third derivatives, only the first derivative and the function itself.

As the expression to be minimized in the corresponding variational problem we take

$$J[\varphi] = \int\limits_a^b [g_2 \varphi''^2 + g_1 \varphi'^2 + g_0 \varphi^2 - 2r\varphi] \, dx + R_a + R_b, \tag{6.13}$$

where

$$R_b = A \varphi_b^2 + 2B \varphi_b \varphi_b' + C \varphi_b'^2 + 2D \varphi_b + 2E \varphi_b' \tag{6.14}$$

and R_a is defined similarly but with b replaced by a and with different constants in place of $A, B, \ldots, E$. The subscript notation here denotes argument values, e.g. $\varphi_b = \varphi(b)$, $\varphi_b' = \varphi'(b)$. The variation $\varphi = y + \varepsilon \eta$ leads to the necessary condition

$$\left. \frac{1}{2} \left(\frac{d}{d\varepsilon} J[y + \varepsilon \eta] \right) \right|_{\varepsilon = 0} \\ = \int\limits_a^b [g_2 y'' \eta'' + g_1 y' \eta' + g_0 y \eta - r\eta] \, dx + R_a^* + R_b^* = 0, \left.\right\} \tag{6.15}$$

where

$$R_b^* = A y_b \eta_b + B (y_b \eta_b' + y_b' \eta_b) + C y_b' \eta_b' + D\eta_b + E \eta_b'$$

and R_a^* is defined correspondingly. Integration by parts then yields

$$0 = \int\limits_a^b \{L[y] - r\} \eta \, dx + S_a + S_b, \tag{6.16}$$

where

$$S_b = g_2 y'' \eta' - g_2' y'' \eta - g_2 y''' \eta + g_1 y' \eta + A y\eta + B y\eta' + \\ + B y'\eta + C y' \eta' + D\eta + E \eta',$$

in which all functions are evaluated at the point $x = b$ (S_a reads similarly). By the usual argument the integrand in (6.16) must vanish; thus y must satisfy (6.12) and we must have $S_a + S_b = 0$. We now rearrange S_a and S_b in the forms

$$S_a = \eta_a W_a + \eta'_a W'_a,$$
$$S_b = \eta_b W_b + \eta'_b W'_b$$

with W_a, W'_a, W_b, W'_b independent of η; for instance

$$\left. \begin{array}{l} W_b = - g_2 y''' - g'_2 y'' + (g_1 + B) y' + A y + D, \\ W'_b = g_2 y'' + C y' + B y + E, \end{array} \right\} \tag{6.17}$$

and there are analogous expressions for W_a, W'_a.

We now consider the prescribed boundary conditions, two at each of the points $x = a$, $x = b$, and distinguish the following cases:

Case I. All the boundary conditions are "essential", i.e. the values of y and y' are prescribed at $x = a$ and $x = b$. Here there are admissible functions such that W_a, W'_a, W_b, W'_b do not vanish, and consequently, for $S_a + S_b$ to vanish, we must demand that $\eta_a = \eta'_a = \eta_b = \eta'_b = 0$, i.e. that the admissible functions satisfy the "essential" boundary conditions.

Case II. All boundary conditions are "suppressible". They may therefore be put into the form

$$\left. \begin{array}{l} y''_b = \beta_1 y_b + \beta_2 y'_b + \beta_3, \\ y'''_b = \beta_4 y_b + \beta_5 y'_b + \beta_6 \end{array} \right\} \tag{6.18}$$

with two corresponding conditions at $x = a$. [We assume that $g_2(a) \neq 0$, $g_2(b) \neq 0$, otherwise the differential equation would be singular at the end points.] For arbitrary values of $\eta_a, \eta'_a, \eta_b, \eta'_b$ we can calculate $\eta''_a, \eta'''_a, \eta''_b, \eta'''_b$ from the homogeneous boundary conditions corresponding to (6.18) and the analogous equations for $x = a$; with these values we can then construct a function η such that $y + \varepsilon \eta$ satisfies all the boundary conditions and is an admissible function. Consequently we must have

$$S_a + S_b = \eta_a W_a + \eta'_a W'_a + \eta_b W_b + \eta'_b W'_b = 0$$

for arbitrary values of $\eta_a, \eta'_a, \eta_b, \eta'_b$, and hence $W_a = W'_a = W_b = W'_b = 0$. These four conditions must therefore be made equivalent to the suppressible boundary conditions at $x = a$ and $x = b$ [(6.18)]. We see immediately that this is not possible in general, for there are six free parameters β_i in (6.18) and only five disposable constants A, B, C, D, E in (6.17). Hence for the equivalence of the boundary conditions to be possible at all a relation must exist between the β_i, namely

$$\beta_1 + \beta_2 \frac{g'_{2b}}{g_{2b}} + \beta_5 - \frac{g_{1b}}{g_{2b}} = 0; \tag{6.19}$$

a similar relation must exist for the boundary conditions at the other boundary $x = a$.

Case III. Mixed boundary conditions. What has been said in case I about essential conditions and in case II about suppressible conditions is combined appropriately; here also only a certain class of suppressible boundary conditions can be dealt with.

Summary. *The fourth-order linear boundary-value problem*

$$\left(g_2(x)\, y''\right)'' - \left(g_1(x)\, y'\right)' + g_0(x)\, y = r(x), \tag{6.20}$$

where

$$g_2(a) \neq 0, \qquad g_2(b) \neq 0,$$

with two linearly independent non-contradictory boundary conditions at each of the points $x = a$, $x = b$, may be identified with the necessary conditions to be satisfied by the solution y of a variational problem $J[\varphi] = $ minimum, where

$$J[\varphi] = \int_a^b \left(g_2\, \varphi''^2 + g_1\, \varphi'^2 + g_0\, \varphi^2 - 2r\, \varphi\right) dx + R_a + R_b, \tag{6.21}$$

under the following conditions:

(i) If all four boundary conditions are essential, then we can put $R_a = R_b = 0$ but we must specifically require that the admissible functions φ shall satisfy all the boundary conditions.

(ii) If the boundary conditions at $x = a$ are suppressible and such that when written in the form

$$\left.\begin{aligned} y_a'' &= \alpha_1\, y_a + \alpha_2\, y_a' + \alpha_3, \\ y_a''' &= \alpha_4\, y_a + \alpha_5\, y_a' + \alpha_6 \end{aligned}\right\} \tag{6.22}$$

the relation

$$(\alpha_1 + \alpha_5)\, g_{2a} - g_{1a} + \alpha_2\, g_{2a}' = 0 \tag{6.23}$$

holds, then we must put

$$\left.\begin{aligned} R_a = &- (\alpha_4\, g_{2a} + \alpha_1\, g_{2a}')\, \varphi_a^2 + 2\alpha_1\, g_{2a}\, \varphi_a\, \varphi_a' + \\ &+ \alpha_2\, g_{2a}\, \varphi_a'^2 - 2(\alpha_6\, g_{2a} + \alpha_3\, g_{2a}')\, \varphi_a + 2\alpha_3\, g_{2a}\, \varphi_a'. \end{aligned}\right\} \tag{6.24}$$

There is no need for φ to satisfy any of the boundary condition at $x = a$.

(iii) Correspondingly, if the boundary conditions at $x = b$ are suppressible and can be written in the form

$$\left.\begin{aligned} y_b'' &= \beta_1\, y_b + \beta_2\, y_b' + \beta_3, \\ y_b''' &= \beta_4\, y_b + \beta_5\, y_b' + \beta_6 \end{aligned}\right\} \tag{6.25}$$

with

$$(\beta_1 + \beta_5)\, g_{2b} - g_{1b} + \beta_2\, g_{2b}' = 0, \tag{6.26}$$

then we must put

$$R_b = (\beta_4 g_{2b} + \beta_1 g'_{2b})\, \varphi_b^2 - 2\beta_1 g_{2b}\,\varphi_b\,\varphi'_b - \beta_2 g_{2b}\,\varphi_b'^2 + \left.\vphantom{\Big|}\right\}$$
$$+ 2(\beta_6 g_{2b} + \beta_3 g'_{2b})\,\varphi_b - 2\beta_3 g_{2b}\,\varphi'_b, \left.\vphantom{\Big|}\right\}$$

$$(6.27)$$

and φ need not satisfy any boundary conditions at $x = b$.

6.3. Example

For the problem of the elastically mounted rail of § 1.3, i.e.

$$[(2 - x^2)\, y'']'' + 40\,y = 2 - x^2, \qquad y''(\pm 1) = y'''(\pm 1) = 0,$$

the corresponding variational problem is

$$J[\varphi] = \int_{-1}^{1} [(2 - x^2)\, \varphi''^2 + 40\varphi^2 - 2(2 - x^2)\,\varphi]\, dx = \text{min.}$$

The function φ need not satisfy any of the boundary conditions, for they are all suppressible. The boundary terms R_a and R_b turn out to be zero in this case since all the α_i and β_i in (6.22), (6.25) are zero.

Because of the symmetry, we take

$$\varphi = \sum_{\nu=1}^{p} a_\nu x^{2\nu-2}, \qquad \text{so that} \qquad \varphi'' = \sum_{\nu=1}^{p} a_\nu (2\nu - 2)(2\nu - 3)\, x^{2\nu-4}.$$

Then for $p = 3$

$$\tfrac{1}{2} J[\varphi] = 40 a_1^2 + \frac{44}{3} a_2^2 + \frac{13\,064}{315} a_3^2 + \frac{40}{3}\, 2 a_1 a_2 + 8 \times 2 a_1 a_3 +$$
$$+ \frac{592}{35}\, 2 a_2 a_3 - \frac{5}{3}\, 2 a_1 - \frac{7}{15}\, 2 a_2 - \frac{9}{35}\, 2 a_3$$

(the check with $a_1 = a_2 = a_3 = a$ is used again here). The necessary conditions for a minimum

$$\frac{\partial J}{\partial a_1} = 0: \quad 40 a_1 + \frac{40}{3} a_2 + \qquad 8 a_3 = \frac{5}{3}$$

$$\frac{\partial J}{\partial a_2} = 0: \quad \frac{40}{3} a_1 + \frac{44}{3} a_2 + \frac{592}{35} a_3 = \frac{7}{15}$$

$$\frac{\partial J}{\partial a_3} = 0: \quad 8 a_1 + \frac{592}{35} a_2 + \frac{13\,064}{315} a_3 = \frac{9}{35}$$

yield the values

$$a_1 = \frac{143\,363}{40 \times 79\,301} = 0{\cdot}045\,195\,8$$

$$a_2 = - \frac{953}{79\,301} = - 0{\cdot}012\,017\,5$$

$$a_3 = \frac{189}{79\,301} = 0{\cdot}002\,383\,3.$$

These give, for example, the approximation

$$y(1) = a_1 + a_2 + a_3 = 0{\cdot}035\,561\,7.$$

6.4. Comparison of RITZ's method with the least-squares process

Consider the boundary-value problem

$$y'' + y + x = 0, \qquad y(0) = y(1) = 0.$$

RITZ's method and the least-squares method both replace the boundary-value problem by a variational problem; here the respective variational problems read

$$J_R[\varphi] = \int_0^1 (\varphi'^2 - \varphi^2 - 2x\varphi)\,dx = \min., \qquad \varphi(0) = \varphi(1) = 0$$

and

$$J_L[\varphi] = \int_0^1 (\varphi'' + \varphi + x)^2\,dx = \min., \qquad \varphi(0) = \varphi(1) = 0.$$

Using the same two-term expression

$$\varphi = a_1(x - x^2) + a_2(x - x^3)$$

for both of these problems, we have

$$J_R = \frac{3}{10}a_1^2 + \frac{9}{10}a_1 a_2 + \frac{76}{105}a_2^2 - \frac{1}{6}a_1 - \frac{4}{15}a_2,$$

$$J_L = \frac{1}{210}(707 a_1^2 + 2121 a_1 a_2 + 2200 a_2^2 - 385 a_1 - 784 a_2 + 70).$$

Then the equations $\partial J/\partial a_1 = \partial J/\partial a_2 = 0$ yield for RITZ's method:

$$a_1 = \frac{8}{369}, \qquad a_2 = \frac{7}{41}.$$

and hence the approximate solution

$$y_R = \frac{1}{369}(71x - 8x^2 - 63x^3);$$

and for the least squares method

$$a_1 = \frac{4448}{101 \times 2437}, \qquad a_2 = \frac{413}{2437},$$

$$y_L = \frac{1}{246137}(46161x - 4448x^2 - 41713x^3).$$

Table III/13 compares some values of y_R and y_L with the exact solution

$$y = \frac{\sin x}{\sin 1} - x.$$

The numerical results and the curves in Fig. III/14 show that of the two methods RITZ's reproduces the function y more accurately while least-squares gives the more accurate values for y''.

Now consider the fourth-order boundary-value problem

$$\frac{d^4 y}{dx^4} = q(x), \qquad y(0) = y''(0) = y(l) = y''(l) = 0$$

Table III/15. *Comparison of results obtained by* Ritz's *method and the least-squares method*

	Exact solution $y(x)$	Approximate solution	
		by Ritz's method: $y_R(x)$	by the least-squares method: $y_L(x)$
$y(\frac{1}{2})$	$0\cdot069\,75$	$\frac{5}{72} = 0\cdot069\,44$ (error $-0\cdot45\%$)	$\frac{134\,035}{1\,969\,096} = 0\cdot068\,07$ (error $-2\cdot4\%$)
$y'(0)$	$0\cdot188\,40$	$\frac{71}{369} = 0\cdot192\,41$ (error $+2\cdot1\%$)	$\frac{46\,161}{246\,137} = 0\cdot187\,54$ (error $-0\cdot46\%$)
$y'(1)$	$-0\cdot357\,91$	$-\frac{134}{369} = -0\cdot363\,14$ (error $-1\cdot5\%$)	$-\frac{87\,874}{246\,137} = -0\cdot357\,01$ (error $+0\cdot25\%$)

for the deflection y of a loaded beam with constant flexural rigidity supported by smooth pin-joints a distance l apart. The corresponding Ritz variational problem reads

$$J_R[\varphi] = \int_0^l (\varphi''^2 - 2q\,\varphi)\,dx$$
$$= \min.,$$
$$\varphi(0) = \varphi(l) = 0.$$

Since $q = y^{IV}$, the integral can be transformed by two integrations by parts to give

$$J_R[\varphi] = \int_0^l (\varphi''^2 - 2\varphi\,y^{IV})\,dx$$
$$= \int_0^l (\varphi''^2 - 2\varphi''\,y'')\,dx$$
$$= \min.$$

This integral differs only by the constant

$$\int_0^l y''^2\,dy = A$$

from the integral in

$$J_R^*[\varphi] = \int_0^l (\varphi'' - y'')^2\,dx$$
$$= \min.,$$

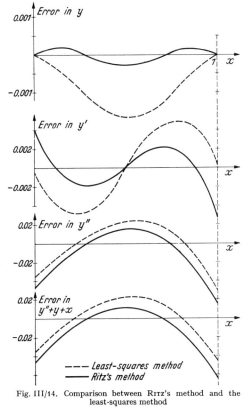

Fig. III/14. Comparison between Ritz's method and the least-squares method

which the Ritz method therefore minimizes with $J_R[\varphi]$. The corresponding least-squares variational problem reads

$$J_L[\varphi] = \int_0^l (\varphi^{IV} - q)^2\,dx = \int_0^l (\varphi^{IV} - y^{IV})^2\,dx = \min.$$

Thus RITZ's method minimizes the error integral of the second derivative while the least-squares method minimizes the error integral of the fourth derivative. Consequently least-squares gives the better approximation to the fourth derivative but its values for the function y itself are less accurate than the corresponding Ritz values; in fact it can happen that they are hopelessly wrong — C. WEBER[1] gives an example in which the approximation to the deflection even has the wrong sign throughout.

§ 7. Series solutions

7.1. Series solutions in general

For many boundary-value problems, especially those with a simple analytic formulation, a solution can be obtained in the form of an infinite series

$$y(x) = \sum_{\varrho=1}^{\infty} c_\varrho \psi_\varrho(x) \tag{7.1}$$

of given functions $\psi_\varrho(x)$ whose coefficients c_ϱ are initially undetermined. *The success of the method depends principally on the nature of the boundary-value problem and on a suitable choice of the system of functions ψ_ϱ.* In the examples of § 7.3 the assumed series converge well and in some cases the series method is actually superior to all other methods; on the other hand, there are cases in which the infinite series either converges only very slowly or does not converge at all.

A general procedure for obtaining a series solution of a linear or non-linear boundary-value problem Ch. I (1.1) with linear boundary conditions Ch. I (1.7) is now described.

Let z be a function satisfying the boundary conditions

$$U_\nu[z] = \gamma_\nu, \qquad (\nu = 1, 2, \ldots, n). \tag{7.2}$$

Then for the ψ_ϱ we choose a system of functions which is complete in the interval considered, $\langle a, b \rangle$ say, (for example, a system of trigonometrical functions or of spherical harmonics) and whose members satisfy the homogeneous boundary conditions

$$U_\nu[\psi_\varrho] = 0 \qquad (\nu = 1, 2, \ldots, n; \ \varrho = 1, 2, \ldots). \tag{7.3}$$

The assumed completeness of the system ψ_ϱ permits us to express the solution y in the form

$$y(x) = z(x) + \sum_{\varrho=1}^{\infty} c_\varrho \psi_\varrho(x), \tag{7.4}$$

which satisfies the boundary conditions (7.2) for any c_ϱ. These coefficients are to be determined so that y satisfies the (linear or non-linear)

---

[1] WEBER, C.: Z. Angew. Math. Mech. **21**, 310−311 (1941).

differential equation (1.1) of Ch. I, i.e. $F(x, y, y', \ldots, y^{(n)}) = 0$. For this
we must assume further that the expansion (7.4) may be differentiated
term by term n times so that it may be substituted in the differential
equation. If y is the required solution, this should yield $F \equiv 0$; therefore
the coefficients in the expansion of F in the form (7.1) (this expansion
also is possible because of the completeness of the ψ_ϱ) must all vanish.
If the ψ_ϱ were also orthogonal, this would lead directly to the equations

$$\int_a^b F \psi_\varrho \, dx = 0 \qquad (\varrho = 1, 2, \ldots); \tag{7.5}$$

otherwise the ψ_ϱ first have to be orthogonalized. Equations (7.5) written
out in detail read

$$\left. \int_a^b F\left(x, z + \sum_{\sigma=1}^{\infty} c_\sigma \psi_\sigma, z' + \sum_{\sigma=1}^{\infty} c_\sigma \psi_\sigma', \ldots, z^{(n)} + \sum_{\sigma=1}^{\infty} c_\sigma \psi_\sigma^{(n)}\right) \psi_\varrho(x) \, dx = 0 \atop (\varrho = 1, 2, \ldots). \right\} \tag{7.6}$$

Thus we have an infinite set of equations, non-linear in general, for
the infinite number of unknowns c_ϱ. For numerical calculation the
series (7.4) is terminated after a finite number of terms (say p); we then
have a finite set of equations for $c_1, \ldots, c_p$:

$$\left. \int_a^b F\left(x, z + \sum_{\sigma=1}^{p} c_\sigma \psi_\sigma, \ldots, z^{(n)} + \sum_{\sigma=1}^{p} c_\sigma \psi_\sigma^{(n)}\right) \psi_\varrho(x) \, dx = 0 \atop (\varrho = 1, 2, \ldots, p). \right\} \tag{7.7}$$

For the special case of a linear second-order problem with essential
boundary conditions these equations are identical with the Ritz-Galerkin
equations (5.23).

A series approach[1] which is occasionally used for problems with one
or more boundary conditions at infinity makes use of two series, a
power series at a finite point $x = x_0$ and an asymptotic series at $x = \infty$;
these expressions for the solution are joined smoothly at a suitable
intermediate point $x = \alpha$.

7.2. Power series solutions

We expand all functions appearing in the differential equation in
powers of x or $x - \xi$ for suitable ξ and assume a corresponding power
series expansion for the solution:

$$y(x) = \sum_{k=0}^{\infty} a_k (x - \xi)^k. \tag{7.8}$$

---

[1] An example of the application of this technique to a non-linear problem can
be found in U. T. Bödewadt: Die Drehströmung über festem Grunde. Z. Angew.
Math. Mech. **20**, 241—253 (1940).

By inserting this series in the differential equation, rearranging in powers of $x - \xi$ and equating to zero the coefficients of $(x - \xi)^k$ (for $k = 0, 1, 2, \ldots$), we obtain in general an infinite set of equations for the a_ν. The n boundary conditions yield n additional equations for the a_ν. This infinite system of equations is solved approximately by retaining only a finite number of equations and the same finite number of unknowns a_ν. In many cases a formula for the successive a_ν is obtained from which convergence of the power series can be inferred and the assumption that the solution can be expressed in the form (7.8) thereby justified (cf. Example I of § 7.3).

7.3. Examples

We will show how the series method is carried out in practice by means of a few examples.

I. The form of the coefficients in the differential equation of the strut problem

$$y'' + (1 + x^2)\, y + 1 = 0, \qquad y(\pm 1) = 0$$

(Example I of § 1.2) suggests that a power series expansion would be convenient here. There are two obvious ways of proceeding. One is to put

$$y = \sum_{\nu=0}^{\infty} a_\nu x^{2\nu}, \tag{7.9}$$

in which we include only the even powers of $x^{2\nu}$ because of symmetry; we then express the a_ν successively in terms of a_0 by inserting the series in the differential equation and equating coefficients to zero; finally we determine a_0 from the boundary condition $y(1) = 0$. The other alternative is to put

$$y = \sum_{\nu=1}^{\infty} b_\nu (1 - x^{2\nu}),$$

which satisfies the boundary conditions already, then determine the b_ν from the infinite set of linear equations obtained by inserting the series in the differential equation and equating the sum of the coefficients of like powers of x to zero. Both methods lead to substantially the same results, so we will describe only the first. Thus using the series (7.9) we have

$$(1 + x^2)\, y = \sum_{\nu=0}^{\infty} a_\nu (x^{2\nu} + x^{2\nu+2}) = a_0 + \sum_{\nu=1}^{\infty} (a_\nu + a_{\nu-1})\, x^{2\nu},$$

$$y'' = \sum_{\nu=0}^{\infty} (2\nu + 2)(2\nu + 1)\, a_{\nu+1} x^{2\nu},$$

and substituting in the differential equation we obtain

$$1 + a_0 + 2a_1 + \sum_{\nu=1}^{\infty} [a_\nu + a_{\nu-1} + (2\nu + 2)(2\nu + 1)\, a_{\nu+1}]\, x^{2\nu} = 0.$$

By the uniqueness theorem for power series, this implies that the coefficients of all powers of x must vanish separately; consequently

$$\left.\begin{array}{l} 1 + a_0 + 2a_1 = 0, \\[2mm] a_{\nu-1} + a_\nu + (2\nu + 2)(2\nu + 1) a_{\nu+1} = 0 \quad (\nu = 1, 2, 3, \ldots). \end{array}\right\} \quad (7.10)$$

From this recursive formula the $a_1, a_2, a_3, \ldots$ can be expressed successively in terms of a_0 (see first column of Table III/14). We can also use this formula to show that

$$|a_\nu| \leq \frac{1 + |a_0|}{(2\nu - 1)\, 2\nu} \quad \text{for} \quad \nu = 1, 2, 3, \ldots .$$

Table III/14. *Calculation of the coefficients in a power series solution*

a_p	p	$\sum_{\nu=0}^{p} a_\nu$	Approximate values of a_0
$a_1 = -\frac{1}{2}(1 + a_0) = -\frac{1}{2!}(1 + a_0)$	1	$\frac{1}{2!}(-1 + a_0)$	1
$a_2 = -\frac{1}{3\times4}(a_0 + a_1) = \frac{1}{4!}(1 - a_0)$	2	$\frac{1}{4!}(-11 + 11 a_0)$	1
$a_3 = -\frac{1}{5\times6}(a_1 + a_2) = \frac{1}{6!}(11 + 13 a_0)$	3	$\frac{1}{6!}(-319 + 343 a_0)$	$\frac{319}{343} = 0{\cdot}93003$
$a_4 = \frac{1}{8!}(-41 + 17 a_0)$	4	$\frac{5}{8!}(-3581 + 3845 a_0)$	$0{\cdot}931339$
$a_5 = \frac{1}{10!}(-575 - 745 a_0)$	5	$\frac{5}{10!}(-322405 + 345901 a_0)$	$0{\cdot}93207304$
$a_6 = \frac{5}{12!}(853 - 157 a_0)$	6	$\frac{5}{12!}(-42556607 + 45658775 a_0)$	$0{\cdot}9320 5757$
$a_7 = \frac{5}{14!}(14327 + 19825 a_0)$	7	$\frac{5}{12!}(-42556528{\cdot}3 + 45658884 a_0)$	$0{\cdot}93205362$

For $\nu = 1, 2$ the validity of this upper bound can be verified from Table III/14; for $\nu > 2$ it follows by complete induction: if the inequality is true for $\nu - 1$ and ν, then

$$|a_{\nu+1}| = \frac{|a_{\nu-1} + a_\nu|}{(2\nu + 1)(2\nu + 2)} \leq \frac{1 + |a_0|}{(2\nu + 1)(2\nu + 2)}\left(\frac{1}{(2\nu - 3)(2\nu - 2)} + \frac{1}{(2\nu - 1)\,2\nu}\right)$$

$$\leq \frac{1 + |a_0|}{(2\nu + 1)(2\nu + 2)} \quad (\nu = 2, 3, \ldots);$$

thus it is also true for $\nu + 1$, and hence for all ν. Since $\sum\limits_{\nu=1}^{\infty} \frac{1}{(2\nu - 1)\, 2\nu}$ converges, the assumed power series converges absolutely and uniformly in the closed interval $-1 \leq x \leq 1$; hence the expansion is justified.

The boundary condition

$$y(1) = \sum_{\nu=0}^{\infty} a_\nu = 0$$

yields an equation for a_0. For numerical calculation the series is truncated after some term, say the term in a_p; the fourth column of Table III/14 gives the sequence of approximations to a_0 for $p = 1, 2, \ldots, 7$.

Error estimation: The error in a_0 due to truncation of the series after the term with $\nu = p$ can be estimated quite easily in this example. We can write

$$a_\nu = \alpha_\nu + a_0 \beta_\nu, \quad \sum_{\nu=0}^{p} a_\nu = A_p + a_0 B_p,$$

where the numbers $\alpha_\nu, \beta_\nu, A_p, B_p$ are independent of a_0 (their values can be found from columns 1 and 3 of Table III/14); then the p-th approximation to a_0 is the quotient $- A_p / B_p$, values of which are given in the fourth column of the table. To illustrate the procedure we will estimate the error in $- A_7 / B_7$. It can be seen from the table that

$$|\alpha_7| < |\alpha_6|, \quad |\beta_7| < |\beta_6|;$$

then from (7.10) we obtain successively the upper bounds

$$|\alpha_8| \leq \frac{2|\alpha_6|}{15 \times 16} = \frac{|\alpha_6|}{120}, \quad |\alpha_9| \leq \frac{|\alpha_6|}{120}, \quad |\alpha_{10}| \leq \frac{|\alpha_6|}{60} \frac{1}{19 \times 20},$$

$$|\alpha_{11}| \leq \frac{|\alpha_6|}{60} \frac{1}{21 \times 22}, \ldots,$$

so that

$$\sum_{\nu=8}^{\infty} |\alpha_\nu| \leq \frac{|\alpha_6|}{60} \left(1 + \frac{1}{19 \times 20} + \frac{1}{21 \times 22} + \frac{1}{23 \times 24} + \cdots\right) = \frac{|\alpha_6|}{60} \times 1 \cdot 027.$$

Precisely similar considerations yield

$$\sum_{\nu=8}^{\infty} |\beta_\nu| \leq \frac{|\beta_6|}{60} \times 1 \cdot 027.$$

Since $|\alpha_6| = \dfrac{4265}{12!}$, the error in A_7 is at most $\dfrac{4265}{60} \times \dfrac{1 \cdot 027}{12!}$; with $- A_7 \approx 213 \times 10^6 / 12!$ the relative error is therefore at most $0 \cdot 35 \times 10^{-6}$. Similarly the relative error in B_7 is at most $\dfrac{785}{60} \times \dfrac{1 \cdot 027}{228 \times 10^6} < 0 \cdot 06 \times 10^{-6}$, and hence the error in a_0 is at most $0 \cdot 42 \times 10^{-6}$. Thus we have the error estimate

$$|a_0 - 0 \cdot 9320536| < 4 \cdot 2 \times 10^{-7}.$$

II. For the non-linear problem (Example II, § 1.2)

$$y'' = \tfrac{3}{2} y^2; \quad y(0) = 4, \quad y(1) = 1$$

we expand $y(x)$ as a power series at $x = 1$, say in powers of $s = 1 - x$:

$$y(s) = 1 + \sum_{\nu=1}^{\infty} a_\nu s^\nu.$$

Substitution in the differential equation yields

$$2a_2 + 6a_3 s + 12a_4 s^2 + 20a_5 s^3 + 30a_6 s^4 + 42a_7 s^5 + \cdots$$
$$= \tfrac{3}{2} + 3a_1 s + (\tfrac{3}{2} a_1^2 + 3a_2) s^2 + (3a_3 + 3a_1 a_2) s^3 + (\tfrac{3}{2} a_2^2 + 3a_1 a_3 + 3a_4) s^4 +$$
$$+ (3a_2 a_3 + 3a_1 a_4 + 3a_5) s^5 + \cdots,$$

and comparing coefficients of like powers of x we find that

$$\tfrac{3}{2} = 2a_2, \quad 6a_3 = 3a_1, \quad 12a_4 = \tfrac{3}{2}a_1^2 + 3a_2, \ldots .$$

By means of these equations we can express $a_3, a_4, \ldots$ successively in terms of a_1, so that the series for y becomes

$$y = 1 + a_1 s + \frac{3}{4} s^2 + \frac{a_1}{2} s^3 + \frac{2a_1^2 + 3}{16} s^4 +$$

$$+ \frac{3a_1}{16} s^5 + \frac{4a_1^2 + 3}{64} s^6 + \frac{a_1^3 + 6a_1}{112} s^7 + \cdots .$$

The boundary condition $y = 4$ at $s = 1$ is then an equation for a_1. If the series is terminated at the term in s^4, we obtain the quadratic equation

$$\frac{a_1^2}{8} + \frac{3}{2}a_1 + \frac{31}{16} = 4$$

with the solutions

$$a_1 = -6 \pm \sqrt{52 \cdot 5} = \begin{cases} 1 \cdot 2457 \\ -13 \cdot 246. \end{cases}$$

If two more terms are retained, we still have only a quadratic to solve, namely

$$\frac{3a_1^2}{16} + \frac{27}{16}a_1 + \frac{127}{64} = 4,$$

from which we obtain the values

$$a_1 = -\tfrac{9}{2} \pm \sqrt{31} = \begin{cases} 1 \cdot 06776 \\ -10 \cdot 068. \end{cases}$$

The corresponding approximations for the everywhere positive solution $y(x)$ for several values of x are given in Table III/15 with their percentage errors in brackets.

The power series for the other (partly negative) solution does not appear to converge since $y(\tfrac{1}{2}) \approx -11$ (cf. Example II, § 1.2).

§8. Some special methods for eigenvalue problems

Eigenvalue problems have already been treated several times in this chapter, but by methods, which are also applicable to ordinary boundary-value problems — actually we used several finite-difference methods and RITZ's method, but other such methods can be used. We now describe some methods designed specifically for eigenvalue problems, giving particular

Table III/15. *Approximate values obtained by applying a power series method to a non-linear problem (Example II)*

| s | x | Approximate values for the everywhere positive solution, obtained by truncating the power series at the term in | | | | other solution | | |
		s^3	s^4	s^5	s^6	s^4	s^5	s^6
0·25	0·75	1·4336 (+9·8%)	1·3696 (+4·9%)	1·3393 (+2·5%)	1·32365 (+1·3%)	−2·282	−2·621	−1·500
0·5	0·5	2·0313 (++14 %)	1·9120 (++11 %)	1·8502 (++4·1%)	1·8168 (++2·2%)	−4·881	−5·443	−3·632
0·75	0·25	2·8633 (+12 %)	2·7395 (++7 %)	2·6656 (++4·1%)	2·6209 (++2·4%)	−4·308	−4·763	−3·496

(percentages in brackets are the relative errors)

prominence to the method of successive approximations or iteration method. In § 8.3 we mention an important bracketing theorem for a special class of eigenvalue problems; in many cases this theorem enables one to obtain limits for the first eigenvalue to an accuracy sufficient for most applications with very little computation.

In the following we make use of a series of results from the theory of eigenvalue problems; for proofs the reader is referred to the literature[1].

8.1. Some concepts and results from the theory of eigenvalue problems

Let the differential equation read

$$M[y] = \lambda N[y], \tag{8.1}$$

where $M[y]$ and $N[y]$ are linear homogeneous ordinary differential expressions [see Ch. I (1.6)] of the form

$$M[y] = \sum_{\nu=0}^{2m} p_\nu(x)\, y^{(\nu)}, \tag{8.2}$$

$$N[y] = \sum_{\nu=0}^{2n} q_\nu(x)\, y^{(\nu)} \tag{8.3}$$

with $m > n$. The $p_\nu(x)$ and $q_\nu(x)$ are given real functions which will here be assumed continuous. With the differential equation we associate $2m$ linear homogeneous boundary conditions [cf. Ch. I (1.7)]

$$U_\mu[y] = 0 \qquad (\mu = 1, 2, \dots, 2m). \tag{8.4}$$

The differential expressions $M[y]$ and $N[y]$ are often assumed to be of the special "self-adjoint" form

$$\left.\begin{aligned} M[y] &= \sum_{\nu=0}^{m} (-1)^\nu\, [f_\nu(x)\, y^{(\nu)}]^{(\nu)}, \\ N[y] &= \sum_{\nu=0}^{n} (-1)^\nu\, [g_\nu(x)\, y^{(\nu)}]^{(\nu)}. \end{aligned}\right\} \tag{8.5}$$

The f_ν and g_ν are given functions which we assume here to be real and to possess continuous derivatives of order ν; further we assume that $f_m \neq 0$ and $g_n \neq 0$. When $n = 0$, the eigenvalue problem, then of the form

$$M[y] = \lambda g_0(x)\, y, \tag{8.6}$$

is called a "special" eigenvalue problem; when $n > 0$, it is called a "general" eigenvalue problem.

---

[1] For example, R. Courant and D. Hilbert: Methods of mathematical physics, Vol. I, 1st English ed. New York: Interscience Publishers, Inc. 1953. — Collatz, L.: Eigenwertaufgaben mit technischen Anwendungen. Leipzig 1949.

For problems in which the eigenvalue λ does not appear in the boundary conditions we make the following definitions:

A *comparison function* is any function other than the zero function which satisfies all the boundary conditions and possesses a continuous derivative of order $2m$.

An *admissible function* is any function other than the zero function which satisfies the essential boundary conditions (see Ch. I, § 1.2) and possesses a continuous derivative of order m.

The eigenvalue problem (8.1), (8.4) is called *self-adjoint* when

$$\left.\begin{aligned} \int_a^b \left(u\,M[v] - v\,M[u]\right) dx = 0, \\ \int_a^b \left(u\,N[v] - v\,N[u]\right) dx = 0 \end{aligned}\right\} \tag{8.7}$$

for any two comparison functions u, v.

The eigenvalue problem (8.1), (8.4) is called *full-definite* when

$$\int_a^b u\,M[u]\,dx > 0 \quad \text{and} \quad \int_a^b u\,N[u]\,dx > 0 \tag{8.8}$$

for any comparison function u.

Whether these conditions of self-adjointness (8.7) and full-definiteness (8.8) are satisfied in any given case can be readily determined by integration by parts[1]; for example, the eigenvalue problem

$$-y'' = \lambda e^x y; \quad y'(0) = 0, \quad y(1) + y'(1) = 0$$

is shown to be self-adjoint and full-definite, respectively, by the following two simple calculations:

$$\int_0^1 \left(u\,v'' - u''\,v\right) dx = \left[u\,v' - v\,u'\right]_0^1$$
$$= u(1)\,v'(1) - v(1)\,u'(1) - u(0)\,v'(0) + v(0)\,u'(0) = 0$$

for any two comparison functions u, v since

$$v'(1) = -v(1), \quad u'(1) = -u(1), \quad u'(0) = v'(0) = 0;$$

and

$$\int_0^1 \left(-u\,u''\right) dx = -\left[u\,u'\right]_0^1 + \int_0^1 u'^2 dx = \left[u(1)\right]^2 + \int_0^1 u'^2 dx > 0$$

for any comparison function u. This last inequality sign could be replaced by an equality sign only for a function with $u' \equiv 0$ and $u(1) = 0$, i.e. for the function $u \equiv 0$, but the zero function is excluded from the domain of comparison functions.

---

[1] See L. COLLATZ: Eigenwertaufgaben, § 4. Leipzig 1949.

In eigenvalue theory an important role is played by "RAYLEIGH's quotient"

$$R[u] = \frac{\int\limits_a^b u\,M[u]\,dx}{\int\limits_a^b u\,N[u]\,dx}. \tag{8.9}$$

For full-definite problems it exists for all comparison functions u and is positive.

A self-adjoint, full-definite eigenvalue problem has the following properties:

It possesses a countably infinite sequence of positive eigenvalues $0 < \lambda_1 \leq \lambda_2 \leq \cdots$ with corresponding eigenfunctions $y_1, y_2, \ldots$. Its first eigenfunction y_1 is the solution of the variational problem of minimizing the Rayleigh quotient (8.9) with respect to the domain of comparison functions; the corresponding minimum value is the first eigenvalue λ_1. More generally, the $(s+1)$-th eigenvalue λ_{s+1} is the minimum value of the Rayleigh quotient (8.9) with respect to the domain of all comparison functions u which are orthogonal to the first s eigenfunctions $y_1, \ldots, y_s$ in a "generalized sense", namely in the sense that they satisfy the "generalized orthogonality relations"

$$\int\limits_a^b u\,N[y_j]\,dx = 0 \qquad (j = 1, 2, \ldots, s).$$

When multiple eigenvalues occur, they must be counted according to their multiplicity.

8.2. The iteration method in the general case

We base the description of the method on the differential equation (8.1) with the boundary conditions (8.4).

In the general case the procedure is as follows: We start with an arbitrary function $F_0(x)$ and from it determine successively a sequence of functions $F_1, F_2, \ldots$ by calculating F_k as the solution of a certain boundary-value problem which depends on F_{k-1}. This boundary-value problem is obtained from the eigenvalue problem (8.1), (8.4) by writing F_{k-1} in place of y in all terms (in the differential equation and the boundary conditions) which are multiplied by λ, writing F_k in place of y in all the remaining terms, and finally putting $\lambda = 1$; thus for the eigenvalue problem

$$y^{IV} + \lambda y'' + y = 0; \quad y'(0) = y''(0) = y'(1) = y(1) - \lambda y'''(1) = 0$$

F_k is determined from F_{k-1} as the solution of the boundary-value problem

$$F_k^{IV} + F_k = -F_{k-1}''; \quad F_k'(0) = F_k''(0) = F_k'(1) = 0, \quad F_k(1) = F_{k-1}'''(1).$$

The absence of λ from the boundary conditions simplifies the description of the procedure: starting from any function $F_0(x)$ we determine successively $F_1, F_2, \ldots$ from the boundary-value problems

$$\left.\begin{array}{l} M[F_k] = N[F_{k-1}], \\ U_\mu[F_k] = 0 \end{array}\right\} \quad (k = 1, 2, \ldots). \tag{8.10}$$

If the ratios of successive terms in the sequence $F_0, F_1, F_2, \ldots$ tend to a constant (independent of x and k) as k increases, say $F_{k-1}/F_k \to K$, in which case we say the sequence converges, then one expects the functions F_k to assume more and more the form of an eigenfunction y_s with eigenvalue $\lambda_s = K$ (an eigenfunction is only determined to within a constant factor). The quotient $M[F_k]/N[F_k]$ should provide an approximation to λ_s but as it is still a function of x we use a more suitable approximate value Λ obtained by taking weighted mean values of the numerator and denominator as in RAYLEIGH'S quotient (8.9):

$$\Lambda = \frac{\int\limits_a^b F_k M[F_k]\, dx}{\int\limits_a^b F_k N[F_k]\, dx} = \frac{\int\limits_a^b F_k N[F_{k-1}]\, dx}{\int\limits_a^b F_k N[F_k]\, dx}.$$

The integrals appearing here in the numerator and denominator are denoted by

$$\left.\begin{array}{l} a_{2k} = \int\limits_a^b F_k N[F_k]\, dx, \\[2mm] a_{2k+1} = \int\limits_a^b F_{k+1} N[F_k]\, dx \end{array}\right\} \quad (k = 0, 1, 2, \ldots)$$

and are called SCHWARZ's constants after H. A. SCHWARZ. Λ is then the quotient of two successive Schwarz constants; if we introduce the Schwarz quotients

$$\mu_{k+1} = \frac{a_k}{a_{k+1}} \quad (k = 0, 1, 2, \ldots), \tag{8.11}$$

Λ is the Schwarz quotient μ_{2k}.

In order to say more about Λ we must make more restrictive assumptions about the eigenvalue problem[1].

8.3. The iteration method for a restricted class of problems

Here we consider only those eigenvalue problems[2] (8.1), (8.4), (8.5) which are self-adjoint and full-definite and whose boundary conditions are independent of the eigenvalue λ.

---

[1] A more general theorem has been established by H. WIELANDT: Das Iterationsverfahren bei nicht selbstadjungierten Eigenwertaufgaben. Math. Z. **50**, 93—143 (1944).

[2] Fewer restrictive assumptions are made by E. STIEFEL and H. ZIEGLER: Natürliche Eigenwertprobleme. Z. Angew. Math. Phys. **1**, 111—138 (1950).

We restrict our choice of starting function $F_0(x)$ slightly by demanding that it shall be continuous, possess continuous derivatives of order up to $2n$ and satisfy just so many of the boundary conditions (not necessarily all) that $a_0 > 0$ and

$$\int_a^b \left(F_0 N[u] - u N[F_0] \right) dx = 0 \qquad (8.12)$$

for all comparison functions u. Integration by parts readily establishes which boundary conditions should be satisfied in any given case.

Consider, for example, a special eigenvalue problem with $N[f] = g_0(x)f$: (8.12) is automatically satisfied and also $a_0 > 0$ since the full-definiteness condition implies that $g_0(x) > 0$. In this case, therefore, F_0 need not satisfy any boundary conditions.

If F_0 satisfies all the boundary conditions, then (8.12) and $a_0 > 0$ are automatically satisfied for any self-adjoint full-definite problem since F_0 is then a comparison function. Usually F_0 will be thus chosen on account of the more accurate results which may be expected; nevertheless it can sometimes be more convenient for calculation to impose a smaller (but sufficient) number of boundary conditions on F_0.

For the sequence of functions $F_0, F_1, F_2, \ldots$ derived from such an F_0 by the iteration process described in § 8.2 the Schwarz constants can be defined generally by

$$a_k = \int_a^b F_i N[F_{k-i}] \, dx \qquad (0 \leq i \leq k, \ k = 0, 1, 2, \ldots). \qquad (8.13)$$

Their dependence on i is only apparent, for using the iterative relation of (8.10) and the self-adjoint condition (8.7) we have

$$a_k = \int_a^b F_i M[F_{k-i+1}] \, dx = \int_a^b F_{k-i+1} M[F_i] \, dx = \int_a^b F_{k-i+1} N[F_{i-1}] \, dx$$
$$= \int_a^b F_{i-1} N[F_{k-i+1}] \, dx,$$

and consequently the integral $\int_a^b F_i N[F_{k-i}] \, dx$ depends only on the sum of the suffices $i + (k-i) = k$; for example,

$$a_2 = \int_a^b F_2 N[F_0] \, dx = \int_a^b F_1 N[F_1] \, dx = \int_a^b F_0 N[F_2] \, dx. \qquad (8.14)$$

The full-definite condition (8.8) implies that $a_1, a_2, \ldots$ are all positive, for

$$\left. \begin{array}{l} a_{2k} = \int_a^b F_k N[F_k] \, dx > 0, \\[2mm] a_{2k-1} = \int_a^b F_k M[F_k] \, dx > 0 \end{array} \right\} \qquad (k = 1, 2, \ldots) \qquad (8.15)$$

and since also $a_0 > 0$, all the Schwarz quotients (8.11) are positive. The quotients μ_{2k} with even suffices may be written as Rayleigh quotients (8.9); for example,

$$\mu_2 = \frac{a_1}{a_2} = \frac{\int\limits_a^b F_1 N[F_0] \, dx}{\int\limits_a^b F_1 N[F_1] \, dx} = \frac{\int\limits_a^b F_1 M[F_1] \, dx}{\int\limits_a^b F_1 N[F_1] \, dx} = R[F_1], \qquad (8.16)$$

and generally we have

$$\mu_{2s} = R[F_s]. \qquad (8.17)$$

Under the assumptions of this section it can be shown that

$$\mu_1 \geq \mu_2 \geq \mu_3 \geq \cdots \geq \lambda_1.$$

It is also possible to prove the following bracketing theorem:

Theorem: *Let the eigenvalue problem (8.1), (8.4), (8.5) be such that*

(i) *the conditions of self-adjointness and full-definiteness are satisfied,*

(ii) *the eigenvalue λ does not appear in the boundary conditions,*

(iii) *the smallest eigenvalue λ_1 is simple,*

and as starting function for the iteration (8.10) take any continuous function $F_0(x)$ with continuous derivatives of order up to $2n$ and which satisfies so many of the prescribed boundary conditions that (8.12) holds and $a_0 > 0$. If μ_k is the k-th Schwarz quotient (8.11) calculated from the iterates F_k via the Schwarz constants a_k (8.13) and if l_2 is a lower bound for the second eigenvalue such that $l_2 > \mu_{k+1}$, then the first eigenvalue λ_1 lies between the limits

$$\mu_{k+1} - \frac{\mu_k - \mu_{k+1}}{\dfrac{l_2}{\mu_{k+1}} - 1} \leq \lambda_1 \leq \mu_{k+1} \qquad (k = 1, 2, \ldots). \qquad (8.18)$$

8.4. Practical application of the method

Starting from a chosen function $F_0(x)$ we have to determine each subsequent function $F_1(x), F_2(x), \ldots$ by solving a boundary-value problem (8.10). If the given differential equation is of very simple form with coefficients given formally by simple expressions, then a number of iterates F_k can be calculated. But for complicated eigenvalue problems the repeated solution of these boundary-value problems presents considerable difficulties, and it is therefore important that for the restricted class of eigenvalue problems of § 8.3 an accuracy sufficient for technical applications is generally achieved by using only $F_0(x)$ and $F_1(x)$. The calculation can be divided into three steps:

1. Find two functions F_0, F_1 with the properties: $M[F_1] = N[F_0]$, F_1 satisfies all the boundary conditions and F_0 satisfies so many that (8.12) holds and $a_0 > 0$.

In many cases two such functions F_0, F_1 can be determined by graphical integrations (when applying the method graphically, one should always integrate and never differentiate); however, we shall not go into the graphical application any further here[1].

For the special eigenvalue problems, for which $N[F_0] \equiv g_0 F_0$, we can choose F_1 arbitrarily from among the comparison functions and calculate F_0 from $g_0 F_0 = M[F_1]$. F_1 can be expressed as an arbitrary linear combination of functions ψ_i which satisfy all the boundary conditions, and the coefficients of the combination will usually be determined so that F_0 also satisfies the boundary conditions. This often leads to substantially better results, for in general the more F_0 and F_1 resemble the first eigenfunction, the more accurate the results turn out to be.

2. Calculate the Schwarz constants and quotients

$$\left.\begin{aligned}
a_0 &= \int_a^b F_0 N[F_0]\, dx, \quad a_1 = \int_a^b F_1 N[F_0]\, dx, \\
a_2 &= \int_a^b F_1 N[F_1]\, dx; \quad \mu_1 = \frac{a_0}{a_1}, \quad \mu_2 = \frac{a_1}{a_2}.
\end{aligned}\right\} \tag{8.19}$$

We then know that

$$\mu_1 \geq \mu_2 \geq \lambda_1.$$

3. Finally we calculate a lower limit for λ_1. For this we need a lower limit l_2 for the second eigenvalue λ_2; it must be greater than μ_2, but as long as this requirement is satisfied it need not be particularly close to λ_2. In many cases such a lower limit l_2 can be obtained by comparison with an eigenvalue problem with constant coefficients (see § 8.5). Then from (8.18) with $k = 1$ we have

$$\mu_2 - \frac{\mu_1 - \mu_2}{\dfrac{l_2}{\mu_2} - 1} \leq \lambda_1 \leq \mu_2. \tag{8.20}$$

For technical applications it is not always essential to have rigorous limits and it will often suffice to have a rough guide to the closeness of μ_2 to λ_1 so that we may know how many decimals in μ_2 can be used for λ_1. In such cases one can replace l_2 in (8.20) by an approximation to λ_2 calculated by the Ritz method or the finite-difference method, or in fact by any other suitable method. Although this procedure is not strictly valid, it can be justified to some extent on the grounds that changes in l_2 have little effect on the lower limit in (8.20) when l_2 is appreciably greater than μ_2.

---

[1] See L. Collatz: Eigenwertaufgaben, § 13. Leipzig 1949.

8.5. An example treated by the iteration method

For the eigenvalue problem

$$- [(1 + x)\, y']' = \lambda (1 + x)\, y, \qquad y'(0) = y(1) = 0$$

of Example III in § 1.2 the conditions to be satisfied by F_0 and F_1 are

$$- [(1 + x)\, F_1']' = (1 + x)\, F_0, \qquad F_1'(0) = F_1(1) = 0.$$

Although F_0 need not satisfy any of the boundary conditions, better results may be expected if it satisfies at least the essential boundary condition $F_0(1) = 0$. We therefore take F_1 to be a polynomial of at least the third degree

$$F_1 = a_0 + a_1 x + a_2 x^2 + a_3 x^3,$$

for we have three conditions to satisfy and a constant factor is indeterminate. The boundary conditions on F_1 eliminate two of the constants; forming F_0 and putting $F_0(1) = 0$ eliminates a third; the fourth is chosen conveniently to give

$$F_1 = 3 - 5 x^2 + 2 x^3, \qquad F_0 = \frac{2 (5 + 4 x - 9 x^2)}{1 + x}.$$

We now calculate the Schwarz constants from (8.19):

$$a_0 = \int_0^1 (1 + x)\, F_0^2\, d x = 256 \ln 2 - 121 = 56{\cdot}445\,68,$$

$$a_1 = \int_0^1 (1 + x)\, F_0 F_1\, d x = \frac{526}{30} \qquad = 17{\cdot}533\,33,$$

$$a_2 = \int_0^1 (1 + x)\, F_1^2\, d x \quad = \frac{572}{105} \qquad = 5{\cdot}447\,6190.$$

These values yield the Schwarz quotients

$$\mu_1 = \frac{a_0}{a_1} = 3{\cdot}219335, \qquad \mu_2 = \frac{a_1}{a_2} = 3{\cdot}218532,$$

which constitute upper limits for λ_1.

To calculate a lower limit from (8.20) we need a lower limit l_2 for the second eigenvalue. This, as has been said before, may be fairly crude and for this example can be obtained by comparison of the given problem with a simpler problem possessing an exact solution. If the function $1 + x$, which occurs on both sides of the differential equation, is replaced by its maximum value of 2 on the right-hand side and by its minimum value of 1 on the left-hand side, we obtain the constant-coefficient eigenvalue problem

$$- y'' = 2 \lambda^* y, \qquad y'(0) = y(1) = 0,$$

whose second eigenvalue λ_2^* is given by

$$2\lambda_2^* = (\tfrac{3}{2}\pi)^2.$$

This new eigenvalue problem has the property[1] that $R^*[u] \leq R[u]$ for all comparison functions u, where $R^*[u]$ and $R[u]$ are the Rayleigh quotients of the new and old problems, respectively. Hence $\lambda_s^* \leq \lambda_s$ for $s = 1, 2, \ldots$ and in particular we can use λ_2^* as a lower limit for λ_2; thus we put

$$l_2 = \lambda_2^* = \tfrac{9}{8}\pi^2.$$

With this value (8.20) yields

$$3.218211 \leq \lambda_1 \leq 3.218532.$$

8.6. The enclosure theorem

The following theorem, which is proved in the theory of eigenvalues and usually known as the enclosure theorem, is often quite useful in numerical work. It is concerned with a certain class of eigenvalue problems, the "single-term" class, for which the operator $N[y]$ consists of only one term:

$$N[y] = (-1)^n [g_n(x)\, y^{(n)}]^{(n)}.$$

The differential equation is consequently of the form

$$M[y] = (-1)^n \lambda [g_n(x)\, y^{(n)}]^{(n)}. \tag{8.21}$$

We impose the usual conditions of self-adjointness and full-definiteness and, as in § 8.1, assume that the function $g_n(x)$ has the same sign throughout the fundamental interval (a, b). We assume further that the given boundary conditions are such that

$$(-1)^n \int_a^b u\, [g\, v^{(n)}]^{(n)}\, dx = \int_a^b g\, u^{(n)}\, v^{(n)}\, dx$$

for any two comparison functions u, v and any function $g(x)$ with continuous derivatives of up to the n-th order (that this condition is satisfied for any given problem can easily be verified by integration by parts).

The theorem states that if we have a function $F_0(x)$ with continuous derivatives of order up to $2n$ and a comparison function $F_1(x)$ such that $M[F_1] = N[F_0]$, cf. (8.10), and if further the function

$$\Phi(x) = \frac{F_0^{(n)}(x)}{F_1^{(n)}(x)} \tag{8.22}$$

[1] Cf. the comparison theorem in L. Collatz: Eigenwertaufgaben, § 8. Leipzig 1949; this deals with the case in which the differential expression M is the same in both problems but the extension to the more general case considered here is immediate.

is bounded and does not change sign in the interval (a, b), then at least one eigenvalue λ_s lies between the maximum and minimum values of Φ in the interval (a, b):

$$\Phi_{\min} \leqq \lambda_s \leqq \Phi_{\max}. \tag{8.23}$$

For special eigenvalue problems ($n = 0$) the conditions to be satisfied by F_0 and F_1 can be formulated more simply[1]. It is sufficient to select any comparison function F_1 and put

$$F_0 = \frac{1}{g_0(x)} M[F_1]. \tag{8.24}$$

Then

$$\Phi(x) = \frac{M[F_1]}{g_0 F_1} = \frac{M[F_1]}{N[F_1]}, \tag{8.25}$$

and we have

$$\left(\frac{M[F_1]}{g_0 F_1}\right)_{\min} \leqq \lambda_s \leqq \left(\frac{M[F_1]}{g_0 F_1}\right)_{\max}. \tag{8.26}$$

The function F_0 need not satisfy any of the boundary conditions but one will usually try to choose F_1 in such a way that F_0 satisfies as many boundary conditions as possible; for by so doing one approximates an eigenfunction more closely and there is a better chance of obtaining reasonably narrow limits in (8.23).

The results can be improved, often quite substantially, by the introduction of a parameter[2] ϱ into the assumed expressions for F_0 and F_1. This parameter is then chosen so as to minimize the difference between the upper and lower limits of (8.23).

Example. Again we consider the vibration problem of § 1.2

$$M[y] = -[(1 + x) y']' = \lambda (1 + x) y, \qquad y'(0) = y(1) = 0$$

but this time we choose for F_1 the expression

$$F_1(x) = (1 + x)^q \sin[a(x - 1)].$$

The boundary condition $F_1(1) = 0$ is already satisfied for all values of the parameters q and a, which are yet to be determined.

To form $M[F_1]$ we evaluate

$$[(1 + x) F_1']' = [q^2 (1 + x)^{q-1} - a^2 (1 + x)^{q+1}] \sin a (x - 1) +$$
$$+ (2q + 1) a (1 + x)^q \cos a (x - 1),$$

---

[1] For the special eigenvalue problems the enclosure theorem for the first eigenvalue was proved by G. Temple: The computation of characteristic numbers and characteristic functions. Proc. Lond. Math. Soc. (2) **29**, 257—280 (1929).

[2] This method is due to F. Kieszling: Eine Methode zur approximativen Berechnung einseitig eingespannter Druckstäbe mit veränderlichem Querschnitt. Z. Angew. Math. Mech. **10**, 594—599 (1930).

and we see that for the quotient function Φ of (8.25) to remain finite the zero of the sine factor in the denominator must be cancelled by a similar zero in the numerator. This can be achieved by choosing $q = -\frac{1}{2}$, for then the cosine term disappears.

We still have to satisfy the other boundary condition $F_1(0) = 0$, so we choose the remaining parameter a to be a root of the equation

$$\tan a = -2a.$$

Let $a_1, a_2, a_3, \ldots$ be the positive roots of this transcendental equation. With these values of a and $q = -\frac{1}{2}$

$$\Phi(x) = \frac{-[(1+x) F_1']'}{(1+x) F_1} = a_k^2 - \frac{1}{4(1+x)^2} \qquad (k = 1, 2, \ldots),$$

and inserting the maximum and minimum values of this function in (8.23) we obtain the limits

$$a_k^2 - \frac{1}{4} \leqq \lambda_s \leqq a_k^2 - \frac{1}{16}.$$

Thus we have obtained limits for infinitely many eigenvalues at one fell swoop. The constant difference between the upper and lower limits means that the percentage error is smaller for higher eigenvalues, and in actual fact this method is probably superior to all others for the higher eigenvalues. Several values are given in Table III/16.

Generalizations of this bracketing theorem have been derived by H. WIELANDT[1]. One formulation, which includes all previously known bracketing theorems as special cases, is based on the Schwarz constants a_k of (8.13). In it, real numbers $b_1, b_2, \ldots, b_p$ are choosen arbitrarily and used, together with the first $p+1$ Schwarz constants $a_0, a_1, \ldots, a_p$, to define two closed sets M_1 and M_2 of real numbers x: M_1 consists of all numbers x for which $\sum\limits_{\nu=1}^{p} b_\nu x^\nu \geqq \dfrac{1}{a_0} \sum\limits_{\nu=1}^{p} a_\nu b_\nu$ and M_2 consists of all

Table III/16. *Bracketing of eigenvalues*

a_k	a_k	Lower	Upper
		limits for λ_s	
1·8366	3·3731	3·123	3·311
4·816	23·19	22·94	23·13
7·917	62·68	62·43	62·62
11·041	121·90	121·65	121·84
14·173	200·86	200·61	200·80
17·308	299·56	299·31	299·50

[1] WIELANDT, H.: Ein Einschließungssatz für charakteristische Wurzeln normaler Matrizen. Arch. Math. **1**, 348—352 (1949), and Fiat-Review, Naturforschung und Medizin in Deutschland 1939—1946, **2**, 98 (1948). An older formulation was given by K. FRIEDRICHS and G. HORVAY: The finite Stieltjes momentum problem. Proc. Nat. Acad. Sci., Wash. **25**, 528—534 (1939); a detailed presentation has been given by H. BÜCKNER: Die praktische Behandlung von Integralgleichungen (Ergebnisse der Angew. Mathematik, H. 1). Berlin-Göttingen-Heidelberg 1952.

numbers x for which $\displaystyle\sum_{\nu=1}^{p} b_\nu x^\nu \leq \frac{1}{a_0} \sum_{\nu=1}^{p} a_\nu b_\nu$. The theorem then states that there is at least one eigenvalue whose reciprocal lies in M_1 and at least one eigenvalue whose reciprocal lies in M_2.

8.7. Three minimum principles

RITZ's method for solving variational problems has already been applied to an eigenvalue problem[1,2] (in § 5) by deriving a variational problem whose Euler conditions can be identified with the eigenvalue problem in question. Now these Euler conditions, although necessary, are not sufficient for an extremum of the variational problem, and hence conclusions about the approximate values obtained cannot be drawn solely from the fact that the eigenvalue problem can be formally identified with them. However, for certain special classes of eigenvalue problem there are other variational problems, based on certain minimum principles of eigenvalue theory, which also have solutions satisfying the eigenvalue problem, but which allow specific statements to be made about the approximate eigenvalues obtained by solving them approximately by RITZ's method. We quote the results without proof.

We now state the above-mentioned minimum principles.

A. RAYLEIGH's minimum principle. For a self-adjoint full-definite eigenvalue problem (8.1), (8.4), (8.5) with boundary conditions independent of the eigenvalue λ the smallest eigenvalue λ_1 is the minimum value of the Rayleigh quotient $R[u]$ (8.9) in the domain of all comparison functions u (cf. § 8.1).

B. KAMKE's minimum principle[3]. Here fewer restrictions are imposed on u at the expense of the generality of the eigenvalue problem; this is often of considerable advantage in practical work. This minimum principle applies to the same eigenvalue problems as RAYLEIGH's except that one additional restriction must be imposed, namely that the eigenvalue problem is "K-definite", a term which we now explain. Firstly the integrals appearing in the Rayleigh quotient (8.9) are put

---

[1] A comprehensive presentation based on the techniques of functional analysis is given by N. ARONSZAJN: Study of eigenvalue problems. The Rayleigh-Ritz and the Weinstein methods for approximation of eigenvalues. Dept. of Math. Oklahoma Agricultural and Mechanical College. Stillwater 1949. 214 pp.

[2] Error estimates for self-adjoint full-definite eigenvalue problems have been obtained by G. BERTRAM: Zur Fehlerabschätzung für das Ritzsche Verfahren bei Eigenwertaufgaben. Diss. 56 pp. Hannover 1950. Other estimates and investigations of the rate of convergence can be found in L. V. KANTOROVICH and V. I. KRYLOV: Näherungsmethoden der Höheren Analysis, pp. 226—329. Berlin 1956.

[3] KAMKE, E.: Über die definiten selbstadjungierten Eigenwertaufgaben IV. Math. Z. **48**, 67—100 (1942).

into "DIRICHLET's form"

$$\left.\begin{array}{l} \int\limits_a^b u\,M[u]\,dx = \int\limits_a^b \sum\limits_{\nu=0}^m f_\nu[u^{(\nu)}]^2\,dx + M_0[u], \\[2mm] \int\limits_a^b u\,N[u]\,dx = \int\limits_a^b \sum\limits_{\nu=0}^n g_\nu[u^{(\nu)}]^2\,dx + N_0[u]; \end{array}\right\} \qquad (8.27)$$

$M_0[u]$ and $N_0[u]$ are called the "Dirichlet boundary expressions". By integrating by parts ν times we transform a general term

$$\int\limits_a^b u\,(-1)^\nu\,[f_\nu u^{(\nu)}]^{(\nu)}\,dx$$

into

$$\int\limits_a^b u^{(\nu)} f_\nu u^{(\nu)}\,dx$$

plus terms involving only boundary values. The Dirichlet boundary expression $M_0[u]$ is then the sum of these boundary terms for all values of ν, and $N_0[u]$ is found similarly. These boundary expressions are quadratic forms in the values of $u, u', u'', \ldots$ at the boundary points $x=a$, $x=b$. If the two forms are positive definite when their variables are restricted to run through all sets of numbers satisfying the relations represented by the boundary conditions and if all the functions $f_\nu(x)$ and $g_\nu(x)$ are non-negative [previously the only restrictions had been on $f_m(x)$ and $g_n(x)$, namely $f_m \neq 0$, $g_n \neq 0$], then the eigenvalue problem is called K-definite.

It can be deduced from the self-adjointness of the eigenvalue problem that the boundary conditions are such that, by suitable linear combination, each boundary derivative of order greater than or equal to m may be expressed in terms of boundary derivatives of order $0, 1, 2, \ldots$, $m-1$. Hence derivatives of order greater or equal to m can be eliminated from $M_0[u]$ and $N_0[u]$, which then become quadratic forms in $u(a)$, $u'(a), \ldots, u^{(m-1)}(a), u(b), u'(b), \ldots, u^{(m-1)}(b)$. RAYLEIGH's quotient (8.9) can therefore be put into the form

$$K[u] = \frac{\int\limits_a^b \sum\limits_{\nu=0}^m f_\nu[u^{(\nu)}]^2\,dx + M_0[u]}{\int\limits_a^b \sum\limits_{\nu=0}^n g_\nu[u^{(\nu)}]^2\,dx + N_0[u]}, \qquad (8.28)$$

in which no boundary derivative of order greater than $m-1$ appears.

KAMKE's theorem then states that λ_1 is the minimum value of the quotient $K[u]$ in the domain of all admissible functions u.

C. A minimum principle for the special eigenvalue problems[1]. For those problems in which $N[y] = g_0(x)\,y$ another minimum principle

---

[1] COLLATZ, L.: Z. Angew. Math. Mech. **19**, 228 (1939).

can be established under the conditions laid down in **A** for the application of Rayleigh's principle; it is based on the iteration method. As mentioned in § 8.4, for the special eigenvalue problems the first iterate $F_1(x)$ in the iteration process may be chosen arbitrarily from among the comparison functions; its predecessor $F_0(x)$ in the iterative sequence is then given by

$$F_0(x) = \frac{1}{g_0(x)} M[F_1].$$ (8.29)

Here the Rayleigh quotient (8.9) may be written as the Schwarz quotient μ_2, as in (8.16):

$$R[u] = R[F_1] = \mu_2,$$

and we know from **A** that the minimum value of this quantity in the domain of comparison functions u is the smallest eigenvalue λ_1. But since the μ_k decrease monotonically, we also have $\mu_1 \geqq \lambda_1$. Now μ_1 may easily be expressed in terms of F_1:

$$\mu_1 = \frac{\int\limits_a^b F_0 N[F_0]\, dx}{\int\limits_a^b F_1 N[F_0]\, dx} = \frac{\int\limits_a^b F_0 M[F_1]\, dx}{\int\limits_a^b F_1 M[F_1]\, dx},$$ (8.30)

where F_0 is given by (8.29), and if we replace F_1 by u to emphasize its arbitrary character, we have

$$\mu_1 = \frac{\int\limits_a^b \frac{1}{g_0(x)} (M[u])^2\, dx}{\int\limits_a^b u\, M[u]\, dx}.$$ (8.31)

The minimum value of this quotient in the domain of comparison functions is the smallest eigenvalue λ_1. In general $R[u]$ is a better approximation to λ_1 than $\mu_1[u]$ for the same function u but $\mu_1[u]$ can often be calculated rather more quickly. The extension of this minimum principle to the general eigenvalue problems is discussed at the end of § 8.8.

8.8. Application of Ritz's method

For any of the three minimum problems just described we can, as in Ch. I (4.1), assume a general expression for u depending on undetermined parameters and then write down the necessary minimum conditions to be satisfied by these parameters. Convenient equations are obtained if the expression for u is chosen to depend linearly on the parameters:

$$u = a_1 v_1(x) + a_2 v_2(x) + \cdots + a_p v_p(x),$$ (8.32)

rather than in a general (non-linear) way. The $v_1, v_2, \ldots, v_p$ will be chosen as fixed comparison functions which are linearly independent in $\langle a, b \rangle$. Then u also will be a comparison function, and will vanish identically only when $a_1 = a_2 = \cdots = a_p = 0$. This case is therefore excluded.

First let us consider RAYLEIGH's principle. Substitution of (8.32) for u in RAYLEIGH's quotient (8.9) yields

$$R[u] = \frac{\int\limits_a^b \sum\limits_{r=1}^p a_r v_r(x) \sum\limits_{s=1}^p a_s M[v_s(x)]\, dx}{\int\limits_a^b \sum\limits_{r=1}^p a_r v_r(x) \sum\limits_{s=1}^p a_s N[v_s(x)]\, dx}. \tag{8.33}$$

For convenience we now introduce quantities m_{rs}, n_{rs} defined by

$$\left.\begin{aligned} m_{rs} &= \int\limits_a^b v_r(x)\, M[v_s(x)]\, dx, \\ n_{rs} &= \int\limits_a^b v_r(x)\, N[v_s(x)]\, dx, \end{aligned}\right\} \tag{8.34}$$

which, on account of the assumed self-adjointness, possess the symmetric properties

$$m_{rs} = m_{sr} \quad \text{and} \quad n_{rs} = n_{sr}. \tag{8.35}$$

Then RAYLEIGH's quotient has the form of a quotient of two quadratic forms Q_1, Q_2 in the parameters a_r:

$$R[u] = \frac{Q_1}{Q_2},$$

where

$$Q_1 = \sum_{r,s=1}^p m_{rs} a_r a_s, \qquad Q_2 = \sum_{r,s=1}^p n_{rs} a_r a_s. \tag{8.36}$$

The assumption of full-definiteness (8.8) means that the numerator and denominator in $R[u]$ can take only positive values (we have excluded the case $a_1 = a_2 = \cdots = a_p = 0$); consequently Q_1 and Q_2 are positive definite quadratic forms.

The necessary conditions for a minimum of $R[u(a_1, \ldots, a_p)]$ read

$$\frac{\partial R}{\partial a_r} = \frac{Q_2 \dfrac{\partial Q_1}{\partial a_r} - Q_1 \dfrac{\partial Q_2}{\partial a_r}}{Q_2^2} = 0 \qquad (r = 1, 2, \ldots, p).$$

The minimum value of R in this restricted class of comparison functions we denote by Λ:

$$\Lambda = \min R[u(a_1, \ldots, a_p)],$$

and use it as an approximate value for the first eigenvalue λ_1. At the same time, of course, Λ is the value of Q_1/Q_2 at the minimum and hence the equations for the a_r may be simplified to

$$\frac{\partial Q_1}{\partial a_r} - \Lambda \frac{\partial Q_2}{\partial a_r} = 0 \qquad (r = 1, 2, \ldots, p). \tag{8.37}$$

Now from (8.36)

$$\frac{\partial Q_1}{\partial a_r} = 2 \sum_{s=1}^{p} m_{rs} a_s, \qquad \frac{\partial Q_2}{\partial a_r} = 2 \sum_{s=1}^{p} n_{rs} a_s,$$

so that the equations (8.37) can be written

$$\sum_{s=1}^{p} a_s (m_{rs} - \Lambda n_{rs}) = 0 \qquad \text{for} \quad r = 1, 2, \ldots, p. \tag{8.38}$$

These equations are called GALERKIN's equations (cf. Ch. I, § 4.3). They are p homogeneous equations for p unknowns a_s and have a non-trivial solution (i.e. one in which at least one a_s is non-zero) if and only if the determinant of their coefficients vanishes:

$$\det(m_{rs} - \Lambda n_{rs}) = \begin{vmatrix} m_{11} - \Lambda n_{11} & m_{12} - \Lambda n_{12} & \ldots & m_{1p} - \Lambda n_{1p} \\ m_{21} - \Lambda n_{21} & m_{22} - \Lambda n_{22} & \ldots & m_{2p} - \Lambda n_{2p} \\ \cdots & \cdots & \cdots & \cdots \\ m_{p1} - \Lambda n_{p1} & m_{p2} - \Lambda n_{p2} & \ldots & m_{pp} - \Lambda n_{pp} \end{vmatrix} = 0. \tag{8.39}$$

This is an algebraic equation for Λ of the p-th degree and has p roots $\Lambda_1, \Lambda_2, \ldots, \Lambda_p$. Under the assumptions which we have made it can be shown that the p roots Λ_k are real and positive, and that, when they are arranged in increasing order of magnitude, they furnish upper limits for the corresponding eigenvalues, i.e.

$$\Lambda_k \geq \lambda_k \qquad \text{for} \quad k = 1, 2, \ldots, p. \tag{8.40}$$

The technique of substituting a linear expression (8.32) for u can be used in exactly the same way for the other two minimum principles of § 8.7. We derive the corresponding equations only for the third minimum principle § 8.7, **C**, but also show how they can be extended to the general eigenvalue problems. We deal first with a special eigenvalue problem, for which the differential equation reads

$$M[y] = \lambda g_0(x) y$$

and (8.31) is to be minimized. Proceeding as for RAYLEIGH's principle we substitute for u an arbitrary linear combination (8.32) of p linearly independent comparison functions $v_r(x)$, which satisfy all the boundary conditions, and obtain

$$\mu_1[u] = \frac{Q_1^{\otimes}}{Q_2^{\otimes}},$$

in which the quadratic forms $Q_1^\otimes$, $Q_2^\otimes$ are here given by

$$Q_1^\otimes = \int_a^b \frac{1}{g_0(x)} \left(\sum_{r=1}^p a_r M[v_r(x)] \right) \left(\sum_{s=1}^p a_s M[v_s(x)] \right) dx = \sum_{r,s=1}^p m_{rs}^\otimes a_r a_s,$$

where

$$m_{rs}^\otimes = \int_a^b \frac{1}{g_0(x)} M[v_r(x)] M[v_s(x)] dx, \tag{8.41}$$

and

$$Q_2^\otimes = \sum_{r,s=1}^p n_{rs}^\otimes a_r a_s,$$

where

$$n_{rs}^\otimes = \int_a^b v_r(x) M[v_s(x)] dx. \tag{8.42}$$

Thus the quantities $n_{rs}^\otimes$ are formed in the same way as the quantities m_{rs} for RAYLEIGH's principle.

The rest of the calculation is identical with the corresponding calculation for RAYLEIGH's principle if the notation Q_1, Q_2, m_{rs}, n_{rs} is modified with a superscript $\otimes$. The equations corresponding to (8.38), i.e.

$$\sum_{s=1}^p a_s (m_{rs}^\otimes - \Lambda n_{rs}^\otimes) = 0 \qquad (r = 1, 2, \dots, p), \tag{8.43}$$

are GRAMMEL's equations[1], and the approximate values Λ_k are the zeros of the determinant

$$\det (m_{rs}^\otimes - \Lambda n_{rs}^\otimes) = \begin{vmatrix} m_{11}^\otimes - \Lambda n_{11}^\otimes & \cdots & m_{1p}^\otimes - \Lambda n_{1p}^\otimes \\ \cdot\cdot\cdot\cdot\cdot\cdot\cdot\cdot\cdot\cdot\cdot\cdot\cdot\cdot\cdot \\ m_{p1}^\otimes - \Lambda n_{p1}^\otimes & \cdots & m_{pp}^\otimes - \Lambda n_{pp}^\otimes \end{vmatrix}. \tag{8.44}$$

In a practical application, particularly one in which the f_r and g_0 are not simple functions, one would normally specify the functions $M[v_r(x)]$ and calculate the $v_r(x)$ by integration, thus avoiding the numerical or graphical differentiation involved in calculating the $M[v_r(x)]$ from chosen $v_r(x)$. The determination of each $v_r(x)$ is equivalent to the first step in the iteration method.

We now turn to the general eigenvalue problem with the differential equation $M[y] = \lambda N[y]$. Clearly we must go back to the original

---

[1] GRAMMEL, R.: Ein neues Verfahren zur Lösung technischer Eigenwertprobleme. Ing.-Arch. 10, 35—46 (1939). GRAMMEL derives the equations in a different way.

Practical examples are worked out by E. MAIER: Biegeschwingungen von spannungslos verwundenen Stäben, insbesondere von Luftschraubenblättern. Ing.-Arch. 11, 73—98 (1940). See also R. GRAMMEL: Über die Lösung technischer Eigenwertprobleme. VDI-Forsch.-Heft, Gebiet Stahlbau 6, 36—42 (1943).

definition of μ_1 in (8.30), which does not depend on the special form of N. Let us put

$$F_1(x) = u(x) = \sum_{r=1}^{p} a_r v_r(x);$$

then we have to determine a function F_0 which is related to F_1 by the differential equation $M[F_1] = N[F_0]$, in accordance with (8.10), and which satisfies those of the boundary conditions sufficient for (8.12) to hold. Corresponding to each of the p functions v_r we determine a function $w_r(x)$ such that $M[v_r] = N[w_r]$ and which satisfies as many of the boundary conditions as are sufficient for (8.12) to hold, i.e. so that

$$\int_a^b \left(w_r N[\varphi] - \varphi N[w_r] \right) d x = 0 \qquad (8.45)$$

for all comparison functions $\varphi(x)$. We can then put

$$F_0(x) = \sum_{r=1}^{p} a_r w_r(x).$$

With this expression substituted in (8.30) we have again a quotient of two quadratic forms $\mu_1 = Q_1^{\otimes}/Q_2^{\otimes}$ but the numerator $Q_1^{\otimes}$ is now defined by

$$Q_1^{\otimes} = \int_a^b \left(\sum_{r=1}^{p} a_r w_r(x) \right) \left(\sum_{s=1}^{p} a_s M[v_s] \right) d x = \sum_{r,s=1}^{p} m_{rs}^{\otimes} a_r a_s,$$

where the $m_{rs}^{\otimes}$ are also defined differently:

$$m_{rs}^{\otimes} = \int_a^b w_r M[v_s] d x.$$

The a_r are determined to minimize μ_1 as before, i.e. from the equations (8.43), and Λ is determined from (8.44), but the new definition of the $m_{rs}^{\otimes}$ is used instead of (8.41); the definition of the $n_{rs}^{\otimes}$ (8.42) is unaltered.

8.9. TEMPLE's quotient

The limits for the first eigenvalue λ_1 given by (8.18) in § 8.3 may be extended to higher eigenvalues, and they then provide an alternative means to the Ritz method for the approximate location of the higher eigenvalues; they can also be used in combination with the Ritz method as in the example below.

Starting with a function $F_0(x)$ which satisfies the same conditions as in § 8.3, we form the next function $F_1(x)$ in the iterative sequence generated by the iteration procedure of § 8.2 and calculate the Schwarz constants a_0, a_1, a_2 and quotients μ_1, μ_2, as in (8.19).

We now assume that $F_0(x)$ may be expanded as a uniformly convergent series of normalized eigenfunctions $y_j(x)$, i.e. normalized in the general sense:

$$\int_a^b y_i N[y_j]\, dx = \begin{cases} 0 & \text{when} \quad i \neq j \\ 1 & \text{when} \quad i = j. \end{cases}$$

We may then write

$$F_0(x) = \sum_{j=1}^{\infty} c_j y_j(x),$$

and assuming further that this series can be differentiated term by term sufficiently often we readily deduce that

$$F_1(x) = \sum_{j=1}^{\infty} c_j \frac{y_j(x)}{\lambda_j} \tag{8.46}$$

and

$$a_k = \sum_{j=1}^{\infty} \frac{c_j^2}{\lambda^k} \quad (k = 0, 1, 2). \tag{8.47}$$

The validity of this expansion can be established[1] when, for example, the eigenvalue problem (8.1), (8.4), (8.5) is self-adjoint and full-definite, $N[y]$ has the special "single-term" form

$$N[y] = (-1)^n [g_n(x) y^{(n)}]^{(n)} \quad \text{with} \quad g_n(x) > 0,$$

the boundary conditions include the equations

$$y(a) = y'(a) = \cdots = y^{(n-1)}(a) = y(b) = y'(b) = \cdots = y^{(n-1)}(b) = 0$$

and $F_0(x)$ is a comparison function.

From the Schwarz constants we now form the "Temple quotient"

$$\left. \begin{aligned} T(t) &= \frac{a_0 - t\, a_1}{a_1 - t\, a_2} = \frac{a_1}{a_2} \cdot \frac{\dfrac{a_0}{a_1} - t}{\dfrac{a_1}{a_2} - t} = \mu_2 \frac{\mu_1 - t}{\mu_2 - t} \\ &= \mu_2 \left(1 + \frac{\mu_1 - \mu_2}{\mu_2 - t}\right) = \mu_2 - \frac{\mu_1 - \mu_2}{\dfrac{t}{\mu_2} - 1}. \end{aligned} \right\} \tag{8.48}$$

Here t is an arbitrary parameter restricted only by the condition that it shall differ from μ_2. This quotient $T(t)$ can easily be expressed in

[1] See L. Collatz: Eigenwertaufgaben. Leipzig 1949. pp. 144 and 191. For formula (8.54) the assumption that $F_0(x)$ is a comparison function can be replaced by the weaker conditions specified for $F_0(x)$ in § 8.3; the proof of this is given by N. J. Lehmann: Beiträge zur numerischen Lösung linearer Eigenwertprobleme. Z. Angew. Math. Mech. **29**, 341—356 (1949); **30**, 1—16 (1950).

terms of the expansion coefficients c_j via (8.47):

$$T(t) = \frac{a_0 - t\,a_1}{a_1 - t\,a_2} = \frac{\sum_{j=1}^{\infty} c_j^2 \left(1 - \frac{t}{\lambda_j}\right)}{\sum_{j=1}^{\infty} c_j^2 \frac{1}{\lambda_j}\left(1 - \frac{t}{\lambda_j}\right)}, \tag{8.49}$$

and hence, for any eigenvalue λ_s,

$$T(t) - \lambda_s = \frac{\sum_{j=1}^{\infty} c_j^2 \left(1 - \frac{\lambda_s}{\lambda_j}\right)\left(1 - \frac{t}{\lambda_j}\right)}{\sum_{=1}^{\infty} c_j^2 \frac{1}{\lambda_j}\left(1 - \frac{t}{\lambda_j}\right)}.$$

Now

$$\mu_2 - t = \frac{a_1 - a_2 t}{a_2} = \frac{\sum_{j=1}^{\infty} c_j^2 \frac{1}{\lambda_j}\left(1 - \frac{t}{\lambda_j}\right)}{\sum_{j=1}^{\infty} \left(\frac{c_j}{\lambda_j}\right)^2},$$

and since the sum of the series which occurs both in the numerator here and in the denominator of the quotient for $T(t) - \lambda_s$ cannot be zero (we have assumed $t \neq \mu_2$), we may multiply the two quotients together and cancel:

$$\left(T(t) - \lambda_s\right)(\mu_2 - t) = \frac{\sum_{j=1}^{\infty} c_j^2 \left(1 - \frac{\lambda_s}{\lambda_j}\right)\left(1 - \frac{t}{\lambda_j}\right)}{\sum_{j=1}^{\infty} \left(\frac{c_j}{\lambda_j}\right)^2}. \tag{8.50}$$

Let the eigenvalues be arranged in ascending order of magnitude, as in § 8.1 (our assumptions imply that $\lambda_1 > 0$ but the present considerations are also valid for $\lambda_1 < 0$). Whether λ_s is simple or multiple, there is a next smaller eigenvalue λ_{s-} and a next larger eigenvalue λ_{s+} (unless $\lambda_s = \lambda_1$, when λ_{s-} does not exist). Thus we have $\lambda_{s-} < \lambda_s < \lambda_{s+}$ and no eigenvalue other than λ_s lies between λ_{s-} and λ_{s+}. When λ_s is simple, $\lambda_{s-} = \lambda_{s-1}$ and $\lambda_{s+} = \lambda_{s+1}$.

Now the numerator of the quotient on the right-hand side of (8.50) cannot be negative if t lies in the interval $\lambda_{s-} \leq t \leq \lambda_{s+}$; moreover, the denominator, whose value is a_2, is always positive since the eigenvalue problem is assumed to be full-definite. Consequently

$$\left(T(t) - \lambda_s\right)(\mu_2 - t) \geq 0 \quad \text{for} \quad \lambda_{s-} \leq t \leq \lambda_{s+}. \tag{8.51}$$

This implies that

and
$$\begin{array}{ll} T(t) \geq \lambda_s & \text{when} \quad t < \mu_2 \\ T(t) \leq \lambda_s & \text{when} \quad t > \mu_2 \end{array} \right\} \tag{8.52}$$

provided always that $\lambda_{s-} \leq t \leq \lambda_{s+}$.

It can be seen immediately that the limits (8.20) for the first eigen-value are special cases of this result: if we know that a number t, greater than μ_2, is a lower limit for λ_2, then (8.52) gives $T(t) \leq \lambda_1$, which is precisely the lower limit in (8.20); if we take the limit as $t \to -\infty$ in (8.48), we obtain $T(-\infty) = \mu_2 \geq \lambda_1$, and hence the result also includes the upper limit in (8.20).

For higher eigenvalues we can summarize thus: If we have an approximation $F_0(x)$ to an eigenfunction y_s corresponding to the eigen-value λ_s which is sufficiently accurate that the Schwarz quotient μ_2 which it generates lies in the interval $\lambda_{s-} < \mu_2 < \lambda_{s+}$, and if also we know an upper limit L_{s-} for λ_{s-} and a lower limit l_{s+} for λ_{s+} which are both so near that

$$\lambda_{s-} \leq L_{s-} < \mu_2 < l_{s+} \leq \lambda_{s+}, \tag{8.53}$$

then λ_s is bracketed by the values $T[l_{s+}]$ and $T[L_{s-}]$:

$$T(l_{s+}) \leq \lambda_s \leq T(L_{s-}). \tag{8.54}$$

Example. As in § 8.5, we again treat the problem

$$-[(1+x)\,y']' = \lambda(1+x)\,y, \quad y'(0) = y(1) = 0$$

by assuming a polynomial expression for $F_1(x)$. This time, however, we use a polynomial of higher degree with an extra arbitrary constant so that in addition to satisfying both boundary conditions on F_1 and the essential boundary condition on F_0, which yields

$$F_0(x) = \frac{2}{1+x} \{c_1(14 + 28\,x - 18\,x^2 - 24\,x^3) + c_2(5 + 4\,x - 9\,x^2)\}$$

$$F_1(x) = \quad c_1(11 - 14\,x^2 + 3\,x^4) \quad\quad + c_2(3 - 5\,x^2 + 2\,x^3),$$

we can determine the ratio $c_1:c_2$ of the remaining free constants to make $F_1(x)$ approximate the second eigenfunction. The Schwarz con-stants generated by this $F_0(x)$ (cf. § 8.5) are

$$a_0 = 4 \quad \{ 204 \cdot 9614\,c_1^2 + 2 \times 53 \cdot 061\,42\,c_1 c_2 + 14 \cdot 111\,42\,c_2^2\}$$

$$a_1 = \frac{2}{21} \{2619 \cdot 8 \quad c_1^2 + 2 \times 693 \cdot 2 \quad c_1 c_2 + 184 \cdot 1 \quad c_2^2\}$$

$$a_2 = \frac{1}{21} \left\{1623 \cdot 3 \quad c_1^2 + 2 \times \frac{1723 \cdot 3}{4} \quad c_1 c_2 + 114 \cdot 4 \quad c_2^2\right\}.$$

One way of determining the ratio $c_1:c_2$ would be to put $c_1 = 1$, calculate the limits (8.54) for different values of c_2 and then interpolate for the value of c_2 which gives the narrowest limits. Here it is con-venient to apply the Rayleigh-Ritz method with $F_1(x)$ as the assumed form of approximate solution; we have already calculated the expres-sions for a_1 and a_2, i.e. the quadratic forms Q_1 and Q_2 of § 8.8, and can write down immediately the Galerkin equations

$$(2 \times 2619 \cdot 8 - \varLambda \times 1623 \cdot 3)\, c_1 + \left(2 \times 693 \cdot 2 - \varLambda \times \frac{1723 \cdot 3}{4}\right) c_2 = 0,$$

$$\left(2 \times 693 \cdot 2 - \varLambda \times \frac{1723 \cdot 3}{4}\right) c_1 + (2 \times 184 \cdot 1 - \varLambda \times 114 \cdot 4)\ c_2 = 0.$$

The determinant of the coefficients is a quadratic expression in $\varLambda$, whose zeros

$$\varLambda = \begin{cases} \varLambda_1 = \ 3 \cdot 218\,505 \\ \varLambda_2 = 23 \cdot 189\,7 \end{cases}$$

are the Rayleigh-Ritz approximations to the first and second eigenvalues. We take $\varLambda = \varLambda_2$ and solve the corresponding set of homogeneous equations to obtain

$$c_1 = 1, \qquad c_2 = -\,3 \cdot 766\,05.$$

With these values we have

$$a_0 = 4 \times 5 \cdot 441\,75, \qquad a_1 = 2 \times 0 \cdot 459\,85, \qquad a_2 = 0 \cdot 039\,664,$$

$$\mu_1 = 23 \cdot 667, \qquad \mu_2 = 23 \cdot 190, \qquad \mu_1 - \mu_2 = 0 \cdot 48.$$

Since μ_2 is equal to the Rayleigh-Ritz approximation $\varLambda_2$, we know immediately that $\mu_2 \geq \lambda_2$. [This is a smaller upper limit than that given by (8.54).]

In order to calculate a lower limit for λ_2 from (8.54) we need a rough lower limit for λ_3. If, as in § 8.5, we compare with a problem having constant coefficients, we obtain a very crude lower limit

$$l_3 = \frac{25}{8}\,\pi^2 \approx 30 \cdot 8.$$

It is greater than μ_2, however, and can therefore be used in (8.54); it yields

$$T(l_3) = 21 \cdot 735 \leq \lambda_2.$$

This limit will be improved if we can deduce a better lower limit for λ_3; for example, $l_3 = 60$ would give

$$T(60) = 22 \cdot 888 \leq \lambda_2.$$

8.10. Some modifications to the iteration method

Theory shows that the ordinary iteration (8 10) will converge to the q-th eigenfunction if $F_0(x)$ is chosen so as to be orthogonal in the generalized sense to the first $q-1$ eigenfunctions. In practice these first $q-1$ eigenfunctions are not known exactly and therefore their components cannot be excluded completely from $F_0(x)$ and subsequent iterates. In a modification of the iteration method given by Koch[1] the unwanted components, which would otherwise become dominant, are suppressed by subtraction at each stage of the iteration of the components of the best available approximations Y_j to the first $q-1$ eigenfunctions. Thus the direct relation (8.10) between F_k and F_{k-1} is replaced by

$$
\left.
\begin{aligned}
M[F_k^*] &= N[F_{k-1}], \\
U_\mu[F_k^*] &= 0, \\
F_k &= F_k^* - \sum_{j=1}^{q-1} \frac{Y_j \int_a^b F_k^* N[Y_j]\, dx}{\int_a^b Y_j N[Y_j]\, dx}
\end{aligned}
\right\} \qquad (k = 1, 2, \ldots).
$$

Another modification[2], which occasionally improves the convergence of both this modified iteration and the ordinary iteration (8.10), consists in replacing $M[F_k]$ by $M[F_k - \vartheta F_{k-1}]$. Thus the iteration (8.10) is modified to

$$
M[F_k - \vartheta F_{k-1}] = N[F_{k-1}], \qquad U_\mu[F_k] = 0 \qquad (k = 1, 2, \ldots).
$$

Assuming that the expansion theorem is applicable we may then write

$$
F_1 = \sum_{j=1}^{\infty} c_j\, y_j \left(\vartheta + \frac{1}{\lambda_j} \right), \qquad F_k = \sum_{j=1}^{\infty} c_j\, y_j \left(\vartheta + \frac{1}{\lambda_j} \right)^k \qquad (k = 1, 2, \ldots)
$$

in place of (8.46).

One can go a step further and use a different ϑ in each cycle[3], in which case the iteration formula becomes

$$
M[F_k - \vartheta_k F_{k-1}] = N[F_{k-1}], \qquad U_\mu[F_k] = 0 \qquad (k = 1, 2, \ldots)
$$

and the expansion formula must be modified to

$$
F_k = \sum_{j=1}^{\infty} c_j\, y_j \left(\vartheta_1 + \frac{1}{\lambda_j} \right)\left(\vartheta_2 + \frac{1}{\lambda_j} \right) \cdots \left(\vartheta_k + \frac{1}{\lambda_j} \right) \qquad (k = 1, 2, \ldots).
$$

[1] Koch, J. J.: Bestimmung höherer kritischer Drehzahlen schnell laufender Wellen. Verh. 2. Internat. Kongr. Techn. Math., Zürich 1926, pp. 213—218.
Another method is given by A. Fraenkle: Ing.-Arch. 1, 499—526 (1930), specifically "Methode II", p. 510 et seq. A method of minimized iterations has been devised by C. Lanczos: An iteration method for the solution of the eigenvalue problem of linear differential and integral operators. J. Res. Nat. Bur. Stand. 45, 255—282 (1950).
[2] Developed for integral equations by G. Wiarda: Integralgleichungen unter besonderer Berücksichtigung der Anwendungen, p. 126. Leipzig 1930.
[3] Developed for integral equations by H. Bückner: Ein unbeschränkt anwendbares Iterationsverfahren für Fredholmsche Integralgleichungen. Math. Nachr. 2, 304—313 (1949).

If approximations Λ_σ are known for, say, the first $p-1$ eigenvalues λ_σ, one can put

$$\left. \begin{aligned} \vartheta_\sigma &= \vartheta_{p+\sigma} = \vartheta_{qp+\sigma} = -\frac{1}{\Lambda_\sigma} \\ \vartheta_p &= \vartheta_{qp} = 0 \end{aligned} \right\} \quad \text{for} \quad \sigma = 1, 2, \ldots, p-1, \quad q = 1, 2, 3, \ldots.$$

This furnishes another iteration method for the p-th eigenfunction provided that the approximations Λ_σ are sufficiently accurate.

8.11. Miscellaneous exercises on Chapter III

1. The stress distribution in a rotating disc (a steam turbine rotor, say) of thickness $\eta(r)$ and radius R may be calculated from the differential equation[1]

$$r^2 S'' + r\left(3 - \frac{r}{\eta}\eta'\right) S' - \frac{m-1}{m}\frac{r}{\eta}\eta' S + \frac{3m+1}{m}\frac{\gamma}{g}\eta r^2 \omega^2 = 0.$$

Here r is the distance from the axis of rotation (Fig. III/15), $S(r) = \eta(r)\sigma_r$, where σ_r is the radial stress, and m, γ, ω, g are constants (usual notation as in reference[1] below). If we consider a thickness profile of the form

$$\eta(r) = \frac{\eta_0}{1 + \left(\dfrac{2r}{R}\right)^2}$$

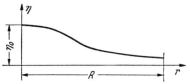

Fig. III/15. Radial section of the disc in Exercise 1

and introduce the dimensionless variables

$$x = \frac{2r}{R}, \quad y = \frac{S}{C}, \quad C = \frac{3m+1}{m}\frac{\gamma}{g}\omega^2\frac{R^2}{4}\eta_0,$$

then the equation with $m = 3$ reads

$$(1 + x^2)y'' + \left(\frac{3}{x} + 5x\right)y' + \frac{4}{3}y + 1 = 0, \tag{8.55}$$

in which the dashes now denote differentiation with respect to x.

Calculate $y(x)$ by the ordinary finite-difference method with interval $h = \frac{1}{2}$, say, for the case in which the boundary conditions are

$$y(\pm 2) = 0\cdot6, \quad y \text{ regular at } x = 0,$$

i.e. for a solid disc with given radial stress at its edge $r = R$ ($x = 2$).

2. Failure of the finite-difference method. Show that for the boundary-value problem

$$x y'' = c y'; \quad y(0) = 0, \quad y(1) = 1$$

the solution given by the ordinary finite-difference method is always partly of the wrong sign when $c > 2$ (when $c = 3$, for example).

---

[1] BIEZENO, C. B., and R. GRAMMEL: Technische Dynamik, 2nd ed., Vol. II, p. 25. Berlin-Göttingen-Heidelberg 1953.

3. Show that the solution of the non-linear problem

$$y'' = \frac{2x^2}{y}, \qquad y(\pm 1) = 1$$

given by the ordinary finite-difference method is exact.

4. The boundary-value problem

$$y'' = y^2; \qquad y(0) = 0, \qquad y(2) = -2$$

possesses only complex solutions. Determine one of these solutions roughly by the ordinary finite-difference method.

5. The boundary-value problem $y'' = 0$; $y(0) = 1$, $y(1) = y'(1)$ has no solution. Can the method of § 5.2 be applied to obtain a variational problem whose Euler conditions are identical with this boundary-value problem?

6. Show that the variational problem corresponding to the boundary-value problem

$$y'' + 4y = 2, \qquad y(\pm 1) = 0$$

according to the formulae of § 5.2 possesses neither a minimum nor a maximum. Can one still apply the Ritz method, say with a polynomial approximation of the form

$$y = \sum_{\nu=1}^{p} a_\nu (1 - x^{2\nu}).$$

7. Failure of the Ritz method. The non-linear boundary-value problem

$$y' y'' = 4x, \qquad y(\pm 1) = 0$$

represents the Euler conditions of the variational problem

$$J[\varphi] = \int_{-1}^{1} [\tfrac{1}{6} \varphi'^3 + 4x\,\varphi]\,dx = \text{extremum}$$

with the auxiliary conditions $\varphi(\pm 1) = 0$. The boundary-value problem has two solutions symmetrical about $x = 0$, namely

$$y_1 = 1 - x^2 \quad \text{and} \quad y_2 = x^2 - 1.$$

Apply RITZ's method with the approximation

$$\varphi = \sum_{\nu=1}^{p} a_\nu (1 - x^{2\nu}) \quad \text{or} \quad \varphi = \sum_{\nu=1}^{p} a_\nu \cos \nu \pi x.$$

8. Bracketing of eigenvalues. Use the enclosure theorem (§ 8.6) to calculate upper and lower limits for some of the smaller eigenvalues of the eigenvalue problem

$$-y'' = \lambda(2 - x^2)\,y; \qquad y(0) = 0, \qquad 2y(1) + y'(1) = 0.$$

(This represents the vibrations of an inhomogeneous string with one end fixed and the other constrained to move transversely under an elastic force.)

9. Use RITZ's method with the two-term approximation

$$\varphi = a_1(1 - x^2) + a_2(1 - x^4)$$

to find the symmetric solutions of the eigenvalue problem

$$-y'' = \lambda(1 + x^2)\,y, \qquad y(\pm 1) = 0,$$

which represents the vibrations of an inhomogeneous string fixed at each end

10. Find upper and lower limits for the first eigenvalue of the problem of the previous exercise by using the iteration method and the theorem of § 8.3

11. Apply RITZ's method to the eigenvalue problem

$$y^{IV} = \lambda (1 + x) y; \quad y(0) = y'(0) = y''(1) = y'''(1) = 0$$

(transverse vibrations of a cantilever of variable cross-section but constant flexural rigidity) with two-term and three-term polynomial approximations.

12. What is the Euler differential equation (6.9) for the variational problem

$$J[\varphi] = \int\limits_a^b f(x) \, \varphi^{(m-1)} \, \varphi^{(m+1)} \, d x = \text{min.},$$

where $m \geq 1$ and $f(x)$ is a given function of x with a continuous second derivative? No boundary conditions need be specified for this.

13. Someone asserts that in the application of RITZ's method to boundary-value problems as described in § 5 the addition of extra boundary terms to the integral, as in (5.6), and the recognition of two quite different types of boundary conditions — essential and suppressible — are quite unnecessary and that all one need do is simply to restrict the admissible functions to the domain of functions which satisfy all the boundary conditions and use merely the integral whose Euler equation coincides with the differential equation in question. Thus for the problem

$$y'' = 0; \quad y(0) = y'(0), \quad y(1) = 1,$$

for example, one would consider the variational problem

$$J^*[\varphi] = \int\limits_0^1 \varphi'^2 d x = \text{min.}$$

subject to the auxiliary conditions $\varphi(0) = \varphi'(0)$, $\varphi(1) = 1$ and argue that the minimizing function $y(x)$ must satisfy the boundary conditions since all admissible functions satisfy them and must also satisfy the Euler equation $y'' = 0$ and therefore it must be the solution of the boundary-value problem. Comment!

8.12. Solutions

1. The singularity at $x = 0$ in (8.55) implies that for a regular solution we must have $y'(0) = 0$. Now $\lim\limits_{x \to 0} \dfrac{y'}{x} = y''(0)$, so that

$$\lim\limits_{x \to 0} \left(\frac{3}{x} + 5x \right) y' = 3 y''(0);$$

hence for the point $x = 0$ the differential equation (8.55) can be replaced by

$$4 y'' + \tfrac{4}{3} y + 1 = 0.$$

With $Y_{-i} = Y_i$ (on account of symmetry) we obtain the difference equations

$$(x = 0) \quad 16(2Y_1 - 2Y_0) + \tfrac{4}{3}Y_0 + 1 = 0,$$
$$(x = \tfrac{1}{2}) \quad 5(Y_2 - 2Y_1 + Y_0) + \tfrac{17}{2}(Y_2 - Y_0) + \tfrac{4}{3}Y_1 + 1 = 0$$

and two more similar equations corresponding to the points $x = 1$, $x = \tfrac{3}{2}$. If we then express $Y_1, Y_2, \dots$ successively in terms of Y_0:

$$Y_1 = 0.958333 Y_0 - 0.03125,$$
$$Y_2 = 0.874486 Y_0 - 0.094136,$$
$$Y_3 = 0.801612 Y_0 - 0.148791,$$
$$Y_4 = 0.742773 Y_0 - 0.192920,$$

the condition $Y_4 = 0.6$ yields

$$Y_0 = 1.067514,$$

and hence

$$Y_1 = 0.991784,$$
$$Y_2 = 0.839390,$$
$$Y_3 = 0.706941.$$

2. The exact solution of the boundary-value problem is $y = x^{1+c}$ and is positive throughout the range $0 < x \leq 1$. With the pivotal interval $h = 1/n$ the difference equation corresponding to $x = h$ is

$$h \frac{Y_2 - 2Y_1 + Y_0}{h^2} = c \frac{Y_2 - Y_0}{2h},$$

which on account of the boundary condition $Y_0 = 0$ reduces to

$$Y_2 = -\frac{4}{c - 2} Y_1.$$

Consequently Y_1 and Y_2 always have opposite signs when $c > 2$. Thus the finite-difference method yields at least one pivotal value Y_i with the wrong sign for all the values of h, however small. For the numerical example $c = 3$, $h = \tfrac{1}{2}$ the exact solution gives $y(\tfrac{1}{2}) = \tfrac{1}{16}$ and the corresponding finite-difference approximation is $Y_1 = -\tfrac{1}{4}$.

3. The exact solution $y = x^2$ is a parabola and for a parabola no error is introduced by the substitution of the second difference quotient for the second derivative. Thus, for example, with $h = \tfrac{2}{3}$ we have $y(\tfrac{1}{3}) = \tfrac{1}{9}$ and this value for Y_1 satisfies the difference equation $\dfrac{1 - 2Y_1 + Y_{-1}}{h^2} =$
$\dfrac{1 - Y_1}{h^2} = \dfrac{2 \cdot \tfrac{1}{9}}{Y_1}$ exactly.

4. Using the pivotal interval $h = 1$ we find that $y(1) \approx -1 \pm i$. With $h = \tfrac{2}{3}$ the approximate solution with positive imaginary part is

$$y(\tfrac{2}{3}) \approx -0.65 + 0.87 i,$$
$$y(\tfrac{4}{3}) \approx -1.45 + 1.24 i$$

and with $h = \tfrac{1}{2}$ it is

$$y\left(\tfrac{1}{2}\right) \approx -0.500 + 0.728\,i,$$
$$y(1) \approx -1.070 + 1.275\,i,$$
$$y\left(\tfrac{3}{2}\right) \approx -1.760 + 1.139\,i;$$

the complex conjugates of these values give the other solution.

The more accurate approximations with $h = 2/n$, $n = 3, 4, \ldots$, which require the solution of systems of two or more equations, are obtained by interpolation. We take several values of $y(h)$ covering a range suggested by the very rough approximation $y(1) \approx -1 + i$, which indicates how the solutions for smaller h run, and for each of these values we calculate $y(2h)$, $y(3h)$, ... from the difference equations. A value of $y(h)$ for which $y(nh) + 2 \approx 0$ is then determined by two-dimensional interpolation.

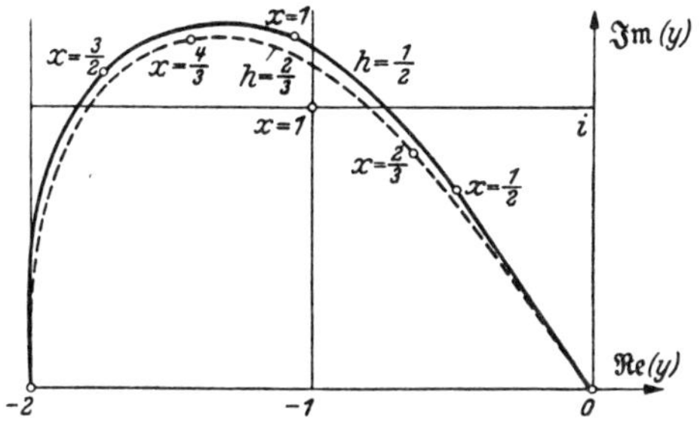

Fig. III/16. The complex solution of a real boundary-value problem

The solution can be regarded as the parametric (parameter x) representation of a curve in the complex y plane which joins the points $y = 0$ and $y = -2$. Fig. III/16 shows the approximations for the points on this curve obtained with $h = \tfrac{1}{2}$ and $h = \tfrac{2}{3}$; the calculated points are shown as small circles.

5. Yes. The required variational problem is

$$J[\varphi] = \int_0^1 \varphi'^2\,dx - [\varphi(1)]^2 = \text{extremum}$$

subject to the auxiliary condition $\varphi(0) = 1$. It also has no solution: with $\varphi = 1 - ax$, $J[\varphi]$ becomes $2a - 1$, which can take all values.

6. From (5.19) the corresponding variational problem is

$$J[\varphi] = \int_{-1}^1 (-\varphi'^2 + 4\varphi^2 - 4\varphi)\,dx = \text{extremum}$$

subject to the auxiliary conditions $\varphi(\pm 1) = 0$. That $J[\varphi]$ has neither a maximum nor a minimum in the domain of continuously differentiable functions $\varphi(x)$ with $\varphi(\pm 1) = 0$ can be shown simply by exhibiting functions in this domain which give the integral arbitrarily large positive and negative values; such functions are

$$\varphi = a(-1 + x^2), \quad \text{which yields} \quad J[\varphi] = 8\left(\tfrac{1}{5}a^2 + \tfrac{2}{3}a\right),$$

and
$$\varphi = a(-x^2 + x^4), \quad \text{which yields} \quad J[\varphi] = -8\left(\tfrac{5}{63} a^2 - \tfrac{2}{15} a\right).$$

However, for $\varphi = y$, where

$$y = \frac{1}{2}\left(1 - \frac{\cos 2x}{\cos 2}\right)$$

is the solution of the boundary-value problem, $J[\varphi]$ has a stationary value; thus the first variation δJ still vanishes when $\varphi = y$ and consequently RITZ's method still gives significant approximations. The stationary character of $J[\varphi]$ for $\varphi = y$ can be seen clearly from the form it takes when we put $\varphi = y(x) + \varepsilon(x)$ with $\varepsilon(\pm 1) = 0$:

$$J[\varphi] = J[y] + \int_{-1}^{1} (4\varepsilon^2 - \varepsilon'^2)\, dx, \quad \text{where} \quad J[y] = \tan 2 - 2.$$

For the two-term Ritz approximation $\varphi = a(1 - x^2) + b(1 - x^4)$ we have

$$\frac{1}{8} J[\varphi] = \frac{1}{5} a^2 + \frac{22}{105} 2ab + \frac{44}{315} b^2 - \frac{2}{3} a - \frac{4}{5} b,$$

and from $\dfrac{\partial J}{\partial a} = 0$, $\dfrac{\partial J}{\partial b} = 0$ we obtain $a \neq \dfrac{7}{3}$, $b = -\dfrac{7}{11}$; these values yield the approximate solution $\varphi = \dfrac{7}{3}(1 - x^2) - \dfrac{7}{11}(1 - x^4)$ and in particular $\varphi(0) = \dfrac{56}{33} = 1{\cdot}69697$ (error $-0{\cdot}3\%$).

7. For both of the suggested Ritz approximations for φ it turns out that $J[\varphi] \equiv 0$; this happens, in fact, for any even function $[\varphi(x) = \varphi(-x)]$, for J is then the integral of an odd function over a symmetric interval. Consequently the a_ν cannot be calculated from the equations $\partial J/\partial a_\varrho = 0$.

8. If we try an approximation of the form

$$F_1 = f(x) \sin(g(x))$$

and calculate

$$F_1', F_1'' \quad \text{and} \quad \Phi = -\frac{F_1''}{(2 - x^2) F_1},$$

we find that the special form

$$F_1 = (1 + bx)^{-\frac{1}{2}} \sin\left[a\left(x + \frac{bx^2}{2}\right)\right]$$

is particularly convenient. This gives

$$\Phi = \frac{a^2 \xi^2 - \dfrac{3}{4} \dfrac{b^2}{\xi^2}}{2 - x^2}, \quad \text{where} \quad \xi = 1 + bx.$$

The boundary condition $2F_1(1) + F_1'(1) = 0$ remains to be satisfied; it implies a relation between a and b, which, if we put $\gamma = a(1 + \tfrac{1}{2}b)$, can be written

$$\frac{\tan \gamma}{\gamma} = \frac{-4(1 + b)^2}{(4 + 3b)(2 + b)}.$$

We now choose a value for b and calculate several of the smaller of the infinitely many corresponding values of γ. Each of these defines a value of a and hence a corresponding $\Phi(x)$, whose upper and lower bounds $\Phi_{\max}$ and $\Phi_{\min}$ give upper and lower bounds for an eigenvalue. By varying b we can easily see when the limits become narrower. Trials with $b = -0.3$ and -0.25 show that the best values are obtained with say $b = -0.285$. The limits for the first three eigenvalues calculated with this value of b are included in Table III/17.

Table III/17. *Bracketing of eigenvalues*

b	γ	a	$\Phi_{\min} \leq \lambda_s \leq \Phi_{\max}$
-0.3	2·409	2·834	$3·24 \leq \lambda_1 \leq 3·98$
	5·190	6·106	$15·2 \leq \lambda_2 \leq 18·61$
-0.25	2·385	2·726	$3·22 \leq \lambda_1 \leq 4·10$
	5·167	5·906	$15·24 \leq \lambda_2 \leq 19·5$
-0.285	2·4028	2·8021	$3·25 \leq \lambda_1 \leq 3·90$
	5·18132	6·04445	$15·25 \leq \lambda_2 \leq 18·56$
	8·16665	9·52336	$37·8 \leq \lambda_3 \leq 46·25$

9. Here we have

$$\frac{1}{2} J[\varphi] = \int_0^1 [\varphi'^2 - \Lambda(1 + x^2)\,\varphi^2]\,dx = 4\left(\frac{1}{3}\,a_1^2 + 2\cdot\frac{2}{5}\,a_1 a_2 + \frac{4}{7}\,a_2^2\right) - $$
$$- 32\Lambda\left(\frac{2}{105}\,a_1^2 + \frac{2}{45}\,a_1 a_2 + \frac{92}{3465}\,a_2^2\right).$$

If $8\Lambda = 3\mu$, the determinant condition for the equations $\partial J/\partial a_\nu = 0$ for $\nu = 1, 2$ reads

$$\begin{vmatrix} 6\mu - 35 & 7\mu - 42 \\ 7\mu - 42 & \dfrac{92}{11}\mu - 60 \end{vmatrix} = 0,$$

which yields the quadratic

$$13\mu^2 - 712\mu + 3696 = 0.$$

The roots of this equation yield upper bounds for the first and third eigenvalues since the second corresponds to an antisymmetric eigenfunction:

$$\Lambda_{1,3} = \frac{3}{26}\left(89 \pm \sqrt{4918}\right) = \begin{cases} 2·177486 \geq \lambda_1 \\ 18·361 \geq \lambda_3. \end{cases}$$

10. According to (8.10) the first iterate is to be calculated here from

$$-F_1'' = (1 + x^2)\,F_0, \qquad F_1(\pm 1) = 0.$$

Starting from $F_0 = 1 - x^2$ we obtain

$$F_1 = \frac{1}{30}\,(14 - 15x^2 + x^6).$$

Then we calculate the Schwarz constants (8.13):

$$\frac{1}{2}\,a_0 = \int_0^1 (1 + x^2)\, F_0^2 \, dx = \frac{64}{105}\,,$$

$$\frac{1}{2}\,a_1 = \frac{3232}{35 \times 330}\,, \qquad \frac{1}{2}\,a_2 = \frac{32 \times 32\,564\cdot8}{900 \times 9009}$$

and from them the Schwarz quotients (8.11):

$$\mu_1 = \frac{a_0}{a_1} = \frac{220}{101} = 2\cdot178\,218\,,$$

$$\mu_2 = \frac{a_1}{a_2} = \frac{177\,255}{81\,412} = 2\cdot177\,259\,.$$

For the application of formula (8.18) we need further a lower limit l_2 for the second eigenvalue λ_2, or here a lower limit l_3 for the third eigenvalue λ_3 since the second eigenfunction is antisymmetric and we are considering here only symmetric modes — this is because we chose a symmetric starting function $F_0(x)$, which can therefore be expanded in a series of symmetric eigenfunctions (cf. § 8.9). The third eigenvalue for the comparable problem with constant coefficients $-y'' = 2\lambda y$, $y(\pm 1) = 0$ is $9\pi^2/8$ and using this value for l_3 in (8.18) we obtain the narrow limits

$$2\cdot177\,034 \leqq \lambda_1 \leqq 2\cdot177\,259\,.$$

11. According to (6.20), (6.21) the corresponding variational problem reads

$$J[\varphi] = \int_0^1 [\varphi''^2 - \lambda(1 + x)\,\varphi^2]\, dx = \text{extremum}\,, \qquad \varphi(0) = \varphi'(0) = 0\,.$$

For the three-term approximation $\varphi = a_1 x^2 + a_2 x^3 + a_3 x^4$ the conditions $\partial J/\partial a_\nu = 0$ ($\nu = 1, 2, 3$) yield three linear homogeneous equations whose determinant reads

$$\begin{vmatrix} 4 - \dfrac{11}{30}\,\Lambda & 6 - \dfrac{13}{42}\,\Lambda & 8 - \dfrac{15}{56}\,\Lambda \\[2mm] 6 - \dfrac{13}{42}\,\Lambda & 12 - \dfrac{15}{56}\,\Lambda & 18 - \dfrac{17}{72}\,\Lambda \\[2mm] 8 - \dfrac{15}{56}\,\Lambda & 18 - \dfrac{17}{72}\,\Lambda & \dfrac{144}{5} - \dfrac{19}{90}\,\Lambda \end{vmatrix}\,,$$

where Λ is an approximation for λ. With $\nu = \Lambda/1008$ the usual determinant condition reduces to

$$5175\,\nu^3 - 44\,634\,\nu^2 + 13\,554\,\nu - 90 = 0\,,$$

from which we obtain

$$\nu = \begin{cases} 0\cdot006\,792\,0 \\ 0\cdot307 \\ 8\cdot33 \end{cases} \qquad \Lambda \approx \begin{cases} 6\cdot8464 \\ 310 \\ 8390\,. \end{cases}$$

For a two-term approximation (φ without the term $a_3 x^4$) the determinant is the sub-determinant of the above determinant indicated by the dotted lines, and if $\Lambda = 840\,\mu$ the determinant condition becomes $425\,\mu^2 - 369\,\mu + 3 = 0$; this yields the values

$$\Lambda \approx \begin{cases} 6\cdot8944 \\ 722\,. \end{cases}$$

12. A possible solution y must satisfy the differential equation

$$2[f(x) y^{(m)}]^{(m)} + [f''(x) y^{(m-1)}]^{(m-1)} = 0,$$

as can be verified by simple manipulations with the aid of the binomial coefficients.

13. The assertion made is untrue. The Ritz method, say with the approximation

$$\varphi(x) = \tfrac{1}{2}(1+x) + \sum_{\nu=1}^{n} c_\nu (1 + x - 2x^{\nu+1}),$$

which satisfies all the boundary conditions, would yield solutions $\varphi_n(x)$ which converge to a limit function $\psi(x)$ as n increases, but this limit function would not be the solution of the given boundary-value problem; for it is determined by the boundary-value problem $\psi'' = 0$; $\psi'(0) = 0$, $\psi(1) = 1$, in which the suppressible boundary condition $\varphi(0) = \varphi'(0)$ has been replaced by the natural (for the suggested variational problem) boundary condition $\psi'(0) = 0$, and for the solution of this boundary-value problem $\psi(0) \neq \psi'(0)$. Thus one could not recognize the approximations as false from the fact that they do not converge; for this reason it is critically important that the variational problem be formulated correctly.

The variational problem suggested in the question, namely $J^*[\varphi] =$ minimum, $\varphi(0) = \varphi'(0)$, $\varphi(1) = 1$ has no solution; with φ subject to these boundary conditions $J^*[\varphi]$ has the lower bound zero, and this is assumed for the function $\varphi \equiv 1$, for which $\varphi(0) \neq \varphi'(0)$, but not for any comparison function. For the solution of the given boundary-value problem, namely $\varphi = y = \tfrac{1}{2}(1+x)$, J^* assumes the value $J^*[y] = \tfrac{1}{4}$, which is not a minimum. The comparison function

$$\varphi(x) = \begin{cases} \dfrac{1}{2\varepsilon + \varepsilon^2}(2\varepsilon + 2\varepsilon x - x^2) & \text{for } 0 \leq x \leq \varepsilon \\ 1 & \text{for } \varepsilon \leq x \leq 1 \end{cases} \quad \text{with} \quad \varepsilon > 0$$

gives J^* the value

$$\frac{\varepsilon}{3}\left(\frac{1}{1+\dfrac{\varepsilon}{2}}\right)^2,$$

and by choosing ε small enough we can make this value as close to zero as we please. As ε decreases the Ritz approximations represented by this family of comparison functions

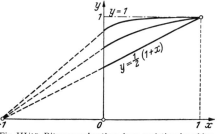

Fig. III/17. Ritz approximations for a variational problem with no solution

approximate more and more closely the function $\varphi \equiv 1$ (see Fig. III/17), yet the solution of the boundary-value problem reads $y = \tfrac{1}{2}(1+x)$.

According to § 5 the correct variational problem corresponding to the given boundary-value problem reads

$$J[\varphi] = \int_0^1 \varphi'^2 dx + [\varphi(0)]^2 = \min.,$$

where φ need satisfy only the essential boundary condition $\varphi(1) = 1$.

Chapter IV

Initial- and initial-/boundary-value problems in partial differential equations

The need for a sound theoretical foundation

In Ch. II, §§ 1.2, 1.3 some fundamental difficulties associated with the treatment of initial-value problems and error estimation for the approximate methods used were discussed with regard to ordinary differential equations. Naturally these difficulties are amplified when partial differential equations are considered; but over and above this, partial differential equations give rise to an extraordinarily large variety of phenomena and types of problem, while such essentials as the existence and uniqueness of solutions and the convergence of approximating sequences are covered by present theory only for a limited number of special classes of problems. These theoretical questions have not yet been settled in a satisfactory manner for many problems which arise in practical work. When confronted with such a problem one may be forced to rely solely on some approximate method, a finite-difference method, for example, and hope that the results obtained will be significant. Naturally such a procedure is not only unsatisfactory but even very questionable, as will be enlarged upon more precisely below; nevertheless, it is often unavoidable when a specific technical problem has to be solved and a theoretical investigation of the corresponding mathematical problem is not asked for. Consequently there is a pressing need for the accumulation of much more practical experience of approximate methods and for research into their theoretical aspects.

That an investigation of the situation is absolutely essential is revealed even by quite simple examples; they show that formal calculation applied to partial differential equations can lead to false results very easily and that approximate methods can converge in a disarmingly innocuous manner to values bearing no relation to the correct solution. ⌐harmlos

Consider, for example, the problem

$$
\frac{\partial^2 u}{\partial x^2} = \frac{\partial^2 u}{\partial y^2};
\quad
\begin{cases}
\left.\begin{array}{l}
u(x, 0) = \cos x \\[4pt]
\dfrac{\partial u(x, 0)}{\partial y} = 0,
\end{array}\right\} \quad \text{for} \quad |x| < \dfrac{\pi}{2}, \\[18pt]
u\left(\pm \dfrac{\pi}{2},\, y\right) = \sin y \quad \text{for} \quad y \geq 0.
\end{cases}
$$

This describes the oscillations of a string of length π which is initially $(y = 0)$ at rest in a displaced position defined by $u(x, 0) = \cos x$ and which is periodically excited at its ends $(x = \pm \pi/2)$.

Let the solution be expanded as a power series in the neighbourhood of the point $x = 0$, $y = 0$:

$$
u(x, y) = \sum_{m, n = 0}^{\infty} a_{mn} x^m y^n.
$$

By inserting the series in the differential equation and equating coefficients we find immediately that the a_{mn} must satisfy the recursive formula

$$
a_{m, n+2} = \frac{(m + 2)(m + 1)}{(n + 2)(n + 1)} a_{m+2, n}.
$$

This relation enables us to express all the a_{mn} with $n \geq 2$ in terms of the $a_{k, 0}$ and $a_{k, 1}$; in particular,

$$
a_{0, 2q} = a_{2q, 0} \quad \text{for} \quad q = 0, 1, 2, \ldots.
$$

To satisfy the initial condition $u_y(x, 0) = 0$ the $a_{k, 1}$, and hence the $a_{0, 2q+1}$, must all be zero. The other initial condition $u(x, 0) = \cos x$ determines the remaining $a_{k, 0}$ ($= a_{0, k}$) with even k:

$$
a_{k, 0} =
\begin{cases}
0 & \text{when } k \text{ is odd} \\[6pt]
\dfrac{(-1)^q}{(2q)!} & \text{when } k = 2q \text{ is even.}
\end{cases}
$$

On the y axis, i.e. when $x = 0$, the power series becomes

$$
\sum_{k=0}^{\infty} a_{0, k} y^k = \sum_{q=0}^{\infty} \frac{(-1)^q}{(2q)!} y^{2q} = \cos y,
$$

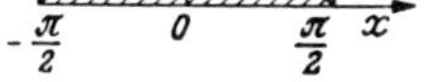

Fig. IV/1. Region outside of which the power series solution breaks down

which converges for all y, and hence represents its sum function $\cos y$ on the whole y axis. Nevertheless, $\cos y$ is not the correct solution along the whole y axis; in fact $u(0, y)$ takes the value $\cos y$ only on that part of the y axis cut off by the characteristics $y = \dfrac{\pi}{2} \pm x$ emanating from the points $\left(\dfrac{\pi}{2}, 0\right)$, $\left(-\dfrac{\pi}{2}, 0\right)$, respectively (see Fig. IV/1).

It is not difficult to construct similar examples for which the power series gives the wrong solution in every interval of the y axis, even though it converges for all values of y. Such a situation arises, for instance, if in the previous example we replace the initial displacement $u(x, 0) = \cos x$ by $u(x, 0) = e^{-(1/x^2)} + c\,x^2$, where c is chosen so that $u\left(\pm \dfrac{\pi}{2},\, 0\right) = 0$, and calculate the coefficients $a_{k,0}$ from the values of the derivatives of u at the point $x = y = 0$ [in this case, of course, even $u(x, 0)$ is not represented by the power series].

It is therefore very desirable, if not essential, that an approximate treatment of a problem in partial differential equations should be coupled with theoretical substantiation of some sort. In this book we cannot develop the theory for each type of problem considered, so the reader must refer to the textbooks on partial differential equations for theoretical details[1]. We shall, however, make use of the results of the theory.

§ 1. The ordinary finite-difference method

Because of its importance for applications, this section is written very comprehensively and much is repeated from earlier sections so that the reader need refer back as little as possible.

The ordinary finite-difference method provides a simple, general method by which one can obtain a reasonable quantitative idea of the solutions of many problems in differential equations. The accuracy achieved is not usually very great but often suffices for technical problems. Refinements of the method will be discussed in § 2.

To simplify the presentation we limit ourselves for the most part to problems with two independent variables x, y, although the method may be applied in exactly the same way to problems with more than two (see the example in § 1.8).

---

[1] Of the numerous textbooks we may mention — COURANT, R., and D. HILBERT: Methoden der mathematischen Physik, 2nd ed., Vol. 1, Berlin 1931, 469 pp.; Vol. 2, Berlin 1937, 549 pp.; English edition, London 1953. — KAMKE, E.: Differentialgleichungen reeller Funktionen, 2nd ed., Leipzig 1944, 442 pp. — Differentialgleichungen, Lösungsmethoden und Lösungen, Vol. 2, Leipzig 1944, 243 pp. — FRANK, PH., and R. v. MISES: Die Differential- und Integralgleichungen der Mechanik und Physik, 2nd ed., Brunswick, Vol. 1, 1930, 916 pp.; Vol. 2, 1935, 1106 pp. — HORN, J.: Partielle Differentialgleichungen, 3rd ed. Berlin and Leipzig 1944, 228 pp. — SOMMERFELD, A.: Partielle Differentialgleichungen der Physik (Vol. 6, consisting of lectures in theoretical physics). Leipzig 1947, 332 pp. — WEBSTER, A. G., and G. SZEGÖ: Partielle Differentialgleichungen der mathematischen Physik. Leipzig and Berlin 1930, 528 pp. — COURANT, R., and K. FRIEDRICHS: Supersonic flow and shock waves, Interscience Publishers Inc. New York 1948, 464 pp. — BERNSTEIN, DOROTHY L.: Existence theorems in partial differential equations (Annals of Mathematics Study 23). Princeton 1950, 228 pp. — SAUER, R.: Anfangswertprobleme bei partiellen Differentialgleichungen. Berlin-Göttingen-Heidelberg: Springer 1952, 229 pp.

1.1. Replacement of derivatives by difference quotients

We base the finite differences on the pivotal points

$$\left.\begin{array}{l} x_i = x_0 + i\,h \\ y_k = y_0 + k\,l \end{array}\right\} \qquad (i, k = 0, \pm 1, \pm 2, \ldots), \qquad (1.1)$$

which may be defined as the nodes of a rectangular mesh made up of the "mesh lines" $x = x_i$, $y = y_k$ ($i, k = 0, \pm 1, \pm 2, \ldots$) displaced by the "mesh widths" h and l, respectively (see Fig. IV/2); (x_0, y_0) is any conveniently chosen origin for the "mesh co-ordinates" i, k. A square mesh is a rectangular mesh with $h = l$.

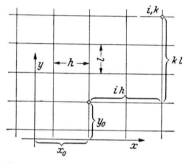

Fig. IV/2. The rectangular finite-difference mesh

Meshes other than rectangular can also be used (see Ch. V, § 2.7) but in general the resulting finite-difference expressions are more complicated and less convenient for numerical work, particularly when the chosen mesh cannot be transformed into itself by a translation[1].

Function values at mesh points, the pivotal values, will be characterized by the appropriate subscripts; thus $u_{i,k}$ will denote the value of the exact solution $u(x, y)$ at the mesh point $x = x_i$, $y = y_k$. $U_{i,k}$ will denote an approximation to the pivotal value $u_{i,k}$.

Just as for an ordinary differential equation (Ch. III, § 1), the partial differential equation is replaced by a difference equation which is derived by the substitution of an appropriate difference quotient for each derivative occurring.

This difference equation is not necessarily unique, for there are often several choices of appropriate difference quotient; for example, the derivative $(\partial u/\partial y)_{i,k}$ can be approximated by any of the following three difference quotients:

the "forward" difference quotient $\qquad \dfrac{U_{i,k+1} - U_{i,k}}{l} = \dfrac{\Delta_y U_{i,k}}{l}$, $\qquad$ (1.2)

the "backward" difference quotient $\qquad \dfrac{U_{i,k} - U_{i,k-1}}{l} = \dfrac{\Delta_y U_{i,k-1}}{l}$, $\qquad$ (1.3)

the "central" difference quotient $\qquad \dfrac{U_{i,k+1} - U_{i,k-1}}{2l} = \dfrac{\Delta_y}{2l}(U_{i,k-1} + U_{i,k})$. $\qquad$ (1.4)

---

[1] For the Tricomi differential equation $k(y)u_{xx} - u_{yy} = f(x, y)$ K. H. BAUERS-FELD: Zum Differenzenverfahren bei Anfangswertaufgaben partieller Differential-gleichungen 2. Ordnung. Diss. Hannover, 1954, uses a rectangular mesh whose mesh width in the x direction is kept constant while its mesh width in the y direction is varied so that the mesh diagonals are chords of the characteristics.

The subscript y to the forward difference operator Δ defined in Ch. I, § 2 signifies that the operand is to be regarded as a function of y alone; thus the effects of Δ_x and Δ_y operating on any function of x and y are

$$\left.\begin{aligned} \Delta_x g(x, y) &= g(x + h, y) - g(x, y), \\ \Delta_y g(x, y) &= g(x, y + k) - g(x, y) \end{aligned}\right\} \tag{1.5}$$

and on a discrete "mesh function" g_{ik}

$$\Delta_x g_{ik} = g_{i+1,k} - g_{ik}, \quad \Delta_y g_{ik} = g_{i,k+1} - g_{ik}. \tag{1.6}$$

It is well known that in general the central difference quotient is a substantially better approximation to the local derivative than either the forward or the backward difference quotient. One might think therefore that it is best to use only central differences in numerical work. In fact, as is demonstrated in § 1.3, this often results in troublesome error propagation (instability).

The second partial derivative $(\partial^2 u / \partial x^2)_{i,k}$ is usually replaced by the second difference quotient

$$\frac{U_{i+1,k} - 2U_{i,k} + U_{i-1,k}}{h^2} = \frac{\Delta_x^2 U_{i-1,k}}{h^2}. \tag{1.7}$$

Generally, we can replace the partial derivative

$$\left(\frac{\partial^{m+n} u}{\partial x^m \partial y^n}\right)_{i,k} \quad \text{by} \quad \frac{\Delta_x^m \Delta_y^n U_{i-r,k-s}}{h^m l^n}, \tag{1.8}$$

in which there is a certain amount of freedom in the choice of r and s; normally r (and s correspondingly) will be $\dfrac{m}{2}$ or $\dfrac{m-1}{2}$ according as m is even or odd, although when m is odd there is also the alternative of replacing $\partial^m u / \partial x^m$ by the symmetrical expression

$$\frac{1}{2h^m}\left[\Delta_x^m\left(U_{i-\frac{m-1}{2},k} + U_{i-\frac{m+1}{2},k}\right)\right]. \tag{1.9}$$

Thus $\left(\dfrac{\partial^3 g}{\partial x^2 \partial y}\right)_{0,0}$, for example, can be replaced either by

$$\frac{1}{h^2 l}(g_{1,1} - 2g_{0,1} + g_{-1,1} - g_{1,0} + 2g_{0,0} - g_{-1,0})$$

or by

$$\frac{1}{2h^2 l}(g_{1,1} - 2g_{0,1} + g_{-1,1} - g_{1,-1} + 2g_{0,-1} - g_{-1,-1}).$$

In this way we can find a finite-difference approximation to any partial derivative, and hence we can set up a difference equation representing the differential equation at each mesh point[1].

---

[1] A method in which only the derivatives in one fixed direction, for example, in the x direction or in the y direction, are replaced by difference quotients is given by D. R. HARTREE and J. R. WOMERSLEY: A method for the numerical or mechanical solution of certain types of partial differential equations. Proc. Roy. Soc. Lond. Ser. A **161**, 353—366 (1937). Consider, for example, the problem $u_{xx} = K u_y$ with given boundary values $u(x, 0)$, $u(0, y)$, $u(a, y)$. If we use a backward

1.2. An example of a parabolic differential equation with given boundary values

If radiation and convection are neglected, the temperature $u(x, t)$ at time t in a thin homogeneous rod at a distance x from one end satisfies the differential equation

$$\frac{\partial^2 u}{\partial x^2} = c \frac{\partial u}{\partial t} \tag{1.10}$$

(the one-dimensional heat equation), where $c = \varrho \sigma / k$ is a function of the density ϱ, specific heat σ and thermal conductivity k. Given the initial temperature distribution $u(x, 0)$ for $0 \leq x \leq a$, where a is the length of the rod, and the end temperatures $u(0, t)$, $u(a, t)$ as functions of time for $t \geq 0$, it is required to calculate the temperature distribution $u(x, t)$ along the rod at subsequent times (cf. Fig. IV/3).

With $y = t$ and $h = a/n$ we cover the region of the (x, y) plane in which we are interested with the mesh defined by

$$x_i = ih, \qquad y_k = kl \qquad (i = 0, 1, 2, \ldots, n; \; k = 0, 1, 2, \ldots).$$

Then if we use first of all the crude approximation with the forward difference quotient in place of the time derivative, the difference equation corresponding to the differential equation (1.10) at the point (x_i, y_k) is

$$\frac{U_{i+1,k} - 2U_{i,k} + U_{i-1,k}}{h^2} = c \frac{U_{i,k+1} - U_{i,k}}{l}. \tag{1.11}$$

This has the following advantages:

1. The mesh width l in the time direction can be chosen so that the term in $U_{i,k}$ does not appear in the difference equation (1.11):

$$l = h^2 \frac{c}{2}. \tag{1.12}$$

This simplifies the calculation considerably, for $U_{i,k+1}$ is then formed merely by taking the arithmetic mean of $U_{i+1,k}$ and $U_{i-1,k}$:

$$U_{i,k+1} = \tfrac{1}{2}(U_{i+1,k} + U_{i-1,k}). \tag{1.13}$$

difference quotient in the y direction, the equation becomes

$$\left(\frac{\partial^2 u}{\partial x^2}\right)_{i,k} = K \frac{u_{i,k} - u_{i,k-1}}{l},$$

and if the values up to the $(k-1)$-th row are already known, we have to solve a boundary-value problem for an ordinary differential equation to obtain the values $u(x, kl)$ on the k-th row. If, on the other hand, we use finite differences in the x direction, so that

$$K \left(\frac{\partial u}{\partial y}\right)_{i,k} = \frac{1}{h^2} \{u_{i+1,k} - 2u_{i,k} + u_{i-1,k}\} \qquad (i = 1, \ldots, n-1),$$

we have to solve an initial-value problem for a system of ordinary differential equations of the first order. Machine methods of solution are particularly suitable here.

Table IV/1. *Finite-difference solution of the heat equation with $\partial u/\partial t$ replaced by the forward difference quotient*

k	$i=0$	$i=1$	$i=2$	$i=3$	$i=4$	$i=5$	$i=6$	$i=7$	$i=8$	Row-sum check S_k	T_k
0	**0**	0	0	0	0	0	0	0	0		
1	**0·5**	0·25	0	0	0	0	0	0	0	0·25	1·1160
2	**0·8660**	0·4955	0·125	0	0	0	0	0	0	0·6205	1·1160
3	**1**	0·6395	0·2790	0·0625	0	0	0	0	0	0·6204	1·6204
4	**0·8660**	0·6316	0·3973	0·1551	0·0312	0	0	0	0	0·9809	1·6204
5	**0·5**	0·4682	0·4365	0·2414	0·0854	0·0156	0	0	0	0·9810	1·8470
6	**0**	0·1896	0·3792	0·2902	0·1440	0·0466	0·0078	0	0	1·2115	1·8470
7	**−0·5**	−0·1318	0·2364	0·2833	0·1874	0·0846	0·0252	0·0039	0	1·2114	1·7114
8	**−0·8660**	−0·4113	0·0434	0·2185	0·2006	0·1180	0·0486	0·0126	0	1·2304	1·7112
9	**−1**	−0·5752	−0·1505	0·1103	0·1772	0·1359	0·0712	0·0243	0	1·2304	1·2304
10	**−0·8660**	−0·5805	−0·2950	−0·0148	0·1209	0·1315	0·0858	0·0356	0	1·0165	1·2304
11	**−0·5**	−0·4264	−0·3528	−0·1252	0·0446	0·1040	0·0872	0·0429	0	1·0164	1·5164
12	**0**	−0·1548	−0·3096	−0·1929	−0·0330	0·0592	0·0738	0·0436	0	0·6126	1·5164
13				−0·2012	−0·0927	0·0075	0·0480	0·0369	0	0·6128	−0·2532
14						−0·0384	0·0158	0·0240	0	0·1152	−0·2532
15								0·0079	0	0·1152	−0·8848
16										−0·3532	−0·8848
17										−0·3532	−1·2192
18										−0·6756	−1·2192
19										−0·6756	−1·1756
20										−0·7732	−1·1756
21										−0·7730	−0·7730
22										−0·6261	−0·7730

−0·1260	−0·1260	0·5864	0·5866	1·1692	1·1692	1·4618	1·4620	1·3828	1·3826	0·9498	0·9496
−0·6260	−0·2794	−0·2796	0·1694	0·1692	0·5959	0·5958	0·8828	0·8828	0·9499	0·9498	0·7768
0	0	0		0	0		0				
−0·0076	−0·0188		−0·0229	−0·0192	−0·0089	0·0050					
−0·0152	−0·0376	−0·0458	−0·0384	−0·0178	0·0100						
−0·0675	−0·0728		−0·0539	−0·0164	0·0290	0·0696					
−0·1198	−0·1082	−0·0620	0·0056	0·0759	0·1293						
−0·1489	−0·0511	0·0652	0·1682	0·2296	0·2324						
−0·1780	0·0060	0·1924	0·3307	0·3833	0·3356						
0·1610	0·4360	0·5962	0·5984	0·4416	0·1678						
0·5	**0·8660**	**1**	**0·8660**	**0·5**	**0**						
30			35								

With a little practice this averaging process can be performed in the head, so that the successive rows can be written down immediately. Build-up of rounding errors can be inhibited effectively simply by rounding to the nearest even end digit; thus, for example, we round

0·43765 to 0·4376,   0·43775 to 0·4378.

2. The difference equation (1.13) relates the values of U at three mesh

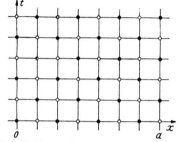

Fig. IV/3. The two interlacing sets of mesh points

points belonging to a sub-set of the whole set of mesh points. There are two such sub-sets (see Fig. IV/3), each complementary to and interlacing the other; they consist of the mesh points with $i+k$ even and odd, respectively, which are marked with black and white spots, respectively, in Fig. IV/3. The calculation need only be carried out for one of these two sub-sets.

Numerical example. Let one end $(x = 2)$ of a rod of length 2 with the initial temperature distribution $u(x, 0) = 0$ be kept at this constant temperature while the temperature at the other end $(x = 0)$ is varied sinusoidally with period $\frac{3}{4} c$, so that $u(2, t) = 0$ and $u(0, t) = \sin \omega t$, where $\omega = \frac{8\pi}{3c}$. Then with $h = \frac{1}{4}$ the boundary temperature $u(0, t)$ reaches its maximum value of 1 when $t = 6l$, i.e. when $k = 6$. Thus we know the boundary values printed in heavy type in Table IV/1. We can immediately proceed to calculate the values of U for one of the two sub-sets of mesh points, starting from the row $k = 0$ and working downwards row by row in accordance with formula (1.13). Thus, for convenience, the time axis in the table points downwards, the opposite direction to that in Fig. IV/3.

An important row-sum check can be derived as follows. Suppose that n is even, say, (as in the numerical example, where $n = 8$) and consider the sub-set of mesh points for which $i + k$ is even. Then for even k we add together the equations obtained from (1.13) by putting $i = 1, 3, \ldots, n - 1$ and find that

$$2 \sum_{i=1}^{\frac{n}{2}} U_{2i-1,\,k+1} = U_{0,\,k} + 2 \sum_{i=1}^{\frac{n}{2}-1} U_{2i,\,k} + U_{n,\,k}; \qquad (1.14)$$

a corresponding formula for odd k can be derived by adding the difference equations with $i = 2, 4, \ldots, n - 2$. If we now introduce quantities S_k and T_k defined by

$$S_k = 2 \sum_{i=1}^{\frac{n}{2}-1} U_{2i,\,k}, \qquad T_k = U_{0,\,k} + S_k + U_{n,\,k}$$

when $k = 2s$ is even and by

$$\sigma_k = U_{1,\,k} + U_{n-1,\,k}, \qquad S_k = \sigma_k + 2 \sum_{i=1}^{\frac{n}{2}-2} U_{2i+1,\,k}, \qquad T_k = \sigma_k + S_k \qquad (1.15)$$

when $k = 2s + 1$ is odd, we find that

$$T_{2s} = T_{2s+1}, \qquad S_{2s+1} = S_{2s+2}. \qquad (1.16)$$

Thus the check consists in forming S_k and T_k for each row and observing alternately whether the value of S_k or T_k is the same as for the previous row. A simpler row-sum check is possible when the calculation is carried out for both sub-meshes simultaneously.

The results obtained show that once the effect of the varying end temperature reaches any fixed point of the rod the temperature at that point oscillates with the exciting frequency but with a phase lag (more than half a wavelength near the end $x = 2$) and that the amplitude decreases towards the end $x = 2$. For small x and large t this behaviour is confirmed by the solution of the heat equation

$$u = e^{-ax} \sin(\omega t - a x),$$

where $a = \sqrt{c\omega/2}$ (see Exercise 1 in § 6.4).

1.3. Error propagation

Now let us try to improve upon the results obtained in § 1.2 by recalculating them from a formula in which the time derivative has been replaced by the more accurate central[1] difference quotient (1.4) instead of the forward difference quotient. Equation (1.11) is now replaced by

$$\frac{U_{i+1,k} - 2U_{i,k} + U_{i-1,k}}{h^2} = c\,\frac{U_{i,k+1} - U_{i,k-1}}{2l}.$$

Hence with the same relation between the mesh widths $l = \frac{1}{2}ch^2$ as in (1.12) we have

$$U_{i,k+1} = U_{i+1,k} - 2U_{i,k} + U_{i-1,k} + U_{i,k-1}. \tag{1.17}$$

With this formula we need starting values on the row $t = l$. We could take over the values already calculated in § 1.2, namely

$$U_{1,1} = U_{2,1} = \cdots = U_{7,1} = 0,$$

but since we are using a more accurate formula, we do not want to lose any accuracy through the propagation of errors due to poor starting values. Consequently we

Table IV/2. *Starting values for the central-difference formula* (1.17)

k'	$i'=0$	$i'=1$	$i'=2$	$i'=3$	$i'=4$	$i'=5$
0	0		0		0	
1		0		0		0
2	0·058 144		0		0	
3		0·029 072		0		0
4	0·116 096		0·014 536		0	
5		0·065 316		0·007 268		0
6	0·173 648		0·036 292		0·003 634	
7		0·104 970		0·019 963		0·001 817
8			0·062 466		0·010 890	
9				0·036 678		

calculate the values on the row $t = l$ by repeating the calculation of § 1.2 with a finer mesh having mesh widths $h' = \frac{1}{3}h$, $l' = \frac{1}{9}l$, which also satisfy the relation (1.12).

This short initial calculation need only be taken as far as is shown in Table IV/2, for we only require the last value:

$$x = 3h', \qquad y = 9l', \qquad U'_{3,9} = 0·036678.$$

We then put $U_{1,1} = U'_{3,9}$, $U_{2,1} = \cdots = U_{7,1} = 0$ and work downwards row by row using formula (1.17), with the results shown in Table IV/3.

---

[1] The third possibility, namely of using a backward difference quotient for the time derivative, leads to the difference equation

$$\sigma(U_{i+1,k} - 2U_{i,k} + U_{i-1,k}) = U_{i,k} - U_{i,k-1}$$

and is followed up by P. LAASONEN: Über eine Methode zur Lösung der Wärmeleitungsgleichung. Acta Math. **81**, 309—317 (1949). With this method the solutions of the finite-difference equations converge to the solution of the differential equation for arbitrary, fixed σ, but the calculation is complicated by the fact that one has to solve a system of linear equations to find the values $U_{i,k}$ on each new row.

Table IV/3. *Solution of the heat equation using the central-difference formula* (1.17)

k	i=0	i=1	i=2	i=3	i=4	i=5	i=6	i=7	i=8
0	0	0	0	0	0	0	0	0	0
1	0·25882	0·03668	0	0	0	0	0	0	0
2	0·5	0·18548	0·03668	0	0	0	0	0	0
3	0·70711	0·20244	0·11210	0·03668	0	0	0	0	0
4	0·86603	0·59979	0·05160	0·03874	0·03668	0	0	0	0
5	0·96593	−0·07951	0·64743	0·04748	−0·03462	0·03668	0	0	0
6	1	2·37217	−1·27529	0·55659	0·19008	−0·10798	0·03668	0	0
7	0·96593	−5·09914	6·12677	−2·15091	0·03383	0·47940	−0·18134	0·03668	0

Just these few rows show already that the calculation is quite useless. If we proceed any further, we obtain numbers with alternating signs and assorted magnitudes which bear no relation whatsoever to the solution of the problem. This is caused by a rapid build-up of

Table IV/4. *Propagation table for the difference equation* (1.17)

	x_{i-1}	x_i	x_{i+1}		
0	0	ε	0	0	0
0	ε	$-2\,\varepsilon$	ε	0	0
ε	$-4\,\varepsilon$	$7\,\varepsilon$	$-4\,\varepsilon$	ε	0
$-6\,\varepsilon$	$17\,\varepsilon$	$-24\,\varepsilon$	$17\,\varepsilon$	$-6\,\varepsilon$	ε
$31\,\varepsilon$	$-68\,\varepsilon$	$89\,\varepsilon$	$-68\,\varepsilon$	$31\,\varepsilon$	$-8\,\varepsilon$
$-144\,\varepsilon$	$273\,\varepsilon$	$-338\,\varepsilon$	$273\,\varepsilon$	$-144\,\varepsilon$	$49\,\varepsilon$
$641\,\varepsilon$	$-1096\,\varepsilon$	$1311\,\varepsilon$	$-1096\,\varepsilon$	$641\,\varepsilon$	$-260\,\varepsilon$

errors, which is a characteristic of formula (1.17). We can analyse such a build-up of errors by assuming that, besides the error already present due to the finite-difference approximation, an additional error ε is made at some stage in the calculation, say at (x_i, y_i). The propagation of this error is easily traced by repeated application of (1.17); thus on the next row $y = y_{k+1}$ it causes additional errors ε, -2ε, ε at $x = x_{i-1}$, x_i, x_{i+1}, which in turn cause additional errors ε, -4ε, 7ε, -4ε, ε at $x = x_{i-2}, \ldots, x_{i+2}$ on the next row, and so on. These additional errors arising from the single error ε can be tabulated in a "propagation table" as in Table IV/4.

Examination of the numbers in the table reveals that the error grows approximately by a factor of four from row to row. Even if the calculation were carried out with so many guarding figures that the propagation of rounding errors could not affect the result, the intrinsic or truncation errors which are inherent in the use of the finite-difference approximations would increase so rapidly in consequence of the unstable character of the difference equation (1.17) that the method would still be quite useless. It could be argued that the presence of the boundaries inhibits the growth of the error, but if we form the propagation table for a single error ε in a boundary value, due to rounding say, we see

Table IV/5. *Propagation of an error on the boundary*

ε	0	0	0	0	0
0	ε	0	0	0	0
0	$-2\,\varepsilon$	ε	0	0	0
0	$6\,\varepsilon$	$-4\,\varepsilon$	ε	0	0
0	$-18\,\varepsilon$	$16\,\varepsilon$	$-6\,\varepsilon$	ε	0
0	$58\,\varepsilon$	$-60\,\varepsilon$	$30\,\varepsilon$	$-8\,\varepsilon$	ε
0	$-194\,\varepsilon$	$224\,\varepsilon$	$-134\,\varepsilon$	$48\,\varepsilon$	$-10\,\varepsilon$

(Table IV/5) that there is no appreciable reduction in the growth of the error, the effect of the boundaries amounting chiefly to a delay of about one row.

In contrast, the propagation table for the difference equation (1.13) (Table IV/6) shows no build-up from an error ε introduced at any point; in fact the propagated errors gradually decrease in magnitude.

Table IV/6. *Stable propagation table for* $U_{i,\,k+1} = \frac{1}{2}\,(U_{i+1,\,k} + U_{i-1,\,k})$

0			ε	0		
0		$0{\cdot}5\,\varepsilon$	$0{\cdot}5\,\varepsilon$	$0{\cdot}5\,\varepsilon$		0
	$0{\cdot}25\,\varepsilon$				$0{\cdot}25\,\varepsilon$	
$0{\cdot}125\,\varepsilon$		$0{\cdot}375\,\varepsilon$	$0{\cdot}5\,\varepsilon$	$0{\cdot}375\,\varepsilon$		$0{\cdot}125\,\varepsilon$
	$0{\cdot}25\,\varepsilon$		$0{\cdot}375\,\varepsilon$		$0{\cdot}25\,\varepsilon$	
$0{\cdot}15625\,\varepsilon$		$0{\cdot}3125\,\varepsilon$	$0{\cdot}3125\,\varepsilon$	$0{\cdot}3125\,\varepsilon$		$0{\cdot}15625\,\varepsilon$

Summary[1]. If we replace the time derivative in the heat equation by the (crude) forward difference quotient and take $l = \frac{1}{2}c\,h^2$, we obtain formula (1.13), whose propagation table shows a gradual decrease in the magnitude of the propagated errors, and the calculation proceeds smoothly (that this method is usable is assured by a convergence proof in § 3.3); if, on the other hand, we replace the time derivative by the central difference quotient, which in itself is more accurate, and use the same mesh relation $l = \frac{1}{2}c\,h^2$, we obtain the formula (1.17), whose propagation table shows a rapid increase in the magnitude of the propagated errors and which is therefore unusable ("unstable").

---

[1] The question of error propagation is discussed further in § 3.4.

1.4. Error propagation and the treatment of boundary conditions

In many applications we are not given the boundary values of u itself; often it is the normal derivative of u or a combination of u and its normal derivative which is specified on the boundary. These types of boundary condition can be dealt with in various ways, which we will now illustrate by an example.

A physical problem. Consider an electromagnet whose core is a solid metal cylinder of specific resistance ϱ and permeability μ. The eddy-current density j and the magnetic field strength H directed along the axis satisfy the equations [1]

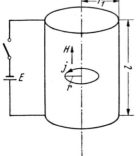

$$4\pi j = -\frac{\partial H}{\partial r}, \qquad \frac{\varrho}{r}\frac{\partial(rj)}{\partial r} = -\mu\frac{\partial H}{\partial t}, \qquad (1.18)$$

where r is the usual cylindrical co-ordinate. Elimination of j yields a second-order parabolic differential equation for H:

$$\frac{\partial^2 H}{\partial r^2} + \frac{1}{r}\frac{\partial H}{\partial r} = \frac{4\pi\mu}{\varrho}\frac{\partial H}{\partial t}. \qquad (1.19)$$

Fig. IV/4. Eddy currents in a metal cylinder

It remains to specify some initial conditions. Let a constant electromotive force E be suddenly switched into the windings circuit at time $t=0$, there being no current in the coil and no magnetic field in the core before this time (Fig. IV/4). The subsequent eddy-current density j and magnetic field strength H are to be calculated as functions of r and t.

If the core is of radius r_1 and length l and the coil has N uniform windings of ohmic resistance R, the boundary conditions [2] can be written

$$\left.\begin{aligned}
t=0:&\quad H=0 \quad\text{for}\quad 0\le r\le r_1,\\
r=r_1:&\quad H+\beta\frac{\partial H}{\partial r}=\frac{1}{\alpha}E \quad\text{for}\quad t>0,\\
r=0:&\quad H,\ \frac{\partial H}{\partial r},\ \frac{\partial^2 H}{\partial r^2} \quad\text{continuous,}
\end{aligned}\right\} \qquad (1.20)$$

where

$$\alpha=\frac{Rl}{4\pi N}, \qquad \beta=\frac{2\pi N^2 r_1\varrho}{Rl}.$$

---

[1] WAGNER, K. W.: Operatorenrechnung, 2nd ed., p. 230 et seq. Leipzig 1950. The equations (1.18) follow from the field equations $\operatorname{curl} H=4\pi j$, $\varrho\operatorname{curl} j=-\mu\dfrac{\partial H}{\partial t}$ (H is the magnetic field strength, j the current vector) by making use of the axial symmetry.

[2] If U is the induced back e.m.f. in the windings, the effective e.m.f. is $E+U$, which must be equal to the potential drop across the windings: $E+U=RI$. Now U can be calculated from the electric field strength at the surface of the core and hence expressed in terms of the eddy-current density: $U=2\pi r_1\varrho N j(r_1)$. Further, the line integral of the magnetic field strength yields $lH(r_1)=4\pi NI$. Elimination of U, I and $j(r_1)$ from these three equations together with (1.18) leads to the boundary condition of (1.20) at $r=r_1$ (cf. K. W. WAGNER: see last footnote).

For our numerical example we consider the case $\beta = \frac{1}{2} r_1$. With the dimensionless variables

$$x = \frac{r}{r_1}, \qquad y = \frac{\varrho t}{4 \pi \mu r_1^2}, \qquad u = \frac{\alpha}{E} H$$

(1.19) becomes[1]

$$\frac{\partial^2 u}{\partial x^2} + \frac{1}{x} \frac{\partial u}{\partial x} = \frac{\partial u}{\partial y}, \tag{1.21}$$

and if the range of the new variable x is extended to cover the whole axial section, the boundary conditions can be put in the form

$$\left.\begin{array}{ll}
y = 0: & u = 0 \quad \text{for} \quad -1 \leq x \leq 1, \\[2mm]
\begin{array}{c} x = 0: \\ \text{(point of symmetry)} \end{array} & u, \quad \dfrac{\partial u}{\partial x}, \quad \dfrac{\partial^2 u}{\partial x^2} \quad \text{continuous}, \\[3mm]
x = 1: & u + \dfrac{1}{2} \dfrac{\partial u}{\partial x} = 1, \\[3mm]
x = -1: & u - \dfrac{1}{2} \dfrac{\partial u}{\partial x} = 1.
\end{array}\right\} \tag{1.22}$$

The last condition may be omitted if we use the symmetry about $x = 0$ and restrict ourselves to the interval $0 \leq x \leq 1$.

Again we employ a mesh (1.1) with $y_0 = 0$ and $2l = h^2$, as in (1.12); and, after our experience in § 1.3, we replace the derivative in the time direction $\left(\dfrac{\partial u}{\partial y}\right)_{i,k}$ by the forward difference quotient $\dfrac{1}{l}\,(U_{i,k+1} - U_{i,k})$. The difference equation does not depend so critically on the radial derivative $\left(\dfrac{\partial u}{\partial x}\right)_{i,k}$ and we therefore replace this derivative by the central difference quotient $\dfrac{1}{2h}\,(U_{i+1,k} - U_{i-1,k})$. The resulting difference

---

[1] A method which is much used, particularly by electrical engineers, for the solution of initial-value and mixed initial-/boundary-value problems with linear differential equations (mostly with constant coefficients and infinite fundamental regions) is the operational method using the Laplace transformation; see, for example, G. DOETSCH: Theorie und Anwendung der Laplace-Transformation. Berlin 1937, 436 pp. — Tabellen zur Laplace-Transformation und Anleitung zum Gebrauch. Berlin and Göttingen 1947, 185 pp. — CHURCHILL, R. V.: Modern operational mathematics in engineering. New York and London 1944, 306 pp. — WAGNER, K. W.: Operatorenrechnung und Laplacesche Transformation nebst Anwendung in Physik und Technik, 2nd ed. Leipzig 1950, 489 pp. — CARSLAW, H. S., and J. C. JAEGER: Conduction of heat in solids. Oxford 1948, 386 pp., here in particular pp. 239—290, 320—338.

From a numerical point of view (for problems not soluble in closed form) the Laplace transform method is only suitable for a restricted range of problems. For some problems it leads to results which can be obtained just as well by other means and for others to series expansions whose numerical evaluation is often merely tedious. Future developments will show whether or not the frequently expressed hopes of the Laplace transform are justified.

Another method for the solution of similar problems is the "mixed" Ritz method, see Ch. V, § 5.9.

equation reads

$$U_{i,k+1} = \frac{1}{2}(U_{i+1,k} + U_{i-1,k}) + \frac{h}{4x_i}(U_{i+1,k} - U_{i-1,k}). \qquad (1.23)$$

The boundary conditions can be taken into account in several ways:

1. Choose a mesh with the boundary $x = 1$ as a mesh line, say $x = x_n$, and replace the derivative in the boundary condition by the backward difference quotient. This gives the "finite" boundary condition

$$U_{n,k} + \frac{1}{2h}(U_{n,k} - U_{n-1,k}) = 1,$$

so that

$$U_{n,k} = \frac{U_{n-1,k} + 2h}{1 + 2h}. \qquad (1.24)$$

2. Again choose a mesh with $x_n = 1$ but replace the boundary derivative by the central difference quotient. This entails keeping a column

Table IV/7. *Propagation tables for various boundary formulae at $x = 1$*

Formula (1.24)			Formula (1.25)			Formula (1.26)		
$x = \frac{1}{3}$	$x = \frac{2}{3}$	$x = 1$	$x = \frac{1}{3}$	$x = \frac{2}{3}$	$x = 1$	$x = \frac{4}{3}$	$x = \frac{2}{5}$	$x = \frac{4}{5}$	$x = \frac{6}{5}$
0	0	ε	0	0	ε	$-1 \cdot 333\,\varepsilon$	0	0	ε
0	$0 \cdot 625\,\varepsilon$	$0 \cdot 375\,\varepsilon$	0	$0 \cdot 625\,\varepsilon$	$-0 \cdot 778\,\varepsilon$	$1 \cdot 662\,\varepsilon$	0	$0 \cdot 625\,\varepsilon$	$0 \cdot 268\,\varepsilon$
$0 \cdot 469\,\varepsilon$	$0 \cdot 234\,\varepsilon$	$0 \cdot 141\,\varepsilon$	$0 \cdot 469\,\varepsilon$	$-0 \cdot 486\,\varepsilon$	$1 \cdot 229\,\varepsilon$	$-2 \cdot 125\,\varepsilon$	$0 \cdot 469\,\varepsilon$	$0 \cdot 167\,\varepsilon$	$0 \cdot 072\,\varepsilon$

of values of $U_{n+1,k}$. If the k-th row is completed, $U_{n,k+1}$ is calculated from (1.23) with $i = n$, then $U_{n+1,k+1}$ is calculated from the "finite" boundary condition

$$U_{n,k+1} + \frac{1}{2}\frac{1}{2h}(U_{n+1,k+1} - U_{n-1,k+1}) = 1,$$

which is used in the form

$$U_{n+1,k+1} = 4h(1 - U_{n,k+1}) + U_{n-1,k+1}. \qquad (1.25)$$

3. Choose a mesh for which the boundary $x = 1$ lies halfway between two mesh lines, say $x = x_n$ and $x = x_{n+1}$, and use $\frac{1}{2}(U_{n,k} + U_{n+1,k})$ as the approximation for u at $x = 1$. Replacing $\partial u / \partial x$ in the boundary condition by the central difference quotient we find that $U_{n+1,k}$ is given by

$$U_{n+1,k} = \frac{2h + U_{n,k}(1 - h)}{1 + h}. \qquad (1.26)$$

The propagation table for a boundary error depends on which of these methods we adopt. Table IV/7 shows the results of using (1.24) and (1.25) with $h = \frac{1}{3}$ and (1.26) with $h = \frac{2}{5}$ [the corresponding homogeneous equations must be used in deriving the error propagation; for example, (1.25) must be used without the term $4h \times 1$ on the right-hand side]. It can be seen that (1.25) compares unfavourably with the crude approximation represented by (1.24). In fact, if we try to calculate the solution using (1.25), the increasing randomness of the results (see Table IV/8) soon convinces us that the use of this apparently reasonable

Table IV/8. *Invalid calculation using the boundary formula* (1.25)

k	$x = 0$	$x = \frac{1}{3}$	$x = \frac{2}{3}$	$x = 1$	$x = \frac{4}{3}$
0	0	0	0	0	0
1	0	0	0	0	1·3333
2	0	0	0	0·7778	0·2963
3	0	0	0·4861	0·1728	1·5890
4		0·3646	0·0907	1·1295	−0·0820
5			0·8427	−0·0100	2·1893

method of dealing with the boundary condition has rendered the calculation quite useless. Of the other two methods (1.24) and (1.26), the latter has the advantage of the accuracy of the central difference quotient.

The other boundary $x = 0$ can also be dealt with in various ways:

1. Using a mesh with $x = 0$ as a mesh line. Since

$$\lim_{x \to 0} \frac{\partial u / \partial x}{x} = \lim_{x \to 0} \frac{\partial^2 u / \partial x^2}{1} = \left(\frac{\partial^2 u}{\partial x^2} \right)_{x=0}, \tag{1.27}$$

the differential equation (1.21) becomes

$$2 \frac{\partial^2 u}{\partial x^2} = \frac{\partial u}{\partial y}$$

as $x \to 0$. We now require a corresponding difference equation which does not lead to instability. From our experience in § 1.3 we try replacing $(\partial u / \partial y)_{0,k}$ by the forward difference quotient $\frac{1}{l} (U_{0,k+1} - U_{0,k})$, but find that the difference equation so obtained, namely

$$U_{0,k+1} = 2 U_{1,k} - U_{0,k} \tag{1.28}$$

(in which we have used the symmetry $U_{1,k} = U_{-1,k}$), propagates a boundary error with increasing magnitude (see Table IV/9). It turns out that the backward difference quotient $\frac{1}{l} (U_{0,k} - U_{0,k-1})$ is the best to use here, as can be seen from the propagation table (Table IV/9) for the resulting difference equation

$$U_{0,k} = \frac{1}{3} (2 U_{1,k} + U_{0,k-1}). \tag{1.29}$$

2. Using a mesh such that the boundary $x = 0$ lies halfway between the two mesh lines $x = -\frac{1}{2} h$ and $x = +\frac{1}{2} h$. With such a mesh we do not need any special

Table IV/9. *Propagation tables for boundary formulae at $x = 0$*

Formula (1.28)			Formula (1.29)		
$x = -h$	$x = 0$	$x = h$	$x = -h$	$x = 0$	$x = h$
0	ε	0	0	ε	0
$0\cdot25\,\varepsilon$	$-\varepsilon$	$0\cdot15\,\varepsilon$	$0\cdot25\,\varepsilon$	$0\cdot333\,\varepsilon$	$0\cdot25\,\varepsilon$
$-0\cdot25\,\varepsilon$	$1\cdot5\,\varepsilon$	$-0\cdot25\,\varepsilon$	$0\cdot083\,\varepsilon$	$0\cdot277\,\varepsilon$	$0\cdot083\,\varepsilon$
$0\cdot445\,\varepsilon$	$-2\,\varepsilon$	$0\cdot445\,\varepsilon$		$0\cdot148\,\varepsilon$	
$2\cdot891\,\varepsilon$					

boundary equations, we merely simplify the difference equation (1.23) by utilizing the symmetry about $x = 0$. This is in fact the method we adopt in the calculation below (Table IV/10) with $h = \frac{1}{4}$.

Having decided on the best equations to use, we can proceed to calculate $U_{i,k}$ row by row from (1.23) for $i = 1, 2, \ldots, n-1$ and from (1.26) for $i = n$.

Table IV/10. *Valid calculation of the growth of a magnetic field in a metal cylinder*

k	y	1 $\swarrow$ $i = 1$ $x = 0\cdot125$	$\frac{1}{3}$ $\frac{2}{3}$ $\searrow$ $\swarrow$ $i = 2$ $x = 0\cdot375$	$\frac{2}{5}$ $\frac{3}{5}$ $\searrow$ $\swarrow$ $i = 3$ $x = 0\cdot625$	$\frac{3}{7}$ $\frac{4}{7}$ $\searrow$ $\swarrow$ $i = 4$ $x = 0\cdot875$	$\frac{3}{5}+0\cdot4$ $\searrow$ $\swarrow$ $i = 5$ $x = 1\cdot125$	Row-sum check ϱ_k	σ_k
0	0	0	0	0	0	0		
1		0	0	0	0	0·4	4·8	
2		0	0	0	0·228 57	0·537 14	8·502 81	−0·000 07
3		0	0	0·137 14	0·306 94	0·584 16	11·829 48	+ 05
4	0·125	0	0·091 43	0·184 16	0·392 58	0·635 55	14·745 09	+ 01
5		0·091 43	0·122 77	0·272 12	0·442 10	0·665 26	17·423 04	+ 03
6		0·122 77	0·211 89	0·314 37	0·496 77	0·698 06	19·838 52	− 04
7		0·211 89	0·250 50	0·382 82	0·533 62	0·720 17	22·077 09	− 07
8	0·250	0·250 50	0·325 84	0·420 37	0·575 59	0·745 35	24·114 12	− 17
9		0·325 84	0·363 75	0·475 69	0·606 07	0·763 64	26·004 93	− 07
10		0·363 75	0·425 74	0·509 14	0·640 23	0·784 14	27·731 76	− 05
11		0·425 74	0·460 68	0·554 43	0·666 28	0·799 77	29·333 55	− 05

These equations are best written out explicitly. For $h = \frac{1}{4}$ they read

$$\bar{u}_1 = u_2, \quad \bar{u}_2 = \tfrac{1}{3}(u_1 + 2u_3), \quad \bar{u}_3 = \tfrac{1}{5}(2u_2 + 3u_4), \quad \bar{u}_4 = \tfrac{1}{7}(3u_3 + 4u_5), \quad \bar{u}_5 = \tfrac{1}{5}(2 + 3\bar{u}_4),$$

where we have used the more concise notation u_i for $U_{i,k}$ and $\bar{u}_i$ for $U_{i,k+1}$. The coefficients can be conveniently recorded at the heads of the columns as in Table IV/10.

As a row-sum check we can form

$$\varrho_k = 3(u_1 + 3u_2 + 5u_3 + 3u_4 + 4u_5)$$

and

$$\sigma_k = \varrho_k - \varrho_{k-1} - 8(1 - u_5)$$

and check that σ_k vanishes.

1.5. Hyperbolic differential equations

Using a rectangular mesh (1.1) and the approximations (1.4), (1.6) we can represent the differential equation

$$a \frac{\partial^2 u}{\partial x^2} + c \frac{\partial^2 u}{\partial y^2} + d \frac{\partial u}{\partial x} + e \frac{\partial u}{\partial y} + f u = r, \qquad (1.30)$$

where a, c, d, e, f, r are given functions of x and y with $a > 0$, $c < 0$, by the difference equation

$$\left. \begin{array}{l} a_{ik} \dfrac{U_{i+1, k} - 2 U_{i, k} + U_{i-1, k}}{h^2} + d_{ik} \dfrac{U_{i+1, k} - U_{i-1, k}}{2h} + \\[2ex] + c_{ik} \dfrac{U_{i, k+1} - 2 U_{i, k} + U_{i, k-1}}{l^2} + e_{ik} \dfrac{U_{i, k+1} - U_{i, k-1}}{2l} + f_{ik} U_{i, k} = r_{ik}. \end{array} \right\} \quad (1.31)$$

For the wave equation

$$\frac{\partial^2 u}{\partial x^2} = \frac{1}{\omega^2} \frac{\partial^2 u}{\partial y^2} \qquad (1.32)$$

and with the mesh relation

$$h = \omega l \qquad (1.33)$$

this difference equation simplifies to

$$U_{i, k+1} = U_{i+1, k} + U_{i-1, k} - U_{i, k-1}. \qquad (1.34)$$

We now consider some examples of the various types of initial- and boundary-value problems which can arise.

1. Values of u and $\partial u / \partial y$ specified on the whole x axis:

$$u = G(x), \qquad \frac{\partial u}{\partial y} = H(x) \quad \text{for} \quad -\infty < x < +\infty, \ y = 0. \quad (1.35)$$

In this case we know immediately the values $U_{i, 0} = G(ih)$ [for a mesh (1.1) with $x_0 = y_0 = 0$]; we can calculate the values $U_{i, 1}$ on the neighbouring row $k = 1$ as follows. Write (1.31) for $k = 0$ in the form

$$U_{i, 1} \left(\frac{c_{i0}}{l^2} + \frac{e_{i0}}{2l} \right) + U_{i, -1} \left(\frac{c_{i0}}{l^2} - \frac{e_{i0}}{2l} \right) = r_i^*, \qquad (1.36)$$

where the r_i^* are known and given by

$$r_i^* = - a_{i0} \frac{U_{i+1, 0} - 2 U_{i, 0} + U_{i-1, 0}}{h^2} - d_{i0} \frac{U_{i+1, 0} - U_{i-1, 0}}{2h} +$$
$$+ 2 c_{i0} \frac{U_{i, 0}}{l^2} - f_{i0} U_{i, 0} + r_{i0};$$

from the boundary condition $\partial u / \partial y = H(x)$ we have

$$U_{i, 1} - U_{i, -1} = 2l H(ih); \qquad (1.37)$$

then elimination of $U_{i,-1}$ between (1.36) and (1.37) yields

$$U_{i,1} = l\,H(i\,h)\left[1 - \frac{l}{2}\frac{e_{i0}}{c_{i0}}\right] + \frac{l^2}{2}\frac{r_i^*}{c_{i0}}. \tag{1.38}$$

We now have starting values for the difference equation (1.31), by means of which we can calculate successively the remaining values of $U_{i,k}$ for $k = 2, 3, 4, \ldots$.

2. Values of u specified on the boundary shown in Fig. IV/5 as a heavy line and values of $\partial u/\partial y$ specified on the part of that boundary which lies along the x axis:

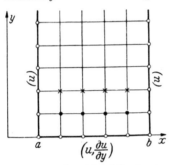

$$\left.\begin{array}{l} u = G(x) \\[4pt] \dfrac{\partial u}{\partial y} = H(x) \end{array}\right\} \quad \begin{array}{l} \text{for } a \le x \le b, \\ y = 0, \end{array}$$

$$u = \varphi(y) \quad \text{for} \quad \begin{array}{l} x = a, \\ y > 0, \end{array}$$

$$u = \psi(y) \quad \text{for} \quad \begin{array}{l} x = b, \\ y > 0. \end{array}$$

$$(1.39)$$

Fig. IV/5. Boundary of the solution domain for the problem (1.30), (1.39)

For a mesh (1.1) with

$$x_0 = a, \quad y_0 = 0, \quad h = \frac{b-a}{n},$$

where n is an integer greater than one, we know the $U_{i,k}$ at the boundary points ("white points" in the figure) and can calculate the values $U_{i,1}$ (at the "black points" in the figure) from (1.38). The remaining points are dealt with using (1.31).

3. Other types of boundary conditions often occur in applications; for example, we could have

$$\alpha u + \beta \frac{\partial u}{\partial x} = \varphi(y) \quad \text{for} \quad x = a, \quad y > 0$$

instead of $u = \varphi(y)$ as in 2. Boundary conditions of this kind can be treated by the methods of § 1.4.

1.6. A numerical example

A uniform circular membrane of radius a is fixed rigidly around its circumference and is deformed[1] into a position defined by the displacement

$$g(r) = v_0\left(1 - \frac{r^2}{a^2}\right),$$

where r is the distance from the centre in the mean plane of the membrane and v_0 is a constant (the maximum displacement of the centre of the membrane). When

---

[1] Cf. K. W. Wagner: Operatorenrechnung, 2nd ed., p. 243. Leipzig 1950.

the membrane is released (at time $t = 0$), vibrations are set up. The displacement $v(r, t)$ at subsequent times is to be calculated. It satisfies the well-known wave equation, which we write down directly in polar co-ordinates:

$$\nabla^2 v - \frac{1}{c^2} \frac{\partial^2 v}{\partial t^2} = \frac{\partial^2 v}{\partial r^2} + \frac{1}{r} \frac{\partial v}{\partial r} - \frac{1}{c^2} \frac{\partial^2 v}{\partial t^2} = 0. \tag{1.40}$$

The constant c denotes the wave velocity.

In the dimensionless co-ordinates

$$u = \frac{v}{v_0}, \qquad x = \frac{r}{a}, \qquad y = \frac{ct}{a}$$

(1.40) reads

$$\frac{\partial^2 u}{\partial x^2} + \frac{1}{x} \frac{\partial u}{\partial x} = \frac{\partial^2 u}{\partial y^2} \tag{1.41}$$

and the boundary conditions become

$$u(x, 0) = 1 - x^2, \frac{\partial u}{\partial y} = 0 \qquad \text{on} \quad y = 0, \quad |x| \leqq 1 \quad \text{(initial conditions)},$$

$$u = 0 \quad \text{(no displacement)} \quad \text{on} \quad x = \pm 1, \quad y > 0 \quad \text{(the fixed boundary)}.$$

The symmetry about the y axis allows us to restrict the calculation to the half region with $x \geqq 0$. Then for a square mesh with $h = l = \frac{1}{5}$ we know the boundary values printed in heavy type in Table IV/12 and can proceed using the equations

$$\left. \begin{aligned} U_{i,1} &= \frac{1}{2} U_{i+1,0} \left(1 + \frac{1}{2i}\right) + \frac{1}{2} U_{i-1,0} \left(1 - \frac{1}{2i}\right), \\ U_{i,k+1} &= U_{i+1,k} \left(1 + \frac{1}{2i}\right) + U_{i-1,k} \left(1 - \frac{1}{2i}\right) - U_{i,k-1} \end{aligned} \right\} \tag{1.42}$$

$$(i = 1, \ldots, n-1; \ k = 1, 2, \ldots).$$

These suffice until we come to deal with values on the y axis; these require special treatment because of the singularity at $x = 0$, or $i = 0$. At this point the differential equation (1.41) becomes

$$2 \frac{\partial^2 u}{\partial x^2} = \frac{\partial^2 u}{\partial y^2}$$

on account of (1.27), and the corresponding difference equation reads

Table IV/11. *Propagation table for* (1.43)

$i = -1$	$i = 0$	$i = 1$
0	ε	0
$\varepsilon/2$	-2ε	$\varepsilon/2$
$-\varepsilon$	7ε	$-\varepsilon$
	-20ε	

$$U_{0,k+1} = 2(U_{1,k} - U_{0,k} + U_{-1,k}) - U_{0,k-1}. \tag{1.43}$$

However, this formula has a very unfavourable propagation table (see Table IV/11) and in fact an attempt at using it yields quite useless results after a few rows.

The simplest way of getting over this difficulty is not to use the differential equation at all, but to put a parabola through the points with $i = \pm 1, \pm 2$ and use its value at $i = 0$ for $U_{0,k}$:

$$U_{0,k} = \frac{4 U_{1,k} - U_{2,k}}{3} = U_{1,k} + \frac{1}{3} (U_{1,k} - U_{2,k}).$$

The last two columns in Table IV/12 provide a simple check which can be applied after each row has been completed. We calculate the

Table IV/12. *Finite-difference approximation for the vibrations of a circular membrane*

β_i		$\beta_0=0.5$	$\beta_1=0.75$	$\beta_2=2.333\,33$	$\beta_3=2.125$	$\beta_4=1.166\,67$		Row-sum check	
$1\pm\dfrac{1}{2i}$		$\frac{1}{2}\quad\frac{3}{2}$	$\frac{3}{4}\quad\frac{5}{4}$	$\frac{5}{6}\quad\frac{7}{6}$	$\frac{7}{8}$				
y	k	$i=0$ $x=0$	$i=1$ $x=0.2$	$i=2$ $x=0.4$	$i=3$ $x=0.6$	$i=4$ $x=0.8$	$i=5$ $x=1$	σ_k	τ_k
0	0	1	0.96	0.84	0.64	0.36	0		2.80
0.2	1	0.92	0.88	0.76	0.56	0.28	0	4.41	2.48
0.4	2	0.68	0.64	0.52	0.32	0.13	0	2.865	1.61
0.6	3	0.28	0.24	0.12	0.025	0	0	0.653 12	0.385
0.8	4	−0.323 75	−0.32	−0.308 75	−0.220	−0.108 13	0	−1.715 94	−0.956 88
1	5	−0.941 67	−0.865 00	−0.635 00	−0.408 44	−0.192 50	0	−3.693 77	−2.100 94
	6	−1.187 60	−1.103 34	−0.850 55	−0.533 74	−0.249 26	0	−4.830 92	−2.736 89
	7	−1.052 95	−1.004 63	−0.859 68	−0.591 15	−0.274 52	0	−4.862 33	−2.729 98
	8	−0.736 26	−0.712 66	−0.641 86	−0.502 93	−0.268 00	0	−3.781 69	−2.125 45
	9	−0.333 89	−0.326 29	−0.303 48	−0.256 40	−0.165 54	0	−1.857 76	−1.051 71
2	10	+0.095 11	0.090 49	0.076 64	0.056 90	0.043 65	0	0.466 09	0.267 68
	11	0.504 25	0.488 81	0.442 47	0.371 19	0.215 33	0	2.691 16	1.517 80
	12	0.849 13	0.825 34	0.753 96	0.563 04	0.281 14	0	4.327 27	2.423 48
	13	1.128 82	1.066 70	0.880 34	0.585 11	0.277 33	0	4.985 47	2.809 48
	14	1.153 62	1.059 58	0.777 45	0.494 13	0.230 83	0	4.504 87	2.561 99
3	15	0.724 38	0.676 29	0.532 01	0.332 06	0.155 03	0		1.695 39

quantities

$$\sigma_k = \sum_{i=0}^{n-1} \beta_i\, U_{i,k}, \qquad \tau_k = \sum_{i=1}^{n-1} U_{i,k},$$

where

$$\beta_0 = \frac{1}{2}, \qquad \beta_1 = \frac{3}{4}, \qquad \beta_i = 2 + \frac{1}{i^2 - 1} \quad \text{for} \quad i = 2, 3, \ldots, n-2,$$

$$\beta_{n-1} = 1 + \frac{1}{2(n-2)},$$

and check that

$$\tau_{k+1} = \sigma_k - \tau_{k-1} \qquad (k = 1, 2, \ldots).$$

The numbers β_i and the coefficients $\left(1 - \dfrac{1}{2i}\right)$ and $\left(1 + \dfrac{1}{2i}\right)$ in (1.42) may be recorded conveniently at the heads of the appropriate columns.

1.7. Graphical treatment of parabolic differential equations by the finite-difference method

It is often convenient to apply the ordinary finite-difference method in graphical form[1]. For instance, the process of forming mean values

---

[1] Although it is not intended that this book should also cover graphical methods, we give here a brief description of this method for parabolic equations because of the many applications.

in the treatment of the heat equation

$$\frac{\partial^2 u}{\partial x^2} = c\,\frac{\partial u}{\partial y} \tag{1.44}$$

by the difference equation (1.13) is very easily performed graphically. The approximations $U_{i,k}$ are plotted against x, the points with equal k defining a sequence of polygons. The $(k+1)$-th polygon is formed from the k-th by joining alternate points of the latter and taking the midpoints of these joins, i.e. the points of intersection with the ordinate lines, as the vertices of the next polygon (see Fig. IV/6) [1].

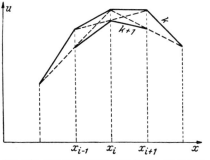

Fig. IV/6. Graphical construction for the solution of the heat equation

This construction is applied to an example in Fig. IV/7; the specific boundary values chosen are

$$\left.\begin{aligned} u(x,0) &= 0 \quad \text{for} \quad -1 \leqq x \leqq 1,\\ u(1,t) &= u(-1,t) = 1 - e^{-\frac{5}{c}t}, \end{aligned}\right\} \tag{1.45}$$

which correspond to a rod being heated from zero temperature by the application of heat to its ends so that their common temperature rises exponentially to the value 1. The graphical solution is carried out with $h = \frac{1}{5}$ and therefore $l = c/50$ in accordance with (1.12). Only the region $x \geqq 0$ is considered on account of symmetry; only the construction lines are shown in Fig. IV/7 since the sides of the polygons, which are shown in Fig. IV/6, would have confused the picture. The values of k are marked at various points; the corresponding time can be calculated from $t = k\,l$.

If we choose a mesh with $2l = c_k h^2 = c(y_k)\,h^2$, the more general differential equation

$$\frac{\partial^2 u}{\partial x^2} + f(x,y)\,\frac{\partial u}{\partial x} = c(y)\,\frac{\partial u}{\partial y} + a(x,y) + b(x,y)\,u \tag{1.46}$$

is represented by the difference equation

$$\left.\begin{aligned} U_{i,k+1} = {}&\frac{1}{2}\left(1 + \frac{h f_{i,k}}{2}\right)U_{i+1,k} +\\ &+ \frac{1}{2}\left(1 - \frac{h f_{i,k}}{2}\right)U_{i-1,k} - \frac{h^2}{2}\,(a_{ik} + b_{ik}\,U_{i,k}). \end{aligned}\right\} \tag{1.47}$$

We can still utilize the chord construction but here we use the ordinate cut off on the line $x = \xi_{i,k} = x_i + \frac{1}{2}h^2 f_{i,k}$ instead of the line $x = x_i$ (see

[1] SCHMIDT, E.: Einführung in die technische Thermodynamik, 2nd ed., p. 282. Berlin 1944.

Fig. IV/8). This ordinate is $\frac{1}{2}(1+\frac{1}{2}hf_{i,k})\,U_{i+1,k}+\frac{1}{2}(1-\frac{1}{2}hf_{i,k})\,U_{i-1,k}$; we can obtain $U_{i,k+1}$ from it by laying off $\frac{1}{2}h^2(a_{ik}+b_{ik}U_{i,k})$. The construction is particularly simple when $f(x, y)$ depends only upon x, for

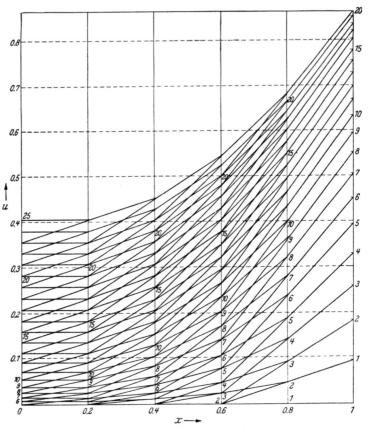

Fig. IV/7. Graphical solution for the varying temperature distribution in a heated rod

then one set of lines $x=\xi_i$ can be used for all k. The variable quantity $c(y)$ does not come into the construction but is needed when we want to find the value of $y=y_k$ corresponding to a particular value of k. This is given by

$$y_k = \frac{h^2}{2}\sum_{\kappa=0}^{k-1} c_\kappa.$$

Fig. IV/9 shows the construction for the example (1.21), (1.22) of § 1.4, which concerned the eddy currents in the metal core of an electromagnet. The special construction for the boundary conditions is described in the next paragraph.

If the values of u are prescribed at a boundary, say $x=\alpha$, then a mesh is chosen with $x=\alpha$ as a mesh line. On the other hand, if the boundary condition at $x=\alpha$

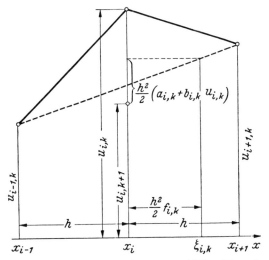

Fig. IV/8. Graphical construction for the more general differential equation (1.46)

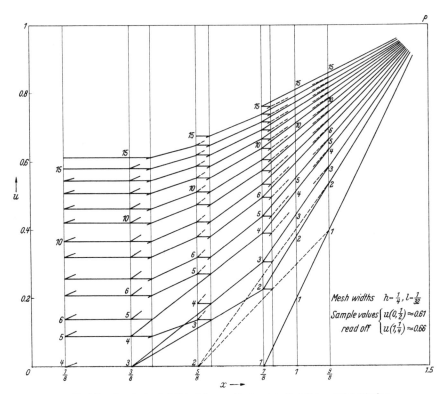

Fig. IV/9. Graphical calculation of the growth of a magnetic field in a metal cylinder

is of the form

$$u + A(y)\frac{\partial u}{\partial x} = B(y) \qquad (1.48)$$

with $A(y) \neq 0$, then a mesh is chosen so that $x = \alpha$ lies halfway between the two mesh lines $x = x_n$ and $x = x_{n+1}$, i.e. so that $\alpha = \frac{1}{2}(x_n + x_{n+1})$; one can then replace the x derivative by the central difference quotient, as in (1.26), which yields

$$\left. \begin{array}{l} U_{n+1,\,k} \\ \quad = \dfrac{2Bh + (2A - h)\,U_{n,\,k}}{2A + h} \end{array} \right\} \quad (1.49)$$

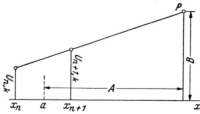

Fig. IV/10. Graphical representation of a boundary condition involving a normal derivative

This equation implies that the point P with co-ordinates $(\alpha + A, B)$ is collinear with the points $(x_n, U_{n,\,k})$ and $(x_{n+1}, U_{n+1,\,k})$, as in Fig. IV/10; hence, knowing $U_{n,\,k}$, we can construct $U_{n+1,\,k}$. If A and B do not depend on y, P is the same point for all k (as for the example in Fig. IV/9).

If u does not occur in the boundary condition (1.48), i.e. $\partial u/\partial x$ is prescribed, then $U_{n+1,\,k}$ can be constructed from $U_{n,\,k}$ merely by drawing a line in the specified direction.

1.8. The two-dimensional heat equation

When we consider the equation of heat flow in two dimensions

$$\frac{\partial^2 u}{\partial x^2} + \frac{\partial^2 u}{\partial y^2} = c\,\frac{\partial u}{\partial z}, \qquad (1.50)$$

we find that an approximate finite-difference solution can still be obtained by an averaging procedure and hence by a simple graphical construction on a plane projection of the (x, y, u) space.

We use a three-dimensional mesh

$$\left. \begin{array}{l} x = x_0 + i\,h_x \\ y = y_0 + k\,h_y \\ z = z_0 + l\,h_z \end{array} \right\} \quad (i, k, l = 0, \pm 1, \pm 2, \ldots) \qquad (1.51)$$

with mesh widths

$$h_x = h_y = h, \qquad h_z = c\,\varrho\,h^2. \qquad (1.52)$$

The corresponding "forward" difference equation reads

$$\left. \begin{array}{l} U_{i,\,k,\,l+1} = \varrho\,(U_{i+1,\,k,\,l} + U_{i-1,\,k,\,l} + U_{i,\,k+1,\,l} + U_{i,\,k-1,\,l}) + \\ \qquad\qquad + (1 - 4\varrho)\,U_{i,\,k,\,l}. \end{array} \right\} \quad (1.53)$$

With $\varrho = \frac{1}{2}$, the value used in (1.13), we have the formula[1]

$$U_{i,\,k,\,l+1} = \tfrac{1}{2}S - U_{i,\,k,\,l}, \qquad (1.54)$$

---

[1] ELSER, K.: Schweizer Arch. Angew. Wiss. Techn. **10**, 341—343 (1944).

$l = 0$

$$
\begin{array}{ccc}
0 & 0 & 0 \\
0 & \boxed{\varepsilon} & 0 \\
0 & 0 & 0 \\
0 & 0 & 0
\end{array}
$$

$l = 1$

$$
\begin{array}{ccc}
0 & \tfrac{1}{2}\varepsilon & 0 \\
\tfrac{1}{2}\varepsilon & \boxed{-\varepsilon} & \tfrac{1}{2}\varepsilon \\
0 & \tfrac{1}{2}\varepsilon & 0 \\
0 & 0 & 0
\end{array}
$$

$l = 2$

$$
\begin{array}{cccc}
\tfrac{1}{2}\varepsilon & -\varepsilon & \tfrac{1}{2}\varepsilon & 0 \\
-\varepsilon & \boxed{2\varepsilon} & -\varepsilon & \tfrac{1}{4}\varepsilon \\
\tfrac{1}{2}\varepsilon & -\varepsilon & \tfrac{1}{2}\varepsilon & 0 \\
0 & \tfrac{1}{4}\varepsilon & 0 & 0
\end{array}
$$

$l = 3$

$$
\begin{array}{ccccc}
-\tfrac{3}{2}\varepsilon & \tfrac{21}{8}\varepsilon & -\tfrac{3}{2}\varepsilon & \tfrac{3}{8}\varepsilon & 0 \\
\tfrac{21}{8}\varepsilon & \boxed{-4\varepsilon} & \tfrac{21}{8}\varepsilon & -\tfrac{3}{4}\varepsilon & \tfrac{1}{8}\varepsilon \\
-\tfrac{3}{2}\varepsilon & \tfrac{21}{8}\varepsilon & -\tfrac{3}{2}\varepsilon & \tfrac{3}{8}\varepsilon & 0 \\
\tfrac{3}{8}\varepsilon & -\tfrac{3}{4}\varepsilon & \tfrac{3}{8}\varepsilon & 0 & 0
\end{array}
$$

$l = 4$

$$
\begin{array}{cccccc}
\tfrac{9}{2}\varepsilon & -\tfrac{13}{2}\varepsilon & \tfrac{9}{2}\varepsilon & -\tfrac{3}{2}\varepsilon & \tfrac{1}{4}\varepsilon & 0 \\
-\tfrac{13}{2}\varepsilon & \boxed{\tfrac{37}{4}} & -\tfrac{13}{2}\varepsilon & \tfrac{5}{2}\varepsilon & -\tfrac{1}{2}\varepsilon & \tfrac{1}{16}\varepsilon \\
\tfrac{9}{2}\varepsilon & -\tfrac{13}{2}\varepsilon & \tfrac{9}{2}\varepsilon & -\tfrac{3}{2}\varepsilon & \tfrac{1}{4}\varepsilon & 0 \\
-\tfrac{3}{2}\varepsilon & \tfrac{5}{2}\varepsilon & -\tfrac{3}{2}\varepsilon & \tfrac{3}{8}\varepsilon & 0 & 0
\end{array}
$$

where

$$
\left.
\begin{aligned}
S = U_{i+1,k,l} + U_{i-1,k,l} + \\
+ U_{i,k+1,l} + U_{i,k-1,l}.
\end{aligned}
\right\} \quad (1.55)
$$

Its error propagation, shown in the sequence of tables above, renders it practically useless; an error ε is approximately doubled at each step in the z direction.

However, putting $\varrho = \tfrac{1}{4}$ in (1.53) yields a usable formula[1] which is also very simple:

$$
U_{i,k,l+1} = \tfrac{1}{4}S, \qquad (1.56)
$$

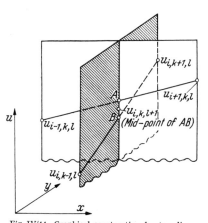

Fig. IV/11. Graphical construction for two-dimensional heat flow

where S is as defined in (1.55). The graphical construction for this formula is shown in Fig. IV/11 and a numerical example is given in Exercise 6 in § 6.4.

---

[1] Mentioned briefly by K. ELSER: see last footnote.

1.9. An indication of further problems

The finite-difference method has been applied with success to many complicated hydrodynamic and aerodynamic problems but to go into details here would be beyond the scope of this book. Suffice it to mention (in addition to the references given in § 5) the application of the finite-difference method to the systems of partial differential equations arising in fluid dynamics[1] and to boundary-value problems for differential equations of "mixed type", i.e. equations which are partly elliptic and partly hyperbolic in the considered region, there being a dividing curve along which the equation is parabolic. These latter problems are important in gas dynamics[2] (the dividing curve corresponds to the transonic region); their numerical solution needs to be treated with particular care and attention.

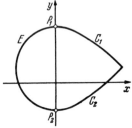

Fig. IV/12.   A boundary-value problem for a differential equation of "mixed type"

As an example which can be formulated very simply (this problem has been treated theoretically) consider the differential equation

$$- u_{xx} + f(x)\,u_{yy} = 0, \qquad (1.57)$$

where $f(x)$ is a non-decreasing function of x with the same sign as x. Then (1.57) is elliptic for $x < 0$, parabolic on the line $x = 0$, and hyperbolic for $x > 0$. Let the boundary Γ consist of an arc E in the half-plane $x \leqq 0$ which has only its end-points P_1 and P_2 in common with the y axis, together with the two characteristics C_1 and C_2 which proceed from P_1 and P_2, respectively, into the half-plane $x \geqq 0$ (Fig. IV/12). Let u be prescribed on the boundaries E and C_1. Then u is determined uniquely inside Γ, and it should be possible to calculate u numerically.

§2. Refinements of the finite-difference method

The finite-difference method can be refined in exactly the same ways for partial differential equations as for ordinary differential equations (Chapter III, § 2).

---

[1] Ways of overcoming difficulties occasioned by the presence of singularities on the boundary are given by H. GÖRTLER: Ein Differenzenverfahren zur Berechnung laminarer Grenzschichten. Ing.-Arch. **16**, 173—187 (1948). — WITTING, H.: Verbesserungen des Differenzenverfahrens von H. Görtler zur Berechnung laminarer Grenzschichten. Z. Angew. Math. Phys. **4**, 376—397 (1953).

[2] The literature on the subject of differential equations of mixed type began with the pioneering work of F. TRICOMI: Sulle equazioni lineari alle derivate parziali di 2° ordino di tipo misto. Mem. Real. Accad. Lincei **14**, 133—247 (1923). It has expanded so rapidly in recent years that we will content ourselves here with mentioning a few arbitrarily selected papers: PROTTER, M. H.: A boundary value problem for an equation of mixed type. Trans. Amer. Math. Soc. **71**, 416—429 (1951). — BERGMAN, ST.: On solutions of linear partial differential equations of mixed type. Amer. J. Math. **74**, 444—474 (1952). — HELLWIG, G.: Anfangs- und Randwertprobleme bei partiellen Differentialgleichungen von wechselndem Typus auf den Rändern. Math. Z. **58**, 337—357 (1953). — LAX, P. D.: Weak Solutions of Nonlinear Hyperbolic Equations and their Numerical Computation. Comm. Pure Appl. Math. **7**, 159—193 (1954).

2.1. The derivation of finite equations

To simplify the description we limit ourselves to linear partial differential equations in two independent variables, x, y. Such an equation of the m-th order can be written

$$L[u] = \sum_{p+q \leq m} A_{pq}(x, y) \frac{\partial^{p+q} u}{\partial x^p \partial y^q} = r(x, y), \qquad (2.1)$$

where the $A_{pq}(x, y)$ and $r(x, y)$ are given functions which we shall assume to be continuous. Let sufficient starting data be given, say on the x axis and certain boundary curves, to determine uniquely a specific solution $u(x, y)$. It is required to calculate the approximations $U_{i, k+1}$ on the $(k+1)$-th row when the calculation based on the rectangular mesh (1.1) has progressed as far as the k-th row. For this purpose we derive a finite equation

$$\sum_{\iota, \varkappa} C_{\iota, \varkappa} U_{\iota, \varkappa} = r_{\iota_0, \varkappa_0} \qquad (2.2)$$

corresponding to (2.1) at (x_0, y_0), in which the coefficients $C_{\iota, \varkappa}$ are determined so that the Taylor expansion of the sum

$$\Phi = \sum_{\iota, \varkappa} C_{\iota, \varkappa} u_{\iota, \varkappa} \qquad (2.3)$$

in terms of u and its partial derivatives at the mesh point $(\iota_0, \varkappa_0)$ agrees with (2.1) at (x_0, y_0) to as high an order as possible, i.e. so that the order of the lowest derivative in the Taylor expansion which does not have the same coefficient as in (2.1) is as high as possible. The summation in (2.2) and (2.3) is to extend over a number of mesh points in the neighbourhood of $(\iota_0, \varkappa_0)$ the choice of which depends on the given differential equation. The number of points may be greater than is absolutely necessary for the Taylor expansion to coincide with the differential equation to some order; for then we have certain free or "surplus" $C_{\iota, \varkappa}$ at our disposal, which can be chosen so as to make the finite equations as stable as possible. A rough indication of the stability is given by the "index" J, which we now define.

Normally we will use a finite equation (2.2) which, besides the term $U_{i, k+1}$ to be calculated, involves only values $U_{\iota, \varkappa}$ with $\varkappa \leq k$. If it is written in the form

$$U_{i, k+1} = - \frac{1}{C_{i, k+1}} \sum_{\varkappa \leq k} C_{\iota, \varkappa} U_{\iota, \varkappa} + \frac{1}{C_{i, k+1}} r_{\iota_0, \varkappa_0}, \qquad (2.4)$$

which expresses $U_{i, k+1}$ in terms of the remaining $U_{\iota, \varkappa}$, its index J is defined as the sum of the absolute values of the coefficients of the remaining $U_{\iota, \varkappa}$:

$$J = \frac{1}{|C_{i, k+1}|} \sum_{\varkappa \leq k} |C_{\iota, \varkappa}|. \qquad (2.5)$$

Apart from the restriction $\varkappa \leq k$, the summation is over all $\iota, \varkappa$ for which $C_{\iota,\varkappa} \neq 0$.

We try to keep the index J as small as possible; this can often be done by suitably choosing surplus coefficients $C_{\iota,\varkappa}$, as suggested above, and also by moving the "centre" $(\iota_0, \varkappa_0)$ of the Taylor expansion "forwards", i.e. by choosing it near the point $(i, k+1)$.

When certain other conditions are satisfied, the condition $J \leq 1$ is sufficient for the stability and convergence of the finite-difference method, see §§ 3.3, 3.4; however, formulae with a greater index can still be usable, for instance, $J = 3$ for equation (1.34).

2.2. Application to the heat equation

To illustrate the derivation of finite equations, let us follow through the procedure for the inhomogeneous heat equation

$$L[u] = -K \frac{\partial u}{\partial y} + \frac{\partial^2 u}{\partial x^2} = r(x, y). \tag{2.6}$$

For simplicity we take the origin of the mesh co-ordinates at the mesh point $(i, k+1)$ and use this point as the centre of the Taylor expansion; thus $\iota_0 = \varkappa_0 = 0$. We extend the summation (2.3) over the points $(\iota, \varkappa)$ with $|\iota| \leq 2, \varkappa = -1, \varkappa = -2$ and $\iota = 0, \varkappa = -3$ and denote the coefficients $C_{\iota,\varkappa}$ by a, b, c, d, e, f, g, j as indicated in Fig. IV/13, taking advantage of symmetry (thus $C_{0,0} = a$, $C_{1,-1} = C_{-1,-1} = c$, etc.). Each $u_{\iota,\varkappa}$ has now to be expanded as a Taylor series centred on the point $(0, 0)$; for example,

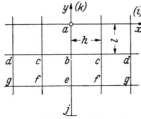

Fig. IV/13. Notation used in setting up a finite expression for the heat equation

$$u_{1,-1} = u_{0,0} + h u_{x,0,0} - l u_{y,0,0} + \frac{h^2}{2} u_{xx,0,0} + \cdots. \tag{2.7}$$

By similarly expanding all the terms which appear in the sum (2.3) we obtain

$$\begin{aligned}
\varPhi = &\, u(a + b + 2c + 2d + e + 2f + 2g + j) + \\
&+ l u_y(-b - 2c - 2d - 2e - 4f - 4g - 3j) + \\
&+ \frac{h^2}{2} u_{xx}(2c + 8d + 2f + 8g) + \\
&+ \frac{l^2}{2} u_{yy}(b + 2c + 2d + 4e + 8f + 8g + 9j) + \\
&+ \frac{h^2 l}{2} u_{xxy}(-2c - 8d - 4f - 16g) + \\
&+ \frac{l^3}{6} u_{yyy}(-b - 2c - 2d - 8e - 16f - 16g - 27j) +
\end{aligned} \tag{2.8}$$

$$+ \frac{h^4}{24} u_{xxxx}(2c + 32d + 2f + 32g) +$$

$$+ \frac{h^2 l^2}{4} u_{xxyy}(2c + 8d + 8f + 32g) +$$

$$+ \frac{l^4}{24} u_{yyyy}(b + 2c + 2d + 16e + 32f + 32g + 81j) +$$

$$+ \text{ terms of the 5th order,}$$

in which all the u, u_y, etc. are evaluated at the point $(0, 0)$.

This expansion is to coincide with the differential expression (2.6) to as high an order as possible, so the second bracketed factor must have the value $-K/l$, and the third must have the value $2/h^2$, while the first, fourth, and as many more as possible, must be zero. For comparison we complete the derivation for formulae correct to several different orders.

1. First-order formulae. Here we equate coefficients of u, u_y, u_{xx} only. This requires three constants, say a, b, c, and from the equations

$$a + b + 2c = 0,$$

$$- b - 2c = - \frac{K}{l},$$

$$2c = \frac{2}{h^2}$$

their values must be

$$a = - \frac{K}{l}, \qquad b = \frac{K}{l} - \frac{2}{h^2}, \qquad c = \frac{1}{h^2}.$$

With the notation

$$\sigma = \frac{l}{Kh^2}$$

we have

$$\left. \begin{aligned} \varPhi &= \frac{K}{l} \left(- u_{0,0} + (1 - 2\sigma) u_{0,-1} + \sigma(u_{1,-1} + u_{-1,-1}) \right) \\ &= \left(- K \frac{\partial u}{\partial y} + \frac{\partial^2 u}{\partial x^2} \right)_{0,0} + \text{remainder term of the 2nd order.} \end{aligned} \right\} \quad (2.9)$$

The finite equation obtained by neglecting the remainder term reads [when expressed in the form (2.4)]

$$U_{0,0} = (1 - 2\sigma) U_{0,-1} + \sigma(U_{1,-1} + U_{-1,-1}) - \sigma h^2 r_{0,0}, \qquad (2.10)$$

which is the formula for the ordinary finite-difference method.

σ is positive, and for all values up to a half the index for equation (2.10) is 1; for $\sigma > \frac{1}{2}$ the index is always greater than 1. In this respect, equation (2.10) is equally favourable for all values of σ in $0 \leq \sigma \leq \frac{1}{2}$;

however, the formula obtained by putting $\sigma = \frac{1}{2}$, i.e.

$$U_{0,0} = \frac{1}{2}(U_{1,-1} + U_{-1,-1}) - \frac{1}{2}h^2 r_{0,0},$$

is more advantageous than a formula resulting from a smaller value of σ, say $\sigma = \frac{1}{3}$, which gives

$$U_{0,0} = \frac{1}{3}(U_{1,-1} + U_{0,-1} + U_{-1,-1}) - \frac{1}{3}h^2 r_{0,0},$$

because a greater value of σ means a greater value of l and hence we progress further in the time direction with the same number of computed values.

2. Second-order formulae. When we try to equate coefficients of all terms of up to the second order inclusive, we see immediately that this cannot be done with the constants a, b, c, d, since the four equations are then inconsistent. With the four constants a, b, c, e the equations are

$$a + b + 2c + e = 0, \qquad -b - 2c - 2e = -\frac{K}{l},$$

$$2c = \frac{2}{h^2}, \qquad b + 2c + 4e = 0,$$

and these can be solved; they yield

$$a = -\frac{3}{2}\frac{K}{l}, \quad b = 2\left(\frac{K}{l} - \frac{1}{h^2}\right), \quad c = \frac{1}{h^2}, \quad e = -\frac{K}{2l}.$$

With these values we have

$$\Phi = \frac{K}{l}\left(-\frac{3}{2}u_{0,0} + (2 - 2\sigma)u_{0,-1} + \sigma(u_{1,-1} + u_{-1,-1}) - \frac{1}{2}u_{0,-2}\right)$$

$$= (L[u])_{0,0} + \text{remainder term of the 3rd order.}$$

The corresponding finite equation reads

$$U_{0,0} = \frac{2}{3}\left[(2 - 2\sigma)U_{0,-1} + \sigma(U_{1,-1} + U_{-1,-1}) - \frac{1}{2}U_{0,-2} - \sigma h^2 r_{0,0}\right]$$

and for $0 \leq \sigma \leq 1$ has the index $J = \frac{5}{3}$. Putting $\sigma = 1$ we obtain the formula

$$U_{0,0} = \frac{1}{3}[2U_{1,-1} + 2U_{-1,-1} - U_{0,-2} - 2h^2 r_{0,0}],$$

which is stable in the sense of § 3.4. As with the ordinary finite-difference formula (1.13), this formula can be used over a sub-set of the whole set of mesh points, but here each row computed takes the solution twice as far as would a row calculated by (1.13).

3. Third-order formulae. We can equate coefficients of all terms of up to the third order inclusive by using the constants a, b, c, e, f.

The solution of the pertinent set of equations yields

$$\Phi = \frac{K}{l}\left\{ -\frac{11}{6}u_{0,0} + 3u_{0,-1} - \frac{3}{2}u_{0,-2} + \frac{1}{3}u_{0,-3} + \right.$$

$$\left. + 2\sigma\left[-2u_{0,-1} + u_{0,-2} + (u_{1,-1} + u_{-1,-1}) - \frac{1}{2}(u_{1,-2} + u_{-1,-2})\right]\right\}$$

$$= -K\frac{\partial u}{\partial y} + \frac{\partial^2 u}{\partial x^2} + \text{remainder term of the 4th order.}$$

The corresponding finite equation has the index $J = \frac{29}{11}$ for $0 \leq \sigma \leq \frac{3}{4}$; this is greater than that for either of the lower order equations and even the inclusion of the values $u_{2,-1}$ and $u_{-2,-1}$ does not improve it. [An expression with a smaller index is given in Z. Angew. Math. Mech. Bd. 16 (1936) p. 245]. A smaller index can be obtained with the Hermitian-type formulae of § 2.3.

2.3. The "Hermitian" methods

The idea of Hermitian-type formulae, which was introduced in Ch. III, § 2.4, for ordinary differential equations, can be readily extended to linear partial differential equations. Consider once more the differential equation (2.1). Instead of deriving a finite equation from the sum in (2.3), we use an expression of the form

$$\Phi = \sum_{\iota, \varkappa} C_{\iota, \varkappa} u_{\iota, \varkappa} + \sum_{\iota, \varkappa} D_{\iota, \varkappa}(L[u])_{\iota, \varkappa}, \tag{2.11}$$

where the $C_{\iota, \varkappa}$ and $D_{\iota, \varkappa}$ are constants to be determined; $(L[u])_{\iota, \varkappa}$ denotes the value taken by the differential expression $L[u]$ at the mesh point $(\iota, \varkappa)$ and the summations are extended over a number of suitably chosen mesh points in the neighbourhood of a specified point $(\iota_0, \varkappa_0)$ with which the equation is associated. As in § 2.1, the mesh points occurring in (2.11) will normally be chosen so that $C_{i,k+1} \neq 0$ and $C_{\iota,k+1} = 0$ for $\iota \neq i$; then with $(L[u])_{\iota, \varkappa} = r_{\iota, \varkappa}$ from (2.1) we use the finite equation

$$\sum_{\iota, \varkappa} C_{\iota, \varkappa} U_{\iota, \varkappa} + \sum_{\iota, \varkappa} D_{\iota, \varkappa} r_{\iota, \varkappa} = 0 \tag{2.12}$$

to calculate the approximation $U_{i,k+1}$ from the $U_{\iota, \varkappa}$ with $\varkappa \leq k$. In order that the approximations obtained by using (2.12) shall be as good as possible, the coefficients $C_{\iota, \varkappa}, D_{\iota, \varkappa}$ are determined so that the Taylor expansion of the expression (2.11) in terms of u and its partial derivatives at the point $(\iota_0, \varkappa_0)$ is zero to as high an order as possible.

Again we use the example of the inhomogeneous heat equation (2.6) to illustrate the derivation. With the same mesh points and the same notation $a, b, c, \ldots$ for the $C_{\iota, \varkappa}$ as in § 2.2 and Fig. IV/13, and with the corresponding capitals A, B, C, ... for the $D_{i,x}$, we write

$$\Phi = a\,u_{0,0} + b\,u_{0,-1} + c\,(u_{1,-1} + u_{-1,-1}) + e\,u_{0,-2} +$$
$$+ f\,(u_{1,-2} + u_{-1,-2}) + A\,(L[u])_{0,0} + B\,(L[u])_{0,-1} + \Bigg\} \quad (2.13)$$
$$+ C\,\{(L[u])_{1,-1} + (L[u])_{-1,-1}\}.$$

Taylor expansion in terms of u and its derivatives at the point $(0, 0)$ yields

$$\Phi = u\,[a + b + 2c + e + 2f] +$$
$$+ u_y\,[l(-b - 2c - 2e - 4f) - K(A + B + 2C)] +$$
$$+ u_{xx}\left[\frac{h^2}{2}(2c + 2f) + A + B + 2C\right] +$$
$$+ u_{yy}\left[\frac{l^2}{2}(b + 2c + 4e + 8f) + Kl(B + 2C)\right] + \Bigg\} \quad (2.14)$$
$$+ u_{xxy}\left[\frac{h^2 l}{2}(-2c - 4f) - l(B + 2C) - h^2 C K\right] +$$
$$+ u_{yyy}\left[\frac{l^3}{6}(-b - 2c - 8e - 16f) - \frac{Kl^2}{2}(B + 2C)\right] +$$
$$+ \text{terms of the 4th order.}$$

If each factor in square brackets is put equal to zero, we have six homogeneous equations for eight unknowns; we therefore express six of the unknowns in terms of the remaining two, say A and f:

$$a = \frac{5}{2}\frac{KA}{l}, \quad b = 2(3\sigma - 1)\frac{KA}{l} + 2f, \quad c = -3\sigma\frac{KA}{l} - f,$$

$$e = -\frac{1}{2}\frac{KA}{l} - 2f, \quad \frac{KB}{l} = 2(1 - \sigma)\frac{KA}{l} + 2f, \quad \frac{KC}{l} = \sigma\frac{KA}{l} - f$$

$$\left(\text{with the usual notation } \sigma = \frac{l}{Kh^2}\right).$$

Thus with $\varepsilon = \frac{l}{KA}f$ we have

$$\frac{5}{2}u_{0,0} + 2(3\sigma - 1 + \varepsilon)\,u_{0,-1} - (3\sigma + \varepsilon)\,(u_{1,-1} + u_{-1,-1}) -$$
$$- \left(\frac{1}{2} + 2\varepsilon\right)u_{0,-2} + \varepsilon\,(u_{1,-2} + u_{-1,-2}) + \frac{l}{K}\{(L[u])_{0,0} + \Bigg\} \quad (2.15)$$
$$+ 2(1 - \sigma + \varepsilon)\,(L[u])_{0,-1} + (\sigma - \varepsilon)\,((L[u])_{1,-1} + (L[u])_{-1,-1})\} +$$
$$+ \text{terms of the 4th order} = 0.$$

It is advisable to check the results of such a Taylor expansion, say by putting $u = x^2$, y^2, y^3, etc.

If the right-hand side of the differential equation (2.6) is a constant, say r, then we obtain a very simple formula by putting $\varepsilon = 0$ and $\sigma = \frac{1}{3}$:

$$u_{0,0} = \frac{2(u_{1,-1} + u_{-1,-1}) + u_{0,-2}}{5} - \frac{6}{5}\frac{l}{K} r + \left.\begin{array}{l} \\ + \text{ remainder term of the 4th order.} \end{array}\right\} \quad (2.16)$$

On account of the larger value of σ, the calculation proceeds "forwards" more rapidly with the formula

$$u_{0,0} = \frac{2(u_{1,-1} + u_{-1,-1}) + u_{1,-2} - u_{0,-2} + u_{-1,-2}}{5} - \frac{6l}{5K} r + \left.\begin{array}{l} \\ + \text{ remainder term of the 4th order,} \end{array}\right\} \quad (2.17)$$

which is obtained by putting $\sigma = \frac{1}{2}$, $\varepsilon = -\frac{1}{2}$. However, (2.16) has the index $J = 1$ while for (2.17) $J = 1\cdot4$, and hence the convergence of the approximate solution to the solution of the initial-value problem is assured by § 3.3 for (2.16) but not for (2.17)[1].

2.4. An example

Let us apply the Hermitian formula (2.16) to the problem

$$\frac{\partial^2 u}{\partial x^2} - K \frac{\partial u}{\partial y} + \beta = 0 \quad \text{for} \quad |x| \le 1, \quad y \ge 0, \quad (2.18)$$

$$u(x, 0) = u(-1, y) = u(1, y) = 0; \quad (2.19)$$

u can be interpreted physically as the temperature at time y of an element of a thin homogeneous rod at a distance x from the centre when heat is generated internally at a constant rate and is conducted away at the ends so as to keep the temperatures there constant at the initial uniform temperature 0 of the whole rod.

To start a calculation based on (2.16) we have to calculate the values $U_{i,1}$ on the first row by some other means. We observe that $u_{xx} = 0$ for the initial temperature distribution $u(x, 0)$; hence, from the differential equation, $u_y = \frac{\beta}{K}$ on $y = 0$, and therefore $u_{xxy} = \frac{d^2}{dx^2}\left(\frac{\beta}{K}\right) = 0$ on $y = 0$. Since the differential equation implies that $u_{xxy} = K u_{yy}$, it follows that $u_{yy} = 0$ on $y = 0$. This means that $U_{i,1} + U_{i,-1} - 2U_{i,0} = 0$ or, since $U_{i,0} = 0$, simply that $U_{i,-1} = -U_{i,1}$. Consequently formula (2.16)

$$U_{i,k+1} = \tfrac{1}{5}(2U_{i+1,k} + 2U_{i-1,k} + U_{i,k-1}) + \alpha, \quad (2.20)$$

where $\alpha = \frac{6}{5}\frac{l\beta}{K}$, becomes

$$U_{i,1} = \tfrac{1}{3}(U_{i+1,0} + U_{i-1,0}) + \tfrac{5}{6}\alpha \quad (2.21)$$

---

[1] Further formulae can be found in J. ALBRECHT: Zum Differenzenverfahren bei parabolischen Differentialgleichungen. Z. Angew. Math. Mech. **37**, 202−212 (1957).

when $k = 0$. This formula gives the required starting values and the calculation can then be continued with (2.20).

The results obtained with a mesh width $h = \frac{1}{4}$ $\left(l \text{ is then determined from } \sigma = \dfrac{l}{K h^2} = \dfrac{1}{3} \right)$ are shown in Table IV/13, which gives the values

Table IV/13. *Solution of the inhomogeneous heat equation by a Hermitian method*

k	$x=-1$	$x=-\frac{3}{4}$	$x=-\frac{1}{2}$	$x=-\frac{1}{4}$	$x=0$	S_k	T_k
0	0		0		0	0	
1		0·833 33		0·833 33		3·333 33	1·666 67
2	0		1·666 67		1·666 67	5	
3		1·833 33		2·5		8·666 67	3·666 67
4	0		3·066 67		3·333 33	9·466 67	
5		2·593 33		4·06		13·306 67	5·186 67
6	0		4·274 67		4·914 67	13·464 01	
7		3·228 53		5·487 74		17·432 54	6·457 06
8	0		5·341 44		6·373 13	17·056 01	
9		3·782 28		6·783 38		21·131 32	7·564 56
10	0		6·294 55		7·701 33	20·290 43	

Steady temperature distribution:

0		15		20		
	8·75		18·75			
0		15		20		

of $U_{i,k}$ (apart from a factor α) for $x \leqq 0$, the other half being symmetrical. The quantities formed in the row-sum check are

$S_\nu = $ sum of all $U_{i,\nu}$ in the ν-th row,

$T_\nu = $ sum of the two outermost U values in the ν-th row;

here the ν-th row means the whole ν-th row and not just the half row $(x \leqq 0)$ reproduced in Table IV/13. Then we should have

$$S_{2k+1} = \frac{1}{5}\left(4 S_{2k} + S_{2k-1}\right) + m\alpha,$$

$$S_{2k} = \frac{1}{5}\left(4 S_{2k-1} + S_{2k-2} - 2 T_{2k-1}\right) + (m-1)\alpha,$$

$$k = 0, 1, 2, \ldots, \quad m = \frac{1}{h} \text{ (here } m = 4\text{)}.$$

The steady temperature distribution (the limit as $y \to \infty$) is given at the bottom of the table; it can be readily determined from the condition $U_{i,k+2} \equiv U_{i,k}$. Since the exact steady temperature distribution is parabolic, the values obtained by the approximate method are exact.

§3. Some theoretical aspects of the finite-difference methods

3.1. Choice of mesh widths

The different mesh relations (1.12) and (1.33) for the parabolic and hyperbolic differential equations, respectively, are intimately connected with the different "domains of dependence" for these equations[1]. Consider, for example, the wave equation

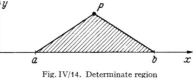

$$\frac{\partial^2 u}{\partial x^2} = \frac{1}{\omega^2}\frac{\partial^2 u}{\partial y^2} \qquad (3.1)$$

Fig. IV/14. Determinate region

with given initial values of u and $\partial u/\partial y$ on the x axis. Then the value $u(x, y)$ at a point P (Fig. IV/14) depends only on the initial data on that part of the x axis, $a \leq x \leq b$, which is cut off by the two "characteristics" through P; this is called the "domain of dependence" of the point P. Reciprocally, the segment $a \leq x \leq b$ of the x axis and the characteristics emanating from its endpoints define a region (shaded in the figures) in which u is determined completely by the initial values of u and $\partial u/\partial y$ on the segment; this may be called the "determinate region" of the initial segment.

If we apply the finite-difference method on a mesh with $h = \nu l$ and with the points $(a, 0)$, $(b, 0)$ among its mesh points, we find that the initial values in $a \leq x \leq b$ determine the $U_{i,k}$ at the mesh points in the triangle formed by the x axis and the straight lines $x \pm \nu y = $ constant

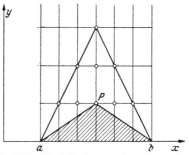

Fig. IV/15. The finite-difference determinate region

which pass through the points $(a, 0)$ and $(b, 0)$, respectively (these mesh points are circled in Fig. IV/15). Now if the number $1/\nu$ is chosen too large, greater than $1/\omega$ in fact, then this finite-difference "determinate region" extends beyond the actual determinate region of the initial segment $a \leq x \leq b$. This means that from initial values in $a \leq x \leq b$ the finite-difference method produces values for u at points where u is influenced by initial values outside of $a \leq x \leq b$. In this case, therefore, we cannot expect the finite-difference approximations to converge to the exact solution of the differential equation as the mesh is refined. However, convergence can be proved for $\nu = \omega$ and also an error estimate can be obtained (see § 3.2).

---

[1] See R. Courant and D. Hilbert: Methoden der mathematischen Physik, Vol. II, p. 307. Berlin 1937, or R. Courant and K. O. Friedrichs: Supersonic flow and shock waves, p. 51. New York: Interscience Publishers Inc. 1948.

Similarly, convergence to the correct solution cannot be expected when the finite-difference method is applied to the heat equation $\partial^2 u/\partial x^2 = \partial u/\partial y$ with a constant mesh ratio $h/l = \nu$. This can be seen by considering the exact solution

$$u = \frac{1}{\sqrt{y}}\, e^{\frac{-x^2}{4y}}, \tag{3.2}$$

which is always positive for $y > 0$ and vanishes on the x axis except at $x = 0$. Now if $0 < a < b$, the finite-difference method would always give zero values in the large triangle of Fig. IV/15; for constant ν this is a fixed triangle and consequently the results would necessarily converge to the wrong solution.

3.2. An error estimate for the inhomogeneous wave equation

We choose here an example admitting of an explicit error estimate from which the convergence[1] of the solution of the finite-difference equations to the solution of the differential equation can be inferred directly.

Consider the initial-value problem for the inhomogeneous wave equation

$$L[u] = \frac{\partial^2 u}{\partial x^2} - \frac{1}{\omega^2}\frac{\partial^2 u}{\partial y^2} = t(x, y) \tag{3.3}$$

in which the values

$$u(x, 0) = G(x) \quad \text{and} \quad \frac{\partial u(x, 0)}{\partial y} = H(x) \tag{3.4}$$

are given on the initial line $y = 0$. The treatment which we shall give here can still be applied when the boundary conditions are more general than this.

First we write down the finite-difference equation corresponding to (3.3) for the mesh (1.1) (with $x_0 = y_0 = 0$):

$$L_{\iota,\varkappa}[U] \equiv \frac{U_{\iota+1,\varkappa} - 2U_{\iota,\varkappa} + U_{\iota-1,\varkappa}}{h^2} - \frac{U_{\iota,\varkappa+1} - 2U_{\iota,\varkappa} + U_{\iota,\varkappa-1}}{\omega^2 l^2} = t_{\iota,\varkappa}. \tag{3.5}$$

---

[1] The proof of convergence may be carried over to more general problems; see R. Courant, K. Friedrichs and H. Lewy: Über die partiellen Differenzengleichungen der mathematischen Physik. Math. Ann. **100**, 32—74 (1928). — Friedrichs, K., and H. Lewy: Das Anfangswertproblem einer beliebigen nichtlinearen hyperbolischen Differentialgleichung beliebiger Ordnung in 2 unabhängigen Variablen; Existenz, Eindeutigkeit und Abhängigkeitsbereich der Lösung. Math. Ann. **99**, 200—221 (1928). — Courant, R., and P. Lax: On nonlinear partial differential equations with two independent variables. Comm. Pure Appl. Math. **2**, 255—273 (1949). — Courant, R., E. Isaacson and Mina Rees: On the solution of non-linear hyperbolic differential equations by finite differences. Comm. Pure Appl. Math. **5**, 243—255 (1952), in which the system

$$\sum_{i=1}^{n}\left(a_{ij}\frac{\partial u_i}{\partial x} + b_{ij}\frac{\partial u_i}{\partial y}\right) = c_j \qquad (j = 1, \ldots, n)$$

is considered, where a_{ij}, b_{ij}, c_j are functions of $x, y, u_1, \ldots, u_n$.

With $h = \omega l$ this simplifies to

$$- h^2 L_{i,\varkappa}[U] = (U_{i,\varkappa+1} - U_{i+1,\varkappa}) - (U_{i-1,\varkappa} - U_{i,\varkappa-1}) = - h^2 t_{i,\varkappa}. \qquad (3.6)$$

On the initial line $y = 0$ we have

$$U_{i,0} = u_{i,0} = u(\iota h, 0) = G(\iota h). \qquad (3.7)$$

The values of U on the line $y = h$ are determined by the other initial condition

$$\frac{U_{i,1} - U_{i,-1}}{2h} = H(\iota h) = \frac{\partial u(\iota h, 0)}{\partial y}: \qquad (3.8)$$

$U_{i,-1}$ can be eliminated between this equation and (3.6) with $\varkappa = 0$, i.e.

$$U_{i,1} - U_{i+1,0} - U_{i-1,0} + U_{i,-1} = - h^2 t_{i,0},$$

to obtain

$$U_{i,1} = \frac{u_{i-1,0} + u_{i+1,0}}{2} + h H(\iota h) - \frac{h^2}{2} t_{i,0}. \qquad (3.9)$$

The solution $U_{i,\varkappa}$ calculated from these starting values by repeated application of (3.6) can be given explicitly. From (3.6) we have

$$U_{i,\varkappa} - U_{i+1,\varkappa-1} = U_{i-1,\varkappa-1} - U_{i,\varkappa-2} - h^2 t_{i,\varkappa-1} = \cdots$$

$$= U_{i-\varkappa+1,1} - U_{i-\varkappa+2,0} - h^2 \sum_{\nu=0}^{\varkappa-2} t_{i-\nu,\varkappa-\nu-1},$$

and similarly

$$U_{i+1,\varkappa-1} - U_{i+2,\varkappa-2} = U_{i-\varkappa+3,1} - U_{i-\varkappa+4,0} - h^2 \sum_{\nu=0}^{\varkappa-3} t_{i-\nu+1,\varkappa-\nu-2}$$

$$\cdots \cdots \cdots \cdots \cdots \cdots \cdots \cdots \cdots$$

$$U_{i+\varkappa-2,2} - U_{i+\varkappa-1,1} = U_{i+\varkappa-3,1} - U_{i+\varkappa-2,0} - h^2 t_{i+\varkappa-2,1}.$$

Addition of these equations then yields

$$U_{i,\varkappa} = \underbrace{\sum_{\nu=1}^{\varkappa} U_{i-\varkappa+2\nu-1,1}}_{\text{sum (i)}} - \underbrace{\sum_{\nu=1}^{\varkappa-1} U_{i-\varkappa+2\nu,0}}_{\text{sum (ii)}} - \underbrace{h^2 \sum_{\mu=1}^{\varkappa-1} \sum_{\nu=1}^{\varkappa-\mu} t_{i-\nu+\mu,\varkappa-\nu-\mu+1}}_{\text{sum (iii)}}. \qquad (3.10)$$

The summations (i) and (ii) extend over the U values at the points circled in Fig. IV/16 and the summation (iii) extends over the values of t at the points marked with crosses.

We now derive an upper bound of the form constant $\times h^2$ for the absolute value of the error in the approximation U at an arbitrary but fixed point in the determinate region for the difference equations; this implies the convergence of the solution $U_{i,\varkappa}$ of the finite-difference equations to the solution (assumed to be differentiable sufficiently often) of the differential equation.

If M_ν is the maximum value of $|\partial^\nu u/\partial x^\nu|$ and $|\partial^\nu u/\partial y^\nu|$ (for $\nu = 3, 4$) in a convex region of the (x, y) plane which includes all mesh points involved in the calculation, then it follows from (3.5) (for remainder term see Table III of the appendix) that

$$L_{\iota, \varkappa}[u] = \frac{\partial^2 u}{\partial x^2} - \frac{1}{\omega^2} \frac{\partial^2 u}{\partial y^2} + \frac{2h^2}{12} \vartheta_{\iota, \varkappa} M_4 = t_{\iota, \varkappa} + \frac{h^2}{6} \vartheta_{\iota, \varkappa} M_4,$$

where $|\vartheta_{\iota, \varkappa}| \leqq 1$; hence the error $\varepsilon = U - u$ satisfies the difference equation

$$L_{\cdot, \varkappa}[\varepsilon] = -\frac{h^2}{6} \vartheta_{\iota, \varkappa} M_4 \quad \text{with} \quad |\vartheta_{\iota, \varkappa}| \leqq 1. \tag{3.11}$$

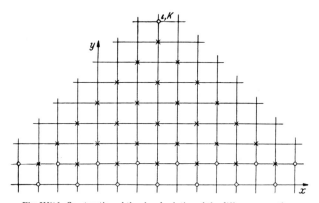

Fig. IV/16. Construction of the closed solution of the difference equations

Since on the "starting line" $y = h$ we have

$$\frac{u_{\iota, 1} - u_{\iota, -1}}{2h} = \frac{\partial u(\iota h, 0)}{\partial y} + \frac{h^2}{6} \vartheta_\iota M_3 \quad \text{with} \quad |\vartheta_\iota| \leqq 1,$$

it follows from (3.8) that

$$\frac{\varepsilon_{\iota, 1} - \varepsilon_{\iota, -1}}{2h} = -\frac{h^2}{6} \vartheta_\iota M_3,$$

and since $\varepsilon_{\iota, 0} = 0$, from (3.11) with $\varkappa = 0$ we also have

$$\varepsilon_{\iota, 1} + \varepsilon_{\iota, -1} = \frac{h^4}{6} \vartheta_{\iota, 0} M_4;$$

we can therefore specify an upper bound for the errors on the starting line $y = h$:

$$|\varepsilon_{\iota, 1}| \leqq \frac{1}{2} |\varepsilon_{\iota, 1} - \varepsilon_{\iota, -1}| + \frac{1}{2} |\varepsilon_{\iota, 1} + \varepsilon_{\iota, -1}| \leqq \frac{h^4}{12} M_4 + \frac{h^3}{6} M_3. \tag{3.12}$$

We now use formula (3.10) to solve (3.11):

$$\varepsilon_{\iota, \varkappa} = \underset{(i)}{\sum} \varepsilon_{\iota + 2\nu + 1, 1} - \underset{(ii)}{\sum} \varepsilon_{\iota + 2\nu + 2, 0} - \frac{h^4}{6} M_4 \underset{(iii)}{\sum \sum} \vartheta_{\iota \ldots, \varkappa \ldots}.$$

The terms in sum (ii) are all zero, the terms of sum (i), $\varkappa$ in number, are each bounded as in (3.12) and the $\frac{1}{2}\varkappa(\varkappa-1)$ terms of sum (iii) are each of absolute value 1 or less; therefore

$$|U_{\iota,\varkappa}-u_{\iota,\varkappa}|=|\varepsilon_{\iota,\varkappa}|\leq\varkappa\left(\frac{h^4}{12}M_4+\frac{h^3}{6}M_3\right)+\frac{h^4}{12}M_4(\varkappa^2-\varkappa).$$

Substituting for $\varkappa=y/h$ we have

$$|U_{\iota,\varkappa}-u_{\iota,\varkappa}|\leq\frac{h^2}{12}(2y\,M_3+y^2\,M_4);\tag{3.13}$$

thus the error at a fixed point (x,y) tends to zero quadratically with h as the mesh is refined.

3.3. The principle of the error estimate for more general problems with linear differential equations

Let the Taylor expansion of the expression (2.3) coincide with the given differential expression in (2.1) up to the terms of the r-th order $(r\geq m)$ inclusive. Under the assumption that in the region considered all the partial derivatives of u of the $(r+1)$-th order exist and are bounded, say by M_{r+1} in absolute value, the remaining terms in the Taylor expansion can be collected together into the single remainder term $\vartheta D l^{r-m+1}M_{r+1}$, where $|\vartheta|\leq 1$ and D depends on the mesh and the differential equation but not on the function u; in fact D is a polynomial in l of at most the m-th degree. The estimation of the quantity M_{r+1} often causes difficulty in practical applications; to get an idea of its order of magnitude one can sometimes calculate approximate values for the partial derivatives from the differences of the numerical solution.

Using this remainder term we can write

$$\sum_{\iota,\varkappa}C_{\iota,\varkappa}u_{\iota,\varkappa}=L[u_{\iota_0\varkappa_0}]+\vartheta D\,l^{r-m+1}M_{r+1}.\tag{3.14}$$

The error $\varepsilon_{\iota,\varkappa}=U_{\iota,\varkappa}-u_{\iota,\varkappa}$ therefore satisfies

$$\sum_{\iota,\varkappa}C_{\iota,\varkappa}\varepsilon_{\iota,\varkappa}=-\vartheta D\,l^{r-m+1}M_{r+1}.\tag{3.15}$$

If we know limits for the errors in the first k rows, this formula enables us to estimate the errors in the $(k+1)$-th row; in principle, therefore, a recursive error estimate is possible. To apply it, we need limits for the errors in the "starting" rows, i.e. the rows with $\varkappa=0$, $1,\ldots,s$ if the values of $\varkappa$ occurring in the finite equation (2.2) differ at most by s. If u is determined uniquely by prescribed initial data on the x axis, so that theoretically we can calculate the value of any partial derivative of u at $y = 0$, then the values of $U_{i,\varkappa}$ in the starting rows could be calculated from the Taylor series

$$U_{i,\varkappa} = u_{i,0} + \sum_{\nu=1}^{r} \frac{l^{\nu} \varkappa^{\nu}}{\nu!} \frac{\partial^{\nu} u_{i,0}}{\partial y^{\nu}} \qquad (\varkappa = 0, 1, 2, \ldots, s - 1)$$

(some other special starting procedure may be possible, of course); we could then use

$$|\varepsilon_{i,\varkappa}| \leq \frac{l^{r+1} \varkappa^{r+1}}{(r+1)!} M_{r+1} \qquad (\varkappa = 0, 1, 2, \ldots, s - 1) \tag{3.16}$$

as the required limits for the starting errors.

If the index J introduced in (2.5) is not greater than one, then in cases with $r > m$ an independent, as opposed to a recursive, error estimate can be given which implies the convergence of the finite-difference approximation to the solution of the initial-value problem. We imagine the differential equation (2.1) to have been multiplied through by such a factor that the coefficient $C_{i,k+1}$ in the finite equation (2.2) is unity; then the finite equation (2.4) reads

$$U_{i,k+1} = -\sum_{\varkappa \leq k} C_{i,\varkappa} U_{i,\varkappa} + r_{i_0, \varkappa_0}. \tag{3.17}$$

Under the assumption $J \leq 1$, the corresponding equation for the error

$$\varepsilon_{i,k+1} = -\sum_{\varkappa \leq k} C_{i,\varkappa} \varepsilon_{i,\varkappa} - \vartheta D l^{r-m+1} M_{r+1}$$

yields the estimate

$$|\varepsilon_{i,k+1}| \leq \max \left(|\varepsilon_k|, |\varepsilon_{k-1}|, \ldots, |\varepsilon_{k-s+1}| \right) + M, \tag{3.18}$$

where $|\varepsilon_{\varkappa}| = \max_{i} |\varepsilon_{i,\varkappa}|$ and

$$M = l^{r-m+1} D_{\max} M_{r+1}.$$

If boundary conditions are imposed which prescribe the values of u on, say, two straight lines which are used as mesh lines, then $\varepsilon = 0$ there. With C denoting the largest of all the limits in (3.16), it follows from (3.18) that for $\varkappa \geq s - 1$

$$|\varepsilon_{i,\varkappa}| \leq Y_{\varkappa} = C + (\varkappa - s + 1) M$$
$$= \left[\frac{[(s-1)l]^{r+1}}{(r+1)!} + (\varkappa - s + 1) D_{\max} l^{r-m+1} \right] M_{r+1}. \left.\right\} \tag{3.19}$$

If $r > m$, these error limits tend to zero with l at a fixed point (x, y) with $y = \varkappa l$; this is still true when $r = m$ provided that the polynomial D has the mesh width l as a factor [this is always the case, for example, when the given differential equation (2.1) has no undifferentiated term in u, i.e. $A_{0,0} = 0$].

The condition $J \leq 1$ and this last condition are satisfied, for example, by formula (2.10) for the inhomogeneous heat equation provided that σ is chosen in the range $0 \leq \sigma \leq \frac{1}{2}$; thus we have shown, in particular, that the finite-difference formula (1.13) yields convergent approximations[1] and that an error estimate can be given for them.

3.4. A more general investigation of error propagation and "stability"

Various courses have been pursued in the quest for systematic means of determining the "stability" characteristics of finite-difference equations, i.e. for criteria which will indicate whether or not the calculation can be rendered useless by unfavourable error propagation. Here we describe first a very simple method[2] by applying it to the initial-/boundary-value problem for the heat equation $u_{xx} = K u_t$ with prescribed values of $u(x, 0)$, $u(0, t)$ and $u(a, t)$.

Let the mesh widths h, l be given by $h = a/N$, $l = K \sigma h^2$, where N is an integer and σ is arbitrary for the present. We assume that the values $U_{j,0}$ on the first row have certain errors ε_j and ask whether the solution of the difference equations

$$\frac{U_{j+1,k} - 2U_{j,k} + U_{j-1,k}}{h^2} - K \frac{U_{j,k+1} - U_{j,k}}{l} = 0$$

remains bounded or not as $k \to \infty$.

Provided that α and β are related by

$$\frac{2(\cos \beta h - 1)}{h^2} - K \frac{e^{\alpha l} - 1}{l} = 0,$$

i.e.

$$e^{\alpha l} = 1 - 4\sigma \sin^2 \frac{\beta h}{2},$$

$u = e^{\alpha t} \sin \beta x$ is a solution of the difference equations.

---

[1] Convergence for the case with u prescribed on the whole x axis and with $\sigma = \frac{1}{2}$ is proved in the paper by R. COURANT, K. FRIEDRICHS and H. LEWY already referred to: Math. Ann. **100**, 32—74 (1928); for the region defined by the strip $0 \leq x \leq a$, $y \geq 0$ and under certain restrictions on the prescribed boundary values convergence is proved for $0 < \sigma \leq \frac{1}{4}$ by W. LEUTERT: J. Math. Phys. **30**, 245—251 (1952), and for $0 < \sigma < \frac{1}{2}$ by F. B. HILDEBRAND: On the convergence of numerical solutions of the heat-flow equation. J. Math. Phys. **31**, 35—41 (1952).

[2] BRIEN, G. O., M. HYMAN and S. KAPLAN: A study of the numerical solutions of partial differential equations. J. Math. Phys. **29**, 223—251 (1951). — HYMAN, M. A.: On the numerical solution of partial differential equations. Proefschrift Techn. Hogeschool, 106 pp. Delft 1953. Cf. also J. TODD: A direct approach to the problem of stability in the numerical solution of partial differential equations. Nat. Bur. Stand. Rep. No. 4260, 1955, 27 pp.

Now if we imagine the initial errors ε_j to be expressed as a finite trigonometrical series of the form

$$U_{j,\,0} = \varepsilon_j = \sum_{\nu=1}^{N-1} A_\nu \sin \frac{\nu j \pi}{N}$$

and replace α, β by α_ν, β_ν with $\beta_\nu = \dfrac{\nu \pi}{h\,N}$, then the required solution of the difference equations is

$$U_{j,\,k} = \sum_{\nu=1}^{N-1} A_\nu e^{\alpha_\nu k l} \sin \frac{\nu j \pi}{N}.$$

We see that the $U_{j,\,k}$ will remain bounded for an arbitrary initial perturbation, i.e. for arbitrary A_ν, if and only if $|e^{\alpha_\nu l}| \leqq 1$ for $\nu = 1, 2, \ldots, N-1$; this is equivalent to the condition

$$4\sigma \sin^2 \frac{\beta_\nu h}{2} \leqq 2,$$

and since $\sin^2 \frac{1}{2} \beta_\nu h \leqq 1$, we obtain as the condition for stability

$$\sigma \leqq \frac{1}{2}, \quad \text{i.e.} \quad l \leqq \frac{K}{2} h^2. \tag{3.20}$$

Equation (1.12) actually specifies the greatest possible value permitted by this condition, namely $l = \frac{1}{2} K h^2$; this value is also recommended in § 2.2 (1st-order formulae).

To avoid any possible misunderstanding, we once more state explicitly that we are concerned here only with the stability behaviour of the solution of the finite-difference equations and do not say anything about the deviation from the solution of the corresponding initial- or initial-/ boundary-value problem.

Similar procedures can be carried out in several other cases. The technique required for dealing with the effects of an isolated disturbance at one mesh point is considerably more complicated than for a disturbance of the form $\sin \beta x$; nevertheless, such isolated disturbances have been investigated for fairly general differential equations[1].

However, for cases in which $J \leqq 1$, J being the index defined in (2.5), *it can be seen immediately that the influence of an isolated disturbance*

---

[1] This theory goes back to JOHN V. NEUMANN; it has been discussed by R. P. EDDY: Stability in the numerical solution of initial value problems in partial differential equations. Naval Ordnance Labor. Memorandum **10**, 232 (1949). — Numerous other papers have been written on the question of stability; suffice it to mention just a few: TODD, J.: A direct Approach to the Problem of Stability in the Numerical Solution of Partial Differential Equations. Comm. Pure Appl. Math. **9**, 597—612 (1956). — LAX, P. D., and R. D. RICHTMYER: Survey of the Stability of Linear Finite Difference Equations. Comm. Pure Appl. Math. **9**, 267—293 (1956). — MLAK, W.: Remarks on the stability problem for parabolic equations. Ann. Polonici Math. **3**, 343—348 (1957).

remains bounded; thus the finite-difference method is always stable when the formula used has $J \leq 1$. On account of the linearity of the differential equation a disturbance $\varepsilon_{i,k}$ superimposed on the solution of the differential equation is propagated according to the homogeneous equation corresponding to (2.4), i.e.

$$\varepsilon_{i,k+1} = -\frac{1}{C_{i,k+1}} \sum_{\varkappa \leq k} C_{\iota,\varkappa} \varepsilon_{\iota,\varkappa}. \tag{3.21}$$

As in § 2.1, the sum extends over all $\iota, \varkappa$ for which $C_{\iota,\varkappa} \neq 0$ except $\varkappa = k+1$. From the definition of J (2.5) and the fact that it does not exceed 1 we see that

$$|\varepsilon_{i,k+1}| \leq \max_{\varkappa \leq k} |\varepsilon_{\iota,\varkappa}|.$$

Therefore, if we introduce an isolated disturbance $\varepsilon_{i_0,k_0} = 1$, so that $\varepsilon_{\iota,\varkappa} = 0$ for $\varkappa < k_0$ and also for $\varkappa = k_0$ except when $\iota = i_0$, we must have $|\varepsilon_{i,k+1}| \leq 1$ for all $k \geq k_0$.

The assumption $J \leq 1$ is satisfied, for example, by equation (2.10) for the heat equation provided that σ is chosen in the range $0 \leq \sigma \leq \frac{1}{2}$; this is the same condition as (3.20).

3.5. An example: The equation for the vibrations of a beam

For illustration we select an example from the theory of the flexural vibrations of thin beams which neglects such subsidiary effects as those due to the changing inclinations of the elements of the beam. In this theory the displacement of the beam satisfies the differential equation

$$\frac{\partial^4 u}{\partial x^4} + K \frac{\partial^2 u}{\partial y^2} = 0. \tag{3.22}$$

Some typical initial and boundary conditions which arise are:
initial conditions (given initial displacement and velocity distributions):

$$u(x, 0) = f_1(x), \qquad \frac{\partial u(x, 0)}{\partial y} = f_2(x) \quad \text{for} \quad 0 \leq x \leq a;$$

boundary conditions (smoothly pinned ends):

$$u(0, y) = \frac{\partial^2 u(0, y)}{\partial x^2} = u(a, y) = \frac{\partial^2 u(a, y)}{\partial x^2} = 0 \quad \text{for} \quad y \geq 0. \tag{3.23}$$

For a rectangular mesh (1.1) with mesh widths $h = a/N$ and l (where N is integral and greater than unity) the ordinary finite-difference method replaces the differential equation (3.22) by the difference equation

$$\left.\begin{aligned}
&\frac{U_{i+2,k} - 4U_{i+1,k} + 6U_{i,k} - 4U_{i-1,k} + U_{i-2,k}}{h^4} + \\
&\qquad + \frac{U_{i,k+1} - 2U_{i,k} + U_{i,k-1}}{l^2} K = 0.
\end{aligned}\right\} \tag{3.24}$$

As we did for the heat equation in (1.12), let us choose a convenient mesh relation; with

$$K\,h^4 = l^2$$

the difference equation simplifies to

$$U_{i,\,k+1} = -\,U_{i+2,\,k} + 4\,U_{i+1,\,k} - 4\,U_{i,\,k} + 4\,U_{i-1,\,k} - U_{i-2,\,k} - U_{i,\,k-1}. \quad (3.25)$$

Once we have found values on two consecutive rows, this formula will give the values on the next row, and it would appear that we could therefore continue the solution as far as required; however, a glance at the propagation table of formula (3.25) shows immediately that it is quite unsuitable for repeated application — just the few rows given in Table IV/14 proclaim its severe instability.

Table IV/14. *The propagation table of formula* (3.25)

0	ε	0	0	0
$4\,\varepsilon$	$-4\,\varepsilon$	$4\,\varepsilon$	$-\,\varepsilon$	0
$-40\,\varepsilon$	$49\,\varepsilon$	$-40\,\varepsilon$	$24\,\varepsilon$	$-8\,\varepsilon$
$496\,\varepsilon$	$-560\,\varepsilon$	$496\,\varepsilon$	$-337\,\varepsilon$	$172\,\varepsilon$
$-6200\,\varepsilon$	$6833\,\varepsilon$	$-6200\,\varepsilon$		

Our choice of mesh relation implies too large a value for the mesh width l for given h. To investigate a better choice of mesh relation, let us introduce the parameter[1]

$$z = \frac{l^2}{K\,h^4}. \quad (3.26)$$

The formula (3.25) just considered arises as the special case $z = 1$.

Suppose that at some stage of the calculation the $U_{j,k}$ values on the last two rows, say $k = p$ and $k = p+1$, have deviated from their true values by errors $\varepsilon_{j,\,p}$ and $\varepsilon_{j,\,p+1}$, respectively. Then we ask again whether the errors $\varepsilon_{j,\,k}$ produced in subsequent rows by using (3.24) remain bounded or not as $k \to \infty$.

It can easily be shown that equation (3.24) possesses the particular solution

$$U_{j,\,k} = e^{(k-p)\,\eta}\sin j\,\xi,$$

if ξ and η are related by

$$2z\,(\cos\xi - 1)^2 = 1 - \cosh\eta; \quad (3.27)$$

therefore with $\xi = \xi_\nu = \dfrac{\nu\,\pi}{N}$ ($\nu = 1, \ldots, N-1$) and corresponding values $\eta = \pm\,\eta_\nu$ from (3.27)

$$v_{j,\,k} = \sum_{\nu=1}^{N-1} \sin\frac{j\,\nu\,\pi}{N}\,\left(A_\nu\,e^{(k-p)\,\eta_\nu} + B_\nu\,e^{-(k-p)\,\eta_\nu}\right)$$

---

[1] COLLATZ, L.: Z. Angew. Math. Mech. **31**, 392–393 (1951).

is also a solution of (3.24). Moreover, it satisfies the finite-difference boundary conditions corresponding to (3.23) and by a suitable choice of A_ν and B_ν can be made to assume the values $\varepsilon_{j,p}$ and $\varepsilon_{j,p+1}$ for $k=p$ and $k=p+1$, respectively: if we obtain quantities α_ν, β_ν from

$$\varepsilon_{j,p} = \sum_{\nu=1}^{N-1} \alpha_\nu \sin \frac{j\nu\pi}{N}, \quad \varepsilon_{j,p+1} = \sum_{\nu=1}^{N-1} \beta_\nu \sin \frac{j\nu\pi}{N}$$

by the formulae of harmonic analysis, then, since $\eta_\nu \neq 0$ ($\cos \xi \neq 1$ and $z \neq 0$ imply that $\cosh \eta \neq 1$), the A_ν and B_ν are uniquely determined by the equations $A_\nu + B_\nu = \alpha_\nu$, $A_\nu e^{\eta\nu} + B_\nu e^{-\eta\nu} = \beta_\nu$.

With these values of A_ν and B_ν we have $v_{j,k} = \varepsilon_{j,k}$, and we see that the $\varepsilon_{j,k}$ will remain bounded for an arbitrary error distribution on rows p and $p+1$, i.e. for arbitrary A_ν, B_ν, if and only if

$$|e^{\eta\nu}| = 1 \quad \text{for} \quad \nu = 1, \ldots, N-1. \tag{3.28}$$

Now let $s = 1 - \cosh \eta$ and $|\operatorname{im} \eta| \leq \pi$. When s is real and positive [as it is here by virtue of (3.27) with $\xi = \nu\pi/N$], the corresponding values of η lie on the imaginary axis for $0 < s \leq 2$ and on the lines $\operatorname{im} \eta = \pm \pi$ for $s > 2$; hence $|e^\eta| = 1$ for $0 < s \leq 2$ and $|e^\eta| > 1$ for $s > 2$. From (3.27), (3.28) our stability condition therefore reads

$$2z(\cos \xi - 1)^2 \leq 2 \quad \text{or} \quad z \leq \frac{1}{\left(\cos \dfrac{\nu\pi}{N} - 1\right)^2} \quad \text{for} \quad \nu = 1, \ldots, N-1.$$

This upper bound for z depends on N, though not strongly; a simple sufficient condition for stability which is valid for all N can be obtained by taking the limit as $N \to \infty$, i.e. by replacing the cosine in the denominator by -1:

$$z \leq \tfrac{1}{4}.$$

With $z = \tfrac{1}{4}$ (3.24) becomes

$$U_{j,k+1} = -U_{j,k-1} + \tfrac{1}{4}(-U_{j+2,k} + 4U_{j+1,k} + \\ + 2U_{j,k} + 4U_{j-1,k} - U_{j-2,k}). \left.\right\} \tag{3.29}$$

The index of this formula is $J = 4$ while that of (3.25) is 15.

§4. Partial differential equations of the first order in one dependent variable

A complete theory for the integration of partial differential equations of the first order in one dependent variable $u(x_1, x_2, \ldots, x_n)$ and n independent variables $x_1, x_2, \ldots, x_n$ has existed for many years. In this theory the integration of the partial differential equation is reduced to the integration of a system of ordinary differential equations, and hence a numerical treatment can be based on the methods described in Chapter II. Our presentation of the theory[1] here will be confined to a summary (in § 4.1) of the results needed for a numerical calculation.

In spite of the completeness of the theory, the numerical integration of the systems of ordinary differential equations which arise is very laborious and there is a need for other approximate methods which will give a quantitative idea of the solution with a moderate amount of computation. We therefore go into other possible methods of treatment, albeit briefly, in §§ 4.3 to 4.5.

Unless otherwise stated, all functions and derivatives which occur are assumed to be continuous.

4.1. Results of the theory in the general case

Consider the first-order partial differential equation

$$F(u, x_1, x_2, \ldots, x_n, p_1, p_2, \ldots, p_n) = 0 \qquad (4.1)$$

for the function $u(x_1, x_2, \ldots, x_n)$ of the independent variables $x_1, x_2, \ldots, x_n$. The p_j denote[2] the partial derivatives of u

$$p_j = \frac{\partial u}{\partial x_j} \qquad (j = 1, 2, \ldots, n), \qquad (4.2)$$

and F is a given function which will be assumed continuous in the arguments specified.

This differential equation is called linear (see Ch. I, § 1.3) when F is linear in u and the p_j, i.e. when it has the form

$$A u + \sum_{j=1}^{n} A_j p_j = B, \qquad (4.3)$$

where A, B and the A_j are given functions of $x_1, x_2, \ldots, x_n$.

It is called quasi-linear (Ch. I, § 1.3) when F is linear only in the p_j, i.e. when it has the form

$$\sum_{j=1}^{n} A_j p_j = B, \qquad (4.4)$$

where now the A_j and B are given functions of $x_1, \ldots, x_n$ and u.

---

[1] Presentations of the theory can be found, for example, in R. Courant and D. Hilbert: Methoden der mathematischen Physik, Vol. II, in particular, pp. 51 to 122. Berlin 1937. — Kamke, E.: Differentialgleichungen reeller Funktionen, 4. Abschnitt, 2nd ed. Leipzig 1944. — Sauer, R.: Anfangswertprobleme bei partiellen Differentialgleichungen, Ch. 2. Berlin-Göttingen-Heidelberg 1952. — Duff, G.: Partial differential equations, Ch.s II, III. Toronto 1956.

[2] In Ch. I, § 1.3 the partial derivatives were denoted by u_j instead of p_j; by using the letter p here we conform to the notation customary in the literature.

Let initial values be prescribed for the unique determination of the function u; in fact let an $(n-1)$-dimensional manifold be defined by prescribing u and the x_j as functions of parameters $t_1, \ldots, t_{n-1}$:

$$x_j = x_j(t_1, \ldots, t_{n-1}) \quad (j = 1, \ldots, n), \qquad u = u(t_1, \ldots, t_{n-1}).$$

This manifold may be extended to a "strip manifold" C by including n functions $p_j(t_1, \ldots, t_{n-1})$ which satisfy the following conditions: the strip conditions $(n-1$ in number)

$$\frac{\partial u}{\partial t_j} = \sum_{k=1}^{n} p_k \frac{\partial x_k}{\partial t_j} \quad (j = 1, \ldots, n-1) \tag{4.5}$$

and the "compatibility" condition

$$F(u, x_1, \ldots, x_n, p_1, \ldots, p_n) = 0. \tag{4.6}$$

To solve the initial-value problem a function $u(x_1, \ldots, x_n)$ must be determined which satisfies the differential equation (4.1) and defines a surface containing the manifold C.

An expression which is critical for the solubility of this initial-value problem is the determinant

$$\varDelta = \begin{vmatrix} F_{p_1} & F_{p_2} & \cdots & F_{p_n} \\ \dfrac{\partial x_1}{\partial t_1} & \dfrac{\partial x_2}{\partial t_1} & \cdots & \dfrac{\partial x_n}{\partial t_1} \\ \cdots & \cdots & \cdots & \cdots \\ \dfrac{\partial x_1}{\partial t_{n-1}} & \dfrac{\partial x_2}{\partial t_{n-1}} & \cdots & \dfrac{\partial x_n}{\partial t_{n-1}} \end{vmatrix}. \tag{4.7}$$

We shall assume that this determinant $\varDelta$ is non-zero at all points of the initial manifold C; then the existence and uniqueness of the solution u is assured (always assuming the continuity of all functions which appear).

The solution u can be built up from the "characteristic strips". These are one-dimensional strip manifolds defined by $(2n+1)$ functions

$$u(s), \quad x_j(s), \quad p_j(s), \tag{4.8}$$

of a parameter s which satisfy the system of "characteristic equations"

$$\frac{du}{ds} = \sum_{k=1}^{n} p_k \frac{\partial F}{\partial p_k}, \quad \frac{dx_j}{ds} = \frac{\partial F}{\partial p_j}, \tag{4.9}$$

$$\frac{dp_j}{ds} = -\frac{\partial F}{\partial x_j} - p_j \frac{\partial F}{\partial u} \qquad (j = 1, \ldots, n). \tag{4.10}$$

To each set of values $t_1, \ldots, t_{n-1}$, i.e. to each point of the initial manifold C, there corresponds one such characteristic strip. The system of $(2n+1)$ ordinary differential equations (4.9), (4.10) for the $(2n+1)$ functions u, x_j, p_j, together with, say, the initial conditions that for $s=0$ the u,

x_j, p_j take the values corresponding to a point of the initial manifold C, determine the functions (4.8) and can be integrated numerically by the methods of Chapter II.

Quasi-linearity of the differential equation (and hence linearity in particular) introduces a considerable simplification: with the equation in the form (4.4) the first $(n+1)$ equations of the characteristic system (4.9), (4.10) become

$$\frac{du}{ds} = B, \qquad \frac{dx_j}{ds} = A_j \qquad (j = 1, 2, \ldots, n), \tag{4.11}$$

in which the p_j no longer appear; thus we need no longer concern ourselves with the p_j and can ignore the equations (4.5), (4.6) and (4.10). Instead of the characteristic strips we have "characteristic curves" $u(s)$, $x_j(s)$, which are solutions of (4.11) and are ordinary space curves in the $(n+1)$-dimensional (u, x_j) space. The determinant (4.7) becomes

$$\Delta = \begin{vmatrix} A_1 & \cdots & A_n \\ \dfrac{\partial x_1}{\partial t_1} & \cdots & \dfrac{\partial x_n}{\partial t_1} \\ \cdots & \cdots & \cdots \\ \dfrac{\partial x_1}{\partial t_{n-1}} & \cdots & \dfrac{\partial x_n}{\partial t_{n-1}} \end{vmatrix} \tag{4.12}$$

and provided that it does not vanish, the solution $u(x_1, \ldots, x_n)$ consists simply of the totality of characteristic curves which pass through the points of the initial manifold.

4.2. An example from the theory of glacier motion

A non-linear differential equation in two independent variables which possesses the simplifying property of quasi-linearity occurs in the theory of glacier motion[1]. It furnishes a mathematical description of the conditions in a glacier, particularly in the glacier tongue, or ablator, and appears in the form

$$[(n + 1)\varkappa u^n - a]\frac{\partial u}{\partial x} + \frac{\partial u}{\partial t} = -a. \tag{4.13}$$

It pertains to a central longitudinal section of a glacier moving down a slightly inclined, straight bed.

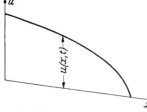

Fig. IV/17. Longitudinal section of a glacier

The variables u and x refer to oblique axes in the plane of the section: $u(x, t)$ is the vertical depth of the ice at time t at a distance x along the bed (Fig. IV/17). The velocity distribution is assumed to be of the form $v = \varkappa u^n$, where $\varkappa$ depends on the slope of the bed and the exponent n is a constant lying between $\frac{1}{4}$ and $\frac{1}{2}$. The remaining symbol a

---

[1] See S. Finsterwalder: Die Theorie der Gletscherschwankungen. Z. Gletscherk. **2**, 81—103 (1907). This gives a detailed treatment of various cases of stationary glaciers, propagation of ridges, glacier shrinkage, the relative slipping of complete ice-blocks, etc.

is an ablation constant; it represents the annual melting on horizontal surfaces. The particular example which we consider here concerns a receding glacier on a very flat bed.

Let the shape of the longitudinal section of the glacier at time $t=0$ be given[1] in dimensionless variables by

$$u = 2\frac{4-x}{5-x} \quad \text{for} \quad 0 \leq x \leq 4 \tag{4.14}$$

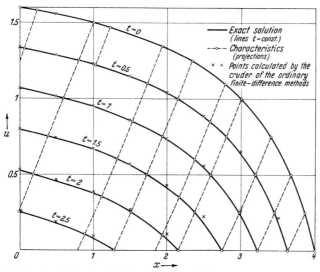

Fig. IV/18. Profile of a glacier at various times as an example of the integration of a first-order partial differential equation

and for numerical values of the parameters take $n = \frac{1}{3}$, $\varkappa = 0.075$, $a = \frac{1}{2}$. Then (4.13) becomes

$$\left[0.1\sqrt[3]{u} - 0.5\right]\frac{\partial u}{\partial x} + \frac{\partial u}{\partial t} = -0.5. \tag{4.15}$$

The characteristic equations (4.11) for equation (4.13) read

$$\frac{du}{ds} = -a, \quad \frac{dx}{ds} = (n+1)\varkappa u^n - a, \quad \frac{dt}{ds} = 1. \tag{4.16}$$

Their general solution is a two-parameter family of curves in the (u, x, t) space; in this case the curves are plane and are given by

$$\left.\begin{array}{l} x + \xi = -\dfrac{\varkappa}{a}u^{n+1} + u = u - 0.15\,u^{\frac{4}{3}}, \\[2mm] u + \eta = -at \qquad = -0.5\,t, \end{array}\right\} \tag{4.17}$$

---

[1] With a more sloping bed (greater $\varkappa$) and sufficiently large values of u the projections of the characteristics on the (x, t) plane slope away from the t axis instead of towards it $(dx/dt < 0)$; then for the complete determination of $u(x, t)$ in the quadrant $x > 0$, $t > 0$, $u(0, t)$ must be prescribed in addition to $u(x, 0)$.

where ξ and η are the parameters. From this two-parameter family we select the one-parameter family of curves which pass through the points of the initial curve (4.14); the surface formed by the totality of these curves constitutes the solution of the problem. We select a set of points on the initial curve and calculate the corresponding characteristic parameters; for example, the point $t=0$, $x=3$, $u=1$ yields $\xi=-2.15$, $\eta=1$. Then the projections of these characteristics may be drawn in a (u, x) plane and graduated at (say equal) intervals in t; curves joining the points with the same values of t give the shape of the longitudinal section at various values of t (Fig. IV/18).

4.3. Power series expansions

If, as will always be assumed, the determinant (4.7) is non-zero over the given initial manifold, it is possible, in principle, to calculate the initial values of the higher partial derivatives of u. The solution can therefore be approximated by a number of terms of its Taylor series and this will give some idea of its behaviour in a neighbourhood of the initial manifold. Naturally nothing can be asserted generally about the extent of the region of convergence. Such a series expansion should always be used circumspectly, for the series can converge in certain regions to a limit function which is not the solution of the initial-value problem; examples which demonstrate this can be constructed quite easily (the example in the introduction to this chapter demonstrates the same phenomenon for a second-order equation).

The higher derivatives are calculated by repeated differentiation of the differential equation and of the initial conditions with respect to the x_j. By differentiating the differential equation and initial conditions in the form (4.5), (4.6) with respect to x_1, for instance, we obtain a system of linear equations for the n quantities $u_{11}, u_{21}, \ldots, u_{n1}$ (with the derivative notation of Ch. I, § 1.3); now the determinant of the coefficients obtained is precisely the determinant (4.7), which is assumed to be non-zero, and we can therefore solve for the derivatives u_{11}, $u_{21}, \ldots, u_{n1}$.

As an illustration consider the example of the last section § 4.2. Let us write

$$f = a - (n+1) \varkappa u^n, \quad u(x,0) = \varphi(x), \quad n(n+1)\varkappa\varphi^{n-1} = \Phi; \quad (4.18)$$

then the differential equation (4.13) reads

$$u_t = -a + f u_x$$

and the initial condition is

$$u(x,0) = \varphi(x) = 2 \cdot \frac{4-x}{5-x}.$$

Differentiation of the differential equation with respect to x and t yields the two equations

$$u_{xt} = -\Phi u_x^2 + f u_{xx},$$
$$u_{tt} = -\Phi u_x u_t + f u_{xt}$$

for the required values of the second derivatives on the initial manifold $t=0$. Two differentiations of the initial condition yield

$$u_x = \varphi' = -\frac{2}{(5-x)^2}, \qquad u_{xx} = \varphi'' = -\frac{4}{(5-x)^3}.$$

Thus we have three linear equations for u_{xx}, u_{xt}, u_{tt}, which in this case can be solved directly by successive substitutions. For u_{tt}, for example, we obtain

$$u_{tt} = f^2 \varphi'' - 2 f \Phi \varphi'^2 + a \Phi \varphi' = -\frac{1}{(5-x)^4} [8 f \Phi + 4(5-x) f^2 + (5-x)^2 \Phi]. \quad (4.19)$$

4.4. Application of the finite-difference method

Of the possible ways of making a rapid quantitative survey of the solution without a laborious integration of the characteristic equations, the finite-difference approach is probably the most suitable. The mesh width in the progressive direction must not be taken too large, of course, for, as in § 3.1, if the method is to make sense, the finite-difference determinate region of any part of the initial manifold must be completely contained within the corresponding determinate region outlined by the characteristics. Consequently the finite-difference method should not be used without first examining the run of the characteristics at least roughly (cf. the example).

There are usually several ways in which the differential equation can be replaced by a difference equation; we illustrate two ways — one more accurate than the other — by applying them to the example of glacier shrinking of § 4.2, i.e. the initial-value problem (4.15), (4.14). We employ the usual rectangular mesh

$$x_i = i h, \qquad y_k \text{ (here } t_k) = k l,$$

and in order to choose l suitably we first estimate the steepness of the characteristics. From (4.16) and using the fact that $u \geq 0$ (the depth of the glacier cannot be negative) and also that $0.5 - 0.1 \sqrt[3]{u} > 0$, we have

$$\left|\frac{dt}{dx}\right| = \frac{1}{|(n+1)\varkappa u^n - a|} = \frac{1}{|0.5 - 0.1 \sqrt[3]{u}|} \geq \frac{1}{0.5} = 2;$$

thus we can safely choose $l \leq 2h$ and for our example we take $l = h = \frac{1}{2}$.

1. The cruder method. Here the derivative with respect to x is replaced by the central difference quotient (1.4) but the derivative with respect to t is approximated crudely by the forward difference quotient

Table IV/15. Calculation of glacier recession by the cruder finite-difference method

t		$i=0$ $x=0$	$i=1$ $x=0.5$	$i=2$ $x=1$	$i=3$ $x=1.5$	$i=4$ $x=2$	$i=5$ $x=2.5$	$i=6$ $x=3$	$i=7$ $x=3.5$	$i=8$ $x=4$
0	$U_{i,0}$	1.6	1.55556	1.5	1.42857	1.33333	1.2	1	0.66667	0
	$\sqrt[3]{U_{i,0}}$	1.1696	1.1587	1.1447	1.1262	1.1006	1.0627	1	0.87358	0
	$\tfrac{1}{2}f(U_{i,0})$	0.19152	0.19207	0.19277	0.19369	0.19497	0.19687	0.20000	0.20632	0.25
	$U_{i+1,0}-U_{i-1,0}$	-0.07778	-0.10000	-0.12698	-0.16667	-0.22857	-0.33333	-0.53333	-1	-1.66667
0.5	$U_{i,1}$	1.33510	1.28635	1.22552	1.14629	1.03877	0.88438	0.64333	0.21035	-0.66667
1	$U_{i,2}$	1.06845	1.01491	0.94800	0.85938	0.73655	0.55450	0.25391	-0.32820	
1.5	$U_{i,3}$	0.79981	0.74085	0.66676	0.56657	0.42410	0.22119	-0.18882		
2	$U_{i,4}$	0.52888	0.46361	0.38080	0.26594	0.10073	-0.16351			
2.5	$U_{i,5}$	0.25520	0.18232	0.08855	-0.04507	-0.24664				
3	$U_{i,6}$	-0.02206	-0.10462	-0.21323						

(1.2). With f defined as in (4.18) we have

$$\frac{U_{i,k+1}-U_{i,k}}{l} = -a + f\frac{U_{i+1,k}-U_{i-1,k}}{2h}, \quad \Bigg\} (4.20)$$

which becomes

$$U_{i,k+1}=U_{i,k}-0.25 + \tfrac{1}{2}f(U_{i,k})[U_{i+1,k}-U_{i-1,k}]$$

when the numerical values we have chosen are inserted.

This formula enables us to calculate immediately the values $U_{i,k}$ above the heavy line in Table IV/15 [for the first row $(k=1)$ the intermediate calculation has been reproduced, but for the subsequent rows only the results are given]. By our choice of mesh ratio we have ensured that this triangle of points lies within the determinate region of the initial segment $0\leq x\leq 4$, $t=0$. Actually we want to cover that part of the quadrant $x\geq 0$, $t\geq 0$ in which $u\geq 0$, so we want to extend the calculation beyond the triangle as far as the boundary $u=0$. This can be done by using unsymmetric difference quotients at the boundaries; for example, the formula

$$U_{0,k+1} = U_{0,k}-0.25+f(U_{0,k})\times \frac{-3U_{0,k}+4U_{1,k}-U_{2,k}}{2} \quad \Bigg\} (4.21)$$

Table IV/16. Calculation of glacier recession by a more accurate finite-difference method

t		$i=0$ $x=0$	$i=1$ $x=0{\cdot}5$	$i=2$ $x=1$	$i=3$ $x=1{\cdot}5$	$i=4$ $x=2$	$i=5$ $x=2{\cdot}5$	$i=6$ $x=3$	$i=7$ $x=3{\cdot}5$	$i=8$ $x=4$
0	$U_{i,0}$	1·6	1·55555	1·5	1·42857	1·33333	1·2	1	0·66667	0
	$\tfrac{1}{2}u_t$ $\tfrac{1}{8}u_{tt}$	−0·26532 −0·00072	−0·26897 −0·00098	−0·27410 −0·00140	−0·28163 −0·00209	−0·29333 −0·00333	−0·31300 −0·00585	−0·35 −0·01188	−0·43340 −0·03121	
0·5	$U_{i,1}$	1·33396	1·28560	1·22451	1·14485	1·03667	0·88115	0·63812	0·20206	−0·66667
	$\sqrt[3]{U_{i,1}}$ $f(U_{i,1})$ $U_{i+1,1}-U_{i-1,1}$	1·1008 0·38992 −0·08399	1·0873 0·39127 −0·10945	1·0698 0·39302 −0·14075	1·0461 0·39539 −0·18784	1·0121 0·39879 −0·26370	0·9587 0·40413 −0·39855	0·8609 0·41391 −0·67909	0·5868 0·44132 −1·30479	
1	$U_{i,2}$	1·06725	1·01273	0·94468	0·85429	0·72817	0·53893	0·21892	−0·40916	
1·5	$U_{i,3}$	0·79597	0·73662	0·66084	0·55714	0·40736	0·16797	−0·27878		
2	$U_{i,4}$	0·52559	0·45737	0·37057	0·24841	0·06243	−0·26628			
2·5	$U_{i,5}$	0·24655	0·17105	0·07137	−0·07756	−0·32957				
3	$U_{i,6}$	−0·02987	−0·12050	−0·24342						

Table IV/17. *Error (in 5th-decimal units) in results obtained for the glacier problem by the various approximate methods*

Method	t	$x=0$	$x=0.5$	$x=1$	$x=1.5$	$x=2$	$x=2.5$	$x=3$	$x=3.5$
Ordinary finite-difference method — 1. Crude approximation using (4.20), (4.21)	0·5	+ 118	81	112	162	245	394	719	1810
	1	235	177	263	388	606	1044	2171	32934
	1·5	358	266	482	724	1187	4014	14679	
	2	512	486	804	1259	2651	10904		
	2·5	724	773	1342	2728	8466			
	3	1152	1625	3316					
Ordinary finite-difference method — 2. Better approximation using (4.22), (4.23)	0·5	4	6	11	18	35	71	− 198	981
	1	− 115	− 41	− 69	− 121	− 232	− 513	− 1328	24838
	1·5	− 26	− 157	− 110	− 219	− 487	− 1308	+ 5683	
	2	− 183	− 138	− 219	− 494	− 1179	+ 627		
	2·5	− 141	+ 354	− 376	− 521	+ 173			
	3	− 371	37	− 297					
$u^{[0]}$ obtained from a constant-coefficient approximation to the given differential equation	0·5	− 497	− 613	− 771	− 1005	− 1359	− 1933	− 2900	− 4225
	1	− 1055	− 1314	− 1680	− 2217	− 3049	− 4406	− 6553	+ 15734
	1·5	− 1682	− 2152	− 2732	− 3660	− 5112	− 7391	− 1439	
	2	− 2376	− 3018	− 3943	− 5335	− 7422	− 6078		
	2·5	− 3129	− 3997	− 5240	− 6654	− 6156			
	3	− 3785	− 4580	− 5361					
$u^{[1]}$ obtained from $u^{[0]}$ by one cycle of the iteration procedure	0·5	− 19	− 26	− 34	− 50	− 75	− 124	− 174	+ 562
	1	− 82	− 114	− 164	− 246	− 392	− 674	− 1228	+ 18863
	1·5	− 201	− 318	− 405	− 617	− 912	− 1496	+ 2410	
	2	− 391	− 556	− 820	− 1284	− 2418	− 654		
	2·5	− 624	− 903	− 1453	− 2717	− 2353			
	3	− 1178	− 1469	− 1840					

Values of the exact solution $u(x,t)$

Method	t	$x=0$	$x=0.5$	$x=1$	$x=1.5$	$x=2$	$x=2.5$	$x=3$	$x=3.5$
Values of the exact solution $u(x,t)$	0·5	1·33392	1·28554	1·22440	1·14467	1·03632	0·88044	0·63614	0·19225
	1	1·06610	1·01314	0·94537	0·85550	0·73049	0·54406	0·23220	−0·65754
	1·5	0·79623	0·73819	0·66194	0·55933	0·41223	0·18105	−0·33561	
	2	0·52376	0·45875	0·37276	0·25335	0·07422	−0·27255		
	2·5	0·24796	0·17459	0·07513	−0·07235	−0·33130			
	3	−0·03358	−0·12087	−0·24639					

can be used at the boundary $i=0$ (there is a similar formula for the other boundary). This is permissible because in the region of interest ($u \geq 0$) the characteristics have $dx/dt < 0$ and the region therefore lies within the determinate region of the initial segment $0 \leq x \leq 4$, $t=0$. Thus we may proceed until the $U_{i,k}$ become negative.

2. The more accurate method. Here we replace the derivatives with respect to x and t both by central difference quotients:

$$U_{i,k+1} = U_{i,k-1} - 0 \cdot 5 + f(U_{i,k})[U_{i+1,k} - U_{i-1,k}]. \qquad (4.22)$$

Starting values are now required on the first *two* rows ($k=0$, $k=1$). The values on the row $k=1$ can be found from

$$U_{i,1} - U_{i,-1} = 2l\,u_t, \qquad U_{i,1} - 2U_{i,0} + U_{i,-1} = l^2 u_{tt},$$

which yield

$$U_{i,1} = U_{i,0} + l\,u_i + \tfrac{1}{2} l^2 u_{tt} \qquad (4.23)$$

(the beginning of the Taylor series). The derivative u_{tt} can be calculated in the same way as for the power series expansion of § 4.3; in fact the required formula has already been derived as (4.19). Since Φ becomes infinite when $x=4$, this formula cannot be used for the value at $x=4$, $t=1$; thus $U_{8,1}$ must be calculated by some other means, and in Table IV/16 the value found by the cruder method is taken over. Here also an unsymmetric formula [corresponding to (4.21)] must be used at the boundary:

$$U_{0,k+1} = U_{0,k-1} - 0 \cdot 5 + f(U_{0,k})[-3U_{0,k} + 4U_{1,k} - U_{2,k}].$$

The errors in the results obtained by these two finite-difference methods are given in Table IV/17 (the exact values were calculated from the characteristics by interpolation). It can be seen that the errors are smaller for the second method than for the first, as was to be expected.

4.5. Iterative methods

An iterative process can be defined by solving the given differential equation (4.1) for one of the partial derivatives, say p_1:

$$p_1 = \frac{\partial u}{\partial x_1} = G(u, x_1, \ldots, x_n, p_2, \ldots, p_n),$$

and replacing u by $u^{[k+1]}$ on the left and by $u^{[k]}$ on the right. Thus the next approximation $u^{[k+1]}$ is determined from the current approximation $u^{[k]}$ by solving the differential equation

$$\frac{\partial u^{[k+1]}}{\partial x_1} = G\left(u^{[k]}, x_1, \ldots, x_n, \frac{\partial u^{[k]}}{\partial x_2}, \ldots, \frac{\partial u^{[k]}}{\partial x_n}\right)$$

for a function $u^{[k+1]}$ satisfying the initial conditions prescribed for u (this method may not always be suitable or even possible). In propagation-type problems the variable corresponding to time will be taken as x_1.

Accordingly, with f defined as in (4.18), the iteration formula for the differential equation (4.13) of the example in § 4.2 reads

$$\frac{\partial u^{[k+1]}}{\partial t} = - a + f(u^{[k]}) \frac{\partial u^{[k]}}{\partial x} .$$

A first approximation can be found by solving the initial-value problem

$$- a \frac{\partial u^{[0]}}{\partial x} + \frac{\partial u^{[0]}}{\partial t} = - a , \qquad u^{[0]}(x, 0) = \varphi(x) = 2 \cdot \frac{4 - x}{5 - x} ,$$

whose differential equation represents a constant-coefficient approximation to (4.13). This simplified equation admits of a general solution expressible very simply in terms of an arbitrary function w:

$$u^{[0]}(x, t) = - a t + w(x + a t).$$

The initial condition implies that $w = \varphi$, so that finally

$$u^{[0]}(x, t) = - a t + \varphi(x + a t) = 2 - \frac{t}{2} - \frac{2}{5 - x - \dfrac{t}{2}} .$$

With

$$u_x^{[0]} = \frac{\partial u^{[0]}}{\partial x} = - \frac{2}{\left(5 - x - \dfrac{t}{2}\right)^2}$$

the equations for the next approximation $u^{[1]}$ read

$$u_t^{[1]} = - a + f(u^{[0]}) u_x^{[0]} , \qquad u^{[1]}(x, 0) = \varphi(x).$$

For a given value of x, $u_t^{[1]}$ can be immediately tabulated at intervals h in t, then $u^{[1]}$ calculated by numerical integration. In Table IV/18 this is done with $h = \frac{1}{2}$ for the values of x used in the finite-difference methods in § 4.4. The values of $u^{[1]}$ for $t = 1, 2, 3, \ldots$ are calculated by SIMPSON's rule and the intermediate values for $t = \frac{1}{2}, \frac{3}{2}, \frac{5}{2}, \ldots$ by the formula

$$\int\limits_0^h g(x) \, d x \approx \frac{h}{12} \left[5 g(0) + 8 g(h) - g(2h) \right]$$

(for the rows with $t \geq 1$ only the results are given). The errors in the values of $u^{[0]}$ and $u^{[1]}$ are also given in Table IV/17.

Table IV/18. *First cycle of the iteration procedure in the iterative solution of the glacier recession problem*

t		$x=0$	$x=0{\cdot}5$	$x=1$	$x=1{\cdot}5$	$x=2$	$x=2{\cdot}5$	$x=3$	$x=3{\cdot}5$	$x=4$
$t=0$	$u^{[0]}$	1·6	1·55556	1·5	1·42857	1·33333	1·2	1	0·66667	0
	$5-x-\frac{t}{2}$	5	4·5	4	3·5	3	2·5	2	1·5	1
	$f(u^{[0]})$	0·38304	0·38413	0·38553	0·38738	0·38994	0·39373	0·4	0·41264	0·5
	$u_x^{[0]}$	−0·08	−0·09877	−0·125	−0·16327	−0·22222	−0·32	−0·5	−0·88889	−2
	$u_t^{[1]}$	−0·53064	−0·53794	−0·54819	−0·56325	−0·58665	−0·62599	−0·7	−0·86679	−2·5
$t=0{\cdot}5$	$u^{[0]}$	1·32895	1·27941	1·21667	1·13462	1·02273	0·86111	0·60714	0·15	
	$5-x-\frac{t}{2}$	4·75	4·25	3·75	3·25	2·75	2·25	1·75	1·25	
	$f(u^{[0]})$	0·39006	0·39144	0·39324	0·39570	0·39925	0·40486	0·41532	0·44687	
	$u_x^{[0]}$	−0·08864	−0·11073	−0·14222	−0·18935	−0·26446	−0·39506	−0·65306	−1·28	
	$u_t^{[1]}$	−0·53458	−0·54334	−0·55593	−0·57493	−0·60558	−0·65994	−0·77123	−1·07199	
	$u^{[1]}$	1·33373	1·28528	1·22406	1·14417	1·03557	0·87920	0·63440	0·19787	
$t=1$	$u^{[0]}$	1·05556	1	0·92857	0·83333	0·7	0·5	0·16667	−0·5	
	$u_t^{[1]}$	−0·53933	−0·55	−0·56571	−0·59020	−0·63159	−0·71032	−0·89553	−1·65874	
	$u^{[1]}$	1·06528	1·01200	0·94373	0·85304	0·72657	0·53732	0·21992	−0·46891	
$t=1{\cdot}5$	$u^{[0]}$	0·77941	0·71667	0·63462	0·52273	0·36111	0·10714	−0·35		
	$u_t^{[1]}$	−0·54517	−0·55838	−0·57840	−0·61092	−0·66940	−0·79551	−1·23020		
	$u^{[1]}$	0·79422	0·73501	0·65789	0·55316	0·40311	0·16609	−0·31151		
$t=2$	$u^{[0]}$	0·5	0·42857	0·33333	0·2	0	−0·33333			
	$u_t^{[1]}$	−0·55258	−0·56932	−0·59570	−0·64129	−0·75	−1·00608			
	$u^{[1]}$	0·51985	0·45319	0·36456	0·24051	0·05004	−0·27909			
$t=2{\cdot}5$	$u^{[0]}$	0·21667	0·13462	0·02273	−0·13889	−0·39286				
	$u_t^{[1]}$	−0·56257	−0·58497	−0·62474	−0·71799	−0·87436				
	$u^{[1]}$	0·24172	0·16556	0·06060	−0·09952	−0·35483				
$t=3$	$u^{[0]}$	−0·07143	−0·16667	−0·3	−0·5					
	$u_t^{[1]}$	−0·58841	−0·62334	−0·68142	−0·78969					
	$u^{[1]}$	−0·04536	−0·13556	−0·26479	−0·47664					

4.6. Application of Hermite's formula

Another method has been suggested by PFLANZ[1]; it utilizes HERMITE's generalization of TAYLOR's formula (Ch. I, § 2.5). For the sake of simplicity we describe the method for a differential equation

$$F(x, y, u, p, q) = 0 \qquad (4.24)$$

in only two independent variables x, y (here $p = \partial u/\partial x$, $q = \partial u/\partial y$). Let U_k, P_k, Q_k be approximations to u, p, q at the point $x = x_k$, $y = y_k$ and U, P, Q corresponding approximations at the point x, y. Then (2.60) of Ch. I with $k = m = 1$ and x_1, y_1, x_0, y_0, h, k replaced by x, y, x_j, y_j, h_j, k_j, where $h_j = x - x_j$, $k_j = y - y_j$, yields

$$U - \frac{h_j}{2} P - \frac{k_j}{2} Q = C_j = U_j + \frac{h_j}{2} P_j + \frac{k_j}{2} Q_j. \qquad (4.25)$$

Now let x_0, y_0, u_0 and x_1, y_1, u_1 be two points on the initial curve; then $U_0, P_0, Q_0, U_1, P_1, Q_1$ are known. If we write down (4.25) for $j = 0$ and $j = 1$, and also demand that U, P, Q satisfy the differential equation (4.24), i.e.

$$F(x, y, U, P, Q) = 0, \qquad (4.26)$$

then we have three equations for the three unknowns U, P, Q. P and Q can be expressed linearly in terms of U by means of (4.25) with $j = 0$ and $j = 1$, so we are left with (4.26) to solve for U; this is usually done iteratively.

As long as the computational labour does not become too great the method may be extended to higher approximations by using (2.60) of Ch. I with $k = m = 2$. Approximations R, S, T to $\dfrac{\partial^2 u}{\partial x^2}$, $\dfrac{\partial^2 u}{\partial x \partial y}$, $\dfrac{\partial^2 u}{\partial y^2}$ must then be included and HERMITE's formula has to be written down for three points on the initial curve:

$$\left. \begin{aligned} 12U - 6(h_j P + k_j Q) + h_j^2 R + 2h_j k_j S + k_j^2 T \\ = 12U_j + 6(h_j P_j + k_j Q_j) + h_j^2 R + 2h_j k_j S_j + k_j^2 T_j \end{aligned} \right\} \quad (j = 0, 1, 2). \qquad (4.27)$$

The system of equations for U, P, Q, R, S, T is completed by (4.26) together with the two equations

$$F_x + F_u P + F_p R + F_q S = 0, \qquad F_y + F_u Q + F_p S + F_q T = 0. \qquad (4.28)$$

§5. The method of characteristics for systems of two differential equations of the first order

In this section we describe briefly the elements of a characteristics method which has been used extensively for hydrodynamic and aerodynamic problems but mostly in graphical mode. For the details, in particular, the exploitation of various advantages afforded by graphical treatment[2], and also for more extensive problems[3], the reader is referred to the literature.

---

[1] PFLANZ, E.: Bemerkungen über die Methode von G. DUFFING zur Integration von Differentialgleichungen. Z. Angew. Math. Mech. **28**, 167—172 (1948). He also gives a numerical example.

[2] MASSAU, J.: Mémoire sur l'intégration graphique des équations aux dérivées partielles. Gent 1900; for a description of the method see in particular §§ 5, 6 Ch. III, pp. 46—58; the method is applied to numerous hydrodynamic problems.

[3] COURANT, R., and D. HILBERT: Methoden der mathematischen Physik, Vol. II, p. 303 et seq. Berlin 1937. — SAUER, R.: Charakteristikenverfahren für die ein-

When carried out numerically, the method has a certain resemblance to the finite-difference method, but unlike that method it has the important advantage (cf. § 3.1) that it builds up an accurate approximation to the true determinate region of a given initial segment.

We shall consider a system of two quasi-linear differential equations in two dependent variables $u(x, y)$ and $v(x, y)$, i.e. a system of the form

$$a_1 u_x + a_2 u_y + a_3 v_x + a_4 v_y = A,$$
$$b_1 u_x + b_2 u_y + b_3 v_x + b_4 v_y = B, \tag{5.1}$$

where a_j, b_j, A, B are given functions of x, y, u, v.

5.1. The characteristics

Let us begin by considering the problem of calculating u and v in the neighbourhood of a curve C on which their values are prescribed. Imagine curvilinear co-ordinates ξ, η introduced in such a way that one of them, say ξ, is constant on the given curve C (Fig. IV/19). We shall assume that in a neighbourhood of C the co-ordinate transformation $(x, y) \to (\xi, \eta)$ is one-one, i.e. the Jacobian of the transformation is non-zero:

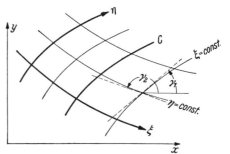

Fig. IV/19. Introduction of the characteristics

$$\Phi = \frac{\partial(\xi, \eta)}{\partial(x, y)} = \begin{vmatrix} \xi_x & \xi_y \\ \eta_x & \eta_y \end{vmatrix} \neq 0. \tag{5.2}$$

u and v are known on C and hence also are their derivatives u_η, v_η along C. We now try to determine the values of the other derivatives u_ξ, v_ξ on C. Substituting

$$u_x = u_\xi \xi_x + u_\eta \eta_x$$

dimensionale instationäre Gasströmung. Ing.-Arch. **13**, 79—89 (1942). — SCHULTZ-GRUNOW, F.: Nichtstationäre, eindimensionale Gasbewegung. Forsch.-Arb. Ing.-Wes. **13**, 125—134 (1942). — SAUER, R.: Charakteristikenverfahren für Kugel- und Zylinderwellen reibungsloser Gase. Z. Angew. Math. Mech. **23**, 29—32 (1943). — Theoretische Einführung in die Gasdynamik, p. 146 et seq. Berlin 1943. — OSWATITSCH, KL.: Über die Charakteristikenverfahren der Hydrodynamik. Z. Angew. Math. Mech. **25**, 195—208 (1947); **27**, 264—270 (1947). (A comprehensive review.) — COURANT, R., and K. FRIEDRICHS: Supersonic flow and shock waves. New York: Interscience Publishers, Inc. 1948. — SAUER, R.: Dreidimensionale Probleme der Charakteristikentheorie partieller Differentialgleichungen. Z. Angew. Math. Mech. **30**, 347—356 (1950). — Anfangswertprobleme bei partiellen Differentialgleichungen, p. 229 et seq. Berlin-Göttingen-Heidelberg 1952. — Hyperbolische Probleme der Gasdynamik mit mehr als zwei unabhängigen Veränderlichen. Z. Angew. Math. Mech. **33**, 331—336 (1953).

and corresponding expressions for u_y, v_x, v_y into (5.1) we obtain

$$
\left.
\begin{aligned}
(a_1 \xi_x + a_2 \xi_y) u_\xi &+ (a_3 \xi_x + a_4 \xi_y) v_\xi \\
&= A - (a_1 \eta_x + a_2 \eta_y) u_\eta - (a_3 \eta_x + a_4 \eta_y) v_\eta, \\
(b_1 \xi_x + b_2 \xi_y) u_\xi &+ (b_3 \xi_x + b_4 \xi_y) v_\xi \\
&= B - (b_1 \eta_x + b_2 \eta_y) u_\eta - (b_3 \eta_x + b_4 \eta_y) v_\eta.
\end{aligned}
\right\} \quad (5.3)
$$

Thus we have two linear equations for the unknown values of u_ξ and v_ξ on C. If the determinant of coefficients

$$
\Delta = \begin{vmatrix} a_1 \xi_x + a_2 \xi_y & a_3 \xi_x + a_4 \xi_y \\ b_1 \xi_x + b_2 \xi_y & b_3 \xi_x + b_4 \xi_y \end{vmatrix}
\tag{5.4}
$$

does not vanish, u_ξ and v_ξ may be calculated. Then in the first instance the values of u and v at neighbouring points $(\xi + \xi', \eta)$ for sufficiently small ξ' can be approximated quite crudely by the first-order terms of their Taylor series.

If, however, the determinant Δ does vanish along C, then the values of u and v at neighbouring points cannot be so calculated. A curve $\xi(x, y) = \text{constant}$ along which Δ vanishes identically is called a characteristic. In general this "characteristic" property depends on the particular solution u, v, so that, unless certain simplifications obtain, we can only speak of a curve as a characteristic with respect to a specific solution u, v. These characteristics form the basis of the approximate method to be described; we therefore derive now several of their properties which will be needed later. It is convenient to introduce a notation for certain determinants which will appear in most of the formulae:

$$
a_{jk} = a_j b_k - a_k b_j, \quad A_j = A b_j - B a_j \quad (j, k = 1, 2, 3, 4); \tag{5.5}
$$

thus the determinant (5.4), for instance, can be expressed in the form

$$
\Delta = a_{13} \xi_x^2 + (a_{14} + a_{23}) \xi_x \xi_y + a_{24} \xi_y^2. \tag{5.6}
$$

Let γ be the angle between the tangent to a characteristic $\xi = \text{constant}$ and the x axis, so that

$$
\tan \gamma = -\frac{\xi_x}{\xi_y}. \tag{5.7}
$$

The equation $\Delta = 0$ yields a quadratic equation for $\tan \gamma$, namely

$$
a_{13} \tan^2 \gamma - (a_{14} + a_{23}) \tan \gamma + a_{24} = 0, \tag{5.8}
$$

from which we obtain the two values

$$
\tan \gamma = \frac{a_{14} + a_{23} \pm D}{2 a_{13}}, \tag{5.9}
$$

where

$$
D^2 = (a_{14} + a_{23})^2 - 4 a_{13} a_{24}. \tag{5.10}
$$

In the following we assume that $D^2 > 0$ so that there are two real values for $\tan \gamma$. The system of differential equations (5.1) is then called hyperbolic with respect to the solution u, v. For a given solution u, v we therefore have two values of $\tan \gamma$, say $\tan \gamma_1$ and $\tan \gamma_2$, at each point (x, y) and hence two corresponding ("characteristic") directions whose direction ratios satisfy the equation $\varDelta = 0$. Thus we have two "direction fields", each of which generates a one-parameter family of characteristic curves. We assume for the remainder that $\xi = $ constant and $\eta = $ constant are the two families of characteristics.

The equations for the characteristics do not depend explicitly on the functions A, B and they depend on u and v only through the functions a_j, b_j. If the latter depend only on x and y, the characteristics can be determined from (5.9) without reference to a particular solution and thus constitute two families of curves which, for a given system (5.1), are fixed once and for all, and are independent of any boundary conditions which would be needed to specify a particular solution.

We now note for later use some algebraic transformations. By expressing the following determinant in terms of the second-order subdeterminants which can be formed from its first two rows, and using the fact that its value must be zero since it has at least one pair of identical rows, we obtain

$$\frac{1}{2} \begin{vmatrix} a_1 & a_2 & a_3 & a_4 \\ b_1 & b_2 & b_3 & b_4 \\ a_1 & a_2 & a_3 & a_4 \\ b_1 & b_2 & b_3 & b_4 \end{vmatrix} = a_{12} a_{34} - a_{13} a_{24} + a_{14} a_{23} = 0. \qquad (5.11)$$

By virtue of this identity, (5.10) can be put in the form

$$D^2 = (a_{23} - a_{14})^2 - 4 a_{12} a_{34}. \qquad (5.12)$$

The expressions (5.9) for the two values of $\tan \gamma$ differ from each other only in the sign of D; we associate them with the two families of characteristics with angles γ_1 and γ_2 (Fig. IV/19) as follows:

$$\left. \begin{aligned} \tan \gamma_1 &= -\frac{\xi_x}{\xi_y} = \frac{2 a_{24}}{a_{14} + a_{23} - D} = \frac{a_{14} + a_{23} + D}{2 a_{13}}, \\ \tan \gamma_2 &= -\frac{\eta_x}{\eta_y} = \frac{2 a_{24}}{a_{14} + a_{23} + D} = \frac{a_{14} + a_{23} - D}{2 a_{13}}. \end{aligned} \right\} \qquad (5.13)$$

The equivalence of the two alternative expressions given here can be verified immediately using (5.10).

5.2. Consistency conditions

Along a characteristic, say $\xi = $ constant, there exists a certain relation between the values of u and v. This relation, which is, of course, a consequence of the characteristic condition, is established as follows.

In a region of the (x, y) plane in which u, v and their partial derivatives are continuous u_ξ and v_ξ have specific finite values satisfying the conditions (5.3), and therefore, since the determinant Δ [(5.4)] vanishes when $\xi = $ constant is a characteristic, the right-hand sides of (5.3) must satisfy the usual consistency conditions along a characteristic; these conditions yield, in fact, a linear relation between the values of u_η and v_η; for example, the condition that the determinant corresponding to u_ξ must vanish, namely

$$\Delta_1 = \begin{vmatrix} A - (a_1 \eta_x + a_2 \eta_y)\, u_\eta - (a_3 \eta_x + a_4 \eta_y)\, v_\eta & a_3 \xi_x + a_4 \xi_y \\ B - (b_1 \eta_x + b_2 \eta_y)\, u_\eta - (b_3 \eta_x + b_4 \eta_y)\, v_\eta & b_3 \xi_x + b_4 \xi_y \end{vmatrix} = 0, \quad (5.14)$$

yields the relation

$$\left. \begin{aligned} A_3 \xi_x + A_4 \xi_y - u_\eta (a_{13} \xi_x \eta_x + a_{14} \xi_y \eta_x + a_{23} \xi_x \eta_y + \\ + a_{24} \xi_y \eta_y) + v_\eta\, a_{34}\, \Phi = 0, \end{aligned} \right\} \quad (5.15)$$

in which we have used the notation of (5.2) and (5.5).

The factor multiplying u_η here can be simplified using (5.7) and (5.8): with $\gamma = \gamma_1$ along $\xi = $ constant we have

$$\xi_y [\eta_x(- a_{13} \tan \gamma_1 + a_{14}) + \eta_y(- a_{23} \tan \gamma_1 + a_{24})]$$
$$= \xi_y [\eta_x(- a_{13} \tan \gamma_1 + a_{14}) + \eta_y \tan \gamma_1 (- a_{13} \tan \gamma_1 + a_{14})]$$
$$= \xi_y \left(\eta_x - \eta_y \frac{\xi_x}{\xi_y} \right) (- a_{13} \tan \gamma_1 + a_{14}) = \Phi(a_{13} \tan \gamma_1 - a_{14}).$$

Then, after division by a_{34}, (5.15) becomes

$$\Phi v_\eta - q_1 \Phi u_\eta + \frac{A_3 \xi_x + A_4 \xi_y}{a_{34}} = 0, \quad (5.16)$$

where [from (5.12), (5.13)]

$$q_1 = \frac{a_{13} \tan \gamma_1 - a_{14}}{a_{34}} = \frac{- a_{14} + a_{23} + D}{2 a_{34}} = \frac{2 a_{12}}{- a_{14} + a_{23} - D}. \quad (5.17)$$

Now let the arc length s be introduced as parameter on the characteristic $\xi = $ constant; then

$$\frac{dv}{ds} = \frac{\partial v}{\partial \eta} \frac{d\eta}{ds} = v_\eta (\eta_x x_s + \eta_y y_s) = v_\eta x_s \left(\eta_x + \eta_y \frac{dy}{dx} \right)$$
$$= v_\eta x_s \left(\eta_x - \eta_y \frac{\xi_x}{\xi_y} \right) = - \frac{v_\eta x_s}{\xi_y}\, \Phi,$$

and similarly

$$\frac{du}{ds} = - \frac{u_\eta x_s}{\xi_y}\, \Phi;$$

also

$$x_s = \cos \gamma_1 \quad \text{and} \quad - \frac{\xi_x}{\xi_y} = \tan \gamma_1,$$

so that from (5.16) we have finally

$$\left(\frac{dv}{ds} - q_1 \frac{du}{ds}\right)_{\xi=\text{const}} = -\frac{A_3}{a_{34}} \sin\gamma_1 + \frac{A_4}{a_{34}} \cos\gamma_1. \qquad (5.18)$$

For a characteristic of the other family $\eta = \text{constant}$, D is to be replaced by $-D$; hence we have (using the same symbol s to denote the new arc length)

$$\left(\frac{dv}{ds} - q_2 \frac{du}{ds}\right)_{\eta=\text{const}} = -\frac{A_3}{a_{34}} \sin\gamma_2 + \frac{A_4}{a_{34}} \cos\gamma_2, \qquad (5.19)$$

where

$$q_2 = \frac{-a_{14} + a_{23} - D}{2a_{34}} = \frac{2a_{12}}{-a_{14} + a_{23} + D}. \qquad (5.20)$$

Thus along the characteristics the values of u and v must satisfy the conditions (5.18), (5.19), respectively, while along curves on which $\varDelta \neq 0$ u and v are not related in this manner.

5.3. The method of characteristics

We suppose now that the values of u and v are given on an arc of a curve K_1 which is nowhere tangent to a characteristic. We may suppose further that the values of u (say) alone are given on a contiguous arc K_2 which is likewise no-where tangent to a character-istic and which makes with K_1 at their intersection P an angle which includes just one of the characteristics passing through P (see Fig. IV/20).

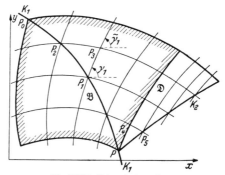

Fig. IV/20. Determinate regions

Let P_1 and P_2 be two points on the curve K_1 and let the characteristic $\xi = \text{constant}$ passing through P_1 intersect the characteristic $\eta = \text{constant}$ passing through P_2 at the point P_3 (Fig. IV/20). We do not yet know the position of P_3, but if P_1 and P_2 are sufficiently close together we can use the point of intersection of the tangents to the characteristics at P_1 and P_2 as an approximation to it. Further, we can approximate to the equations (5.18), (5.19), which hold along the respective characteristics, by replacing the deriv-atives by difference quotients for the steps $s_1 = P_1 P_3$ and $s_2 = P_2 P_3$, respectively. In this way we obtain two equations for u_3 and v_3:

$$(v_3 - v_1) - q_1(u_3 - u_1) = -\frac{s_1}{a_{34}} (A_3 \sin\gamma_1 - A_4 \cos\gamma_1), \qquad (5.21)$$

$$(v_3 - v_2) - q_2(u_3 - u_2) = -\frac{s_2}{a_{34}} (A_3 \sin\gamma_2 - A_4 \cos\gamma_2). \qquad (5.22)$$

In similar fashion we can calculate values for u and v at all points of the approximate characteristic mesh within the determinate region of the arc K_1 (the curvilinear "parallelogram" shaded in Fig. IV/20), or rather within the polygonal approximation to this region outlined by the approximate characteristics through P and P_0.

If, as mentioned at the beginning, we also have values of u prescribed on the curve K_2, we can start say at P_5 in Fig. IV/20 and use the equation corresponding to (5.22) for P_4, P_5 to calculate the missing value v_5 from the known values u_4, v_4, u_5; the next interior point can then be treated as above, and in this way, using just the one equation whenever we have to deal with a boundary point, we can proceed to fill in the region between K_2 and the enclosed characteristic through P.

We can improve somewhat on the approximation represented by (5.21), (5.22) by using one of the methods of the second chapter for the integration of (5.18), (5.19), such as one of the simple methods of § 1.5 of that chapter. We can also improve on the approximate position of P_3 by using, for example, the principle of the Euler-Cauchy method: firstly, a provisional approximation is calculated by the method described above, i.e. the intersection of the tangents at the points P_1, P_2 is used as a provisional position $\tilde{P}_3$ for the point P_3 of the characteristic mesh and corresponding values $\tilde{u}_3$, $\tilde{v}_3$ for u_3, v_3 are calculated from (5.21), (5.22); from these are calculated corresponding approximations $\tilde{\gamma}_1$, $\tilde{\gamma}_2$ to the characteristic directions at P_3 (see Fig. IV/20); then the inter-section of the new straight lines through P_1 and P_2 making the angles $\frac{1}{2}(\gamma_1+\tilde{\gamma}_1)$ and $\frac{1}{2}(\gamma_2+\tilde{\gamma}_2)$, respectively, with the x axis may be expected to give a better approximation to the position of P_3. A simple method from which better values for u and v may be expected is to use now the mean values $\frac{1}{2}(u_1+\tilde{u}_3)$, $\frac{1}{2}(v_1+\tilde{v}_3)$ for u and v in the right-hand side of (5.21) and correspondingly for (5.22); effectively, we write down (5.18) for the mid-point of $\overline{P_1 P_3}$ with these mean-value approximations for the local values of u and v, and correspondingly for (5.19), and replace the derivatives by central difference quotients.

5.4. Example

We choose a particularly simple example with differential equations admitting of predetermined characteristics and for which the exact solution is known so that the error in any approximation is also known.

The current $J(x, t)$ and potential $V(x, t)$ in an electric cable satisfy the equations

$$C\,\frac{\partial V}{\partial t} + S V = -\frac{\partial J}{\partial x}, \qquad L\,\frac{\partial J}{\partial t} + R J = -\frac{\partial V}{\partial x}, \tag{5.23}$$

where t is the time, x is measured along the cable and C, R, L, S are the usual symbols for capacity, resistance, self-induction and leakage (per unit length of cable).

Let us consider the case $S = 0$. Then with $u = J, v = -CV, t = y$, $LC = \alpha^2, RC = \beta$ the differential equations become

$$\left.\begin{array}{c} u_x - v_y = 0 \\ -\alpha^2 u_y + v_x = \beta u. \end{array}\right\} \quad (5.24)$$

Here the matrix of the coefficients a_j, b_j, A, B in (5.1) reads

$$\begin{pmatrix} 1 & 0 & 0 & -1 & 0 \\ 0 & -\alpha^2 & +1 & 0 & \beta u \end{pmatrix}$$

and the values of the determinants (5.5) are

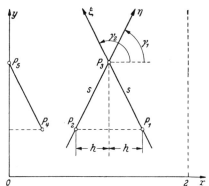

Fig. IV/21. The fixed straight-line characteristics of the example of § 5.4

$$a_{12} = -\alpha^2, \quad a_{13} = 1, \quad a_{14} = a_{23} = 0, \quad a_{24} = -\alpha^2, \quad a_{34} = 1,$$
$$A_3 = 0, \quad A_4 = \beta u.$$

From (5.10), (5.13) we have

$$D^2 = 4\alpha^2, \quad D = 2\alpha, \quad \tan \gamma_1 = \alpha, \quad \tan \gamma_2 = -\alpha,$$

so that the characteristics are the straight lines $y \pm \alpha x = \text{constant}$; they are independent of boundary conditions and particular solutions.

If, for convenience, we measure the arc length s from the initial line $y = 0$, we have $\cos \gamma_1 = -\cos \gamma_2 = (1 + \alpha^2)^{-\frac{1}{2}}$; then, since $q_1 = \alpha, q_2 = -\alpha$ from (5.17), (5.20), the relations (5.18), (5.19) which hold along the characteristics read

$$\left.\begin{array}{c} \left(\dfrac{dv}{ds} - \alpha \dfrac{du}{ds}\right)_{\xi=\text{const}} = \dfrac{1}{\sqrt{1+\alpha^2}} \beta u, \\[2ex] \left(\dfrac{dv}{ds} + \alpha \dfrac{du}{ds}\right)_{\eta=\text{const}} = -\dfrac{1}{\sqrt{1+\alpha^2}} \beta u. \end{array}\right\} \quad (5.25)$$

If we write down these equations for two points P_1, P_2 lying on a parallel to the x axis a distance $2h$ apart (Fig. IV/21) and replace the derivatives by forward difference quotients, we obtain (with function values at the points P_j in Fig. IV/21 denoted by u_j, v_j)

$$v_3 - v_2 - \alpha(u_3 - u_2) = h\beta u_2, \quad v_3 - v_1 + \alpha(u_3 - u_1) = -h\beta u_1 \quad (5.26)$$

and hence the first crude approximations

$$\left.\begin{array}{c} v_3 = \frac{1}{2}\left[(v_1 + v_2) + (\alpha - \beta h)(u_1 - u_2)\right], \\[2ex] u_3 = \dfrac{1}{2\alpha}\left[(v_1 - v_2) + (\alpha - \beta h)(u_1 + u_2)\right]. \end{array}\right\} \quad (5.27)$$

We can approximate equations (5.25) rather more accurately by writing them down for the mid-points of $\overline{P_1 P_3}$ and $\overline{P_2 P_3}$, using the arithmetic means of the end-point values for the function values at the mid-points and replacing the derivatives by central difference quotients:

$$\left.\begin{aligned}
v_3 - v_2 - \alpha\,(u_3 - u_2) &= h\beta\,\frac{u_2 + u_3}{2}\,, \\
v_3 - v_1 + \alpha\,(u_3 - u_1) &= -\,h\beta\,\frac{u_1 + u_3}{2}\,.
\end{aligned}\right\} \tag{5.28}$$

Solving these two equations for u_3, v_3 we obtain the approximations

$$\left.\begin{aligned}
v_3 &= \frac{1}{2}\left[(v_1 + v_2) + \left(\alpha - \frac{\beta\,h}{2}\right)(u_1 - u_2)\right], \\
u_3 &= \frac{1}{2\alpha + h\beta}\left[(v_1 - v_2) + \left(\alpha - \frac{\beta\,h}{2}\right)(u_1 + u_2)\right].
\end{aligned}\right\} \tag{5.29}$$

We now consider a particular problem with the initial and boundary values

$$\left.\begin{aligned}
u\,(x, 0) &= \sin\frac{\pi}{2}\,x \\
v\,(x, 0) &= 0
\end{aligned}\right\} \quad \text{for} \quad 0 \le x \le 2,$$

$$u\,(0, y) = u\,(2, y) = 0 \quad \text{for} \quad y \ge 0.$$

From (5.27) and an appropriate boundary formula, u and v can be calculated row by row for $y = \alpha h$, $2\alpha h$, $3\alpha h$, ... Since u is given on the boundary, the appropriate formula to be used to complete each row at the boundaries is the second formula of (5.26); for the boundary at $x = 0$, for instance, we use

$$v_5 = v_4 - \alpha\,(u_5 - u_4) - h\beta\,u_4,$$

in which the subscripts refer to the points P_4, P_5 situated as in Fig. IV/21 and the value of u_5 is given (zero in the present example). Since symmetry (or antisymmetry in the case of v) exists about the line $x = 1$, we can restrict the calculation to the half $0 \le x \le 1$. The results obtained with $h = \frac{1}{4}$ for the case $\alpha = 2$, $\beta = 1$ are exhibited in Table IV/19. The value of u for each mesh point is given with the associated error (in fifth-decimal units) in brackets and the corresponding value of v immediately below. The error was obtained by comparison with the exact solution

$$u\,(x, y) = e^{-\frac{1}{4}y}\sin\frac{\pi}{2}\,x\left\{\cos\frac{\nu\,y}{8} - \frac{1}{\nu}\sin\frac{\nu\,y}{8}\right\}, \quad \text{where} \quad \nu = \sqrt{4\pi^2 - 1}\,.$$

The last column of the table is a check column containing row sums, the summations being extended over the whole row $0 \le x \le 2$ for u but only over the half row $0 \le x \le 1$ for v (the sum of all v values for $0 \le x \le 2$

Table IV/19. *Values of* $\begin{Bmatrix} u \\ v \end{Bmatrix}$ *obtained by using* (5·26), (5·27) *with* $h = \frac{1}{4}$

x =	0	0·25	0·5	0·75	1	1·25	Row sums
y = 0	0 0		**0·70711** **0**		**1** **0**		2·41422 0
y = 0·5		0·30936 (−154) 0·61872		0·74686 (−374) 0·25628		0·74686 −0·25628	2·11244 0·875
y = 1	**0** 1·16010		0·37149 (−374) 0·82031		0·52536 (−529) 0		1·26834 1·98041
y = 1·5		0·07758 (−126) 0·31526		0·18729 (−306) 0·54479			0·52974 1·86005
y = 2	**0** 1·45102		−0·07674 (+99) 1·02602		−0·10852 (+141) 0		−0·26200 2·47704
y = 2·5		−0·13982 (+293) 1·17137		−0·33756 (+707) 0·48520			−0·95476 1·65657
y = 3	**0**		−0·38040 (981)		−0·53797 (1387)		−1·29877

Table IV/20. *Values of* $\begin{cases} u \\ v \end{cases}$ *obtained by using* (5.26), (5.27) *with* $h = \frac{1}{8}$

$x =$	0	0·125	0·25	0·375	0·5	0·625	0·75	0·875	1	Row sums
$y = 0$	0 / 0		**0.38268** / **0**		**0.70711** / **0**		**0.92388** / **0**		**1** / **0**	5·02734 / 0
$y = 0·25$		0·17938 / 0·35876		0·51084 / 0·30415		0·76453 / 0·20322		0·90182 / 0·07136		4·71314 / 0·93750
$y = 0·5$	0		0·30989 (−101)		0·57260 (−188)		0·74814 (−246)		0·80978 (−266)	4·07104

Table IV/21. *Values of* $\begin{cases} u \\ v \end{cases}$ *obtained by using* (5.28), (5.29) *with* $h = \frac{1}{4}$

$x =$	0	0·25	0·5	0·75	1
$y = 0$	**0** / **0**		**0.70711** / **0**		**0** / **1**
$y = 0·5$		0·31196 (+106) / 0·62692		0·75314 (+254) / 0·27458	
$y = 1$	**0** / 1·24784		0·37853 (330) / 0·88235		0·53532 (467) / **0**
$y = 1·5$		0·08100 (216)		0·19556 (521)	

is zero and therefore does not provide a check). The check consists in verifying the relation

$$8 S_{m+1} = 7(S_m - u_a) - 4v_a$$

(where u_a and v_a are the first values of u and v appearing in the m-th row), which should hold between successive row sums S_m, S_{m+1} of the u values.

Table IV/20 gives the beginning of a similar calculation based on the finer characteristic mesh with $h = \frac{1}{8}$ and Table IV/21 gives the beginning of a calculation based on the coarser mesh with $h = \frac{1}{4}$ but employing the more accurate formulae (5.28), (5.29). Again the errors in the u values are given. The gain in accuracy over the first rough calculation is slight.

§6. Supplements

In §§ 6.1 and 6.2 we prove theorems which permit the deduction of bounds for the error in approximate solutions of various problems in parabolic differential equations. These theorems play a role for parabolic equations similar to that played by the boundary-maximum theorem for elliptic equations (Ch. V, § 3), which likewise provides a possible basis for the estimation of the error in approximate solutions. Rather less is known about hyperbolic differential equations in this respect; a theorem has been established[1] which states that under certain conditions the maximum of a function satisfying a differential inequality is assumed only on a boundary curve, but the cases covered by the theorem do not yet possess the degree of generality which has been achieved for elliptic and parabolic equations. Nevertheless, it is worthy of note that an error estimate is possible for the "mixed-type" problem with equation (1.57), which was mentioned in § 1.9 (the Tricomi problem).

6.1. Monotonic character of a wide class of initial-/boundary-value problems in non-linear parabolic differential equations

A very general estimation theorem which nevertheless admits of an elementary proof has been established by WESTPHAL[2]. Let B be the open region

$$0 < y < Y, \qquad x_0(y) < x < x_1(y),$$

---

[1] AGMON, S., L. NIRENBERG and M. H. PROTTER: A maximum principle for a class of hyperbolic equations and applications to equations of mixed elliptic-hyperbolic type. Comm. Pure Appl. Math. **6**, 455−470 (1953). − PROTTER, M. H.: A Boundary Value Problem for an Equation of Mixed Type. Trans. Amer. Math. Soc. **71**, 416−429 (1951).

Some applications to error estimation for hyperbolic differential equations will be published shortly by the present author; cf. also L. COLLATZ: Fehlermaß-prinzipien in der praktischen Analysis. Proc. Internat. Congr. Math. Vol. I, Amsterdam 1954.

[2] WESTPHAL, H.: Zur Abschätzung der Lösungen nichtlinearer parabolischer Differentialgleichungen. Math. Z. **51**, 690−695 (1949). − Similar and more general theorems can be found in the following papers: − PICONE, M.: Sul problema della

where Y is a positive constant and $x_0(y)$, $x_1(y)$ are two given continuous functions of y such that $x_0(y) < x_1(y)$ for $0 \leq y \leq Y$ (see Fig. IV/22).

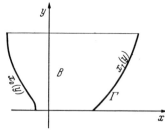

We shall need to distinguish between that part of the boundary of B which lies along the line $y = Y$ and the remaining part; we denote them by Γ_Y and Γ, respectively. Now let T be the operator defined by

$$T u = u_y - f(x, y, u, u_x, u_{xx}) \qquad (6.1)$$

Fig. IV/22. Boundaries for a class of boundary-value problems with non-linear parabolic differential equations

for functions $u(x, y)$ which are continuous in the closed region $\overline{B} = B + \Gamma + \Gamma_Y$ and for which u_y, u_x, u_{xx} exist in B. If we restrict the given function $f(x, y, u, u_x, u_{xx})$ to be monotonic non-decreasing with respect to u_{xx} for any fixed values of x, y, u, u_x, then the following theorem holds.

Theorem: *If, with the above definitions and restrictions, two functions u and v are such that $Tv < Tu$ in $B + \Gamma_Y$ and $v < u$ on Γ, then $v < u$ throughout $\overline{B}$.*

Proof: Let $w = u - v$; then $w > 0$ on Γ and we wish to show that this is true also in $B + \Gamma_Y$. Let M be the set of values of y in $0 \leq y \leq Y$ for which there is at least one value of x in $x_0(y) < x < x_1(y)$ with $w \leq 0$; we wish to show that this set is null. Suppose that it is not. If it is an infinite set, it will have a lower limit $\tilde{y}$ and there will be a sequence of values y_ν with corresponding values x_ν such that $y_\nu \to \tilde{y}$ and $w(x_\nu, y_\nu) \leq 0$; now these x_ν form a bounded infinite set, which must have a limit $\tilde{x}$, and we can choose a sub-sequence x_ν', with corresponding values y_ν', such that the points (x_ν', y_ν') converge to the limit point $\tilde{P} = (\tilde{x}, \tilde{y})$; then by continuity $w(\tilde{P}) \leq 0$. If M is finite, we take the smallest member of M as $\tilde{y}$ and a corresponding x value as $\tilde{x}$. Now $\tilde{P}$ cannot lie on Γ, so we can choose a sequence of points $(\tilde{x}, y)$ with $w > 0$ which tend to $\tilde{P}$ from below (i.e. with $y < \tilde{y}$); hence, by continuity, $w(\tilde{P}) \geq 0$ and we must have $w(\tilde{P}) = 0$; we see also that $w_y(\tilde{P}) \leq 0$ (w_y is assumed to exist

propagazione del calore in un messo di frontiera conduttore, isotrope e anogenes. Math. Ann. **101**, 701—712 (1929). — Marasimhan: On the Asymptotic Stability of Solutions of Parabolic Differential Equations. J. Rational Mech. Anal. **3**, 303—313 (1954). — Nirenberg, L. A.: A Strong Maximum Principle for Parabolic Equations. Comm. Pure Appl. Math. **6**, 167—177 (1953). — Szarski, L.: Sur la limitation et unicité des solution d'un système non-linéaire d'équations paraboliques aux derivées partielles du second ordre. Ann. Polonoci Math. **2**, 237—249 (1955). — Collatz, L.: Fehlerabschätzungen für Näherungslösungen parabolischer Differentialgleichungen. Anais Acad. Brasiliera de Ciéncias **28**, 1—9 (1956).

in B). Similarly, $w \geq 0$ on the whole line segment $y = \tilde{y}$, $x_0(\tilde{y}) \leq x \leq x_1(\tilde{y})$ and therefore, since w_x and w_{xx} exist at $\tilde{P}$,

$$w_x(\tilde{P}) = 0, \qquad w_{xx}(\tilde{P}) \geq 0.$$

In terms of u and v the situation at $\tilde{P}$ is that $u = v$, $u_x = v_{,x}$ $u_y \leq v_y$, $u_{xx} \geq v_{xx}$; consequently

$$f(x, y, u, u_x, u_{xx}) \geq f(x, y, v, v_x, v_{xx}) \quad \text{and} \quad T u \leq T v,$$

which contradicts our definition of u and v. Thus the set M must be null and the theorem is proved.

6.2. Estimation theorems for the solutions

In the reference cited, WESTPHAL makes among others the two following applications of the theorem in § 6.1.

Theorem 1. *With the definitions of § 6.1 let there be constants ε_1, ε_2, δ such that*

$$|T u| \leq \varepsilon_1, \quad |T v| \leq \varepsilon_2 \quad in \quad B + \Gamma_Y \quad and \quad |u - v| \leq \delta \quad on \ \Gamma; \qquad (6.2)$$

further, in addition to the monotonic condition of § 6.1, let f satisfy

$$|f(x, y, u_1, u_x, u_{xx}) - f(x, y, u_2, u_x, u_{xx})| \leq C^*, \qquad (6.3)$$

where C^* is a constant, for all values of x, y, u, u_x, u_{xx} which come into consideration. Then

$$|u - v| \leq \delta + (C^* + \varepsilon_1 + \varepsilon_2) y \qquad (6.4)$$

in the whole of $\overline{B}$.

Proof. Again let $w = u - v$, and apply to it the operator T^* defined for fixed v by

$$T^* \varphi = \varphi_y - f(x, y, v, v_x + \varphi_x, v_{xx} + \varphi_{xx}) + f(x, y, v, v_x, v_{xx}):$$

we have

$$\begin{aligned} T^* w &= u_y - v_y - f(x, y, v, u_x, u_{xx}) + f(x, y, v, v_x, v_{xx}) \\ &= T u - T v + f(x, y, u, u_x, u_{xx}) - f(x, y, v, u_x, u_{xx}) \\ &\leq \varepsilon_1 + \varepsilon_2 + C^* \end{aligned}$$

in $B + \Gamma_Y$. Now the function defining $T^* \varphi$ is a monotonic non-decreasing function of φ_{xx}, so that T^* has the same monotonic property as was prescribed for T in § 6.1; moreover, the function $W = \delta + \varepsilon + (C^* + \varepsilon_1 + \varepsilon_2 + \varepsilon) y$, for which

$$T^* W = C^* + \varepsilon_1 + \varepsilon_2 + \varepsilon,$$

is such that $T^* w < T^* W$ in $B + \Gamma_Y$ and $w < W$ on Γ for any positive constant ε; consequently we can use the theorem of § 6.1 to show that $w < W$ in the whole of $\overline{B}$. Similarly $-w < W$, and hence $|w| < W$, in the whole of $\overline{B}$. (6.4) follows by letting ε tend to zero.

Theorem 2. *Let the region B of § 6.1 be such that*

$$x_0(y) \leq -A, \qquad x_1(y) \geq A > 0$$

and in addition to the monotonic condition of § 6.1 let f satisfy a Lipschitz condition of the form

$$\left. \begin{aligned} & |f(x, y, u', u'_x, u'_{xx}) - f(x, y, u'', u''_x, u''_{xx})| \\ & \qquad \leq M_0 |u' - u''| + M_1 |u'_x - u''_x| + M_2 |u'_{xx} - u''_{xx}|. \end{aligned} \right\} \tag{6.5}$$

Then for two functions u, v satisfying the differential equation $Tu = Tv = 0$ in $B + \Gamma_Y$ and which are equal on that part of the boundary Γ along $y = 0$ and differ by at most δ on the remainder of Γ, we have

$$|u - v| \leq \delta e^{(M_0 + M_1 + 2M_2) y - A} \left(e^{|x|} + \tfrac{1}{2} \right) \tag{6.6}$$

in the whole of $\overline{B}$. By putting $\delta = 0$ we obtain the uniqueness theorem for the first boundary-value problem.

Proof. Again let $w = u - v$, but this time consider the operator

$$T^* \varphi = \varphi_y - M_0 |\varphi| - M_1 |\varphi_x| - M_2 \Phi(\varphi_{xx}),$$

where

$$\Phi(z) = \tfrac{1}{2} (z + |z|) = \begin{cases} z & \text{for} \quad z \geq 0 \\ 0 & \text{for} \quad z < 0; \end{cases}$$

$\Phi(z)$ is a monotonic non-decreasing function of z, so that T^* again satisfies the monotonic condition of § 6.1.

Now

$$\begin{aligned} u_y - v_y &= f(x, y, u, u_x, u_{xx}) - f(x, y, v, v_x, v_{xx}) \\ &= f(x, y, u, u_x, v_{xx}) - f(x, y, v, v_x, v_{xx}) + \\ & \quad + f(x, y, u, u_x, u_{xx}) - f(x, y, u, u_x, v_{xx}) \\ &\leq M_0 |w| + M_1 |w_x| + M_2 \Phi(w_{xx}), \end{aligned}$$

where we have used the fact that f is a monotonic non-decreasing function of its last argument. Hence $T^* w \leq 0$.

For the function

$$W = \delta e^{(M_0 + M_1 + 2M_2 + \varepsilon) y - A} \left(e^{|x|} + \tfrac{1}{2} e^{-2|x|} \right),$$

where ε is any positive constant, a simple calculation yields

$$\begin{aligned} T^* W = \delta\, e^{(M_0 + M_1 + 2M_2 + \varepsilon) y - A} \times \\ \times \left[\varepsilon \left(e^{|x|} + \tfrac{1}{2} e^{-2|x|} \right) + \tfrac{3}{2} M_1 e^{-2|x|} + M_2 \left(e^{|x|} - e^{-2|x|} \right) \right]. \end{aligned}$$

Consequently $T^* W > 0 \geq T^* w$. Since we also have $W > w$ on Γ, it follows from the theorem of § 6.1 that $w < W$ in the whole of $\overline{B}$. By interchanging u and v we can show also that $-w < W$, and hence $|w| < W$, in the whole of $\overline{B}$. (6.6) follows by letting ε tend to zero.

GÖRTLER[1] has considered such estimation theorems generalized for the solutions of the system of partial differential equations

$$u\, u_x + v\, u_y = f(x, u, u_y, u_{yy})$$
$$u_x + v_y = 0,$$

which is of great importance in fluid dynamics.

Example. Consider the example of § 2.4 with $K = \beta = 1$:

$$T u = u_y - u_{xx} - 1 = 0 \quad \text{in} \quad |x| \leq 1, \quad y \geq 0 \quad (B + \Gamma)$$
$$u(x, 0) = 0 \quad \text{for} \quad |x| \leq 1, \quad u(\pm 1, y) = 0 \quad \text{for} \quad y \geq 0 \quad (\text{boundary } \Gamma).$$

The function

$$v(x, y) = \tfrac{1}{2}(1 - x^2) + \sum_{n=1}^{N} a_n \cos(c_n x)\, e^{-c_n^2 y},$$

where $c_n = (n - \tfrac{1}{2})\pi$, satisfies the differential equation $Tv = 0$ and the boundary conditions $v(\pm 1, y) = 0$ (for $y \geq 0$) for any integer N and for arbitrary constants a_n. From Theorem 1 with $\varepsilon_1 = \varepsilon_2 = C^* = 0$ it follows that $|u - v| \leq \delta$ in the whole of $\overline{B}$, where δ is an upper bound for $|u - v|$ along the initial line segment $y = 0$, $-1 \leq x \leq 1$. We therefore choose the a_n so that

$$\left| \tfrac{1}{2}(1 - x^2) + \sum_{n=1}^{N} a_n \cos(c_n x) \right| \leq \delta$$

in the interval $-1 \leq x \leq 1$ with δ as small as possible. For $N = 3$ we can take δ as small as $0\cdot002$ with the constants $a_1 = -0\cdot516$, $a_2 = 0\cdot0193$, $a_3 = -0\cdot005$. The Hermitian finite-difference method of § 2.4 yields the approximate value $u(0, \tfrac{5}{24}) \approx \frac{7\cdot70133}{40} = 0\cdot19253$ at the point $i = 0$, $k = 10$ (since $l = \tfrac{1}{48}$ and $\alpha = \tfrac{1}{40}$). Here we obtain the approximation $v(0, \tfrac{5}{24}) = 0\cdot19158$ and can assert with certainty that $|u(0, \tfrac{5}{24}) - 0\cdot19158| \leq 0\cdot002$. This error estimate can be improved somewhat by the addition of a suitable constant to v.

6.3. Reduction to boundary-value problems

As for ordinary differential equations (Ch. II, § 5.9), initial- and initial-/boundary-value problems in partial differential equations may also be transformed into boundary-value problems, often in several different ways, and then dealt with by one of the several methods

---

[1] GÖRTLER, H.: Über die Lösungen nichtlinearer partieller Differentialgleichungen vom Reibungsschichttypus. Z. Angew. Math. Mech. **30**, 265−267 (1950).

available for the treatment of boundary-value problems (Ch. V). The transformation is effected by raising the order of the differential equation and adding further boundary conditions, which can often be chosen quite arbitrarily.

Such a treatment of the heat-flow problem

$$\frac{\partial u}{\partial t} = K \frac{\partial^2 u}{\partial x^2},$$

$$u(x, 0) = u(0, t) = 0, \qquad u(L, t) = 100$$

is given in detail by ALLEN and SEVERN[1].

They put $u = \dfrac{\partial w}{\partial t} + K \dfrac{\partial^2 w}{\partial x^2}$, so that $\dfrac{\partial^2 w}{\partial t^2} = K^2 \dfrac{\partial^4 w}{\partial x^4}$, and as additional boundary conditions take

$$w(0, t) = w(L, t) = \frac{\partial w}{\partial t}(x, T) = 0,$$

where T is a suitably chosen time. This boundary-value problem for the rectangular region $0 \leq x \leq L$, $0 \leq t \leq T$ they then solve by the finite-difference method in conjunction with the relaxation technique (Ch. V, § 1.6).

6.4. Miscellaneous exercises on Chapter IV

1. The numerical example of § 1.2 concerned the heat flow in a rod at one end of which the temperature was varied sinusoidally. After a long time (for large t) the temperature distribution settles down to a regular oscillation with the imposed frequency ω. Calculate a half-cycle of this oscillating distribution by the finite-difference method [formula (1.13) with mesh width $h = \frac{1}{4}$ as in § 1.2] by introducing the values $U_{i,q}$ on some row $k = q$ (q assumed to be large and for convenience a multiple of 12) as unknowns

$$U_{2,q} = a, \qquad U_{4,q} = b, \qquad U_{6,q} = c,$$

expressing the $U_{i,k}$ in the strip $q < k \leq q + 12$ in terms of a, b and c by means of the formula (1.13) and the boundary conditions, say $U_{0,q+2s} = \sin\left[\pi\left(1 + \dfrac{s}{6}\right)\right]$ and $U_{8,q+2s} = 0$ for $s = 0, 1, \ldots, 5$, and then determining a, b, c from the requirement that

$$U_{2r,q} = -U_{2r,q+12} \quad \text{for} \quad r = 1, 2, 3.$$

2. Let there be initially a linear fall in temperature from $100°$ C to $0°$ C along a thin homogeneous rod of length 10. Then let the hot end be cooled rapidly while the other end is kept at $0°$ C. Calculate approximately the temperature distribution $u(x, y)$ at several subsequent times y from the initial-/boundary-value problem

$$\frac{\partial^2 u}{\partial x^2} = 2 \frac{\partial u}{\partial y}; \qquad u(x, 0) = 10x, \qquad u(0, y) = 0, \qquad u(10, y) = 100 e^{-0.1 y}$$

by means of the finite-difference method of § 1 with the mesh widths $h = 1$ and $\frac{1}{2}$.

3. In dealing with the heat flow in a thin homogeneous rod in § 1.2 we neglected radiation and convection effects. If we take into consideration a heat loss to the surrounding medium across the surface of the rod with the rate of loss of heat

---

[1] ALLEN, D. N. DE G., and R. T. SEVERN: The application of relaxation methods to the solution of non-elliptic partial differential equations. Quart. J. Mech. Appl. Math. 4, 209−222 (1951).

proportional to the temperature difference (or to the temperature u of the rod if we take the temperature of the surroundings as zero), then the differential equation reads

$$\frac{\partial^2 u}{\partial x^2} = \frac{\varrho\,\sigma}{K}\frac{\partial u}{\partial t} + \lambda u,$$

where ϱ, σ, K, λ are physical constants. If the rod (of length $2a$) is initially heated to a uniform temperature $u = 1$ and then allowed to cool by itself, the initial and boundary conditions are

$$u(x, 0) = 1 \quad \text{for} \quad -a \leq x \leq a,$$

$$u \pm K\frac{\partial u}{\partial x} = 0 \quad \text{for} \quad x = \pm a \quad \text{and} \quad t > 0.$$

Calculate the solution of this problem for $\lambda = \frac{1}{2}$, $a = 1$, $K = 1$. The introduction of the new variable $y = \dfrac{Kt}{\varrho\sigma}$ reduces the differential equation to

$$\frac{\partial^2 u}{\partial x^2} = \frac{\partial u}{\partial y} + \frac{1}{2}u.$$

4. Use the finite-difference method to calculate an approximate solution of the cable problem which was treated by the method of characteristics in § 5.4, reducing the system of two equations for u and v to a single equation for u and similarly eliminating v from the boundary conditions. Compare the approximate solution so obtained with the exact solution given in § 5.4 and with the other approximate solutions obtained in § 5.4.

5. Apply the ordinary finite-difference method to the problem

$$J_{tt} + \tfrac{1}{2}J_t = J_{xx}; \quad J = 0 \quad \text{for} \quad x = 0 \quad \text{and} \quad x = 1,$$

$$\left.\begin{array}{l} J = 0 \\ \dfrac{\partial J}{\partial t} = -\dfrac{dq_0}{dx} = e^{-x} \end{array}\right\} \quad \text{for} \quad t = 0, \quad \text{and} \quad 0 < x < 1$$

[transient current produced by an initially non-uniform charge distribution $q_0(x) = e^{-x}$ in an open circuit consisting of a thin uniform conductor] and incorporate a running check.

6. The free cooling of a long square prism from a uniform temperature in excess of that of the surroundings can be reduced to the problem

$$u_{xx} + u_{yy} = u_t \quad \text{in the region} \quad |x| \leq 1, \quad |y| \leq 1, \quad t \geq 0,$$

$$u - \frac{\partial u}{\partial \nu} = 0 \quad \text{on the boundary} \quad |x| = 1, \quad |y| = 1, \quad t \geq 0,$$

and $u = 1$ for $t = 0$. Apply the ordinary finite-difference method in the form of equation (1.56) of § 1.8, incorporating a current check.

7. In § 1.8 we applied the ordinary finite-difference method to the heat equation (1.50) with two space co-ordinates x, y, and with the mesh widths (1.52) we obtained the formula (1.53), which depends on the parameter ϱ. The formula was shown to be unstable for $\varrho = 1$, and the value $\varrho = \frac{1}{4}$ was recommended. Is the formula stable for $\varrho = \frac{1}{3}$?

8. Let the two-dimensional wave equation

$$u_{xx} + u_{yy} = C u_{tt}$$

be approximated on the mesh (1.51) with the mesh widths $h_x = h_y = h$, $h_t = (c\varrho)^{\frac{1}{2}} h_x$ by the difference equation

$$U_{i,k,l+1} = 2U_{i,k,l} - U_{i,k,l-1} + \varrho(S - 4U_{i,k,l}),$$

where S is as defined in (1.55). Is the formula stable for $\varrho = 1$ and $\varrho = \frac{1}{2}$, respectively?

9. In § 2.2 several finite expressions of a higher approximation were derived for the equation $u_{xx} - K u_y = r(x, y)$, K being a non-zero constant. With the same mesh notation as used there (mesh widths h, l with $l = \sigma K h^2$) determine the quantities c and σ in the expression

$$\Phi = U_{0,1} - U_{0,0} - \sigma(U_{1,0} - 2U_{0,0} + U_{-1,0}) + c$$

so that Φ is equal to a remainder term of the sixth order by expanding Φ by TAYLOR's theorem and using the relations

$$K^2 u_{yy} = \frac{\partial^4 u}{\partial x^4} - S_1, \qquad K^3 u_{yyy} = \frac{\partial^6 u}{\partial x^6} - S_2,$$

where

$$S_1 = r_{xx} + K r_y, \qquad S_2 = \frac{\partial^4 r}{\partial x^4} + K r_{xxy} + K^2 r_{yy}.$$

10. Derive in the same way as in Exercise 9 a finite expression with a sixth-order remainder term for the equation

$$V^2 u = u_{xx} + u_{yy} = K u_t + r(x, y, t).$$

6.5. Solutions

1. The equations for the a, b, c read

$$
\begin{aligned}
1131a + 140b + 91c &= 229 + 92\sqrt{3} = 388.35 \\
140a + 1222b + 140c &= 88 + 60\sqrt{3} = 191.92 \\
91a + 140b + 1131c &= 29 + 12\sqrt{3} = 49.785,
\end{aligned}
$$

Table IV/22. *Half-cycle of the oscillating temperature distribution*

k	$i=0$	$i=1$	$i=2$	$i=3$	$i=4$	$i=5$	$i=6$	$i=7$	$i=8$
q	0		0·3284		0·1191		0·0028		0
$q+1$		0·1642		0·2238		0·0610		0·0014	
$q+2$	−0·5		0·1940		0·1424		0·0312		0
		−0·1530		0·1682		0·0868		0·0156	
	−0·8660		0·0076		0·1275		0·0512		0
$q+5$		−0·4292		0·0676		0·0894		0·0256	
	−1		−0·1808		0·0785		0·0575		0
		−0·5904		−0·0512		0·0680		0·0288	
	−0·8660		−0·3208		0·0084		0·0484		0
		−0·5934		−0·1562		0·0284		0·0242	
$q+10$	−0·5		−0·3748		−0·0639		0·0263		0
		−0·4374		−0·2194		−0·0188		0·0132	
$q+12$	0		−0·3284		−0·1191		−0·0028		0

and since the matrix of coefficients has strongly dominant diagonal elements, they may be solved conveniently by single-step iteration; we obtain

$$a = 0.3284, \qquad b = 0.1191, \qquad c = 0.0028.$$

Table IV/22 shows the corresponding temperature distribution calculated from (1.13).

2. With the mesh relation (1.12) the finite-difference formula reduces to (1.13) and the calculation consists solely in forming arithmetic means. The results for $h = 1$ and $h = 0·5$ are given in Tables IV/23, IV/24 and the temperature distribution is shown for various times in Fig. IV/23.

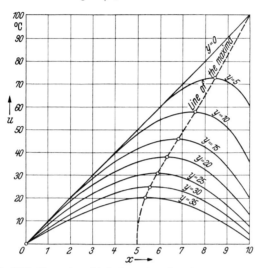

Fig. IV/23. Temperature distribution u along the rod at various times y

Table IV/23. *Finite-difference approximation for the mesh widths $h = l = 1$*

$k=y$	$x=0$	$x=1$	$x=2$	$x=3$	$x=4$	$x=5$	$x=6$	$x=7$	$x=8$	$x=9$	$x=10$
0	0		20		40		60		80		100
1		10		30		50		70		90	
2	0		20		40		60		80		81·873
3		10		30		50		70		80·937	
4	0		20		40		60		75·468		67·032
5		10		30		50		67·734		71·250	
6	0		20		40		58·867		69·492		54·881
7		10		30		49·434		64·180		62·187	
8	0		20		39·717		56·807		63·183		44·933
9		10		29·858		48·262		59·995		54·058	
10	0		19·929		39·060		54·129		57·027		36·788
11		9·965		29·495		46·594		55·578		46·907	
12	0		19·730		38·044		51·086		51·242		30·119
16	0		18·870		35·173		44·323		40·550		20·190
20	0		17·460		31·522		37·993		32·607		13·534
24	0		15·711		27·722		32·217		26·120		9·072
28	0		13·854		24·010		27·091		21·440		6·081
32	0		12·026		20·581		22·768		16·978		4·076
36	0		10·326		17·477		18·934		13·578		2·732

Table IV/24. *A section of the results for the mesh widths* $h = \tfrac{1}{2}$, $l = \tfrac{1}{4}$

k	y	x=3·5	x=4	x=4·5	x=5	x=5·5	x=6	x=6·5	x=7	x=7·5	x=8	x=8·5	x=9	x=9·5	x=10
0	0		40		50		60		70		80		90		100
1		35		45		55		65		75		85		95	
2			40		50		60		70		80		90		95·1:
3		35		45		55		65		75		85		92·562	
4	1		40		50		60		70		80		88·781		90·4·
5		35		45		55		65		75		84·391		89·633	
6			40		50		60		70		79·695		87·012		86·0
7		35		45		55		65		74·848		83·354		86·542	
8	2		40		50		60		69·924		79·101		84·948		81·8
12	3		40		49·996		59·926		69·432		77·249		80·397		74·0
16	4		39·995		49·945		59·658		68·481		74·802		75·654		67·0

3. We use the mesh

$$x_i = (i - \tfrac{1}{2})\, h, \qquad y_k = k\, l$$

with the mesh widths $h = \tfrac{1}{5}$ and $l = \tfrac{1}{50}$ [in accordance with (1.12)]. Then the finite-difference equation reads

$$U_{i,k+1} = \tfrac{1}{2}(U_{i+1,k} + U_{i-1,k}) - 0\cdot01\, U_{i,k}$$

and the boundary condition (with $n = 5$)

$$\frac{U_{n,k} + U_{n+1,k}}{2} + \frac{U_{n+1,k} - U_{n,k}}{h} = 0, \quad \text{so that} \quad U_{n+1,k} = \frac{9}{11}\, U_{n,k}.$$

We need to use this boundary condition on the initial row $y = 0$ in order to calculate the value $U_{6,0}$ which is required for the calculation of $U_{5,1}$. If we had simply put $U_{6,0} = 0$, say, this discontinuity would have made its presence felt in large fluctuations in the U values, and the unevenness of the U distribution would

Table IV/25. *Temperature in a cooling rod by the finite-difference method*

k	y	i=1 x=0·1	i=2 x=0·3	i=3 x=0·5	i=4 x=0·7	i=5 x=0·9	i=6 x=1·1	Row-sum check σ_k	$U_{4,k-1}$ minus $U_{4,k}$
0	0	1	1	1	1	1	0·81818	5	
1	0·02	0·99	0·99	0·99	0·99	0·89909	0·73562	4·85909	0·01000
2	0·04	0·9801	0·9801	0·9801	0·93465	0·85382	0·69858	4·72877	0·05535
3	0·06	0·97030	0·97030	0·94757	0·90761	0·80808	0·66115	4·60386	0·02704
		0·96060	0·94923	0·92948	0·86875	0·77630	0·63515	4·48436	0·03886
5	0·10	0·94531	0·93555	0·89970	0·84420	0·74419	0·60889	4·36895	0·02455
		0·93098	0·91315	0·88088	0·81350	0·71910	0·58835	4·25761	0·03070
		0·91276	0·89680	0·85452	0·79186	0·69373	0·56759	4·14967	0·02164
		0·89565	0·87467	0·83578	0·76621	0·67279	0·55047	4·04510	0·02565
		0·87620	0·85697	0·81208	0·74662	0·65161	0·53313	3·94348	0·01959
10	0·20	0·85782	0·83557	0·79367	0·72438	0·63336	0·51820	3·84480	0·02224
		0·83812	0·81739	0·77204	0·70627	0·61496	0·50315	3·74878	0·01811
		0·81937	0·79691	0·75411	0·68644	0·59856	0·48974	3·65538	0·01983
		0·79995	0·77877	0·73413	0·66947	0·58210	0·47626	3·56442	0·01697
14	0·28	0·78136	0·75925	0·71678	0·65142	0·56704	0·46394	3·47585	0·01805

certainly have been tar greater than that shown by the values obtained with no discontinuity (Table IV/25). The differences of the column $i = 4$, which are given in the last column of the table, show up this unevenness; it is quite considerable initially but gradually dies away. The row-sum check to be satisfied by the row-sums recorded in the penultimate column is

$$\sigma_{k+1} = (1 - 0.01)\,\sigma_k - \tfrac{1}{9}U_{n+1,k}, \quad \text{where} \quad \sigma_k = \sum_{i=1}^{n} U_{i,k}.$$

4. With $\alpha = 2$, $\beta = 1$ the problem reads

$$u_{xx} = 4u_{yy} + u_y,$$

$$
\left.
\begin{aligned}
u(x, 0) &= \sin\frac{\pi}{2}x \\[4pt]
u_y(x, 0) &= -\frac{1}{4}\sin\frac{\pi}{2}x
\end{aligned}
\right\} \quad \text{for} \ \ 0 \leq x \leq 2,
$$

$$u(0, y) = u(2, y) = 0 \qquad \text{for} \ \ 0 \leq y.$$

With $h = \tfrac{1}{4}$, $l = \tfrac{1}{2}$ the difference equation reads

$$U_{i,k+1} = \tfrac{1}{17}[16(U_{i+1,k} + U_{i-1,k}) - 15 U_{i,k-1}] \qquad (k = 0, 1, 2, \ldots).$$

To find starting values we must proceed as in (1.35) to (1.38); we obtain the formula

$$U_{i,1} = \tfrac{1}{32}(16 U_{i+1,0} - \tfrac{15}{4}U_{i,0} + 16 U_{i-1,0}).$$

The first few rows of the calculation are given in Table IV/27, which includes a row-sum check in the last two columns. If $S_m = \sum_{j=1}^{7} U_{j,m}$ is the sum of the U values in the m-th row, then

$$\tau_m = 17 S_{m+1} - 32(S_m - U_{1,m}) + 15 S_{m-1}$$

should be zero. The τ values are very sensitive to small variations in the U values. The errors in the U values are given in brackets in fifth-decimal units. Initially their magnitude is substantially greater than that of the corresponding errors obtained by the method of characteristics in § 5.4.

Table IV/26. *Current produced by a non-uniform charge distribution*

t	$x=0$	$x=0.2$	$x=0.4$	$x=0.6$	$x=0.8$	$x=1$	σ_k	τ_k
0	0	0	0	0	0	0	0	0
0.2	0	0.15556	0.12736	0.10427	0.08537	0	0.23163	0.47257
0.4	0	0.12130	0.24746	0.20260	0.09931	0	0.45006	0.67067
0.6	0	0.09493	0.19325	0.23591	0.11571	0	0.42916	0.63980
0.8	0	0.07430	0.09120	0.11094	0.13483	0	0.20214	0.41126
1	0	0.00096	0.00158	0.00182	0.00096	0	0.00339	0.00532
1.2	0	−0.06572	−0.07986	−0.09796	−0.12026	0	−0.17782	−0.36380
1.4	0	−0.07693	−0.15731	−0.19223	−0.09416	0	−0.34954	−0.52064
1.6	0	−0.09036	−0.18409	−0.15087	−0.07428	0	−0.33496	−0.49959
2.2	0	0.09296	0.07532	0.06114	0.05059	0	0.13647	0.28001
2.8	0	0.04604	0.05705	0.06886	0.08289	0	0.12590	0.25483
3.4	0	−0.04615	−0.09481	−0.11598	−0.05673	0	−0.21079	−0.31367
4	0	−0.00182	−0.00343	−0.00297	−0.00182	0	−0.00640	−0.01004

Table IV/27. Finite-difference approximation for the problem treated by the method of characteristics in § 5.4

$x=$	0	0·25	0·5	0·75	1	Row sums S_m	τ_m
$y=$							
−0·5	0	0·40438	0·74719	0·97625	1·05669	5·31233	−0·00050
0	0	**0·38268**	**0·70711**	**0·92388**	**1**	5·02734	
0·5	0	0·30871 (−219)	0·57042 (−406)	0·74528 (−532)	0·80669 (−575)	4·05551	+0·00014
1	0	0·19921 (−386)	0·36807 (−716)	0·48092 (−934)	0·52053 (−1012)	2·61692	+0·00030
2	0	0·07403 (−481)	0·13681 (−888)	0·17873 (−1162)	0·19347 (−1256)	0·97261	−0·00030
	0	−0·04701 (+494)	−0·08688 (−915)	−0·11349 (−1193)	−0·12286 (−1293)	−0·61762	−0·00009
3	0	−0·14709 (−434)	−0·27177 (−800)	−0·35511 (−1048)	−0·38434 (−1131)	−1·93228	−0·00002
	0	−0·21430 (−312)	−0·39600 (−579)	−0·51738 (−755)	−0·56004 (−820)	−2·81540	
	0	−0·24292					

Table IV/28. Cooling of a square prism

l	$U_{0,0,l}$ (1/8)	$U_{0,1,l}$ (7/8)	$U_{0,2,l}$ (2/8)	$U_{0,3,l}$ (1/8)	$U_{1,1,l}$ (3/8)	$U_{1,2,l}$ (9/8)	$U_{1,3,l}$ (1/8)	$U_{2,2,l}$ (1/8)	$U_{2,3,l}$ (2/8)	σ_l
1	1	1	1	0·66667	1	1	0·66667	1	0·66667	5·66667
2	1	1	0·91667	0·61111	1	0·91667	0·61111	0·83333	0·55556	5·37500
3	1	0·97917	0·86111	0·57407	0·95833	0·84028	0·56019	0·73611	0·49074	5·08623
4	0·97917	0·94444	0·80845	0·53897	0·90973	0·77894	0·51929	0·66551	0·44367	4·80527
5	0·94444	0·90177	0·76032	0·50688	0·86169	0·72575	0·48383	0·61131	0·40754	4·53354
6	0·90177	0·85704	0·71504	0·47669	0·81376	0·67929	0·45286	0·56664	0·37776	4·27497
7	0·85704	0·81108	0·67308	0·44872	0·76817	0·63708	0·42472	0·52853	0·35235	4·02861
8	0·81108	0·76661	0·63349	0·42233	0·72408	0·59863	0·39909	0·49472	0·32981	3·79602
9	0·76661	0·72318	0·59655	0·39770	0·68262	0·56285	0·37523	0·46422	0·30948	3·57575
10	0·72318	0·68210	0·56164	0·37443	0·64301	0·52966	0·35310	0·43616	0·29077	3·36834
11	0·68210	0·64271	0·52896	0·35264	0·60588	0·49848	0·33232	0·41021	0·27347	3·17241
12	0·64271	0·60571	0·49808	0·33205	0·57060	0·46934	0·31289	0·38598	0·25732	2·98805
13	0·60571	0·57050	0·46911	0·31274	0·53752	0·44189	0·29459	0·36333	0·24222	2·81411
14	0·57050	0·53747	0·44176	0·29450	0·50620	0·41614	0·27743	0·34206	0·22804	2·65044
15	0·53747	0·50616	0·41606	0·27737	0·47680	0·39186	0·26124	0·32209	0·21473	2·49611
16	0·50616	0·47678	0·39181	0·26121	0·44901	0·36905	0·24603	0·30330	0·20220	2·35088

5. With $h = 0.2$ the finite-difference equations read

$$J_{i,1} = l\left(1 - \frac{l}{4}\right)e^{-ih},$$

$$J_{i,k+1} = \frac{1}{1.05}\,[J_{i+1,k} + J_{i-1,k} - 0.95\,J_{i,k-1}] \qquad (k = 1, 2, \ldots),$$

for which we deduce the check $1.05\,\tau_{k+1} = \sigma_k + \tau_k - 0.95\,\tau_{k-1}$, where $\sigma_k = J_{2,k} + J_{3,k}$ and $\tau_k = \sum\limits_{i=1}^{4} J_{i,k}$. The results are given in Table IV/26, where only every third row is reproduced after the row $k = 8$.

6. With the mesh widths $h_x = h_y = \frac{2}{5}$, $h_z = \frac{1}{4}h_x^2 = \frac{1}{25}$ (and $x_0 = y_0 = 0$) the boundary condition is represented by $U_{i,3,l} = \frac{2}{3}U_{i,2,l}$. A convenient arrangement of the calculation is shown in Table IV/28, which includes a column for the quantities $\sigma_l = \sum\limits_{i,k}\beta_{i,k}U_{i,k,l}$, where the $\beta_{i,k}$ are the coefficients $\frac{1}{4}, \frac{7}{4}, \ldots$ encircled at the heads of the columns and the summation extends over all mesh points (i, k) which occur in the table; these quantities are needed for a check, which consists in verifying that

$$\tau_{l+1} = \sigma_l,$$

where $\tau_l = \sum\limits_{i,k}^{*} U_{i,k,l}$ over all interior mesh points occurring in the table.

7. No. One should not be deceived by the fact that the absolute maximum of the propagated errors (set out below in the manner of § 1.8) decreases at first:

$l = 0$	$l = 1$	$l = 2$	$l = 3$
$0 \quad 0$	$\frac{1}{3}\varepsilon \quad 0$	$-\frac{2}{9}\varepsilon \quad \frac{2}{9}\varepsilon \quad 0$	$\frac{4}{9}\varepsilon \quad -\frac{2}{9}\varepsilon \quad \frac{1}{9}\varepsilon \quad 0$
$\boxed{\varepsilon} \quad 0$	$\boxed{-\frac{1}{3}\varepsilon} \quad \frac{1}{3}\varepsilon$	$\boxed{\frac{5}{9}\varepsilon} \quad -\frac{2}{9}\varepsilon \quad \frac{1}{9}\varepsilon$	$\boxed{-\frac{13}{27}\varepsilon} \quad \frac{4}{9}\varepsilon \quad -\frac{1}{9}\varepsilon \quad \frac{1}{27}\varepsilon$

8. For $\varrho = 1$ the formula is unstable[1], while for $\varrho = \frac{1}{2}$ we obtain the simple formula $U_{i,k,l+1} = \frac{1}{2}S - U_{i,k,l-1}$, which is stable.

A more accurate formula using more mesh points is given by MILNE[2] on p. 144 of his book.

9. The Taylor expansion yields

$$\varPhi = c - \frac{l}{K}r - \frac{l^2}{2K^2}S_1 - \frac{l^3}{6K^3}S_2 + u_{xx}\left(\frac{l}{K} - \sigma h^2\right) +$$

$$+ \frac{\partial^4 u}{\partial x^4}\left(\frac{l^2}{2K^2} - \sigma\frac{h^4}{12}\right) + \frac{l^3}{6K^3}\frac{\widetilde{\partial^6 u}}{\partial x^6} - \frac{\sigma h^6}{360}\frac{\widetilde{\partial^6 u}}{\partial x^6},$$

where all values are taken to be at $x = y = 0$ except those distinguished by wavy lines, which are to be taken at certain intermediate points. The factor multiplying u_{xx} vanishes, and so also does the factor multiplying $\partial^4 u/\partial x^4$ if we choose $\sigma = \frac{1}{6}$,

---

[1] LJUSTERNIK, L. A.: Dokl. Akad. Nauk SSSR. (N. S.) **89**, 613—616 (1953) [Russian], gives the critical value ϱ_0 (stability for $\varrho < \varrho_0$, instability for $\varrho > \varrho_0$) for several finite-difference representations of the form (1.54) for the heat equation (1.50); ϱ_0 depends on the smallest eigenvalue for the corresponding difference operator.

[2] MILNE, W. E.: Numerical solution of differential equations. New York and London 1953.

i.e. $l = \frac{1}{6} K h^2$. For this value of σ we therefore obtain the finite equation[1]

$$U_{0,1} = \frac{1}{6} (U_{-1,0} + 4 U_{0,0} + U_{1,0}) - \frac{h^2}{6} r_{0,0} - \frac{h^4}{72} (r_{xx} + K r_y)_{0,0}.$$

10. With the notation for the u values in the plane $t = t_0 = \text{constant}$ defined in Table VIII of the appendix, and using the results of that table, we find that for the mesh widths $h_x = h_y = h$, $h_t = \sigma K h^2$

$$Q = a u_a + b \sum u_b + e \sum u_e + p u(x_0, y_0, t_0 + h_t) + q$$

$$= \left\{ q - p \sigma r h^2 - \frac{p}{2} \sigma^2 h^4 (K r_t + V^2 r) \right\} + u_a [p + a + 20] + h^2 V^2 u_a [p \sigma + 6] +$$

$$+ \frac{h^4}{2} V^4 u_a [p \sigma^2 + 1] + \frac{p \sigma^3 h^6}{6} (V^6 u_a - K^2 r_{tt} - K V^2 r_t - V^4 r) +$$

$$+ \frac{h^6}{60} (V^6 u_a + 2 V^2 D u_a) + \cdots.$$

Here we have put $b = 4$, $e = 1$ in order that no term in $D u_a$ shall appear. If we put the three quantities in square brackets equal to zero, we obtain $p = -36$, $\sigma = \frac{1}{6}$, $a = 16$. Then the equation $Q = 0$ yields[2]

$$36 u(x_0, y_0, t_0 + \tfrac{1}{6} K h^2) = 16 u_a + 4 \sum u_b + \sum u_e - 6 h^2 r_{0,0} -$$

$$- \tfrac{1}{2} h^4 (K r_t + V^2 r)_{0,0} + \text{remainder term of the 6th order.}$$

This formula has index $J = 1$ and is therefore stable.

Chapter V

Boundary-value problems
in partial differential equations

Many of the methods described for boundary-value problems in ordinary differential equations in Chapter III carry over without difficulty to partial differential equations. What has already been said in the introduction to Chapter IV about the need for more theoretical investigation and more practical experience applies particularly to boundary-value problems for partial differential equations and bears repeating here. For the solutions of these problems we do not possess existence and uniqueness theorems covering anything like the range desirable from the standpoint of technical applications; moreover, the diversity of problems which arise in applications is continually increasing. There is also an urgent need for existing approximate methods to be subjected to extensive practical tests and thorough theoretical investigations on a much larger scale than hitherto.

§1. The ordinary finite-difference method

The finite-difference method is applicable to boundary-value problems generally. The equations are easily set up and, if a coarse mesh is used, the solution is usually obtained to an accuracy sufficient for technical purposes

---

[1] MILNE (see last footnote) p. 122. For $r(x, y) = \text{constant}$ this formula includes the formula (2.10) with $\sigma = \frac{1}{6}$.

[2] This formula is given by MILNE p. 137 for the case $r \equiv 0$.

with a relatively short calculation. Moreover, for partial differential equations the finite-difference method may be the only practicable means of solution; for boundaries can occur of such shapes that other methods cannot cope with the boundary conditions at all, or at least, not without difficulty.

If, however, the solution is required to a greater accuracy, the ordinary finite-difference method has the disadvantage of slow convergence and it is usually better to use one of the refinements of the method described in § 2 rather than repeat the calculation with a smaller mesh width.

1.1. Description of the method[1]

Although the method may be applied to problems in three or more independent variables (cf. § 1.7), we will simplify the description, as in Ch. IV, § 1, by restricting ourselves to two variables x, y.

As in Ch. IV, § 1, we introduce a rectangular mesh

$$\left. \begin{array}{l} x_i = x_0 + i h \\ y_k = y_0 + k l \end{array} \right\} \qquad (i, k = 0, \pm 1, \pm 2, \ldots) \qquad (1.1)$$

and characterize function values at mesh points by corresponding subscripts (for instance, $u_{i,k}$ denotes $u(x_i, y_k)$ and $U_{i,k}$ denotes an approximation to it). As in Ch. IV (1.8), we can associate with every partial derivative $\dfrac{\partial^{m+n} u}{\partial x^m \partial y^n}$ a difference quotient which involves only function values at mesh points; thus, given any differential equation for u, we can derive in this way a corresponding difference equation, which provides a relation between the approximate values $U_{i,k}$. Such an equation is written down for every mesh point in the interior of the considered region. In a similar manner we derive difference equations corresponding to the boundary conditions, until we have as many equations as unknown values $U_{i,k}$. We can say nothing about the solubility of this system of equations without being more specific about the original boundary-value problem (cf. § 1.2).

---

[1] The following is a selection of the extensive literature on the subject: RUNGE, C.: Über eine Methode, die partielle Differentialgleichungen $V^2 u =$ Constans numerisch zu integrieren. Z. Math. Phys. **56**, 225—232 (1908). — RICHARDSON, L. F.: The approximate arithmetical solution by finite differences of physical problems involving differential equations with an application to the stresses in a masonry dam. Phil. Trans. Roy. Soc. Lond., Ser. A **210**, 308—357 (1911). — LIEBMANN, H.: Die angenäherte Ermittlung harmonischer Funktionen und konformer Abbildung. S.-B. Bayer. Akad. Wiss., Math.-phys. Kl. **1918**, 385—416. — HENCKY, H.: Die numerische Bearbeitung von partiellen Differentialgleichungen in der Technik. Z. Angew. Math. Mech. **2**, 58—66 (1922). — COURANT, R.: Über Randwertaufgaben bei partiellen Differentialgleichungen. Z. Angew. Math. Mech. **6**, 322—325 (1926). — COURANT, R., K. FRIEDRICHS and H. LEWY: Über die partiellen Differenzengleichungen der mathematischen Physik. Math. Ann. **100**, 32—74 (1928). — MARCUS, H.: Die Theorie elastischer Gewebe und ihre Anwendung auf die Berechnung biegsamer Platten, 2nd ed. Berlin 1932.

Since the practical derivation of the difference equations has already been treated in detail for similar cases (§ 1 of Ch. III, §§ 1, 2 of Ch. IV), there is little point in pursuing it further here; suffice it to explain the application of the method by means of the simple examples in §§ 1.5, 1.6. In §§ 1.2 to 1.4 we consider a special class of problems for which precise assertions can be made concerning convergence and error estimation.

1.2. Linear elliptic differential equations of the second order

Consider the differential equation

$$L[u] \equiv a \frac{\partial^2 u}{\partial x^2} + c \frac{\partial^2 u}{\partial y^2} + d \frac{\partial u}{\partial x} + e \frac{\partial u}{\partial y} - g u = r \qquad (1.2)$$

and let boundary values $\bar{u}$ be prescribed for $u(x, y)$ on a simple, closed, piecewise-smooth curve Γ in the (x, y) plane (Fig. V/1). This is the

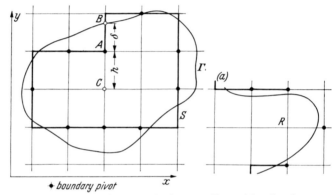

Fig. V/1. The boundary curve Γ and corresponding mesh boundary S

so-called "first" boundary-value problem, or the boundary-value problem of the first kind, for the differential equation (1.2) and the boundary Γ. We assume that the given functions a, c, d, e, g, r are continuous, and further that a, c are positive and g non-negative in a region B^* of the (x, y) plane containing the boundary Γ. The region actually bounded by Γ we denote by B and usually take this to be the closed region including the boundary Γ. B^* is as large an extension of B as is needed for the problem under consideration, and in some cases may be the improper extension $B^* = B$ (cf. § 2.1).

We first introduce some useful terminology concerning the mesh points of the basic finite-difference mesh. Two mesh points will be referred to as "neighbouring points" if they are consecutive mesh points on a mesh line. The term "pivot" will be used for a mesh point which lies within or on Γ. The pivots can be divided into "interior pivots" and "boundary pivots": an interior pivot is a pivot for which all four neighbouring points are also pivots; a boundary pivot, on the other hand, is a pivot for which at least one of its neighbouring points is not a pivot[1]. We assume that the interior pivots are "connected", i.e. given any two non-neighbouring interior pivots P and P^*, we can find further interior pivots $P_1, P_2, \ldots, P_n$ ($n \geq 1$) such that consecutive members of the sequence $P\,P_1\,P_2 \ldots P_n\,P^*$ are neighbouring points. (An assumption of "simple connectivity" is not needed for our purposes.) The boundary curve Γ may be associated with a "mesh boundary" S (Fig. V/1); this is a polygon with sides consisting only of sections of mesh lines and having the property that all mesh points which lie on them are either boundary pivots or isolated corners projecting beyond Γ[2].

We now derive the equations for the approximate pivotal values $U_{i,k}$. For every interior pivot we write down a difference equation corresponding to (1.2):

$$a_{i,k} \frac{U_{i+1,k} - 2U_{i,k} + U_{i-1,k}}{h^2} + c_{i,k} \frac{U_{i,k+1} - 2U_{i,k} + U_{i,k-1}}{l^2} +$$

$$+ d_{i,k} \frac{U_{i+1,k} - U_{i-1,k}}{2h} + e_{i,k} \frac{U_{i,k+1} - U_{i,k-1}}{2l} - g_{i,k} U_{i,k} = r_{i,k}.$$

If a square mesh ($h = l$) is used, this can be written

where
$$\left.\begin{aligned}
&l_{ik}[U] = h^2 r_{ik}, \\
&l_{ik}[U] \equiv l^*_{ik}[U] - (2a_{ik} + 2c_{ik} + g_{ik}h^2)\, U_{ik}, \\
&l^*_{ik}[U] \equiv \left(a_{ik} - \frac{h}{2} d_{ik}\right) U_{i-1,k} + \left(a_{ik} + \frac{h}{2} d_{ik}\right) U_{i+1,k} + \\
&\qquad + \left(c_{ik} - \frac{h}{2} e_{ik}\right) U_{i,k-1} + \left(c_{ik} + \frac{h}{2} e_{ik}\right) U_{i,k+1}.
\end{aligned}\right\} \quad (1.3)$$

With every boundary pivot A we associate a boundary point B (i.e. a point on Γ) which lies on a mesh line through A such that the distance δ from A to B is less than the mesh width h ($0 \leq \delta < h$). We assume that h is chosen so small that C, the neighbouring point to A furthest from B (Fig. V/1), is an interior pivot. Then for the pivot A we write down the equation

$$U(A) = \frac{\delta U(C) + h\,\bar{u}(B)}{\delta + h}, \quad (1.4)$$

---

[1] These definitions of "interior" and "boundary" pivots are convenient for the theory, at least, as far as it is developed in the following. In a practical calculation, however, it is often desirable to use function values at mesh points lying near, though outside of, the boundary curve Γ [for example, the heavily marked points in the subsidiary figure (a) of Fig. V/1] and it is then convenient to include these points as boundary pivots. As a consequence, some points which were originally boundary pivots, such as the point R in the figure, become interior pivots.

[2] Actually we do not need this "mesh boundary" for the theory presented here; for our purposes the classification of the pivots into boundary pivots and connected interior pivots is sufficient.

in which the modified notation has the obvious significance; the equation implies that graphically the values $\bar{u}(B)$, $U(A)$, $U(C)$ are collinear. If A lies on Γ, then $B=A$, $\delta=0$ and $U(A)$ is given its required boundary value $\bar{u}(A)$.

Another approach for curved boundaries is to derive difference equations corresponding to (1.3) which make use of actual boundary points as pivotal points instead of only mesh points. For example, if $a_1 h$, $a_2 h$, $a_3 h$, $a_4 h$ are the distances from a mesh point $P_0 = (i, k)$ of four neighbouring pivotal points $P_1 = (i + a_1, k)$, $P_2 = (i, k + a_2)$, $P_3 = (i - a_3, k)$, $P_4 = (i, k - a_4)$, as in Fig. V/2, then

$$\frac{2u(P_1)}{a_1(a_1 + a_3)} + \frac{2u(P_2)}{a_2(a_2 + a_4)} + \frac{2u(P_3)}{a_3(a_3 + a_1)} + \frac{2u(P_4)}{a_4(a_4 + a_2)} -$$

$$- 2\left(\frac{1}{a_1 a_3} + \frac{1}{a_2 a_4}\right) u(P_0) = h^2 (u_{xx} + u_{yy})(P_0) + O(h^3)$$

as can be verified by substituting the Taylor expansions centred on P_0 for the values of u which appear. If the differential equation is of the form $u_{xx} + u_{yy} = gu - r$, this formula without the remainder term, and accordingly with $u(P_j)$ replaced by the approximations $U(P_j)$, provides a difference equation corresponding to the point P_0. A similar application of TAYLOR's theorem can also be used to derive corresponding difference equations for the more general differential equation (1.2).

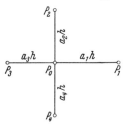

Fig. V/2. Unsymmetrical point pattern for curved boundaries

In many applications the boundary condition for the whole of the boundary Γ or some part of it Γ^* does not specify u but instead prescribes the value of a linear combination of u and its derivative $\partial u / \partial \nu$ in the direction of the inward normal ν, i.e. we can have a condition of the form

$$A_1(s) u + A_2(s) \frac{\partial u}{\partial \nu} = A_3(s) \quad \text{on} \quad \Gamma^*, \tag{1.5}$$

where s is the boundary parameter. When $\Gamma^* = \Gamma$, the problem of determining u is called the "second" boundary-value problem if $A_1(s) \equiv 0$ and the "third" boundary-value problem if $A_1(s) \not\equiv 0$ (cf. Ch. I, § 3.3); when Γ^* is properly a part of Γ, we speak of a mixed boundary-value problem. One way of deriving a difference equation to simulate a boundary condition of the type (1.5) is the following[1,2].

[1] See BATSCHELET, E.: Über die numerische Auflösung von Randwertproblemen bei elliptischen partiellen Differentialgleichungen. Z. Angew. Math. Phys. **3**, 165—193 (1952), where error estimates for the finite-difference method are also given; see also F. B. HILDEBRAND: Methods of applied mathematics, p. 307. New York 1952. — F. S SHAW: An introduction to relaxation methods, p. 120 et seq. New York 1953.

[2] More accurate formulae can be derived by bringing in more mesh points. A formula using 11 points is given by W. E. MILNE: Numerical solution of dif-

Consider a boundary pivot A of a sufficiently fine square mesh and let the local normal through A meet the boundary Γ in a point B which belongs to Γ^* (Fig. V/3). The line BA produced will cut a mesh line at a first point C between two neighbouring mesh points D, E. Let the associated distances be denoted as follows:

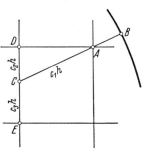

$$\overline{AC} = c_1 h, \quad \overline{DC} = c_2 h, \quad \overline{EC} = c_3 h,$$

where

$$1 \leq c_1 \leq \sqrt{2}, \quad c_2 \geq 0, \quad c_3 \geq 0,$$

$$c_2 + c_3 = 1.$$

Fig. V/3. One way of dealing with the third boundary-value problem

Then for the boundary pivot A we can write down the difference equation

$$A_1(s)\, U(A) + A_2(s) \frac{1}{c_1 h}\left[\frac{c_3\, U(D) + c_2\, U(E)}{c_2 + c_3} - U(A)\right] = A_3(s), \quad (1.6)$$

where s is to be given the value corresponding to the boundary point B.

Iterative solution of the system of difference equations. When a fine mesh is used, the system of difference equations becomes rather large, and in consequence an iterative solution is frequently resorted to. Thus, starting from approximations $U_{ik}^{[0]}$, we determine successive iterates $U_{ik}^{[1]}, U_{ik}^{[2]}, \ldots$ from the general formulae

$$(2a_{ik} + 2c_{ik} + g_{ik} h^2)\, U_{ik}^{[\nu+1]} = l_{ik}^*[U^{[\nu]}] - h^2 r_{ik}, \quad (1.7)$$

$$U^{[\nu+1]}(A) = \frac{\delta\, U^{[\nu]}(C) + h\, \bar{u}(B)}{\delta + h} \qquad (\nu = 0, 1, 2, \ldots). \quad (1.8)$$

For LAPLACE's equation

$$\nabla^2 u = \frac{\partial^2 u}{\partial x^2} + \frac{\partial^2 u}{\partial y^2} = 0$$

the iteration formula (1.7) reduces to

$$U_{ik}^{[\nu+1]} = \tfrac{1}{4}(U_{i+1,k}^{[\nu]} + U_{i,k+1}^{[\nu]} + U_{i-1,k}^{[\nu]} + U_{i,k-1}^{[\nu]})$$
$$= \tfrac{1}{4} \times (\text{Sum of the } \nu\text{-th iterates at the neighbouring points}); \quad (1.9)$$

ferential equations, p. 150. New York and London 1953. — Further formulae are given by J. ALBRECHT and W. UHLMANN: Differenzenverfahren für die erste Randwertaufgabe mit krummlinigen Rändern bei $\Delta u(x, y) = r(x, y, u)$. Z. Angew. Math. Mech. **37**, 212—224 (1957). — UHLMANN, W.: Differenzenverfahren für die 1. Randwertaufgabe mit krummflächigen Rändern bei $\Delta u(x, y, z) = r(x, y, z, u)$. Z. Angew. Math. Mech. **38**, 130—139 (1958). — Differenzenverfahren für die 2. und 3. Randwertaufgabe mit krummlinigen Rändern bei $\Delta u(x, y) = r(x, y, u)$. Z. Angew. Math. Mech. **38**, 236—251 (1958).

thus at each stage the value at the interior pivot (i, k) is to be replaced by the mean of the four neighbouring values (LIEBMANN's averaging procedure[1]).

Existence and uniqueness of the solution of the system of difference equations. The system of equations (1.3), (1.4) for the U_{ik} will always possess a uniquely determined solution if the mesh width h is chosen so small that $a_{ik} \pm \frac{1}{2} h d_{ik}$ and $c_{ik} \pm \frac{1}{2} h e_{ik}$ are positive for all pivots (i, k). This statement follows directly from Theorem 2 of Ch. I, § 5.5, for we can easily verify that the conditions of that theorem are satisfied (the a_{ik} in § 5.5 do not have the same meaning as here). Firstly, the equations can be written down with the coefficients $2a_{ik} + 2c_{ik} + h^2 g_{ik}$ or 1 (corresponding to interior or boundary equations, respectively) on the leading diagonal; then the diagonal coefficients are positive and the non-zero off-diagonal coefficients, which can only be $-a_{ik} \pm \frac{1}{2} h d_{ik}$, $-c_{ik} \pm \frac{1}{2} h e_{ik}$ or $-\dfrac{\delta}{\delta + h}$, are all negative. Secondly, all row sums of coefficients are non-negative and at least one row sum is positive: when they occur, equations of the form (1.4) give positive row sums since $1 - \dfrac{\delta}{\delta + h} > 0$; when no such equations occur, i.e. when all boundary pivots lie on the boundary curve Γ, a positive row sum will arise from one of the equations (1.3) corresponding to an interior pivot with at least one of its neighbouring pivots, say P, on Γ, for in such an equation $u(P)$ is known and is therefore part of the inhomogeneous term, so that its coefficient is not to be included in the row sum. Thirdly, the non-decomposition of the matrix of coefficients follows from the assumption that the interior pivots are connected.

The weaker conditions of Theorem 3 of Ch. I, § 5.5 are therefore also satisfied, so that the convergence of the total-step and single-step iteration procedures[2] is assured at the same time.

1.3. Principle of an error estimate for the finite-difference method

An error estimate[3] for the class of problems considered in § 1.2 has been given by GERSCHGORIN *; existence of the solution $u(x, y)$ is

---

* GERSCHGORIN, S.: Fehlerabschätzung für das Differenzenverfahren zur Lösung partieller Differentialgleichungen. Z. Angew. Math. Mech. **10**, 373—382 (1930).

[1] LIEBMANN, H.: Die angenäherte Ermittlung harmonischer Funktionen und konformer Abbildung. S.-B. Bayer. Akad. Wiss., Math.-phys. Kl. **1918**, 385—416. — WOLF, F.: Über die angenäherte numerische Berechnung harmonischer und biharmonischer Funktionen. Z. Angew. Math. Mech. **6**, 118—150 (1926).

[2] A different convergence proof based on subharmonic functions has been given by J. B. DIAZ and R. C. ROBERTS: On the numerical solution of the Dirichlet problem for Laplace's difference equation. Quart. Appl. Math. **9**, 355—360 (1952).

[3] Error estimation can be looked at in another way, which is suggested by the inexactness of applied problems. This has been discussed by R. v. MISES:

assumed and also knowledge of upper bounds M_n for the maxima in B of the absolute values of the partial derivatives $\partial^n u / \partial x^n$ and $\partial^n u / \partial y^n$ up to $n = 4$:

$$\left| \frac{\partial^n u}{\partial x^n} \right| \leq M_n, \quad \left| \frac{\partial^n u}{\partial y^n} \right| \leq M_n \quad \text{in } B \text{ for } n = 1, 2, 3, 4. \tag{1.10}$$

It will often be difficult to establish strict upper bounds M_n, and sometimes, for want of something better, one will use instead upper bounds for the corresponding difference quotients of the calculated approximations U_{ik}. Since the partial derivatives may exceed these bounds, one clearly cannot claim that the error limits so obtained are rigorous; however, they should yield an indication of the size of the error, and sometimes one may be content with this[1].

The derivation of the error estimate follows the same general pattern as for ordinary differential equations (cf. Ch. III, § 3). We write down the difference expressions corresponding to (1.3) for the exact solution $u(x, y)$ and expand them by means of TAYLOR's theorem into differential

---

On network methods in conformal mapping and in related problems. Nat. Bur. Stand., Appl. Math. Ser. **18**, 1—7 (1952). He maintains that from a physical point of view it is more logical to look for an approximate solution U such that

$$|L[U]| \leq \delta, \quad |\Phi_\mu[U]| \leq \delta$$

for a given tolerance δ ($L[u] = 0$ being the given differential equation and $\Phi_\mu[u] = 0$ the associated boundary conditions) than to investigate the error $\varepsilon = U - u$ in a given approximate solution U. Thus the idea of convergence is replaced by the idea of simultaneous approximation of the differential equation and the boundary conditions.

Application of this idea to the finite-difference method is a little troublesome; having calculated approximate values at the mesh points, one still has to construct from them a function U for which $L[U]$ can be formed. If, for example, we are treating a boundary-value problem with a second-order differential equation in two independent variables x, y by the finite-difference method and have obtained approximate values U_{ik} for a function $u(x, y)$ at the mesh points (x_i, y_k), then the next step is to construct a set of polynomials, one for each cell of the mesh, which join together smoothly to form one continuous function $U(x, y)$ with continuous first and second derivatives with respect to each independent variable.

[1] This difficulty in establishing or estimating the quantities M_n which embarrasses the application of the formulae (1.26) to (1.28) can sometimes be overcome by using special properties of the problem in question; for instance, GERSCHGORIN (loc. cit.) has shown that M_4 can be calculated exactly for the torsion problem for a rectangular bar. W. WASOW [J. Res. Nat. Bur. Stand. **48** (1952) Res. Paper 2321] employs a different approach in his derivation of error estimates for the plane Laplace equation $u_{xx} + u_{yy} = 0$ in a rectangle. He considers the first boundary-value problem with $u = f(s)$ on the boundary, where $f(s)$ is a "reasonable" function, and starts by reducing the problem to one in which $f(s) = 0$ at the four corners of the rectangle; this can be done by subtracting out a suitable harmonic polynomial of the form $a_0 + a_1 x + a_2 y + a_3 x y$. Then u can be written as the sum of four functions u_j satisfying $\nabla^2 u_j = 0$ ($j = 1, 2, 3, 4$) each of which also satisfies

expressions plus remainder terms. For instance, we have

$$u_{i+1,k} + u_{i-1,k} - 2u_{ik} = h^2 \left(\frac{\partial^2 u}{\partial x^2}\right)_{ik} + \frac{h^4}{24}\left(\frac{\partial^4 u_{\overline{i,i-1},k}}{\partial x^4} + \frac{\partial^4 u_{\overline{i,i+1},k}}{\partial x^4}\right),$$

in which we have used the notation $\partial^n u_{\overline{i,j},k}/\partial x^n$ to denote a value of $\partial^n u/\partial x^n$ at a point on the section of mesh line between the mesh points (i, k) and (j, k). Similarly expanding the other difference expressions in (1.3), we obtain

$$\left.\begin{aligned}
\frac{1}{h^2} l_{ik}[u] &= a_{ik}\left\{\left(\frac{\partial^2 u}{\partial x^2}\right)_{ik} + \frac{h^2}{24}\left(\frac{\partial^4 u_{\overline{i,i-1},k}}{\partial x^4} + \frac{\partial^4 u_{\overline{i,i+1},k}}{\partial x^4}\right)\right\} + \\
&\quad + c_{ik}\{\cdots\} + d_{ik}\{\cdots\} + e_{ik}\{\cdots\} - g_{ik} u_{ik} \\
&= (L[u])_{ik} + \frac{1}{h^2} R_{ik}[u] = r_{ik} + \frac{1}{h^2} R_{ik}[u],
\end{aligned}\right\} \quad (1.11)$$

in which the remainder term $R_{ik}[u]$ can be expressed in the form

$$\left.\begin{aligned}
R_{ik}&[u] \\
&= \frac{h^4}{24}\left\{a_{ik}\left(\frac{\partial^4 u_{\overline{i,i-1},k}}{\partial x^4} + \frac{\partial^4 u_{\overline{i,i+1},k}}{\partial x^4}\right) + c_{ik}\left(\frac{\partial^4 u_{i,\overline{k,k-1}}}{\partial y^4} + \frac{\partial^4 u_{i,\overline{k,k+1}}}{\partial y^4}\right)\right\} + \\
&\quad + \frac{h^4}{12}\left\{d_{ik}\left(\frac{\partial^3 u_{\overline{i,i-1},k}}{\partial x^3} + \frac{\partial^3 u_{\overline{i,i+1},k}}{\partial x^3}\right) + e_{ik}\left(\frac{\partial^3 u_{i,\overline{k,k-1}}}{\partial y^3} + \frac{\partial^3 u_{i,\overline{k,k+1}}}{\partial y^3}\right)\right\}.
\end{aligned}\right\} \quad (1.12)$$

Accordingly, since the approximations U_{ik} satisfy (1.3), the errors

$$\varepsilon_{ik} = U_{ik} - u_{ik} \quad (1.13)$$

at interior pivots (i, k) satisfy the difference equations

$$l_{ik}[\varepsilon] = - R_{ik}[u]. \quad (1.14)$$

Assuming that the derivatives of u are bounded as in (1.10) we have

$$\left|\frac{1}{h^2} R_{ik}[u]\right| \leq \frac{h^2}{12}\{M_4(a_{ik} + c_{ik}) + 2M_3(|d_{ik}| + |e_{ik}|)\} = h^2 C_1, \text{ say}. \quad (1.15)$$

---

the modified boundary condition on just one of the four sides of the rectangle and is zero on the other three.

With a suitable choice of co-ordinate system u_1, for example, may be expressed as a series of the form

$$u_1 = \sum_{n=1}^{\infty} d_n \sin b_n x \sinh c_n y.$$

Now the finite-difference solution, which in itself is only defined at the mesh points, can also be represented by such a series, and this enables one to estimate the difference between the two solutions. If the mesh width is h, the upper bound for the absolute error finally arrived at is $(A_2 M_2^* + A_3 M_3^*) h^2$, where A_2, A_3 depend only on the size of the rectangle and M_2^*, M_3^* are upper bounds for the absolute values of the second and third derivatives, respectively, of the given boundary values along the sides of the rectangle. All these quantities can be computed from the data.

Using TAYLOR's theorem in a similar manner at the boundary pivots, we have (for a boundary pivot A and points B, C as in Fig. V/1 on a line parallel the x axis, for instance)

$$u(C) = u(B) + \frac{\partial u(B)}{\partial x} (h + \delta) + \frac{\partial^2 u(\bar{B})}{\partial x^2} \cdot \frac{(h + \delta)^2}{2}$$

$$u(A) = u(B) + \frac{\partial u(B)}{\partial x} \delta + \frac{\partial^2 u(\bar{\bar{B}})}{\partial x^2} \cdot \frac{\delta^2}{2},$$

where $\bar{B}, \bar{\bar{B}}$ are intermediate points. Elimination of $\partial u/\partial x$ yields

$$u(A) = \frac{h}{h + \delta} u(B) + \frac{\delta}{h + \delta} u(C) - R_A, \tag{1.16}$$

where

$$R_A = \frac{\delta(h + \delta)}{2} \cdot \frac{\partial^2 u(\bar{B})}{\partial x^2} - \frac{\delta^2}{2} \cdot \frac{\partial^2 u(\bar{\bar{B}})}{\partial x^2}.$$

Comparison with (1.4) reveals that the error ε is given by

$$\varepsilon(A) = \frac{\delta}{h + \delta} \varepsilon(C) + R_A \tag{1.17}$$

at the boundary pivots, and hence, using the fact that $\delta < h$, we have

$$|\varepsilon(A)| \leq \tfrac{1}{2} |\varepsilon(C)| + \tfrac{3}{2} h^2 M_2. \tag{1.18}$$

To estimate the error ε from (1.14), (1.15), (1.18), GERSCHGORIN introduces an auxiliary function W defined by

$$L[W] = -1 \quad \text{in } B, \quad W = 0 \quad \text{on } \Gamma.$$

Since we can hardly determine this function W in general, or even find limits for it, we will restrict ourselves for the remainder to a special case for which such an auxiliary function can be majorized and thereby a simple error estimate obtained. This special case is nevertheless still sufficiently general to cover many technically important problems. The additional assumption which we make is that we can find an ellipse E with axes parallel to the co-ordinate axes which has the following properties: no point of E lies inside of Γ and its semi-axes p and q are such that

$$Q = \frac{a}{p^2} + \frac{c}{q^2} - \frac{|d|}{p} - \frac{|e|}{q} > 0 \tag{1.19}$$

everywhere in B.

Let (x_c, y_c) be the centre of this ellipse and consider the auxiliary function

$$Z(x, y) = \mu + \beta \left[1 - \left(\frac{x - x_c}{p} \right)^2 - \left(\frac{y - y_c}{q} \right)^2 \right] = \mu + \beta \cdot \varphi(x, y), \tag{1.20}$$

in which μ and β are positive constants to be determined. Firstly β is chosen so that

$$-\frac{1}{h^2} l_{ik}[Z] \geq h^2 C_1 \quad \text{in } B. \tag{1.21}$$

From (1.11), (1.12) we have

$$-\frac{1}{h^2} l_{ik}[Z] = -\left(L[Z]\right)_{ik} - \frac{1}{h^2} R_{ik}[Z] = -\left(L[Z]\right)_{ik} = \mu g + 2\beta \times$$

$$\times \left\{ \frac{a}{p^2} + \frac{c}{q^2} + d\,\frac{(x - x_c)}{p^2} + e\,\frac{(y - y_c)}{q^2} + \frac{g}{2}\left[1 - \left(\frac{x - x_c}{p}\right)^2 - \left(\frac{y - y_c}{q}\right)^2 \right] \right\}.$$

Now for the ellipse E, (1.19) holds and also

$$|x - x_c| \leq p, \quad |y - y_c| \leq q, \quad 1 - \left(\frac{x - x_c}{p}\right)^2 - \left(\frac{y - y_c}{q}\right)^2 \geq 0 \quad \text{in } B;$$

consequently

$$-\frac{1}{h^2} l_{ik}[Z] \geq 2\beta\, Q \geq h^2 C_1, \tag{1.22}$$

in which the last inequality determines a lower bound for β. Thus we can choose

$$\beta = \frac{h^2}{2}\left(\frac{C_1}{Q}\right)_{\text{max in } B}. \tag{1.23}$$

We now choose μ so that the result corresponding to (1.21) for boundary pivots also holds. In fact we verify that $\mu = \beta + 3h^2 M_2$ is a suitable choice. For a typical boundary pivot A and its associated interior pivot C, we have

$$Z(A) - \frac{\delta}{h + \delta} Z(C) = \mu\,\frac{h}{h + \delta} + \beta\left(\varphi(A) - \frac{\delta}{h + \delta}\,\varphi(C)\right)$$

$$= \beta\left[\frac{h}{h + \delta} + \varphi(A) - \frac{\delta}{h + \delta}\,\varphi(C)\right] + 3h^2 M_2\,\frac{h}{h + \delta}.$$

Here, the last term is at least $\frac{3}{2}h^2 M_2$ and, since $0 \leq \delta \leq h$, $0 \leq \varphi(A) \leq 1$, $0 \leq \varphi(C) \leq 1$, the factor multiplying β is non-negative (in fact it lies between 0 and 2); thus the right-hand side cannot be less than the upper bound $\frac{3}{2}h^2 M_2$ of (1.17) for $|R_A|$ and we have

$$\left| \varepsilon(A) - \frac{\delta}{h + \delta}\,\varepsilon(C) \right| \leq \frac{3}{2}\,h^2 M_2 \leq Z(A) - \frac{\delta}{h + \delta}\,Z(C). \tag{1.24}$$

This corresponds to

$$\left| -\frac{1}{h^2} l_{ik}[\varepsilon] \right| \leq -\frac{1}{h^2} l_{ik}[Z]$$

for interior pivots, which follows from (1.14), (1.15), (1.22).

The matrix of the system of these inequalities written down for every pivot satisfies the conditions of Theorem 2 of Ch. I, § 5.5, as we have already seen in § 1.2; we can therefore use the monotonic property which is assured by that theorem to obtain the error estimate

$$|\varepsilon(A)| \leq Z(A) \quad \text{and} \quad |\varepsilon_{ik}| \leq Z_{ik} \quad \text{in } B, \tag{1.25}$$

or, using the maximum value of Z from (1.20),

$$|U_{ik} - u_{ik}| \leq \mu + \beta, \tag{1.26}$$

where

$$\beta = \frac{h^2}{24} \left(\frac{M_4(a+c) + 2M_3(|d| + |e|)}{\dfrac{a}{p^2} + \dfrac{c}{q^2} - \dfrac{|d|}{p} - \dfrac{|e|}{q}} \right)_{\text{max in } B} \tag{1.27}$$

and

$$\mu = \beta + 3h^2 M_2. \tag{1.28}$$

When all boundary pivots actually lie on the boundary curve Γ, no equations of the type (1.4) are needed and the estimate (1.26), (1.27) is valid with $\mu = 0$.

The special case with the differential equation

$$a(x, y) \frac{\partial^2 u}{\partial x^2} + c(x, y) \frac{\partial^2 u}{\partial y^2} = r(x, y) \tag{1.29}$$

includes many important problems. Since $d = e = 0$ for this case, the expression (1.27) for β simplifies considerably to

$$\beta = h^2 \frac{M_4 \varrho^2}{24} \tag{1.30}$$

if we choose $p = q = \varrho$ as the radius of a circle enclosing B.

1.4. An error estimate for the iterative solution of the difference equations

For the class of problems considered in § 1.2 a very simple error estimate for the iteration procedure described there can be derived by using again the monotonic property assured by Theorem 2 of Ch. I, § 5.5. In § 1.2 it has already been established that the matrix of coefficients associated with the system of equations (1.3), (1.4) satisfies the conditions of that theorem. The practical application of the consequent possibility of bracketing the solution of the difference equations is described by means of an example in § 1.6, where it is used for error estimation in connection with the relaxation method. We use it here in a different way to derive an error estimate of more general type.

Suppose that we have continued the iteration process until the numbers have settled to the accuracy of the number of decimals carried, say m. We then have approximations $U_{ik}^{[\nu]}$ for which no further iterations improve the last decimal. This, however, is no guarantee that the last decimal is correct; in fact, as is well illustrated by an example given by WOLF[1], it can still differ considerably from the corresponding figure in the exact solution U_{ik} of the system of equations (1.3), (1.4). Consequently one is interested in having an indication of how great this departure $\zeta_{ik}^{[\nu]} = U_{ik}^{[\nu]} - U_{ik}$ can be for known changes $\sigma_{ik}^{[\nu]} = U_{ik}^{[\nu+1]} - U_{ik}^{[\nu]}$.

From (1.3) and (1.7) it follows that

$$
\left.
\begin{aligned}
\sigma_{ik}^{[\nu]} &= \frac{h^2}{2a_{ik} + 2c_{ik} + g_{ik}h^2} \left[\frac{l_{ik}^{*}[U^{[\nu]}]}{h^2} - r_{ik} \right] - U_{ik}^{[\nu]} \\
&= \frac{h^2}{2a_{ik} + 2c_{ik} + g_{ik}h^2} \left[\frac{l_{ik}[U^{[\nu]}]}{h^2} - r_{ik} \right] \\
&= \frac{h^2}{2a_{ik} + 2c_{ik} + g_{ik}h^2} \left[\frac{l_{ik}[\zeta^{[\nu]}]}{h^2} \right];
\end{aligned}
\right\} \tag{1.31}
$$

similarly from (1.4) and (1.8) we have

$$
\left.
\begin{aligned}
\sigma^{[\nu]}(A) &= U^{[\nu+1]}(A) - U^{[\nu]}(A) = \frac{\delta}{h+\delta} U^{[\nu]}(C) + \\
&+ \frac{h}{h+\delta} \bar{u}(B) - U^{[\nu]}(A) = \frac{\delta}{h+\delta} \zeta^{[\nu]}(C) - \zeta^{[\nu]}(A).
\end{aligned}
\right\} \tag{1.32}
$$

Thus the errors $\zeta_{ik}^{[\nu]}$, $\zeta^{[\nu]}(A)$ satisfy a system of equations with the same matrix of coefficients as the system (1.14), (1.17) but with different right-hand sides. The $-\frac{1}{h^2} R_{ik}[u]$ and $-R_A$, which are bounded by $\pm h^2 C_1$ and $\pm \frac{3}{2}h^2 M_2$, respectively, are replaced here by $\frac{2a_{ik} + 2c_{ik} + g_{ik}h^2}{h^2} \sigma_{ik}^{[\nu]}$ and $\sigma^{[\nu]}(A)$, respectively. Since all the changes $\sigma_{ik}^{[\nu]}$, $\sigma^{[\nu]}(A)$ are known, we can find bounds for the new right-hand sides, and we have precisely the same situation as in §1.3. Consequently the error estimate given there [(1.26) to (1.28)] can be carried over immediately with the substitutions just mentioned:

$$
|\zeta_{ik}^{[\nu]}| = |U_{ik}^{[\nu]} - U_{ik}| \leq \mu^{*} + \beta^{*}, \tag{1.33}
$$

where

$$
\beta^{*} = \frac{1}{h^2} \left(\frac{a + c + \frac{1}{2}g h^2}{\frac{a}{p^2} + \frac{c}{q^2} - \frac{|d|}{p} - \frac{|e|}{q}} \right)_{\text{max in } B} \cdot |\sigma_{ik}^{[\nu]}|_{\text{max}} \tag{1.34}
$$

and

$$
\mu^{*} = \beta^{*} + 2 |\sigma^{[\nu]}(A)|_{\text{max}}. \tag{1.35}
$$

In particular, for the differential equation (1.29) we can use

$$
\beta^{*} = \frac{\varrho^2 |\sigma_{ik}^{[\nu]}|_{\text{max}}}{h^2}, \tag{1.36}
$$

where ϱ has the same significance as in (1.30).

---

[1] See pp. 130—131 of the paper by F. WOLF already cited in § 1.2 [Z. Angew. Math. Mech. 6, 118—150 (1926)].

If this iterative calculation is carried out with m decimals, then for the stage at which the numbers settle we can put $|\sigma^{[\nu]}|_{max} = \frac{1}{2} \times 10^{-m}$. If, for instance, $\varrho = 4$, $h = 0.1$ and $m = 4$, β^* can be as large as 0.08 and we cannot even be sure of the first decimal. This situation worsens as h decreases, and often it is not allowed for sufficiently.

1.5. Examples of the application of the ordinary finite-difference method

I. A problem in plane potential flow. We consider the flow of an incompressible "ideal" fluid through a two-dimensional channel in which there is a right-angle bend. Let ABC and DEF be the walls of the channel as in Fig. V/4. We are to calculate the velocity distribution and streamlines within the region $ABCDEF$ assuming that the fluid flows in with unit velocity across the entrance AF and out with unit velocity across the exit CD. The velocity components v_x and v_y can be written as the partial derivatives of a stream-function Ψ or of a potential Φ:

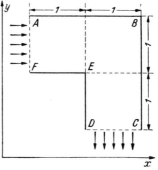

Fig. V/4. Flow through a bent channel

$$v_x = \frac{\partial \Phi}{\partial x} = \frac{\partial \Psi}{\partial y}, \quad v_y = \frac{\partial \Phi}{\partial y} = -\frac{\partial \Psi}{\partial x};$$

$$\nabla^2 \Phi = \nabla^2 \Psi = 0.$$

If we work in terms of Φ, we have the "second boundary-value problem" of potential theory, for the normal derivative $\partial \Phi / \partial \nu$, representing the normal component of the velocity (ν directed inwards), is prescribed on the whole of the boundary. Thus we have $\nabla^2 \Phi = 0$ within the region $ABCDEF$ and

$$\frac{\partial \Phi}{\partial \nu} = \begin{cases} 0 \text{ along } ABC \text{ and } DEF, \\ 1 \text{ along } AF, \\ -1 \text{ along } CD. \end{cases}$$

The second boundary-value problem of potential theory may be reduced to the first by the introduction of the conjugate potential function, which in this case coincides with the stream-function Ψ ($= -u$ say). (The present problem can also be formulated directly in terms of the stream-function.)

The required streamlines are therefore the lines $u = $ constant, where u is the harmonic function determined by the boundary values

$$u = \begin{cases} 0 \text{ along } ABC, \\ 1 \text{ along } DEF, \\ \text{linearly increasing along } AF \text{ and } CD. \end{cases}$$

(a) First of all a rough idea of the distribution of u values is obtained by using a coarse mesh with $h = \frac{1}{3}$. If we make use of the symmetry about BE, the number of unknowns reduces to nine; we denote them by $a, b, \ldots, i$ at the points indicated in Fig. V/5. We use the difference equation (1.3) to express $c, d, e, f, \ldots, i$ successively in terms of a and b:

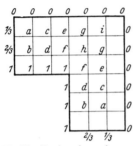

Fig. V/5. The boundary-value problem for the conjugate potential function

$$c = 4a - b - \tfrac{1}{3},$$
$$d = -a + 4b - \tfrac{5}{3},$$
$$e = 16a - 8b + \tfrac{1}{3},$$
$$\ldots\ldots\ldots\ldots\ldots;$$

then the difference equations for the two pivots on the line of symmetry yield a pair of simultaneous equations for a and b with the solution

$$a = \frac{5021}{15\,525} \quad \text{and} \quad b = \frac{10\,177}{15\,525},$$

which, substituted back, gives the approximate values

0	0	0	0	0	0	0
0·333 33	0·323 41	0·304 80	0·263 77	0·182 54	0·091 27	0
0·666 67	0·655 52	0·632 01	0·567 73	0·375 14	0·182 54	0
1	1	1	1	0·567 73	0·263 77	0

(b) We next illustrate LIEBMANN's iteration method on the equations obtained using a slightly finer mesh with $h = \tfrac{1}{4}$. The iteration is started with the following values, which were obtained by graphical interpolation from the values calculated in (a) with $h = \tfrac{1}{3}$:

0	0	0	0	0	0	0	0	0
0·25	0·25	0·24	0·22	0·19	0·15	0·10	0·05	0
0·50	0·50	0·49	0·46	0·39	0·30	0·20		0
0·75	0·75	0·74	0·71	0·65	0·48			0
1	1	1	1	1				0

We calculate new values at each interior pivot from the four neighbouring values by the averaging procedure of (1.9). The results are set out below: beneath each new value we record in brackets the change from the old value (in fourth-decimal units), which may be regarded as a correction.

0	0	0	0	0	0	0	0	0
0·25	0·2475	0·24	0·2225	0·19	0·1475	0·10	0·05	0
	(−25)	(0)	(+25)	(0)	(−25)	(0)	(0)	
0·50	0·4975	0·4850	0·4525	0·40	0·3050	0·20		0
	(−25)	(−50)	(−75)	(+100)	(+50)	(0)		
0·75	0·7475	0·7375	0·7125	0·6450	0·4750			0
	(−25)	(−25)	(+25)	(−50)	(−50)			
1	1	1	1	1				0

By applying the same averaging procedure to these changes we obtain the changes for the next cycle, which in turn yield the changes for the following cycle (in brackets):

```
0    0          0          0          0          0         0          0         0
0 -  6(-10)  -13(+14)  -19(+ 9)  +25(- 6)  +13(+ 6)  -  6(+10)  0(-3)  0
0 -25(-12)  -31(- 6)  +25(-27)  -19(+19)  +  6(+ 5)  +25(0)              0
0 -13(- 9)  -13(-20)  -38(+ 8)  +19(-14)     0(+13)                      0
0    0          0          0          0                                   0
```

One can continue averaging the changes, then calculate "best" values by extrapolation and repeat the whole process again with these as starting values. In general the convergence of the process is slow and many suggestions for improving it have been made[1]. These do not always have the desired effect; for example, an extrapolation based on a geometric series can, in fact, make matters worse. (A different type of correction calculation using relaxation is described in § 1.6.) In our example here a number of further iterations yields the following results, for which the next iteration would produce a maximum change of 0·00003:

```
0       0         0         0         0         0         0         0         0
0·25   0·24404   0·23517   0·21948   0·19217   0·15028   0·10135   0·05068   0
0·50   0·49110   0·47716   0·45064   0·39898   0·30756   0·20447             0
0·75   0·74323   0·73185   0·70701   0·64564   0·47660                       0
1       1         1         1         1                                       0
```

One normally increases the accuracy by using a smaller mesh width, but here the accuracy can also be increased by subtracting out the singularity[2]. This can be a useful expedient in many other situations in which singularities appear.

II. An equation of more general type (1.2). If a single turn of a helical spring of small angle α and radius R is deformed into a plane ring under the influence of an axial load, the stress-function Φ can be shown to satisfy the dif-

---

[1] WELLER, R., G. H. SHORTLEY and B. FRIED: J. Appl. Phys., Lancaster Pa. **9**, 334—344 (1939); **11**, 283—290 (1940).

[2] This device is used by S. GERSCHGORIN: Z. Angew. Math. Mech. **10**, 373—382 (1930); see also W. E. MILNE: Numerical solution of differential equations, p. 221. New York and London 1953. — WOODS, L. C.: The relaxation treatment of singular points in Poissons equation. Quart. J. Mech. Appl. Math. **6**, 163—185 (1953), deals with three types of singularities: 1. A logarithmic singularity in u at a point P, which can be subtracted out. 2. A discontinuity in u at P; for instance, there may be a jump in the given boundary values at a point P on the boundary [for plane potential problems one can subtract out the function $a\varphi$, where (r, φ) are polar co-ordinates centred at P and a is suitably chosen]. 3. The derivatives of u are unbounded at P although u is continuous there (this is the case in the above example). With suitable α, β, m one can subtract the function $(\alpha \sin m\varphi + \beta \cos m\varphi) r^m$.

ferential equation[1]

$$\frac{\partial^2 \Phi}{\partial x^2} + \frac{\partial^2 \Phi}{\partial y^2} + \frac{3}{R-y}\frac{\partial \Phi}{\partial y} - 2G\lambda = 0$$

and the boundary condition

$$\Phi = 0 \quad \text{on } \Gamma,$$

where Γ is the boundary of the cross-section in the (x, y) plane (containing the axis of the spring) (Fig. V/6). G is the modulus of rigidity and $\lambda = \dfrac{\sin\alpha\cos\alpha}{R}$.

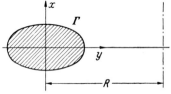

Fig. V/6. Cross-section of a coil of a helical spring

The shear-stress components are given by

$$\tau_y = \left(\frac{R}{R-y}\right)^2 \frac{\partial \Phi}{\partial x},$$

$$\tau_x = -\left(\frac{R}{R-y}\right)^2 \frac{\partial \Phi}{\partial y}.$$

We consider the special case with the rectangular cross-section

$$|x| \leq \tfrac{1}{2}, \quad |y| \leq 1$$

and $R = 5$. Then, in terms of a new dependent variable u defined by

$$\Phi = -2G\lambda u,$$

the problem reads

$$u_{xx} + u_{yy} + \frac{3}{5-y} u_y + 1 = 0, \quad u = 0 \text{ on } \Gamma.$$

(a) For a coarse mesh with $h = \tfrac{1}{2}$ there are only the three unknown u values a, b, c at the points indicated in Fig. V/7. They satisfy the difference equations

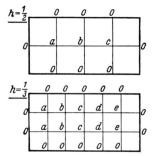

Fig. V/7. Notation for pivotal values

$$4(b - 4a) + \frac{3}{5+\tfrac{1}{2}}\frac{b-0}{1} + 1 = 0,$$

$$4(c + a - 4b) + \frac{3}{5}\frac{c-a}{1} + 1 = 0,$$

$$4(b - 4c) + \frac{3}{5-\tfrac{1}{2}}\frac{0-b}{1} + 1 = 0$$

with the solution

$$a = \frac{1379}{16 \times 929} \fallingdotseq 0.09277,$$

$$b = \frac{99}{929} = 0.10657,$$

$$c = \frac{1259}{16 \times 929} = 0.08470.$$

(b) For a mesh with $h = \tfrac{1}{4}$ we have, on account of symmetry, just five unknown values a, b, c, d, e as in Fig. V/7. If we express four of these unknowns in terms of the other, say e, by means of the difference equations, then the last equation determines e:

$$d = 3.391304\,e - \tfrac{1}{8} \times 1.130435,$$
$$c = 10.154783\,e - \tfrac{1}{8} \times 4.918261,$$
$$b = 29.70435\,e - \tfrac{1}{8} \times 16.12367,$$
$$a = 86.07586\,e - \tfrac{1}{8} \times 48.54253,$$
$$7680.677\,e - \tfrac{1}{8} \times 4388.762 = 0.$$

[1] Biezeno, C. B., and R. Grammel: Technische Dynamik, 2nd ed., Vol. I, p. 351. 1953.

Therefore

$$a = 0.07128,$$
$$b = 0.09439,$$
$$c = 0.09825,$$
$$d = 0.08971,$$
$$e = 0.0634892,$$
$$\tfrac{9}{8}c = 0.11053.$$

The last value $\tfrac{9}{8}c$ corresponds to the centre of the rectangle.

(c) For the finer mesh with $h = \tfrac{1}{4}$ the difference equations are best solved by iteration. With mesh points $(x_i, y_k) = (ih, kh)$ the equations can be put in the form

$$U_{i,k} = \frac{1}{4}\,(U_{i+1,k} + U_{i-1,k} + U_{i,k+1} + U_{i,k-1}) + \frac{3}{32}\,\frac{1}{5-y}\,(U_{i+1,k} - U_{i-1,k}) + \frac{1}{64}\,.$$

As in Example I, we estimate starting values for the U_{ik} from graphs of the results of the calculation with $h = \tfrac{1}{3}$. For convenience we iterate only on the changes. The results are exhibited in Table V/1, whose last row gives the factors $\dfrac{3}{32}\cdot\dfrac{1}{5-y}$, which can be calculated once and for all.

Table V/1. *Results of the iterative solution of the difference equations with $h = \tfrac{1}{4}$*

i	x	$k=-4$ $y=-1$	$k=-3$	$k=-2$	$k=-1$	$k=0$ $y=0$	$k=1$	$k=2$	$k=3$	$k=4$ $y=1$
-2	-0.6	0	0	0	0	0	0	0	0	0
-1	-0.25	0	0.05244	0.07447	0.08289	0.08420	0.08010	0.06912	0.04622	0
0	0	0	0.06794	0.09798	0.10968	0.11151	0.10576	0.09050	0.05936	0
$\dfrac{3}{32}\dfrac{1}{5-y}$			0.016304	0.017046	0.017857	0.018750	0.019736	0.020833	0.022059	

III. A differential equation of the fourth order.

If a non-uniformly loaded square plate of side $2A$ has two opposite sides firmly clamped and the other two smoothly hinged (Fig. V/8), the deflection u will satisfy

$$\nabla^4 u = \frac{p}{N} = \left(\frac{x}{A}\right)^2 \frac{p_0}{N}\,,$$

$$u = \frac{\partial u}{\partial x} = 0 \quad \text{for} \quad |x| = A,$$

$$u = \nabla^2 u = 0 \quad \text{for} \quad |y| = A.$$

Let p_0 and N be constants.

Using the symmetry of the problem, we can reduce the number of unknown values for the mesh width $h = \tfrac{1}{2}A$ to four, namely the a, b, c, d as shown in the figure. In setting up the difference equations, we find that values outside of the square are needed. As indicated in the figure, these can be expressed in terms of the values inside by means of the boundary conditions; for example, the values $-c$, $-d$ follow from the difference equations representing the boundary condition $V^2 u = 0$. Let us put

$$k = \frac{h^4}{2}\frac{p_0}{N} = \frac{q}{128};$$

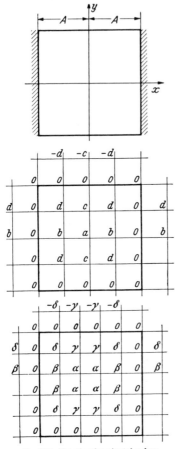

then, using Table VI of the appendix, we can immediately write down the set of difference equations:

$$20a - 16b - 16c + 8d \qquad = 0,$$
$$20b - 8a - 16d + 4c + 2b = 2k,$$
$$20c - 8a - 16d + 4b \qquad = 0,$$
$$20d - 8b - 8c + 2a + 2d = 2k.$$

The solution is

$$a = \frac{310}{497}k = 0\cdot00487\,q,$$

$$b = \frac{257}{497}k = 0\cdot00405\,q,$$

$$c = \frac{227}{497}k = 0\cdot00357\,q,$$

$$d = \frac{193}{497}k = 0\cdot00304\,q.$$

If we choose $h = \tfrac{2}{5}A$, we obtain somewhat more accurate results with still only four unknowns (see Fig. V/8). Thus from the difference equations

$$6\alpha - 5\beta - 5\gamma + 2\delta = s,$$
$$-5\alpha + 13\beta + 2\gamma - 7\delta = 9s,$$
$$-5\alpha + 2\beta + 11\gamma - 7\delta = s,$$
$$2\alpha - 7\beta - 7\gamma + 20\delta = 9s,$$

Fig. V/8. Notation for pivotal values

where $s = \dfrac{h^4}{25}\cdot\dfrac{p_0}{N} = \dfrac{16}{15625}q$, we obtain the values

$$\alpha = \frac{6755}{1522}s = 0\cdot00454\,q, \qquad \beta = \frac{4693}{1522}s = 0\cdot00315\,q,$$

$$\gamma = \frac{4383}{1522}s = 0\cdot00294\,q, \qquad \delta = \frac{3186}{1522}s = 0\cdot00214\,q.$$

1.6. Relaxation with error estimation

The practical application of the ordinary finite-difference method in conjunction with the relaxation procedure[1] will be explained by means

---

[1] See, for example, G. SHORTLEY, R. WELLER, P. DARBY and E. H. GAMBLE: Numerical solution of axisymmetrical problems, with applications to electrostatics

of an example. We consider the problem of determining the steady temperature distribution $u(x, y)$ in a homogeneous square plate of side A whose edges are kept at the constant temperatures $u=0$ and $u=1$ as specified in Fig. V/9; u satisfies the potential equation $\nabla^2 u = 0$.

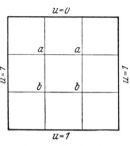

As usual we begin with a coarse mesh so as to get a rough idea of the solution from which we can estimate starting values for the next step. With $h = \frac{1}{3}A$ we have two unknown values a, b (see Fig. V/9); they satisfy the difference equations

$$4a = a + b + 1, \qquad 4b = a + b + 1 + 1,$$

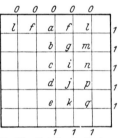

from which we obtain $a = \frac{5}{8}, b = \frac{7}{8}$. These values, rounded to 0·63 and 0·88, are taken over for the first approximation on the finer mesh with $h = \frac{1}{6}A$. Thus we have first approximations to the values g, j in Fig. V/9; we estimate the remaining values by graphical interpolation, dealing with f, i, k first. These starting values are recorded in the upper-left quadrants at the intersections in Table V/2, Stage (i); each intersection corresponds to a pivot. The numbers in the upper-right quadrants are the arithmetic means

Fig. V/9. Notation for the temperature distribution in a square plate

of the values in the four neighbouring upper-left quadrants, i.e. the next approximation in the iteration of § 1.2; for the point a, for instance, this is

$$0\cdot25 \times (0\cdot39 + 0 + 0\cdot39 + 0\cdot59) = 0\cdot3425.$$

In the lower-right quadrants we enter the changes thus produced, expressing them in fourth-decimal units (for example, at a we have $0\cdot3425 - 0\cdot32 = 225 \times 10^{-4}$).

Now if the difference equation for a particular pivot is satisfied by the starting values, the corresponding change must be zero[1]; con-

and torsion. J. Appl. Phys. **18**, 116—129 (1947). A detailed description, which also describes many little artifices, is given by L. Fox: Some improvements in the use of relaxation methods for the solution of ordinary and partial differential equations. Proc. Roy. Soc. Lond., Ser. A **190**, 31—59 (1947). Comprehensive treatments are to be found in the books by SOUTHWELL which were cited earlier and also in F. S. SHAW: An introduction to relaxation methods. New York: Dover Publ., Inc. 1953, 396 pp. A theory is developed by G. TEMPLE: The general theory of relaxation methods applied to linear systems. Proc. Roy. Soc. Lond., Ser. A **169**, 476—500 (1939).

[1] Translator's note: These changes produced in the starting values by one step of the iteration procedure are, in fact, identical with (or at least proportional to) what would in the customary relaxational parlance be called the "residuals" corresponding to the starting values; see Ch. III, § 1.4.

Table V/2. *Various stages in the solution of the difference equations by relaxation*

Stage (i)

	0	0	0·39	0·56	1
0·32	0·3425		0·3775	0·5375	1
		+225	−125	−225	
0·59	0·5825	0·63	0·7625		
	−75	0	+25		
0·75	0·7755	0·78	0·865		
	+50	0	+50		
0·87	0·8625	0·88	0·925		
	−75	0	+50		
0·94	0·9425	0·945	0·9675		
	+25	−50	+75		
1					1

Stage (ii)

	0	0·3313	0·3716	0·3714	0·5316	0·5317
						1
0·5820	−3/0·5819	0·6224	−2/0·6222	0·7552	+1/0·7550
0·7512	−1/0·7512	0·78	−2/0·7799	0·8660	−2/0·8660
0·8628	0/0·8627	0·88	−1/0·8797	0·9288	0/0·9288
0·9396	−1/0·9393	0·9472	−3/0·9472	0·9692	0/0·9690
1	−3		0	1	−2

Stage (iii)

	0	0·33064	0·37082	0·37084	0·53132	0·53134
						1
0·58088	0/0·58085	0·62132	−2/0·62130	0·75452	+2/0·75452
0·75012	−3/0·75011	0·77896	−2/0·77894	0·86544	0/0·86543
0·86164	−1/0·86164	0·87888	−2/0·87885	0·92824	−1/0·92825
0·93868	0/0·93869	0·94656	−3/0·94656	0·96868	+1/0·96870
1	+1		0	1	+2

Table V/3. *Numerical details of the relaxation procedure (all numbers expressed in fourth-decimal units)*

(1) Relaxation of the starting values

Corrections		Changes		
+200	−120	−240		
			[+225] −200 −60	[−125] +50 +60 +120	[−225] +240 −30
			[−75] +50 +15	[−125] +50 −60 +120	[−225] +240 −30
+60			[+50] −60 −15	[0] +15 +15	[+50] −60 +15
−60			[−75] +15 +60	[0] −15 +15 −5	[+50] +15 −60 +20
	−20	+80	[+25] −15 −10	[−50] +20 +20	[+75] +15 −80

(2) Relaxation of the improved values

Corrections		Changes		
−60	−60	−60	[−35] +15 +24 −2 −5	[−15] +15 −6 +4 −1	[−15] +30 −16 +3 +1
−24	−4	+16			
−60	−60	−60	[−10] +15 −6 −8 +20 −12	[−30] +15 +1 −16 −5	[−20] +30 +4 −12 −1
−20	−16	+12			
−48			[−25] −15 −5 +48 −3	[+30] −15 +1 −4 −12	[+5] −15 +7 +3
−12		+28	[0] −1 −12 +12	[0] −1 +2 −3	[+25] −28 +3
−4	−8	+12	[0] +4 −4	[−10] +3 +8	[+5] +7 −12

sequently we relax the starting values, i.e. make corrections to them, so as to make the changes as small as possible. If at the pivot a, for example, we had taken the value 0·34 instead of 0·32, a correction of $+200 \times 10^{-4}$, the change at a would have been only 25 instead of 225 (in 10^{-4} units), but the changes at the four neighbouring pivots would have been greater by $+50$. The procedure is most conveniently carried out in two relaxation tables, one for the corrections and the other for the changes, as in Table V/3. The numbers which are framed in the arrays of changes are the changes current at the start of each table, the other numbers being the alterations produced by the corrections.

For the relaxation of the starting values we note that the correction of $+200$ at a alters the changes by the amounts underlined in the table, i.e. -200 at a and $+50$ at b and f. Similarly we make corrections at all pivots where particularly large changes appear, recording these corrections and the corresponding amounts by which they alter the changes in the appropriate columns. In doing this we must remember that on account of symmetry we are correcting two values at once, except on the line of symmetry, so that any change which is affected by both must be altered accordingly; thus the correction of -120 at f alters the changes at g and l by -30, but the change at a by -60.

It is convenient to review the situation now by calculating the net effect on the changes so far and start again in a fresh table with these new changes as starting changes; these are framed in the table and it can be seen that their maximum absolute value is now 35 instead of 225. We continue the relaxation from these improved values as in the table, making corrections which eventually bring the maximum absolute value of the changes down to 3×10^{-4}; at this stage we must increase the number of decimals in order to be able to make further improvements. The current values of u are easily calculated by adding the corrections to the starting values; for example, for a we have

$$0·32 + 0·0200 - 0·0084 = 0·3316.$$

As a check we carry out the averaging process of the Liebmann iteration on these improved u values and thus make an independent calculation of the new changes [Table V/2, Stage (ii)]. The corresponding results after repeating the procedure with an extra decimal are given in Table V/2, Stage (iii)[1].

---

[1] An "over-relaxation" in which, for the solution of the system of linear equations

$$\sum_{k=1}^{n} a_{jk} x_k = r_j \qquad (j = 1, \ldots, n),$$

one determines a sequence of approximations $x_j^{[\nu]}$ generally by

$$x_j^{[\nu+1]} = x_j^{[\nu]} - \omega \left\{ \sum_{k=1}^{j-1} a_{jk} x_k^{[\nu+1]} + \sum_{k=j}^{n} a_{jk} x_k^{[\nu]} - r_j \right\}$$

is investigated by D. M. YOUNG: Iterative methods for solving partial difference

A point of computational technique may be mentioned here. It is a fact that the magnitudes of the changes can be reduced easily in a region where the changes alternate in sign, but not in a region where they have the same sign. A useful technique for dealing with this latter situation is provided by the so-called "block relaxation", in which one makes identical corrections at a block of points[1]. The general effect achieved by this is typified by the following example, in which corrections of 4α are made at points in a rectangular block:

Corrections						Effect on the changes				
						α	α	α	α	α
4α	4α	4α	4α	4α	α	-2α	$-\alpha$	$-\alpha$	$-\alpha$	-2α	α
4α	4α	4α	4α	4α	α	$-\alpha$	0	0	0	$-\alpha$	α
4α	4α	4α	4α	4α	α	-2α	$-\alpha$	$-\alpha$	$-\alpha$	-2α	α
						α	α	α	α	α

With this characteristic pattern in mind it is easy to determine for any particular case which points to include in the block and what value to take for the correction 4α. An example occurs in Table V/3, (2), in which the changes at the pivots a, b, f, g, l, m are all negative; the block correction of -60 produces the underlined alterations in the table of changes.

A comparison of Stages (ii) and (iii) of Table V/2 reveals that in the values of u the third decimal has altered even though the changes in Stage (ii) are confined to the fourth decimal. The possibility of this "ill-conditioning" has already been mentioned in § 1.4; it can become dangerously bad, particularly when a large number of pivots are used.

Error estimation with relaxation. From § 1.4 and Theorem 2 of Ch. I, § 5.5 we have the following bracketing rule for the solutions of the difference equations of § 1.2:

---

equations of elliptic type. Trans. Amer. Math. Soc. **76**, 92—111 (1954). — On the Solution of Linear Systems by Iteration. Proc. Symp. Appl. Math. **6**, 283—298 (1956). He gives a rule for the suitable choice of the "relaxation factor" ω. Previously L. F. RICHARDSON: Phil. Trans. Roy. Soc. Lond., Ser. A **210**, 307—357 (1911), had used a relaxation factor which even varied from step to step. — See also L. F. RICHARDSON: Phil. Trans. Roy. Soc. Lond., Ser. A **242**, 439—491 (1950). Over-relaxation is also considered by D. N. DE G. ALLEN: La méthode de libération des liaisons … . Colloques Internat. Centre Nat. Rech. Sci. **14** (Méthodes de calcul dans les problèmes de méchaniques) 11—34. Marseilles and Paris 1949.

[1] For a description of block relaxation see R. V. SOUTHWELL: Relaxation methods in theoretical physics, p. 55. Oxford 1946. Numerous applications of block relaxation are to be found in the literature; see, for example, D. C. GILLES: The use of interlacing nets for the application of relaxation methods to problems involving two dependent variables, with a foreword by W. G. BICKLEY. Proc. Roy. Soc. Lond., Ser. A **193**, 407—433 (1948). — DUSINBERRE, G. M.: Numerical analysis of heat flow, 227 pp. New York-Toronto-London 1949; on p. 65 of this book a large number of point patterns for block relaxation are reproduced.

If we have a set of approximate values for the solution of the difference equations of § 1.2 for which the changes are everywhere non-positive (or everywhere non-negative), then each of these values is greater than or equal to (or, respectively, less than or equal to) the corresponding exact value of the solution of the difference equations. (Whether they are greater or less than the corresponding values of the exact solution of the boundary-value problem is another matter).

A set of changes with no variations in sign can usually be achieved quite quickly with block relaxation. Table V/4 shows the relaxations required to bring about all non-positive and all non-negative changes, respectively, for the approximations obtained in our present example at Stage (iii) (Table V/2, upper-left quadrants); all numbers are expressed in fifth-decimal units. In the first half A. the effect of the corrections is given in detail: the framed numbers are the starting changes taken from Table V/2, Stage (iii); the underlined numbers are the alterations produced by the block of five corrections of $+4$ in the third column; combination of all the alterations yields the non-positive new changes. The corresponding approximate solution, which is everywhere greater than or equal to the exact solution of the difference equations, is exhibited in Table V/5 (same arrangement as in Table V/2). The second half B. of Table V/4 gives the corrections necessary to produce all non-negative changes; the corresponding approximate values, which are consequently lower limits, are exhibited in Table V/6. Thus for the value b, for example, we have the limits

$$0.58064 \leqq b \leqq 0.58088.$$

In actual fact, although we have not proved it here, the numbers in the upper-right quadrants of Tables V/5 and V/6, i.e. the next Liebmann iterates, also provide upper and lower limits, respectively, so that the more accurate result

$$0.58066 \leqq b \leqq 0.58085$$

also holds[1]

A generalized form of block relaxation[2] which is usually more effective than the ordinary form for large blocks is obtained by choosing the values of the corrections so that they represent a discrete sub-harmonic function in the sense that the value at each point of the block is greater than or equal to the arithmetic mean of the values at the four neighbouring points, the values outside of the block being zero.

---

[1] For proof see L. COLLATZ: Einschließungssätze bei Iteration und Relaxation. Z. Angew. Math. Mech. **32**, 76—84 (1952).

[2] The "Scheibenrelaxation" of E. STIEFEL: Über einige Methoden der Relaxationsrechnung. Z. Angew. Math. Phys. **3**, 1—33 (1952).

Table V/4. *Relaxation of an approximate solution to obtain changes with no variations in sign*

A. Non-positive changes (for an upper bound)									B. Non-negative changes				
Corrections			Alterations produced			New changes			Corrections			New changes	
0	0	+4	[0]	[−2]+1	[+2]−3	0	−1	−1	−12	−12	− 4	0	+1
0	0	+4	[−3]	[−2]+1	[0]−2	−3	−1	−2	−24	−20	−12	+2	0 +
0	0	+4	[−1]+1	[−2]+1	[−1]−2	0	−1	−3	−24	−24	−12	0	+3
+4	0	+4	[0]−3 (+1 +1)	[−3]+1	[+1]−2	−3	0	−1	−20	−20	− 8	+2	+2
+4	+4	+4	[+1]−3 (+1 +1)	[0]+1 (+1 −4)	[+2]−3 (+1)	0	−2	0	− 8	− 8	− 4	0	0 +

Table V/5. *Upper bounds for the solution of the difference equations*

0		0		0		
0·33064	0·33064	0·37084	0·37083	0·53136	0·53135	1
0·58088	(0) 0·58085	0·62132	(−1) 0·62131	0·75456	(−1) 0·75454	1
0·75012	(−3) 0·75012	0·77896	(−1) 0·77895	0·86548	(−2) 0·86545	1
0·86168	(0) 0·86165	0·87888	(−1) 0·87888	0·92828	(−3) 0·92827	1
0·93872	(−3) 0·93872	0·94560	(0) 0·94558	0·96872	(−1) 0·96872	1
	(0) 1	1	(−2) 1		(0)	

Table V/6. *Lower bounds for the solution of the difference equations*

0		0		0		
0·33052	0·33052	0·37072	0·37073	0·53128	0·53128	1
0·58064	(0) 0·58066	0·62112	(+1) 0·62112	0·75440	(0) 0·75443	1
0·74988	(+2) 0·74988	0·77872	(0) 0·77875	0·86532	(+3) 0·86532	1
0·86144	(0) 0·86146	0·87868	(+3) 0·87870	0·92816	(0) 0·92816	1
0·93860	(+2) 0·93860	0·94648	(+2) 0·94648	0·96864	(0) 0·96866	1
	(0) 1	1	(0) 1		(+2)	

This leaves considerable freedom of choice, and a variety of such group relaxations can easily be constructed; the following is a typical example:

Corrections					Effect on the changes			
					α	2α	2α	α
4α	8α	8α	4α	α	0	−2α	−2α	0	α
8α	12α	12α	8α	2α	−2α	−2α	−2α	−2α	2α
8α	12α	12α	8α	2α	−2α	−2α	−2α	−2α	2α
4α	8α	8α	4α	α	0	−2α	−2α	0	α
					α	2α	2α	α

We conclude this section on relaxation by mentioning a phenomenon which manifests itself in systematic relaxation procedures such as the modification of the Liebmann iteration (1.9) in which one deals with the pivots in a fixed cycle and always uses the latest improved values at the neighbouring pivots. To simplify the description we confine ourselves to two equations in two unknowns:

$$
\left.\begin{array}{l}
a_{11}x_1 + a_{12}x_2 = r_1 \\
a_{21}x_1 + a_{22}x_2 = r_2.
\end{array}\right\} \quad (1.37)
$$

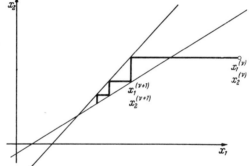

Consider the iteration procedure defined by

$$x_1^{[v+1]} = \frac{1}{a_{11}}\left(r_1 - a_{12}x_2^{[v]}\right),$$

$$x_2^{[v+1]} = \frac{1}{a_{22}}\left(r_2 - a_{21}x_1^{[v+1]}\right)$$

$$(v = 0,1,2,\ldots)$$

Fig. V/10. A drawback of systematic relaxation procedures due to the occurence of "cages"

(the single-step iteration method of Ch. I, § 5.5) and assume that it converges. The procedure admits of a simple geometrical representation if the two lines defined by (1.37) are drawn in the (x_1, x_2) plane, as in Fig. V/10. The zigzag line connects the sequence of points $(x_1^{[v]}, x_2^{[v]})$, $(x_1^{[v+1]}, x_2^{[v]})$, $(x_1^{[v+1]}, x_2^{[v+1]})$, ... and we see immediately that for k greater than a certain value k_0 (here $k_0 = 0$) the points $(x_1^{[k]}, x_2^{[k]})$ all remain in a fixed "wedge", which STIEFEL (see last footnote) calls a "Käfig", i.e. a "cage". When the angle of the wedge is small, the convergence of the sequence of points to the solution, i.e. to the point of intersection, is slow. Similar "cages" occur also with larger systems of equations; they cause the situation in which substantial improvements are obtained from the first stage of a systematic relaxation procedure (getting into the cage) but only moderate improvements from subsequent stages.

1.7. Three independent variables (spatial problems)

For most problems with more than two independent variables solution by finite differences is very tedious, for as soon as one starts to refine the mesh the number of unknown pivotal values increases very rapidly. Nevertheless one can easily think of examples for which the finite-difference method is the only really feasible numerical method. It may be remarked that the extension to more than two independent variables introduces no new fundamental difficulties.

We choose for illustration an example possessing a high degree of symmetry; this permits a considerable reduction in the number of unknowns. Suppose that a homogeneous material occupies the cube $|x| \leq 1$, $|y| \leq 1$, $|z| \leq 1$ and that two opposite vertices are maintained at the temperatures $+1$ and -1, respectively, while the rest of the surface is insulated. Then the steady temperature distribution satisfies the boundary-value problem

$\nabla^2 u = 0$ inside the cube,

$\dfrac{\partial u}{\partial v} = 0$ for $|x| = 1$, $|y| = 1$, $|z| = 1$,

$u = 1$ for $x = y = z = 1$ and $u = -1$ for $x = y = z = -1$.

For a three-dimensional square mesh with mesh width $h = \frac{2}{3}$ we have the nine unknown pivotal values $a, b, \ldots, k$ as shown in Fig. V/11. The boundary condition $\partial u/\partial v = 0$ is satisfied by introducing symmetrical function values at neighbouring points outside of the cube, as shown in Fig. V/11 for a few typical points. It is expedient in setting up the difference equations corresponding to $\nabla^2 u = 0$ (formula in Table VI of the appendix) to make a sketch as in Fig. V/11. We obtain the equations

$$
\begin{aligned}
-6a + b + 4d + 1 &= 0, \\
a - 6b + c + 4e &= 0, \\
b - 3c + 2f &= 0, \\
a - 3d + e + i &= 0, \\
b + d - 6e + f + g + 2k &= 0, \\
c + 2e - 7f - 2g &= 0, \\
e - f - 3g - k &= 0, \\
d - 2i + k &= 0, \\
2e - g + i - 8k &= 0,
\end{aligned}
$$

which yield

$a = \dfrac{1367}{5147} = 0{\cdot}2656,$ $b = \dfrac{415}{5147} = 0{\cdot}0807,$ $c = \dfrac{199}{5147} = 0{\cdot}0387,$

$d = \dfrac{660}{5147} = 0{\cdot}1282,$ $e = \dfrac{231}{5147} = 0{\cdot}0450,$ $f = \dfrac{91}{5147} = 0{\cdot}0177,$

$g = \dfrac{12}{5147} = 0{\cdot}0023,$ $i = \dfrac{382}{5147} = 0{\cdot}0743,$ $k = \dfrac{104}{5147} = 0{\cdot}0202.$

In some cases it is possible to transform the spatial problem into a plane problem. For example, TRANTER[1] reduces the first boundary-value problem for the equation

$$\nabla^2 u + f(x, y, z) = 0 \qquad (1.38)$$

in a cylinder, i.e. with the boundary conditions

$u = h_1(x, y)$ for $z = 0$
$u = h_2(x, y)$ for $z = \pi$
$u = g(x, y, z)$ on the curved surface of a cylinder parallel to the z axis,

to a set of first boundary-value problems in the region of the (x, y) plane representing the cross-section of the cylinder by means of finite Fourier transforms

$$v(n) = v(n, x, y) = \int_0^\pi u \sin nz \, dz$$

$$(n = 1, 2, 3, \ldots).$$

If (1.38) is multiplied by $\sin nz$ and integrated with respect to z from 0 to π, integration by parts yields a two-dimensional partial differential equation for $v(n)$:

$$\left.
\begin{aligned}
&\left(\frac{\partial^2}{\partial x^2} + \frac{\partial^2}{\partial y^2} - n^2\right) v(n) + \\
&+ n[h_1 - (-1)^n h_2] + \\
&+ \int_0^\pi f \sin nz \, dz = 0.
\end{aligned}
\right\} \quad (1.39)$$

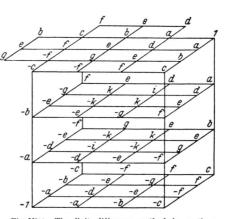

Fig. V/11. The finite-difference method for a three-dimensional temperature distribution

On the boundary of the cross-section $v(n)$ must take the value

$$\int_0^\pi g \sin nz \, dz.$$

This boundary-value problem is solved approximately for several values of n by the finite-difference method and relaxation, then u is determined by Fourier synthesis.

TRANTER also considers the mixed boundary-value problem with the normal derivative prescribed on the plane ends of the cylinder; for this he uses the cosine transform

$$v(n) = \int_0^\pi u \cos nz \, dz.$$

The appropriate transforms for the corresponding problems with a semi-infinite cylinder are

$$v(\xi) = \int_0^\infty u \begin{Bmatrix} \sin \\ \cos \end{Bmatrix} \xi z \, dz.$$

This technique of reducing the number of independent variables by means of integral transforms (Fourier, Laplace and allied transforms all have their particular applications) can be used in a variety of problems.

---

[1] TRANTER, C. J.: The combined use of relaxation methods and Fourier transforms in the solution of some three-dimensional boundary value problems. Quart. J. Mech. Appl. Math. 1, 281—286 (1948); a numerical example is given.

1.8. Arbitrary mesh systems

A formula which approximates a differential equation to a certain order of accuracy by a relation between pivotal values (i.e. function values at mesh points) generally requires more of these pivotal values when the mesh is chosen to be other than rectangular[1] (here we are again restricting ourselves to the plane). For example, in the case of the second-order differential equation

$$L[u] = A u_{xx} + 2B u_{xy} + C u_{yy} + 2D u_x + 2E u_y + 2F u = r(x, y) \qquad (1.40)$$

(with constant $A, B, \ldots, F$) the corresponding relation, centred on a mesh point P_0 say, must in general involve also the values at five neighbouring points $P_1, \ldots, P_5$; only in special cases does the number reduce to four. If we denote the expression in the pivotal values by

$$Q = \sum_{\nu=0}^{5} a_\nu u_\nu,$$

where u_ν is the value of u at the point P_ν with co-ordinates x_ν, y_ν, and expand these function values by TAYLOR's theorem at the point P_0 (which, for simplicity, we take to be the origin $x_0 = 0, y_0 = 0$):

$$Q = u_0 \sum_{\nu=0}^{5} a_\nu + \left(\frac{\partial u}{\partial x}\right)_0 \sum_{\nu=0}^{5} a_\nu x_\nu + \left(\frac{\partial u}{\partial y}\right)_0 \sum_{\nu=0}^{5} a_\nu y_\nu + \frac{1}{2}\left(\frac{\partial^2 u}{\partial x^2}\right)_0 \sum_{\nu=0}^{5} a_\nu x_\nu^2 + \cdots,$$

then by comparing with the given differential expression $L[u]$ we obtain (with an arbitrary constant ϱ) the six equations

$$\sum_{\nu=0}^{5} a_\nu x_\nu^2 = \varrho A, \qquad \sum_{\nu=0}^{5} a_\nu x_\nu y_\nu = \varrho B, \qquad \sum_{\nu=0}^{5} a_\nu y_\nu^2 = \varrho C,$$

$$\sum_{\nu=0}^{5} a_\nu x_\nu = \varrho D, \qquad \sum_{\nu=0}^{5} a_\nu y_\nu = \varrho E, \qquad \sum_{\nu=0}^{5} a_\nu = \varrho F$$

for the a_ν. Since a_0 appears in the last equation only, the other equations provide five linear equations for $a_1, \ldots, a_5$. We now ask in what circumstances just four neighbouring values, say $u_1, \ldots, u_4$ (so that $a_5 = 0$), will be sufficient. With $a_5 = 0$ the first five equations constitute five homogeneous equations for $a_1, a_2, a_3, u_4, \varrho$ and will possess a non-trivial solution if and only if the determinant vanishes:

$$\begin{vmatrix} x_1^2 & x_2^2 & x_3^2 & x_4^2 & A \\ x_1 y_1 & x_2 y_2 & x_3 y_3 & x_4 y_4 & B \\ y_1^2 & y_2^2 & y_3^2 & y_4^2 & C \\ x_1 & x_2 & x_3 & x_4 & D \\ y_1 & y_2 & y_3 & y_4 & E \end{vmatrix} = 0.$$

There are therefore five constants $\alpha, \beta, \gamma, \delta, \varepsilon$, not all zero, such that

$$\alpha x_j^2 + \beta x_j y_j + \gamma y_j^2 + \delta x_j + \varepsilon y_j = 0 \quad \text{for} \quad j = 1, 2, 3, 4$$

and

$$\alpha A + \beta B + \gamma C + \delta D + \varepsilon E = 0. \qquad (1.41)$$

This means that the four points P_1, P_2, P_3, P_4 lie with P_0 on the conic

$$\alpha x^2 + \beta x y + \gamma y^2 + \delta x + \varepsilon y = 0,$$

where $\alpha, \beta, \gamma, \delta, \varepsilon$ must satisfy the last equation of (1.41).

---

[1] See R. v. MISES: On network methods in conformal mapping and in related problems. Nat. Bur. Stand., Appl. Math. Ser. **18**, 1−7 (1952).

For the equation $\nabla^2 u = r(x, y)$, in which $A = C$, $B = D = E = 0$, we have $\gamma = -\alpha$ and the conic is a rectangular hyperbola; the degenerate case of a pair of orthogonal straight lines is the conic used with a rectangular mesh.

1.9. Solution of the difference equations by finite sums

We present the method[1] by describing its application to the first boundary, value problem of potential theory for a rectangular region: $\nabla^2 u = 0$ in $0 \leq x \leq a$ $0 \leq y \leq b$ and u prescribed on the boundary. For a mesh with the points

$$x_j = j h, \quad y_k = k l, \quad h = \frac{a}{M + 1}, \quad l = \frac{b}{N + 1}$$

$$(j = 0, 1, \ldots, M + 1; \; k = 0, 1, \ldots, N + 1)$$

the difference equations [(1.3) with $a_{i,k} = c_{i,k} = 1$, $d_{i,k} = e_{i,k} = g_{i,k} = r_{i,k} = 0$] have particular solutions of the form

$$\lambda^k \sin \frac{n \pi j h}{a},$$

where λ, $\varrho = \frac{n \pi h}{a}$ and $r = \frac{l}{h}$ are related by the equation

$$2 r^2 (\cos \varrho - 1) + \left(\lambda + \frac{1}{\lambda} - 2 \right) = 0.$$

For given r and ϱ this yields two values for λ:

$$\left.\begin{matrix} \lambda_1 \\ \lambda_2 \end{matrix}\right\} = \left\{\begin{matrix} \lambda_{1,n} \\ \lambda_{2,n} \end{matrix}\right\} = \mu \pm \sqrt{\mu^2 - 1},$$

where $\mu = 1 - r^2 (\cos \varrho - 1)$.

Consequently the finite sum

$$u' = \sum_{n=1}^{M} \sin \frac{n \pi j h}{a} (P_n' \lambda_{1,n}^k + Q_n' \lambda_{2,n}^k), \tag{1.42}$$

in which P_n', Q_n' are constants to be determined, is a solution of the difference equations. It vanishes on the boundaries $x = 0$ and $x = a$ and by suitable choice of P_n', Q_n' can be made to take the prescribed boundary values of u on the boundaries $y = 0$ and $y = b$. If u'' is the corresponding finite sum which vanishes on $y = 0$ and $y = b$ and takes the prescribed boundary values on $x = 0$ and $x = a$, then $u' + u''$ will be the required solution of the difference equations.

To determine P_n', Q_n' we first put $y = 0$ and obtain from (1.42)

$$u(j h, 0) = \sum_{n=1}^{M} (P_n' + Q_n') \sin \frac{n \pi j h}{a}.$$

Thus the quantities $P_n' + Q_n'$ are the coefficients in a harmonic analysis of the given function $u(j h, 0)$ and can be determined by any of the well-known methods (RUNGE's scheme, etc.). Similarly, putting $y = b$, we have

$$u(j h, b) = \sum_{n=1}^{M} (P_n' \lambda_{1,n}^{N+1} + Q_n' \lambda_{2,n}^{N+1}) \sin \frac{n \pi j h}{a},$$

and the quantities $P_n' \lambda_{1,n}^{N+1} + Q_n' \lambda_{2,n}^{N+1}$ can be determined in the same way. We then have two equations for P_n' and Q_n'. The u values at the individual mesh points can be calculated from the difference equations once the values on two consecutive rows have been obtained by evaluating the finite sums.

[1] HYMAN, M. A.: Non-iterative numerical solution of boundary value problems. Appl. Sci. Res. B **2**, 325–351 (1952).

Similar finite-sum solutions can also be used for the second and third boundary-value problems and for differential equations of higher order with constant coefficients (the biharmonic equation, for example). The method can be carried over to problems in three or more dimensions by using multiple sums.

1.10. Simplification of the calculation by decomposition of the finite-difference equations[1]

Solution of the difference equations corresponding to a linear boundary-value problem with a differential equation of the form (1.2) can often be simplified as follows. Firstly, in addition to assuming the existence of a solution, we assume that the linear difference equation associated with a boundary pivot P as a representation of the boundary condition involves, in addition to the value $U(P)$ (which must occur), only the approximate values U at neighbouring pivots. Those pivots P_i $(i = 1, 2, \ldots, n)$ for which the value of U is not given immediately by the boundary condition are now divided (chequer-wise) into two sets A $(i = 1, 2, \ldots, p$; we can take $p \leq \frac{1}{2}n)$ and B $(i = p + 1, \ldots, n)$ such that no two neighbouring points belong to the same set [the values $U(P_i)$ are classified accordingly as A values or B values]. Then, multiplying each of the difference equations by a suitable constant if need be, we can put the system of equations into the form

$$-\alpha U_i + \sum_{k=p+1}^{n} b_{ik} U_k + s_i = 0 \quad (i = 1, 2, \ldots, p), \tag{1.43}$$

$$\sum_{k=1}^{p} b_{ik} U_k - \alpha U_i + s_i = 0 \quad (i = p + 1, \ldots, n). \tag{1.44}$$

In matrix notation this can be written

$$\left(\begin{array}{c|c} -\alpha I_p & B_1 \\ \hline B_2 & -\alpha I_{n-p} \end{array} \right) \left(\begin{array}{c} x_A \\ x_B \end{array} \right) + \left(\begin{array}{c} s_A \\ s_B \end{array} \right) = 0,$$

in which the matrix and vectors are partitioned into sub-matrices and sub-vectors, respectively, by lines between the p-th and $(p + 1)$-th rows and columns, and obvious notations are used for these partitions; for instance, I_p is the p-rowed unit matrix and x_A is the vector of unknowns in set A, etc. We now eliminate the unknowns of set B by pre-multiplying the system by the matrix $\left(\begin{array}{c|c} \alpha I_p & B_1 \\ \hline 0 & I_{n-p} \end{array} \right)$; this transforms the original matrix into the "decomposed" matrix $\left(\begin{array}{c|c} -\alpha^2 I_p + B_1 B_2 & 0 \\ \hline B_2 & -\alpha I_{n-p} \end{array} \right)$.
In this way we obtain a "reduced system" for the unknowns in set A:

$$-\alpha^2 U_i + \sum_{k=1}^{p} \beta_{ik} U_k + \sigma_i = 0, \quad \beta_{ik} = \sum_{j=p+1}^{n} b_{ij} b_{jk}, \quad \sigma_i = \alpha s_i + \sum_{j=p+1}^{n} b_{ij} s_j$$
$$(i = 1, 2, \ldots, p),$$

i.e.

$$(-\alpha^2 I_p + B_1 B_2) x_A + \sigma_A = 0, \quad \sigma_A = \alpha s_A + B_1 s_B. \tag{1.45}$$

Thus we only have to solve p equations; the remaining unknowns are expressed explicitly in terms of the solution of these equations by (1.44). Another advantage

---

[1] The material for this section was kindly made available to me by Herr J. Schröder [for a detailed presentation see J. Schröder: Z. Angew. Math. Mech. **34**, 241—253 (1954)].

of this method is that far fewer iteration cycles are needed to solve the reduced system by single-step (or total-step) iteration than to solve similarly the original system to the same accuracy in the A values [1].

For differential equations of the form $\nabla^2 u - g u = r$, where g is a non-negative constant, use of a square mesh (mesh width h) usually permits a simple, direct derivation of the reduced system which does not involve the matrix multiplications $B_1 B_2$, $B_1 s_B$. One way of doing this is to use special stencils [2] which can be constructed by superimposing ordinary stencils. However, if U is known at all the boundary pivots, the reduced system can be written down immediately, for the coefficients are given by the following simple formulae:

$$\left.\begin{aligned}
&\alpha = 4 + g h^2 \\
&\beta_{ik} = \text{number of pivots } P_j \text{ which are neighbours to both} \\
&\qquad P_i \text{ and } P_k \\
&\text{(in particular, } \beta_{ii} = \text{number of pivots } P_j \text{ which are neighbours to } P_i) \\
&\sigma_i = \alpha s_i + \sum_j s_j, \text{ where the sum extends over all } j \text{ for} \\
&\qquad \text{which } P_i \text{ and } P_j \text{ are neighbouring points.}
\end{aligned}\right\} \quad (1.46)$$

Clearly the matrix of the system is symmetric in this case.

The technique can also be applied with advantage to eigenvalue problems. If, for instance, the differential equation reads $\nabla^2 u - g u + \lambda u = 0$, where g is again a non-negative constant, and if with a square mesh (mesh width h) $U = 0$ at all boundary pivots, then by the same procedure as above we can put the difference equations into the form (1.43), (1.44) [with $s_i = 0$ and $\alpha = 4 + (g - \Lambda) h^2$, where Λ is an approximate eigenvalue] and set up the corresponding reduced system (1.45) for which the β_{ik} may be found from the formula of (1.46). By virtue of the equation

$$\det \left(\frac{-\alpha I_p \mid B_1}{B_2 \mid -\alpha I_{n-p}} \right) = (-1)^{n-p} \alpha^{n-2p} \det(-\alpha^2 I_p + B_1 B_2), \quad (1.47)$$

the "latent roots" α of the matrix $B = \left(\dfrac{0 \mid B_1}{B_2 \mid 0} \right)$ may be calculated using the determinant of the reduced system. They are obviously symmetrically placed about $\alpha = 0$, and if (x_A/x_B) is a latent vector corresponding to α, then $(x_A/-x_B)$ is a latent vector corresponding to $-\alpha$. In the iterative calculation of the greatest latent root and its corresponding latent vector the same accuracy is achieved in half the number of cycles if the matrix of the reduced system is used rather than that of the original system.

Example: Let us find a function $u(x, y)$ which satisfies the equation $\nabla^2 u + \lambda u = 0$ inside the region illustrated in Fig. V/12 and vanishes on the boundary. Using the heavily lined square mesh with mesh width $h = 1$, we have seven pivots P_i (marked 1, 2, ..., 7 in the figure); set A comprises the pivots P_1, P_2, P_3 and set B the remainder ($n = 7$, $p = 3$). The numbers β_{ik} ($i, k = 1, 2, 3$) and b_{ik} ($i = 4, 5, 6, 7$;

---

[1] See SCHRÖDER's paper.

[2] Following MILNE we apply the term "stencil" to an array of coefficients set out in a pattern corresponding to the points whose associated approximate values they are to multiply. Thus the ordinary stencil here is
$$\begin{array}{|c|c|c|} \hline & 1 & \\ \hline 1 & -\alpha & 1 \\ \hline & 1 & \\ \hline \end{array}.$$

$k = 1, 2, 3$) are found from the formula of (1.46) (for example: P_1 and P_2 have two neighbouring pivots in common so that $\beta_{12} = \beta_{21} = 2$) and from the ordinary difference equations, respectively:

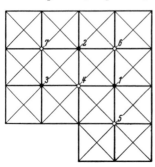

Fig. V/12. Reduction of the system of difference equations

$\beta_{i,k}$	β_i	3	2	1	6
		2	3	2	7
		1	2	2	5
$b_{i,k}$	b_i	$=$	1	1	1	3
		1	0	0	1
		1	1	0	2
		0	1	1	2

As a check one may calculate the numbers

$$\beta_i = \sum_{k=1}^{3} \beta_{ik} \text{ and } b_i = \sum_{k=1}^{3} b_{ik} \text{ and verify that } \beta_i = \sum_{j=4}^{7} b_{ji} b_j.$$

From (1.47) the latent roots α are given by $\alpha(\alpha^6 - 8\alpha^4 + 12\alpha^2 - 3) = 0$. One pair of roots is $\alpha_2 = 1\cdot2523$ and $\alpha_6 = -\alpha_2$; these give $\Lambda_2 = 2\cdot7477$ and $\Lambda_6 = 5\cdot2523$. A solution of the reduced system corresponding to α_2 and α_6 is

$$U_1 = -1\cdot3168, \qquad U_2 = 0\cdot4426, \qquad U_3 = 1.$$

The ordinary difference equations (1.44) (with $s_i = 0$, $\alpha = \alpha_2, \alpha_6$) then yield the following values for the remaining U_i:

$$U_4 = \pm 0\cdot1004, \qquad U_5 = \mp 1\cdot0515, \qquad U_6 = \mp 0\cdot6981, \qquad U_7 = \pm 1\cdot1519.$$

Mesh refinement: For the type of eigenvalue problem just considered ($\nabla^2 u - gu + \lambda u = 0$, $g = \text{constant} \geq 0$, $U = 0$ at the boundary pivots for a square mesh) one may seek to improve a calculated eigenvector x and corresponding approximate eigenvalue $\Lambda = g + \dfrac{1}{h^2}(4 - \alpha)$ by a perturbation calculation starting from the known quantities x and α (this method of improving calculated values by means of a perturbation calculation may also be applied to the corresponding inhomogeneous boundary-value problem). We make two further assumptions: the latent root α is simple and non-zero, and the boundary Γ consists of complete sides or diagonals of the squares of the "basic mesh", i.e. the mesh with which the known approximations x and α were calculated. The diagonals of the squares determine a new mesh, the "diagonal mesh", which has at least $2n$ pivots and which takes over directly the old pivots $P_1, \ldots, P_n$ as the new set A^* of non-neighbouring pivots. Since the Laplace operator is unaltered by a rotation of the co-ordinate axes, the coefficients in the difference equations are similar except for the smaller mesh width $h/\sqrt{2}$ and the corresponding reduced system can be written

$$(-\alpha^{*2} I_n + B_1^* B_2^*) x_{A^*}^* = 0, \qquad \alpha^* = 4 + \frac{h^2}{2}(g - \Lambda^*),$$

where asterisked quantities refer to the diagonal mesh. It can be shown that

$$A_0 = \tfrac{1}{2} B^2 + 2B$$

does not differ greatly from $B_1^* B_2^*$, so we put

$$B_1^* B_2^* = A_0 + \varepsilon A_1,$$

in which we shall take $\varepsilon = 1$, and solve the equation $(-\alpha^{*2} I_n + A_0 + \varepsilon A_1) x_{A*}^* = 0$ by a perturbation method: we put $x_{A*}^* = \varphi_0 + \varepsilon \varphi_1 + \varepsilon^2 \varphi_2 + \cdots$ with $\varphi_0 = x$, $\alpha^{*2} = \mu_0 + \varepsilon \mu_1 + \varepsilon^2 \mu_2 + \cdots$ with $\mu_0 = \frac{1}{2}\alpha^2 + 2\alpha$ and equate coefficients of powers of ε. Using the fact that A_0 is symmetrical we find that

$$\mu_0 + \mu_1 = 2\alpha + r, \quad \text{where} \quad r = \frac{(R\,\varphi_0, \varphi_0)}{(\varphi_0, \varphi_0)}, \quad R = B_1^* B_2^* - 2B = \left(\begin{array}{c|c} R_1 & 0 \\ \hline 0 & R_2 \end{array}\right);$$

here (y, z) denotes the inner product of two n-dimensional vectors y and z.

For the above example the diagonal mesh provides twenty-one pivots as against the seven pivots of the basic mesh. We have

$$R_1 = \begin{pmatrix} 4 & 1 & 0 \\ 1 & 4 & 1 \\ 0 & 1 & 4 \end{pmatrix}, \quad R_2 = \begin{pmatrix} 4 & 1 & 1 & 1 \\ 1 & 4 & 0 & 0 \\ 1 & 0 & 4 & 0 \\ 1 & 0 & 0 & 4 \end{pmatrix},$$

and for $\alpha = \alpha_2$

$$\mu_0 + \mu_1 = 6 \cdot 4363, \quad \mu_0 + \mu_1 + \mu_2 = 6 \cdot 4746 \quad \text{and} \quad \mu_0 + \mu_1 + \mu_2 + \mu_3 = 6 \cdot 4670.$$

These three numbers lead to the approximate values $\Lambda_2^* \approx 2 \cdot 9260$, $2 \cdot 9110$, $2 \cdot 9139$, respectively (the exact solution for the calculation with twenty-one pivots is $\Lambda_2^* = 2 \cdot 9135$), as compared with the value $\Lambda_2 = 2 \cdot 7477$ from the basic mesh.

§2. Refinements of the finite-difference method

The derivation and application of the formulae for the methods to be described here are for the most part so similar to the corresponding considerations in Ch. III, § 2 and Ch. IV, § 2 that we can be brief and place more emphasis on the generality of the methods.

2.1. The finite-difference method to a higher approximation in the general case

Let a function $u(x_1, x_2, \ldots, x_n)$ be defined in a simply connected, closed region B^* of the n-dimensional co-ordinate space by an m-th-order linear differential equation

$$L[u] = \sum_{\alpha_1 + \cdots + \alpha_n \leq m} A_{\alpha_1, \alpha_2, \ldots, \alpha_n} \frac{\partial^{\alpha_1 + \cdots + \alpha_n} u}{\partial x_1^{\alpha_1} \ldots \partial x_n^{\alpha_n}} = t(x_1, \ldots, x_n) \qquad (2.1)$$

together with linear boundary conditions. For example, we might prescribe on certain hypersurfaces Γ_μ values of u or of $\partial u/\partial x_\varrho$ or, more generally, of a linear combination of the partial derivatives of u of up to the $(m-1)$-th order. These surfaces need not lie on the boundary of the region of definition B^*; it is often convenient to take into consideration values of the function u outside of the "boundary" surfaces on which the boundary conditions are given.

For the numerical calculation of u from the given data we introduce a system of pivotal points P_j $(j = 1, 2, \ldots, N)$ which all lie in B^* (they need not be arranged in a regular pattern) and seek approximations U_j to the values $u(P_j)$ of the exact solution u at the points P_j. For the solution of this problem we set up a system of linear equations for the U_j. Such an equation is obtained by forming the sum

$$\sum_{\varrho=1}^{N} C_{\varrho} u (P_{\varrho})$$

and determining the constants C_{ϱ} so that it approximates the differential expression of (2.1) at a point P_j. We say the equation is written for, or corresponds to, the point P_j.

Each term $C_{\varrho} u (P_{\varrho})$ of the sum is expanded by TAYLOR's theorem about the point P_j:

$$
\left.
\begin{aligned}
u (P_{\varrho}) &= u (P_j) + \sum_{\nu=1}^{n} \left(x_\nu (P_{\varrho}) - x_\nu (P_j) \right) \frac{\partial u (P_j)}{\partial x_\nu} + \\
&+ \frac{1}{2!} \sum_{\nu, \mu=1}^{n} \left(x_\nu (P_{\varrho}) - x_\nu (P_j) \right) \left(x_\mu (P_{\varrho}) - x_\mu (P_j) \right) \frac{\partial^2 u (P_j)}{\partial x_\nu \partial x_\mu} + \\
&+ \cdots + \text{terms of the } r\text{-th order} + \text{remainder term}
\end{aligned}
\right\} \quad (2.2)
$$

and the whole sum $\sum C_{\varrho} u (P_{\varrho})$ then rearranged in terms of $u (P_j)$ and the partial derivatives of u at the point P_j:

$$\sum_{\varrho=1}^{N} C_{\varrho} u (P_{\varrho}) = \sum_{\alpha_1 + \cdots + \alpha_n \leq r} B_{\alpha_1, \ldots, \alpha_n} \frac{\partial^{\alpha_1 + \cdots + \alpha_n} u (P_j)}{\partial x_1^{\alpha_1} \ldots \partial x_n^{\alpha_n}} + \text{remainder term.} \quad (2.3)$$

The new coefficients $B_{\alpha_1, \ldots, \alpha_n}$ depend linearly on the C_{ϱ}. If h is the smallest non-zero number among the values of the quantities $|x_j (P_k) - x_j (P_j)|$, and M_{r+1} is the maximum in B^* of the absolute values of all the $(r+1)$-th partial derivatives of u, then the remainder term can be expressed in the form $\vartheta D_j h^{r+1} M_{r+1}$, where $|\vartheta| \leq 1$ and D_j depends on the positions of the points P_{ϱ} but not on u; when the disposition of the points P_j is known, numerical limits can be given for the D_j.

Our object is to make the expression on the left-hand side of (2.3) as accurate a representation of the differential expression $L[u]$ in (2.1) as possible; we can then write approximately

$$\sum_{\varrho=1}^{N} C_{\varrho} u (P_{\varrho}) \approx L[u (P_j)] = t (P_j)$$

and take

$$\sum_{\varrho=1}^{N} C_{\varrho} U_{\varrho} = t (P_j) \quad (2.4)$$

as one of the equations for the U_j.

To achieve our object we try to make the coefficients of $\frac{\partial^{\alpha_1 + \cdots + \alpha_n} u}{\partial x_1^{\alpha_1} \ldots \partial x_n^{\alpha_n}}$ in (2.1) and (2.3) agree for all derivatives of up to as high an order as possible. First of all we must have

$$A_{\alpha_1, \ldots, \alpha_n} = B_{\alpha_1, \ldots, \alpha_n} (C_{\varrho}), \quad (2.5)$$

i.e. we must have agreement for $\alpha = \alpha_1 + \alpha_2 + \cdots + \alpha_n = 0, 1, 2, \ldots, m$ at least. This minimum requirement yields the special case of the ordinary finite-difference method, for which (2.3) becomes

$$\sum_{\varrho=1}^{N} C_\varrho u(P_\varrho) = L[u(P_j)] + \vartheta D_j h^{m+1} M_{m+1}.$$

The linear equations given by (2.5) admit of simple solutions[1] in which many of the C_ϱ are zero and the few non-zero C_ϱ correspond to points P_ϱ lying near P_j.

Here we are more interested in higher approximations, i.e. those for which the order of the remainder term is greater than $m + 1$:

$$\sum_{\varrho=1}^{N} C_\varrho u(P_\varrho) = L[u(P_j)] + \vartheta D_j h^{r+1} M_{r+1}, \tag{2.6}$$

where $r > m$. To this end, we add to the equations (2.5) for the C_ϱ the further equations

$$B_{\alpha_1, \ldots, \alpha_n} = 0 \quad \text{for} \quad m < \alpha \leqq r.$$

Insertion in (2.4) of a set of numbers C_ϱ which satisfy these equations yields one of the desired equations for the U_j.

Further equations of a higher approximation can be derived by finding a different set of numbers C_ϱ which satisfy the necessary equations[2] or by using a different point P_i as "centre" of the Taylor expansions. In addition, linear equations for the U_j may be derived in the same way from the boundary conditions.

One endeavours to set up as many equations for the U_j as will yield a system of linear equations with a non-zero determinant.

2.2. A general principle for error estimation

If we have such a system of equations for the U_j, then in principle we can obtain estimates for the errors

$$\varepsilon_j = U_j - u(P_j).$$

If we assume that the solution u possesses partial derivatives of up to and including the $(r + 1)$-th order, then insertion of u into (2.6) yields

$$\sum_{\varrho=1}^{N} C_\varrho u(P_\varrho) = L[u(P_j)] + \vartheta D_j h^{r+1} M_{r+1} = t(P_j) + \vartheta D_j h^{r+1} M_{r+1};$$

---

[1] Provided that the distribution of pivotal points in B^* is sufficiently dense the number of C_ϱ at our disposal will be far greater than the number of equations to be satisfied.

[2] Strictly we ought to write $C_\varrho^{(j, \varkappa)}$, $\vartheta^{(j, \varkappa)}$, $D^{(j, \varkappa)}$, say, for even with the same point P_j several sets of values for the C_ϱ are possible. Since misunderstanding is unlikely, we use the simpler notation with fewer indices.

then subtracting from (2.4) we find that the errors ε_j satisfy the equation

$$\sum_{\varrho=1}^{N} C_\varrho \varepsilon_\varrho = -\vartheta D_j h^{r+1} M_{r+1}. \tag{2.7}$$

Thus, apart from the different right-hand sides, the errors ε_j satisfy the same system of equations as the approximate values U_j. Since the determinant of the coefficients was assumed to be non-zero, we can solve the system of equations (2.7) for the ε_ϱ.

For the finite-difference methods described here for the numerical calculation of the solution u of a linear boundary-value problem to a higher approximation, the theoretical possibility of being able to obtain error estimates follows from the possibility of being able to apply the method in the first place, provided that u is assumed to be differentiable sufficiently often in the region B^*.

Uniqueness of the solution u need not be assumed: what we estimate is the departure of our approximate pivotal values from the values of a solution whose partial derivatives of the $(r+1)$-th order are bounded absolutely in B^* by the constant M_{r+1} in (2.7). As regards the difficulties which attend the estimation of M_{r+1}, cf. the remarks made in Ch. IV, § 3.3.

2.3. Derivation of finite expressions

The Taylor expansion method of § 2.1 provides a technique for setting up a finite expression to represent any given differential expression. Here we describe an operator method which shows that there exist finite expressions of an arbitrarily high order of approximation, i.e. expressions with remainder terms of order $r+1$ for any prescribed r.

We select a typical term

$$\frac{\partial^{\alpha_1+\alpha_2+\cdots+\alpha_n} u}{\partial x_1^{\alpha_1} \partial x_2^{\alpha_2} \ldots \partial x_n^{\alpha_n}} \tag{2.8}$$

and take our pivotal points at the nodes of a square mesh of mesh width h. We can then use the displacement operator E_i which transforms a function $u(x_1, x_2, \ldots, x_n)$ into $u(x_1, x_2, \ldots, x_{i-1}, x_i+h, x_{i+1}, \ldots, x_n)$. This operator[1] obeys the laws of ordinary algebra and also commutes with the differential operators $\partial^{\alpha_i}/\partial x_i^{\alpha_i}$. Raised to an integral power p (positive, negative or zero) it has the effect

$$E_i^p u(x_1, x_2, \ldots, x_i, \ldots, x_n) \equiv u(x_1, x_2, \ldots, x_{i-1}, x_i+ph, x_{i+1}, \ldots, x_n). \tag{2.9}$$

In the equations between operators which we write in the following we imagine the operations to be performed on rational integral functions

[1] STEFFENSEN, J. F.: Interpolation, p. 4 et seq. and p. 178 et seq. Baltimore 1927. — BRUWIER, L.: Sur une équation aux dérivées et aux différences mêlées. Mathesis **47**, 103—104 (1933).

$v(x_1, x_2, \ldots, x_n)$ which are of degree not greater than r in each of the independent variables. This avoids having to take the remainder terms into account each time.

According to § 2.2 of Chapter III there exists for each individual operator $\partial^{\alpha_i}/\partial x_i^{\alpha_i}$ a finite operator[1]

$$\frac{\partial^{\alpha_i}}{\partial x_i^{\alpha_i}} = \sum_{\varrho = \left[-\frac{r}{2}\right]}^{\left[\frac{r}{2}\right]} A_\varrho^{(i)} E_i^\varrho \tag{2.10}$$

for any given integer r.

If we multiply this equation by $\partial^{\alpha_k}/\partial x_k^{\alpha_k}$ $(k \neq i)$, then, since $E_i^\varrho v$ is also a rational integral function of $x_1, x_2, \ldots, x_n$ of degree not greater than r in each of the independent variables, we can replace

$$\frac{\partial^{\alpha_k}}{\partial x_k^{\alpha_k}} E_i^\varrho v \quad \text{by} \quad \sum_\sigma A_\sigma^{(k)} E_k^\sigma E_i^\varrho v;$$

consequently

$$\frac{\partial^{\alpha_i}}{\partial x_i^{\alpha_i}} \frac{\partial^{\alpha_k}}{\partial x_k^{\alpha_k}} = \left\{ \sum_\varrho A_\varrho^{(i)} E_i^\varrho \right\} \left\{ \sum_\sigma A_\sigma^{(k)} E_k^\sigma \right\}$$

and

$$\prod_{i=1}^n \left[\sum_{\varrho = \left[-\frac{r}{2}\right]}^{\left[\frac{r}{2}\right]} A_\varrho^{(i)} E_i^\varrho \right] - \frac{\partial^{\alpha_1 + \cdots + \alpha_n}}{\partial x_1^{\alpha_1} \ldots \partial x_n^{\alpha_n}} \equiv 0 \tag{2.11}$$

for any one of our functions $v(x_1, \ldots, x_n)$.

If the product in (2.11) is now multiplied out and applied to a function u, and then each term expanded by TAYLOR's theorem with a remainder term of the $(r+1)$-th order, exactly as in § 2.1, we have

$$\prod_{i=1}^n \left[\sum_\varrho A_\varrho^{(i)} E_i^\varrho \right] u - \frac{\partial^{\alpha_1 + \cdots + \alpha_n} u}{\partial x_1^{\alpha_1} \ldots \partial x_n^{\alpha_n}}$$
$$= \sum_{\beta_1 + \cdots + \beta_n \leq r} B_{\beta_1, \ldots, \beta_n} \frac{\partial^{\beta_1 + \cdots + \beta_n} u}{\partial x_1^{\beta_1} \ldots \partial x_n^{\beta_n}} + \vartheta D_j h^{r+1} M_{r+1}$$

(notation D_j, M_{r+1} as in § 2.1). The constants $B_{\beta_1, \beta_2, \ldots, \beta_n}$ are the same not only for all our rational integral functions, but also for all functions u which possess partial derivatives of up to and including the $(r+1)$-th order. Now take u to be any rational integral function of (total) degree not greater than r; then the right-hand side must be zero since these functions are included among those for which (2.11) is valid. Further, $M_{r+1} = 0$ for these functions, so if we put $u = $ constant, we deduce that $B_{0, 0, \ldots, 0} = 0$; similarly we can put $u = x_i$, $u = x_i x_k$,

---

[1] The customary notation $[x]$ is used for the greatest integer which is not greater than x; for example, $[2.5] = 2$, $[2] = 2$, $[-2.5] = -3$.

$u = x_i x_k x_l, \ldots$, from which it follows that all $B_{\beta_1, \ldots, \beta_n} = 0$ for $\beta_1 + \beta_2 + \cdots + \beta_n \leq r$. Thus for all functions u with continuous partial derivatives of the $(r + 1)$-th order we have

$$\prod_{i=1}^{n} \left[\sum_{\varrho = \left[-\frac{r}{2}\right]}^{\left[\frac{r}{2}\right]} A_{\varrho}^{(i)} E_i^{\varrho} \right] u = \frac{\partial^{\alpha_1 + \cdots + \alpha_n} u}{\partial x_1^{\alpha_1} \ldots \partial x_n^{\alpha_n}} + \vartheta D_j \, h^{r+1} M_{r+1}. \qquad (2.12)$$

Although this operator method[1] shows the existence of finite expressions of an arbitrarily high order of approximation, the Taylor expansion method of § 2.1 is often more profitable in practice, for it can often be used to derive simpler expressions involving fewer points[2]. Several finite expressions for the frequently occurring differential expressions $\nabla^2 u$ and $\nabla^4 u$ are given for two independent variables in Table VI of the appendix (also for three independent variables for $\nabla^2 u$).

2.4. Utilization of function values at exterior mesh points

With the finite-difference methods of a higher approximation it often happens that a finite equation written down for a mesh point near the boundary involves approximate values at mesh points which lie outside

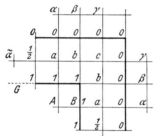

Fig. V/13. Utilization of exterior mesh
points

of the fundamental region, i.e. the region enclosed by the boundaries on which the boundary values are prescribed. To render possible the elimination of these values one usually employs finite equations of a lower approximation on the boundary. This is illustrated in the following examples.

I. Consider the problem of Example I, § 1.5, which was to determine a potential function with given values on the boundaries shown in Fig. V/4. If we take $h = \frac{1}{2}$ and make use of the symmetry of the problem, we have just three unknown values a, b, c at the points indicated in Fig. V/13, but finite equations of a higher approximation written down for these points will involve function values outside of the fundamental region — in the simplest case, the values $\alpha, \beta, \gamma; \tilde{\alpha}, A, B$ as in the figure.

The unknown value B is included because of the singular behaviour at the re-entrant corner. The section of the u surface in the vertical plane containing the line $\beta b B$ will have a discontinuous slope at the

---

[1] Operators are used by W. G. Bickley: Finite-difference formulae for the square lattice. Quart. J. Mech. Appl. Math. 1, 35—42 (1948).

[2] Examples for comparison of the methods can be found in L. Collatz: Schr. Math. Sem. u. Inst. Angew. Math. Univ. Berlin 3, 18 (1935).

corner as shown by the heavy line in Fig. V/14. Consequently the true value of 1 at B is not consistent with a polynomial representation, whereas a fictitious extrapolated value can be made so.

Using the finite expression for $V^2 u$ given in Table VI of the appendix we obtain the equations

$$-60a + 16(b + 1 + \tfrac{1}{2} + 0) -$$
$$- (c + A + \tilde{\alpha} + \alpha) = 0,$$
$$-60b + 16(c + 1 + a + 0) -$$
$$- (0 + B + \tfrac{1}{2} + \beta) = 0,$$
$$-60c + 16(0 + b + b + 0) -$$
$$- (\gamma + a + a + \gamma) = 0.$$

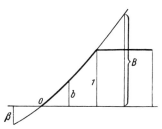

Fig. V/14. Introduction of a fictitious value near a singularity

To eliminate the extra unknowns we express them in terms of a, b and c by means of the ordinary difference equations corresponding to the boundary points:

$$-4 \times 0 + 0 + a + 0 + \alpha = 0, \quad \text{yields} \quad \alpha = -a,$$
$$\text{and similarly we obtain} \quad \beta = -b, \quad \gamma = -c,$$
$$-4 \times \tfrac{1}{2} + a + 1 + \tilde{\alpha} + 0 = 0, \quad \text{yields} \quad \tilde{\alpha} = 1 - a,$$
$$-4 \times 1 + 1 + A + 1 + a = 0, \quad \text{yields} \quad A = 2 - a.$$

Similarly we put $B = 2 - b$ as though the value 1 on the line G in Fig. V/13 extended beyond the re-entrant corner.

With these substitutions we find that

$$-57a + 16b - c = -21,$$
$$32a - 116b + 32c = -27,$$
$$-a + 16b - 29c = 0,$$

and hence

$$a = \frac{2571}{5308} = 0.48436, \qquad b = \frac{2265}{5308} = 0.42671, \qquad c = \frac{1161}{5308} = 0.21873.$$

If more accurate values are needed, the calculation can be repeated with a smaller mesh width h; the larger system of equations which arises can be solved iteratively.

II. A boundary-value problem whose physical background is sketched in § 4.2 is the following:

$$V^2 u = -1, \quad \frac{\partial u}{\partial \nu} = u \quad \text{on the boundary of the square} \quad |x| \leq 1, \ |y| \leq 1,$$

where ν is the inward normal. To illustrate the method, we use the rather large mesh width $h = \frac{2}{3}$; using the symmetry we have the five unknowns a, b, c, d, e (Fig. V/15). We describe three methods.

A. The ordinary finite-difference method.

The ordinary difference equations for a, b and c are

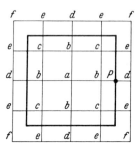

Fig. V/15. Notation for the example treated by several finite-difference methods

$$-4a + 4b + h^2 = 0,$$
$$-4b + a + 2c + d + h^2 = 0,$$
$$-4c + 2b + 2e + h^2 = 0;$$

to these we add the equations corresponding to the boundary conditions:

$$\frac{b-d}{h} = \frac{b+d}{2},$$

i.e.

$$b(2-h) = d(2+h),$$

and similarly

$$c(2-h) = e(2+h).$$

Solving these equations we obtain

$$a = \frac{53}{28}h^2 = \frac{53}{63} = 0{\cdot}8413, \qquad b = \frac{46}{63} = 0{\cdot}7302, \qquad c = \frac{40}{63} = 0{\cdot}6349,$$

$$d = \frac{1}{2}b = 0{\cdot}3651, \qquad e = \frac{1}{2}c = 0{\cdot}3175.$$

B. A finite equation of higher approximation could have been written down for the point a. If we therefore replace the first of the equations in A by the equation

$$-60a + 64b - 4d + 12h^2 = 0$$

and use the remaining equations as they are, the solution of the new system of equations is

$$a = \frac{388}{459} = 0{\cdot}8453, \qquad b = \frac{336}{459} = 0{\cdot}7102, \qquad c = \frac{292}{459} = 0{\cdot}6362,$$

$$d = \frac{1}{2}b, \qquad e = \frac{1}{2}c.$$

These values are worse than those obtained in A, which is rather surprising at first sight. This demonstrates the unadvisability of using a very accurate equation at one point when very crude approximations are retained at others (here at the boundary). We therefore approximate the boundary conditions more accurately in the next method.

C. Method of a higher approximation.

Here we approximate the boundary condition $\partial u/\partial v = u$ at the point P, say, in Fig. V/15 by putting a parabola through the points a, b, d.

At P this parabola has the ordinate and derivative values

$$u_P = \frac{3d + 6b - a}{8} \quad \text{and} \quad -(u_v)_P = \frac{d - b}{h},$$

which substituted in the boundary condition yield

$$\frac{3d + 6b - a}{8} + \frac{d - b}{h} = 0.$$

With $h = \frac{2}{3}$ we have the more accurate finite boundary condition

$$15d = 6b + a, \quad \text{and similarly} \quad 15e = 6c + b. \tag{2.13}$$

From these two equations, together with the main equations used in B:

$$-60a + 64b - 4d + 12h^2 = 0,$$
$$-4b + a + 2c + d + h^2 = 0,$$
$$-4c + 2b + 2e + h^2 = 0,$$

we obtain

$$a = \frac{5787}{7092} = 0.8160, \quad b = \frac{4983}{7092} = 0.7026, \quad c = \frac{4307}{7092} = 0.6073,$$

$$d = 0.3354, \quad e = 0.2898.$$

These values are much better than those found in A and B.

III. In Example II of § 1.5 the boundary-value problem

$$u_{xx} + u_{yy} + \frac{3}{5 - y} u_y + 1 = 0,$$

$u = 0$ for $|x| = \frac{1}{2}$ and for $|y| = 1$

was treated by the ordinary finite-difference method for the mesh widths $h = \frac{1}{2}, \frac{1}{3}, \frac{1}{4}$. Here we apply a method of higher approximation for $h = \frac{1}{2}$. With the notation of Fig. V/16 the finite equations for the three interior points read

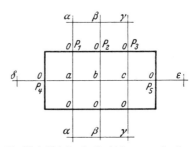

Fig. V/16. Notation for the higher approximation solution of Example III

$$\frac{-c + 16b - 30a - \delta}{3} + \frac{-2\alpha - 30a}{3} + \frac{6}{11} \frac{-c + 8b + \delta}{6} + 1 = 0,$$

$$\left.\frac{16c - 30b + 16a}{3} + \frac{-2\beta - 30b}{3} + \frac{3}{5} \frac{8c - 8a}{6} + 1 = 0, \right\} \tag{2.14}$$

$$\frac{-\varepsilon - 30c + 16b - a}{3} + \frac{-2\gamma - 30c}{3} + \frac{2}{3} \frac{-\varepsilon - 8b + a}{6} + 1 = 0,$$

in which appear the unknown values $\alpha, \beta, \gamma, \delta, \varepsilon$ outside of the rectangle.

These extra values are eliminated by using again the ordinary difference equations corresponding to the differential equation for the points $P_1, \ldots, P_5$:

$$\frac{\alpha + a}{h^2} + 1 = \frac{\beta + b}{h^2} + 1 = \frac{\gamma + c}{h^2} + 1 = 0,$$

$$\frac{a + \delta}{h^2} + \frac{1}{2}(a - \delta) + 1 = \frac{c + \varepsilon}{h^2} + \frac{3}{4}(\varepsilon - c) + 1 = 0,$$

by means of which the $\alpha, \ldots, \varepsilon$ can be expressed in terms of a, b, c.

Substituting these values in (2.14) we obtain three linear equations for a, b, c which have strongly dominant diagonal coefficients and are best solved by iteration:

$$-4394a + 1400b - 98c + 285 \cdot 5 = 0,$$
$$136a - 580b + 184c + 35 = 0,$$
$$-76a + 1520b - 6508c + 431 = 0.$$

Their solution is

$$a = 0 \cdot 098\,78,$$
$$b = 0 \cdot 112\,49,$$
$$c = 0 \cdot 091\,34.$$

2.5. Hermitian finite-difference methods (Mehrstellenverfahren)

These methods employ formulae of the Hermitian type discussed in Ch. III, § 2 and Ch. IV, § 2.3; their derivation here is very similar, though rather more general.

We consider a boundary-value problem as in § 2.1, and with the same notation as used there seek approximate values U_j for the values $u(P_j)$ of the exact solution at the pivotal points P_j distributed (not necessarily in a regular pattern) throughout the region B^*.

We write the differential equation (not necessarily linear) in the form

$$f\left(x_1, \ldots, x_n, u, L_1[u], L_2[u], \ldots, L_p[u]\right) = 0, \qquad (2.15)$$

where $L_1[u], L_2[u], \ldots, L_p[u]$ are given linear differential expressions in u as in (2.1). For each of these differential expressions, and also for any further linear differential expressions $L_{p+1}[u], \ldots, L_q[u]$ which may occur in the boundary conditions, we write down expressions of the form

$$\Phi_s[u] = \sum_{\varrho=1}^{N} C_{\varrho, s}\, u(P_\varrho) + \sum_{\varrho=1}^{N} D_{\varrho, s}\left(L_s[u]\right)_{P_\varrho} \quad (s = 1, \ldots, q) \qquad (2.16)$$

and expand them by TAYLOR's theorem in terms of the values of u and its partial derivatives at the point P_j. The constants $C_{\varrho, s}$ and $D_{\varrho, s}$ are to be chosen so that the expansion vanishes identically to as high an order as possible. For most practical calculations one will naturally choose the pivotal points in a regular pattern and use expressions Φ_s for which only the $C_{\varrho, s}$ and $D_{\varrho, s}$ corresponding to "mesh points" near

P_j are non-zero. The Taylor expansion and determination of the coefficients are illustrated in detail in Ch. IV, §§ 2.2 and 2.3 for the differential expression

$$L[u] = \frac{\partial^2 u}{\partial x^2} - k \frac{\partial u}{\partial y}.$$

Expressions of the form (2.16) are given for the individual derivatives

$$\frac{\partial^\nu u}{\partial x_\sigma^\nu}$$

(for small ν) in Table III of the appendix and for the operators V^2 and V^4 in Table VI [1].

An approximate equation

$$\Phi_s[U_j] = \sum_{\varrho=1}^N C_{\varrho,s} U_\varrho + \sum_{\varrho=1}^N D_{\varrho,s} U_{\varrho,s} = 0 \qquad \left(\begin{matrix} j = 1, \ldots, N \\ s = 1, \ldots, q \end{matrix} \right), \qquad (2.17)$$

where $U_{\varrho,s}$ denotes an approximate value of $(L_s[u])_{P_\varrho}$, is now written down for each point P_j and for each expression $L_s[u]$, making $N \times q$ equations in all. We have $N \times (q+1)$ unknowns $U_\varrho, U_{\varrho,s}$, so N further equations are required. These are obtained by writing down the boundary conditions for those points P_j lying on or near the boundary and the differential equation (2.15) with $(L_s[u])_{P_j}$ replaced by $U_{j,s}$ for those points P_j lying "further inside". Whether the differential equation or a boundary condition is written down for any specific point will be decided on the basis of the particular problem under consideration.

For a linear differential equation $L[u] = r(x_1, \ldots, x_n)$ a more general form [2] of the expression (2.16) can be used, namely

$$\Phi[u] = \sum_{\varrho=1}^N C_\varrho u(P_\varrho) + \sum_{\varrho=1}^N D_\varrho (L[u])_{P_\varrho} + \sum_{\varrho=1}^N \sum_{\mu=1}^M E_{\varrho,\mu} (M_\mu [L[u]])_{P_\varrho},$$

where the M_μ are chosen operators and the $E_{\varrho,\mu}$ are constants to be determined in the usual way (formally the second sum is included in the double sum and could therefore be omitted). In the corresponding approximate equation we replace $(M_\mu[L[u]])_{P_\varrho}$ by $(M_\mu[r])_{P_\varrho}$, which is known. The inclusion of these extra terms offers the possibility in many cases of achieving a higher order approximation without involving any extra pivotal points. Naturally a high accuracy formula of this type will only be used when the solution possesses continuous derivatives of

---

[1] See also L. COLLATZ: Das Mehrstellenverfahren bei Plattenaufgaben. Z. Angew. Math. Mech. **30**, 385−388 (1950) and R. ZURMÜHL: Behandlung der Plattenaufgabe nach dem verbesserten Differenzenverfahren. Z. Angew. Math. Mech. **37**, 1−16 (1957).

[2] COLLATZ, L.: Z. Angew. Math. Mech. **31**, 232 (1951).

a sufficiently high order. As an example we mention the use of the expression

$$\Phi[u] = 20u_a - 4\sum u_b - \sum u_c + \frac{h^2}{5}\left(34\,V^2 u_a - \sum V^2 u_b\right) + $$
$$+ \frac{h^4}{30}\left(17\,V^4 u_a + \sum V^4 u_b\right) = O\left(h^8\right)$$

for the two-dimensional Poisson equation, in which $L[u] = u_{xx} + u_{yy}$; it is based on a square mesh with mesh width h and makes use of just one operator $M_1 = V^2$. The notation used is as follows: a is an arbitrary mesh point P, $u_a = u(P)$, $\sum u_b$ is the sum of the u values at the four neighbouring points and $\sum u_c$ is the sum of the u values at the four mesh points at a distance $\sqrt{2}\,h$ from P; and similarly for $\sum V^2 u_b$, etc.[1].

Other ways of improving on the accuracy of the ordinary finite-difference method are considered by WOODS[2]. For the differential equation $V^2 u = r(x, y)$, which he treats in detail, he suggests a method which is effectively an iterative solution of the system of equations obtained from the first Hermitian formula in Table VI of the appendix. This formula is easily derived by Taylor expansion, and with the same notation as above can be written in the form

$$4u_a - \sum u_b + h^2 r_a = D[u] + O(h^6),$$

where

$$D[u] = \frac{1}{6}\left(\sum u_c - 2\sum u_b + 4u_a\right) - \frac{h^2}{12}\left(\sum r_b - 4r_a\right).$$

First of all, approximations U' are obtained from the ordinary difference equations

$$4U'_a - \sum U'_b + h^2 r_a = 0$$

(together with the boundary conditions). With these values U' the quantities $D[U']$ are calculated and then corrections v' obtained from the inhomogeneous difference equations

$$4v'_a - \sum v'_b = D[U']$$

---

[1] A special formula of this type, namely

$$20u_a - 4\sum u_b - \sum u_c + 6h^2 V^2 u_a - \frac{h^4}{2} V^4 u_a = O(h^6),$$

was given as early as 1934 by SH. MIKELADZE: Sur l'intégration numérique d'équations différentielles aux dérivées partielles. Bull. Acad. Sci. URSS. **6**, 819—841 (1934) (Summaries in French and Russian); cf. also equation (2.21) of § 2.7. Further formulae of this type can be found in SH. E. MIKELADZE: Über die numerische Lösung der Differentialgleichung $u_{xx} + u_{yy} + u_{zz} = \varphi(x, y, z)$. C. R. (Doklady) Acad. Sci. URSS. **14**, 177—179 (1937), also 181—182.

[2] WOODS, L. C.: Improvements to the accuracy of arithmetical solutions to certain two-dimensional field problems. Quart. J. Mech. Appl. Math. **3**, 349—363 (1950).

(with homogeneous boundary conditions), so that the new approximations $U'' = U' + v'$ satisfy the equations

$$4 U_a'' - \sum U_b'' + h^2 r_a = D[U'].$$

If need be, further corrections v'', ... can be calculated in the same way.

2.6. Examples of the use of Hermitian formulae

In order that the methods may be compared, we use the same examples as in § 2.4 (and in the same order).

I. From Table VI of the appendix the Hermitian equations for $\nabla^2 u = 0$ centred on the points a, b, c (notation as in Fig. V/13) can be written down immediately:

$$40a - 8(b + 1 + \tfrac{1}{2} + 0) - 2(1 + 1 + 0 + 0) = 0,$$
$$40b - 8(c + 1 + a + 0) - 2(b + 1 + 0 + 0) = 0,$$
$$40c - 8(0 + b + b + 0) - 2(0 + 1 + 0 + 0) = 0.$$

Here no exterior pivotal values appear, in contrast to § 2.4. We obtain

$$a = \frac{40}{83} = 0.48193, \qquad b = \frac{34}{83} = 0.40964, \qquad c = \frac{71}{332} = 0.21386.$$

II. With the notation of Fig. V/15 the Hermitian equations read

$$40a - 32b - 8c = \frac{16}{3},$$
$$40b - 8(a + 2c + d) - 4(b + e) = \frac{16}{3},$$
$$40c - 16(b + e) - 2(a + 2d + f) = \frac{16}{3};$$

to these we must add the equations (2.13), corresponding to the boundary conditions, and also the analogous equation

$$15f = 6e + d.$$

Solution of this system of equations yields

$$a = \frac{147177}{176484} = 0.8339,$$

$$b = 0.7199, \qquad c = 0.6234, \qquad d = 0.3436, \qquad e = 0.2974, \qquad f = 0.1418.$$

III. For the boundary-value problem

$$u_{xx} + u_{yy} + \frac{3}{5 - y} u_y + 1 = 0 \tag{2.18}$$

with $u = 0$ on the sides of the rectangle $|x| \leq \frac{1}{2}$, $|y| \leq 1$ we use a different notation from § 2.4. With the mesh width $h = \frac{1}{2}$ we have the unknown function values U_2, U_3, U_4 (see Fig. V/17) and the unknown values U_1', U_2', U_3', U_4', U_5' of the partial derivative in the y direction; the corresponding values of $V^2 u$, which we denote by V_1^2, ..., V_5^2, can be expressed in terms of the U_j' by means of (2.18):

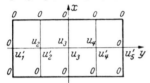

Fig. V/17. Notation for an example treated by the Hermitian method

$$V_1^2 = -1 - \tfrac{1}{2} U_1', \qquad V_2^2 = -1 - \tfrac{6}{11} U_2',$$

$$V_3^2 = -1 - \tfrac{3}{5} U_3', \qquad V_4^2 = -1 - \tfrac{2}{3} U_4',$$

$$V_5^2 = -1 - \tfrac{3}{4} U_5',$$

while for $|x| = \frac{1}{2}$ we have $V^2 u = -1$ since $u_y = 0$. Thus the Hermitian equations read

$$
\left.
\begin{aligned}
40\,U_2 - 8\,U_3 - 2 - \tfrac{12}{11} U_2' + \tfrac{1}{4}(-4 - \tfrac{3}{5} U_3' - \tfrac{1}{2} U_1') &= 0, \\
40\,U_3 - 8\,U_2 - 8\,U_4 - 2 - \tfrac{6}{5} U_3' + \tfrac{1}{4}(-4 - \tfrac{2}{3} U_4' - \tfrac{6}{11} U_2') &= 0, \\
40\,U_4 - 8\,U_3 - 2 - \tfrac{4}{3} U_4' + \tfrac{1}{4}(-4 - \tfrac{3}{4} U_5' - \tfrac{3}{5} U_3') &= 0.
\end{aligned}
\right\}
\qquad (2.19)
$$

According to the general procedure described in § 2.5 one would now write down a set of Hermitian equations for the U_j'; from Table III of the appendix such an equation centred on $y = \frac{1}{2}$, for example, would be

$$-U_3 - \tfrac{1}{6}(U_5' + 4\,U_4' + U_3') = 0.$$

However, in the present simple case it is more convenient, and at the same time more accurate, to put a fourth-order interpolation polynomial through the five function values U_i and use its derivatives at the points $y = -1$, $-\frac{1}{2}$, 0, $\frac{1}{2}$, 1 to express the U_i' in terms of the U_i. Thus, evaluating the derivative of the polynomial

$$U(y) = \left[-\tfrac{4}{3} U_2\, y\,(2y - 1) + U_3(4y^2 - 1) - \tfrac{4}{3} U_4\, y\,(2y + 1)\right](y^2 - 1)$$

for these values of y, we obtain

$$U_1' = 8\,U_2 - 6\,U_3 + \tfrac{8}{3} U_4, \qquad U_2' = -\tfrac{5}{3} U_2 + 3\,U_3 - U_4, \qquad U_3' = -\tfrac{4}{3} U_2 + \tfrac{4}{3} U_4,$$

$$U_4' = U_2 - 3\,U_3 + \tfrac{5}{3} U_4, \qquad U_5' = -\tfrac{8}{3} U_2 + 6\,U_3 - 8\,U_4.$$

Substitution of these expressions in (2.19) yields a system of equations for the U_i:

$$27072\,U_2 - 6945\,U_3 + 368\,U_4 - 1980 = 0,$$

$$-3138\,U_2 + 19845\,U_3 - 4822\,U_4 - 1485 = 0,$$

$$-228\,U_2 - 1845\,U_3 + 14068\,U_4 - 1080 = 0$$

which may be solved conveniently by iteration since it has strongly dominant diagonal coefficients. The solution is

$$U_2 = 0 \cdot 10098, \qquad U_3 = 0 \cdot 11346, \qquad U_4 = 0 \cdot 09329.$$

IV. We mention here an interesting theorem due to PÓLYA[1]. Let a bounded, simply connected region B in the (x, y) plane be covered by a square mesh of mesh width h and let B' be a region contained within B and bounded only by sections

---

[1] PÓLYA, G.: Sur une interprétation de la méthode des différences finies, qui peut fournir des bornes supérieures ou inférieures. C. R. Acad. Sci., Paris **235**, 995–997 (1952).

of mesh lines. If λ_k is an eigenvalue of $\nabla^2 u + \lambda u = 0$ in B with $u = 0$ on the boundary Γ of B, λ_k' the corresponding eigenvalue for the region B' and Λ_k the corresponding eigenvalue of the particular difference equations

$$U_{i+1,j} + U_{i,j+1} + U_{i-1,j} + U_{i,j-1} - 4U_{i,j} +$$

$$+ \frac{\lambda h^2}{12}(6U_{i,j} + U_{i+1,j} + U_{i+1,j+1} + U_{i,j+1} + U_{i-1,j} + U_{i-1,j-1} + U_{i,j-1}) = 0$$

for the region B', then $\lambda_k \leq \lambda_k' \leq \Lambda_k$ for all k for which Λ_k exists. The corresponding result for the ordinary difference equations does not necessarily hold.

2.7. Triangular and hexagonal mesh systems

There are many types of boundary for which it is expedient to use a non-rectangular mesh. For illustration we select a mesh of equilateral triangles (Fig. V/18) and describe the derivation of finite expressions for the operators ∇^2 and ∇^4.

Let O be a general mesh point. We start by expanding the function values at the neighbouring mesh points $P, Q, \ldots$ into Taylor series centred on O; for example, with the notation $\alpha = \frac{1}{2}h$, $\beta = \frac{1}{2}\sqrt{3}\,h$, where h is the mesh width (here the length of the sides of the equilateral triangles), the value at P (Fig. V/18) has the expansion

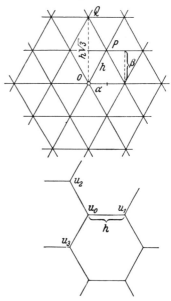

Fig. V/18. Triangular and hexagonal meshes

$$u_P = u_0 + \alpha u_x + \beta u_y +$$

$$+ \frac{1}{2!}(\alpha^2 u_{xx} + 2\alpha\beta u_{xy} + \beta^2 u_{yy}) +$$

$$+ \frac{1}{3!}(\alpha^3 u_{xxx} + 3\alpha^2\beta u_{xxy} + 3\alpha\beta^2 u_{xyy} + \beta^3 u_{yyy}) + \quad\quad (2.20)$$

$$+ \frac{1}{4!}(\alpha^4 u_{xxxx} + 4\alpha^3\beta u_{xxxy} + 6\alpha^2\beta^2 u_{xxyy} + 4\alpha\beta^3 u_{xyyy} +$$

$$+ \beta^4 u_{yyyy}) + \cdots.$$

If we write down the corresponding expansions for the other five mesh points at a distance h from O and add all six together, most terms cancel out because of the symmetry and we are left with the result

$$\sum u_P = 6u_0 + \frac{3h^2}{2}\nabla^2 u_0 + \frac{3h^4}{32}\nabla^4 u_0 + \cdots. \quad\quad (2.21)$$

Similarly, addition of the corresponding Taylor expansions for the six function values at the mesh points at a distance $\sqrt{3}\,h$ from O yields the formula

$$\Sigma\,u_Q = 6\,u_0 + \frac{9\,h^2}{2}\,\nabla^2 u_0 + \frac{27\,h^4}{32}\,\nabla^4 u_0 + \cdots, \qquad (2.22)$$

which can also be derived immediately from (2.21) by replacing h by $\sqrt{3}\,h$.

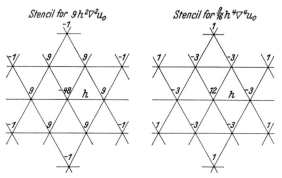

Fig. V/19. Stencils for $\nabla^2 u$ and $\nabla^4 u$

Thus as a first approximation for $\nabla^2 u_0$ we have the formula

$$\Sigma\,u_P - 6\,u_0 = \frac{3\,h^2}{2}\,\nabla^2 u_0 + O(h^4) \qquad (2.23)$$

and as a second approximation the formula

$$-48\,u_0 + 9\,\Sigma\,u_P - \Sigma\,u_Q = 9\,h^2\,\nabla^2 u_0 + O(h^6), \qquad (2.24)$$

which is obtained by eliminating $\nabla^4 u_0$ between (2.21) and (2.22).

Elimination of $\nabla^2 u_0$ between (2.21) and (2.22) yields an approximation for $\nabla^4 u_0$:

$$12\,u_0 - 3\,\Sigma\,u_P + \Sigma\,u_Q = \tfrac{9}{16}\,h^4\,\nabla^4 u_0 + O(h^6). \qquad (2.25)$$

These results are best visualized as stencils of coefficients laid over the corresponding pattern of mesh points as in Fig. V/19; such stencils are also helpful in setting up the system of difference equations.

For a mesh of regular hexagons ("honeycomb pattern") we obtain in the same way

$$u_1 + u_2 + u_3 = 3\,u_0 + \frac{3\,h^2}{4}\,\nabla^2 u_0 + O(h^3), \qquad (2.26)$$

where the function values u_0, u_1, u_2, u_3 correspond to the points indicated in Fig. V/18.

2.8. Applications to membrane and plate problems

Example I. Consider a uniform plate in the shape of a trapezium (sides of lengths $A, A, A, 2A$ and angles of $60°$ and $120°$ as in Fig. V/20) with its long edge clamped and its other edges freely supported. If it is subjected to a uniformly distributed trans-
verse load of intensity p
per unit area and its
flexural rigidity is N,
the transverse displace-
ment u satisfies the differential equation

$$\nabla^4 u = \frac{p}{N}$$

and the boundary con-
ditions

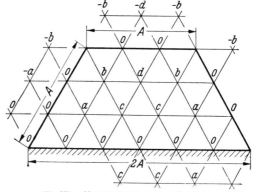

Fig. V/20. Notation for the trapezoidal plate

$u = \dfrac{\partial u}{\partial v} = 0$ along the clamped edge (v in the direction
of the inward normal),

$u = \nabla^2 u = 0$ along the freely supported edges.

From (2.25) the finite equations for a triangular mesh with mesh width $h = \frac{1}{3}A$ read

$$12a - 3b - 3c + d = k = \frac{A^4}{144} \cdot \frac{p}{N},$$

$$-3a + 10b - 2c - 3d = k,$$

$$-3a - 2b + 10c - 3d = k,$$

$$2a - 6b - 6c + 11d = k,$$

where a, b, c, d are the approximate function values as indicated in Fig. V/20.

In these equations we have used the values at exterior mesh points inferred from the boundary conditions (symmetrical values about the long clamped edge and antisymmetrical values about the freely supported edges — see Fig. V/20).

Solution of the equations yields

$$a = \frac{58}{241} k, \quad b = c = \frac{95}{241} k, \quad d = \frac{115}{241} k,$$

i.e.

$$a = 0{\cdot}001\,671 \frac{p A^4}{N}, \quad b = c = 0{\cdot}002\,737 \frac{p A^4}{N}, \quad d = 0{\cdot}003\,314 \frac{p A^4}{N}.$$

A finer mesh with $h = \frac{1}{4}A$ is used in Exercise 9 of § 6.7.

Example II. The normal modes of vibration of a homogeneous membrane stretched over a frame in the shape of a regular hexagon H with sides of length L (Fig. V/21) are given by the solutions of the eigenvalue problem

$$\nabla^2 u = - \lambda u \text{ in } H,$$

$$u = 0 \text{ on the boundary of } H.$$

A. Triangular mesh with $h = \frac{1}{3}L$. Here we have nineteen mesh points inside of H. For the corresponding function values we adopt the nota-

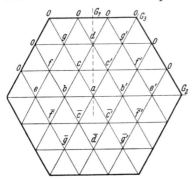

Fig. V/21. Hexagonal membrane

tion $a, b, c, \ldots, g, b', \ldots, \bar{g}'$ as in Fig. V/21, in which the same letter is associated with points lying symmetrically about the lines G_1 and G_2 (Fig. V/21) and dashes and bars distinguish on which sides of the lines they lie; for example, the image of c in G_1 is c' and in G_2 is $\bar{c}$, and the image of $\bar{c}$ in G_1 is $\bar{c}'$. A general calculation would lead to an algebraic equation of the nineteenth degree for the approximation Λ to the eigenvalue λ, but by postulating the various symmetries which can occur we can achieve our object with equations of a very much lower degree.

A 1. Symmetry about G_1 and G_2. All bars and dashes can be omitted since $c = c' = \bar{c} = \bar{c}'$ etc. and we have just seven equations for $a, b, \ldots, g$ and Λ. With $\nu = 6 - \frac{3}{2} h^2 \Lambda$ these read [from (2.23)]

$$\nu a = 2b + 4c$$
$$\nu b = a + 2c + e + 2f,$$
$$\nu c = a + b + c + d + f + g,$$
$$\nu d = 2c + 2g,$$
$$\nu e = b + 2f,$$
$$\nu f = b + c + e + g,$$
$$\nu g = c + d + f.$$

We can take out the factors $(\nu + 1)$ and $(\nu^2 - 3)$ from the determinant of coefficients $D(\nu)$ by combining several rows and columns, and the condition on ν for a non-trivial solution becomes

$$D(\nu) = (\nu + 1)(\nu^2 - 3)(\nu^4 - 2\nu^3 - 15\nu^2 + 24) = 0.$$

Table V/7. *The normal modes of vibration of a hexagonal membrane*

Meshes		Triangular meshes				Various hexagonal mesh systems							
Number of interior mesh points		7		19		6		12		24		4	
Mesh width		$h=\frac{1}{2}L$		$h=\frac{1}{3}L$		$h=\frac{1}{2}L$		$h=\frac{1}{3}L$		$h=\frac{1}{4}L$		$h=\frac{1}{\sqrt{3}}L$	
Symmetry about	Nodal lines	ν	ΛL^2	ν	ΛL^2	ν	ΛL^2	ν	ΛL^2	ν	ΛL^2	ν	ΛL^2
G_1, G_2 full symm.		$1+\sqrt{7}$	6·28	4·8715	6·77	2	5·33	$1+\sqrt{2}$	7·030	2·6750	6·933	$\sqrt{3}$	5·072
G_1 / G_2		1	13·3	3·355	15·87	1	10·67	$\frac{1+\sqrt{5}}{2}$	16·58	2·214	16·76	0	12
G_1, G_2 / —		−1	18·7	$\sqrt{3}$	25·6	−1	21·3	$\frac{\sqrt{5}-1}{2}$	28·58	1·68	28·2		
G_1, G_2 full symm.		$1-\sqrt{7}$	20·4	1·23	28·6			$1-\sqrt{2}$	41·0	1·55	31·0	$-\sqrt{3}$	18·9
G_2		−2	21·3	$\sqrt{2}-1$	33·5	−2	26·7	$-1+\sqrt{2}$	31·0	1·21	38·2		
G_1				0	36								
G_1 / G_2				−0·476	38·8			$\frac{1-\sqrt{5}}{2}$	43·4	1	42·7		
G_1, G_2 / —				−1	42			$\frac{-1-\sqrt{5}}{2}$	55·4	0·54	52·5		
G_1, G_2 full symm.				−1·63	45·8					−1·21	90		

The largest root of this equation gives the "gravest" mode (smallest value of Λ); this corresponds to the special case of full symmetry (six axes of symmetry), in which the displacements are all of the same sign and nodal lines are absent. The corresponding values are given in the first row of Table V/7. The subsequent rows show the corresponding results for the higher eigenvalues in increasing order of magnitude; sketches of the corresponding nodal lines are also given. These lines of zero displacement can be obtained approximately from the

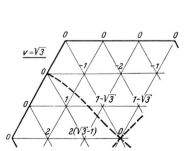

Fig. V/22. A higher mode with nodal line

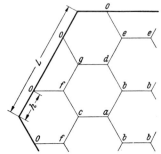

Fig. V/23. Hexagonal mesh for the calculation of normal modes

corresponding approximate mode form given by the ratios $a:b:c:d:e:f:g$; these are easily calculated for each ν value from the homogeneous equations, say by putting one of the non-zero values equal to 1. This is shown in Fig. V/22 for $\nu = \sqrt{3}$.

A2. Symmetry about G_1, antisymmetry about G_2. Here the dashes can be omitted, as before, but omission of the bars is to be accompanied by a reversal of sign; thus, for example, $c = c' = -\bar{c} = -\bar{c}'$. Further $a = b = e = 0$, and we are left with just four equations for c, d, f, g:

$$\nu c = c + d + f + g,$$
$$\nu d = 2c + 2g,$$
$$\nu f = c + g,$$
$$\nu g = c + d + f.$$

The determinantal condition now leads to the quartic

$$\nu\left(\nu^3 - \nu^2 - 7\nu - 3\right) = 0.$$

A3, A4. Antisymmetry about G_1, symmetry and antisymmetry, respectively, about G_2. The procedure is the same as in A1 and A2; the number of unknowns is reduced to five and three, respectively, with the results shown in Table V/7. For the higher modes some uncertainty exists in the ordering of the eigenvalues; this can only be removed by increasing the accuracy of the calculation.

B. Hexagonal mesh. The results for several different mesh widths are exhibited in Table V/7. It will suffice to describe the beginning of the calculation with $h = \frac{1}{4}L$, for which the mesh system and the notation $a, b, \ldots, g$ for the function values are indicated in Fig. V/23. Again we postulate the various symmetries, and taking the case of symmetry about G_1 and G_2 as an example we have from (2.26) the equations

$$\nu a = 2b + c,$$
$$\nu b = a + b + d,$$
$$\dots\dots\dots\dots\dots,$$

where $\nu = 3 - \tfrac{3}{4}h^2 \Lambda$. The values $c, d, \dots, g$ can easily be expressed successively in terms of a and b:

$$c = 2\nu\, \frac{a}{2} - 2b,$$

$$f = (\nu^2 - 1)\, \frac{a}{2} - \nu b,$$

$$g = (\nu^3 - 3\nu)\, \frac{a}{2} + (-\nu^2 + 2)\, b,$$

$$\dots\dots\dots\dots\dots\dots\dots\dots\dots$$

and the last two homogeneous equations then lead directly to the equation for ν:

$$(\nu + 1)(\nu^3 - 3\nu^2 - \nu + 5)(\nu^3 - 4\nu + 2) = 0.$$

A hexagonal mesh can be fitted into the hexagon H in various other ways, as indicated in Table V/7.

Example III. If the membrane of Example II is replaced by a homogeneous plate, the differential equation governing the vibrations becomes

$$\nabla^4 u = \lambda u;$$

the simplest associated boundary conditions are $u = 0$, $\nabla^2 u = 0$ at a freely supported edge and $u = 0$, $\partial u/\partial \nu = 0$ at a clamped edge. If all the edges are freely supported, the eigenfunctions, or mode forms, are the same as for the membrane, for a solution ν of $\nabla^2 v = -\lambda v$ with $v = 0$ on the boundary satisfies $\nabla^4 v = \lambda^2 v$ and also the boundary conditions $v = \nabla^2 v = 0$. For a clamped plate, on the other hand, a new calculation is required. We have to use mesh points outside of the hexagon, and because of the clamping, the associated function values are related to the interior values in symmetrical fashion about the edges, as in Example I. With $h = \tfrac{1}{2}L$, $\nu = 12 - \tfrac{9}{16}h^4 \Lambda$ and the function values as in Fig. V/24, we have from (2.25) the equations

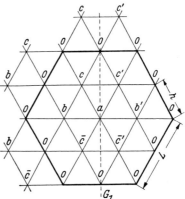

Fig. V/24. Notation for a clamped hexagonal plate

$$\nu a = 3(b + c + c' + b' + \bar{c}' + \bar{c}),$$
$$\nu b = 3(c + a + \bar{c}) - 2b - c' - \bar{c}',$$
$$\dots\dots\dots\dots\dots\dots\dots\dots$$

As before, these can be simplified very considerably by postulating the various symmetries that can occur. Symmetry about the line G_1 leads to the values $\nu = 1$, $\pm \sqrt{55}$, 2, -4 and antisymmetry to the values $\nu = 2$, -4, -10; the corresponding values of ΛL^4 are 104, 284, 284, 455, 455, 524, 626.

§ 3. The boundary-maximum theorem and the bracketing of solutions

The boundary-maximum theorem for second-order elliptic differential equations owes its importance to the fact that for many cases of first and third boundary-value problems it provides a basis from which one may be able to obtain in a simple manner rigorous error limits for approximate solutions, and hence also upper and lower bounds for the exact solution. The simple nature of the calculations which are involved is demonstrated by the examples in § 3.4.

3.1. The general boundary-maximum theorem

The error estimates of the following sections are based on the boundary-maximum theorem. This theorem was first established for the potential equation but is valid in much more general cases and will therefore be presented generally here.

Consider the second-order linear differential equation

with
$$\left.\begin{array}{l} L[u] = r \\ L[u] = - \sum_{j,k=1}^{n} a_{jk} u_{jk} - \sum_{j=1}^{n} b_j u_j + c u, \end{array}\right\} \tag{3.1}$$

whose dependent variable $u(x_1, \ldots, x_n)$ is a function of the n independent variables $x_1, \ldots, x_n$ in a closed region $B + \Gamma$ of the n-dimensional space, B being an open region and Γ its set of boundary points. The differential expression may be written more simply as

$$L[u] = - a_{jk} u_{jk} - b_j u_j + c u \tag{3.2}$$

if we make use of the summation convention which is customary in tensor calculus; in this convention the summation signs are omitted and summation from 1 to n understood over each Latin subscript which appears twice in a term. The subscripts attached to a and b serve merely to distinguish different functions but those attached to u (and v, w, z in the following) denote as in (1.11) the partial derivatives

$$u_j = \frac{\partial u}{\partial x_j}, \qquad u_{jk} = \frac{\partial^2 u}{\partial x_j \partial x_k}.$$

We assume here that the given functions a_{jk}, b_j, c, r are continuous in $B + \Gamma$.

The differential equation is said to be elliptic when the matrix a_{jk} (which can be taken as symmetric: $a_{jk} = a_{kj}$) is positive definite in $B + \Gamma$. The theorem applies to elliptic differential equations with $c \geq 0$. Let the open region B be bounded (and connected) and let its boundary Γ form a closed, connected, piecewise smooth[1], $(n-1)$-dimensional hypersurface. Then we can state

---

[1] Courant, R., and D. Hilbert: Methoden der mathematischen Physik, Vol. II, p. 228. Berlin 1937.

The boundary-maximum theorem: *If a non-constant function* $w(x_1, \ldots, x_n)$ *with continuous partial derivatives of up to and including the second order is such that* $L[w] \leq 0$ *in* B, *where* L *is the differential operator of* (3.2) *with* (a_{jk}) *positive definite and* $c \geq 0$, *and if further (this condition may be omitted for the case* $c = 0$) w *is not everywhere negative in* $B + \Gamma$, *then the maximum value* M *attained by* w *in* $B + \Gamma$ *is assumed only on the boundary* Γ.

Proof (indirect)[1]: We derive a contradiction from the assumption that there is a point P_1 inside B with $w(P_1) = M$. Since w is not constant, there must be a second point P_2 in B with $w(P_2) < M$. There is a curve C joining P_1 and P_2 and a positive number $\varrho > 0$ such that every hypersphere of radius ϱ with a point P on C as centre lies entirely within B. The continuity of w implies the existence of a point P_3 on C such that $w(P_3) = M$ and $w(P) < M$ for all points P on C lying between P_2 and P_3. We now choose a point P_4 on C lying between P_2 and P_3 and sufficiently near P_3 that the hypersphere S_1 with P_4 as centre and with P_3 on its surface lies entirely within B. We therefore have a hypersphere S_1 with $w(P_4) < M$ at its centre P_4 and $w(P_3) = M$ at a point P_3 on its surface. We can now shrink this hypersphere until we arrive at a hypersphere S_2, centre P_4, with $w < M$ at all points inside S_2 and with at least one point P_5 on its surface where $w(P_5) = M$. Then the hypersphere S_3 with $\overline{P_4 P_5}$ as a diameter will have $w = M$ at only one point P_5 on its surface; elsewhere on its surface and in its interior $w < M$. Let P_6 be the centre of S_3 and R its radius. We now define a fourth hypersphere S_4 with centre P_5 and radius $\varrho' < R$ and also so small that S_4 lies completely within B. Then, with the origin of the co-ordinates $x_1, \ldots, x_n$ transferred temporarily to the point P_6 and with the distance from it denoted by r $(r^2 = x_j x_j)$, positive constants δ and α can be chosen so that the auxiliary function $z = w + \delta(e^{-\alpha r^2} - e^{-\alpha R^2})$ is less than M everywhere on the surface Φ of S_4; for on the part of Φ which lies outside of S_3 the bracketed factor is negative and on the remainder, which is a closed set, $w \leq M' < M$, so that, for given α, δ can be chosen so small that

---

[1] Cf. E. Hopf: Elementare Bemerkungen über die Lösungen partieller Differentialgleichungen 2. Ordnung vom elliptischen Typus. S.-B. Preuss. Akad. Wiss., Phys.-Math. Kl., pp. 147—152, Berlin 1927. — Courant, R., and D. Hilbert: Methoden der mathematischen Physik, Vol. II, p. 275. Berlin 1937. — Bateman, H.: Partial differential equations of mathematical physics, p. 135. New York 1944. — The theorem can be proved very simply for the special case of the potential equation — one way makes use of Poisson's integral, cf. for example Ph. Frank and R. v. Mises: Die Differentialgleichungen und Integralgleichungen der Mechanik und Physik, 2nd ed., Vol. I, pp. 691—692. Brunswick 1930. "A priori" limits for the solutions of boundary-value problems are derived by G. Fichera: Methods of functional linear analysis in mathematical physics. Proc. Internat. Math. Congr. Amsterdam 1954.

$z < M$ there also. Further, $z(P_5) = M$ from the definition of z. Consequently we have a function z which takes the value M at the centre of S_4 and is less than M everywhere on the surface of S_4; it must therefore have a relative maximum with $z \geq M$ for at least one point P_7 inside of S_4.

So far α has been an arbitrary positive number; we now choose it so that $L[z] < 0$ at P_7. We have

$$L[z] = L[w] + \delta \{ L[e^{-\alpha r^2}] - c\, e^{-\alpha R^2} \}$$

and

$$e^{\alpha r^2} \{ L[e^{-\alpha r^2}] - c\, e^{-\alpha R^2} \} = -4\alpha^2 a_{jk} x_j x_k + 2\alpha (a_{jj} + b_j x_j) + c - c\, e^{\alpha (r^2 - R^2)}.$$

Here α^2 is multiplied by a negative factor and the last term is non-positive since $c \geq 0$; therefore by choosing α large enough we can ensure that the expression in curly brackets takes only negative values (excluding zero) in the whole of S_4. Since $L[w] \leq 0$, we also have $L[z] < 0$ everywhere in S_4, in particular at P_7.

Now at P_7, z has a relative maximum; at P_7, therefore, $z_j = 0$ ($j = 1, \dots, n$) and the negatives of the second partial derivatives $-z_{jk}$ form a positive definite or semi-definite matrix:

$$-z_{jk} \xi_j \xi_k \geq 0$$

for arbitrary real numbers $\xi_1, \dots, \xi_n$. Consequently

$$L[z]_{P_7} = -a_{jk} z_{jk} + c z,$$

where both the a_{jk} and the $-z_{jk}$ are the coefficients of positive (semi-) definite quadratic forms, so that

$$-a_{jk} z_{jk} \geq 0$$

from a result due to Fejer[1]. This implies immediately that $L[z]_{P_7} \geq 0$ for the case $c = 0$; and since the conditions of the theorem require that $M \geq 0$ when $c \geq 0$, we have $z(P_7) \geq M \geq 0$, and hence $L[z]_{P_7} \geq 0$, also for the general case $c \geq 0$. Since z was chosen above so that $L[z]_{P_7} < 0$, we have arrived at the desired contradiction.

Corollary[2]. *The corresponding boundary-minimum theorem, in which the inequalities satisfied by $L[w]$ and M are reversed (i.e. have the signs*

---

[1] Fejer, L.: Über die Eindeutigkeit der Lösung der linearen partiellen Differentialgleichungen zweiter Ordnung. Math. Z. **1**, 70—79 (1918).

[2] The analogous theorem to the boundary-maximum theorem for the corresponding difference equations has been proved by T. S. Motzkin and W. Wasow: On the approximation of linear elliptic differential equations by difference equations with positive coefficients. J. Math. Phys. **31**, 253—259 (1953); this result can be proved very easily, cf. R. Courant and D. Hilbert: Methoden der mathematischen Physik, Vol. II, p. 275. 1937.

$\leqq$, $\geqq$ *interchanged) and* M *is the minimum value attained by* w, *is also valid (the boundary-maximum theorem has only to be applied to the function* $-w$).

A further consequence of the boundary-maximum theorem is the following:

If two functions w_1 *and* w_2 *with continuous partial derivatives of up to and including the second order both vanish on the boundary* Γ *and are such that* $|L[w_1]|\leqq L[w_2]$ *in* B, *then* $|w_1|\leqq w_2$ *in* B.

This can be seen by writing $w=w_1-w_2$, $w^*=-w_1-w_2$, so that $L[w]\leqq0$ and $L[w^*]\leqq0$. The maximum of w cannot be negative since there are points where $w=0$ (on the boundary Γ); thus w satisfies the conditions of the boundary-maximum theorem, and hence $M=0$ since $w\equiv0$ on Γ; consequently $w\leqq0$. Similarly $w^*\leqq0$, and therefore $|w_1|\leqq w_2$.

3.2. General error estimation for the first boundary-value problem

Consider now the first boundary-value problem associated with the equation (3.1), in which u takes prescribed values $\tilde{u}$ on the boundary Γ:

$$u=\tilde{u}\quad\text{on }\Gamma;\tag{3.3}$$

for simplicity we will assume that the given function $\tilde{u}$ is continuous on Γ. The tilde will be used to denote boundary values for other functions also; for example, $\tilde{v}$ will denote the value of v on the boundary.

Let v be an approximate solution of the boundary-value problem which satisfies at least one of the equations (3.1), (3.3). We distinguish two cases (cf. Ch. I, § 4.1).

Case 1. v satisfies the differential equation $L[u]=r$ exactly but need not take the prescribed boundary values. The error function $w=v-u$ then satisfies the boundary-value problem $L[w]=0$, $\tilde{w}=\tilde{v}-\tilde{u}$, in which the boundary values $\tilde{w}$ are known for a given approximate solution v. Since $L[w]=0$, w satisfies both of the conditions $L[w]\leqq0$ and $L[w]\geqq0$, which appear in the boundary-maximum theorem and its corollary, respectively. If $w\geqq0$ somewhere in $B+\Gamma$, then by the boundary-maximum theorem w assumes its maximum value only on the boundary and $w_{\max}=\tilde{w}_{\max}\geqq0$. Similarly, if $w\leqq0$ somewhere in $B+\Gamma$, then by the corollary w assumes its minimum value only on the boundary and $w_{\min}=\tilde{w}_{\min}\leqq0$. Consequently, if $w=0$ somewhere in $B+\Gamma$, then we have $\tilde{w}_{\min}\leqq w\leqq\tilde{w}_{\max}$, where $\tilde{w}_{\min}\leqq0$, $\tilde{w}_{\max}\geqq0$. If $w\neq0$ in $B+\Gamma$, i.e. if w does not change sign in $B+\Gamma$ (w being a continuous function), then for $w>0$, say, we have $0<w\leqq\tilde{w}_{\max}$.

We can distinguish these various cases according to the behaviour of $\tilde{w}$ instead of w; this is more convenient for application and leads to the

Error estimate. *For the boundary-value problem* (3.1), (3.3) *with* $c \geq 0$ *and* (a_{jk}) *positive definite let* v *be an approximate solution which satisfies the differential equation* $L[u] = r$ *exactly. The boundary values* $\tilde{w} = \tilde{v} - \tilde{u}$ *of the error function* $w = v - u$ *are determinable. If* $\tilde{w}$ *takes the value zero somewhere on* Γ, *then*

$$\tilde{w}_{\min} \leq w \leq \tilde{w}_{\max} \tag{3.4}$$

in B. *If* $\tilde{w}$ *does not take the value zero somewhere on* Γ, *i.e.* $\tilde{w}$ *is everywhere positive or everywhere negative on* Γ, *then* w *is everywhere positive or everywhere negative, respectively, in* B *and*

$$|w| \leq |\tilde{w}|_{\max}. \tag{3.5}$$

In the case $c = 0$, (3.4) *holds independently of the behaviour of* $\tilde{w}$.

From the point of view of practical analysis the situation described here deserves mention as being particularly favourable for error estimation. Error estimates for other problems involving differential equations often turn out to be far more intricate and much less precise than those obtained here. If in fact we were specifically aiming to find constant error bounds valid over the whole of the region B, we could not do better than the estimates (3.4) and (3.5), for the error actually reaches the specified limits on the boundary Γ and the constants are therefore best-possible.

Case 2. The approximate solution v satisfies the boundary condition (3.3) but not necessarily the differential equation (3.1). This situation arises, for instance, when RITZ's method has been used (cf. § 5) and is therefore of particular interest. The residual function $L[v] - r = L[v] - L[u]$ is now not necessarily zero. We suppose that we can construct two auxiliary functions q_1 and q_2 such that

$$L[v] + L[q_1] \leq r \quad \text{and} \quad L[v] + L[q_2] \geq r \tag{3.6}$$

everywhere in $B + \Gamma$. One way of doing this is to evaluate $L[v]$, then choose suitable functions q_1^* and q_2^*, form $L[q_1^*]$ and $L[q_2^*]$ and try to find constants c_1, c_2 such that the inequalities are satisfied in $B + \Gamma$ by $q_1 = c_1 q_1^*$ and $q_2 = c_2 q_2^*$. Such functions q_1, q_2 can always be specified when a function z is known with $L[z] \leq -A < 0$ in B; for the Laplace expression

$$L[u] = -\sum_{j=1}^{n} \frac{\partial^2 u}{\partial x_j^2}$$

we can use, for example,

$$z = \frac{A}{2n} \sum_{j=1}^{n} x_j^2 = q_1^* = q_2^*.$$

The boundary values of the functions $v + q_1 - u$ and $v + q_2 - u$ are known; let M_1 be the maximum of $v + q_1 - u$ and M_2 the minimum of $v + q_2 - u$ on the boundary. Now put

$$\mu_1 = \begin{cases} M_1 & \text{for the case} \quad c = 0 \\ \max(M_1, 0) & \text{for the case} \quad c \geq 0, \end{cases}$$

$$\mu_2 = \begin{cases} M_2 & \text{for the case} \quad c = 0 \\ \min(M_2, 0) & \text{for the case} \quad c \geq 0. \end{cases}$$

Then, since $L[v + q_1 - u] \leq 0$ and $L[v + q_2 - u] \geq 0$ in $B + \Gamma$, the boundary-maximum theorem and its corollary yield

$$v + q_1 - u \leq \mu_1, \qquad v + q_2 - u \geq \mu_2,$$

from which we obtain upper and lower bounds for u:

$$v + q_1 - \mu_1 \leq u \leq v + q_2 - \mu_2. \tag{3.7}$$

A somewhat simpler estimate can be derived if a function Z can be found with

$$L[Z] \geq A > 0 \quad \text{in } B, \qquad Z = 0 \quad \text{on } \Gamma. \tag{3.8}$$

Then with

$$\varrho = \max_{\text{in } B} |L[v] - r|$$

we have

$$|L[v - u]| = |L[v] - r| \leq \varrho \leq \frac{\varrho}{A} L[Z] = L\left[\frac{\varrho}{A} Z\right],$$

and by the consequence of the boundary-maximum theorem mentioned at the end of § 3.1

$$|v - u| \leq \frac{Z}{A} \max_{\text{in } B} |L[v] - r|. \tag{3.9}$$

3.3. Error estimation for the third boundary-value problem

We consider a mixed boundary-value problem in which Γ is composed of two parts Γ_1 and Γ_2, but do not exclude the degenerate cases in which one part coincides with Γ and the other is null. Let the boundary value $u = A_4$ be prescribed on Γ_1 and let the boundary condition on Γ_2 be of the form[1]

$$A_1 u + A_2 L^*[u] = A_3. \tag{3.10}$$

---

[1] GRÜNSCH, H. J.: Eine Fehlerabschätzung bei der 3. Randwertaufgabe der Potentialtheorie. Z. Angew. Math. Mech. **32**, 279—281 (1952). — See also C. PUCCI: Maggiorizazione della suluzione di un problema al contorno, di tipo misto, relativo a una equazione a derivate parzioli, lineare, del secondo ordine. Atti Acad. Naz. Lincei, Ser. VIII **13**, 360—366 (1953). — Bounds for Solutions of Laplace's Equation Satisfying Mixed Conditions. J. Rational Mech. Anal. **2**, 299—302 (1953).

where A_4 and A_1, A_2, A_3 are given functions on Γ_1 and Γ_2, respectively, and

$$L^*[u] = A\,\frac{\partial u}{\partial \sigma},$$

as defined in Ch. I, § 3.3, with $A > 0$ and σ denoting the conormal; further let

$$\frac{A_2}{A_1} < 0 \quad \text{on } \Gamma_2.$$

The other assumptions made in Ch. I, § 3.3 for the formulation of the third boundary-value problem are also made here.

Now let v be an approximate solution which satisfies the differential equation (3.1) exactly but the boundary conditions only approximately:

$$\left.\begin{array}{ll} v = A_4^* & \text{on } \Gamma_1, \\[2mm] A_1 v + A_2 A\,\dfrac{\partial v}{\partial \sigma} = A_3^* & \text{on } \Gamma_2. \end{array}\right\} \tag{3.11}$$

We now define a boundary function γ along the whole of the boundary Γ as the defect in the satisfaction of the boundary conditions, with the mixed condition normalized so that the coefficient of u is unity. Thus the error function $w = v - u$, which satisfies the homogeneous differential equation corresponding to (3.1):

$$L[w] = 0,$$

is subject to the boundary conditions

$$\left.\begin{array}{ll} w = A_4^* - A_4 = \gamma & \text{on } \Gamma_1, \\[2mm] w + \dfrac{A_2 A}{A_1}\,\dfrac{\partial w}{\partial \sigma} = \dfrac{A_3^* - A_3}{A_1} = \gamma & \text{on } \Gamma_2. \end{array}\right\} \tag{3.12}$$

Again w satisfies both of the conditions $L[w] \leq 0$ and $L[w] \geq 0$ of § 3.1 since $L[w] = 0$. If $w \geq 0$ somewhere in $B + \Gamma_1 + \Gamma_2$, then by the boundary-maximum theorem w assumes its non-negative maximum value M only on the boundary $\Gamma = \Gamma_1 + \Gamma_2$, say at a point Q.

If Q is on Γ_1, then $w(Q) = \gamma(Q) = M$, and consequently $M \leq \gamma_{\max}$, where $\gamma_{\max}$ denotes the maximum value of γ on Γ; if Q is on Γ_2, then $w(Q) \leq \gamma(Q)$, for the maximum property of $w(Q) = M$ implies that

$$\frac{\partial w}{\partial \sigma} \leq 0$$

at Q (it was shown in Ch. I, § 3.3 that for an elliptic differential equation the conormal points into the interior): in both cases $w \leq \gamma_{\max}$.

Similarly, if $w \leq 0$ somewhere in $B + \Gamma$, then $w \geq \gamma_{\min}$. In the case where the coefficient c in the differential equation (3.1) is zero these inequalities hold irrespective of the behaviour of w. Summarizing we have the

Error estimate. *For the (mixed) boundary-value problem* (3.1), (3.10) *with* $c \geq 0$, (a_{jk}) *positive definite and* $\dfrac{A_2}{A_1} < 0$, *let* v *be an approximate solution which satisfies the differential equation. The boundary function* γ *associated with the error function* $w = v - u$ *can then be determined from* (3.12). *If* γ *takes both positive and negative values, or if* γ *takes the value zero, then*

$$\gamma_{\min} \leq w \leq \gamma_{\max} \tag{3.13}$$

in B. *If* γ *has a fixed sign, then* w *has the same fixed sign and*

$$|w| \leq |\gamma|_{\max} \tag{3.14}$$

in B. *For the case* $c = 0$, (3.13) *holds independently of whether the sign of* γ *varies or not.*

3.4. Examples

It is often essential to know particular solutions of the differential equation if an error estimate is to be readily worked out. The derivation of particular solutions usually presents no difficulty when the differential equation is separable, i.e. when solutions in Bernoulli product form can be obtained by solving ordinary differential equations [error estimation and the derivation of particular solutions are treated in more detail in an example given by the author in Z. Angew. Math. Mech. **32**, 207 (1952)]. There are also methods of obtaining particular solutions in general cases[1].

I. The torsion problem for a beam of square cross-section (with sides of length 2) leads to the boundary-value problem

$$\nabla^2 u = 0 \text{ in } B, \qquad \text{i.e. for} \quad |x| \leq 1, \ |y| \leq 1,$$
$$u = x^2 + y^2 \text{ on } \Gamma, \qquad \text{i.e. for} \quad |x| = 1 \text{ and for } |y| = 1.$$

Let us assume for v an expression of the form

$$v = a_0 + a_1 v_1 + a_2 v_2,$$

where

$$v_1 = x^4 - 6x^2 y^2 + y^4 = \operatorname{re}(x + iy)^4,$$
$$v_2 = x^8 - 28 x^6 y^2 + 70 x^4 y^4 - 28 x^2 y^6 + y^8 = \operatorname{re}(x + iy)^8.$$

It is expedient to make fairly accurate sketches of the behaviour on the boundary of the functions v_1, v_2, u, and also of several approximate solutions v — here, on account of symmetry, we need only consider the

---

[1] BERGMAN, St.: Operatorenmethoden in der Gasdynamık. Z. Angew. Math. Mech. **32**, 33—45 (1952).

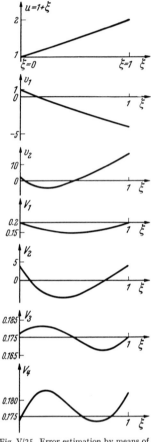

Fig. V/25. Error estimation by means of the boundary-maximum theorem

line $y = 1$ (cf. Fig. V/25, in which $\xi = x^2$); then it is easy[1] to see how v can be adjusted by the addition of $c_1 v_1$ or $c_2 v_2$, respectively, so as to approximate u more closely (aiming at a minimum maximum-error) and to see what would be suitable values for c_1, c_2. Each adjustment is followed up numerically with the aid, say, of a table of function values at equidistant points, in which can be included a simple row-sum check as in Table V/8. If only a_0 and $a_1 v_1$ are used, we obtain

$$\varphi_1 = 1 \cdot 175 - 0 \cdot 2 v_1 (x, y),$$

and for the error along $y = 1$ we have

$$|1 \cdot 175 - 0 \cdot 2 v_1 - (x^2 + 1)| \le 0 \cdot 025;$$

then from the error estimate of §3.2, Case I

$$|u - \varphi_1| \le 0 \cdot 025$$

holds in B also. If v_2 is used as well, we obtain

$$\varphi_2 = 1 \cdot 1786 - 0 \cdot 2 v_1 + 0 \cdot 006 V_2 + 0 \cdot 0019 v_1,$$

where $V_2 = v_2 + 3 v_1$, with $|\varphi_2 - u| \le 0 \cdot 005$ on the boundary, and hence $|\varphi_2 - u| \le 0 \cdot 005$ also in B; in particular, we obtain the limits

$$1 \cdot 1736 \le u(0, 0) \le 1 \cdot 1836$$

Table V/8. *Tabular values of various approximating profiles*

	$\xi=0$	$\xi=0{\cdot}2$	$\xi=0{\cdot}3$	$\xi=0{\cdot}5$	$\xi=0{\cdot}7$	$\xi=0{\cdot}8$	$\xi=1$	Row-sum check
$u=1+\xi$ (with $\xi=x^2$)	1	1·2	1·3	1·5	1·7	1·8	2	10·5
$v_1=1-6\,\xi+\xi^2$	1	−0·16	−0·71	−1·75	−2·71	−3·16	−4	−11·49
$v_2=1-28\xi+70\xi^2-28\xi^3+\xi^4$	1	−2·0224	−1·8479	1·0625	6·3361	9·4736	16	30·001
$V_1=u+0{\cdot}2v_1-1$	0·2	0·168	0·158	0·150	0·168	0·168	0·2	1·202
$V_2=v_2+3\,v_1$	4	−2·5024	−3·9779	−4·1875	−1·7939	−0·0064	4	−4·468
$V_3=V_1-0{\cdot}006\,V_2$	0·176	0·18301	0·18187	0·175125	0·16876	0·16804	0·176	1·228
$V_4=V_3-0{\cdot}0019\,v_1$	0·1741	0·18331	0·18322	0·17845	0·17391	0·17404	0·1836	1·250

[1] A fairly experienced computer will usually achieve the desired result very quickly in this way; an automatic procedure could be used, of course, and the a_0 determined by TREFFTZ's method (cf. § 6 — the present example is in fact treated by this method in § 6.5) or one of the error distribution principles, but much more computation would be involved.

for the value at the centre of the square (the error in the mean value $1 \cdot 1786$ is less than $\frac{1}{2}\%$).

II. For the heat conduction problem (cf. § 4.2)

$$\nabla^2 u = -1 \text{ in } B, \quad \text{i.e. for} \quad |x| \leq 1, \ |y| \leq 1$$

$$u = \frac{\partial u}{\partial v} \text{ on } \Gamma, \quad \text{i.e. for} \quad |x| = 1 \text{ and for } |y| = 1,$$

we use an approximate solution of the form

$$v = -\frac{x^2 + y^2}{4} + a_0 + a_1 v_1 + a_2 v_2$$

with the same v_1 and v_2 as in the last example.

Then on $y = 1$ we have (with $\xi = x^2$)

$$\gamma = \left(v + \frac{\partial v}{\partial y} \right)_{y=1} = -\frac{3 + \xi}{4} + a_0 + a_1 V_1 + a_2 V_2,$$

where

$$V_1 = 5 - 18\xi + \xi^2, \quad V_2 = 9 - 196\xi + 350\xi^2 - 84\xi^3 + \xi^4.$$

As in the previous example, we try to choose a_0, a_1, a_2 so as to make the absolute value of the error $|\gamma|$ as small as possible. As in Example I, the following results were obtained with the aid of graphs and tabulated values of the various functions involved.

With $a_2 = 0$, i.e. using v_1 only, we obtain

$$\left| -\frac{3 + \xi}{4} + 0 \cdot 8217 - 0 \cdot 0147 V_1 \right| \leq 0 \cdot 0019,$$

while if we include v_2, we effectively gain another figure:

$$\left| -\frac{3 + \xi}{4} + 0 \cdot 821\,66 - 0 \cdot 014436 V_1 + 0 \cdot 000063\,1 V_2 \right| \leq 0 \cdot 000\,19.$$

According to (3.13) we obtain from this last result the error estimate

$$\left| -\frac{x^2 + y^2}{4} + 0 \cdot 821\,66 - 0 \cdot 014436 v_1 + 0 \cdot 000063\,1 v_2 - u \right| \leq 0 \cdot 000\,19,$$

which is valid in the whole of $B + \Gamma$; in particular, we have the useful limits

$$0 \cdot 821\,47 \leq u(0, 0) \leq 0 \cdot 821\,85,$$

$$0 \cdot 557\,10 \leq u(0, 1) \leq 0 \cdot 557\,48,$$

$$0 \cdot 380\,22 \leq u(1, 1) \leq 0 \cdot 380\,60.$$

3.5. Upper and lower bounds for solutions of the biharmonic equation

Here we describe briefly, making use of the definitions and results already given in Ch. I, § 3.3, a fundamental way[1] of deriving limits for the solution u of the "clamped plate problem"

$$\nabla^4 u = p \text{ in } B,$$
$$u = f, \quad u_\nu = g \text{ on } \Gamma$$

with region B, boundary Γ and inward normal ν as in Ch. I, §§ 3.2, 3.5. The existence of a solution to such a boundary-value problem will be assumed.

Then the problem

$$\nabla^4 u^* = 0 \text{ in } B,$$
$$u^* = -\varrho, \quad u_\nu^* = -\varrho_\nu \text{ on } \Gamma$$

will also possess a solution u^*; here we select an arbitrary interior point (x_0, y_0) of B and take ϱ to be the "fundamental solution" which is singular at (x_0, y_0) as defined in Ch. I (3.40). According to Ch. I (3.41) we have

$$8\pi u(x_0, y_0) = J_1 + J_2,$$

where

$$J_1 = \iint_B \varrho \nabla^4 u \, dx \, dy + \int_\Gamma [u_\nu \nabla^2 \varrho - u(\nabla^2 \varrho)_\nu] \, ds$$

is known but

$$J_2 = \int_\Gamma [\varrho(\nabla^2 u)_\nu - \varrho_\nu \nabla^2 u] \, ds \tag{3.15}$$

is not (at least, not until the problem has been solved).

Now let v, w, v^*, w^* be functions satisfying

$$\left.\begin{array}{ll} \nabla^4 v = p, \quad \nabla^4 v^* = 0 \text{ in } B, \\ w = f, \quad w_\nu = g, \quad w^* = -\varrho, \quad w_\nu^* = -\varrho_\nu \text{ on } \Gamma. \end{array}\right\} \tag{3.16}$$

Then from Ch. I (3.36), (3.39) we have

$$(D[u - w, u^* - w^*])^2 \le D[u - w] D[u^* - w^*] \le D[v - w] D[v^* - w^*],$$

---

[1] DIAZ, J. B., and H. J. GREENBERG: Upper and lower bounds for the solution of the first biharmonic boundary-value problem. J. Math. Phys. **27**, 193—201 (1948). Another way, which is based almost entirely on mechanical concepts, is described by P. FUNK and E. BERGER: Eingrenzung für die größte Durchbiegung einer gleichmäßig belasteten quadratischen Platte. Federhofer-Girkmann Festschrift 1952, pp. 199—204. — Limits for further problems from the theory of elasticity, including limits for the derivatives of the solution, are given by J. L. SYNGE: Upper and lower bounds for the solutions of problems of elasticity. Proc. Roy. Irish Acad., Sect. A **53**, 41—46 (1950). — Pointwise bounds for the solution of certain boundary value problems. Proc. Roy. Soc. Lond., Ser. A **208**, 170—175 (1951). — MAPLE, C. G.: The Dirichlet Problem: Bounds at a point for the Solution and its Derivatives. Quart. Appl. Math. **8**, 213—228 (1950). — COOPERMAN, PH.: An extension of the method of Trefftz for finding local bounds on the solutions of boundary value problems and on their derivatives. Quart. Appl. Math. **10**, 359—373 (1953). — NICOLOVIUS, R.: Abschätzung der Lösung der ersten Platten-Randwertaufgabe nach der Methode von Maple-Synge. Z. Angew. Math. Mech. **37**, 344—349 (1957).

where D is the integral operator defined in Ch. I (3.34). Now the function appearing to the second power on the left-hand side can be expressed in terms of $u(x_0, y_0)$ and known functions. We have

$$D[u - w, \; u^* - w^*] = D[w, w^*] - D[u, w^*] + D[u - w, u^*],$$

in which the last two terms can be evaluated by the formula (3.37) of Ch. I; with $\varphi = u^*, \; \psi = u - w$ we see that

$$D[u - w, u^*] = 0$$

and with $\varphi = u, \; \psi = w^*$ that

$$D[u, w^*] = \iint\limits_{B} w^* \, V^4 u \, dx \, dy - J_2.$$

Substituting for J_2 from the expression for $8\pi u(x_0, y_0)$ [Ch. I (3.41)] we obtain

$$D[u - w, \; u^* - w^*] = D[w, w^*] - \iint\limits_{B} w^* \, V^4 u \, dx \, dy + 8\pi u(x_0, y_0) - J_1.$$

With

$$W = J_1 + \iint\limits_{B} w^* \, V^4 u \, dx \, dy - D[w, w^*]$$

we therefore have the estimate

$$\left(8\pi u(x_0, y_0) - W\right)^2 \leqq D[v - w] \, D[v^* - w^*]. \tag{3.17}$$

To use this estimate, we have to find four functions v, w, v^*, w^* satisfying (3.15). A suitable function for w^* is probably the most difficult to find. DIAZ and GREENBERG recommend that outside of a small circle C, centre (x_0, y_0), w^* be chosen as $-\varrho$, and inside of C as a polynomial in r such that w^* possesses continuous second derivatives everywhere, including on C.

Another fundamental approach has been demonstrated by MIRANDA[1]. If the boundary (possibly consisting of several separate curves) possesses a continuously turning tangent, and if the boundary functions f, g and also $f' = \dfrac{df}{ds}$ (s denoting the arc length along the boundary) are continuous, then the following estimates hold for the homogeneous equation ($p = 0$):

$$\sqrt{u_x^2 + u_y^2} \leqq Q = K_1[\max_\Gamma |g| + \max_\Gamma |f'|] + K_2 \max_\Gamma |f|,$$
$$|u(P)| \leqq \delta Q + \max_\Gamma |f|,$$

in which P is an interior point of B, δ is the distance of P from the boundary Γ and K_1, K_2 are constants depending only on the geometry of the region B; the calculation of numerical values for these constants is rather laborious. This yields a fundamental error estimate for an approximate solution of the clamped plate problem which satisfies the differential equation exactly.

§4. Some general methods

Various general methods have already been discussed in Ch. I, §§ 4 and 5; it suffices here to refer to these methods and give examples. More general boundary-value problems for partial differential equations were formulated in Ch. I, § 1.3.

---

[1] MIRANDA, C.: Formule di maggiorazione e teorema di esistenza per le funzioni biarmoniche di due variabili. Giorn. Mat. Battaglini **78**, 97−118 (1948/49).

4.1. Boundary-value problems of monotonic type for partial differential equations of the second and fourth orders

As in the case of ordinary differential equations, problems of monotonic type can be treated effectively by making use of their monotonic property. Let the differential equation read

$$T[u] = L[u] + F(x_j, u) = 0, \tag{4.1}$$

as in Ch. I (5.52). We consider first equations of the second order and focus attention on those equations for which $L[u]$ has the form (3.2) with coefficients satisfying the conditions laid down in § 3.1. Let the boundary condition be

$$\text{either (case I)} \qquad\qquad u = \gamma \quad \text{on } \Gamma \tag{4.2}$$

$$\text{or (case II)} \quad A_1 u + A_2 L^*[u] = A_3 \quad \text{on } \Gamma, \tag{4.3}$$

where γ, A_1, A_2, A_3 are continuous functions of position on Γ with $\dfrac{A_2}{A_1} < 0$, and I^* has the same significance as in Ch. I, § 3.3. We can then state the following

Theorem. *If the function $F(x_j, u)$ in the differential equation (4.1) is such that $\partial F/\partial u$ exists and is non-negative in a domain H of the $(x_1, \ldots, x_n, u)$ space which is convex with respect to u, then the boundary-value problems (4.1), (4.2) and (4.1), (4.3) with the differential expression $L[u]$ of (3.2) [satisfying the conditions of § 3.1] and the boundary conditions (4.2) and (4.3), respectively, [satisfying the above assumptions] are of monotonic type. This implies that if a solution u exists and lies in H, then it is bracketed by any two functions u_1 and u_2 which satisfy the boundary conditions and are such that $T[u_1] \leq 0 \leq T[u_2]$:*

$$u_1 \leq u \leq u_2 \quad \text{in } B. \tag{4.4}$$

The theorem still holds when, as in § 3.3, (4.2) is prescribed on a part Γ_1 of the boundary Γ and (4.3) is prescribed on the remaining part Γ_2.

To prove the theorem we have only to show (see Ch. I, § 5.6) that

$$L[\varepsilon] + \varepsilon A(x_j) \geq 0 \quad \text{in } B,$$

where $A(x_j) \geq 0$, together with the appropriate homogeneous boundary condition (case I: $\varepsilon = 0$ on Γ, case II: $A_1 \varepsilon + A_2 L^*[\varepsilon] = 0$), implies that $\varepsilon \geq 0$. For this we use the boundary-maximum theorem of § 3.1.

Suppose that the minimum value of ε is negative and let $x_j = \xi_j$ be a point at which this minimum value is attained.

1. If ξ_j is an interior point of B, then since ε is continuous, there must be a neighbourhood of ξ_j in which $\varepsilon < 0$; in this neighbourhood we

therefore have $L[\varepsilon] \geq 0$. But according to the boundary-maximum theorem this means that there cannot be a negative minimum value in the interior. Consequently ξ_j cannot be an interior point of B.

2. Neither can it be a boundary point; for in case I $\varepsilon = 0$ on the boundary and in case II a negative minimum on the boundary leads to a contradiction as follows. At ξ_j on Γ we must have $\dfrac{\partial \varepsilon}{\partial \sigma} \geq 0$, for σ, being in the direction of the conormal, points inwards (for elliptic equations). Therefore $\dfrac{A_2}{A_1} L^*[\varepsilon] \leq 0$ at ξ_j, which, since we have assumed that $\varepsilon(\xi_j) < 0$, is incompatible with the boundary conditions.

Thus we must have $\varepsilon \geq 0$ and the theorem is proved. It is applied to a numerical example in § 4.3.

The monotonic character of some fourth-order boundary-value problems can be demonstrated by means of this theorem. Consider, for example, the following simple problem which arises in the standard treatment of the flexure of a freely supported flat plate:

$$\nabla^4 u \equiv \frac{\partial^4 u}{\partial x^4} + 2 \frac{\partial^4 u}{\partial x^2 \partial y^2} + \frac{\partial^4 u}{\partial y^4} = r(x, y) \text{ in } B, \tag{4.5}$$

$$u = \gamma_1(s), \qquad \nabla^2 u = \gamma_2(s) \text{ on } \Gamma, \tag{4.6}$$

where s is the arc length along the boundary curve Γ of a simply-connected plane region B and r, γ_1, γ_2 are given functions. Here we define the operator T by

$$T v = \nabla^4 v - r(x, y)$$

and restrict its domain of operation to those functions v in B which possess fourth-order partial derivatives and satisfy the boundary conditions. This operator also is monotonic, for it can be readily shown that

$$\nabla^4 \varepsilon \geq 0 \text{ in } B \quad \text{and} \quad \nabla^2 \varepsilon = \varepsilon = 0 \text{ on } \Gamma$$

implies $\varepsilon \geq 0$: we put $-\nabla^2 \varepsilon = \zeta$, so that $-\nabla^2 \zeta \geq 0$ in B, $\zeta = 0$ on Γ; then, according to the above theorem, $\zeta \geq 0$ and consequently $-\nabla^2 \varepsilon \geq 0$ in B, $\varepsilon = 0$ on Γ; using the above theorem again we have $\varepsilon \geq 0$.

4.2. Error distribution principles. Boundary and interior collocation

For these methods (see Ch. I, § 4) we use as an approximation to the required solution $u(x_1, \ldots, x_n)$ some function $w(x_1, \ldots, x_n, a_1, \ldots, a_p)$ which depends on p parameters a_ϱ and which, for arbitrary values of these parameters, satisfies either the differential equation or the boundary conditions, whichever is the more convenient. Insertion of w into the boundary conditions or the differential equation, respectively, yields residual error functions $\varepsilon(x_j, a_\varrho)$, and by choosing the a_ϱ suitably we make these error functions approximate the zero function as closely as possible in the sense of one of the principles mentioned in Ch. I, § 4. We recall that methods of this sort may be conveniently classified into boundary, interior and mixed methods, as in Ch. I, § 4.1. For several important classes of boundary-value problems we have means for deriving bounds for the error of an approximation function in a quite simple manner, cf. § 3 and § 4.1; it is therefore very useful to possess in the procedures based on the error distribution principles methods of a parallel simplicity for the calculation of approximation functions.

Example. Consider the problem

$$\nabla^2 u = \frac{\partial^2 u}{\partial x^2} + \frac{\partial^2 u}{\partial y^2} = -1 \quad \text{in the interior of } B,$$

$$u = \frac{\partial u}{\partial v} \quad \text{on the boundary of } B,$$

where B is the square $|x| \leq 1, |y| \leq 1$ (Fig. V/26) and v is in the direction of the inward normal; u could be interpreted physically as the tem-

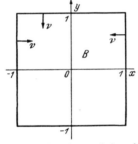

perature distribution (measured as the excess over the temperature of the surroundings) in the cross-section of a square wire carrying a current and for which the amount of heat lost to the surroundings is proportional to the excess temperature at the surface.

For our approximation function we can assume an expression [Ch. I (4.10)] which satisfies either the differential equation or the boundary conditions. One way of obtain-ing an expression which satisfies the boundary conditions is to write down a polynomial

Fig. V/26. Boundary and inward normals for a "third" boundary-value problem

expression satisfying any symmetry conditions that exist but whose coefficients $a_{\mu\nu}$ are otherwise arbitrary:

$$w = a_{00} + a_{20}(x^2 + y^2) + a_{22}x^2y^2 + a_{40}(x^4 + y^4) + a_{42}(x^4y^2 + x^2y^4) + \cdots,$$

and then determine these coefficients from the boundary conditions (cf. § 4.3). In this way we obtain, for example,

$$\left. \begin{aligned} w = a_1 v_1 + a_2 v_2 &= a_1 [9 - 3(x^2 + y^2) + x^2y^2] + \\ &+ a_2 [30 - 5(x^2 + y^2) - 3(x^4 + y^4) + x^2y^2(x^2 + y^2)]. \end{aligned} \right\} \quad (4.7)$$

For an expression which satisfies the differential equation we require a particular solution, together with a series of functions satisfying the homogeneous equation (and also any symmetries which exist). In our present case the latter can be obtained, for example, as the real parts of the functions $(x+iy)^{4\varrho}$; then with $-\frac{1}{4}(x^2+y^2)$ satisfying the inhomogeneous equation we have

$$w = -\tfrac{1}{4}(x^2 + y^2) + a_0 + a_1(x^4 - 6x^2y^2 + y^4) + \\ + a_2(x^8 - 28x^6y^2 + 70x^4y^4 - 28x^2y^6 + y^8). \Bigg\} \quad (4.8)$$

We now illustrate several methods for determining the a_ϱ.

I. Collocation[1]. A. We use first the expression (4.7) to illustrate collocation as an "interior" method. Substitution in the differential expression yields

$$\nabla^2 w = a_1[-12 + 2(x^2 + y^2)] + \\ + a_2[-20 - 36(x^2 + y^2) + 2(x^4 + y^4) + 24x^2y^2], \Bigg\} \quad (4.9)$$

which is to be put equal to -1 at the collocation points.

The choice of collocation points is a matter of some uncertainty; although normally some attempt at a uniform distribution would be made, we have a virtually unrestricted choice, but no reliable guide to the effect this choice has on the results. In the case of a single-term expression (only $a_1 \neq 0$) the coefficient a_1 is given by

$$a_1 = \frac{1}{2(6 - x^2 - y^2)}$$

for an arbitrary position (x, y) of the single collocation point, and will vary between $\frac{1}{12}$ (for $x = y = 0$) and $\frac{1}{8}$ (for $x = y = 1$) as the collocation point moves about the square; thus we can obtain values for $w(0, 0)$ lying anywhere in the range from 0·75 to 1·125, which is a rather large variation.

For a two-term expression with a_1 and a_2 we have to dispose two collocation points. On account of the symmetries which exist, collocation actually occurs in general at sixteen points; we try to choose the two arbitrary points so that the full set of sixteen collocation points are distributed as uniformly as possible. If, for example, we choose the points $x = \frac{1}{2}$, $y = \frac{1}{4}$ and $x = 1$, $y = \frac{1}{2}$, we obtain the two equations

$$a_1 \frac{91}{8} + a_2 \frac{3935}{128} = 1, \qquad a_1 \frac{19}{2} + a_2 \frac{455}{8} = 1,$$

which yield the values

$$a_1 = \frac{446}{6057} = 0·07363, \qquad a_2 = \frac{32}{6057} = 0·005283.$$

For $u(0, 0)$ these give the approximate value $w(0, 0) = \dfrac{4974}{6057} = 0·82120.$

Table V/9 exhibits the results for various positions of the collocation points.

---

[1] See also R. A. Frazer, W. P. Jones and Sylvia W. Skan: Approximations to functions and to the solutions of differential equations. Rep. and Mem. No. 1799 (2913), Aero. Res. Comm. 1937, 33 pp.

Table V/9. *Collocation for the temperature distribution in a square wire carrying a current*

	Expression	Co-ordinates of the collocation points	Position	Results
Interior collocation Expression (4.7)	only a_1	$x = y = \dfrac{1}{2}$		$a_1 = \dfrac{1}{11}$ $w(0,0) = \dfrac{9}{11} = 0.8182$
	only a_1	$x = y = \dfrac{2}{3}$		$a_1 = \dfrac{9}{92}$ $w(0,0) = \dfrac{81}{92} = 0.8804$
	only a_1	$x = \dfrac{3}{4},\; y = \dfrac{1}{4}$		$a_1 = \dfrac{4}{43}$ $w(0,0) = \dfrac{36}{43} = 0.8372$
	with a_1, a_2	$x = \dfrac{1}{2},\; y = \dfrac{1}{4}$ $x = \dfrac{3}{4},\; y = \dfrac{1}{2}$		$a_1 = 0.07400$ $a_2 = 0.005148$ $w(0,0) = \dfrac{4080}{4973} = 0.82043$
	with a_1, a_2	$x = \dfrac{1}{2},\; y = \dfrac{1}{4}$ $x = 1,\; y = \dfrac{1}{2}$		$a_1 = 0.073634$ $a_2 = 0.005283$ $w(0,0) = \dfrac{4974}{6057} = 0.82120$
Boundary collocation Expression (4.8)	only a_0	$x = 1,\; y = \dfrac{1}{2}$		$a_0 = \dfrac{13}{16}$ $w(0,0) = a_0 = 0.8125$
	with a_0, a_1	$x = 1,\; y = \dfrac{1}{4}$ $x = 1,\; y = \dfrac{3}{4}$		$a_0 = w(0,0) = \dfrac{14615}{17792}$ $= 0.821431$ $a_1 = -\dfrac{2}{139} = -0.01439$
	with a_0, a_1, a_2	$x = 1,\; y = 0$ $x = 1,\; y = \sqrt{\dfrac{1}{2}}$ $x = 1,\; y = 1$		$a_0 = w(0,0) = \dfrac{12846}{15635}$ $= 0.821618$ $a_1 = -0.014400$ $a_2 = 0.000064$

B. Greater convenience and less uncertainty are afforded by the use of the other expression (4.8), which satisfies the differential equation. This is because the collocation points now lie only on the boundary and it is much easier to ensure some degree of uniformity in their distribution.

On account of symmetry we need only consider the part of the boundary along $x=1$, where $\dfrac{\partial w}{\partial v}=-\dfrac{\partial w}{\partial x}$; we have

$$\left(w+\frac{\partial w}{\partial x}\right)_{x=1}=-\frac{3+y^2}{4}+$$
$$+\,a_0+a_1(5-18y^2+y^4)+a_2(9-196y^2+350y^4-84y^6+y^8),$$

which in accordance with the boundary conditions is to approximate the zero function as closely as possible.

First of all, let us put $a_1=a_2=0$; then for a general boundary collocation point $(1, y)$

$$a_0=\frac{3+y^2}{4}.$$

Now $|y|$ lies between 0 and 1, so a_0 can vary from 0·75 to 1; for $y=\frac{1}{2}$ we have $a_0=w(0, 0)=0\cdot8125$.

If we include the term in a_1, we can demand that the boundary condition be satisfied at $y=\frac{1}{4}$ and $y=\frac{3}{4}$, for example; then from

$$a_0+a_1\frac{993}{256}=\frac{49}{64}, \qquad a_0-a_1\frac{1231}{256}=\frac{57}{64}$$

we obtain

$$a_0=w(0, 0)=0\cdot821\,43,$$

$$a_1=-\,0\cdot014\,39.$$

The results for a three-parameter expression are also given in Table V/9.

II. GALERKIN's method. We use the same example to illustrate GALERKIN's method as an interior method; thus we take the expression $w=a_1v_1+a_2v_2$ of (4.7) as an approximation function. Here we determine a_1 and a_2 from the equations [see Ch. I (4.13)]

$$\int\limits_{0}^{1}\int\limits_{0}^{1} v_\varrho(\nabla^2 w+1)\,dx\,dy=0 \qquad (\varrho=1, 2), \qquad (4.10)$$

where $\nabla^2 w$ is given by (4.9). The evaluation of the integrals leads to the equations

$$\frac{6}{5}a_1+\frac{6782}{1575}a_2=\frac{1}{9}, \qquad \frac{6782}{1575}a_1+\frac{74\,144}{4725}a_2=\frac{2}{5}.$$

The first approximation ($a_2=0$) can be found from the first equation by putting $a_2=0$:

$$a_1 = \frac{5}{54}, \qquad w(0,0) = \frac{5}{6} = 0\cdot 833\ldots.$$

If a_2 is taken into account, we obtain

$$a_1 = \frac{39305}{536397} = 0\cdot 073\,2760, \qquad a_2 = 0\cdot 005\,383\,1, \qquad w(0,0) = 0\cdot 820978.$$

4.3. The least-squares method as an interior and a boundary method

After the general description in Ch. I, § 4.2 we restrict ourselves here to just two examples.

I. Interior method. Let us consider again the example of § 4.2:

$$\left.\begin{array}{ll} V^2 u = -1 & \text{in the interior} \\[2mm] \dfrac{\partial u}{\partial v} = u & \text{on the boundary} \end{array}\right\} \text{ of the square } |x|\leq 1,\ |y|\leq 1. \quad (4.11)$$

We start with the simplest case of a one-parameter approximation. We need a function $\varphi(x,y)$ which satisfies the boundary conditions. Taking into account the symmetries of the problem, we put

$$\varphi = c_1 + c_2(x^2+y^2) + c_3 x^2 y^2;$$

then on the part of the boundary $x=1$, $|y|\leq 1$, for example, we have

$$-\frac{\partial \varphi}{\partial v} = \frac{\partial \varphi}{\partial x} = 2c_2 x + 2c_3 x y^2,$$

and the boundary condition yields

$$\left(\varphi - \frac{\partial \varphi}{\partial v}\right)_{x=1} = c_1 + 3c_2 + (c_2 + 3c_3)y^2 = 0;$$

consequently

$$c_2 = -3c_3, \qquad c_1 = -3c_2 = 9c_3,$$

and we take

$$\varphi = a\left(9 - 3(x^2+y^2) + x^2 y^2\right).$$

Then

$$V^2\varphi = a\left(-12 + 2(x^2+y^2)\right),$$

and inserting this in the expression to be minimized according to the minimum principle of Ch. I (4.5), namely

$$J[\varphi] = \iint_B (V^2\varphi + 1)^2\, dx\, dy, \qquad (4.12)$$

and evaluating the integral, we obtain

$$J = 4\left(1 - \frac{64}{3}a + 112 \times \frac{46}{45}a^2\right).$$

The condition

$$\frac{1}{4}\frac{\partial J}{\partial a} = -\frac{64}{3} + 224 \times \frac{46}{45}\, a = 0$$

yields $a = \frac{15}{161}$ and hence the approximate solution

$$\varphi = \frac{15}{161}\left(9 - 3\left(x^2 + y^2\right) + x^2 y^2\right)$$

This gives the following values at the key points of the cross-section:

$$\varphi(0, 0) = \frac{135}{161} = 0\cdot8385 \quad (\text{error } +2\cdot1\%),$$

$$\varphi(0, 1) = \frac{90}{161} = 0\cdot5590 \quad (\text{error } +0\cdot3\%),$$

$$\varphi(1, 1) = \frac{60}{161} = 0\cdot3727 \quad (\text{error } -1\cdot4\%).$$

Proceeding to the second approximation, we put

$$\varphi = c_1 + c_2\left(x^2 + y^2\right) + c_3 x^2 y^2 + c_4\left(x^4 + y^4\right) + c_5\left(x^2 y^4 + x^4 y^2\right).$$

With the relations between the c_i required by the boundary conditions the expression takes the form

$$\varphi = 9\alpha + (-3\alpha + 15\beta)\left(x^2 + y^2\right) + (\alpha - 10\beta)\, x^2 y^2 - 9\beta\left(x^4 + y^4\right) + 3\beta\left(x^2 y^4 + x^4 y^2\right);$$

this function satisfies the boundary conditions for arbitrary values of α and β and therefore provides a two-parameter approximation function. With $2\alpha = a_1 = a$, $2\beta = a_2 = b$ it yields

$$\nabla^2\varphi + 1 = 1 + a A_1 + b A_2,$$

where

$$A_1 = -6 + x^2 + y^2, \qquad A_2 = 30 - 64\left(x^2 + y^2\right) + 3\left(x^4 + y^4\right) + 36 x^2 y^2.$$

From the conditions $\partial J/\partial a_i = 0$ (for $i = 1, 2$) we obtain the equations

$$\iint_B A_1\, dx\, dy + a \iint_B A_1^2\, dx\, dy + b \iint_B A_1 A_2\, dx\, dy = 0,$$

$$\iint_B A_2\, dx\, dy + a \iint_B A_1 A_2\, dx\, dy + b \iint_B A_2^2\, dx\, dy = 0,$$

where

$$\alpha_{10} = \iint_B A_1\, dx\, dy = -4 \times \frac{16}{3}, \qquad \alpha_{20} = \iint_B A_2\, dx\, dy = -4 \times \frac{112}{15},$$

$$\alpha_{11} = \iint_B A_1^2\, dx\, dy = 4 \times \frac{1288}{45}, \qquad \alpha_{12} = \alpha_{21} = \iint_B A_1 A_2\, dx\, dy = 4 \times \frac{9776}{315},$$

$$\alpha_{22} = \iint_B A_2^2\, dx\, dy = 4 \times \frac{788192}{1575}.$$

We can check the calculation of these integrals by evaluating $J[\varphi]$ for $a = b = 1$: we have

$$\nabla^2\varphi + 1 = 25 - 63\left(x^2 + y^2\right) + 36 x^2 y^2 + 3\left(x^4 + y^4\right)$$

and hence

$$\iint_B (\nabla^2 \varphi + 1)^2 \, dx \, dy = 4 \times \frac{99\,143}{175} \, ;$$

this value should be equal to

$$\iint_B (1 + A_1 + A_2)^2 \, dx \, dy = 4 + 2\alpha_{10} + 2\alpha_{20} + \alpha_{11} + 2\alpha_{12} + \alpha_{22}.$$

The equations

$$\alpha_{10} + \alpha_{11} a + \alpha_{12} b = 0,$$
$$\alpha_{20} + \alpha_{21} a + \alpha_{22} b = 0$$

read

$$-\frac{16}{3} + \frac{1288}{45} a + \frac{9776}{315} b = 0,$$

$$-\frac{112}{15} + \frac{9776}{315} a + \frac{788\,192}{1575} b = 0,$$

i.e.

$$-210 + 1127 a + 1222 b = 0,$$
$$-735 + 3055 a + 49262 b = 0,$$

and yield

$$a = \frac{175}{4} \frac{2999}{719237} = 0\cdot182424,$$

$$b = \frac{175}{4} \frac{593}{719237} = 0\cdot003\,6071.$$

The approximate values at the key points of the cross-section are now

$$\varphi(0, 0) = \tfrac{9}{2} a \qquad = 0\cdot820909 \ (\text{error} \ -0\cdot08\%),$$
$$\varphi(0, 1) = 3(a + b) = 0\cdot558094 \ (\text{error} \ +0\cdot16\%),$$
$$\varphi(1, 1) = 2a + 4b = 0\cdot379277 \ (\text{error} \ -0\cdot19\%).$$

Upper and lower bounds for the solution. It may be mentioned here that for this problem upper and lower bounds for the solution, though admittedly fairly coarse, can be obtained very simply by direct use of its monotonic property — which is assured by the theorem of § 4.1. If a function φ satisfying the boundary conditions is such that $\nabla^2\varphi + 1 \leq 0$ or ≥ 0 throughout the whole region, then φ is respectively an upper or lower bound for the solution u; thus, for instance, in the approximation function which we have just been using we want to choose the constants a, b so that $\nabla^2\varphi + 1$ does not change sign (naturally we will also try to keep its magnitude as small as possible). First of all we construct a linear combination of the functions A_1 and A_2 which is as near constant as possible. It can be seen by evaluating these functions at the "key" points that the best combination (up to a constant factor) is $43 A_1 + A_2$; we have

$$-246 \leq 43 A_1 + A_2 \leq -228$$

and hence

$$\frac{1}{228}(43 A_1 + A_2) + 1 \leq 0 \leq \frac{1}{246}(43 A_1 + A_2) + 1.$$

Consequently φ is an upper bound for u when $\alpha = \dfrac{43}{456}$, $\beta = \dfrac{1}{456}$ and a lower bound when $\alpha = \dfrac{43}{492}$, $\beta = \dfrac{1}{492}$. The mean of these two bounds gives an approximation with known error limits; at the centre of the square, for example, we have $u(0, 0) = 0\cdot8177 \pm 0\cdot0311$.

II. Boundary method. We choose a different problem here, namely

$$\nabla^2 u = x^2 - 1, \qquad u = 0 \quad \text{for} \quad |x| = 1 \quad \text{and for} \quad |y| = \tfrac{1}{2};$$

u can be interpreted as the transverse displacement of a membrane stretched over a rectangular frame and distorted by a non-uniformly distributed transverse load.

Table V/10. *Values of the individual contributions a_{mn}*

$a_{00} = \dfrac{3}{2}$	$a_{10} = \dfrac{5}{6}$	$a_{20} = \dfrac{7}{10}$	$a_{30} = \dfrac{9}{14}$	$a_{40} = \dfrac{11}{18}$
$a_{01} = \dfrac{7}{24}$	$a_{11} = \dfrac{1}{8}$	$a_{21} = \dfrac{11}{120}$	$a_{31} = \dfrac{13}{168}$	$a_{41} = \dfrac{5}{72}$
$a_{02} = \dfrac{11}{160}$				

The function $\varphi = w_0 + \displaystyle\sum_{\nu=1}^{p} a_\nu w_\nu$ with

$$w_0 = \frac{x^4}{12} - \frac{x^2}{2}, \qquad w_\nu = \mathrm{re}\,(x + i\,y)^{2\nu}$$

satisfies the differential equation exactly for arbitrary a_ν and exhibits the symmetries possessed by the problem. We restrict attention to the first quadrant and denote

$$\int_0^{\frac{1}{2}} F(1,\,y)\,dy + \int_0^1 F(x,\,\tfrac{1}{2})\,dx \qquad \text{by} \qquad \int_\Gamma F\,ds.$$

Then the least-squares requirement reads

$$J[\varphi] = \int_\Gamma \varphi^2\,ds = \text{minimum}, \tag{4.13}$$

or

$$\frac{1}{2}\frac{\partial J}{\partial a_\nu} = 0 = \int_\Gamma \varphi\,w_\nu\,ds \qquad (\nu = 1, 2, \ldots, p). \tag{4.14}$$

To expedite the derivation of the equations for the a_ν from (4.14), we construct a table of values of the quantities

$$a_{mn} = \int_\Gamma x^{2m}\,y^{2n}\,ds;$$

Table V/10 gives sufficient values for the case $p = 2$, for which we obtain the equations

$$\frac{3}{2}\,a_1 + \frac{13}{24}\,a_2 = \frac{43}{120}, \qquad \frac{13}{24}\,a_1 + \frac{83}{160}\,a_2 = \frac{2435}{120 \times 84}$$

with the solution

$$a_1 = \frac{16643}{60 \times 2443} = 0{\cdot}1135, \qquad a_2 = \frac{848}{2443} = 0{\cdot}3471.$$

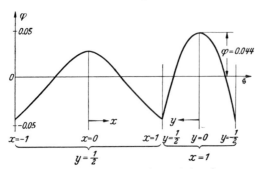

Fig. V/27. The error distribution on the boundary

The boundary values of the corresponding approximation φ are shown graphically in Fig. V/27; because of the zero boundary condition this graph represents the error distribution on the boundary. Since $u - \frac{1}{12} x^4 + \frac{1}{2} x^2$ is a harmonic function, we infer from the boundary-maximum theorem of § 3.1 that the maximum absolute error in φ on the boundary, about $0{\cdot}044$, is not exceeded in the interior: $|\varphi - u| \leqq 0{\cdot}044$ in the whole rectangle.

4.4. Series solutions

A method which is often used for the solution of boundary-value problems is to find a series expansion for the solution u of the form

$$u = \sum_{m=1}^{\infty} a_m \Psi_m \quad \text{say, or} \quad u = \sum_{m,n=1}^{\infty} a_{mn} \Psi_{mn}, \tag{4.15}$$

where Ψ_m or Ψ_{mn} are chosen as known functions of the independent variables and the a_m or a_{mn}, respectively, are constants to be determined. Naturally triple sums and higher multiple sums can also be used as occasion demands.

What was said in Ch. III, § 7.1 concerning series solutions for ordinary differential equations is obviously also valid here; for partial differential equations the effectiveness of a series solution depends to a particularly large extent on a propitious choice of the functions Ψ and on what is involved in carrying out the method.

Ordinarily we choose the functions Ψ so that they satisfy either

(i) the differential equation,

or (ii) the boundary conditions,

or (iii) at least some of the boundary conditions,

then seek to determine the expansion coefficients a_m (or a_{mn}) so that the series satisfies the differential equation or the boundary conditions when inserted. This usually leads to an infinite system of equations for infinitely many unknowns a_m; such a system is normally solved in an approximate manner by putting all the unknowns after the first N equal to zero and determining the remaining unknowns from the first N equations. A thorough treatment of the subject of series solutions would have to cover such a wide range of phenomena that we content ourselves here with a few examples which demonstrate the procedure involved in the application of the method; for the rest, reference can be made to the literature, where numerous examples[1] of series expansions are to be found[2].

4.5. Examples of the use of power series and related series

I. For the heat conduction problem (cf. § 4.2)

$$\left. \begin{array}{l} \nabla^2 u = -1 \text{ in the interior} \\ \partial u/\partial v = u \text{ on the boundary} \end{array} \right\} \begin{array}{c} \text{of the square } |x| \leqq 1, \quad |y| \leqq 1, \\ (v = \text{inward normal}), \end{array} \left. \begin{array}{l} \\ \end{array} \right\} (4.16)$$

which has already been treated by many different methods, the power series method is superior to all others in shortness and accuracy. We make use of the symmetries of the problem and instead of the general double series

$$u = \sum_{m,n=0}^{\infty} a_{mn} x^m y^n$$

---

[1] For instance, in A. SOMMERFELD: Partielle Differentialgleichungen der Physik. Leipzig 1947. — FRANK, PH., and R. v. MISES: Differential- und Integralgleichungen der Mechanik und Physik. Brunswick 1930 and 1935. See also our Ch. IV, introduction and § 4.3.

[2] A method for the solution of boundary-value problems based on the Bergman "kernel function" associated with the given region and the given differential equation is given by ST. BERGMAN: Kernel functions and conformal mapping. Surveys of the Amer. Math. Soc. 5 (1950); see also ST. BERGMAN and M. SCHIFFER: Kernel functions and elliptic differential equations in mathematical physics, Part B. New York: Academic Press Inc. 1953. The kernel function can be obtained by expansion in terms of the functions of a complete orthonormal system. For the potential equation in a simply-connected region of the complex z plane, for example the system of polynomials obtained from the powers of z by orthogonalization with respect to the given region is suitable. The kernel function is defined as the difference between the corresponding Green and Neumann functions. There are simple rules for modifying the kernel function when the differential equation or the region is altered. This method has been used at the National Bureau of Standards, Washington, to solve boundary-value problems for the potential equation in doubly-connected regions bounded by concentric rectangles. Experience must show whether it is best to work in terms of orthonormal systems or to tabulate the kernel function for several different regions and a given differential equation, say the potential equation.

use the simple series

$$u = -\tfrac{1}{4}(x^2 + y^2) + \sum_{\varrho=0}^{\infty} a_{2\varrho}\, \mathrm{re}\,(x + i\,y)^{4\varrho}$$
$$= -\tfrac{1}{4}(x^2 + y^2) + a_0 + a_2(x^4 - 6x^2 y^2 + y^4) +$$
$$+ a_4(x^8 - 28x^6 y^2 + 70 x^4 y^4 - 28 x^2 y^6 + y^8) +$$
$$+ a_6(x^{12} - 66 x^{10} y^2 + 495 x^8 y^4 - 924 x^6 y^6 + \cdots) + \cdots.$$

The first term $-\tfrac{1}{4}(x^2 + y^2)$ satisfies the inhomogeneous potential equation and the subsequent terms, being the real parts of analytic functions, satisfy the homogeneous equation; therefore $\nabla^2 u = -1$ for arbitrary a_ϱ. Only powers $(x + iy)^{4\varrho}$ which satisfy all the symmetries of the problem have been used. The a_ϱ will now be determined so that the boundary condition $\partial u/\partial\nu = u$ is satisfied. We form

$$(u)_{x=1} = a_0 - \tfrac{1}{4}(1 + y^2) + a_2(1 - 6y^2 + y^4) +$$
$$+ a_4(1 - 28 y^2 + 70 y^4 - 28 y^6 + y^8) +$$
$$+ a_6(1 - 66 y^2 + 495 y^4 - 924 y^6 + \cdots) + \cdots,$$
$$\left(\frac{\partial u}{\partial x}\right)_{x=1} = -\tfrac{1}{2} + a_2(4 - 12 y^2) + a_4(8 - 168 y^2 + 280 y^4 - 56 y^6) +$$
$$+ a_6(12 - 660 y^2 + 3960 y^4 - 5544 y^6 + \cdots) + \cdots.$$

The sum of these two expressions is to vanish identically, so the coefficient multiplying each power y^{2k} in the sum must be put equal to zero; this yields the following infinite system of equations for the unknowns a_ϱ:

$$
\begin{aligned}
a_0 + 5a_2 + \quad 9a_4 + \quad 13a_6 + \cdots &= \tfrac{3}{4} \quad \text{(coefficient of } y^0 \text{)},\\
18a_2 + 196a_4 + \quad 726a_6 + \cdots &= -\tfrac{1}{4} \quad (\quad ,, \quad \quad ,, \;\; y^2),\\
a_2 + 350a_4 + 4455a_6 + \cdots &= 0 \quad (\quad ,, \quad \quad ,, \;\; y^4),\\
84a_4 + 6468a_6 + \cdots &= 0 \quad (\quad ,, \quad \quad ,, \;\; y^6).
\end{aligned}
$$

We solve these equations approximately in the customary fashion: for the ϱ-th approximation we solve the first ϱ equations for the first ϱ unknowns with the remaining unknowns put equal to zero. In the second approximation, for example, we solve the equations

$$a_0 + 5a_2 = \tfrac{3}{4}, \qquad 18a_2 = -\tfrac{1}{4},$$

and obtain

$$a_0 = \frac{59}{72}, \qquad a_2 = -\frac{1}{72}.$$

Table V/11 gives the first four approximations for the coefficients of the power series, together with the values calculated from them for the key points of the square. The convergence of the successive approximations is good.

Table V/11. *First four approximations by the power series method*

	1st approximation	2nd approximation	3rd approximation	4th approximation
a_0	$\dfrac{3}{4}=0{\cdot}75$	$\dfrac{59}{72}= 0{\cdot}81944$	$\dfrac{20053}{24416}= 0{\cdot}82131$	$\dfrac{1283427}{1562176}=0{\cdot}821564$
a_2	0	$-\dfrac{1}{72}=-0{\cdot}01389$	$-\dfrac{25}{1744}=-0{\cdot}01432$	$-\dfrac{22495}{1562176}=-0{\cdot}01440$
a_4	0	0	$\dfrac{1}{24416}= 0{\cdot}000041$	$\dfrac{77}{1562176}=0{\cdot}0000493$
a_6	0	0	0	$\dfrac{-1}{1562176}=-0{\cdot}00000064$
$u(0,0)$	$0{\cdot}75$	$0{\cdot}819444$	$\dfrac{20053}{24416}=0{\cdot}821306$	$\dfrac{1283427}{1562176}=0{\cdot}8215636$
$u(0,1)$	$0{\cdot}50$	$0{\cdot}555555$	$\dfrac{13600}{24416}=0{\cdot}557012$	$\dfrac{870464}{1562176}=0{\cdot}5572125$
$u(1,1)$	$0{\cdot}25$	$0{\cdot}375$	$\dfrac{9261}{24416}=0{\cdot}379300$	$\dfrac{593615}{1562176}=0{\cdot}3799924$

II. We now illustrate the opposite procedure on the plate problem of § 1.5

$$\left.\begin{array}{l} \nabla^4 u = x^2, \\ u = u_x = 0 \quad \text{for} \quad |x| = 1, \quad u = \nabla^2 u = 0 \quad \text{for} \quad |y| = 1, \end{array}\right\} \quad (4.17)$$

i.e. we use a series which already satisfies all the boundary conditions:

$$u = (1 - x^2)^2 (1 - y^2) \sum_{m,n=0}^{\infty} a_{mn}\, x^{2m}\, \eta_n(y)$$

with

$$\eta_n(y) = (4n+5)\, y^{2n} - (4n+1)\, y^{2n+2}.$$

Again we choose a series which also satisfies the symmetries of the problem.

The expressions $\nabla^4 \Psi_{mn}$ are now worked out for the first few quantities $\Psi_{mn} = (1-x^2)^2(1-y^2)\, x^{2m}\, \eta_n(y)$, but only as far as the highest powers which we intend to consider. We will illustrate the procedure with a quite crude three-term approximation involving only a_{00}, a_{01}, a_{10}. For this we need only those terms of $\nabla^4 \Psi_{mn}$ which are $O(x^2+y^2)$, and by substituting in the differential equation and equating coefficients we obtain:

$$\text{absolute term:} \qquad 10 a_{00} - 20 a_{01} - 12 a_{10} = 0,$$

$$\text{factor multiplying } x^2: \quad -14 a_{00} + 46 a_{01} + 100 a_{10} = \frac{1}{24},$$

$$\text{factor multiplying } y^2: \quad -10 a_{00} + 140 a_{01} + 14 a_{10} = 0.$$

Solution of this system of equations yields

$$a_{00} = \frac{35}{24 \times 2487}, \qquad a_{01} = \frac{-1}{48 \times 2487}, \qquad a_{10} = \frac{15}{12 \times 2487}.$$

The approximate value

$$u(0,0) \approx 5 a_{00} = 0{\cdot}00293$$

corresponding to this three-term approximation is still very crude and more terms must be taken into consideration if any real accuracy is to be attained.

4.6. Eigenfunction expansions

The set of expansion functions Ψ is often chosen as the set of eigenfunctions of the associated eigenvalue problem. Consider a linear inhomogeneous boundary-value problem of the form

$$\left. \begin{aligned} L[u^*] &= r^* \quad \text{in } B, \\ U_\mu[u^*] &= v_\mu \quad \text{on } \Gamma \quad (\mu = 1, \ldots, k), \end{aligned} \right\} \tag{4.18}$$

where, as usual, Γ is the boundary of a two(or more)-dimensional region B, L and the U_μ are given linear differential expressions, r^* and the v_μ are given functions of position, and u^* is to be determined. We suppose that the boundary conditions $U_\mu[u^*] = v_\mu$ render the determination unique. First of all we reduce the problem to one with homogeneous boundary conditions: if w^* is any function which satisfies the inhomogeneous boundary conditions, and also possesses derivatives of a sufficiently high order (so that $L[w^*]$ exists), then the function $u = u^* - w^*$ satisfies a linear boundary-value problem with homogeneous boundary conditions, namely

$$\left. \begin{aligned} L[u] &= r = r^* - L[w^*] \quad \text{in } B, \\ U_\mu[u] &= 0 \quad\quad\quad\quad\quad \text{on } \Gamma \quad (\mu = 1, \ldots, k). \end{aligned} \right\} \tag{4.19}$$

We now suppose that the eigenvalue problem

$$\left. \begin{aligned} L[\psi] &= \lambda \psi \quad \text{in } B, \\ U_\mu[\psi] &= 0 \quad\quad \text{on } \Gamma \quad (\mu = 1, \ldots, k) \end{aligned} \right\} \tag{4.20}$$

possesses a complete system of eigenfunctions $\psi_1, \psi_2, \ldots$ corresponding to the eigenvalues $\lambda_1, \lambda_2, \ldots$, and expand the function r in (4.19) into a uniformly convergent series of these eigenfunctions (we assume that this is possible):

$$r = \sum_{\varrho=1}^{\infty} c_\varrho \psi_\varrho. \tag{4.21}$$

Then

$$u = \sum_{\varrho=1}^{\infty} \frac{c_\varrho}{\lambda_\varrho} \psi_\varrho \tag{4.22}$$

is the solution of the inhomogeneous boundary-value problem (4.19), assuming that $L[u]$ may be formed term by term for this series (4.22).

Example. The boundary-value problem which arises in the problem of the torsion of prisms (cf. Example I, § 3.4) is customarily reduced to one with homogeneous boundary conditions by putting $w^* = x^2 + y^2$; apart from a constant factor we obtain

$$\nabla^2 u = -1 \quad \text{in } B,$$
$$u = 0 \quad\quad \text{on } \Gamma.$$

Let us consider a prism of rectangular section bounded by the lines $x=0$, $x=a$, $y=0$, $y=b$.

The eigenfunctions of the associated eigenvalue problem

$$\nabla^2 u = \lambda u \quad \text{in } B, \quad u=0 \quad \text{on } \Gamma$$

are given by

$$\Psi_{m,n} = \sin \frac{m\pi x}{a} \sin \frac{n\pi y}{b} \quad (m, n = 1, 2, \ldots)$$

and the corresponding eigenvalues by

$$\lambda_{m,n} = -\left(\frac{m^2}{a^2} + \frac{n^2}{b^2}\right)\pi^2.$$

The eigenfunction expansion of the function $r=-1$ is here a double Fourier expansion

$$-1 = \sum_{m,n=1}^{\infty} c_{m,n} \sin \frac{m\pi x}{a} \sin \frac{n\pi y}{b} \quad \text{for} \quad 0<x<a, \ 0<y<b, \quad (4.23)$$

so that the coefficients $c_{m,n}$ are Fourier coefficients. They are obtained by the well-known method of multiplying by $\Psi_{p,q}$ and integrating over the fundamental region:

$$-\int_0^a \sin \frac{p\pi x}{a} dx \int_0^b \sin \frac{q\pi y}{b} dy = c_{p,q} \int_0^a \sin^2 \frac{p\pi x}{a} dx \int_0^b \sin^2 \frac{q\pi y}{b} dy = c_{p,q} \frac{ab}{4}.$$

This yields

$$c_{p,q} = \begin{cases} 0 & \text{when at least one of } p, q \text{ is even,} \\ -\dfrac{16}{\pi^2 p q} & \text{when both } p \text{ and } q \text{ are odd.} \end{cases}$$

The solution (4.22) therefore reads

$$u = \frac{16}{\pi^4} \sum_{p,q \text{ odd}} \frac{\sin \dfrac{p\pi x}{a} \sin \dfrac{q\pi y}{b}}{pq\left(\dfrac{p^2}{a^2} + \dfrac{q^2}{b^2}\right)}, \quad (4.24)$$

where p, q run through all pairs of odd positive integers.

Let us evaluate the value at the centre point for the special case of a square section. From (4.24)

$$u\left(\frac{a}{2}, \frac{a}{2}\right) = \frac{16a^2}{\pi^4} \sum_{p,q \text{ odd}} \frac{(-1)^{\frac{p+q}{2}-1}}{pq(p^2+q^2)}$$

$$= \frac{8a^2}{\pi^4} \sum_{p,q \text{ odd}} \frac{(-1)^{\frac{p+q}{2}-1}}{\left(\dfrac{p+q}{2}\right)^4 - \left(\dfrac{p-q}{2}\right)^4} = \frac{8a^2}{\pi^4} S,$$
(4.25)

where

$$S = \sum_{k=1}^{\infty} (-1)^{k+1} A_k \quad \text{with} \quad A_k = \sum_{\substack{j=1,3,5,\ldots \\ j \text{ odd}}}^{2k-1} \frac{1}{k^4 - (k-j)^4};$$

for example,

$$A_4 = \frac{2}{4^4 - 3^4} + \frac{2}{4^4 - 1^4} \quad \text{and} \quad A_5 = \frac{2}{5^4 - 4^4} + \frac{2}{5^4 - 2^4} + \frac{1}{5^4}.$$

This rearrangement of the double series (permissible because of the absolute convergence), in which terms with the same $p + q$, and therefore with the same sign, are grouped together into the finite sums A_k converts it into an alternating

Table V/12. *Evaluation of the alternating series for the value at the centre*

k	Terms $(-1)^{k+1} A_k$	Partial sums s_k	First smoothing s'_k	Second smoothing s''_k
1	1	1		
2	−0·13333333	0·86666667	0·91077873	
3	0·04311491	0·90978158	0·89418493	0·89926310
4	−0·01927171	0·89050987	0·89790383	0·89667191
5	0·01030412	0·90081399	0·89669503	0·89711961
6	−0·00617112	0·89464227	0·89718455	0·89700470
7	0·00399741	0·89863968	0·89695466	0·89704204
8	−0·00274267	0·89589701	0·89707429	0·89702754
9	0·00196644	0·89786345	0·89700691	0·89703385
10	−0·00145973	0·89640372	0·89704729	0·89703084
11	0·00111455	0·89751827	0·89702187	0·89703239
12	−0·00087105	0·89664722	0·89703853	
13	0·00069417	0·89734139		

simple series S in the A_k; moreover, these A_k are easy to calculate, since the denominators are the differences of the fourth powers 1, 16, 81, ... Table V/12 gives the calculated values for k up to 13 of the quantities A_k, the partial sums $s_k = \sum_{i=1}^{k} (-1)^{i+1} A_i$ and the quantities s'_k, s''_k which are obtained from the s_k by smoothing:

$$s'_k = \tfrac{1}{4}(s_{k-1} + 2s_k + s_{k+1}), \qquad s''_k = \tfrac{1}{4}(s'_{k-1} + 2s'_k + s'_{k+1})$$

(this smoothing process gains three more decimals). We deduce from these figures that to six decimals

$$S = 0·897032,$$

and hence

$$u\left(\frac{a}{2}, \frac{a}{2}\right) = 0·0736713 a^2.$$

For the special case of a square of side $a = 2$ (cf. Example I, § 3.4) we have

$$u(1, 1) = 0·294685.$$

For this example the series method would appear to be far superior to all other methods.

§5. The Ritz method

5.1. The Ritz method for linear boundary-value problems of the second order

As in the case of ordinary differential equations (Ch. III, § 5), it is also possible for many boundary-value problems of even order in partial differential equations to specify an integral expression $J[\varphi]$ which can be formed for a certain class of functions φ and which assumes its minimum value just for that function u which satisfies the boundary-value problem, and therefore to find an approximate solution of the boundary-value problem by inserting an approximation function for φ in $J[\varphi]$ and determining the parameters to make J a minimum for this restricted class of functions.

For a boundary-value problem of the second order in two independent variables we consider the variational problem [corresponding to (5.6) in Ch. III]

$$J[\varphi] = \iint_B [A\,\varphi_x^2 + 2B\,\varphi_x\varphi_y + C\,\varphi_y^2 + F\,\varphi^2 - 2r\,\varphi]\,dx\,dy + \left.{\vphantom{\int}}\right\}$$
$$+ \int_\Gamma (S\,\varphi^2 + 2T\,\varphi)\,ds = \text{extremum}, \left.\right\} \quad (5.1)$$

and investigate the range of boundary-value problems which can be covered by it. B is assumed to be a bounded, simply-connected, closed region with a piecewise-smooth boundary curve Γ, along which the arc length s is measured anti-clockwise from some fixed point. The extremum, which we may suppose for definiteness to be a minimum, shall be relative to the set of values of J obtained by letting φ run through the domain of continuous functions which possess continuous partial derivatives of up to and including the second order and which also satisfy on Γ a certain linear boundary condition of the form

$$a_1(s)\,\varphi + a_2^*(s)\,\varphi_x + a_3^*(s)\,\varphi_y = a_4(s), \quad (5.2)$$

which will be specified more precisely later. We assume that there is a smallest value of J and a corresponding function u among the admissible functions φ for which J assumes this smallest value. The question of the existence of a minimum of J will not be pursued further here. Thus we suppose that

$$J[u] \leq J[\varphi]. \quad (5.3)$$

If $\eta(x, y)$ is any function with continuous second partial derivatives which satisfies the homogeneous boundary condition corresponding to (5.2), i.e.

$$a_1\eta + a_2^*\,\eta_x + a_3^*\,\eta_y = 0 \quad \text{on } \Gamma, \quad (5.4)$$

then $\varphi = u + \varepsilon\eta$ is an admissible function for any value of ε, and $J[u + \varepsilon\eta] = \Phi(\varepsilon)$ is a function of ε which assumes its minimum value for $\varepsilon = 0$ and whose derivative with respect to ε must therefore vanish at $\varepsilon = 0$:

$$\Phi'(0) = \left(\frac{\partial J[u + \varepsilon \eta]}{\partial \varepsilon}\right)_{\varepsilon=0} = 0. \tag{5.5}$$

By inserting $\varphi = u + \varepsilon \eta$ in (5.1) and calculating the derivative in (5.5) we obtain

$$\tfrac{1}{2}\Phi'(0) = J[u, \eta] + \iint_B r\, u\, dx\, dy - \int_\Gamma T u\, ds = 0,$$

where $J[u, \eta]$ is the integral of the corresponding bilinear expression:

$$J[\varphi, \psi] = \iint_B \left[A\, \varphi_x \psi_x + B\, (\varphi_x \psi_y + \varphi_y \psi_x) + C\, \varphi_y \psi_y + F\, \varphi \psi - \atop - r(\varphi + \psi)\right] dx\, dy + \int_\Gamma [S\, \varphi \psi + T(\varphi + \psi)]\, ds; \left.\right\} \tag{5.6}$$

transformation of the double integral by means of GREEN's formula [Ch. I (3.4)] then yields

$$\tfrac{1}{2}\Phi'(0) = \iint_B \eta\, \{L[u] - r\}\, dx\, dy - \int_\Gamma \eta\, L^*[u]\, ds + \int_\Gamma \eta\, (S u + T)\, ds = 0, \tag{5.7}$$

in which $L[u]$, $L^*[u]$ are the linear differential expressions defined in Ch. I, (3.3), (3.5):

$$L[u] = -\frac{\partial}{\partial x}(A\, u_x + B\, u_y) - \frac{\partial}{\partial y}(B\, u_x + C\, u_y) + F\, u,$$

$$L^*[u] = (A\, u_x + B\, u_y) \cos(\nu, x) + (B\, u_x + C\, u_y) \cos(\nu, y).$$

Now (5.7) can hold for all functions $\eta(x, y)$ satisfying the present conditions only if the factor $L[u] - r$ vanishes identically in B. This follows by the same argument as was used in Ch. III, § 5.2: if $L[u] - r$ were not equal to zero, but were positive say, at some interior point (x_0, y_0) of B, then by continuity it would also be positive in a small neighbourhood $(x - x_0)^2 + (y - y_0)^2 \leq \delta^2$ of (x_0, y_0), and for any η function which satisfies (5.7), say $\eta = \eta^*(x, \nu)$, we could specify another, $\eta = \eta^{**}$, which does not:

$$\eta^{**} = \begin{cases} \eta^* + [(x - x_0)^2 + (y - y_0)^2 - \delta^2]^3 & \text{for } (x - x_0)^2 + (y - y_0)^2 \leq \delta^2 \\ \eta^* & \text{otherwise.} \end{cases}$$

Consequently u must satisfy the differential equation

$$L[u] = r(x, y), \tag{5.8}$$

called the Euler equation of the variational problem, and (5.7) reduces to

$$\int_\Gamma \eta\, L^{**}[u]\, ds = 0, \tag{5.9}$$

where

$$L^{**}[u] = Su + T - L^* = Su + T - (A u_x + B u_y) \cos(v, x) - \\ - (B u_x + C u_y) \cos(v, y). \quad (5.10)$$

If our assumption of the existence of a minimizing function u is not to be contradicted, this boundary condition (5.9) must be compatible with the boundary condition (5.2) (with φ replaced by u); we see immediately that this is not always possible, and proceed to discuss the conditions under which it is.

5.2. Discussion of various boundary conditions

We distinguish two cases according as u is or is not prescribed on the boundary Γ.

Case I. First boundary-value problem. Here the boundary values of u are prescribed:

$$u(s) = g(s) \quad \text{on } \Gamma. \quad (5.11)$$

The corresponding homogeneous boundary condition for η is therefore $\eta = 0$ on Γ and the necessary condition (5.9) is automatically satisfied; thus no condition is imposed on $L^{**}[u]$ and we may choose $S = T = 0$.

Case II. Third boundary-value problem. Here the value of a linear combination of u and its first partial derivatives as in (5.2) is prescribed on the boundary Γ. Consequently there are functions η which are not zero on Γ, and therefore in (5.9) we must have $L^{**}[u] = 0$; this condition is therefore to be identifiable with the boundary condition (5.2). Let us denote the direction cosines of the inward normal v by

$$\alpha = \cos(v, x) = -\cos(s, y), \qquad \beta = \cos(v, y) = \cos(s, x) \quad (5.12)$$

and express φ_x and φ_y in terms of the normal and tangential derivatives φ_v and φ_s:

$$\varphi_x = \alpha \varphi_v + \beta \varphi_s, \\ \varphi_y = \beta \varphi_v - \alpha \varphi_s. \quad (5.13)$$

Then the two boundary conditions (5.2) and $L^{**}[u] = 0$, which are to represent the same condition, read

$$a_1 \varphi + (-a_3^* \alpha + a_2^* \beta) \varphi_s + (a_2^* \alpha + a_3^* \beta) \varphi_v - a_4 = 0, \quad (5.14)$$

$$S u + [(C-A) \alpha \beta + B(\alpha^2 - \beta^2)] u_s - [A \alpha^2 + 2 B \alpha \beta + C \beta^2] u_v + T = 0. \quad (5.15)$$

If we regard the differential equation as prescribed, so that A, B, C are fixed, we have only two quantities S and T at our disposal in (5.15), and it is clear that they cannot be chosen to make (5.15) equivalent to (5.14) unless the coefficients in the given boundary condition are such that the ratio of the coefficients of the partial derivatives in each of these two equations are the same:

$$\frac{a_2}{a_3} = \frac{-a_3^* \alpha + a_2^* \beta}{a_2^* \alpha + a_3^* \beta} = \frac{(A - C)\alpha\beta - B(\alpha^2 - \beta^2)}{A\alpha^2 + 2B\alpha\beta + C\beta^2} \qquad (5.16)$$

We exclude the case when the denominator vanishes identically (on the boundary), for then the boundary condition provides a relation between u and u_s, i.e. a differential equation for the boundary value $u(s)$, and by integration of this differential equation the boundary-value problem can be reduced to a first boundary-value problem. The denominator can still vanish on parts of the boundary Γ without vanishing everywhere on it. As this case will not be pursued any further, we will assume here that $A\alpha^2 + 2B\alpha\beta + C\beta^2 \neq 0$ on Γ. The results of the last two sections can then be summarized in the following

Theorem: *The boundary-value problem*

$$L[u] = -\frac{\partial}{\partial x}(A u_x + B u_y) - \frac{\partial}{\partial y}(B u_x + C u_y) + F u = r(x, y) \text{ in } B, \quad (5.17)$$

$$a_1(s)\, u + a_2(s)\, u_s + a_3(s)\, u_\nu = a_4(s) \quad \text{on } \Gamma \qquad (5.18)$$

with $a_3 \neq 0$ may be written as the necessary Euler conditions for a variational problem of the form

$$J[\varphi] = \iint_B [A \varphi_x^2 + 2B \varphi_x \varphi_y + C \varphi_y^2 + F \varphi^2 - 2r\varphi]\, dx\, dy + \left. \right\} \qquad (5.19)$$
$$+ \int_\Gamma (S \varphi^2 + 2T \varphi)\, ds = \text{extremum}$$

if and only if the coefficients satisfy the condition

$$\frac{a_2}{a_3} = \frac{(A - C)\alpha\beta - B(\alpha^2 - \beta^2)}{A\alpha^2 + 2B\alpha\beta + C\beta^2}, \qquad (5.20)$$

where

$$\alpha = \cos(\nu, x), \qquad \beta = \cos(\nu, y).$$

S and T are necessarily related to the other coefficients by

$$S = -(A\alpha^2 + 2B\alpha\beta + C\beta^2)\frac{a_1}{a_3}, \left. \right\} \qquad (5.21)$$
$$T = +(A\alpha^2 + 2B\alpha\beta + C\beta^2)\frac{a_4}{a_3},$$

but the functions φ admitted for comparison need not be restricted to satisfy any boundary conditions.

For the corresponding first boundary-value problem, i.e. for the case $a_2 = a_3 = 0$, $a_1 \neq 0$, the values of S and T are irrelevant, and may be taken to be zero, but it is now essential that the admissible functions φ all satisfy the boundary condition: $\varphi(s) = a_4/a_1$ on Γ.

5.3. A special class of boundary-value problems

We specialize the results of the last section by taking $A = C$ and $B = 0$; this leaves the class of differential equations of the form

$$-\frac{\partial}{\partial x}[A(x, y) u_x] - \frac{\partial}{\partial y}[A(x, y) u_y] + F(x, y) u = r(x, y) \text{ in } B, \quad (5.22)$$

which occur frequently in applications. To satisfy the condition (5.19) we must also take $a_2 = 0$; thus we restrict ourselves to boundary conditions of the form

$$a_1(s) u + a_3(s) u_v = a_4(s) \quad \text{on } \Gamma, \quad (5.23)$$

which also occur frequently in practice.

Then for the case $\alpha_3 \neq 0$ the corresponding variational problem reads (since $\alpha^2 + \beta^2 = 1$)

$$\left.\begin{aligned}
J[\varphi] = \iint_B [A(\varphi_x^2 + \varphi_y^2) + F \varphi^2 - 2r \varphi] \, dx \, dy + \\
+ \int_\Gamma \frac{A}{a_3}(-a_1 \varphi^2 + 2a_4 \varphi) \, ds = \text{extremum},
\end{aligned}\right\} \quad (5.24)$$

for which φ need not satisfy any boundary conditions.

For the case $a_3 = 0$, $a_1 \neq 0$ the boundary integral in (5.24) may be omitted, but the admissible functions φ must all take the boundary value a_4/a_1 on Γ.

5.4. Example

Once more we consider the heat conduction problem

$$\left.\begin{aligned}
\nabla^2 u = -1 & \quad \text{in the interior} \\
\partial u/\partial v = u & \quad \text{on the boundary}
\end{aligned}\right\} \text{ of the square } |x| \leq 1, \; |y| \leq 1. \quad (5.25)$$

From § 5.3 the corresponding variational problem reads

$$J[\varphi] = \iint_B [\varphi_x^2 + \varphi_y^2 - 2\varphi] \, dx \, dy + \int_\Gamma \varphi^2 \, ds = \text{extremum}, \quad (5.26)$$

for which φ need not satisfy any boundary conditions. Using polynomial approximation functions which already satisfy the symmetry conditions, we take for our first approximation the two-parameter expression

$$\varphi = a_1 + a_2(x^2 + y^2).$$

With

$$\varphi_x = 2a_2 x, \qquad \varphi_y = 2a_2 y$$

we obtain

$$\iint_B [\varphi_x^2 + \varphi_y^2 - 2\varphi]\, dx\, dy = -8a_1 - \frac{16}{3} a_2 + \frac{32}{3} a_2^2,$$

$$\int_\Gamma \varphi^2\, ds = 8a_1^2 + \frac{64}{3} a_1 a_2 + \frac{224}{15} a_2^2;$$

hence

$$J[\varphi] = 8\left[a_1^2 + \frac{8}{3} a_1 a_2 + \frac{16}{5} a_2^2 - a_1 - \frac{2}{3} a_2\right]. \qquad (5.27)$$

Table V/13. *Results for the heat conduction problem obtained by Ritz's method*

Point	Two-parameter expression		Three-parameter expression	
	Approximate value	Error	Approximate value	Error
$x=0,\ y=0$	$\dfrac{13}{16} \approx 0{\cdot}8125$	$-1{\cdot}1\%$	$\dfrac{139}{168} \approx 0{\cdot}827\,38$	$+0{\cdot}7\%$
$x=0,\ y=1$	$\dfrac{37}{64} \approx 0{\cdot}5781$	$+3{\cdot}8\%$	$\dfrac{47}{84} \approx 0{\cdot}559\,52$	$+0{\cdot}4\%$
$x=1,\ y=1$	$\dfrac{11}{32} \approx 0{\cdot}3438$	$-9{\cdot}5\%$	$\dfrac{8}{21} \approx 0{\cdot}380\,95$	$+0{\cdot}25\%$

From the two extremum conditions

$$\frac{1}{8}\frac{\partial J}{\partial a_1} = 0 = 2a_1 + \frac{8}{3} a_2 - 1,$$

$$\frac{1}{8}\frac{\partial J}{\partial a_2} = 0 = \frac{8}{3} a_1 + \frac{32}{5} a_2 - \frac{2}{3}$$

we obtain $a_1 = \dfrac{13}{16}$, $a_2 = -\dfrac{15}{64}$, and hence the approximate solution

$$\varphi = \frac{1}{64}\left[52 - 15\,(x^2 + y^2)\right].$$

Several numerical values, together with their errors, are given in Table V/13.

For a better approximation we use the three-parameter expression

$$\varphi = a_1 + a_2\,(x^2 + y^2) + a_3\,x^2 y^2.$$

J is now given by

$$\frac{1}{8} J[\varphi] = a_1^2 + \frac{8}{3} a_1 a_2 + \frac{16}{5} a_2^2 + \frac{2}{3} a_1 a_3 + \frac{88}{45} a_2 a_3 + \frac{7}{15} a_3^2 - a_1 - \frac{2}{3} a_2 - \frac{1}{9} a_3$$

and the extremum conditions

$$\frac{1}{8}\frac{\partial J}{\partial a_1} = 2\,a_1 + \frac{8}{3} a_2 + \frac{2}{3} a_3 - 1 = 0,$$

$$\frac{1}{8}\frac{\partial J}{\partial a_2} = \frac{8}{3} a_1 + \frac{32}{5} a_2 + \frac{88}{45} a_3 - \frac{2}{3} = 0,$$

$$\frac{1}{8}\frac{\partial J}{\partial a_3} = \frac{2}{3} a_1 + \frac{88}{45} a_2 + \frac{14}{15} a_3 - \frac{1}{9} = 0$$

yield
$$a_1 = \frac{139}{168}, \qquad a_2 = -\frac{15}{56}, \qquad a_3 = \frac{5}{56}.$$

The new approximate solution is

$$\varphi = \frac{1}{168}\,[139 - 45\,(x^2 + y^2) + 15\,x^2\,y^2];$$

numerical values at key points are compared with the first approximation in Table V/13.

In this example J has been worked out as a quadratic function of the a_ϱ for illustration; this is not essential, for the differentiations with respect to the parameters can be performed before the integral is evaluated, i.e. under the integral sign, and in fact the calculation can often be shortened thereby.

5.5. A differential equation of the fourth order

For the plate problem[1] of § 1.5

$$\nabla^4 u = \frac{p}{N} = x^2 \quad \text{in } B: \; |x| \leq 1, \; |y| \leq 1, \tag{5.28}$$

$$u = u_x = 0 \quad \text{for } |x| = 1, \qquad u = \nabla^2 u = 0 \quad \text{for } |y| = 1 \tag{5.29}$$

we consider an integral expression of the form[2]

$$J[\varphi] = \iint_B \left[(\nabla^2 \varphi)^2 - 2\,\frac{p}{N}\,\varphi \right] dx\,dy. \tag{5.30}$$

To investigate the associated variational problem we proceed exactly as in §5.1, and can therefore be brief. We put $\varphi = u + \varepsilon \eta$ and form $\Phi(\varepsilon) = J[u + \varepsilon \eta]$; then (5.5), which here reads

$$\Phi'(0) = \left(\frac{\partial J}{\partial \varepsilon}\right)_{\varepsilon=0} = 2 \iint_B \left(\nabla^2 u\, \nabla^2 \eta - \frac{p}{N}\,\eta \right) dx\,dy = 0, \tag{5.31}$$

provides a necessary condition to be satisfied by u if it is to minimize $J[\varphi]$.

If we apply GREEN's formula [Ch. I (3.8)] with $\varphi = \nabla^2 u$, $\psi = \eta$, i.e.

$$\iint_B (\nabla^2 u\, \nabla^2 \eta - \eta\, \nabla^4 u)\, dx\,dy = \int_\Gamma \left(\eta\, \frac{\partial \nabla^2 u}{\partial \nu} - \nabla^2 u\, \frac{\partial \eta}{\partial \nu} \right) ds, \tag{5.32}$$

to the first term in the double integral of (5.31), the necessary extremum

---

[1] More general problems of stressed elastic bodies have been considered by WEBER; he obtains bounds for the displacements by making use of two minimum principles. C. WEBER: Eingrenzung von Verschiebungen mit Hilfe der Minimalsätze. Z. Angew. Math. Mech. **22**, 126—130 (1942). — Eingrenzung von Verschiebungen und Zerrungen mit Hilfe der Minimalsätze. Z. Angew. Math. Mech. **22**, 130—136 (1942).

[2] By way of further literature on variational problems for double integrals involving second derivatives we may mention A. R. FORSYTH: Calculus of variations, 656 pp. Cambridge 1927, in particular Ch. XI, pp. 567—600.

condition (5.31) becomes

$$\iint\limits_{B} \eta \left(V^4 u - \frac{p}{N}\right) dx\, dy + \int\limits_{\Gamma} \left(\eta \frac{\partial V^2 u}{\partial \nu} - V^2 u \frac{\partial \eta}{\partial \nu}\right) ds = 0. \quad (5.33)$$

The definitions of boundary curve Γ, arc length s and inward normal ν follow those in Ch. I, § 3.1.

As before, the factor multiplying η in the double integral must vanish, and hence the minimizing function u must satisfy the given differential equation. Thus the double integral vanishes. In order that the boundary integral shall also vanish we must demand that $\eta = 0$ everywhere on the boundary and that $\partial\eta/\partial\nu = 0$ on the parts of the boundary where the plate is clamped; the second boundary condition on the rest of the boundary, $V^2\eta = 0$, may be suppressed, since the fact that we do not demand the vanishing of η_ν on this part of the boundary automatically ensures that the minimizing function u satisfies $V^2 u = 0$ here; we say that $V^2 u = 0$ is a "natural" boundary condition for the variational problem $J[\varphi] = \min$. Thus, as in the case of ordinary differential equations, the admissible functions φ must satisfy the essential boundary conditions, but need not satisfy the suppressible boundary conditions. Unless the latter are also natural boundary conditions, as in the present example, $J[\varphi]$ must be modified by the addition of suitable boundary integral expressions so as to obtain a variational problem for which the suppressible boundary conditions are natural, otherwise they will not be satisfied by the minimizing function u.

As a family of admissible functions for the present example we could use, for instance,

$$\varphi = (1 - x^2)^2 (1 - y^2) (a_1 + a_2 x^2 + a_3 y^2 + a_4 x^4 + a_5 x^2 y^2 + a_6 y^4);$$

these satisfy all the essential boundary conditions and also the appropriate symmetry conditions. To illustrate the method we will use the one-parameter family obtained by putting all parameters but a_1 equal to zero; the integral in (5.30) becomes

$$J[\varphi] = \frac{128}{35} \left(\frac{64}{5} a_1^2 - \frac{a_1}{9}\right), \quad (5.34)$$

and the condition $\partial J/\partial a_1 = 0$ leads to

$$a_1 = \varphi(0, 0) = \frac{5}{1152} \approx 0 \cdot 00435.$$

5.6. Direct proof of two minimum principles for a biharmonic boundary-value problem

Consider the boundary-value problem

$$V^4 u = p(x, y) \quad \text{in } B, \quad (5.35)$$

$$u = f(s), \quad u_\nu = g(s) \quad \text{on } \Gamma \quad (5.36)$$

with region B, boundary Γ, inward normal ν as in § 3.5, and assume the existence of a solution u.

Now let v and w be functions satisfying the differential equation and boundary conditions, respectively:

$$V^4 v = p \quad \text{in } B; \quad w = f, \quad w_\nu = g \quad \text{on } \Gamma. \tag{5.37}$$

Then from (3.38) of Ch. I we have

$$D[v - w] = D[v - u] + D[u - w], \tag{5.38}$$

and hence

$$D[v - u] \leq D[v - w] \tag{5.39}$$

and

$$D[u - w] \leq D[v - w]. \tag{5.40}$$

Each of these two inequalities contains a minimum principle.

I. Consider first (5.39). On account of (5.38) we have equality if and only if $D[u - w] = 0$. Now from the definition of $D[\varphi]$, Ch. I (3.34), $D[\varphi] = 0$ implies that $V^2 \varphi = 0$; further $V^2(u - w) = 0$ in B with $u - w = 0$ on the boundary Γ implies that $u - w \equiv 0$ in B. Consequently

$$D[v - u] = D[v - w] \quad \text{if and only if} \quad u = w.$$

With each side expanded as in (3.35) of Ch. I, (5.39) here reads

$$D[v] - 2D[u, v] + D[u] \leq D[v] - 2D[v, w] + D[w]. \tag{5.41}$$

If we put $\varphi = v$, $\psi = u - w$ in (3.37) of Ch. I, the boundary integral vanishes on account of (5.37) and we have

$$D[u - w, v] = D[u, v] - D[w, v] = \iint_B (u - w) V^4 v \, dx \, dy$$
$$= \iint_B (u - w) V^4 u \, dx \, dy.$$

Hence (5.41) may be written

$$\iint_B [(V^2 u)^2 - 2u V^4 u] \, dx \, dy \leq \iint_B [(V^2 w)^2 - 2w V^4 u] \, dx \, dy. \tag{5.42}$$

This result can be stated as the following

Theorem: *If w runs through the class of functions which possess continuous partial derivatives of the second order and satisfy the boundary conditions in (5.37), i.e. $w = f$, $w_\nu = g$ on Γ, then the integral*

$$J[w] = \iint_B [(V^2 w)^2 - 2w \, p] \, dx \, dy \tag{5.43}$$

assumes its minimum value when and only when $w = u$, where u is the solution of the boundary-value problem (5.35), (5.36).

II. The treatment of the second inequality (5.40) is quite analogous. On account of (5.38) we have equality if and only if $D[u-v]=0$, i.e. $V^2[u-v]=0$. If we put $\varphi = u - v$, $\psi = w$ in (3.37) of Ch. I, the double integral on the right-hand side disappears and we have

$$D[u-v, w] = D[u, w] - D[v, w] = \int_\Gamma \left[w\{V^2(u-v)\}_\nu - w_\nu V^2(u-v) \right] ds.$$

The inequality (5.40), which can be written in the form

$$D[u] - 2D[u, w] + D[w] \leqq D[v] - 2D[v, w] + D[w],$$

therefore yields

$$\iint_B (V^2u)^2\, dx\, dy - 2\int_\Gamma [w(V^2u)_\nu - w_\nu V^2u]\, ds$$
$$\leqq \iint_B (V^2v)^2\, dx\, dy - 2\int_\Gamma [w(V^2v) - w_\nu V^2v]\, ds.$$

Thus we may state the

Theorem[1]. *If v runs through the class of functions which possess continuous partial derivatives of the fourth order and satisfy the differential equation $V^4v = p$ of (5.37), then the integral*

$$J[v] = \iint_B (V^2v)^2\, dx\, dy + 2\int_\Gamma [g\, V^2v - f(V^2v)_\nu]\, ds \qquad (5.44)$$

assumes a minimum value for all functions of the form $u + \varphi$, where u is the solution of the boundary-value problem (5.35), (5.36) and φ is any admissible function with $V^2\varphi = 0$.

Example. We choose an example similar to one considered by WEGNER (loc. cit.). For a uniformly loaded rectangular plate whose edges are clamped and have lengths in the ratio $1:2$ we have

$$V^4u = 1 \text{ in } B,$$
$$u = \frac{\partial u}{\partial \nu} = 0 \text{ on } \Gamma,$$

where Γ is the boundary of the rectangle B with $|x| \leqq 2$, $|y| \leqq 1$.

Let us assume for V^2v an expression of the form

$$V^2v = \frac{x^2 + y^2}{4} + \sum_{\nu=0}^{p} c_\nu \varphi_\nu \quad \text{with} \quad \varphi_\nu = \mathrm{re}\,(x + i\,y)^{2\nu};$$

---

[1] Established for homogeneous boundary conditions $f = g = 0$ by U. WEGNER: Ein neues Verfahren zur Berechnung der Spannungen in Scheiben. Forsch.-Arb. Ing.-Wes. **13**, 114—149 (1942), and for inhomogeneous boundary conditions by J. B. DIAZ and H. J. GREENBERG: Upper and lower bounds for the solution of the first biharmonic boundary-value problem. J. Math. Phys. **27**, 193—201 (1948). WEGNER also gives numerical examples.

then $\nabla^4 v = 1$ for arbitrary c_ν. Since the boundary conditions are homogeneous, the boundary integral in (5.44) does not appear, and the conditions $\partial J/\partial c_\nu = 0$ (for $\nu = 0, 1, \ldots, p$) read for $p = 1$

$$2c_0 + 2\, c_1 + \frac{5}{6} = 0,$$

$$2c_0 + \frac{226}{45}\, c_1 + \frac{3}{2} = 0,$$

with the solution

$$c_0 = -\frac{40}{204} \approx -0{\cdot}196, \qquad c_1 = -\frac{45}{204} \approx -0{\cdot}221;$$

for $p = 2$ the calculation yields

$$c_0 = -\frac{1850}{3 \times 3671} \approx -0{\cdot}168, \qquad c_1 = -\frac{3945}{4 \times 3671} \approx -0{\cdot}269,$$

$$c_2 = \frac{100}{3671} \approx 0{\cdot}0272.$$

In this way we obtain directly an approximation for $\nabla^2 u$ instead of for u. Consequently the method is very suitable when the stress distribution is of most interest, but less so when the displacement has to be calculated.

5.7. More than two independent variables

We procede along the same lines[1] as in §§ 5.1—5.3, but because of the greater length and complexity of the formulae, we restrict ourselves to a simple case, namely that for which the differential equation is of the self-adjoint form (5.49); and further we present this simple case more concisely.

Corresponding to (5.1) we consider the variational problem

$$\left. \begin{aligned} J[\varphi] = \int_B \left(\sum_{i,k=1}^{m} A_{ik}\, \frac{\partial \varphi}{\partial x_i}\, \frac{\partial \varphi}{\partial x_k} + q\,\varphi^2 - 2r\,\varphi \right) d\tau\ + \\ + \int_\Gamma (K\,\varphi^2 + 2M\,\varphi)\, dS = \text{extremum}, \end{aligned} \right\} \tag{5.45}$$

where B is a given, finite, simply-connected, (say) closed region of the $(x_1, x_2, \ldots, x_m)$ space bounded by a closed surface Γ which is made up of a finite number of "faces", each with a continuously turning tangent plane; $d\tau = dx_1\, dx_2 \ldots dx_m$ denotes, as in Ch. I, § 3.2, the volume element in B, and dS the surface element on Γ; A_{ik}, q, r and K, M are given continuous functions of position in B and on Γ, respectively; further the A_{ik} are symmetric ($A_{ki} = A_{ik}$) and possess continuous first partial derivatives. The domain of admissible functions will be restricted to

---

[1] See also R. WEINSTOCK: Calculus of variations, 326 pp. New York-Toronto-London 1952.

those functions $\varphi(x_1, x_2, \ldots, x_m)$ which are continuous and have continuous partial derivatives of up to and including the second order in B and which satisfy such boundary conditions on the boundary Γ as are found to be necessary. Again we assume that there is an admissible function $u(x_1, x_2, \ldots, x_m)$ for which J attains its minimum value, and then derive conditions which must be satisfied by this function. Such a condition is that for any family of admissible functions of the form $\varphi = u + \varepsilon\eta(x_1, x_2, \ldots, x_m)$, where ε is a parameter, the function $\Phi(\varepsilon) = J[u + \varepsilon\eta]$ must have a minimum value when $\varepsilon = 0$; hence the condition $\Phi'(0) = 0$ of (5.5) must be satisfied.

Here we have

$$\left.\begin{aligned}\frac{1}{2}\, \Phi'(0) &= \frac{1}{2}\left\{\frac{\partial}{\partial\varepsilon}\, J[u + \varepsilon\eta]\right\}_{\varepsilon=0} \\ &= \int_B \left\{\sum_{i,k=1}^m A_{ik}\frac{\partial u}{\partial x_i}\frac{\partial\eta}{\partial x_k} + qu\eta - r\eta\right\}d\tau + \int_\Gamma \{Ku\eta + M\eta\}\,dS = 0.\end{aligned}\right\} \quad (5.46)$$

Now according to the formulae (3.13) to (3.15) of Ch. I

$$\int_B \left\{\sum_{i,k=1}^m A_{ik}\frac{\partial u}{\partial x_i}\frac{\partial\eta}{\partial x_k} + qu\eta\right\}d\tau = \int_B \eta\, L[u]\,d\tau - \int_\Gamma \eta\, L^*[u]\,dS, \quad (5.47)$$

where $L[u]$ and $L^*[u]$ are defined, as in Ch. I (3.12), (3.15), by

$$L[u] = -\sum_{i,k=1}^m \frac{\partial}{\partial x_i}\left(A_{ik}\frac{\partial u}{\partial x_k}\right) + qu,$$

$$L^*[u] = \sum_{i,k=1}^m A_{ik}\frac{\partial u}{\partial x_k}\cos(\nu, x_i) = A\frac{\partial u}{\partial\sigma};$$

we can therefore rewrite (5.46) in the form

$$\int_B \eta\{L[u] - r\}d\tau + \int_\Gamma \eta\{Ku + M - L^*[u]\}\,dS = 0. \quad (5.48)$$

By the usual argument it can be deduced from the arbitrariness of η that the factor multiplying η in the space integral must vanish inside B; thus as the Euler differential equation we obtain

$$L[u] = -\sum_{i,k=1}^m \frac{\partial}{\partial x_i}\left(A_{ik}\frac{\partial u}{\partial x_k}\right) + qu = r. \quad (5.49)$$

The necessary condition (5.48) then reduces to

$$\int_\Gamma \eta\{Ku + M - L^*[u]\}\,dS = 0. \quad (5.50)$$

For further discussion we select two special cases (not all boundary conditions can be dealt with by this method anyway).

5.8. Special cases

Case I. First boundary-value problem. Here, since the value of u is prescribed on the boundary Γ, say $u = g$, we must demand that $\eta = 0$ on Γ, i.e. that all admissible functions φ satisfy the boundary condition $\varphi = g$ on Γ. Then the condition (5.50) is satisfied for arbitrary K and M; in particular, we can put $K = M = 0$, i.e. we can omit the boundary integral in (5.45).

Case II. Third boundary-value problem. Suppose that we require

$$L^*[u] + qu + w = 0 \quad \text{on } \Gamma,$$

where q, w are given functions of position. Here we minimize J for functions φ which are not prescribed on the boundary, so that the minimizing function u must satisfy $Ku + M - L^*[u] = 0$ on Γ. Thus if we choose $K = -q$, $M = -w$, the required boundary condition will be satisfied, and the admissible functions φ need not be restricted by any boundary conditions.

For the particular differential equations with

$$A_{ik} = \delta_{ik} p = \begin{cases} p & \text{for} \quad i = k \\ 0 & \text{for} \quad i \neq k, \end{cases}$$

where p is a given function of position in B, the above boundary condition reads

$$p u_\nu + q u + w = 0,$$

where u_ν is the derivative in the direction of the inward normal.

Example. For the three-dimensional problem

$$u_{xx} + u_{yy} + u_{zz} = 0,$$

$$\frac{\partial u}{\partial \nu} = 0 \quad \text{for} \quad |x| = 1, \quad |y| = 1, \quad |z| = 1.$$

$$u = \sigma \quad \text{for} \quad x = y = z = \sigma \quad \text{with} \quad \sigma = \pm 1$$

(temperature distribution in a cube C as in § 1.7) the volume integral in (5.45) reduces to the Dirichlet integral

$$J[\varphi] = \iiint\limits_C (\text{grad } \varphi)^2 \, dx \, dy \, dz \tag{5.51}$$

and φ need only satisfy the boundary condition $\varphi = \sigma$ at the corners $x = y = z = \sigma$ with $\sigma = \pm 1$. The simplest family of admissible functions which also satisfy the symmetries of the problem may be defined by

$$\varphi = a(x + y + z) + (1 - 3a) x y z.$$

Using this expression we have

$$\varphi_x = a + (1 - 3a) y z,$$
$$\tfrac{1}{8} J[\varphi] = 6a^2 - 2a + \tfrac{1}{3},$$

and from $\partial J/\partial a = 0$ we obtain $a = \tfrac{1}{6}$; the approximate solution therefore reads

$$\varphi = \tfrac{1}{6}(x + y + z) + \tfrac{1}{2} x y z.$$

5.9. The mixed Ritz expression

As in § 5.7, let $J[\varphi]$ be an integral expression which is minimized with respect to a certain domain of admissible functions φ by the function u which is the solution of a given (say) linear boundary-value problem with the differential equation

$$L[x_1, x_2, \ldots, x_n, u, u_{x_1}, u_{x_2}, \ldots] = 0$$

and the boundary conditions $U_\varkappa[u] = \gamma_\varkappa$ ($\varkappa = 1, \ldots, k$). The admissible functions φ may have to satisfy certain boundary conditions (cf. § 5.7, for example), which we will denote by

$$V_\mu[\varphi] = v_\mu \qquad (\mu = 1, \ldots, m). \tag{5.52}$$

The basis of the ordinary Ritz method is the minimization of J with respect to a p-parameter family of admissible functions defined by an expression of the form

$$\varphi(x_\varrho) = w_0(x_\varrho) + \sum_{\nu=1}^{p} a_\nu w_\nu(x_\varrho), \tag{5.53}$$

where w_0 satisfies the inhomogeneous boundary conditions $V_\mu[w_0] = v_\mu$ and the w_ν for $1 \leq \nu \leq p$ satisfy the corresponding homogeneous conditions $V_\mu[w_\nu] = 0$. Now if there is an independent variable, x_n say, derivatives with respect to which do not occur in the boundary conditions, then φ, as defined in (5.53), would still satisfy the boundary conditions even if the a_ν were allowed to depend on x_n, and were thus regarded as functions of x_n to be determined[1]. This defines a wider class of admissible functions, which lies somewhere between the p-parameter family (5.53) (an ordinary Ritz expression, leading to a system of linear equations for the a_ν) and the class of all admissible functions [with the a_ν depending on all the x_ϱ, so that the sum in (5.53) can be replaced by a single unknown function $a(x_\varrho)$; this leads to the Euler differential equation of the variational problem].

If we insert this "mixed" expression, in which the a_ν depend on a single variable x_n, the integral expression J becomes a functional of the p functions $a_\nu(x_n)$; we write

$$J = \tilde{J}[a_\nu(x_n)].$$

---

[1] KANTOROVICH, L.: Sur une méthode directe de la solution approximative du problème du minimum d'un intégral double [Russian]. Leningrad Bull. Ac. Sci. 7, 647—652 (1933) (Jb. Fortschr. Math. **59**, 1149). The method has been used by HILLEL PORITZKY: The reduction of the solution of certain partial differential equations to ordinary differential equations. Trans. 5th Intern. Congr. Appl. Mech., Cambridge (Mass.) 1938, pp. 700—707, by N. S. SEMENOW: Biegung von Rechtecksplatten. Zbl. Mech. **11**, 12 (1939) and also by E. METTLER: Allgemeine Theorie der Stabilität erzwungener Schwingungen elastischer Körper. Ing.-Arch. **17**, 418—449 (1949), in particular p. 420 et seq. and p. 445 et seq.

The requirement that $\tilde{J}$ be minimized leads to a system of Euler equations consisting of p ordinary differential equations for the p unknown functions $a_\nu(x_n)$.

The method is also applicable to initial-/boundary-value problems (Ch. IV); if the time t is one of the independent variables, it can be used as x_n, and sometimes one can then draw conclusions about the behaviour of the solution as t increases indefinitely[1].

Examples. I. Consider the first boundary-value problem

$$V^2 u = r(x, y, z) \quad \text{in } B,$$

$$u = 0 \quad \text{on the boundary } \Gamma \text{ of } B.$$

According to § 5.3 the corresponding variational problem reads

$$J[\varphi] = \int_B \{(\mathrm{grad}\,\varphi)^2 + 2r\,\varphi\}\,d\tau = \min.$$

Let $\varphi = a(x)\,w(x, y, z)$, where w (or the product aw if w does not depend on x) vanishes on the boundary Γ. Then $J[\varphi] = J[a(x)\,w] = \tilde{J}[a]$. We derive an Euler differential equation for the minimizing function $A(x)$ by the usual procedure of putting $a(x) = A(x) + \varepsilon\eta(x)$. We must have

$$\frac{1}{2}\left(\frac{\partial \tilde{J}[A + \varepsilon\eta]}{\partial\varepsilon}\right)_{\varepsilon=0} = \int_B \{(\mathrm{grad}\,(A\,w),\ \mathrm{grad}\,(\eta\,w)) + r\,\eta\,w\}\,d\tau = 0.$$

By using formula (3.17) of Ch. I we can separate out η as a factor:

$$\int_B \eta\,w\{-V^2(A\,w) + r\}\,d\tau - \int_\Gamma \eta\,w\,\frac{\partial(A\,w)}{\partial\nu}\,dS = 0.$$

Now w (or the product aw), and hence also ηw, vanishes on the boundary, so that the boundary integral is zero. The remaining volume integral can be written as a repeated integral of the form

$$\int \eta(x)\left\lceil \iint_{B^*(x)} w\{-V^2(A(x)\,w) + r\}\,dy\,dz\ \right. dx,$$

where $B^*(x)$ is the cross-section of B in the plane $x = $ constant. We see that, on account of the arbitrariness of $\eta(x)$, the double integral over the cross-section $B^*(x)$ must vanish for each x in B. This yields an ordinary differential equation of the second order for $A(x)$:

$$A'' \iint_{B^*} w^2\,dy\,dz + 2A' \iint_{B^*} w\,\frac{\partial w}{\partial x}\,dy\,dz + A \iint_{B^*} w\,V^2 w\,dy\,dz - \iint_{B^*} r\,w\,dy\,dz = 0,$$

where dashes denote derivatives with respect to x.

In the general formulation with $w = 0$ everywhere on Γ the boundary conditions for $A(x)$ are that it remains finite at the end-points, and the uniqueness depends on the fact that the differential equation is singular at the end-points; it may be more convenient, however, to choose w so that A must vanish at the end-points. With $A(x)$ so determined $A(x)\,w(x, y, z)$ is the desired approximate solution.

---

[1] An example is given by F. WEIDENHAMMER: Der eingespannte, axial pulsierend belastete Stab als Stabilitätsproblem. Z. Angew. Math. Mech. **30**, 235—237 (1950).

II. Example in which the fundamental region extends to infinity:

$$\nabla^2 u = 0 \text{ in } B; \quad u(x, \pm 1) = 0, \quad u(0, y) = 1 - y^2,$$

where B is the region $x \geq 0$, $|y| \leq 1$. Here we have to minimize the Dirichlet integral:

$$J = \iint_B (\varphi_x^2 + \varphi_y^2) \, dx \, dy = \min.$$

We first put $\varphi(x, y) = (1 - y^2) f(x)$; then with dashes denoting differentiation with respect to x the new functional $J = \tilde{J}[f]$ is given by

$$\frac{1}{2} \tilde{J}[f] = \int_0^\infty \left(\frac{8}{15} f'^2 + \frac{4}{3} f^2 \right) dx.$$

If $f = F(x)$ minimizes this integral, we have by the usual linear variation method

$$0 = \frac{1}{4} \left(\frac{\partial \tilde{J}[F + \varepsilon \eta]}{\partial \varepsilon} \right)_{\varepsilon=0} = \int_0^\infty \left(\frac{8}{15} F' \eta' + \frac{4}{3} F \eta \right) dx$$

$$= \int_0^\infty \eta \left(-\frac{8}{15} F'' + \frac{4}{3} F \right) dx + \left[\frac{8}{15} \eta F' \right]_0^\infty.$$

By the usual arguments depending on the arbitrariness of η the factor multiplying η in the integral must vanish, and also we must have $\eta(0) = F'(\infty) = 0$. Thus $F(x)$ must satisfy the Euler equation

$$-\frac{8}{15} F'' + \frac{4}{3} F = 0$$

with the boundary conditions $F(0) = 1$, $F'(\infty) = 0$. Hence $F(x) = e^{-kx}$, where $k = \sqrt{2.5}$, and

$$\varphi(x, y) = (1 - y^2) e^{-\sqrt{2.5} x}.$$

Table V/14. *Approximations for* $u(x, 0)$

x	One-term expression	Two-term expression
0·25	0·6735	0·687 44
0·5	0·4536	0·467 63
0·75	0·3055	0·316 73
1	0·2057	0·214 14

Table V/14 gives a few sample values for comparison with the next approximation.

A two-term expression

$$\varphi(x, y) = (1 - y^2) f_1(x) + (1 - y^2)^2 f_2(x)$$

leads in exactly the same way to the pair of simultaneous ordinary differential equations

$$-\frac{8}{15} F_1'' - \frac{16}{35} F_2'' + \frac{4}{3} F_1 + \frac{16}{15} F_2 = 0,$$

$$-\frac{16}{35} F_1'' - \frac{128}{315} F_2'' + \frac{16}{15} F_1 + \frac{128}{105} F_2 = 0$$

with the boundary conditions $F_1(0) = 1$, $F_2(0) = 0$, $F_1'(\infty) = F_2'(\infty) = 0$. It follows that

$$F_1(x) = a\,e^{k_1 x} + (1-a)\,e^{k_2 x}$$
$$F_2(x) = b\,e^{k_1 x} - b\,e^{k_2 x},$$

where k_1 and k_2 are the negative roots of $k^4 - 28k^2 + 63 = 0$; if we put $k^2 = \varrho$, we have

$$\varrho_{1,2} = 14 \pm \sqrt{133} = \begin{cases} 25 \cdot 533 \\ 2 \cdot 4674 \end{cases}$$

and $a = \dfrac{7 - \varrho_2}{2\sqrt{133}}$, $b = \dfrac{7 - \varrho_1}{16}\,a$. Some values of the approximation function when $y = 0$ are given in Table V/14.

§6. The Trefftz[1] method

While the Ritz method is, in the sense of Ch. I, §4.1, an interior method, the Trefftz method[2] belongs to the category of boundary methods, and therefore possesses over the Ritz method the advantages already mentioned in Ch. I, §4.1. Moreover, error estimation by means of the boundary-maximum theorem (§3) is in many ways more simple for the Trefftz method than for the Ritz method; see §6.3 and the example in §6.5.

6.1. Derivation of the Trefftz equations

Consider the first boundary-value problem with the differential equation

$$L[u] = -\frac{\partial}{\partial x}(A\,u_x + B\,u_y) - \frac{\partial}{\partial y}(B\,u_x + C\,u_y) + F\,u = r(x,y) \text{ in } B \quad (6.1)$$

---

[1] ERICH TREFFTZ, born 21 February 1888, son of a Leipzig merchant, studied in Aachen, Göttingen and Strasbourg. He recieved much stimulation from his uncle, CARL RUNGE, whom he assisted at Göttingen and accompanied to New York when RUNGE went there as exchange professor in the Columbia University. He obtained his doctor's degree at Strasbourg in 1913 with work instigated by v. MISES. After being wounded in the first World War he went to Aachen, where he became ordinary professor of applied mathematics in 1919. In 1922 he accepted a professorship in applied mechanics at Dresden, a post which he held until his death from a malignant disease on the 21 January 1937. In his obituary [Z. Angew. *Nachruf* Math. Mech. **17**, 1 (1937)] L. PRANDTL wrote of him: "His domestic happiness, the pleasure he took in his work, the devotion of his students and the affection of his friends made him a happy man." His work in the field of mechanics was concerned chiefly with hydrodynamics, the theory of vibrations and elasticity [see R. GRAMMEL: Das wissenschaftliche Werk von Erich Trefftz. Z. Angew. Math. Mech. **18**, 1−11 (1938)].

[2] TREFFTZ, E.: Ein Gegenstück zum Ritzschen Verfahren. Verh. Kongr. Techn. Mech., Zürich 1926, pp. 131−137. − Konvergenz und Fehlerschätzung beim Ritzschen Verfahren. Math. Ann. **100**, 503−521 (1928). In these papers the method is presented for the potential equation and the biharmonic equation; we shall restrict ourselves to second-order equations but of the more general type (6.1) and (6.21).

and the boundary condition

$$u = g(s) \text{ on } \Gamma, \tag{6.2}$$

where s is the arc length along the boundary Γ of a region B as in § 5.1, F, r, g are given continuous functions, A, B, C are given functions with continuous first partial derivatives and $A > 0, C > 0, AC - B^2 > 0, F \geq 0$.

Now suppose that we know a particular solution $\overline{w}$ of the inhomogeneous equation (6.1) and a number of linearly independent solutions $w_1, w_2, \ldots, w_m$ of the corresponding homogeneous equation:

$$L[\overline{w}] = r, \quad L[w_1] = L[w_2] = \cdots = L[w_m] = 0. \tag{6.3}$$

Then

$$W = \overline{w} + \sum_{\sigma=1}^{m} a_\sigma w_\sigma \tag{6.4}$$

is also a solution of (6.1), and the a_σ are at our disposal for making W approximate the solution u of the boundary-value problem as closely as possible.

The obvious way of choosing the a_σ would be to employ one of the error distribution principles of Ch. I, § 4.2; by the least-squares method, for example, we would demand that

$$J^* = \int_\Gamma [W(s) - g(s)]^2 \, ds = \min., \tag{6.5}$$

which on the insertion of the expression (6.4) for W yields the following symmetric system of linear equations for the a_ϱ:

$$\sum_{\varrho=1}^{m} a_\varrho \int_\Gamma w_\varrho w_\sigma \, ds = \int_\Gamma (g - \overline{w}) w_\sigma \, ds \qquad (\sigma = 1, 2, \ldots, m). \tag{6.6}$$

TREFFTZ's method of determining the a_ϱ, however, is based on quite a different principle and makes further use of the differential equation. As with RITZ's method, we define the integral expression

$$J[\varphi, \psi] = \iint_B [A \, \varphi_x \psi_x + B (\varphi_x \psi_y + \varphi_y \psi_x) + C \, \varphi_y \psi_y + F \varphi \psi] \, dx \, dy \tag{6.7}$$

and put

$$J[\varphi, \varphi] = J[\varphi]. \tag{6.8}$$

As addition law we have

$$J[\varphi + \psi] = J[\varphi] + 2J[\varphi, \psi] + J[\psi]. \tag{6.9}$$

Now $J[\varphi]$ is a measure of the total deviation of the function φ from a constant (from zero in cases where $F > 0$ somewhere in B); for $J[\varphi] \geq 0$, and $J[\varphi] = 0$ if and only if φ_x, φ_y and $F\varphi^2$ vanish identically in B. TREFFTZ's method is to demand that the error ε in the linear combination (6.4)

$$\varepsilon = W - u = \overline{w} + \sum_{\sigma=1}^{m} a_\sigma w_\sigma - u \tag{6.10}$$

shall be as small as possible over the region B when assessed in terms of the measure J, i.e.

$$J[\varepsilon] = J\left[\overline{w} + \sum_{\sigma=1}^{m} a_\sigma w_\sigma - u\right] = \min. \tag{6.11}$$

In the cases when J measures the deviation from a constant which need not necessarily be zero, for example, when L is the Laplace operator (with $A = C = 1$, $B = F = 0$), this minimization leaves an additive constant free in W, which has to be determined by some other means; see the example in § 6.5.

The necessary minimum conditions

$$\frac{\partial J\left[\overline{w} + \sum_{\varrho=1}^{m} a_\varrho w_\varrho - u\right]}{\partial a_\sigma} = 2 J\left[\overline{w} + \sum_{\varrho=1}^{m} a_\varrho w_\varrho - u, w_\sigma\right] = 0 \tag{6.12}$$

$$(\sigma = 1, 2, \ldots, m)$$

yield a system of linear equations for the a_ϱ whose coefficients are double integrals over the region B. These integrals may be transformed into boundary integrals by means of GREEN's formula (3.4) of Ch. I. With $\varphi = \overline{w} + \sum_{\varrho=1}^{m} a_\varrho w_\varrho - u$, $\psi = w_\sigma$ this formula reads

$$J\left[\overline{w} + \sum_\varrho a_\varrho w_\varrho - u, w_\sigma\right] - \iint_B \left(\overline{w} + \sum_\varrho a_\varrho w_\varrho - u\right) L[w_\sigma] \, dx \, dy +$$

$$+ \int_\Gamma \left(\overline{w} + \sum_\varrho a_\varrho w_\varrho - u\right) L^*[w_\sigma] \, ds = 0 \qquad (\sigma = 1, 2, \ldots, m).$$

Here the first term vanishes by virtue of (6.12) and the second by virtue of (6.3); only the boundary integral remains, and since $u(s) = g(s)$ is known on the boundary, we have

$$\sum_{\varrho=1}^{m} a_\varrho \int_\Gamma w_\varrho L^*[w_\sigma] \, ds = \int_\Gamma (g - \overline{w}) L^*[w_\sigma] \, ds \qquad (\sigma = 1, 2, \ldots, m). \tag{6.13}$$

These are TREFFTZ's equations.

6.2. A maximum property

$\overline{w}$, chosen as a solution of the inhomogeneous differential equation (6.1), is a fixed function independent of the a_ϱ; consequently so also is $\overline{w} - u$. If the a_ϱ could be chosen ideally we would have $J\left[\sum_{\varrho=1}^{m} a_\varrho w_\varrho\right] = J[\overline{w} - u]$; we show that by TREFFTZ's method $J\left[\sum_{\varrho=1}^{m} a_\varrho w_\varrho\right]$ approximates

$J[\overline{w} - u]$ from below. From (6.10) and (6.9) we have

$$J\left[\sum_\varrho a_\varrho w_\varrho\right] = J[\varepsilon + u - \overline{w}] = J[\overline{w} - u - \varepsilon]$$
$$= J[\overline{w} - u] - 2J[\overline{w} - u, \varepsilon] + J[\varepsilon].$$

Now $\overline{w} - u = \varepsilon - \sum_\varrho a_\varrho w_\varrho$, so that, since $J[\varphi, \psi]$ is a bilinear functional,

$$J[\overline{w} - u, \varepsilon] = J[\varepsilon] - \sum_\varrho a_\varrho J[w_\varrho, \varepsilon];$$

but from (6.12) $J[w_\varrho, \varepsilon] = J[\varepsilon, w_\varrho] = 0$, so that

$$J[\overline{w} - u, \varepsilon] = J[\varepsilon].$$

Consequently

$$J\left[\sum_\varrho a_\varrho w_\varrho\right] = J[\overline{w} - u] - J[\varepsilon], \tag{6.14}$$

and since $J[\varepsilon]$ is non-negative,

$$J\left[\sum_{\varrho=1}^m a_\varrho w_\varrho\right] \leq J[\overline{w} - u]. \tag{6.15}$$

If Ritz's method is used to find an approximation for $\overline{w} - u$, the integral expression $J[\overline{w} - u]$ is approximated from above; for $\zeta = \overline{w} - u$ is defined by the homogeneous differential equation $L[\zeta] = 0$ and the known boundary values

$$\zeta(s) = \overline{w}(s) - g(s),$$

and is the solution of the variational problem

$$J[\varphi] = \text{min.}, \qquad \varphi = \overline{w}(s) - g(s) \quad \text{on } \Gamma$$

(assuming, as in § 5.1, that a unique solution of this problem exists). Thus by using a Ritz expression

$$\zeta \approx \vartheta = \vartheta_0 + \sum_{\varrho=1}^p b_\varrho \vartheta_\varrho,$$

where $\vartheta_0(s) = \overline{w}(s) - g(s)$, $\vartheta_\varrho(s) = 0$ $(1 \leq \varrho \leq p)$ on the boundary, but otherwise the ϑ_ϱ are chosen arbitrarily from the class of continuous functions with continuous partial derivatives, we can determine from the necessary minimum conditions $\dfrac{\partial J[\vartheta]}{\partial b_\varrho} = 0$ $(\varrho = 1, \ldots, p)$ an approximation ϑ such that $J[\vartheta] \geq J[\zeta]$. By using both Trefftz's and Ritz's method we can therefore bracket[1] the value of the integral $J[\overline{w} - u]$:

$$J\left[\sum_{\varrho=1}^m a_\varrho w_\varrho\right] \leq J[\overline{w} - u] \leq J[\vartheta]. \tag{6.16}$$

---

[1] Another estimate is given by J. B. Diaz: On the estimation of torsional rigidity and other physical quantities. Proc. 1st Nat. Congr. Appl. Mech. 1953, pp. 259—263, and a further estimate in connection with the finite-difference method by G. Pólya: C. R. Acad. Sci., Paris **235**, 995—997 (1952).

For many problems — the torsion problem[1] for a shaft of constant cross-section for example, — the value of this integral can be of greater interest than values of the dependent variable.

6.3. Special case of the potential equation

For the first boundary-value problem for the inhomogeneous potential equation

$$-V^2 u = r(x, y) \text{ in } B, \quad u = g(s) \text{ on } \Gamma \tag{6.17}$$

the differential expression $L[u]$ of (6.1) has

$$A \equiv C \equiv 1, \quad B \equiv F \equiv 0$$

and the boundary expression $L^*[u]$ reduces to

$$L^*[\psi] = \frac{\partial \psi}{\partial \nu}.$$

If $\overline{w}$ and $w_1, \ldots, w_m$ are solutions of the inhomogeneous and homogeneous differential equation, respectively, i.e.

$$-V^2 \overline{w} = r; \quad V^2 w_1 = V^2 w_2 = \cdots = V^2 w_m = 0,$$

the unknowns a_ϱ in the approximate solution

$$u \approx \overline{w} + \sum_{\sigma=1}^{m} a_\sigma w_\sigma$$

are determined in TREFFTZ's method from the equations

$$\sum_{\varrho=1}^{m} a_\varrho \int_{\Gamma} w_\varrho \frac{\partial w_\sigma}{\partial \nu} \, ds = \int_{\Gamma} (g - \overline{w}) \frac{\partial w_\sigma}{\partial \nu} \, ds \qquad (\sigma = 1, 2, \ldots, m). \tag{6.18}$$

In this special case the integral expressions (6.7), (6.8) read

$$J[\varphi, \psi] = \iint_{B} (\varphi_x \psi_x + \varphi_y \psi_y) \, dx \, dy = \iint_{B} \operatorname{grad} \varphi \operatorname{grad} \psi \, dx \, dy,$$

$$J[\varphi] = \iint_{B} (\varphi_x^2 + \varphi_y^2) \, dx \, dy = \iint_{B} \operatorname{grad}^2 \varphi \, dx \, dy,$$

and we see that TREFFTZ's method requires that the mean square of the gradient of the error function, grad ε, where

$$\varepsilon = \overline{w} + \sum_{\sigma=1}^{m} a_\sigma w_\sigma - u, \tag{6.19}$$

shall be as small as possible.

---

[1] The bracketing of the dependent variable is considered by C. WEBER, who makes use of auxiliary problems [Z. Angew. Math. Mech. **22**, 126—136 (1942)].

Error estimate. With the a_ϱ calculated from (6.18) inserted in (6.19) we form the error function $\varepsilon(s)$ along the boundary Γ. If ε_{min} and ε_{max} are the smallest and largest values of ε on the boundary Γ, then from the boundary-maximum theorem of § 3.1 we know that ε lies within these limits also in B:

$$\varepsilon_{min} \leqq \overline{w} + \sum_{\sigma=1}^{m} a_\sigma w_\sigma - u \leqq \varepsilon_{max}. \tag{6.20}$$

6.4. More than two independent variables

As with RITZ's method, we can also extend TREFFTZ's method to self-adjoint equations of the form (5.49) in higher dimensions:

$$L[u] = - \sum_{i,\,k=1}^{m} \frac{\partial}{\partial x_i} \left(A_{ik} \frac{\partial u}{\partial x_k} \right) + q u = r \tag{6.21}$$

with $A_{ik} = A_{ki}$, but here we must also demand that the quadratic form

$$Q = \sum_{i,\,k=1}^{m} A_{ik} z_i z_k \tag{6.22}$$

shall be positive definite for all x_r in B and that $q \geqq 0$. Apart from this, the assumptions we make concerning the region B, boundary Γ and functions A_{ik}, q, r are the same as in § 5.7. For the present we restrict ourselves to the first boundary-value problem with $u = g$ prescribed on the boundary Γ.

The derivation of the Trefftz equations follows exactly the same lines as in § 6.1. From a particular solution $\overline{w}$ of the inhomogeneous differential equation (6.21) and p linearly independent solutions $w_1, \ldots, w_p$ of the corresponding homogeneous equation we form the function

$$W = \overline{w} + \sum_{\varrho=1}^{p} a_\varrho w_\varrho, \tag{6.23}$$

which satisfies the inhomogeneous equation (6.21) for arbitrary a_ϱ.

Now we introduce the integral expression [(3.14) of Ch. I]

$$J[\varphi, \psi] = \int_B \left\{ \sum_{i,\,k=1}^{m} A_{ik} \frac{\partial \varphi}{\partial x_i} \frac{\partial \psi}{\partial x_k} + q \varphi \psi \right\} d\tau \tag{6.24}$$

and put

$$J[\varphi] = J[\varphi, \varphi]. \tag{6.25}$$

From our assumptions concerning the coefficients A_{ik} and q we have $J[\varphi] \geqq 0$, and $J[\varphi] = 0$ only if $\partial \varphi / \partial x_i \equiv 0$, i.e. $\varphi = \text{constant}$ (and only if $\varphi \equiv 0$ when $q > 0$ somewhere in B). Thus, as before, we can use $J[\varepsilon]$

as a measure of the deviation of the error function

$$\varepsilon = W - u = \overline{w} + \sum_{\varrho=1}^{p} a_\varrho w_\varrho - u \qquad (6.26)$$

from a constant (which may be zero) and demand that this measure shall be as small as possible:

$$J[\varepsilon] = J\left[\overline{w} + \sum_{\varrho=1}^{p} a_\varrho w_\varrho - u\right] = \min. \qquad (6.27)$$

This yields the necessary conditions

$$\frac{\partial J}{\partial a_\mu} = 2J\left[\overline{w} + \sum_{\varrho=1}^{p} a_\varrho w_\varrho - u, w_\mu\right] = 2J[\varepsilon, w_\mu] = 0 \quad (\mu = 1, \ldots, p), \quad (6.28)$$

which may be transformed into

$$\int_B \varepsilon L[w_\mu] d\tau - \int_\Gamma \varepsilon L^*[w_\mu] dS = 0 \qquad (\mu = 1, \ldots, p) \qquad (6.29)$$

by means of the formulae (3.13) to (3.15) of Ch. I. Since $L[w_\mu] = 0$, we obtain the system of linear equations

$$\sum_{\varrho=1}^{p} a_\varrho \int_\Gamma w_\varrho L^*[w_\mu] dS = \int_\Gamma (g - \overline{w}) L^*[w_\mu] dS \qquad (\mu = 1, \ldots, p), \qquad (6.30)$$

from which the a_ϱ are to be determined.

If $q = 0$ and the matrix A_{ik} in (6.21) is the unit matrix (δ_{ik}), we have $L[u] = -\nabla^2 u$ and $L^*[\psi] = \partial\psi/\partial\nu$.

The argument in § 6.2 leading to the result (6.15) applies equally well for the higher-dimensional measure J of (6.24), (6.25), so that here also the Trefftz approximation function is such that

$$J\left[\sum_{\varrho=1}^{p} a_\varrho w_\varrho\right] \leq J[\overline{w} - u]. \qquad (6.31)$$

6.5. Example

The torsion problem for a shaft of square cross-section leads to the boundary-value problem

$$\nabla^2 u = -1 \quad \text{for} \quad |x| \leq 1, \; |y| \leq 1,$$
$$u = 0 \quad \text{for} \quad |x| = 1 \quad \text{and for} \quad |y| = 1.$$

Here solutions $\overline{w}, w_\mu$ satisfying the inhomogeneous and homogeneous differential equation, respectively, and also satisfying the symmetries of the problem, are

$$\overline{w} = -\tfrac{1}{4}(x^2 + y^2)$$
$$w_1 = x^4 - 6x^2 y^2 + y^4 = \mathrm{re}\,(x + iy)^4,$$
$$w_2 = x^8 - 28x^6 y^2 + 70x^4 y^4 - 28x^2 y^6 + y^8 = \mathrm{re}\,(x + iy)^8.$$

The function $w =$ constant, which also satisfies the homogeneous equation, may be disregarded for the present (cf. the remarks in § 6.1 about $J[\varphi]$ measuring the deviation from a constant). The equations (6.18) for determining the constants a_μ can be written down immediately; because of the symmetries we need only take the boundary integral over a half-side, say $x = 1$, $0 \leq y \leq 1$, and the equations for $p = 2$ read

$$\sum_{\varrho=1}^{2} a_\varrho \int_0^1 \left(w_\varrho \frac{\partial w_\mu}{\partial x} \right)_{x=1} dy = - \int_0^1 \left(\overline{w} \frac{\partial w_\mu}{\partial x} \right)_{x=1} dy \qquad (\mu = 1, 2).$$

With the integrals evaluated, for example,

$$\int_0^1 \left(\overline{w} \frac{\partial w_1}{\partial x} \right)_{x=1} dy = - \int_0^1 (1 + y^2)\,(1 - 3y^2)\,dy = \frac{4}{15},$$

they read

$$\frac{48}{7} a_1 - \frac{1664}{99} a_2 + \frac{1}{3} = 0,$$

$$-208 a_1 + \frac{15232}{13} a_2 - 11 = 0,$$

from which we obtain

$$a_1 = -\frac{3619}{79856},$$

$$a_2 = \frac{429}{319424}.$$

As mentioned previously, we have a disposable additive constant, so the Trefftz approximation can be written

$$w = C - \frac{1}{4}(x^2 + y^2) - \frac{14476}{319424}(x^4 - 6x^2 y^2 + y^4) +$$

$$+ \frac{429}{319424}(x^8 - 28x^6 y^2 + 70x^4 y^4 - 28x^2 y^6 + y^8);$$

C will be determined so that the boundary values of w approximate the zero boundary value as closely as possible. On the boundary we have

$$(w)_{x=1} - C = -\gamma + f(y), \qquad \text{where} \qquad \gamma = \frac{93903}{319424}$$

and

$$319424 f(y) = -5012 y^2 + 15554 y^4 - 12012 y^6 + 429 y^8,$$

so that to determine C we need to find the variation σ in the function $f(y)$ over the interval $0 \leq y \leq 1$; for then, with $C = \gamma + \frac{1}{2}\sigma - f_{max}$, the boundary values of w will lie between $\frac{1}{2}\sigma$ and $-\frac{1}{2}\sigma$. With the aid of the graph in Fig. V/28 we find that $\sigma = \dfrac{1150}{319424}$, and hence $C = 0\cdot295\,434$.

According to the error estimate of §6.3, the absolute error in w cannot exceed $\frac{1}{2}\sigma$ anywhere in the square. In particular, for $u(0, 0)$ we have the strict limits

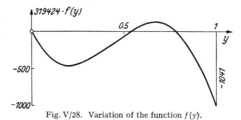

Fig. V/28. Variation of the function $f(y)$.

$$C - \frac{\sigma}{2} = 0\cdot2936 \leq u(0, 0)$$

$$\leq 0\cdot2972 = C + \frac{\sigma}{2}.$$

6.6. Generalization to the second and third boundary-value problems

Here we describe a different way[1] of deriving the Trefftz equations (6.30) which is applicable to the second and third boundary-value problems also. Consider again the differential equation (6.21) but let the boundary condition be of the more general type

$$A_1 u + A_2 L^*[u] = A_3 \qquad (6.32)$$

on the boundary Γ, where A_1, A_2, A_3 are given functions of position on Γ and $L^*[u] = A\dfrac{\partial u}{\partial \sigma}$ is as defined by (3.15) of Ch. I.

As in §6.1, we determine an approximation of the form

$$u \approx W = \overline{w} + \sum_{\varrho=1}^{p} a_\varrho w_\varrho,$$

where $\overline{w}$ and $w_1, \ldots, w_p$ are solutions of the inhomogeneous and homogeneous differential equation, respectively, as in (6.3), but here we apply GREEN's theorem in the form (3.16) of Ch. I directly to the functions $\varphi = u - \overline{w}$ and $\psi = w_\varrho$. Since $L[u - \overline{w}] = L[w_\varrho] = 0$, the volume integrals disappear and we have

$$\int_\Gamma \{(u - \overline{w}) L^*[w_\varrho] - w_\varrho L^*[u - \overline{w}]\}\,dS = 0 \qquad (\varrho = 1, \ldots, p). \qquad (6.33)$$

[1] A similar approach is considered by M. PICONE: Sur le calcul de la déformation d'un solide élastique encastré. 7th Internat. Congr. Appl. Mech., London 1948, and G. FICHERA: Risultati concernenti la risoluzioni delle equazioni funzionali lineari dovuti all' Insituto Nazionale per le applicazioni del calcolo. Mem. Accad. Naz. Lincei, Sci. Fis. Math., Ser. VIII 3, Sez. I, Fasc. 1, Rome 1950. — See also J. ALBRECHT: Eine einheitliche Herleitung der Gleichungen von TREFFTZ und GALERKIN. Z. Angew. Math. Mech. 35, 193—195 (1955).

First of all let us consider the first boundary-value problem again. Here $u - \overline{w}$ is known on the boundary, but not $L^*[u - \overline{w}]$. We replace $u - \overline{w}$ in $L^*[u - \overline{w}]$ by the approximation $\sum\limits_{\varrho=1}^{p} a_\varrho w_\varrho$ and obtain thereby a system of p linear equations for the a_ϱ:

$$\int\limits_{\Gamma} (u - \overline{w}) L^*[w_\varrho] \, dS - \sum\limits_{\tau=1}^{p} a_\tau \int\limits_{\Gamma} w_\varrho L^*[w_\tau] \, dS = 0 \qquad (\varrho = 1, \dots, p).$$

On account of the relation

$$\int\limits_{\Gamma} \{w_\varrho L^*[w_\tau] - w_\tau L^*[w_\varrho]\} \, dS = 0,$$

which follows from (3.16) of Ch. I with $\varphi = w_\varrho$, $\psi = w_\tau$, this system of linear equations is precisely the same as (6.30).

Now consider the general boundary condition (6.32), firstly with $A_1 \neq 0$ on Γ. Then on Γ we have

$$u - \overline{w} = - \frac{A_2}{A_1} L^*[u - \overline{w}] + A_3^*,$$

where

$$A_3^* = \frac{A_3}{A_1} - \overline{w} - \frac{A_2}{A_1} L^*[\overline{w}],$$

and (6.33) can be written

$$\int\limits_{\Gamma} \left(L^*[u - \overline{w}] \left\{ \frac{A_2}{A_1} L^*[w_\varrho] + w_\varrho \right\} - A_3^* L^*[w_\varrho] \right) dS = 0 \qquad (\varrho = 1, \dots, p). \quad (6.34)$$

Here we can again replace $L^*[u - \overline{w}]$ by $\sum\limits_{\tau=1}^{p} a_\tau L^*[w_\tau]$ to obtain a system of linear equations for the p unknowns a_τ. For the cases in which $A_2 \neq 0$ we could also have expressed $L^*[u - \overline{w}]$ in terms of $u - \overline{w}$; this would have led to a different system of equations for the a_τ.

If $A_1 \equiv 0$ on Γ, then $L^*[u]$ is known on Γ; by replacing $u - \overline{w}$ by $\sum\limits_{\tau=1}^{p} a_\tau w_\tau$ in the first term of the integrand in (6.33) we obtain the equations

$$\sum\limits_{\tau=1}^{p} a_\tau \int\limits_{\Gamma} w_\tau L^*[w_\varrho] \, dS = \int\limits_{\Gamma} w_\varrho L^*[u - \overline{w}] \, dS \qquad (\varrho = 1, \dots, p). \quad (6.35)$$

If Γ consists of two parts Γ_1 and Γ_2 on which $A_1 \neq 0$ and $A_1 \equiv 0$, respectively, we obtain a system of equations for the a_ϱ by splitting up the integral of (6.33) into one along Γ_1 and one along Γ_2, and using the transformation which leads to (6.34) in the former and that which leads to (6.35) in the latter. Again no volume integrals have to be evaluated; all the coefficients are expressed in terms of boundary integrals.

6.7. Miscellaneous exercises on Chapter V

1. **Transformer field.** Calculate a solution of LAPLACE's equation $\nabla^2 u = 0$ in the annular region between two concentric squares (Fig. V/29) on the boundaries of which u is constant at a different value on each, say 0 on the outer boundary and 1 on the inner. This is a problem for which the finite-difference method is particularly suitable; with other methods, series expansions, for instance, difficulty is experienced in dealing with the boundary conditions.

2. **Torsion of an I-section girder.** Let the cross-section of a girder be composed of seven squares of side a arranged as in Fig. V/30 to form an I-shaped region.

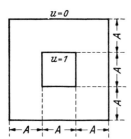

Fig. V/29. Transformer field

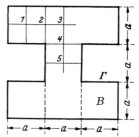

Fig. V/30. Torsion of a girder of I section

The Prandtl stress function satisfies $\nabla^2 u = -2$ inside of this region and on the boundary $u = 0$. Solve this problem approximately by the finite-difference method with

(a) the mesh width $h = a/2$ (cf. Fig. V/30),
(b) the mesh width $h = a/4$ (cf. Fig. V/35),
(c) the mesh width $h = a/2$ again, but by the Hermitian method of § 2.5.

3. Solve the torsion problem of § 6.5, i.e.

$$\nabla^2 u = -1 \quad \text{for} \quad |x| \le 1, \ |y| \le 1,$$
$$u = 0 \quad \text{for} \quad |x| = 1 \quad \text{and for} \quad |y| = 1,$$

by assuming a power series solution of the form

$$u = \sum_{j,k=0}^{\infty} a_{jk} x^j y^k.$$

Calculate in particular $u(0, 0)$.

4. For comparison, solve the problem of the last exercise again, this time using a series expansion of the form

$$u = \sum_{\substack{m,n=0 \\ m \ge n}}^{\infty} (1 - x^2)(1 - y^2) b_{mn}(x^{2m} y^{2n} + x^{2n} y^{2m}),$$

which already satisfies the boundary conditions and symmetries of the problem. Substitute this series in the differential equation and equate coefficients.

5. For the same problem as in 3. and 4. compare the ordinary finite-difference method with the Hermitian method for $h = \frac{1}{2}$.

6. Normal modes of vibration of the air in a room. The amplitude of the vibration satisfies the equation[1]

$$\frac{\partial^2 v}{\partial x^2} + \frac{\partial^2 v}{\partial y^2} + \frac{\partial^2 v}{\partial z^2} = -\varkappa v,$$

in which the constant $\varkappa$ is related to the natural frequency and represents an eigenvalue parameter to be determined. The boundary conditions are that $v = 0$ on fixed walls and $\partial v/\partial v = 0$ (v being the inward normal) on the boundaries of the room across which the air can move freely.

Consider here a rectangular room which has fixed walls with dimensions $a = 4\,\mathrm{m}$, $b = 6\,\mathrm{m}$, $c = 4\,\mathrm{m}$ as in Fig. V/31 and with two open

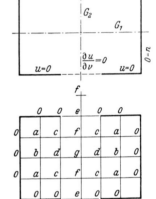

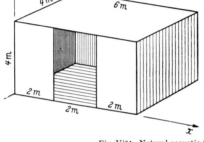

Fig. V/31. Natural acoustic frequencies of a room

doorways, one in the centre of each of the long walls and measuring $4\,\mathrm{m} \times 2\,\mathrm{m}$. If the system of co-ordinates x, y, z is chosen as in Fig. V/31, the problem can be reduced to

$$\nabla^2 u = \frac{\partial^2 u}{\partial x^2} + \frac{\partial^2 u}{\partial y^2} = -\lambda u$$

by assuming solutions of the form $v = u(x, y) \sin(n z/c)$; λ is a new eigenvalue parameter.

Determine the first few eigenvalues λ approximately by means of the ordinary finite-difference method with $h = 1\,\mathrm{m}$.

7. One end-face of a solid cylinder of radius 1 (one unit of length) and length 2 is kept at the temperature $u = 1$ and the other at the temperature $u = -1$ (Fig. V/32), while from the curved surface heat is lost to the surroundings at a rate given by the boundary condition

$$\frac{\partial u}{\partial v} = \alpha u.$$

Use the finite-difference method to calculate approximately the steady temperature distribution ($\nabla^2 u = 0$) in the cylinder for $\alpha = 1$ and $\alpha = 5$.

---

[1] See, for instance, A. G. WEBSTER and G. SZEGÖ: Partielle Differentialgleichungen der mathematischen Physik, p. 36. Leipzig and Berlin 1930.

8. Find Ritz approximations, using one term and two terms, respectively, for the solution of the problem

$$u_{xx} + u_{yy} + \frac{3}{5 - y} u_y + 1 = 0, \quad u = 0 \quad \text{for} \quad |x| = \frac{1}{2} \quad \text{and for} \quad |y| = 1$$

of Example II in § 1.5 (shear stress in a helical spring).

9. Work out Example I of § 2.8 (concerning a loaded trapezoidal plate) with the smaller mesh width $h = A/4$.

10. A homogeneous membrane stretched over a square frame Γ (length of side 1) has an elastic thread running through it along a diagonal D, which divides the square into two triangles B_1, B_2 (Fig. V/33). The amplitude distribution $u(x, y)$ for a normal mode of vibration satisfies the equations[1]

$$V^2 u + \lambda u = 0 \text{ in } B_1 \text{ and } B_2,$$
$$u = 0 \text{ on } \Gamma,$$
$$\frac{\partial u}{\partial v_1} + \frac{\partial u}{\partial v_2} + \alpha \frac{\partial^2 u}{\partial s^2} + \lambda \varrho u = 0 \text{ on } D,$$

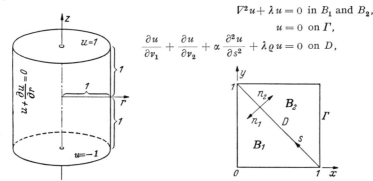

Fig. V/32. A potential problem with axial symmetry   Fig. V/33. Membrane containing an elastic thread

where s denotes the arc length along D and v_1, v_2 are the inward normals from D into B_1, B_2, respectively; α is proportional to the tension in the thread and ϱ to its density. With $\alpha = 6$, $\varrho = 1$ determine the fundamental mode approximately by means of

(a) the ordinary finite-difference method,
(b) the Ritz method or the collocation method.

11. In Example II of Ch. III, § 1.4 the steady temperature distribution in a rod in which heat was generated according to an exponential law was obtained by relaxation. A corresponding two-dimensional problem can be formulated as follows:

$$V^2 u + \frac{1 + e^u}{2} = 0 \quad \text{for} \quad |x| \leq 1, |y| \leq 1,$$

$$u = 0 \quad \text{for} \quad |x| = 1 \quad \text{and for} \quad |y| = 1.$$

(a) Find an approximate temperature distribution by the ordinary finite-difference method.

(b) How do other methods such as RITZ's, GALERKIN's, least-squares and collocation compare when applied to this non-linear problem?

---

[1] COURANT, R.: Über die Anwendung der Variationsrechnung in der Theorie der Eigenschwingungen und über neue Klassen von Funktionalgleichungen. Acta Math. **49**, 1—68 (1926), here p. 60.

6.8. Solutions

1. Because of the symmetry we need only consider that part of the annular region which lies in one half-quadrant. The ordinary finite-difference method yields the following values:

for $h = A/2$

0	0	0
0·4167	0·2083	0
1	0·4167	0
1	0·4583	0

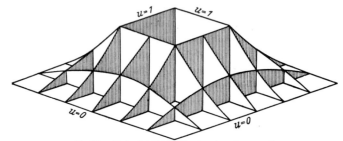

Fig. V/34. Potential distribution in a transformer coil

for $h = A/4$

	0	0	0
	0·099 86	0·049 93	0
0·303 53	0·201 70	0·099 86	0
0·472 05	0·303 53	0·147 82	0
1 0·640 57	0·392 56	0·187 89	0
1 0·697 65	0·438 27	0·211 17	0
1 0·711 75	0·451 70	0·218 51	0

Assuming that the error in these results is $O(h^2)$ we can extrapolate to obtain the new approximate values

0	0	0
0·3845	0·1995	0
1	0·3845	0
1	0·4495	0

Values obtained by the higher approximation method of §§ 2.3, 2.4 with $h = A/2$ are worse at as many points as they are better than those obtained by the ordinary finite-difference method with the same mesh width (Fig. V/34):

0	0	0
0·405 16	0·212 03	0
1	0·405 16	0
1	0·462 03	0

The more accurate formulae are based on the inclusion of higher order terms in the Taylor expansion, whose validity is restricted by the presence of the singularities at the corners of the inner square.

2. (a) Mesh width $h = a/2$. The five unknown function values $U_1, U_2, \ldots, U_5$ (numbered as in Fig. V/30) satisfy the difference equations

$$2h^2 = 4U_1 - U_2 = 4U_2 - U_1 - U_3 = 4U_3 - 2U_2 - U_4 = 4U_4 - U_3 - U_5 = 4U_5 - 2U_4,$$

from which we obtain

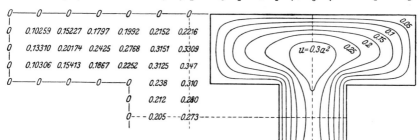

$$U_1 = \frac{251}{334}h^2 = 0.18787\,a^2,$$

$$U_2 = \frac{336}{334}h^2 = 0.25150\,a^2,$$

$$U_3 = \frac{425}{334}h^2 = 0.31811\,a^2,$$

$$U_4 = \frac{360}{334}h^2 = 0.26946\,a^2,$$

$$U_5 = \frac{347}{334}h^2 = 0.25973\,a^2.$$

Fig. V/35. Numbering of the points of the finer mesh

(b) Mesh width $h = a/4$. Let the unknown function values $U_1, U_2, \ldots$ correspond to the points marked 1, 2, ... in Fig. V/35. The calculation can easily be carried out exactly (without iteration) by putting $U_1 = \alpha$, $U_2 = \beta$, $U_3 = \gamma$ and expressing

Fig. V/36. Results of the finite-difference approximation for the shear stress in a twisted girder

the remaining U_ν in terms of these first three unknowns. This is straightforward as far as U_{17}. Then the difference equation for the point 16

$$4U_{16} - 2U_{13} - U_{17} = 2h^2$$

yields a relation between α, β and γ. This relation can be used to reduce the size of the coefficients of α, β, γ at each step in the continuation of the calculation down the web, i.e. in the successive expressions for U_{18} to U_{26}. With the further relations implied by the conditions $U_{21} = U_{25}$, $U_{22} = U_{26}$ we have a system of equations for α, β, γ, which yields

$$\alpha = 0.8207286 \times 2h^2,$$

$$\beta = 1.0647835 \times 2h^2,$$

$$\gamma = 0.8244528 \times 2h^2.$$

Fig. V/36 shows the approximate values of U_ν/a^2, together with lines of shearing stress calculated from them by interpolation.

(c) The equations for the Hermitian method read

$$24\,h^2 = 40U_1 - 8\,U_2 = 40\,U_2 - 8\,U_1 - 8\,U_3 - 2\,U_4$$
$$= 40U_3 - 16\,U_2 - 8\,U_4 = 40\,U_4 - 8\,U_3 - 8\,U_5 - 4\,U_2 = 40\,U_5 - 16\,U_4,$$

and have the solution

$$U_1 = \frac{1544\cdot7}{7582}\,a^2 = 0\cdot203\,73\,a^2, \qquad U_4 = \frac{2226}{7582}\,a^2 = 0\cdot293\,59\,a^2,$$

$$U_2 = \frac{2037}{7582}\,a^2 = 0\cdot268\,66\,a^2, \qquad U_5 = \frac{2027\cdot7}{7582}\,a^2 = 0\cdot267\,44\,a^2.$$

$$U_3 = \frac{2397\cdot3}{7582}\,a^2 = 0\cdot316\,18\,a^2,$$

Most of the (c) values lie nearer to the (b) values than to the (a), i.e. nearer to the values which are to be regarded as the more accurate; since the amount of computation involved in (c) is only a little greater than that in (a), the Hermitian method appears to be quite favourable in this case.

3. We retain only those terms in the assumed power series solution which satisfy the symmetries of the problem, and arrange them in the form

$$u = \sum_{n=0}^{\infty} \sum_{\varrho=0}^{n} b_{n,\varrho}\,x^{2n-2\varrho}\,y^{2\varrho} \quad \text{with} \quad b_{n,\varrho} = b_{n,n-\varrho}.$$

Substituting in the differential equation and equating coefficients, we obtain

for $n = 1$: $\quad 2b_{1,0} + 2b_{1,1} = -1$, which (since $b_{1,0} = b_{1,1}$) yields $b_{1,0} = -\tfrac{1}{4}$,

and for $n > 1$:

$$b_{n,\sigma+1} = -\,b_{n,\sigma}\,\frac{(2n - 2\sigma)(2n - 2\sigma - 1)}{(2\sigma + 2)(2\sigma + 1)}$$

i.e.

$$b_{n,\sigma} = (-1)^{\sigma}\binom{2n}{2\sigma}\,b_{n,0} \qquad (\sigma = 0, 1, \ldots, n).$$

Thus $b_{n,\sigma} = 0$ for n odd, and with $b_{2m,0} = c_m$ we have

$$u(x, y) = -\frac{1}{4}\,(x^2 + y^2) + \sum_{m=0}^{\infty} c_m \sum_{\sigma=0}^{2m} (-1)^{\sigma}\binom{4m}{2\sigma}\,x^{4m-2\sigma}\,y^{2\sigma},$$

which satisfies the differential equation for arbitrary c_m [the factors multiplying the c_m are the real parts of the functions $(x + iy)^{4m}$].

For $x = 1$ we must have $u(1, y) \equiv 0$, i.e. all the coefficients of y^0, y^2, y^4, ... in $u(1, y)$ must vanish. This yields an infinite system of linear equations for the c_m:

coefficient of y^0 :	$\tfrac{1}{4} = c_0 + c_1 +$	$c_2 +$	$c_3 +$	$c_4 +$	$c_5 + \cdots$
„ „ y^2 :	$-\tfrac{1}{4} =$	$6c_1 + 28c_2 +$	$66c_3 +$	$120c_4 +$	$190c_5 + \cdots$
„ „ y^4 :	$0 =$	$c_1 + 70c_2 +$	$495c_3 +$	$1820c_4 +$	$4845c_5 + \cdots$
„ „ y^6 :	$0 =$	$+ 28c_2 +$	$924c_3 +$	$8008c_4 +$	$38760c_5 + \cdots$
„ „ y^8 :	$0 =$		$c_2 + 495c_3 +$	$12870c_4 +$	$125970c_5 + \cdots$
„ „ y^{10}:	$0 =$		$66c_3 +$	$8008c_4 +$	$184756c_5 + \cdots$

which is solved approximately as in § 4.5. Table V/15 shows the results of the first few approximations up to the sixth; the approximate value of $u(0, 0)$ is given by c_0. The convergence is seen to be very good.

Table V/15. *Results of the approximations of order $\nu = 1, \ldots, 6$*

ν	c_0	c_1	c_2	c_3	c_4	c_5
1	0·25					
2	0·291 667	−0·041 667				
3	0·294 005 1	−0·044 6429	0·000 638			
4	0·294 4328	−0·045 2303	0·000 8224	−0·000 024 92		
5	0·294 5656	−0·045 4185	0·000 8911	−0·000 0396	0·000 001 5	
6	0·294 6195	−0·045 4962	0·000 9216	−0·000 0476	0·000 002 77	−0·000 000 1029

4. It is a help in forming $\nabla^2 u$ to make a short table of the Laplacians $\nabla^2 s_{m,n}$ of the symmetric functions $s_{m,n} = x^{2m} y^{2n} + x^{2n} y^{2m}$ which occur (Table V/16); the product multiplying $b_{m,n}$ can be expanded as $s_{m,n} - s_{m+1,n} - s_{m,n+1} + s_{m+1,n+1}$. Such a table can be very useful for various other purposes also.

Table V/16. *Laplacians of various symmetric functions*

φ	$\nabla^2 \varphi$
$x^2 + y^2$	4
$x^4 + y^4$ $x^2 y^2$	$12(x^2 + y^2)$ $2(x^2 + y^2)$
$x^6 + y^6$ $x^4 y^2 + x^2 y^4$	$30(x^4 + y^4)$ $2(x^4 + y^4) + 24 x^2 y^2$
$x^8 + y^8$ $x^6 y^2 + x^2 y^6$ $x^4 y^4$	$56(x^6 + y^6)$ $2(x^6 + y^6) + 30(x^4 y^2 + x^2 y^4)$ $12(x^4 y^2 + x^2 y^4)$
$x^{10} + y^{10}$ $x^8 y^2 + x^2 y^8$ $x^6 y^4 + x^4 y^6$	$90(x^8 + y^8)$ $2(x^8 + y^8) + 56(x^6 y^2 + x^2 y^6)$ $12(x^6 y^2 + x^2 y^6) + 60 x^4 y^4$

Substituting the series in the differential equation and equating coefficients, we obtain with $2b_{mm} = \beta_m$ the following infinite system of equations:

coefficient of $x^0 y^0$: $\quad -1 = -4\beta_0 + 4b_{10}$

,,   ,, $(x^2 + y^2)$: $\quad 0 = 2\beta_0 - 16b_{10} + 12b_{20} + 2\beta_1$

,,   ,, $(x^4 + y^4)$: $\quad 0 = 2b_{10} - 32b_{20} - 2\beta_1 + 30b_{30} + 2b_{21}$

,,   ,, $x^2 y^2$: $\quad 0 = 24b_{10} - 24b_{20} - 24\beta_1 + 24b_{21}$

,,   ,, $(x^6 + y^6)$: $\quad 0 = 2b_{20} - 58b_{30} - 2b_{21}$

,,   ,, $(x^4 y^2 + x^2 y^4)$: $\quad 0 = 30b_{20} + 12\beta_1 - 30b_{30} - 54b_{21}$

The dotted lines mark off the successive finite systems from which the successive approximations are calculated (as in § 4.5). The corresponding approximate values

Table V/17. *Results of the first four approximations by a series method*

Approximation	β_0	Error in β_0	b_{10}	b_{20}	β_1	b_{30}	b_{21}
1st	$\frac{1}{4} = 0\cdot25$	-15 %	—	—	—	—	—
2nd	$\frac{2}{7} = 0\cdot28571$	$-3\cdot0$ %	$0\cdot03571$	—	—	—	—
3rd	$\frac{7}{24} = 0\cdot291667$	$-1\cdot0$ %	$0\cdot041667$	0	$0\cdot041667$	—	—
4th	$\frac{164}{559} = 0\cdot293381$	$-0\cdot44$%	$0\cdot043381$	$-0\cdot000447$	$0\cdot056351$	$-0\cdot000447$	$0\cdot012522$

are given in Table V/17, which also gives the error in β_0, the approximation for $u(0, 0)$.

5. The approximations for the function values a, b, c as defined in Fig. V/37 are exhibited in Table V/18. Here the trivial increase in computational work entailed by the Hermitian method is handsomely repaid by the considerable gain in accuracy.

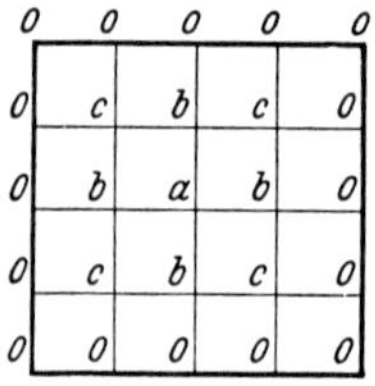

Fig. V/37. Notation for the torsion problem for a square shaft

6. For the amplitude distribution which is symmetric about both axes of symmetry G_1, G_2 we have seven unknown values $a, b, \ldots, g$ as in Fig. V/33. By expressing all unknowns in terms of a, b and $\nu = 4 - \Lambda h^2$ we obtain finally

$$\nu^7 - 14\nu^5 + 37\nu^3 - 16\nu = 0,$$

which has the roots

$$\nu = 0, \quad \pm 0\cdot7332, \quad \pm 1\cdot6699, \quad \pm 3\cdot2671.$$

The corresponding approximate eigenvalues Λ_i are given in Table V/19, together with the corresponding results for the antisymmetric modes.

Table V/18. *Comparison of the ordinary and Hermitian finite-difference methods*

Ordinary finite-difference method		Hermitian method	
Equations	Results	Equations	Results
$4a - 4b = h^2 = \frac{1}{4}$	$a = \frac{9}{32} = 0\cdot28125$	$40a - 32b - 8c = 12h^2 = 3$	$a = \frac{1272}{4312} = 0\cdot29499$
$-a + 4b - 2c = \frac{1}{4}$	$b = \frac{7}{32} = 0\cdot21875$	$-8a + 36b - 16c = 3$	$b = \frac{990}{4312} = 0\cdot22959$
$-2b + 4c = \frac{1}{4}$	$c = \frac{11}{64} = 0\cdot1719$	$-2a - 16b + 40c = 3$	$c = \frac{783}{4312} = 0\cdot18159$
	Error in a: $-4\cdot5\%$		Error in a: $+0\cdot1\%$

7. Take cylindrical co-ordinates r, z as in Fig. V/32 and use a mesh in the (r, z) plane as in Fig. V/38 with $h = \frac{2}{5}$. With the notation of Fig. V/38 the boundary condition $\partial u/\partial r = -u$ corresponds to

$$g = e\,\frac{2-h}{2+h} = \frac{2}{3}\,e, \qquad k = \frac{2}{3}\,f.$$

Table V/19. *Results for the natural acoustic frequencies of a room*

Mode symmetric about	G_1, G_2	G_1	G_2	—
and antisymmetric about	—	G_2	G_1	G_1, G_2
Approximate eigenvalues Λ	$\Lambda_1 = 0{\cdot}7329$ $\Lambda_4 = 2{\cdot}330$ $\Lambda_6 = 3{\cdot}267$ $\Lambda_9 = 4$ $\Lambda_{12} = 4{\cdot}733$ $\Lambda_{14} = 5{\cdot}670$ $\Lambda_{17} = 7{\cdot}27$	$\Lambda_2 = 1{\cdot}586$ $\Lambda_8 = 3{\cdot}586$ $\Lambda_{10} = 4{\cdot}414$ $\Lambda_{16} = 6{\cdot}414$	$\Lambda_3 = 1{\cdot}864$ $\Lambda_7 = 3{\cdot}338$ $\Lambda_{11} = 4{\cdot}662$ $\Lambda_{15} = 6{\cdot}136$	$\Lambda_5 = 3$ $\Lambda_{13} = 5$

As $r \to 0$ the differential equation

$$\nabla^2 u = u_{rr} + \frac{1}{r}\, u_r + u_{zz} = 0$$

tends to $2u_{rr} + u_{zz} = 0$. See Table V/20 for equations and results.

In the case with greater heat loss at the boundary, i.e. with $\partial u/\partial r = -5u$, we obtain temperatures which depart much more from the linear distribution $u = z$.

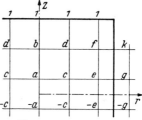

Fig. V/38. Finite-difference mesh in a cylinder

Table V/20. *Approximate temperature distribution in a cylinder*

Equations for the finite-difference method for the case $\dfrac{\partial u}{\partial r} = -u$	Results for the case $\dfrac{\partial u}{\partial r} = -u$	Results for the case $\dfrac{\partial u}{\partial r} = -5u$
$7a-\ b-\ 4c\qquad\qquad = 0$ $-a+6b\qquad -4d\qquad = 1$ $-a\qquad +10c-2d-\ 3e\ = 0$ $-\ b-\ 2c+8d\qquad -\ 3f = 2$ $-\ 9c\qquad +50e-12f = 0$ $-9d-12e+38f = 12$	$\uparrow z$ $-1\!-\!\!-\!\!-\!1\!-\!\!-\!\!-\!1-$ $\ $ $0{\cdot}5621\quad 0{\cdot}5480\quad 0{\cdot}4927$ $\ $ $0{\cdot}1760\quad 0{\cdot}1720\quad 0{\cdot}1492$ $-\mid-\!\cdot\!-\!\cdot\!-\!\cdot\!-\!\cdot\!-\mid\!\to^{\;r}$	$\uparrow z$ $-1\!-\!\!-\!\!-\!1\!-\!\!-\!\!-\!1-$ $\ $ $0{\cdot}514\quad 0{\cdot}487\quad 0{\cdot}365$ $\ $ $0{\cdot}154\quad 0{\cdot}141\quad 0{\cdot}094$ $-\mid-\!\cdot\!-\!\cdot\!-\!\cdot\!-\!\cdot\!-\mid\!\to^{\;r}$

8. First of all the differential equation must be put into the self-adjoint form (5.17):

$$(5-y)^{-3}\frac{\partial^2 u}{\partial x^2} + \frac{\partial}{\partial y}\left[(5-y)^{-3}\frac{\partial u}{\partial y}\right] + (5-y)^{-3} = 0.$$

Then from (5.20) the corresponding variational problem reads

$$J[\varphi] = \iint\limits_{B} (5-y)^{-3}\,[\varphi_x^2 + \varphi_y^2 - 2\varphi]\,dx\,dy = \min.,$$

where B is the rectangle given in the problem and the admissible functions φ are to vanish on the boundary. We assume the two-parameter expression

$$\varphi = (1-y^2)\,(1-4x^2)\,(5-y)^3\,(a+by),$$

Table V/21. *Shear stress in a helical spring by Ritz's method*

One-parameter expression	Two-parameter expression		y	Approximate values for $u(0, y)$	
Solution	Equations	Solution		1-param. exp.	2-param. exp.
$a = \dfrac{7}{7264}$ $= 0 \cdot 000\,963\,7$	$17\,025\,a - 1931\,b = \dfrac{525}{32}$ $1931\,a + 4065\,b = 0$	$a = 0 \cdot 001\,018\,5$ $b = 0 \cdot 000\,483\,8$	$-0 \cdot 5$ 0 $0 \cdot 5$	$0 \cdot 120\,25$ $0 \cdot 120\,46$ $0 \cdot 065\,86$	$0 \cdot 096\,91$ $0 \cdot 127\,32$ $0 \cdot 086\,14$

in which b is to be put equal to zero for a one-parameter approximation; the factor $(5 - y)^3$ serves to simplify the calculation of the integrals which occur. See Table V/21 for the results.

In Fig. V/39 these results are compared with those obtained by the ordinary finite-difference method with mesh width $h = \frac{1}{4}$. The finite-difference method

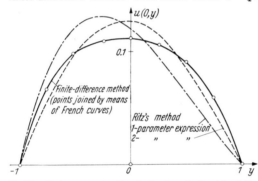

yields a curve for $u(0, y)$ which has a noticeably flat peak; further terms are needed in the Ritz approximation in order to be able to reproduce a curve of this shape.

9. With function values denoted by $a, b, \ldots$ as in Fig. V/40 and

$$\frac{p A^4}{N} = \varrho, \qquad k = \frac{9}{16} h^4 \frac{p}{N}$$

Fig. V/39. Various approximations by Ritz's method and the finite-difference method

the equations in matrix form read $A\,x = r$ with

$$A = \begin{Bmatrix} 12 & -3 & 0 & -3 & 1 & 0 & 0 & 0 \\ -3 & 11 & -3 & -3 & -3 & 1 & 1 & 0 \\ 0 & -3 & 10 & 1 & -3 & -3 & 0 & 1 \\ -3 & -3 & 1 & 13 & -3 & 0 & -3 & 1 \\ 1 & -3 & -3 & -3 & 12 & -2 & -2 & -3 \\ 0 & 1 & -3 & 0 & -2 & 8 & 1 & -3 \\ 0 & 1 & 0 & -3 & -2 & 1 & 10 & -3 \\ 0 & 0 & 2 & 2 & -6 & -6 & -6 & 12 \end{Bmatrix}, \quad x = \begin{Bmatrix} a \\ b \\ c \\ d \\ e \\ f \\ g \\ i \end{Bmatrix}, \quad r = \begin{Bmatrix} k \\ k \\ k \\ k \\ k \\ k \\ k \\ k \end{Bmatrix}.$$

Solution of these equations yields the following pivotal values, which are set out in their mesh positions:

$$c = 0 \cdot 001\,840\,\varrho, \qquad f = 0 \cdot 002\,419\,\varrho,$$
$$b = 0 \cdot 001\,665\,\varrho, \qquad e = 0 \cdot 002\,820\,\varrho, \qquad i = 0 \cdot 003\,131\,\varrho,$$
$$a = 0 \cdot 000\,711\,\varrho, \qquad d = 0 \cdot 001\,385\,\varrho, \qquad g = 0 \cdot 001\,730\,\varrho.$$

10. (a) First take $h = \frac{1}{3}$. Let the function values be denoted by a, b as in Fig. V/41, let Λ be an approximate value for λ and put $\mu = \Lambda h^2$; then corresponding

to the differential equation we have

$$2b - 4a + \mu a = 0.$$

To approximate the condition along the diagonal D we could replace $\partial u/\partial \nu_1$ at the point b by $\dfrac{0-b}{h\sqrt{2}}$, but this would give a very crude approximation; instead

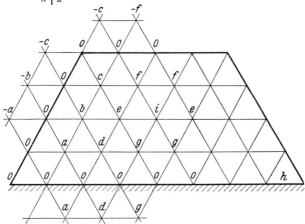

Fig. V/40. Notation for the finer mesh on the trapezoidal plate

we introduce an intermediate point c (Fig. V/41) and use for $\partial u/\partial \nu_1$ the derivative at b of the parabola through the ordinates $0, c, b$, along the line G_2. The ordinate value c can be obtained by parabolic interpolation along the line G_1; thus $c = \frac{3}{4}a$ and we have

$$\left(\frac{\partial u}{\partial \nu_1}\right)_b = \frac{-3b + 4c - 0}{h\sqrt{2}} = \frac{-3b + 3a}{\frac{1}{3}\sqrt{2}}.$$

This yields

$$2 \cdot \frac{-3b + 3a}{\frac{1}{3}\sqrt{2}} + 6 \cdot \frac{b - 2b + 0}{\left(\frac{1}{3}\sqrt{2}\right)^2} + \Lambda b = 0$$

as an approximation to the condition along D.

As characteristic equation we have

$$\begin{vmatrix} -4+\mu & 2 \\ \sqrt{2} & -3 - \sqrt{2} + \mu \end{vmatrix} = 0,$$

and from it we obtain

$$\mu = \begin{cases} 2 \cdot 513 \\ 5 \cdot 902 \end{cases} \quad \text{and} \quad \Lambda = \begin{cases} 22 \cdot 61 \\ 53 \cdot 11. \end{cases}$$

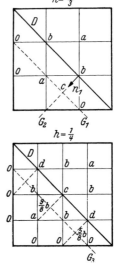

Fig. V/41. Notation and extra pivotal points

The corresponding solutions of the homogeneous equations are respectively

$$a = 1, \qquad b = 0 \cdot 744,$$

and

$$a = 1, \qquad b = -0 \cdot 951.$$

Now take $h = \frac{1}{4}$, let the function values be denoted by a, b, c, d as in Fig. V/41 and introduce intermediate values along the line G_3 (calculated by parabolic interpolation as for $h = \frac{1}{3}$). In matrix form the equations read

$$A\,x = \varkappa\,x$$

with $\Lambda h^2 = \mu = \varkappa + 4$ and

$$
x = \left\{ \begin{array}{c} a \\ b \\ c \\ d \end{array} \right\}, \qquad
A = \left\{ \begin{array}{cccc}
0 & 2 & 0 & 0 \\
1 & 0 & 1 & 1 \\
-\dfrac{\sqrt{2}}{4} & \dfrac{9}{8}\sqrt{2} & -2 - \dfrac{3}{4}\sqrt{2} & 6 \\
0 & \dfrac{5}{8}\sqrt{2} & 3 & -2 - \dfrac{3}{4}\sqrt{2}
\end{array} \right\}.
$$

Here it is convenient to determine the first few eigenvalues by means of bracketing theorems[1]; this avoids having to set up and solve an algebraic equation of the fourth degree. We obtain

$$\varkappa_1 = 2\cdot483; \quad \Lambda_1 = 24\cdot27; \quad a = 4\cdot025, \quad b = 5, \quad c = 4\cdot93, \quad d = 3\cdot465,$$
$$\varkappa_2 = 0\cdot62; \quad \Lambda_2 = 54\cdot1; \quad a = 10, \quad b = 3\cdot1, \quad c = -4\cdot9, \quad d = -3\cdot23.$$

The fundamental mode is depicted in Fig. V/42, in which the restraining influence of the thread can be discerned. The thread has this effect because $\alpha = 6$ is large compared with $\varrho = 1$; if α had been small and ϱ large it would have had the opposite effect.

(b) The Ritz variational problem reads

Fig. V/42. Fundamental mode of vibration of a membrane containing a thread

$$J = \iint\limits_{B_1 + B_2} (\varphi_x^2 + \varphi_y^2 - \lambda \varphi^2)\, dx\, dy +$$

$$+ 6 \int\limits_{D} \left\{ \left(\frac{d\varphi}{ds} \right)^2 - \lambda \varphi^2 \right\} ds = \text{extremum}$$

with φ restricted to be zero on the boundary and continuous along D. To illustrate the method we find a very rough approximation using the one-parameter expressions $\varphi = a\,xy$ in B_1, $\varphi = a(1-x)(1-y)$ in B_2. From $\partial J/\partial a = 0$ we obtain the approximation $\Lambda_1 = \dfrac{15\,(35 - 3\sqrt{2})}{17} = 27\cdot14$.

11. (a) With the function values denoted by a, α as in Fig. V/43 we obtain the approximate values of Table V/22.

We use these approximations to find rough starting values for an iterative solution of the equations for a finer mesh with $h = \frac{1}{2}$ and function value notation a, b, c as in Fig. V/43.

The equations for the ordinary finite-difference method read

$$
\left. \begin{array}{rl}
8(\;4a - 4b\;\;\;\;\;\;) &= 1 + e^a, \\
8(-a + 4b - 2c) &= 1 + e^b, \\
8(\;\;\;\;\;\; - 2b + 4c) &= 1 + e^c.
\end{array} \right\} \tag{6.36}
$$

[1] See, for instance, L. Collatz: Eigenwertaufgaben mit technischen Anwendungen, p. 291 et seq. Leipzig 1949.

Table V/22. *Rough approximations for a non-linear heat flow problem*

Ordinary finite-difference method		Hermitian method	
$h=1$	$h=\tfrac{2}{3}$	$h=1$	$h=\tfrac{2}{3}$
$8a = 1 + e^a$ $a = \begin{cases} 0\cdot2925 \\ 3\cdot204 \end{cases}$	$9\alpha = 1 + e^\alpha$ $\alpha = \begin{cases} 0\cdot2544 \\ 3\cdot382 \end{cases}$	$10a = 2 + e^a$ $a = \begin{cases} 0\cdot3406 \\ 3\cdot495 \end{cases}$	$99\alpha = 14 + 10e^\alpha$ $\alpha = \begin{cases} 0\cdot2743 \\ 3\cdot506 \end{cases}$

For the "cool" solution (with the small u values) a suitable iteration process is to solve these equations with the current approximations substituted in the right-hand side. The calculation is best performed by inverting the matrix of coefficients

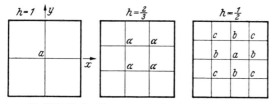

Fig. V/43. Finite-difference meshes for a non-linear problem

on the left-hand side, so that the $(n + 1)$-th approximation is given explicitly in terms of the n-th approximation by

$$\left.\begin{aligned} a_{n+1} &= 6A + 2B + 2C, \\ b_{n+1} &= 2A + 2B + 2C, \\ c_{n+1} &= A + B + 3C, \end{aligned}\right\} \quad \text{where} \quad \left\{\begin{aligned} 128A &= 1 + e^{a_n}, \\ 32B &= 1 + e^{b_n}, \\ 64C &= 1 + e^{c_n}. \end{aligned}\right.$$

Table V/23. *Iterative solution of the non-linear finite-difference equations*

n	a_n	b_n	c_n	e^{a_n}	e^{b_n}	e^{c_n}	$2A$	$2B$	$2C$
0	0·34	0·26	0·19	1·405	1·297	1·209	0·03758	0·1436	0·06903
1	0·325	0·250	0·194	1·384	1·284	1·214	0·03725	0·14275	0·06919
2	0·3237	0·2492	0·1938	1·3822	1·2830	1·2138	0·037222	0·142688	0·069181
3	0·32353	0·24909	0·19373						

Successive stages of the iteration are shown in Table V/23. We infer that to four decimals the solution of the finite-difference equations is $a = 0\cdot3235$, $b = 0\cdot2491$, $c = 0\cdot1937$.

The h^2-extrapolation of Ch. III (1.8) yields

$$u(0, 0) \approx 0\cdot334$$

as a better approximation for the temperature at the centre.

For the "hot" solution (with the much larger u values) we start from rough values approximately ten times as great: $a_0 = 3\cdot4$, $b_0 = 2\cdot6$, $c_0 = 1\cdot9$. In this case, neither the above iteration process nor the corresponding inverse process, in which the current approximation is inserted in the left-hand sides of (6.36) instead of the right-hand sides and the next approximation obtained by solving for the exponentials, is found to be suitable. We therefore employ the relaxation method of § 1.6, using the change produced by the inverse iteration just mentioned as the measure of the residual error, i.e. we define the residuals for approximations a, b, c as the differences between these values and those obtained as the next approximations a', b', c' by inserting a, b, c in the left-hand sides of (6.36), solving for the exponentials and taking natural logarithms. We make corrections to the starting values so as to reduce the magnitude of the changes $a - a', \ldots$. The decrease in the sum S of the absolute values of these changes is a rough measure of the effectiveness of each step. The calculation is shown in Table V/24, which is set out in the same way as Table III/9 for the corresponding one-dimensional problem, with the addition of a column for S. So many steps were needed because the original starting values a_0, b_0, c_0 were rather poor approximations. To accelerate the process a group relaxation was used in the last line of the table; an appropriate group correction α, β, γ can be calculated fairly accurately by virtue of the fact that the exponentials can be approximated closely by linear functions for small α, β, γ. We want to choose these corrections so that the values $a + \alpha, b + \beta, c + \gamma$ satisfy (6.36). Now $32(a + \alpha) - 32(b + \beta) = 1 + e^{a + \alpha}$, for instance, can be written $e^{a'} + 32\alpha - 32\beta = e^{a'} \cdot e^{(a - a') + \alpha}$; since $(a - a')$ and α may be expected to be of the same order of magnitude, we expand the last exponential to obtain $32\alpha - 32\beta = e^{a'}[(a - a') + \alpha]$, and similarly for the other equations. With the values from the penultimate row of the table we have

$$32\alpha - 32\beta \qquad\quad = 80 \cdot 28 (\quad 0 \cdot 015 + \alpha),$$
$$- 8\alpha + 32\beta - 16\gamma = \quad 6 \cdot 52 (- 0 \cdot 015 + \beta),$$
$$- 16\beta + 32\gamma = \quad 2 \cdot 84 (\quad 0 \cdot 006 + \gamma)$$

and hence

$$\alpha = - 0 \cdot 0163,$$
$$\beta = - 0 \cdot 0131,$$
$$\gamma = - 0 \cdot 0066.$$

(b) Application of the Ritz method does not introduce any fundamental difficulties. A corresponding variational problem can be specified:

$$J[\varphi] = \iint_Q [\varphi_x^2 + \varphi_y^2 - \varphi - e^\varphi] \, dx \, dy = \text{extremum}$$

where Q is the square $|x| \leq 1$, $|y| \leq 1$ and φ is to vanish on the boundary of Q; and also an appropriate approximation function:

$$\varphi = (1 - x^2)(1 - y^2)[a_0 + a_1(x^2 + y^2) + a_2 x^2 y^2 + \cdots].$$

Our difficulties start, however, when we come to evaluate the integrals which arise — even with a one-parameter approximation (only $a_0 \neq 0$) they are quite formidable. By and large, one may say that any method which entails integration over the square is undesirable here. A method which does not suffer from this drawback is the collocation method, but in this method one is faced with the uncertainty regarding the choice of collocation points.

Applying this method with the approximation

$$u \approx w = (1 - x^2)(1 - y^2)[a_0 + a_1(x^2 + y^2)],$$

we determine the parameters so that

$$\nabla^2 w = a_0[-4 + 2(x^2 + y^2)] + a_1[4 - 16(x^2 + y^2) + 24 x^2 y^2 + 2(x^4 + y^4)]$$

is equal to $- \tfrac{1}{2}(1 + e^w)$ at the collocation points.

Table V/24. Relaxational treatment of the non-linear finite-difference equations for the "hot" temperature distribution

	Starting values			$32(a-b)-1=$			New values			Changes			Absolute sum S
	a	b	c	$e^{a'}$	$e^{b'}$	$e^{c'}$	a'	b'	c'	$a-a'$	$b-b'$	$c-c'$	
	3·4	2·6	1·9	24·6	24·6	18·2	3·20	3·20	2·90	0·20	−0·60	−1·00	1·80
Corrections	0·1 / 3·5	2·6	1·9	3·2 / 27·8	−0·8 / 23·8	18·2	3·33	3·17	2·90	0·17	−0·57	−1·00	1·74
Corrections	3·5	2·6	−0·1 / 1·8	27·8	1·6 / 25·4	−3·2 / 15·0	3·33	3·24	2·71	0·17	−0·64	−0·91	1·72
Corrections	3·5	−0·2 / 2·4	−0·3 / 1·5	6·4 / 34·2	−1·6 / 23·8	−6·4 / 8·6	3·53	3·17	2·15	−0·03	−0·77	−0·65	1·45
Corrections	0·6 / 4·1	2·4	1·5	19·2 / 53·4	−4·8 / 19·0	8·6	3·98	2·94	2·15	0·12	−0·54	−0·65	1·31
Corrections	4·1	−0·2 / 2·2	−0·2 / 1·3	6·4 / 59·8	−3·2 / 15·8	−3·2 / 5·4	4·09	2·76	1·69	0·01	−0·56	−0·39	0·96
Corrections	0·2 / 4·3	2·2	1·3	6·4 / 66·2	−1·6 / 14·2	5·4	4·19	2·65	1·69	0·11	−0·45	−0·39	0·95
Corrections	4·3	−0·2 / 2·0	−0·2 / 1·1	6·4 / 72·6	−3·2 / 11·0	−3·2 / 2·2	4·29	2·40	0·79	0·01	−0·40	+0·31	0·72
Corrections	4·3	−0·05 / 1·95	1·1	1·6 / 74·2	−1·6 / 9·4	0·8 / 3·0	4·31	2·24	1·10	−0·01	−0·29	0	0·30
Corrections	0·1 / 4·4	1·95	1·1	3·2 / 77·4	−0·8 / 8·6	3·0	4·35	2·15	1·10	0·05	−0·20	0	0·25
Corrections	4·4	−0·1 / 1·85	−0·05 / 1·05	3·2 / 80·6	−2·4 / 6·2	3·0	4·39	1·82	1·10	0·01	+0·03	−0·05	0·09
Corrections	4·4	0·01 / 1·86	1·05	−0·32 / 80·28	+0·32 / 6·52	−0·16 / 2·84	4·385	1·875	1·044	0·015	−0·015	0·006	0·036
Corrections	α 4·3837	β 1·8469	γ 1·0434	80·1776	6·3368	2·8384	4·3842	1·8464	1·0433	−0·0005	0·0005	0·0001	0·0011

For the first approximation (only $a_0 \neq 0$) we have one collocation point to choose. One would normally choose it roughly in the middle of the half-quadrant, say at $x = \frac{1}{2}$, $y = \frac{1}{4}$ (see Fig. V/44); this yields

$$e^{\beta} = 9 \cdot 6\beta - 1 \quad \text{with} \quad \beta = \frac{45}{64} a_0,$$

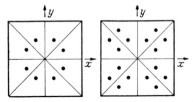

Fig. V/44. Collocation points for a non-linear heat flow problem

from which we obtain the values

$$\beta = \begin{cases} 0 \cdot 2360 \\ 3 \cdot 475 \end{cases}$$

and

$$w(0, 0) = a_0 = \begin{cases} 0 \cdot 3356 \\ 4 \cdot 94. \end{cases}$$

How strongly dependent the results are on the choice of collocation points and how uncertain they are in consequence is shown in Table V/25; for comparison, results are given for some positions which obviously would not normally be chosen.

Table V/25. *Uncertainty in the results*

Collocation point	Value for a_0
$x = 0$, $\quad y = 0$,	$0 \cdot 29$ and 3.2
$x = 1$, $\quad y = 0$,	$0 \cdot 5$ $\qquad$ —
$x = \frac{2}{3}$, $\quad y = \frac{1}{3}$,	$0 \cdot 382$ and $7 \cdot 61$
$x = 1$, $\quad y = 1$,	∞ $\qquad$ —

For the two-parameter approximation (including a_1) with $x = \frac{1}{2}$, $y = \frac{1}{4}$ and $x = \frac{3}{4}$, $y = \frac{1}{2}$ as collocation points (Fig. V/44) it is convenient to write $a_0 = 13\mu - 5\nu$, $a_1 = 16(\mu - \nu)$, so that the collocation equations become

$$36\mu - 9\nu = \frac{1}{2}\left(1 + e^{\frac{45}{8}\mu}\right),$$

$$-47\mu + 66\nu = \frac{1}{2}\left(1 + e^{\frac{21}{8}\nu}\right).$$

The first equation can be solved immediately for ν in terms of μ, and the second for μ in terms of ν, so that it is a simple matter to represent them graphically; their points of intersection are

$$\begin{cases} \mu = 0 \cdot 0438 \\ \nu = 0 \cdot 0485, \end{cases} \text{ so that } \begin{cases} a_0 = 0 \cdot 327 \\ a_1 = 0 \cdot 075, \end{cases} \text{ and } \begin{cases} \mu = 0 \cdot 64 \\ \nu = 0 \cdot 50, \end{cases} \text{ so that } \begin{cases} a_0 = \quad 5 \cdot 8 \\ a_1 = -2 \cdot 2. \end{cases}$$

Chapter VI

Integral and functional equations

§ 1. General methods for integral equations

1.1. Definitions

An equation for a function $u(x_1, x_2, \ldots, x_n)$ of n independent variables $x_1, x_2, \ldots, x_n$, in the simplest case for a function $y(x)$, is called an integral equation when it involves an integral with the function u appearing in its integrand and with at least one of the arguments of u among its variables of integration. When the equation also involves somewhere a derivative of u, it is called an integro-differential equation.

A special role is played by the linear integral equations

$$\lambda \int_a^b K(x, \xi)\, y(\xi)\, d\xi = f(x) \tag{1.1}$$

and

$$y(x) - \lambda \int_a^b K(x, \xi)\, y(\xi)\, d\xi = f(x) \tag{1.2}$$

of the first and second kinds, respectively, (written down here for the special case of one independent variable x). In these equations the "kernel" $K(x, \xi)$ is a given, say continuous function of x and ξ, $f(x)$ a given continuous function of x, and $y(x)$ the function to be determined. The interval of integration (a, b) may extend to infinity in one or both directions.

The theorems and methods described in the following for integral equations with one independent variable x are all equally valid for integral equations with several (finitely many) independent variables, in which the integration, instead of being taken over the fundamental interval $a \leq x \leq b$, is taken over the corresponding fundamental region.

If $f(x) \equiv 0$, the integral equation (1.2) is called homogeneous; otherwise inhomogeneous. For the inhomogeneous integral equation λ is to be regarded as a given parameter, while for the homogeneous equation it is essentially an eigenvalue parameter since the integral equation then presents an eigenvalue problem, in which ordinarily the object is to determine those values of λ, the *eigenvalues*, for which the integral equation possesses "non-trivial" solutions (i.e. solutions which do not vanish identically), correspondingly called the *eigenfunctions*.

The integral equation (1.2) is said to be "regular" when the interval (a, b) is finite and the kernel $K(x, \xi)$ is bounded and continuous[1];

---

[1] The theorems for regular integral equations may easily be carried over to cases in which fewer assumptions are made about the kernel. The definitions of regular and singular integral equations used here follow those in Ph. Frank and R. v. Mises: Differential- und Integralgleichungen der Mechanik und Physik, 2nd ed., Vol. 1, p. 535. Brunswick 1930.

"singular" integral equations, for which the fundamental interval is infinite or the kernel violates the conditions of boundedness and continuity, also occur frequently in applications (cf. § 3).

If the upper limit of integration in (1.1) or (1.2) is replaced by the variable x, we obtain a "Volterra integral equation"; corresponding to (1.2), for example, we have the Volterra integral equation of the second kind

$$y(x) - \lambda \int\limits_a^x K(x, \xi)\, y(\xi)\, d\xi = f(x).$$

For regular integral equations there exists a complete theory[1].

A variety of applied problems give rise also to integro-differential equations; we shall frequently consider the following type of linear integro-differential equation:

$$M[y(x)] - \lambda \int\limits_a^b K(x, \xi)\, N[y(\xi)]\, d\xi = f(x), \tag{1.3}$$

with boundary conditions

$$U_\mu[y] = \gamma_\mu \qquad (\mu = 1, 2, \ldots, m). \tag{1.4}$$

Here $M[y]$ and $N[y]$ are linear differential expressions in y as in Ch. III, (8.2), (8.3), $M[y]$ being of order $m \geq 0$, and $U_\mu[y] = \gamma_\mu$ are m given linear boundary conditions for y of the form (1.7), (1.8) of Ch. I. If M and N are both of order zero, so that no derivatives appear in equation (1.3), then we have an ordinary integral equation of the form (1.2) and the boundary conditions drop out.

Occasionally initial- and boundary-value problems in differential equations are reduced to integral equations[2], cf. Examples I, II of § 1.3, but the importance of the connection between the two kinds of formulation would appear to lie primarily in the theoretical field; for the numerical solution of a differential equation problem one generally prefers to apply the methods of Ch.s II to V directly rather than to first transform the problem into an integral equation.

---

[1] See, for instance, PH. FRANK and R. v. MISES: (see last footnote). — COURANT, R., and D. HILBERT: Methods of mathematical physics, 1st English ed., Vol. I. New York: Interscience Publishers, Inc. 1953. — HAMEL, G.: Integralgleichungen, 2nd ed. Berlin 1949. — SCHMEIDLER, W.: Integralgleichungen mit Anwendungen in Physik und Technik, Vol. I, Lineare Integralgleichungen, 611 pp. Leipzig 1950.

[2] See, for instance, W. SCHMEIDLER: (see last footnote) pp. 328—360. — COLLATZ, L.: Eigenwertaufgaben, pp. 90—109. Leipzig 1949.

1.2. Replacement of the integrals by finite sums

As for ordinary and partial differential equations, the finite-difference method is also a very generally applicable method for integral and integro-differential equations. In this method the derivatives are replaced by finite expressions as in Ch. III, § 1 and Ch. IV, § 1 and the integrals by sums derived from a quadrature formula of a suitable kind.

We explain the method with reference to a not necessarily linear integral equation of the form

$$\int_a^b \Phi\big(x, \xi, y(x), y(\xi)\big) \, d\xi = 0 \tag{1.5}$$

with finite interval (a, b).

Let us use a quadrature formula

$$\int_a^b F(x) \, dx = \sum_{\nu=1}^n A_\nu F(x_\nu) + R, \tag{1.6}$$

where the x_ν are chosen pivotal points in $\langle a, b \rangle$, the A_ν are appropriate weighting factors and R denotes the remainder term. Let Y_j be an approximation for $y(x_j)$. If we write down (1.5) for $x = x_j$ and replace the integral by a finite sum in accordance with the quadrature formula (1.6) without the remainder term, we obtain an equation for the approximate values Y_j:

$$\sum_{\nu=1}^n A_\nu \Phi(x_j, x_\nu, Y_j, Y_\nu) = 0 \qquad (j = 1, 2, \ldots, n).$$

If such an equation is written down for $j = 1, 2, \ldots, n$, we obtain a system of n (in general non-linear) equations for the same number of unknowns Y_j.

For the linear integral equation (1.2), for instance, the finite equations read

$$Y_j - \lambda h \{ \tfrac{1}{2} K_{j,0} Y_0 + \sum_{k=1}^{n-1} K_{j,k} Y_k + \tfrac{1}{2} K_{j,n} Y_n \} = f_j \qquad (j = 0, 1, \ldots, n) \tag{1.7}$$

when the quadrature formula used is the trapezium rule with the pivotal points $x_j = a + jh$ $\left(\text{pivotal interval } h = \dfrac{b-a}{n}\right)$. Here we have written $K_{j,k}$ and f_j for $K(x_j, x_k)$ and $f(x_j)$, respectively. Thus we have $n+1$ linear equations for the Y_j.

If the integral equation is homogeneous, i.e. $f(x) \equiv 0$, then the linear equations (1.7) are also homogeneous and have a non-trivial solution if and only if the determinant of the coefficients vanishes. This yields an algebraic equation for λ of (in general) the $(n+1)$-th degree. The roots $\Lambda_1, \Lambda_2, \ldots, \Lambda_{n+1}$ of this algebraic equation are used as approximate values[1] for the first $n+1$ eigenvalues λ_ν.

If SIMPSON's rule or some other more accurate quadrature formula[2] is used, one should take care that it is applied only to intervals in which the integrand is differentiable sufficiently often; for an integrand with, say, a discontinuous derivative — as is the case, for example, with certain GREEN's functions for differential equation problems — SIMPSON's rule can naturally yield worse results than the trapezium rule if applied indiscriminately.

1.3. Examples

I. Inhomogeneous linear integral equation of the second kind. In the first boundary-value problem of potential theory a solution $u(x, y)$ of the potential equation is to be determined which takes given boundary values $g(t)$ on the boundary Γ of a region B. Let the boundary curve Γ be defined in terms of the boundary parameter t by

$$x = \xi(t), \qquad y = \eta(t).$$

The solution may[3] be written in the form

$$u(x, y) = \int_\Gamma \mu(t) \frac{d\vartheta}{dt} dt, \qquad (1.8)$$

---

[1] WIELANDT, H.: Proc. Internat. Congr. Math. Amsterdam 1954, Vol. II, p. 391, and: Error bounds for eigenvalues of symmetric integral equations. Proc. Symp. Appl. Math. VI, New York-Toronto-London 1956, pp. 261—282, gives an error estimate for the approximate values $\Lambda_\nu^{(n)}$ (for the eigenvalues λ_ν) obtained by using the quadrature formula (1.6) in the case of a real, symmetric, square-integrable (cf. § 2.3) kernel $K(x, \xi)$. If we imagine the $\Lambda_\nu^{(n)}$ as the first $n+1$ members of an infinite sequence whose remaining members are all zero, and if the λ_ν are numbered suitably, then there is a number C depending only on the kernel and on the quadrature formula such that (for the interval $\langle a, b \rangle = \langle 0, 1 \rangle$) the estimate

$$\left| \frac{1}{\lambda_\nu} - \frac{1}{\Lambda_\nu^{(n)}} \right| \leqq C$$

holds for all ν; for the trapezium rule we can put

$$C = \frac{0.54}{n-1} \sup_{x, \xi} \left| \frac{\partial K(x, \xi)}{\partial x} \right|$$

and for SIMPSON's rule

$$C = \frac{0.75}{(n-1)^2} \sup_{x, \xi} \left| \frac{\partial^2 K(x, \xi)}{\partial x^2} \right|.$$

[2] GAUSS's and CHEBYSHEV's quadrature formulae are recommended by E. J. NYSTRÖM: Über die praktische Auflösung von linearen Integralgleichungen und Anwendungen auf Randwertaufgaben der Potentialtheorie. Commentationes physico-mathematicae. Acta Soc. Sci. Fenn. 4, Nr. 15, 1—52. Helsingfors 1928. Error estimates are given by L. V. KANTOROVICH and V. I. KRYLOV: Näherungsmethoden der höheren Analysis, pp. 94—155. Berlin 1956.

[3] NYSTRÖM, E. J.: (see last footnote).

where ϑ is the angle (see Fig. VI/1) defined by

$$\vartheta = \tan^{-1} \frac{\eta(t) - y}{\xi(t) - x}$$

and μ satisfies the integral equation of the second kind

$$\pi\mu(s) + \int_{\Gamma} K(s, t)\,\mu(t)\,dt = g(s) \quad \text{with} \quad K(s, t) = \frac{\partial}{\partial t}\tan^{-1}\frac{\eta(t) - \eta(s)}{\xi(t) - \xi(s)}. \quad (1.9)$$

If, for example, Γ is the ellipse

$$x = a\cos t, \quad y = b\sin t, \quad a \geq b, \quad (1.10)$$

the formula for the kernel yields

$$K(s, t) = \frac{ab}{a^2 + b^2 - (a^2 - b^2)\cos(s + t)}. \quad (1.11)$$

Let us consider the particular example of the steady temperature distribution in a long cylinder of elliptical cross-section (Fig. VI/2) at the surface of which

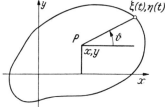

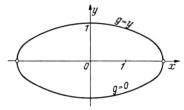

Fig. VI/1. The first boundary-value problem of potential theory

Fig. VI/2. Cross-section of an elliptical cylinder with the prescribed values for the temperature on the boundary

[the boundary of the ellipse (1.10) with $a = 2$, $b = 1$] the temperature $g(s)$ is prescribed as follows: $g(s) = 0$ for $y \leq 0$, $g(s) = y$ for $y \geq 0$ (linearly increasing temperature). Suppose that we are required to calculate approximately the temperature $u(0, 0)$ at the centre of the cross-section. According to (1.9) we have to solve the integral equation

$$\pi\mu(s) + \int_{-\pi}^{\pi} \frac{2}{5 - 3\cos(s + t)}\,\mu(t)\,dt = \begin{cases} \sin s & \text{for} & 0 \leq s \leq \pi, \\ 0 & \text{for} & -\pi \leq s \leq 0 \end{cases}$$

(there are, of course, numerous other methods for the solution of this problem, cf. Ch. V).

With the pivotal interval $h = \dfrac{\pi}{4}$ there are nine unknown pivotal values $\mu\left(\dfrac{\pi}{4}j\right)$ in the interval $-\pi \leq s \leq \pi$. We can reduce this number by using the symmetry of the problem; if μ_j denotes the corresponding approximate pivotal value, we have

$$\mu_j = \mu_{4-j}.$$

If the integral is evaluated approximately by the trapezium rule, we obtain for the approximations $\mu_2, \mu_1, \mu_0, \mu_{-1}, \mu_{-2}$ the linear equations

$$4\mu_2 + 0.25 \quad \mu_2 + 0.5617\mu_1 + 0.8 \quad \mu_0 + 1.3895\mu_{-1} + \quad\quad \mu_{-2} = \frac{4}{\pi} \approx 1.2732,$$

$$4\mu_1 + 0.2808\mu_2 + 0.65 \quad \mu_1 + 0.9756\mu_0 + 1.4 \quad \mu_{-1} + 0.6948\mu_{-2} = \frac{4}{\pi\sqrt{2}} \approx 0.9003,$$

$$4\mu_0 + 0.4 \quad \mu_2 + 0.9756\mu_1 + 1.25 \quad \mu_0 + 0.9756\mu_{-1} + 0.4 \quad \mu_{-2} = 0,$$

$$4\mu_{-1} + 0.6948\mu_2 + 1.4 \quad \mu_1 + 0.9756\mu_0 + 0.65 \quad \mu_{-1} + 0.2808\mu_{-2} = 0,$$

$$4\mu_{-2} + \quad \mu_2 + 1.3895\mu_1 + 0.8 \quad \mu_0 + 0.5617\mu_{-1} + 0.25 \quad \mu_{-2} = 0,$$

with the solution (obtainable conveniently by iteration)

$$\mu_2 = 0.3422, \quad \mu_1 = 0.2329, \quad \mu_0 = -0.0396, \quad \mu_{-1} = -0.1048, \quad \mu_{-2} = -0.1354.$$

These values yield the approximation

$$u(0,0) = \int_{-\pi}^{\pi} \frac{\mu(t)}{2\cos^2 t + \frac{1}{2}\sin^2 t}\, dt \approx \frac{\pi}{4}\left[2(\mu_2 + \mu_{-2}) + 1.6(\mu_1 + \mu_{-1}) + \mu_0\right] = 0.4547$$

for the temperature at the centre of the cross-section.

II. An eigenvalue problem. In Exercise 11 of Ch. III, § 8.11 the problem representing the flexural vibrations of a cantilever, namely

$$y^{IV} = \lambda(1+x)\,y, \quad y(0) = y'(0) = y''(1) = y'''(1) = 0, \tag{1.12}$$

was treated by RITZ's method. Here it will first be transformed into an integral equation with the aid of its GREEN's function $G(x, \xi)$. Since the problem

$$y^{IV}(x) = r(x), \quad y(0) = y'(0) = y''(1) = y'''(1) = 0$$

is equivalent to

$$y(x) = \int_0^1 G(x, \xi)\, r(\xi)\, d\xi, \quad \text{where} \quad G(x, \xi) = \begin{cases} \dfrac{x^2}{6}(3\xi - x) & \text{for} \quad x \leq \xi, \\ \dfrac{\xi^2}{6}(3x - \xi) & \text{for} \quad x \geq \xi, \end{cases}$$

(1.12) may be replaced by the homogeneous integral equation of the second kind

$$y(x) = \lambda \int_0^1 K(x, \xi)\, y(\xi)\, d\xi \quad \text{with} \quad K(x, \xi) = G(x, \xi)(1 + \xi).$$

This kernel is unsymmetric, but it could easily be made symmetric by introducing $s(x) = y(x)\sqrt{1+x}$; however, for the following calculation it is more convenient to work with the unsymmetric kernel and so avoid the square roots.

We illustrate several ways of replacing the integral by a finite sum by using different quadrature formulae.

1. Using the trapezium rule. This is the simplest method, but also the crudest, and accordingly the results are not at all accurate.

First of all let us use just three pivotal points $x = 0, \frac{1}{2}, 1$. It is expedient to record the values of the kernel neatly in an array as in Table VI/1.

If Λ, y_1, y_2 are approximate values for λ, $y(\frac{1}{2})$, $y(1)$, and if $\Lambda = 6/\varrho$, the approximate equations corresponding to the integral equation read

$$\text{(for } x = \tfrac{1}{2}) \qquad \varrho\, y_1 = \frac{1}{4}\left[0 + 2 \times \frac{3}{8}\, y_1 + \frac{5}{4}\, y_2\right],$$

$$\text{(for } x = 1) \qquad \varrho\, y_2 = \frac{1}{4}\left[0 + 2 \times \frac{15}{16}\, y_1 + 4\, y_2\right].$$

Table VI/1. *Values of* $6K(x, \xi)$

	$x=0$	$x=\frac{1}{2}$	$x=1$
$\xi = 1$	0	$\frac{5}{4}$	4
$\xi = \frac{1}{2}$	0	$\frac{3}{8}$	$\frac{15}{16}$
$\xi = 0$	0	0	0

Table VI/2. *Values of* $6K(x, \xi)$

	$x=0$	$x=\frac{1}{3}$	$x=\frac{2}{3}$	$x=1$
$\xi = 1$	0	$\frac{48}{81}$	$\frac{168}{81}$	$\frac{324}{81} = 4$
$\xi = \frac{2}{3}$	0	$\frac{25}{81}$	$\frac{80}{81}$	$\frac{140}{81}$
$\xi = \frac{1}{3}$	0	$\frac{8}{81}$	$\frac{20}{81}$	$\frac{32}{81}$
$\xi = 0$	0	0	0	0

With $4\varrho = \tau$ the determinant of this homogeneous system of equations for y_1, y_2 reads

$$\begin{vmatrix} \frac{3}{4} - \tau & \frac{5}{4} \\ \frac{15}{8} & 4 - \tau \end{vmatrix} = \tau^2 - \frac{19}{4}\, \tau + \frac{21}{32};$$

its zeros are $\tau = \frac{1}{8}\left(19 \pm \sqrt{319}\right)$, so that $\lambda \approx \Lambda = \dfrac{24}{\tau} = \begin{cases} 5 \cdot 21 \\ 168 \cdot 5. \end{cases}$

Although the method using the trapezium rule is not to be recommended on account of its low accuracy, we apply it again with the smaller pivotal interval $h = \frac{1}{3}$ and $x = 0, \frac{1}{3}, \frac{2}{3}, 1$; in any case we shall need the corresponding values of the kernel (Table VI/2) when we apply the three-eights rule in 2. below and also we introduce a matrix notation which will be needed later.

If y_j is an approximation for $y(\frac{1}{3}j)$, the corresponding homogeneous equations may be written concisely in the matrix form

$$A\, z = \sigma z, \tag{1.13}$$

where

$$z = \begin{pmatrix} y_1 \\ y_2 \\ y_3 \end{pmatrix}, \quad A = \begin{pmatrix} 8 & 25 & 24 \\ 20 & 80 & 84 \\ 32 & 140 & 162 \end{pmatrix}, \quad \sigma = 243\varrho = 243 \times \frac{6}{\Lambda}.$$

The required values of σ are the latent roots of the matrix A; we obtain

$$\sigma = \begin{cases} 242 \cdot 3 \\ 6 \cdot 78 \\ 0 \cdot 95, \end{cases} \quad \text{and hence} \quad \Lambda = \begin{cases} 6 \cdot 01 \\ 215 \\ 1530. \end{cases}$$

2. **Using more accurate formulae.** By using better quadrature formulae we can obtain more accurate values with the same amount of computation as in 1. not counting the calculation of the kernel values, which is the same in each case.

With the pivotal points $x = 0, \frac{1}{2}, 1$ we use SIMPSON's rule. Then the approximate equations corresponding to the integral equation become (with $\varrho = 6/\varLambda$)

$$(\text{for } x = \tfrac{1}{2}) \qquad \varrho\, y_1 = \frac{1}{6}\left[1 \times 0 + 4 \times \frac{3}{8}\, y_1 + \frac{5}{4}\, y_2\right],$$

$$(\text{for } x = 1) \qquad \varrho\, y_2 = \frac{1}{6}\left[1 \times 0 + 4 \times \frac{15}{16}\, y_1 + 4 y_2\right].$$

From the usual determinant condition, which here reads

$$\begin{vmatrix} \dfrac{3}{2} - \tau & \dfrac{5}{4} \\[2mm] \dfrac{15}{4} & 4 - \tau \end{vmatrix} = \tau^2 - \frac{11}{2}\,\tau + \frac{21}{16} = 0,$$

where $\tau = 6\varrho$, we obtain

$$\tau = \begin{cases} \dfrac{21}{4} \\[2mm] \dfrac{1}{4} \end{cases} \qquad \text{and} \qquad \varLambda = \frac{36}{\tau} = \begin{cases} \dfrac{48}{7} = 6\text{·}8571 \\[2mm] 144; \end{cases}$$

the value $\varLambda_1 = \dfrac{48}{7}$ in particular is a very good approximation.

With the pivotal points $x = 0, \frac{1}{3}, \frac{2}{3}, 1$ one can use the "three-eighths rule"

$$\int\limits_0^{3h} f(x)\, dx \approx \frac{3h}{8}\,[f(0) + 3f(h) + 3f(2h) + f(3h)];$$

this yields a matrix equation of the form (1.13) with

$$A = \begin{pmatrix} 8 & 25 & 16 \\ 20 & 80 & 56 \\ 32 & 140 & 108 \end{pmatrix}, \qquad \sigma = 216\varrho = 216\,\frac{6}{\varLambda},$$

and we obtain

$$\sigma = \begin{cases} 189\text{·}16 \\ 5\text{·}91 \\ 0\text{·}93, \end{cases} \qquad \varLambda = \begin{cases} 6\text{·}851 \\ 219 \\ 1390. \end{cases}$$

Naturally with so few pivotal points we cannot expect much accuracy from the approximations for the higher eigenvalues.

III. An eigenvalue problem for a function of two independent variables. Consider the integral equation

$$u(x, y) = \lambda \int\limits_{-1}^{1} \int\limits_{-1}^{1} |x\xi + y\eta|\, u(\xi, \eta)\, d\xi\, d\eta. \tag{1.14}$$

Here, instead of an ordinary integral, we have a double integral to approximate by a finite sum. For this there are again crude formulae and more accurate formulae at our disposal. For illustration of the method we confine ourselves to the simple formula

$$\int_{-1}^{1}\int_{-1}^{1}\varphi(\xi,\eta)\,d\xi\,d\eta \approx \frac{4}{n^2}\sum_{\nu,\mu=1}^{n}\varphi(a_\nu,a_\mu),\qquad(1.15)$$

in which the a_ν are a set of distinct points in the interval $(-1, 1)$.

If we first take $n=2$ and $a_1=-k$, $a_2=k$, and denote the pivotal values of u by a, b, c, d as in Fig. VI/3, we obtain the four approximate equations

$$\nu a = 2a + 2d = \nu d,$$

$$\nu b = 2b + 2c = \nu c,$$

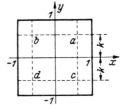

Fig. VI/3. Notation for the eigenvalue problem of Example III

where $\nu = 1/(\Lambda k^2)$, Λ being an approximation for λ.

The determinant condition reduces to $\nu^2(4-\nu)^2=0$, so if we exclude $\nu=0$ (which would give $\Lambda=\infty$) we have the double root $\nu=4$; the corresponding eigenfunction is characterized by $a=d$, $b=c$, with a, b otherwise arbitrary, and the approximate eigenvalue Λ is $1/(4k^2)$, where k is yet to be chosen. The choice $k=\frac{1}{2}$ gives $\Lambda=1$; but a better choice is $k=\frac{1}{3}\sqrt{3}$, for (1.15) then becomes CHEBYSHEV's quadrature formula; the corresponding value of Λ is 0·75.

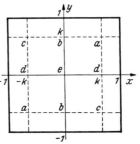

Fig. VI/4. Notation for nine pivotal points (assuming central symmetry)

For the larger number of pivotal points defined by $n=3$, $a_1=-k$, $a_2=0$, $a_3=k$ we exclude the case $1/\Lambda=0$ from the start and make use of the consequent central symmetry about the point $x=y=0$. We are left with five unknown function values a, b, c, d, e as indicated in Fig. VI/4; these must satisfy the equations

$$\varrho a = 2a + b \quad\;\;\; + d$$
$$\varrho b = \;\;\; a + b + \;\; c$$
$$\varrho c = \quad\quad\;\; b + 2c + d$$
$$\varrho d = \;\;\; a \quad\;\; + \;\; c + d$$
$$\varrho e = 0,$$

where $\varrho = \dfrac{9}{8\Lambda k^2}$. Fig. VI/5 shows the corresponding eigenvalues and eigenfunctions for the choice $k=\frac{2}{3}$; the simple calculation involved can

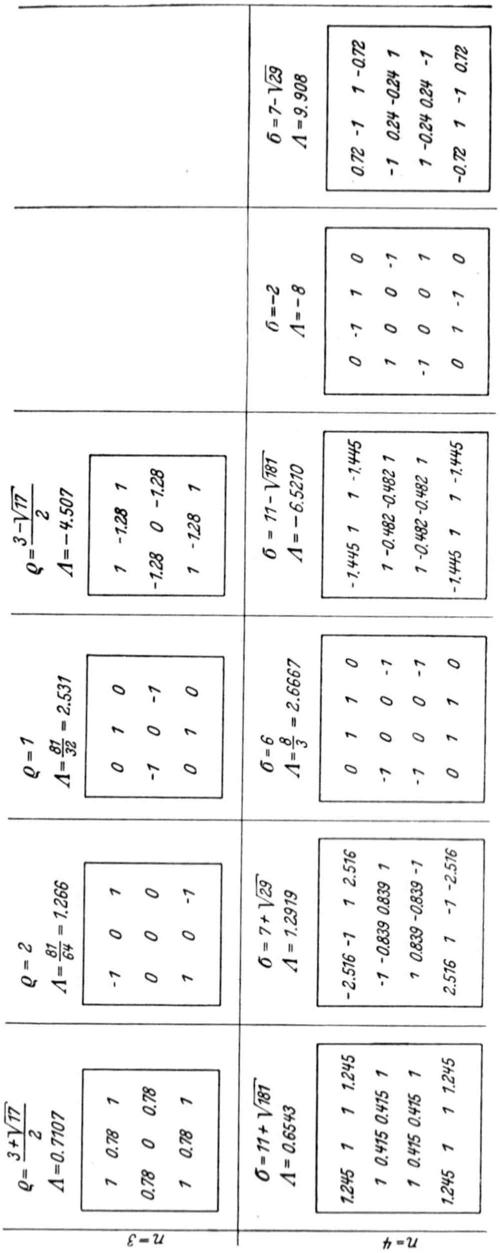

Fig. VI/5. Approximate eigenvalues and eigenfunctions for the eigenvalue problem with two independent variables (Example III)

be performed very quickly. Also shown are the approximations for $n=4$, $a_1=-3k$, $a_2=-k$, $a_3=k$, $a_4=3k$ with $k=\frac{1}{4}$ [here $\sigma=1/(\varLambda k^2)$], which are likewise obtainable with only a short calculation. (The values are still fairly crude; cf. the values calculated in § 1.5.)

If k is chosen so that the error in the quadrature formula (1.15) is of as high an order as possible, it turns out that the approximations for the lower eigenvalues are improved, but that those for the higher eigenvalues are worsened. For $n=3$ the choice $k=\frac{1}{2}\sqrt{2}$ (CHEBYSHEV's quadrature formula) yields for Λ the approximations $\frac{9}{16}(\sqrt{17}-3)=0\cdot632$, $\frac{9}{8}=1\cdot125$, $-\frac{9}{16}(\sqrt{17}+3)=-4\cdot007$; with $n=4$ and $k=\frac{1}{15}\sqrt{15}$ we obtain for Λ the values $\frac{1}{4}(\sqrt{181}-11)=0\cdot613$, $1\cdot211$, $2\cdot5$, $-6\cdot113$, $-7\cdot5$, $9\cdot289$.

As in Ch. V, § 2.8, Example II, the computational work is greatly reduced if we postulate the various symmetries and anti-symmetries which can exist.

IV. A non-linear integral equation. Consider the equation

$$\int_0^1 \frac{[y(x)+y(\xi)]^2}{1+x+\xi}\,d\xi = 1. \tag{1.16}$$

If we first make a very rough approximation to the integral by replacing the integrand by a constant, say its value at the midpoint $\xi=\frac{1}{2}$, we find that for $x=\frac{1}{2}$

$$\tfrac{1}{2}[2y(\tfrac{1}{2})]^2 \approx 1; \quad \text{so that} \quad y(\tfrac{1}{2}) \approx \pm\sqrt{\tfrac{1}{2}} \approx \pm 0\cdot707.$$

If $y(x)$ is a solution of the integral equation, then obviously so is $-y(x)$.

By using the trapezium rule for the integral we obtain the two equations

(for $x=0$) $\qquad 4y_0^2 + \tfrac{1}{2}(y_0+y_1)^2 = 2$,

(for $x=1$) $\qquad \tfrac{1}{2}(y_0+y_1)^2 + \tfrac{4}{3}y_1^2 = 2$

for the approximate values y_0, y_1 of $y(0)$, $y(1)$. From these equations it follows immediately that $3y_0^2=y_1^2$; for the remainder we restrict attention to the solution corresponding to $y_1=\sqrt{3}\,y_0$ with the positive sign. For this solution we obtain

$$y_0 = \sqrt{\frac{2}{33}(6-\sqrt{3})} = 0\cdot5086 \quad \text{and} \quad y_1 = y_0\sqrt{3} \approx 0\cdot8809.$$

By using SIMPSON's rule (with three approximate pivotal values y_0, y_1, y_2 at the points $x=0,\frac{1}{2},1$) we obtain the three non-linear equations

(for $x=0$) $\qquad 4y_0^2 + \tfrac{8}{3}(y_0+y_1)^2 + \tfrac{1}{2}(y_0+y_2)^2 = 6$,

(for $x=\tfrac{1}{2}$) $\qquad \tfrac{2}{3}(y_0+y_1)^2 + 8y_1^2 \qquad + \tfrac{2}{5}(y_1+y_2)^2 = 6$,

(for $x=1$) $\qquad \tfrac{1}{2}(y_0+y_2)^2 + \tfrac{8}{5}(y_1+y_2)^2 + \tfrac{4}{3}y_2^2 \qquad = 6$.

These are best solved iteratively by expressing the principal term (underlined) in each equation in terms of the remaining ones, i.e. using the iterative scheme defined by

$$y_0^{[\nu+1]} + y_1^{[\nu+1]} = [\tfrac{3}{8}\{6 - 4 y_0^{[\nu]\,2} - \tfrac{1}{2}(y_0^{[\nu]} + y_2^{[\nu]})^2\}]^{\frac{1}{2}}$$

for the first equation and similarly for the others, the bracketed superscripts $[\nu]$ and $[\nu+1]$ characterizing as usual the values obtained by the ν-th and $(\nu+1)$-th cycles of the iteration procedure.

The way in which the calculation is carried out is evident from Table VI/3, in which several cycles of the iteration are reproduced. The starting values

Table VI/3. *Iterative solution of the equations obtained by using* SIMPSON'S *rule*

ν	$y_0^{[\nu]}$	$y_1^{[\nu]}$	$y_2^{[\nu]}$	$y_0^{[\nu]}+y_1^{[\nu]}$	$y_0^{[\nu]}+y_2^{[\nu]}$	$y_1^{[\nu]}+y_2^{[\nu]}$	$y_0^{[\nu+1]}+y_1^{[\nu+1]}$	$y_1^{[\nu+1]}$	$y_1^{[\nu+1]}+y_2^{[\nu+1]}$
0	0·5	0·7	0·9	1·2	1·4	1·6	1·228	0·7085	1·569
1	0·519	0·709	0·861	1·228	1·380	1·569	1·221	0·7080	1·594
2	0·513	0·708	0·886	1·221	1·399	1·594	1·220	0·7062	1·576
3	0·514	0·706	0·870						
4	0·514	0·706	0·877	1·220	1·391	1·583	1·2210	0·7076	1·5825
5	0·5134	0·7076	0·8749						

y_0, y_1, y_2 are taken from the results obtained above with a smaller number of pivotal points. As shown in the table, the calculation can be shortened by estimating new starting values after a few cycles of the iteration.

1.4. The iteration method

With integral equations of the form

$$y(x) = \int_a^b G\big(x, \xi, y(x), y(\xi)\big)\, d\xi \qquad (1.17)$$

one can choose arbitrarily a function $F_0(x)$ and from it calculate a sequence of functions $F_1, F_2, \ldots$ according to the iterative formula

$$F_{n+1}(x) = \int_a^b G\big(x, \xi, F_n(x), F_n(\xi)\big)\, d\xi \qquad (n = 0, 1, 2, \ldots). \qquad (1.18)$$

For the linear integral equation (1.2) this formula[1] reads

$$F_{n+1}(x) = f(x) + \lambda \int_a^b K(x, \xi) F_n(\xi)\, d\xi \qquad (n = 0, 1, 2, \ldots), \qquad (1.19)$$

and the sequence $F_n(x)$ converges to the solution of the integral equation if the kernel $K(x, \xi)$ satisfies the conditions laid down in § 2.3 and

---

[1] For cases in which the variation of $K(x, \xi)$ with ξ is small C. WAGNER: On the numerical evaluation of Fredholm integral equations with the aid of the Liouville-Neumann series. J. Math. Phys. **30**, 232—234 (1952), gives a correction which can be used to improve the values at the current stage of the iteration.

$|\lambda| \leq |\lambda_1|$, where λ_1 is the smallest in absolute value of the eigenvalues of the homogeneous integral equation[1] corresponding to (1.2).

For eigenvalue problems [equation (1.2) with $f(x) \equiv 0$] the method parallels the iteration method in Ch. III, § 8 for eigenvalue problems in ordinary differential equations. Here the iterative procedure is defined by

$$F_{n+1}(x) = \int_a^b K(x, \xi) F_n(\xi) \, d\xi \qquad (n = 0, 1, 2, \ldots), \qquad (1.20)$$

and from the successive iterates we calculate the Schwarz constants and Schwarz quotients

$$a_k = \int_a^b F_0(x) F_k(x) \, dx, \qquad \mu_{k+1} = \frac{a_k}{a_{k+1}} \qquad (k = 0, 1, 2, \ldots). \qquad (1.21)$$

For symmetric kernels $K(x, \xi)$ we can also write

$$a_k = \int_a^b F_j(x) F_{k-j}(x) \, dx \qquad (0 \leq j \leq k; \; k = 0, 1, 2, \ldots). \qquad (1.22)$$

The μ_k then provide successive approximations to one of the eigenvalues (care should be exercised in dealing with homogeneous integral equations with unsymmetric kernels, for they need not possess real eigenvalues). Error estimates have been established[2] for problems with symmetric kernels; if, in addition, the kernels are positive definite, the estimates for the smallest eigenvalue of Ch. III (8.18) are also valid here, cf. § 2.3.

1.5. Examples of the iteration method

I. An eigenvalue problem. Consider again the integral equation (1.14), for which approximate solutions were obtained in § 1.3; these solutions suggest that

$$F_0(x, y) = |x| + |y|$$

should be a reasonable starting function.

According to (1.20) we have now to calculate

$$F_1(x, y) = \int_{-1}^1 \int_{-1}^1 |x\xi + y\eta| \, (|\xi| + |\eta|) \, d\xi \, d\eta. \qquad (1.23)$$

---

[1] A detailed and comprehensive presentation can be found in H. BÜCKNER: Die praktische Behandlung von Integralgleichungen. (Ergebnisse der Angew. Math., H. 1.) Berlin-Göttingen-Heidelberg 1952.

[2] COLLATZ, L.: Schrittweise Näherungen bei Integralgleichungen und Eigenwertproblemen. Math. Z. **46**, 692—708 (1940). — IGLISCH, R.: Bemerkungen zu einigen von Herrn COLLATZ angegebenen Eigenwertabschätzungen bei linearen Integralgleichungen. Math. Ann. **118**, 263—275 (1941).

In evaluating this integral we first suppose that $y \geq x > 0$; in this half-quadrant we have

$$\frac{1}{2} F_1 = \int_0^1 \int_0^1 (x\xi + y\eta)(\xi + \eta)\, d\xi\, d\eta + \int_{-1}^0 \int_{-\frac{x\xi}{y}}^1 (x\xi + y\eta)(-\xi + \eta)\, d\xi\, d\eta +$$

$$+ \int_0^1 \int_{-\frac{x\xi}{y}}^0 (x\xi + y\eta)(\xi - \eta)\, d\xi\, d\eta = \frac{7}{6} y + \frac{x^2}{4y} + \frac{x^3}{12 y^2}.$$

F_1 is determined in the other seven half-quadrants of the fundamental region by symmetry considerations. Then from (1.21) we have

$$a_0 = \int_{-1}^1 \int_{-1}^1 F_0^2\, dx\, dy = 4 \int_0^1 \int_0^1 (x+y)^2\, dx\, dy = 8 \times \frac{7}{12},$$

$$a_1 = \int_{-1}^1 \int_{-1}^1 F_0 F_1\, dx\, dy = 8 \int_0^1 \int_0^y 2(x+y)\left(\frac{7}{6} y + \frac{x^2}{4y} + \frac{x^3}{12 y^2}\right) dx\, dy = 8 \times \frac{29}{30},$$

and similarly from (1.22)

$$a_2 = \int_{-1}^1 \int_{-1}^1 F_1^2\, dx\, dy = 8 \times \frac{2713}{1680}.$$

Forming the corresponding Schwarz quotients, we obtain for the first eigenvalue the approximations

$$\mu_1 = \frac{a_0}{a_1} = \frac{35}{58} \approx 0\cdot 603\,45 \quad \text{and} \quad \mu_2 = \frac{a_1}{a_2} = \frac{1624}{2713} \approx 0\cdot 598\,599.$$

II. A non-linear integral equation. In order to apply the iteration method to the integral equation

$$\int_0^1 \frac{[y(x) + y(\xi)]^2}{1 + x + \xi}\, d\xi = 1, \tag{1.24}$$

we must first put it in the form (1.17). Naturally a general rule which will always be effective in producing the required form cannot be laid down for non-linear cases, but here we can derive a quadratic equation for $y(x)$ by taking it outside of the integral sign:

$$[y(x)]^2 \varphi_0(x) + 2y(x)\,\varphi_1(x) + \varphi_2(x) = 1, \tag{1.25}$$

where

$$\varphi_\nu(x) = \int_0^1 \frac{[y(\xi)]^\nu}{1 + x + \xi}\, d\xi \qquad (\nu = 0, 1, 2).$$

If we define

$$\varphi_{\nu,n} = \int_0^1 \frac{[F_n(\xi)]^\nu}{1+x+\xi}\, d\xi,$$

where $F_n(x)$ is the n-th approximation, and substitute it for $\varphi_\nu(x)$ in the solution of (1.25), we obtain as the next approximation

$$F_{n+1}(x) = \frac{1}{\varphi_{0,n}}\left[-\varphi_{1,n} + \sqrt{\varphi_{1,n}^2 + \varphi_{0,n}(1-\varphi_{2,n})}\right]. \tag{1.26}$$

As starting function we choose $F_0(x) = \text{constant} = \delta$; then with $L(x) = \ln\dfrac{2+x}{1+x}$ we obtain $F_1(x) = \dfrac{1}{\sqrt{L(x)}} - \delta$. We now determine δ so that $F_0(x)$ and $F_1(x)$ are of the same order of magnitude, say by demanding that they take the same value at the mid-point $x = \frac{1}{2}$; this yields $\delta = \dfrac{1}{2\sqrt{L(\frac{1}{2})}} \approx 0\cdot 6996$. Values of $F_1(x)$ for this value of δ are given in Table VI/4.

Table VI/4. *Specimen values of the iterates*

x	$F_0(x)$	$F_1(x)$	$F_1^*(x)$	$F_2^*(x)$
0	0·6996	0·5015	0·5	0·518 52
0·5	0·6996	0·6996	0·7	0·711 35
1	0·6996	0·8708	0·9	0·879 69

It is now convenient to round off $F_1(x)$ to $F_1^*(x) = 0\cdot 5 + 0\cdot 4x$ and apply the next iteration step to $F_1^*(x)$, for which the integrations are much simpler:

$$\varphi_{0,1}^*(x) = L(x),$$

$$\varphi_{1,1}^*(x) = 0\cdot 4 + (0\cdot 1 - 0\cdot 4x)\, L(x),$$

$$\varphi_{2,1}^*(x) = 0\cdot 16(2-x) + (0\cdot 1 - 0\cdot 4x)^2\, L(x).$$

We then calculate $F_2^*(x)$ from (1.26); specimen values are reproduced in Table VI/4.

III. An error estimate for a non-linear equation. As an example of an equation for which the existence of and bounds for the solution can be deduced directly from the general iteration theorem of Ch. I, § 5.2, consider the non-linear equation

$$\int_0^1 \sqrt{x+y(\xi)}\, d\xi = y(x)$$

(the application of the theorem does not always proceed so smoothly as for this equation). The operator T is here the integral operator defined by the left-hand side of the equation, and the equation is already of the form $Ty(x) = y(x)$ to which the theorem applies.

We must first define a norm, then determine a Lipschitz constant K as in Ch. I (5.8) for the operator T. By TAYLOR's theorem we have

$$T f_1(x) - T f_2(x) = \int_0^1 \frac{[f_1(\xi) - f_2(\xi)] \, d\xi}{2 \sqrt{x + f_1(\xi) + \vartheta(\xi) [f_2(\xi) - f_1(\xi)]}},$$

where the values of ϑ lie in the range $0 < \vartheta < 1$. Now let us introduce as norm

$$\|f(x)\| = \max_{\langle 0, 1 \rangle} |f(x)|$$

for a continuous function $f(x)$ in $\langle 0, 1 \rangle$, and as the sub-space F of Ch. I, § 5.2 let us choose the sub-space of functions $f(x)$ with $0 < m \le f(x)$ for fixed m; then

$$\|T f_1 - T f_2\| \le K \|f_1 - f_2\| \quad \text{if} \quad K = \max_{\langle 0, 1 \rangle} \int_0^1 \frac{d\xi}{2 \sqrt{x + m}} = \frac{1}{2 \sqrt{m}}.$$

Next we use one of the finite-sum methods of §§ 1.2, 1.3 to get a rough quantitative idea of a possible solution. The approximate values y_0, y_1 for $y(0)$, $y(1)$ given by the approximate equations $\frac{1}{2} (\sqrt{y_0} + \sqrt{y_1}) = y_0$, $\frac{1}{2} (\sqrt{1 + y_0} + \sqrt{1 + y_1}) = y_1$ are $y_0 = 1.17$, $y_1 = 1.53$. We therefore try $m = 1$ and choose $u_0(x) = a + bx$, where $a = 1.17$ and $b = 0.36$, as starting function. One step of the iteration procedure yields

$$u_1(x) = T u_0(x) = \frac{2}{3b} [(x + a + b)^{\frac{3}{2}} - (x + a)^{\frac{3}{2}}];$$

in particular, $u_1(0) = 1.1610$. By determining the maximum difference between u_0 and u_1 we find that $\|u_1 - u_0\| \le 0.01$, and hence, since the choice $m = 1$ gives $K = \frac{1}{2}$ and $\frac{K}{1 - K} = 1$, the "sphere" Σ of Ch. I (5.13) consists of those functions $h(x)$ with $|h - u_1| \le 0.01$; since this implies that $h(x) \ge 1$, the sphere certainly lies in F, and consequently, by the general theorem of Ch. I, § 5.2, there exists precisely one function $y(x)$ which satisfies the integral equation and the condition $y(x) \ge 1$, and this function lies in the strip $|y(x) - u_1(x)| \le 0.01$.

1.6. Error distribution principles

The error distribution principles of Ch. I, § 4, employed for the approximate solution of differential equations in Ch. III, § 4 and Ch. V, § 4, can be employed similarly for the approximate solution of integral equations.

We shall consider general integral equations in the form (1.5) and linear integral equations in the form (1.2). Let us assume for $y(x)$ the approximation

$$y(x) \approx w(x) = \sum_{j=1}^{p} a_j v_j(x). \tag{1.27}$$

Insertion of the approximation function w into the integral equation yields a residual error function $\varepsilon(x)$; for the general equation (1.5)

$$\varepsilon(x) = \varepsilon(x, a_1, \ldots, a_p) = \int_a^b \Phi(x, \xi, w(x), w(\xi)) \, d\xi \tag{1.28}$$

and for the linear equation (1.2)

$$\varepsilon(x) = \varepsilon(x, a_1, \ldots, a_p) = w(x) - \lambda \int_a^b K(x, \xi) \, w(\xi) \, d\xi - f(x).$$

Again there are several principles on which the determination of the a_j can be based; these were listed and described in detail in Ch. I, § 4.2. Here we mention only the least-squares method and collocation.

1. Least-squares method. Here we demand that

$$J = J(a_1, a_2, \ldots, a_p) = \int_a^b \varepsilon^2 \, dx = \min., \tag{1.29}$$

and obtain in the necessary minimum conditions

$$\frac{\partial J}{\partial a_\varrho} = 0 \qquad (\varrho = 1, 2, \ldots, p)$$

p (in general non-linear) equations for the a_ϱ.

For the linear integral equation (1.2) we obtain the system of linear equations

$$\sum_{k=1}^{p} b_{jk} a_k = r_j \qquad (j = 1, \ldots, p)$$

with

$$b_{jk} = \int_a^b \left[v_j(x) - \lambda \int_a^b K(x, \xi) \, v_j(\xi) \, d\xi \right] \left[v_k(x) - \lambda \int_a^b K(x, \xi) \, v_k(\xi) \, d\xi \right] dx$$

and

$$r_j = \int_a^b f(x) \left[v_j(x) - \lambda \int_a^b K(x, \xi) \, v_j(\xi) \, d\xi \right] dx.$$

2. Collocation. Here we demand that ε shall vanish at p points x_ϱ in the interval $\langle a, b \rangle$:

$$\varepsilon(x_\varrho) = 0 \qquad (\varrho = 1, 2, \ldots, p; \ a \le x_1 < x_2 < \cdots < x_p \le b),$$

which similarly yields p (in general non-linear) equations for the a_ϱ.

Example. The non-linear integral equation

$$\int_0^1 \sqrt{x + \xi} \, [y(\xi)]^2 \, d\xi = y(x)$$

has the trivial solution $y(x) \equiv 0$, but also possesses a non-trivial solution. For this non-trivial solution we assume an approximation of the form

$$y \approx w = \sqrt{a_1 + a_2\, x}\,,$$

thus departing somewhat from the usual procedure represented by (1.27); it is, however, a more obvious choice in this case. The residual error function which arises is given by

$$\varepsilon(x) = \int_0^1 \sqrt{x+\xi}\,[a_2(x+\xi) + a_1 - a_2\,x]\,d\xi - \sqrt{a_1+a_2\,x}$$

$$= \tfrac{2}{5} a_2 \left(\sqrt{x+1}^5 - \sqrt{x}^5\right) + \tfrac{2}{3}(a_1 - a_2\,x)\left(\sqrt{x+1}^3 - \sqrt{x}^3\right) - \sqrt{a_1+a_2\,x}\,.$$

If we use first of all a single-parameter approximation with $a_2 = 0$, and determine a_1 in accordance with the collocation principle from $\varepsilon = 0$ at $x = \dfrac{1}{2}$, we find that $\sqrt{a_1} = \dfrac{3\sqrt{2}}{3\sqrt{3}-1} \approx 1{\cdot}011$.

If we incorporate both parameters a_1 and a_2, and use $x = \tfrac{1}{4}$ and $x = \tfrac{3}{4}$ as collocation points, a_1 and a_2 are to be determined from the system of non-linear equations

$$\varepsilon\!\left(\tfrac{1}{4}\right) = a_2\,\frac{25\sqrt{5}-1}{80} + \left(a_1 - \frac{a_2}{4}\right)\frac{5\sqrt{5}-1}{12} - \sqrt{a_1 + \tfrac{1}{4} a_2} = 0,$$

$$\varepsilon\!\left(\tfrac{3}{4}\right) = a_2\,\frac{49\sqrt{7}-9\sqrt{3}}{80} + \left(a_1 - \frac{3a_2}{4}\right)\frac{7\sqrt{7}-3\sqrt{3}}{12} - \sqrt{a_1 + \tfrac{3}{4} a_2} = 0.$$

To solve these equations we put $a_2 = a_1\gamma$. Then after dividing them through by $\sqrt{a_1}$ we can immediately eliminate $\sqrt{a_1}$ between them and obtain an equation for γ:

$$\sqrt{\frac{4+\gamma}{4+3\gamma}} = \frac{0{\cdot}848\,362 + 0{\cdot}474\,181\gamma}{1{\cdot}110\,342 + 0{\cdot}592\,910\gamma}\,.$$

This has the solution $\gamma = 1{\cdot}8643$, which yields $a_1 = 0{\cdot}4886$, $a_2 = 0{\cdot}9108$; specimen values of the corresponding approximation for $y(x)$ are

x	0	0·5	1
$w(x)$	0·6990	0·9716	1·1830

1.7. Connection with variational problems

In order to extend the technique of Ch. III, §§ 5, 6 and Ch. V, § 5 to integral equations, we must be able to find a variational problem whose Euler equation can be identified with the given integral equation. First of all we derive the Euler equation for a general class of variational problems, namely

$$J = J[u] = \int_a^b\!\!\int_a^b F\big(x, \xi, u(x), u(\xi)\big)\,dx\,d\xi + \int_a^b G\big(\xi, u(\xi)\big)\,d\xi = \text{extr.} \quad (1.30)$$

with respect to the domain of continuous functions $u(x)$ in the interval $\langle a, b \rangle$; we shall assume here that the given functions $F(x, \xi, u_1, u_2)$ and $G(\xi, u)$ are continuous in all their arguments, x, ξ, u_1, u_2 and ξ, u, respectively. We derive here only the necessary Euler equation which must be satisfied by a solution $y(x)$ if it exists, and assume that the variational problem (1.30) does, in fact, possess such a solution with the property that

$$J[y] \leq J[u].$$

In the usual way we consider any one-parameter family of admissible functions of the form

$$u = y + \varepsilon \eta,$$

where ε is the parameter and η an admissible function. For this family of functions $J[y + \varepsilon \eta] = \Phi(\varepsilon)$ is a function of ε which must have a minimum value for $\varepsilon = 0$. A necessary condition that y shall minimize J is therefore that

$$\left(\frac{d\Phi}{d\varepsilon} \right)_{\varepsilon=0} = \left(\frac{d}{d\varepsilon} J[y + \varepsilon \eta] \right)_{\varepsilon=0} = 0$$

for any admissible function η.

Let us denote the partial derivatives of F with respect to u_1 and u_2 and of G with respect to u by subscripts thus

$$F_j = \frac{\partial F}{\partial u_j} \quad (j = 1, 2), \qquad G_u = \frac{\partial G}{\partial u}.$$

Taylor expansion of F yields

$$F\big(x, \xi, y(x) + \varepsilon \eta(x), y(\xi) + \varepsilon \eta(\xi)\big) = F\big(x, \xi, y(x), y(\xi)\big) + \\ + \varepsilon \eta(x) F_1\big(x, \xi, y(x), y(\xi)\big) + \varepsilon \eta(\xi) F_2\big(x, \xi, y(x), y(\xi)\big) + \varepsilon^2 \ldots,$$

and hence

$$\left(\frac{d}{d\varepsilon} J[y + \varepsilon \eta] \right)_{\varepsilon=0} = \int\limits_a^b \int\limits_a^b [\eta(x) F_1 + \eta(\xi) F_2] \, dx \, d\xi + \int\limits_a^b \eta(\xi) G_u \, d\xi.$$

Now in the second term in the integrand of the first integral we can interchange x and ξ, so that $\eta(x)$ will appear as a factor in both terms; further we can write x in place of ξ in the second integral. Our necessary condition can therefore be written

$$\int\limits_a^b \eta(x) \, S \, dx = 0, \tag{1.31}$$

where

$$S = \int\limits_a^b [F_1(x, \xi, y(x), y(\xi)) + F_2(\xi, x, y(\xi), y(x))] \, d\xi + G_u\big(x, y(x)\big). \tag{1.32}$$

Since (1.31) is to hold for any continuous function $\eta(x)$, S must vanish identically; thus $S=0$ is the necessary condition (Euler equation) for the variational problem (1.30).

We now consider the inverse problem of finding a suitable variational problem for a given integral equation. We try to bring the integral equation

$$g\left(x, y(x)\right) + \int_a^b h\left(x, \xi, y(x), y(\xi)\right) d\xi = 0 \qquad (1.33)$$

into the form $S=0$. First of all we can put

$$g(x, y) = G_u(x, y),$$

so that

$$G(x, y) = \int_{y_0}^y g(x, t) dt, \qquad (1.34)$$

in which y_0 can be chosen arbitrarily. Sometimes it may be expedient to multiply the integral equation (1.33) through by a suitable function $s(x)$ beforehand.

In order to bring (1.33) into the form $S=0$ we have yet to satisfy

$$h\left(x, \xi, y(x), y(\xi)\right) = F_1\left(x, \xi, y(x), y(\xi)\right) + F_2\left(\xi, x, y(\xi), y(x)\right). \quad (1.35)$$

The question of the existence of a function F related to a given function h by (1.35) will not be pursued further here; we content ourselves with listing in Table VI/5 some typical possibilities obtained by calculating h from some simple forms for F.

Table VI/5. *Some corresponding integrand functions*

$F(x, \xi, y(x), y(\xi))$	$h(x, \xi, y(x), y(\xi))$
$\frac{1}{2}K(x, \xi)\, y(x)\, y(\xi)$	$\frac{1}{2}[K(x, \xi) + K(\xi, x)]\, y(\xi)$
$K(x, \xi)\, y(x)$	$K(x, \xi)$
$K(x, \xi)\int_0^{y(x)} \varphi(u)\, du$	$K(x, \xi)\, \varphi(y(x))$
$K(x, \xi)\, \varphi(y(x)\, y(\xi))$	$[K(x, \xi) + K(\xi, x)]\, \varphi'(y(x)\, y(\xi))\, y(\xi)$
$K(x, \xi)\, \varphi(y(x) + y(\xi))$	$[K(x, \xi) + K(\xi, x)]\, \varphi'(y(x) + y(\xi))$

By referring to this table, which can easily be extended, and using (1.34) one can find a corresponding variational problem (1.30) for many cases of integral equations of the form (1.33); for example, for the linear integral equation

$$y(x) - \lambda \int_a^b K(x, \xi)\, y(\xi)\, d\xi = f(x) \qquad (1.36)$$

with a symmetric kernel $K(x, \xi) = K(\xi, x)$ we obtain the variational problem

$$
\left.
\begin{aligned}
J[u] &= \tfrac{1}{2} \lambda \int_a^b \int_a^b K(x, \xi)\, u(x)\, u(\xi)\, dx\, d\xi + \\
&+ \int_a^b \left(f(\xi)\, u(\xi) - \tfrac{1}{2}[u(\xi)]^2 \right) d\xi = \text{extremum}.
\end{aligned}
\right\}
\tag{1.37}
$$

Examples. I. A linear inhomogeneous integral equation arising in a measurement problem.

Suppose that when an attempt is made to measure an intensity distribution $y(x)$, say along the section $-1 \leq x \leq 1$ of the x axis, the reading $f(x)$ differs from the true value $y(x)$ because of the influence of neighbouring elements. Let the influence at the point x of an element $d\xi$ at the point ξ be $K(x, \xi)\, y(\xi)\, d\xi$; normally $K(x, \xi)$ will depend only on the distance $|x - \xi|$, and here we will assume that

$$
K(x, \xi) = \begin{cases} 1 - (x - \xi)^2 & \text{for } |x - \xi| \leq 1 \\ 0 & \text{for } |x - \xi| \geq 1. \end{cases}
\tag{1.38}
$$

Then the reading taken at the point x will be

$$
y(x) + \int_{-1}^{1} K(x, \xi)\, y(\xi)\, d\xi = f(x).
$$

Thus we have an integral equation for the required true distribution $y(x)$ in terms of the observed (measured) distribution $f(x)$. For this example we will take

$$
f(x) = \frac{1}{1 + x^2}.
$$

(1.36), (1.37) yield immediately the corresponding variational problem

$$
J[u] = \frac{1}{2} \int_{-1}^{1} \int_{-1}^{1} K(x, \xi)\, u(x)\, u(\xi)\, dx\, d\xi + \frac{1}{2} \int_{-1}^{1} u^2(x)\, dx - \int_{-1}^{1} \frac{u(x)}{1 + x^2}\, dx = \text{extr.}
$$

Taking account of the symmetry, we assume the two-term approximation

$$
u = a + b x^2,
\tag{1.39}
$$

and instead of first forming J, we form the expressions $\partial J / \partial a$, $\partial J / \partial b$ directly; differentiation of the double integral with respect to a, for instance, yields

$$
\int_{-1}^{1} \int_{-1}^{1} K(x, \xi)\, (2a + b(x^2 + \xi^2))\, dx\, d\xi
$$

$$
= 2 \int_{0}^{1} \int_{\xi-1}^{1} (2a + b(x^2 + \xi^2))\, (1 - (x - \xi)^2)\, dx\, d\xi = \frac{13}{3} a + \frac{6}{5} b
$$

We obtain for a and b the equations

$$
\frac{\partial J}{\partial a} = \frac{25}{6} a + \frac{19}{15} b - \frac{\pi}{2} = 0,
$$

$$
\frac{\partial J}{\partial b} = \frac{19}{15} a + \frac{73}{60} b - \frac{4 - \pi}{2} = 0,
$$

with the solution

$$
a = 0{\cdot}4428,
$$

$$
b = -0{\cdot}2164.
$$

This parabolic approximation is quite crude, however; in Fig. VI/6 it is compared with the results obtained by the summation methods of § 1.2 with $h = \frac{1}{2}$ and $h = \frac{1}{3}$; these results show that the curve of the solution $y(x)$ is bell-shaped, and it cannot therefore be followed closely by a parabola, which has no points of inflexion.

II. A non-linear integral equation (already treated by collocation in § 1.6).

There are various ways of setting up a corresponding variational problem for the equation

$$\int_0^1 \sqrt{x + \xi}\, y^2(\xi)\, d\xi = y(x);$$

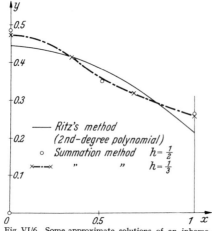

Fig. VI/6. Some approximate solutions of an inhomogeneous integral equation obtained by Rɪᴛᴢ's method and summation methods

one could, for example, multiply the equation through by $y(x)$, thereby bringing the integrand into the form given in Table VI/5, right-hand column, last row but one; we see from the left-hand column that this is generated by (1.35) from the function

$$F = \tfrac{1}{4}\sqrt{x + \xi}\, y^2(x)\, y^2(\xi).$$

Here, however, it is somewhat simpler to write the integral equation as

$$\int_0^1 \sqrt{x + \xi}\, z(\xi)\, d\xi = \sqrt{z(x)}$$

by introducing $z(x) = y^2(x)$; then according to (1.33), (1.34), (1.36), (1.37) a corresponding variational expression reads

$$J[u] = \int_0^1\!\!\int_0^1 \tfrac{1}{2}\sqrt{x + \xi}\, u(x)\, u(\xi)\, dx\, d\xi - \tfrac{2}{3}\int_0^1 u^{\frac{3}{2}}(x)\, dx.$$

For the simple linear approximation $u(x) = a + bx$ the necessary minimum conditions read

$$\frac{\partial J}{\partial a} = \int_0^1\!\!\int_0^1 \tfrac{1}{2}\sqrt{x + \xi}\{2a + b(x + \xi)\}\, dx\, d\xi - \int_0^1 \sqrt{a + bx}\, dx = 0,$$

$$\frac{\partial J}{\partial b} = \int_0^1\!\!\int_0^1 \tfrac{1}{2}\sqrt{x + \xi}\{a(x + \xi) + 2b\xi x\}\, dx\, d\xi - \int_0^1 x\sqrt{a + bx}\, dx = 0.$$

On evaluating the integrals we obtain the two non-linear equations

$$\alpha a + \beta b = \frac{2}{3b}\left(\sqrt{a+b}^3 - \sqrt{a}^3\right)$$

$$\beta a + \gamma b = \frac{2}{5b^2}\left(\sqrt{a+b}^5 - \sqrt{a}^5\right) - \frac{2a}{3b^2}\left(\sqrt{a+b}^3 - \sqrt{a}^3\right),$$

where

$$\alpha = \frac{8}{15}\left(2\sqrt{2} - 1\right) \approx 0.975\,15,$$

$$\beta = \frac{4}{35}\left(4\sqrt{2} - 1\right) \approx 0.532\,21,$$

$$\gamma = \frac{16}{135}\left(\sqrt{2} + 1\right) \approx 0.286\,13,$$

and these can easily be reduced to a single equation for the ratio $c = b/a$:

$$\frac{\alpha + \beta c}{\beta + \gamma c} = \frac{\frac{2}{3}c\left(\sqrt{1+c}^3 - 1\right)}{\frac{2}{5}\left(\sqrt{1+c}^5 - 1\right) - \frac{2}{3}\left(\sqrt{1+c}^3 - 1\right)};$$

this has the solution $c = 1.8861$, yielding $a = 0.4861$, $b = 0.9168$. Specimen values of the corresponding approximate solution u are

x	0	0.5	1
u	0.6972	0.9719	1.1844

which agree well with the results of § 1.6.

1.8. Integro-differential equations and variational problems

Here we can be brief because of the similarity with the last section and with Ch. III, § 6.1.

Suppose that the expression

$$
\begin{aligned}
J[u] = &\int_a^b\!\!\int_a^b F\left(x, \xi, u(x), u(\xi), u'(x), u'(\xi), \ldots, u^{(n)}(x), u^{(n)}(\xi)\right) dx\,d\xi + \\
&+ \int_a^b G\left(\xi, u(\xi), u'(\xi), \ldots, u^{(n)}(\xi)\right) d\xi + \sum_{\nu=0}^{n-1}\left(a_\nu u^{(\nu)}(a) + \right. \\
&\left. + b_\nu u^{(\nu)}(b)\right) + \sum_{\mu,\nu=0}^{n-1}\left(a_{\mu\nu}u^{(\mu)}(a)u^{(\nu)}(a) + b_{\mu\nu}u^{(\mu)}(b)u^{(\nu)}(b)\right)
\end{aligned}
\tag{1.40}
$$

is to be minimized. F and G are given functions continuous in each of their arguments and with as many continuous partial derivatives as are needed in the following; and $a_\nu, b_\nu, a_{\mu\nu}, b_{\mu\nu}$ are given constants.

Here we can prescribe certain boundary conditions

$$U_\mu[u(a), u(b), u'(a), u'(b), \ldots, u^{(n-1)}(a), u^{(n-1)}(b)] = \gamma_\mu \atop (\mu = 1, 2, \ldots, k;\ k \leq 2n),} \tag{1.41}$$

which we will assume to be linear in the boundary values specified. Let there exist a solution $y(x)$ of this variational problem. Then the usual procedure of considering a one-parameter family of admissible functions $u = y + \varepsilon\eta$ leads to the necessary condition

$$\left.\begin{aligned} 0 &= \left(\frac{d}{d\varepsilon} J[y + \varepsilon\eta]\right)_{\varepsilon=0} \\ &= \int_a^b\int_a^b \sum_{v=0}^n \left(F_{1v}\,\eta^{(v)}(x) + F_{2v}\,\eta^{(v)}(\xi)\right) dx\, d\xi + \int_a^b \sum_{v=0}^n G_v\eta^{(v)}(\xi)\, d\xi + \Psi \end{aligned}\right\} \tag{1.42}$$

for any function η satisfying the homogeneous boundary conditions

$$U_\mu[\eta] = 0 \qquad (\mu = 1, 2, \ldots, k), \tag{1.43}$$

where in (1.42) we have introduced the notation

$$F_{1v} = \frac{\partial F}{\partial u^{(v)}(x)}, \qquad F_{2v} = \frac{\partial F}{\partial u^{(v)}(\xi)}, \qquad G_v = \frac{\partial G}{\partial u^{(v)}}, \tag{1.44}$$

$$\left.\begin{aligned} \Psi &= \sum_{v=0}^{n-1}[a_v\eta^{(v)}(a) + b_v\eta^{(v)}(b)] + \sum_{\mu,v=0}^{n-1}[a_{\mu v}(y^{(\mu)}(a)\,\eta^{(v)}(a) + \\ &\quad + y^{(v)}(a)\,\eta^{(\mu)}(a)) + b_{\mu v}(y^{(\mu)}(b)\,\eta^{(v)}(b) + y^{(v)}(b)\,\eta^{(\mu)}(b))]. \end{aligned}\right\} \tag{1.45}$$

Further, let a tilde embellishing a function symbol signify that x and ξ have been interchanged at every place where they occur in that function, so that, for example,

$$\tilde{F}_{1v} = F_{1v}(\xi, x, u(\xi), u(x), u'(\xi), u'(x), \ldots, u^{(n)}(\xi), u^{(n)}(x)),$$

and put

$$F_{1v} + \tilde{F}_{2v} = \Phi_v.$$

Then by interchanging x and ξ in the single integral, and also in the F_2 terms of the double integral, we can write (1.42) in the form

$$0 = \int_a^b\int_a^b \sum_{v=0}^n \eta^{(v)}(x)\,\Phi_v\,dx\,d\xi + \int_a^b \sum_{v=0}^n \eta^{(v)}(x)\,G_v\,dx + \Psi. \tag{1.46}$$

We can now transform it further by integration by parts. We have, for example,

$$\int_a^b\int_a^b \eta'(x)\,\Phi_1\,dx\,d\xi = \left[\eta(x)\int_a^b \Phi_1\,d\xi\right]_a^b - \int_a^b \eta(x)\left(\int_a^b \frac{d}{dx}\,\Phi_1\,d\xi\right)dx;$$

and similarly transforming all other terms (integrating the term in $\eta^{(\nu)}$ by parts ν times), we obtain

$$
\begin{aligned}
)=\int_a^b \eta(x) &\left\{ \int_a^b \left(\Phi_0 - \frac{d}{dx}\Phi_1 + \frac{d^2}{dx^2}\Phi_2 - \cdots \right) d\xi + \left(G_0 - \frac{d}{dx}G_1 + \frac{d^2}{dx^2}G_2 - \cdots \right) \right\} dx + \\
&+ \left[\eta(x) \left\{ \int_a^b \left(\Phi_1 - \frac{d}{dx}\Phi_2 + \frac{d^2}{dx^2}\Phi_3 - \cdots \right) d\xi + \left(G_1 - \frac{d}{dx}G_2 + \frac{d^2}{dx^2}G_3 - \cdots \right) \right\} \right]_a^b + \\
&+ \left[\eta'(x) \left\{ \int_a^b \left(\Phi_2 - \frac{d}{dx}\Phi_3 + \frac{d^2}{dx^2}\Phi_4 - \cdots \right) d\xi + \left(G_2 - \frac{d}{dx}G_3 + \frac{d^2}{dx^2}G_4 - \cdots \right) \right\} \right]_a^b + \\
&+ \left[\eta''(x) \left\{ \cdots \right\} \right]_a^b + \cdots + \left[\eta^{(n-1)}(x) \left\{ \int_a^b \Phi_n \, d\xi + G_n \right\} \right]_a^b + \Psi .
\end{aligned}
\qquad (1.47)
$$

On account of the arbitrariness of η, the factor multiplying $\eta(x)$ in the integral must vanish; this yields the integro-differential equation

$$
\int_a^b \left(\sum_{\nu=0}^n (-1)^\nu \frac{d^\nu}{dx^\nu}\, \Phi_\nu \right) d\xi + \sum_{\nu=0}^n (-1)^\nu \frac{d^\nu}{dx^\nu} \left(\frac{\partial G}{\partial u^{(\nu)}(x)} \right) = 0; \qquad (1.48)
$$

The integral with respect to x therefore drops out of (1.47), leaving a sum which is linear and homogeneous in the boundary values $\eta(a)$, $\eta'(a), \ldots, \eta^{(n-1)}(a), \eta(b), \eta'(b), \ldots, \eta^{(n-1)}(b)$. In general, k of these $2n$ boundary values can be expressed in terms of the remaining ones by means of the boundary conditions (1.43). These $2n - k$ remaining boundary values may be called "free boundary values", and will be denoted in any order by $\eta_1, \eta_2, \ldots, \eta_{2n-k}$. If all boundary values of η appearing in (1.47) are now expressed in terms of the free boundary values, the sum remaining in (1.47) assumes the form

$$
\sum_{\nu=1}^{2n-k} \eta_\nu W_\nu [y] = 0. \qquad (1.49)
$$

Since the η_ν may be chosen arbitrarily, we must have

$$
W_\nu [y] = 0 \qquad (\nu = 1, 2, \ldots, 2n - k). \qquad (1.50)
$$

These equations constitute $2n - k$ further boundary conditions to be satisfied necessarily by the solution y of the variational problem.

Thus y must satisfy the integro-differential equation (1.48), which in general is of the $2n$-th order, together with the $2n$ boundary conditions (1.41), (1.50). In general this determines y uniquely.

The following two possibilities should be noted: that integrals over the fundamental interval can occur in the boundary conditions (1.50) derived from (1.47), and that boundary values of y can appear in the integro-differential equation (1.48) if integration by parts is applied to the integral.

Example. The vertical oscillations of a suspension bridge satisfy the integro-differential equation[1]

$$(EIy'')'' - Hy'' + k \int_{-l}^{l} y(\xi)\,d\xi = m\omega^2 y$$

together with certain boundary conditions, which, if the bridge is assumed to be pinned at each end, read

$$y(\pm l) = y''(\pm l) = 0.$$

Here, EI is the flexural rigidity of the stiffening girder, H the horizontal component of the tension in the cable, m the mass per unit length of the oscillating parts, ω the frequency of the oscillations, y the deflection due to the oscillations, $2l$ the span and k a certain constant.

For this problem a corresponding variational problem reads[2]

$$J[u] = \int_{-l}^{l}(EIu''^2 + Hu'^2 - m\omega^2 u^2)\,dx + k\left(\int_{-l}^{l} u(\xi)\,d\xi\right)^2 = \text{extr.}$$

with respect to the domain of admissible functions u satisfying the essential boundary conditions $u(\pm 1) = 0$.

1.9. Series solutions

Integral equations frequently arise whose form suits them for treatment by a series expansion method of one kind or another — a power series or trigonometric series method, for example —, the appropriate series to use depending on the particular form of the integral equation. For instance, trigonometric series are very effective for linear integral equations of the form (1.2) with "convolution-type kernels", i.e. integral equations of the form

$$y(x) = f(x) + \lambda \int_{a}^{b} K(x - \xi)\,y(\xi)\,d\xi, \tag{1.51}$$

in which the kernel is a function only of the difference $x - \xi$, provided also that K is periodic with period $b - a$:

$$K(x - \xi) = K(x - \xi + b - a).$$

We imagine the functions $K(x)$, $f(x)$ and $y(x)$ to be expanded as trigonometric series written in complex form:

$$K(x) = \sum_{\nu=-\infty}^{\infty} k_\nu e^{2\pi i \frac{\nu x}{b-a}}, \quad f(x) = \sum_{\nu=-\infty}^{\infty} f_\nu e^{2\pi i \frac{\nu x}{b-a}}, \left.\begin{array}{}\\\\\\\\\end{array}\right\} \tag{1.52}$$

$$y(x) = \sum_{\nu=-\infty}^{\infty} y_\nu e^{2\pi i \frac{\nu x}{b-a}}$$

---

[1] Klöppel, K., and H. Lie: Lotrechte Schwingungen von Hängebrücken. Ing.-Arch. **13**, 211−266 (1942).

[2] A numerical example can be found in L. Collatz: Eigenwertaufgaben, pp. 244, 377. Leipzig 1949.

and insert these series into the integral equation (1.51); assuming their convergence to be sufficiently strong that they may be multiplied together and integrated term by term we obtain

$$\sum_{\nu=-\infty}^{\infty} (y_\nu - f_\nu)\, e^{2\pi i \frac{\nu x}{a-b}} = \lambda \sum_{\nu=-\infty}^{\infty} k_\nu e^{2\pi i \frac{\nu x}{b-a}} \sum_{\mu=-\infty}^{\infty} y_\mu \int_a^b e^{2\pi i \frac{\mu-\nu}{b-a}\xi}\, d\xi.$$

Now

$$\int_a^b e^{2\pi i \frac{k\xi}{b-a}}\, d\xi = \begin{cases} b-a & \text{for} \quad k=0 \\ 0 & \text{for} \quad \text{integral } k \neq 0, \end{cases}$$

so that the only non-zero terms in the double sum are those with $\mu = \nu$, and by equating coefficients of $e^{2\pi i \frac{\nu x}{b-a}}$ we find that

$$y_\nu - f_\nu = \lambda (b-a)\, k_\nu\, y_\nu,$$

or

$$y_\nu = \frac{f_\nu}{1 - \lambda (b-a)\, k_\nu}. \tag{1.53}$$

Thus, provided that the denominator in (1.53) does not vanish, i.e. provided that λ is not one of the eigenvalues $\dfrac{1}{(b-a)\, k_\nu}$, we know the Fourier coefficients y_ν and hence also, from (1.52), the solution $y(x)$.

When the fundamental interval is infinite, we use FOURIER's integral theorem instead of the Fourier expansions (1.52) (see textbooks on integral equations)[1].

For linear integral equations of the form (1.2) the solution $y(x)$ is sometimes[2] expanded as a series of functions which are orthogonal over the fundamental interval, i.e. functions $z_1(x), z_2(x), \ldots$ with

$$\int_a^b z_j(x)\, z_k(x)\, dx = \begin{cases} 0 & \text{for} \quad j \neq k, \\ s_j & \text{for} \quad j = k. \end{cases} \tag{1.54}$$

The series

$$y(x) = \sum_{k=1}^{\infty} a_k z_k(x) \tag{1.55}$$

---

[1] An application to the integro-differential equation

$$y(x) = \int_{-\infty}^{+\infty} K(|x - \xi|)\, (p(\xi) - c\, y^{\mathrm{IV}}(\xi))\, d\xi$$

[for which boundary conditions are replaced by the condition that y shall be integrable in $(-\infty, \infty)$], which arises in the investigation of the effects of elasticity in railway track mountings, can be found in M. E. REISSNER: On the theory of beams resting on a yielding foundation. Proc. Nat. Acad. Sci., Wash. **23**, 328—333 (1937).

[2] Various series expansions are used by D. ENSKOG: Eine allgemeine Methode zur Auflösung von linearen Integralgleichungen. Math. Z. **24**, 670—683 (1926).

is inserted into the integral equation (1.2) and the resulting equation

$$\sum_{k=1}^{\infty} a_k z_k(x) - \lambda \int_a^b \sum_{k=1}^{\infty} a_k K(x, \xi) z_k(\xi)\, d\xi = f(x)$$

multiplied by $z_j(x)$, then integrated over the fundamental interval $\langle a, b \rangle$. Assuming that term-by-term integration is permissible (which has to be verified separately in individual cases), we obtain the following infinite system of equations for a countably infinite number of unknowns a_j:

$$a_j s_j - \lambda \sum_{k=1}^{\infty} \varkappa_{jk} a_k = r_j \qquad (j = 1, 2, \ldots), \tag{1.56}$$

where

$$\varkappa_{jk} = \int_a^b \int_a^b K(x, \xi)\, z_j(x)\, z_k(\xi)\, d x\, d\xi, \qquad r_j = \int_a^b f(x)\, z_j(x)\, d x. \tag{1.57}$$

This infinite system is usually solved approximately by retaining only the first p equations and solving them for the first p unknowns $a_1, \ldots, a_p$ with the remaining $a_r (r > p)$ put equal to zero. The values so obtained may be called the p-th approximation and will be denoted by $a_1^{(p)}, \ldots, a_p^{(p)}$; they are calculated from the finite system of equations

$$a_j^{(p)} s_j - \lambda \sum_{k=1}^{p} \varkappa_{jk} a_k^{(p)} = r_j \qquad (j = 1, 2, \ldots, p). \tag{1.58}$$

For the corresponding eigenvalue problem, i.e. (1.2) with $f(x) \equiv 0$, we have $r_j \equiv 0$ and we calculate the p-th approximations $\Lambda_1^{(p)}, \Lambda_2^{(p)}, \ldots, \Lambda_p^{(p)}$ to the eigenvalues λ_j as the roots of the algebraic equation obtained by putting the determinant of (1.58) equal to zero[1]:

$$\begin{vmatrix} s_1 - \lambda \varkappa_{11} & - \lambda \varkappa_{12} \ldots & - \lambda \varkappa_{1p} \\ - \lambda \varkappa_{21} & s_2 - \lambda \varkappa_{22} \ldots & - \lambda \varkappa_{2p} \\ \cdots \cdots \cdots \cdots \cdots \cdots \cdots \\ - \lambda \varkappa_{p1} & - \lambda \varkappa_{p2} \ldots s_p - \lambda \varkappa_{pp} \end{vmatrix} = 0. \tag{1.59}$$

---

[1] Lösch, F.: Zur praktischen Berechnung der Eigenwerte linearer Integralgleichungen. Z. Angew. Math. Mech. **24**, 35—41 (1944). Under the assumption that $K(x, \xi)$ is continuous and can be expanded uniformly as a series of the form

$$K(x, \xi) = \sum_{\nu=1}^{\infty} \varphi_\nu(x)\, z_\nu(\xi) \quad \text{with} \quad \varphi_\nu(x) = \int_a^b K(x, \xi)\, z_\nu(\xi)\, d\xi,$$

Lösch proves that the $\Lambda_j^{(p)}$ tend to the eigenvalues λ_j of the integral equation in "position and order" and that when the eigenvalues of the integral equation are all simple, the suitably normalized approximating functions

$$\sum_{j=1}^{p} a_j^{(p)} z_j(x)$$

converge uniformly in $a \le x \le b$ to the corresponding eigenfunction of the integral equation. Symmetry of the kernel is not assumed.

The method is also applicable to linear integro-differential equations of the form (1.3) with the boundary conditions (1.4). Here we use a set of functions $z_0(x)$, $z_1(x)$, $z_2(x)$, ... with $z_0(x)$ satisfying the inhomogeneous boundary conditions (1.4) and the other $z_j(x)$ satisfying the corresponding homogeneous conditions:

$$U_\mu[z_0] = \gamma_\mu \qquad \begin{cases} \mu = 1, 2, \ldots, m \\ j = 1, 2, \ldots \end{cases}$$
$$U_\mu[z_j] = 0$$

Inserting

$$y = z_0 + \sum_{k=1}^{\infty} a_k z_k \tag{1.60}$$

into the integro-differential equation (1.3), we obtain (again assuming that the various operations may be performed term by term)

$$\left. \begin{aligned} \sum_{k=1}^{\infty} a_k M[z_k] - \lambda \int_a^b \sum_{k=1}^{\infty} a_k K(x, \xi)\, N[z_k(\xi)]\, d\xi = \varphi(x) \\ = f(x) - M[z_0(x)] + \lambda \int_a^b K(x, \xi)\, N[z_0(\xi)]\, d\xi. \end{aligned} \right\} \tag{1.61}$$

The constants a_k could be determined approximately by one of the general methods of § 1.6 — collocation, for example — but we can also use a method allied to the method described above for the special case $M[y] = N[y] = y$. Let $w_1(x)$, $w_2(x)$, ... be a system of functions which is complete over the interval $\langle a, b \rangle$. Then multiplication of (1.61) by $w_j(x)$ and integration over the interval $\langle a, b \rangle$ yields

$$\sum_{k=1}^{\infty} (m_{jk} - \lambda n_{jk})\, a_k = r_j \qquad (j = 1, 2, \ldots), \tag{1.62}$$

where

$$\left. \begin{aligned} m_{jk} &= \int_a^b w_j(x)\, M[z_k(x)]\, dx, \\ n_{jk} &= \int_a^b \int_a^b K(x, \xi)\, w_j(x)\, N[z_k(\xi)]\, dx\, d\xi, \\ r_j &= \int_a^b \varphi(x)\, w_j(x)\, dx. \end{aligned} \right\} \tag{1.63}$$

Again we have [in (1.62)] an infinite system of equations for the unknowns a_j and can obtain approximate solutions as above by retaining only the first p equations and the first p unknowns.

For the corresponding eigenvalue problem, in which $f(x)$, γ_μ, $z_0(x)$, $\varphi(x)$, r_j are all identically zero, it is to be recommended that the $z_k(x)$ and $w_j(x)$ be so chosen that $m_{jk} = 0$ for $j \neq k$; then with $\lambda = 1/\varkappa$ the equation for the approximate eigenvalues reduces to the secular equation

$$
\begin{vmatrix}
n_{11} - \varkappa\, m_{11} & n_{12} & \cdots\, n_{1p} \\
n_{21} & n_{22} - \varkappa\, m_{22} \cdots n_{2p} \\
\cdots\cdots\cdots\cdots\cdots\cdots\cdots \\
n_{p1} & n_{p2} & \cdots n_{pp} - \varkappa\, m_{pp}
\end{vmatrix} = 0.
$$

This can be achieved, for example, by choosing the $z_k(x)$ as the eigen-functions of the eigenvalue problem

$$
M[z_k] = \lambda z_k, \qquad U_\mu[z_k] = 0
$$

and putting

$$
w_j(x) = z_j(x).
$$

1.10. Examples

I. An inhomogeneous integro-differential equation. Consider the equation

$$
\frac{1+x}{2}\, y'' + 10 \int_0^1 \frac{y(\xi)\, d\xi}{1+x+\xi} + 1 = 0
$$

with the boundary conditions

$$
y(0) = y(1) = 0.
$$

For the series satisfying the boundary conditions we take

$$
y(x) = \sum_{k=1}^{\infty} a_k(x - x^{k+1}).
$$

For a two-parameter approximation we retain only the first two terms, putting $a_k = 0$ for $k \geq 3$; then insertion into the equation yields

$$
a_1(-1-x) + a_2(-3x - 3x^2) + 10 \int_0^1 \frac{a_1(\xi - \xi^2) + a_2(\xi - \xi^3)}{1+x+\xi}\, d\xi + 1 = 0. \qquad (1.64)
$$

Here, suitable functions for the $w_j(x)$ are the powers $x^{j-1}\,(j = 1, 2, \ldots)$; thus one equation for a_1, a_2 is obtained by integrating (1.64) over the interval $\langle 0, 1\rangle$ and the other by first multiplying (1.64) by x, then integrating over $\langle 0, 1\rangle$. Before writing these equations down, we evaluate separately the more complicated integrals which occur:

$$
J_{11} = \int_0^1\!\!\int_0^1 \frac{(\xi - \xi^2)\, dx\, d\xi}{1+x+\xi} = \frac{1}{6}\,(8 + 32\ln 2 - 27\ln 3) \qquad = 0\!\cdot\!0863630,
$$

$$
J_{12} = \int_0^1\!\!\int_0^1 \frac{(\xi - \xi^3)\, dx\, d\xi}{1+x+\xi} = \frac{1}{4}\left(-\frac{23}{6} - 8\ln 2 + 9\ln 3\right) \qquad = 0\!\cdot\!1272500,
$$

$$J_{21} = \int_0^1\int_0^1 \frac{x(\xi - \xi^2)\,dx\,d\xi}{1 + x + \xi} = \frac{1}{4}\left(\frac{9}{2} + 8\ln 2 - 9\ln 3\right) \qquad = 0{\cdot}0394167,$$

$$J_{22} = \int_0^1\int_0^1 \frac{x(\xi - \xi^3)\,dx\,d\xi}{1 + x + \xi} = \frac{1}{60}\left(-\frac{49}{2} - 88\ln 2 + 81\ln 3\right) = 0{\cdot}0581774$$

(a useful check on the order of magnitude of the value obtained in each case can be made by considering the range of values assumed by the integrand).

The two equations for a_1 and a_2 read

$$a_1\left(-\tfrac{3}{2} + 10 J_{11}\right) + a_2\left(-\tfrac{5}{2} + 10 J_{12}\right) + 1 = 0,$$
$$a_1\left(-\tfrac{5}{6} + 10 J_{21}\right) + a_2\left(-\tfrac{7}{4} + 10 J_{22}\right) + \tfrac{1}{2} = 0,$$

and have the solution

$$a_1 = 2{\cdot}7135, \qquad a_2 = -0{\cdot}59206.$$

As an approximation for the mid-point value, for example, we obtain

$$y\left(\tfrac{1}{2}\right) = 0{\cdot}4563.$$

II. A non-linear integral equation. For the integral equation

$$\int_0^1 \frac{[y(x) + y(\xi)]^2}{1 + x + \xi}\,d\xi = 1,$$

already considered in §§ 1.3, 1.5, it is expedient to move the origin to the centre of the fundamental interval before using power series expansions; we introduce the new variables $u = x - \tfrac{1}{2}$, $v = \xi - \tfrac{1}{2}$, $z(u) = y(u + \tfrac{1}{2})$, thus transforming the integral equation into

$$\int_{-\frac{1}{2}}^{\frac{1}{2}} \frac{[z(u) + z(v)]^2}{2 + u + v}\,dv = 1,$$

and then insert the power series

$$z(u) = a_0 + a_1 u + a_2 u^2 + \cdots, \qquad \frac{1}{2 + u + v} = \frac{1}{2} - \frac{u + v}{4} + \frac{(u + v)^2}{8} - + \cdots.$$

If only terms of at most the second degree in u and v after multiplying out the series in the integrand are retained, the integration yields

$$1 = \int_{-\frac{1}{2}}^{\frac{1}{2}} \ldots dv \approx 2 a_0^2 + (2 a_0 a_1 - a_0^2) u + \left(\frac{a_0^2}{2} + \frac{a_1^2}{2} + 2 a_0 a_2 - a_0 a_1\right)\left(u^2 + \frac{1}{12}\right).$$

By equating coefficients we obtain the three equations

$$1 = 2 a_0^2,$$
$$0 = 2 a_0 a_1 - a_0^2,$$
$$0 = \frac{a_1^2}{2} + 2 a_0 a_2 - a_0 a_1 + \frac{a_0^2}{2},$$

with the solution

$$a_1 = \frac{1}{2}\,a_0, \qquad a_2 = -\frac{1}{16}\,a_0, \qquad a_0 = \pm\sqrt{\frac{1}{2}};$$

and hence

$$z \approx \pm\sqrt{\frac{1}{2}}\left(1 + \frac{1}{2}\,u - \frac{1}{16}\,u^2\right), \qquad y \approx \pm\sqrt{\frac{1}{2}}\left(\frac{47}{64} + \frac{9}{16}\,x - \frac{1}{16}\,x^2\right).$$

For the positive solution, for example, we obtain the approximations $0\cdot5192$, $0\cdot7071$, $0\cdot8729$ for $y(0)$, $y(\frac{1}{2})$, $y(1)$, respectively.

If we had retained only the linear terms in the integrand, we would still have obtained a fairly useful approximation, namely

$$z \approx \pm\sqrt{\frac{1}{2}}\left(1 + \frac{1}{2}\,u\right), \qquad y \approx \pm\sqrt{\frac{1}{2}}\left(\frac{3}{4} + \frac{1}{2}\,x\right).$$

§2. Some special methods for linear integral equations

2.1. Approximation of kernels by degenerate kernels

The method to be described now relates to linear integral equations of the form (1.2) or to the more general case of linear integro-differential equations of the form (1.3) with the boundary conditions (1.4).

In this method the kernel $K(x, \xi)$ is approximated by kernels $K_n(x, \xi)$ of the form

$$K_n(x, \xi) = \sum_{j=1}^{n} A_j(x)\,B_j(\xi). \tag{2.1}$$

A kernel which can be written in this way as a finite sum of functions in which the variables x and ξ can be separated, i.e. confined to separate factors $A_j(x)$ and $B_j(\xi)$, is called a "degenerate kernel". Any continuous kernel may be approximated to any degree of accuracy by a degenerate kernel. In practice one often uses polynomials or trigonometric functions for the functions $A_j(x)$, $B_j(\xi)$. A degenerate approximation $K_n(x, \xi)$ to a kernel $K(x, \xi)$ can also be determined[1] from the equation

$$\begin{vmatrix} K_n(x, \xi) & K(x, \xi_1) & K(x, \xi_2) & \dots K(x, \xi_n) \\ K(x_1, \xi) & K(x_1, \xi_1) & K(x_1, \xi_2) & \dots K(x_1, \xi_n) \\ \cdot & \cdot & \cdot & \cdot \cdot \cdot \cdot \\ K(x_n, \xi) & K(x_n, \xi_1) & K(x_n, \xi_2) & \dots K(x_n, \xi_n) \end{vmatrix} = 0, \tag{2.2}$$

where the x_j and ξ_j are selected points of the fundamental interval; the degenerate kernel K_n then coincides with the given kernel K along the $2n$ lines $x = x_j$ and $\xi = \xi_j$:

$$K_n(x, \xi_j) = K(x, \xi_j) \quad \text{and} \quad K_n(x_j, \xi) = K(x_j, \xi) \quad \text{for} \quad j = 1, 2, \dots, n.$$

---

[1] BATEMAN, H.: On the numerical solution of linear integral equations. Proc. Roy. Soc. Lond., Ser. A **100**, 441—449 (1922).

For a degenerate kernel the solution of the integro-differential equation (1.3) with the boundary conditions (1.4) may be reduced to the solution of a set of ordinary boundary-value problems and of a system of linear equations.

We therefore replace the kernel $K(x, \xi)$ in the integro-differential equation (1.3) by the degenerate approximation $K_n(x, \xi)$. Then by interchanging the order of the summation and integration, and taking the functions $A_j(x)$ outside of the integral sign, we obtain

$$M[y(x)] = f(x) + \lambda \sum_{j=1}^{n} c_j A_j(x), \qquad (2.3)$$

where

$$c_j = \int_a^b B_j(\xi) N[y(\xi)] \, d\xi. \qquad (2.4)$$

Now let $z(x)$ be the solution of the boundary-value problem

$$M[y(x)] = f(x), \qquad U_\mu[y] = \gamma_\mu \qquad (\mu = 1, 2, \ldots, m)$$

and $z_k(x)$ (for $k = 1, \ldots, n$) the solution of the boundary-value problem

$$M[y(x)] = A_k(x), \qquad U_\mu[y] = 0 \qquad (\mu = 1, 2, \ldots, m)$$

(we assume that these $n + 1$ boundary-value problems possess unique solutions); then the boundary-value problem (2.3), (1.4) has the solution

$$y(x) = z(x) + \lambda \sum_{k=1}^{n} c_k z_k(x). \qquad (2.5)$$

We now insert this expression for y into (2.4), and obtain a system of equations for the c_j:

$$c_j = \int_a^b B_j(\xi) N[z(\xi)] \, d\xi + \lambda \sum_{k=1}^{n} c_k \int_a^b B_j(\xi) N[z_k(\xi)] \, d\xi.$$

If we introduce the Kronecker symbol

$$\delta_{jk} = \begin{cases} 0 & \text{for} \quad j \neq k, \\ 1 & \text{for} \quad j = k \end{cases}$$

and define

$$a_{jk} = \int_a^b B_j(\xi) N[z_k(\xi)] \, d\xi, \qquad r_j = \int_a^b B_j(\xi) N[z(\xi)] \, dx, \qquad (2.6)$$

these equations can be written more compactly as

$$\sum_{k=1}^{n} (\delta_{jk} - \lambda a_{jk}) c_k = r_j \qquad (j = 1, 2, \ldots, n). \qquad (2.7)$$

Thus the c_k can be calculated provided that the determinant

$$\Delta = \det\left(\delta_{jk} - \lambda a_{jk}\right) \tag{2.8}$$

does not vanish, and with these c_k values (2.5) gives the solution of the integro-differential equation with the degenerate kernel (2.1).

If we have a completely homogeneous problem with $f(x) \equiv 0$, $\gamma_\mu \equiv 0$ (an eigenvalue problem), then $z(x) \equiv 0$ and $r_j \equiv 0$, and the zeros of the determinant provide approximations Λ_j for the eigenvalues λ_j.

If the degenerate kernel $K_n(x, \xi)$ defined by (2.2) is used, the solution of the integral equation (1.2) of the second kind can be expressed in the form[1]

$$y(x) = f(x) + \lambda \int_a^b G(x, \xi, \lambda) f(\xi)\, d\xi,$$

where the solving kernel $G(x, \xi, \lambda)$ is to be determined from the equation

$$\begin{vmatrix} G(x, \xi, \lambda) & K(x, \xi_1) & K(x, \xi_2) \dots K(x, \xi_n) \\ K(x_1, \xi) & A_{11} & A_{12} & \dots A_{1n} \\ \cdot\ \cdot\ \cdot\ \cdot\ \cdot\ \cdot\ \cdot\ \cdot\ \cdot\ \cdot\ \cdot\ \cdot\ \cdot \\ K(x_n, \xi) & A_{n1} & A_{n2} & \dots A_{nn} \end{vmatrix} = 0$$

with

$$A_{jk} = K(x_j, \xi_k) - \lambda \int_a^b K(x_j, t) K(t, \xi_k)\, dt.$$

2.2. Example

For the problem

$$\frac{1+x}{2} y'' + 10 \int_0^1 \frac{y(\xi)\, d\xi}{1 + x + \xi} + 1 = 0, \qquad y(0) = y(1) = 0, \tag{2.9}$$

considered in § 1.10, let the kernel $K(x, \xi) = \dfrac{1}{1 + x + \xi}$ be approximated first of all quite crudely by a constant A; the solution $\eta(x)$ of the approximate integro-differential equation can then be given in closed form. The equation reads

$$\eta'' = \frac{C}{1+x} \qquad \text{with} \qquad C = -2 - 20A \int_0^1 \eta(\xi)\, d\xi$$

and can be integrated immediately; using the boundary conditions $\eta(0) = \eta(1) = 0$, we obtain

$$\eta = C[(1 + x) \ln(1 + x) - 2x \ln 2].$$

---

[1] BATEMAN, H.: (see last footnote) p. 443.

To determine C we calculate

$$\int_0^1 \eta(\xi)\, d\xi = C(\ln 2 - \tfrac{3}{4}) = -C \times 0.056853 = -C\,\delta, \text{ say},$$

insert this value for the integral into the equation defining C and find that

$$C = \frac{2}{20A\,\delta - 1} \approx \frac{-2}{1 - 1.1371A}.$$

If we take $A = \tfrac{1}{2}$ as being a reasonable average value for the kernel, we have $C = -4.6353$, and hence $\eta(x)$ is known; in particular, $\eta(\tfrac{1}{2}) = 0.3938$.

Now let us approximate the kernel $K(x, \xi) = \dfrac{1}{1 + x + \xi}$ more accurately by using a linear function $A + B(x + \xi)$. The corresponding approximate solution $\eta(x)$ is now to be determined from

$$(1 + x)\,\eta'' = -2 - 20 \int_0^1 [A + B(x + \xi)]\,\eta(\xi)\, d\xi = C + D(1 + x),$$

$$\eta(0) = \eta(1) = 0,$$

and again we have a problem which can be solved in closed form:

$$\eta = C[(1 + x)\ln(1 + x) - 2x \ln 2] + D\,\frac{x^2 - x}{2}.$$

To determine C and D we calculate

$$\int_0^1 \eta(\xi)\, d\xi = -C\,\delta - \frac{1}{12}\,D, \qquad \int_0^1 \xi\,\eta(\xi)\, d\xi = -\frac{1}{36}\,C - \frac{1}{24}\,D$$

(with δ as defined above) and insert these values in the equation defining C and D; then by equating coefficients (of powers of x) we obtain two linear equations for C and D:

$$(1 - 20A\,\delta - \tfrac{5}{9}\,B)\,C + (1 - \tfrac{5}{3}A - \tfrac{5}{6}\,B)\,D = -2,$$

$$-20B\,\delta\,C + (1 - \tfrac{5}{3}\,B)\,D = 0.$$

The values of C and D depend on the choice of values for A and B. Suitable values for A and B could be determined by the method of least squares, but here we use values $\left(A = 0.8,\ B = -\dfrac{4}{15}\right)$ which were obtained simply by means of a graph; they give $C = -7.6386$, $D = 1.6035$ and $\eta(\tfrac{1}{2}) = 0.4485$.

Naturally one could use still more accurate approximations for the kernel $K(x, \xi)$.

2.3. The iteration method for eigenvalue problems

The theory of the method of successive approximations, or iteration method, of Ch. III, § 8.3 applies also to the linear homogeneous integral equation

$$y(x) = \lambda \int_a^b K(x, \xi) \, y(\xi) \, d\xi \tag{2.10}$$

with symmetric kernel $K(x, \xi)$ when this kernel is square-integrable and continuous in the mean[1]. Starting from an arbitrarily chosen, continuous function $F_0(x)$, one determines further functions $F_n(x)$ from the iteration formula (1.20) and calculates the corresponding Schwarz constants a_n and quotients μ_n defined by (1.21), (1.22). One can then make use of the results stated below.

We arrange the eigenvalues of (2.10) in order of increasing absolute value, treating each multiple eigenvalue, say of multiplicity m, as a set of m coincident eigenvalues:

$$0 \leq |\lambda_1| \leq |\lambda_2| \leq |\lambda_3| \leq \cdots,$$

and for the following assume that $0 < |\lambda_1| < |\lambda_2|$.

Then we may state:

1. *If, besides being symmetric, square-integrable and continuous in the mean, the kernel is "positive definite", i.e. the eigenvalues are all positive, then the μ_n decrease monotonically:*

$$\mu_1 \geq \mu_2 \geq \cdots \geq \lambda_1 > 0, \tag{2.11}$$

and

$$0 \leq \mu_{n+1} - \lambda_1 \leq \frac{\mu_n - \mu_{n+1}}{\dfrac{l_2}{\mu_{n+1}} - 1}, \tag{2.12}$$

where l_2 is a lower limit for the second eigenvalue, but is greater than μ_{n+1}:

$$\lambda_2 \geq l_2 > \mu_{n+1}.$$

2. *In the more general cases for which the condition of positive definiteness of the kernel is relaxed, the μ_n need not form a monotonic sequence;*

---

[1] That is when the integrals

$$\int K(x, \xi) \, dx, \quad \int K^2(x, \xi) \, dx, \quad \iint K(x, \xi) \, dx \, d\xi, \quad \iint K^2(x, \xi) \, dx \, d\xi$$

over the fundamental interval all exist and are bounded (square-integrability) and

$$\lim_{x \to x_1} \int [K(x, \xi) - K(x_1, \xi)]^2 \, d\xi = 0$$

(continuity in the mean); both of these conditions are satisfied when $K(x, \xi)$ is continuous. See G. HAMEL: Integralgleichungen, 2nd ed., p. 68. Berlin 1949.

now, however, the products $\mu_{2r-1}\mu_{2r}$ decrease monotonically:

$$\mu_1\mu_2 \geqq \mu_3\mu_4 \geqq \mu_5\mu_6 \geqq \cdots \geqq \lambda_1^2 \geqq 0, \tag{2.13}$$

and the corresponding limits for the first eigenvalue[1] take the form

$$0 \leqq \mu_{2n+1}\mu_{2n+2} - \lambda_1^2 \leqq \frac{\mu_{2n-1}\mu_{2n} - \mu_{2n+1}\mu_{2n+2}}{\dfrac{l_2^2}{\mu_{2n+1}\mu_{2n+2}} - 1}, \tag{2.14}$$

where l_2 is a number such that

$$\lambda_2^2 \geqq l_2^2 > \mu_{2n+1}\mu_{2n+2}.$$

3. *The enclosure theorem also holds[2]: If the function*

$$G(x) = \frac{F_0(x)}{F_1(x)} \tag{2.15}$$

lies between finite limits $G_{\min}$ and $G_{\max}$ and does not change sign in the fundamental interval, then $G_{\min}$ and $G_{\max}$ enclose at least one eigenvalue λ_k of (2.10):

$$G_{\min} \leqq \lambda_k \leqq G_{\max}. \tag{2.16}$$

Sometimes a suitable value for l_2 or a direct estimate for λ_1 can be obtained from the relation[3]

$$\sum_{\nu=1}^{\infty} \frac{1}{\lambda_\nu^2} = \int_a^b\int_a^b [K(x,\xi)]^2\, dx\, d\xi = k;$$

for example, we can deduce that

$$\frac{1}{\lambda_2^2} \leqq k - \frac{1}{\lambda_1^2}, \quad \text{i.e.} \quad \lambda_2 \geqq \left(k - \frac{1}{\lambda_1^2}\right)^{-\frac{1}{2}}.$$

Occasionally one also uses the "iterated kernels", in terms of which the above relation and allied results can be expressed; these "iterated kernels" are defined in § 3.1, where another use for them is described.

The various modifications to the iteration method which were described in Ch. III, §§ 8.9, 8.10 can be carried over immediately to integral equations; in the literature, in fact, they are often established for integral equations first.

---

[1] Similar results for the higher eigenvalues and other generalizations are given by L. COLLATZ: Math. Z. **46**, 692—708 (1940), and R. IGLISCH: Math. Ann. **118**, 263—275 (1941).

[2] COLLATZ, L.: Math. Z. **47**, 395—398 (1941).

[3] See, for instance, G. HAMEL: Integralgleichungen, 2nd ed., p. 68. Berlin 1949.

§3. Singular integral equations

Applied problems frequently lead to integral and integro-differential equations for which the fundamental region is infinite or for which the kernel $K(x, \xi)$ for equations of the form (1.2), (1.3) is unbounded. A numerical treatment of such singular equations must, of course, take into account the nature of the singularity, but so many different kinds of situation can arise that a unified description seems hardly possible; it must suffice here to select a few typical cases and use them to illustrate the application of some of the methods which have been used hitherto.

Two of the most frequently occurring types of singular kernel $K(x, \xi)$ can be written in the forms

$$\frac{H(x, \xi)}{(x - \xi)^\alpha}, \qquad H(x, \xi) \ln |x - \xi|,$$

where $H(x, \xi)$ is continuous in the fundamental region $a \leq (x, \xi) \leq b$.

3.1. Smoothing of the kernel

For an integral equation of the form (1.2) a "smoother" kernel can be obtained by utilizing the smoothing property of integration in the following way[1]. We multiply the equation by $K(v, x)$ and integrate with respect to x over the fundamental interval, then make use of the original equation to obtain

$$\frac{y(v) - f(v)}{\lambda} = \int_a^b K(v, x)\, y(x)\, dx$$

$$= \lambda \int_a^b \int_a^b K(v, x)\, K(x, \xi)\, y(\xi)\, dx\, d\xi + \int_a^b K(v, x)\, f(x)\, dx;$$

this also is an integral equation of the form (1.2), and can be written

$$y(v) = \lambda^2 \int_a^b K_2(v, \xi)\, y(\xi)\, d\xi + f_2(v), \tag{3.1}$$

where

$$K_2(v, \xi) = \int_a^b K(v, x)\, K(x, \xi)\, dx, \quad f_2(v) = f(v) + \lambda \int_a^b K(v, x)\, f(x)\, dx. \tag{3.2}$$

In general the new kernel K_2 will be "smoother" than K. In particular, for a singular symmetric kernel of the form

$$K(x, \xi) = \frac{H(x, \xi)}{(x - \xi)^\alpha} \tag{3.3}$$

---

[1] See G. HAMEL: Integralgleichungen, 2nd ed., p. 140. Berlin 1949; or P. MORSE and H. FESHBACH: Methods of theoretical physics, Part I, p. 922. New York-Toronto-London: McGraw-Hill 1953.

with $0 < \alpha \leq \frac{1}{2}$ the new kernel K_2 will no longer be singular. For the case $\frac{1}{2} < \alpha < 1$, K_2 will still be singular, but we can repeat the process on equation (3.1) and so on; after a finite number of such "smoothing steps" we will have a kernel K_n which is no longer singular. These kernels $K_2, K_3, \ldots$ are known as "iterated kernels".

It should be noted that for the case with $\alpha = 1$ in (3.3) this smoothing process does not help, for the iterated kernels are again of the form (3.3). This important case will be dealt with in greater detail in § 3.2.

3.2. Singular equations with Cauchy-type integrals

A number of applied problems give rise to singular integral and integro-differential equations of the first and second kinds of the forms (the integral notation will be explained later)

$$\lambda \oint_a^b \frac{H(x,\xi)}{x-\xi} N[y(\xi)] \, d\xi = f(x) \tag{3.4}$$

and

$$y(x) - \lambda \oint_a^b \frac{H(x,\xi)}{x-\xi} N[y(\xi)] \, d\xi = f(x), \tag{3.5}$$

respectively[1], where $H(x, \xi)$ and $f(x)$ are given continuous functions and $N[y]$ is a linear differential expression in y of the n-th order. In the simplest case $n = 0$ we have an equation of the form

$$\lambda \oint_a^b \frac{H(x,\xi) \, y(\xi)}{x-\xi} \, d\xi = f(x).$$

An integral of the type considered here, usually known as a Cauchy integral, is improper and in general does not converge. In order to attach a meaning to it, we define its Cauchy principal value, distinguished by a C superimposed on the integral sign, as the limit

$$\oint_a^b \frac{H(x,\xi)}{x-\xi} \, d\xi = \lim_{\varepsilon \to 0} \left\{ \int_a^{x-\varepsilon} \frac{H(x,\xi)}{x-\xi} \, d\xi + \int_{x+\varepsilon}^b \frac{H(x,\xi)}{x-\xi} \, d\xi \right\},$$

provided that this limit exists. This is the value to be understood in the above equations.

The difficulties which attend the solution of integral equations with these "Cauchy kernels" can often be surmounted by making use of certain singular integrals whose Cauchy principal values are known; the formulae (3.8), (3.15) below, for example, are used to this end.

---

[1] For the one-dimensional singular Fredholm integral equation with Cauchy-type kernel a complete theory has been given by N. J. MUSKHELISHVILI: Singular integral equations, 447 pp. Groningen 1953.

Consider the integral equation

$$\frac{1}{2\pi} \oint_{-a}^{a} \frac{y(\xi)\,d\xi}{\xi - x} = f(x). \tag{3.6}$$

First we transform it by the introduction of the new variables

$$x = -a\cos\varphi, \qquad \xi = -a\cos\psi$$

into

$$\frac{1}{2\pi} \oint_{0}^{\pi} \frac{y(-a\cos\psi)\sin\psi}{\cos\varphi - \cos\psi}\,d\psi = f(-a\cos\varphi). \tag{3.7}$$

If we now assume for the function $y(-a\cos\psi)\sin\psi = g(\psi)$ appearing in the numerator of the integrand an expansion of the form

$$g(\psi) = y(-a\cos\psi)\sin\psi = \tfrac{1}{2}b_0 + \sum_{n=1}^{\infty} b_n \cos n\psi,$$

we can evaluate the integral in (3.7) by using the formula[1]

$$\oint_{0}^{\pi} \frac{\cos n\psi}{\cos\psi - \cos\varphi}\,d\psi = \pi \frac{\sin n\varphi}{\sin\varphi} \qquad (n = 0, 1, 2, \ldots;\ 0 < \varphi < \pi); \tag{3.8}$$

(3.7) then becomes

$$-\frac{1}{2} \sum_{n=1}^{\infty} b_n \frac{\sin n\varphi}{\sin\varphi} = f(-a\cos\varphi). \tag{3.9}$$

We now assume that the function $f(-a\cos\varphi)\sin\varphi$ may be expanded as a uniformly convergent Fourier series of the form

$$f(-a\cos\varphi)\sin\varphi = -\tfrac{1}{2}\sum_{n=1}^{\infty} b_n \sin n\varphi \qquad (0 < \varphi < \pi); \tag{3.10}$$

the b_n (except b_0) are thereby determined, and in

$$y(\xi) = \frac{1}{\sqrt{1 - \left(\frac{\xi}{a}\right)^2}} \left(\frac{1}{2}b_0 + \sum_{n=1}^{\infty} b_n \cos n\psi \right), \tag{3.11}$$

where the $\cos n\psi$ can be expressed in known fashion as polynomials in $\cos\psi$ (and hence in ξ), we have a family of solutions of the integral equation (3.6). We note that the constant b_0 can be chosen arbitrarily and that the solution $y(\xi)$ can become infinite like $(a^2 - \xi^2)^{-\frac{1}{2}}$ at the end-points $\xi = \pm a$.

---

[1] See, for example, G. HAMEL: Integralgleichungen, 2nd ed., p. 145. Berlin 1949.

Other singular integral equations, such as

$$\oint\limits_{-a}^{+a} \frac{y(\xi)\,d\xi}{x^2-\xi^2} = f(x), \tag{3.12}$$

for example, may be solved in a similar way[1].

On account of its importance in the theory of conformal transformations, we will consider one more particular integral equation, namely

$$f(\varphi) = -\frac{1}{2\pi} \oint\limits_{0}^{2\pi} y(\vartheta) \cot \frac{\vartheta-\varphi}{2}\,d\vartheta. \tag{3.13}$$

If we substitute the Fourier series

$$y(\vartheta) = \frac{1}{2} a_0 + \sum_{n=1}^{\infty} (a_n \cos n\,\vartheta + b_n \sin n\,\vartheta) \tag{3.14}$$

for $y(\vartheta)$ in the integral and use the formulae

$$\left.\begin{array}{l}
\displaystyle\oint\limits_{0}^{2\pi} \cos n\,\vartheta \cot \frac{\vartheta-\varphi}{2}\,d\vartheta = -2\pi \sin n\,\varphi \\[2em]
\displaystyle\oint\limits_{0}^{2\pi} \sin n\,\vartheta \cot \frac{\vartheta-\varphi}{2}\,d\vartheta = 2\pi \cos n\,\varphi
\end{array}\right\} \quad (n = 0, 1, 2, \ldots), \tag{3.15}$$

the integral equation (3.13) becomes

$$f(\varphi) = \sum_{n=1}^{\infty} (a_n \sin n\,\varphi - b_n \cos n\,\varphi). \tag{3.16}$$

If we assume that the given function $f(\varphi)$ may be expanded as a uniformly convergent Fourier series of the form (3.16) — this implies in particular that $f(\varphi)$ must satisfy the condition

$$\int\limits_{0}^{2\pi} f(\varphi)\,d\varphi = 0$$

—, then the unknowns a_n and b_n (except a_0) are determined immediately by Fourier analysis of $f(\varphi)$. With these values in (3.14) we again obtain solutions $y(\vartheta)$ in which a constant a_0 may be chosen arbitrarily. With the aid of numerical tables and formulae for trigonometric interpolation

---

[1] HAMEL, G.: Integralgleichungen, 2nd ed., p. 148. Berlin 1949.

WITTICH reduces this solution[1] to a form suitable for numerical evaluation.

For singular integral equations which cannot be solved in closed form or by series expansions one tries to remove the singularity by some means or other in order that a quadrature formula may be used to evaluate the integral. A device which is often used for this purpose with kernels of Cauchy type is the following[2]. Consider the Cauchy integral

$$J(x) = \oint_a^b \frac{f(x,\xi)}{x-\xi} \, d\xi$$

and assume that $f(x, \xi)$ possesses a continuous derivative with respect to ξ. When x lies nearer a than b, say, so that $a < x < b$, $h = x - a < b - x$, we can write

$$J = J_1 + J_2,$$

where

$$J_1 = \int_a^x \frac{f(x,\xi)}{x-\xi} \, d\xi + \int_x^{x+h} \frac{f(x, 2x - \xi)}{x-\xi} \, d\xi$$

and

$$J_2 = \int_x^{x+h} \frac{f(x, \xi) - f(x, 2x - \xi)}{x-\xi} \, d\xi + \int_{x+h}^b \frac{f(x,\xi)}{x-\xi} \, d\xi.$$

---

[1] THEODORSEN, T., and J. E. GARRICK: General potential theory. Nat. Adv. Comm. Aeronaut. Rep. Nr. 452 (1933) 1—35. Practical methods for conformal transformation are not dealt with in the present book because the subject lies at the very edge of the field with which the book is concerned; however, we can at least give a selection of references: TREFFTZ, E.: Eine neue Methode zur Lösung der Randwertaufgabe partieller Differentialgleichungen. Math. Ann. **79** (1919). — FRANK, PH., and R. v. MISES: Differential- und Integralgleichungen der Mechanik und Physik, Vol. I, pp. 729—734. Brunswick 1930. — HEINHOLD, J.: Ein Schmiegungsverfahren der konformen Abbildung. S.-B. Bayer. Akad. Wiss., Math.-nat. Kl. **1948**, 203—222 (where further literature is mentioned), also Z. Angew. Math. Mech. **30**, 286—287 (1950). — WITTICH, H.: Konforme Abbildung einfach zusammenhängender Gebiete. Z. Angew. Math. Mech. **25/27**, 131—132 (1947). — LÖSCH, F.: Auftrieb und Moment eines unsymmetrischen Doppelflügels. Luftf.-Forschg. **17**, 22—31 (1940) (doubly-connected region). — WITTICH, H.: Bemerkungen zur Druckverteilungsrechnung nach THEODORSEN-GARRICK. Jb. dtsch. Luftf.-Forschung **1941**. — WARSCHAWSKI, S. E.: Recent results in numerical methods of conformal mapping. Proc. Symp. Appl. Math. VI, pp. 219—250. New York-Toronto-London 1956. — Experiments in the computation of conformal maps (edited by JOHN TODD). Nat. Bur. Stand., Appl. Math. Ser. **42**, 61 pp. (1955). — KANTOROVICH, L. V., and V. I. KRYLOV: Näherungsmethoden der Höheren Analysis, pp. 330—508. Berlin 1956.

[2] A clear graphical explanation of the device and an application to a technical problem can be found in C. WEBER: Randverformung der Halbebene durch eine Normalbelastung, „Ein Ei des Kolumbus?". Z. Angew. Math. Mech. **30**, 240—242 (1950).

Now the integrals in J_1 combine to form one integral over the interval $a \leq \xi \leq 2x - a$ with an integrand whose numerator is symmetric and denominator antisymmetric about the mid-point $\xi = x$; hence, by the definition of the Cauchy principal value, $J_1 = 0$. The evaluation of J_2 presents no difficulties since the two integrands now remain bounded, the limit

$$\lim_{\xi \to x} \frac{f(x, \xi) - f(x, 2x - \xi)}{x - \xi} = -2 \left(\frac{\partial f(x, \xi)}{\partial \xi} \right)_{\xi = x}$$

existing by virtue of the assumption that $f(x, \xi)$ is differentiable with respect to ξ.

Example. For the integral equation

$$\frac{1}{2\pi} \oint_{-1}^{1} \frac{y(\xi)\, d\xi}{\xi - x} = |x|$$

(3.10) requires that the function $|\cos \varphi| \sin \varphi$ be expanded as a Fourier sine series:

$$-2 |\cos \varphi| \sin \varphi = \sum_{n=1}^{\infty} b_n \sin n\, \varphi;$$

by the usual rules of Fourier analysis we obtain

$$b_{2k} = 0, \qquad b_{2k-1} = (-1)^{k-1} \frac{8}{\pi(2k - 3)(2k + 1)} \qquad (k = 1, 2, \ldots).$$

From (3.11) it follows that

$$y(\xi) = \frac{1}{\sqrt{1 - \xi^2}} \left(\frac{1}{2} b_0 + \sum_{n=1}^{\infty} b_n \cos n\, \psi \right), \quad \text{where} \quad \xi = -\cos \psi,$$

is a solution of the integral equation. The constant b_0 is arbitrary and $(1 - \xi^2)^{-\frac{1}{2}}$ must be a solution of the corresponding homogeneous equation. If we omit the arbitrary multiples of this function which may be added to any solution, we are left with the particular solution

$$y(\xi) = \frac{1}{\sqrt{1 - \xi^2}} \frac{8}{\pi} \left(\frac{\cos \psi}{(-1) \cdot 3} - \frac{\cos 3\psi}{1 \cdot 5} + \frac{\cos 5\psi}{3 \cdot 7} - \frac{\cos 7\psi}{5 \cdot 9} + - \cdots \right)$$

$$= \frac{8}{\pi} \frac{1}{\sqrt{1 - \xi^2}} g, \text{ say.}$$

The sum function of this Fourier series, which we have denoted by g, may be given in closed form[1]. If, however, graphical accuracy will suffice, one can quickly plot a few terms of the Fourier series to obtain

[1] By putting $e^{i\psi} = z$, $\cos n\psi = \operatorname{re} z^n$ and summing the power series which results, one can show that

$$g(\psi) = -\frac{\cos \psi}{2} \left(1 + \sin \psi \ln \tan \left(\frac{\pi}{4} - \frac{\psi}{2} \right) \right) \quad \text{for} \quad 0 \leq \psi \leq \frac{\pi}{2}.$$

a graph of g against ψ, transfer this graph onto a different abscissa scale to represent g as a function of ξ, then scale up the ordinate values in the appropriate manner to obtain finally a graph of $y(\xi)$ against ξ (see Fig. VI/7).

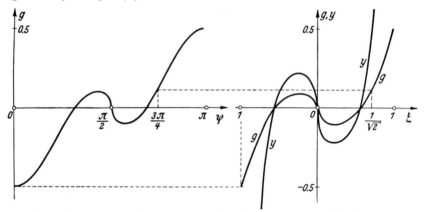

Fig. VI/7. Graphical evaluation of a Fourier series solution of a singular integral equation

3.3. Closed-form solutions

Some singular integral equations may be solved in closed form with the aid of inversion formulae[1]. Several examples of such formulae are listed in Table VI/6, in which for simplicity $f(x)$, say, is regarded as a

Table VI/6. *Inversion formulae for some well-known singular integral equations*

Integral equation	Solution
$f(x) = \dfrac{1}{2\pi} \displaystyle\oint_{-\pi}^{\pi} \left(1 + \cot\dfrac{1}{2}(x-\xi)\right) y(\xi)\, d\xi$	$y(\xi) = \dfrac{1}{2\pi} \displaystyle\oint_{-\pi}^{\pi} \left(1 + \cot\dfrac{1}{2}(x-\xi)\right) f(x)\, dx$
$f(x) = \sqrt{\dfrac{2}{\pi}} \displaystyle\int_{0}^{\infty} \cos(x\xi)\, y(\xi)\, d\xi$	$y(\xi) = \sqrt{\dfrac{2}{\pi}} \displaystyle\int_{0}^{\infty} \cos(x\xi)\, f(x)\, dx$
$f(x) = \displaystyle\int_{0}^{x} \dfrac{y(\xi)\, d\xi}{\sqrt{x-\xi}}$	$y(\xi) = \dfrac{1}{\pi} \dfrac{d}{d\xi} \displaystyle\int_{0}^{\xi} \dfrac{f(x)\, dx}{\sqrt{\xi-x}}$
$f(x) = \dfrac{1}{2\pi} \displaystyle\oint_{-1}^{1} \dfrac{y(\xi)\, d\xi}{x-\xi}$	$\sqrt{1-\xi^2}\,\pi y(\xi) = 2\displaystyle\oint_{-1}^{1} \dfrac{f(x)\sqrt{1-x^2}}{x-\xi}\, dx - \displaystyle\int_{-1}^{1} y(x)\, dx$

[1] See, for example, W. Magnus and F. Oberhettinger: Formeln und Sätze für die speziellen Funktionen der mathematischen Physik, 2nd ed., p. 186. Berlin 1948.

given continuous function; actually, continuity may be replaced by a weaker condition, but for this and other details we must refer the reader to the literature.

3.4. Approximation of the kernel by degenerate kernels

The method of § 2.1 can also be applied to equations with singular kernels if these kernels have known expansions in terms of their eigenfunctions, for we can then truncate these expansions to obtain degenerate approximations to the given singular kernels.

Thus, for example, the kernel

$$K(\varphi, \psi) = \ln |\cos \varphi - \cos \psi|,$$

with a logarithmic singularity, has the expansion

$$K(\varphi, \psi) = \sum_{n=0}^{\infty} \frac{y_n(\varphi)\, y_n(\psi)}{\lambda_n}$$

with eigenfunctions $y_n(\varphi)$ given by

$$y_0 = \frac{1}{\sqrt{\pi}}, \quad y_n(\varphi) = \sqrt{\frac{2}{\pi}} \cos n\varphi \quad (n = 1, 2, \ldots)$$

(these form an orthonormal system) and eigenvalues λ_n by

$$\lambda_0 = -\frac{1}{\pi \ln 2}, \quad \lambda_n = -\frac{n}{\pi} \quad (n = 1, 2, \ldots).$$

For PRANDTL's circulation equation

$$\alpha_i(y) = \frac{1}{4\pi v} \int_{-\frac{b}{2}}^{\frac{b}{2}} \frac{d\Gamma}{dy'} \frac{dy'}{y - y'}$$

(an explanation of the notation can be found in the literature[1]), which

---

[1] The following is only a short selection from the extensive literature on the subject: K. SCHRÖDER: Über die Prandtlsche Integro-Differentialgleichung der Tragflügeltheorie. S.-B. Preuss. Akad. Wiss., Math.-nat. Kl. **16**, 1–35 (1939). — SÖHNGEN, H.: Die Lösungen der Integralgleichung $g(x) = \frac{1}{2\pi} \int_{-a}^{a} \frac{f(\xi)\, d\xi}{x - \xi}$ und deren Anwendung in der Tragflügeltheorie. Math. Z. **45**, 245–264 (1939). — MULTHOPP, H.: Die Berechnung der Auftriebsverteilung von Tragflügeln. Luftf.-Forschg. **15**, 153–169 (1938). — SCHWABE, M.: Luftf.-Forschg. **15**, 170–180 (1938), gives a computing scheme for MULTHOPP's method for the calculation of the lift distribution over a wing.

occurs in wing theory, this method of replacing the kernel by a degenerate kernel leads to a system of equations first derived by MULTHOPP, but in a different way[1].

§4. Volterra integral equations

4.1. Preliminary remarks

If the upper limit in the integral equation is variable, it is called a Volterra integral equation; in particular,

$$f(x) = \int_a^x K(x, \xi)\, y(\xi)\, d\xi \tag{4.1}$$

is called a Volterra integral equation of the first kind and

$$y(x) = f(x) + \int_a^x K(x, \xi)\, y(\xi)\, d\xi \tag{4.2}$$

a Volterra integral equation of the second kind. $y(x)$ is to be determined for given functions $f(x)$ and $K(x, \xi)$ which we assume to be continuous for $a \le x \le b$ and $a \le x \le \xi \le b$, respectively.

In general, differentiation with respect to x transforms a Volterra integral equation of the first kind into one of the second kind; thus we obtain

$$f'(x) = K(x, x)\, y(x) + \int_a^x \frac{\partial K(x, \xi)}{\partial x}\, y(\xi)\, d\xi,$$

which can be written in the form (4.2) by dividing through by $K(x, x)$:

$$y(x) = \frac{f'(x)}{K(x, x)} - \int_a^x \frac{\partial K(x, \xi)}{\partial x}\, \frac{y(\xi)}{K(x, x)}\, d\xi. \tag{4.3}$$

For this transformation to be legitimate we must assume that $f(x)$ and $K(x, \xi)$ are differentiable with respect to x and that $K(x, x) \neq 0$.

Under the assumptions of continuity made above, it may be shown that the Volterra integral equation of the second kind (4.2) always possesses one and only one continuous solution $y(x)$[2].

---

[1] That MULTHOPP's equations could be derived by using the principle mentioned here was recognized by K. JAECKEL: Praktische Auswertung singulärer Integralgleichungen. Lecture. Hannover 1949.

[2] See, for example, G. HAMEL: Integralgleichungen, 2nd ed., p. 32 et seq. Berlin 1949.

Volterra integral equations which occur in applications often have kernels of convolution type, i.e. kernels which are functions of the difference $\delta = x - \xi$ alone: $K(x, \xi) = K(x - \xi) = K(\delta)$. For such cases WHITTAKER[1] recommends that $K(\delta)$ be approximated by a sum of kernel functions of similar form, i.e. of convolution type, but for which the integral equation can be solved in closed form. He exhibits several classes of such kernel functions. Here we mention only the class of kernels of the form $K(\delta) = \sum_{\nu=1}^{n} q_\nu e^{p_\nu \delta}$; for this kernel the solution of (4.2) (with $a = 0$) reads

$$y(x) = f(x) + \int_0^x \widetilde{K}(x - s) f(s)\, ds,$$

where

$$\widetilde{K}(x) = \sum_{\nu=1}^{n} \frac{(\alpha_\nu - p_1)(\alpha_\nu - p_2) \dots (\alpha_\nu - p_n)}{(\alpha_\nu - \alpha_1)(\alpha_\nu - \alpha_2) \dots (\alpha_\nu - \alpha_{\nu-1})(\alpha_\nu - \alpha_{\nu+1}) \dots (\alpha_\nu - \alpha_n)} e^{p_\nu x},$$

the $\alpha_1, \alpha_2, \dots, \alpha_n$ being the roots (assumed to be distinct) of the equation

$$\sum_{\nu=1}^{n} \frac{q_\nu}{x - p_\nu} = 1.$$

4.2. Step-by-step numerical solution

Perhaps the most obvious method for the numerical solution of (4.2) is to replace the integral by a finite sum[2]. Let

$$x_i = \xi_i = a + ih, \qquad y_i = y(x_i), \qquad f_i = f(x_i), \qquad K_{il} = K(x_i, \xi_l), \qquad (4.4)$$

where h is a suitably chosen pivotal interval, and let Y_i be an approximation to be calculated for y_i. Then from (4.2) it follows first of all that $f_0 = y_0 = Y_0$. For $x = x_1$ we evaluate the integral by the trapezium rule (the use of a more accurate quadrature formula is described in § 4.3) and obtain the equation

$$Y_1 = f_1 + \frac{h}{2}(K_{10} Y_0 + K_{11} Y_1);$$

thus we can calculate

$$Y_1 = \frac{f_1 + \dfrac{h}{2} K_{10} Y_0}{1 - \dfrac{h}{2} K_{11}}. \qquad (4.5)$$

[1] WHITTAKER, E. T.: On the numerical solution of integral equations. Proc. Roy. Soc. Lond., Ser. A **94**, 367—383 (1918).

[2] Cf. J. R. CARSON: Electrical circuit theory and the operational calculus, p. 145. New York: McGraw-Hill 1926. A similar method is used by A. HUBER: Eine Näherungsmethode zur Auflösung Volterrascher Integralgleichungen. Mh. Math. Phys. **47**, 240—246 (1939); he replaces the function to be determined by piecewise-linear functions. Cf. also V. I. KRYLOV: Application of the Euler-Laplace formula to the approximate solution of integral equations of Volterra type. Trudy Mat. Inst. Steklov **28**, 33—72 (1949) [Russian; reviewed in Zbl. Math. **41**, 79 (1952), also in Math. Rev. **12**, 540 (1951)].

For $x = x_n$ we obtain in the same way the equation

$$Y_n = f_n + \frac{h}{2}(K_{n0} Y_0 + 2 K_{n1} Y_1 + 2 K_{n2} Y_2 + \cdots + 2 K_{n,n-1} Y_{n-1} + K_{nn} Y_n),$$

and hence we can calculate

$$Y_n = \frac{1}{1 - \dfrac{h}{2} K_{nn}} \left(f_n + \frac{h}{2} K_{n0} Y_0 + h \sum_{i=1}^{n-1} K_{ni} Y_i \right) \qquad (4.6)$$

in terms of the values $Y_0, Y_1, \ldots, Y_{n-1}$ calculated in the previous steps.

Example. Linear transmission systems. Suppose that we have an arbitrary linear transmission system (whose physical nature we may leave unspecified), i.e. we imagine a system with an input and output for which the associated input and output quantities (functions of time) are related by a linear transformation. If we denote the input function by $y(t)$ and the output function by $S(t)$, this transformation may be expressed in the form[1]

$$S(t) = y(t) A(0) - \int_0^t y(\xi) \frac{d}{d\xi} A(t - \xi) \, d\xi \qquad (4.7)$$

or, equivalently,

$$S(t) = \frac{d}{dt} \int_0^t y(t - \xi) A(\xi) \, d\xi; \qquad (4.8)$$

here $A(t)$ is the response to a unit step function, i.e. the output $S(t)$ produced when $y(t) = 0$ for $t < 0$ and $y(t) = 1$ for $t > 0$. The physical nature of the input and output quantities may or may not be the same; in a mechanical system, for example, the input might be a force and the output a displacement, and in an electrical system both quantities might be currents.

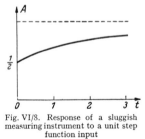

Fig. VI/8. Response of a sluggish measuring instrument to a unit step function input

For our example we will consider a sluggish measuring instrument whose immediate response to a unit step function input is a deflection of only a half a unit, which then gradually increases to the correct unit deflection according to the law (see Fig. VI/8)

$$A(t) = 1 + \frac{e^{-t} - 1}{2t}. \qquad (4.9)$$

The question arises what is the time distribution of the quantity to be measured $y(t)$ if a specific time distribution of the reading $S(t)$ has

[1] See, for example, K. W. WAGNER: Operatorenrechnung, 2nd ed., p. 14. Leipzig 1950.

been recorded. Equation (4.7) is then a Volterra integral equation of the second kind for $y(t)$. We will consider the case $S(t) \equiv 1$; thus we ask how the quantity to be measured $y(t)$ must vary in order that the deflection may remain constant.

Integration of (4.8) from 0 to t yields

$$t = \int_0^t y(t - \xi) A(\xi) \, d\xi; \tag{4.10}$$

this is a Volterra integral equation of the first kind in a form which can be obtained by a trivial change in the variable of integration for

Table VI/7. *Step-by-step solution of a Volterra integral equation*

t	$A(t)$	Calculation with $h=0 \cdot 2$		Calculation with $h=0 \cdot 1$	
		Approximation Y	Error	Approximation Y	Error
0	0·5	2		2	
0·1	0·524 187 0			1·903 252	
0·2	0·546 827 0	1·812 692	−0·009	1·822 052	+0·0005
0·3	0·568 030 3			1·744 497	
0·4	0·587 900 0	1·683 48	+0·003	1·680 822	+0·0009
0·5	0·606 530 7			1·618 255	
0·6	0·624 009 7	1·558 94	−0·008	1·568 44	+0·0012
0·7	0·640 418 1			1·517 57	
0·8	0·655 830 6	1·483 37	+0·006	1·478 77	+0·0015
0·9	0·670 316 5			1·437 00	
1	0·683 939 7	1·396 3	−0·009	1·406 98	+0·0018

any kernel of convolution type. If $A_n = A(nh)$ and Y_n is the approximation for $y(nh)$, the equation derived by using the trapezium rule to evaluate the integral reads

$$t_n = nh = \frac{h}{2} (Y_0 A_n + 2 Y_1 A_{n-1} + 2 Y_2 A_{n-2} + \cdots + 2 Y_{n-1} A_1 + Y_n A_0);$$

hence

$$Y_n = \frac{1}{A_0} \left(2n - Y_0 A_n - 2 \sum_{\nu=1}^{n-1} Y_\nu A_{n-\nu} \right). \tag{4.11}$$

The values obtained by means of this formula with the two different pivotal intervals $h = 0 \cdot 2$ and $h = 0 \cdot 1$ are given in Table VI/7. Better values can be obtained by using a more accurate quadrature formula — SIMPSON's rule, for example, cf. § 4.3.

4.3. Method of successive approximations (iteration method)

Starting from a continuous function $y_0(x)$ we construct a sequence of functions[1] $y_n(x)$ according to the iteration formula

$$y_{n+1}(x) = f(x) + \int_a^x K(x, \xi)\, y_n(\xi)\, d\xi \qquad (n = 0, 1, 2, \ldots). \qquad (4.12)$$

Under the continuity assumptions made in §4.1, the sequence converges uniformly to the solution $y(x)$ of (4.2) in every finite interval $\langle a, b^* \rangle$ with $b^* < b$. If we denote pivotal values of $y_n(x)$ by

$$y_{(n)k} = y_n(x_k)$$

and evaluate the integral by SIMPSON's rule, the iteration formula (4.12) reads

$$
\left.
\begin{aligned}
y_{(n+1), 2k} = f_{2k} + \frac{h}{3}\, (&K_{2k, 0}\, y_{(n)0} + 4K_{2k, 1}\, y_{(n)1} + \\
&+ 2K_{2k, 2}\, y_{(n)2} + 4K_{2k, 3}\, y_{(n)3} + 2K_{2k, 4}\, y_{(n)4} + \cdots + \\
&+ K_{2k, 2k}\, y_{(n)2k}) \qquad (k = 1, 2, 3, \ldots;\ n = 0, 1, 2, \ldots).
\end{aligned}
\right\} \quad (4.13)
$$

With this formula we can calculate the values $y_{(n+1)\nu}$ for even ν, but for the next cycle of the iteration we also need the values for odd ν; these can be obtained by interpolating between the "even" values, say by putting a parabola through three consecutive points:

$$y_{(n)\nu} = \tfrac{1}{8}\, (3\, y_{(n), \nu-1} + 6\, y_{(n), \nu+1} - y_{(n), \nu+3}) \qquad (4.14)$$

or, rather more accurately, by putting a cubic through four consecutive points:

$$y_{(n)1} = \frac{1}{16}\, (5\, y_{(n)0} + 15\, y_{(n)2} - 5\, y_{(n)4} + y_{(n)6}), \qquad (4.15)$$

$$
\left.
\begin{aligned}
y_{(n)\nu} = \frac{1}{16}\, (&- y_{(n), \nu-3} + 9\, y_{(n), \nu-1} + 9\, y_{(n), \nu+1} - y_{(n), \nu+3}) \\
&(\nu = 3, 5, 7, \ldots).
\end{aligned}
\right\} \quad (4.16)
$$

If the solution is taken as far as the point $x = x_{2p}$, the value $y_{(n), 2p-1}$, must, of course, be calculated from the formula corresponding to (4.15) and not from (4.16).

Example. We consider the same problem as in §4.2, i.e. the problem (4.7), (4.9) concerning a sluggish measuring instrument. With $S(t) \equiv 1$, $A(0) = \tfrac{1}{2}$ and $A'(t) = \tfrac{1}{2} t^{-2}[1 - (1 + t)\, e^{-t}]$ the iteration formula (4.12) reads

$$y_{n+1}(t) = 2 - 2 \int_0^1 y_n(\xi)\, A'(t - \xi)\, d\xi. \qquad (4.17)$$

For the calculation implied by (4.13) it is convenient to make a separate table of the various multiples of the pivotal values of $A'(t)$ which will actually be needed. These are reproduced in Table VI/8 for the chosen pivotal interval $h = 0.1$.

---

[1] An application can be found in K. ZOLLER: Die Entzerrung bei linearen physikalischen Systemen. Ing.-Arch. **15**, 1—18 (1944).

Table VI/8. *Multiples of $A'(t)$ needed for* SIMPSON'S *rule*

t	$A'(t)$	$2A'(t)$	$4A'(t)$
0	0·25		
0·1	0·233 9420		0·935 7680
0·2	0·219 0388	0·438 077 5	
0·3	0·205 201 8		0·820 807 0
0·4	0·192 3498	0·384 6996	
0·5	0·180 4080		0·721 6321
0·6	0·169 307 5	0·338 6149	
0·7	0·158 9847		0·635 9387
0·8	0·149 381 2	0·298 762 3	
0·9	0·140 4430		0·561 7720
1	0·132 1206		

Table VI/9. *Iterative solution with integrals evaluated by* SIMPSON'S *rule with $h = 0·1$*

t	Simpson sum	$y_1(t)$	$y_2(t)$	$y_3(t)$	$y_4(t)$	$y_5(t)$
0		2	2	2	2	2
0·1		1·903 27	1·905 70	1·905 678	1·905 680 7	1·905 6798
0·2	2·8096	1·812 69	1·821 848	1·821 544	1·821 5509	1·821 5506
0·3		1·7279	1·747 58	1·746 590	1·746 625 3	1·746 6246
0·4	5·2740	1·6484	1·682 00	1·679 808	1·679 9182	1·679 9139
0·5		1·5739	1·6242	1·620 08	1·620 334	1·620 323
0·6	7·4406	1·504 0	1·573 4	1·566 72	1·567 216	1·567 186
0·7		1·438 3	1·529 0	1·518 84	1·519 711	1·519 650
0·8	9·3499	1·376 7	1·490 4	1·475 90	1·477 304	1·477 192
0·9		1·318 8	1·456 9	1·437 3	1·439 40	1·439 207
1	11·0364	1·264 2	1·428 0	1·402 3	1·405 39	1·405 090

Table VI/10. *Iterative solution with integrals evaluated by* SIMPSON'S *rule with $h = 0·2$*

t	$y_1(t)$	$y_2(t)$	$y_3(t)$	$y_4(t)$	$y_5(t)$	$y_6(t)$
0	2	2	2	2	2	2
0·2	1·8130	1·822 52	1·822 411	1·822 416	1·822 404	
0·4	1·6484	1·681 97	1·679 733	1·679 821	1·679 8168	1·679 8184
0·6	1·5038	1·573 0	1·566 16	1·566 639	1·566 618	
0·8	1·3767	1·490 3	1·475 88	1·477 292	1·477 183	1·477 191
1	1·264	1·427 7	1·401 98	1·405 10	1·404 80	
1·2	1·165	1·382 4	1·341 98	1·347 80	1·347 12	1·347 187
1.4	1·076	1·351	1·292 3	1·302 02	1·300 69	
1·6	0·998	1·329	1·249 3	1·264 39	1·262 08	1·262 379

We have $y_n(0) = 2$. Starting from the quite crude approximation $y_0(x) \equiv 2$ we obtain the values shown in Table VI/9 for the first five cycles of the iteration procedure. For the first iterate $y_1(t)$ we also give the value of the integral calculated by SIMPSON's rule (the Simpson sum), and indicate the interpolated values by indenting them.

For comparison we give in Table VI/10 the values obtained by a calculation performed with the double step $h = 0.2$.

4.4. Power series solutions

If the kernel $K(x, \xi)$ and the inhomogeneous term $f(x)$ are analytic functions of simple form, it is often convenient to calculate the solution $y(x)$ for small values of $x - a$ by means of its Taylor series

$$y(x) = \sum_{\nu=0}^{\infty} \frac{y^{(\nu)}(a)}{\nu!} (x - a)^{\nu}. \tag{4.18}$$

We will illustrate the procedure for the integral equation of the second kind (4.2). First we express each derivative $y^{(\nu)}(x)$ in terms of lower derivatives by repeated differentiation of (4.2); for the first two derivatives, for example, we obtain

$$\left. \begin{aligned} y'(x) &= f'(x) + K(x, x)\, y(x) + \int_a^x \frac{\partial K(x, \xi)}{\partial x}\, y(\xi)\, d\xi, \\[2mm] y''(x) &= f''(x) + \frac{dK(x, x)}{dx}\, y(x) + K(x, x)\, y'(x) + \\[2mm] &\quad + \frac{\partial K(x, x)}{\partial x}\, y(x) + \int_a^x \frac{\partial^2 K(x, \xi)}{\partial x^2}\, y(\xi)\, d(\xi), \end{aligned} \right\} \tag{4.19}$$

where we have used the convention that

$$\frac{dK(x, x)}{dx} = \left(\frac{\partial K(x, \xi)}{\partial x} + \frac{\partial K(x, \xi)}{\partial \xi} \right)_{\xi = x}$$

and

$$\frac{\partial K(x, x)}{\partial x} = \left(\frac{\partial K(x, \xi)}{\partial x} \right)_{\xi = x}.$$

We now put $x = a$ in (4.2) and (4.19) and obtain a set of equations from which we can calculate successively the derivatives of $y(x)$ at $x = a$ which are needed in (4.18):

$$\left. \begin{aligned} y(a) &= f(a), \\[2mm] y'(a) &= f'(a) + K(a, a)\, y(a), \\[2mm] y''(a) &= f''(a) + K(a, a)\, y'(a) + \\[2mm] &\quad + \left[2\frac{\partial K(x, \xi)}{\partial x} + \frac{\partial K(x, \xi)}{\partial \xi} \right]_{\substack{x=a \\ \xi=a}} \times y(a), \end{aligned} \right\} \tag{4.20}$$

$$\cdot \; \cdot \; \cdot \; \cdot \; \cdot \; \cdot \; \cdot \; \cdot \; \cdot \; \cdot \; \cdot \; \cdot \; \cdot \; \cdot \; \cdot \; \cdot \; \cdot \; \cdot$$

§5. Functional equations

Functional equations can appear in so many different forms that there would be little sense in setting out to describe generally applicable methods; the more fruitful approach here is to look at particular equations to see what methods suggest themselves. In any case most of the methods mentioned in §§ 1 and 2 for the solution of integral equations can be readily adapted for the solution of more general equations, and we therefore limit ourselves to a few examples.

5.1. Examples of functional equations

Any equation which expresses a property possessed by one or more functions or by a class of functions may be called a functional equation[1]. In such an equation might appear, for example, a function $u(x, y)$, its partial derivatives, its values at points other than (x, y), say $u(x+h, y+k)$, integrals with integrands involving u, etc. Thus differential equations, integral equations, integro-differential equations, difference equations, in fact all the equations dealt with in this book, are functional equations. Sometimes, however, the term functional equation is used in a more restricted sense applying only to those cases in which the arguments of the function to be determined are not the same throughout the equation. If we restrict the meaning of the term in this way, we may, for example, refer to an equation in which the arguments of the unknown function are not the same throughout and in which derivatives of the unknown function appear as a functional-differential equation.

We now select just a few examples from the very many different types of functional equation which can occur; at the same time we indicate possible ways of dealing with them to show that there is a corresponding diversity of methods for their solution.

A very simple functional-differential equation is

$$y'_{(x)} = y_{(x-1)}$$

(whenever there is a possibility of the argument being misread as a factor, as here, we write it as a subscript). Here, $y(x)$ in the interval $\langle 0, 1 \rangle$, say, may be chosen arbitrarily from among the differentiable functions with $y'(1) = y(0)$, then $y(x)$ in the intervals $\langle 1, 2 \rangle, \langle 2, 3 \rangle, \dots$ determined successively by integration in accordance with the differential equation. Thus further conditions are required to determine a unique solution. In applied problems such conditions will usually arise naturally in the formulation of the particular problem. For example, in the

---

[1] Cf. S. PINCHERLE: Encyklopädie der mathematischen Wissenschaften, Vol. II, Part I, 2nd half, pp. 788—817. Leipzig 1904—1916. — KAMKE, E.: Differential-gleichungen, Lösungsmethoden und Lösungen, Vol. I, pp. 630—636. Leipzig 1942.

theory of structures[1] problems occur in which the additional conditions necessary to determine a unique solution of difference equations of the form

$$\sum_{k=0}^{n} a_{r,k}\, y_{(x_0 + (k+r)\,h)} = c_r, \qquad (r = 0, 1, \ldots, p)$$

(with given $a_{r,k}$, c_r, x_0, h) are provided by certain boundary conditions; one then has a system of linear equations for a finite number of values of y, which can be solved by one of the usual methods.

Another very simple functional equation occurs in spectroscopy. If the width of the slit in a spectrometer is s, the measured (known) energy distribution $E(x)$ is related to the "true" (required) energy distribution $J(x)$ by the equation[2]

$$\frac{dE(x)}{dx} = J_{\left(x + \frac{s}{2}\right)} - J_{\left(x - \frac{s}{2}\right)}.$$

This equation is solved with the aid of summation symbols[3].

A problem in kinetics leads to the difference equation[4]

$$y_{(x)} + y_{\left(x + \frac{2\pi}{3}\right)} + y_{\left(x + \frac{4\pi}{3}\right)} = h = \text{constant}.$$

The corresponding homogeneous equation, to which it can be reduced by putting $y_{(x)} = g_{(x)} + \frac{1}{3}h$, is satisfied by any trigonometric series of the form

$$g_{(x)} = \sum_{n=1}^{\infty} (a_n \cos n\,x + b_n \sin n\,x)$$

with $a_n = b_n = 0$ for all values of n which are divisible by 3. Thus there are infinitely many possible "guide" curves.

Another type of functional equation involves the "iterates" $\varphi_n(x)$ of a function $y = \varphi(x)$. These are defined successively for $n = 1, 2, \ldots$ by the iterative formula $\varphi_{n+1}(x) = \varphi(\varphi_n(x))$, $\varphi_1 = \varphi(x)$ [for example, $\varphi_2(x) = \ln \ln x$ if $\varphi(x) = \ln x$]. Such sequences of iterates are considered

[1] See, for instance, P. FUNK: Die linearen Differenzengleichungen und ihre Anwendung in der Theorie der Baukonstruktionen. Berlin 1920. — BLEICH, F., and E. MELAN: Die gewöhnlichen und partiellen Differenzengleichungen in der Baustatik. Berlin 1927. See also W. E. MILNE: Numerical calculus, 393 pp. Princeton 1949, in particular pp. 324—348.

[2] MEYER-EPPLER, W.: Z. Instrumentenkunde **60**, 198 (1940).

[3] See N. E. NÖRLUND: Differenzenrechnung, Berlin 1924, or L. M. MILNE-THOMSON: Calculus of finite-differences, Ch. VIII, London 1933.

[4] FISCHER, H. J.: Kurven, in denen ein Drei- oder Vieleck so herumbewegt werden kann, daß seine Ecken die Kurve durchlaufen. Dtsch. Math. **1**, 485—498 (1936).

in various contexts; for example, the repeated application of NEWTON's formula for the approximate determination of a zero of an algebraic or transcendental equation $f(x) = 0$ yields the iterates of the function $\varphi(x) = x - f(x)/f'(x)$. A question of interest is whether the sequence $\varphi_n(x)$ converges as $n \to \infty$, or, more precisely, for what values of x does it converge. Many investigations in function theory concern this iteration of functions, especially the case in which the functions are rational; these investigations are rather complicated in parts, and we cannot go into them here.

Functional-differential equations arise in the analysis of control processes in which time-lags are taken into account. Probably the simplest case is that of an oscillatory system of one degree of freedom which is acted upon by a force $P(t)$ whose magnitude at time t depends on the displacement of the system at a previous time $t - \tau$, where τ is a constant time-lag. If $x_{(t)}$ is the displacement of the system at time t, so that $P(t)$ is a function of $x_{(t-\tau)}$, the equation of the system might read

$$m\ddot{x}_{(t)} + k\dot{x}_{(t)} + c\,x_{(t)} = a + b\,x_{(t-\tau)}.$$

More general systems give rise to equations of the form

$$\sum_{r=0}^{n} \sum_{k=0}^{p} a_{r,k}\, y^{(r)}_{(t-\tau_k)} = r(t).$$

Several investigations[1] concern those solutions which, with their first $n-1$ derivatives, grow no faster than some power of t as $|t| \to \infty$. The question of stability, i.e. what conditions on the coefficients yield systems for which all continuous solutions of the functional-differential equation remain bounded as $t \to +\infty$, is particularly important for control processes. Again one can first remove the inhomogeneous term from the equation. One can then obtain an indication of the stability by assuming a solution of the form $y(t) = e^{st}$; this leads to a transcendental equation for s:

$$\sum_{r=0}^{n} \sum_{k=0}^{p} a_{r,k}\, s^r e^{-s\tau_k} = 0.$$

If $s_1, s_2, \ldots$ are the roots of this equation, then $y(t) = \sum_{\sigma=1}^{\infty} c_\sigma e^{s_\sigma t}$, with constants c_σ such that the series converges, but otherwise arbitrary, is

---

[1] SCHMIDT, ERHARD: Über eine Klasse linearer funktionaler Differentialgleichungen. Math. Ann. **70**, 499−524 (1911). — HILB, E.: Lineare funktionale Differentialgleichungen. Math. Ann. **78**, 137−170 (1918), among others.

also a solution; and if any one of the roots s_σ has a positive real part, then the control system is unstable. The theory of nomograms is often made use of in stability investigations[1].

5.2. Examples of analytic, continuous and discontinuous solutions of functional equations

For many functional equations a geometrical interpretation of the equation can yield a graphical method for the construction of solutions.

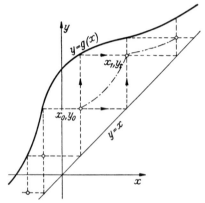

Fig. VI/9. Construction of solutions of the functional equation $y[y(x)] = g(x)$

Example. Consider the functional equation

$$y\big(y(x)\big) = g(x). \qquad (5.1)$$

Thus we are required to find a function $y(x)$ whose second iterate $y_2(x)$ coincides with a given function $g(x)$.

Let the given function $g(x)$ be real and continuous, say, and assume for the present that

1. $g(x) > x$ for all x,

2. $g(x)$ increases monotonically with x.

Let (x_0, y_0) be a point through which a curve $y(x)$ which satisfies (5.1) is to pass. Then a countably finite number of points on this curve can be computed successively by the formula

$$\left.\begin{array}{l} x_{n+1} = y(x_n) \\[4pt] y_{n+1} = g(x_n) \end{array}\right\} \quad (n = 0, 1, 2, \ldots);$$

alternatively these points may be located graphically by the self-evident construction shown in Fig. VI/9. This sequence of points can also be extended "backwards" $(n = -1, -2, \ldots)$. It can be seen from the construction that if the points (x_0, y_0) and (x_1, y_1) are joined by any curve C which is not crossed more than once by each parallel to the

---

[1] See R. C. Oldenbourg and H. Sartorius: Dynamik selbsttätiger Regelungen, 2nd ed., 258 pp. Munich 1951. — Engel, F. V. A. in collaboration with R. C. Oldenbourg: Mittelbare Regler und Regelanlagen. Berlin: VDI-Verlag 1944. — Hahn, W.: Bericht über Differential-Differenzengleichungen mit festen und veränderlichen Spannen. Jber. Dtsch. Math. Ver. **57**, 55—84 (1954). Another applied problem giving rise to a functional-differential equation is treated by C. Meissner (Zürich): Bestimmung des Profils einer Seilbahn, auf der unter Mitberücksichtigung des Gewichtes des Drahtseiles gleichförmige Bewegung möglich sein soll. Schweiz. Bauztg. **54**, No. 7, 96—98 (1909).

co-ordinate axes, then the construction can be repeated for all points of this curve; in this way we can complete a particular solution of (5.1) for each such curve C. This procedure can still be carried out if there are discontinuities in the curve C; hence we can also construct discontinuous solutions of (5.1) in the same way.

If $g(x)$ possesses a power series expansion:

$$g(x) = \sum_{n=0}^{\infty} g_n x^n,$$

we may expect there to be solutions possessing power series expansions:

$$y(x) = \sum_{n=0}^{\infty} a_n x^n.$$

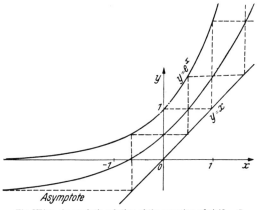

Fig. VI/10. An analytic solution of the equation $y[y(x)] = e^x$

We set out to find them by substituting the assumed form of solution into the functional equation and equating coefficients of powers of x. This yields an infinite system of non-linear equations for the unknowns a_ν:

$$\left.\begin{aligned}
g_0 &= a_0 & & + a_0 a_1 + a_0^2 a_2 + a_0^3 a_3 & & + \cdots \\
g_1 &= & & a_1^2 + 2 a_0 a_1 a_2 + 3 a_0^2 a_1 a_3 & & + \cdots \\
g_2 &= 2 a_0 a_2^2 + a_1 a_2 + a_1^2 a_2 & & + 3 a_0 a_3 (a_1^2 + a_0 a_2) & & + \cdots \\
g_3 &= 2 a_0 a_2 a_3 + 2 a_1 a_2^2 + a_1 a_3 + a_1^3 a_3 & & + 3 a_0^2 a_3^2 + 6 a_0 a_1 a_2 a_3 & & + \cdots.
\end{aligned}\right\} \quad (5.2)$$

Normally an approximate solution of such a system of equations will be determined by solving a finite system obtained from the infinite system by truncation, i.e. by ignoring all but the first p (say) equations and putting $a_\nu = 0$ for $\nu \geq p$. In (5.2) the successive truncated systems are indicated by dotted lines.

We now consider some specific forms for $g(x)$.

1. $g(x) = e^x$. The above procedure for obtaining a power series solution[1] here yields

$$y \approx \tfrac{1}{2} + x$$

---

[1] The methods of function theory have been used to demonstrate the existence of an analytic solution for this case by H. Kneser: Reelle analytische Lösungen der Gleichung $\varphi(\varphi(x)) = e^x$ und verwandter Funktionalgleichungen. J. Reine Angew. Math. **187**, 56—67 (1949). Mathematicians' attention was directed to this equation by its occurrence in practical industrial problems.

as a first approximation ($p = 2$) and

$$y \approx 0.4979 + 0.8781\,x + 0.2618\,x^2$$

as a second approximation ($p = 3$). These curves can be extended for large $|x|$ by means of the graphical construction (see Fig. VI/10); the construction shows that the solution has an asymptote parallel to the x axis.

2. $g(x) = 1 - x^2$. This function violates the two conditions that $g(x)$ shall increase monotonically with x and always be greater than x.

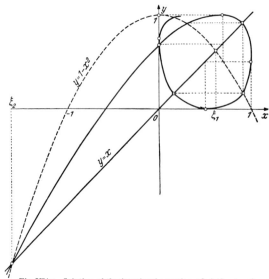

There are now two points where $g(x) = x$; their abscissae are $x = \xi_1$ and $x = \xi_2$, where $\xi_{1,2} = \frac{1}{2}(-1 \pm \sqrt{5})$. No real differentiable solution can pass through the point $x = y = \xi_1 = \frac{1}{2}(-1 + \sqrt{5})$, for differentiation of the functional equation yields

$$y'\big(y(x)\big) \cdot y'(x) = -2x,$$

from which it follows that $y'(\xi_1) = \sqrt{-2\xi_1}$ for $x = y = \xi_1$, and this value is imaginary. There is, however, an analytic solution through the

Fig. VI/11. Solution of the functional equation $y\,[y(x)] = 1 - x^2$

point $x = y = \xi_2 = -\frac{1}{2}(1 + \sqrt{5})$, and for this solution we can assume a power series expansion; for the second approximation ($a_\nu = 0$ for $\nu > 2$) (5.2) reduces to three equations, which with $g_0 = 1$, $g_1 = 0$, $g_2 = -1$ lead to the quartic $a_1^2(a_1^2 + 2a_1) = 4$ for a_1 and thence to the two real approximations

$$y \approx \quad 0.648 + 1.090\,x - 0.842\,x^2,$$
$$y \approx -6.244 - 2.320\,x - 0.186\,x^2.$$

Another way of obtaining a power series solution is to calculate the derivatives of $y(x)$ at the point $x = \xi_2$ from repeated differentiations of the equation and then write down the Taylor series for $y(x)$ at $x = \xi_2$. Thus we have

$$y''(y)\,[y'(x)]^2 + y'(y)\,y''(x) = -2, \quad \text{so that} \quad y''(\xi_2) = \frac{-2}{y'(\xi_2)\,(1 + y'(\xi_2))};$$

similarly

$$y'''(\xi_2) = -3\frac{[y''(\xi_2)]^2}{1+[y'(\xi_2)]^2} \quad \text{etc.},$$

and we obtain

$$y = \xi_2 + 1\cdot799\,(x-\xi_2) - 0\cdot1986\,(x-\xi_2)^2 - 0\cdot0187\,(x-\xi_2)^3 + \cdots.$$

The curve corresponding to this solution is shown in Fig. VI/11, where the initial part calculated from the power series has been continued by means of the graphical construction. Once inside the square $0\le(x,y)\le1$, the curve cannot get out again and "inscribes" it an infinite number of times, clinging more and more closely to the sides of the square with each revolution.

5.3. Example of a functional-differential equation from mechanics

Consider the small oscillations of a mechanical system consisting of a homogeneous string of length l fixed at one end $x=l$ and attached at the other $x=0$ to the centre of a transversely mounted spring as in Fig. VI/12.

The displacement of the string $u(x,t)$ at the point x at time t satisfies the wave equation

$$\frac{\partial^2 u}{\partial t^2} = C^2\frac{\partial^2 u}{\partial x^2}, \qquad (5.3)$$

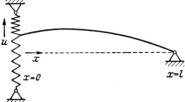

Fig. VI/12. A non-linear oscillatory system

where C is a given constant. A displacement $u(0,t)=u_0(t)$ at the point $x=0$ produces an opposing force $H(u_0)$ in the spring, $H(u)$ being a given, in general non-linear, function of u; we will assume also that $H(u)$ is an odd function of u. The boundary conditions therefore read

$$\left.\begin{array}{r}\dfrac{\partial u(0,t)}{\partial x} = G\left(u(0,t)\right), \\[2mm] u(l,t) = 0,\end{array}\right\} \qquad (5.4)$$

where $G(u)=kH(u)$ is likewise a given, non-linear, odd function of u (k is a constant depending on the tension in the string). Our object is to investigate solutions which are periodic in time; the period T will depend on the maximum displacement. Let us look for a solution which passes through the position of equilibrium at time $t=0$, say, i.e. a solution for which

$$u(x,0) = 0 \quad \text{for} \quad 0\le x\le l. \qquad (5.5)$$

Now the general solution of (5.3) can be expressed in the form

$$u = f(x+Ct) - g(x-Ct),$$

where f and g are arbitrary functions. The initial condition (5.5) then becomes

$$f(x) = g(x) \quad \text{for} \quad 0\le x\le l,$$

and since the values of $f(x)$ for $x<0$ and of $g(x)$ for $x>l$ do not affect the solution for $t>0$, which is what interests us, we can put $f(x)=g(x)$ for all x. The boundary conditions (5.4) then reduce to

$$f(l + y) = f(l - y), \qquad (5.6)$$
$$f'(y) - f'(-y) = G(f(y) - f(-y)), \qquad (5.7)$$

where $y = C\, t$.

It remains to express the condition of periodicity in terms of f: we require that

$$f(x + \tau) = f(x - \tau)$$

for some real finite number τ; then $f(x)$ will have the period 2τ and $u(x, t)$, as a function of t, the period $T = 2\tau/C$.

Thus for given $G(u)$ we seek periodic solutions of the functional-differential equation (5.7) which are symmetric about the point $x = l$.

A simple way of obtaining a solution of this problem is to choose a value for the period 2τ and determine a solution of (5.7) with this period, say by the finite-

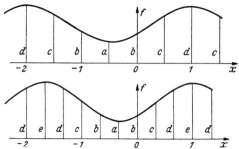

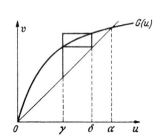

Fig. VI/13. Notation for the finite-difference solution of the functional equation

Fig. VI/14. Solution of the transcendental equations

difference method; naturally for given $G(u)$ there may be values of 2τ for which no solution exists.

We illustrate the procedure with $l = 1$, and choose first $2\tau = 3$. We need only consider a half-cycle, and if we divide this interval into three, i.e. use a finite-difference step $h = \frac{1}{3}\tau = \frac{1}{2}$, we have four unknown f values a, b, c, d as indicated in Fig. VI/13. The finite-difference equations read

$$\text{(for } x = \tfrac{1}{2}) \qquad d - b = G(c - a), \quad \text{or} \quad \beta = G(\alpha),$$
$$\text{(for } x = 1) \qquad c - a = G(d - b), \quad \text{or} \quad \alpha = G(\beta)$$

(no new equations are obtained for $x = \frac{3}{2}$ and $x = 2$); the α and β introduced here are corresponding values of $u(0, t)$:

$$\alpha = c - a = f\left(\tfrac{1}{2}\right) - f\left(-\tfrac{1}{2}\right) = u\left(0, \tfrac{T}{6}\right), \quad \beta = d - b = f(1) - f(-1) = u\left(0, \tfrac{T}{3}\right).$$

For a normal spring $G(u)$ is monotonic, and the values of α and β are given by the intersections of the curve $v = G(u)$ with the straight line $v = u$ (Fig. VI/14).

Let us now choose another period, say $2\tau = \frac{8}{3}$, and this time use more pivotal points, say with $h = \frac{1}{4}\tau = \frac{1}{3}$. We now have five unknown f values a, b, c, d, e (as in Fig. VI/13) and three corresponding values of $u(0, t)$: $u\left(0, \tfrac{T}{8}\right) = f\left(\tfrac{1}{3}\right) - f\left(-\tfrac{1}{3}\right) = c - a = \gamma$, $u\left(0, \tfrac{T}{4}\right) = d - b = \delta$, $u\left(0, \tfrac{3T}{8}\right) = e - c = \varepsilon$. If we put

$G^* = 2hG = \tfrac{2}{3}G$, the difference equations read

$$\begin{array}{lr}
(\text{for } x = \tfrac{1}{3}) & \delta = G^*(\gamma), \\
(\text{for } x = \tfrac{2}{3}) & \gamma + \varepsilon = G^*(\delta), \\
(\text{for } x = 1) & \delta = G^*(\varepsilon)
\end{array}$$

(no new equations are obtained for $x = \tfrac{4}{3}, \tfrac{5}{3}, \ldots$).

If $G(u)$ is monotonic, we must have $\gamma = \varepsilon$ and the equations simplify to the pair of equations

$$\delta = G^*(\gamma), \qquad 2\gamma = G^*(\delta).$$

These also may easily be solved graphically [see Fig. VI/14; in this sketch, which is only for illustration, we have not drawn a new curve for $G^*(u)$].

5.4. Miscellaneous exercises on Chapter VI

1. Consider a luminous, line "object" whose intensity of illumination is a function $z(\xi)$ of the distance ξ along the line, and let its image formed by an optical system (Fig. VI/15) be another line, say the x axis, illuminated with the intensity $y(x)$; further let all parts of the lines outside of the sections $|x| \leq 1$ and $|\xi| \leq 1$ be shielded by blinds. Then the intensity distributions $y(x)$ for $|x| \leq 1$ and $z(\xi)$ for $|\xi| \leq 1$ are related by an integral equation[1] of the form

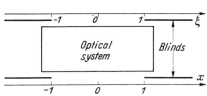

$$y(x) = \int\limits_{-1}^{1} K(x, \xi)\, z(\xi)\, d\xi;$$

Fig. VI/15. Optical system with line object and image

the kernel $K(x, \xi)$ depends on the optical system used, but may be approximated by

$$K(x, \xi) = \begin{cases} 1 + \cos \pi (x - \xi) & \text{for} \quad |x - \xi| \leq 1, \\ 0 & \text{for} \quad |x - \xi| \geq 1. \end{cases}$$

For what intensity distributions are the object and image distributions similar, i.e. such that $z(x) = \lambda y(x)$? Calculate approximations for the first few eigenfunctions by the finite-sum method of § 1.2.

2. Apply the enclosure theorem (2.15), (2.16) of § 2.3 to the integral equation

$$y(x) = \lambda \int\limits_{0}^{1} e^{x\xi} y(\xi)\, d\xi.$$

3. Determine an approximate solution of the non-linear integral equation

$$\int\limits_{0}^{1} \frac{d\xi}{y(x) + y(\xi)} = 1 + x$$

by the finite-sum method of § 1.2. May the Ritz method of § 1.7 also be applied to this equation?

[1] FRANK, PH., and R. v. MISES: Die Differential- und Integralgleichungen der Mechanik und Physik, Vol. I, p. 473. Brunswick 1930.

4. Determine a real analytic solution of the functional equation

$$y_{(x + y(x))} = 1 - x^2$$

(where the argument is again written as a subscript so that it cannot be read as a factor) under the assumption that such a solution exists.

5. Use the finite-difference method to obtain approximate solutions of the eigenvalue problem presented by the functional-differential equation

$$- y''_{(x)} = \lambda y_{(1-x)}$$

with the boundary conditions

$$y(0) = y'(1) = 0.$$

6. Apply (a) the Ritz method and (b) the power series method to the problem of the last exercise.

7. Let us end by applying the two well-tried methods

(a) the finite-difference method

(b) the Ritz method,

which have been used repeatedly throughout this book, to the partial functional-differential eigenvalue problem

$$\nabla^2 u(x, y) + \lambda u(- x, - y) = 0$$

with the boundary conditions

$$u = 0 \quad \text{for} \quad x = 1 \quad \text{and for} \quad y = 1,$$

$$\frac{\partial u}{\partial v} = 0 \quad \text{for} \quad x = -1 \quad \text{and for} \quad y = -1.$$

Calculate approximations for the first few eigenvalues.

5.5. Solutions

1. (a) Three pivotal points $x_j = j$ with $j = 0, \pm 1$. Let y_j be the approximate pivotal value of $y(x)$ at $x = x_j$. We must remember that there are discontinuities in the derivatives of the kernel and, as mentioned in § 1.2, must choose our quadrature formulae accordingly; if we evaluate the integral by the trapezium rule for $x = \pm 1$ and by SIMPSON's rule for $x = 0$, we obtain from the equations

$$y_{-1} = \Lambda y_{-1}, \quad y_0 = \Lambda \tfrac{1}{3} 8 y_0, \quad y_1 = \Lambda y_1$$

the three approximate values

$$\Lambda = \tfrac{3}{8}, 1, 1$$

for the eigenvalue λ.

(b) Five pivotal points $x_j = \tfrac{1}{2} j$ with $j = 0, \pm 1, \pm 2$. As usual with symmetric eigenvalue problems we can reduce the number of unknowns by treating the symmetric and antisymmetric solutions separately; thus if y_j is an approximation for $y(x_j)$, we postulate $y_j = y_{-j}$ for symmetric solutions and $y_j = - y_{-j}$ for antisymmetric. For the symmetric solutions we obtain the equations

$$y_2 = \frac{\Lambda}{6} (2 y_2 + 4 y_1),$$

$$y_1 = \frac{\Lambda}{8} \frac{3}{2} (y_2 + 6 y_1 + 3 y_0),$$

$$y_0 = \frac{\Lambda}{6} (8 y_1 + 4 y_0)$$

(it is a help in setting them up to record the values of the kernel $K(x, \xi)$ in an array as in Fig. VI/16); we have taken account of the points at which the kernel has discontinuous higher derivatives by using SIMPSON's rule for $x = 0$, 1 and the "three-eighths" rule for $x = \frac{1}{2}$.

For $\varkappa = 1/\Lambda$ we obtain the equation

$$- 24\varkappa^3 + 51\varkappa^2 - \frac{34}{3}\varkappa - 2 = 0,$$

which yields the approximate eigenvalues

$$\Lambda = 0.541, \quad 2.53, \quad -8.73.$$

By postulating antisymmetry we obtain similarly

$$\varkappa = \frac{1}{48}\left(35 \pm \sqrt{649}\right); \quad \Lambda = 0.794, \quad 5.04.$$

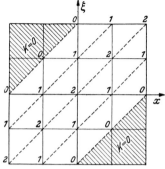

Fig. VI/16. Values of the kernel

The points given by the corresponding solutions of the homogeneous equations are plotted in Fig. VI/17 and joined by straight lines, so that the approximate eigenfunctions are represented by piecewise-linear functions; smooth approximations could be obtained by rounding off the corners.

2. An obvious choice for $F_0(x)$ is e^{ax}; then

$$F_1(x) = \int_0^1 e^{x\xi} e^{a\xi}\,d\xi = \frac{e^{a+x} - 1}{a + x}$$

and

$$\Phi(x) = \frac{F_0(x)}{F_1(x)} = \frac{e^{ax}(a + x)}{e^{a+x} - 1}.$$

Curves of $\Phi(x)$ against x are drawn in Fig. VI/18 for several values of a; the difference $\Phi_{max} - \Phi_{min}$ appears to be smallest for a value of a about 0.59; for this value we obtain the limits

$$\Phi_{min} = 0.7338 \leq \lambda \leq 0.7417 = \Phi_{max}.$$

The mean value, which can be in error by at most 0.53 %, gives $\lambda \approx 0.7377$.

At the same time $F_1(x)$ with $a = 0.59$ provides an approximation for the corresponding eigenfunction.

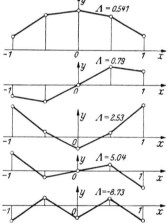

Fig. VI/17. Approximations obtained by the summation method for distributions of illumination which undergo pure magnification in an optical system

3. (a) First of all let us find a very crude approximation by taking only two pivotal points $x = 0$ and $x = 1$ and using the trapezium rule; then for y_0 and y_1, the approximate values of $y(0)$ and $y(1)$, we obtain the equations

$$\frac{1}{2y_0} + \frac{1}{y_0 + y_1} = 2, \qquad \frac{1}{y_0 + y_1} + \frac{1}{2y_1} = 4.$$

These yield a quadratic for the ratio $\eta = y_1/y_0$, from which we calculate the two values

$$\eta = \frac{y_1}{y_0} = -\frac{3}{4} \pm \frac{1}{4}\sqrt{17} = \begin{cases} 0\cdot2808 \\ -1\cdot7808; \end{cases}$$

we therefore obtain two solutions $\left(8\,y_0 = 1 \pm \sqrt{17},\; 16\,y_1 = 7 \mp \sqrt{17}\right)$:

$$\text{1st solution:} \qquad y_0 = 0\cdot640, \qquad y_1 = 0\cdot180$$

$$\text{2nd solution:} \qquad y_0 = -0\cdot390, \quad y_1 = 0\cdot695.$$

For the second solution, which changes sign, the integral equation is singular and the integral must be regarded as a Cauchy principal value (cf. § 3.2); con-

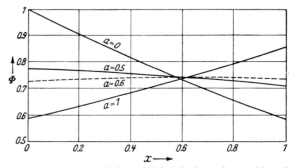

Fig. VI/18. A family of functions $\Phi(x)$ from which is to be chosen the one with smallest variation

sequently, indiscriminate use of the finite-sum method should not be carried any further in this case, and we will proceed to higher accuracy only for the solution with constant sign.

(b) Let y_0, y_1, y_2 be approximations for $y(0)$, $y(\tfrac{1}{2})$, $y(1)$. Evaluating the integral by SIMPSON's rule we obtain

$$\frac{1}{2\,y_0} + \frac{4}{y_0 + y_1} + \frac{1}{y_0 + y_2} = 6,$$

$$\frac{1}{y_0 + y_1} + \frac{4}{2\,y_1} + \frac{1}{y_1 + y_2} = 9,$$

$$\frac{1}{y_0 + y_2} + \frac{4}{y_1 + y_2} + \frac{1}{2\,y_2} = 12.$$

If we write $a = y_0 + y_1$, $b = y_1$, $c = y_1 + y_2$, we have a system of non-linear equations for a, b, c for which an iterative solution is suitable; thus we calculate the $(n+1)$-th approximation $a_{n+1}, b_{n+1}, c_{n+1}$ from the n-th by the formulae

$$\left.\begin{aligned} \frac{4}{a_{n+1}} &= 6 - \frac{1}{2\,(a_n - b_n)} - \frac{1}{a_n + c_n - 2b_n}, \\[4pt] \frac{2}{b_{n+1}} &= 9 - \frac{1}{a_{n+1}} - \frac{1}{c_n}, \\[4pt] \frac{4}{c_{n+1}} &= 12 - \frac{1}{a_{n+1} + c_n - 2b_{n+1}} - \frac{1}{2\,(c_n - b_{n+1})} \end{aligned}\right\} \qquad (n = 0, 1, 2, \ldots).$$

This iteration converges well, and yields the values

$$a = 0.990, \qquad b = 0.332, \qquad c = 0.505,$$

from which we obtain the approximate solution

$$y_0 = 0.658, \qquad y_1 = 0.332, \qquad y_2 = 0.173.$$

(c) In Ritz's method we have to find approximate solutions of the variational problem

$$J[u] = \tfrac{1}{2} \int_0^1 \int_0^1 \ln\left(u(x) + u(\xi)\right) dx\, d\xi - \int_0^1 (1+x)\, u(x)\, dx = \text{extremum.}$$

In the first approximation with $u = a$ we have $J = \tfrac{1}{2} \ln 2a - \tfrac{3}{2} a$, and $\partial J/\partial a = 0$ yields the value $a = \tfrac{1}{3}$; however, even for the second approximation with $u = a + b x$ the amount of calculation involved is already disproportionately large in comparison with the finite-sum method.

4. (a) Let us see first whether $y(x)$ can vanish at some point $x = \xi$; if, for the present, we consider only single-valued solutions, the functional equation at such a point would read $0 = 1 - \xi^2$, so that we must have $\xi = \pm 1$. Let us now determine the derivatives at $\xi = \pm 1$ of the solutions which vanish at these points. By repeated differentiation of the functional equation we obtain

$$y'_{(x+y(x))} [1 + y'(x)] = -2x,$$

$$y''_{(x+y(x))} [1 + y'(x)]^2 + y''(x)\, y'_{(x+y(x))} = -2;$$

now for $x = \xi = \pm 1$ and $y(\xi) = 0$ the first equation reduces to a quadratic for $y'(\xi)$:

$$[y'(\xi)]^2 + y'(\xi) + 2\xi = 0;$$

since this has real solutions only when $\xi = -1$ we need no longer consider the point $\xi = 1$. For $\xi = -1$ we have $y'(\xi) = 1$ or -2, and for each of these two values the higher derivatives at $\xi = -1$ are determined uniquely. To calculate the two solutions $y_1(x), y_2(x)$ through the point $(-1, 0)$ we put $x + 1 = u$ and insert the power series

$$y_{(-1+u)} = \sum_{\nu=1}^{\infty} a_\nu u^\nu$$

into the functional equation, which then reads

$$y_{(-1+u+y(-1+u))} = 2u - u^2;$$

for $a_1 = -2$ and $a_1 = 1$ we obtain the respective expansions

$$y_{1(-1+u)} = -2u + u^2 - \frac{2}{3} u^3 + \frac{1}{3} u^4 - \cdots,$$

$$y_{2(-1+u)} = u - \frac{1}{5} u^2 - \frac{4}{225} u^3 - \frac{11}{3825} u^4 - \cdots.$$

(b) To pursue the solution $y_1(x)$ further, let us calculate sequences of points (x_j, y_j) from the equations

$$x_{j+1} = x_j + y_j, \qquad y_{j+1} = 1 - x_j^2 \qquad (j = 0, 1, 2, \ldots)$$

with $x_0 = 0$ and y_0 as a parameter; we choose several values of y_0 covering a range in which we expect to find the value associated with $y_1(x)$. Each of these sequences lies on a solution of the functional equation, and as can be seen in Fig. VI/19, which shows several points of the sequences for $y_0 = -1.4, -1.38, -1.397$, these curves

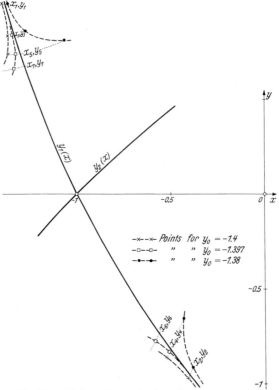

-x--x- Points for $y_0 = -1.4$
-□--□-   ,,   ,, $y_0 = -1.397$
-•--•-   ,,   ,, $y_0 = -1.38$

Fig. VI/19. Solution of the functional equation $y_{(x+y(x))} = 1 - x^2$

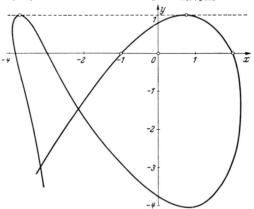

Fig. VI/20. A many-valued solution of the functional equation

diverge in varying degrees either side; by interpolation the initial value corresponding to the non-diverging curve is found to be $y_0 = -1.39635$.

(c) By continuing the solution $y_2(x)$ by means of the functional equation we
obtain a many-valued function; a part of it is reproduced in Fig. VI/20.

5. With the pivotal value notation a, b, c as in Fig. VI/21 and Λ as an approximate value of λ the finite-difference equations read

$$9(b - 2a) + \Lambda b = 9(c - 2b + a) + \Lambda a = 9(2b - 2c) = 0,$$

and for non-trivial solutions we must have $\Lambda = 9(-1 \pm \sqrt{2})$. These values are expressed in decimals in Table VI/11, together with the values obtained for $h = \frac{1}{4}$ and $h = \frac{1}{5}$.

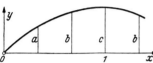

Fig. VI/21. Pivotal value notation for the functional-differential eigenvalue problem

Table VI/11. *Successive approximations for the eigenvalues*

Pivotal interval	Λ_{-2}	Λ_{-1}	Λ_1	Λ_2
$h = \frac{1}{3}$		$-21{\cdot}73$	$3{\cdot}728$	
$h = \frac{1}{4}$		$-22{\cdot}26$	$3{\cdot}634$	$50{\cdot}6$
$h = \frac{1}{5}$	$-87{\cdot}25$	$-22{\cdot}29$	$3{\cdot}591$	$56{\cdot}0$

The problem possesses infinitely many positive and infinitely many negative eigenvalues.

6. (a) The corresponding variational problem reads

$$J[u] = \int\limits_0^1 [u'^2_{(x)} - \lambda u_{(x)} u_{(1-x)}] \, dx = \text{extremum}$$

with respect to the domain of functions u possessing continuous derivatives and satisfying the condition $u(0) = 0$. Thus if we put $u = y + \varepsilon\eta$, where $\eta(0) = 0$, in $J[u]$, the condition $(\partial J/\partial \varepsilon)_{\varepsilon=0}$, which is necessary if y is to minimize J, leads to

$$\int\limits_0^1 \eta(x) \{-y''_{(x)} - \lambda y_{(1-x)}\} \, dx + [\eta \, y']_0^1 = 0$$

on integration by parts, and the usual arguments show that y, which already satisfies $y(0) = 0$, must also satisfy the functional-differential equation and the boundary condition $y'(1) = 0$.

Let us use the two-parameter Ritz expression

$$u = c_1(2x - x^2) + c_2(3x - x^3),$$

which satisfies both boundary conditions [u need only satisfy the condition $u(0) = 0$, but here it is no more difficult to satisfy the other boundary condition at the same time].

The necessary conditions $\partial J/\partial c_1 = \partial J/\partial c_2 = 0$ yield two homogeneous linear equations for the c_j; their determinant

$$\begin{vmatrix} \dfrac{4}{3} - \Lambda\,\dfrac{11}{30} & \dfrac{5}{2} - \Lambda\,\dfrac{2}{3} \\[2ex] \dfrac{5}{2} - \Lambda\,\dfrac{2}{3} & \dfrac{24}{5} - \Lambda\,\dfrac{169}{140} \end{vmatrix}$$

vanishes for the values

$$\Lambda = \begin{cases} 3{\cdot}51984 \\ -23{\cdot}346. \end{cases}$$

A one-parameter expression ($c_2 = 0$) would have lead to the value $\Lambda = \dfrac{40}{11} \approx 3{\cdot}636$, while for a two-parameter expression satisfying the boundary condition $u(0) = 0$ only, i.e. $u = c_1 x + c_2 x^2$, we would have obtained the values

$$\Lambda = 4(-8 \pm \sqrt{79}) = \begin{cases} 3{\cdot}553 \\ -67{\cdot}55. \end{cases}$$

(b) At the point $x = 1$ we have $y'(1) = 0$, and from the functional equation $y''(1) = -\lambda y(0) = 0$. We therefore assume the power series solution

$$y_{(1-x)} = a_0 + a_3 x^3 + a_4 x^4 + a_5 x^5 \dots .$$

By inserting the series in the functional equation and equating coefficients we obtain an infinite system of equations for the a_j; with $\lambda = 20\mu$ they read

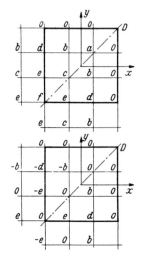

$$a_0 + a_3 + a_4 + a_5 + a_6 + \dots = 0,$$
$$10\mu\, a_0 + 3a_3 + 6a_4 + 10a_5 + 15a_6 + \dots = 0,$$
$$a_3 + 4a_4 + 10a_5 + 20a_6 + \dots = 0,$$
$$a_4 + 5a_5 + 15a_6 + \dots = 0,$$
$$-\mu a_3 \qquad\quad + a_5 + 6a_6 + \dots = 0,$$
$$\cdot\;\cdot\;\cdot\;\cdot\;\cdot\;\cdot\;\cdot\;\cdot\;\cdot\;\cdot\;\cdot\;\cdot\;\cdot\;\cdot\;\cdot\;\cdot$$

Apart from a missing line and the terms involving μ, the coefficients are binomial coefficients.

If we truncate the system by retaining only the first five equations and putting $a_j = 0$ for $j \geq 7$, the zeros of the truncated determinant yield the approximate eigenvalues

$$\Lambda = \frac{1}{2}(-5 \pm \sqrt{145}) = \begin{cases} 3 \cdot 5208 \\ -8 \cdot 52. \end{cases}$$

7. (a) For the pivotal interval $h = \frac{2}{3}$ and with pivotal values $a, b, \dots, f$ as in Fig. VI/22 for solutions symmetrical about the diagonal D, the finite-difference equations read

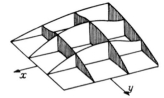

Fig. VI/22. An eigenvalue problem for a partial functional-differential equation

$$-4a + \qquad 2b + \mu c \qquad\qquad = 0,$$
$$a + (\mu - 4)b + \quad c + d \qquad\qquad = 0,$$
$$\mu a + \qquad 2b - 4c \qquad + 2e \qquad = 0,$$
$$2b \qquad - 4d + e \qquad = 0,$$
$$2c + \quad d - 4e + \quad f = 0,$$
$$4e - 4f = 0,$$

where $\mu = \Lambda h^2$ and Λ is an approximate value of λ.

For $\nu = \frac{1}{2}\mu$ we obtain the equation $11\nu^3 - 19\nu^2 - 41\nu + 26 = 0$, which yields

$$\nu = \begin{cases} -1 \cdot 580, \\ 0 \cdot 54100, \\ 2 \cdot 7659, \end{cases} \text{ and hence } \begin{cases} \Lambda_{-1} = -7 \cdot 11, \\ \Lambda_1 = 2 \cdot 434, \\ \Lambda_2 = 12 \cdot 45; \end{cases}$$

the eigenfunction corresponding to Λ_1 is depicted in Fig. VI/22. The value obtained for λ_1 by the simpler calculation with $h = 1$ is $\Lambda_1 = \frac{8}{3} = 2 \cdot 667$.

A corresponding calculation (with $h = \frac{2}{3}$) for the antisymmetric solutions (cf. Fig. VI/22) yields $\Lambda_{-2} = -7 \cdot 8$.

(b) A corresponding variational problem reads

$$J[\varphi] = \iint\limits_{Q} [\varphi_x^2 + \varphi_y^2 - \lambda\varphi(x, y)\,\varphi(-x, -y)]\,dx\,dy = \text{extremum}$$

with $\varphi(1, y) = \varphi(x, 1) = 0$; Q is the square $|x| \leq 1$, $|y| \leq 1$.

This can be verified in the usual way by putting $\varphi = u + \varepsilon\eta$, where $\eta(1, y) = \eta(x, 1) = 0$, and calculating

$$\left(\frac{\partial J(u + \varepsilon\eta)}{\partial \varepsilon}\right)_{\varepsilon=0}$$

$$= \iint_Q [2(u_x\eta_x + u_y\eta_y) - \lambda\{u(x, y)\eta(-x, -y) + u(-x, -y)\eta(x, y)\}]\,dx\,dy.$$

To find the conditions on u necessary for this expression to be zero for any η we transform the first term by GREEN's formula (3.7) of Ch. I:

$$\iint_Q (u_x\eta_x + u_y\eta_y)\,dx\,dy = -\iint_Q \eta\,\nabla^2 u\,dx\,dy - \int_\Gamma \eta\,u_\nu\,ds,$$

and change the signs of x and y in the term in $\eta(-x, -y)$, obtaining the necessary condition

$$\iint_Q \eta\{\nabla^2 u(x, y) + \lambda u(-x, -y)\}\,dx\,dy + \int_\Gamma \eta\,u_\nu\,ds = 0.$$

By the usual arguments we deduce from the arbitrariness of η that the factor multiplying η in the double integral must vanish, i.e. u must satisfy the functional equation, and that on the part of the boundary where we have not required that $\eta = 0$ the normal derivative u_ν must also vanish, so that u must also satisfy all the given boundary conditions.

Let us now assume for φ the expression

$$\varphi = (1 - x)(1 - y)(\beta + x)(\beta + y)$$

(non-linear in the parameter β). Then J takes the form $J = f_1(\beta) - \Lambda f_2(\beta)$. In the usual way $\dfrac{\partial J}{\partial \beta} = f_1' - \Lambda f_2' = 0$ yields at the same time the value $\Lambda = \left(\dfrac{f_1}{f_2}\right)_{\text{extr.}}$

$\left[\text{from } \left(\dfrac{f_1}{f_2}\right)' = \dfrac{f_1'f_2 - f_1f_2'}{f_2^2} = 0 \text{ it follows that } \dfrac{f_1}{f_2} = \dfrac{f_1'}{f_2'}\right].$ Here

$$\Lambda = 10\,\frac{(7 - 6\beta + 3\beta^2)(2 - 5\beta + 5\beta^2)}{(1 - 5\beta^2)^2},$$

and from a graph we find that the approximate position of the minimum is at the point $\beta \approx 2$ and the corresponding approximate value for λ_1 is $\Lambda = 2 \cdot 326$.

Appendix

Table I. *Approximate methods for ordinary differential equations of the first order*

$$y' = f(x, y)$$

Notation: $y_s =$ approximation for $y(x_s)$; $f_s = f(x_s, y_s)$

(Explanations in the text of Chapter II)

	Name	Formulae
Formulae of Runge-Kutta type	1st order (polygon method)	$y_{r+1} = y_r + h f_r$
	2nd order (improved polygon method)	$f^*_{r+\frac{1}{2}} = f\left(x_{r+\frac{1}{2}}, y_r + \frac{1}{2} h f_r\right);\quad y_{r+1} = y_r + h f^*_{r+\frac{1}{2}}$
		$y^*_{r+1} = y_r + h f_r;\quad y_{r+1} = y_r + \frac{1}{2} h\left[f_r + f(x_{r+1}, y^*_{r+1})\right]$
	3rd order (HEUN)	$k_1 = h f_r;\quad k_2 = h f\left(x_{r+\frac{1}{2}}, y_r + \frac{1}{2} k_1\right);$ $k_3 = h f(x_{r+1}, y_r - k_1 + 2 k_2);\quad y_{r+1} = y_r + \frac{1}{6}(k_1 + 4 k_2 + k_3)$
		$k_1 = h f_r;\ k_2 = h f\left(x_{r+\frac{1}{3}}, y_r + \frac{1}{3} k_1\right);\ k_3 = h f\left(x_{r+\frac{2}{3}}, y_r + \frac{2}{3} k_2\right)$ $y_{r+1} = y_r + \frac{1}{4} k_1 + \frac{3}{4} k_3$
	4th order (KUTTA)	$k_1 = h f_r;\ k_2 = h f\left(x_{r+\frac{1}{2}}, y_r + \frac{1}{2} k_1\right);\ k_3 = h f\left(x_{r+\frac{1}{2}}, y_r + \frac{1}{2} k_2\right)$ $k_4 = h f(x_{r+1}, y_r + k_3);\quad y_{r+1} = y_r + \frac{1}{6}(k_1 + 2 k_2 + 2 k_3 + k_4)$
	DUFFING's formula	$y_{r+1} = y_r + \frac{1}{6} h\left[4 f_r + 2 f_{r+1} + h\left(\frac{d}{dx} f\right)_{x=x_r}\right]$

		Name	Formulae
Formulae for finite-difference methods	**Extrapolation**	ADAMS	$y_{r+1} = y_r + h\left[f_r + \frac{1}{2} \nabla f_r + \frac{5}{12} \nabla^2 f_r + \frac{3}{8} \nabla^3 f_r + \frac{251}{720} \nabla^4 f_r + \right.$ $\left. + \frac{95}{288} \nabla^5 f_r + \frac{19087}{60480} \nabla^6 f_r + \cdots\right]$
		NYSTRÖM	$y_{r+1} = y_{r-1} + h\left[2 f_r + \frac{1}{3} \nabla^2 f_r + \frac{1}{3} \nabla^3 f_r + \frac{29}{90} \nabla^4 f_r + \frac{14}{45} \nabla^5 f_r + \cdots\right]$
	Interpolation	ADAMS	$y_{r+1} = y_r + h\left[f_{r+1} - \frac{1}{2} \nabla f_{r+1} - \frac{1}{12} \nabla^2 f_{r+1} - \frac{1}{24} \nabla^3 f_{r+1} - \right.$ $\left. - \frac{19}{720} \nabla^4 f_{r+1} - \frac{3}{160} \nabla^5 f_{r+1} - \frac{863}{60480} \nabla^6 f_{r+1} - \cdots\right]$
		Central differences	$y_{r+1} = y_{r-1} + h\left[2 f_r + \frac{1}{3} \nabla^2 f_{r+1} - \frac{1}{90} \nabla^4 f_{r+2} + \frac{1}{756} \nabla^6 f_{r+3} - \cdots\right]$
		MILNE's formulae	$y^*_{r+1} = y_{r-3} + 4 h\left[f_{r-1} + \frac{2}{3} \nabla^2 f_r\right];\ f^*_{r+1} = f(x_{r+1}, y^*_{r+1})$ $y_{r+1} = y_{r-1} + h\left[2 f_r + \frac{1}{3} \nabla^2 f^*_{r+1}\right]$
		QUADE's formula	$y_{r+1} = \frac{8}{19}(y_r - y_{r-2}) + y_{r-3} + \frac{6 h}{19}(f_{r+1} + 4 f_r + 4 f_{r-2} + f_{r-3})$

Table II. *Approximate methods for ordinary differential equations of the second order*

$$y'' = f(x, y, y')$$

Notation: y_s, y_s' approximations for $y(x_s)$, $y'(x_s)$, resp.; $f_s = f(x_s, y_s, y_s')$. (When required, y_s' is to be calculated from the corresponding formula for a differential equation of the first order).

Name	Formulae
RUNGE-KUTTA-NYSTRÖM	$k_1 = \dfrac{h^2}{2} f(x_r, y_r, y_r')$ $k_2 = \dfrac{h^2}{2} f\left(x_{r+\frac{1}{2}}, y_r + \dfrac{h}{2} y_r' + \dfrac{1}{4} k_1, y_r' + \dfrac{1}{h} k_1\right)$ $k_3 = \dfrac{h^2}{2} f\left(x_{r+\frac{1}{2}}, y_r + \dfrac{h}{2} y_r' + \dfrac{1}{4} k_1, y_r' + \dfrac{1}{h} k_2\right)$ $k_4 = \dfrac{h^2}{2} f\left(x_{r+1}, y_r + h y_r' + k_3, y_r' + \dfrac{2}{h} k_3\right)$ $y_{r+1} = y_r + h y_r' + \dfrac{1}{3} (k_1 + k_2 + k_3)$ $y_{r+1}' = y_r' + \dfrac{1}{3h} (k_1 + 2k_2 + 2k_3 + k_4)$
ADAMS extrapolation	$y_{r+1} = y_r + h y_r' + h^2\left(\dfrac{1}{2} f_r + \dfrac{1}{6} \nabla f_r + \dfrac{1}{8} \nabla^2 f_r + \dfrac{19}{180} \nabla^3 f_r + \dfrac{3}{32} \nabla^4 f_r + \cdots\right)$
STÖRMER-NYSTRÖM extrapolation	$y_{r+1} = 2y_r - y_{r-1} + h^2\left(f_r + \dfrac{1}{12} \nabla^2 f_r + \dfrac{1}{12} \nabla^3 f_r + \dfrac{19}{240} \nabla^4 f_r + \right.$ $\left. + \dfrac{3}{40} \nabla^5 f_r + \dfrac{863}{12\,096} \nabla^6 f_r + \cdots\right)$
ADAMS interpolation	$y_{r+1} = y_r + h y_r' + h^2\left(\dfrac{1}{2} f_{r+1} - \dfrac{1}{3} \nabla f_{r+1} - \dfrac{1}{24} \nabla^2 f_{r+1} - \right.$ $\left. - \dfrac{7}{360} \nabla^3 f_{r+1} - \dfrac{17}{1440} \nabla^4 f_{r+1} - \cdots\right)$
Central-difference method	$y_{r+1} = 2y_r - y_{r-1} +$ $+ h^2\left(f_r + \dfrac{1}{12} \nabla^2 f_{r+1} - \dfrac{1}{240} \nabla^4 f_{r+2} + \dfrac{31}{60\,480} \nabla^6 f_{r+3} - \cdots\right)$
MILNE extrapolation[1]	$y_{r+1} = y_r + y_{r-2} - y_{r-3} + h^2\left(3f_{r-1} + \dfrac{5}{4} \nabla^2 f_r\right)$ $y_{r+1} = y_r + y_{r-4} - y_{r-5} + \dfrac{h^2}{48} (67f_r - 8f_{r-1} + 122f_{r-2} - 8f_{r-3} + 67f_{r-4})$
MILNE interpolation	$y_{r+1} = y_r + y_{r-2} - y_{r-3} +$ $+ \dfrac{h^2}{240} (17f_{r+1} + 232f_r + 222f_{r-1} + 232f_{r-2} + 17f_{r-3})$

[1] MILNE, W. E.: Amer. Math. Monthly **40**, 322—327 (1933). — HARTREE, D. R.: Mem. Manchester **76**, 91—107 (1932).

Table III. *Finite-difference expressions for ordinary differential equations*

		Formulae [Notation: $y_j = y(jh)$, $y'_j = y'(jh)$, etc.]	The next non-vanishing term of the Taylor expansion
Formulae for y'	symmetric	$y'_0 = \dfrac{1}{2h}(-y_{-1} + y_1) +$	$-\dfrac{1}{6} h^2 y'''_0 - \cdots$
		$y'_0 = \dfrac{1}{12h}(y_{-2} - 8y_{-1} + 8y_1 - y_2) +$	$+\dfrac{1}{30} h^4 y^{V}_0 + \cdots$
		$y'_0 = \dfrac{1}{60h}(-y_{-3} + 9y_{-2} - 45y_{-1} + 45y_1 - 9y_2 + y_3) +$	$-\dfrac{1}{140} h^6 y^{VII}_0 + \cdots$
		$y'_{-1} + 4y'_0 + y'_1 + \dfrac{3}{h}(y_{-1} - y_1) = 0 +$	$+\dfrac{1}{30} h^4 y^{V}_0 + \cdots$
		$y'_{-1} + 3y'_0 + y'_1 + \dfrac{1}{12h}(y_{-2} + 28y_{-1} - 28y_1 - y_2) = 0 +$	$-\dfrac{1}{420} h^6 y^{VII}_0 - \cdots$
		$y'_{-2} + 16y'_{-1} + 36y'_0 + 16y'_1 + y'_2 +$ $+ \dfrac{5}{6h}(5y_{-2} + 32y_{-1} - 32y_1 - 5y_2) = 0 +$	$+\dfrac{1}{630} h^8 y^{IX}_0 + \cdots$
		$7y'_{-2} + 32y'_{-1} + 12y'_0 + 32y'_1 + 7y'_2 + \dfrac{45}{2h}(y_{-2} - y_2) = 0 +$	$+\dfrac{4}{21} h^6 y^{VII}_0 + \cdots$
		$y'_{-2} + 4y'_{-1} + 4y'_1 + y'_2 +$ $+ \dfrac{1}{6h}(19y_{-2} - 8y_{-1} + 8y_1 - 19y_2) = 0 +$	$+\dfrac{1}{35} h^6 y^{VII}_0 + \cdots$
	unsymmetric	$y'_0 = \dfrac{1}{h}(-y_0 + y_1) +$	$-\dfrac{1}{2} h\, y''_0 - \cdots$
		$y'_0 = \dfrac{1}{2h}(-3y_0 + 4y_1 - y_2) +$	$+\dfrac{1}{3} h^2 y'''_0 + \cdots$
		$y'_0 = \dfrac{1}{12h}(-3y_{-1} - 10y_0 + 18y_1 - 6y_2 + y_3) +$	$-\dfrac{1}{20} h^4 y^{V}_0 + \cdots$
		$y'_0 = \dfrac{1}{60h}(2y_{-2} - 24y_{-1} - 35y_0 + 80y_1 - 30y_2 +$ $+ 8y_3 - y_4) +$	$+\dfrac{1}{105} h^6 y^{VII}_0 + \cdots$
		$y'_0 + y'_1 + \dfrac{2}{h}(y_0 - y_1) = 0 +$	$+\dfrac{1}{6} h^2 y'''_0 + \cdots$
		$y'_{-1} + 9y'_0 + 9y'_1 + y'_2 +$ $+ \dfrac{1}{3h}(11y_{-1} + 27y_0 - 27y_1 - 11y_2) = 0 +$	$+\dfrac{1}{140} h^6 y^{VII}_0 + \cdots$
Formulae for y''	symmetric	$y''_0 = \dfrac{1}{h^2}(y_{-1} - 2y_0 + y_1) +$	$-\dfrac{1}{12} h^2 y^{IV}_0 + \cdots$
		$y''_0 = \dfrac{1}{12h^2}(-y_{-2} + 16y_{-1} - 30y_0 + 16y_1 - y_2) +$	$+\dfrac{1}{90} h^4 y^{VI}_0 + \cdots$
		$y''_0 = \dfrac{1}{180h^2}(2y_{-3} - 27y_{-2} + 270y_{-1} - 490y_0 +$ $+ 270y_1 - 27y_2 + 2y_3) +$	$-\dfrac{1}{560} h^6 y^{VIII}_0 + \cdots$
		$y''_{-1} + 10y''_0 + y''_1 - \dfrac{12}{h^2}(y_{-1} - 2y_0 + y_1) = 0 +$	$+\dfrac{1}{20} h^4 y^{VI}_0 + \cdots$
		$2y''_{-1} + 11y''_0 + 2y''_1 - \dfrac{3}{4h^2}(y_{-2} + 16y_{-1} - 34y_0 +$ $+ 16y_1 + y_2) = 0 +$	$-\dfrac{23}{5040} h^6 y^{VIII}_0 + \cdots$

Table III (continued)

		Formulae [Notation: $y_j = y(jh)$, $y'_j = y'(jh)$, etc.]	The next non-vanishing term of the Taylor expansion
Formulae for y''	symmetric	$23 y''_{-2} + 688 y''_{-1} + 2358 y''_0 + 688 y''_1 + 23 y''_2 -$ $- \frac{15}{h^2}(31 y_{-2} + 128 y_{-1} - 318 y_0 + 128 y_1 + 31 y_2) = 0 +$	$+\dfrac{79}{1260} h^8 y_0^{X} + \cdots$
		$y''_{-1} - 8 y''_0 + y''_1 + \frac{9}{h}(y'_{-1} - y'_1) +$ $+ \frac{24}{h^2}(y_{-1} - 2 y_0 + y_1) = 0 +$	$+\dfrac{1}{2520} h^6 y_0^{VIII} + \cdots$
		$y''_{-1} - y''_1 + \frac{1}{h}(7 y'_{-1} + 16 y'_0 + 7 y'_1) +$ $+ \frac{15}{h^2}(y_{-1} - y_1) = 0 +$	$-\dfrac{1}{315} h^5 y_0^{VII} + \cdots$
	unsymmetric	$y''_0 = \frac{1}{h^2}(2 y_0 - 5 y_1 + 4 y_2 - y_3) +$	$+\dfrac{11}{12} h^2 y_0^{IV} + \cdots$
		$y''_0 = \frac{1}{12 h^2}(11 y_{-1} - 20 y_0 + 6 y_1 + 4 y_2 - y_3) +$	$+\dfrac{1}{12} h^3 y_0^{V} + \cdots$
		$y''_0 = \frac{1}{180 h^2}(-13 y_{-2} + 228 y_{-1} - 420 y_0 + 200 y_1 +$ $+ 15 y_2 - 12 y_3 + 2 y_4) +$	$-\dfrac{1}{90} h^5 y_0^{VII} + \cdots$
Formulae for y'''	symmetric	$y'''_0 = \frac{1}{2 h^3}(-y_{-2} + 2 y_{-1} - 2 y_1 + y_2) +$	$-\dfrac{1}{4} h^2 y_0^{V} + \cdots$
		$y'''_0 = \frac{1}{8 h^3}(y_{-3} - 8 y_{-2} + 13 y_{-1} - 13 y_1 + 8 y_2 - y_3) +$	$+\dfrac{7}{120} h^4 y_0^{VII} + \cdots$
		$y'''_{-1} + 2 y'''_0 + y'''_1 + \frac{2}{h^3}(y_{-2} - 2 y_{-1} + 2 y_1 - y_2) = 0 +$	$-\dfrac{1}{60} h^4 y_0^{VII} + \cdots$
		$y'''_{-2} + 56 y'''_{-1} + 126 y'''_0 + 56 y'''_1 + y'''_2 +$ $+ \frac{120}{h^3}(y_{-2} - 2 y_{-1} + 2 y_1 - y_2) = 0 +$	$-\dfrac{1}{252} h^6 y_0^{IX} + \cdots$
	unsymmetric	$y'''_0 = \frac{1}{2 h^3}(-3 y_{-1} + 10 y_0 - 12 y_1 + 6 y_2 - y_3) +$	$+\dfrac{1}{4} h^2 y_0^{V} + \cdots$
		$y'''_0 = \frac{1}{8 h^3}(-y_{-2} - 8 y_{-1} + 35 y_0 - 48 y_1 + 29 y_2$ $- 8 y_3 + y_4) +$	$-\dfrac{1}{15} h^4 y_0^{VII} + \cdots$
Formulae for y^{IV}	symmetric	$y_0^{IV} = \frac{1}{h^4}(y_{-2} - 4 y_{-1} + 6 y_0 - 4 y_1 + y_2) +$	$-\dfrac{1}{6} h^2 y_0^{VI} + \cdots$
		$y_0^{IV} = \frac{1}{6 h^4}(-y_{-3} + 12 y_{-2} - 39 y_{-1} +$ $+ 56 y_0 - 39 y_1 + 12 y_2 - y_3) +$	$+\dfrac{7}{240} h^4 y_0^{VIII} + \cdots$
		$y_{-1}^{IV} + 4 y_0^{IV} + y_1^{IV} - \frac{6}{h^4}(y_{-2} - 4 y_{-1} + 6 y_0 - 4 y_1 + y_2)$ $= 0 +$	$+\dfrac{1}{120} h^4 y_0^{VIII} + \cdots$
		$y_{-2}^{IV} - 124 y_{-1}^{IV} - 474 y_0^{IV} - 124 y_1^{IV} + y_2^{IV} +$ $+ \frac{720}{h^4}(y_{-2} - 4 y_{-1} + 6 y_0 - 4 y_1 + y_2) = 0 +$	$+\dfrac{5}{21} h^6 y_0^{X} + \cdots$

Table IV. *Euler expressions for functions of one independent variable*

(To facilitate the setting up of variational problems corresponding to given ordinary differential equations.)

$F(x, y, y', y'', ..., y^{(n)})$	$\dfrac{\partial F}{\partial y} - \dfrac{d}{dx}\left(\dfrac{\partial F}{\partial y'}\right) + \dfrac{d^2}{dx^2}\left(\dfrac{\partial F}{\partial y''}\right) - \cdots + (-1)^n \dfrac{d^n}{dx^n}\left(\dfrac{\partial F}{\partial y^{(n)}}\right)$
$\frac{1}{2} h_n(x)\, [y^{(n)}]^2$	$(-1)^n\, [h_n(x)\, y^{(n)}]^{(n)}$
special case $\frac{1}{2} h_0(x)\, y^2$	$h_0(x)\, y$
,, ,, $\frac{1}{2} h_1(x)\, y'^2$	$-[h_1(x)\, y']'$
,, ,, $\frac{1}{2} h_2(x)\, y''^2$	$[h_2(x)\, y'']''$
$-h(x, y)\, y'^n$	$y'^{n-2}[n(n-1)\, h\, y'' + n\, h_x\, y' + (n-1)\, h_y\, y'^2]$
special case $h(x, y)$	$h_y(x, y)$
,, ,, $h(x, y)\, y'$	$-h_x$
,, ,, $f(y)\, y'$	0
,, ,, $-h(x, y)\, y'^2$	$2h\, y'' + 2h_x\, y' + h_y\, y'^2$
,, ,, $-f(y)\, y'^2$	$2f\, y'' + f'\, y'^2$
$h(x, y)\, y''^n$	$\begin{cases} y''^{n-3}[n(n-1)\, h\, y''\, y^{IV} + n(n-1)\, y''' \{2h_x\, y'' + \\ + 2h_y\, y'\, y'' + (n-2)\, h\} + y''^3(n+1)\, h_y + \\ + y''^2\, n\{h_{xx} + 2h_{xy}\, y' + h_{yy}\, y'^2\}] \end{cases}$
special case $h(x, y)\, y''$	$2h_y\, y'' + h_{xx} + 2h_{xy}\, y' + h_{yy}\, y'^2$
,, ,, $h(x, y)\, y''^2$	$\begin{cases} 2h\, y^{IV} + 4(h_x + h_y\, y')\, y''' + 3h_y\, y''^2 + \\ + 2y''(h_{xx} + 2h_{xy}\, y' + h_{yy}\, y'^2) \end{cases}$
$(-1)^{n+1} f(x)\, [y^{(n-1)} y^{(n+1)} - y^{(n)2}]$	$4\, [f\, y^{(n)}]^{(n)} + [f''\, y^{(n-1)}]^{(n-1)}$
special case $f(x)\, [y\, y'' - y'^2]$	$4\, [f\, y']' + f''\, y$
$(-1)^{n+1} f(x)\, [y^{(n-1)} y^{(n+1)} + y^{(n)2}]$	$[f''\, y^{(n-1)}]^{(n-1)}$
special case $f(x)\, [y\, y'' + y'^2]$	$f''\, y$
,, ,, $(a + bx) \times$ $\times [y^{(n-1)} y^{(n+1)} + y^{(n)2}]$	0
$h(y')$	$-h''(y')\, y''$
$h(y'')$	$h'''(y'')\, y^{IV} + h^{IV}(y'')\, y'''^2$
special case $\dfrac{1}{24}\, y''^4$	$y''\, y^{IV} + y'''^2$
$h(x, y, y')\, y''$	$y''(2h_y + h_{xy'} + h_{yy'}\, y') + h_{xx} + 2h_{xy}\, y' + h_{yy}\, y'^2$
special case $p(x)\, q(y')\, y''$	$\dfrac{d}{dx}\, [p'(x)\, q(y')] = p'\, q'\, y'' + p''\, q$

Table V. *Euler expressions for functions of two independent variables*

(To facilitate the setting up of variational problems corresponding to given partial differential equations.)

$F(x, y, u, u_x, u_y, u_{xx}, u_{xy}, u_{yy})$	$F_u - \dfrac{\partial}{\partial x} F_{u_x} - \dfrac{\partial}{\partial y} F_{u_y} + \dfrac{\partial^2}{\partial x^2} F_{u_{xx}} + \dfrac{\partial^2}{\partial x \partial y} F_{u_{xy}} + \dfrac{\partial^2}{\partial y^2} F_{u_{yy}}$
$h(x, y) g(u)$	$h(x, y) g'(u)$
special case $h(x, y) u$	$h(x, y)$
,, ,, $\dfrac{1}{2} h(x, y) u^2$	$h(x, y) u$
$h(x, y) u_x$	$-h_x$
$-\dfrac{1}{2} h(x, y) u_x^2$	$\dfrac{\partial}{\partial x} (h u_x)$
$h(x, y) u u_x$	$-h_x u$
$-h(x, y) u_x u_y$	$\dfrac{\partial}{\partial x} (h u_y) + \dfrac{\partial}{\partial y} (h u_x) = 2 h u_{xy} + h_x u_y + h_y u_x$
special case $-\dfrac{1}{2} u_x u_y$	u_{xy}
$-\dfrac{1}{2} h(x, y) u_y^2$	$\dfrac{\partial}{\partial y} (h u_y)$
$-\dfrac{1}{2} (u_x^2 + u_y^2)$	$\nabla^2 u \equiv u_{xx} + u_{yy}$
$h(x, y) u u_{xx}$	$h u_{xx} + \dfrac{\partial^2}{\partial x^2} (h u) = 2 h u_{xx} + 2 h_x u_x + h_{xx} u$
$\dfrac{1}{2} h(x, y) u_{xx}^2$	$\dfrac{\partial^2}{\partial x^2} (h u_{xx})$
$\dfrac{1}{2} h(x, y) u_{xy}^2$	$\dfrac{\partial^2}{\partial x \partial y} (h u_{xy})$
$h(x, y) u_{xx} u_{yy}$	$\dfrac{\partial^2}{\partial x^2} (h u_{yy}) + \dfrac{\partial^2}{\partial y^2} (h u_{xx})$
u_{xy}^2	$\left.\vphantom{\begin{matrix}a\\b\end{matrix}}\right\} 2 u_{xxyy}$
$u_{xx} u_{yy}$	
$\dfrac{1}{2} (\nabla^2 u)^2 = \dfrac{1}{2} (u_{xx} + u_{yy})^2$	$\left.\vphantom{\begin{matrix}a\\b\\c\end{matrix}}\right\} \nabla^4 u \equiv u_{xxxx} + 2 u_{xxyy} + u_{yyyy}$
$\dfrac{1}{2} (u_{xx}^2 + 2 u_{xy}^2 + u_{yy}^2)$	
$u_{xx} u_{yy} - u_{xy}^2$	0
$\displaystyle\int_0^u f(x, y, v)\, dv$	$f(x, y, u)$
$g(x, y, u) (u_x^2 + u_y^2)$	$-2g \nabla^2 u - g_u (u_x^2 + u_y^2) - 2 g_x u_x - 2 g_y u_y$

Table VI. *Stencils for the differential operators ∇^2 and ∇^4 (for meshes formed from squares, equilateral triangles and cubes)*

Formula relates	Stencil	Stencil to be read as	Further terms of the Taylor expansion. Notation: $\zeta_{j,k} = \left(\dfrac{\partial^{j+k} u}{\partial x^j \partial y^k} + \dfrac{\partial^{j+k} u}{\partial x^k \partial y^j}\right)_{\substack{x=0 \\ y=0}}$
$h^2\nabla^2 u_{0,0}$ with values of u	$\begin{array}{ccc} & 1 & \\ 1 & -4 & 1 \\ & 1 & \end{array}$	$h^2\nabla^2 u_{0,0} = -4u_{0,0} + u_{1,0} + u_{0,1} + u_{-1,0} + u_{0,-1} +$	$-\dfrac{h^4}{12}\zeta_{0,4} - \dfrac{h^6}{360}\zeta_{0,6} - \cdots$
$12h^2\nabla^2 u_{0,0}$ with values of u	$\begin{array}{ccccc} & & -1 & & \\ & & 16 & & \\ -1 & 16 & -60 & 16 & -1 \\ & & 16 & & \\ & & -1 & & \end{array}$	$12h^2\nabla^2 u_{0,0} = -60u_{0,0} + 16(u_{1,0}+\cdots) - (u_{2,0}+\cdots) +$	$\dfrac{2}{15}h^6\zeta_{0,6} + \dfrac{h^8}{84}\zeta_{0,8} + \cdots$
values of $\boxed{h^2\nabla^2 u}$ and values of u	$\begin{array}{ccc} -2 & -8① & -2 \\ -8① & 40① & -8① \\ -2 & -8① & -2 \end{array}$	$8h^2\nabla^2 u_{0,0} + h^2\nabla^2 u_{1,0} + h^2\nabla^2 u_{0,1} + h^2\nabla^2 u_{-1,0} + h^2\nabla^2 u_{0,-1} + 40u_{0,0} - 8(u_{1,0}+u_{0,1}+\cdot+\cdot) - 2(u_{1,1}+u_{1,-1}+\cdot+\cdot) = 0 +$	$\dfrac{h^6}{60}(3\zeta_{0,6} - 5\zeta_{2,4}) + \cdots$
values of $\boxed{\dfrac{1}{12}h^2\nabla^2 u}$ and values of u	$\begin{array}{ccc} -1① & -4④ & -1① \\ -4④ & 20\,\boxed{52} & -4④ \\ -1① & -4④ & -1① \end{array}$	$\dfrac{1}{12}h^2[52\nabla^2 u_{0,0} + 4(\nabla^2 u_{1,0} + \nabla^2 u_{0,1} + \cdot+\cdot)] + (\nabla^2 u_{1,1} + \nabla^2 u_{1,-1} + \cdot+\cdot)] + 20u_{0,0} - 4(u_{1,0}+u_{0,1}+\cdot+\cdot) - (u_{1,1}+u_{1,-1}+\cdot+\cdot) = 0 +$	$\dfrac{h^6}{120}(3\zeta_{0,6} + 5\zeta_{2,4}) + \cdots$

values of $\dfrac{2}{3}h^2\nabla^2 u$ and values of u		$\dfrac{2}{3}h^2[46\nabla^2 u_{0,0} + 10(\nabla^2 u_{1,0} + \nabla^2 u_{0,1} + \cdot + \cdot) + (\nabla^2 u_{1,1} + \nabla^2 u_{1,-1} + \cdot + \cdot)] + 140 u_{0,0} - 16(u_{1,0} + u_{1,1} + \cdots) - 3(u_{2,0} + u_{0,2} + \cdot + \cdot) = 0 +$	$\dfrac{h^8}{1260}(-23\zeta_{0,8} + 42\zeta_{2,6}) + \cdots$
$\dfrac{1}{10}h^4\nabla^4 u_{0,0}$ with values of $\dfrac{1}{15}h^2\nabla^2 u$ and values of u		$\dfrac{3}{10}h^4\nabla^4 u_{0,0} + \dfrac{1}{15}h^2[82\nabla^2 u_{0,0} + \nabla^2 u_{1,0} + \cdot + \cdot + \cdot + \cdot] + \nabla^2 u_{1,1} + \cdot + \cdot + \cdot + \cdot + \cdot) + 20 u_{0,0} - 4(u_{1,0} + u_{0,1} + \cdot + \cdot) - (u_{1,1} + u_{1,-1} + \cdot + \cdot) = 0 +$	$\dfrac{h^8}{50400}(13\zeta_{0,8} + 168\zeta_{2,6} + 105\zeta_{4,4}) + \cdots$
$h^4\nabla^4 u_{0,0}$ with values of u		$h^4\nabla^4 u_{0,0} = 20 u_{0,0} - 8(u_{1,0} + u_{0,1} + \cdot + \cdot) + 2(u_{1,1} + u_{1,-1} + \cdot + \cdot) + (u_{2,0} + u_{0,2} + \cdot + \cdot) +$	$-\dfrac{h^6}{6}(\zeta_{0,6} + \zeta_{2,4}) - \cdots$
$6h^4\nabla^4 u_{0,0}$ with values of u		$6h^4\nabla^4 u_{0,0} = 184 u_{0,0} - 77(u_{1,0} + u_{0,1} + \cdot + \cdot) + 20(u_{1,1} + u_{1,-1} + \cdot + \cdot) + 14(u_{2,0} + u_{0,2} + \cdot + \cdot) - (u_{0,3} + u_{1,2} + u_{2,1} + \cdots) +$	$\dfrac{h^8}{120}(21\zeta_{0,8} + 16\zeta_{2,6} + 5\zeta_{4,4}) + \cdots$

Table VI continued

Formula relates	Stencil	Stencil to be read as	Further terms of the Taylor expansion. Notation: $u_{j,k} = \left(\dfrac{\partial^{j+k}u}{\partial x^j\,\partial y^k}\right)_{\substack{z=0\\y=0}}$
values of $\dfrac{1}{2}h^4\nabla^4 u$ and values of u	$\begin{array}{ccccc} & -1 & -1 & -1 & \\ -1 & 2 & 10① & 2 & -1 \\ -1 & 10① & -36② & 10① & -1 \\ -1 & 2 & 10① & 2 & -1 \\ & -1 & -1 & -1 & \end{array}$	$\dfrac{1}{2}h^4[2\nabla^4 u_{0,0} + (\nabla^4 u_{1,0} + \nabla^4 u_{0,1} + \cdots)] - 36 u_{0,0} + 10(u_{1,0} + u_{0,1} + \cdots) + 2(u_{1,1} + u_{1,-1} + \cdots) - (u_{2,0} + u_{2,1} + u_{1,2} + \cdots) = 0 +$	$\dfrac{h^8}{240}(\zeta_{0,8} - 24\zeta_{2,6} - 15\zeta_{4,4}) + \cdots$
values of $\dfrac{h^4}{20}\nabla^4 u$ and values of u	$\begin{array}{ccccc} 1 & 8 & 18① & 8 & 1 \\ 8 & -8㉒ & -74④ & -8㉒ & 8 \\ 18① & -74④ & 468㉟ & -74④ & 18① \\ 8 & -8㉒ & -74④ & -8㉒ & 8 \\ 1 & 8 & 18① & 8 & 1 \end{array}$	$\dfrac{h^4}{20}\big[-332\nabla^4 u_{0,0} - 72(\nabla^4 u_{1,0} + \nabla^4 u_{0,1} + \cdots) - 26(\nabla^4 u_{1,1} + \nabla^4 u_{1,-1} + \cdots) + (\nabla^4 u_{2,0} + \nabla^4 u_{0,2} + \cdots)\big] + 468 u_{0,0} - 144(u_{1,0} + u_{0,1} + \cdots) - 8(u_{1,1} + \cdots) + 18(u_{2,0} + \cdots) + 8(u_{1,2} + u_{2,1} + \cdots) + (u_{2,2} + \cdots) = 0 +$	$\dfrac{h^{10}}{420}(5\zeta_{0,10} + 12\zeta_{2,8} + 21\zeta_{4,6}) + \cdots$
$\dfrac{3}{2}h^2\nabla^2 u_{0,0}$ with values of u	$\begin{array}{ccc} & 1 & 1 \\ 1 & -6 & 1 \\ 1 & 1 & \end{array}$	$\dfrac{3}{2}h^2\nabla^2 u_{0,0} = -6u_{0,0} + (u_{1,0} + \cdots) +$	$-\dfrac{3h^4}{32}(\nabla^4 u)_{0,0} - \dfrac{h^6}{3840}(11u_{60} + 15u_{42} + 45u_{24} + 9u_{06}) + \cdots$

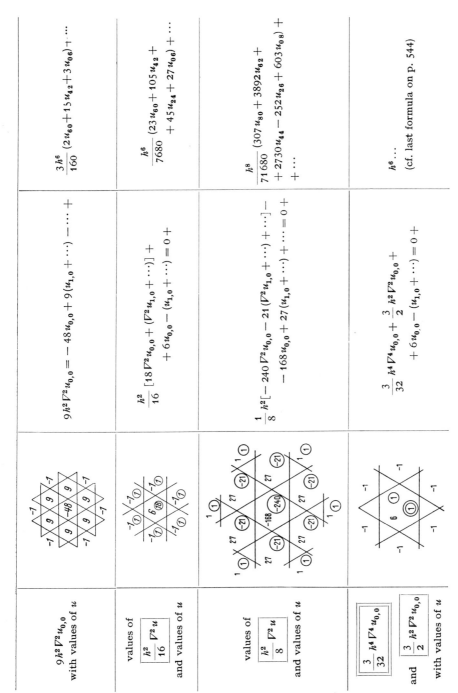

$9h^2 \nabla^2 u_{0,0}$ with values of u

$$9h^2\nabla^2 u_{0,0} = -48u_{0,0} + 9(u_{1,0} + \cdots) - \cdots +$$

$$\frac{3h^6}{160}\left(2u_{60} + 15u_{42} + 3u_{06}\right) + \cdots$$

values of $\boxed{\dfrac{h^2}{16}\nabla^2 u}$ and values of u

$$\frac{h^2}{16}\left[18\nabla^2 u_{0,0} + (\nabla^2 u_{1,0} + \cdots)\right] +$$
$$+ 6u_{0,0} - (u_{1,0} + \cdots) = 0 +$$

$$\frac{h^6}{7680}\left(23u_{60} + 105u_{42} + 27u_{06}\right) +$$
$$+ 45u_{24} + 27u_{06}) + \cdots$$

values of $\boxed{\dfrac{h^2}{8}\nabla^2 u}$ and values of u

$$\frac{1}{8}h^2\left[-240\nabla^2 u_{0,0} - 21(\nabla^2 u_{1,0} + \cdots) + \cdots\right] -$$
$$- 168u_{0,0} + 27(u_{1,0} + \cdots) + \cdots = 0 +$$

$$\frac{h^8}{71680}\left(307u_{80} + 3892u_{62} +\right.$$
$$+ 2730u_{44} - 252u_{26} + 603u_{08}) +$$
$$+ \cdots$$

$\boxed{\dfrac{3}{32}h^4\nabla^4 u_{0,0}}$ and $\boxed{\dfrac{3}{2}h^2\nabla^2 u_{0,0}}$ with values of u

$$\frac{3}{32}h^4\nabla^4 u_{0,0} + \frac{3}{2}h^2\nabla^2 u_{0,0} +$$
$$+ 6u_{0,0} - (u_{1,0} + \cdots) = 0 +$$

$$h^6 \cdots.$$
(cf. last formula on p. 544)

Table VI continued

Formula relates	Stencil	Stencil to be read as	Further terms of the Taylor expansion. Notation: $$u_{jk}=\left(\frac{\partial^{j+k} u}{\partial x^j \partial y^k}\right)_{\substack{z=0\\y=0}}$$
$\dfrac{9}{16} h^4 \nabla^4 u_{0,0}$ with values of u		$\dfrac{9}{16} h^4 \nabla^4 u_{0,0} = 12 u_{0,0} - 3(u_{1,0} + \cdots) + \cdots +$	$-\dfrac{h^6}{128}(7 u_{60} + 39 u_{42} + 9 u_{24} + 9 u_{06}) + \cdots$
values of $\boxed{\dfrac{3}{16} h^4 \nabla^4 u}$ and values of u		$\dfrac{3}{16} h^4 [6 \nabla^4 u_{0,0} + (\nabla^4 u_{1,0} + \cdots)] -$ $- 42 u_{0,0} + 10(u_{1,0} + \cdots) - \cdots = 0 +$	$\dfrac{h^8}{5120}(u_{80} + 76 u_{62} + 30 u_{44} -$ $- 36 u_{26} + 9 u_{08}) + \cdots$

Cubical mesh (notation $u_m, \Sigma u_s, \ldots, \nabla^2 u_m, \Sigma \nabla^2 u_s, \ldots$ as in Table X) [from J. ALBRECHT: Z. Angew. Math. Mech. **33**, 48 (1953). where further formulae can be found]

$\Sigma u_s - 6 u_m - h^2 \nabla^2 u_m =$

$\Sigma u_k + 2 \Sigma u_s - 24 \Sigma u_m - \frac{1}{2} h^2 [\Sigma \nabla^2 u_s + 6 \nabla^2 u_m] =$

$\Sigma u_k + 2 \Sigma u_s - 24 u_m - \frac{1}{12} h^2 [\Sigma \nabla^2 u_k + 2 \Sigma \nabla^2 u_s + 48 \nabla^2 u_m] =$

$\Sigma u_e + 3 \Sigma u_k + 14 \Sigma u_s - 128 u_m - \frac{1}{3} h^2 [\Sigma \nabla^2 u_k - \Sigma \nabla^2 u_s + 84 \nabla^2 u_m] - \frac{3}{2} h^4 \nabla^4 u_m =$

remainder term of the 4th order
,, ,, ,, ,, 6th ,,
,, ,, ,, ,, 6th ,,
,, ,, ,, ,, 8th ,,

Table VII. Catalogue of examples treated

Abbreviations for the methods used: Bm = boundary-maximum theorem, Ch = method of characteristics, Cl = collocation, D = ordinary finite-difference method, Di = improved finite-difference methods, Ds = finite-difference step-by-step method, Es = exact solution, Et = enclosure theorem, Fs = finite-sum method, G = GALERKIN's method, It = iteration method, Ls = least-squares method, Mp = method based on the monotonic property, Pt = perturbation method, Ps = power series, Pl = polygon method, R = RITZ's method, RK = Runge-Kutta method, Se = Series expansion, T = TREFFTZ's method.

No.	Type	Equation(s)	Initial or boundary conditions	Possible physical applications and other remarks	Methods used and page numbers
		Chapter II. Initial-value problems in ordinary differential equations			
1	Differential equation of the first order	$y' = y$	$y(0) = 1$		Ls 138
2		$y' = x + y$	$y(0) = 0$		Ds 82
3		$y' = y - \dfrac{2x}{y}$	$y(0) = 1$		Es 55, Pl 55, RK 73, Ps 80, Ds 92, It 115
4		$y' = x^2 + y^2$	$y(0) = -1$		Ps 137, Ds 137
5		$y' = -y - 2y^3 + \sin 2x$	$y(0) = 0$	Transient effects when an electromagnet is switched on	Ds 91
6	System of first order equations	$\dot{u}_1 = u_2\, u_3 - 0{\cdot}6\, u_1$ $\dot{u}_2 = -u_1\, u_3 - 0{\cdot}3\, u_2$ $\dot{u}_3 = \dfrac{1}{3}\, u_1\, u_2 - 0{\cdot}2\, u_3$	$u_1(0) = u_2(0) = 1$ $u_3(0) = 0$	Motion of a body about a fixed point (EULER's equations)	RK 75
7	Diff. equ. of higher order	$y'' = -y^3$	$y(0) = 0{\cdot}2, \ y'(0) = 0$	Oscillations of a mass attached to a spring	RK 137
8		$y''' = -xy$	$y(0) = 0, \ y'(0) = 1, \ y''(0) = 0$	Boundary layer along a flat plate	Ds 138
9		$y''' = -yy''$	$y(0) = y'(0) = 0, \ y''(0) = 1$		RK 75, Ds 130, RK 138, Ps 130, 138
		Chapter III. Boundary-value problems in ordinary differential equations			
1	Linear differential equation of the second order	$y'' = 0$	$y(0) = 1, \ y(1) = y'(1)$	No solution	R 252
2		$y'' + 4y = 2$	$y(\pm 1) = 0$		R 252, Es 256
3		$y'' + y + x = 0$	$y(0) = y(1) = 0$		R 220, Ls 220
4		$y'' - \dfrac{1}{x^2}\, y = -\dfrac{1}{x}$	$y(2) = y(3) = 0$		Es 178, D 178
5		$y'' + (1 + x^2) y + 1 = 0$	$y(\pm 1) = 0$	Bending of a beam	D 143, 155, Di 163, 168, Cl 182, Pt 188, It 195, R 209, Ps 224
6		$y'' = \dfrac{1 + x}{2 + x}\, y$	$y(0) = 1, \ y(\infty) = 0$	Temperature distribution in an infinitely long rod	D 150, Cl 182
7		$(1 + x^2)\, y'' + \left(\dfrac{3}{x} + 5x\right) y' + \dfrac{4}{3}\, y + 1 = 0$	$y(\pm 2) = 0{\cdot}6, \ y$ at $x = 0$ regular	Stress distribution in a steam turbine rotor	D 251
8	Eigen-value problem	$y'' + \lambda\, (1 + x^2)\, y = 0$	$y(\pm 1) = 0$	Vibration of an inhomogeneous string	R 252, It 253
9		$y'' + \lambda\, (2 - x^3)\, y = 0$	$y(0) = 0, \ 2y(1) + y'(1) = 0$	Longitudinal vibrations of a rod	Et 252
10		$-[(1 + x)\, y']' = \lambda\, (1 + x)\, y$	$y'(0) = y(1) = 0$		D 147, Di 163, 170, R 210, It 235, Et 237, R 248

Table VII (continued)

No.	Type	Equation	Initial or boundary conditions	Possible physical applications and other remarks	Methods used and page numbers
11	Non-linear differential equation of the second order	$y'' = \dfrac{2x^2}{\,}$	$y(\pm1)=1$		Es 252, D 252
12		$y'y''=\dfrac{y}{4x}$	$y(\pm1)=0$	Only complex solutions	R 252
13		$y''=y^3$	$y(0)=0,\ y(2)=-2$		D 252
14		$y''=\dfrac{3}{2}y^2$	$y(0)=4,\ y(1)=1$		Es 145, D 145, 180, Ls 184, R 212, Ps 226
15		$y''+(1+e^y)=0$	$y(0)=0,\ y(1)=1$		D 159
16		$y''+6y+y^2+\dfrac{3}{2}\cos x=0$	$y(0)-y(2\pi)=y'(0)-y'(2\pi)=0$	Forced oscillations of a system with a non-linear restoring force	It 198
17		$y''=6xy^2$	$y(0)=y(1)=1$		Mp 201
18	Diff. equ. of the fourth order	$[(2-x^2)y'']''+40\,y=2-x^2$	$y''(\pm1)=y'''(\pm1)=0$	Transversely loaded, elastically embedded rail	D 152, R 219
19		$y^{IV}=\lambda(1+x)\,y$	$y(0)=y'(0)=y''(1)=y'''(1)=0$	Flexural vibrations of a rod with variable cross-section	R 253, Fs 472

Chapter IV. Initial and initial-/boundary-value problems in partial differential equations

No.	Type	Equation	Initial or boundary conditions	Possible physical applications and other remarks	Methods used and page numbers
1	Differential equation of the first order	$(0.1\sqrt[3]{u}-0.5)u_x+u_y=-0.5$	$u(x,0)=2\dfrac{4-x}{5-x}$ for $0\le x\le4$	Motion of a glacier	Ch 309, Ps 310, D 311, It 316
2		$\begin{aligned}u_x-v_y&=0\\ -4u_y+v_x&=u\end{aligned}$	$\begin{cases}u(x,0)=\sin\frac{\pi}{2}x,\ v(x,0)=0 & \text{for } 0\le x\le2\\ u(0,y)=u(2,y)=0 & \text{for } y\ge0\end{cases}$	Transient currents in an electric cable	Ch 324, Es 326, D 335
3	Parabolic type	$u_{xx}=c\,u_y$	$\begin{cases}u(x,0)=0 & \text{for } 0\le x\le2\\ u(0,y)=\sin\frac{8\pi}{3c}y,\ u(2,y)=0 & \text{for } y\ge0\end{cases}$	Heat conduction in a rod	D 267, 268, 334
4		$u_{xx}=2u_y$	$\begin{cases}u(\pm1,y)=1-e^{-\frac{5}{c}y} & \text{for } y\ge0\\ u(x,0)=10x & \text{for } 0\le x\le10\end{cases}$	Heat conduction in a rod	D 281
5		$u_{xx}-Ku_y+\beta=0$	$\begin{cases}u(0,y)=0,\ u(10,y)=100\,e^{-0.1y} & \text{for } y\ge0\\ u(x,0)=u(\pm1,y)=0\end{cases}$	Heat conduction with constant heat generation	D 334
6		$u_{xx}=u_y+\dfrac{1}{2}u$	$\begin{cases}u(x,0)=1 & \text{for } |x|\le1\\ u\pm u_x=0 & \text{for } x=\pm1 \text{ and } y>0\end{cases}$	Heat conduction with heat loss to the surroundings	D 334
7		$u_{xx}+\dfrac{1}{x}u_x=u_y$	$\begin{cases}u\pm\frac{1}{2}u_x=1 & \text{for } x=\pm1 \text{ and } y>0\\ u(x,0)=0 & \text{for } |x|\le1\end{cases}$	Eddy current density in a metal cylinder	D 273, 282
8			$u(x,0)=0$ for $|x|\le1$		
9	Hyperbolic type	$u_{xx}=u_{yy}$	$\begin{cases}u_y(x,0)=0,\ u(x,0)=\cos x & \text{for } |x|\le\frac{\pi}{2}\\ u\!\left(\pm\frac{\pi}{2},y\right)=\sin y & \text{for } y\ge0\end{cases}$		Ps 261, Es 261
10		$u_{xx}=u_{yy}+\dfrac{1}{x}\,u_y$	$u(x,0)=0,\ u_y(x,0)=e^{-x}$ for $0\le x\le1$	Transient current produced by an initially non-uniform charge distribution	D 335

No.	Type	Differential equation	Boundary conditions	Problem	References
11	Hyperb.	$u_{xx} + \dfrac{1}{x} u_x = u_{yy}$	$\begin{cases} u(x,0)=1-x^2,\ u_y(x,0)=0 & \text{for } |x|\leq 1 \\ u(\pm 1,y)=0 & \text{for } y\geq 0 \end{cases}$	Vibrations of a circular membrane	D 279
12	Parabol.	$u_{xx} + u_{yy} = u_t$	$u = \dfrac{\partial u}{\partial \nu}$ for $|x|=1,\ |y|=1$; $u=1$ for $t=0$	Cooling of a long square prism	D 335
13	Parabol.	$u_{zzzz} + K u_{yy} = 0$	$\begin{cases} u(x,0),\ u_y(x,0) \text{ prescribed for } 0\leq x\leq a, \\ u(0,y)=u_{zz}(0,y)=u(a,y)=u_{zz}(a,y)=0 \text{ for } y\geq 0 \end{cases}$	Flexural vibrations of a rod	D 303

Chapter V. *Boundary-value problems in partial differential equations*

No.	Type	Differential equation	Boundary conditions	Problem	References
1	Laplace's equation in two independent variables	$\nabla^2 u = u_{xx} + u_{yy} = 0$	$u=1$ on three sides of a square; $u=0$ on the fourth side	Temperature distribution in a plate	D 361–366
2			$u=x^2+y^2$ on all four sides of a square	Torsion of a square-sectioned girder	Bm 403
3			$\dfrac{\partial u}{\partial \nu}$ prescribed on all six sides of an L-shaped region	Fluid flow through a sharply bent channel	D 355, Di 380, 387
4			Two concentric squares: $u=1$ on inner boundary, $u=0$ on outer boundary	Transformer field	D 451
5	Poisson's equation in two independent variables	$\nabla^2 u = u_{xx} + u_{yy} = -1$	$u(x,\pm 1)=0,\ u(0,y)=1-y^2$	Infinite fundamental region	R 440
6			$u=0$ for $|x|=1$ and for $|y|=1$ (Square section); $u=0$ on the boundary (I section)	Torsion problem for a girder — Square section / I section	Se 422, T 447, Ps 451, Se 451, D 451, Di 451
7		$\nabla^2 u = u_{xx} + u_{yy} = x^2-1$	$u=\dfrac{\partial u}{\partial \nu}$ for $|x|=1$ and for $|y|=1$	Temperature in a square-sectioned wire carrying a current	D 381, Di 381, 387, Bm 405, Cl 411, Gl 413, Ls 414, Ps 419, R 429
8			$u=0$ for $|x|=1$ and for $|y|=\dfrac{1}{2}$	Distortion of a rectangular membrane by a given pressure distribution	Ls 417
9	Further elliptic differential equations	$u_{xx} + u_{yy} + \dfrac{3}{5-y}\,u_y = -1$	$u=0$ for $|x|=\dfrac{1}{2}$ and for $|y|=1$	Stresses in a helical spring	D 358, Di 383, 387, R 453
10		$\nabla^2 u = u_{rr} + \dfrac{1}{r}\,u_r + u_{zz} = 0$	$u=1$ for $z=1$, $u=-1$ for $z=-1$; $u=\dfrac{\partial u}{\partial \nu}$ for $r=1$	Temperature distribution in a cylinder	D 452
11		$\nabla^2 u = u_{xx} + u_{yy} + u_{zz} = 0$	$u_\nu=0$ over each face of cube; $u=\pm 1$ at two opposite corners	Temperature distribution in a cube	D 368, R 437

Table VII (continued)

No.	Type	Equation	Initial or boundary conditions	Possible physical applications and other remarks	Methods used and page numbers
12	Eigenvalue problems with elliptic differential equations	$\nabla^2 u = u_{xx} + u_{yy} = -\lambda u$	$u=0$ on the boundary (of a regular hexagon)	Normal modes of vibration of a hexagonal membrane	D 392
13			$u=0$ along the sides of a square; special condition along the diagonal	Normal modes of vibration of a membrane containing an elastic thread	D 453, R 453
14		$\nabla^2 u = u_{xx} + u_{yy} + u_{zz} = -\lambda u$	$u=0$ on the boundary		D 373
15			$u_\nu=0$ on part of the boundary $u=0$ on the remainder	Natural acoustic frequencies of a room	D 452
16	Non-lin. ell. d.e.	$\nabla^2 u = u_{xx} + u_{yy} = -\tfrac{1}{2}(1 + e^u)$	$u=0$ for $|x|=1$ and for $|y|=1$	Temperature distribution in a plate	D 453, R 453, Cl 453
17	Biharmonic problems	$\nabla^4 u = \dfrac{p}{N}$	$u=u_\nu=0$ along the long side of a trapezium $\nabla^2 u=0$ along the other three sides	Uniformly loaded trapezoidal plate with its long edge clamped and the others freely supported	D 391, 453
18		$\nabla^4 u = 1$	$u=u_\nu=0$ on the boundary (of the rectangle $|x|\leq 1,\ |y|\leq 2$)	Uniformly loaded rectangular plate clamped along its edges	R 434
19		$\nabla^4 u = x^2$	$u=u_x=0$ for $|x|=1$ $u=u_{yy}=0$ for $|y|=1$	Non-uniformly loaded square plate with the two pairs of opposite sides clamped and freely supported, resp.	D 359, Se 421, R 431
20		$\nabla^4 u = \lambda u$	$u=u_\nu=0$ on the boundary (of a regular hexagon)	Normal modes of vibration of a clamped hexagonal plate	D 395

Chapter VI. *Integral and functional equations*

No.	Type	Equation		Possible physical applications and other remarks	Methods used and page numbers
1	Inhomogeneous linear integral equations	$\pi y(x) + \displaystyle\int_0^{2\pi} \frac{2y(\xi)}{5 - 3\cos(x+\xi)}\, d\xi = \begin{cases} \sin x & \text{for } 0 \leq x \leq \pi \\ 0 & \text{for } -\pi \leq x \leq 0 \end{cases}$		First boundary-value problem of potential theory; steady temperature distribution in an elliptic cylinder	Fs 471
2		$y(x) + \displaystyle\int_{-1}^{1} K(x,\xi)\, y(\xi)\, d\xi = \frac{1}{1+x^2}$	with $K(x,\xi) = \begin{cases} 1-(x-\xi)^2 & \text{for } |x-\xi| \leq 1 \\ 0 & \text{for } |x-\xi| \geq 1 \end{cases}$	Calculation of true intensity from readings affected by interference	R 487
3		$\displaystyle\int_0^x y(x-\xi)\left\{1 + \frac{e^{-\xi}-1}{2\xi}\right\} d\xi = x$		Linear transmission system, sluggish measuring instrument, Volterra integral equation	Fs 514, It 516
4		$\dfrac{1}{2\pi}\displaystyle\oint_{-1}^{1} \frac{y(\xi)\, d\xi}{\xi - x} = |x|$		Singular integral equation	Se 509

No.	Type	Equation	Side conditions	Application	References
5	Eigen-value problem	$y(x)=\lambda\int_0^1 e^{x\xi}y(\xi)\,d\xi$			Et 527, It 527
6	Eigenvalue problems	$y(x)=\lambda\int_{-1}^1 K(x,\xi)y(\xi)\,d\xi$ with $K(x,\xi)=\begin{cases}1+\cos\pi(x-\xi)&\text{for }|x-\xi|\le1\\0&\text{for }|x-\xi|\ge1\end{cases}$		Distortion of intensity (of illumination) distribution by an optical system	Fs 527
7		$y(x)=\lambda\int_0^1 K(x,\xi)y(\xi)\,d\xi$ with $K(x,\xi)=\begin{cases}\dfrac16 x^2(3\xi-x)(1+\xi)&\text{for }x\le\xi\\[4pt]\dfrac16\xi^2(3x-\xi)(1+\xi)&\text{for }x\ge\xi\end{cases}$		Flexural vibrations of a beam of variable cross-section	R 253, Fs 472
8		$u(x,y)=\lambda\int_{-1}^1\int_{-1}^1|x\xi+y\eta|\,u(\xi,\eta)\,d\xi\,d\eta$			Fs 474, It 479
9	Non-linear integral equations	$\int_0^1\dfrac{[y(x)+y(\xi)]^2}{1+x+\xi}\,d\xi=1$			Fs 477, It 480, Se 497
10		$\int_0^1\dfrac{d\xi}{y(x)+y(\xi)}=1+x$			Fs 527
11		$\int_0^1\sqrt{x+\xi}\,[y(\xi)]^2\,d\xi=y(x)$			Cl 483, R 488
		$\int_{-1}^1\sqrt{x+y(\xi)}\,d\xi=y(x)$			It 481
12	Integro-differential equations	$\dfrac{1+x}{2}y''+10\int_0^1\dfrac{y(\xi)}{1+x+\xi}\,d\xi+1=0$	$y(0)=y(1)=0$		Se 496, Degenerate kernel 500
13		$(EIy'')''-Hy''+k\int_{-l}^l y(\xi)\,d\xi=m\omega^2 y$	$y(\pm l)=y''(\pm l)=0$	Vertical oscillations of a suspension bridge	R 492
14	Functional equations	$y(y(x))=e^x$	—		Ps 523
15		$y(y(x))=1-x^2$	—		Ps 523
16		$y'(x)-y'(-x)=G[y(-x)-y(-x)]$	—		D 526
17		$y(x+y(x))=1-x^2$		Non-linear oscillations of a stretched string attached to a spring	Ps 528
18		$-y'_{(x)}=\lambda y(1-x)$	$y(0)=y'(1)=0$		D 528, R 528, Ps 528
19		$\nabla^2 u(x,y)+\lambda u(-x,-y)=0$	$\{u(1,y)=u(x,1)=0;\ u_z(-1,y)=u_y(x,-1)=0\}$		D 528, R 528

Table VIII. *Taylor expansion of a general finite expression involving the differential operators* ∇^2 *and* ∇^4 *for a square mesh.* (All the formulae of Table VI for a square mesh may be derived from this Taylor expansion.)

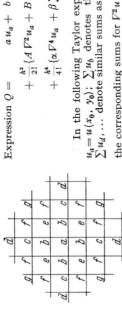

Expression $Q =$

$$au_a + b\sum u_b + c\sum u_c + d\sum u_d + e\sum u_e + f\sum u_f + g\sum u_g$$
$$+ \frac{h^2}{2!}\{A\nabla^2 u_a + B\sum\nabla^2 u_b + C\sum\nabla^2 u_c + D\sum\nabla^2 u_d + E\sum\nabla^2 u_e + F\sum\nabla^2 u_f + G\sum\nabla^2 u_g\}$$
$$+ \frac{h^4}{4!}\{\alpha\nabla^4 u_a + \beta\sum\nabla^4 u_b + \gamma\sum\nabla^4 u_c + \delta\sum\nabla^4 u_d + \varepsilon\sum\nabla^4 u_e + \zeta\sum\nabla^4 u_f + \eta\sum\nabla^4 u_g\}.$$

In the following Taylor expansion all derivatives are evaluated at the point $x = a$; $h =$ mesh width; $u_a = u(x_0, y_0)$; $\sum u_b$ denotes the sum $u(x_0 + h, y_0) + u(x_0, y_0 + h) + u(x_0 - h, y_0) + u(x_0, y_0 - h)$; $\sum u_c$, $\sum u_d$, ... denote similar sums as indicated in the diagram; $\nabla^2 u_a$, $\sum\nabla^2 u_b$, ... and $\nabla^4 u_a$, $\sum\nabla^4 u_b$, ... denote the corresponding sums for $\nabla^2 u$ and $\nabla^4 u$; $\nabla^2 = \dfrac{\partial^2}{\partial x^2} + \dfrac{\partial^2}{\partial y^2}$; $\mathsf{D} = \dfrac{\partial^4}{\partial x^2\,\partial y^2}$.

Then $Q =$

$$u_a\,[4\{b + c + d + e + 2f + g\} + a]$$
$$+ \frac{h^2}{2!}\nabla^2 u_a\,[2\{b + 4c + 9d + 2e + 10f + 8g\} + 4\{B + C + D + E + 2F + G\} + A]$$
$$+ \frac{h^4}{4!}\nabla^4 u_a\,\Big[2\{b + 16c + 81d + 2e + 34f + 32g\} + \binom{4}{2}2\{B + 4C + 9D + 2E + 10F + 8G\} +$$
$$+ 4\{\beta + \gamma + \delta + \varepsilon + 2\zeta + \eta\} + \alpha\Big]$$

$$- \frac{h^4}{4!}\mathsf{D}u_a\,[4\{b + 16c + 81d - 4e - 14f - 64g\}]$$
$$+ \frac{h^6}{6!}\nabla^6 u_a\,\Big[2\{b + 64c + 729d + 2e + 130f + 128g\} + \binom{6}{2}2\{B + 16C + 81D + 2E + 34F + 32G\} +$$
$$+ \binom{6}{4}2\{\beta + 4\gamma + 9\delta + 2\varepsilon + 10\zeta + 8\eta\}\Big]$$

$$- \frac{h^6}{6!}\nabla^2\mathsf{D}u_a\,\Big[6\{b + 64c + 729d - 8e - 70f - 512g\} + \binom{6}{2}4\{B + 16C + 81D - 4E - 14F - 64G\}\Big]$$
$$+ \frac{h^8}{8!}\nabla^8 u_a\,\Big[2\{b + 256c + 6561d + 2e + 514f + 512g\} + \binom{8}{2}2\{B + 64C + 729D + 2E + 130F + 128G\} +$$
$$+ \binom{8}{4}2\{\beta + 16\gamma + 81\delta + 2\varepsilon + 34\zeta + 32\eta\}\Big]$$

$$- \frac{h^8}{8!}\nabla^4\mathsf{D}u_a\,\Big[8\{b + 256c + 6561d - 12e - 438f - 3072g\} + \binom{8}{2}6\{B + 64C + 729D - 8E - 70F - 512G\} +$$
$$+ \binom{8}{4}4\{\beta + 16\gamma + 81\delta - 4\varepsilon - 14\zeta - 64\eta\}\Big]$$

$$+ \frac{h^8}{8!}\mathsf{D}^2 u_a\,[4\{b + 256c + 6561d + 16e - 1054f + 4096g\}]$$

$$+ \text{terms of the 10th and higher orders.}$$

Table IX. *Taylor expansion of a general finite expression involving the differential operators* ∇^2 *and* ∇^4 *for a triangular mesh.* (All formulae of Table VI for a triangular mesh may be derived from this Taylor expansion.) Explanation of notation as for Table VIII, together with

$$u_{kl} = \left(\frac{\partial^{k+l}u}{\partial x^k \partial y^l}\right)_{x_0,y_0} \quad \text{and}$$

$$\nabla^6_* = u_{60} - 15u_{42} + 15u_{24} - u_{06},$$
$$\nabla^8_* = u_{80} - 14u_{62} + 14u_{26} - u_{08},$$
$$\nabla^{10}_* = u_{10,0} - 13u_{82} - 14u_{64} + 14u_{46} + 13u_{28} - u_{0,10}.$$

Expression $Q =$
$$a u_a + b \Sigma u_b + c \Sigma u_c + d \Sigma u_d +$$
$$+ \frac{h^2}{2!}\{A \nabla^2 u_a + B \Sigma \nabla^2 u_b + C \Sigma \nabla^2 u_c + D \Sigma \nabla^2 u_d\} +$$
$$+ \frac{h^4}{4!}\{\alpha \nabla^4 u_a + \beta \Sigma \nabla^4 u_b + \gamma \Sigma \nabla^4 u_c + \delta \Sigma \nabla^4 u_d\}.$$

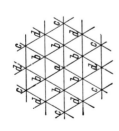

$$Q = u_a\left[\, 6 \{b + c + d\} + a\right]$$
$$+ \frac{h^2}{2!}\nabla^2 u_a\left[\, 3 \{b + 4c + 3d\} + 6 \{B + C + D\} + A\right]$$
$$+ \frac{h^4}{4!}\nabla^4 u_a\left[\, \frac{9}{4} \{b + 16c + 9d\} + \binom{4}{2} 3 \{B + 4C + 3D\} + 6 \{\beta + \gamma + \delta\} + \alpha\right]$$
$$+ \frac{h^6}{6!}\nabla^6 u_a\left[\, \frac{15}{8} \{b + 64c + 27d\} + \binom{6}{2} \frac{9}{4} \{B + 16C + 9D\} + \binom{6}{4} 3 \{\beta + 4\gamma + 3\delta\}\right]$$
$$+ \frac{h^6}{6!}\nabla^6_* u_a\left[\, \frac{3}{16} \{b + 64c - 27d\}\right]$$
$$+ \frac{h^8}{8!}\nabla^8 u_a\left[\, \frac{105}{64} \{b + 256c + 81d\} + \binom{8}{2} \frac{15}{8} \{B + 64C + 27D\} + \binom{8}{4} \frac{9}{4} \{\beta + 16\gamma + 9\delta\}\right]$$
$$+ \frac{h^8}{8!}\nabla^8_* u_a\left[\, \frac{3}{8} \{b + 256c - 81d\} + \binom{8}{2} \frac{3}{16} \{B + 64C - 27D\}\right]$$
$$+ \frac{h^{10}}{10!}\nabla^{10} u_a\left[\, \frac{189}{128} \{b + 1024c + 243d\} + \binom{10}{2} \frac{105}{64} \{B + 256C + 81D\} + \binom{10}{4} \frac{15}{8} \{\beta + 64\gamma + 27\delta\}\right]$$
$$+ \frac{h^{10}}{10!}\nabla^{10}_* u_a\left[\, \frac{135}{256} \{b + 1024c - 243d\} + \binom{10}{2} \frac{3}{8} \{B + 256C - 81D\} + \binom{10}{4} \frac{3}{16} \{\beta + 64\gamma - 27\delta\}\right]$$

$+$ terms of the 12th and higher orders.

Table X. *Taylor expansion of a general finite expression involving the differential operators ∇^2 and ∇^4 for a cubical mesh*
[from J. ALBRECHT: Z. Angew. Math. Mech. **33**, 41–48 (1953)]

In the following Taylor expansion all derivatives are evaluated at the point $m = (x_0, y_0, z_0)$; h = mesh width; $u_m = u(x_0, y_0, z_0)$; $\sum u_s$, $\sum u_k$, $\sum u_e$ denote the sums over the u values at all mesh points whose distances from m are h, $\sqrt{2}\,h$, $\sqrt{3}\,h$, respectively; a corresponding notation is used for the values of $\nabla^2 u$ and $\nabla^4 u$; further

$$\nabla^{2n} = \left(\frac{\partial^2}{\partial x^2} + \frac{\partial^2}{\partial y^2} + \frac{\partial^2}{\partial z^2}\right)^n; \quad u_{jkl} = \frac{\partial^{j+k+l}u}{\partial x^j\,\partial y^k\,\partial z^l}; \quad \nabla_3^{2k}u = u_{kk0} + u_{k0k} + u_{0kk} \quad \text{for } k > 0.$$

Expression $Q =$
$$m\,u_m + s\sum u_s + k\sum u_k + e\sum u_e$$
$$+\; \frac{h^2}{2!}\{M\nabla^2 u_m + S\sum\nabla^2 u_s + K\sum\nabla^2 u_k + E\sum\nabla^2 u_e\}$$
$$+\; \frac{h^4}{4!}\{\mu\nabla^4 u_m + \sigma\sum\nabla^4 u_s + \varkappa\sum\nabla^4 u_k + \varepsilon\sum\nabla^4 u_e\}.$$

$$
\begin{aligned}
Q = \quad & u_m && [[\; 6s + 12k + 8e + m]] \\[2pt]
+\; \frac{h^2}{2!}\quad & \nabla^2 u_m && [\{\; 2s + 8k + 8e\} + \{\; 6S + 12K + 8E + M\}] \\[2pt]
+\; \frac{h^4}{4!}\quad & \nabla^4 u_m && [\{\; 2s + 8k + 8e\} + \tbinom{4}{2}\{\; 2S + 8K + 8E\} + \{\; 6\sigma + 12\varkappa + 8\varepsilon + \mu\}] \\[2pt]
+\; \frac{h^4}{4!}\quad & \nabla_3^4 u_m && [\{-4s + 8k + 32e\}] \\[2pt]
+\; \frac{h^6}{6!}\quad & \nabla^6 u_m && [\{\; 2s + 8k + 8e\} + \tbinom{6}{2}\{\; 2S + 8K + 8E\} + \tbinom{6}{4}\{\; 2\sigma + 8\varkappa + 8\varepsilon\}] \\[2pt]
+\; \frac{h^6}{6!}\quad & \nabla^2\nabla_3^4 u_m && [\{-6s + 36k + 96e\} + \tbinom{6}{2}\{-4S + 8K + 32E\}] \\[2pt]
+\; \frac{h^6}{6!}\quad & u_{222m} && [\{\; 6s - 156k + 384e\}] \\[2pt]
+\; \frac{h^8}{8!}\quad & \nabla^8 u_m && [\{\; 2s + 8k + 8e\} + \tbinom{8}{2}\{\; 2S + 8K + 8E\} + \tbinom{8}{4}\{\; 2\sigma + 8\varkappa + 8\varepsilon\}] \\[2pt]
+\; \frac{h^8}{8!}\quad & \nabla^4\nabla_3^4 u_m && [\{-8s + 80k + 192e\} + \tbinom{8}{2}\{-6S + 36K + 96E\} + \tbinom{8}{4}\{-4\sigma + 8\varkappa + 32\varepsilon\}] \\[2pt]
+\; \frac{h^8}{8!}\quad & \nabla^2 u_{222m} && [\{16s - 496k + 2304e\} + \tbinom{8}{2}\{\; 6S - 156K + 384E\}] \\[2pt]
+\; \frac{h^8}{8!}\quad & \nabla_3^8 u_m && [\{\; 4s + 72k + 128e\}]
\end{aligned}
$$

+ terms of the 10th and higher orders.

Author Index

Subject Index

CIVIL CODE OF THE RUSSIAN FEDERATION

CIVIL CODE OF THE RUSSIAN FEDERATION

Parts One, Two, and Three

Edited, Compiled, and
Translated from the Russian,
with an Introduction by

WILLIAM E. BUTLER

*Professor of Comparative Law in the University of London
Foreign Member, Russian Academy of Natural Sciences and
National Academy of Sciences of Ukraine
Director, The Vinogradoff Institute, University College London
M. M. Speranskii Professor of International and Comparative Law,
Moscow Higher School of Social and Economic Sciences*

OXFORD

UNIVERSITY PRESS

OXFORD

UNIVERSITY PRESS

Great Clarendon Street, Oxford OX2 6DP

Oxford University Press is a department of the University of Oxford.
It furthers the University's objective of excellence in research, scholarship,
and education by publishing worldwide in

Oxford New York

Auckland Bangkok Buenos Aires Cape Town Chennai
Dar es Salaam Delhi Hong Kong Istanbul Karachi Kolkata
Kuala Lumpur Madrid Melbourne Mexico City Mumbai Nairobi
São Paulo Shanghai Taipei Tokyo Toronto

Oxford is a registered trade mark of Oxford University Press
in the UK and in certain other countries

Published in the United States
by Oxford University Press Inc., New York

Status juris: 20 January 2003

British Library Cataloguing in Publication Data
Data available

Library of Congress Cataloging in Publication Data
Data available

ISBN 0–19–926153–9

1 3 5 7 9 10 8 6 4 2

Typeset by Hope Services (Abingdon) Ltd.
Printed in Great Britain
on acid-free paper by
Biddles Ltd., Guildford and King's Lynn

William E. Butler is Professor of Comparative Law in the University of London; Academician of the Russian Academy of Natural Sciences and of the National Academy of Sciences of Ukraine; Director of The Vinogradoff Institute, University College London. He was the founder and Dean (1993–98) of the Faculty of Law, Moscow Higher School of Social and Economic Sciences and continues to teach there as the M. M. Speranskii Professor of International and Comparative Law, and in spring 2002 as professor at the Chair of Civil and Family Law, Moscow State Legal Academy. In 1999 he was elected to the Russian Academy of Legal Sciences. He read law at Harvard University, where he has been Visiting Professor of Law, and holds a Master of Russian Law (LL.M.) from the School of Law, Institute of State and Law, Russian Academy of Sciences, and higher doctorates from The Johns Hopkins University (Ph.D.) and the University of London (LL.D.). He has advised the Governments of the USSR, Belarus, Russian Federation, Republic Azerbaidzhan, Republic Kazakhstan, Kyrgyz Republic, Tadzhikistan, Turkmenistan, Ukraine, Uzbekistan, and Lithuania on aspects of law reform and participated in the drafting of a number of key legislative acts. He served as a member of the EC/IS Joint Task Force on Law Reform in the Independent States and has advised the World Bank, the European Bank of Reconstruction and Development, the European Commission, the United Kingdom Know-How Fund, and specialised agencies of the United Nations on law reform in the Independent States. In 1995 he was elected a member of the Russian Court of International Commercial Arbitration, and in 1999 to the Court of Economic Arbitration attached to the Urals Chamber of Trade and Industry (Ekaterinburg). In 2002 he co-founded Phoenix Law Associates, a law firm based in Moscow and specialising in the legal systems of the CIS.

CONTENTS

CIVIL CODE OF THE RUSSIAN FEDERATION

PART ONE

SECTION I. GENERAL PROVISIONS

Subsection 1. Basic Provisions

Subsection 2. Persons

Subsection 3. Objects of Civil Rights

Subsection 4. Transactions and Representation

Subsection 5. Periods. Limitations

SECTION II. THE RIGHT OF OWNERSHIP AND OTHER RIGHTS TO THING

SECTION III. GENERAL PART OF THE LAW OF OBLIGATIONS

Subsection 1. General Provisions on Obligations

Subsection 2. General Provisions on Contract

PART TWO

SECTION IV. INDIVIDUAL TYPES OF OBLIGATIONS

PART THREE

SECTION V. INHERITANCE LAW

SECTION VI. INTERNATIONAL PRIVATE LAW

Contents

PREFACE

This volume contains in a specially prepared translation the Civil Code of the Russian Federation, incorporating all amendments, changes, and additions, and the Federal Laws introducing each Part of the Civil Code into operation, as amended. A Russian/English glossary of civil-law terminology is given, drawn from the translation and from my *Russian-English Legal Dictionary* (2001). The subject index gives the Russian language term for each index entry.

Although the volume stands independently as a resource on Russian law, it accompanies and complements my treatise on the subject, *Russian Law*, the second edition of which is appearing simultaneously, and a collection of laws and decrees, many specifically called for by the Civil Code, which elaborate and clarify Civil Code provisions, *Russian Commercial and Company Law*. Both are published by Oxford University Press.

Status juris: 20 January 2003

W. E. Butler
Bicester, Oxon.

ON THE CIVIL CODE

The Civil Code of the Russian Federation is the only Civil Code to have been accorded the honour of three academic translations within the space of five years by foreign legal scholars,[1] together with an uncountable number of commercial translations of the most dubious quality. The reasons for this accolade are, in my view, several. The Civil Code represents a unique blend of theory, structure, and norms which collectively represent the transitional era that the Code is intended to both serve and facilitate. The transition is one from a socialist planned economy to a species of market economy whose ultimate configuration is not wholly in evidence. It is a transition without precedent and without benchmarks; no one can predict for certain whether the Civil Code will be successful or not in this mission. This is an inconsummately fascinating transition, legally speaking, for it places law at the forefront of introducing and implementing social, economic, and institutional changes designed to take Russian from a Soviet past to a non-Soviet, democratic future founded on the rule of law.

An authentically Russian creation, integrating prerevolutionary Russian civil-law principles, institutions and terminology with those of the Soviet era which either proved themselves to be useful or which, in the opinion of the draftsmen, cannot yet be dispensed with entirely, the Civil Code also draws upon modern European models, relevant Anglo-American experience, and international treaties. The principal 'schools' of civil law in Russia—Moscow, St. Petersburg, Ekaterinburg (formerly Sverdlovsk)—all took an active role in the drafting and thoroughly aired their respective conceptual approaches to the Code.[2]

The importance of the Russian Civil Code is further enhanced by its stature as a model for other members of the Commonwealth of Independent States (CIS), a role fulfilled as a written text and as the object of discussions within the Inter-Parliamentary Assembly of the CIS amongst expert civilists from throughout the

---

[1] My own translations have appeared in W. E. Butler, *Civil Code of the Russian Federation (Part I)* (1995); *Civil Code of the Russian Federation (Parts I and II)* (1997; 2nd edn, 1998); *Russian Civil Legislation* (1999); *Civil Code of the Russian Federation. Parts 1 and 2* (Moscow: Iurinfor, 2001), in parallel Russian and English texts as of 15 September 2001. Professor Peter B. Maggs and A. N. Zhiltsov published a translation with a strong emphasis upon American legal equivalents in *The Civil Code of the Russian Federation. Parts 1 and 2* (1997), accompanied by a comprehensive introduction prepared by A. L. Makovsky and S. A. Khokhlov. Christopher Osakwe introduced his translation with a substantial commentary in C. Osakwe, *Russian Civil Code Annotated* (Moscow: Moscow University Press, 2000).

[2] See on the Sverdlovsk school, Антология уральской цивилистики. *1925–1989:* сборник статей (2001).

former Soviet Union. And there are practitioner considerations, for the influx of foreign investors into Russia during the 1990s required professional legal advice in which the place of the Civil Code was crucial.

Civil-law relations were not drastically affected by the disappearance of the former Soviet Union in and of itself on or about 25 December 1991, for the Russian Federation had in place the Civil Code of the Russian Socialist Federated Soviet Republic (RSFSR), confirmed by a Law of the RSFSR on 11 June 1964, which replicated in all material respects the 1961 Fundamental Principles of Civil Legislation of the USSR and Union Republics (FPCivL). The inappropriateness of those enactments as products of the Soviet era was ameliorated by introducing into force on the territory of the Russian Federation from 3 August 1992 the 1991 Fundamental Principles of Civil Legislation of the USSR and Republics (scheduled to enter into force as from 1 January 1992, but not doing so because of the extinction of the USSR), together with a host of individual laws based on market relations: ownership, lease, enterprises, pledge, intellectual property, and the like. The last have overridden the respective articles of the Civil Code, and other provisions have simply become redundant in the course of time. And Parts One and Two of the Civil Code of the Russian Federation have in turn been the subject of clarifying decrees by the Federal Assembly, the Constitutional Court of the Russian Federation, the Supreme Court of the Russian Federation, and the Supreme Arbitrazh Court of the Russian Federation.

Although the Russian Federation was not the first Independent State to replace the 'General Part' of its Civil Code (Estonia did so on 28 June 1994), the text of the Civil Code of the Russian Federation became the standard point of reference, functioning veritably as a model code, for most of the other Independent States, together with a more rudimentary Model Civil Code developed within the Commonwealth of Independent States. The Belarus Civil Code remained the paradigm for an interim stage of reworking the 1964 generation of civil codes until its full-scale replacement entered into force from 1 July 1999.[3] The Republic Kazakhstan adopted the Civil Code of the Republic Kazakhstan in two parts on 27 December 1994 and 1 July 1999 based principally on the Russian model.[4] The Republic Uzbekistan adopted a full Civil Code in two parts, the first on 21 December 1995 and the second on 28 August 1996, with effect from 1 January 1997.[5] New civil codes also have been enacted by Kyrgyzia, Georgia, Armenia, Tadzhikistan, and Turkmenistan.[6]

The accompanying Federal Laws on the procedure for introducing the Civil Code into force will repay careful reading. It should be noted that various portions

[3] For translations, see W. E. Butler, *Civil Code of the Republic Belarus* (1994; 2nd edn, 1998); Butler, *Civil Code of the Republic Belarus* (2000).

[4] See W. E. Butler, *Civil Code of the Republic Kazakhstan* (1995; 2nd edn, 1996; 3rd edn, 1997).

[5] Translated in W. E. Butler, *Civil Code of the Republic Uzbekistan* (1997; 2nd edn, 1998; 3rd edn, 1999).

[6] Translated in W. E. Butler, *Turkmenistan Civil Code of Saparmurat Turkmenbashi* (1999).

of the Civil Code enter into force at different times. Whereas 1 January 1995 is the general date of introduction into force, the provisions affecting the creation and operation of juridical persons become effective from the date of official publication. The stipulation that the Chapter on land relations only enters into force when the Land Code of the Russian Federation enters into force was altered by an amendment of 16 April 2001 (in anticipation of the Land Code, which was adopted 25 October 2001 and entered into force on 30 October 2001). Part Two of the Civil Code entered into force from 1 March 1996, and Part Three as from 1 March 2002. The application of all the Parts is subject to decisions of the Constitutional Court of the Russian Federation, decrees of the plenums of the Supreme Court of the Russian Federation and Supreme Arbitrazh Court of the Russian Federation, sometimes adopted jointly, and decrees of the last two courts in individual cases. And given that Russian doctrine with respect to the role of judicial precedent in the Russian legal system has changed, the published decisions of all Russian courts must be taken into account.

The present text contains Parts One, Two, and Three of the Civil Code, or what might be loosely called the 'General Part' and the greater portion of the 'Special Part'. Part Three did not, in the end, include provisions on intellectual property; it remains to be seen whether they will one day be included, or whether the present version will be treated as definitive.

Parts One and Two of the Civil Code have been amended on eleven occasions. On 20 February 1996 Article 64(1) was amended, and on 12 August 1996 two separate laws were enacted, one to amend Article 185(4) and the second to amend Article 855(2). On 24 October 1997 Article 855(2) was added to yet again. Amendments also were introduced by Laws of 23 December 1997, 8 July 1999, 17 December 1999, 16 April 2001 (the footnote to the heading of Chapter 17 was deleted), 15 May 2001 (Article 292 was changed), 21 March 2002 (affecting Articles 51, 54, 61, 62, and 63), 14 November 2002 (affecting Articles 48, 54, 113, 114, 115, and 300), 26 November 2002 (affecting Articles 318, 1086, 1087, and 1091), and twice on 10 January 2003 (affecting Articles 358, 789, 855, 908, and 1063).

The translation of the Civil Code of the Russian Federation is based on the official text published in СЗ РФ (1994), no. 32, item 3301; (1996), no. 9, item 773; no. 34, items 4025 and 4026; (1997), no. 43, item 4903; (1999), no. 28, item 3471; no. 51, item 6288; (2001), no. 17, item 1644; no. 21, item 2063; no. 49, item 4552; (2002), no. 12, item 1093; no. 48, items 4737 and 4746; (2003), no. 2, items 160 and 167. The sources of all other documents translated in the present volume are given in the headnote to each.

W. E. Butler

DETAILED CONTENTS LIST

CIVIL CODE OF THE RUSSIAN FEDERATION (PARTS ONE, TWO, AND THREE)

PART ONE

SECTION I. GENERAL PROVISIONS

Subsection 1. Basic Provisions

Chapter 1. Civil Legislation

Chapter 2. Origin of Civil Rights and Duties, and Effectuation and Defence of Civil Rights

Subsection 2. Persons

Chapter 3. Citizens (Natural Persons)

Chapter 4. Juridical Persons

§1. Basic Provisions

§2. Economic Partnerships and Societies
1. General Provisions

2. Full Partnership

Chapter 18. Right of Ownership and Other Rights to Thing in Dwelling Premises

Chapter 19. Right of Economic Jurisdiction, Right of Operative Management

Chapter 20. Defence of Right of Ownership and Other Rights to Thing

SECTION III. GENERAL PART OF THE LAW OF OBLIGATIONS

Subsection 1. General Provisions on Obligations

Chapter 21. *Concept of and Parties to Obligation*

Chapter 22. *Performance of Obligations*

Chapter 25. *Responsibility for Violation of Obligations*

Chapter 26. *Termination of Obligations*

Subsection 2. General Provisions on Contract

Chapter 27. Concept and Conditions of Contract

Chapter 28. Conclusion of Contract

PART TWO

SECTION IV. INDIVIDUAL TYPES OF OBLIGATIONS

Chapter 30. Purchase-Sale

§1. General Provisions on Purchase-Sale

Chapter 34. Lease

§1. General Provisions on Lease

§2. Rental

Chapter 49. Commission

Chapter 52. *Agency*

Chapter 53. *Trust Management of Property*

Chapter 54. *Commercial Concession*

PART THREE

SECTION V. INHERITANCE LAW

SECTION VI. INTERNATIONAL PRIVATE LAW

Chapter 66. General Provisions

Chapter 67. Law Subject to Application When Determining Legal Status of Persons

ABBREVIATIONS

CIS	Commonwealth of Independent States
FPCivL	Fundamental Principles of Civil Legislation of the USSR and Republics
RSFSR	Russian Soviet Federated Socialist Republic
СЗ РФ	Собрание законодательства Российской Федерации [Collection of Legislation of the Russian Federation]
USSR	Union of Soviet Socialist Republics

CIVIL CODE OF THE RUSSIAN FEDERATION

PART ONE

SECTION I. GENERAL PROVISIONS

Subsection 1. Basic Provisions

Chapter 1. Civil Legislation

Article 1. Basic Principles of Civil Legislation

1. Civil legislation shall be based on recognition of the equality of the participants of the relations regulated by it, the inviolability of ownership, the freedom of contract, the inadmissibility of arbitrary interference by anyone in private matters, and the necessity for the unobstructed effectuation of civil rights, ensuring the restoration of rights violated, and the judicial defence thereof.

2. Citizens (natural persons) and juridical persons shall acquire and effectuate their civil rights by their own will and in their own interest. They shall be free in establishing their rights and duties on the basis of contract and in determining any conditions of a contract which are not contrary to legislation.

Civil rights may be limited on the basis of a federal law and only to the extent that this is necessary for the purposes of defending the fundamental principles of the constitutional system, morality, public health, and the rights and legal interests of other persons, and to ensure the defence of the country and the security of the State.

3. Goods, services, and financial assets shall move freely throughout the entire territory of the Russian Federation.

Limitations on the free movement of goods and services may be introduced in accordance with a federal law if this is necessary in order to ensure safety, defence of human life and health, and the protection of nature and cultural valuables.

Article 2. Relations Regulated by Civil Legislation

1. Civil legislation shall determine the legal status of the participants of civil turnover, the grounds for the origin and the procedure for the effectuation of the right of ownership and other rights to a thing and exclusive rights to the results of intellectual activity (intellectual property), shall regulate contractual and other obligations, and also other property and related personal nonproperty relations based on equality, autonomy of will, and the property autonomy of the participants thereof.

Citizens and juridical persons shall be the participants of relations regulated by civil legislation. The Russian Federation, subjects of the Russian Federation, and municipal formations also may participate in relations regulated by civil legislation (Article 124).

Civil legislation shall regulate relations between persons effectuating entrepreneurial activity or with the participation thereof, proceeding from the fact that autonomous activity effectuated at one's own risk directed towards the systematic obtaining of profit from the use of property, the sale of goods, the fulfilment of work, or the rendering of services by persons registered in this capacity in the procedure established by law shall be entrepreneurial.

The rules established by civil legislation shall apply to relations with the participation of foreign citizens, stateless persons, and foreign juridical persons unless provided otherwise by a federal law.

2. Inalienable rights and freedoms of man and other nonmaterial benefits shall be defended by civil legislation insofar as does not follow otherwise from the essence of these nonmaterial benefits.

3. Civil legislation shall not apply to property relations based on administrative or other power subordination of one party to another, including to tax and other financial and administrative relations, unless provided otherwise by legislation.

Article 3. Civil Legislation and Other Acts Which Contain Norms of Civil Law

1. In accordance with the Constitution of the Russian Federation civil legislation shall be within the jurisdiction of the Russian Federation.

2. Civil legislation shall consist of the present Code and other federal laws (hereinafter: laws) adopted in accordance therewith regulating the relations specified in Article 2(1) and (2) of the present Code.

Norms of civil law contained in other laws must conform to the present Code.

3. Relations specified in Article 2(1) and (2) of the present Code may also be regulated by edicts of the President of the Russian Federation, which must not be contrary to the present Code and other laws.

4. On the basis of and in execution of the present Code and other laws, and of edicts of the President of the Russian Federation, the Government of the Russian Federation shall have the right to adopt decrees containing norms of civil law.

5. In the event an edict of the President of the Russian Federation or a decree of the Government of the Russian Federation is contrary to the present Code or other law, the present Code or respective law shall apply.

6. The operation and application of norms of civil law contained in edicts of the President of the Russian Federation and decrees of the Government of the Russian Federation (hereinafter: other legal acts) shall be determined by the rules of the present Chapter.

7. Ministries and other federal agencies of executive power may issue acts containing norms of civil law in the instances and within the limits provided for by the present Code, other laws, and other legal acts.

Article 4. Operation of Civil Legislation in Time

1. Acts of civil legislation shall not have retroactive force and shall apply to relations which arose after the introduction thereof into operation.

The operation of a law shall extend to relations which arose before the introduction thereof into operation only in instances when this is expressly provided for by a law.

2. With regard to relations which arose before the introduction into operation of an act of civil legislation, it shall apply to the rights and duties which arose after the introduction thereof into operation. Relations of parties regarding a contract concluded before the introduction into operation of an act of civil legislation shall be regulated in accordance with Article 422 of the present Code.

Article 5. Customs of Business Turnover

1. A custom of business turnover shall be deemed to be a rule of behaviour which has been formed and extensively applied in any domain of entrepreneurial activity and is not provided for by legislation irrespective of whether it has been fixed in any document.

2. Customs of business turnover which are contrary to provisions of legislation or a contract binding upon the participants of the respective relation shall not be applied.

Article 6. Application of Civil Legislation by Analogy

1. In instances when relations provided for by Article 2(1) and (2) of the present Code have not been expressly regulated by legislation or by agreement of the parties and there is no custom of business turnover applicable to them, civil legislation regulating similar relations (analogy of lex) shall apply to such relations if this is not contrary to the essence thereof.

2. When it is impossible to use analogy of lex, the rights and duties of the parties shall be determined by proceeding from the general principles and sense of civil legislation (analogy of jus) and the requirements of good faith, reasonableness, and justness.

Article 7. Civil Legislation and Norms of International Law

1. Generally-recognised principles and norms of international law and international treaties of the Russian Federation shall, in accordance with the Constitution of the Russian Federation, be an integral part of the legal system of the Russian Federation.

2. International treaties of the Russian Federation shall apply to the relations specified in Article 2(1) and (2) of the present Code directly except for instances when it follows from the international treaty that the publication of a domestic act is required in order to apply it.

If other rules have been established by an international treaty of the Russian Federation than those which have been provided for by civil legislation, the rules of the international treaty shall apply.

Chapter 2. Origin of Civil Rights and Duties, and Effectuation and Defence of Civil Rights

Article 8. Grounds for Civil Rights and Duties to Arise

1. Civil rights and duties shall arise from the grounds provided for by a law and other legal acts, and also from the actions of citizens and juridical persons, which although not provided for by a law or such acts but by virtue of the basic principles and sense of civil legislation give rise to civil rights and duties.

In accordance therewith civil rights and duties shall arise:

(1) from contracts and other transactions provided for by a law, and also from contracts and other transactions which, although not provided for by a law, are not contrary thereto;

(2) from acts of State agencies and agencies of local self-government which have been provided for by a law as a ground for civil rights and duties to arise;

(3) from a judicial decision which has established civil rights and duties;

(4) as a result of the acquisition of property on the grounds permitted by a law;

(5) as a result of the creation of works of science, literature, art, inventions, and other results of intellectual activity;

(6) as a consequence of the causing of harm to another person;

(7) as a consequence of unfounded enrichment;

(8) as a consequence of other actions of citizens and juridical persons;

(9) as a consequence of events with which a law or other legal act connects the ensuing of civil-law consequences.

2. The rights to property subject to State registration shall arise from the moment of registration of the respective rights thereto, unless established otherwise by a law.

Article 9. Effectuation of Civil Rights

1. Citizens and juridical persons shall effectuate the civil rights belonging to them at their discretion.

2. The refusal of citizens and juridical persons to effectuate rights belonging to them shall not entail the termination of those rights except for instances provided for by a law.

Article 10. Limits of Effectuation of Civil Rights

1. The actions of citizens and juridical persons effectuated exclusively with the intention to cause harm to another person, and also abuse of right in other forms, shall not be permitted.
The use of civil rights for the purposes of limiting competition, and also abuse of a dominant position in the market, shall not be permitted.

2. In the event of the failure to comply with the requirements provided for by point 1 of the present Article, a court, arbitrazh court, or arbitration court may refuse to defend the right belonging to the person.

3. In instances when a law makes the defence of civil rights dependent upon whether these rights have been effectuated reasonably and in good faith, the reasonableness of the actions and the good faith of the participants of civil law relations shall be presupposed.

Article 11. Judicial Defence of Civil Rights

1. A court, arbitrazh court, or arbitration court (hereinafter: court) shall effectuate the defence of violated or contested civil rights in accordance with the jurisdiction of cases established by procedural legislation.

2. The defence of civil rights in an administrative procedure shall be effectuated only in the instances provided for by a law. A decision adopted in an administrative procedure may be appealed to a court.

Article 12. Means of Defence of Civil Rights

The defence of civil rights shall be effectuated by means of:
recognition of the right;
restoration of the situation which existed before the violation of the right and the suppression of actions violating a right or creating a threat to violate it;

recognition of a contested transaction to be invalid and the application of the consequences of the invalidity thereof, and the application of the consequences of the invalidity of a null transaction;

recognition of an act of a State agency or agency of local self-government to be invalid;

self-defence of right;

awarding performance of a duty in kind;

compensation of losses;

recovery of a penalty;

contributory compensation for moral harm;

termination or change of legal relation;

non-application by court of act of a State agency or agency of local self-government which is contrary to a law;

other means provided for by a law.

Article 13. Deeming Act of State Agency or Agency of Local Self-Government to be Invalid

A non-normative act of a State agency or agency of local self-government, and in instances provided for by a law, also a normative act, not corresponding to a law or other legal acts and violating civil rights and interests of a citizen or juridical person protected by a law may be deemed by a court to be invalid.

In the event that an act is deemed by a court to be invalid, the violated right shall be subject to restoration or to defence by other means provided for by Article 12 of the present Code.

Article 14. Self-Defence of Civil Rights

The self-defence of civil rights shall be permitted.

Means of self-defence must be commensurate to the violation and not exceed the limits of actions necessary to suppress it.

Article 15. Compensation of Losses

1. A person whose right has been violated may demand full compensation of losses caused to him unless compensation of losses in a lesser amount has been provided for by a law or contract.

2. By losses is understood expenses which the person whose right has been violated made or must make in order to restore the violated right, loss or damage of his property (real damage), and also revenues not received which this person would have received under ordinary conditions of civil turnover if his right had not been violated (lost advantage).

If the person who has violated a right has received revenues as a consequence thereof, the person whose right was violated shall have the right to demand compensation together with the other losses, for lost advantage in an amount of not less than such revenues.

Article 16. Compensation of Losses Caused by State Agencies and Agencies of Local Self-Government

Losses caused to a citizen or juridical person as a result of the illegal actions (or failure to act) of State agencies, agencies of local self-government, or officials of these agencies, including the issuance of the act of a State agency or agency of local self-government which does not correspond to a law or other legal act, shall be subject to compensation by the Russian Federation, respective subject of the Russian Federation, or municipal formation.

Subsection 2. Persons

Chapter 3. Citizens (Natural Persons)

Article 17. Legal Capacity of Citizen

1. The capacity to have civil rights and to bear duties (civil legal capacity) shall be recognised in equal measure for all citizens.

2. The legal capacity of a citizen shall arise at the moment of his birth and be terminated by death.

Article 18. Content of Legal Capacity of Citizens

Citizens may have property by right of ownership, inherit and bequeath property, engage in entrepreneurial and any other activity not prohibited by a law; create juridical persons autonomously or jointly with other citizens and juridical persons; conclude any transactions which are not contrary to a law and participate in obligations; select the place of residence; have the rights of authors to works of science, literature and art, inventions and other results of intellectual activity protected by a law; and have other property and personal nonproperty rights.

Article 19. Name of Citizen

1. A citizen shall acquire and effectuate rights and duties under his own name, including surname and own forename, and also patronymic unless it arises otherwise from a law or national custom.

A citizen may used a pseudonym (fictitious name) in the instances and in the procedure provided for by a law.

2. A citizen shall have the right to change his name in the procedure established by a law. The change of name by a citizen shall not be grounds for the termination or change of his rights and duties acquired under the previous name.

A citizen shall be obliged to take necessary measures in order to inform his debtors and creditors about the change of his name and shall bear the risk of the consequences caused by these persons lacking information about the change of his name.

A citizen who has changed name shall have the right to demand the making, at his expense, of respective changes in documents formalised in his previous name.

3. The name received by a citizen at birth, and also a change of name, shall be subject to registration in the procedure established for the registration of acts of civil status.

4. The acquisition of rights and duties under the name of another person shall not be permitted.

5. Harm caused to a citizen as a result of the incorrect use of his name shall be subject to compensation in accordance with the present Code.

In the event of the distortion or of the use of the name of a citizen by means or in a form affecting his honour, dignity, or business reputation, the rules provided for by Article 152 of the present Code shall apply.

Article 20. Place of Residence of Citizen

1. The place where a citizen permanently or primarily resides shall be deemed to be the place of residence.

2. The place of residence of minors who have not attained fourteen years of age or citizens under trusteeship shall be deemed to be the place of residence of their legal representatives—parents, adoptive parents, or trustees.

Article 21. Dispositive Legal Capacity of Citizen

1. The capacity of a citizen by his own actions to acquire and effectuate civil rights, to create civil duties for himself, and to perform them (civil dispositive legal capacity) shall arise in full with the ensuing of majority, that is, upon attaining eighteen years of age.

2. When by law entry into marriage is permitted before attaining eighteen years of age, a citizen who has not attained eighteen years of age shall acquire civil dispositive legal capacity in full from the time of entering into marriage.

Dispositive legal capacity acquired as a result of concluding a marriage shall be retained in full also in the event of dissolution of the marriage before attaining eighteen years of age.

When deeming a marriage to be invalid a court may adopt a decision concerning the loss by the minor spouse of full dispositive legal capacity from the moment determined by the court.

Article 22. Inadmissibility of Deprivation and Limitation of Legal Capacity and Dispositive Legal Capacity of Citizen

1. No one may be limited in legal capacity and dispositive legal capacity other than in the instances and in the procedure established by a law.

2. The failure to comply with the conditions and procedure established by a law for the limitation of dispositive legal capacity of citizens or of their right to engage in entrepreneurial or other activity shall entail the invalidity of the act of the State or other agency establishing the respective limitation.

3. The full or partial renunciation by a citizen of legal capacity or dispositive legal capacity and other transactions directed towards a limitation of legal capacity or dispositive legal capacity shall be void except for instances when such transactions are permitted by a law.

Article 23. Entrepreneurial Activity of Citizen

1. A citizen shall have the right to engage in entrepreneurial activity without the formation of a juridical person from the moment of State registration as an individual entrepreneur.

2. The head of a peasant (or farmer) economy effectuating activity without the formation of a juridical person (Article 257) shall be deemed to be an entrepreneur from the moment of State registration of the peasant (or farmer) economy.

3. The rules of the present Code which regulate the activity of juridical persons that are commercial organisations shall apply respectively to the entrepreneurial activity of citizens effectuated without the formation of a juridical person unless it arises otherwise from a law, other legal acts, or the essence of a legal relation.

4. A citizen effectuating entrepreneurial activity without the formation of a juridical person in violation of the requirements of point 1 of the present Article shall not have the right to refer with respect to transactions concluded by him to the fact that he is not an entrepreneur. A court may apply to such transactions the rules of the present Code on obligations connected with the effectuation of entrepreneurial activity.

Article 24. Property Responsibility of Citizen

A citizen shall be liable for his obligations with all of the property belonging to him except for property against which execution may not be levied in accordance with a law.

A List of the property of citizens against which execution may not be levied shall be established by civil procedure legislation.

Article 25. Insolvency (or Bankruptcy) of Individual Entrepreneur

1. An individual entrepreneur who is not in a state to satisfy the demands of creditors connected with the effectuation by him of entrepreneurial activity may be deemed to be insolvent (or bankrupt) by decision of a court. From the moment of the rendering of such decision his registration as an individual entrepreneur shall lose force.

2. When effectuating the procedure for deeming an individual entrepreneur to be bankrupt his creditors with regard to obligations not connected with the effectuation of entrepreneurial activity by him also shall have the right to present their demands. The demands of the said creditors not declared by them in this procedure shall retain force after the completion of the bankruptcy proceeding of an individual entrepreneur.

3. The demands of creditors of an individual entrepreneur in the event of him being deemed bankrupt shall be satisfied at the expense of the property belonging to him against which execution may be levied in the following priority:

in first priority shall be satisfied the demands of citizens to whom an entrepreneur bears responsibility for causing harm to life or health by means of the capitalisation of the respective time payments, and also demands relating to the recovery of alimony;

in second priority the settlement of accounts shall be made for the payment of severance benefits and the payment of labour with persons working under a labour contract [*dogovor*], including a *kontrakt* and payment of remuneration with regard to author's contracts;

in third priority shall be satisfied the demands of creditors secured by a pledge of property belonging to an individual entrepreneur;

in fourth priority the indebtedness for obligatory payments to the budget and to extrabudgetary funds shall be repaid;

in fifth priority the settlement of accounts shall be made with other creditors in accordance with a law.

4. After completing the settlement of accounts with creditors, the individual entrepreneur deemed to be bankrupt shall be relieved from the performance of residual obligations connected with his entrepreneurial activity and other demands presented for performance and taken into account when deeming the entrepreneur to be bankrupt.

The demands of citizens to whom the citizen declared to be bankrupt bears responsibility for causing harm to life or health, and also other demands of a personal character, shall retain force.

5. The grounds and procedure for a court to deem an individual entrepreneur to be bankrupt or the declaration by him of his own bankruptcy shall be established by a law on insolvency (or bankruptcy).

Article 26. Dispositive Legal Capacity of Minors from Fourteen to Eighteen Years of Age

1. Minors in age from fourteen to eighteen years shall conclude transactions, except those named in point 2 of the present Article, with the written consent of their legal representatives—parents, adoptive parents, or guardian.

A transaction concluded by such a minor shall be valid also in the event of the subsequent written approval thereof by his parents, adoptive parents, or guardian.

2. Minors in age from fourteen to eighteen years shall have the right autonomously, without the consent of the parents, adoptive parents, and guardian, to:

(1) dispose of their earnings, stipends, and other revenues;

(2) effectuate the rights of an author of a work of science, literature or art, invention, or other result of his intellectual activity protected by a law;

(3) in accordance with a law, make deposits in credit institutions and dispose of them;

(4) conclude petty domestic transactions and other transactions provided for by Article 28(2) of the present Code.

Upon attaining sixteen years of age minors also shall have the right to be members of cooperatives in accordance with the laws on cooperatives.

3. Minors in age from fourteen to eighteen years shall autonomously bear property responsibility for transactions concluded by them in accordance with points 1 and 2 of the present Article. For the harm caused by them such minors shall bear responsibility in accordance with the present Code.

4. When there are sufficient grounds a court may, upon the petition of parents, adoptive parents, or guardian or the agency of trusteeship and guardianship, limit or deprive a minor from fourteen to eighteen years of age of the right to autonomously dispose of his earnings, stipend, or other revenues, except for instances when this minor acquired dispositive legal capacity in full in accordance with Article 21(2) or with Article 27 of the present Code.

Article 27. Emancipation

1. A minor who has attained sixteen years of age may be declared to have full dispositive legal capacity if he works under a labour contract [*dogovor*], including a *kontrakt* or with the consent of parents, adoptive parents, or guardian engages in entrepreneurial activity.

A minor shall be declared to have full dispositive legal capacity (emancipation) by decision of an agency of trusteeship and guardianship with the consent of both parents, adoptive parents, or the guardian, or in the absence of such consent, by decision of a court.

2. Parents, adoptive parents, and a guardian shall not bear responsibility for obligations of an emancipated minor, in particular, obligations which arose as a consequence of the causing of harm by him.

Article 28. Dispositive Legal Capacity of Youth

1. Transactions, except those specified in point 2 of the present Article, may be concluded for minors who have not attained fourteen years of age (youth) only in their names by the parents, adoptive parents, or trustees.

The rules provided for by Article 37(2) and (3) of the present Code shall apply to transactions of the legal representatives of a minor with his property.

2. Youth in age from six to fourteen years shall have the right autonomously to conclude:
(1) petty domestic transactions;
(2) transactions directed towards receiving advantages without compensation which do not require notarial certification or State registration;
(3) transactions relating to the disposition of assets granted by a legal representative or with the consent of the last by a third person for a specified purpose or for free disposition.

3. The parents, adoptive parents, or trustees shall bear property responsibility for transactions of a youth, including transactions concluded by him autonomously, unless it is proved that the obligation was violated not through his fault. These persons in accordance with a law also shall be liable for harm caused by a youth.

Article 29. Deeming of Citizen to Lack Dispositive Legal Capacity

1. A citizen who as a consequence of mental disturbance cannot understand the significance of his actions or direct them may be deemed by a court to lack dispositive legal capacity in the procedure established by the civil procedure legislation. A trusteeship shall be established over him.

2. Transactions in the name of a citizen deemed to lack dispositive legal capacity shall be concluded by his trustee.

3. If the grounds by virtue of which a citizen was deemed to lack dispositive legal capacity have disappeared, the court shall deem him to have dispositive legal capacity. The trusteeship established over him shall be vacated on the basis of the decision of a court.

Article 30. Limitation of Dispositive Legal Capacity of Citizens

1. A citizen who as a consequence of abusing alcoholic beverages or narcotic means places his family in a grave material position may be limited in dispositive legal capacity by a court in the procedure established by civil procedure legislation. A guardianship shall be established over him.
He shall have the right autonomously to conclude petty domestic transactions.
He may conclude other transactions, and also receive earnings, pension, and other revenues and dispose of them only with the consent of the guardian. However, such citizen shall autonomously bear property responsibility for transactions concluded by him and for harm caused by him.

2. If the grounds by virtue of which a citizen was limited in dispositive legal capacity have disappeared, a court shall vacate the limitation of his dispositive

legal capacity. The guardianship established over the citizen shall be vacated on the basis of the decision of a court.

Article 31. Trusteeship and Guardianship

1. Trusteeship and guardianship shall be established in order to defend the rights and interests of persons who lack dispositive legal capacity or who are not citizens with full dispositive legal capacity. Trusteeship and guardianship over minor citizens also shall be established for the purpose of nurturing them. The rights and duties of trustees and guardians corresponding thereto shall be determined by legislation on marriage and the family.

2. Trustees and guardians shall act in defence of the rights and interests of their wards in relations with any persons, including in courts, without a special power.

3. A trusteeship and guardianship over minors shall be established when they have no parents or adoptive parents, the deprivation of parents by a court of parental rights, and also in instances when such citizens for other reasons are left without parental care, in particular, when the parents evade nurturing them or in defence of their rights and interests.

Article 32. Trusteeship

1. Trusteeship shall be established over youths, and also over citizens deemed by a court to lack dispositive legal capacity as a consequence of mental disturbance.

2. Trustees shall be representatives of the wards by virtue of a law and conclude all necessary transactions in their name and in their interests.

Article 33. Guardianship

1. Guardianship shall be established over minors from fourteen to eighteen years of age, and also over citizens limited in dispositive legal capacity by a court as a consequence of abusing alcoholic beverages or narcotic means.

2. Guardians shall give consent to the conclusion of those transactions which citizens under guardianship do not have the right to conclude autonomously.

Guardians shall render assistance to wards in the effectuation by them of their rights and the performance of duties, and also protect them against abuses on the part of third persons.

Article 34. Agencies of Trusteeship and Guardianship

1. Agencies of local self-government shall be agencies of trusteeship and guardianship.

2. A court shall be obliged within three days from the time of a decision entering into legal force concerning a citizen being deemed to lack dispositive legal

capacity or to have limited dispositive legal capacity, to notify the trusteeship and guardianship agency thereof at the place of residence of this citizen in order to establish trusteeship or guardianship over him.

3. The agency of trusteeship and guardianship at the place of residence of the wards shall effectuate supervision over the activity of their trustees and guardians.

Article 35. Trustees and Guardians

1. A trustee or guardian shall be appointed by an agency of trusteeship and guardianship at the place of residence of the person needing trusteeship or guardianship within a month from the moment when the said agencies become aware of the need to establish the trusteeship or guardianship over the citizen. When there are circumstances deserving attention the trustee or guardian may be appointed by the agency of trusteeship and guardianship at the place of residence of the trustee (or guardian). If a trustee or guardian has not been appointed within a month for a person who needs a trusteeship or guardianship, the performance of the duties of trustee or guardian shall be entrusted to a trusteeship or guardianship agency.

The appointment of the trustee or guardian may be appealed to a court by interested persons.

2. Only citizens who have reached majority and who have dispositive legal capacity may be appointed trustees and guardians. Citizens deprived of parental rights may not be appointed as trustees and guardians.

3. A trustee or guardian may be appointed only with his consent. His moral and other personal qualities, ability to fulfil the duties of trustee or guardian, relations existing between him and the person needing trusteeship or guardianship, and, if possible, the wish of the ward, must be taken into account in so doing.

4. Trustees and guardians for citizens needing trusteeship or guardianship and situated or placed in respective nurturing or treatment institutions, institutions of social defence of the populace, or other analogous institutions shall be those institutions.

Article 36. Performance by Trustees and Guardians of Their Duties

1. Duties relating to trusteeship and guardianship shall be performed without compensation except for instances provided for by a law.

2. Trustees and guardians of minor citizens shall be obliged to reside jointly with their wards. Separate residence of a guardian and ward who has attained sixteen years of age shall be permitted with the authorisation of an agency of trusteeship and guardianship on condition that this does not reflect unfavourably on the nurturing and defence of the rights and interests of the ward.

Trustees and guardians shall be obliged to notify agencies of trusteeship and guardianship concerning any change of place of residence.

3. Trustees and guardians shall be obliged to be concerned for the maintenance of their wards, providing them with care and treatment and defending their rights and interests.

Trustees and guardians of minors must be concerned about their education and nurturing.

4. The duties specified in point 3 of the present Article shall not be entrusted to guardians over citizens who have reached majority and who have been limited in dispositive legal capacity by a court.

5. If the grounds by virtue of which a citizen was deemed to lack dispositive legal capacity or to have limited dispositive legal capacity as a consequence of the abuse of alcoholic beverages or narcotic means have disappeared, the trustee or guardian shall be obliged to petition the court to deem the ward to have dispositive legal capacity and to remove the trusteeship or guardianship from him.

Article 37. Disposition of Property of Ward

1. Revenues of a citizen under wardship, including revenues due to the ward from the management of his property, except for revenues which the ward has the right to dispose of autonomously, shall be expended exclusively by the trustee or guardian in the interests of the ward and with the prior authorisation of the agency of trusteeship and guardianship.

Without the prior authorisation of the agency of trusteeship and guardianship the trustee or guardian shall have the right to make expenses necessary to maintain the ward at the expense of amounts which are due to the ward as his revenue.

2. A trustee shall not have the right without the prior authorisation of agencies of trusteeship and guardianship to conclude, and a guardian—to give consent to the conclusion of transactions relating to the alienation, including exchange or gift, of the property of the ward, hire (or lease) thereof, use free of charge, or pledge, or transactions entailing a waiver of rights which belong to the ward, the separation of his property or an apportionment of a participatory share from it, and also any other transactions entailing a reduction of the property of the ward.

The procedure for the management by a trustee or guardian of the property of a ward shall be determined by a law.

3. A trustee and guardian, their spouses and close relatives shall not have the right to conclude transactions with a ward except for the transfer of property to the ward as a gift or for use free of charge, and also to represent the ward when concluding transactions or conducting judicial cases between the ward and a spouse of the trustee or guardian and their close relatives.

Article 38. Trust Management of Property of Ward

1. When the permanent management of immovable and valuable moveable property of a ward is necessary, the agency of trusteeship and guardianship shall

conclude with the manager determined by this agency a contract concerning trust management of such property. In this event the trustee or guardian shall retain his powers with respect to that property of the ward which is not transferred to trust management.

In the event of the effectuation by the manager of powers relating to the management of the property of the ward the operation of the rules provided for by Articles 37(2) and (3) of the present Code shall extend to the manager.

2. The trust management of the property of a ward shall be terminated on the grounds provided for by a law for the termination of a contract on the trust management of property, and also in the instances of termination of the trusteeship and guardianship.

Article 39. Relieving and Removing Trustees and Guardians from Performance by Them of Their Duties

1. An agency of trusteeship and guardianship shall relieve a trustee or guardian from the performance of their duties in instances of the return of minors to their parents or the adoption thereof.

When a ward is placed in a respective nurturing and treatment institution, institution of social defence of the populace, or other analogous institution, the trusteeship and guardianship agency shall relieve the trustee or guardian previously appointed from the performance of his duties unless this is contrary to the interests of the ward.

2. When there are important reasons (illness, change of property position, lack of mutual understanding with the ward, etc.), a trustee or guardian may be relieved from the performance of his duties at his request.

3. In instances of the improper fulfilment by a trustee or guardian of duties placed on him, including his using the trusteeship or guardianship for mercenary purposes or leaving the ward without supervision and necessary assistance, the agency of trusteeship and guardianship may remove the trustee or guardian from the performance of these duties and to take necessary measures to bring the guilty citizen to the responsibility established by a law.

Article 40. Termination of Trusteeship and Guardianship

1. Trusteeship and guardianship over citizens who have reached majority shall be terminated in the instances of a court rendering a decision to deem the ward to have dispositive legal capacity or vacating limitations on his dispositive legal capacity upon the application of the trustee, guardian, or trusteeship and guardianship agency.

2. Upon a youth under wardship attaining fourteen years of age the trusteeship over him shall be terminated, and the citizen effectuating the duties of trustee shall become the guardian of the minor without an additional decision thereon.

3. Upon a minor ward attaining eighteen years of age, the guardianship over the minor shall terminate without a special decision, and also when he enters into a marriage and in other instances of the acquisition by him of full dispositive legal capacity before attaining majority (Article 21[2] and Article 27).

Article 41. Home Visiting of Citizens Who Have Dispositive Legal Capacity

1. At the request of a citizen who has reached majority and has dispositive legal capacity and who by reason of state of health cannot autonomously effectuate and defend his rights and perform duties, a guardianship in the form of home visiting may be established over him.

2. The guardian (or assistant) of a citizen who has reached majority and who has dispositive legal capacity may be appointed by an agency of trusteeship and guardianship only with the consent of this citizen.

3. The disposition of property belonging to a ward who has reached majority and who has dispositive legal capacity shall be effectuated by the guardian (or assistant) on the basis of a contract of commission or trust management concluded with the ward. The conclusion of domestic transactions and other transactions directed towards maintenance and the satisfaction of domestic requirements of the ward shall be effectuated by his guardian (or assistant) with the consent of the ward.

4. Home visiting over a citizen who has reached majority and who has dispositive legal capacity established in accordance with point 1 of the present Article shall be terminated at the demand of the citizen who is under home visits.

The guardian (or assistant) of a citizen who is under home visits shall be relieved from the fulfilment of the duties placed on him in the instances provided for by Article 39 of the present Code.

Article 42. Deeming of Citizen to be Missing

A citizen may, upon the application of interested persons, be deemed by a court to be missing if within one year there is no information at his place of residence concerning his whereabouts.

When it is impossible to establish the date of receipt of the last information concerning the absent person, the commencement of calculation of the period for deeming a person to be missing shall be considered to be the first date of the month following that in which the last information was received concerning the missing person, and when it is impossible to establish this month, the first of January of the following year.

Article 43. Consequences of Deeming Citizen to be Missing

1. The property of a citizen deemed to be missing shall, when permanent management is necessary, on the basis of the decision of a court be transferred to

the person who is determined by the agency of trusteeship and guardianship and shall operate on the basis of a contract on trust management concluded with this agency.

From this property maintenance shall be issued to citizens whom the missing person is obliged to maintain, and indebtedness shall be repaid with regard to other obligations of the missing person.

2. An agency of trusteeship and guardianship may also before the expiry of a year from the date of receiving information about the whereabouts of the missing citizen appoint a manager of his property.

3. The consequences of deeming a person to be missing which are not provided for by the present Article shall be determined by a law.

Article 44. Vacating Decision on Deeming Citizen to be Missing

In the event of the appearance or discovery of the whereabouts of a citizen deemed to be missing a court shall vacate the decision deeming him to be missing. The management of the property of this citizen shall be vacated on the basis of the decision of the court.

Article 45. Declaration of Citizen to be Deceased

1. A citizen may be declared by a court to be deceased if there is no information at his place of residence concerning his whereabouts in the course of five years, and if he was missing under circumstances threatening death or giving grounds to suppose he perished from a specific accident, within six months.

2. A military serviceman or other citizen who is missing in connection with military actions may be declared by a court to be deceased not earlier than upon the expiry of two years from the date the military actions end.

3. The date of death of a citizen declared to be deceased shall be considered to be the date of entry into legal force of the decision of the court declaring him to be deceased. In the event a citizen is declared to be deceased who is missing under circumstances threatening death or giving grounds to suppose he perished in a specific accident, the court may deem the date of death of this citizen to be the date of the presupposed perishing.

Article 46. Consequences of Appearance of Citizen Declared to be Deceased

1. In the event of the appearance or discovery of the whereabouts of a citizen declared to be deceased, the court shall vacate the decision declaring him to be deceased.

2. Irrespective of the time of his appearance, a citizen may demand from any person the return of property preserved which passed to this person without

compensation after the declaration of the citizen to be deceased, except for instances provided for by Article 302(3) of the present Code.

Persons to whom property of a citizen declared to be deceased has passed according to transactions for compensation shall be obliged to return this property to him if it is proved that, in acquiring the property, they knew that the citizen declared to be deceased is alive. If it is impossible to return such property in kind, the value thereof shall be compensated.

Article 47. Registration of Acts of Civil Status

1. The following acts of civil status shall be subject to registration:
(1) birth;
(2) conclusion of marriage;
(3) dissolution of marriage;
(4) adoption;
(5) establishment of paternity;
(6) change of name;
(7) death of a citizen.

2. Acts of civil status shall be registered by agencies for the registry of acts of civil status by means of making respective entries in the book for the registration of acts of civil status (documentary books) and the issuance to citizens of certificates on the basis of these entries.

3. The correction and the changing of entries of acts of civil status shall be done by the agency for the registry of acts of civil status when there are sufficient grounds and the absence of a dispute between interested persons.

When there is a dispute between interested persons, or the refusal of an agency for the registry of acts of civil status to correct or change an entry, the dispute shall be settled by a court.

The annulment and restoration of entries of acts of civil status shall be done by the agency for the registry of acts of civil status on the basis of the decision of a court.

4. Agencies effectuating the registration of acts of civil status, the procedure for the registration of these acts, the procedure for the change, restoration, and annulment of entries of acts of civil status, the forms of the documentary books and certificates, and also the procedure and periods for keeping the documentary books, shall be determined by a law on acts of civil status.

Chapter 4. Juridical Persons
§1. Basic Provisions

Article 48. Concept of Juridical Person

1. A juridical person shall be deemed to be an organisation which has solitary property in ownership, economic jurisdiction, or operative management and is

liable for its obligations with such property and may in its own name acquire and effectuate property and personal nonproperty rights, bear duties, and be a plaintiff or defendant in court.

Juridical persons must have an autonomous balance sheet or estimate.

2. In connection with participation in the formation of the property of a juridical person its founders (or participants) may have rights of obligations with respect to this juridical person or rights to a thing with respect to its property.

To juridical persons with respect to which the participants thereof have rights of obligations shall be relegated: economic partnerships and societies; production and consumer cooperatives.

To juridical persons with respect to whose property the founders thereof have the right of ownership or other right to a thing shall be relegated: State and municipal unitary enterprises, and also institutions financed by the owner [as amended by Law of 14 November 2002. СЗ РФ (2002), no. 48, item 4746].

3. To juridical persons with respect to which the founders (or participants) thereof do not have property rights shall be relegated: social and religious organisations (or associations); philanthropic and other foundations; associations of juridical persons (associations and unions).

Article 49. Legal Capacity of Juridical Person

1. A juridical person may have civil rights corresponding to the purposes of the activity provided for in its constitutive documents and shall bear the duties connected with such activity.

Commercial organisations may, except for unitary enterprises and other types of organisations provided for by a law, have civil rights and bear civil duties necessary for the effectuation of any types of activity not prohibited by a law.

A juridical person may engage in individual types of activity, a List of which is determined by a law, only on the basis of a special authorisation (or license).

2. A juridical person may be limited in rights only in the instances and in the procedure provided for by a law. A decision concerning limitation of rights may be appealed by a juridical person to a court.

3. The legal capacity of a juridical person shall arise at the moment of its creation (Article 51[2]) and shall terminate at the moment of the completion of its liquidation (Article 63[8]).

The right of a juridical person to effectuate activity to engage in which it is necessary to obtain a license shall arise from the moment of the receipt of such license or within the period specified therein and shall terminate upon the expiry of the period of its operation, unless established otherwise by a law or other legal acts.

Article 50. Commercial and Noncommercial Organisations

1. Organisations pursuing the deriving of profit as the principal purpose of their activity (commercial organisations) or not having the deriving of profit as such purpose and not distributing profit received among the participants (noncommercial organisations) may be juridical persons.

2. Juridical persons which are commercial organisations may be created in the form of economic partnerships and societies, production cooperatives, and State and municipal unitary enterprises.

3. Juridical persons which are noncommercial organisations may be created in the form of consumer cooperatives, social or religious organisations (or associations) financed by the owner of institutions, philanthropic and other foundations, and also in other forms provided for by a law.

Noncommercial organisations may effectuate entrepreneurial activity only insofar as this serves the attainment of the purposes for which they have been created and corresponds to such purposes.

4. The creation of associations of commercial and(or) noncommercial organisations in the form of associations and unions shall be permitted.

Article 51. State Registration of Juridical Persons

1. A juridical person shall be subject to State registration in an empowered State agency in the procedure determined by the Law on the State Registration of Juridical Persons. The data of State registration shall be included in the unified State register of juridical persons open for general familiarisation.

A refusal of State registration of a juridical person shall be permitted only in the instances established by a law.

The refusal of State registration of a juridical person, and also the evasion of such registration, may be appealed to a court [as amended by Law of 21 March 2002. СЗ РФ (2002), no. 12, item 1093].

2. A juridical person shall be considered to be created from the day of the making of the respective entry in the unified State register of juridical persons [as amended by Law of 21 March 2002. СЗ РФ (2002), no. 12, item 1093].

Article 52. Constitutive Documents of Juridical Person

1. A juridical person shall operate on the basis of a charter, or a constitutive contract and a charter, or only a constitutive contract. In the instances provided for by a law, a juridical person which is not a commercial organisation may operate on the basis of a General Statute on organisations of the particular type.

The constitutive contract of a juridical person shall be concluded, and the charter confirmed, by its founders (or participants).

A juridical person created in accordance with the present Code by one founder shall operate on the basis of a charter confirmed by that founder.

2. The name of the juridical person, its location, the procedure for the management of the activity of the juridical person must be determined in the constitutive documents of the juridical person, and also they shall contain other information provided for by the law for juridical persons of the respective type. The subject and purpose of activity of the juridical person must be determined in the constitutive documents of noncommercial organisations and unitary enterprises, and in the instances provided for by a law, also other commercial organisations. The subject and specified purposes of the activity of a commercial organisation may be provided for by the constitutive documents of other commercial organisations also in instances when according to a law this is not obligatory.

In a constitutive contract the founders shall be obliged to create a juridical person and determine the procedure for joint activity relating to its creation and the conditions for the transfer of their property to it and participation in its activity. The conditions and procedure for distribution among the participants of profit and losses, the management activity of the juridical person, and the withdrawal of founders (or participants) from the composition thereof shall also be determined by the contract.

3. Changes of constitutive documents shall acquire force for third persons from the moment of their State registration, and in the instances established by a law, from the moment of informing the agency effectuating State registration of such changes. However, juridical persons and their founders (or participants) shall not have the right to refer to the lack of registration of such changes in relations with third persons who have acted by taking these changes into account.

Article 53. Organs of Juridical Person

1. A juridical person shall acquire civil rights and assume civil duties through its organs operating in accordance with a law, other legal acts, and the constitutive documents.

The procedure for the appointment or election of the organs of a juridical person shall be determined by a law and the constitutive documents.

2. In the instances provided for by a law a juridical person may acquire civil rights and assume civil duties through its participants.

3. A person who by virtue of a law or constitutive documents of a juridical person acts in its name must operate in the interests of the juridical person represented by him in good faith and reasonably. He shall be obliged at the demand of the founders (or participants) of the juridical person, unless provided otherwise by a law or contract, to compensate losses caused by him to the juridical person.

Article 54. Name and Location of Juridical Person

1. A juridical person shall have its own name containing an indication of its organisational-legal form. The names of noncommercial organisations, and in the instances provided for by a law, the names of commercial organisations, must contain an indication of the character of activity of the juridical person [as amended 14 November 2002. СЗ РФ (2002), no. 48, item 4746].

2. The location of a juridical person shall be determined by the place of its State registration. The State registration of a juridical person shall be effectuated at the location of its permanently operating executive organ, and in the absence of a permanently operating executive organ—other organ or person having the right to act in the name of the juridical person without a power of attorney [as amended 21 March 2002. СЗ РФ (2002), no. 12, item 1093].

3. The name and location of a juridical person shall be specified in its constitutive documents.

4. A juridical person which is a commercial organisation must have a firm name.

A juridical person whose firm name has been registered in the established procedure shall have the exclusive right to the use thereof.

A person who unlawfully uses another's registered firm name shall be obliged, at the demand of the possessor of the right to the firm name, to terminate the use thereof and compensate the losses caused.

The procedure for registration and use of firm names shall be determined by a law and other legal acts in accordance with the present Code.

Article 55. Representations and Branches

1. A solitary subdivision of a juridical person situated outside the location thereof which represents the interests of the juridical person and effectuates the defence thereof shall be a representation.

2. A solitary subdivision of a juridical person situated outside the location thereof and effectuating all or part of its functions, including the function of representation, shall be a branch.

3. Representations and branches shall not be juridical persons. They shall be endowed with property by the juridical person which created them and shall operate on the basis of Statutes confirmed by them.

The directors of representations and branches shall be appointed by the juridical person and operate on the basis of the power of attorney thereof.

Representations and branches must be specified in the constitutive documents of the juridical person which created them.

Article 56. Responsibility of Juridical Person

1. Juridical persons, except those being financed by the owner of institutions, shall be liable for their obligations with all of the property belonging to them.

2. A treasury enterprise and an institution being financed by the owner shall be liable for its obligations in the procedure and on the conditions established by Article 113(5) and Articles 115 and 120 of the present Code.

3. The founder (or participant) of a juridical person or the owner of its property shall not be liable for obligations of the juridical person, and the juridical person shall not be liable for the obligations of the founder (or participant) or owner, except for the instances provided for by the present Code or by the constitutive documents of the juridical person.

If the insolvency (or bankruptcy) of a juridical person has been caused by the founders (or participants) or by the owner of the property of the juridical person or by other persons who have the right to give instructions which are binding upon this juridical person or otherwise have the possibility to determine its actions, subsidiary responsibility for its obligations may be placed upon such persons in the event the property of the juridical person is insufficient.

Article 57. Reorganisation of Juridical Person

1. The reorganisation of a juridical person (merger, accession, division, separation, transformation) may be effectuated by decision of its founders (or participants) or the organ of the juridical person empowered therefor by the constitutive documents.

2. In the instances established by a law the reorganisation of a juridical person in the form of its division or the separation from its composition of one or several juridical persons therefrom shall be effectuated by decision of empowered State agencies or by decision of a court.

If the founders (or participants) of a juridical person, agency empowered by them, or organ of the juridical person empowered to reorganise it by its constitutive documents does not effectuate the reorganisation of the juridical person within the period specified in the decision of the empowered State agency, the court upon the suit of the said State agency shall appoint an external administrator of the juridical person and charge him with effectuating the reorganisation of this juridical person. From the moment of appointment of the external administrator to him shall pass the powers relating to managing the affairs of the juridical person. The external administrator shall act in the name of the juridical person in court, draw up the separation balance sheet and transfer it for consideration of the court together with the constitutive documents of the juridical persons arising as a result of the reorganisation. The confirmation by the court of the said documents shall be the grounds for State registration of the newly arisen juridical persons.

3. In the instances established by a law the reorganisation of juridical persons in the form of merger, accession, or transformation may be effectuated only with the consent of empowered State agencies.

4. The juridical person shall be considered to be reorganised, except for instances of reorganisation in the form of accession, from the moment of State registration of the juridical persons which have arisen anew.

In the event of the reorganisation of a juridical person in the form of accession thereto of another juridical person, the first of them shall be considered to be reorganised from the moment of making an entry in the unified State register of juridical persons concerning the termination of the activity of the juridical person which has acceded.

Article 58. Legal Succession in Event of Reorganisation of Juridical Persons

1. In the event of the merger of juridical persons the rights and duties of each of them shall pass to the juridical person which newly arises in accordance with the act of transfer.

2. In the event of the accession of a juridical person to another juridical person, to the last shall pass the rights and duties of the acceding juridical person in accordance with the act of transfer.

3. In the event of the division of a juridical person its rights and duties shall pass to the juridical persons which newly arise in accordance with the division balance sheet.

4. In the event of separation from a juridical person of one or several juridical persons, to each of them shall pass the rights and duties of the reorganised juridical person in accordance with the separation balance sheet.

5. In the event of the transformation of a juridical person of one type into a juridical person of another type (change of organisational-legal form), to the juridical person which arises anew shall pass the rights and duties of the reorganised juridical person in accordance with the act of transfer.

Article 59. Act of Transfer and Division Balance Sheet

1. The act of transfer and division balance sheet must contain provisions concerning legal succession regarding all obligations of the reorganised juridical person with respect to all creditors and debtors thereof, including obligations being contested by the parties.

2. The act of transfer and division balance sheet shall be confirmed by the founders (or participants) of the juridical person or by the agency which adopted the decision concerning reorganisation of the juridical persons and shall be submitted together with the constitutive documents for the State registration of the

juridical persons which newly arose or the making of changes in the constitutive documents of existing juridical persons.

The failure to submit with the constitutive documents the act of transfer or division balance sheet respectively, and also the absence therein of provisions concerning legal succession with regard to the obligations of the reorganised juridical person, shall entail a refusal of State registration of the juridical persons which newly arose.

Article 60. Guarantees of Rights of Creditors of Juridical Person in Event of its Reorganisation

1. The founders (or participants) of the juridical person or agency which adopted the decision concerning the reorganisation of a juridical person shall be obliged in writing to inform the creditors of the juridical person being reorganised thereof.

2. The creditor of the juridical person being reorganised shall have the right to demand the termination or the performance before time of the obligation, the debtor with regard to which this juridical person is, and compensation of losses.

3. If the division balance sheet does not make it possible to determine the legal successor of the reorganised juridical person, the juridical persons which newly arose shall bear joint and several responsibility for the obligations of the reorganised juridical person to its creditors.

Article 61. Liquidation of Juridical Person

1. The liquidation of a juridical person shall entail the termination thereof without the transfer of rights and duties by way of legal succession to other persons.

2. A juridical person may be liquidated by:

decision of its founders (or participants) or the organ of the juridical person empowered by the constitutive documents, including in connection with the expiry of the period for which the juridical person was created, achievement of the purpose for which it was created [as amended 21 March 2002. СЗ РФ (2002), no. 12, item 1093];

decision of a court in the event of flagrant violations of a law permitted in the creation thereof if these violations are of an ineradicable character, or of the effectuation of activity without proper authorisation (or license), or activity prohibited by a law, or with other repeated or flagrant violations of a law or other legal acts, or in the event of the systematic effectuation by a social or religious organisation (or association) or by a philanthropic or other foundation of activity which is contrary to its charter purposes, and also in the other instances provided for by the present Code [as amended 21 March 2002. СЗ РФ (2002), no. 12, item 1093].

3. A demand concerning liquidation of a juridical person on the grounds specified in point 2 of the present Article may be presented to a court by a State agency or agency of local self-government to which the right to present such demand has been granted by a law.

Duties relating to the effectuation of the liquidation of a juridical person may be placed by decision of a court concerning the liquidation of a juridical person on its founders (or participants), or organ empowered to liquidate the juridical person by its constitutive documents.

4. A juridical person which is a commercial organisation or operating in the form of a consumer cooperative, philanthropic or other foundation shall also be liquidated in accordance with Article 65 of the present Code as a consequence of its being deemed to be insolvent (or bankrupt).

If the value of the property of such a juridical person is insufficient to satisfy the demands of creditors, it may be liquidated only in the procedure provided for by Article 65 of the present Code.

The provisions on the liquidation of juridical persons as a consequence of insolvency (or bankruptcy) shall not extend to treasury enterprises.

Article 62. Duties of Person Which Adopted Decision Concerning Liquidation of Juridical Person

1. The founders (or participants) of a juridical person or agency which adopted the decision concerning liquidation of the juridical person shall be obliged immediately in writing to notify the empowered State agency thereof to place in the unified State register of juridical persons information concerning the fact that the juridical person is in the process of liquidation [as amended 21 March 2002. СЗ РФ (2002), no. 12, item 1093].

2. The founders (or participants) of a juridical person or agency which has adopted the decision concerning the liquidation of a juridical person shall appoint a liquidation commission (or liquidator) and establish in accordance with the present Code and other laws the procedure and periods of liquidation [as amended 21 March 2002. СЗ РФ (2002), no. 12, item 1093].

3. From the moment of the appointment of the liquidation commission the powers relating to the management of the affairs of the juridical person shall pass to it. The liquidation commission shall act in court in the name of the juridical person being liquidated.

Article 63. Procedure for Liquidation of Juridical Person

1. The liquidation commission shall place in press organs in which data concerning the State registration of a juridical person is published a publication concerning its liquidation and the procedure and periods for declaring demands

by its creditors. This period may not be less than two months from the moment of the publication concerning liquidation.

The liquidation commission shall take measures to elicit creditors and receive debtor indebtedness, and also inform creditors in writing concerning the liquidation of a juridical person.

2. After the period ends for the presentation of demands by creditors, the liquidation commission shall draw up the interim liquidation balance sheet, which shall contain information concerning the composition of the property of the juridical person being liquidated, a list of demands presented by creditors, and also the results of their consideration.

The interim liquidation balance sheet shall be confirmed by the founders (or participants) of the juridical person or agency which adopted the decision concerning the liquidation of the juridical person. In instances established by a law the interim liquidation balance sheet shall be confirmed by agreement with the empowered State agency [as amended 21 March 2002. СЗ РФ (2002), no. 12, item 1093].

3. If a juridical person being liquidated (except institutions) has monetary means which are insufficient to satisfy the demands of creditors, the liquidation commission shall effectuate the sale of the property of the juridical person at public sales in the procedure established for the execution of judicial decisions.

4. The payment of monetary amounts to creditors of a juridical person being liquidated shall be by the liquidation commission in the order of priority established by Article 64 of the present Code and in accordance with the interim liquidation balance sheet, commencing from the date of its confirmation, except for creditors of the fifth priority, payment to whom shall be made upon the expiry of a month from the date of confirmation of the interim liquidation balance sheet.

5. After the completion of the settlement of accounts with creditors, the liquidation commission shall draw up the liquidation balance sheet, which shall be confirmed by the founders (or participants) of the juridical person or by the agency which adopted the decision concerning the liquidation of the juridical person. In instances established by a law the interim liquidation balance sheet shall be confirmed by agreement with the empowered State agency [as amended 21 March 2002. СЗ РФ (2002), no. 12, item 1093].

6. In the event that the property of a treasury enterprise being liquidated is insufficient, and the institution being liquidated has monetary means to satisfy the demands of creditors, the last shall have the right to bring suit in a court to satisfy the remaining part of the demands at the expense of the owner of the property of this enterprise or institution.

7. The property of a juridical person remaining after the satisfaction of the demands of creditors shall be transferred to the founders (or participants) thereof having rights to a thing in this property or rights of obligations with respect to this juridical person, unless provided otherwise by a law, other legal acts or the constitutive documents of the juridical person.

8. The liquidation of a juridical person shall be considered to be completed and the juridical person to have terminated existence after the making of entries thereof in the unified State register of juridical persons.

Article 64. Satisfaction of Demands of Creditors

1. When liquidating a juridical person the demands of its creditors shall be satisfied in the following priority:

in first priority the demands of citizens to whom the juridical person being liquidated bears responsibility for the causing of harm to life or health by means of capitalising the respective time payments shall be satisfied;

in second priority accounts shall be settled with regard to the payment of severance benefits and the payment for labour with persons who work under a labour contract [*договор*], including a *kontrakt* [*контрак*] and with regard to the payment of remuneration under authors' contracts;

in third priority the demands of creditors relating to obligations secured by the pledge of property of the juridical person being liquidated shall be satisfied;

in fourth priority the indebtedness relating to obligatory payments to the budget and extrabudgetary funds shall be paid;

in fifth priority accounts shall be settled with other creditors in accordance with a law.

In the event of the liquidation of banks or other credit institutions which attract the means of citizens, in first priority shall be satisfied the demands of citizens who are the creditors of the banks or other credit institutions which attract the means of citizens [added 20 February 1996. СЗ РФ (1996), no. 9, item 773].

2. The demands of each priority shall be satisfied after the full satisfaction of demands of the preceding priority.

3. In the event the property of a juridical person being liquidated is insufficient, it shall be distributed among the creditors of the respective priority in proportion to the amounts of demands subject to satisfaction, unless established otherwise by a law.

4. In the event of the refusal of the liquidation commission to satisfy the demands of a creditor or the evasion of the consideration thereof, a creditor shall have the right before confirmation of the liquidation balance sheet of the juridical person to bring suit against the liquidation commission in court. By decision of the court the demands of the creditor may be satisfied at the expense of the residual property of the juridical person being liquidated.

5. Demands of a creditor declared after the expiry of the period established by the liquidation commission for the presentation thereof shall be satisfied from the property of the juridical person being liquidated remaining after the satisfaction of the demands of creditors declared within the period.

6. Demands of creditors not satisfied because of the insufficiency of the property of the juridical person being liquidated shall be considered to be paid. Demands of creditors not recognised by the liquidation commission, if the creditor has not brought suit in court, and also demands whose satisfaction has been refused to a creditor by decision of a court, also shall be considered to be paid.

Article 65. Insolvency (or Bankruptcy) of Juridical Person

1. A juridical person which is a commercial organisation, except for a treasury enterprise, and also a juridical person operating in the form of a consumer cooperative or philanthropic or other foundation, may, by decision of a court, be deemed to be insolvent (or bankrupt) if it is not in a state to satisfy the demands of creditors.
   The deeming of a juridical person to be bankrupt by a court shall entail its liquidation.

2. A juridical person which is a commercial organisation, and also a juridical person operating in the form of a consumer cooperative or a philanthropic or other foundation, may adopt, jointly with creditors, a decision concerning the declaration of its bankruptcy and voluntary liquidation.

3. The grounds for deeming a juridical person to be bankrupt by a court, or declaring its own bankruptcy, and also the procedure for the liquidation of such juridical person, shall be established by a law on insolvency (or bankruptcy). Demands of creditors shall be satisfied in the priority provided for by Article 64(1) of the present Code.

§2. Economic Partnerships and Societies

1. General Provisions

Article 66. Basic Provisions on Economic Partnerships and Societies

1. Commercial organisations with a charter (or contributed) capital divided into participatory shares (or contributions) of the founders (or participants) shall be deemed to be economic partnerships and societies. The property created at the expense of contributions of the founders (or participants), and also produced and acquired by the economic partnership or society in the process of its activity, shall belong to it by right of ownership.
   In the instances provided for by the present Code an economic society may be created by one person, who shall become its sole participant.

2. Economic partnerships may be created in the form of a full partnership and limited partnership (kommandit partnership).

3. Economic societies may be created in the form of a joint-stock society, limited responsibility society, or additional responsibility society.

4. Individual entrepreneurs and (or) commercial organisations may be participants of full partnerships and may be full partners in limited partnerships.

Citizens and juridical persons may be participants of economic societies and contributors to limited partnerships.

State agencies and agencies of local self-government shall not have the right to act as participants of economic societies and contributors to limited partnerships unless established otherwise by a law.

Institutions financed by owners may be participants of economic societies and contributors to partnerships with the authorisation of the owner unless established otherwise by a law.

The participation of individual categories of citizens in economic partnerships and societies, except for open joint-stock societies, may be prohibited or limited by a law.

5. Economic partnerships and societies may be the founders (or participants) of other economic partnerships and societies except for instances provided for by the present Code and by other laws.

6. Money, securities, other things or property rights, or other rights having monetary value may be a contribution to the property of an economic partnership or society.

The monetary value of the contribution of a participant of an economic society shall be made by agreement between the participants (or founders) of the society and in the instances provided for by laws shall be subject to independent expert verification.

7. Economic partnerships, and also limited responsibility or additional responsibility societies, shall not have the right to issue stocks.

Article 67. Rights and Duties of Participants of Economic Partnership or Society

1. The participants of an economic partnership or society shall have the right to:

participate in the management of the affairs of the partnership or society, except for instances provided for by Article 84(2) of the present Code and by the law on joint-stock societies;

receive information concerning the activity of the partnership or society and familiarise themselves with its bookkeeping records and other documentation in the procedure established by the constitutive documents;

take part in the distribution of profit;

receive in the event of the liquidation of the partnership or society the part of the property remaining after the settlement of accounts with creditors or the value thereof.

The participants of an economic partnership or society also may have other rights provided for by the present Code, laws on economic societies, or constitutive documents of a partnership or society.

2. The participants of an economic partnership or society shall be obliged to:

make contributions in the procedure and amounts, by the means, and within the periods provided for by the constitutive documents;

not divulge confidential information concerning the activity of the partnership or society.

The participants of an economic partnership or society may also bear other duties provided for by its constitutive documents.

Article 68. Transformation of Economic Partnerships and Societies

1. Economic partnerships and societies of one type may be transformed into economic partnerships and societies of another type or into production cooperatives by decision of the general meeting of participants in the procedure established by the present Code.

2. In the event of the transformation of a partnership into a society, each full partner who has become a participant (or stockholder) of the society shall for two years bear subsidiary responsibility with all of its property for obligations which passed to the society from the partnership. The alienation by a former partner of participatory shares (or stocks) belonging to it shall not relieve it from such responsibility. The rules set out in the present point shall apply respectively when transforming a partnership into a production cooperative.

2. Full Partnership

Article 69. Basic Provisions on Full Partnership

1. A partnership whose participants (full partners) in accordance with a contract concluded between them engage in entrepreneurial activity in the name of the partnership and bear responsibility for its obligations with all of the property belonging to them, shall be deemed to be a full [partnership].

2. A person may be a participant only of one full partnership.

3. The firm name of a full partnership must contain either the names of all its participants, and the words 'full partnership', or the name of one or several participants with the addition of the words 'and company', and the words 'full partnership'.

Article 70. Constitutive Contract of Full Partnership

1. A full partnership shall be created and operate on the basis of a constitutive contract. The constitutive contract shall be signed by all of its participants.

2. The constitutive contract of a full partnership must contain, in addition to the information specified in Article 52(2) of the present Code, the conditions concerning the amount and composition of the contributed capital of the partnership; the amount and procedure for changing the participatory shares of each of the participants in the contributed capital; the amount, composition, periods, and procedure for making contributions by them; the responsibility of the participants for a violation of the duties relating to making contributions.

Article 71. Management in Full Partnership

1. The management of the activity of a full partnership shall be effectuated by the common consent of all the participants. Instances when a decision is adopted by a majority vote of the participants may be provided for by the constitutive contract of the partnership.

2. Each participant of a full partnership shall have one vote unless another procedure for determining the number of votes of its participants has been provided for by the constitutive contract.

3. Each participant of a partnership, irrespective of whether he is empowered to conduct the affairs of the partnership, shall have the right to familiarise himself with all of the documentation relating to conducting the affairs. A waiver of this right or limitation thereof, including by agreement of the participants of the partnership, shall be void.

Article 72. Conducting the Affairs of Full Partnership

1. Each participant of a full partnership shall have the right to operate in the name of the partnership unless it has been established by the constitutive contract that all of its participants conduct the affairs jointly, or the conducting of affairs has been entrusted by them to individual participants.

In the event of the joint conducting of the affairs of the partnership by its participants, the consent of all participants of the partnership shall be required in order to conclude each transaction.

If the conduct of the affairs of the partnership is entrusted by its participants to one or certain of them, the remaining participants must, in order to conclude transactions in the name of the partnership, have a power of attorney from the participant(s) to whom the conducting of the affairs of the partnership has been entrusted.

In relations with third persons the partnership shall not have the right to refer to provisions of the constitutive contract limiting the powers of the participants of the partnership, except for instances when the partnership proves that the third person at the moment of concluding the transaction knew or knowingly should have known about the lack of the participant of the partnership's right to operate in the name of the partnership.

2. The powers to conduct the affairs of the partnership granted to one or several participants may be terminated by a court upon the demand of one or several other participants of the partnership when there are serious grounds, in particular, as a consequence of a flagrant violation by an empowered person(s) of his duties or of discovering that he is not capable of reasonably conducting affairs. Necessary changes shall be made in the constitutive contract of the partnership on the basis of a judicial decision.

Article 73. Duties of Participant of Full Partnership

1. The participant of a full partnership shall be obliged to participate in its activity in accordance with the conditions of the constitutive contract.

2. The participant of a full partnership shall be obliged to make not less than half of his contribution to the contributed capital of the partnership at the moment of registration thereof. The remaining portion must contributed by the participant within the periods established by the constitutive contract. In the event of the failure to fulfil the said duty the participant shall be obliged to pay ten percent annual interest on the uncontributed portion of the contribution and to compensate the losses caused unless other consequences have been established by the constitutive contract.

3. The participant of a full partnership shall not have the right without the consent of the remaining participants to conclude in its name and in its interests or in the interests of third persons a transaction of the same nature as that which comprises the subject of activity of the partnership.

In the event of a violation of this rule, the partnership shall have the right at its choice to demand from such participant either compensation for losses caused to the partnership, or the transfer to the partnership of all advantage acquired under such transactions.

Article 74. Distribution of Profit and Losses of Full Partnership

1. Profit and losses of a full partnership shall be distributed between the participants thereof in proportion to their participatory shares in the contributed capital unless provided otherwise by the constitutive contract or other agreement of the participants. An agreement concerning the elimination of any of the participants of the partnership from participation in the profits or in the losses shall not be permitted.

2. If as a consequence of losses incurred by the partnership, the value of its net assets becomes less than the amount of its contributed capital, the profit received

by the partnership shall not be distributed among the participants so long as the value of the net assets does not exceed the amount of contributed capital.

Article 75. Responsibility of Participants of Full Partnership for its Obligations

1. The participants of a partnership shall jointly and severally bear subsidiary responsibility with all of their property for the obligations of the partnership.

2. The participant of a full partnership who is not a founder thereof shall be liable equally with the other participants also for obligations which arose before his joining the partnership.

A participant who has withdrawn from the partnership shall be liable for the obligations of the partnership which arose before the moment of his withdrawal equally with the remaining participants for two years from the date of confirmation of the report on the activity of the partnership for the year in which he withdrew from the partnership.

3. An agreement of the participants of the partnership concerning the limitation or elimination of the responsibility provided for in the present Article shall be void.

Article 76. Change of Composition of Participants of Full Partnership

1. In instances of the withdrawal or death of any of the participants of a full partnership, the deeming of one of them to be missing, to lack dispositive legal capacity, or to have limited dispositive legal capacity, or to be insolvent (or bankrupt), the opening with respect to one of the participants of reorganisation procedures by decision of a court, the liquidation of a juridical person participating in the partnership, or the levy of execution by a creditor of one of the participants against the part of the property corresponding to his participatory share in the contributed capital, the partnership may continue its activity if this has been provided for by the constitutive contract of the partnership or by agreement of the remaining participants.

2. The participants of a full partnership shall have the right to demand in a judicial proceeding the expulsion of any of the participants from the partnership upon the unanimous decision of the remaining participants and when there are serious grounds, in particular, as a consequence of a flagrant violation by this participant of his duties or the discovered inability of him to reasonably conduct affairs.

Article 77. Withdrawal of Participant from Full Partnership

1. The participant of a full partnership shall have the right to withdraw therefrom, having declared his refusal to participate in the partnership.

A refusal to participate in a full partnership founded without specification of the period must be declared by the participant not less than six months before the actual withdrawal from the partnership. A refusal to participate in a full partnership before time which is founded for a specified period shall be permitted only for a justifiable reason.

2. An agreement between the participants of a partnership to waive the right to withdraw from the partnership shall be void.

Article 78. Consequences of Withdrawal of Participant from Full Partnership

1. A participant who has withdrawn from a full partnership shall be paid the value of the portion of the property of the partnership corresponding to the participatory share of this participant in the contributed capital unless provided otherwise by the constitutive contract. By agreement of the participant who is withdrawing, the payment of the value of the property may be replaced by the issuance of property in kind.

Except for the instance provided for in Article 80 of the present Code, the part of the property of the partnership due to the withdrawing participant or the value thereof shall be determined according to the balance sheet drawn up at the moment of his withdrawal.

2. In the event of the death of the participant of a full partnership his heir may join the full partnership only with the consent of the other participants.

A juridical person which is the legal successor of a reorganised juridical person which participated in a full partnership shall have the right to join the partnership with the consent of its other participants unless provided otherwise by the constitutive contract of the partnership.

The settlement of accounts with the heir (or legal successor) who has not joined the partnership shall be made in accordance with point 1 of the present Article. The heir (or legal successor) of a participant of a full partnership shall bear responsibility for the obligations of the partnership to third persons, with regard to which in accordance with Article 75(2) of the present Code the withdrawn participant would have been liable, within the limits of the property of the withdrawn participant of the partnership which passed to him.

3. If one of the participants withdrew from the partnership, the participatory shares of the remaining participants in the contributed capital of the partnership shall be increased respectively, unless provided otherwise by the constitutive contract or other agreement of the participants.

Article 79. Transfer of Participatory Share of Participant in Contributed Capital of Full Partnership

The participant of a full partnership shall have the right with the consent of the remaining participants thereof to transfer his participatory share in the contributed capital or part thereof to another participant of the partnership or to a third person.

In the event of the transfer of the participatory share (or part of the participatory share) to another person, to him shall pass wholly or in respective part the rights which belong to the participant who has transferred the participatory share (or portion of the participatory share). The person to whom the participatory share (or portion of the participatory share) is transferred shall bear responsibility for the obligations of the partnership in the procedure established by Article 75(2), paragraph one, of the present Code.

The transfer by the participant of a partnership of the entire participatory share to another person shall terminate his participation in the partnership and entail the consequences provided for by Article 75(2) of the present Code.

Article 80. Levying Execution on Participatory Share of Participant in Contributed Capital of Full Partnership

Levying execution on the participatory share of a participant in the contributed capital of a full partnership regarding the own debts of the participant shall be permitted only when his other property is insufficient to cover the debts. Creditors of such a participant shall have the right to demand of the full partnership the apportionment of the part of the property of the partnership corresponding to the participatory share of the debtor in the contributed capital for the purpose of levying execution on this property. The part of the property of the partnership subject to apportionment or the value thereof shall be determined according to the balance sheet drawn up at the moment of the creditors presenting the demand concerning the apportionment.

The levy of execution on property corresponding to the participatory share of a participant in the contributed capital of a full partnership shall terminate his participation in the partnership and entail the consequences provided for by Article 75(2), paragraph two, of the present Code.

Article 81. Liquidation of Full Partnership

A full partnership shall be liquidated on the grounds specified in Article 61 of the present Code, and also when a sole participant remains in the partnership. Such participant shall have the right within six months from the moment when he became the sole participant of the partnership to transform such partnership into an economic society in the procedure established by the present Code.

A full partnership shall be liquidated also in the instances specified in Article 76(1) of the present Code unless it has been provided by the constitutive contract of the partnership or agreement of the remaining participants that the partnership shall continue its activity.

3. Limited Partnership

Article 82. Basic Provisions on Limited Partnership

1. A limited partnership (kommandit partnership) shall be deemed to be a partnership in which together with participants effectuating entrepreneurial

37

activity in the name of the partnership and liable for the obligations of the partnership with their property (full partners) there are one or several participant-contributors (kommanditists) who shall bear the risk of losses connected with the activity of the partnership within the limits of the amounts of the contributions made by them and shall not take part in the effectuation by the partnership of entrepreneurial activity.

2. The status of full partners participating in a limited partnership and their responsibility for the obligations of the partnership shall be determined by the rules of the present Code concerning the participants of a full partnership.

3. A person may be a full partner only in one limited partnership.

A participant of a full partnership may not be a full partner in a limited partnership.

A full partner in a limited partnership may not be a participant of a full partnership.

4. The firm name of a limited partnership must contain either the name(s) of all of the full partners and the words 'limited partnership' or 'kommandit partnership', or the name of not less than one full partner with the addition of the words 'and company' and the words 'limited partnership' or 'kommandit partnership'.

If the name of a contributor is included in the firm name of a limited partnership, such contributor shall become a full partner.

5. The rules of the present Code on the full partnership shall apply to a limited partnership insofar as this is not contrary to the rules of the present Code on the limited partnership.

Article 83. Constitutive Contract of Limited Partnership

1. A limited partnership shall be created and shall operate on the basis of a constitutive contract. The constitutive contract shall be signed by all of the full partners.

2. The constitutive contract of a limited partnership must contain, besides the information specified in Article 52(2) of the present Code, the conditions concerning the amount and composition of the contributed capital of the partnership; the amount and procedure for changing the participatory shares of each of the full partners in the contributed capital; the amount, composition, periods, and procedure for the making of contributions by them, their responsibility for a violation of the duties relating to the making of contributions; the aggregate amount of the contributions to be made by the contributors.

Article 84. Management in Limited Partnership and Conducting of its Affairs

1. The management of the activity of a limited partnership shall be effectuated by the full partners. The procedure for the management and conducting of the

affairs of such partnership by its full partners shall be established by them according to the rules of the present Code concerning the full partnership.

2. The contributors shall not have the right to participate in the management and conducting of the affairs of a limited partnership, nor to act in its name, other than under a power of attorney. They shall not have the right to contest the actions of the full partners relating to the management and the conducting of the affairs of the partnership.

Article 85. Rights and Duties of Contributor to Limited Partnership

1. The contributor to a limited partnership shall be obliged to make a contribution to the contributed capital. The making of a contribution shall be certified by a certificate of participation issued to the contributor by the partnership.

2. The contributor to a limited partnership shall have the right to:
(1) receive part of the profit of the partnership due for his participatory share in the contributed capital in the procedure provided for by the constitutive contract;
(2) familiarise himself with the annual reports and balance sheets of the partnership;
(3) at the end of the financial year withdraw from the partnership and receive his contribution in the procedure provided for by the constitutive contract;
(4) transfer his participatory share in the contributed capital or part thereof to another contributor or to a third person. The contributors shall enjoy a preferential right of purchase against third persons of the participatory share (or part thereof) according to the conditions and procedure provided for by Article 93(2) of the present Code. The transfer by the contributor of his entire participatory share to another person shall terminate his participation in the partnership.

Other rights of a contributor also may be provided for by the constitutive contract of a limited partnership.

Article 86. Liquidation of Limited Partnership

1. A limited partnership shall be liquidated in the event of the withdrawal of all the contributors who participated therein. However, the full partners shall have the right instead of liquidation to transform the limited partnership into a full partnership.

A limited partnership shall be liquidated also on the grounds of liquidation of a full partnership (Article 81). However, a limited partnership shall be preserved if at least one full partner and one contributor remain therein.

2. In the event of the liquidation of a limited partnership, including in the event of bankruptcy, contributors shall have the preferential right before the full partners to receive contributions from the property of the partnership remaining after the satisfaction of the demands of its creditors.

The property of the partnership remaining thereafter shall be distributed between the full partners and the contributors in proportion to their participatory shares in the contributed capital of the partnership unless another procedure has been established by the constitutive contract or by agreement of the full partners and the contributors.

4. Limited Responsibility Society

Article 87. Basic Provisions on Limited Responsibility Society

1. A limited responsibility society shall be deemed to be a society founded by one or several persons, the charter capital of which is divided into participatory shares of amounts determined by the constitutive documents; the participants of a limited responsibility society shall not be liable for its obligations and shall bear the risk of losses connected with the activity of the society within the limits of the value of the contributions made by them.

The participants of a society who have not made contributions in full shall bear joint and several responsibility for its obligations within the limits of the value of the unpaid portion of the contribution of each of the participants.

2. The firm name of a limited responsibility society must contain the name of the society, and also the words 'limited responsibility'.

3. The legal status of a limited responsibility society and the rights and duties of its participants shall be determined by the present Code and by the law on limited responsibility societies.

The peculiarities of the legal status of credit organisations created in the form of limited responsibility societies and the rights and duties of the participants thereof shall be determined by laws regulating the activity of credit organisations [added by Law of 8 July 1999. СЗ РФ (1999), no. 28, item 3471].

Article 88. Participants of Limited Responsibility Society

1. The number of participants of a limited responsibility society must not exceed the limit established by the law on limited responsibility societies. Otherwise it shall be subject to transformation into a joint-stock society within a year, and upon the expiry of this period, liquidation in a judicial proceeding if the number of its participants is not reduced to up to the limit established by a law.

2. A limited responsibility society may not have as a sole participant another economic society consisting of one person.

Article 89. Constitutive Documents of Limited Responsibility Society

1. The constitutive documents of a limited responsibility society shall be the constitutive contract signed by its founders and the charter confirmed by them. If the society is founded by one person, the charter shall be its constitutive document.

2. The constitutive documents of a limited responsibility society must contain, besides the information specified in Article 52(2) of the present Code, conditions concerning the amount of the charter capital of the society; concerning the amount of the participatory shares of each participant; the amount, composition, periods, and procedure for the making of contributions by them, the responsibility of the participants for a violation of the duties relating to making the contributions; the composition and competence of the management organs of the society, and the procedure for the adoption of decisions by them, including on questions with regard to which decisions shall be adopted unanimously or by a qualified majority of votes, and also other information provided for by the law on limited responsibility societies.

Article 90. Charter Capital of Limited Responsibility Society

1. The charter capital of a limited responsibility society shall be comprised of the value of the contributions of its participants.

The charter capital shall determine the minimum amount of the property of the society guaranteeing the interests of its creditors. The amount of charter capital of the society may not be less than the amount determined by the law on limited responsibility societies.

2. The relieving of a participant of a limited responsibility society from the duty to make a contribution to the charter capital of the society, including by means of setting off demands against the society, shall not be permitted, except for the instances provided for by a law [as amended 8 July 1999. СЗ РФ (1999), no. 28, item 3471].

3. The charter capital of a limited responsibility society must at the moment of registration of the society be paid up by its participants by not less than half. The remaining unpaid portion of the charter capital of the society shall be subject to being paid up by its participants within one year of the activity of the society. In the event of a violation of this duty the society must either declare a reduction of its charter capital and register this reduction in the established procedure, or terminate its activity by way of liquidation.

4. If at the end of the second or each subsequent financial year the value of net assets of a limited responsibility society proves to be less than the charter capital, the society shall be obliged to declare a reduction of its charter capital and to register the reduction in the established procedure. If the value of the said means of the society becomes less than the minimum amount of charter capital determined by a law, the society shall be subject to liquidation.

5. A reduction of the charter capital of a limited responsibility society shall be permitted after informing all of its creditors. The last shall have the right in this event to demand the termination or performance before time of the respective obligations of the society and the compensation of losses to them.

The rights and duties of creditors of credit organisations created in the form of limited responsibility societies shall also be determined also by laws regulating the activity of credit organisations [added by Law of 8 July 1999. СЗ РФ (1999), no. 28, item 3471].

6. An increase of charter capital of the society shall be permitted after the making of contributions in full by all of its participants.

Article 91. Management in Limited Responsibility Society

1. The highest organ of a limited responsibility society shall be the general meeting of its participants.

An executive organ (collegial and/or one-man) effectuating the current direction of its activity and accountable to the general meeting of its participants shall be created in a limited responsibility society. A one-man management organ of the society also may be elected not from among its participants.

2. The competence of the management organs of the society, and also the procedure for the adoption of decisions by them and acting in the name of the society, shall be determined in accordance with the present Code by the law on limited responsibility societies and by the charter of the society.

3. There shall be relegated to the exclusive competence of the general meeting of participants of a limited responsibility society:

(1) change of the charter of the society, change of the amount of its charter capital;

(2) formation of executive organs of the society and the termination of their powers before time;

(3) confirmation of the annual reports and bookkeeping balance sheets of the society and distribution of its profits and losses;

(4) decision concerning the reorganisation or liquidation of the society;

(5) election of the audit commission (or internal auditor) of the society.

The deciding of other questions may be relegated also by the law on limited responsibility societies to the exclusive competence of the general meeting.

Questions relegated to the exclusive competence of the general meeting of participants of the society may not be transferred by it for decision of the executive organ of the society.

4. In order to verify and confirm the correctness of the annual financial report of the limited responsibility society, it shall have the right annually to recruit a professional auditor not connected by property interests with the society or its participants (external audit). The audit verification of the yearly financial report of the society may also be carried out at the demand of any of its participants.

The procedure for conducting audit verifications of the activity of the society shall be determined by a law and by the charter of the society.

5. The publication by the society of information concerning the results of the conducting of its affairs (public accountability) shall not be required except for instances provided for by the law on limited responsibility societies.

Article 92. Reorganisation and Liquidation of Limited Responsibility Society

1. A limited responsibility society may be voluntarily reorganised or liquidated by the unanimous decision of its participants.

Other grounds for the reorganisation and liquidation of the society, and also the procedure for its reorganisation and liquidation, shall be determined by the present Code and other laws.

2. A limited responsibility society shall have the right to transform itself into a joint-stock society or into a production cooperative.

Article 93. Transfer of Participatory Share in Charter Capital of Limited Responsibility Society to Another Person

1. The participant of a limited responsibility society shall have the right to sell or otherwise assign its participatory share in the charter capital of the society or part thereof to one or several participants of the particular society.

2. The alienation by the participant of a society of his participatory share (or part thereof) to third persons shall be permitted unless provided otherwise by the charter of the society.

The participants of the society shall enjoy a preferential right to purchase the participatory share of the participant (or part thereof) in proportion to the amounts of their participatory shares unless the charter of the society or an agreement of its participants has provided a different procedure for effectuating this right. If the participants of the society do not take advantage of their preferential right within a month from the date of notification or within another period provided for by the charter of the society or by agreement of the participants thereof, the participatory share of the participant may be alienated to a third person.

3. If in accordance with the charter of a limited responsibility society the alienation of the participatory share of a participant (or part thereof) to third persons is impossible, and the other participants of the society decline to purchase it, the society shall be obliged to pay the participant its actual value or to issue the property in kind to him which corresponds to such value.

4. The participatory share of the participant of a limited responsibility society may be alienated until the payment thereof in full only in that part which has been already paid up.

5. In the event of the acquisition of the participatory share of a participant (or part thereof) by the limited responsibility society itself, it shall be obliged to realise

it to the other participants or to third persons within the periods and in the procedure provided for by the law on limited responsibility societies and by the constitutive documents of the society, or to reduce its charter capital in accordance with Article 90(4) and (5) of the present Code.

6. The participatory shares in the charter capital of a limited responsibility society shall pass to the heirs of citizens and to the legal successors of juridical persons who are participants of the society unless the constitutive documents of the society have provided that such transfer shall be permitted only with the consent of the other participants of the society. A refusal to consent to the transfer of a participatory share shall entail the duty of the society to pay to the heirs (or legal successors) of the participant its real value or to issue property in kind to them for such value in the procedure and on the conditions provided for by the law on limited responsibility societies and the constitutive documents of the society.

Article 94. Withdrawal of Participant of Limited Responsibility Society from the Society

A participant of a limited responsibility society shall have the right at any time to withdraw from the society irrespective of the consent of its other participants. In so doing he must be paid the value of the part of the property corresponding to his participatory share in the charter capital of the society in the procedure, by the means, and within the periods provided for by the law on limited responsibility societies and the constitutive documents of the society.

5. Additional Responsibility Society

Article 95. Basic Provisions on Additional Responsibility Societies

1. An additional responsibility society shall be deemed to be a society founded by one or several persons, whose charter capital has been divided into participatory shares of the amounts determined by the constitutive documents; the participants of such a society shall bear subsidiary responsibility jointly and severally for its obligations with their property in a multiple identical for all of the value of their contributions determined by the constitutive documents of the society. In the event of the bankruptcy of one of the participants, his responsibility for the obligations of the society shall be distributed among the remaining participants in proportion to their contributions unless another procedure for the distribution of responsibility has been provided for by the constitutive documents of the society.

2. The firm name of an additional responsibility society must contain the name of the society and the words 'additional responsibility'.

3. The rules of the present Code on the limited responsibility society shall apply to the additional responsibility society insofar as the present Article does not provide otherwise.

6. Joint-Stock Society

Article 96. Basic Provisions on the Joint-Stock Society

1. A joint-stock society shall be deemed to be a society whose charter capital has been divided into a determined number of stocks; the participants of a joint-stock society (stockholders) shall not be liable for its obligations and shall bear the risk of losses connected with the activity of the society within the limits of the value of the stocks belonging to them.

Stockholders who have not fully paid up stocks shall bear joint and several responsibility for the obligations of the joint-stock society within the limits of the unpaid portion of the value of the stocks belonging to them.

2. The firm name of a joint-stock society must contain its name and an indication that the society is a joint-stock [society].

3. The legal status of a joint-stock society and the rights and duties of the stockholders shall be determined in accordance with the present Code and by the law on joint-stock societies.

The peculiarities of the legal status of joint-stock societies created by means of the privatisation of State and municipal enterprises shall be determined also by laws and other legal acts on the privatisation of these enterprises.

The peculiarities of the legal status of credit organisations created in the form of joint-stock societies and the rights and duties of the stockholders thereof shall be determined also by laws retulating the activity of credit organisations [added by Law of 8 July 1999. СЗ РФ (1999), no. 28, item 3471].

Article 97. Open and Closed Joint-Stock Societies

1. A joint-stock society whose participants may alienate the stocks belonging to them without the consent of other stockholders shall be deemed to be an open joint-stock society. Such a joint-stock society shall have the right to conduct an open subscription for the stocks issued by it and the free sale thereof on the conditions established by a law and other legal acts.

An open joint-stock society shall be obliged annually to publish for general information a yearly report, the bookkeeping balance sheet, and the profits and losses account.

2. A joint-stock society whose stocks are distributed only among its founders or other previously determined group of persons shall be deemed to be a closed joint-stock society. Such society shall not have the right to conduct an open subscription for the stocks to be issued by it or otherwise offer them for acquisition to an unlimited group of persons.

The stockholders of a closed joint-stock society shall have a preferential right to acquire stocks being sold by other stockholders of this society.

The number of participants of a closed joint-stock society must not exceed the number established by the law on joint-stock societies; otherwise it shall be subject to transformation into an open joint-stock society within a year, and upon the expiry of this period, to liquidation in a judicial proceeding if the number thereof is not reduced up to the limit established by the law.

In the instances provided for by the law on joint-stock societies a closed joint-stock society may be obliged to publish the documents for general information specified in point 1 of the present Article.

Article 98. Formation of Joint-Stock Society

1. The founders of a joint-stock society shall conclude between themselves a contract determining the procedure for the effectuation by them of joint activity relating to the creation of the society; the amount of the charter capital of the society; the categories of stocks to be issued and the procedure for placing them, and also other conditions provided for by the law on joint-stock societies.

The contract concerning the creation of a joint-stock society shall be concluded in writing.

2. The founders of a joint-stock society shall bear joint and several responsibility for the obligations which arose before the registration of the society.

The society shall bear responsibility for the obligations of the founders connected with its creation only in the event of subsequent approval of their actions by the general meeting of stockholders.

3. The constitutive document of a joint-stock society shall be its charter, confirmed by the founders.

The charter of a joint-stock society must, besides the information specified in Article 52(2) of the present Code, contain conditions concerning the categories of stocks to be issued by the society, their par value and quantity; the amount of the charter capital of the society; the rights of stockholders; the composition and competence of the management organs of the society and the procedure for the adoption of decisions by them, including questions the decisions regarding which shall be adopted unanimously or by a qualified majority of votes. The charter of the joint-stock society also must contain other information provided for by the law on joint-stock societies.

4. The procedure for the performance of other actions relating to the creation of a joint-stock society, including the competence of the constitutive meeting, shall be determined by the law on joint-stock societies.

5. The peculiarities of creating joint-stock societies in the event of the privatisation of State and municipal enterprises shall be determined by the laws and other legal acts on the privatisation of these enterprises.

6. A joint-stock society may be created by one person or consist of one person in the event of the acquisition by one stockholder of all the stocks of the society. Information thereon must be contained in the charter of the society, be registered, and be published for general information.

A joint-stock society may not have another economic society consisting of one person as the sole participant.

Article 99. Charter Capital of Joint-Stock Society

1. The charter capital of a joint-stock society shall comprise the par value of the stocks of the society acquired by the stockholders.

The charter capital of the society shall determine the minimum amount of the property of the society, guaranteeing the interests of its creditors. It may not be less than the amount provided for by the law on joint-stock societies.

2. Relieving a stockholder from the duty to pay up the stocks of the society, including relieving it of this duty by means of setting off demands against the society, shall not be permitted.

3. An open subscription for the stocks of a joint-stock society shall not be permitted until the charter capital is fully paid up. When a joint-stock society is founded, all of its stocks must be distributed among the founders.

4. If at the end of the second and each subsequent financial year the value of the net means of the society proves to be less than the charter capital, the society shall be obliged to declare and register in the established procedure a reduction of its charter capital. If the value of the said means of the society becomes less than the minimum amount of charter capital (point 1 of the present Article) determined by a law, the society shall be subject to liquidation.

5. Limitations on the number, total par value of stocks, or maximum number of votes belonging to one stockholder may be established by a law or by the charter of the society.

Article 100. Increase of Charter Capital of Joint-Stock Society

1. A joint-stock society shall have the right by decision of the general meeting of stockholders to increase the charter capital by means of increasing the par value of the stocks or the issuance of additional stocks.

2. An increase of the charter capital of a joint-stock society shall be permitted after it is paid up in full. An increase of charter capital of the society in order to cover losses incurred by it shall not be permitted.

3. In the instances provided for by the law on joint-stock societies a preferential right of stockholders who possess common or other voting stocks to purchase stocks additionally to be issued by the society may be established by the charter of the society.

Article 101. Reduction of Charter Capital of Joint-Stock Society

1. A joint-stock society shall have the right by decision of the general meeting of stockholders to reduce the charter capital by means of reducing the par value of the stocks, or by means of buying up part of the stocks for the purposes of reducing the total quantity thereof.

A reduction of charter capital of a society shall be permitted after informing all of its creditors in the procedure determined by the law on joint-stock societies. In this connection the creditors of the society shall have the right to demand the termination before time and the performance of the respective obligations of the society and compensation of losses to them.

The rights and duties of creditors of credit organisations created in the form of joint-stock societies shall be determined also by laws regulating the activity of credit organisations [added by Law of 8 July 1999. СЗ РФ (1999), no. 28, item 3471].

2. A reduction of the charter capital of a joint-stock society by means of the buying up and redemption of part of the stocks shall be permitted if such possibility has been provided for in the charter of the society.

Article 102. Limitations on Issuance of Securities and Payment of Dividends of Joint-Stock Society

1. The participatory share of preferred stocks in the overall amount of the charter capital of a joint-stock society must not exceed 25%.

2. A joint-stock society shall have the right to issue bonds for an amount not exceeding the charter capital or the amount of security granted to the society for these purposes by third persons after the charter capital has been paid up in full. In the absence of security, the issuance of bonds shall be permitted not earlier than the third year of existence of the joint-stock society and on condition of the proper confirmation for this period of two yearly balance sheets of the society.

3. A joint-stock society shall not have the right to declare and pay a dividend:
until all of the charter capital is paid up in full;
if the value of net assets of the joint-stock society is less than its charter capital and reserve fund or becomes less than the amount thereof as a result of the payment of dividends.

Article 103. Management in Joint-Stock Society

1. The highest management organ of a joint-stock society shall be the general meeting of its stockholders.

There shall be relegated to the exclusive competence of the general meeting of stockholders:

(1) change of the charter of the society, including change of the amount of its charter capital;

(2) election of members of the council of directors (or supervisory council) and audit commission (or internal auditor) of the society, and the termination of their powers before time;

(3) formation of the executive organs of the society and the termination of their powers before time, unless the deciding of these questions has been relegated by the charter of the society to the competence of the council of directors (or supervisory council);

(4) confirmation of the yearly reports, bookkeeping balance sheets, profits and losses accounts of the society, and distribution of its profits and losses;

(5) decision concerning reorganisation or liquidation of the society;

The deciding of other questions also may be relegated to the exclusive competence of the general meeting of stockholders by the law on joint-stock societies.

Questions relegated by law to the exclusive competence of the general meeting of stockholders may not be transferred by them for decision of the executive organs of the society.

2. In a society with more than 50 stockholders a council of directors (or supervisory council) shall be created.

In the event a council of directors (or supervisory council) is created, its exclusive competence must be determined by the charter of the society in accordance with a law on joint-stock societies. Questions relegated by the charter to the exclusive competence of the council of directors (or supervisory council) may not be transferred by them for decision of executive organs of the society.

3. The executive organ of a society may be collegial (board, directorate) and/or one-man (director, director-general). It shall effectuate current direction over the activity of the society and be accountable to the council of directors (or supervisory council) and general meeting of stockholders.

To the competence of the executive organ of the society shall be relegated the deciding of all questions which do not constitute the exclusive competence of other management organs of the society specified by a law or by the charter of the society.

By decision of the general meeting of stockholders the powers of an executive organ of the society may be transferred under a contract to another commercial organisation or individual entrepreneur (manager).

4. The competence of the management organs of a joint-stock society, and also the procedure for the adoption of decisions by them and of acting in the name of the society, shall be determined in accordance with the present Code by the law on joint-stock societies and by the charter of the society.

5. A joint-stock society obliged in accordance with the present Code or the law on joint-stock societies to publish for general information the documents specified in Article 97(1) of the present Code must, in order to verify and confirm the

correctness of the yearly financial reports, annually recruit a professional auditor not connected by property interests with the society or the participants thereof.

The audit verification of the activity of a joint-stock society, including those not obliged to publish the said documents for general information, must be conducted at any time at the demand of the stockholders, the aggregate participatory share of which in the charter capital comprises ten or more percent.

The procedure for conducting audit verifications of the activity of a joint-stock society shall be determined by a law and by the charter of the society.

Article 104. Reorganisation and Liquidation of Joint-Stock Society

1. A joint-stock society may be voluntarily reorganised or liquidated by decision of the general meeting of stockholders.

Other grounds and the procedure for reorganisation and liquidation of the joint-stock society shall be determined by the present Code and other laws.

2. A joint-stock society shall have the right to be transformed into a limited responsibility society or a production cooperative, and also in a noncommercial organisation in accordance with a law [as amended 8 July 1999. СЗ РФ (1999), no. 28, item 3471].

7. Subsidiary and Dependent Societies

Article 105. Subsidiary Economic Society

1. An economic society shall be deemed to be a subsidiary if another (principal) economic society or partnership by virtue of predominant participation in its charter capital, or in accordance with a contract concluded between them, or otherwise has the possibility to determine the decisions adopted by that society.

2. A subsidiary society shall not be liable for the debts of its principal society (or partnership).

The principal society (or partnership) which has the right to give to the subsidiary society instructions binding upon it, including under a contract with it, shall be liable jointly and severally with the subsidiary society with regard to transactions concluded by the last in performance of such instructions.

In the event of the insolvency (or bankruptcy) of the subsidiary society through the fault of the principal society (or partnership) the last shall bear subsidiary responsibility for its debts.

3. The participants (or stockholders) of a subsidiary society shall have the right to demand compensation by the principal society (or partnership) for losses caused through its fault to the subsidiary society unless established otherwise by the laws on economic societies.

Article 106. Dependent Economic Society

1. An economic society shall be deemed to be dependent if another (predominant, participating) society has more than 20% of the voting stocks of the joint-stock society or 20% of the charter capital of a limited responsibility society.

2. The economic society which has acquired more than 20% of the voting stocks of a joint-stock society or 20% of the charter capital of a limited responsibility society shall be obliged immediately to publish information concerning this in the procedure provided for by laws on economic societies.

3. The limits of mutual participation of economic societies in the charter capital of one another and the number of votes which one of such societies may use at the general meeting of participants or stockholders of the other society shall be determined by a law.

§3. Production Cooperatives

Article 107. Concept of Production Cooperative

1. A production cooperative (or artel) shall be deemed to be a voluntary association of citizens on the basis of membership for the purpose of joint production or other economic activity (production, processing, and sale of industrial, agricultural, and other products, fulfilment of work, trade, domestic servicing, rendering of other services) based on their personal labour and other participation and the combining of property share contributions by the members (or participants) thereof. The participation of juridical persons in its activity may be provided for by a law and by the constitutive documents of the production cooperative. A production cooperative shall be a commercial organisation.

2. Members of a production cooperative shall bear subsidiary responsibility for the obligations of the cooperative within the amounts and in the procedure provided for by the law on production cooperatives and by the charter of the cooperative.

3. The firm name of a cooperative must contain its name and the words 'production cooperative' or 'artel'.

4. The legal status of production cooperatives and the rights and duties of their members shall be determined in accordance with the present Code by the laws on the production cooperatives.

Article 108. Formation of Production Cooperatives

1. The constitutive document of a cooperative shall be its charter, confirmed by the general meeting of its members.

2. The charter of a cooperative must contain, besides the information specified in Article 52(2) of the present Code, the conditions concerning the amount

of the share contributions of the members of the cooperative; the composition and procedure for the making of share contributions by members of the cooperative and their responsibility for a violation of the obligation to make share contributions; the character and procedure of labour participation of its members in the activity of the cooperative and their responsibility for a violation of the obligation relating to personal labour participation; the procedure for the distribution of profits and losses of the cooperative; the amount and conditions of subsidiary responsibility of its members for the debts of the cooperative, the composition and competence of the management organs of the cooperative and the procedure for the adoption of decisions by them, including on questions the decisions regarding which shall be adopted unanimously or by a qualified majority of votes.

3. The number of members of a cooperative must be not less than five.

Article 109. Property of Production Cooperative

1. Property in the ownership of a production cooperative shall be divided into shares of its members in accordance with the charter of the cooperative.

It may be established by the charter of the cooperative that a determined part of the property belonging to the cooperative shall comprise the indivisible funds to be used for the purposes determined by the charter.

A decision concerning the formation of the indivisible funds shall be adopted by the members of the cooperative unanimously unless provided otherwise by the charter of the cooperative.

2. The member of a cooperative shall be obliged to contribute not less than 10% of his share contribution at the moment of registration of the cooperative, and the remaining portion within a year from the moment of registration.

3. A cooperative shall not have the right to issue stocks.

4. The profit of a cooperative shall be distributed among its members in accordance with their labour participation unless a different procedure has been provided by a law and by the charter of the cooperative.

The property remaining after the liquidation of the cooperative and the satisfaction of the demands of its creditors shall be distributed in the same procedure.

Article 110. Management in Production Cooperative

1. The highest management organ of a cooperative shall be the general meeting of its members.

A supervisory council which shall effectuate control over the activity of executive organs of the cooperative may be created in a cooperative with more than 50 members.

The executive organs of a cooperative shall be the board and/or its chairman. They shall effectuate current direction over the activity of the cooperative and shall be accountable to the supervisory council and the general meeting of the members of the cooperative.

Only members of the cooperative may be members of the supervisory council and board of the cooperative, and also the chairman of the cooperative. A member of the cooperative may not simultaneously be a member of the supervisory council and a member of the board or the chairman of the cooperative.

2. The competence of the management organs of a cooperative and the procedure for the adoption of decisions by them shall be determined by a law and by the charter of the cooperative.

3. There shall be relegated to the exclusive competence of the general meeting of members of the cooperative:

(1) change of the charter of the cooperative;

(2) formation of the supervisory council and termination of the powers of its members, and also the formation and termination of the powers of the executive organs of the cooperative unless this right according to the charter of the cooperative has been transferred to the supervisory council;

(3) admission and expulsion of members of the cooperative;

(4) confirmation of the yearly reports and bookkeeping balance sheets of the cooperative and distribution of its profits and losses;

(5) decision concerning the reorganisation and liquidation of the cooperative;

The deciding of other questions also may be relegated to the exclusive competence of the general meeting by the law on production cooperatives and by the charter of the cooperative.

Questions relegated to the exclusive competence of the general meeting or supervisory council of a cooperative may not be transferred by them for decision of the executive organs of the cooperative.

4. A member of a cooperative shall have one vote when decisions are adopted by the general meeting.

Article 111. Termination of Membership in Production Cooperative and Transfer of Share

1. The member of a cooperative shall have the right at his discretion to withdraw from the cooperative. In this event the value of the share must be paid or the property corresponding to his share issued to him, and also other payments provided for by the charter of the cooperative effectuated.

The payment of the value of the share or the issuance of other property to a withdrawing member of a cooperative shall be done at the end of the financial year and after confirmation of the bookkeeping balance sheet of the cooperative unless provided otherwise by the charter of the cooperative.

2. A member of a cooperative may be expelled from the cooperative upon the decision of the general meeting in the event of the failure to perform or the improper performance of the duties placed on him by the charter of the cooperative, and also in other instances provided for by a law and by the charter of the cooperative.

A member of the supervisory council or executive organ may be expelled from the cooperative by decision of the general meeting in connection with his membership in an analogous cooperative.

A member of a cooperative expelled therefrom shall have the right to receive the share and other payments provided for by the charter of the cooperative in accordance with point 1 of the present Article.

3. The member of a cooperative shall have the right to transfer his share or part thereof to another member of the cooperative unless provided otherwise by a law and by the charter of the cooperative.

The transfer of a share (or part thereof) to a citizen who is not a member of the cooperative shall be permitted only with the consent of the cooperative. In this event the other members of the cooperative shall enjoy the preferential right to purchase such share (or part thereof).

4. In the event of the death of a member of a production cooperative his heirs may be admitted as members of the cooperative unless provided otherwise by the charter of the cooperative. Otherwise the cooperative shall pay to the heirs the value of the share of the deceased member of the cooperative.

5. The levying of execution on the share of a member of a production cooperative shall be permitted with regard to the own debts of a member of the cooperative only if his other property is insufficient to cover such debts in the procedure provided for by a law and by the charter of the cooperative. Execution may not be levied for debts of a member of a cooperative on the indivisible funds of the cooperative.

Article 112. Reorganisation and Liquidation of Production Cooperatives

1. A production cooperative may be voluntarily reorganised or liquidated by decision of the general meeting of its members.

Other grounds and the procedure for the reorganisation and liquidation of a cooperative shall be determined by the present Code and by other laws.

2. A production cooperative may, by unanimous decision of its members, be transformed into an economic partnership or society.

§4. State and Municipal Unitary Enterprises

Article 113. Unitary Enterprise

1. A commercial organisation not endowed with the right of ownership to property consolidated to it by the owner shall be deemed to be a unitary enterprise. The property of a unitary enterprise shall be indivisible and may not be

distributed according to contributions (or participatory shares, shares), including among the workers of the enterprise.

The charter of a unitary enterprise must contain, besides the information specified in Article 52(2) of the present Code, information concerning the subject and purposes of activity of the enterprise, and also the amount of the charter fund of the enterprise and the procedure and sources for forming it, except for treasury enterprises [as amended by Law of 14 November 2002. СЗ РФ (2002), no. 48, item 4746].

Only State and municipal enterprises may be created in the form of unitary enterprises.

2. The property of the State or municipal unitary enterprise shall respectively be in State or municipal ownership and shall belong to that enterprise by right of economic jurisdiction or operative management.

3. The firm name of a unitary enterprise must contain an indication of the owner of its property.

4. The organ of a unitary enterprise shall be the director, who shall be appointed by the owner or by an agency empowered by the owner and accountable to it.

5. A unitary enterprise shall be liable for its obligations with all of the property belonging to it.

A unitary enterprise shall not bear responsibility for the obligations of the owner of its property.

6. The legal status of State and municipal unitary enterprises shall be determined by the present Code and by the law on State and municipal unitary enterprises.

Article 114. Unitary Enterprise Based on Right of Economic Jurisdiction

1. A unitary enterprise based on the right of economic jurisdiction shall be created by decision of the State agency or agency of local self-government empowered to do so.

2. The constitutive document of an enterprise based on the right of economic jurisdiction shall be its charter, confirmed by the empowered State agency or agency of local self-government.

3. The amount of the charter fund of an enterprise based on the right of economic jurisdiction may not be less than the amount determined by the law on State and municipal unitary enterprises.

4. The procedure for forming the charter fund of the enterprise based on the right of economic jurisdiction or operative management shall be determined by the Law on State and Municipal Unitary Enterprises [as amended by Law of 14 November 2002. СЗ РФ (2002), no. 48, item 4746].

5. If at the end of the financial year the value of the net assets of an enterprise based on the right of economic jurisdiction proves to be less than the amount of the charter fund, the agency empowered to create such enterprises shall be obliged to reduce the charter fund in the established procedure. If the value of net assets becomes less than the amount determined by a law, the enterprise may be liquidated by decision of a court.

6. In the event of the adoption of a decision to reduce the charter fund, the enterprise shall be obliged in writing to inform its creditors thereof.

The creditor of the enterprise shall have the right to demand the termination or the performance before time of an obligation, the debtor with regard to which this enterprise is, and compensation of losses [repealed by Law of 14 November 2002. СЗ РФ (2002), no. 48, item 4746].

7. The owner of the property of an enterprise based on the right of economic jurisdiction shall not be liable for the obligations of the enterprise, except for instances provided for by Article 56(3) of the present Code. This rule also shall apply to the responsibility of the enterprise which founded the subsidiary enterprise with regard to the obligations of the last [renumbered by Law of 14 November 2002. СЗ РФ (2002), no. 48, item 4746].

Article 115. Unitary Enterprise Based on Right of Operative Management

1. In the instances and in the procedure provided for by the Law on State and Municipal Unitary Enterprises' a unitary enterprise may be created by right of operative management (treasury enterprise) on the base of State or municipal property.

2. The constitutive document of a treasury enterprise shall be its charter confirmed by an empowered State agency or agency of local self-government.

3. The firm name of a unitary enterprise based on the right of operative management must contain an indication that this enterprise is a treasury [enterprise].

4. The rights of a treasury enterprise to the property consolidated to it shall be determined in accordance with Articles 296 and 297 of the present Code and the Law on State and Municipal Unitary Enterprises.

5. The owner of the property of a treasury enterprise shall bear subsidiary responsibility for the obligations of such enterprise if its property is insufficient.

6. A treasury enterprise may be reorganised or liquidated in accordance with the Law on State and Municipal Enterprises [Article 115 in the version of the Law of 14 November 2002. СЗ РФ (2002), no. 48, item 4746].

§5. Noncommercial Organisations

Article 116. Consumer Cooperative

1. A voluntary association of citizens and juridical persons on the basis of membership for the purpose of satisfying material and other requirements of participants to be effectuated by means of combining the property share contributions of the members thereof shall be deemed to be a consumer cooperative.

2. The charter of a consumer cooperative must contain, besides the information specified in Article 52(2) of the present Code, conditions concerning the amount of share contributions of the members of the cooperative; the composition and procedure for the making of share contributions by members of the cooperative and their responsibility for a violation of the obligation relating to the making of the share contributions; the composition and competence of the management organs of the cooperative and the procedure for the adoption of decisions by them, including on questions, the decisions regarding which are adopted unanimously or by a qualified majority of votes; and the procedure for the covering by members of the cooperative of losses incurred by it.

3. The name of the consumer cooperative must contain an indication of the principal purpose of its activity, and also either the word 'cooperative' or the words 'consumer cooperative' or 'consumer society'.

4. The members of the consumer cooperative shall be obliged within three months after confirmation of the annual balance sheet to cover losses which have formed by means of additional contributions. In the event of the failure to fulfil this duty, the cooperative may be liquidated in a judicial proceeding at the demand of the creditors.

The members of a consumer cooperative shall bear jointly and severally subsidiary responsibility for its obligations within the limits of the uncontributed portion of the additional contribution of each member of the cooperative.

5. The revenues received by the consumer cooperative from entrepreneurial activity effectuated by the cooperative in accordance with a law and the charter shall be distributed among its members.

6. The legal status of consumer cooperatives, and also the rights and duties of their members, shall be determined in accordance with the present Code by the laws on consumer cooperative societies.

Article 117. Social and Religious Organisations (or Associations)

1. Social and religious organisations (or associations) shall be deemed to be voluntary associations of citizens who have combined in the procedure

established by law on the basis of a community of their interests in order to satisfy spiritual and other nonmaterial requirements.

Social and religious organisations shall be noncommercial organisations. They shall have the right to effectuate entrepreneurial activity only for the achievement of the purposes for which they were created and the purposes corresponding thereto.

2. The participants (or members) of social and religious organisations shall not retain rights to property transferred in ownership by them to such organisations, including membership dues. They shall not be liable for the obligations of social and religious organisations in which they participate as members, and the said organisations shall not be liable for the obligations of their members.

3. The peculiarities of the legal status of social and religious organisations as participants of relations regulated by the present Code shall be determined by a law.

Article 118. Foundations

1. For the purposes of the present Code a foundation shall be deemed to be a noncommercial organisation not having membership, founded by citizens and/or juridical persons on the basis of voluntary property contributions, and pursuing social, philanthropic, cultural, educational, and other socially-useful purposes.

The property transferred to the foundation by its founders shall be the ownership of the foundation. The founders shall not be liable for the obligations of the foundation created by them, and the foundation shall not be liable for the obligations of its founders.

2. The foundation shall use property for the purposes determined in its charter. The foundation shall have the right to engage in entrepreneurial activity necessary to attain the socially-useful purposes for which the foundation was created and corresponding to those purposes. In order to effectuate entrepreneurial activity foundations shall have the right to create economic societies or to participate in them.

The foundation shall be obliged annually to publish reports on the use of its property.

3. The procedure for the management of the foundation and the procedure for forming its organs shall be determined by its charter confirmed by the founders.

4. The charter of a foundation must, besides the information specified in Article 52(2) of the present Code, contain: the name of the foundation, including the word 'foundation'; information concerning the purposes of the foundation; indications about the foundation organs, including the trusteeship council effectuating supervision over the activity of the foundation; the procedure for the appointment of officials of the foundation and of relieving them; the location of

the foundation; and the fate of the property of the foundation in the event of its liquidation.

Article 119. Change of Charter and Liquidation of Foundation

1. The charter of a foundation may be changed by organs of the foundation if the possibility of its change has been provided for by the charter in such a procedure.

If retention of the charter in unchanged form entails consequences which were impossible to foresee when founding the foundation and the possibility of changing the charter has not been provided for therein or the charter is not changed by empowered persons, the right to make changes shall belong to a court upon the application of organs of the foundation or agency empowered to effectuate supervision over its activity.

2. A decision concerning liquidation of a foundation may be adopted only by a court upon the application of the interested persons.

A foundation may be liquidated:

(1) if the property of the foundation is insufficient for the effectuation of its purposes and the likelihood of receiving the necessary property is unrealistic;

(2) if the purposes of the foundation cannot be achieved and necessary changes in the purposes of the foundation cannot be made;

(3) in the event the foundation evades in its activity the purposes provided for by the charter;

(4) in other instances provided for by a law.

3. In the event of the liquidation of the foundation its property remaining after satisfaction of the demands of the creditors shall be directed to the purposes specified in the charter of the foundation.

Article 120. Institutions

1. An institution shall be deemed to be an organisation created by the owner in order to effectuate management, socio-cultural, or other functions of a non-commercial character and financed by it wholly or partially.

The rights of an institution to property consolidated to it shall be determined in accordance with Article 296 of the present Code.

2. An institution shall be liable for its obligations with the monetary means at its disposition. The owner of the respective property shall bear subsidiary responsibility for its obligations in the event of their insufficiency.

3. The peculiarities of the legal status of individual types of State and other institutions shall be determined by a law and other legal acts.

Article 121. Associations of Juridical Persons (Associations [*ассоциация*] and Unions)

1. Commercial organisations may for the purposes of coordinating their entrepreneurial activity, and also of representing and defending common property interests, under a contract between themselves create associations in the form of associations (or unions) which are noncommercial organisations.

If by decision of the participants the conducting of entrepreneurial activity is placed on an association (or union), such association (or union) shall be transformed into an economic society or partnership in the procedure provided for by the present Code, or may create an economic society in order to effectuate entrepreneurial activity or participate in such society.

2. Social and other noncommercial organisations, including institutions, may voluntarily unite into associations (or unions) of these organisations.

An association (or union) of noncommercial organisations shall be a noncommercial organisation.

3. Members of an association (or union) shall retain their autonomy and rights of a juridical person.

4. An association (or union) shall not be liable for the obligations of its members. Members of an association (or union) shall bear subsidiary responsibility for its obligations in the amount and in the procedure provided for by the constitutive documents of the association.

5. The name of an association (or union) must contain an indication of the principal subject of the activity of its members with inclusion of the word 'association' or 'union'.

Article 122. Constitutive Documents of Associations and Unions

1. The constitutive documents of an association (or union) shall be the constitutive contract signed by its members and the charter confirmed by them.

2. The constitutive documents of an association must contain, besides the information specified in Article 52(2) of the present Code, conditions concerning the composition and competence of management organs of the association (or union) and the procedure for the adoption of decisions by them, including questions, the decisions with regard to which shall be adopted unanimously or by a qualified majority of votes of members of the association (or union); and the procedure for the distribution of property remaining after liquidation of the association (or union).

Article 123. Rights and Duties of Members of Associations and Unions

1. The members of an association (or union) shall have the right to use its services free of charge.

2. The member of an association (or union) shall have the right at its discretion to withdraw from the association (or union) at the end of the financial year. In this event it shall bear subsidiary responsibility for obligations of the association (or union) in proportion to its contribution for two years from the moment of withdrawal.

The member of an association (or union) may be expelled therefrom by decision of the remaining participants in the instances and in the procedure established by the constitutive documents of the association (or union). Rules relating to withdrawal from an association (or union) shall apply with respect to the responsibility of the expelled member of the association (or union).

3. With the consent of the members of the association (or union), a new participant may join. The joining of an association (or union) by a new member may be conditioned by its subsidiary responsibility for obligations of the association (or union) which arose before its joining.

Chapter 5. Participation of the Russian Federation, Subjects of the Russian Federation, and Municipal Formations in Relations Regulated by Civil Legislation

Article 124. Russian Federation, Subjects of Russian Federation, and Municipal Formations—Subjects of Civil Law

1. The Russian Federation, subjects of the Russian Federation—republics, territories, regions, cities of federal significance, autonomous region, autonomous national areas, and also city and rural settlements and other municipal formations shall act in relations regulated by civil legislation on equal principles with other participants of these relations—citizens and juridical persons.

2. Unless it arises otherwise from a law or the peculiarities of the particular subjects, the norms determining the participation of juridical persons in relations regulated by civil legislation shall apply to the subjects of civil law specified in point 1 of the present Article.

Article 125. Procedure for Participation of Russian Federation, Subjects of the Russian Federation, and Municipal Formations in Relations Regulated by Civil Legislation

1. Agencies of State power within the frameworks of their competence established by acts determining the status of these agencies may by their actions acquire and effectuate property and personal nonproperty rights and duties and act in court in the name of the Russian Federation and of subjects of the Russian Federation.

2. Agencies of local self-government within the frameworks of their competence established by acts determining the status of these agencies may by their actions acquire and effectuate the rights and duties specified in point 1 of the present Article in the name of the municipal formations.

3. In the instances and in the procedure provided for by federal laws, edicts of the President of the Russian Federation, and decrees of the Government of the Russian Federation, and normative acts of subjects of the Russian Federation and municipal formations, State agencies, agencies of local self-government, and also juridical persons, and citizens may act upon their special commission in their names.

Article 126. Responsibility for Obligations of Russian Federation, Subject of Russian Federation, and Municipal Formation

1. The Russian Federation, subject of the Russian Federation, and municipal formation shall be liable for its obligations with the property belonging to it by right of ownership, except for property which was consolidated by it to juridical persons created by them by right of economic jurisdiction or operative management, and also property which may be only in State or municipal ownership.

The levy of execution on land and other natural resources in State or municipal ownership shall be permitted in the instances provided for by a law.

2. Juridical persons created by the Russian Federation, subjects of the Russian Federation, or municipal formations shall not be liable for their obligations.

3. The Russian Federation, subjects of the Russian Federation, and municipal formations shall not be liable for obligations of juridical persons created by them except for instances provided for by a law.

4. The Russian Federation shall not be liable for obligations of subjects of the Federation and municipal formations.

5. Subjects of the Russian Federation and municipal formations shall not be liable for obligations of one another, nor for obligations of the Russian Federation.

6. The rules of points 2–5 of the present Article shall not extend to instances when the Russian Federation has assumed a guarantee (or suretyship) for the obligations of a subject of the Russian Federation, municipal formation, or juridical person, and the said subjects have undertaken a guarantee (or suretyship) with regard to obligations of the Russian Federation.

Article 127. Peculiarities of Responsibility of Russian Federation and Subjects of Russian Federation in Relations Being Regulated by Civil Legislation with Participation of Foreign Juridical Persons, Citizens, and States

The peculiarities of the responsibility of the Russian Federation and subjects of the Russian Federation in relations regulated by civil legislation with the participation of foreign juridical persons, citizens, and States shall be determined by a law on the immunity of the State and its ownership.

Subsection 3. Objects of Civil Rights

Chapter 6. General Provisions

Article 128. Types of Objects of Civil Rights

To objects of civil rights shall be relegated things, including money and securities, other property, including property rights; work and services; information; the results of intellectual activity, including exclusive rights thereto (intellectual property); and nonmaterial benefits.

Article 129. Circulability of Objects of Civil Rights

1. Objects of civil rights may be freely alienated or be transferred from one person to another by way of universal legal succession (inheritance, reorganisation of juridical person) or by other means unless they have been withdrawn from turnover or are limited in turnover.

2. Types of objects of civil rights whose being in turnover is not permitted (objects withdrawn from turnover) must be expressly specified in a law.

Types of objects of civil rights which may belong only to determined participants of turnover or whose being in turnover is permitted by a special authorisation (objects of limited circulability) shall be determined in the procedure established by a law.

3. Land and other natural resources may be alienated or pass from one person to another by other means to the extent that their turnover is permitted by laws on land and other natural resources.

Article 130. Immovable and Movable Things

1. To immovable things (immovable property, immovable) shall be relegated land plots, subsoil plots, solitary water objects, and all that is firmly connected with the land, that is, objects whose movement without incommensurate damage to the purpose thereof is impossible, including forests, perennial plantings, buildings, and installations.

To immovable things also shall be relegated aircraft and sea-going vessels subject to State registration, vessels of internal navigation, and space objects. Other property also may be relegated by a law to immovable things.

2. Things which are not relegated to immovable, including money and securities, shall be deemed to be movable property. The registration of rights to movable things shall not be required except for instances specified in a law.

Article 131. State Registration of Immovable

1. The right of ownership and other rights to a thing in immovable things, limitations of these rights, the origin thereof, transfer, and termination shall be subject to State registration in a unified State register by justice institutions. There shall be subject to registration: right of ownership, right of economic jurisdiction, right of operative management, right of inheritable possession for life, right of permanent use, mortgage, servitudes, and also other rights in the instances provided for by the present Code and other laws.

2. In the instances provided for by a law a special registration or recording of individual types of immovable property may be effectuated together with the State registration.

3. The agency effectuating State registration of rights to an immovable and transactions with it shall be obliged upon the petition of the rightholder to certify the registration made by means of the issuance of a document concerning the registered right or transaction or by making inscriptions on the document submitted for registration.

4. The agency effectuating State registration of rights to an immovable and transactions with it shall be obliged to provide information concerning the registration made and the rights registered to any person.

The information shall be provided in any agency effectuating the registration of an immovable irrespective of the place of performing the registration.

5. A refusal of State registration of the right to an immovable or transaction with it or the evading of registration by the respective agency may be appealed to a court.

6. The procedure for State registration and the grounds for refusal of registration shall be established in accordance with the present Code by the law on registration of the rights to immovable property and transactions with it.

Article 132. Enterprise

1. An enterprise as an object of rights shall be deemed to be a property complex used to effectuate entrepreneurial activity.

An enterprise as a whole, as a property complex, shall be deemed to be an immovable.

2. An enterprise as a whole or part thereof may be the object of a purchase-sale, pledge, lease, and other transactions connected with the establishment, change, and termination of rights to a thing.

Within an enterprise as a property complex shall be all types of property designated for its activity, including land plots, buildings, installations, equipment, tools, raw material, products, rights of demand, debts, and also rights to designation which individualise the enterprise, its products, work, and services (firm name, trademarks, service marks), and other exclusive rights, unless provided otherwise by a law or contract.

Article 133. Indivisible Things

A thing, the division of which in kind is impossible without changing the purpose thereof, shall be deemed to be indivisible.

The peculiarities of the partition of a participatory share in the right of ownership to an indivisible thing shall be determined by Articles 252 and 258 of the present Code.

Article 134. Complex Things

If various kinds of thing form a single whole presupposing the use thereof for a common purpose, they shall be considered to be one thing (complex thing).

The operation of a transaction concluded with regard to a complex thing shall extend to all of its constituent parts unless provided otherwise by a contract.

Article 135. Principal Thing and Appurtenance

A thing designated to serve another, principal thing and connected therewith by a common purpose (appurtenance) shall follow the fate of the principal thing unless provided otherwise by contract.

Article 136. Fruits, Products, and Revenues

Proceeds received as a result of the use of property (fruits, product, revenues) shall belong to the person using this property on legal grounds unless provided otherwise by a law, other legal acts, or by the contract concerning the use of this property.

Article 137. Fauna

To fauna shall apply the general rules on property insofar as not established otherwise by a law or other legal acts.

When effectuating rights the cruel treatment of fauna contrary to the principles of humaneness shall not be permitted.

Article 138. Intellectual Property

In the instances and in the procedure established by the present Code and by other laws an exclusive right (intellectual property) of a citizen or juridical person shall be recognised to the results of intellectual activity and the means of individualisation of the juridical person equated to them or the individualisation of a product or the work fulfilled or services (firm name, trademark, service mark, and others).

The use of the results of intellectual activity and means of individualisation which are the object of exclusive rights (intellectual property) may be effectuated by third persons only with the consent of the possessor of the right.

Article 139. Employment and Commercial Secret

1. Information shall constitute an employment or commercial secret when the information has real or potential commercial value by virtue of its being unknown to third persons, there is no free access to it on legal grounds, and the possessor of the information takes measures to protect its confidentiality. Information which cannot constitute an employment or commercial secret shall be determined by a law and other legal acts.

2. Information constituting an employment or commercial secret shall be defended by the means provided for by the present Code and other laws.

Persons who have received information which constitutes an employment or commercial secret by illegal methods shall be obliged to compensate the losses caused. The same duty shall be placed on workers who, contrary to the labour contract, including a *kontrakt*, have divulged an employment or commercial secret and on contracting parties who have done so despite a civil-law contract.

Article 140. Money (Currency)

1. The ruble shall be the legal means of payment obligatory for acceptance at face value throughout the entire territory of the Russian Federation.

Payments on the territory of the Russian Federation shall be effectuated in the form of cash and noncash settlement accounts.

2. The instances, procedure, and conditions for the use of foreign currency on the territory of the Russian Federation shall be determined by a law or in the procedure established by it.

Article 141. Currency Valuables

The types of property deemed to be currency valuables and the procedure for concluding transactions with them shall be determined by the law on currency regulation and currency control.

The right of ownership in currency valuables shall be defended in the Russian Federation on the general grounds.

Chapter 7. Securities

Article 142. Security

1. A security shall be a document certifying, in compliance with the established form and obligatory requisites, property rights whose effectuation or transfer shall be possible only when presenting it.

With the transfer of the security shall pass all rights in aggregate which are certified by it.

2. In the instances provided for by a law or in the procedure established by it in order to effectuate and transfer the rights certified by a security evidence of its being consolidated in a special register (ordinary or computerised) shall be sufficient.

Article 143. Types of Securities

There shall be relegated to securities: State bond, bond, bill of exchange, cheque, deposit and savings certificates, bank bearer savings book, bill of lading, stock, privatisation securities, and other documents which have been relegated to securities by the laws on securities or in the procedure established by them.

Article 144. Requirements for Security

1. The types of rights which shall be certified by securities, the obligatory requisites of securities, the requirements for the form of a security, and other necessary requirements shall be determined by a law or in the procedure established by it.

2. The absence of obligatory requisites of a security or the failure of the security to conform to the form established for it shall entail the nullity thereof.

Article 145. Subjects of Rights Certified by Security

1. Rights certified by a security may belong to:
(1) the bearer of the security (bearer security), or
(2) the person named on the security (inscribed security), or
(3) the person named on the security, who may himself effectuate these rights or appoint by his instruction (order) another empowered person (order security).

2. The possibility of the issuance of securities of a specified type as inscribed, or as order, or as bearer securities may be excluded by a law.

Article 146. Transfer of Rights Relating to Security

1. In order to transfer to another person the rights certified by a bearer security it shall be sufficient to hand over the security to this person.

2. The rights certified by an inscribed security shall be transferred in the procedure established for the assignment of demands (cession). In accordance with Article 390 of the present Code a person who has transferred the right relating to a security shall bear responsibility only for the invalidity of the respective demand, but not for the failure to perform it.

3. The rights relating to an order security shall be transferred by means of making an inscription of transfer on this security—an endorsement. The endorser shall bear responsibility not only for the existence of the right, but also for the effectuation thereof.

An endorsement made on a security shall pass all the rights certified by the security to the person to whom or to the order of whom the rights relating to the security are transferred—the endorsee. An endorsement may be in blank (without specifying the person to whom performance must be made) or to order (specifying the person to whom or to the order of whom performance must be made).

An endorsement may be limited only to a commission to effectuate the rights certified by the security without the transfer of these rights to the endorsee (endorsement of entrustment). In this event the endorsee shall act as a representative.

Article 147. Performance Relating to Security

1. The person who has issued a security and all the persons who have endorsed it shall be liable to the legal possessor thereof jointly and severally. In the event of the satisfaction of the demand of the legal possessor of the security concerning the performance of the obligation certified by it by one or several persons from among those who are obliged before him according to the security, they shall acquire the right of a demand for indemnification (regression) against the other persons who are obliged with regard to the security.

2. A refusal to perform an obligation certified by a security by referring to the absence of grounds of the obligation or to its invalidity shall not be permitted.

The possessor of a security who has discovered a forgery or a counterfeit security shall have the right to present to the person who transferred the security to him a demand for proper performance of the obligation certified by the security and for compensation of losses.

Article 148. Reinstatement of Security

The rights relating to lost bearer securities and order securities shall be reinstated by a court in the procedure provided for by procedural legislation.

Article 149. Paperless Securities

1. In the instances determined by a law or in the procedure established by it, a person who has received a special license may fix the rights consolidated by an inscribed or order security, including in paperless form (with the assistance of electronic computer technology, and the like). The rules established for securities shall apply to this form of fixation of rights unless it arises otherwise from the peculiarities of the fixation.

The person who has effectuated the fixation of the right in paperless form shall be obliged upon the demand of the possessor of the right to issue a document to him certifying the consolidated right.

The rights certified by means of the said fixation, the procedure for the official fixation of rights and rightholders, the procedure for the documentary confirmation of entries, and the procedure for performing operations with paperless securities shall be determined by a law or in the procedure established by it.

2. Operations with paperless securities may be performed only by having recourse to the person who has officially performed the entry of the rights. The

transfer, granting, and limitation of rights must be officially fixed by this person, who shall bear responsibility for the preservation of official entries, ensuring their confidentiality, the submission of correct data concerning such entries, and the performance of official entries concerning operations carried on.

Chapter 8. *Nonmaterial Benefits and the Defence Thereof*

Article 150. Nonmaterial Benefits

1. Life and health, the dignity of the person, personal inviolability, honour and good name, business reputation, inviolability of private life, personal and family secrecy, the right of free movement, and choice of place of sojourn and residence, the right to name, the right of authorship, other personal nonproperty rights and other nonmaterial benefits which belong to a citizen from birth or by virtue of a law shall be inalienable and not transferable by other means. In the instances and in the procedure provided for by a law, personal nonproperty rights and other nonmaterial benefits which belonged to a deceased person may be effectuated and defended by other persons, including heirs of the possessor of the right.

2. Nonmaterial benefits shall be defended in accordance with the present Code and other laws in the instances and procedure provided for by them, and also in those instances and within those limits in which the use of the means of the defence of civil rights (Article 12) arises from the essence of the violated non-material right and the character of the consequences of this violation.

Article 151. Contributory Compensation of Moral Harm

If moral harm has been caused to a citizen (physical or moral suffering) by actions violating his personal nonproperty rights or infringing on other non-material benefits belonging to a citizen, and also in other instances provided for by a law, the court may impose on the offender the duty of monetary contributory compensation of the said harm.

When determining the amounts of contributory compensation of moral harm the court shall take into account the degree of fault of the offender and other cir-cumstances deserving of attention. The court must also take into account the extent of physical and moral suffering connected with the individual peculiarities of the person to whom harm was caused.

Article 152. Defence of Honour, Dignity, and Business Reputation

1. A citizen shall have the right to demand through a court the refutation of information defaming his honour, dignity, or business reputation, unless the dis-seminator of such information proves that it corresponds to reality.

Upon the demand of interested persons the defence of the honour and dignity of a citizen shall also be permitted after his death.

2. If information defaming the honour, dignity, or business reputation of the citizen has been disseminated in the mass media, it must be refuted in the same mass media.

If the said information is contained in a document emanating from an organisation, the document shall be subject to replacement or recall.

The procedure for refutation in other instances shall be established by a court.

3. A citizen with respect to whom the mass media have published information impinging upon his rights or interests protected by a law shall have the right to publication of his reply in the same mass media.

4. If the decision of a court has not been fulfilled, the court shall have the right to impose a fine on the offender to be recovered in the amount and in the procedure provided for by procedural legislation to the revenue of the Russian Federation. The payment of the fine shall not relieve the offender from the duty to fulfil the action provided for by the court decision.

5. A citizen with respect to whom information has been disseminated which defames his honour, dignity or business reputation shall have the right, together with the refutation of such information, to demand compensation of losses and moral harm caused by the dissemination thereof.

6. If it is impossible to establish the person who has disseminated information defaming the honour, dignity, or business reputation of a citizen, the person with respect to whom such information has been disseminated shall have the right to apply to a court with a statement concerning recognition that the information being disseminated does not correspond to reality.

7. The rules of the present Article concerning the defence of business reputation of a citizen respectively shall apply to the defence of the business reputation of a juridical person.

Subsection 4. Transactions and Representation

Chapter 9. Transactions

§1. Concept, Types, and Form of Transactions

Article 153. Concept of Transaction

The actions of citizens and juridical persons directed towards the establishment, change, or termination of civil rights and duties shall be deemed to be transactions.

Article 154. Contracts and Unilateral Transactions

1. Transactions may be bilateral or multilateral (contracts), and unilateral.

2. A transaction for the conclusion of which in accordance with a law, other legal acts, or by agreement of the parties the expression of the will of one party is necessary and sufficient shall be considered to be unilateral.

3. The expression of the concordant will of two parties (bilateral transaction), or three or more parties (multilateral transaction) shall be necessary for the conclusion of a contract.

Article 155. Duties Regarding Unilateral Transaction

A unilateral transaction shall create duties for the person who has concluded the transaction. It may create duties for other persons only in the instances established by a law or by agreement with those persons.

Article 156. Legal Regulation of Unilateral Transactions

The general provisions on obligations and on contracts shall apply to unilateral transactions respectively insofar as this is not contrary to a law or to the unilateral character and essence of the transaction.

Article 157. Transactions Concluded Under a Condition

1. A transaction shall be considered to be concluded under a condition subsequent if the parties have made the arising of rights and duties dependent upon a circumstance relative to which it is unknown as to whether this will ensue or not.

2. A transaction shall be considered to be concluded under a condition precedent if the parties have made the termination of rights and duties dependent upon a circumstance relative to which it is unknown whether this will ensue or not.

3. If the ensuing of the condition is obstructed by a party not in good faith for whom the ensuing of the condition is disadvantageous, the condition shall be deemed to have ensued.

If the ensuing of the condition is facilitated by a party not in good faith for whom the ensuing of the condition is advantageous, the condition shall be deemed not to have ensued.

Article 158. Form of Transactions

1. Transactions shall be concluded orally or in written form (simple or notarial).

2. A transaction which may be concluded orally shall be considered to be concluded also when from the behaviour of the person his will to conclude the transaction is manifest.

3. Silence shall be deemed to be an expression of will to conclude a transaction in the instances provided for by a law or by agreement of the parties.

Article 159. Oral Transactions

1. A transaction for which the written (simple or notarial) form has not been established by a law or by agreement of the parties may be concluded orally.

2. Unless otherwise established by agreement of the parties, all transactions to be performed by those who concluded them themselves may be concluded orally, except for transactions for which the notarial form has been established and transactions the failure to comply with the simple written form of which entails their invalidity.

3. Transactions in performance of a contract concluded in written form may, by agreement of the parties, be concluded orally unless this is contrary to a law, other legal acts, and a contract.

Article 160. Written Form of Transaction

1. A transaction in written form must be concluded by means of drawing up a document reflecting the content thereof and signed by the person or persons concluding the transaction, or by persons duly empowered by them.

Bilateral (or multilateral) transactions may be concluded by the means established by Article 434(2) and (3) of the present Code.

Additional requirements to which the form of the transaction must conform may be established by a law, other legal acts, and by agreement of the parties (conclusion on a letterhead of a specified form; affixing of seal, and others), and consequences provided for the failure to comply with these requirements. If such consequences have not been provided, the consequences of the failure to comply with the simple written form of a transaction shall apply (Article 162[1]).

2. The use when concluding a transaction of a facsimile reproduction of a signature with the assistance of mechanical or other means of copying, electronic-cypher signature, or other analogue of a signature in one's own hand shall be permitted in the instances and procedure provided for by a law, other legal acts, or by agreement of the parties.

3. If a citizen as a consequence of physical defect, illness, or illiteracy cannot sign in his own hand, then at his request another citizen may sign the transaction. The signature of the last must be certified by a notary or by another official having the right to perform such a notarial action, specifying the reasons by virtue of which the person concluding the transaction could not sign it in his own hand.

However, when concluding the transactions specified in Article 185(4) of the present Code and powers of attorney to conclude them, the signature of he who signs the transaction may be certified also by the organisation in which the citizen who cannot sign in his own hand works or by the administration of the inpatient treatment institution in which he is situated for care.

Article 161. Transactions Concluded in Simple Written Form

1. There must be concluded in simple written form, except for transactions requiring notarial certification:
   (1) transactions of juridical persons between themselves and with citizens;
   (2) transactions of citizens between themselves for an amount exceeding not less than ten times the minimum amount of payment of labour, and in the instances provided for by a law, irrespective of the amount of the transaction.

2. Compliance with the simple written form shall not be required for transactions which in accordance with Article 159 of the present Code may be concluded orally.

Article 162. Consequences of Failure to Comply with Simple Written Form of Transaction

1. The failure to comply with the simple written form of a transaction shall deprive the parties of the right in the event of a dispute to refer in confirmation of the transaction and its conditions to witness testimony, but shall not deprive them of the right also to cite written and other evidence.

2. In the instances expressly specified in a law or in the agreement of the parties the failure to comply with the simple written form of a transaction shall entail its invalidity.

3. The failure to comply with the simple written form of a foreign economic transaction shall entail the invalidity of the transaction.

Article 163. Notarially Certified Transactions

1. The notarial certification of a transaction shall be effectuated by means of the performing on the document of an endorsement of certification corresponding to the requirements of Article 160 of the present Code by a notary or other official having the right to perform such a notarial action.

2. Notarial certification of transactions shall be obligatory:
   (1) in the instances specified in a law;
   (2) in the instances provided for by agreement of the parties, although according to a law for transactions of the particular type this form is not required.

Article 164. State Registration of Transactions

1. Transactions with land and other immovable property shall be subject to State registration in the instances and in the procedure provided for by Article 131 of the present Code and by the law on the registration of rights to immovable property and transactions with it.

2. State registration of transactions with movable property of determined types may be established by a law.

Article 165. Consequences of Failure to Comply with Notarial Form of Transaction and Requirements for Registration Thereof

1. The failure to comply with the notarial form, and in the instances established by a law, the requirements concerning State registration of a transaction shall entail its invalidity. Such a transaction shall be considered to be void.

2. If one of the parties wholly or partially has performed a transaction requiring notarial certification, and the other party has evaded such certification of the transaction, the court shall have the right at the demand of the party who performed the transaction to deem the transaction to be valid. In this event subsequent notarial certification of the transaction shall not be required.

3. If a transaction requiring State registration has been concluded in the proper form but one of the parties evades the registration thereof, a court shall have the right at the demand of the other party to render a decision concerning registration of the transaction. In this event the transaction shall be registered in accordance with the decision of the court.

4. In the instances provided for by points 2 and 3 of the present Article the party who unjustifiably evades notarial certification or State registration of a transaction must compensate the other party for losses caused by the delay in concluding or registering the transaction.

§2. Invalidity of Transactions

Article 166. Contested and Void Transactions

1. A transaction shall be invalid on the grounds established by the present Code by virtue of being deemed such by a court (contested transaction) or irrespective of such deeming (void transaction).

2. A demand to deem a contested transaction to be invalid may be brought by the persons specified in the present Code.

A demand concerning the application of the consequences of the invalidity of a void transaction may be presented by any interested person. The court shall have the right to apply such consequences at its own initiative.

Article 167. General Provisions on Consequences of Invalidity of Transaction

1. An invalid transaction shall not entail legal consequences, except for those which are connected with its invalidity, and shall be invalid from the moment of its conclusion.

2. In the event of the invalidity of a transaction, each of the parties shall be obliged to return to the other everything received according to the transaction, and if it is impossible to return that received in kind (including when that received is expressed in the use of property, work fulfilled, or service provided), to compensate its value in money, unless other consequences of the invalidity of the transaction have been provided for by a law.

3. If it follows from the content of a contested transaction that it may be only terminated at a future time, the court deeming the transaction to be invalid shall terminate its operation at the future time.

Article 168. Invalidity of Transaction Not Conforming to a Law or Other Legal Acts

A transaction not corresponding to the requirements of a law or other legal acts shall be void unless the law establishes that such a transaction is contestable or provides other consequences for the violation.

Article 169. Invalidity of Transaction Concluded for Purpose Contrary to Fundamental Principles of Legal Order and Morality

A transaction concluded for a purpose knowingly contrary to the fundamental principles of legal order or morality shall be void.

When both parties to such a transaction have intent—in the event of performance of the transaction by both parties—everything received by them under the transaction shall be recovered to the revenue of the Russian Federation, and in the event of the performance of the transaction by one party, everything received shall be recovered for the revenue of the Russian Federation from the other party and everything due from it to the first party in compensation of that received.

When only one party of such a transaction has intent, everything received by it under the transaction must be returned to the other party, and everything received by the last or due to it in compensation of that performed shall be recovered to the revenue of the Russian Federation.

Article 170. Invalidity of Fictitious and Sham Transactions

1. A fictitious transaction, that is, a transaction concluded only for form, without the intention to create legal consequences corresponding to it, shall be void.

2. A sham transaction, that is, a transaction which is concluded for the purpose of concealing another transaction, shall be void. To the transaction which the parties actually had in view, taking into account the essence of the transaction, shall apply the rules relevant thereto.

Article 171. Invalidity of Transaction Concluded by Citizen Deemed to Lack Dispositive Legal Capacity

1. A transaction concluded by a citizen deemed to lack dispositive legal capacity as a consequence of mental disturbance shall be void.

Each of the parties to such a transaction shall be obliged to return to the other everything received in kind, and if it is impossible to return that received in kind, to compensate its value in money.

The party having dispositive legal capacity shall be obliged, in addition, to compensate to the other party the real damage incurred by it if the party having dispositive legal capacity knew or should have known about the lack of dispositive legal capacity of the other party.

2. In the interests of a citizen deemed to lack dispositive legal capacity as a consequence of mental disturbance, the transaction concluded by him may, at the demand of his trustee, be deemed by a court to be valid if it was concluded to the advantage of this citizen.

Article 172. Invalidity of Transaction Concluded by Minor Who Has Not Attained Fourteen Years of Age

1. A transaction concluded by a minor who has not attained fourteen years of age (youth) shall be void. To such transaction shall apply the rules provided for by Article 171(1), paragraphs two and three, of the present Code.

2. In the interests of the youth, a transaction concluded by him may at the demand of his parents, adoptive parents, or trustee be deemed by a court to be valid if it was concluded to the advantage of the youth.

3. The rules of the present Article shall not extend to petty domestic and other transactions of youth which they have the right to conclude autonomously in accordance with Article 28 of the present Code.

Article 173. Invalidity of Transaction of Juridical Person Exceeding the Limits of Its Legal Capacity

A transaction concluded by a juridical person which is contrary to the purposes of the activity specifically limited in its constitutive documents, or by a juridical person not having a license to engage in the respective activity, may be deemed by a court to be invalid upon the suit of this juridical person, its founder (or participant), or the State agency effectuating control or supervision over the activity of the juridical person if it is proved that the other party to the transaction knew or knowingly should have known about the illegality thereof.

Article 174. Consequences of Limitation of Powers to Conclude Transaction

If the powers of a person to conclude a transaction have been limited by a contract or the powers of the organ of a juridical person, by its constitutive documents, in comparison with those as determined in a power of attorney or in a law, or which may be considered to be obvious from the situation in which the transaction was concluded, and when concluding it such person or organ exceeded the limits of these limitations, the transaction may be deemed by a court to be invalid upon the suit of the person in whose interests the limitations were established only in instances when it is proved that the other party to the transaction knew or knowingly should have known about the said limitations.

Article 175. Invalidity of Transaction Concluded by Minor from Fourteen to Eighteen Years of Age

1. A transaction concluded by a minor from fourteen to eighteen years of age without the consent of his parents, adoptive parents, or guardian in the instances when such consent is required in accordance with Article 26 of the present Code may be deemed by a court to be invalid upon the suit of the parents, adoptive parents, or guardian.

If such transaction is deemed to be invalid, the rules provided for by Article 171(1), paragraphs two and three, of the present Code shall apply respectively.

2. The rules of the present Article shall not extend to transactions of minors who have come to have full dispositive legal capacity.

Article 176. Invalidity of Transaction Concluded by Citizen Limited by Court in Dispositive Legal Capacity

1. A transaction relating to the disposition of property concluded without the consent of the guardian by a citizen limited by a court in dispositive legal capacity as a consequence of the abuse of alcoholic beverages or narcotic means may be deemed by a court to be invalid upon the suit of the guardian.

If such transaction has been deemed to be invalid, the rules provided for by Article 171(1), paragraphs two and three, of the present Code shall apply respectively.

2. The rules of the present Article shall not extend to petty domestic transactions which a citizen limited in dispositive legal capacity has the right to conclude autonomously in accordance with Article 30 of the present Code.

Article 177. Invalidity of Transaction Concluded by Citizen Not Capable of Understanding the Significance of His Actions or Guiding Them

1. A transaction concluded by a citizen, although having dispositive legal capacity but at the moment of concluding it was in such a state that he was not capable of understanding the significance of his actions or guiding them, may be deemed by a court to be invalid upon the suit of this citizen or other persons whose rights or interests protected by law have been violated as a result of the conclusion thereof.

2. A transaction concluded by a citizen who is subsequently deemed to lack dispositive legal capacity may be deemed by a court to be invalid upon the suit of his trustee if it is proved that at the moment of concluding the transaction the citizen was not capable of understanding the significance of his actions or guiding them.

3. If a transaction is deemed to be invalid on the grounds of the present Article, the rules provided for by Article 171(1), paragraphs two and three, of the present Code shall apply respectively.

Article 178. Invalidity of Transaction Concluded Under Influence of Delusion

1. A transaction concluded under the influence of delusion having material significance may be deemed to be invalid by a court upon the suit of the party who acted under the influence of delusion.

Delusion relative to the nature of the transaction, the identity or such qualities of its subject which significantly reduce the possibility of using it for its purpose shall have material significance. Delusion relative to the motives for the transaction shall not have material significance.

2. If the transaction is deemed to be invalid as concluded under the influence of delusion, the rules provided for by Article 167(2) of the present Code shall apply respectively.

In addition, the party at whose suit the transaction was deemed to be invalid shall have the right to demand from the other party compensation for real damage caused to it if it is proved that the delusion arose through the fault of the other party. If this is not proved, the party at whose suit the transaction was deemed to be invalid shall be obliged to compensate the other party at its demand for real damage caused to it even if the delusion arose through circumstances not dependent upon the deluded party.

Article 179. Invalidity of Transaction Concluded Under Influence of Fraud, Coercion, Threat, or Ill-Intentioned Agreement of Representative of One Party with Other Party or Confluence of Grave Circumstances

1. A transaction concluded under the influence of fraud, coercion, threat, or ill-intentioned agreement of a representative of one party with the other party, and also a transaction which a person was forced to conclude as a consequence of the confluence of grave circumstances on conditions extremely disadvantageous for himself which the other party took advantage of (cabalistic transaction) may be deemed by a court to be invalid upon the suit of the victim.

2. If the transaction was deemed invalid on one of the grounds specified in point 1 of the present Article, then the other party shall return to the victim everything received by it under the transaction, and if it is impossible to return everything received in kind, the value thereof shall be compensated in money. Property received under a transaction by the victim from the other party, and also that due it in compensation transferred to the other party, shall go to the revenue of the Russian Federation. If it is impossible to transfer the property to the revenue of the State in kind the value thereof in money shall be recovered. In addition, the victim shall be compensated by the other party for real damage caused to it.

Article 180. Consequences of Invalidity of Part of Transaction

The invalidity of part of the transaction shall not entail the invalidity of its other parts if it is possible to suppose that the transaction would have been concluded also without including the invalid part thereof.

Article 181. Periods of Limitations Regarding Invalid Transactions

1. A suit concerning the application of the consequences of the invalidity of a void transaction may be brought within ten years from the date when the performance thereof commenced.

2. A suit to deem a contested transaction to be invalid and concerning the application of the consequences of its invalidity may be brought within a year from the date of the termination of the coercion or threat under whose influence the transaction was concluded (Article 179[1]), or from the date when the plaintiff knew or should have known about other circumstances which are the grounds for deeming the transaction to be invalid.

Chapter 10. Representation. Power of Attorney

Article 182. Representation

1. A transaction concluded by one person (representative) in the name of another person (person represented) by virtue of a power based on a power

of attorney, specification of a law, or act of an empowered State agency or agency of local self-government shall directly create, change, and terminate civil rights and duties of the person represented.

A power also may be manifest from the situation in which the representative acts (seller in retail trade, cashier, and others).

2. Persons acting, although in the interests of another but in their own name (commercial intermediaries, bankruptcy administrators, executors in the event of inheritance, and others), and also persons empowered to enter into negotiations relative to possible future transactions, shall not be representatives.

3. A representative may not conclude transactions in the name of the person represented with respect to himself personally. He also may not conclude such transactions with respect to another person whose representative he is simultaneously, except for instances of commercial representation.

4. The conclusion through a representative of a transaction which by its character may be concluded only personally, and likewise other transactions specified in a law, shall not be permitted.

Article 183. Conclusion of Transaction by Unempowered Person

1. In the absence of powers to act in the name of another person or in the event of exceeding such powers, the transaction shall be considered to be concluded in the name of and in the interests of the person who concluded it unless the other person (person represented) subsequently approves expressly the particular transaction.

2. Subsequent approval of a transaction by the person represented shall create, change, and terminate civil rights and duties for him with regard to the particular transaction from the moment of its conclusion.

Article 184. Commercial Representation

1. A person who permanently and autonomously is representing in the name of entrepreneurs when they conclude contracts in the sphere of entrepreneurial activity shall be a commercial representative.

2. The simultaneous commercial representation of various parties in a transaction shall be permitted with the consent of these parties and in other instances provided for by a law. In so doing the commercial representative shall be obliged to perform the commissions given to him with the care of an ordinary entrepreneur.

A commercial representative shall have the right to demand the payment of stipulated remuneration and compensation for costs incurred by him when performing the commission from the parties to the contract in equal participatory shares unless provided otherwise by agreement between them.

3. A commercial representation shall be effectuated on the basis of a contract concluded in written form and containing an indication of the powers of the representative, and in the absence of such indications, also a power of attorney.

A commercial representative shall be obliged to preserve the secrecy of the information made known to him concerning trade transactions also after the performance of the commission given to him.

4. The peculiarities of commercial representation in individual spheres of entrepreneurial activity shall be established by a law and other legal acts.

Article 185. Power of Attorney

1. A power of attorney shall be deemed to be a written power issued by one person to another person for representation to third persons. The written power to conclude a transaction by a representative may be presented by the person represented directly to the respective third person.

2. A power of attorney to conclude transactions requiring the notarial form must be notarially certified, except for instances provided for by a law.

3. There shall be equated to notarially certified powers of attorney:
(1) powers of attorney of military servicemen and other persons being treated in military hospitals, sanatoriums, and other military treatment institutions certified by the head of such institution, his deputy for medical affairs, and the senior or duty doctor;
(2) powers of attorney of military servicemen, and in centres for the stationing of military units, formations, institutions, and military training institutions where there are no notarial offices nor other agencies which perform notarial actions, also the powers of attorney of workers and employees, members of their families and members of the families of military servicemen certified by the commander (or head) of his unit, formation, institution, or educational institution;
(3) powers of attorney of persons in places of deprivation of freedom certified by the head of the respective place of deprivation of freedom;
(4) powers of attorney of citizens who have reached majority and have dispositive legal capacity and who are in institutions for the social defence of the populace which are certified by the administration of this institution or the director (or his deputy) of the respective agency of social defence of the populace.

4. A power of attorney to receive earnings and other payments connected with labour relations, to receive the remuneration of authors and inventors, pensions, benefits, and stipends, deposits of citizens in banks and to receive correspondence, including monetary and parcel, may be certified also by the organisation in which the principal works or studies, housing-operations organisation at his place of residence, and the administration of an inpatient treatment institution in which he is being treated.

A power of attorney for the representative of a citizen to receive his deposit in a bank, monetary means from his bank account, correspondence addressed to him in communications organisations, and also to conclude other transactions in the name of the citizen specified in paragraph one of the present point, may be certified by the respective bank or communications organisation. Such a power of attorney shall be certified free of charge [Added 12 August 1996. СЗ РФ (1996), no. 34, item 4026].

5. A power of attorney in the name of a juridical person shall be issued over the signature of its director or other person empowered by the constitutive documents to do so, with the seal of this organisation affixed.

A power of attorney in the name of a juridical person based on State or municipal ownership for the receipt or issuance of money and other property valuables must be signed also by the chief (or senior) bookkeeper of this organisation.

Article 186. Period of Power of Attorney

1. The period of operation of a power of attorney may not exceed three years. If the period has not been specified in the power of attorney, it shall retain force for a year from the date of the conclusion thereof.

A power of attorney in which the date of its conclusion has not been specified shall be void.

2. A power of attorney certified by a notary and intended for the performance of actions abroad and not containing an indication of the period of its operation shall retain force until the revocation thereof by the person who has issued the power of attorney.

Article 187. Transfer of Power of Attorney

1. A person to whom the power of attorney was issued must personally perform those actions for which he is empowered. He may transfer the power of attorney to perform them to another person if empowered to do so by the power of attorney or forced by virtue of circumstances in order to protect the interests of the person who issued the power of attorney.

2. The person who transferred the power of attorney to another person must notify the person who issued the power of attorney thereof and communicate to him necessary information about the person to whom the powers have been transferred. The failure to perform this duty shall place on the person who transferred the power responsibility for the actions of the person to whom he transferred the power as they were his own.

3. A power of attorney issued by way of transfer must be notarially certified, except for instances provided for by Article 185[4] of the present Code.

4. A period of operation of a power of attorney issued by way of transfer may not exceed the period of operation of the power of attorney on the basis of which it was issued.

Article 188. Termination of Power of Attorney

1. The operation of a power of attorney shall terminate as a consequence of:

(1) expiry of the period of the power of attorney;

(2) revocation of the power of attorney by the person who issued it;

(3) renunciation by the person to whom the power of attorney was issued;

(4) termination of the juridical person in whose name the power of attorney was issued;

(5) termination of the juridical person to whom the power of attorney was issued;

(6) death of the citizen who issued the power of attorney, deeming him to lack dispositive legal capacity, limited dispositive legal capacity, or to be missing;

(7) death of the citizen to whom the power of attorney was issued, deeming him to lack dispositive legal capacity, limited dispositive legal capacity, or to be missing.

2. A person who issued the power of attorney may at any time revoke the power of attorney or the transfer of the power of attorney, and the person to whom the power of attorney was issued, to renounce it. An agreement concerning the waiver of these rights shall be void.

3. The transfer of a power of attorney shall lose force with the termination of the power of attorney.

Article 189. Consequences of Termination of Power of Attorney

1. A person who has issued a power of attorney and subsequently revoked it shall be obliged to notify the person to whom the power of attorney was issued about the revocation thereof, as well as third persons known to him with respect to whom the power of attorney was issued for representation. The same duty shall be placed on the legal successors of the person who issued a power of attorney in instances of the termination thereof on the grounds provided for in Article 188(1), subpoints (4) and (6), of the present Code.

2. The rights and duties which arose as a result of the actions of the person to whom a power of attorney is issued before this person knew or should have known about its termination shall retain force for the person who issued the power of attorney and his legal successors with respect to third persons. This rule shall not apply if the third person knew or should have known that the operation of the power of attorney had terminated.

3. With regard to the termination of a power of attorney the person to whom it was issued or his legal successors shall be obliged immediately to return the power of attorney.

Subsection 5. Periods. Limitations

Chapter 11. Calculation of Periods

Article 190. Determination of Period

A period established by a law, other legal acts, a transaction, or designated by a court shall be determined by a calendar date or by the expiry of a period of time which shall be calculated by years, months, weeks, days, or hours.

A period may also be determined by specifying an event which must inevitably ensue.

Article 191. Commencement of Period Determined by Period of Time

The running of a period determined by a period of time shall commence on the following day after the calendar date or ensuing of the event by which its commencement is determined.

Article 192. Ending of Period Determined by Period of Time

1. A period calculated by years shall expire in the corresponding month and date of the last year of the period.

The rules for periods calculated by months shall apply to a period calculated by a half-year.

2. To a period calculated by quarters of a year shall apply the rules for periods calculated by months. The quarter shall be considered to be equal to three months, and the quarters shall be calculated from the beginning of the year.

3. A period calculated by months shall expire on the corresponding date of the last month of the period.

A period determined as a half-month shall be considered as a period calculated by days and shall be considered to be equal to fifteen days.

If the ending of a period calculated by months comes in such month in which there is no corresponding date, the period shall expire on the last day of that month.

4. A period calculated by weeks shall expire on the corresponding day of the last week of the period.

Article 193. Ending of Period on Non-Work Day

If the last day of a period comes on a non-work day, the next work day following shall be considered to be the day of ending of the period.

Article 194. Procedure for Performing Actions on Last Day of Period

1. If a period has been established for performing any action whatever, it may be fulfilled up to 24:00 hours of the last day of the period.

However, if this action had to be performed in an organisation, the period shall expire at that hour when in this organisation the respective operations terminate according to the established rules.

2. Written applications and notifications handed in to a communications organisation before 24:00 hours of the last day of the period shall be considered to be made within the period.

Chapter 12. Limitations

Article 195. Concept of Limitations

The period for the defence of a right upon the suit of a person whose right has been violated shall be deemed to be a limitation.

Article 196. General Period of Limitations

The general period of limitations shall be established as three years.

Article 197. Special Periods of Limitations

1. Special periods of limitations may be established for individual types of demands by a law which are reduced or are longer in comparison with the general period.

2. The rules of Articles 195, 198–207 of the present Code shall also extend to special periods of limitation unless established otherwise by a law.

Article 198. Invalidity of Agreement to Change Periods of Limitations

Periods of limitations and the procedure for calculating them may not be changed by agreement of the parties.

The grounds for the suspension and interruption of the running of periods of limitations shall be established by the present Code and by other laws.

Article 199. Application of Limitations

1. A demand concerning the defence of a violated right shall be accepted for consideration by a court irrespective of the expiry of the period of limitations.

2. Limitations shall be applied by a court only upon the application of the party to the dispute which was made before the rendering of the decision by the court.

The expiry of the period of limitations concerning the application of which a party to the dispute has applied for shall be grounds for the court to render a decision to reject the suit.

Article 200. Commencement of Running of Period of Limitations

1. The running of the period of limitations shall commence from the day when the person knew or should have known about the violation of his right.

Exceptions from this rule shall be established by the present Code and by other laws.

2. With regard to obligations with a specified period of performance, the running of the limitations shall commence upon the end of the period of performance.

With regard to obligations the period for the performance of which has not been determined or has been determined by the moment of demand, the running of limitations shall commence from the moment when the right arises with the creditor to present a demand concerning the performance of the obligation, and if the debtor is granted an exemption period for performance of such demand, the calculation of limitations shall commence upon the ending of the said period.

3. With regard to regressive obligations the running of the limitations shall commence from the moment of performance of the principal obligation.

Article 201. Period of Limitations in Event of Change of Persons in Obligation

The change of persons in an obligation shall not entail a change of the period of limitations and the procedure for calculating it.

Article 202. Suspension of Running of Period of Limitations

1. The running of the period of limitations shall be suspended:
(1) if the bringing of suit has been hindered by an extraordinary and unavoidable circumstance under the particular conditions (insuperable force);
(2) if the plaintiff or defendant is in the Armed Forces which are transferred to a military situation;
(3) by virtue of a deferral for the performance of obligations established on the basis of a law by the Government of the Russian Federation (moratorium);
(4) by virtue of the suspension of the operation of a law or other legal act regulating the respective relation.

2. The running of the period of limitations shall be suspended on condition that the circumstances specified in the present Article arose or have continued to exist during the last six months of the period of limitations, and if this period is equal to six months or less than six months, the running of the period of limitations.

3. The running of the period shall continue from the date of termination of the circumstances serving as the grounds for suspension of the limitation. The remaining part of the period shall be lengthened up to six months, and if the period of limitations is equal to six months or less than six months, up to the period of limitations.

Article 203. Interruption of Running of Period of Limitations

The running of the period of limitations shall be interrupted by bringing suit in the established procedure, and also by the performance of actions by the obliged person which testify to recognition of the debt.

After the interruption, the running of the period of limitations shall commence anew; the time which elapsed before interruption shall not be calculated in the new period.

Article 204. Running of Period of Limitations in Event of Leaving Suit Without Consideration

If a suit has been left without consideration by a court, then the running of the period of limitations which commenced before bringing suit shall continue in the general procedure.

If a suit brought in a criminal case has been left by a court without consideration, then the running of the period of limitations which commenced before bringing suit shall be suspended until the entry into legal force of the judgment by which the suit was left without consideration; the time during which the limitation was suspended shall not be calculated in the period of limitations. In so doing if the remaining portion of the period is less than six months, it shall be lengthened up to six months.

Article 205. Restoration of Period of Limitations

In exceptional instances, when the court deems the reason for the lapsing of the period of limitations to be justifiable with regard to circumstances connected with the person of the plaintiff (grave illness, helpless state, illiteracy, and the like), the violated right of the citizen shall be subject to defence. The reasons for the lapse of a period of limitations may be deemed to be justifiable if they occurred in the last six months of the period of limitations, and if this period is equal to six months or to less than six months, the running of the period of limitations.

Article 206. Performance of Duty Upon Expiry of Period of Limitations

A debtor or other obliged person who has performed a duty upon the expiry of the period of limitations shall not have the right to demand back that which has been performed, even though at the moment of performance the said person did not know about the expiry of the limitations.

Article 207. Application of Period of Limitations to Additional Demands

The period of limitations regarding additional demands (penalty, pledge, suretyship, and the like) shall expire with the expiry of the period of limitations regarding the principal demand.

Article 208. Demands to Which Limitations Do Not Extend

Limitations shall not extend to:

demands concerning the defence of personal nonproperty rights and other nonmaterial benefits, except for instances provided for by a law;

demands of depositors against a bank concerning the issuance of deposits;

demands concerning compensation of harm caused to the life or health of a citizen. However, demands presented upon the expiry of three years from the moment of the arising of the right to compensation for such harm shall be satisfied for the lapsed time for not more than three years preceding the bringing of suit;

demands of the owner or other possessor concerning the elimination of any violations of his right, even though these violations were not connected with a deprivation of possession (Article 304);

other demands in the instances established by a law.

SECTION II. RIGHT OF OWNERSHIP AND OTHER RIGHTS TO THING

Chapter 13. General Provisions

Article 209. Content of Right of Ownership

1. The rights of possession, use, and disposition of his property shall belong to the owner.

2. The owner shall have the right at his discretion to perform with respect to property belonging to him any actions which are not contrary to a law and other legal acts and do not violate the rights and the interests protected by a law of other persons, including to alienate his property in ownership to other persons, to transfer to them while remaining the owner the rights of possession, use, and disposition of the property, to pledge out property and to encumber it by other means, and to otherwise dispose of it.

3. The possession, use, and disposition of land and other natural resources to the extent that their turnover is permitted by a law (Article 129) shall be effectuated by the owner thereof freely unless this causes damage to the environment and violates the rights and legal interests of other persons.

4. The owner may transfer his property in trust management to another person (trust manager). The transfer of property to trust management shall not entail the transfer of the right of ownership to the trust manager, who shall be obliged to effectuate the management of the property in the interests of the owner or a third person specified by him.

Article 210. Burden of Maintenance of Property

The owner shall bear the burden of maintenance of the property belonging to him unless provided otherwise by law or contract.

Article 211. Risk of Accidental Perishing of Property

The risk of accidental perishing or accidental damaging of property shall be borne by the owner thereof unless provided otherwise by law or by contract.

Article 212. Subjects of Right of Ownership

1. Private, State, municipal, and other forms of ownership shall be recognised in the Russian Federation.

2. Property may be in the ownership of citizens and juridical persons, and also of the Russian Federation, subjects of the Russian Federation, and municipal formations.

3. The peculiarities of the acquisition and termination of the right of ownership to property and the possession, use, and disposition of it depending upon whether the property is in the ownership of a citizen or juridical person or in the ownership of the Russian Federation, subject of the Russian Federation, or municipal formation may be established only by a law.

The types of property which may be only in State or municipal ownership shall be determined by a law.

4. The rights of all owners shall be defended equally.

Article 213. Right of Ownership of Citizens and Juridical Persons

1. Any property may be in the ownership of citizens and juridical persons, except for individual types of property which in accordance with a law may not belong to citizens or juridical persons.

2. The quantity and value of property in the ownership of citizens and juridical persons shall not be limited, except for instances when such limitations have been established by a law for the purposes provided for by Article 1(2) of the present Code.

3. Commercial and noncommercial organisations, except State and municipal enterprises, and also institutions financed by the owner, shall be the owners of property transferred to them as contributions (or dues) of their founders (or participants, members), and also property acquired by these juridical persons on other grounds.

4. Social and religious organisations (or associations), philanthropic and other foundations shall be the owners of property acquired by them and may use it only

in order to achieve the purposes provided for by the constitutive documents thereof. The founders (or participants, members) of these organisations shall lose the right to property transferred by them to the ownership of the respective organisation. In the event of the liquidation of such organisation the property thereof remaining after satisfaction of the demands of creditors shall be used for the purposes specified in its constitutive documents.

Article 214. Right of State Ownership

1. Property which belongs by right of ownership to the Russian Federation (federal ownership) and property belonging by right of ownership to subjects of the Russian Federation—republics, territories, regions, cities of federal significance, autonomous region, and autonomous national areas (ownership of subject of the Russian Federation)—shall be State ownership in the Russian Federation.

2. Land and other natural resources which are not in the ownership of citizens, juridical persons, or municipal formations shall be State ownership.

3. The agencies and persons specified in Article 125 of the present Code shall effectuate the rights of owner in the name of the Russian Federation and subjects of the Russian Federation.

4. Property in State ownership shall be consolidated to State enterprises and institutions in possession, use, and disposition in accordance with the present Code (Articles 294, 296).

Means of the respective budget and other State property not consolidated to State enterprises and institutions shall constitute the State treasury of the Russian Federation, the treasury of a republic within the Russian Federation, or the treasury of a territory, region, city of federal significance, autonomous region, or autonomous national area.

5. The relegation of State property to federal ownership and the ownership of subjects of the Russian Federation shall be effectuated in the procedure established by a law.

Article 215. Right of Municipal Ownership

1. Property belonging by right of ownership to city and rural settlements, and also to other municipal formations, shall be municipal ownership.

2. Agencies of local self-government and the persons specified in Article 125 of the present Code shall effectuate the rights of owner in the name of the municipal formation.

3. Property in municipal ownership shall be consolidated to municipal enterprises and institutions in possession, use, and disposition in accordance with the present Code (Articles 294, 296).

Means of the local budget and other municipal property not consolidated to municipal enterprises and institutions shall constitute the municipal treasury of the respective city or rural settlement or other municipal formation.

Article 216. Rights to Thing of Persons Who Are Not Owners

1. The rights to a thing, in addition to the right of ownership, are, in particular:

the right of inheritable possession for life of a land plot (Article 265);

the right of permanent (or perpetual) use of a land plot (Article 268);

servitudes (Articles 274, 275);

the right of economic jurisdiction over property (Article 294) and the right of operative management of property (Article 296).

2. The rights to a thing in property may belong to persons who are not the owners of this property.

3. The transfer of the right of ownership in property to another person shall not be grounds for the termination of other rights to a thing in this property.

4. The rights to a thing of a person who is not an owner shall be defended against the violation thereof by any person in the procedure provided for by Article 305 of the present Code.

Article 217. Privatisation of State and Municipal Property

Property in State or municipal ownership may be transferred by its owner to the ownership of citizens and juridical persons in the procedure provided for by laws on the privatisation of State and municipal property.

When privatising State and municipal property the provisions provided for by the present Code which regulate the procedure for the acquisition and termination of the right of ownership shall apply unless the laws on privatisation provide otherwise.

Chapter 14. Acquisition of Right of Ownership

Article 218. Grounds for Acquisition of Right of Ownership

1. The right of ownership to a new thing manufactured or created by a person for himself in compliance with a law and other legal acts shall be acquired by this person.

The right of ownership to fruits, products, and revenues received as a result of the use of property shall be acquired on the grounds provided for by Article 136 of the present Code.

2. The right of ownership to property which an owner has may be acquired by another person on the basis of a contract of purchase/sale, barter, gift, or other transaction concerning the alienation of this property.

In the event of the death of a citizen the right of ownership to the property belonging to him shall pass by inheritance to other persons in accordance with the will or a law.

In the event of the reorganisation of a juridical person the right of ownership to the property belonging to it shall pass to the juridical person's legal successors of the reorganised juridical person.

3. In the instances and in the procedure provided for by the present Code, a person may acquire the right of ownership in property which has no owner, in property whose owner is unknown, or in property which the owner has renounced or to which he has lost the right of ownership on other grounds provided for by a law.

4. The member of a housing, housing-construction, dacha, garage, or other consumer cooperative, other persons having the right to share-accumulation who have fully made their share contribution to an apartment, dacha, garage, or other premise granted to these persons by the cooperative shall acquire the right of ownership to the said property.

Article 219. Arising of Right of Ownership in Newly Created Immovable Property

The right of ownership in a building, installation, and other newly created immovable property subject to State registration shall arise from the moment of such registration.

Article 220. Converting

1. Unless provided otherwise by a contract, the right of ownership to a new movable thing manufactured by the person by means of converting materials which do not belong to him shall be acquired by the owner of the materials.

However, if the value of the converting materially exceeds the value of the materials, the right of ownership to a new thing shall be acquired by the person who, acting in good faith, has effectuated the converting for himself.

2. Unless provided otherwise by a contract, the owner of materials who has acquired the right of ownership in a thing manufactured from them shall be obliged to compensate the value of the converting to the person who has effectuated it, and in the event of the acquisition of the right of ownership to the new thing by this person, the last shall be obliged to compensate the owner of the materials for the value thereof.

3. The owner of the materials who has lost them as a result of the actions not in good faith of the person who effectuated the converting shall have the right to demand the transfer of the new thing to his ownership and compensation of losses caused to him.

Article 221. Conversion to Ownership of Things Generally-Accessible for Gathering

In the instances when in accordance with a law, general authorisation by a particular owner, or in accordance with local custom the gathering of berries, catching of fish, and gathering or extraction of other generally-accessible things and fauna is permitted in forests, waters, or on other territories, the right of ownership in the respective things shall be acquired by the person effectuating the gathering or extraction thereof.

Article 222. Arbitrary Structure

1. A dwelling house, other structure, installation, or other immovable property created on a land plot not allotted for these purposes in the procedure established by a law and other legal acts, or created without obtaining the necessary authorisations for this, or with a material violation of urban construction and construction norms and rules, shall be an arbitrary structure.

2. A person who has effectuated an arbitrary structure shall not acquire the right of ownership in it. He shall not have the right to dispose of the structure—to sell, give, lease out, or conclude other transactions.

The arbitrary structure shall be subject to demolition by the person who effectuated it or at his expense, except for the instances provided for by point 3 of the present Article.

3. The right of ownership to an arbitrary structure may be recognised by a court for the person who effectuated the building on a land plot not belonging to him on condition that the particular plot will be granted to this person in the established procedure under the structure erected.

The right of ownership in an arbitrary structure may be recognised by a court for the person in whose ownership, inheritable possession for life, permanent (or perpetual) use the land plot is where the structure was effectuated. In this event the person whose right of ownership to the structure has been recognised shall compensate the person who effectuated it for the expenses for the structure in the amount determined by the court.

The right of ownership in the arbitrary structure may not be recognised for the said persons if retention of the structure violates the rights and the interests protected by a law of the other persons or creates a threat to the life and health of citizens.

Article 223. Moment of Origin of Right of Ownership in Acquirer Under Contract

1. The right of ownership in the acquirer of a thing under a contract shall arise from the moment of the transfer thereof unless provided otherwise by a law or by a contract.

2. In instances when the alienation of property is subject to State registration, the right of ownership in the acquirer shall arise from the moment of such registration, unless established otherwise by a law.

Article 224. Transfer of Thing

1. The handing over of a thing to the acquirer shall be deemed to be the transfer, and likewise the handing over to a carrier for despatch to the acquirer, or to a communications organisation for the sending to the acquirer of a thing alienated without the obligation of delivery.

A thing shall be considered to be handed over to the acquirer from the moment of its actual receipt in the possession of the acquirer or person specified by him.

2. If at the moment of concluding a contract on the alienation of a thing it is already in the possession of the acquirer, the thing shall be deemed to be transferred to him from this moment.

3. The transfer of a bill of lading or other goods-disposition document therefor shall be equated to the transfer of a thing.

Article 225. Masterless Things

1. A thing which has no owner or whose owner is unknown, or a thing, the right of ownership of which has been renounced by the owner, shall be masterless.

2. Unless this is excluded by the rules of the present Code on the acquisition of the right of ownership to a thing which the owner has renounced (Article 226), on find (Articles 227–228), on neglected animals (Articles 230–231), and treasure (Article 233), the right of ownership in masterless movable things may be acquired by virtue of acquisitive prescription.

3. Masterless immovable things shall be accepted for recording by the agency effectuating State registration of the right to immovable property upon the application of the agency of local self-government on whose territory they are situated.

Upon the expiry of a year from the day of the masterless immovable thing being recorded, the agency empowered to manage municipal property may apply to a court with a demand to recognise the right of municipal ownership in this thing.

A masterless immovable thing not deemed by decision of a court to have entered into municipal ownership may be accepted anew into the possession, use, and disposition of the owner who left it, or acquired in ownership by virtue of acquisitive prescription.

Article 226. Movable Things Which the Owner Has Renounced

1. Movable things thrown away by the owner or otherwise left by him (discarded things) with a view to renouncing the right of ownership may be converted

by other persons into their ownership in the procedure provided for by point 2 of the present Article.

2. The person in whose ownership, possession, or use a land plot, water, or other object is where the discarded thing is situated, the value of which is clearly lower than an amount corresponding to five times the minimum amount of payment of labour, or discarded scrap metal, defective products, melted alloys, slagheaps and discharges formed when extracting minerals, production wastes, and other wastes shall have the right to convert these things to his ownership, having commenced to use them or having performed other actions testifying to the conversion of the thing into ownership.

Other discarded things shall enter into the ownership of the person who has entered into possession of them if they have been deemed to be masterless by a court upon the application of this person.

Article 227. Find

1. The finder of a lost thing shall be obliged immediately to inform the person who lost it thereof or the owner of the thing or some other person known to him as having the right to receive it and shall return the found thing to this person.

If a thing has been found in a premise or on transport it shall be subject to being handed over to a person representing the possessor of this premise or means of transport. In this event the person to whom the find has been handed over shall acquire the rights and bear the duties of the person who found the thing.

2. If the person having the right to demand the return of a found thing or his whereabouts are unknown, the finder of the thing shall be obliged to declare the find to the police or agency of local self-government.

3. The finder of a thing shall have the right to keep it or to hand it over for keeping to the police or agency of local self-government or to a person specified by them.

A perishable thing or a thing whose costs for keeping are incommensurately great in comparison with its value may be realised by the finder of the thing, receiving written evidence certifying the amount of the receipts. The money derived from the sale of a found thing shall be subject to return to the person empowered to receive it.

4. The finder of a thing shall be liable for its loss or damage only in the event of intent or gross negligence and within the limits of the value of the thing.

Article 228. Acquisition of Right of Ownership in Find

1. If in the course of six months from the moment of declaring the find to the police or agency of local self-government (Article 227[2]) the person empowered to receive the found thing has not been established or does not himself declare

his right to the thing to the person who found it, or to the police or agency of local self-government, the finder of the thing shall acquire the right of ownership in it.

2. If the finder of a thing refuses to acquire the found thing in ownership, it shall enter municipal ownership.

Article 229. Compensation for Expenses Connected with Find and Remuneration to Finder of Thing

1. The finder and returner of a thing to the person empowered to receive it shall have the right to receive from this person, and in instances of the thing passing into municipal ownership, from the respective agency of local self-government, compensation of necessary expenses connected with keeping, handing over, or realisation of the thing, and also expenditures for the discovery of the person empowered to receive the thing.

2. The finder of a thing shall have the right to demand from the person empowered to receive the thing remuneration for the find in the amount of up to 20% of the value of the thing. If a found thing is of value only to the person empowered to receive it, the amount of remuneration shall be determined by agreement with this person.

The right to remuneration shall arise unless the finder of the thing has not declared the find or has attempted to conceal it.

Article 230. Neglected Animals

1. A person who has detained neglected or stray livestock or other neglected domestic animals shall be obliged to return them to their owner, and if the owner of the animals or his whereabouts is unknown, not later than within three days from the moment of detention to declare the discovered animals to the police or agency of local self-government, which shall take measures to seek the owner.

2. During the search for the owner of the animals they may be left by the person who detained them with himself for maintenance and for use, or handed over for maintenance and use to another person having the necessary conditions for this. At the request of the person who detained neglected animals, the seeking of a person who has the necessary conditions for their maintenance and use and the transfer of the animals thereto shall be effectuated by the police or agency of local self-government.

3. The person who has detained neglected animals and the person to whom they have been transferred for maintenance and use shall be obliged to maintain them properly and shall be liable for the perishing and spoilage to the animals if there is fault and within the limits of the value thereof.

Article 231. Acquisition of Right of Ownership to Neglected Livestock

1. If within six months from the moment of declaring the detention of neglected domestic animals their owner is not discovered or he himself does not declare his right to them, the right of ownership to them shall be acquired by the person with whom they are situated for maintenance and use.

If this person refuses to acquire in ownership animals maintained by him they shall enter into municipal ownership and be used in the procedure determined by the agency of local self-government.

2. In the event of the appearance of the former owner of the animals after their transfer into the ownership of another person, the owner shall have the right when there are circumstances testifying to the retention of an attachment on the part of these animals towards him or the cruel or other improper treatment of them by the new owner to demand their return on the conditions determined by an agreement with the new owner, and in the event of the failure to reach agreement, by a court.

Article 232. Compensation of Expenses for Maintenance of Neglected Animals and Remuneration for Them

The person who detained animals and the person with whom they are situated for maintenance and use shall, in the event of the return of neglected domestic animals to the owner, have the right to compensation by their owner of necessary expenses connected with the maintenance of the animals, setting off the advantages derived from the use thereof.

The person who has detained neglected domestic animals shall have the right to remuneration in accordance with Article 229(2) of the present Code.

Article 233. Treasure

1. Treasure, that is, money or valuable objects buried in the earth or concealed by other means whose owner cannot be established or by virtue of a law has lost the right to them shall enter into the ownership of the person to whom the property belongs (land plot, structure, etc.) where the treasure was concealed, and the person who discovered the treasure, in equal participatory shares, unless established otherwise by an agreement between them.

When a treasure is discovered by a person who has made diggings or searches for the valuables without the consent of the owner of the land plot or other property where the treasure was concealed, the treasure shall be subject to transfer to the owner of the land plot or other property where the treasure was discovered.

2. In the event of the discovery of a treasure containing a thing relegated to monuments of history or culture, it shall be subject to transfer to State ownership. In so doing the owner of the land plot or other property where the treasure was

concealed and the person who discovered the treasure shall have the right to receive remuneration in the amount of 50% of the value of the treasure. The remuneration shall be distributed between these persons in equal participatory shares unless established otherwise by an agreement between them.

In the event of the discovery of such treasure by a person who has made diggings or searches for valuables without the consent of the owner of the property where the treasure was concealed, the remuneration shall not be paid to this person and shall go to the owner in full.

3. The rules of the present Article shall not apply to persons within whose labour or employment duties are the conducting of diggings and searches directed towards the discovery of treasure.

Article 234. Acquisitive Prescription

1. A person—citizen or juridical person—who is not the owner of property but in good faith, openly, and uninterruptedly possesses as his own immovable property for fifteen years or other property for five years shall acquire the right of ownership in such property (acquisitive prescription).

The right of ownership in immovable and other property subject to State registration shall arise in the person who acquired this property by virtue of acquisitive prescription from the moment of such registration.

2. Until the acquisition of the right of ownership in property by virtue of acquisitive prescription, the person possessing the property as his own shall have the right to defend his possession against third persons who are not the owners of the property, and also who do not have the rights to possession thereof by virtue of another ground provided for by a law or contract.

3. A person referring to the prescription of possession may join to the time of his possession all of the time during which this property was possessed by he whom this person is the legal successor of.

4. The running of the period of acquisitive prescription with respect to things situated with a person from whose possession they may be demanded and obtained in accordance with Articles 301 and 305 of the present Code shall commence not earlier than the expiry of the period of limitations with respect to the corresponding demands.

Chapter 15. *Termination of Right of Ownership*

Article 235. Grounds for Termination of Right of Ownership

1. The right of ownership shall terminate in the event of the alienation by the owner of his property to other persons, renunciation of the right of ownership by the owner, the perishing or destruction of the property, and the loss of the right of ownership to property in other instances provided for by a law.

2. The compulsory withdrawal from the owner of property shall not be permitted except for instances when on the grounds provided for by a law, it shall be done by:

(1) levying execution on property for obligations (Article 237);

(2) alienation of property which by virtue of a law cannot belong to the particular person (Article 238);

(3) alienation of immovable property in connection with the withdrawal of a plot (Article 239);

(4) purchase of improvidently maintained cultural valuables and domestic livestock (Articles 240 and 241);

(5) requisition (Article 242);

(6) confiscation (Article 243);

(7) alienation of property in the instances provided for by Article 252(4), Article 272(2), and Articles 282, 285, and 293 of the present Code.

By decision of the owner in the procedure provided for by laws on privatisation the property in State or municipal ownership shall be alienated into the ownership of citizens and juridical persons.

Property in the ownership of citizens and juridical persons shall be converted into State ownership (nationalisation) on the basis of a law, with compensation of the value of this property and other losses in the procedure established by Article 306 of the present Code.

Article 236. Renunciation of Right of Ownership

A citizen or juridical person may renounce the right of ownership in property belonging to him, so declaring or having performed other actions which specifically testify to the elimination of the possession, use, and disposition of the property without the intention to preserve any rights whatever to such property.

The renunciation of the right of ownership shall not entail the termination of the rights and duties of the owner with respect to the corresponding property until the moment of the acquisition of the right of ownership thereto by another person.

Article 237. Levy of Execution on Property for Obligations of Owner

1. The withdrawal of property by means of levying execution on it with regard to obligations of the owner shall be done on the basis of the decision of a court, unless another procedure for levy of execution has been provided for by a law or by contract.

2. The right of ownership in property on which execution is levied shall terminate in the owner from the moment that the right of ownership arises in the withdrawn property in the person to whom this property passes.

Article 238. Termination of Right of Ownership of Person in Property Which Cannot Belong to Him

1. If on the grounds permitted by a law property which by virtue of law cannot belong to him turns out to be in the ownership of a person, this property must be alienated by the owner within a year from the moment the right of ownership arises in the property unless another period has been established by a law.

2. In instances when the property has not been alienated by the owner within the periods specified in point 1 of the present Article, such property, taking into account its character and purpose, shall by decision of a court rendered upon the application of a State agency or agency of local self-government be subject to compulsory sale, with transfer to the former owner of the amounts received, or the transfer to State or municipal ownership with compensation of the value of the property to the former owner determined by a court. In so doing the expenditures for alienation of the property shall be deducted.

3. If in the ownership of a citizen or juridical person on the grounds permitted by a law there turns out to be a thing for whose acquisition a special authorisation is necessary, the issuance of which has been refused to the owner, this thing shall be subject to alienation in the procedure established for property which cannot belong to the particular owner.

Article 239. Alienation of Immovable Property in Connection with Withdrawal of Plot on Which It Is Situated

1. When the withdrawal of a land plot for State or municipal needs or in view of the improper use of land is impossible without the termination of the right of ownership in a building, installation, or other immovable property situated on the particular plot, this property may be withdrawn from the owner by means of the purchase by the State or sale at public sale in the procedure provided for respectively by Articles 279–282 and 284–286 of the present Code.

The demand to withdraw immovable property shall not be subject to satisfaction if the State agency or agency of local self-government which has applied to a court with this demand does not prove that the use of the land plot for the purposes for which it is being withdrawn is impossible without terminating the right of ownership to the particular immovable property.

2. The rules of the present Article respectively shall apply in the event of termination of the right of ownership to immovable property in connection with the withdrawal of mining allotments, aquatory plots, and other plots on which property is situated.

Article 240. Purchase of Improvidently Maintained Cultural Valuables

In instances when the owner of cultural valuables relegated in accordance with a law to [the category of] specially valuable and protected by the State improvidently maintains these valuables, which threatens them with their losing their significance, such valuables may by decision of a court be withdrawn from the owner through purchase by the State or sale at a public sale.

In the event of the purchase of cultural valuables the owner shall be compensated their value in the amount established by agreement of the parties, and in the event of a dispute, by a court. In the event of the sale at a public sale the amount received from the sale shall be transferred to the owner less the expenses for holding the public sale.

Article 241. Purchase of Domestic Animals In Event of Improper Treatment Thereof

When the owner of domestic animals treats them in clear contravention of the requirements of the rules established on the basis of a law and the norms accepted in society for the humane attitude towards animals, these animals may be withdrawn from the owner by means of the purchase thereof by the person who has brought the respective demand in court. The price of the purchase shall be determined by agreement of the parties, and in the event of a dispute, by a court.

Article 242. Requisition

1. In instances of natural disasters, wrecks, epidemics, epizootic, and other circumstances of an extraordinary character property may, in the interests of society by decision of State agencies, be withdrawn from the owner in the procedure and on the conditions established by a law with payment of the value of the property to him (requisition).

2. The valuation according to which the value of the requisitioned property is compensated to the owner may be contested by it in a court.

3. The person whose property was requisitioned shall have the right when the operation of the circumstances terminates in connection with which the requisition was made to demand in court the return to him of the property which has been preserved.

Article 243. Confiscation

1. In the instances provided for by a law property may be withdrawn without compensation from the owner by decision of a court in the form of a sanction for the commission of a crime or other violation of law (confiscation).

2. In the instances provided for by a law, confiscation may be made in an administrative procedure. The decision concerning confiscation adopted in an administrative procedure may be appealed to a court.

101

Chapter 16. Common Ownership

Article 244. Concept and Grounds of Origin of Common Ownership

1. Property in the ownership of two or several persons shall belong to them by right of common ownership.

2. Property may be in common ownership with the determination of the participatory share of each of the owners in the right of ownership (participatory share ownership) or without the determination of such participatory shares (joint ownership).

3. Common ownership to property shall be participatory share except for instances when the formation of joint ownership to this property has been provided for by a law.

4. Common ownership shall arise when two or several persons enter into the ownership of property which cannot be divided without changing its purpose (indivisible things), or is not subject to division by virtue of a law.

Common ownership in divisible property shall arise in the instances provided for by a law or contract.

5. By agreement of the participants of joint ownership, and in the event of not achieving consent, by decision of a court, the participatory share ownership of these persons may be established for common property.

Article 245. Determination of Participatory Share in Right of Participatory Share Ownership

1. If the participatory shares of the participants of participatory share ownership cannot be determined on the basis of a law and was not established by agreement of all of its participants, the participatory shares shall be considered to be equal.

2. The procedure for the determination and change of their participatory shares may be established by agreement of all the participants of participatory share ownership depending upon the contribution of each of them in the formation and growth of the common property.

3. A participant of participatory share ownership who has effectuated at his own expense while complying with the established procedure for the use of common property indivisible improvements of this property shall have the right to a corresponding increase of his participatory share in the right to common property.

Divisible improvements of common property, unless provided otherwise by agreement of the participants of the participatory share ownership, shall enter into the ownership of those participants who produced them.

Article 246. Disposition of Property in Participatory Share Ownership

1. The disposition of property in participatory share ownership shall be effectuated by agreement of all of its participants.

2. A participant of participatory share ownership shall have the right at his discretion to sell, give, bequeath, or pledge his participatory share or to otherwise dispose of it while complying, in the event of alienation for payment, with the rules provided for by Article 250 of the present Code.

Article 247. Possession and Use of Property in Participatory Share Ownership

1. The possession and use of property in participatory share ownership shall be effectuated by the agreement of all of its participants, and if consent is not achieved, in the procedure established by a court.

2. A participant of participatory share ownership shall have the right to grant part of the common property commensurate to his participatory share for possession and use, and if this is impossible, shall have the right to demand of the other participants who possess and use the property respective contributory compensation approximating his participatory share.

Article 248. Fruits, Products, and Revenues from Use of Property in Participatory Share Ownership

The fruits, products, and revenues from the use of property in participatory share ownership shall become part of the common property and be distributed among the participants of participatory ownership commensurately with their participatory shares unless provided otherwise by an agreement between them.

Article 249. Expenses for Maintenance of Property in Participatory Share Ownership

Each participant of participatory share ownership shall be obliged in proportion to their participatory shares to participate in the payment of taxes, charges, and other payments relating to the common property, and also the costs for its maintenance and preservation.

Article 250. Preferential Right of Purchase

1. In the event of the sale of a participatory share in the right of common ownership to an outside person the remaining participants of participatory share ownership shall have the preferential right of purchase of the participatory share being sold at the price for which it is being sold and on other equal conditions, except for sales at a public sale.

The public sale for the sale of the participatory share in the right of common ownership may, in the absence of consent thereto of all the participants of participatory share ownership, be held in the instances provided for by Article 255, paragraph two, of the present Code, and in the other instances provided for by a law.

2. The seller of a participatory share shall be obliged to notify in written form the remaining participants of participatory share ownership of the intention to sell his participatory share to an outside person, specifying the price and other conditions on which he is selling it. If the other participants of participatory share ownership refuse to purchase or do not acquire the participatory share being sold in the right of ownership to immovable property within a month, and in the right of ownership to movable property, within ten days from the date of notification, the seller shall have the right to sell his participatory share to any person.

3. In the event of the sale of a participatory share in violation of the preferential right of purchase, any other participant of participatory share ownership shall have the right within three months to demand in a judicial proceeding the transfer to him of the rights and duties of the purchaser.

4. The assignment of a preferential right of purchase of the participatory share shall not be permitted.

5. The rules of the present Article shall apply also when alienating a participatory share under a contract of barter.

Article 251. Moment of Transfer of Participatory Share in Right of Common Ownership to Acquirer Under Contract

The participatory share in the right of common ownership shall pass to the acquirer under a contract from the moment of conclusion of the contract unless provided otherwise by an agreement of the parties.

The moment of transfer of a participatory share in the right of common ownership under a contract subject to State registration shall be determined in accordance with Article 223(2) of the present Code.

Article 252. Division of Property in Participatory Share Ownership and Partition of Participatory Share Therefrom

1. Property in common participatory share ownership may be divided between its participants by an agreement between them.

2. A participant of common participatory share ownership shall have the right to demand the partition of his participatory share from the common property.

3. If the participants of participatory share ownership do not reach agreement concerning the means and conditions for the division of common property or partition of the participatory share ownership of one of them, the participant of participatory share ownership shall have the right in a judicial proceeding to demand the partition of his participatory share in kind from the common property.

If the partition of a participatory share in kind is not permitted by a law or is impossible without incommensurate damage to property in common ownership, the partitioning owner shall have the right to payment to him of the value of his participatory share by the other participants of participatory share ownership.

4. The incommensurateness of property partitionable in kind to a participant of participatory share ownership on the basis of the present Article to his participatory share in the right of ownership shall be eliminated by the payment of a corresponding monetary amount or other contributory compensation.

The payment to a participant of participatory share ownership by the remaining co-owners of contributory compensation in place of the partition of his participatory share in kind shall be permitted with his consent. In instances when the participatory share of the respective owner is insignificant and cannot be truly partitioned and he does not have a material interest in the use of the common property, a court may also, in the absence of the consent of this owner, oblige the other participants of participatory ownership to pay contributory compensation to him.

5. With the receipt of the contributory compensation in accordance with points 3 and 4 of the present Article the owner shall lose the right to the participatory share in the common property.

Article 253. Possession, Use, and Disposition of Property in Joint Ownership

1. The participants of joint ownership, unless provided otherwise by agreement between them, shall possess and use common property in common.

2. The disposition of property in joint ownership shall be effectuated by the consent of all the participants, which shall be presupposed irrespective as to which of the participants has concluded the transaction with regard to disposition of the property.

3. Each of the participants of joint ownership shall have the right to conclude transactions relating to the disposition of common property unless it arises otherwise from the agreement of all the participants. A transaction concluded by one of the participants of joint ownership which is connected with the disposition of common property may be deemed to be invalid at the demand of the remaining participants for reasons that the participant who concluded the transaction lacked the necessary powers only if it is proved that the other party to the transaction knew or knowingly should have known about this.

4. The rules of the present Article shall apply insofar as not established otherwise for individual types of joint ownership by the present Code or other laws.

Article 254. Division of Property in Joint Ownership and Partition of Participatory Share Therefrom

1. The separation of common property between participants of joint ownership, and also the partition of the participatory share of one of them, may be effectuated after the preliminary determination of the participatory share of each of the participants in the right to common property.

2. In the event of the separation of common property and the partition of a participatory share therefrom, unless provided otherwise by law or by agreement of the participants, their participatory shares shall be deemed to be equal.

3. The grounds and procedure for the separation of common property and the partition of the participatory share therefrom shall be determined according to the rules of Article 252 of the present Code insofar as has not been established otherwise for individual types of joint ownership by the present Code and other laws and does not arise from the essence of the relations of the participants of joint ownership.

Article 255. Levying Execution on Participatory Share in Common Property

The creditor of a participant of participatory share or joint ownership shall, in the event that the owner's other property is insufficient, have the right to present a demand concerning partition of the participatory share of the debtor in the common property in order to levy execution on it.

If in such instances the apportionment of the participatory share in kind is impossible or the other participants of participatory share or joint ownership object to this, the creditor shall have the right to demand the sale by the debtor of his participatory share to the other participants of common ownership at a price commensurate with the market value of this participatory share, with the assets received from the sale being applied to repayment of the debt.

In the event of the refusal of the remaining participants of common ownership to acquire the participatory share of the debtor, the creditor shall have the right to demand in court the levy of execution on the participatory share of the debtor in the right of common ownership by means of the sale of this participatory share at a public sale.

Article 256. Common Ownership of Spouses

1. Property acquired by spouses during marriage shall be their joint ownership unless another regime for this property has been established by a contract between them.

2. The property which belonged to each of the spouses before entering into marriage, and also received by one of the spouses during marriage as a gift or by way of inheritance, shall be his ownership.

Things of individual use (clothing, footwear, and others), except for jewellery and other articles of embellishment, although acquired during the marriage at the expense of the common assets of the spouses, shall be deemed to be the ownership of that spouse who used them.

The property of each spouse may be deemed to be their joint ownership if it is established that during the marriage investments were made at the expense of the common property of the spouses or the personal property of the other spouse which significantly increased the value of this property (capital repair, reconstruction, re-equipping, and others). The present rule shall not apply if a contract between the spouses has provided otherwise.

3. Execution may be levied with regard to the obligations of one spouse on property in his ownership, and also on his participatory share in the common property of the spouses which would have been due to him in the event this property is separated.

4. The rules for determining the participatory shares of spouses in common property in the event of the separation thereof and the procedure for such separation shall be established by legislation on marriage and the family.

Article 257. Ownership of Peasant (or Farmer's) Economy

1. The property of a peasant (or farmer's) economy shall belong to its members by right of joint ownership unless established otherwise by a law or a contract between them.

2. The land plot, plantings, economic and other structures, soil conservation and other installations, productive and working livestock, poultry, agricultural and other technology and equipment, means of transport, tools, and other property acquired for the economy for common assets of its members granted in ownership to this economy shall be in the joint ownership of the members of the peasant (or farmer's) economy.

3. The fruits, products, and revenues received as a result of the activity of the peasant (or farmer's) economy shall be the common property of the members of the peasant (or farmer's) economy and shall be used by agreement between them.

Article 258. Separation of Property of Peasant (or Farmer's) Economy

1. In the event of the termination of a peasant (or farmer's) economy in connection with the withdrawal therefrom of all of its members or on other grounds, the common property shall be subject to separation according to the rules provided for by Articles 252 and 254 of the present Code.

The land plot in such instances shall be divided according to the rules established by the present Code and by land legislation.

2. The land plot and means of production belonging to the peasant (or farmer's) economy shall not be subject to separation in the event of the withdrawal of one of its members from the economy. The person withdrawing from the economy shall have the right to receive monetary contributory compensation commensurate with his participatory share in the common ownership in such property.

3. In the instances provided for by the present Article the participatory shares of members of a peasant (or farmers's) economy in the right of joint ownership to the property of the economy shall be deemed to be equal unless established otherwise by an agreement between them.

Article 259. Ownership of Economic Partnership or Cooperative Formed on the Base of the Property of Peasant (or Farmer's) Economy

1. An economic partnership or production cooperative may be created by the members of a peasant (or farmer's) economy on the base of the property of the economy. Such economic partnership or cooperative shall, as a juridical person, possess the right of ownership in the property transferred to it in the form of contributions and other payments by members of the farmer economy, and also to property received as a result of its activity and acquired on other grounds permitted by a law.

2. The amount of the contributions of the participants of the partnership or members of the cooperative created on the base of property of a peasant (or farmer's) economy shall be established by proceeding from their participatory shares in the right of common ownership in the property of the economy determined in accordance with Article 258(3) of the present Code.

Chapter 17. Right of Ownership and Other Rights to Thing in Land
[as amended 16 April 2001. СЗ РФ (2001),
no. 17, item 1644]

Article 260. General Provisions on the Right of Ownership in Land

1. Persons having a land plot in ownership shall have the right to sell it, give, pledge, or lease out, and otherwise dispose of it (Article 209) insofar as the respective land has not been excluded from turnover on the basis of a law or is not limited in turnover.

2. Lands of agricultural or other designation, the use of which for other purposes is not permitted or is limited, shall be determined on the basis of a law and in the procedure established by it. The use of a land plot relegated to such lands may be effectuated within the limits determined by its designation.

Article 261. Land Plot as Object of Right of Ownership

1. The territorial boundaries of a land plot shall be determined in the procedure established by land legislation on the basis of documents issued to the owner by State agencies for land resources and land tenure.

2. Unless established otherwise by a law, the right of ownership to the land plot shall extend to the surface (or soil) layer and enclosed waters situated within the boundaries of this plot and the forest and plants situated thereon.

3. The owner of a land plot shall have the right to use at his discretion everything that is situated above and below the surface of this plot, unless provided otherwise by the laws on the subsoil, on the use of airspace, and by other laws, and does not violate the rights of other persons.

Article 262. Land Plots of Common Use. Access to Land Plot

1. Citizens shall have the right freely, without any authorisations whatever, to be on land plots not closed off for general access which are in State or municipal ownership and to use natural objects on these plots within the limits permitted by a law and other legal acts, and also by the owner of the respective land plot.

2. If a land plot is not fenced off or the owner thereof has not clearly designated by other means that entry to the plot is not permitted without his authorisation, any person may traverse this plot on condition that this does not cause damage or disturbance to the owner.

Article 263. Building on Land Plot

1. The owner of a land plot may erect buildings and installations thereon, effectuate the restructuring or demolition thereof, and authorise construction on this plot to other persons. These rights shall be effectuated on condition of compliance with city construction and construction norms and rules, and also requirements concerning the designation of the land plot (Article 260[2]).

2. Unless provided otherwise by a law or a contract, the owner of a land plot shall acquire the right of ownership in a building, installation, and other immovable property erected or created by him for himself on the plot belonging to him.

The consequences of an arbitrary structure made by the owner on a land plot belonging to him shall be determined by Article 222 of the present Code.

Article 264. Rights to Land of Persons Who Are Not Owners of Land Plots

1. Land plots and immovable property situated on them may be granted by their owners to other persons for permanent or fixed-term use, including on lease.

2. The person who is not the owner of a land plot shall effectuate the rights of possession and use of the plot belonging to him on the conditions and within the limits established by a law or by a contract with the owner.

3. Unless provided otherwise by a law or by contract, the possessor of a land plot who is not the owner shall not have the right to dispose of this plot.

Article 265. Grounds for Acquisition of Right of Inheritable Possession for Life of Land Plot

The right of inheritable possession for life of a land plot which is in State or municipal ownership shall be acquired by citizens on the grounds and in the procedure which has been provided for by land legislation.

Article 266. Possession and Use of Land Plot by Right of Inheritable Possession for Life

1. A citizen who possesses the right of inheritable possession for life (possessor of land plot) shall have the right of possession and use of the land plot transferable by inheritance.

2. Unless it arises otherwise from the conditions of use of the land plot established by a law, the possessor of the land plot shall have the right to erect buildings and installations thereon and to create other immovable property, acquiring the right of ownership therein.

Article 267. Disposition of Land Plot in Inheritable Possession for Life

1. The possessor of a land plot may transfer it to other persons on lease or fixed-term use without compensation.

2. The sale or pledge of the land plot and the conclusion of other transactions by the possessor thereof which entail or might entail the alienation of the land plot shall not be permitted.

Article 268. Grounds for Acquisition of Right of Permanent (Perpetual) Use of Land Plot

1. The right of permanent (perpetual) use of a land plot which is in State or municipal ownership shall be granted to citizens and juridical persons on the basis of a decision of the State or municipal agency empowered to grant land plots for such use.

2. The right of permanent use of a land plot also may be acquired by the owner of the building, installation, and other immovable property in the instances provided for by Article 271(1) of the present Code.

3. In the event of the reorganisation of a juridical person, the right of permanent use of a land plot belonging to it shall pass by way of legal succession.

Article 269. Possession and Use of Land by Right of Permanent Use

1. The person to whom a land plot has been granted for permanent use shall effectuate the possession and use of this plot within the limits established by a law, other legal acts, and by the act on granting the plot for use.

2. Unless provided otherwise by a law, the person to whom the land plot has been granted for permanent use shall have the right autonomously to use the plot for the purposes for which it has been granted, including the erection for such purposes on the plot of buildings, installations, and other immovable property. The buildings, installations, and other immovable property created by this person for himself shall be his ownership.

Article 270. Disposition of Land Plot in Permanent Use

The person to whom a land plot has been granted for permanent use shall have the right to transfer this plot on lease or for fixed-term use without compensation only with the consent of the owner of the plot.

Article 271. Right to Use Land Plot by Owner of Immovable

1. The owner of a building, installation, and other immovable on a land plot which belongs to another person shall have the right of use of the portion of the land plot under this immovable granted by such person.

Unless it arises otherwise from a law, the decision on granting the land in State or municipal ownership, or contract, the owner of the building or installation shall have the right of permanent use of the part of the land plot (Articles 268–270) on which this immovable property is located.

2. In the event of the transfer of the right of ownership to the immovable on another's land plot to another person he shall acquire the right of use of the respective part of the land plot on the same conditions and in the same amount as the previous owner of the immovable.

The transfer of the right of ownership in a land plot shall not be grounds for the termination or change of the right to use this plot which belongs to the owner of the immovable.

3. The owner of an immovable on another's land plot shall have the right to possess, use, and dispose of this immovable at his discretion, including the demolition of the respective buildings and installations insofar as this is not contrary to the conditions of use of the particular plot established by a law or by contract.

Article 272. Consequences of Loss by Owner of Immovable of the Right to Use Land Plot

1. In the event of the termination of the right of use of a land plot granted to the owner of immovable property situated on this plot (Article 271), the rights to an immovable left by its owner on the land plot shall be determined in accordance with an agreement between the owner of the plot and the owner of the respective immovable property.

2. In the absence of or the failure to reach an agreement specified in point 1 of the present Article, the consequences of the termination of the right of use of the

land plot shall be determined by a court at the demand of the owner of the land plot or the owner of the immovable.

The owner of a land plot shall have the right to demand in court that the owner of an immovable shall, after termination of the right of use of a plot, free it of the immovable and bring the plot into its initial state.

In instances when the demolition of a building or installation on a land plot has been prohibited in accordance with a law and other legal acts (dwelling houses, monuments of history and culture, and so forth), or is not subject to effectuation in view of the value of the building or installation clearly exceeding the value of the land allotted beneath it, taking into account the grounds for the termination of the right of use of the land plot and upon the respective demands of the parties being presented, the court may:

recognise the right of the owner of the immovable to acquire the land plot on which this immovable is situated in ownership, or the right of the owner of the land plot to acquire the immovables left thereon, or

establish conditions for the use of the land plot by the owner of the immovable for a new period.

3. The rules of the present Article shall not apply when a land plot is withdrawn for State or municipal needs (Article 283), and also when the rights to a land plot are terminated in view of its improper use (Article 286).

Article 273. Transfer of Right to Land Plot in Event of Alienation of Buildings or Installations Situated on It

In the event of the transfer of the right of ownership to a building or installation which belonged to the owner of a land plot on which it is situated, the rights to the land plot determined by agreement of the parties shall pass to the acquirer of the building (or installation).

Unless otherwise provided for by a contract on alienation of the building or installation, to the acquirer shall pass the right of ownership in that part of the land plot which is occupied by the building (or installation) and is necessary for its use.

Article 274. Right of Limited Use of Another's Land Plot (Servitude)

1. The owner of immovable property (land plot, other immovable) shall have the right to demand from the owner of the neighbouring land plot, and, when necessary, also from the owner of another land plot (neighbouring plot) the granting of a right of limited use of the neighbouring plot (servitude).

A servitude may be established in order to ensure passage through the neighbouring land plot, the laying and operation of transmission and communications lines and pipelines, ensuring of water supply and soil conservation, and also other needs of the owner of the immovable property which cannot be ensured without establishing a servitude.

2. The encumberment of a land plot by a servitude shall not deprive the owner of the plot of the rights of possession, use, and disposition of this plot.

3. A servitude shall be established by agreement between the person requiring the establishment of the servitude and the owner of the neighbouring plot and shall be subject to registration in the procedure established for registration of the rights to immovable property. In the event of the failure to reach agreement concerning the establishment or the conditions of a servitude, upon the suit of the person requiring the establishment of the servitude the dispute shall be settled by a court.

4. On the conditions and in the procedure provided for by points 1 and 3 of the present Article a servitude may be established also in the interests and upon the demand of the person to whom the plot was granted by right of inheritable possession for life or right of permanent use.

5. The owner of a plot encumbered by a servitude shall have the right, unless otherwise provided by a law, to demand of the persons in whose interests the servitude was established commensurate payment for the use of the plot.

Article 275. Preservation of Servitude in Event of Transfer of Rights to Land Plot

1. A servitude shall be preserved in the event of the transfer of the rights to a land plot which is encumbered by this servitude to another person.

2. A servitude may not be an autonomous subject of purchase-sale or pledge and may not be transferred by any means whatever to persons who are not the owners of the immovable property to ensure the use of which the servitude was established.

Article 276. Termination of Servitude

1. Upon the demand of the owner of a land plot encumbered by a servitude, the servitude may be terminated in view of the grounds for which the servitude was established having disappeared.

2. In instances when a land plot belonging to a citizen or juridical person cannot be used as a result of being encumbered with a servitude in accordance with the designation of the plot, the owner shall have the right to demand in court the termination of the servitude.

Article 277. Encumberment of Buildings and Installations by Servitude

Buildings, installations, and other immovable property, the limited use of which is necessary which is not connected with the use of the land plot, may be encumbered by a servitude in accordance with the rules provided for by Articles 274–276 of the present Code.

Article 278. Levy of Execution on Land Plot

Levy of execution on the land plot with regard to obligations of its owner shall be permitted only on the basis of the decision of a court.

Article 279. Purchase of Land Plot for State and Municipal Needs

1. A land plot may be withdrawn from the owner for State or municipal needs by means of purchase.

Depending upon for whose needs the land is withdrawn, the purchase shall be effectuated by the Russian Federation, respective subject of the Russian Federation, or municipal formation.

2. The decision concerning withdrawal of the land plot for State or municipal needs shall be adopted by federal agencies of executive power and by agencies of executive power of the subjects of the Russian Federation.

State agencies empowered to adopt decisions concerning the withdrawal of land plots for State or municipal needs and the procedure for the preparation and adoption of these decisions shall be determined by federal land legislation.

3. The owner of a land plot must be informed in writing not later than a year before the forthcoming withdrawal of the land plot by the agency which adopted the decision concerning the withdrawal. The purchase of the land plot before the expiry of a year from the date of receipt by the owner of such notification shall be permitted only with the consent of the owner.

4. The decision of the State agency to withdraw the land plot for State or municipal needs shall be subject to State registration at the agency effectuating the registration of the rights to the land plot. The owner of the land plot must be notified about the registration made, specifying the date thereof.

5. The purchase for State or municipal needs of part of the land plot shall be permitted not other than with the consent of the owner.

Article 280. Rights of Owner of Land Plot Subject to Withdrawal for State or Municipal Needs

The owner of a land plot subject to withdrawal for State or municipal needs may, from the moment of State registration of the decision concerning withdrawal until the reaching of agreement or the adoption by a court of the decision concerning the purchase of the land plot, possess, use, and dispose of it at its discretion and make necessary expenditures ensuring the use of the plot in accordance with its designation. However, the owner shall bear the risk of taking upon himself when the purchase price of the land plot is determined (Article 281) the expenditures and losses connected with the new construction, expansion, and reconstruction of buildings and installations on the land plot within the said period.

Article 281. Purchase Price of Land Plot Withdrawn for State or Municipal Needs

1. Payment for a land plot withdrawn for State or municipal needs (purchase price) and the periods and other conditions of the purchase shall be determined by agreement with the owner of the plot. An agreement shall include the obligation of the Russian Federation, subject of the Russian Federation, or municipal formation to pay the purchase price for the plot being withdrawn.

2. When determining the purchase price, the market value of the land plot and of immovable property situated thereon shall be included therein, and also all losses caused to the owner by the withdrawal of the land plot, including losses which he bears in connection with the termination before time of his obligations to third persons, including lost advantage.

3. By agreement with the owner another land plot may be granted to him in place of the plot withdrawn for State or municipal needs, setting off the value thereof in the purchase price.

Article 282. Purchase of Land Plot for State or Municipal Needs by Decision of Court

If the owner does not agree with the decision concerning the withdrawal of a land plot from him for State or municipal needs, or agreement has not been reached with him concerning the purchase price or other conditions of the purchase, the State agency which adopted this decision may bring suit to purchase the land plot in a court. The suit concerning the purchase of the land plot for State or municipal needs may be filed within two years from the moment of sending the notice specified in Article 279(3) of the present Code to the owner of the plot.

Article 283. Termination of Rights of Possession and Use of Land Plot in Event of its Withdrawal for State or Municipal Needs

In the instances when a land plot being withdrawn for State or municipal needs is in possession and use by right of inheritable possession for life or permanent use, the termination of these rights shall be effectuated according to the rules provided for by Articles 279–282 of the present Code.

Article 284. Withdrawal of Land Plot Which is Not Used in Accordance with Its Designation

A land plot may be withdrawn from the owner in instances when the plot is earmarked for agricultural production or for housing or other construction and is not used for the respective purpose within three years unless a longer period has been established by a law. The time needed to exploit the plot, and also the time during which the plot cannot be used for the designation because of natural calamities or in view of other circumstances precluding such use, shall not be included within this period.

Article 285. Withdrawal of Land Plot Being Used in Violation of Legislation

A land plot may be withdrawn from the owner if the use of the plot is effectuated in flagrant violation of the rules for the rational use of land established by land legislation, in particular, if the plot is not used in accordance with its special-purpose designation or its use leads to a material reduction of the fertility of agricultural land or to a significant worsening of the ecological situation.

Article 286. Procedure for Withdrawal of Land Plot in View of its Improper Use

1. The agency of State power or local self-government empowered to adopt decisions concerning the withdrawal of land plots on the grounds provided for by Articles 284 and 285 of the present Code, and also the procedure for the obligatory timely warning of owners of the plots about violations permitted shall be determined by land legislation.

2. If the owner of the land plot informs in writing the agency which adopted the decision to withdraw the land plot of his consent to execute this decision, the plot shall be subject to sale at a public sale.

3. If the owner of the land plot does not agree with the decision to withdraw the plot from him, the agency which has adopted the decision to withdraw the plot may present a demand in court concerning the sale of the plot.

Article 287. Termination of Rights to Land Plot Belonging to Persons Who Are Not Owners Thereof

The termination of the rights to a land plot belonging to lessees and other persons who are not the owners thereof shall, in view of the improper use of the plot by these persons, be effectuated on the grounds and in the procedure which has been established by land legislation.

Chapter 18. Right of Ownership and Other Rights to Thing in Dwelling Premises

Article 288. Ownership in Dwelling Premise

1. The owner shall effectuate the rights of possession, use, and disposition of a dwelling premise belonging to him in accordance with its designation.

2. Dwelling premises are intended for the residence of citizens.

A citizen-owner of a dwelling premise may use it for personal residence and the residence of members of his family.

Dwelling premises may be leased by their owners for residence on the basis of a contract.

3. The siting in dwelling premises of industrial entities shall not be permitted.

The siting by the owner in a dwelling premise belonging to him of enterprises, institutions, and organisations shall be permitted only after the transfer of such premise to nonliving. The transfer of premises from living to nonliving shall be in the procedure determined by housing legislation.

Article 289. Apartment as Object of Right of Ownership

A participatory share in the right of ownership to the common property of a house (Article 290) also shall belong to the owner of an apartment in an apartment house together with the premises occupied by the apartment which belongs to him.

Article 290. Common Property of Owners of Apartments in Apartment House

1. The common premises of the house, supporting construction of the house, mechanical, electrical, sanitary-technical, and other equipment beyond the limits or within the apartment servicing more than one apartment shall belong by right of common participatory share ownership to the owners of the apartments in the apartment house.

2. The owner of an apartment shall not have the right to alienate his participatory share in the right of ownership to common property of a dwelling house, nor to perform other actions which entail the transfer of this participatory share separately from the right of ownership to the apartment.

Article 291. Partnership of Owners of Housing

1. The owners of apartments shall, in order to ensure the operation of the apartment house and the use of the apartments and the common property thereof, form a partnership of the owners of the apartments (or housing).

2. The partnership of owners of the housing shall be a noncommercial organisation created and operating in accordance with the law on partnerships of owners of housing.

Article 292. Rights of Members of Family of Owners of Dwelling Premise

1. The members of the family of the owner residing in a dwelling premise belonging to him shall have the right to use this premise on the conditions provided for by housing legislation.

Members of a family of the owner who have dispositive legal capacity and reside in a dwelling premise belonging to him shall bear joint and several responsibility with the owner for obligations arising from the use of the dwelling premise [added 15 May 2001. СЗ РФ (2001), no. 21, item 2063].

2. The transfer of the right of ownership to a dwelling house or apartment to another person shall not be grounds for termination of the right of use of the

dwelling premise by members of the family of the former owner, unless established otherwise by a law [as amended 15 May 2001. СЗ РФ (2001), no. 21, item 2063].

3. The members of the family of the owner of the dwelling premise may demand the elimination of the violations of their rights to a dwelling premise from any persons, including the owner of the premise.

4. The alienation of the dwelling premise in which minors, persons lacking dispositive legal capacity, or limited in dispositive legal capacity who are members of the family of the owner reside, if in so doing the rights or interests of the said persons protected by a law are affected, shall be permitted with the consent of the agency of trusteeship and guardianship [as amended 15 May 2001. СЗ РФ (2001), no. 21, item 2063].

Article 293. Termination of Right of Ownership in Improvidently Maintained Dwelling Premise

If the owner of a dwelling premise uses it not for the designation or systematically violates the rights and interests of neighbours, or treats the dwelling improvidently, allowing the destruction thereof, the agency of local self-government may warn the owner about the need to eliminate violations, and if they entail destruction of the premise, also to designate a commensurate period for repair of the premise to the owner.

If the owner continues, after a warning, to violate the rights and interests of the neighbours or to use the dwelling premise not for its designation, or without justifiable reasons does not make the necessary repair, the court may at the suit of the agency of local self-government adopt a decision concerning the sale at a public sale of this dwelling premise with payment to the owner of means derived from the sale, deducting expenses for execution of the judicial decision.

Chapter 19. Right of Economic Jurisdiction, Right of Operative Management

Article 294. Right of Economic Jurisdiction

A State or municipal unitary enterprise to which property belongs by right of economic jurisdiction shall possess, use, and dispose of this property within the limits determined in accordance with the present Code.

Article 295. Rights of Owner with Respect to Property in Economic Jurisdiction

1. The owner of property in economic jurisdiction shall, in accordance with a law, decide questions of the creation of the enterprise and determination of the subject and purposes of its activity, its reorganisation, and liquidation, appoint the director of the enterprise, and shall effectuate control over use according to designation and preservation of property belonging to the enterprise.

The owner shall have the right to receive part of the profit from the use of the property in the economic jurisdiction of the enterprise.

2. An enterprise shall not have the right to sell immovable property belonging to it by right of economic jurisdiction, lease it out, pledge it, contribute it as a contribution to the charter (or contributed) capital of economic societies and partnerships or by other means dispose of this property without the consent of the owner.

The remaining property belonging to the enterprise it shall dispose of autonomously, except for the instances established by a law or other legal acts.

Article 296. Right of Operative Management

1. A treasury enterprise, and also an institution, shall effectuate with respect to property consolidated to it within the limits established by a law and in accordance with the purposes of its activity, the planning tasks of the owner, and the designation of the property, the rights of possession, use, and disposition thereof.

2. The owner of the property consolidated to a treasury enterprise or institution shall have the right to withdraw surplus or unused property or property not used according to its designation, and to dispose of it according to his discretion.

Article 297. Disposition of Property of Treasury Enterprise

1. A treasury enterprise shall have the right to alienate or by other means to dispose of property consolidated to it only with the consent of the owner of this property.

A treasury enterprise autonomously shall realise the product produced by it unless established otherwise by a law or other legal acts.

2. The procedure for the distribution of revenues of the treasury enterprise shall be determined by the owner of its property.

Article 298. Disposition of Property of Institution

1. An institution shall not have the right to alienate or by other means to dispose of property consolidated to it and property acquired at the expense of assets allotted to it under the estimate.

2. If in accordance with the constitutive documents an institution has been granted the right to effectuate activity which brings revenues, the revenues received from such activity and the property acquired at the expense of such revenues shall be at the autonomous disposition of the institution and shall be taken into account on a separate balance sheet.

Article 299. Acquisition and Termination of Right of Economic Jurisdiction and Right of Operative Management

1. The right of economic jurisdiction or the right of operative management of property with respect to which a decision has been adopted by the owner

concerning the consolidation to a unitary enterprise or institution shall arise at this enterprise or institution from the moment of the transfer of the property, unless established otherwise by a law and other legal acts or by decision of the owner.

2. The fruits, products, and revenues from the use of the property in economic jurisdiction or operative management, and also the property acquired by a unitary enterprise or institution under contract or other grounds, shall enter the economic jurisdiction or the operative management of the enterprise or institution in the procedure established by the present Code, other laws, and other legal acts for acquisition of the right of ownership.

3. The right of economic jurisdiction and the right of operative management of property shall terminate upon the grounds and in the procedure provided for by the present Code, other laws, and other legal acts for the termination of the right of ownership, and also in instances of the lawful withdrawal of the property from the enterprise or institution by decision of the owner.

Article 300. Retention of Rights to Property in Event of Transfer of Enterprise or Institution to Another Owner

1. In the event of the transfer of the right of ownership in a State or municipal enterprise as a property complex to another owner of State or municipal property, this enterprise shall retain the right of economic jurisdiction or the right of operative management in the property belonging to it [as amended by the Law of 14 November 2002. СЗ РФ (2002), no. 48, item 4746].

2. In the event of the transfer of the right of ownership in an institution to another person, this institution shall retain the right of operative management to the property belonging to it.

Chapter 20. Defence of Right of Ownership and Other Rights to Thing

Article 301. Demanding and Obtaining Property from Another's Illegal Possession

The owner shall have the right to demand and obtain his property from another's illegal possession.

Article 302. Demanding and Obtaining Property from a Good-Faith Acquirer

1. If property has been acquired for compensation from a person who did not have the right to alienate it, of which the acquirer did not know and could not have known (good-faith acquirer), then the owner shall have the right to demand and obtain this property from the acquirer when the property has been lost by the owner or person to whom the property was transferred by the owner in possession, or stolen from one or the other, or left the possession thereof by means other than the will thereof.

2. If property was acquired without compensation from a person who did not have the right to alienate it, the owner shall have the right to demand and obtain the property in all instances.

3. Money, and also bearer securities, may not be demanded and obtained from a good-faith acquirer.

Article 303. Settlement of Accounts in Event of Return of Property from Illegal Possession

When demanding and obtaining property from another's illegal possession the owner also shall have the right to demand and obtain from a person who knew or should have known that his possession is illegal (possessor not in good faith) the return or compensation of all revenues which this person derived or should have been derived throughout the entire period of possession; and from a good-faith possessor, the return of compensation of all revenues which he derived or should have derived from the time when he knew or should have known about the unlawfulness of the possession or received a writ relating to the suit of the owner for the return of the property.

A possessor, both in good faith and not in good faith, in turn shall have the right to demand from the owner compensation for necessary expenditures made by him on the property for that time from which revenues from the property are due to the owner.

A possessor in good faith shall have the right to retain for himself improvements made by him if they can be separated without damaging the property. If such a separation of the improvements is impossible, a good-faith possessor shall have the right to demand compensation for expenditures made for the improvement, but not more than the amount of the increase of the value of the property.

Article 304. Defence of Rights of Owner Against Violations Not Connected with Deprivation of Possession

An owner may demand the elimination of any violations of his right, even though these violations are not combined with deprivation of possession.

Article 305. Defence of Rights of Possessor Who is Not the Owner

The rights provided for by Articles 301–304 of the present Code shall belong also to a person who, although not the owner, is the possessor of the property by right of inheritable possession for life, economic jurisdiction or operative management or on another grounds provided for by a law or contract. This person shall have the right to the defence of his possession against the owner as well.

Article 306. Consequences of Termination of Right of Ownership by Virtue of Law

In the event of the adoption by the Russian Federation of a law terminating the right of ownership the losses caused to the owner as a result of the adoption of this act, including the value of the property, shall be compensated by the State. Disputes concerning compensation of losses shall be settled by a court.

SECTION III. GENERAL PART OF THE LAW OF OBLIGATIONS

Subsection 1. General Provisions on Obligations

Chapter 21. Concept of and Parties to Obligation

Article 307. Concept of Obligation and Grounds of its Origin

1. By virtue of an obligation one person (the debtor) shall be obliged to perform to the benefit of another person (the creditor) a determined action, that is, to transfer property, fulfil work, pay money, and the like or to refrain from a determined action, and the creditor shall have the right to demand from the debtor the performance of his duty.

2. Obligations shall arise from a contract, as a consequence of the causing of harm, and from other grounds specified in the present Code.

Article 308. Parties to Obligation

1. One or simultaneously several persons may participate in an obligation as each of its parties—creditor or debtor.

The invalidity of the demands of the creditor with respect to one of the persons participating in the obligation as a debtor, and likewise the expiry of the period of limitations with regard to the demand against such person, shall not in and of itself affect his demands against the other such persons.

2. If each of the parties to a contract bears a duty in favour of the other party, it shall be considered to be the debtor of the other party in that which he is obliged to do to its benefit, and simultaneously the creditor thereof in that which it has the right to demand from it.

3. An obligation shall not create duties for persons who are not participating therein as parties (for third persons).

In the instances provided for by a law, other legal acts, or by agreement of the parties an obligation may create rights for third persons with respect to one or both parties to the obligation.

Chapter 22. Performance of Obligations

Article 309. General Provisions

Obligations must be performed duly in accordance with the conditions of the obligation and the requirements of a law and other legal acts, and in the absence of such conditions and requirements—in accordance with the customs of business turnover or other usually presented requirements.

Article 310. Inadmissibility of Unilateral Refusal to Perform Obligation

A unilateral refusal to perform an obligation and a unilateral change of its conditions shall not be permitted except for instances provided for by a law. A unilateral refusal to perform an obligation connected with the effectuation by its parties of entrepreneurial activity, and a unilateral change of conditions of such obligation shall also be permitted in the instances provided for by the contract unless it follows otherwise from a law or the essence of the obligation.

Article 311. Performance of Obligation in Parts

A creditor shall have the right not to accept the performance of an obligation in parts unless provided otherwise by a law, other legal acts, or by the conditions of the obligation and does not arise from the customs of business turnover or the essence of the obligation.

Article 312. Performance of Obligation to Proper Person

Unless provided otherwise by agreement of the parties and does not arise from the customs of business turnover or the essence of the obligation, the debtor shall have the right when performing the obligation to demand evidence that performance is accepted by the creditor himself or by a person empowered by him and shall bear the risk of the consequences of the failure to present such demand.

Article 313. Performance of Obligation by Third Person

1. Unless the duty of the debtor to perform an obligation personally arises from a law, other legal acts, the conditions of the obligation, or the essence thereof, the performance of the obligation may be placed by the debtor on a third person. In this event the creditor shall be obliged to accept the performance offered by a third person for the debtor.

2. A third person who is subject to the danger of losing his right to the property of the debtor (right of lease, pledge, or others) as a consequence of execution being levied by the creditor on this property may at his own expense satisfy the demand of the creditor without the consent of the debtor. In this event the rights of the creditor regarding the obligation shall pass to the third person in accordance with Articles 382–387 of the present Code.

Article 314. Period for Performance of Obligation

1. If an obligation provides for or enables the date of its performance or the period of time during which it must be performed to be determined, the obligation shall be subject to performance on that day or, respectively, at any moment within the limits of that period.

2. When an obligation does not provide for a period for its performance and does not contain conditions enabling this period to be determined, it must be performed within a reasonable period after the origin of the obligation.

An obligation not performed within a reasonable period, and likewise an obligation whose period of performance has been determined by the moment of demand, the debtor shall be obliged to perform within a seven-day period from the day of presentation by the creditor of the demand concerning the performance thereof unless the duty to perform within another period arises from a law, other legal acts, the conditions of the obligation, the customs of business turnover, or the essence of the obligation.

Article 315. Performance of Obligation Before Time

The debtor shall have the right to perform an obligation before the period unless provided otherwise by a law, other legal acts, or the conditions of the obligation, or does not arise from the essence thereof. However, the performance of an obligation before time connected with the effectuation of entrepreneurial activity by the parties thereof shall be permitted only in instances when the possibility to perform an obligation before the period provided for by a law, other legal acts, or by the conditions of the obligation or arises from the customs of business turnover or the essence of the obligation.

Article 316. Place of Performance of Obligation

If the place of performance has not been determined by a law, other legal acts, or by contract and is not manifest from customs of business turnover or the essence of the obligation, performance must be made:

with regard to an obligation to transfer a land plot, building, installation, or other immovable property: at the location of the property;

with regard to an obligation to transfer a good or other property providing for the carriage thereof: at the place of handing over the property to the first carrier for delivery to the creditor;

with regard to other obligations of an entrepreneur to transfer a good or other property: at the place of manufacture or keeping of the property if this place was known to the creditor at the moment of origin of the obligation;

with regard to a monetary obligation: at the place of residence of the creditor at the moment of origin of the obligation, and if the creditor is a juridical person— at its location at the moment of origin of the obligation; if the creditor at the moment of performance of the obligation has changed place of residence or location and notified the debtor thereof—at the new place of residence or location of the creditor, relegating the expenses connected with the change of place of performance to the account of the creditor;

with regard to all other obligations: at the place of residence of the debtor, and if the debtor is a juridical person—at its location.

Article 317. Currency of Monetary Obligations

1. Monetary obligations must be expressed in rubles (Article 140).

2. It may be provided in a monetary obligation that it shall be subject to payment in rubles in an amount equivalent to a determined amount in foreign currency or in prearranged monetary units (ecu, special borrowing rights, and others). In this event the amount subject to payment in rubles shall be determined according to the official exchange rate of the respective currency or prearranged monetary units on the day of payment unless other exchange rate or another date for determining it has been established by a law or by agreement of the parties.

3. The use of foreign currency, and also of payment documents in foreign currency, shall be permitted when effectuating the settlement of accounts on the territory of the Russian Federation with regard to obligations in the instances and on the conditions determined by a law or in the procedure established by it.

Article 318. Increase of Amounts Payable for Maintenance of Citizen

An amount payable with regard to a monetary obligation directly for the maintenance of a citizen: in compensation of harm caused to life or health, under a contract of maintenance for life, and in other instances—shall be indexed by taking into account the level of inflation in the proceedure and instances which have been provided for by a law [as amended by Law of 26 November 2002. СЗ РФ (2002), no. 48, item 4737].

Article 319. Priority of Payment of Demands Relating to Monetary Obligation

The amount of payment made which is insufficient for the performance of a monetary obligation in full shall in the absence of another agreement pay first of all the costs of the creditor with regard to the receipt of performance and then, interest, and last, the principal amount of the debt.

Article 320. Performance of Alternative Obligation

The right of choice, unless it arises otherwise from a law, other legal acts, or the conditions of the obligation, shall belong to the debtor who is obliged to transfer to the creditor one property or another or to perform one of two or several actions.

Article 321. Performance of Obligation in Which Several Creditors or Several Debtors Participate

If several creditors or several debtors participate in an obligation, each of the creditors shall have the right to demand performance, and each of the debtors shall be obliged to perform the obligation in equal participatory share with the others insofar as it does not arise otherwise from a law, other legal acts, or the conditions of the obligation.

Article 322. Joint and Several Obligations

1. A joint and several duty (or responsibility) or a joint and several demand shall arise if the joint-and-severalness of the duty or demand has been provided for by a contract or established by a law, in particular, when the subject of the obligation is indivisible.

2. The duties of several debtors relating to an obligation connected with entrepreneurial activity, and likewise also demands of several creditors to such obligation, shall be joint and several unless provided otherwise by a law, other legal acts, or by the conditions of the obligation.

Article 323. Rights of Creditor in Event of Joint and Several Duty

1. In the event of the joint and several duty of debtors, the creditor shall have the right to demand performance both from all of the debtors jointly or from any of them individually, either for the whole or for part of the debt.

2. A creditor who has not received full satisfaction from one of the joint and several debtors shall have the right to demand that which has not been received from the other joint and several debtors.

Joint and several debtors shall remain obliged so long as the obligation is not performed in full.

Article 324. Objections to Demands of Creditor in Event of Joint and Several Duty

In the event of the joint and several duty a debtor shall not have the right to advance objections against the demand of a creditor based on those relations of other debtors with the creditor in which the said debtor does not participate.

Article 325. Performance of Joint and Several Duty by One of Debtors

1. The performance of a joint and several duty in full by one of the debtors shall relieve the other debtors from performance to the creditor.

2. Unless it arises otherwise from relations between the joint and several debtors:

(1) the debtor who has performed the joint and several duty shall have the right of a regressive demand against the other debtors in equal participatory shares, deducting the participatory share falling on himself;

(2) that unpaid by one of the joint and several debtors to the debtor who has performed the joint and several duty shall fall in equal participatory shares on this debtor and on the remaining debtors.

3. The rules of the present Article shall apply respectively when terminating the joint and several obligation by setting-off the counter-demand of one of the debtors.

Article 326. Joint and Several Demands

1. In the event of the joint and severalness of a demand, any of the joint and several creditors shall have the right to present the demand to the debtor in full.

Until the demand is presented by one of the joint and several creditors, the debtor shall have the right to perform the obligation to any of them at his discretion.

2. The debtor shall not have the right to advance against the demand of one of the joint and several creditors objections based on those relations of the debtor with another joint and several creditor in which the particular creditor did not participate.

3. The performance of the obligation in full to one of the joint and several creditors shall relieve the debtor from performance to the other creditors.

4. A joint and several creditor who has received performance from the debtor shall be obliged to compensate that which is due to the other creditors unless it arises otherwise from the relations between them in the equal participatory shares.

Article 327. Performance of Obligation by Placing Debt on Deposit

1. A debtor shall have the right to place money due from him or securities on deposit with a notary, and in the instances established by a law, on deposit with a court if the obligation cannot be performed by the debtor as a consequence of:

(1) the absence of the creditor or person empowered by him to accept performance at the place where the obligation must be performed;

(2) the lack of dispositive legal capacity of the creditor and absence of his representative;

(3) the evident lack of specificity as to who is the creditor with regard to the obligation, in particular, in connection with a dispute in this regard between the creditor and other persons;

(4) avoidance by the creditor of accepting performance or other delay on his part.

2. The placing of a monetary amount or securities on deposit with a notary or court shall be considered to be performance of the obligation.

The notary or court with whom the deposit has been made of money or securities shall notify the creditor thereof.

Article 328. Counter Performance of Obligations

1. The performance of an obligation by one of the parties which in accordance with a contract is conditioned by the performance by the other party of its obligations shall be deemed to be counter performance.

2. In the event of the failure of an obliged party to provide the performance of an obligation stipulated by the contract or of the existence of circumstances

obviously testifying that such performance will not be made within the established period, the party on whom counter performance lies shall have the right to suspend the performance of this obligation or to waive the performance of this obligation and demand compensation of losses.

If the performance of an obligation stipulated by a contract was not made in full, the party on whom the counter performance lies shall have the right to suspend the performance of this obligation or to waive performance in that part corresponding to the performance not provided.

3. If counter performance of an obligation has been made notwithstanding the failure of the other party to provide performance of his obligation stipulated by the contract, this party shall be obliged to provide such performance.

4. The rules provided for by points 2 and 3 of the present Article shall apply unless provided otherwise by the contract or by a law.

Chapter 23. Securing Performance of Obligations

§1. General Provisions

Article 329. Means of Securing the Performance of Obligations

1. The performance of obligations may be secured by a penalty, pledge, withholding of property from the debtor, suretyship, bank guarantee, deposit, and other means provided for by a law or by a contract.

2. The invalidity of an agreement to secure the performance of an obligation shall not entail the invalidity of this obligation (principal obligation).

3. The invalidity of the principal obligation shall entail the invalidity of the obligation securing it unless established otherwise by a law.

§2. Penalty

Article 330. Concept of Penalty

1. A penalty (or fine, forfeit) shall be deemed to be a monetary amount determined by a law or by contract which the debtor shall be obliged to pay to the creditor in the event of the failure to perform or the improper performance of an obligation, in particular in the event of the delay of performance. With regard to a demand concerning payment of a penalty the creditor shall not be obliged to prove the causing of losses to him.

2. A creditor shall not have the right to demand payment of a penalty unless the debtor bears responsibility for the failure to perform or the improper performance of the obligation.

Article 331. Form of Agreement on Penalty

An agreement concerning a penalty must be concluded in written form irrespective of the form of the principal obligation.

The failure to comply with the written form shall entail the invalidity of the agreement concerning the penalty.

Article 332. Legal Penalty

1. A creditor shall have the right to demand the payment of a penalty specified by a law (legal penalty) irrespective of whether the duty to pay it has been provided for by agreement of the parties.

2. The amount of a legal penalty may by agreement of the parties be increased if the law does not so prohibit.

Article 333. Reduction of Penalty

If a penalty subject to payment is clearly incommensurate to the consequences of the violation of the obligation, a court shall have the right to reduce the penalty.

The rules of the present Article shall not affect the rights of the debtor to a reduction of the amount of his responsibility on the basis of Article 404 of the present Code and the rights of the creditor to compensation of losses in the instances provided for by Article 394 of the present Code.

§3. Pledge

Article 334. Concept and Grounds for Arising of Pledge

1. By virtue of a pledge the creditor with regard to an obligation secured by a pledge (pledgeholder) shall have the right, in the event of the failure of the debtor to perform this obligation, to receive satisfaction from the value of the pledged property preferentially before other creditors of the person to whom this property (pledgor) belongs with the exceptions established by a law.

The pledgeholder shall have the right to receive on the same principles satisfaction from insurance compensation for loss or damage of the pledged property irrespective of to whose benefit it was insured unless the loss or damage occurred for reasons for which the pledgeholder is liable.

2. A pledge of land plots, enterprises, buildings, installations, apartments, and other immovable property (mortgage) shall be regulated by a law on mortgage. The general rules on pledge contained in the present Code shall apply to mortgage in the instances when other rules have not been established by the present Code or by the law on mortgage.

3. A pledge shall arise by virtue of a contract. Pledge also shall arise on the basis of a law when the circumstances specified therein ensue if it is provided in the law

which property and for securing the performance of which obligation there is deemed to be a pledge.

The rules of the present Code on pledge arising by virtue of contract shall respectively apply to a pledge arising on the basis of a law unless established otherwise by a law.

Article 335. Pledgor

1. Either the debtor himself or a third person may be a pledgor.

2. The pledgor of a thing may be the owner thereof or a person having the right of economic jurisdiction over it.

The person to whom a thing belongs by right of economic jurisdiction shall have the right to pledge it without the consent of the owner in the instances provided for by Article 295(2) of the present Code.

3. The pledgor of a right may be the person to whom the pledged right belongs.

The pledge of the right of lease or other right to another's thing shall be permitted without the consent of its owner or the person having the right of economic jurisdiction over it unless the alienation of this right without the consent of the said persons has been prohibited by a law or by contract.

Article 336. Subject of Pledge

1. Any property, including things and property rights (or demands), may be the subject of a pledge except for property withdrawn from turnover, demands inextricably connected with the person of the creditor, in particular, demands concerning alimony, compensation for harm caused to life or health, and other rights whose assignment to another person has been prohibited by a law.

2. The pledge of individual types of property, in particular the property of citizens against which levy of execution is not permitted, may be prohibited or limited by a law.

Article 337. Demand Secured by Pledge

Unless provided otherwise by a contract, a pledge shall secure a demand in that amount which it has at the moment of satisfaction, in particular, interest, penalty, compensation of losses caused by delay of performance, and also compensation of necessary expenses of the pledgeholder for maintenance of the pledged thing and expenses relating to recovery.

Article 338. Pledge With and Without Transfer of Pledged Property to Pledgeholder

1. Pledged property shall remain with the pledgor unless provided otherwise by contract.

Property on which a mortgage is established, and also pledged goods in turnover, shall not be transferred to the pledgeholder.

2. The subject of a pledge may be left with the pledgor under lock and seal of the pledgeholder.

The subject of a pledge may be left with the pledgor with the imposition of marks testifying to the pledge (firm pledge).

3. The subject of a pledge transferred by the pledgor for a time in possession and use to a third person shall be considered to be a pledge while leaving it with the pledgor.

4. In the event of the pledge of a property right certified by a security, it shall be transferred to the pledgeholder or on deposit to a notary unless provided otherwise by the contract.

Article 339. Contract on Pledge, Its Form and Registration

1. In the contract on pledge must be specified the subject of pledge and its valuation, the essence, amount, and period for performance of the obligation secured by the pledge. It also must contain an indication of with which of the parties the pledged property is situated.

2. A contract on pledge must be concluded in written form.

A contract on mortgage, and also contracts on the pledge of movable property or rights to property to secure obligations under a contract which must be notarially certified, shall be subject to notarial certification.

3. A contract on mortgage must be registered in the procedure established for the registration of transactions with the respective property.

4. The failure to comply with the rules contained in points 2 and 3 of the present Article shall entail the invalidity of the contract on pledge.

Article 340. Property to Which Rights of Pledgeholder Extends

1. The rights of the pledgeholder (right of pledge) to a thing which is the subject of pledge shall extend to its appurtenances unless provided otherwise by contract.

The right of pledge shall extend to fruits, products, and revenues received as a result of the use of the pledged property in the instances provided for by contract.

2. In the event of the mortgage of an enterprise or other property complex as a whole the right of pledge shall extend to all of its property, movable and immovable, including the rights of demand and exclusive rights, amongst them those acquired during the period of the mortgage, unless provided otherwise by a law or contract.

3. The mortgage of a building or installation shall be permitted only with the simultaneous mortgage under the same contract of the land plot on which this building or installation is situated, or the part of this plot functionally ensuring the pledged object, or the right of lease of this plot or corresponding part thereof which belongs to the pledgor.

4. In the event of the mortgage of a land plot, the right of pledge shall not extend to buildings and installations of the pledgor situated or erected on this plot unless a condition provides otherwise in the contract.

In the absence in the contract of such a condition the pledgor shall in the event execution is levied on the pledged land plot retain the right of limited use (servitude) of that part thereof which is necessary for the use of the building or installation in accordance with its designation. Conditions for the use of this part of the plot shall be determined by agreement of the pledgor with the pledgeholder, and in the event of a dispute, by a court.

5. If a mortgage has been established on a land plot on which a building or installation is situated which belongs not to the pledgor, but to another person, then when the pledgeholder levies execution on this plot and in the event of its sale at a public sale, to the acquirer of the plot shall pass the rights and duties which the pledgor had with respect to this person.

6. The pledge of things and property rights which the pledgor acquires in future may be provided for by the contract on pledge, and with respect to a pledge arising on the basis of a law, by the law.

Article 341. Origin of Right of Pledge

1. Unless provided otherwise by the contract on pledge, the right of pledge shall arise from the moment of conclusion of the contract on pledge, and with respect to the pledge of property which is subject to transfer to the pledgeholder, from the moment of transfer of this property.

2. The right of pledge to goods in turnover shall arise in accordance with the rules of Article 357(2) of the present Code.

Article 342. Subsequent Pledge

1. If property under pledge becomes the subject of yet another pledge to secure other demands (subsequent pledge), the demands of the subsequent pledgeholder shall be satisfied from the value of this property after the demands of preceding pledgeholders.

2. A subsequent pledge shall be permitted if it is not prohibited by the preceding contracts on pledge.

3. The pledgor shall be obliged to communicate to each subsequent pledgeholder information about all of the existing pledges of the particular property

provided for by Article 339(1) of the present Code and shall be liable for losses caused to pledgeholders by the failure to fulfil this duty.

Article 343. Content and Preservation of Pledged Property

1. The pledgor or pledgeholder, depending upon which of them has the pledged property (Article 338), shall be obliged, unless provided otherwise by law or contract, to:

(1) insure pledged property at its full value at the expense of the pledgor against risks of loss and damage, and if the full value of the property exceeds the amount of the demand secured by the pledge, for an amount not lower than the amount of the demand;

(2) take measures necessary in order to ensure the preservation of the pledged property, including to defend it against infringements and demands on the part of third persons;

(3) inform immediately the other party concerning the arising of the threat of loss or damage to the pledged property.

2. The pledgeholder and the pledgor shall have the right to verify by documents the actual existence, quantity, state, and conditions of keeping the pledged property situated with the other party.

3. In the event of a flagrant violation by the pledgeholder of the duties specified in point 1 of the present Article which create a threat of loss or damage of the pledged property, the pledgor shall have the right to demand the termination of the pledge before time.

Article 344. Consequences of Loss of or Damage to Pledged Property

1. The pledgor shall bear the risk of accidental perishing or accidental damage to pledged property unless provided otherwise by the contract on pledge.

2. The pledgeholder shall be liable for the full or partial loss of or damage to the subject of pledge transferred to him unless it is proved that he may be relieved of responsibility in accordance with Article 401 of the present Code.

A pledgeholder shall be liable for the loss of the subject of pledge in the amount of its real value, and for the damaging thereof—the amount by which the value was reduced irrespective of the amount at which the subject of pledge was valued when transferring it to the pledgeholder.

If as a result of damage to the subject of pledge it has so changed that it cannot be used for its express designation, the pledgor shall have the right to reject it and to demand compensation for the loss thereof.

The duty of the pledgeholder to compensate the pledgor for other losses caused by the loss of or damage to the subject of pledge may be provided for by a contract.

The pledgor who is a debtor with regard to the obligation secured by a pledge shall have the right to set off the demand against the pledgeholder concerning compensation of losses caused by the loss of or damage to the subject of pledge in repaying the obligation secured by the pledge.

Article 345. Substitution and Restoration of Subject of Pledge

1. The replacement of the subject of pledge shall be permitted with the consent of the pledgeholder unless provided otherwise by law or by contract.

2. If the subject of pledge has perished or is damaged or the right of ownership therein or the right of economic jurisdiction has been terminated on the grounds established by a law, the pledgor shall have the right within a reasonable period to restore the subject of pledge or substitute other property of equal value unless provided otherwise by the contract.

Article 346. Use and Disposition of Subject of Pledge

1. The pledgor shall have the right, unless provided otherwise by contract and it follows from the essence of the pledge, to use the subject of pledge in accordance with its designation, including to derive fruits and revenues from it.

2. Unless provided otherwise by a law or contract and it does not arise from the essence of the pledge, the pledgor shall have the right to alienate the subject of pledge, to transfer it on lease or for use without compensation to another person, or otherwise to dispose of it only with the consent of the pledgeholder.

An agreement limiting the right of the pledgor to bequeath pledged property shall be void.

3. The pledgeholder shall have the right to use the subject of pledge transferred to him only in the instances provided for by contract, regularly submitting a report on use to the pledgor. The duty may be placed under a contract on the pledgeholder to derive fruits and revenues from the subject of pledge for the purposes of repaying the principal obligation or in the interests of the pledgor.

Article 347. Defence by Pledgeholder of His Rights to Subject of Pledge

1. The pledgeholder with whom the pledged property is situated or should be situated shall have the right to demand and obtain it from another's illegal possession, including from the possession of the pledgor (Articles 301, 302, 305).

2. In instances when according to the conditions of the contract the pledgeholder has been granted the right to use a subject of pledge transferred to him, he may demand from other persons, including from the pledgor, the elimination of any violations of his right, even though these violations were not linked with deprivation of possession (Articles 304, 305).

Article 348. Grounds for Levying Execution on Pledged Property

1. Execution may be levied on pledged property in order to satisfy the demands of the pledgeholder (creditor) in the event of the failure to perform or

the improper performance by the debtor of an obligation secured by the pledge for which he is liable.

2. Levy of execution on pledged property may be refused if the violation of the obligation secured by the pledge which was permitted by the debtor is extremely insignificant and the amount of the demands of the pledgeholder as a consequence thereof is clearly incommensurate with the value of the pledged property.

Article 349. Procedure for Levy of Execution on Pledged Property

1. The demands of the pledgeholder (creditor) shall be satisfied from the value of pledged immovable property by decision of a court.

The satisfaction of the demand of a pledgeholder at the expense of the pledged immovable property without recourse to a court shall be permitted on the basis of a notarially certified agreement of the pledgeholder with the pledgor concluded after the grounds arise to levy execution on the subject of pledge. Such an agreement may be deemed by a court to be invalid upon the suit of the person whose rights were violated by such agreement.

2. The demands of a pledgeholder at the expense of pledged movable property shall be satisfied by decision of a court unless provided otherwise by agreement of the pledgor with the pledgeholder. However, execution may be levied on the subject of a pledge transferred to the pledgeholder in the procedure established by the contract on pledge unless a different procedure has been established by a law.

3. Execution may be levied on the subject of pledge only by decision of a court in instances when:

(1) in order to conclude a contract on pledge the consent or authorisation of another person or agency is required;

(2) property having significant historical, artistic, or other cultural value for society is the subject of pledge;

(3) the pledgor is absent and it is impossible to establish his whereabouts.

Article 350. Realisation of Pledged Property

1. Pledged property on which execution is levied in accordance with Article 349 of the present Code shall be realised (or sold) by means of sale at public sale in the procedure established by procedural legislation unless a different procedure has been established by a law.

2. At the request of the pledgor the court shall have the right in deciding to levy execution on pledged property to defer the sale thereof at public sale for a period of up to one year. The deferral shall not affect the rights and duties of the parties with regard to an obligation secured by the pledge of this property and shall not relieve the debtor from compensating the losses of the creditor and the penalty which grew during the period of deferral.

3. The initial sale price of pledged property at which the public sale begins shall be determined by decision of a court in the instances of levying execution on property in a judicial proceeding or by agreement of the pledgeholder with the pledgor in other instances.

The pledged property shall be sold to the person who offered the highest price at the public sale.

4. When a public sale is declared to be unconstituted, a pledgeholder shall have the right by agreement with the pledgor to acquire the pledged property and to set off against the purchase price his demands secured by the pledge. The rules concerning the contract of purchase/sale shall apply to such an agreement.

In the event a second public sale is declared to be unconstituted, the pledgeholder shall have the right to retain the subject of pledge, valuing it in an amount of not more than 10% less than the initial sale price at the second public sale.

If the pledgeholder does not take advantage of the right to retain the subject of pledge within a month from the date of announcement of the second public sale being unconstituted, the contract on pledge shall terminate.

5. If the amount received when realising the pledged property is insufficient to cover the demand of the pledgeholder, he shall have the right in the absence of another indication in a law or the contract to receive the amount in arrears from other property of the debtor which does not enjoy a preference based on pledge.

6. If the amount received in the event of the realisation of the pledged property exceeds the amount of the demand of the pledgeholder secured by the pledge, the difference shall be returned to the pledgor.

7. A debtor and pledgor who are a third person shall have the right at any time before the sale of the subject of pledge to terminate the levy of execution on it and the realisation thereof, having performed the obligation secured by the pledge or that part thereof whose performance has been delayed. An agreement limiting this right shall be void.

Article 351. Performance of Obligation Before Time Secured by Pledge and Levy of Execution on Pledged Property

1. A pledgeholder shall have the right to demand the performance before time of an obligation secured by pledge in instances of:

(1) the subject of pledge departed from the possession of the pledgor with whom it was left not in accordance with the conditions of the contract on pledge;

(2) a violation by the pledgor of the rules concerning substitution of the subject of pledge (Article 345);

(3) loss of the subject of pledge under circumstances for which the pledgeholder is not liable if the pledgor has not taken advantage of the right provided for by Article 345(2) of the present Code.

2. The pledgeholder shall have the right to demand the performance before time of the obligation secured by pledge and, if this demand is not satisfied, to levy execution against the subject of pledge in instances of:

(1) a violation by the pledgor of rules concerning subsequent pledge (Article 342);

(2) the failure of the pledgor to fulfil the duties provided for by Article 343(1), subpoints (1) and (2) and Article 343(2) of the present Code;

(3) a violation by the pledgor of the rules concerning the disposition of pledged property (Article 346[2]).

Article 352. Termination of Pledge

1. A pledge shall terminate:

(1) with the termination of the obligation secured by the pledge;

(2) upon the demand of the pledgor when there exist the grounds provided for by Article 343(3) of the present Code;

(3) in the event of the perishing of the pledged thing or termination of the pledged right unless the pledgor has taken advantage of the right provided for by Article 345(2) of the present Code;

(4) in the event of the sale of pledged property at public sale, and also when the realisation thereof has proved to be impossible (Article 350[4]).

2. A notation in the register in which the contract on mortgage was registered must be made concerning the termination of the mortgage.

3. In the event of the termination of pledge as a consequence of the performance of the obligation secured by the pledge or at the demand of the pledgor (Article 343[3]), the pledgeholder with whom the pledged property is situated shall be obliged immediately to return it to the pledgor.

Article 353. Preservation of Pledge in Event of Transfer of Right to Pledged Property to Another Person

1. In the event of the transfer of the right of ownership to pledged property or the right of economic jurisdiction from the pledgor to another person as a result of the alienation of this property with or without compensation or by way of universal legal succession, the right of pledge shall retain force.

The legal successor of the pledgor shall take the place of the pledgor and shall bear all the duties of the pledgor, unless established otherwise by an agreement with the pledgeholder.

2. If the property of the pledgor which is the subject of pledge has passed by way of legal succession to several persons, each of the legal successors (or acquirers of the property) shall bear the consequences arising from the pledge for the failure to perform the obligation secured by the pledge commensurate with the part of the said property which has passed to him. However, if the subject of the pledge is

indivisible or on other grounds remains in the common joint ownership of the legal successors, they shall become joint and several pledgors.

Article 354. Consequences of Compulsory Withdrawal of Pledged Property

1. If the right of ownership of a pledgor to property which is the subject of pledge terminates on the grounds and in the procedure established by a law as a consequence of the withdrawal (or purchase) for State or municipal needs, requisition, or nationalisation, and the pledgor is granted other property or respective compensation, the right of pledge shall extend to the property granted instead or, respectively, the pledgeholder shall acquire the right of preferential satisfaction of his demand from the amount of compensation due to the pledgor. The pledgeholder also shall have the right to demand the performance before time of the obligation secured by the pledge.

2. In the instances when property which is the subject of pledge is withdrawn from the pledgor in the procedure established by a law on the grounds that in reality the owner of this property is another person (Article 301), or in the form of a sanction for the commission of a crime or another violation of law (Article 243), the pledge with respect to this property shall terminate. In these instances the pledgeholder shall have the right to demand performance before time of the obligation secured by the pledge.

Article 355. Assignment of Rights Regarding Contract on Pledge

A pledgeholder shall have the right to transfer his rights under a contract on pledge to another person while complying with the rules on the transfer of the rights of a creditor by means of the assignment of a demand (Articles 382–390).

The assignment by the pledgeholder of his rights under a contract on pledge to another person shall be valid if the rights of demand against the debtor regarding the principal obligation secured by pledge have been assigned.

Unless proved otherwise, the assignment of rights under a contract on mortgage shall also mean the assignment of rights relating to the obligation secured by the mortgage.

Article 356. Transfer of Debt Regarding Obligation Secured by Pledge

With the transfer of a debt regarding an obligation secured by a pledge to another person, the pledge shall terminate unless the pledgor has given consent to the creditor to be liable for the new debtor.

Article 357. Pledge of Goods in Turnover

1. A pledge of goods while leaving them with the pledgor and granting to the pledgor the right to change the composition and natural form of the pledged property (goods reserves, raw material, materials, semi-fabricates, finished products,

and the like) on condition that their total value does not become less than that specified in the contract on pledge, shall be deemed to be a pledge of goods in turnover.

A reduction of the value of the pledged goods in turnover shall be permitted commensurately with the performed part of the obligation secured by pledge unless provided otherwise by the contract.

2. Goods in turnover alienated by a pledgor shall cease to be the subject of pledge from the moment of their transfer into ownership, economic jurisdiction, or operative management of the acquirer, and goods acquired by the pledgor which are specified in the contract on pledge shall become the subject of pledge from the moment the right of the pledgor arises to ownership or economic jurisdiction.

3. The pledgor of goods in turnover shall be obliged to keep a register book of pledges in which entries are made concerning the conditions of the pledge of goods and all operations entailing a change of the composition or natural form of the pledged goods, including the processing thereof, on the day of the last operation.

4. In the event of a violation by the pledgor of the conditions of a pledge of goods in turnover, the pledgeholder shall have the right, by placing his marks and seals on the pledged goods, to suspend operations with them until the violation is eliminated.

Article 358. Pledge of Things in Pawnshop

1. The acceptance from citizens on pledge of movable property intended for personal consumption to secure short-term credits may be effectuated as entrepreneurial activity by specialised organisations—pawnshops [as amended by Federal Law No. 15-ФЗ, 10 January 2003].

2. A contract on the pledge of things in a pawnshop shall be formalised by the issuance of a pledge ticket by the pawnshop.

3. The pledged things shall be transferred to the pawnshop.

The pawnshop shall be obliged to insure to the benefit of the pledgor at his expense things accepted on pledge for the full amount of their value as established in accordance with prices for things of this type and quality ordinarily established in trade at the moment of their acceptance on pledge.

The pawnshop shall not have the right to use and dispose of pledged things.

4. The pawnshop shall bear responsibility for loss of and damage to pledged things unless it is proved that the loss or damage occurred as a consequence of insuperable force.

5. In the event of the failure to return the amount of the credit within the established period secured by the pledge of things to the pawnshop, the pawnshop

shall have the right on the basis of an endorsement of execution of a notary upon the expiry of a month's grace period to sell this property in the procedure established for the realisation of pledged property (Article 350[3][4][6] and [7]). After this demand of the pawnshop against the pledgor (debtor) is paid, even if the amount received when realising the pledged property is insufficient for the full satisfaction thereof.

6. The rules for granting credits to citizens by pawnshops under pledge of things belonging to citizens shall be established in accordance with the present Code by a law.

7. The conditions of a contract on pledge of things in a pawnshop limiting the rights of the pledgor in comparison with the rights granted him by the present Code and by other laws shall be void. In place of such conditions the respective provisions of the law shall be applied.

§4. Withholding

Article 359. Grounds of Withholding

1. A creditor with whom a thing is situated which is subject to transfer to a debtor or to a person specified by the debtor shall have the right in the event of the failure of the debtor to perform the obligation relating to payment for this thing within the period or to compensate the creditor for expenses and other losses connected therewith to withhold it so long as the respective obligation is not performed.

The withholding of a thing may also secure demands that, although not connected with the payment for the thing or compensation of expenses therefor and other losses, but arose from an obligation, the parties to which act as entrepreneurs.

2. A creditor may withhold a thing situated with him notwithstanding the fact that after this thing came into the possession of the creditor the rights thereto have been acquired by a third person.

3. The rules of the present Article shall apply unless provided otherwise by a contract.

Article 360. Satisfaction of Demands at Expense of Withheld Property

The demands of a creditor who is withholding a thing shall be satisfied from the value thereof in the amount and procedure provided for the satisfaction of demands secured by a pledge.

§5. Suretyship

Article 361. Contract of Suretyship

Under a contract of suretyship the surety shall be obliged to the creditor of another person to be liable for the performance by the last of his obligation in full or in part.

A contract of suretyship may be concluded also in order to secure an obligation which arises in future.

Article 362. Form of Contract of Suretyship

A contract of suretyship must be concluded in written form. The failure to comply with the written form shall entail the invalidity of the contract of suretyship.

Article 363. Responsibility of Surety

1. In the event of the failure to perform or the improper performance by the debtor of an obligation secured by a suretyship, the surety and the debtor shall be liable to a creditor jointly and severally unless subsidiary responsibility of the surety has been provided for by a law or by the contract of suretyship.

2. The surety shall be liable to the creditor in the same amount as the debtor, including the payment of interest, compensation for court costs relating to recovery of the debt, and other losses of the creditor caused by the failure to perform or the improper performance of the obligation by the debtor unless provided otherwise by the contract of suretyship.

3. Persons who have given a suretyship jointly shall be liable to the creditor jointly and severally unless provided otherwise by the contract of suretyship.

Article 364. Right of Surety to Objections Against Demand of Creditor

The surety shall have the right to advance against the demand of the creditor the objections which the debtor could present unless it arises otherwise from the contract of suretyship. The surety shall not lose the right to these objections even if the debtor waived them or has acknowledged his debt.

Article 365. Rights of Surety Who Has Performed Obligation

1. To a surety who has performed an obligation shall pass the rights of the creditor with regard to this obligation and the rights which belonged to the creditor as pledgeholder in that amount in which the surety satisfied the demand of the creditor. The surety also shall have the right to demand from the debtor the payment of interest on the amount paid to the creditor and compensation for other losses incurred in connection with responsibility for the debtor.

2. Upon performance by the surety of the obligation the creditor shall be obliged to hand over to the surety the documents certifying the demand against the debtor and to transfer the rights which secure this demand.

3. The rules established by the present Article shall be applied unless provided otherwise by a law, other legal acts, or by the contract of surety with the debtor and does not arise from the relations between them.

Article 366. Notification of Surety About Performance of Obligation by Debtor

The debtor who has performed an obligation secured by a suretyship shall be obliged immediately to notify the surety thereof. Otherwise the surety who in turn has performed the obligation shall have the right to recover from the creditor that unjustifiably received or to present a regressive demand against the debtor. In the last instance the debtor shall have the right to recover from the creditor only that unjustifiably received.

Article 367. Termination of Suretyship

1. A suretyship shall terminate with the termination of the obligation secured by it, and also in the event of a change thereof entailing an increase of responsibility or other unfavourable consequences for the surety without the consent of the last.

2. A suretyship shall terminate with the transfer to another person of the debt relating to the obligation secured by the suretyship unless the surety has given consent to the creditor to be liable for the new debtor.

3. A suretyship shall terminate if the creditor has refused to accept proper performance offered by the debtor or the surety.

4. A suretyship shall terminate upon the expiry of the period for which it was given specified in the contract of suretyship. If such period has not been established, it shall terminate if the creditor within a year from the date of ensuing of the period of performance of the obligation secured by the suretyship a suit against the surety has not been brought. When the period of performance of the principal obligation has not been specified and cannot be determined or is determined by the moment of demand, the suretyship shall terminate if the creditor does not bring suit against the surety within two years from the date of concluding the contract of suretyship.

§6. Bank Guarantee

Article 368. Concept of Bank Guarantee

By virtue of a bank guarantee the bank, other credit institution, or insurance organisation (guarantor) gives at the request of another person (principal) a written obligation to pay to the creditor of the principal (beneficiary) a monetary amount in accordance with the conditions of the obligation given by the guarantor upon the presentation by the beneficiary of a written demand concerning the payment thereof.

Article 369. Securing Obligation of Principal by Bank Guarantee

1. A bank guarantee shall secure the proper performance by the principal of his obligation to the beneficiary (principal obligation).

2. The principal shall pay remuneration to the guarantor for the issuance of a bank guarantee.

Article 370. Independence of Bank Guarantee from Principal Obligation

The obligation of the guarantor to the beneficiary provided for by the bank guarantee is not dependent in the relations between them upon the principal obligation, to secure the performance of which it was issued even if the guarantee contains a reference to this obligation.

Article 371. Irrevocability of Bank Guarantee

A bank guarantee may not be revoked by the guarantor unless provided otherwise therein.

Article 372. Nontransferability of Rights Under Bank Guarantee

The right of demand against the guarantor which belongs to the beneficiary under the bank guarantee may not be transferred to another person unless provided otherwise in the guarantee.

Article 373. Entry of Bank Guarantee into Force

A bank guarantee shall enter into force from the date of its issuance unless provided otherwise in the guarantee.

Article 374. Presentation of Demand Under Bank Guarantee

1. The demand of a beneficiary concerning the payment of a monetary amount under a bank guarantee must be submitted to the guarantor in written form with the documents specified in the guarantee appended. In the demand or in the annex thereto the beneficiary must specify what the violation by the principal of the principal obligation consists of, to secure which the guarantee was issued.

2. The demand of the beneficiary must be submitted to the guarantor before the period specified in the guarantee for which it was issued ends.

Article 375. Duties of Guarantor When Demand of Beneficiary is Considered

1. Upon receipt of the demand of the beneficiary the guarantor must without delay inform the principal thereof and transfer to him a copy of the demand with all documents relevant thereto.

2. The guarantor must consider the demand of the beneficiary with the documents appended thereto within a reasonable period and manifest reasonable care so as to establish whether this demand and the documents appended thereto correspond to the conditions of the guarantee.

Article 376. Refusal of Guarantor to Satisfy Demand of Beneficiary

1. The guarantor shall refuse to the beneficiary satisfaction of his demand if this demand or the documents appended thereto do not correspond to the conditions of the guarantee or have been submitted to the guarantor at the end of the period specified in the guarantee.

The guarantor must immediately notify the beneficiary about the refusal to satisfy his demand.

2. If it becomes known to the guarantor before satisfaction of the demand of the beneficiary that the principal obligation secured by the bank guarantee has been performed fully or in respective part, has been terminated on other grounds, or is invalid, he must immediately communicate this to the beneficiary and the principal.

A second demand of the beneficiary received by the guarantor after such notification shall be subject to satisfaction by the guarantor.

Article 377. Limits of Obligation of Guarantor

1. An obligation of the guarantor to a beneficiary provided for by a bank guarantee shall be limited to payment of the amount for which the guarantee was issued.

2. Unless provided otherwise in the guarantee, the responsibility of the guarantor to the beneficiary for the failure to fulfil or the improper fulfilment by the guarantor of the obligation under the guarantee shall not be limited to the amount for which the guarantee was issued.

Article 378. Termination of Bank Guarantee

1. The obligation of the guarantor to a beneficiary under a guarantee shall be terminated:

(1) by payment to the beneficiary of the amount for which the guarantee was issued;

(2) ending of period specified in the guarantee for which it was issued;

(3) as a consequence of the waiver by the beneficiary of his rights under the guarantee and the return thereof to the guarantor;

(4) as a consequence of the waiver by the beneficiary of his rights under the guarantee by a written statement to release the guarantor from his obligations.

The termination of the obligation of the guarantor on the grounds specified in subpoints 1, 2, and 4 of the present point shall not depend upon whether the guarantee is returned to him.

2. A guarantor to whom the termination of a guarantee has become known must without delay notify the principal thereof.

Article 379. Regressive Demands of Guarantor Against Principal

1. The right of the guarantor to demand from the principal by way of regression the compensation of amounts paid to the beneficiary under a bank guarantee shall be determined by agreement of the guarantor with the principal for the performance of which the guarantee was issued.

2. The guarantor shall not have the right to demand from the principal the compensation of amounts paid to the beneficiary not in accordance with the conditions of the guarantee or for a violation of the obligation of the guarantor to the beneficiary unless provided otherwise by agreement of the guarantor with the principal.

§7. Deposit

Article 380. Concept of Deposit. Form of Agreement on Deposit

1. A deposit shall be deemed to be a monetary amount issued by one of the contracting parties on the account under a contract of payment due from him to the other party as evidence of the conclusion of the contract and to secure its performance.

2. An agreement on deposit, irrespective of the amount of the deposit, must be concluded in written form.

3. In the event of doubt as to whether the amount is paid on the account of payments due from a party under the contract as a deposit, in particular as a consequence of the failure to comply with the rule established by point 2 of the present Article, this amount shall be considered to be paid as an advance unless proved otherwise.

Article 381. Consequences of Termination and Failure to Perform Obligation Secured by Deposit

1. In the event of the termination of the obligation before the commencement of the performance thereof by agreement of the parties or as a consequence of the impossibility of performance (Article 416) the deposit must be returned.

2. If for the failure to perform the contract the party which gave the deposit is responsible, it shall remain with the other party. If for the failure to perform the contract the party which received the deposit is responsible, it shall be obliged to pay the other party twice the amount of the deposit.

In addition the party responsible for the failure to perform the contract shall be obliged to compensate the other party for losses, setting off the amount of the deposit unless provided otherwise in the contract.

Chapter 24. Change of Persons in Obligation

§1. Transfer of Rights of Creditor to Another Person

Article 382. Grounds and Procedure for Transfer of Rights of Creditor to Another Person

1. The right (or demand) belonging to a creditor on the basis of an obligation may be transferred by him to another person with regard to the transaction (assignment of demand) or pass to another person on the basis of a law.

The rules concerning the transfer of the rights of a creditor to another person shall not apply to regressive demands.

2. The consent of the debtor, unless provided otherwise by a law or by contract, shall not be required for the transfer of the rights of the creditor to another person.

3. If the debtor was not informed in writing about the transfer of the rights of the creditor to another person, the new creditor shall bear the risk of the unfavourable consequences caused by this for him. In this event the performance of the obligation to the initial creditor shall be deemed to be performance to a proper creditor.

Article 383. Rights Which May Not Pass to Other Persons

The transfer to another person of the rights inextricably connected with the person of the creditor, in particular, demands concerning alimony and compensation of harm caused to life or health, shall not be permitted.

Article 384. Amount of Rights of Creditor Passing to Another Person

Unless provided otherwise by a law or by contract, the right of the initial creditor shall pass to the new creditor in that amount and on those conditions which existed at the moment of transfer of the right. In particular, to the new creditor shall pass the rights securing the performance of the obligation, and also other rights connected with the demand, including the right to unpaid interest.

Article 385. Evidence of Rights of New Creditor

1. The debtor shall have the right not to perform the obligation to a new creditor until the submission to him of evidence of the transfer of the demand to this person.

2. The creditor who has assigned a demand to another person shall be obliged to transfer to him the documents certifying the right of demand and to communicate information having significance for the effectuation of the demand.

Article 386. Objections of Debtor Against Demand of New Creditor

The debtor shall have the right to advance against the demand of a new creditor objections which he had against the initial creditor at the moment of receipt of notification concerning the transfer of the rights regarding the obligation to the new creditor.

Article 387. Transfer of Rights of Creditor to Another Person on Basis of Law

The rights of a creditor relating to an obligation shall pass to another person on the basis of a law and ensuing of the circumstances specified therein:

as a result of universal legal succession to the rights of the creditor;

by decision of a court concerning the transfer of the rights of the creditor to another person when the possibility of such transfer has been provided for by a law;

as a consequence of the performance of the obligation of the debtor by his surety or by a pledgor who is not the debtor with regard to this obligation;

in the event of the subrogation to an insurer of the rights of the creditor against the debtor who is responsible for the ensuing of an insured event;

in the other instances provided for by a law.

Article 388. Conditions of Assignment of Demand

1. The assignment of a demand by the creditor to another person shall be permitted if this is not contrary to a law, other legal acts, or a contract.

2. The assignment of the demand regarding an obligation in which the person of the creditor has material significance for the debtor without the consent of the debtor shall not be permitted.

Article 389. Form of Assignment of Demand

1. The assignment of a demand based on a transaction concluded in simple written or notarial form must be concluded in the respective written form.

2. The assignment of a demand relating to a transaction requiring State registration must be registered in the procedure established for registration of this transaction unless established otherwise by a law.

3. The assignment of a demand regarding an order security shall be concluded by means of an endorsement on this security (Article 146[3]).

Article 390. Responsibility of Creditor Who Has Assigned Demand

The initial creditor who has assigned a demand shall be liable to the new creditor for the invalidity of the demand transferred to him but shall not be liable for the failure to perform this demand by the debtor except when the initial creditor assumed a suretyship for the debtor to the new creditor.

§2. Transfer of Debt

Article 391. Condition and Form of Transfer of Debt

1. The transfer by the debtor of his debt to another person shall be permitted only with the consent of the creditor.

2. The rules contained in Article 389(1) and (2) of the present Code shall apply respectively to the form of the transfer of the debt.

Article 392. Objections of New Debtor Against Demand of Creditor

The new debtor shall have the right to advance against the demand of the creditor the objections based on relations between the creditor and the initial debtor.

Chapter 25. Responsibility for Violation of Obligations

Article 393. Duty of Debtor to Compensate Losses

1. A debtor shall be obliged to compensate the creditor for losses caused by the failure to perform or improper performance of an obligation.

2. Losses shall be determined in accordance with the rules provided for by Article 15 of the present Code.

3. Unless provided otherwise by a law, other legal acts, or by contract, when determining losses the prices which existed in that place where the obligation was to have been performed on the date of voluntary satisfaction by the debtor of the demand of the creditor shall be taken into account, and if the demand was not voluntarily satisfied, on the day of bringing the suit. Proceeding from the circumstances, a court may satisfy the demand to compensate losses by taking into account the prices existing on the day of rendering the decision.

4. When determining lost advantage the measures undertaken by the creditor to receive it and the preparations made for this purpose shall be taken into account.

Article 394. Losses and Penalty

1. If a penalty has been established for the failure to perform or for improper performance of an obligation, the losses shall be compensated in the part not covered by the penalty.

Instances may be provided for by a law or by contract: when the recovery of only a penalty is permitted, but not of losses; when losses may be recovered in full above the penalty; when at the choice of the creditor either a penalty or losses may be recovered.

2. When limited responsibility (Article 400) was established for the failure to perform or improper performance, the losses subject to compensation in the part not covered by the penalty or above or in place of it may be recovered up to the limits established by such limitation.

Article 395. Responsibility for Failure to Perform Monetary Obligation

1. For the use of another's monetary means as a consequence of unlawful withholding, avoidance of the return thereof, other delay in the payment thereof or the

unjustified receipt or savings thereof at the expense of another person interest shall be subject to payment on the amount of these means. The amount of interest shall be determined as the rate of bank interest on the day of performance of the monetary obligation or respective part thereof which existed at the place of residence of the creditor, and if the creditor is a juridical person, at the place of its location. In the event of the recovery of a debt in a judicial proceeding the court may satisfy the demand of the creditor by proceeding from the bank interest rate on the date of presenting the suit or on the date of rendering the decision. These rules shall apply unless another amount of interest has been established by a law or by the contract.

2. If the losses caused to a creditor by the unlawful use of his monetary means exceed the amount of interest due him on the basis of point 1 of the present Article, he shall have the right to demand from the debtor compensation of losses in that part exceeding this amount.

3. Interest for the use of another's means shall be recovered as of the day of payment of the amount of these means to the creditor unless a shorter period has been established for calculating interest by a law, other legal acts, or by contract.

Article 396. Responsibility and Performance of Obligation in Kind

1. The payment of a penalty and compensation of losses in the event of the improper performance of an obligation shall not relieve the debtor from performance of the obligation in kind unless provided otherwise by a law or by contract.

2. The compensation of losses in the event of the failure to perform an obligation and the payment of a penalty for the failure to perform shall relieve the debtor from the performance of the obligation in kind unless provided otherwise by a law or by contract.

3. The refusal of a creditor to accept performance which as a consequence of delay has lost interest for him (Article 405[2]), and also the payment of a penalty established as release-money (Article 409), shall relieve the debtor from performance of the obligation in kind.

Article 397. Performance of Obligation at Expense of Debtor

In the event of the failure to perform an obligation by a debtor to manufacture and transfer a thing in ownership, economic jurisdiction, or operative management, or to transfer a thing for use to a creditor, or to fulfil specified work or to render a service for him, the creditor shall have the right within a reasonable period to commission the fulfilment of the obligation to third persons for a reasonable price or to fulfil it by his own efforts unless it follows otherwise from a law, other legal acts, the contract, or the essence of the obligation, and to demand from the debtor compensation for necessary expenses and other losses incurred.

Article 398. Consequences of Failure to Perform Obligation to Transfer Individually-Specified Thing

In the event of the failure to perform an obligation to transfer an individually-specified thing in ownership, economic jurisdiction, operative management, or use for compensation to a creditor, the last shall have the right to demand that this thing be taken away from the debtor and the transfer thereof to the creditor on the conditions provided for by the obligation. This right shall disappear if the thing already has been transferred to a third person having the right of ownership, economic jurisdiction, or operative management. If the thing has not yet been transferred, those creditors to whose benefit the obligation arose earlier shall have priority, and if this is impossible to establish, then he who has brought suit earlier.

The creditor shall have the right to demand compensation of losses instead of a demand to transfer the thing to him which is the subject of the obligation.

Article 399. Subsidiary Responsibility

1. Until demands are brought against the person who in accordance with a law, other legal acts, or with the conditions of an obligation bears responsibility additionally to the responsibility of another person who is the principal debtor (subsidiary responsibility), the creditor must present the demand to the principal debtor.

If the principal debtor refused to satisfy the demand of the creditor or the creditor has not received a reply from him within a reasonable period to the demand presented, this demand may be presented to the person who bears subsidiary responsibility.

2. A creditor shall not have the right to demand satisfaction of his demand against the principal debtor from a person bearing subsidiary responsibility if this demand can be satisfied by means of the set-off of the counter demand against the principal debtor, or by the uncontested recovery of assets from the principal debtor.

3. The person who bears subsidiary responsibility must until satisfaction of the demand presented to him by the creditor warn the principal debtor thereof, and if suit is brought against such person, to involve the principal debtor in the case. Otherwise the principal debtor shall have the right to advance against the regressive demand of the person subsidiarily liable objections which he had against the creditor.

Article 400. Limitation of Amount of Responsibility for Obligations

1. The right to full compensation of losses may be limited (limited responsibility) by a law with regard to individual types of obligations and with regard to obligations connected with the determined nature of activity.

2. An agreement concerning the limitation of the amount of responsibility of a debtor under a contract of adhesion or other contract in which the citizen acting as a consumer is a creditor shall be void if the amount of responsibility for the particular type of obligations or for the particular violation has been determined by a law and if the agreement was concluded before the circumstances ensue which entail responsibility for the failure to perform or the improper performance of the obligation.

Article 401. Grounds of Responsibility for Violation of Obligation

1. A person who has not performed an obligation or who performed it improperly shall bear responsibility when there is fault (intent or negligence) except for instances when by a law or by contract other grounds of responsibility have been provided for.

The person shall be deemed to be not at fault if with that degree of concern and attentiveness which is required of him according to the character of the obligation and conditions of turnover he has taken all measures for proper performance of the obligation.

2. The absence of fault shall be proved by the person who violated the obligation.

3. Unless provided otherwise by a law or by contract, the person who has not performed or who has improperly performed an obligation shall, when effectuating entrepreneurial activity, bear responsibility unless it is proved that proper performance proved to be impossible as a consequence of insuperable force, that is, extraordinary and unavertable circumstances under the particular conditions. There shall not be relegated to such circumstances, in particular, a violation of duties on the part of the contracting parties of the debtor, the absence in the market of goods necessary for performance, and the lack of necessary monetary means on the part of the debtor.

4. An agreement concluded beforehand concerning the elimination or limitation of responsibility for an intentional violation of an obligation shall be void.

Article 402. Responsibility of Debtor for His Workers

The actions of the workers of a debtor relating to the performance of his obligation shall be considered to be the actions of the debtor. The debtor shall be liable for these actions if they entailed the failure to perform or the improper performance of the obligation.

Article 403. Responsibility of Debtor for Actions of Third Persons

A debtor shall be liable for the failure to perform or the improper performance of the obligation by third persons on whom performance has been placed unless it has been established by a law that responsibility shall be borne by the third person who is the immediate performer.

Article 404. Fault of Creditor

1. If the failure to perform or improper performance of an obligation has occurred through the fault of both parties, a court shall respectively reduce the amount of responsibility of the debtor. A court also shall have the right to reduce the amount of responsibility of the debtor if the creditor has intentionally or through negligence facilitated the increase of the amount of losses caused by the failure to perform or improper performance, or has not taken reasonable measures to reduce them.

2. The rules of point 1 of the present Article respectively shall also apply in instances when the debtor, by virtue of a law or contract, bears responsibility for the failure to perform or improper performance of an obligation irrespective of his fault.

Article 405. Delay of Debtor

1. A debtor who has delayed performance shall be liable to the creditor for losses caused by the delay and for the consequences which incidentally ensued during the delay of the impossibility of performance.

2. If as a consequence of delay of the debtor the performance has lost interest for the creditor, he may refuse to accept performance and demand compensation of losses.

3. The debtor shall not be considered to be delayed if the obligation cannot be performed as a consequence of delay by the creditor.

Article 406. Delay of Creditor

1. The creditor shall be considered to have delayed if he refused to accept proper performance offered by the debtor or did not perform actions provided for by a law, other legal acts, or by contract or arising from the customs of business turnover or from the essence of the obligation, until the performance of which the debtor could not perform his obligation.
    A creditor shall be considered to have delayed also in the instances specified in Article 408(2) of the present Code.

2. The delay of a creditor shall give the debtor the right to compensation of losses caused by the delay if the creditor does not prove that the delay occurred through circumstances for which neither he himself nor those persons on whom by virtue of a law, other legal acts, or the commission of the creditor acceptance of performance was placed are liable.

3. With regard to a monetary obligation the debtor shall not be obliged to pay interest for the time of delay by the creditor.

Chapter 26. Termination of Obligations

Article 407. Grounds for Termination of Obligations

1. An obligation shall terminate wholly or in part on the grounds provided for by the present Code, other laws, other legal acts, or contract.

2. The termination of an obligation relating to the demand of one of the parties shall be permitted only in the instances provided for by a law or by contract.

Article 408. Termination of Obligation by Performance

1. Proper performance shall terminate an obligation.

2. A creditor accepting performance shall be obliged at the demand of the debtor to issue him a receipt of receiving performance fully or in respective part.

If a debtor has issued a debt document to a creditor in certification of the obligation, then the creditor, in accepting performance, must return this document, and if it is impossible to return, specify this on the receipt issued by him. A receipt may be replaced by an inscription on the returned debt document. The debt document being with the debtor shall certify, unless proven otherwise, the termination of the obligation.

If the creditor refuses to issue the receipt, to return the debt document, or to note on the receipt the impossibility of the return thereof, the debtor shall have the right to withhold performance. In these instances the creditor shall be considered to have delayed.

Article 409. Release-Money

By agreement of the parties an obligation may be terminated by granting release-money in place of performance (payment of money, transfer of property, etc.). The amount, periods, and procedure for granting release-money shall be established by the parties.

Article 410. Termination of Obligation by Set-Off

An obligation shall be terminated wholly or in part by the set-off of a counter demand of the same type, the period for which has ensued or the period of which has not been specified or is determined by the moment of demand. The statements of one party are sufficient for set-off.

Article 411. Instances of the Inadmissibility of Set-Off

The set-off of demands shall not be permitted:
if according to the statement of the other party the period of limitations is subject to application to the demand and this period has expired;
concerning compensation of harm caused to life or health;
concerning the recovery of alimony;

concerning maintenance for life;

in other instances provided for by a law or by a contract.

Article 412. Set-Off in Event of Assignment of Demand

In the event of the assignment of a demand the debtor shall have the right to set-off against the demand of the new creditor his counter demand against the initial creditor.

The set-off shall be made if the demand arose on a grounds which existed at the moment of receipt by the debtor of notification concerning the assignment of the demand and the period of the demand ensued before receiving it or this period was not specified or is determined by the moment of demand.

Article 413. Termination of Obligation by Coincidence of Debtor and Creditor in One Person

An obligation shall terminate by the coinciding of the debtor and the creditor in one person.

Article 414. Termination of Obligation by Novation

1. An obligation shall be terminated by agreement of the parties concerning the substitution of the initial obligation which existed between them by another obligation between the same persons providing for another subject or means of performance (novation).

2. Novation shall not be permitted with respect to obligations relating to compensation for harm caused to life or health and relating to the payment of alimony.

3. Novation shall terminate additional obligations connected with the initial one unless provided otherwise by agreement of the parties.

Article 415. Forgiveness of Debt

An obligation shall be terminated by the release of the debtor by the creditor from the duties on him unless this violates the rights of other persons with respect to the property of the creditor.

Article 416. Termination of Obligation for Impossibility of Performance

1. An obligation shall be terminated by the impossibility of performance if it was caused by a circumstance for which none of the parties is liable.

2. In the event of the impossibility of performance by a debtor of an obligation caused by the guilty actions of the creditor, the last shall not have the right to demand the return of that which has been performed under the obligation.

Article 417. Termination of Obligation on Basis of Act of State Agency

1. If as a result of the issuance of an act of a State agency the performance of an obligation becomes impossible wholly or partially, the obligation shall terminate

wholly or in respective part. The parties who have incurred losses as a result thereof shall have the right to demand compensation thereof in accordance with Articles 13 and 16 of the present Code.

2. In the event of the act of the State agency being deemed in the established procedure to be invalid on the basis of which an obligation has terminated, the obligation shall be restored unless it arises otherwise from an agreement of the parties or the essence of the obligation and performance has not lost interest for the creditor.

Article 418. Termination of Obligation by Death of Citizen

1. An obligation shall be terminated by the death of the debtor if performance cannot be made without the personal participation of the debtor or the obligation is otherwise inextricably connected with the person of the debtor.

2. An obligation shall be terminated by the death of the creditor if the performance was intended personally for the creditor or the obligation was otherwise inextricably connected with the person of the creditor.

Article 419. Termination of Obligation by Liquidation of Juridical Person

An obligation shall be terminated by the liquidation of a juridical person (debtor or creditor) except for instances when by a law or other legal acts the performance of an obligation of a liquidated juridical person is placed on another person (relating to demands concerning compensation of harm caused to life and health, and others).

Subsection 2. General Provisions on Contract

Chapter 27. Concept and Conditions of Contract

Article 420. Concept of Contract

1. A contract shall be considered to be an agreement of two or several persons concerning the establishment, change, or termination of civil rights and duties.

2. The rules concerning bilateral and multilateral transactions provided for by Chapter 9 of the present Code shall apply to contracts.

3. The general provisions on obligations (Articles 307–419) shall apply to obligations which arose from a contract unless provided otherwise by rules of the present Chapter and by rules on individual types of contracts contained in the present Code.

4. The general provisions on contract shall apply to contracts concluded by more than two parties unless this is contrary to the multilateral character of such contracts.

Article 421. Freedom of Contract

1. Citizens and juridical persons shall be free in concluding a contract.

Coercion to conclude a contract shall not be permitted except for instances when the duty to conclude a contract has been provided for by the present Code, by a law, or by an obligation voluntarily accepted.

2. The parties may conclude a contract which is either provided for or is not provided for by a law or other legal acts.

3. The parties may conclude a contract which contains elements of various contracts provided for by a law or other legal acts (mixed contract). The rules on contracts whose elements are contained in a mixed contract shall apply to the relations of the parties under a mixed contract unless it arises otherwise from the agreement of the parties or the essence of the mixed contract.

4. The conditions of a contract shall be determined by discretion of the parties except for instances when the content of the respective condition has been prescribed by a law or other legal acts (Article 422).

In instances when a condition of a contract has been provided for by a norm which applies insofar as not established otherwise by agreement of the parties (dispositive norm), the parties may by their agreement exclude the application thereof or establish a condition which differs from that provided therein. In the absence of such agreement, the condition of the contract shall be determined by the dispositive norm.

5. If the condition of a contract has not been determined by the parties or by a dispositive norm, the respective conditions shall be determined by the customs of business turnover applicable to the relations of the parties.

Article 422. Contract and Law [закон]

1. A contract must correspond to rules established by a law and other legal acts (imperative norms) which are binding upon the parties and prevailing at the moment of its conclusion.

2. If after the conclusion of a contract a law is adopted which establishes rules binding upon the parties other than those by which were prevailing when concluding the contract, the conditions of the contract concluded shall retain force except for instances when it is established in a law that its operation extends to relations which arose from the contracts previously concluded.

Article 423. Contracts For and Without Compensation

1. A contract under which a party must receive payment or other counter-giving for the performance of his duties shall be for compensation.

2. A contract under which one party is obliged to grant something to the other party without receiving payment from him or other counter-giving shall be deemed to be without compensation.

3. A contract shall be presupposed to be for compensation unless it arises otherwise from a law, other legal acts, or the content or essence of the contract.

Article 424. Price

1. The performance of a contract shall be paid for at the price established by agreement of the parties.

In the instances provided for by a law the prices (or tariffs, price scales, rates, etc.) established or regulated by duly empowered State agencies shall be applied.

2. A change of the price after the conclusion of a contract shall be permitted in the instances and on the conditions provided for by contract, by a law, or in the procedure established by a law.

3. In instances when in a contract for compensation the price has not been provided for and may not be determined by proceeding from the conditions of the contract, the performance of the contract must be paid for at the price which under comparable circumstances usually is recovered for analogous goods, work, or services.

Article 425. Operation of Contract

1. A contract shall enter into force and become binding for the parties from the moment of its conclusion.

2. The parties shall have the right to establish that the conditions of a contract concluded by them shall apply to their relations which arose before the conclusion of the contract.

3. It may be provided by a law or by contract that the end of the period of operation of a contract shall entail the termination of the obligations of the parties under the contract.

A contract which lacks such a condition shall be deemed to be in force until the moment of the ending of performance of the obligation by the parties determined therein.

4. The ending of the period of operation of a contract shall not relieve the parties from responsibility for the violation thereof.

Article 426. Public Contract

1. A contract concluded by a commercial organisation and establishing its duties relating to the sale of goods, fulfilment of work, or rendering of services which this organisation by the character of its activity must effectuate with respect

to everyone who has recourse to it (retail trade, carriage by common-use transport, communications services, electric power supply, medical, hotel servicing, and so forth) shall be deemed to be a public contract.

A commercial organisation shall not have the right to prefer one person to others with respect to the conclusion of a public contract except for the instances provided by a law or other legal acts.

2. The price of goods, work, and services, and also other conditions of a public contract, shall be established identically for all consumers except for instances when the granting of privileges for individual categories of consumers is permitted by a law and other legal acts.

3. A refusal of a commercial organisation to conclude a public contract when it is possible to grant the respective goods or services to the consumer or to fulfil the respective work for him shall not be permitted.

In the event of an unjustified evasion by a commercial organisation from the concluding of a public contract the provisions provided for by Article 445(4) of the present Code shall apply.

4. In instances provided for by a law the Government of the Russian Federation may issue rules binding upon the parties when concluding and performing public contracts (standard contracts, statutes, etc.).

5. The conditions of a public contract which do not correspond to the requirements established by points 2 and 4 of the present Article shall be void.

Article 427. Model Conditions of Contract

1. It may be provided in a contract that its individual conditions shall be determined by model conditions worked out for contracts of the respective type and published in the press.

2. In instances when a reference to the model conditions is not contained in the contract, such model conditions shall apply to the relations of the parties as customs of business turnover if they correspond to the requirements established by Article 5 and by Article 421(5) of the present Code.

3. Model conditions may be set out in the form of a model contract or other document containing these conditions.

Article 428. Contract of Adhesion

1. A contract of adhesion shall be deemed to be a contract whose conditions have been determined by one of the parties in records or other standard forms and which can be accepted by the other party not other than by means of adhering to the contract as a whole being offered.

2. A party which has adhered to the contract shall have the right to demand dissolution or change of the contract if the contract of adhesion, although not

contrary to a law and other legal acts, deprives this party of the rights usually granted under contracts of that type, excludes or limits the responsibility of the other party for a violation of obligations, or contains other conditions clearly burdensome for the adhering party which it, proceeding from its own reasonably understandable interests, would not accept if it had the opportunity to participate in determining the conditions of the contract.

3. When the circumstances provided for in point 2 of the present Article are present, the demand concerning dissolution or change of the contract presented by a party which has adhered to a contract in connection with the effectuation of its entrepreneurial activity shall not be subject to satisfaction, if the adhering party knew or should have known on what conditions the contract is concluded.

Article 429. Preliminary Contract

1. Under a preliminary contract the parties shall be obliged to conclude in future a contract concerning the transfer of property, fulfilment of work, or rendering of services (principal contract) on the conditions provided for by the preliminary contract.

2. The preliminary contract shall be concluded in the form established for the principal contract, and if the form of the principal contract has not been established, then in written form. The failure to comply with the rules concerning the form of the preliminary contract shall entail its being void.

3. A preliminary contract must contain conditions enabling the establishment of the subject, and also the other material conditions, of the principal contract.

4. A preliminary contract shall specify the period in which the parties are obliged to conclude the principal contract.

If such period has not been determined in the preliminary contract, the principal contract shall be subject to conclusion within a year from the moment of concluding the preliminary contract.

5. In instances when a party which has concluded a preliminary contract evades the conclusion of the principal contract, the provisions provided for by Article 445(4) of the present Code shall apply.

6. The obligations provided for by a preliminary contract shall terminate if before the ending of the period in which the parties must conclude the principal contract it has not been concluded or one of the parties does not send to the other party an offer to conclude this contract.

Article 430. Contract to Benefit of Third Person

1. A contract to the benefit of a third person shall be deemed to be a contract in which the parties have established that a debtor is obliged to make performance

not to the creditor, but to a third person specified or not specified in the contract and having the right to demand performance of the obligation to his benefit from the debtor.

2. Unless provided otherwise by a law, other legal acts, or a contract, from the moment of the expression by the third person to the debtor of an intention to take advantage of his right under the contract, the parties may not dissolve or change the contract concluded by them without the consent of the third person.

3. The debtor in the contract shall have the right to advance against the demand of the third person objections which he could have advanced against the creditor.

4. When a third person waives a right granted to him under the contract, the creditor may take advantage of this right unless this is contrary to a law, other legal acts, and the contract.

Article 431. Interpretation of Contract

In the event of the interpretation of the conditions of a contract by a court the literal meaning of the words and expressions contained therein shall be taken into account. The literal meaning of the condition of a contract in the event of its ambiguity shall be established by means of comparing with the other conditions and with the sense of the contract as a whole.

If the rules contained in paragraph one of the present Article do not enable the content of the contract to be determined, the true common will of the parties must be elicited by taking into account the purpose of the contract. In so doing all the respective circumstances, including negotiations preceding the contract and correspondence, practice being established in the mutual relations of the parties, the customs of business turnover, and the subsequent conduct of the parties, shall be taken into account.

Chapter 28. Conclusion of Contract

Article 432. Basic Provisions on Conclusion of Contract

1. A contract shall be considered to be concluded when agreement regarding all the material conditions of the contract has been reached in the form required in appropriate instances.

Conditions concerning the subject of the contract, conditions which are named in a law or other legal acts as material or necessary for contracts of the particular type, and also all those conditions relative to which agreement must be reached according to the statement of one of the parties, shall be material.

2. A contract shall be concluded by means of sending an offer (proposal to conclude a contract) by one party and its acceptance (acceptance of the proposal) by the other party.

Article 433. Moment of Conclusion of Contract

1. A contract shall be deemed to be concluded at the moment of receipt by the person who has sent an offer of its acceptance.

2. If in accordance with a law the transfer of property also is necessary in order to conclude a contract, the contract shall be considered to be concluded from the moment of the transfer of the respective property (Article 224).

3. A contract subject to State registration shall be considered to be concluded from the moment of the registration thereof, unless established otherwise by a law.

Article 434. Form of Contract

1. A contract may be concluded in any form provided for in order to conclude transactions unless a determined form has been established by a law for contracts of the particular type.

If the parties have agreed to conclude a contract in a determined form, it shall be considered to be concluded after imparting the stipulated form to it even though such form has not been required by a law for contracts of the particular type.

2. A contract in written form may be concluded by means of drawing up one document signed by the parties, and also by means of the exchange of documents by means of postal, telegraph, teletype, telephone, electronic, or other communications, enabling it to be reliably established that the document emanates from a party under the contract.

3. The written form of the contract shall be considered to be complied with if the written offer to conclude a contract has been accepted in the procedure provided for by Article 438(3) of the present Code.

Article 435. Offer

1. A proposal addressed to one or several specific persons which is sufficiently definite and expresses the intention of the person who has made the proposal to consider himself to have concluded a contract with the addressee who will accept the proposal shall be deemed to be an offer.

An offer must contain the material conditions of the contract.

2. An offer shall bind the person who sent it from the moment of its receipt by the addressee.

If notice of the revocation of an offer has been received earlier than or simultaneously with the offer itself, the offer shall be considered to be not received.

Article 436. Irrevocability of Offer

An offer received by the addressee may not be revoked within the period established for its acceptance unless otherwise stipulated in the offer itself or does not arise from the essence of the proposal or situation in which it was made.

Article 437. Invitation to Make Offers. Public Offer

1. An advertisement and other proposals addressed to an indefinite group of persons shall be regarded as an invitation to make offers unless expressly specified otherwise in the proposal.

2. A proposal containing all the material conditions of a contract from which the will of the person making the proposal is seen to conclude a contract on the conditions specified in the proposal with anyone who responds shall be deemed to be an offer (public offer).

Article 438. Acceptance

1. The reply of a person to whom an offer has been addressed concerning the acceptance thereof shall be deemed to be an acceptance.
An acceptance must be full and unconditional.

2. Silence shall not be an acceptance unless it arises otherwise from a law, custom of business turnover, or former business relations of the parties.

3. The performance by a person who has received an offer within the period established for its acceptance of actions relating to the fulfilment of the conditions of the contract specified therein (shipment of goods, provision of services, fulfilment of work, payment of respective amount, etc.) shall be considered to be an acceptance unless provided otherwise by a law, other legal acts, or specified in the offer.

Article 439. Revocation of Acceptance

If notice of the revocation of an acceptance has been received by the person to whom the offer was sent earlier than or simultaneously with the acceptance itself, the acceptance shall be considered to be not received.

Article 440. Conclusion of Contract on Basis of Offer Determining Period for Acceptance

When a period for acceptance is determined in an offer, a contract shall be considered to be concluded if the acceptance was received by a person who has sent the offer within the limits of the period specified therein.

Article 441. Conclusion of Contract on Basis of Offer Not Determining Period for Acceptance

1. When a written offer does not determine the period for acceptance, the contract shall be considered to be concluded if the acceptance was received by the person who has sent the offer before the end of the period established by a law or other legal acts, and if such a period has not been established, during the time normally necessary for this.

2. When an offer has been made orally without specifying the period for acceptance, the contract shall be considered to be concluded if the other party immediately declared its acceptance thereof.

Article 442. Acceptance Received Late

In the instances when a notification of acceptance sent in good time has been received late, the acceptance shall not be considered to be late if a party who has sent the offer immediately does not inform the other party about the late receipt of the acceptance.

If a party who has sent an offer immediately communicates to the other party the accepting of its acceptance which has been received late, the contract shall be considered to be concluded.

Article 443. Acceptance on Other Conditions

A reply concerning consent to conclude a contract on conditions other than those proposed in the offer shall not be an acceptance.

Such a reply shall be deemed to be a refusal to accept and, at the same time, a new offer.

Article 444. Place of Conclusion of Contract

Unless the place of conclusion has been specified in the contract, a contract shall be deemed to be concluded at the place of residence of a citizen or location of the juridical person which sent the offer.

Article 445. Conclusion of Contract in Obligatory Procedure

1. In instances when in accordance with the present Code or other laws the conclusion of a contract for one of the parties to whom an offer (or draft contract) is obligatory, this party must send to the other party a notification concerning acceptance or rejection of acceptance or acceptance of the offer (or draft contract) on other conditions (protocol of disagreements to draft contract) within 30 days from the date of receipt of the offer.

A party who has sent an offer and received from the party for whom the conclusion of the contract is obligatory a notification of its acceptance on other conditions (protocol of disagreements to draft contract) shall have the right to transfer disagreements which have arisen when concluding the contract for consideration of a court within 30 days from the date of receipt of such notification or the expiry of the period for acceptance.

2. In instances when in accordance with the present Code or other laws the conclusion of a contract is obligatory for the party who has sent an offer (or draft contract) and a protocol of disagreements to the draft contract is sent to it within 30 days, this party shall be obliged within 30 days from the date of receipt of the protocol of disagreements to notify the other party about acceptance of the

contract in the wording thereof or about rejection of the protocol of disagreements.

In the event the protocol of disagreements is rejected or of the failure to receive notice concerning the results of the consideration thereof within the specified period, the party which sent the protocol of disagreements shall have the right to transfer the disagreements which arose when concluding the contract for the consideration of a court.

3. The rules concerning the periods provided for by points 1 and 2 of the present Article shall apply unless other periods have been established by a law, other legal acts, or have not been agreed by the parties.

4. If a party for which in accordance with the present Code or other laws the conclusion of the contract is obligatory evades the conclusion thereof, the other party shall have the right to apply to a court with a demand to compel the contract to be concluded.

The party which has unjustifiably evaded the conclusion of the contract must compensate the other party for losses caused by this.

Article 446. Precontractual Disputes

In the instances of the transfer of disagreements which arose when concluding a contract for the consideration of a court on the basis of Article 445 of the present Code or by agreement of the parties, the conditions of the contract with regard to which the parties had disagreements shall be determined in accordance with the decision of the court.

Article 447. Conclusion of Contract at Public Sale

1. A contract may, unless it arises otherwise from the essence thereof, be concluded by means of holding public sales. The contract shall be concluded with the person who has won the public sale.

2. The owner of the thing or the possessor of the property right or a specialised organisation may act as the organiser of a public sale. A specialised organisation shall operate on the basis of a contract with the owner of the thing or the possessor of the property right and shall act in its name or in his own name.

3. In the instances specified in the present Code or in another law, contracts concerning the sale of a thing or property right may be concluded only by means of holding public sales.

4. A public sale shall be held in the form of an auction or a competition.

The person who has won the public sale at an auction shall be deemed the person who has proposed the highest price, and at a competition—the person who in accordance with the opinion of a competition commission previously appointed by the organiser of the public sale proposed the best conditions.

The form of public sale shall be determined by the owner of the thing being sold or the possessor of the property right being realised unless provided otherwise by a law.

5. An auction or competition in which only one participant took part shall be deemed to be unconstituted.

6. The rules provided for by Articles 448 and 449 of the present Code shall apply to a public sale held by way of execution of the decision of a court unless provided otherwise by procedural legislation.

Article 448. Organisation and Procedure for Holding Public Sale

1. Auctions and competitions may be open and closed.
Any person may participate in an open auction or open competition. Only persons specially invited for this purpose shall participate in a closed auction or closed competition.

2. Unless provided otherwise by a law, a notice concerning the conducting of a public sale must be made by the organiser not less than thirty days before conducting it. The notice must contain in any event information concerning the time, place, and form of the public sale, the subject thereof and the procedure for conducting it, including the formalisation of participation in the public sale, determination of the person who wins the public sale, and also information concerning the starting price.
If the subject of the public sale is only the right to conclude a contract, the period granted for this must be specified in the notice of the forthcoming public sale.

3. Unless provided otherwise in a law or in the notice on conducting the public sale, the organiser of an open public sale who has given notice shall have the right to refuse to hold an auction at any time, but not later than three days before the date ensues for conducting it, and to hold a competition, not later than thirty days before conducting the competition.
In instances when the organiser of an open public sale refuses to conduct it in violation of the specified periods, he shall be obliged to compensate participants for real damage incurred.
The organiser of a closed auction or closed competition shall be obliged to compensate the invited participants for real damage irrespective of precisely when the sending of the notice followed the refusal to hold the public sale.

4. The participants of a public sale shall make a deposit in the amount, within the periods, and in the procedure specified in the notice on conducting the public sale. If the public sale is unconstituted, the deposit shall be subject to return. A deposit shall also be returned to persons who participated in the public sale, but did not win it.

When concluding a contract with a person who has won a public sale, the amount of the deposit made by him shall be deducted from the performance of the obligations under the contract concluded.

5. A person who has won a public sale and the organiser of the public sale shall sign a protocol on the day of conducting the auction or competition concerning the results of the public sale, which shall have the force of a contract. In the event of the person who won the public sale evading the signature of the contract, he shall lose the deposit made by him. In the event of the organiser of a public sale evading the signature of the protocol, he shall be obliged to return the deposit in double the amount, and also to compensate the person who won the public sale for losses caused by participation in the public sale in that part exceeding the amount of the deposit.

If only the right to conclude a contract was the subject of a public sale, such contract must be signed by the parties not later than twenty days or other period specified in the notice after the completion of the public sale and formalisation of the protocol. In the event of one of them evading the conclusion of the contract, the other party shall have the right to apply to a court with a demand to compel the contract to be concluded, and also for compensation of losses caused by evading the conclusion thereof.

Article 449. Consequences of Violation of Rules for Conducting Public Sale

1. A public sale conducted in violation of the rules established by a law may be deemed by a court to be invalid upon the suit of an interested person.

2. The deeming of a public sale to be invalid shall entail the invalidity of the contract concluded with the person who won the public sale.

Chapter 29. Change and Dissolution of Contract

Article 450. Grounds for Change and Dissolution of Contract

1. A change and dissolution of a contract shall be possible by agreement of the parties unless provided otherwise by the present Code, by other laws, or by contract.

2. Upon the demand of one of the parties a contract may be changed or dissolved by decision of a court only:
(1) in the event of a material violation of the contract by the other party;
(2) in other instances provided for by the present Code, by other laws, or by contract.
A violation of a contract by one of the parties which entails for the other party such damage that it is deprived in significant degree of that which it had the right to count on when concluding the contract shall be deemed to be material.

3. In the event of a unilateral refusal to perform a contract wholly or partially when such refusal is permitted by a law or by agreement of the parties, the contract shall be considered to be dissolved or changed respectively.

Article 451. Change and Dissolution of Contract in Connection with Material Change of Circumstances

1. A material change of circumstances from which the parties proceeded when concluding a contract shall be a grounds for the change or dissolution thereof unless provided otherwise by the contract or it arises from the essence thereof.

A change of circumstances shall be deemed to be material when they have changed such that if the parties could reasonably foresee this, the contract would not have been concluded at all by them or it would have been concluded on significantly differing conditions.

2. If the parties have not reached agreement concerning the bringing of the contract into conformity with the materially changed circumstances or the dissolution thereof, the contract may be dissolved, and on the grounds provided for by point 4 of the present Article, changed by a court at the demand of the interested party when the following conditions simultaneously exist:

(1) at the moment of concluding the contract the parties proceeded from the fact that such a change of circumstances would not occur;

(2) the change of circumstances has been caused by reasons which the interested party could not overcome after they arose with that degree of concern and care which are required of him by the character of the contract and the conditions of turnover;

(3) the performance of the contract without a change of its conditions would so violate the correlation of property interests of the parties which correspond to the contract and entail for the interested party such damage that it would be deprived to a significant degree of that which it had the right to count on when concluding the contract;

(4) it does not arise from the customs of business turnover or the essence of the contract that the risk of the change of circumstances is borne by the interested party.

3. In the event of the dissolution of a contract as a consequence of a material change of circumstances the court at the demand of any of the parties shall determine the consequences of the dissolution of the contract by proceeding from the need for a just distribution between the parties of the expenses incurred by them in connection with the performance of this contract.

4. The change of a contract in connection with a material change of circumstances shall be permitted by decision of a court in exceptional instances when dissolution of the contract is contrary to social interests or entails damage for the parties which significantly exceeds the expenditures needed to perform the contract on the conditions changed by the court.

Article 452. Procedure for Change of and Dissolution of Contract

1. An agreement concerning a change of or dissolution of a contract shall be concluded in the same form as the contract unless it arises otherwise from a law, other legal acts, the contract, or the customs of business turnover.

2. A demand concerning a change of or dissolution of a contract may be made by a party in court only after receipt of a refusal of the other party to the proposal to change or dissolve the contract or of not receiving a reply within the period specified in the proposal or established by law or by contract, and in the absence thereof, within a thirty-day period.

Article 453. Consequences of Change of and Dissolution of Contract

1. In the event of the change of a contract, the obligations of the parties shall be preserved in the changed form.

2. In the event of the dissolution of the contract the obligations of the parties shall terminate.

3. In the event of the change of or the dissolution of a contract, the obligations shall be considered to be changed or terminated from the moment of the conclusion of an agreement of the parties concerning the change of or dissolution of the contract unless it follows otherwise from the agreement or the character of the change of the contract, and in the event of the change of or the dissolution of a contract in a judicial proceeding, from the moment of the entry into legal force of the decision of the court on the change of or the dissolution of the contract.

4. The parties shall not have the right to demand the return of that which was performed by them under the obligation before the moment of change or dissolution of the contract, unless established otherwise by a law or by agreement of the parties.

5. If a material violation of the contract by one of the parties has served as the grounds for change of or the dissolution of a contract, the other party shall have the right to demand compensation of losses caused by the change of or the dissolution of the contract.

PART TWO

SECTION IV. INDIVIDUAL TYPES OF OBLIGATIONS

Chapter 30. Purchase-Sale

§1. General Provisions on Purchase-Sale

Article 454. Contract of Purchase-Sale

1. Under a contract of purchase-sale one party (seller) shall be obliged to transfer a thing (or good) to the ownership of the other party (purchaser), and the purchaser shall be obliged to accept this good and to pay for it the determined monetary amount (price).

2. Unless special rules have been established for the purchase-sale thereof by a law, the provisions provided for by the present paragraph shall apply to the sale of securities and currency valuables.

3. In the instances provided for by the present Code or other law, the peculiarities of the purchase and sale of goods of individual types shall be determined by laws and other legal acts.

4. The provisions provided for by the present paragraph shall apply to the sale of property rights unless it arises otherwise from the content or character of these rights.

5. Unless provided otherwise by rules of the present Code concerning these types of contracts, the provisions provided for by the present paragraph shall apply to individual types of contract for purchase-sale (retail purchase-sale, delivery of goods, delivery of goods for State needs, agricultural procurement contract, electric power supply, sale of immovable, sale of enterprise).

Article 455. Condition of Contract on Good

1. Any things may be a good under a contract of purchase-sale in compliance with the rules provided for by Article 129 of the present Code.

2. A contract may be concluded for the purchase-sale of a good which the seller has at the moment of concluding the contract, and also a good which will be created or acquired by the seller in the future, unless established otherwise by a law or unless it so arises from the character of the good.

3. The condition of a contract of purchase-sale concerning a good shall be considered to be agreed if the contract enables the name and quantity of the good to be determined.

Article 456. Duties of Seller Relating to Transfer of Good

1. The seller shall be obliged to transfer to the purchaser the good provided for by the contract of purchase-sale.

2. Unless provided otherwise by the contract of purchase-sale, the seller shall be obliged simultaneously with the transfer of the thing to transfer to the purchaser the appurtenances thereto, as well as the documents relating thereto (technical passport, certificate of quality, instructions relating to operation, and so forth) provided for by a law, other legal acts, or the contract.

Article 457. Period for Performance of Duty to Transfer Good

1. The period for the performance by the seller of the duty to transfer the good to the purchaser shall be determined by the contract of purchase-sale, and if the contract does not enable this period to be determined, in accordance with the rules provided for by Article 314 of the present Code.

2. A contract of purchase-sale shall be deemed to be concluded with a condition of the performance thereof within a strictly determined period if it clearly arises from the contract that in the event of a violation of the period for the performance thereof the purchaser loses interest in the contract.

The seller shall have the right to perform such contract before the ensuing or after the expiry of the period determined therein only with the consent of the purchaser.

Article 458. Moment of Performance of Duty of Seller to Transfer Good

1. Unless provided otherwise by the contract of purchase-sale, the duty of the seller to transfer the good to the purchaser shall be considered to be performed at the moment of:

handing over the good to the purchaser or person specified by him if the duty of the seller relating to the delivery of the good has been provided for by the contract;

placing the good at the disposition of the purchaser if the good must be transferred to the purchaser or person specified by him at the location of the good. A good shall be considered to be placed at the disposition of the purchaser when within the period provided for by the contract the good is ready for transfer to the proper place and the purchaser in accordance with the conditions of the contract has knowledge of the readiness of the good for transfer. A good shall not be deemed to be ready for transfer if it is not identified for the purposes of the contract by means of marking or in other form.

2. In the instances when the duty of the seller relating to delivery of a good or transfer of a good at its location to the purchaser does not arise from a contract of purchase-sale, the duty of the seller to transfer the good to the purchaser shall be considered to be performed at the moment of handing over the good to the carrier or communications organisation for delivery to the purchaser, unless provided otherwise by the contract.

Article 459. Transfer of Risk of Accidental Perishing of Good

1. Unless provided otherwise by the contract of purchase-sale, the risk of accidental perishing or accidental damage of a good shall pass to the purchaser from the moment when in accordance with a law or the contract the seller is considered to have performed his duty relating to transfer of the good to the purchaser.

2. The risk of accidental perishing or accidental damage of a good sold while en route shall pass to the purchaser from the moment of the conclusion of the contract of purchase-sale unless provided otherwise by such contract or the customs of business turnover.

The condition of the contract that the risk of accidental perishing or accidental damage of the good passes to the purchaser from the moment of handing over the good to the first carrier may, at the demand of the purchaser, be deemed by a court to be invalid if at the moment of concluding the contract the seller knew or should have known that the good has been lost or damaged and did not communicate this to the purchaser.

Article 460. Duty of Seller to Transfer Good Free of Rights of Third Persons

1. The seller shall be obliged to transfer a good to the purchaser free from any rights of third persons, except for the instance when the purchaser has agreed to accept the good encumbered by the rights of third persons.

The failure of the seller to perform this duty shall give the purchaser the right to demand a reduction of the price of the good or dissolution of the contract of purchase-sale unless it is proved that the purchaser knew or should have known about the rights of third persons to this good.

2. The rules provided for by point 1 of the present Article shall apply respectively also when with respect to the good at the moment of its transfer to the purchaser there are claims of third persons, which were known to the seller if these claims subsequently were deemed to be lawful in the established procedure.

Article 461. Responsibility of Seller in Event of Seizure of Good from Purchaser

1. In the event of the seizure of the good from the purchaser by third persons on grounds which arose before the performance of the contract of purchase-sale,

the seller shall be obliged to compensate the purchaser for losses incurred by him unless it is proved that the purchaser knew or should have known about the existence of these grounds.

2. An agreement of the parties concerning relieving the seller from responsibility in the event a good acquired is demanded and obtained from the purchaser by third persons or a limitation thereof shall be invalid.

Article 462. Duties of Purchaser and Seller in Event of Filing of Suit Concerning Seizure of Good

If a third person on grounds which arose before performance of the contract of purchase-sale brings suit against the purchaser concerning seizure of the good, the purchaser shall be obliged to involve the seller in the case, and the seller shall be obliged to appear in this case on the side of the purchaser.

The failure of the purchaser to involve the seller in the case shall relieve the seller from responsibility to the purchaser if the seller proves that, having accepted to participate in the case, he could have averted the seizure of the good sold from the purchaser.

The seller involved by the purchaser in a case but who has not taken part therein shall be deprived of the right to prove the incorrectness of the conducting of the case by the purchaser.

Article 463. Consequences of Failure to Perform Duty to Transfer Good

1. If the seller refuses to transfer the good sold to the purchaser, the purchaser shall have the right to refuse to perform the contract of purchase-sale.

2. In the event of the refusal of the seller to transfer an individually-specified thing, the purchaser shall have the right to present the demand to the seller provided for by Article 398 of the present Code.

Article 464. Consequences of Failure to Perform Duty to Transfer Appurtenances and Documents Relating to Good

If the seller does not transfer or refuses to transfer to the purchaser appurtenances or documents relating to the good which he should transfer in accordance with a law, other legal acts, or the contract of purchase-sale (Article 456[2]), the purchaser shall have the right to designate a reasonable period to him for the transfer thereof.

When appurtenances or documents relating to a good have not been transferred by the seller within the specified period, the purchaser shall have the right to refuse the good unless provided otherwise by the contract.

Article 465. Quantity of Good

1. The quantity of a good subject to transfer to a purchaser shall be provided for by the contract of purchase-sale in respective units of measurement or in mon-

etary expression. The condition concerning the quantity of a good may be agreed by means of establishing in the contract the procedure for determining it.

2. If the contract of purchase-sale does not enable the quantity of the good subject to transfer to be determined, the contract shall not be considered to be concluded.

Article 466. Consequences of Violation of Condition Concerning Quantity of Good

1. If the seller has transferred to the purchaser in violation of the contract of purchase-sale a lesser quantity of the good that has been determined by the contract, the purchaser shall have the right, unless provided otherwise by the contract, either to demand to transfer the short quantity of the good or to refuse the good transferred and the payment thereof, and if the good has been paid for, to demand the return of the monetary amount paid.

2. If the seller has transferred to the purchaser a good in a quantity exceeding that specified in the contract of purchase-sale, the purchaser shall be obliged to notify the seller thereof in the procedure provided for by Article 483(1) of the present Code. When within a reasonable period after receipt of the attention of the purchaser the seller does not dispose of the respective part of the good, the purchaser shall have the right, unless provided otherwise by the contract, to accept the entire good.

3. In the event of the acceptance of the good by the purchaser in a quantity exceeding that specified in the contract of purchase-sale (point 2 of the present Article), the good additionally accepted shall be paid for at the price determined for the good accepted in accordance with the contract unless another price has been determined by agreement of the parties.

Article 467. Assortment of Goods

1. If under the contract of purchase-sale goods are subject to transfer in a specified correlation with regard to types, models, dimensions, colours, or other indicia (assortment), the seller shall be obliged to transfer the goods to the purchaser in the assortment agreed by the parties.

2. If the assortment in a contract of purchase-sale is not determined and the procedure for the determination thereof has not been established in the contract but it arises from the essence of the obligation that the goods must be transferred to the purchaser in an assortment, the seller shall have the right to transfer the goods to the purchaser in an assortment by proceeding from the requirements of the purchaser which were known to the seller at the moment of conclusion of the contract, or to refuse to perform the contract.

Article 468. Consequences of Violation of Condition on Assortment of Goods

1. In the event of the transfer by the seller of goods provided for by the contract of purchase-sale in an assortment which does not correspond to the contract, the purchaser shall have the right to refuse to accept and to pay for them, and if they have been paid for, to require the return of the monetary amount paid.

2. If the seller has transferred to the purchaser, together with goods whose assortment corresponds to the contract of purchase-sale, goods in violation of the conditions concerning the assortment, the purchase shall have the right at his choice to:

accept the goods corresponding to the conditions concerning the assortment and refuse the other goods;

refuse all of the goods transferred;

require the replacement of the goods which do not correspond to the condition concerning the assortment with goods in the assortment provided for by the contract;

accept all the goods transferred.

3. In the event of the refusal of goods whose assortment does not correspond to the condition of the contract of purchase-sale, or of presenting a demand concerning replacement of the goods which do not correspond to the condition concerning the assortment, the purchaser shall have the right also to refuse to pay for these goods, and if they have been paid for, to require the return of the monetary amount paid.

4. Goods which do not correspond to the condition of the contract of purchase-sale concerning assortment shall be considered to be accepted if the purchaser within a reasonable period after the receipt thereof does not notify the seller concerning his refusal of the goods.

5. If the purchaser has not refused goods whose assortment does not correspond to the contract of purchase-sale, he shall be obliged to pay for them at the price agreed with the seller. When necessary measures have not been taken by the seller to agree the price within a reasonable period, the purchaser shall pay for the goods at the price which at the moment of concluding the contract under comparable circumstances are usually recovered for analogous goods.

6. The rules of the present Article shall apply unless provided otherwise by the contract of purchase-sale.

Article 469. Quality of Good

1. The seller shall be obliged to transfer a good to the purchaser whose quality corresponds to the contract of purchase-sale.

2. In the absence in a contract of purchase-sale of conditions concerning the quality of a good the seller shall be obliged to transfer to the purchaser a good fit for the purposes for which a good of such nature is usually used.

If the seller when concluding the contract was given notice by the purchaser of the specific purposes for the acquisition of the good, the seller shall be obliged to transfer to the purchaser a good fit for use in accordance with these purposes.

3. In the event of the sale of a good according to a sample and/or description, the seller shall be obliged to transfer to the purchaser a good which corresponds to the sample and/or description.

4. If obligatory requirements for the quality of a good being sold have been provided by a law or in the procedure established by it, then the seller effectuating entrepreneurial activity shall be obliged to transfer a good to the purchaser which corresponds to these obligatory requirements [as amended 17 December 1999. СЗ РФ (1999), no. 51, item 6288].

By agreement between the seller and the purchaser a good may be transferred which corresponds to higher quality requirements in comparison with the obligatory requirements in the procedure provided for by a law or established by it [as amended 17 December 1999. СЗ РФ (1999), no. 51, item 6288].

Article 470. Guarantee of Quality of Good

1. A good which a seller is obliged to transfer to a purchaser must correspond to the requirements provided for by Article 469 of the present Code at the moment of the transfer to the purchaser unless another moment of determining the conformity of the good to these requirements has been provided for by the contract of purchase-sale and within the limits of a reasonable period must be fit for the purposes for which the goods of such nature are usually used.

2. When by a contract of purchase-sale the granting by the seller of a guarantee of the quality of a good has been provided for, the seller shall be obliged to transfer to the purchaser a good which must correspond to the requirements provided for by Article 469 of the present Code during the determined time established by the contract (guarantee period).

3. A guarantee of the quality of a good also shall extend to all parts comprising it (sets of manufacture) unless provided otherwise by the contract of purchase-sale.

Article 471. Calculation of Guarantee Period

1. A guarantee period shall begin to run from the moment of the transfer of the good to the purchaser (Article 457) unless provided otherwise by the contract of purchase-sale.

2. If a purchaser is deprived of the possibility to use the good with respect to which a guarantee period has been established by the contract under

circumstances within the control of the seller, the guarantee period shall not run until the elimination of the respective circumstances by the seller.

Unless provided otherwise by the contract, a guarantee period shall be extended for the time during which the good could not be used because of defects discovered therein, on condition of the notification of the seller about the defects of the good in the procedure established by Article 483 of the present Code.

3. Unless provided otherwise by the contract of purchase-sale, a guarantee period for a set of manufactures shall be considered to be equal to the guarantee period for the principal manufacture and shall begin to run simultaneously with the guarantee period on the principal manufacture.

4. For a good (or set of manufactures) transferred by the seller in place of a good (or set of manufactures) in which during the guarantee period defects were discovered (Article 476), a guarantee period shall be established of the same duration as that replaced, unless provided otherwise by the contract of purchase-sale.

Article 472. Period of Fitness of Good

1. The duty to determine a period upon the expiry of which a good shall be considered to be unfit for use for designation (period of fitness) may be provided for by a law or in the procedure established by it [as amended 17 December 1999. СЗ РФ (1999), no. 51, item 6288].

2. A good for which a period of fitness has been established the seller shall be obliged to transfer to the purchaser so calculated that it may be used for its designation until the expiry of the period of fitness, unless provided otherwise by the contract [as amended 17 December 1999. СЗ РФ (1999), no. 51, item 6288].

Article 473. Calculation of Period of Fitness of Good

The period of fitness of a good shall be determined by the period of time calculated from the date of its manufacture during which the good is fit for use or by the date until the ensuing of which the good is fit for use.

Article 474. Verification of Quality of Good

1. The verification of the quality of a good may be provided for by a law, other legal acts, obligatory requirements of State standards, or by the contract of purchase-sale.

The procedure for verification of the quality of a good shall be established by a law, other legal acts, obligatory requirements of State standards, or by the contract.

In instances when the procedure for verification has been established by a law, other legal acts, or obligatory requirements of State standards, the procedure for the verification of the quality of goods determined by the contract must correspond to these requirements.

2. If the procedure for the verification of the quality of a good is not established in accordance with point 1 of the present Article, the quality of a good shall be verified in accordance with the customs of business turnover or other ordinarily used conditions for the verification of a good subject to transfer under a contract of purchase-sale.

3. If the duty of the seller to verify the quality of a good transferred to the purchaser (test, analysis, inspection, and so forth) has been provided for by a law, other legal acts, obligatory requirements of State standards, or by the contract of purchase-sale, the seller must grant to the purchaser evidence of the effectuation of the verification of the quality of the good.

4. The procedure, and also other conditions for the verification of the quality of a good to be made both by the seller and by the purchaser, must be one and the same.

Article 475. Consequences of Transfer of Good of Improper Quality

1. If the defects of a good were not stipulated by the seller, the purchaser to whom the good of improper quality was transferred shall have the right at his choice to require from the seller:

commensurate reduction of the purchase price;

elimination without compensation of the defects of the good within a reasonable period;

compensation of his expenses to eliminate the defects of the good.

2. In the event of a material violation of the requirements for the quality of a good (discovery of unremovable defects, defects which cannot be eliminated without incommensurate expenses or expenditures of time or are elicited repeatedly, or manifest themselves anew after their elimination, and other similar defects), the purchaser shall have the right at his choice to:

refuse to perform the contract of purchase-sale and to require the return of the monetary amount paid for the good;

require the replacement of the good of improper quality by a good corresponding to the contract.

3. Demands concerning the elimination of defects or replacement of a good specified in points 1 and 2 of the present Article may be presented by the purchaser unless it arises otherwise from the character of the good or the essence of the obligation.

4. In the event of the improper quality of part of the goods which are part of a set (Article 479), the purchaser shall have the right to effectuate with respect to this part of the goods the rights provided for by points 1 and 2 of the present Article.

5. The rules provided for by the present Article shall apply unless established otherwise by the present Code or another law.

Article 476. Defects of Good for Which Seller is Liable

1. The seller shall be liable for the defects of a good if the purchaser proves that the defects of the good arose before the transfer thereof to the purchaser or for reasons which arose before this moment.

2. With respect to a good for which a guarantee of quality has been granted by the seller the seller shall be liable for defects of the good unless it is proved that the defects of the good arose after the transfer thereof to the purchaser as a consequence of a violation by the purchaser of the rules for the use of the good or the keeping thereof, or the actions of third persons, or insuperable force.

Article 477. Periods for Discovery of Defects in Transferred Good

1. Unless established otherwise by a law or by a contract of purchase-sale, the purchaser shall have the right to present demands connected with the defects of a good on condition that they were discovered within the periods established by the present Article.

2. If a guarantee period or a period of fitness has not been established for a good, demands connected with defects of a good may be presented by the purchaser on condition that the defects in the good sold were discovered within a reasonable period, but within the limits of two years from the date of the transfer of the good to the purchaser, or a longer period when such period was established by a law or by a contract of purchase-sale. The period for eliciting the defects of a good subject to carriage or to despatch by post shall be calculated from the date of delivery of the good at the place of its destination.

3. If a guarantee period has been established for a good, the purchaser shall have the right to present demands connected with defects of the good in the event the defects are discovered during the guarantee period.

When a guarantee period has been established in a contract of purchase-sale for sets of manufactures which is less in duration than the principal manufacture, the purchaser shall have the right to present demands connected with the defects of the sets of manufactures in the event of their discovery during the guarantee period for the principal manufacture.

If a guarantee period has been established in a contract for a set of manufactures which is of lesser duration than for the principal manufacture, the purchaser shall have the right to present demands connected with defects of the set of manufactures when they are discovered during the guarantee period for it irrespective of the expiry of the guarantee period for the principal manufacture.

4. With respect to a good for which a period of fitness has been established, the purchaser shall have the right to present demands connected with defects of the good if they were discovered within the period of fitness of the good.

5. In instances when the guarantee period provided for by a contract constitutes less than two years and the defects of a good are discovered by the purchaser upon the expiry of the guarantee period but within the limits of two years from the day of transfer of the good to the purchaser, the seller shall bear responsibility if the purchaser proves that the defects of the good arose before the transfer of the good to the purchaser or for reasons which arose before this moment.

Article 478. Completeness of Good

1. The seller shall be obliged to transfer to the purchaser a good which corresponds to the conditions of the contract of purchase-sale concerning completeness.

2. When the completeness of a good has not been determined by a contract of purchase-sale, the seller shall be obliged to transfer to the purchaser a good whose completeness is determined by the customs of business turnover or other ordinarily presented requirements.

Article 479. Complete Set of Goods

1. If the duty of the seller to transfer to the purchaser a determined selection of goods in a complete set (complete set of goods) has been provided for by a contract of purchase-sale, the obligation shall be considered to be performed from the moment of the transfer of all goods included in the complete set.

2. Unless otherwise provided for by the contract of purchase-sale and does not arise from the essence of the obligation, the seller shall be obliged to transfer simultaneously to the purchaser all goods which comprise the complete set.

Article 480. Consequences of Transfer of Incomplete Good

1. In the event of the transfer of an incomplete good (Article 478) the purchaser shall have the right at his choice to require from the seller:
commensurate reduction of the purchase price;
making the good complete within a reasonable period.

2. If the seller within a reasonable period has not fulfilled the demand of the purchaser to make the good complete, the purchaser shall have the right at his choice to:
require replacement of the incomplete good by a complete [one];
refuse to perform the contract of purchase-sale and require the return of the monetary amount paid.

3. The consequences provided for by points 1 and 2 of the present Article shall also apply in the event of a violation by the seller of the duty to transfer a complete

set of goods (Article 479) to the purchaser unless provided otherwise by the contract of purchase-sale and does not arise from the essence of the obligation.

Article 481. Packaging and Packing

1. Unless provided otherwise by the contract of purchase-sale and does not arise from the essence of the obligation, the seller shall be obliged to transfer the good to the purchaser in packaging and/or packing, except for a good which by its character does not require packaging and/or packing.

2. Unless the requirements for packaging and packing have been determined by the contract of purchase-sale, the good must be packaged and/or packed by the means usual for such good, and in the absence of such, by a means ensuring the preservation of goods of such nature under the usual conditions for keeping and transportation.

3. If in the procedure established by a law obligatory requirements for packaging and/or packing have been provided for, the seller who effectuates entrepreneurial activity shall be obliged to transfer to the purchaser the good in packaging and/or packing which corresponds to these obligatory requirements.

Article 482. Consequences of Transfer of Good Without Packaging and/or Packing or in Improper Packaging and/or Packing

1. In instances when a good which is subject to packaging and/or packing is transferred to the purchaser without packaging and/or packing or in improper packaging and/or packing, the purchaser shall have the right to demand of the seller to package and/or to pack the good or to replace the improper packaging and/or packing unless it arises otherwise from the essence of the obligation or the character of the good.

2. In the instances provided for in point 1 of the present Article, the purchaser shall have the right instead of presenting demands to the seller which are specified in this point to present to him demands arising from the transfer of a good of improper quality (Article 475).

Article 483. Notification of Seller About Improper Performance of Contract of Purchase-Sale

1. The purchaser shall be obliged to notify the seller about a violation of the conditions of the contract of purchase-sale concerning quantity, assortment, quality, completeness, packaging and/or packing of the good within the period provided for by a law, other legal acts, or the contract, and if such period is not established, within a reasonable period after such violation of the respective condition of the contract should have been discovered by proceeding from the character and designation of the good.

2. In the event of the failure to fulfil the rules provided for by point 1 of the present Article, the seller shall have the right to refuse satisfaction wholly or partially of the demands of the purchaser concerning the transfer to him of the short quantity of the good, replacement of the good which does not correspond to the conditions of the contract of purchase-sale concerning quality or assortment, elimination of the defects of a good, to make the good complete or to replace a good which is not complete with a complete [one], the packaging and/or packing of a good or the replacement of improper packaging and/or packing of a good, if it is proved that the failure to fulfil these rules by the purchaser has entailed the impossibility of satisfying his demands or entails for the seller incommensurate expenses in comparison with those which he would bear if he had been notified in good time about the violation of the contract.

3. If the seller knew or should have known that the goods transferred to the purchaser do not correspond to the conditions of the contract of purchase-sale, he shall not have the right to refer to the provisions provided for by points 1 and 2 of the present Article.

Article 484. Duty of Purchaser to Accept Good

1. The purchaser shall be obliged to accept a good transferred to him except for instances when he has the right to demand the replacement of the good or to refuse performance of the contract of purchase-sale.

2. Unless provided otherwise by a law, other legal acts, or the contract of purchase-sale, the purchaser shall be obliged to perform actions which in accordance with requirements usually presented are necessary on his part in order to ensure the transfer and receipt of the respective good.

3. In instances when the purchaser in violation of a law, other legal acts, or contract of purchase-sale does not accept a good or refuses to accept it, the seller shall have the right to demand that the purchaser accept the good or to refuse performance of the contract.

Article 485. Price of Good

1. The purchaser shall be obliged to pay for a good at the price provided for by the contract of purchase-sale or, if it is not provided for by the contract and may not be determined by proceeding from the conditions thereof, at the price determined in accordance with Article 424(3) of the present Code, and also to perform at his expense actions which in accordance with a law, other legal acts, the contract, or requirements usually presented are necessary in order to effectuate payment.

2. When the price has been established depending on the weight of the good, it shall be determined according to net weight unless provided otherwise by the contract of purchase-sale.

3. If the contract of purchase-sale provides that the price of a good is subject to change depending upon indicators which condition the price of the good (cost of production, expenditures, and so forth) but in so doing does not determine the means for revision of the price, the price shall be determined by proceeding from the correlation of these indicators at the moment of conclusion of the contract and at the moment of transfer of the good. In the event of delay by the seller of performance of the duty to transfer the good, the price shall be determined by proceeding from the correlation of these indicators at the moment of conclusion of the contract and at the moment of transfer of the good provided for by the contract, and if it is not provided for by the contract, at the moment determined in accordance with Article 314 of the present Code.

The rules provided for by the present point shall apply unless established otherwise by the present Code, another law, other legal acts, or the contract and does not arise from the essence of the obligation.

Article 486. Payment for Good

1. The purchaser shall be obliged to pay for the good directly before and after the transfer of the good to him by the seller unless provided otherwise by the present Code, another law, other legal acts, or by the contract of purchase-sale and does not arise from the essence of the obligation.

2. Unless payment by instalment for a good has been provided by the contract of purchase-sale, the purchaser shall be obliged to pay to the seller the price of the transferred good in full.

3. If the purchaser does not pay in a timely way for the good transferred in accordance with the contract of purchase-sale, the seller shall have the right to require payment for the good and the payment of interest in accordance with Article 395 of the present Code.

4. If the purchaser in violation of the contract of purchase sale refuses to accept and to pay for the good, the seller shall have the right at his choice to require payment for the good or to refuse performance of the contract.

5. In instances when the seller in accordance with the contract of purchase-sale is obliged to transfer to the purchaser not only goods which are not paid for by the purchaser, but also other goods, the seller shall have the right to suspend the transfer of these goods until the full payment for all goods previously transferred unless provided otherwise by a law, other legal acts, or contract.

Article 487. Preliminary Payment for Good

1. In instances when the duty of the purchaser to pay for a good wholly or partially before the transfer of the good by the seller (preliminary payment) has been provided for by the contract of purchase-sale, the purchaser must make payment

within the period provided for by the contract, and if such period has not been provided by the contract, within the period determined in accordance with Article 314 of the present Code.

2. In the event of the failure of the purchaser to perform the duty to pay in advance for the good, the rules provided for by Article 328 of the present Code shall apply.

3. In an instance when a seller who has received the amount of preliminary payment does not perform the duty to transfer the good within the established period (Article 457), the purchaser shall have the right to require the transfer of the good paid for or the return of the amount of preliminary payment for the good not transferred by the seller.

4. Unless provided otherwise by the contract of purchase-sale, in an instance when the seller does not perform the duty relating to the transfer of the good paid for in advance the amount of preliminary payment shall be subject to the payment of interest in accordance with Article 395 of the present Code from the date when under the contract the transfer of the good should have been made up to the date of the transfer of the good to the purchaser or the return to him of the amount paid in advance by him. The duty of the seller to pay interest on the amount of the preliminary payment, commencing from the date of receipt of this amount from the purchaser, may be provided for by the contract.

Article 488. Payment for Good Sold on Credit

1. In an instance when payment for a good over a determined time after the transfer thereof to the purchaser (sale of good on credit) has been provided for by the contract of purchase-sale, the purchaser must make payment within the period provided for by the contract, and if such period is not provided for by the contract, within the period determined in accordance with Article 314 of the present Code.

2. In the event of the failure of the seller to perform the duty to transfer the good, the rules provided for by Article 328 of the present Code shall apply.

3. In an instance when the purchaser which received the good does not perform the duty with regard to payment within the period established by the contract of purchase-sale, the seller shall have the right to require payment of the good transferred or return of the goods not paid for

4. Unless provided otherwise by the present Code or by the contract of purchase-sale, in an instance when the purchaser does not perform the duty with regard to payment for the good transferred within the period established by the contract, interest shall be subject to payment on the delayed amount in accordance with Article 395 of the present Code from the date when under the contract

the good should have been paid for up to the date of payment for the good by the purchaser.

The duty of the purchaser to pay interest on the amount corresponding to the price of the good, commencing from the date of transfer of the good by the seller, may be provided for by the contract.

5. Unless provided otherwise by the contract of purchase-sale, from the moment of transfer of the good to the purchaser and up to the payment for it a good sold on credit shall be deemed to be under pledge to the seller in order to secure performance by the purchaser of its duty with regard to payment for the good.

Article 489. Payment for Good by Instalment

1. Payment for a good by instalment may be provided for by a contract concerning the sale of a good on credit.

A contract concerning the sale of a good on credit with a condition concerning instalment payment shall be considered to be concluded if, together with the other material conditions of the contract of purchase-sale, the price of the good and the procedure, periods, and amounts of payments have been specified.

2. When a purchaser does not make a regular payment within the period established by the contract for a good sold by instalment and transferred to him, the seller shall have the right to refuse performance of the contract and to require the return of the good sold, except for instances when the amount of payments received from the purchaser exceeds half of the price of the good unless provided otherwise by the contract.

3. The rules provided for by Article 488 (2), (4) and (5) of the present Code shall apply to a contract concerning the sale of a good on credit with a condition concerning instalment payment.

Article 490. Insurance of Good

The duty of the seller or purchaser to insure a good may be provided for by the contract of purchase-sale.

In an instance when a party obliged to insure a good does not effectuate insurance in accordance with the conditions of the contract, the other party shall have the right to insure the good and require from the obliged party compensation of expenses for the insurance or to refuse performance of the contract.

Article 491. Preservation of Right of Ownership for Seller

In instances when it is provided by a contract of purchase-sale that the right of ownership in a good transferred to the purchaser shall be preserved for the seller until payment for the good or the ensuing of other circumstances, the purchaser shall not have the right before transfer of the right of ownership to him to alienate the good or otherwise dispose of it unless provided otherwise by a law or contract or does not arise from the designation and properties of the good.

In instances when within the period provided for by the contract the transferred good is not paid for and other circumstances do not ensue under which the right of ownership passes to the purchaser, the seller shall have the right to require from the purchaser to return the good to him unless provided otherwise by the contract.

§2. Retail Purchase-Sale

Article 492. Contract of Retail Purchase-Sale

1. Under a contract of retail purchase-sale the seller effectuating entrepreneurial activity relating to the sale of goods on retail shall be obliged to transfer to the purchaser a good intended for personal, family, home, or other use which is not connected with entrepreneurial activity.

2. A contract of retail purchase-sale shall be a public contract (Article 426).

3. The laws concerning the defence of the rights of consumers and other legal acts adopted in accordance with them shall apply to relations under a contract of retail purchase-sale with the participation of a citizen-purchaser which are not regulated by the present Code.

Article 493. Form of Contract of Retail Purchase-Sale

Unless provided otherwise by a law or a contract of retail purchase-sale, including the conditions of standard contracts or other standard forms to which the purchaser adheres (Article 428), a contract of retail purchase-sale shall be considered to be concluded in proper form from the moment of the issuance by the seller to the purchaser of a cashier or goods cheque or other document confirming payment of the good. The absence of the said documents with the purchaser shall not deprive him of the possibility to refer to witness testimony in confirmation of the conclusion of the contract and its conditions.

Article 494. Public Offer of Good

1. The proposal of a good in its advertisement, catalogues, and descriptions of the goods addressed to an indefinite group of persons shall be deemed to be a public offer (Article 437[2]) if it contains all the material conditions of the contract of retail purchase-sale.

2. Exhibition at the place of sale (on counters, in showcases, and so forth) of goods, the demonstration of models thereof, or the granting of information concerning goods being sold (descriptions, catalogues, photographs, and so forth) at the place of sale thereof shall be deemed to be a public offer irrespective of whether the price and other material conditions of the contract of retail purchase-sale have

been specified, except for an instance when the seller visibly determined that the respective goods are not intended for sale.

Article 495. Granting of Information Concerning Good to Purchaser

1. The seller shall be obliged to grant to the purchaser necessary and reliable information concerning the good proposed for sale which corresponds to the requirements established by a law, other legal acts, and those usually presented in retail trade for the content and means of granting such information.

2. The purchaser shall have the right before the conclusion of a contract of retail purchase-sale to inspect the good and require the conducting in his presence of a verification of the properties or demonstration of the use of the good unless this is precluded in view of the character of the good and is not contrary to the rules accepted in retail trade.

3. If the purchaser has not been granted the possibility at once to receive information at the place of sale concerning the good specified in points 1 and 2 of the present Article, he shall have the right to require from the seller the compensation of losses caused by unfounded avoidance of conclusion of the contract of retail purchase-sale (Article 445[4]), and if the contract has been concluded, within a reasonable period to refuse performance of the contract and to require the return of the amount paid for the good and compensation of other losses.

4. A seller who has not granted the purchaser the possibility to receive respective information concerning a good shall bear responsibility also for defects of the good which arose after its transfer to the purchaser with respect to which the purchaser proves that they arose in connection with the absence of such information.

Article 496. Sale of Good with Condition Concerning Acceptance Thereof by Purchaser Within Determined Period

A contract of retail purchase-sale may be concluded with a condition concerning acceptance by the purchaser of the good within a period determined by the contract during which this good may not be sold to another purchaser.

Unless provided otherwise by the contract, the failure of the purchaser to appear or the failure to perform other necessary actions in order to accept the good within the period determined by the contract may be considered by the seller as the refusal of the purchaser to perform the contract.

Additional expenses of the seller relating to ensuring the transfer of the good to the purchaser within the period determined by the contract shall be included in the price of the good unless provided otherwise by a law, other legal acts, or contract.

Article 497. Sale of Goods by Samples

1. A contract of retail purchase-sale may be concluded on the basis of famil-iarising the purchaser with a sample of the good (or description thereof, catalogue of goods, and so forth) proposed by the seller.

2. Unless provided otherwise by a law, other legal acts, or the contract, a con-tract of purchase-sale for a good by a sample shall be considered to be performed from the moment of delivery of the good to the place specified in the contract, and if the place of transfer of the good is not determined by the contract, from the moment of delivery of the good to the purchaser at the place of residence of the citizen or location of the juridical person.

3. A purchaser before the transfer of the good shall have the right to refuse per-formance of a contract of retail purchase-sale on condition of compensation to the seller for necessary expenses incurred in connection with the performance of actions relating to fulfilment of the contract.

Article 498. Sale of Goods With Use of Automatic Machines

1. In the instances when goods are sold with the use of automatic machines, the possessor of the automatic machines shall be obliged to bring information to purchasers concerning the seller of the goods by means of placing on the auto-matic machine or granting to the purchasers by other means information con-cerning the name (or firm name) of the seller, location, work regime, and also actions which are necessary to be performed by the purchaser in order to receive the good.

2. A contract of retail purchase-sale with the use of automatic machines shall be considered to be concluded from the moment of the performance of actions by the purchaser which are necessary in order to receive the good.

3. If the purchaser has not been granted a good paid for, the seller shall be obliged at the demand of the purchaser to grant the good at once to the purchaser or to return the amount paid by him.

4. In instances when an automatic machine is used to change money or the acquisition of banknotes for payment or the exchange of currency, the rules con-cerning retail purchase-sale shall apply unless it arises otherwise from the essence of the obligation.

Article 499. Sale of Good With Condition of Delivery Thereof to Purchaser

1. In an instance when a contract of retail purchase-sale has been concluded with a condition concerning delivery of the good to the purchaser, the seller shall be obliged within the period established by the contract to deliver the good at the place specified by the purchaser, and if the place of delivery is not specified by the

purchaser, at the place of residence of the citizen or location of the juridical person which is the purchaser.

2. A contract of retail purchase-sale shall be considered to be performed from the moment of handing of the good over to the purchaser, and in the event of his absence, to any person presenting the receipt or other document testifying to the conclusion of the contract or formalisation of the delivery of the good, unless provided otherwise by a law, other legal acts, or the contract or does not arise from the essence of the obligation.

3. In an instance when the time for delivery of a good for handing over to its purchaser is not determined by the contract, the good must be delivered within a reasonable period after receipt of the demand of the purchaser.

Article 500. Price and Payment for Good

1. The purchaser shall be obliged to pay for a good at the price declared by the seller at the moment of concluding the contract of retail purchase-sale unless provided otherwise by a law, other legal acts, or does not arise from the essence of the obligation.

2. In an instance when preliminary payment for a good (Article 487) has been provided for by the contract of retail purchase-sale, the failure of the purchaser to pay for the good within the period established by the contract shall be deemed to be repudiation of the contract by the purchaser unless provided otherwise by agreement of the parties.

3. The rules provided for by Article 488(4), paragraph one, of the present Code shall not be subject to application to contracts of retail purchase-sale on credit, including with a condition for payment by the purchaser for the goods by instalment.

The purchaser shall have the right to pay up for the good at any time within the limits of the instalment period for payment for the good established by the contract.

Article 501. Contract of Hire-Sale

It may be provided by a contract that before the transfer of the right of ownership to a good to the purchaser (Article 491) the purchaser shall be the hirer (or lessee) of the good transferred to him (contract of hire-sale).

Unless provided otherwise by the contract, the purchaser shall become the owner of the good from the moment of paying up the good.

Article 502. Exchange of Good

1. The purchaser shall have the right within fourteen days from the moment of the transfer to him of a non-foodstuff good, unless a longer period is declared

by the seller, to exchange a good bought at the place of purchase and other places declared by the seller for an analogous good of another size, form, dimension, style, colour, or complement, making the necessary settlement with the seller in the event of a difference in price.

If the seller does not have the good necessary for the exchange, the purchaser shall have the right to return the acquired good to the seller and to receive the monetary amount paid for it.

The demand of the purchaser concerning an exchange or return of the good shall be subject to satisfaction if the good was not used, has retained its consumer properties, and there is evidence of its acquisition from the particular seller.

2. A list of goods not subject to exchange or return on the grounds specified in the present Article shall be determined in the procedure established by a law or other legal acts.

Article 503. Rights of Purchaser in Event of Sale to Him of Good of Improper Quality

1. A purchaser to whom a good of improper quality was sold, unless its defects were stipulated by the seller, shall have the right at his choice to require:
replacement of the poor-quality good by a good of proper quality;
commensurate reduction of the purchase price;
immediate elimination without compensation of the defects of the good;
compensation of expenses for elimination of the defects of the good.
The purchaser shall have the right to demand the replacement of a technically complex or expensive good in the event of a material violation of the requirements for quality (Article 475[2]).

2. In the event of the discovery of defects of a good whose properties do not enable them to be eliminated (foodstuffs, domestic sundries, and so forth) the purchaser shall have the right at his choice to require the replacement of such good by a good of proper quality or commensurate reduction of the purchase price.

3. Instead of presenting the demands provided for by points 1 and 2 of the present Article, the purchaser shall have the right to refuse performance of the contract of retail purchase-sale and to require the return of the monetary amount paid for the good.

In so doing the purchaser must at the demand of the seller and at his expense return the good of improper quality received.

In the event of the return to the purchaser of the monetary amount paid for the good, the seller shall not have the right to withhold from it the amount by which the value of the good was lowered because of the full or partial use of the good, loss of goods appearance, or other similar circumstances.

Article 504. Compensation of Difference in Price in Event of Replacement of Good, Reduction of Purchase Price, and Return of Good of Improper Quality

1. In the event of the replacement of a poor-quality good not corresponding to the contract of retail purchase-sale with a good of proper quality, the seller shall not have the right to demand compensation of the difference between the price of the good established by the contract and the price of the good existing at the moment of replacement of the good or rendering by a court of a decision concerning replacement of the good.

2. In the event of the replacement of a poor-quality good by an analogous good but different in dimension, style, quality, or other indicia, the good of proper quality shall be subject to compensation of the difference between the price of the good being replaced at the moment of replacement and the price of the good being transferred in place of the good of improper quality.

If the demand of the purchaser is not satisfied by the seller, the price of the good being replaced and the price of the good being transferred in place of it shall be determined at the moment of the rendering by a court of the decision concerning replacement of the good.

3. In the event a demand is presented concerning a commensurate reduction of the purchase price of the good in settlement, the price of the good at the moment of presenting the demand concerning the price reduction shall be used, and if the demand of the purchase is not satisfied voluntarily, at the moment of the rendering by a court of a decision concerning the commensurate reduction of the price.

4. In the event of the return to the seller of a good of improper quality the purchaser shall have the right to require compensation for the difference between the price of the good established by the contract of retail purchase-sale and the price of the respective good at the moment of voluntary satisfaction of his demand, and if the demand is not satisfied voluntarily, at the moment of the rendering of a decision by the court.

Article 505. Responsibility of Seller and Performance of Obligation in Kind

In the event of the failure of the seller to perform an obligation under the contract of retail purchase-sale, compensation of losses and the payment of a penalty shall not relieve the seller from performance of the obligation in kind.

§3. Delivery of Goods

Article 506. Contract of Delivery

Under a contract of delivery the supplier-seller effectuating entrepreneurial activity shall be obliged to transfer goods produced or purchased by him within the stipulated period or periods to the purchaser for use in entrepreneurial activity or for other purposes not connected with personal, family, home, and other analogous use.

Article 507. Settlement of Disagreements When Concluding Contract of Delivery

1. In an instance when in concluding a contract of delivery disagreements arose between the parties with regard to individual conditions of the contract, the party proposing to conclude the contract and receiving a proposal from the other party to agree these conditions must within thirty days from the date of receipt of this proposal, unless another period has been established by a law or has been agreed by the parties, take measures relating to agreeing the respective conditions of the contract or inform the other party in writing of his refusal to conclude it.

2. The party who received the proposal relating to respective conditions of the contract but did not take measures relating to agreeing the conditions of the contract of delivery and did not inform the other party about the refusal to conclude the contract within the period provided for by point 1 of the present Article shall be obliged to compensate the losses caused by the avoidance of agreeing the conditions of the contract.

Article 508. Periods for Delivery of Goods

1. In an instance when the delivery of goods has been provided for by the parties within the period of operation of the contract of delivery by individual lots and the periods of delivery of the individual lots (or delivery periods) have not been determined therein, the goods must be delivered in lots of equal measure monthly unless it arises otherwise from a law, other legal acts, the essence of the obligation or the customs of business turnover.

2. Together with the determination of the periods of delivery a delivery schedule for the goods (ten-day, twenty-four hour, hour, and so forth) may be established in the contract of delivery.

3. Goods may be delivered before time with the consent of the purchaser.
Goods delivered before time and accepted by the purchaser shall be credited to the quantity of goods subject to delivery within the following period.

Article 509. Procedure for Delivery of Goods

1. The delivery of goods shall be effectuated by the supplier by means of shipment (or transfer) of the goods to the purchaser who is a party to the contract of delivery or to the person specified in the contract as the recipient.

2. In an instance when the right of the purchaser has been provided for by the contract of delivery to give an instruction to the supplier concerning the shipment

(or transfer) of the goods to the recipients (shipment order), the shipment (or transfer) of the goods shall be effectuated by the supplier to the recipients specified in the shipment order.

The content of a shipment order and the period for sending it by the purchaser to the supplier shall be determined by the contract. If the period for sending the shipment order is not provided for by the contract, it must be sent to the supplier not later than thirty days before the ensuing of the delivery period.

3. The failure of the purchaser to present the shipment order within the established period shall give the supplier the right either to refuse performance of the contract of delivery or to require payment for the goods from the recipient. In addition, the supplier shall have the right to require compensation of losses caused in connection with the failure to present the shipping order.

Article 510. Delivery of Goods

1. The delivery of goods shall be effectuated by the supplier by means of the shipment thereof by transport provided for by the contract of delivery and on the conditions determined in the contract.

In instances when what type of transport or on what conditions the delivery shall be effectuated is not determined in the contract, the right of choice of the type of transport or determination of the conditions of delivery of the goods shall belong to the supplier unless it arises otherwise from a law, other legal acts, the essence of the obligation, or customs of business turnover.

2. The receipt of goods by the purchaser (or recipient) at the location of the supplier (selection of goods) may be provided for by the contract of delivery.

If the period of selection is not provided for by the contract, the selection of goods by the purchaser (or recipient) must be made within a reasonable period after the receipt of the notice of the supplier concerning the readiness of the goods.

Article 511. Augmentation of Short Deliveries of Goods

1. The supplier who has permitted a short delivery of goods in an individual period of delivery shall be obliged to augment the short-delivered quantity of goods within the following period(s) within the limits of the period of operation of the contract of delivery unless provided otherwise by the contract.

2. In an instance when goods have been shipped by the supplier to several recipients specified in the contract of delivery or shipment order of the recipient, the goods delivered to one recipient in excess of the quantity provided for in the contract or shipment order shall not be credited to cover the short delivery to other purchasers unless provided otherwise by the contract.

3. The purchaser shall have the right, having informed the supplier, to refuse acceptance of the goods whose delivery has been delayed unless provided other-

wise in the contract. The goods delivered before receipt of the notice by the supplier the purchaser shall be obliged to accept and to pay for.

Article 512. Assortment of Goods When Augmenting Short Delivery

1. The assortment of goods whose short delivery is subject to augmentation shall be determined by agreement of the parties. In the absence of such agreement the supplier shall be obliged to augment the short-delivered quantity of the goods in the assortment established for that period in which the short delivery was permitted.

2. The delivery of goods of one name in a larger quantity than is provided for the contract of delivery shall not be credited towards covering the short delivery of goods of another name within the same assortment and shall be subject to augmentation except for instances when such delivery was made with the prior written consent of the purchaser.

Article 513. Acceptance of Goods by Purchaser

1. The purchaser (or recipient) shall be obliged to perform all necessary actions ensuring the acceptance of goods delivered in accordance with the contract of delivery.

2. Goods accepted by the purchaser (or recipient) must be inspected by him within the period determined by a law, other legal acts, the contract of delivery, or the customs of business turnover.

The purchaser (or recipient) shall be obliged within the same period to verify the quantity and quality of the goods accepted in the procedure established by a law, other legal acts, the contract, or the customs of business turnover and to inform the supplier about elicited nonconformities or short deliveries of goods immediately in writing.

3. In the event of the receipt of goods delivered from a transport organisation, the purchaser (or recipient) shall be obliged to verify the conformity of the goods to the information specified in the transport and accompanying documents, and also to accept these goods from the transport organisation in compliance with the rules provided for by laws and other legal acts regulating the activity of transport.

Article 514. Custody of Good Not Accepted by Purchaser

1. When the purchaser (or recipient) in accordance with a law, other legal acts, or contract of delivery refuses a good transferred by the supplier, he shall be obliged to ensure the preservation of this good (custody) and immediately notify the supplier.

2. The supplier shall be obliged to remove the good accepted by the purchaser (or recipient) for custody or to dispose of it within a reasonable period.

If the supplier within this period does not dispose of the good, the purchaser shall have the right to realise the good or to return it to the supplier.

3. Necessary expenses incurred by the purchaser in connection with the acceptance of the good for custody, realisation of the good, or the return thereof to the seller shall be subject to compensation by the supplier.

In so doing that received from the realisation of the good shall be transferred to the supplier, deducting that due to the purchaser.

4. In instances when the purchaser does not accept a good from the supplier or refuses acceptance thereof without grounds established by a law, other legal acts, or the contract, the supplier shall have the right to require payment for the good from the purchaser.

Article 515. Selection of Goods

1. When the selection of goods by the purchaser (or recipient) has been provided for by the contract of delivery at the location of the supplier (Article 510[2]), the purchaser shall be obliged to effectuate the inspection of the goods to be transferred to the place of their transfer unless provided otherwise by a law, other legal acts, or it does not arise from the essence of the obligation.

2. The failure of the purchase (or recipient) to select goods within the period established by the contract of delivery, and in the absence thereof within a reasonable period after receipt of the notice of the supplier concerning the readiness of the goods shall give the supplier the right to refuse performance of the contract or to require payment for the goods from the recipient.

Article 516. Settlement of Accounts for Goods to be Delivered

1. The purchaser shall pay for delivered goods in compliance with the procedure and form for the settlement of accounts provided for by the contract of delivery. If the procedure and form for the settlement of accounts has not been determined by agreement of the parties, the settlement of accounts shall be effectuated by payment orders.

2. If it is provided by a contract of delivery that the payment for goods is effectuated by the recipient (payer) and the last unfoundedly has refused to pay or did not pay for the goods within the period established by the contract, the supplier shall have the right to require payment for the goods delivered from the purchaser.

3. In an instance when the delivery of goods by individual parts within a set is provided for in a contract of delivery, payment for the goods by the purchaser shall be made after the shipment (or selection) of the last part within the set unless established otherwise by the contract.

Article 517. Packaging and Packing

Unless established otherwise by the contract of delivery, the purchaser (or recipient) shall be obliged to return to the supplier nondisposable packaging and means of packing in which the good arrived in the procedure and within the periods established by a law, other legal acts, obligatory rules adopted in accordance therewith, or by the contract.

Other packaging, and also packing, of the good shall be subject to return to the supplier only in the instances provided for by the contract.

Article 518. Consequences of Delivery of Goods of Improper Quality

1. The purchaser (or recipient) to whom the goods of improper quality are delivered shall have the right to present demands to the supplier provided for by Article 475 of the present Code except for the instance when the supplier who received the notice of the purchaser concerning defects of the delivered goods replaces without delay the goods delivered with goods of the proper quality.

2. The purchaser (or recipient) effectuating the sale of goods delivered to him at retail shall have the right to demand the replacement of a good of improper quality within a reasonable period returned by a consumer unless provided otherwise by the contract of delivery.

Article 519. Consequences of Delivery of Incomplete Goods

1. The purchaser (or recipient) to whom goods have been delivered in violation of the conditions of a contract of delivery, the requirements of a law, other legal acts, or requirements usually presented for completeness shall have the right to present the demands to the supplier provided for by Article 480 of the present Code, except for the instance when the supplier who has received notice of the purchaser concerning the incompleteness of the delivered goods without delay makes the goods complete or replaces them with complete goods.

2. The purchaser (or recipient) effectuating the sale of goods at retail shall have the right to demand replacement of the incomplete goods within a reasonable period returned by a consumer with complete ones unless provided otherwise by the contract of delivery.

Article 520. Rights of Purchaser in Event of Failure to Deliver Goods, Failure to Fulfil Requirements Concerning Elimination of Defects of Goods or Completing Sets of Goods

1. If the supplier does not deliver the quantity of goods provided for by the contract of delivery or did not fulfil the demands of the purchaser concerning the replacement of poor-quality goods or concerning the completion of the sets of goods within the established period, the purchaser shall have the right to acquire

the nondelivered goods from other persons, relegating to the supplier all necessary and reasonable expenses for the acquisition thereof.

The calculation of the expenses of the purchaser for the acquisition of goods from other persons in instances of the short delivery thereof by the supplier or the failure to fulfil the demands of the purchaser concerning elimination of the defects of the goods or concerning the completion of sets of the goods shall be made according to the rules provided for by Article 524(1) of the present Code.

2. The purchaser (or recipient) shall have the right to refuse to pay for goods of improper quality and incomplete goods, and if such goods have been paid for, to require the return of the paid amounts until elimination of the defects and completion of the sets of the goods or the replacement thereof.

Article 521. Penalty for Short Delivery or Delay of Delivery of Goods

The penalty established by a law or contract of delivery for the short delivery or delay of delivery of goods shall be recovered from the supplier until the actual performance of the obligation within the limits of his duty to augment the short-delivered quantity of goods within subsequent delivery periods unless another procedure for payment of the penalty has been established by a law or the contract.

Article 522. Cancellation of Obligations of Same Kind Under Several Contracts of Delivery

1. In instances when the delivery of goods of the same name is effectuated by the supplier to the purchaser simultaneously under several contracts of delivery and the quantity of goods delivered is insufficient in order to cancel the obligations of the supplier under all the contracts, the goods delivered must be credited to the performance of the contract specified by the supplier when effectuating delivery or without delay after the delivery.

2. If the purchaser paid the supplier for goods of one name received under several contracts of delivery and the amounts of payment are insufficient to cancel the obligations of the purchaser under all the contracts, the amount paid must be credited to performance of the contract specified by the purchaser when effectuating payment for the goods or without delay after payment.

3. If the supplier or purchaser did not take advantage of the rights granted respectively by points 1 and 2 of the present Article, the performance of the obligation shall be credited to cancellation of the obligations under the contract whose period of performance ensued earlier. If the period of performance of obligations under several contracts has ensued simultaneously, the performance granted shall be credited in proportion to the cancellation of obligations under all the contracts.

Article 523. Unilateral Refusal of Performance of Contract of Delivery

1. A unilateral refusal of performance of a contract of delivery (wholly or partially) or a unilateral change thereof shall be permitted in the event of a material violation of the contract by one of the parties (Article 450[2], paragraph four).

2. A violation of a contract of delivery by the supplier shall be presupposed to be material in instances of:

the delivery of goods of improper quality with defects which cannot be eliminated within a period acceptable to the purchaser;

the repeated violation of the periods for delivery of goods.

3. The violation of a contract of delivery by the purchaser shall be presupposed to be material in instances of:

the repeated violation of the periods for the payment of goods;

the repeated failure to select goods.

4. A contract of delivery shall be considered to be changed or dissolved from the moment of receipt by one party of notice by the other party concerning the unilateral refusal of performance of the contract wholly or partially unless another period for dissolution or change of the contract is provided for in the notice or has been determined by agreement of the parties.

Article 524. Calculation of Losses in Event of Dissolution of Contract

1. If within a reasonable period after dissolution of the contract as a consequence of a violation of an obligation by the seller the purchaser has bought from another person a good at a higher but reasonable price in place of that provided for by the contract, the purchaser may present a demand to the seller concerning compensation of losses in the form of the difference between the price established in the contract and the price of the transaction concluded in lieu.

2. If within a reasonable period after dissolution of the contract as a consequence of a violation of an obligation by the purchaser the seller has sold the good to another person at a lower but reasonable price than that provided for by the contract, the seller may present a demand to the purchaser concerning compensation of losses in the form of the difference between the price established in the contract and the price for the transaction concluded in lieu.

3. If after the dissolution of a contract on the grounds provided for by points 1 and 2 of the present Article a transaction has not been concluded in place of the dissolved contract and for the particular good there is a current price, a party may present a demand concerning compensation of losses in the form of the difference between the price established in the contract and the current price at the moment of dissolution of the contract.

The current price shall be deemed to be the price usually recovered under comparable circumstances for an analogous good at the place where the transfer of the good should have been effectuated. If a current price does not exist at that place, the current price applied in another place may be used which may serve as a reasonable replacement, taking into account the difference in expenses for the transportation of the good.

4. Satisfaction of the demands provided for by points 1, 2, and 3 of the present Article shall not relieve the party which did not perform or which improperly performed the obligation from compensation of other losses caused to the other party on the basis of Article 15 of the present Code.

§4. Delivery of Goods for State Needs

Article 525. Grounds for Delivery of Goods for State Needs

1. The delivery of goods for State needs shall be effectuated on the basis of a State contract for the delivery of goods for State needs, and also contracts for delivery of goods concluded in accordance therewith for State needs (Article 530[2]). State needs shall be deemed to be requirements of the Russian Federation or subjects of the Russian Federation determined in the procedure established by a law which are ensured at the expense of assets of the budgets and extrabudgetary sources of financing.

2. The rules concerning the contract of delivery (Articles 506–523) shall apply to relations relating to the delivery of goods for State needs unless provided otherwise by the rules of the present Code.

The laws on the delivery of goods for State needs shall apply to relations relating to the delivery of goods for State needs in the part not regulated by the present paragraph.

Article 526. State Contract for Delivery of Goods for State Needs

Under a State contract for the delivery of goods for State needs (hereinafter: State contract), the supplier (or executor) shall be obliged to transfer the goods to the State customer or at his instruction to another person, and the State customer shall be obliged to ensure payment for the goods delivered.

Article 527. Grounds for Conclusion of State Contract

1. A State contract shall be concluded on the basis of the order of a State customer for the delivery of goods for State needs accepted by the supplier (or executor).

The conclusion of a State contract shall be obligatory for the State customer which placed the order accepted by the supplier (or executor).

2. The conclusion of a State contract shall be obligatory for the supplier (or executor) only in the instances established by a law and on condition that the State

customer will compensate all losses which may be caused to the supplier (or executor) in connection with the fulfilment of the State contract.

3. The condition concerning compensation of losses provided for by point 2 of the present Article shall not apply with respect to a treasury enterprise.

4. If the order for the delivery of goods for State needs is placed according to a competition, the conclusion of a State contract with the supplier (or executor) who is declared the winner of the competition shall be obligatory for the State customer.

Article 528. Procedure for Conclusion of State Contract

1. The draft State contract shall be worked out by the State customer and sent to the supplier (or executor) unless provided otherwise by agreement between them.

2. The party who has received the draft State contract shall sign it within not later than a thirty-day period and return one example of the State contract to the other party, and if there are disagreements regarding the conditions of the State contract shall draw up within the same period a protocol of disagreements and send it together with the signed State contract to the other party or inform it of the refusal to conclude the State contract.

3. The party who has received the State contract with the protocol of disagreements must within thirty days consider the disagreements, take measures to agree them with the other party, and inform the other party about the acceptance of the State contract in the wording thereof or reject the protocol of disagreements.

In the event of the rejection of the protocol of disagreements or expiry of this period, the unsettled disagreements with regard to the State contract whose conclusion is obligatory for one of the parties may be transferred by the other party within not later than thirty days for the consideration of a court.

4. In an instance when a State contract is concluded according to the results of a competition for the placement of an order for the delivery of goods for State needs, the State contract must be concluded not later than twenty days from the date of conducting the competition.

5. If the party for which the conclusion of a State contract is obligatory evades the conclusion thereof, the other party shall have the right to apply to a court with the demand to compel this party to conclude the State contract.

Article 529. Conclusion of Contract for Delivery of Goods for State Needs

1. If it is provided by a State contract that the delivery of goods shall be effectuated by a supplier (or executor) determined by the State customer to the purchaser under contracts for delivery of goods for State needs, the State customer

199

shall not later than a thirty-day period from the date of signature of the State contract send notice to the supplier (or executor) and to the purchaser concerning the attachment of the purchaser to the supplier (or executor).

The notice concerning the attachment of the purchaser to the supplier (or executor) issued by the State customer in accordance with the State contract shall be grounds for the conclusion of the contract for delivery of goods for State needs.

2. The supplier (or executor) shall be obliged to send the draft contract for delivery of goods for State needs to the purchaser specified in the notice concerning the attachment not later than thirty days from the date of receipt of the notice from the State customer unless another procedure for the preparation of the draft contract is provided for by the State contract or the draft contract is not submitted by the purchaser.

3. The party who has received the draft contract for delivery of goods for State needs shall sign it and return one example to the other party within thirty days from the date of receipt of the draft, and if there are disagreements with regard to the contract to draw up within this same period a protocol of disagreements and send it together with the signed contract to the other party.

4. The party who has received the signed draft contract for delivery of goods for State needs with the protocol of disagreements must within thirty days consider the disagreements, take measures to agree the conditions of the contract with the other party, and notify the other party about the acceptance of the contract in the wording thereof—or reject the protocol of disagreements. The unsettled disagreements within a thirty-day period may be transferred by the interested party for consideration of a court.

5. If the supplier (or executor) evades conclusion of the contract for delivery of goods for State needs, the purchaser shall have the right to apply to a court with a demand concerning compelling the supplier (or executor) to conclude the contract on the conditions of the draft contract worked out by the purchaser.

Article 530. Refusal of Purchaser to Conclude Contract for Delivery of Goods for State Needs

1. The purchaser shall have the right wholly or partially to refuse goods specified in the notice concerning attachment and to conclude the contract for the delivery thereof.

In this event the supplier (or executor) must immediately inform the State customer and shall have the right to demand from him notice concerning attachment to another purchaser.

2. The State customer shall not later than thirty days from the date of receipt of the notification of the supplier (or executor) either issue the notice concerning

attachment of another purchaser to him or send to the supplier (or executor) a shipment order specifying the recipient of the goods or communicate his consent to accept and to pay for the goods.

3. In the event of the failure to fulfil the duties provided for by point 2 of the present Article by the State customer, the supplier (or executor) shall have the right either to require the State customer to accept and pay for the good or to realise the good at his discretion, relegating reasonable expenses connected with the realisation thereof to the State customer.

Article 531. Performance of State Contract

1. In instances when in accordance with the conditions of the State contract the delivery of goods is effectuated directly to the State customer or at his instruction (or shipment order) to another person (recipient), the relations of the parties with regard to the performance of the State contract shall be regulated by the rules provided for by Articles 506–523 of the present Code.

2. In instances when the delivery of goods for State needs is effectuated by the recipient specified in the shipment order, the goods shall be paid for by the State customer unless another procedure for the settlement of accounts is provided for by the State contract.

Article 532. Payment for Good Under Contract for Delivery of Goods for State Needs

In the event of the delivery of goods to purchasers under contracts for delivery of goods for State needs, the goods shall be paid for by the purchasers at the prices determined in accordance with a State contract unless another procedure for determining the prices and settlement of accounts has been provided for by the State contract.

In the event of payment by the purchaser for goods under a contract for delivery of goods for State needs, the State customer shall be deemed to be a surety of the purchaser with regard to this obligation (Articles 361–367).

Article 533. Compensation of Losses Caused in Connection with Fulfilment or Dissolution of State Contract

1. Unless provided otherwise by laws on the delivery of goods for State needs or by the State contract, losses which were caused to the supplier (or executor) in connection with the fulfilment of the State contract (Article 527[2]) shall be subject to compensation by the State customer not later than thirty days from the date of transfer of the good in accordance with the State contract.

2. In an instance when losses caused to the supplier (or executor) in connection with the fulfilment of a State contract are not compensated in accordance with the State contract, the supplier (or executor) shall have the right to refuse

performance of the State contract and require compensation of losses caused by the dissolution of the State contract.

3. In the event of the dissolution of the State contract with regard to the grounds specified in point 2 of the present Article, the supplier shall have the right to refuse performance of the contract for delivery of the good for State needs.

Losses caused to the purchaser by such refusal of the supplier shall be compensated by the State customer.

Article 534. Refusal by State Customer of Goods Delivered Under State Contract

In the instances provided for by a law, the State customer shall have the right wholly or partially to refuse goods whose delivery is provided for by the State contract on condition of the compensation to the supplier of losses caused by such refusal.

If the refusal by a State customer of goods whose delivery is provided for by a State contract entails the dissolution or change of the contract for delivery of goods for State needs, the losses caused to the purchaser by such dissolution or change shall be compensated by the State customer.

§5. Agricultural Procurement Contract

Article 535. Contract of Agricultural Procurement

1. Under a contract of agricultural procurement the producer of agricultural products shall be obliged to transfer the agricultural products grown (or produced) by him to the procurer—the person effectuating the purchase of such products for processing or sale.

2. The rules concerning the contract of delivery (Articles 506–524), and in respective instances concerning the delivery of goods for State needs (Articles 525–534), shall apply to relations under the contract of agricultural procurement which are not regulated by the rules of the present paragraph.

Article 536. Duties of Procurer

1. Unless provided otherwise by the contract of agricultural procurement, the procurer shall be obliged to accept the agricultural product from the producer at the location thereof and to ensure the carriage thereof.

2. In an instance when the acceptance of an agricultural product is effectuated at the location of the procurer or other place specified by him, the procurer shall not have the right to refuse to accept the agricultural product which corresponds to the conditions of the contract of agricultural procurement and transferred to the procurer within the period stipulated by the contract.

3. The duty of the procurer effectuating the processing of agricultural products may be provided for by the contract of agricultural procurement to return to the producer at his demand the wastes from the processing of agricultural products with payment at the price determined by the contract.

Article 537. Duties of Producer of Agricultural Product

The producer of an agricultural product shall be obliged to transfer the agricultural product grown (or produced) to the procurer in the quantity and assortment provided for by the contract of agricultural procurement.

Article 538. Responsibility of Producer of Agricultural Product

The producer of an agricultural product who has not performed the obligation or has improperly performed the obligation shall bear responsibility when his fault is present.

§6. Electric Power Supply

Article 539. Contract of Electric Power Supply

1. Under a contract of electric power supply the electric power supply organisation shall be obliged to provide electric power to the subscriber (or consumer) through the connected network, and the subscriber shall be obliged to pay for the electric power accepted, and also to comply with the consumption regime provided for by the contract, ensure the safety of operation of the electric power networks within its jurisdiction and the good repair of the instruments and equipment connected with the consumption of electric power used by it.

2. The contract of electric power supply shall be concluded with the subscriber when he has an electric power receiving device which meets the established technical requirements and is connected to the networks of the electric power supply organisation, and other necessary equipment, and also when the recording of the consumption of electric power is ensured.

3. Laws and other legal acts concerning electric power supply, and also obligatory rules adopted in accordance with them, shall apply to relations under the contract of electric power supply not regulated by the present Code.

Article 540. Conclusion and Extension of Contract of Electric Power Supply

1. In an instance when a citizen who is using electric power for domestic consumption is a subscriber under the contract of electric power supply, the contract shall be considered to be concluded from the moment of first actual connection of the subscriber in the established procedure to the connecting network.

Unless provided otherwise by agreement of the parties, such a contract shall be considered to be concluded for an indefinite period and may be changed or dissolved on the grounds provided for by Article 546 of the present Code.

2. The contract of electric power supply concluded for a determined period shall be considered to be extended for the same period and on the same conditions unless one of the parties declares the termination or change thereof before the end of the period of its operation or the conclusion of a new contract.

3. If one of the parties before the end of the period of operation of the contract has made a proposal to conclude a new contract, the relations of the parties until the conclusion of the new contract shall be regulated by the previously concluded contract.

Article 541. Quantity of Electric Power

1. The electric power supply organisation shall be obliged to provide the subscriber with electric power through the connecting network in the quantity provided for by the contract of electric power supply and in compliance with the provision regime agreed by the parties. The quantity provided by the electric power supply organisation and used by the subscriber of electric power shall be determined in accordance with the record data concerning the actual consumption thereof.

2. The right of the subscriber to change the quantity of electric power used by him as determined by the contract may be provided for by the contract of electric power supply on condition of compensation by him of expenses incurred by the electric power supply organisation in connection with ensuring the provision of electric power in a quantity not stipulated by the contract.

3. In an instance when a citizen is the subscriber under the contract of electric power supply who uses electric power for domestic consumption, he shall have the right to use electric power in the quantity necessary for him.

Article 542. Quality of Electric Power

1. The quality of electric power provided by the electric power supply organisation must correspond to the requirements established by State standards and other obligatory rules or provided for by the contract of electric power supply.

2. In the event of a violation by the electric power supply organisation of the requirements for quality of electric power, the subscriber shall have the right to refuse payment for such electric power. In so doing the electric power supply organisation shall have the right to demand compensation by the subscriber for the cost of that which the subscriber saved unfoundedly as a consequence of the use of this electric power (Article 1105[2]).

Article 543. Duties of Purchaser for Maintenance and Operation of Networks, Instruments, and Equipment

1. The subscriber shall be obliged to ensure the proper technical state and safety of electric power networks being operated, instruments, and equipment, to comply with the established consumption regime for electric power, and also immediately to notify the electric power supply organisation about accidents, fires, the disrepair of instruments for recording electric power, and other violations arising when using electric power.

2. In an instance when a citizen is a subscriber under a contract of electric power supply who is using electric power for domestic consumption, the duty to ensure the proper technical state and safety of the electric power networks, and also the instruments for recording the consumption of electric power, shall be placed on the electric power supply organisation unless established otherwise by a law or other legal acts.

3. The requirements for the technical state and operation of electric power networks, instruments, and equipment, and also the procedure for effectuating control over compliance therewith, shall be determined by a law, other legal acts, and the obligatory rules adopted in accordance therewith.

Article 544. Payment for Electric Power

1. Payment for electric power shall be made for the quantity of electric power actually accepted by the subscriber in accordance with the data of recording electric power unless provided otherwise by a law, other legal acts, or agreement of the parties.

2. The procedure for the settlement of accounts for electric power shall be determined by a law, other legal acts, or agreement of the parties.

Article 545. Sub-subscriber

A subscriber may transfer electric power accepted by him from the electric power supply organisation through the connecting network to another person (sub-subscriber) only with the consent of the electric power supply organisation.

Article 546. Change and Dissolution of Contract of Electric Power Supply

1. In an instance when a citizen is a subscriber under a contract of electric power supply who is using electric power for domestic consumption, he shall have the right to dissolve the contract unilaterally on condition of informing the electric power supply organisation thereof and of payment in full for the electric power used.

In an instance when a juridical person is a subscriber under the contract of electric power supply, the electric power supply organisation shall have the right to refuse performance of the contract unilaterally on the grounds provided for by Article 523 of the present Code, except for instances established by a law or other legal acts.

2. An interruption in provision or the termination or limitation of provision of electric power shall be permitted by agreement of the parties, except for instances when the unsatisfactory state of electric power devices of the subscriber certified by a State electric power supervision agency threatens an accident or creates a threat to the life and safety of citizens. The electric power supply organisation must warn the subscriber about an interruption in provision or the termination or limitation of provision of electric power.

3. An interruption in provision or termination or limitation of provision of electric power without the agreement of the subscriber and without a respective warning thereof shall be permitted when necessary to take urgent measures relating to the prevention or liquidation of an accident in the system of the electric power supply organisation on condition of immediate notification of the subscriber thereof.

Article 547. Responsibility Under Contract of Electric Power Supply

1. In instances of the failure to perform or the improper performance of obligations under a contract of electric power supply, the party who has violated the obligation shall be obliged to compensate the real damage caused by this (Article 15[2]).

2. If as a result of the regulation of the consumption regime for electric power effectuated on the basis of a law or other legal acts an interruption in the provision of electric power to the subscriber is permitted, the electric power supply organisation shall bear responsibility for the failure to perform or the improper performance of contractual obligations when there is fault thereof.

Article 548. Application of Rules on Electric Power Supply to Other Contracts

1. The rules provided for by Articles 539–547 of the present Code shall apply to relations connected with the supply of heating electric power through a connecting network unless established otherwise by a law or other legal acts.

2. The rules concerning the contract of electric power supply (Articles 539–547) shall apply to relations connected with supply through a connecting network of gas, oil and oil products, water, and other goods unless established otherwise by a law, other legal acts, or does not arise from the essence of the obligation.

§7. Sale of Immovable

Article 549. Contract of Sale of Immovable

1. Under a contract of purchase-sale of immovable property (contract of sale of immovable) the seller shall be obliged to transfer to the ownership of the purchaser a land plot, building, installation, apartment, or other immovable property (Article 130).

2. The rules provided for by the present paragraph shall apply to the sale of enterprises insofar as not provided otherwise by the rules concerning the contract of sale of an enterprise (Articles 559–566).

Article 550. Form of Contract of Sale of Immovable

A contract of sale of an immovable shall be concluded in written form by means of drawing up one document signed by the parties (Article 434[2]).

The failure to comply with the form of the contract of sale of an immovable shall entail its invalidity.

Article 551. State Registration of Transfer of Right of Ownership to Immovable

1. The transfer of the right of ownership to an immovable under a contract of sale of an immovable to a purchaser shall be subject to State registration.

2. The performance of a contract of sale of an immovable by the parties before State registration of the transfer of the right of ownership shall not be grounds for a change of their relations with third persons.

3. In an instance when one of the parties evades State registration of the transfer of the right of ownership to an immovable a court shall have the right at the demand of the other party to render a decision concerning State registration of the transfer of the right of ownership. The party which unfoundedly evaded State registration of the transfer of the right of ownership must compensate the other party for losses caused by the delay of registration.

Article 552. Rights to Land Plot in Event of Sale of Building, Installation, or Other Immovable Situated Thereon

1. Under a contract for the sale of a building, installation, or other immovable the rights shall pass to the purchaser simultaneously with the transfer of the right of ownership to such immovable to that part of the land plot which is occupied by this immovable and necessary in order to use it.

2. In an instance when the seller is an owner of the land plot on which an immovable being sold is situated, the right of ownership shall be transferred to the

purchaser or the right of lease or other right provided for by the contract of sale of an immovable to the corresponding part of the land plot shall be granted.

If the right to the corresponding land plot is not determined by the contract for the immovable being transferred to the purchaser, to the purchaser shall pass the right of ownership to that part of the land plot which is occupied by the immovable and necessary in order to use it.

3. The sale of an immovable situated on a land plot which does not belong to the seller by right of ownership shall be permitted without the consent of the owner of this plot if this is not contrary to the conditions of use of such plot established by a law or by the contract.

In the event of the sale of such immovable, the purchaser shall acquire the right of use of the respective part of the land plot on the same conditions as the seller of the immovable.

Article 553. Right to Immovable in Event of Sale of Land Plot

In instances when the land plot on which a building, installation, or other immovable belonging to the seller is situated is sold without the transfer of this immovable to the ownership of the purchaser, the right of use of this land plot which is occupied by the immovable and necessary for its use shall be retained for the seller on the conditions determined by the contract of sale.

If the conditions for the use of the respective part of the land plot have not been determined by the contract for the sale thereof, the seller shall retain the right of limited use (servitude) of that part of the land plot which is occupied by the immovable and is necessary for its use in accordance with its designation.

Article 554. Determination of Subject in Contract of Sale of Immovable

Data must be specified in a contract of sale of an immovable enabling the immovable property subject to transfer to the purchaser under the contract to be established definitely, including data determining the location of the immovable on the respective land plot or as part of other immovable property.

In the absence of such data in the contract, the condition concerning immovable property subject to transfer shall be considered to be not agreed by the parties, and the respective contract shall not be considered to be concluded.

Article 555. Price in Contract of Sale of Immovable

1. The contract of sale of an immovable must provide for the price of this property.

In the absence in the contract of a condition concerning the price of the immovable agreed by the parties in written form, the contract concerning the sale thereof shall be considered to be not concluded. In so doing the rules for determining the prices provided for by Article 424(3) of the present Code shall not apply.

2. Unless provided otherwise by a law or contract for the sale of an immovable, the price established therein of the building, installation, or other immovable property situated on the land plot shall include the price of the immovable property transferred therewith of the respective part of the land plot or rights thereto.

3. In instances when the price of the immovable in a contract of sale of an immovable is established per unit of the space thereof or other indicator of its dimension, the total price of such immovable property subject to payment shall be determined by proceeding from the actual dimension of the immovable property transferred to the purchaser.

Article 556. Transfer of Immovable

1. The transfer of an immovable by the seller and the acceptance thereof by the purchaser shall be effectuated under the act of transfer or other document concerning the transfer signed by the parties.

Unless provided otherwise by a law or contract, the obligation of the seller to transfer the immovable to the purchaser shall be considered to be performed after the handing over of this property to the purchaser and the signature by the parties of the respective document concerning transfer.

The evasion by one of the parties of signature of the document concerning transfer of the immovable on the conditions provided for by the contract shall be considered to be a refusal respectively of the seller to perform the duty to transfer the property and of the purchaser, the duty to accept the property.

2. The acceptance by the purchaser of an immovable which does not correspond to the conditions of the contract of sale of an immovable, including when such nonconformity is stipulated in the document concerning the transfer of the immovable, shall not be grounds for relieving the seller of responsibility for improper performance of the contract.

Article 557. Consequences of Transfer of Immovable of Improper Quality

In the event of the transfer by the seller to the purchaser of an immovable which does not correspond to the conditions of the contract of sale of an immovable concerning the quality thereof, the rules of Article 475 of the present Code shall apply except for the provisions concerning the right of the purchaser to require replacement of the good of improper quality for a good corresponding to the contract.

Article 558. Peculiarities of Sale of Dwelling Premises

1. A material condition of the contract of sale of a dwelling house, apartment, part of a dwelling house or apartment in which persons reside who retain in accordance with a law the right of use of this dwelling premise after its acquisition by the purchaser shall be the list of such persons specifying their rights to use the dwelling premise being sold.

2. The contract of sale of a dwelling house, apartment, or part of a dwelling house or apartment shall be subject to State registration and shall be considered to be concluded from the moment of such registration.

§8. Sale of Enterprise

Article 559. Contract of Sale of Enterprise

1. Under a contract of sale of an enterprise the seller shall be obliged to transfer the enterprise as a whole as a property complex (Article 132) to the ownership of the purchaser, except for the rights and duties which the seller does not have the right to transfer to other persons.

2. The rights to a firm name, trademark, service mark, and other means of individualisation of the seller and its goods, work, or services, and also the license of the right of use of such means of individualisation belonging to it on the basis of a license, shall pass to the purchaser unless provided otherwise by the contract.

3. The rights of the seller received by it on the basis of an authorisation (or license) to engage in the respective activity shall not be subject to transfer to the purchaser of the enterprise unless established otherwise by a law or other legal acts. The transfer to the purchaser as part of the enterprise of obligations whose performance by the purchaser is impossible when such authorisation (or license) is lacking shall not relieve the seller from the respective obligations to creditors. The seller and the purchase shall bear joint and several responsibility to creditors for the failure to perform such obligations.

Article 560. Form and State Registration of Contract of Sale of Enterprise

1. The contract of sale of an enterprise shall be concluded in written form by means of drawing up one document signed by the parties (Article 434[2]) with the obligatory appending thereto of the documents specified in Article 561(2) of the present Code.

2. The failure to comply with the form of the contract of sale of an enterprise shall entail its invalidity.

3. The contract of sale of an enterprise shall be subject to State registration and shall be considered to be concluded from the moment of such registration.

Article 561. Certification of Composition of Enterprise Being Sold

1. The composition and value of the enterprise being sold shall be determined in the contract of sale of an enterprise on the basis of the full inventorisation of the enterprise conducted in accordance with the established rules of such inventorisation.

2. Before signature of the contract of sale of an enterprise there must be drawn up and considered by the parties: the act of inventorisation, the bookkeeping bal-

ance sheet, the opinion of an independent auditor concerning the composition and value of the enterprise, and also a list of all the debts (or obligations) included within the enterprise, specifying the creditors, the character, amount, and periods of their demands.

The property, rights, and duties specified in the named documents shall be subject to transfer by the seller to the purchaser unless it follows otherwise from the rules of Article 559 of the present Code and is not established by agreement of the parties.

Article 562. Rights of Creditors in Event of Sale of Enterprise

1. The creditors with regard to obligations included within the enterprise being sold must be informed in writing before the transfer thereof about its sale by one of the parties of the contract of sale of the enterprise.

2. The creditor who in writing has not communicated to the seller or purchaser his consent to the transfer of the debt shall have the right within three months from the date of receipt of the notice concerning the sale of the enterprise to require either the termination or performance before time of the obligation and compensation by the seller of the losses caused thereby or recognition of the contract of sale of the enterprise to be invalid wholly or in respective part.

3. The creditor who was not informed about the sale of the enterprise in the procedure provided for by point 1 of the present Article may bring suit concerning satisfaction of the demands provided for by point 2 of the present Article within a year from the date when he knew or should have known about the transfer of the enterprise by the seller to the purchaser.

4. After the transfer of the enterprise to the purchaser, the seller and the purchaser shall bear joint and several responsibility for debts included within the enterprise being transferred which were transferred to the purchaser without the consent of the creditor.

Article 563. Transfer of Enterprise

1. The transfer of an enterprise by the seller to the purchaser shall be effectuated under the act of transfer, in which shall be specified data concerning the composition of the enterprise and notice of the creditors concerning the sale of the enterprise, and also information concerning defects elicited in the property being transferred and list of property, the duties with regard to the transfer of which have not been performed by the seller in view of the loss thereof.

The preparation of the enterprise for transfer, including the drawing up and submission for signature of the act of transfer, shall be the duty of the seller and shall be effectuated at his expense unless provided otherwise by the contract.

2. An enterprise shall be considered to be transferred to the purchaser from the date of signature of the act of transfer by both parties.

From this moment the risk of accidental perishing or accidental damage of property being transferred within the enterprise shall pass to the purchaser.

Article 564. Transfer of Right of Ownership to Enterprise

1. The right of ownership to an enterprise shall pass to the purchaser from the moment of State registration of this right.

2. Unless provided otherwise by the contract of sale of an enterprise, the right of ownership to an enterprise shall pass to the purchaser and shall be subject to State registration directly after the transfer of the enterprise to the purchaser (Article 563).

3. In instances when the retention by the seller of the right of ownership to an enterprise transferred to the purchaser has been provided for by the contract before payment for the enterprise or before the ensuing of other circumstances, the purchaser shall have the right until the transfer of the right of ownership to it to dispose of the property and of the rights which are part of the enterprise being transferred to that extent as is necessary for the purposes for which the enterprise was acquired.

Article 565. Consequences of Transfer and Acceptance of Enterprise with Defects

1. The consequences of the transfer by the seller and acceptance by the purchaser under an act of transfer of an enterprise the composition of which does not correspond to the sale of the enterprise provided for by the contract, including with respect to the quality of the property being transferred, shall be determined on the basis of the rules provided for by Articles 460–462, 466, 469, 475, and 479 of the present Code unless it arises otherwise from the contract and is not provided for by points 2–4 of the present Article.

2. In an instance when an enterprise has been transferred and accepted under the act of transfer in which information concerning elicited defects of the enterprise and lost property have been specified (Article 563[1]), the purchaser shall have the right to demand a corresponding reduction of the purchase price of the enterprise unless the right to present other demands in such instances has been provided for by the contract of sale of the enterprise.

3. The purchaser shall have the right to demand a reduction of the purchase price in the event of the transfer to it within the enterprise of debts (or obligations) of the seller which were not specified in the contract of sale of the enterprise or act of transfer unless the seller proves that the purchaser knew of such debts (or obligations) during the conclusion of the contract and transfer of the enterprise.

4. The seller in the event of receiving notice of the purchaser concerning defects of the property being transferred within the enterprise, or the absence of

individual types of property in such composition which are subject to transfer, may without delay replace the property of improper quality or grant short-delivered property to the purchaser.

5. The purchaser shall have the right in a judicial proceeding to demand dissolution or change of the contract of sale of an enterprise and the return of that which was performed by the parties under the contract if it is established that the enterprise, in view of the defects for which the seller is liable, is unfit for the purposes named in the contract of sale and these defects have not been eliminated by the seller on the conditions, in the procedure, and within the periods which have been established in accordance with the present Code, other laws, other legal acts, or the contract, or the elimination of such defects is impossible.

Article 566. Application to Contract of Sale of Enterprise of Rules Concerning Consequences of Invalidity of Transactions and Concerning Change or Dissolution of Contract

The rules of the present Code concerning the consequences of the invalidity of transactions and a change or dissolution of the contract of purchase-sale providing for the return or recovery in kind of that received under the contract from one party or from both parties shall apply to the contract of sale of an enterprise unless such consequences materially violate the rights and the interests protected by a law of the creditors of the seller and purchaser, of other persons, and are not contrary to social interests.

Chapter 31. Barter

Article 567. Contract of Barter

1. Under a contract of barter, each of the parties shall be obliged to transfer to the ownership of the other party one good in exchange for another.

2. To the contract of barter shall apply respectively the rules concerning purchase-sale (Chapter 30) if this is not contrary to the rules of the present Chapter and the essence of the barter. In so doing each of the parties shall be deemed to be the seller of the good which he is obliged to transfer and the purchaser of the good which he is obliged to accept in exchange.

Article 568. Prices and Expenses Under Contract of Barter

1. Unless it arises otherwise from the contract of barter, goods subject to exchange shall be offered of equal value and expenses for the transfer thereof and acceptance shall be effectuated in each instance by that party which bears the respective duties.

2. In an instance when in accordance with the contract of barter the goods to be exchanged are deemed not to be of equal value, the party obliged to transfer a

good whose price is lower than the price of the good being granted in exchange must pay for the difference in prices directly before or after the performance of its duty to transfer the good unless another procedure for payment has been provided for by the contract.

Article 569. Counter Performance of Obligation to Transfer Good Under Contract of Barter

In an instance when in accordance with a contract of barter the periods for the transfer of the goods to be exchanged do not coincide, the rules concerning counter performance of obligations (Article 328) shall apply to the performance of the obligation to transfer the good by the party which must transfer the good after the transfer of the good by the other party.

Article 570. Transfer of Right of Ownership in Goods to be Exchanged

Unless provided otherwise by a law or by the contract of barter, the right of ownership in the goods to be exchanged shall pass to the parties acting under the contract of barter as purchasers simultaneously after the performance of the obligations to transfer the respective goods by both parties.

Article 571. Responsibility for Seizure of Good Acquired Under Contract of Barter

The party from whom a good is seized by a third person which was acquired under a contract of barter shall have the right when there are grounds provided for by Article 461 of the present Code to require from the other party the return of the good received by the last in the exchange and/or compensation of losses.

Chapter 32. Gift

Article 572. Contract of Gift

1. Under a contract of gift one party (donor) transfers without compensation or is obliged to transfer to the other party (donee) a thing in ownership or a property right (or demand) against himself or against a third person or relieves or is obliged to relieve him of a property duty to himself or to a third person.

When there exists a counter transfer of a thing or right or a counter obligation the contract shall not be deemed to be a gift. The rules provided for by Article 170(2) of the present Code shall apply to such contract.

2. The promise to transfer without compensation any thing or property right whatever or to relieve someone of a property duty (promise of gift) shall be deemed to be a contract of gift and shall bind the promisor if the promise is made in the proper form (Article 574[2]) and contains a clearly expressed intention to perform in the future the transfer of the thing or right without compensation to a specific person or to relieve him from a property duty.

The promise to give all of his property or part of his property without specifying the specific subject of gift in the form of a thing, right, or relief from a duty shall be void.

3. A contract providing for the transfer of a gift to the donee after the death of the donor shall be void.

The rules of civil legislation concerning inheritance shall apply to such kind of gift.

Article 573. Refusal of Donee to Accept Gift

1. The donee shall have the right at any time before the transfer of the gift to him to refuse it. In this event the contract of gift shall be considered to be dissolved.

2. If the contract of gift is concluded in written form, the refusal of the gift must also be concluded in written form. In an instance when the contract of gift has been registered (Article 574[3]), the refusal to accept the gift also shall be subject to State registration.

3. If the contract of gift was concluded in written form, the donor shall have the right to demand from the donee compensation of real damage caused by the refusal to accept the gift.

Article 574. Form of Contract of Gift

1. A gift accompanied by the transfer of the gift to the donee may be concluded orally, except for instances provided for by points 2 and 3 of the present Article.

The transfer of the gift shall be effectuated by means of handing it over, by symbolic transfer (handing over keys, and so forth) or by handing over right-establishing documents.

2. A contract of gift of moveable property must be concluded in written form in instances when:

the donor is a juridical person and the value of the gift exceeds five minimum amounts of payment for labour established by a law;

the contract contains a promise of a gift in the future.

In the instances provided for in the present point a contract of gift concluded orally shall be void.

3. A contract of gift of immovable property shall be subject to State registration.

Article 575. Prohibition of Gift

A gift shall not be permitted except for ordinary presents whose value does not exceed five minimum amounts of payment for labour established by a law:

(1) in the name of youths and citizens deemed to lack dispositive legal capacity, by their legal representatives;

(2) to workers of treatment and nurturing institutions, institutions of social defence, and other analogous institutions by citizens situated therein for treatment, maintenance, or nurturing, by the spouses and relatives of these citizens;

(3) to State employees and employees of agencies of municipal formations in connection with their official position or in connection with the performance by them of their official duties;

(4) in relations between commercial organisations.

Article 576. Limitations of Gift

1. A juridical person to whom a thing belongs by right of economic jurisdiction or operative management shall have the right to give it with the consent of the owner unless provided otherwise by a law. This limitation shall not extend to ordinary gifts of small value.

2. The gift of property in common joint ownership shall be permitted by consent of all the participants of joint ownership in compliance with the rules provided for by Article 253 of the present Code.

3. The gift of a right of demand belonging to the donor against a third person shall be effectuated in compliance with the rules provided for by Articles 382–386, 388, and 389 of the present Code.

4. A gift by means of the performance for the donee of his duty to a third person shall be effectuated in compliance with the rules provided for by Article 313(1) of the present Code.

A gift by means of the transfer by the donor to himself of the debt of the donee to a third person shall be effectuated in compliance with the rules provided for by Articles 391 and 392 of the present Code.

5. A power of attorney to perform a gift by a representative in which the donee is not named and the subject of the gift is not specified shall be void.

Article 577. Refusal of Performance of Contract of Gift

1. The donor shall have the right to refuse performance of a contract which contains a promise to transfer a thing or right in the future to a donee, or to relieve the donee from a property duty, if after conclusion of the contract the property or family position or the state of health of the donor has so changed that performance of the contract in the new conditions leads to a material reduction of his standard of living.

2. The donor shall have the right to refuse performance of a contract which contains a promise to transfer a thing or right to the donee in the future or to relieve the donee from a property duty on the grounds giving him the right to revoke the gift (Article 578[1]).

3. The refusal of the donor of performance of the contract of gift on the grounds provided for by points 1 and 2 of the present Article shall not give a right of demand for compensation of losses to the donee.

Article 578. Revocation of Gift

1. The donor shall have the right to revoke a gift if the donee has committed an attempt on his life, the life of another member of his family or close relatives, or has intentionally caused bodily injuries to the donor.

In the event of the intentional deprivation of life of the donor by the donee, the right to demand revocation of the gift in court shall belong to the heirs of the donor.

2. The donor shall have the right to require revocation of the gift in a judicial proceeding if the treatment by the donee of the thing given was of great nonproperty value to the donor and creates a threat of its irretrievable loss.

3. Upon the demand of an interested person a court may revoke a gift made by an individual entrepreneur or juridical person in violation of the provisions of the law on insolvency (or bankruptcy) at the expense of means connected with its entrepreneurial activity within six months which preceded the declaration of such person to be insolvent (or bankrupt).

4. The right of the donor to revoke the gift in the event he outlives the donee may be stipulated in the contract of gift.

5. In the event of revocation of a gift the donee shall be obliged to return the thing given if it has been preserved in kind at the moment of revocation of the gift.

Article 579. Instances in Which Refusal of Performance of Contract of Gift and Revocation of Gift are Impossible

The rules concerning the refusal of performance of a contract of gift (Article 577) and revocation of a gift (Article 578) shall not apply to ordinary presents of small value.

Article 580. Consequences of Causing Harm as Consequence of Defects of Thing Given

Harm caused to the life, health, or property of a donee citizen as a consequence of defects of the thing given shall be subject to compensation by the donor in accordance with the rules provided for by Chapter 59 of the present Code if it is proved that these defects arose before the transfer of the thing to the donee and were not obvious and the donor, although he knew about them, did not warn the donee about them.

Article 581. Legal Succession in Event of Promise of Gift

1. The rights of a donee to whom a gift was promised under a contract of gift shall not pass to his heirs (or legal successors) unless provided otherwise by the contract of gift.

2. The duties of the donor who promised the gift shall pass to his heirs (or legal successors) unless provided otherwise by the contract of gift.

Article 582. Donations

1. The gift of a thing or right for generally useful purposes shall be deemed to be a donation.

Donations may be made to citizens, treatment and nurturing institutions, institutions of social defence, and other analogous institutions, philanthropic, scientific, and educational institutions, foundations, museums, and other institutions of culture, social and religious organisations, and also to the State and other subjects of civil law specified in Article 124 of the present Code.

2. No authorisations or consent whatever shall be required to accept donations.

3. The donation of property to a citizen must be, and to juridical persons may be, conditioned by the donor by the use of this property for a determined designation. In the absence of such a condition the donation of the property to a citizen shall be considered to be an ordinary gift, and in other instances the donated property shall be used by the donee in accordance with the designation of the property.

A juridical person which has accepted a donation for whose use a determined designation has been established must carry out a solitary recording of all operations relating to the use of the donated property.

4. If the use of donated property in accordance with the designation specified by the donor becomes as a consequence of changed circumstances impossible, it may be used for another designation only with the consent of the donor, and in the event of the death of a citizen-donor or the liquidation of a juridical person-donor, by decision of a court.

5. The use of donated property not in accordance with the designation specified by the donor or a change of such designation in violation of the rules provided for by point 4 of the present Article shall give the right to the donor, his heirs, or other legal successor to demand revocation of the donation.

6. Articles 578 and 581 of the present Code shall not apply to donations.

Chapter 33. Rent and Maintenance of Dependent for Life

§1. General Provisions on Rent and Maintenance of Dependent for Life

Article 583. Contract of Rent

1. Under a contract of rent one party (recipient of the rent) shall transfer property in ownership to the other party (payer of rent), and the payer of rent shall be obliged in exchange for the property received periodically to pay to the recipient of rent in the form of a determined monetary amount or of granting assets for his maintenance in other form.

2. Under a contract of rent the establishment of the duty to pay rent in perpetuity (permanent rent) or for the period of life of the recipient of the rent (rent for life) shall be permitted. Rent for life may be established on conditions of the maintenance for life of a dependent citizen.

Article 584. Form of Contract of Rent

The contract of rent shall be subject to notarial certification, and a contract providing for the alienation of immovable property under payment of rent shall also be subject to State registration.

Article 585. Alienation of Property Under Payment of Rent

1. Property which is alienated under payment of rent may be transferred by the recipient of the rent to the ownership of the payer of rent for payment or free of charge.

2. In an instance when the transfer of property for payment is provided for by the contract of rent, the rules concerning purchase-sale (Chapter 30) shall apply to relations of the parties relating to the transfer and payment, and in an instance when such property is transferred free of charge, the rules concerning the contract of gift (Chapter 32) insofar as not established otherwise by the rules of the present Chapter and is not contrary to the essence of the contract of rent.

Article 586. Encumberment of Immovable Property by Rent

1. Rent shall encumber a land plot, enterprise, building, installation, or other immovable property transferred under payment thereof. In the event of the alienation of such property by the payer of the rent, his obligations under the contract of rent shall pass to the acquirer of the property.

2. The person who has transferred immovable property encumbered by rent to the ownership of another person shall bear subsidiary responsibility with him (Article 399) with regard to the demands of the recipient of the rent which arose in connection with a violation of the contract of rent unless joint and several

responsibility for this obligation has been provided for by the present Code, another law, or the contract.

Article 587. Securing Payment of Rent

1. In the event of the transfer of a land plot or other immovable property under payment of rent, the recipient of the rent shall acquire the right of pledge on this property to secure the obligation of the payer of the rent.

2. The condition establishing the duty of the payer of rent to grant security for the performance of his obligations (Article 329) or to insure to the benefit of the recipient of rent the risk of responsibility for the failure to perform or the improper performance of these obligations shall be a material condition of a contract providing for the transfer of a monetary amount or other moveable property as payment of rent.

3. In the event of the failure of the payer of the rent to fulfil the duties provided for by point 2 of the present Article, and also in the event of the loss of the security or worsening of the conditions thereof under circumstances for which the recipient of rent is not liable, the recipient of rent shall have the right to dissolve the contract of rent and require compensation of losses caused by the dissolution of the contract.

Article 588. Responsibility for Delay of Payment of Rent

For a delay of payment of rent the payer of rent shall pay interest to the recipient of rent provided for by Article 395 of the present Code unless another amount of interest has been established by the contract of rent.

§2. Permanent Rent

Article 589. Recipient of Permanent Rent

1. Only citizens, and also noncommercial organisations, may be recipients of permanent rent unless this is contrary to a law and corresponds to the purposes of their activity.

2. The rights of the recipient of rent under a contract of permanent rent may be transferred to the persons specified in point 1 of the present Article by means of assignment of a demand and transfer by inheritance or by way of legal succession in the event of the reorganisation of juridical persons, unless provided otherwise by a law or by a contract.

Article 590. Form and Amount of Permanent Rent

1. Permanent rent shall be paid in money in the amount established by the contract.

The payment of rent by means of the granting of things, fulfilment of work, or rendering of services which correspond to the value of the monetary amount of rent may be provided for by the contract of permanent rent.

2. Unless provided otherwise by the contract of permanent rent, the amount of rent to be paid shall be increased in proportion to the increase of the minimum amount of payment for labour established by a law.

Article 591. Periods for Payment of Permanent Rent

Unless provided otherwise by the contract of permanent rent, permanent rent shall be paid at the end of each calendar quarter.

Article 592. Right of Payer to Purchase Permanent Rent

1. The payer of permanent rent shall have the right to refuse the further payment of rent by means of purchasing it.

2. Such refusal shall be valid on condition that it is declared by the payer of rent in written form not later than three months before the termination of the payment of rent or longer period provided for by the contract of permanent rent. In so doing the obligation relating to the payment of rent shall not terminate until the receipt of the entire amount of the purchase by the recipient of the rent unless a different procedure of purchase has been provided for by the contract.

3. The condition of a contract of permanent rent concerning the waiver of the payer of permanent rent to purchase it shall be void.

It may be provided by a contract that the right of purchase of permanent rent may not be effectuated during the life of the recipient of rent or during another period not exceeding thirty years from the moment of conclusion of the contract.

Article 593. Purchase of Permanent Rent at Demand of Recipient of Rent

The recipient of permanent rent shall have the right to demand the purchase of rent by the payer in instances when:

the payer of rent delayed the payment thereof by more than one year, unless provided otherwise by the contract of permanent rent;

the payer of rent violated his obligations relating to securing the payment of rent (Article 587);

the payer of rent is deemed to be insolvent or other obligations have arisen which obviously testify that the rent will not be paid in the amount and within the periods which have been established by the contract;

the immovable property transferred under payment of rent has ensued to common ownership or been divided among several persons;

and in other instances provided for by the contract.

Article 594. Purchase Price of Permanent Rent

1. The purchase of permanent rent in the instances provided for by Articles 592 and 593 of the present Code shall be made at the price determined by the contract of permanent rent.

2. In the absence of a condition concerning the purchase price in a contract of permanent rent under which property has been transferred for payment under payment of permanent rent, the purchase shall be effectuated at the price corresponding to the yearly amount of rent subject to payment.

3. In the absence of a condition concerning the purchase price in a contract of permanent rent under which property has been transferred under payment of rent free of charge, the price of the transferred property determined according to the rules provided for by Article 424(3) of the present Code shall be included in the purchase price together with the yearly amount of rent payments.

Article 595. Risk of Accidental Perishing of Property Transferred Under Payment of Permanent Rent

1. The risk of accidental perishing or accidental damage of property transferred free of charge under payment of permanent rent shall be borne by the payer of rent.

2. In the event of the accidental perishing or accidental damaging of property transferred for payment under payment of permanent rent, the payer shall have the right to demand respectively termination of the obligation relating to the payment of rent or a change of the conditions for the payment thereof.

§3. Rent for Life

Article 596. Recipient of Rent for Life

1. Rent for life may be established for the period of life of a citizen transferring property under payment of rent or for the period of life of another citizen specified by him.

2. The establishment of rent for life to the benefit of several citizens whose participatory share in the right to receive rent shall be considered to be equal shall be permitted unless provided otherwise by the contract of rent for life.

In the event of the death of one of the recipients of rent his participatory share in the right to receive rent shall pass to the recipients of rent who survive him unless provided otherwise by the contract of rent for life, and in the event of the death of the last recipient of rent, the obligation to pay the rent shall terminate.

3. The contract establishing rent for life to the benefit of a citizen who is deceased at the moment of concluding the contract shall be void.

Article 597. Amount of Rent for Life

1. Rent for life shall be determined in the contract as a monetary amount periodically to be paid to the recipient of rent during his life.

2. The amount of rent for life determined in the contract calculated per month must be not less than the minimum amount of payment for labour established by a law, and in the instances provided for by Article 318 of the present Code, shall be subject to increase.

Article 598. Period for Payment of Rent for Life

Unless provided otherwise by the contract of rent for life, the rent for life shall be paid at the end of each calendar month.

Article 599. Dissolution of Contract of Rent for Life at Demand of Recipient of Rent

1. In the event of a material violation of a contract of rent for life by the payer of rent, the recipient of rent shall have the right to demand from the payer of rent the purchase of rent on the conditions provided for by Article 594 of the present Code or dissolution of the contract and compensation of losses.

2. If an apartment, dwelling house, or other property under payment of rent for life has been alienated free of charge, the recipient of the rent shall have the right in the event of a material violation of the contract by the payer of the rent to require the return of this property, setting off the value thereof at the expense of the purchase price of the rent.

Article 600. Risk of Accidental Perishing of Property Transferred Under Payment of Rent for Life

The accidental perishing or accidental damaging of property transferred under payment of rent for life shall not relieve the payer of rent from the obligation to pay it on the conditions provided for by the contract of rent for life.

§4. Maintenance of Dependent for Life

Article 601. Contract of Maintenance of Dependent for Life

1. Under a contract of maintenance of a dependent for life the recipient-citizen of rent shall transfer a dwelling house, apartment, land plot, or other immovable belonging to him to the ownership of the payer of rent, who shall be obliged to effectuate the maintenance of the dependent-citizen for life and/or third person(s) specified by him.

2. The rules concerning rent for life shall apply to the contract of maintenance of a dependent for life unless provided otherwise by the rules of the present paragraph.

Article 602. Duty Relating to Granting of Maintenance of Dependent

1. The duty of the payer of rent relating to granting maintenance of a dependent may include provision of the requirements for housing, nourishment, and clothing, and if the state of health of the citizen so requires, also care for him. The payment for ritual services by the payer of rent also may be provided for by the contract for the maintenance of a dependent for life.

2. The value of the entire amount of maintenance of a dependent must be determined in the contract of maintenance of a dependent for life. In so doing the value of the total amount of maintenance per month may not be less than two minimum amounts of payment for labour established by a law.

3. When settling a dispute between the parties concerning the amount of maintenance which is granted or should be granted to a citizen, the court must be guided by the principles of good faith and reasonableness.

Article 603. Replacement of Maintenance for Life by Periodic Payments

The possibility of replacing the granting of maintenance of a dependent in kind by the payment during the life of the citizen of periodic payments in money may be provided for by a contract of maintenance of a dependent for life.

Article 604. Alienation and Use of Property Transferred to Secure Maintenance for Life

The payer of rent shall have the right to alienate, pledge, or by other means encumber immovable property transferred to him as security for maintenance for life only with the prior consent of the recipient of rent.

The payer of rent shall be obliged to take necessary measures so that during the period of granting maintenance of a dependent for life the use of the said property does not lead to a reduction of the value of this property.

Article 605. Termination of Maintenance of Dependent for Life

1. The obligation of maintenance of a dependent for life shall terminate with the death of the recipient of rent.

2. In the event of a material violation by the payer of rent of his obligations, the recipient of rent shall have the right to require the return of immovable property transferred to secure the maintenance for life or the payment to him of the purchase price on the conditions established by Article 594 of the present Code. In so doing the payer of rent shall not have the right to demand contributory compensation of the expenses incurred in connection with the maintenance of the recipient of rent.

Chapter 34. Lease

§1. General Provisions on Lease

Article 606. Contract of Lease

Under a contract of lease (or property rental) the lessor (or renter) shall be obliged to grant to the lessee (or tenant) property for payment in temporary possession and use or temporary use.

The fruits, products, and revenues received by the lessee as a result of the use of the leased property in accordance with the contract shall be his ownership.

Article 607. Objects of Lease

1. Land plots and other solitary natural objects, enterprises, and other property complexes, buildings, installations, equipment, means of transport, and other things which do not lose their natural properties in the process of their use (nonconsumable things) may be transferred on lease.

Types of property, the leasing out of which is not permitted or is limited, may be established by a law.

2. The peculiarities of leasing out land plots and other solitary natural objects may be established by a law.

3. Data enabling to definitely establish the property which is subject to transfer to the lessee as the object of a lease must be specified in the contract of lease. In the absence of such data in the contract, the condition concerning the object which is subject to transfer on lease shall be considered to be not agreed by the parties and the respective contract shall not be considered to be concluded.

Article 608. Lessor

The right to lease out property shall belong to its owner. Persons empowered by a law or by the owner to lease out property also may be lessors.

Article 609. Form and State Registration of Contract of Lease

1. A contract of lease for a period of more than a year, and if one of the parties of the contract is a juridical person, irrespective of the period, must be concluded in written form.

2. A contract of lease for immovable property shall be subject to State registration unless established otherwise by a law.

3. A contract of lease for property which provides for the transfer thereafter of the right of ownership to this property to the lessee shall be concluded in the form provided for the contract of purchase-sale of such property.

Article 610. Period of Contract of Lease

1. A contract of lease shall be concluded for the period determined by the contract.

2. If the period of the lease has not been determined in the contract, the contract of lease shall be considered to be concluded for an indefinite period.

In this event each of the parties shall have the right at any time to repudiate the contract, having warned the other party thereof one month in advance, and in the event of the lease of immovable property, three months. Another period may be established by a law or by a contract for warning of the termination of a contract of lease concluded for an indefinite period.

3. Maximum periods for a contract for individual types of lease may be established by a law, as well as for the lease of individual types of property. In these instances if the period of the lease is not determined in the contract and neither of the parties has repudiated the contract upon the expiry of the maximum period established by a law, the contract shall terminate upon the expiry of the maximum period.

A contract of lease concluded for a period exceeding the maximum period established by a law shall be considered to be concluded for the period equal to the maximum.

Article 611. Granting Property to Lessee

1. The lessor shall be obliged to grant property to the lessee in a state corresponding to the conditions of the contract of lease and designation of the property.

2. Property shall be leased out together with all of its appurtenances and documents relating thereto (technical passport, certificate of quality, and so forth), unless provided otherwise by the contract.

If such appurtenances and documents were not transferred and without them the lessee cannot use the property in accordance with its designation or is to a significant degree deprived of that which he had the right to count upon when concluding the contract, he may require the granting to him by the lessor of such appurtenances and documents or dissolution of the contract, and also compensation of losses.

3. If the lessor did not grant to the lessee property hired out within the period specified in the contract of lease, and in an instance when in the contract such period is not specified, within a reasonable period, the lessee shall have the right to demand and obtain this property from him in accordance with Article 398 of the present Code and to require compensation of losses caused by the delay of performance or to require dissolution of the contract and compensation of losses caused by his failure to perform.

Article 612. Responsibility of Lessor for Defects of Property Leased Out

1. The lessor shall be liable for defects of property leased out which wholly or partially obstruct the use thereof even if during the conclusion of the contract of lease he did not know about these defects.

In the event of the discovery of such defects the lessee shall have the right at his choice to:

require of the lessor either the elimination of the defects of the property without compensation or a commensurate reduction of the lease payment, or compensation of his expenses for elimination of the defects of the property;

directly withhold the amount of expenses incurred by him for elimination of the said defects from the lease payment, having informed the lessor in advance thereof;

require the dissolution of the contract before time.

The lessor who is notified about the demands of the lessee or about his intention to eliminate the defects of the property at the expense of the lessor may without delay replace the property granted to the lessee with other analogous property in a proper state or eliminate the defects of the property without compensation.

If satisfaction of the demands of the lessee or withholding of expenses by him for elimination of defects from the lease payment does not fully cover the losses caused to the lessee, he shall have the right to require compensation of the uncovered part of the losses.

2. The lessor shall not be liable for defects of property leased out which were stipulated by him when concluding the contract of lease or were previously known to the lessee or should have been discovered by the lessee during the inspection of the property or verification of its good repair when concluding the contract or transferring the property on lease.

Article 613. Rights of Third Persons to Property Leased Out

The transfer of property on lease shall not be grounds for termination or change of the rights of third persons to such property.

When concluding a contract of lease the lessor shall be obliged to warn the lessee about all the rights of third persons to the property being leased out (servitude, right of pledge, and so forth). The failure of the lessor to perform this duty shall give the lessee the right to demand a reduction of lease payment or dissolution of the contract and compensation of losses.

Article 614. Lease Payment

1. The lessee shall be obliged to make payment in a timely way for the use of property (lease payment).

The procedure, conditions, and periods of making the lease payment shall be determined by the contract of lease. In an instance when they have not been determined by the contract, it shall be considered that the procedure, conditions, and periods have been established which are usually applied when leasing analogous property under comparable circumstances.

2. The lease payment shall be established for the entire leased property as a whole or individually for each of the constituent parts in the form of:

(1) payments determined in a fixed amount made periodically or one time;

(2) the established participatory share of the products, fruits, or revenues received as a result of the use of the leased property;

(3) the granting of determined services by the lessee;

(4) the transfer by the lessee to the lessor of things stipulated by the contract in ownership or on lease;

(5) imposing on the lessee expenditures stipulated by the contract for the improvement of the leased property.

The parties may provide in the contract of lease for combining the said forms of lease payment or other forms of payment of the lease.

3. Unless provided otherwise by the contract, the amount of lease payment may be changed by agreement of the parties within the periods provided for by the contract, but not more often than once a year. Other minimum periods for revision of the amount of lease payment for individual types of lease, and also for the lease of individual types of property, may be provided for by a law.

4. Unless provided otherwise by a law, the lessee shall have the right to require a respective reduction of the lease payment if by virtue of circumstances for which he is not liable the conditions of use provided for by the contract of lease or the state of the property materially have worsened.

5. Unless provided otherwise by the contract of lease, in the event of a material violation by the lessee of the periods for making the lease payment the lessor shall have the right to require the making of lease payment by him before time within the period established by the lessor. In so doing the lessor shall not have the right to demand the making of lease payment before time for more than two periods in succession.

Article 615. Use of Leased Property

1. The lessee shall be obliged to use leased property in accordance with the conditions of the contract of lease, and if such conditions have not been determined in the contract, in accordance with the designation of the property.

2. The lessee shall have the right with the consent of the lessor to sublease the leased property (or sub-hire) and to transfer his rights and duties under the contract of lease to another person (re-hire), to grant the leased property for use without compensation, and also to pledge lease rights and contribute them as a contribution to the charter capital of economic partnerships and societies or a

share contribution to a production cooperative unless established otherwise by the present Code, another law, or other legal acts. In the said instances, except for re-hire, the lessee shall remain responsible under the contract to the lessor.

The contract of sublease may not be concluded for a period exceeding the period of the contract of lease.

The rules concerning contracts of lease, unless established otherwise by a law or other legal acts, shall apply to contracts of sublease.

3. If a lessee uses property not in accordance with the conditions of the contract of lease or the designation of the property, the lessor shall have the right to require dissolution of the contract and compensation of losses.

Article 616. Duties of Parties With Regard to Maintenance of Leased Property

1. The lessor shall be obliged to perform at his own expense capital repair to the property transferred on lease unless provided otherwise by a law, other legal acts, or the contract of lease.

Capital repair must be performed within the period established by the contract, and if it is not determined by the contract or called for by urgent necessity, within a reasonable period.

A violation by the lessor of the duty with regard to the performance of capital repair shall give the lessee the right at his choice to:

perform the capital repair provided for by the contract or caused by urgent necessity and to recover from the lessor the value of the repair or to set it off from the account of the lease payment;

require a corresponding reduction of lease payment;

require dissolution of the contract and compensation of losses.

2. The lessee shall be obliged to maintain the property in a proper state, to perform current repair at his own expense, and to bear expenses for the maintenance of the property unless otherwise established by a law or by the contract of lease.

Article 617. Preservation of Contract of Lease in Force in Event of Change of Parties

1. Transfer of the right of ownership (or economic jurisdiction, operative management, inheritable possession for life) to property leased out to another person shall not be grounds for a change of or dissolution of the contract of lease.

2. In the event of the death of a citizen leasing immovable property, his rights and duties under the contract of lease shall pass to the heir unless provided otherwise by a law or by the contract.

The lessor shall not have the right to refuse such heir from entering into the contract for the remaining period of its operation except for an instance when the conclusion of the contract was conditioned by the personal qualities of the lessee.

Article 618. Termination of Contract of Sublease in Event of Termination of Contract of Lease Before Time

1. Unless provided otherwise by the contract of lease, the termination of the contract of lease before time shall entail the termination of the contract of sublease concluded in accordance with it. The sublessee shall have the right in this event to conclude a contract of lease with him for property in his use in accordance with the contract of sublease within the limits of the remaining period of sublease on the conditions corresponding to the conditions of the terminated contract of lease.

2. If the contract of lease is void on the grounds provided for by the present Code, the contracts of sublease concluded in accordance with it also shall be void.

Article 619. Dissolution of Contract Before Time Upon Demand of Lessor

Upon the demand of the lessor the contract of lease may be dissolved before time by a court in instances when the lessee:

(1) uses the property with material violation of the conditions of the contract or the designation of the property or with repeated violations;

(2) materially worsens the property;

(3) does not make lease payment more than two times in succession upon the expiry of the period established by the contract;

(4) does not make capital repair of the property within the periods established by the contract of lease, and in the absence thereof in the contract within reasonable periods in those instances when in accordance with a law, other legal acts, or the contract the performance of capital repair is the duty of the lessee.

Other grounds for the dissolution of the contract before time also may be established by the contract of lease upon the demand of the lessor in accordance with Article 450(2) of the present Code.

The lessor shall have the right to demand dissolution of the contract before time only after sending a written warning to the lessee about the necessity to perform the obligation within a reasonable period.

Article 620. Dissolution of Contract Before Time Upon Demand of Lessee

Upon the demand of the lessee the contract of lease may be dissolved before time by a court in instances when:

(1) the lessor does not grant property for use to the lessee or creates obstacles to the use of the property in accordance with the conditions of the contract or the designation of the property;

(2) the property transferred to the lessee has defects which obstruct its use by him which were not stipulated by the lessor when concluding the contract, were not known to the lessee in advance, and should not have been discovered by the

lessee during the inspection of the property or verification of its good condition when concluding the contract;

(3) the lessor does not make capital repair of the property which is his duty within the periods established by the contract of lease, and in the absence thereof in the contract, within reasonable periods;

(4) the property by virtue of circumstances for which the lessee is not liable turns out to be in a state not fit for use.

Other grounds of dissolution of the contract before time also may be established by the contract of lease upon the demand of the lessee in accordance with Article 450(2) of the present Code.

Article 621. Preferential Right of Lessee to Conclude Contract of Lease for New Period

1. Unless provided otherwise by a law or by the contract of lease, the lessee who has properly performed his duties shall have upon the expiry of the period of the contract a preferential right under equal conditions before other persons to conclude a contract of lease for a new period. The lessee shall be obliged to inform the lessor in writing about the wish to conclude such contract within the period specified in the contract of lease, and if such period is not specified in the contract, within a reasonable period before the end of the operation of the contract.

When concluding a contract of lease for a new period, the conditions of the contract may be changed by agreement of the parties.

If the lessor has refused the lessee to conclude a contract for a new period, but within a year from the date of expiry of the period of the contract with him concluded a contract of lease with another person, the lessee shall have the right at his choice to demand in court the transfer to himself of the rights and duties under the contract concluded and compensation of losses caused by the refusal to renew the contract of lease with him or only compensation of such losses.

2. If the lessee continues to use the property after the expiry of the period of the contract in the absence of objections on the part of the lessor, the contract shall be considered to be renewed on the same conditions for an indefinite period (Article 610).

Article 622. Return of Leased Property to Lessor

In the event of the termination of the contract of lease the lessee shall be obliged to return property to the lessor in that state in which he received it, taking into account normal wear and tear, or in the state stipulated by the contract.

If the lessee has not returned the leased property or returned it not in good time, the lessor shall have the right to demand the making of lease payment for the entire time of the delay. In an instance when the said payment does not cover the losses caused to the lessor, he may demand the compensation thereof.

In an instance when a penalty has been provided for by the contract for the untimely return of the leased property, losses may be recovered in the full amount above the penalty unless provided otherwise by the contract.

Article 623. Improvement of Leased Property

1. Separable improvements of leased property made by the lessee shall be his ownership unless provided otherwise by the contract of lease.

2. In an instance when a lessee has made improvements of leased property at the expense of own assets and with the consent of the lessor which are not separable without harm to the property, the lessee shall have the right after termination of the contract to compensation of the cost of these improvements unless provided otherwise by the contract of lease.

3. The cost of inseparable improvements of leased property made by the lessee without the consent of the lessor shall not be subject to compensation unless provided otherwise by a law.

4. Improvements of leased property, both separable and inseparable, made at the expense of amortisation deductions from this property shall be the ownership of the lessor.

Article 624. Purchase of Leased Property

1. It may be provided in a law or contract of lease that leased property passes to the ownership of the lessee upon the expiry of the period of lease or before the expiry thereof on condition of the lessee making the entire purchase price stipulated by the contract.

2. If the condition concerning the purchase of leased property is not provided for in the contract of lease, it may be established by an additional agreement of the parties, who in so doing shall have the right to agree on the set-off of the lease payment previously paid against the purchase price.

3. Instances of the prohibition of the purchase of leased property may be established by a law.

Article 625. Peculiarities of Individual Types of Lease and Lease of Individual Types of Property

The provisions provided for by the present paragraph shall apply to individual types of contract of lease and to contracts of lease of individual types of property (rental, lease of means of transport, lease of buildings and installations, lease of enterprises, finance lease) unless established otherwise by the rules of the present Code concerning these contracts.

§2. Rental

Article 626. Contract of Rental

1. Under a contract of rental the lessor effectuating the lease of property as a permanent entrepreneurial activity shall be obliged to grant moveable property to the lessee in temporary possession and use for payment.

Property granted under a contract of rental shall be used for consumption purposes unless provided otherwise by the contract or does not arise from the essence of the obligation.

2. A contract of rental shall be concluded in written form.

3. A contract of rental shall be a public contract (Article 426).

Article 627. Period of Contract of Rental

1. A contract of rental shall be concluded for a period of up to one year.

2. The rules concerning renewal of a contract of lease for an indefinite period and concerning the preferential right of the lessee to renewal of the contract of lease (Article 621) shall not apply to the contract of rental.

3. The lessee shall have the right to repudiate the contract of rental at any time, having warned the lessor in writing about his intention by not less than ten days.

Article 628. Granting of Property to Lessee

A lessor who has concluded a contract of rental shall be obliged in the presence of the lessee to verify the good state of the property being leased, and also to familiarise the lessee with the rules for the operation of the property or to issue him written instructions concerning the use of this property.

Article 629. Elimination of Defects of Property Leased

1. In the event of the discovery by the lessee of defects in the property leased which wholly or partially obstruct the use thereof, the lessor shall be obliged within a ten-day period from the date of the statement of the lessee concerning the defects, unless a shorter period has been established by the contract of rental, to eliminate the defects of the property without compensation on the site or to replace the particular property with other analogous property which is in a proper state.

2. If the defects of the leased property were a consequence of a violation by the lessee of the rules for operation and maintenance of the property, the lessee shall pay the lessor the cost of the repair and transportation of the property.

Article 630. Lease Payment Under Contract of Rental

1. Lease payment under a contract of rental shall be established in the form of payments determined in a lump sum to be made periodically or at one time.

233

2. In the event of the return of the property before time by the lessee, the lessor shall return to him the corresponding part of the lease payment received, calculating it from the day following the day of the actual return of the property.

3. The recovery of indebtedness from the lessee with regard to a lease payment shall be made in an uncontested proceeding on the basis of the endorsement of execution of a notary.

Article 631. Use of Leased Property

1. The capital and current repair of property leased under a contract of rental shall be the duty of the lessor.

2. The sublease of property granted to a lessee under a contract of rental, the transfer by him of his rights and duties under the contract of rental to another person, the granting of this property for uncompensated use, the pledge of lease rights, and the contribution thereof as a property contribution to economic partnerships and societies or as a share contribution to production cooperatives shall not be permitted.

§3. Lease of Means of Transport

1. Lease of Means of Transport With Provision of Services Relating to Management and Technical Operation

Article 632. Contract of Lease of Means of Transport with Crew

Under a contract of lease (time-charter) of a means of transport with a crew the lessor shall grant the means of transport to the lessee in temporary possession and use for payment and render services with his own forces with regard to the management and technical operation thereof.

The rules concerning renewal of a contract of lease for an indefinite period and concerning the preferential right of the lessee to conclude a contract of lease for a new period (Article 621) shall not apply to the contract of lease of means of transport with a crew.

Article 633. Form of Contract of Lease of Means of Transport with Crew

A contract of lease of means of transport with a crew must be concluded in written form irrespective of the period thereof. The rules concerning the registration of contracts of lease provided for by Article 609(2) of the present Code shall not apply to such a contract.

Article 634. Duty of Lessor Relating to Maintenance of Means of Transport

The lessor throughout the entire period of the contract of lease of means of transport with a crew shall be obliged to maintain the proper state of the means of transport leased, including the effectuation of current and capital repair and the granting of necessary appurtenances.

Article 635. Duties of Lessor With Regard to Management and Technical Operation of Means of Transport

1. Services with regard to management and technical operation of means of transport granted by the lessor to the lessee must ensure its normal and safe operation in accordance with the purposes of the lease specified in the contract. A more extensive group of services to be granted to the lessee may be provided for by the contract of lease of means of transport with a crew.

2. The composition of the crew of means of transport and the skills thereof must correspond to the rules and conditions of the contract which are binding upon the parties, and if such requirements have not been established by rules and conditions of the contract which are binding upon the parties, the requirements of usual practice for the operation of means of transport of the particular type and the conditions of the contract.

Members of the crew shall be workers of the lessor. They shall be subordinate to the regulations of the lessor with respect to the management and technical operation and to the regulations of the lessee affecting the commercial operation of the means of transport.

Unless provided otherwise by the contract of lease, expenses relating to the payment for services of members of the crew, and also expenses for the maintenance thereof, shall be borne by the lessor.

Article 636. Duty of Lessee Relating to Payment of Expenses Connected with Commercial Operation of Means of Transport

Unless provided otherwise by the contract of lease of means of transport with a crew, the lessee shall bear expenses arising in connection with the commercial operation of the means of transport, including expenses for the payment of fuel and other materials expended in the process of operation and for the payment of fees.

Article 637. Insurance of Means of Transport

Unless provided otherwise by a contract of lease of means of transport with a crew, the duty to insure the means of transport and/or to insure responsibility for damage which may be caused by it or in connection with its operation shall be placed on the lessor in those instances when such insurance is obligatory by virtue of a law or contract.

Article 638. Contracts With Third Persons Concerning Use of Means of Transport

1. Unless provided otherwise by the contract of lease of means of transport with a crew, the lessee shall have the right without the consent of the lessor to sublease the means of transport.

2. The lessee within the framework of effectuating commercial operation of leased means of transport shall have the right without the consent of the lessor to conclude in his own name with third persons contracts of carriage and other contracts unless they are contrary to the purposes of the use of means of transport specified in the contract of lease, and if such purposes have not been established, the designation of the means of transport.

Article 639. Responsibility for Harm Caused to Means of Transport

In the event of the perishing or damage of leased means of transport, the lessee shall be obliged to compensate the lessor for losses caused if the last proves that the perishing or damage of the means of transport occurred with regard to circumstances for which the lessee is liable in accordance with a law or the contract of lease.

Article 640. Responsibility for Harm Caused by Means of Transport

Responsibility for harm caused to third persons by leased means of transport, mechanisms thereof, devices, and equipment shall be borne by the lessor in accordance with the rules provided for by Chapter 59 of the present Code. He shall have the right to present a regressive demand against the lessee concerning compensation of amounts paid to third persons if it is proved that the harm arose through the fault of the lessee.

Article 641. Peculiarities of Lease of Individual Types of Means of Transport

Other peculiarities of the lease of individual types of means of transport with the granting of services relating to management and technical operation may be established, besides those provided for by the present paragraph, by transport charters and codes.

2. Lease of Means of Transport Without Provision of Services Relating to Management and Technical Operation

Article 642. Contract of Lease of Means of Transport Without Crew

Under a contract of lease of means of transport without a crew the lessor shall grant means of transport to the lessee in temporary possession and use for payment without rendering services relating to the management and technical operation thereof.

The rules concerning renewal of the contract of lease for an indefinite period and concerning the preferential right of the lessee to conclude a contract of lease for a new period (Article 621) shall not apply to the contract of lease of means of transport without a crew.

Article 643. Form of Contract of Lease of Means of Transport Without Crew

A contract of lease of means of transport without crew must be concluded in written form irrespective of the period thereof. The rules concerning the registration of contracts of lease provided for by Article 609(2) of the present Code shall not apply to such a contract.

Article 644. Duty of Lessee With Regard to Maintenance of Means of Transport

The lessee during the entire period of the contract of lease of means of transport without crew shall be obliged to maintain the proper state of leased means of transport, including the effectuation of current and capital repair.

Article 645. Duties of Lessee With Regard to Management of Means of Transport and Its Technical Operation

The lessee shall effectuate the management of the leased means of transport with his own forces and the operation thereof, both commercial and technical.

Article 646. Duty of Lessee Relating to Payment of Expenses for Maintenance of Means of Transport

Unless provided otherwise by the contract of lease of means of transport without crew, the lessee shall bear the expenses for the maintenance of the leased means of transport, the insurance thereof, including insurance of his responsibility, and also expenses arising in connection with its operation.

Article 647. Contracts with Third Persons Concerning Use of Means of Transport

1. Unless provided otherwise by the contract of lease of means of transport, the lessee shall have the right without the consent of the lessor to sublease the leased means of transport on conditions of the contract of lease of means of transport with or without crew.

2. The lessee shall have the right without the consent of the lessor to conclude in his own name with third persons contracts of carriage and other contracts unless they are contrary to the purposes of the use of the means of transport specified in the contract of lease, and if such purposes have not been established, to the designation of the means of transport.

Article 648. Responsibility for Harm Caused by Means of Transport

Responsibility for harm caused to third persons by a means of transport, its mechanisms, devices, and equipment shall be borne by the lessee in accordance with the rules of Chapter 59 of the present Code.

Article 649. Peculiarities of Lease of Individual Types of Means of Transport

Other peculiarities of the lease of individual types of means of transport without granting services with regard to management and technical operation besides those provided for by the present paragraph may be established by transport charters and codes.

§4. Lease of Buildings and Installations

Article 650. Contract of Lease of Building or Installation

1. Under the contract of lease of a building or installation the lessor shall be obliged to transfer a building or installation in temporary possession and use or temporary use to the lessee.

2. The rules of the present paragraph shall apply to the lease of enterprises unless provided otherwise by the rules of the present Code concerning the lease of an enterprise.

Article 651. Form and State Registration of Contract of Lease of Building or Installation

1. A contract of lease of a building or installation shall be concluded in written form by means of drawing up one document signed by the parties (Article 434[2]).

The failure to comply with the form of the contract of lease of a building or installation shall entail its invalidity.

2. The contract of lease of a building or installation concluded for a period of not less than one year shall be subject to State registration and shall be considered to be concluded from the moment of such registration.

Article 652. Rights to Land Plot in Event of Lease of Building or Installation Situated Thereon

1. Under a contract of lease of a building or installation, to the lessee shall pass simultaneously with the transfer of the rights of possession and use of such immovable the rights to that part of the land plot which is occupied by this immovable and is necessary for its use.

2. In instances when the lessor is the owner of the land plot on which the building or installation to be leased is situated, the right of lease shall be granted to the lessee or other right to the respective part of the land plot provided for by the contract of lease of the building or installation.

Unless the right to the respective land plot to be transferred to the lessee has been determined by the contract, to him shall pass for the period of the lease of the building or installation the right to use that part of the land plot which is occupied by the building or installation and is necessary for its use in accordance with its designation.

3. The lease of a building or installation situated on the land plot which does not belong to the lessor by right of ownership shall be permitted with the consent of the owner of this plot unless this is contrary to the conditions of the use of this plot established by a law or by contract with the owner of the land plot.

Article 653. Preservation by Lessee of Building or Installation of Right to Use Land Plot in Event of Sale Thereof

In instances when the land plot on which the leased building or installation is situated is sold to another person, the right of use of the part of the land plot which is occupied by the building or installation and is necessary for its use shall be preserved for the lessee of this building or installation on the conditions which operated before the sale of the land plot.

Article 654. Amount of Lease Payment

1. The contract of lease of a building or installation must provide for the amount of lease payment. In the absence of the condition concerning the amount of lease payment agreed by the parties in written form, the contract of lease of a building or installation shall be considered to be not concluded. In so doing the rules for determining the price provided for by Article 424(3) of the present Code shall not apply.

2. The payment established in the contract of lease of a building or installation for the use of the building or installation shall include payment for use of the land plot on which it is situated or the respective part of the plot transferred together with it, unless provided otherwise by a law or the contract.

3. In instances when the payment for lease of a building or installation has been established in the contract per unit of space of the building (or installation) or other indicator of the dimension thereof, the lease payment shall be determined by proceeding from the actual dimension of the building or installation transferred to the lessee.

Article 655. Transfer of Building or Installation

1. The transfer of a building or installation by the lessor and the acceptance thereof by the lessee shall be effectuated under an act of transfer or other document concerning the transfer to be signed by the parties.

Unless provided otherwise by a law or the contract of lease of the building or installation, the obligation of the lessor to transfer the building or installation to the lessee shall be considered to be performed after the granting thereof to the lessee in possession or use and signature by the parties of the respective document concerning transfer.

The avoidance by one party of signature of the document concerning the transfer of the building or installation on the conditions provided for by a contract shall be considered as a refusal respectively of the lessor to perform the duty with regard to the transfer of property and of the lessee to accept the property.

2. In the event of the termination of the contract of lease of a building or installation, the leased building or installation must be returned to the lessor in compliance with the rules provided for by point 1 of the present Article.

§5. Lease of Enterprises

Article 656. Contract of Lease of Enterprise

1. Under the contract of lease of an enterprise as a whole as a property complex used in order to effectuate entrepreneurial activity the lessor shall be obliged to grant to the lessee for payment in temporary possession and use land plots, buildings, installations, equipment, and other basic means within the composition of the enterprise, transfer in the procedure, on the conditions, and within the limits determined by the contract the stocks of raw material, fuel, materials, and other circulating means, the rights of use of the land, water, and other natural resources, buildings, installations, and equipment, and other property rights of the lessor connected with the enterprise, the rights to designation individualising the activity of the enterprise, and other exclusive rights, and also to assign to it the rights of demand and transfer debts to it which relate to the enterprise. The transfer of the rights of possession and use of property which is in the ownership of other persons, including land and other natural resources, shall be in the procedure provided for by a law and other legal acts.

2. The rights of the lessor received by him on the basis of an authorisation (or license) to engage in a respective activity shall not be subject to transfer to the lessee unless established otherwise by a law or other legal acts. The inclusion in the composition of the enterprise to be transferred under the contract of obligations whose performance by the lessee is impossible in the absence of such authorisation (or license) shall not relieve the lessor from the respective obligations to creditors.

Article 657. Rights of Creditors in Event of Lease of Enterprise

1. Creditors with regard to obligations included as part of the enterprise must be notified before the transfer thereof to the lessee by the lessor in writing concerning the transfer of the enterprise on lease.

2. A creditor who has not communicated to the lessor in writing his consent to the transfer of the debt shall have the right within three months from the date of receiving notification concerning the transfer of an enterprise on lease to demand the termination or performance before time of the obligation and compensation of the losses caused by this.

3. A creditor who has not been informed about the transfer of an enterprise on lease in the procedure provided for by point 1 of the present Article may bring suit concerning the satisfaction of the demands provided for by point 2 of the present Article within a year from the date when he knew or should have known about the transfer of the property on lease.

4. After the transfer of an enterprise on lease the lessor and the lessee shall bear joint and several responsibility for the debts included as part of the transferred enterprise which were transferred to the lessee without the consent of the creditor.

Article 658. Form and State Registration of Contract of Lease of Enterprise

1. The contract of lease of an enterprise shall be concluded in written form by means of drawing up one document signed by the parties (Article 434[2]).

2. The contract of lease of an enterprise shall be subject to State registration and shall be considered to be concluded from the moment of such registration.

3. The failure to comply with the form of the contract of lease of an enterprise shall entail its invalidity.

Article 659. Transfer of Leased Enterprise

The transfer of an enterprise to the lessee shall be effectuated under an act of transfer.

The preparation of the enterprise for transfer, including the drawing up and granting for signature of the act of transfer, shall be the duty of the lessor and shall be effectuated at his expense unless provided otherwise by the contract of lease of the enterprise.

Article 660. Use of Property of Leased Enterprise

Unless provided otherwise by the contract of lease of an enterprise, the lessee shall have the right without the consent of the lessor to sell, exchange, grant for temporary use, or loan material valuables which are part of the property of the leased enterprise, sublease them, and transfer their rights and duties under the contract of lease with respect to such valuables to another person on condition that this does not entail a reduction of the value of the enterprise and does not violate other provisions of the contract of lease of the enterprise. The said procedure shall not apply with respect to land and other natural resources, nor in other instances provided for by a law.

Unless provided otherwise by the contract of lease of an enterprise, the lessee shall have the right without the consent of the lessor to make changes in the composition of the leased property complex and conduct the conversion, expansion, or technical re-equipping which increases its value.

Article 661. Duties of Lessee with Regard to Maintenance of Enterprise and Payment of Expenses for Operation Thereof

1. The lessee of an enterprise shall be obliged within the entire period of operation of the contract of lease of the enterprise to maintain the enterprise in a proper technical state, including to effectuate the current and capital repair thereof.

2. Expenses connected with the operation of the leased enterprise shall be placed on the lessee unless provided otherwise by the contract, and also the payment of payments for insurance of the leased property.

Article 662. Making of Improvements by Lessee in Leased Enterprise

The lessee of an enterprise shall have the right to compensation for the cost to him of inseparable improvements of the leased property irrespective of the authorisation of the lessor for such improvements unless provided otherwise by the contract of lease of the enterprise.

The lessor may be relieved by a court from the duty to compensate the lessee for the cost of such improvements if it is proved that the costs of the lessee for such improvements exceeds the value of the leased property incommensurately to the improvement of its quality and/or operational properties or when effectuating such improvements the principles of good faith and reasonableness were violated.

Article 663. Application to Contract of Lease of Enterprise of Rules Concerning Consequences of Invalidity of Transactions and Changes and Dissolution of Contract

The rules of the present Code concerning the consequences of the invalidity of the transactions and changes and dissolution of a contract providing for the return or the recovery in kind of that received under the contract from one or from both parties shall apply to a contract of lease of an enterprise if such consequences do not violate materially the rights and the interests protected by a law of creditors of the lessor and lessee and other persons and are not contrary to social interests.

Article 664. Return of Leased Enterprise

In the event of the termination of a contract of lease of an enterprise, the leased property complex must be returned to the lessor in compliance with the rules provided for by Articles 656, 657, and 659 of the present Code. The preparation of the enterprise for transfer to the lessor, including the drawing up and submission for signature of the act of transfer, shall in this event be the duty of the lessee and shall be effectuated at his expense unless provided otherwise by the contract.

§6. Finance Lease (Finance Leasing)

Article 665. Contract of Finance Lease

According to a contract of finance lease (contract of finance leasing) the lessor shall be obliged to acquire in ownership property specified by the lessee from a seller determined by him and to grant this property in temporary possession and use for entrepreneurial purposes to the lessee for payment. The lessor in this event shall not bear responsibility for the choice of the subject of lease and the seller.

It may be provided by the contract of finance lease that the choice of seller or property to be acquired shall be effectuated by the lessor.

Article 666. Subject of Contract of Finance Lease

Any non-consumer things to be used for entrepreneurial activity, except land plots and other natural objects, may be the subject of a contract of finance lease.

Article 667. Informing Seller About Leasing Out of Property

The lessor, in acquiring property for a lessee, must inform the seller that the property is intended for the transfer thereof on lease to a determined person.

Article 668. Transfer to Lessee of Subject of Contract of Finance Lease

1. Unless provided otherwise by the contract of finance lease, the property which is the subject of this contract shall be transferred by the seller directly to the lessee at the location of the last.

2. In an instance when property which is the subject of a contract of finance lease is not transferred to the lessee within the period specified in this contract, and if such period has not been specified in this contract, within a reasonable period, the lessee shall have the right if the delay was permitted under circumstances for which the lessor is liable to demand dissolution of the contract and compensation of losses.

Article 669. Transfer to Lessee of Risk of Accidental Perishing or Accidental Spoilage of Property

The risk of accidental perishing or accidental spoilage of leased property shall pass to the lessee at the moment of the transfer of the leased property to him unless provided otherwise by the contract of finance lease.

Article 670. Responsibility of Seller

1. The lessee shall have the right to present demands directly to the seller of the property which is the subject of the contract of finance lease and which arise from the contract of purchase-sale concluded between the seller and the lessor, in particular, with respect to the quality and completeness of the property, the periods for the delivery thereof, and in other instances of improper performance

of the contract by the seller. In so doing the lessee shall have the rights and bear the duties provided for by the present Code for a purchaser except the duties to pay for the property acquired as if he were a party to the contract of purchase-sale of the said property. However, the lessee may not dissolve the contract of purchase-sale with the seller without the consent of the lessor.

In relations with the seller the lessee and lessor shall act as joint and several creditors (Article 326).

2. Unless provided otherwise by the contract of finance lease, the lessor shall not be liable to the lessee for the fulfilment by the seller of demands arising from the contract of purchase-sale, except for instances when responsibility for the choice of the seller lies on the lessor. In the last instance the lessee shall have the right at his choice to present demands arising from the contract of purchase-sale both directly to the seller of the property and to the lessor, who shall bear joint and several responsibility.

Chapter 35. Rental of Dwelling Premise

Article 671. Contract of Rental of Dwelling Premise

1. Under a contract of rental of a dwelling premise one party—the owner of the dwelling premise or person empowered by him (landlord)—shall be obliged to grant to the other party (tenant) a dwelling premise for payment in possession and use in order to reside therein.

2. A dwelling premise may be granted to juridical persons in possession and/or use on the basis of a contract of lease or other contract. A juridical person may use the dwelling premise only for the residing of citizens.

Article 672. Contract of Rental of Dwelling Premise in State and Municipal Housing Fund of Social Use

1. Dwelling premises shall be granted to citizens by a contract of social rental of dwelling premise in the State and municipal dwelling fund of social use.

2. Members of the family residing jointly with the tenant under a contract of social rental of a dwelling premise shall enjoy all the rights and bear all the duties under the contract of rental of a dwelling premise equally with the tenant.

At the demand of the tenant and members of his family the contract may be concluded with one of the members of the family. In the event of the death of the tenant or of his departure from the dwelling premise the contract shall be concluded with one of the members of the family residing in the dwelling premise.

3. The contract of social rental of a dwelling premise shall be concluded on the grounds, on the conditions, and in the procedure provided for by housing legislation. The rules of Articles 674, 675, 678, 680, 681, and 685(1)-(3) of the present Code shall apply to such contract. Other provisions of the present Code shall

apply to the contract of social rental of dwelling premise unless provided otherwise by housing legislation.

Article 673. Object of Contract of Rental of Dwelling Premise

1. An isolated dwelling premise fit for permanent residing (apartment, dwelling house, part of apartment or dwelling house) may be the object of a contract of rental of a dwelling premise.

The fitness of a dwelling premise for residing shall be determined in the procedure provided for by housing legislation.

2. The tenant of a dwelling premise in an apartment house shall have the right, together with the use of the dwelling premise, to use the property specified in Article 290 of the present Code.

Article 674. Form of Contract of Rental of Dwelling Premise

A contract of rental of dwelling premise shall be concluded in written form.

Article 675. Preservation of Contract of Rental of Dwelling Premise in Event of Transfer of Right of Ownership to Dwelling Premise

The transfer of the right of ownership to a dwelling premise occupied under a contract of rental of dwelling premise shall not entail dissolution or change of the contract of rental of the dwelling premise. The new owner shall become the landlord on the conditions of the contract of rental previously concluded.

Article 676. Duties of Landlord of Dwelling Premise

1. The landlord shall be obliged to transfer to the tenant a vacant dwelling premise in the state fit for residing.

2. The landlord shall be obliged to effectuate the proper operation of the dwelling house in which the rented dwelling premise is situated, to grant or to ensure the granting of the necessary municipal services to the tenant for payment, ensure the conducting of the repair of the common property of an apartment house and devices for rendering municipal services situated in the dwelling premise.

Article 677. Tenant and Citizens Permanently Residing Together With Him

1. Only a citizen may be a tenant under a contract of rental of a dwelling premise.

2. The citizens permanently residing in the dwelling premise together with the tenant must be specified in the contract. In the absence of such indications in the contract these citizens shall be evicted in accordance with the rules of Article 679 of the present Code.

Citizens permanently residing jointly with the tenant shall have equal rights with him with regard to the use of the dwelling premise. Relations between the tenant and such citizens shall be determined by a law.

3. The tenant shall bear responsibility to the landlord for the actions of citizens permanently residing jointly with him which violate the conditions of the contract of rental of the dwelling premise.

4. Citizens permanently residing together with the tenant may, having notified the landlord, conclude with the tenant a contract that all citizens permanently residing in the dwelling premise shall bear jointly with the tenant joint and several responsibility to the landlord. In this event such citizens shall be co-tenants.

Article 678. Duties of Tenant of Dwelling Premise

The tenant shall be obliged to use the dwelling premise only for residing, ensure the preservation of the dwelling premise, and maintain it in a proper state.

The tenant shall not have the right to rebuild and convert a dwelling premise without the consent of the landlord.

The tenant shall be obliged to make payment in good time for the dwelling premise. Unless established otherwise by the contract, the tenant shall be obliged autonomously to make municipal payments.

Article 679. Moving In of Citizens Permanently Residing With Tenant

With the consent of the landlord, tenant, and citizens permanently residing with him, other citizens may move in the dwelling premise as permanently residing with the tenant. In the event of moving in minor children, such consent shall not be required.

Moving in shall be permitted on condition of compliance with the requirements of legislation on the norm of dwelling space per person except in the event of the moving in of minor children.

Article 680. Temporary Inhabitants

The tenant and citizens permanently residing with him shall have the right by common consent and with prior notification of the landlord to authorize the uncompensated residing in the dwelling premise of temporary inhabitants (users). The landlord may prohibit the residing of temporary inhabitants on condition of the failure to comply with the requirements of legislation concerning the norm of dwelling space per person. The period of the residing of temporary inhabitants may not exceed six months.

Temporary inhabitants shall not possess an autonomous right of use of the dwelling premise. The tenant shall bear responsibility to the landlord for their actions.

Temporary inhabitants shall be obliged to free the dwelling premise upon the expiry of the period of residing agreed with them, and if the period is not agreed, not later than seven days from the date of the respective demand being presented by the tenant or by any citizen permanently residing with them.

Article 681. Repair of Rented Dwelling Premise

1. The current repair of a rented dwelling premise shall be the duty of the tenant unless established otherwise by the contract of rental of the dwelling premise.

2. The capital repair of a rented dwelling premise shall be the duty of the tenant unless established otherwise by the contract of rental of the dwelling premise.

3. The re-equipping of a dwelling house in which a rented dwelling premise is situated, if such re-equipping materially changes the conditions of use of the dwelling premise, shall not be permitted without the consent of the tenant.

Article 682. Payment for Dwelling Premise

1. The amount of payment for a dwelling premise shall be established by agreement of the parties in the contract of rental of a dwelling premise. If in accordance with a law the maximum amount of payment for the dwelling premise is established, the payment established in the contract must not exceed this amount.

2. A unilateral change of the amount of payment for the dwelling premise shall not be permitted except for instances provided for by a law or by the contract.

3. Payment for the dwelling premise must be made by the tenant within the periods provided for by the contract of rental of the dwelling premise. Unless the periods have been provided for by the contract, the payment must be made by the tenant monthly in the procedure established by the Housing Code of the Russian Federation.

Article 683. Period in Contract of Rental of Dwelling Premise

1. The contract of rental of a dwelling premise shall be concluded for a period not exceeding five years. If the period is not determined in the contract, the contract shall be considered to be concluded for five years.

2. The rules provided for by Article 677(2), Articles 680 and 684–686, and Article 687(2), paragraph four, of the present Code shall not apply to the contract of the rental of a dwelling premise concluded for a term of up to one year (short-term rental) unless provided otherwise by the contract.

Article 684. Preferential Right of Tenant to Conclusion of Contract for New Period

Upon the expiry of the period of a contract of rental of a dwelling premise the tenant shall have a preferential right to conclude a contract of rental of the dwelling premise for a new period.

Not later than three months before the expiry of the period of the contract of rental of a dwelling premise the landlord must propose to the tenant to conclude a contract on the same or other conditions or to warn the tenant about the refusal to prolong the contract in connection with a decision not to rent out the dwelling premise for not less than a year. If the landlord has not fulfilled this duty, and the tenant has not refused to prolong the contract, the contract shall be considered to be prolonged on the same conditions and for the same period.

In the event of agreeing the conditions of the contract the tenant shall not have the right to demand an increase of the number of persons permanently residing with him under the contract of rental of the dwelling premise.

If the landlord has refused to prolong the contract in connection with a decision not to rent out the premise but within a year from the date of the expiry of the period of the contract with the tenant has concluded a contract of rental of the dwelling premise with another person, the tenant shall have the right to demand the deeming of this contract to be invalid and/or compensation of losses caused by the refusal to renew the contract with him.

Article 685. Sub-Rental of Dwelling Premise

1. Under a contract of sub-rental of a dwelling premise the tenant shall with the consent of the landlord transfer for a period part or all of the premise rented by him to the use of the subtenant. The subtenant shall not acquire autonomously the right of use of the dwelling premise. The tenant shall remain responsible to the landlord under the contract of rental of the dwelling premise.

2. The contract of sub-rental of a dwelling premise may be concluded on condition of compliance with the requirements of legislation concerning the norm of dwelling space per person.

3. The contract of sub-rental of a dwelling premise shall be for compensation.

4. The period of the contract of sub-rental of a dwelling premise may not exceed the period of the contract of rental of a dwelling premise.

5. In the event of the termination before time of the contract of rental of a dwelling premise the contract of sub-rental of the dwelling premise shall terminate simultaneously with it.

6. The rules concerning the preferential right to conclude a contract for a new period shall not extend to the contract of sub-rental of a dwelling premise.

Article 686. Replacement of Tenant in Contract of Rental of Dwelling Premise

1. Upon the demand of the tenant and other citizens permanently residing with him and with the consent of the landlord, the tenant in a contract of rental of a dwelling premise may be replaced by one of the citizens who has reached majority and is permanently residing with the tenant.

2. In the event of the death of the tenant or of his departure from the dwelling premise, the contract shall continue to operate on those same conditions and one of the citizens permanently residing with the previous tenant shall become the tenant by general consent among them. If such consent is not reached, all the citizens permanently residing in the dwelling premise shall become co-tenants.

Article 687. Dissolution of Contract of Rental of Dwelling Premise

1. The tenant of a dwelling premise shall have the right with the consent of other citizens permanently residing with him at any time to dissolve the contract of rental with a warning in writing to the landlord three months beforehand.

2. The contract of rental of a dwelling premise may be dissolved in a judicial proceeding upon the demand of the landlord in instances of:
the failure of the tenant to make payment for the dwelling premise for six months unless a longer period has been established by the contract, and in the event of a short-term rental in the event of the failure to make payment more than twice upon the expiry of the period of payment established by the contract;
destruction or spoilage of the dwelling premise by the tenant or other citizens for whose actions he is liable.

By decision of a court the tenant may be granted a period of not more than a year in order to eliminate violations which serve as the grounds for dissolution of a contract of rental of a dwelling premise. If during the period determined by a court the tenant does not eliminate violations permitted or does not take all necessary measures in order to eliminate them, the court upon a second application of the landlord shall adopt a decision concerning dissolution of the contract of rental of a dwelling premise. In so doing at the request of the tenant the court in the decision concerning dissolution of the contract may defer execution of the decision for a period of not more than a year.

3. The contract of rental of a dwelling premise may be dissolved in a judicial proceeding at the demand of any of the parties to the contract:
if the premise ceases to be fit for permanent habitation, and also in the event of the damaged state thereof;
in other instances provided for by housing legislation.

4. If the tenant of a dwelling premise or other citizens for whose actions he is liable uses the dwelling premise not for the designation or systematically violates the rights and interests of neighbours, the landlord may warn the tenant about the need to eliminate the violation.

If the tenant or other citizens for whose actions he is liable continues after a warning to use the dwelling premise not for the designation or to violate the rights and interests of neighbours, the landlord shall have the right in a judicial proceeding to dissolve the contract of rental of the dwelling premise. In this event the rules provided for by point 2, paragraph four, of the present Article shall apply.

Article 688. Consequences of Dissolution of Contract of Rental of Dwelling Premise

In the event of the dissolution of a contract of rental of a dwelling premise the tenant and other citizens residing in the dwelling premise at the moment of dissolution of the contract shall be subject to eviction from the dwelling premise on the grounds of the decision of a court.

Chapter 36. Uncompensated Use

Article 689. Contract of Uncompensated Use

1. Under a contract of uncompensated use (or contract of loan) one party (the lender) is obliged to transfer or transfers a thing for temporary uncompensated use to the other party (loan recipient), and the last is obliged to return this thing in the same state in which it was received, taking into account normal wear and tear, or in a state stipulated by the contract.

2. The rules provided for by Article 607, Article 610(1) and (2), paragraph one, Article 615(1) and (3), Article 621(2), and Article 623(1) and (3) of the present Code shall apply respectively to the contract of uncompensated use.

Article 690. Lender

1. The right to transfer a thing for uncompensated use shall belong to the owner thereof and to other persons empowered to do so by a law or by the owner.

2. A commercial organisation shall not have the right to transfer property for uncompensated use to a person who is the founder, participant, director, or member of the management or control organs thereof.

Article 691. Granting Thing for Uncompensated Use

1. The lender shall be obliged to grant a thing in a state corresponding to the conditions of the contract of uncompensated use and the designation thereof.

2. A thing shall be granted for uncompensated use with all of its appurtenances and documents relating thereto (instruction manuals for use, technical passport, and so forth) unless provided otherwise by the contract.

If such appurtenances and documents were not transferred and without them the thing cannot be used for the designation or its use to a significant degree loses value for the loan recipient, the last shall have the right to require such appurtenances and documents be granted to him or to dissolution of the contract and compensation for real damage incurred by him.

Article 692. Consequences of Failure to Grant Thing for Uncompensated Use

If the lender does not transfer a thing to the loan recipient, the last shall have the right to require dissolution of the contract of uncompensated use and compensation for real damage incurred by him.

Article 693. Responsibility for Defects of Thing Transferred for Uncompensated Use

1. The lender shall be liable for defects of a thing which he intentionally or through gross negligence did not stipulate when concluding the contract of uncompensated use.

In the event of the discovery of such defects the loan recipient shall have the right at his choice to require of the lender the uncompensated elimination of the defects of the thing or compensation of his expenses for elimination of the defects of the thing or dissolution before time of the contract and compensation for real damage incurred by him.

2. The lender notified about the demands of the loan recipient or of his intention to eliminate the defects of the thing at the expense of the lender may without delay replace the thing in disrepair with another analogous thing which is in a proper state.

3. The lender shall not be liable for defects of a thing which were stipulated by him when concluding the contract or were known to the loan recipient beforehand or should have been discovered by the loan recipient during the inspection of the thing or verification of its good repair when concluding the contract or when transferring the thing.

Article 694. Rights of Third Persons to Thing Being Transferred for Uncompensated Use

The transfer of a thing for uncompensated use shall not be grounds for the change or termination of the rights of third persons to this thing.

When concluding a contract of uncompensated use the lender shall be obliged to warn the loan recipient about all the rights of third persons to this thing (servitude, right of pledge, and so forth). The failure to perform this duty shall give the loan recipient the right to demand dissolution of the contract and compensation for real damage incurred by him.

Article 695. Duties of Loan Recipient With Regard to Maintenance of Thing

The loan recipient shall be obliged to maintain the thing received for compensated use in a correct state of repair, including the effectuation of current and capital repair, and bear all expenses for the maintenance thereof unless provided otherwise by the contract of uncompensated use.

Article 696. Risk of Accidental Perishing or Accidental Damaging of Thing

The loan recipient shall bear the risk of accidental perishing or accidental damage of a thing received for uncompensated use if the thing perished or was spoiled in connection with the fact that he used it not in accordance with the contract of uncompensated use or designation of the thing or transferred it to a third person without the consent of the lender. The loan recipient also shall bear the risk of accidental perishing or accidental damage of the thing if, taking into account the actual circumstances, he could prevent the perishing or spoilage thereof by having sacrificed his thing but preferred to preserve his thing.

Article 697. Responsibility for Harm Caused to Third Person as Result of Use of Thing

The lender shall be liable for harm caused to a third person as a result of the use of the thing unless it is proved that the harm was caused as a consequence of the intent or gross negligence of the loan recipient or person from whom this thing is situated with the consent of the lender.

Article 698. Dissolution Before Time of Contract of Uncompensated Use

1. The lender shall have the right to require dissolution before time of a contract of uncompensated use in instances when the loan recipient:
    uses the thing not in accordance with the contract or designation of the thing;
    does not fulfil duties with regard to maintenance of the thing in a proper state of repair or the maintenance thereof;
    materially worsens the state of the thing;
    has transferred the thing without the consent of the lender to a third person.

2. The loan recipient shall have the right to demand the dissolution before time of the contract of uncompensated use:
    in the event of the discovery of defects making the normal use of the thing impossible or encumbered, concerning the existence of which he did not know and could not have known at the moment of concluding the contract;
    if the thing by virtue of circumstances for which he is not liable proves to be in a state not fit for use;
    if when concluding the contract the lender did not warn him about the rights of third persons to the thing being transferred;
    in the event of the failure of the lender to perform the duty to transfer the thing or appurtenances thereof and documents relating to it.

Article 699. Repudiation of Contract of Uncompensated Use

1. Each of the parties shall have the right at any time to repudiate the contract of uncompensated use concluded without specification of the period, having notified the other party thereof one month beforehand unless a different period of notice has been provided for by the contract.

2. Unless provided otherwise by the contract, the loan recipient shall have the right at any time to repudiate the contract concluded with specification of the period in the procedure provided for by point 1 of the present Article.

Article 700. Change of Parties in Contract of Uncompensated Use

1. The lender shall have the right to alienate the thing or transfer it for compensated use to a third person. In so doing to the new owner or user shall pass the rights relating to the previously concluded contract of uncompensated use, and his rights with respect to the thing shall be encumbered by the rights of the loan recipient.

2. In the event of the death of a citizen-lender or the reorganisation or liquidation of a juridical person-lender the rights and duties of the lender under the contract of uncompensated use shall pass to the heir (or legal successor) or to another person to whom the right of ownership has passed to the thing or other right on the basis of which the thing was transferred in uncompensated use.

In the event of the reorganisation of a juridical person-loan recipient its rights and duties under the contract shall pass to the juridical person which is the legal successor thereof unless provided otherwise by the contract.

Article 701. Termination of Contract of Uncompensated Use

A contract of uncompensated use shall terminate in the event of the death of the citizen-loan recipient or liquidation of the juridical person-loan recipient unless provided otherwise by the contract.

Chapter 37. Independent-Work

§1. General Provisions on Independent-Work

Article 702. Contract of Independent-Work

1. Under a contract of independent-work one party (the independent-work contractor) shall be obliged to fulfil according to the planning task of the other party (customer) determined work and to hand over the result thereof to the customer, and the customer shall be obliged to accept the result of the work and to pay for it.

2. The provisions provided for by the present paragraph shall apply to individual types of contract of independent-work (domestic independent-work,

construction independent-work, independent-work for the fulfilment of design and survey work, independent-work for State needs) unless established otherwise by the rules of the present Code concerning these types of contracts.

Article 703. Work to be Fulfilled Under Contract of Independent-Work

1. A contract of independent-work shall be concluded for the manufacture or converting (or processing) of a thing or for the fulfilment of other work with the transfer of its results to the customer.

2. Under a contract of independent-work concluded for the manufacture of a thing the independent-work contractor shall pass the rights thereto to the customer.

3. Unless provided otherwise by the contract, the independent-work contractor autonomously shall determine the means of fulfilment of the planning task of the customer.

Article 704. Fulfilment of Work by the Means of Independent-Work Contractor

1. Unless provided otherwise by the contract of independent-work, the work shall be fulfilled by the means of the independent-work contractor—from his materials and by his forces and means.

2. The independent-work contractor shall bear responsibility for the improper quality of materials and equipment granted by him, and also for granting materials and equipment encumbered by the rights of third persons.

Article 705. Distribution of Risks Among Parties

1. Unless provided otherwise by the present Code, other laws, or by the contract of independent-work:
the risk of accidental perishing or accidental damage of materials, equipment, things transferred for conversion (or processing), or other property used to perform the contract shall be borne by the party granting them;
the risk of accidental perishing or accidental damage of the result of the work fulfilled shall be borne by the independent-work contractor until the acceptance thereof by the customer.

2. In the event of the delay of the transfer or acceptance of the result of the work the risks provided for by point 1 of the present Article shall be borne by the party who permitted the delay.

Article 706. General Independent-Work Contractor and Subindependent-Work Contractor

1. Unless the duty of the independent-work contractor to fulfil the work provided for in the contract personally arises from a law or contract of independent-

work, the independent-work contractor shall have the right to recruit other persons to perform his obligations (subindependent-work contractor). In this event the independent-work contractor shall act in the role of general independent-work contractor.

2. The independent-work contractor who has recruited a subindependent-work contractor to perform the contract of independent-work in violation of the provisions of point 1 of the present Article or the contract shall bear responsibility to the customer for losses caused by the participation of the subindependent-work contractor in performance of the contract.

3. The general independent-work contractor shall bear responsibility to the customer for the consequences of the failure to perform or the improper performance of obligations by the subindependent-work contractor in accordance with the rules of Article 313(1) and Article 403 of the present Code, and to the subindependent-work contractor, responsibility for the failure to perform or the improper performance by the customer of obligations under the contract of independent-work.

Unless provided otherwise by a law or contract, the customer and the subindependent-work contractor shall not have the right to present to one another demands connected with a violation of the contracts concluded by each of them with the general independent-work contractor.

4. With the consent of the general independent-work contractor the customer shall have the right to conclude contracts for the fulfilment of individual work with other persons. In this event the said persons shall bear responsibility for the failure to perform or the improper performance of work directly to the customer.

Article 707. Participation in Performance of Work of Several Persons

1. If two persons or more act on the side of the independent-work contractor, in the event of the indivisibility of the subject of the obligation they shall be deemed with respect to the customer to be joint and several debtors and respectively joint and several creditors.

2. In the event of the divisibility of the subject of the obligation, and also in other instances provided for by a law, other legal acts, or the contract, each of the persons specified in point 1 of the present Article shall acquire the rights and bear the duties with respect to the customer within the limits of his participatory share (Article 321).

Article 708. Periods for Fulfilment of Work

1. The beginning and ending periods of fulfilment of the work shall be specified in the contract of independent-work. By agreement between the parties

periods for completing individual stages of work (intermediate periods) also may be provided for in the contract.

Unless established otherwise by a law, other legal acts, or provided for by the contract, the independent-work contractor shall bear responsibility for a violation of both the beginning and ending, and also the intermediate, periods for the fulfilment of work.

2. The beginning, ending, and intermediate periods for the fulfilment of work specified in the contract of independent-work may be changed in the instances and in the procedure provided for by the contract.

3. The consequences of delay of performance specified in Article 405(2) of the present Code shall ensue in the event of a violation of the ending of the period for the fulfilment of the work, and also other periods established by the independent-work contract [as amended 17 December 1999. СЗ РФ (1999), no. 51, item 6288].

Article 709. Price of Work

1. In the contract of independent-work the price of the work subject to fulfilment and the means of determining it shall be specified. In the absence of such indications in the contract the price shall be determined in accordance with Article 424(3) of the present Code.

2. The price in the contract of independent-work shall include contributory compensation for the costs of the independent-work contractor and the remuneration due to it.

3. The price of the work may be determined by means of drawing up an estimate.

In instances when work is fulfilled in accordance with the estimate drawn up by the independent-work contractor, the estimate shall acquire force and become part of the contract of independent-work from the moment of the confirmation thereof by the customer.

4. The price of the work (or estimate) may be approximate or firm. In the absence of other indications in the contract of independent-work the price of the work shall be considered to be firm.

5. If the necessity arises for carrying out additional work and for this reason the price of the work determined approximately materially increases, the independent-work contractor shall be obliged to warn the customer thereof in good time. The customer who has not agreed to the increase of the price of the work specified in the contract of independent-work shall have the right to repudiate the contract. In this event the independent-work contractor may demand payment from the customer to him of the price for the part of the work fulfilled.

An independent-work contractor who has not warned the customer in good time about the necessity of increasing the price of the work specified in the contract shall be obliged to fulfil the contract, retaining the right to payment for the work at the price determined in the contract.

6. An independent-work contractor shall not have the right to demand an increase of a firm price, nor the customer to reduce it, including in an instance when at the moment of concluding the contract of independent-work the possibility to provide for the full amount of work subject to fulfilment or the expenses necessary for this was excluded.

In the event of the material growth of the cost of materials and equipment granted by the independent-work contractor, and also of services rendered to him by third persons which could not be provided for when concluding the contract, the independent-work contractor shall have the right to demand an increase of the established price, and in the event of the refusal of the customer to fulfil this demand, dissolution of the contract in accordance with Article 451 of the present Code.

Article 710. Economies of Independent-Work Contractor

1. In instances when the actual expenses of the independent-work contractor prove to be less than those which were taken into account when determining the price of the work, the independent-work contractor shall retain the right to payment for the work at the price provided for by the contract of independent-work unless the customer proves that the economies received by the independent-work contractor influenced the quality of the work fulfilled.

2. A distribution of the economies received by the independent-work contractor between the parties may be provided for in the contract of independent-work.

Article 711. Procedure for Payment of Work

1. Unless preliminary payment for work fulfilled or individual stages thereof has been provided for by the contract of independent-work, the customer shall be obliged to pay the independent-work contractor the stipulated price after the final handing over of the results of the work, on condition that the work is properly fulfilled and within the agreed period, or with the consent of the customer before time.

2. The independent-work contractor shall have the right to demand the payment to him of an advance or a deposit only in the instances and in the amount specified in a law or in the contract of independent-work.

Article 712. Right of Independent-Work Contractor to Withholding

In the event of the failure of the customer to perform the duty to pay the established price or other amount due to the independent-work contractor in connection with the fulfilment of the contract of independent-work, the independent-work contractor shall have the right to withholding in accordance with Articles 359 and 360 of the present Code the result of the work, and also the equipment belonging to the customer transferred for converting (or processing) the thing, the balance of the unused material, and other property of the customer which proves to be with him until the payment by the customer of the respective amounts.

Article 713. Fulfilment of Work with Use of Material of Customer

1. The independent-work contractor shall be obliged to use the material granted by the customer economically and carefully, after the ending of the work to submit a report to the customer concerning the expenditure of the material, and also to return the balance thereof or with the consent of the customer to reduce the price of the work by taking into account the cost of the unused material remaining with the independent-work contractor.

2. If the result of the work was not achieved or the result achieved turned out to have defects which make it unfit for the use provided for in the contract of independent-work, and in the absence in the contract of a respective condition of unfitness for ordinary use for reasons caused by the defects of the material granted by the customer, the independent-work contractor shall have the right to require payment for the work fulfilled by him.

3. The independent-work contractor may effectuate the right specified in point 2 of the present Article if it is proved that the defects of the material could not be discovered when the independent-work contractor properly received this material.

Article 714. Responsibility of Independent-Work Contractor for Non-Preservation of Property Granted by Customer

The independent-work contractor shall bear responsibility for the non-preservation of material, equipment, the thing transferred for conversion (or processing), or other property granted by the customer which turned out to be in the possession of the independent-work contractor in connection with the performance of the contract of independent-work.

Article 715. Rights of Customer During Fulfilment of Work by Independent-Work Contractor

1. The customer shall have the right at any time to verify the course and quality of the work to be fulfilled by the independent-work contractor without interfering in his activity.

2. If the independent-work contractor does not embark upon the timely performance of the contract of independent-work or fulfills the work so slowly that

the ending thereof within the period clearly becomes impossible, the customer shall have the right to repudiate performance of the contract and to require compensation of losses.

3. If during the fulfilment of work it becomes evident that it will not be properly fulfilled, the customer shall have the right to designate a reasonable period to the independent-work contractor for elimination of the defects and in the event of the failure to perform this demand by the independent-work contractor within the designated period to repudiate the contract of independent-work or to commission rectification of the work to another person at the expense of the independent-work contractor, and also to require compensation of losses.

Article 716. Circumstances of Which Independent-Work Contractor is Obliged to Warn Customer

1. The independent-work contractor shall be obliged immediately to warn the customer and until receiving instructions from him to suspend the work in the event of discovering:

the unfitness or poor quality of the material, equipment, technical documentation, or thing transferred for conversion (or processing) granted by the customer;

possible consequences unfavourable for the customer of the fulfilment of his instructions concerning the means of performance of the work;

other circumstances beyond the control of the independent-work contractor which threaten the fitness or stability of the results of the work to be fulfilled or create the impossibility of completing it within the period.

2. An independent-work contractor who has not warned the customer about the circumstances specified in point 1 of the present Article or who has continued the work without awaiting the expiry of the period specified in the contract, and in the absence thereof of a reasonable period for a reply to the warning or notwithstanding the timely instruction of the customer concerning termination of the work, shall not have the right to refer to the said circumstances in the event of the presentation to him or by him to the customer of respective demands.

3. If the customer, notwithstanding a timely and substantiated warning on the part of the independent-work contractor concerning the circumstances specified in point 1 of the present Article, within a reasonable period does not replace unfit or poor-quality material, equipment, technical documentation, or thing transferred for conversion (or processing), does not change instructions concerning the means of fulfilling the work, or does not take other necessary measures in order to eliminate the circumstances threatening the fitness thereof, the independent-work contractor shall have the right to refuse performance of the contract of independent-work and to require compensation of losses caused by its termination.

Article 717. Refusal of Customer of Performance of Contract of Independent-Work

Unless provided otherwise by the contract of independent-work, the customer may at any time before the handing over to him of the result of the work refuse to perform the contract, having paid the independent-work contractor the part of the established price in proportion to the part of the work fulfilled before receiving the notification concerning the refusal of the customer of performance of the contract. The customer also shall be obliged to compensate the independent-work contractor for losses caused by the termination of the contract of independent-work within the limits of the difference between the price determined for all of the work and the part of the price paid for the work fulfilled.

Article 718. Assistance of Customer

1. The customer shall be obliged in the instances, in the amount, and in the procedure provided for by the contract of independent-work to render assistance to the independent-work contractor in the fulfilment of the work.

In the event of the failure of the customer to perform this duty the independent-work contractor shall have the right to demand compensation of losses caused, including additional costs caused by idle-time, or to carry over the periods for performance of the work, or to increase the price of the work specified in the contract.

2. In instances when performance of the work under the contract of independent-work has become impossible as a consequence of actions or omissions of the customer, the independent-work contractor shall retain the right to payment to him of the price specified in the contract taking into account the part of the work fulfilled.

Article 719. Failure of Customer to Perform Counter Duties Under Contract of Independent-Work

1. The independent-work contractor shall have the right not to embark upon the work, and to suspend work begun, in instances when a violation by the customer of his duties under the contract of independent-work, in particular the failure to grant material, equipment, technical documentation, or proper conversion (or processing) of the thing prevents the performance of the contract by the independent-work contractor, and also when there are circumstances obviously testifying to the fact that performance of the said duties will not be made within the established period (Article 328).

2. Unless provided otherwise by the contract of independent-work, the independent-work contractor shall have the right, when the circumstances speci-

fied in point 1 of the present Article exist, to refuse performance of the contract and to require compensation of losses.

Article 720. Acceptance by Customer of Work Fulfilled by Independent-Work Contractor

1. The customer shall be obliged within the periods and in the procedure which are provided for by the contract of independent-work to inspect and accept, with the participation of the independent-work contractor, the work fulfilled (or result thereof), and in the event of discovering deviations from the contract which worsen the result of the work or other defects in the work, to immediately declare this to the independent-work contractor.

2. The customer who has discovered defects in the work when receiving it shall have the right to refer to them in instances when in the act or in the other document certifying acceptance these defects or the possibility of subsequent presentation of a demand concerning their elimination were stipulated.

3. Unless provided otherwise by the contract of independent-work, the customer who has accepted work without verification shall be deprived of the right to refer to defects of the work which could have been established by ordinary means of the acceptance thereof (obvious defects).

4. The customer who has discovered after acceptance of the work deviations therein from the contract of independent-work or other defects which could not be established by ordinary means of acceptance (latent defects), including those which were intentionally concealed by the independent-work contractor, shall be obliged to notify the independent-work contractor thereof within a reasonable period with regard to the discovery thereof.

5. In the event a dispute arises between the customer and the independent-work contractor on the occasion of defects of the work fulfilled or the reasons therefor, an expert examination must be designated at the demand of any of the parties. The expenses with regard to the expert examination shall be borne by the independent-work contractor except for instances when the expert examination established the absence of violations by the independent-work contractor of the contract of independent-work or a causal link between the actions of the independent-work contractor and the defects discovered. In the said instances the expenses for the expert examination shall be borne by the party which required the designation of the expert examination, and if it was designated by agreement between the parties, both parties equally.

6. Unless provided otherwise by the contract of independent-work, in the event the customer evades accepting the work fulfilled, the independent-work contractor shall have the right upon the expiry of a month from the date when according to the contract the result of the work should have been transferred to

the customer, and on condition of a subsequent warning twice of the customer, to sell the result of the work, and to deposit the amount received less all payments due to the independent-work contractor in the name of the customer in the procedure provided for by Article 327 of the present Code.

7. If the evasion of the customer of acceptance of the work fulfilled entailed delay in handing over the work, the risk of accidental perishing of the manufactured (or converted or processed) thing shall be deemed to have passed to the customer at the moment when the transfer of the thing should have happened.

Article 721. Quality of Work

1. The quality of work fulfilled by the independent-work contractor must correspond to the conditions of the contract of independent-work, and in the absence or incompleteness of the conditions of the contract, to the requirements ordinarily presented for work of the respective nature. Unless provided otherwise by a law, other legal acts, or contract, the result of work fulfilled must at the moment of transfer to the customer possess the properties specified in the contract or determined requirements usually presented, and within the limits of a reasonable period be fit for the use established by the contract, and if such use is not provided for by the contract, for ordinary use of the result of work of this nature.

2. If obligatory requirements for work to be fulfilled under a contract of independent-work have been provided for by a law, other legal acts, or in the procedure established by them, the independent-work contractor acting as an entrepreneur shall be obliged to fulfil the work while complying with these obligatory requirements.

The independent-work contractor may accept the duty to fulfil work under the contract which meets the requirements for quality that are higher in comparison with the established requirements which are obligatory for the parties.

Article 722. Guarantee of Quality of Work

1. In an instance when a guarantee period is provided for the result of work by a law, other legal acts, the contract of independent-work, or the customs of business turnover, the result of the work must correspond during the entire guarantee period to the conditions of the contract concerning quality (Article 721[1]).

2. The guarantee of quality of the result of work, unless provided otherwise by the contract of independent-work, shall extend to everything comprising the result of the work.

Article 723. Responsibility of Independent-Work Contractor for Improper Quality of Work

1. In instances when work is fulfilled by an independent-work contractor with deviations from the contract of independent-work which worsen the result of the

work, or with other defects which make it unfit for the use provided for in the contract or in the event of the absence in the contract of a corresponding condition of unfitness for ordinary use, the customer shall have the right, unless established otherwise by a law or the contract, at his choice to require from the independent-work contractor:

elimination of the defects without compensation within a reasonable period;

commensurate reduction of the price established for the work;

compensation of his expenses for elimination of the defects when the right of the customer to eliminate them has been provided for in the contract of independent-work (Article 397).

2. The independent-work contractor shall have the right instead of eliminating the defects for which he is liable to fulfil the work without compensation anew with compensation to the customer of losses caused by the delay of performance. In this event the customer shall be obliged to return the result of the work previously transferred to him to the independent-work contractor if by the character of the work such return is possible.

3. If deviations in the work from the conditions of the contract of independent-work or other defects of the result of the work are not eliminated within a reasonable period established by the customer or are material and ineradicable, the customer shall have the right to refuse performance of the contract and to require compensation of losses caused.

4. The condition of the contract of independent-work concerning relieving the independent-work contractor of responsibility for determined defects shall not relieve him from responsibility if it is proved that such defects arose as a consequence of guilty actions or the failure to act of the independent-work contractor.

5. The independent-work contractor who has granted material to fulfil work shall be liable for its quality according to the rules concerning responsibility of the seller for goods of improper quality (Article 475).

Article 724. Periods of Discovery of Improper Quality of Result of Work

1. Unless established otherwise by a law or the contract of independent-work, the customer shall have the right to present demands connected with the improper quality of the result of work on condition that it is elicited within the periods established by the present Article.

2. In an instance when a guarantee period is not established for the result of the work, demands connected with defects of the result of the work may be presented by the customer on condition that they were discovered within a reasonable period, but within the limits of two years from the date of transfer of the result of the work, unless other periods have been established by a law, the contract, or the customs of business turnover.

3. The customer shall have the right to present demands connected with defects of the result of the work discovered within the guarantee period.

4. In an instance when the guarantee period provided for by the contract constitutes less than two years and the defects of the result of the work are discovered by the customer upon the expiry of the guarantee period but within the limits of two years from the moment provided for by point 5 of the present Article, the independent-work contractor shall bear responsibility if the customer proves that the defects arose before the transfer of the result of the work to the customer or for reasons which arose before that moment.

5. Unless provided otherwise by the contract of independent-work, the guarantee period (Article 722[1]) shall commence to run from the moment when the result of the work fulfilled was accepted or should have been accepted by the customer.

6. The rules contained in Article 471(2) and (4) of the present Code respectively shall apply to calculating the guarantee period under the contract of independent-work unless provided otherwise by a law, other legal acts, agreement of the parties, or does not arise from the peculiarities of the contract of independent-work.

Article 725. Limitation With Regard to Suits Concerning Improper Quality of Work

1. The period of limitations for demands presented in connection with improper quality of work to be fulfilled under the contract of independent-work shall constitute one year, and with respect to buildings and installations shall be determined according to the rules of Article 196 of the present Code.

2. If in accordance with the contract of independent-work the result of the work was accepted by the customer in parts, the running of the period of limitations shall commence from the date of acceptance of the result of the work as a whole.

3. If a guarantee period has been established by a law, other legal acts, or the contract of independent-work and the statement with regard to defects of the result of the work was made within the limits of the guarantee period, the running of the period of limitations specified in point 1 of the present Article shall commence from the date of the statement concerning the defects.

Article 726. Duty of Independent-Work Contractor to Transfer Information to Customer

The independent-work contractor shall be obliged to transfer to the customer together with the result of the work information affecting the operation or other use of the subject of the contract of independent-work if this is provided for by the contract or the character of the information such that without it the use of the result of the work is impossible for the purposes specified in the contract.

Article 727. Confidentiality of Information Received by Parties

If a party thanks to the performance of his obligation under the contract of independent-work has received information from the other party concerning new solutions and technical knowledge, including not protected by a law, and also information which might be considered as a commercial secret (Article 139), the party who has received such information shall not have the right to communicate it to third persons without the consent of the other party.

The procedure and conditions for the use of such information shall be determined by agreement of the parties.

Article 728. Return by Independent-Work Contractor of Property Transferred by Customer

In instances when the customer on the basis of Article 715(2) or Article 723(3) of the present Code dissolves a contract of independent-work, the independent-work contractor shall be obliged to return the materials, equipment, thing transferred for conversion (or processing), and other property granted by the customer or to transfer them to the person specified by the customer, and if this proves to be impossible, to compensate the cost of the materials, equipment, and other property.

Article 729. Consequences of Termination of Contract of Independent-Work Before Acceptance of Result of Work

In the event of the termination of a contract of independent-work on the grounds provided for by a law or the contract before the acceptance by the customer of the result of the work to be fulfilled by the independent-work contractor (Article 720[1]), the customer shall have the right to demand the transfer to him of the result of the uncompleted work with contributory compensation to the independent-work contractor of expenditures made.

§2. Domestic Independent-Work

Article 730. Contract of Domestic Independent-Work

1. Under a contract of domestic independent-work the independent-work contractor effectuating the respective entrepreneurial activity shall be obliged to fulfil according to the planning task of a citizen (customer) the work determined which is intended to satisfy the domestic or other personal requirements of the customer, and the customer shall be obliged to accept and to pay for the work.

2. The contract of domestic independent-work shall be a public contract (Article 426).

3. The laws on defence of the rights of consumers and other legal acts adopted in accordance with them shall apply to relations under the contract of domestic independent-work not regulated by the present Code.

Article 731. Guarantees of Rights of Customer

1. An independent-work contractor shall not have the right to bind the customer to include additional work or services in the contract of domestic independent-work. The customer shall have the right to refuse payment for work or services not provided for by the contract.

2. The customer shall have the right at any time before the work is handed over to him to refuse performance of the contract of domestic independent-work, having paid to the independent-work contractor the part of the established price in proportion to the part of the work fulfilled before informing of the refusal of performance of the contract and having compensated the independent-work contractor for expenses made until this moment for the purposes of performance of the contract unless they are within the said part of the price of the work. Conditions of a contract depriving the customer of this right shall be void.

Article 732. Granting Information to Customer About Proposed Work

1. The independent-work contractor shall be obliged before conclusion of the contract of domestic independent-work to grant to the customer necessary and reliable information concerning the proposed work, the types thereof, and peculiarities, the price and form of payment, and also notify the customer at his request of other information relating to the contract and respective work. If by the character of the work this has significance, the independent-work contractor must indicate to the customer the specific person who will fulfil it.

2. If a customer has not been granted the possibility to receive immediately at the place of conclusion of a contract of domestic independent-work information concerning the work specified in point 1 of the present Article, he shall have the right to demand from the independent-work contractor losses caused by an unsubstantiated evasion of conclusion of the contract (Article 445[4]) [as amended 17 December 1999. СЗ РФ (1999), no. 51, item 6288].

The customer shall have the right to demand dissolution of a concluded contract of domestic independent-work without payment for the work fulfilled, and also compensation of losses, in instances when as a consequence of the incompleteness or unreliability of information received from the independnent-work contractor the contract was concluded for the fulfilment work not possessing the properties which the customer had in view.

An independent-work contractor who had not granted for the customer information concerning the work specified in point 1 of the present Article shall bear responsibility also for those defects of work which arose after the transfer thereof to the customer as a consequence of him lacking such information [added by Law of 17 December 1999. СЗ РФ (1999), no. 51, item 6288].

Article 733. Fulfilment of Work from Material of Independent-Work Contractor

1. If work under the contract of domestic independent-work is fulfilled from material of the independent-work contractor, the material shall be paid for by the customer when concluding the contract wholly or in the part specified in the contract, with the final settlement of accounts when the work fulfilled by the independent-work contractor is received by the customer.

In accordance with the contract the material may be granted by the independent-work contractor on credit, including with a condition of payment for the material by the customer on instalment.

2. A change of price of the material granted by the independent-work contractor after the conclusion of a contract of domestic independent-work shall not entail a resettlement of accounts.

Article 734. Fulfilment of Work from Material of Customer

If work under the contract of domestic independent-work is fulfilled from the material of the customer, the precise name, description, and price of the material determined by agreement of the parties must be specified on the receipt or other document issued by the independent-work contractor to the customer when concluding the contract. A valuation of the material on the receipt or other analogous document may be subsequently contested by the customer in a court.

Article 735. Price and Payment for Work

The price of work in a contract of domestic independent-work shall be determined by agreement of the parties and may not be higher than that established or regulated by respective State agencies. The work shall be paid for by the customer after the final handing over thereof by the independent-work contractor. With the consent of the customer the work may be paid for by him when concluding the contract in full or by means of issuance of an advance.

Article 736. Warning of Customer About Conditions of Use of Work Fulfilled

When handing work over to the customer, the independent-work contractor shall be obliged to notify him about the requirements which it is necessary to comply with for the effective and safe use of the result of the work, and also possible consequences for the customer himself and other persons of the failure to comply with the respective requirements.

Article 737. Consequences of Discovery of Defects in Work Fulfilled

1. In the event of the discovery of defects during the acceptance of the result of the work or after the acceptance thereof within the guarantee period, and if it was not established, a reasonable period but not later than two years (for immoveable property, five years) from the day of acceptance of the result of the work, the customer shall have the right at his choice to effectuate one of the rights provided for in Article 723 of the present Code or to require fulfilment of the work a second time without compensation or compensation of expenses incurred by him for rectification of the defects by his own means or by third persons [as amended 17 December 1999. СЗ РФ (1999), no. 51, item 6288].

2. In the event of the discovery of material defects of the result of the work the customer shall have the right to present a demand to the independent-work contractor concerning the elimination of such defects without compensation if it is proved that they arose before acceptance of the result of the work by the customer or for reasons which arose before that moment. This demand may be presented by the customer if the said defects were discovered upon the expiry of two years (or for immoveable property, five years) from the day of acceptance of the result of the work by the customer but within the limits of the service period established for the result of the work or within ten years from the day of acceptance of the result of the work by the customer if a service period has not been established [as amended 17 December 1999. СЗ РФ (1999), no. 51, item 6288].

3. In the event of the failure of the independent-work contractor to fulfil the demand specified in point 2 of the present Article, the customer shall have the right within this same period to require either the return of part of the price paid for the work or compensation of expenses incurred in connection with the elimination of defects by the customer by his forces or with the assistance of third persons, or to refuse performance of the contract and demand compensation of losses caused [as amended 17 December 1999. СЗ РФ (1999), no. 51, item 6288].

Article 738. Consequences of Failure of Customer to Appear for Receipt of Result of Work

In the event of the failure of the customer to appear for the receipt of the result of the work fulfilled or other evasion of the customer from acceptance thereof, the independent-work contractor shall have the right, having warned the customer in writing, upon the expiry of two months from the date of such warning to sell the result of the work for a reasonable price and to place the amount received, less all payments due to the independent-work contractor, on deposit in the procedure provided for by Article 327 of the present Code.

Article 739. Rights of Customer in Event of Improper Fulfilment or Failure to Fulfil Work Under Contract of Domestic Independent-Work

In the event of the improper fulfilment or the failure to fulfil work under a contract of domestic independent-work, the customer may take advantage of the rights granted to the purchaser in accordance with Articles 503–505 of the present Code.

§3. Construction Independent-Work

Article 740. Contract of Construction Independent-Work

1. Under a contract of construction independent-work the independent-work contractor shall be obliged within the period established by the contract to build a determined object according to the planning task of the customer or to fulfil other construction work, and the customer shall be obliged to create necessary conditions for the independent-work contractor in order to fulfil the work, accept the result thereof, and to pay the stipulated price.

2. The contract of construction independent-work shall be concluded for the construction or reconstruction of an enterprise, building (including dwelling house), installation, or other object, and also for the fulfilment of assembly, launching, and other work inextricably connected with the object being built. The rules concerning the contract of construction independent-work shall also apply to work relating to the capital repair of buildings and installations unless provided otherwise by the contract.

In the instances provided for by the contract, the independent-work contract shall assume the duty to ensure the operation of the object after its acceptance by the customer during the period specified in the contract.

3. In instances when under the contract of construction independent-work work is fulfilled in order to satisfy domestic or other personal requirements of a citizen (or customer), the rules of paragraph 2 of the present Chapter concerning the rights of the customer under the contract of domestic independent-work shall apply respectively to such contract.

Article 741. Distribution of Risk Between Parties

1. The risk of accidental perishing or accidental damaging of the object of construction comprising the subject of a contract of construction independent-work before the acceptance of this object by the customer shall be borne by the independent-work contractor.

2. If the object of construction before acceptance by the customer has perished or been damaged as a consequence of the poor-quality of the material granted by the customer (or parts, construction design) or equipment or the performance of

erroneous instructions of the customer, the independent-work contractor shall have the right to demand the payment of the entire cost of the work provided for by the estimate on condition that the duties provided for by Article 716(1) of the present Code were fulfilled by him.

Article 742. Insurance of Object of Construction

1. The duty of the party on whom the risk of accidental perishing or accidental damaging of the object of construction, material, equipment, and other property being used in the construction, or the responsibility for the causing of harm to other persons when effectuating the construction, to insure the respective risks may be provided for by the contract of construction independent-work.

The party on whom the duty is placed with regard to insurance must grant to the other party evidence of the conclusion by it of the contract of insurance on the conditions provided for by the contract of construction independent-work, including data concerning the insurer, the amount of the insured amount, and the risks insured.

2. Insurance shall not relieve the respective party from the duty to take necessary measures in order to prevent the ensuing of the insured event.

Article 743. Technical Documentation and Estimate

1. The independent-work contractor shall be obliged to effectuate construction and work connected therewith in accordance with the technical documentation determining the amount and content of the work and other requirements presented to him and with the estimate determining the price of the work.

In the absence of other instructions in the contract of construction independent-work it shall be presupposed that the independent-work contractor is obliged to fulfil all the work specified in the technical documentation and in the estimate.

2. The composition and content of the technical documentation must be determined by the contract of construction independent-work, and it also must be provided which of the parties and within what period must grant the respective documentation.

3. An independent-work contractor who has discovered in the course of construction work which is not clarified in the technical documentation and in connection therewith the need to conduct additional work and increase the estimate cost of the construction shall be obliged to inform the customer thereof.

In the event of the failure to receive a reply from the customer to his communication within ten days, unless another period has been provided by a law or by the contract of construction independent-work, the independent-work contractor shall be obliged to suspend the respective work with relegation of the losses caused by the idleness to the account of the customer. The customer shall be relieved from compensation of such losses if the absence of the necessity to conduct the additional work is proved.

4. An independent-work contractor who has not fulfilled the duties established by point 3 of the present Article shall be deprived of the right to demand payment from the customer for the additional work fulfilled by him and compensation of losses caused thereby unless it is proved that the necessity for immediate actions in the interests of the customer, in particular in connection with the fact that the suspension of work could lead to the perishing or damaging of the object of construction.

5. In the event of the consent of the customer to carry on and pay for the additional work, the independent-work contractor shall have the right to refuse the fulfilment thereof only in instances when such is not within the sphere of the professional activity of the independent-work contractor or may not be fulfilled by the independent-work contractor for reasons beyond his control.

Article 744. Making Changes in Technical Documentation

1. The customer shall have the right to make changes in the technical documentation on condition that the additional work caused by this does not exceed in cost 10% of the total cost of the construction specified in the estimate and does not change the character of the work provided for in the contract of construction independent-work.

2. The making of changes in the technical documentation greater than those specified in point 1 of the present Article shall be effectuated on the basis of an additional estimate agreed by the parties.

3. The independent-work contractor shall have the right to demand in accordance with Article 450 of the present Code a revision of the estimate if for reasons beyond his control the cost of the work exceeded the estimate by not less than 10%.

4. The independent-work contractor shall have the right to demand compensation of reasonable expenses which were incurred by him in connection with the establishment and elimination of defects in the technical documentation.

Article 745. Provision of Construction with Materials and Equipment

1. The duty relating to provision of the construction with materials, including parts and construction designs or equipment, shall be borne by the independent-work contractor unless it is provided by the contract of construction independent-work that the provision of construction as a whole or in determined part shall be effectuated by the customer.

2. The party within those duty is the provision of the construction shall bear responsibility for the impossibility being discovered of using the materials or the

equipment granted by it without worsening the quality of the work to be fulfilled unless it is proved that the impossibility of the use arose through circumstances for which the other party is liable.

3. In the event of the impossibility being discovered of the use of the materials or equipment granted by the customer without worsening the quality of the work to be fulfilled and of the refusal of the customer to replace them, the independent-work contractor shall have the right to repudiate the contract of construction independent-work and to require from the customer the payment of the price of the contract in proportion to the fulfilment of part of the work.

Article 746. Payment for Work

1. Payment for the work fulfilled by the independent-work contractor shall be made by the customer in the amount provided for by the estimate, within the periods, and in the procedure which has been established by a law or by the contract of construction independent-work. In the absence of respective instructions in a law or in the contract the payment for work shall be made in accordance with Article 711 of the present Code.

2. Payment for work at one time or in full after acceptance of the object by the customer may be provided for by the contract of construction independent-work.

Article 747. Additional Duties of Customer Under Contract of Construction Independent-Work

1. The customer shall be obliged in a timely way to grant a land plot for construction. The expanse and state of the land plot granted must correspond to the conditions contained in the contract of construction independent-work, and in the absence of such conditions, ensure the timely commencement of the work, normal conducting thereof, and completion within the period.

2. The customer shall be obliged in the instances and procedure provided for by the contract of construction independent-work to transfer for use to the independent-work contractor buildings and installations necessary in order to effectuate the work, ensure transportation of goods to his address, the temporary laying of the network for electric power supply, water, and steam and to render other services.

3. Payment for services granted by the customer specified in point 2 of the present Article shall be effectuated in the instances and on the conditions provided for by the contract of construction independent-work.

Article 748. Control and Supervision of Customer Over Fulfilment of Work Under Contract of Construction Independent-Work

1. The customer shall have the right to effectuate control and supervision over the course and quality of work being fulfilled, compliance with the periods

thereof for the fulfilment (or schedule), the quality of materials granted by the independent-work contractor, and also the correctness of the use of the materials of the customer by the independent-work contractor without interfering in so doing in the operational-economic activity of the independent-work contractor.

2. The customer who has discovered when effectuating control and supervision over the fulfilment of work deviations from the conditions of the contract of construction independent-work which may worsen the quality of work or other defects thereof shall be obliged immediately to state this to the independent-work contractor. The customer who has not made such statement shall lose the right thereafter to refer to the defects discovered by him.

3. The independent-work contractor shall be obliged to perform instructions of the customer received in the course of the construction if such instructions are not contrary to the conditions of the contract of construction independent-work and do not represent interference in the operational-economic activity of the independent-work contractor.

4. The independent-work contractor who has improperly fulfilled work shall not have the right to refer to the fact that the customer did not effectuate control and supervision over the fulfilment thereof except for instances when the duty to effectuate such control and supervision is placed on the customer by a law.

Article 749. Participation of Engineer (or Engineering Organisation) in Effectuation of Rights and Fulfilment of Duties of Customer

The customer for the purpose of effectuating control and supervision over construction and the taking of decisions in his name in mutual relations with the independent-work contractor may conclude autonomously without the consent of the independent-work contractor a contract concerning the rendering of services of such type to the customer with a respective engineer (or engineering organisation). In this event the functions of such engineer (or engineering organisation) connected with the consequences of his actions for the independent-work contractor shall be determined in the contract of construction independent-work.

Article 750. Cooperation of Parties in Contract of Construction Independent-Work

1. If when fulfilling construction and work connected therewith obstacles are discovered to the proper performance of the contract of construction independent-work, each of the parties shall be obliged to take all reasonable measures within its control to eliminate such obstacles. The party who does not perform this duty shall lose the right to compensation of losses caused by the fact that the respective obstacles were not eliminated.

2. The expenses of the party connected with the performance of the duties specified in point 1 of the present Article shall be subject to compensation by the other party in the instances when this is provided for by the contract of construction independent-work.

Article 751. Duties of Independent-Work Contractor Relating to Protection of Environment and Ensuring Safety of Construction Work

1. The independent-work contractor when effectuating construction and work connected therewith shall be obliged to comply with the requirements of a law and other legal acts concerning protection of the environment and the safety of construction work.

The independent-work contractor shall bear responsibility for a violation of the said requirements.

2. The independent-work contractor shall not have the right to use materials and equipment in the course of effectuating work provided by the customer or to fulfil his instructions if this may lead to a violation of requirements for the protection of the environment and safety of construction work which are obligatory for the parties.

Article 752. Consequences of Shutting Down of Construction

If for reasons beyond the control of the parties the work under the contract of construction independent-work is suspended and the object of construction shut down, the customer shall be obliged to pay the independent-work contractor in full for the fulfilment of work up to the moment of the shutting down, and also to compensate expenses caused by the necessity to terminate work and shut down the construction, setting off advantages which the independent-work contractor received or could receive as a consequence of the termination of the work.

Article 753. Handing Over and Acceptance of Work

1. The customer who has received the communication of an independent-work contractor concerning the readiness for handing over the result of the work fulfilled under the contract of construction independent-work shall be obliged or, if this is provided for by the contract, stage of work fulfilled, immediately to embark upon the acceptance thereof.

2. The customer shall organise and effectuate the acceptance of the result of work at his expense unless provided otherwise by the contract of construction independent-work.

In the instances provided for by a law or other legal acts the representatives of State agencies and agencies of local self-government must participate in the acceptance of the result of the work.

3. The customer, having accepted the result of an individual stage of work in advance, shall bear the risk of the consequences of the perishing or damaging of the result of the work which occurred not through the fault of the independent-work contractor.

4. The handing over of the result of the work by the independent-work contractor and its acceptance by the customer shall be formalised by an act signed by both parties. In the event of the refusal of one of the parties to sign the act, a notation shall be made thereon concerning this and the act shall be signed by the other party.

A unilateral act of handing over and acceptance of the result of work may be deemed by a court to be invalid only if the reasons for the refusal to sign the act are deemed by it to be substantiated.

5. In instances when this has been provided for by a law or contract of construction independent-work or arises from the character of the work to be fulfilled under the contract, preliminary tests must precede acceptance of the result of the work. In these instances acceptance may be effectuated only in the event of a positive result of the preliminary tests.

6. The customer shall have the right to refuse acceptance of the result of the work in the event of the discovery of defects which exclude the possibility of using it for the purpose specified in the contract of construction independent-work and cannot be eliminated by the independent-work contractor or by the customer.

Article 754. Responsibility of Independent-Work Contractor for Quality of Work

1. The independent-work contractor shall bear responsibility to the customer for deviations permitted from the requirements provided for in the technical documentation and construction norms and rules obligatory for the parties, and also for the failure to achieve the indicators of the object of construction specified in the technical documentation, including those such as the production capacity of the enterprise.

In the event of the reconstruction (renewal, restructuring, restoration, and so forth) of a building or installation, the responsibility shall be placed on the independent-work contractor for a reduction or the loss of firmness, stability, or reliability of the building, installation, or part thereof.

2. The independent-work contractor shall not bear responsibility for petty deviations permitted by him without the consent of the customer from the technical documentation if it is proved that they did not influence the quality of the object of construction.

Article 755. Guarantees of Quality in Contract of Construction Independent-Work

1. The independent-work contractor, unless provided otherwise by the contract of construction independent-work, shall guarantee the achievement by the object of construction of the indicators specified in the technical documentation and the possibility of operation of the object in accordance with the contract of construction independent-work throughout the extent of the guarantee period. The guarantee period established by a law may be increased by agreement of the parties.

2. The independent-work contractor shall bear responsibility for defects discovered within the limits of the guarantee period unless it is proved that they occurred as a consequence of normal wear and tear of the object or parts thereof, the incorrect operation thereof or the incorrectness of the instructions for its operation worked out by the customer himself or by third persons involved by him, or improper repair of the object made by the customer himself or third persons involved by him.

3. The running of the guarantee period shall be interrupted for the entire time during the extent of which the object could not be operated as a consequence of the defects for which the independent-work contractor is liable.

4. In the event of the discovery during the guarantee period of the defects specified in Article 754(1) of the present Code, the customer must state them to the independent-work contractor within a reasonable period of the discovery thereof.

Article 756. Periods of Discovery of Improper Quality of Construction Work

In the event of the presentation of demands connected with the improper quality of the result of work, the rules provided for by Article 724(1)-(5) of the present Code shall be applied.

In so doing the maximum period for the discovery of defects in accordance with Article 724(2) and (4) of the present Code shall comprise five years.

Article 757. Elimination of Defects at Expense of Customer

1. The duty of the independent-work contractor to eliminate defects at the demand of the customer and at his expense for which the independent-work contractor does not bear responsibility may be provided for by the contract of construction independent-work.

2. The independent-work contractor shall have the right to refuse fulfilment of the duty specified in point 1 of the present Article in instances when the elimination of the defects is not connected directly with the subject of the contract or

cannot be effectuated by the independent-work contractor for reasons beyond his control.

§4. Independent-Work Contract for Fulfilment of Design and Survey Work

Article 758. Contract of Independent-Work for Fulfilment of Design and Survey Work

Under a contract of independent-work for the fulfilment of design and survey work the independent-work contractor (or designer, surveyor) shall be obliged to work out technical documentation according to the planning task of the customer and/or to fulfil the survey work, and the customer shall be obliged to accept and to pay for the result thereof.

Article 759. Base Data for Fulfilment of Design and Survey Work

1. Under a contract of independent-work for the fulfilment of design and survey work the customer shall be obliged to transfer to the independent-work contractor a planning task for the design, and also other base data necessary in order to draw up the technical documentation. The planning task for the fulfilment of design work may be prepared on behalf of the customer by the independent-work contractor. In this event the planning task shall become obligatory for the parties from the moment of its confirmation by the customer.

2. The independent-work contractor shall be obliged to comply with the requirements contained in the planning task and other base data for the fulfilment of the design and survey work and shall have the right to depart from such only with the consent of the customer.

Article 760. Duties of Independent-Work Contractor

1. Under a contract of independent-work for the fulfilment of design and survey work the independent-work contractor shall be obliged to:

fulfil work in accordance with the planning task and other base data for the design and with the contract;

agree the finished technical documentation with the customer, and, when necessary, together with the customer, with competent State agencies and agencies of local self-government;

transfer the finished technical documentation to the customer and the results of the survey work.

An independent-work contractor shall not have the right to transfer technical documentation to third persons without the consent of the customer.

2. The independent-work contractor under a contract of independent-work for the fulfilment of design and survey work shall guarantee to the customer the absence in third persons of the right to obstruct the fulfilment of the work or limit

the fulfilment thereof on the basis of the technical documentation prepared by the independent-work contractor.

Article 761. Responsibility of Independent-Work Contractor for Improper Fulfilment of Design and Survey Work

1. The independent-work contractor under a contract of independent-work for the fulfilment of design and survey work shall bear responsibility for the improper drawing up of technical documentation and the fulfilment of survey work, including defects discovered subsequently in the course of construction, and also in the process of the operation of the object created on the basis of technical documentation and the data of survey work.

2. In the event of the discovery of defects in the technical documentation or in the survey work the independent-work contractor shall be obliged at the demand of the customer to redo the technical documentation without compensation and respectively perform necessary additional survey work, and also to compensate the customer for losses caused unless established otherwise by a law or by the contract of independent-work for the fulfilment of design and survey work.

Article 762. Duties of Customer

Under an independent-work contract for the fulfilment of design and survey work the customer shall be obliged, unless provided otherwise by the contract, to:

pay the independent-work contractor the established price in full after the completion of all work or to pay for it in parts after the completion of individual stages of work;

use technical documentation received from the independent-work contractor only for the purposes provided for by the contract, not transfer the technical documentation to third persons, and not divulge the data contained therein without the consent of the independent-work contractor;

render assistance to the independent-work contractor in the fulfilment of design and survey work to the extent and on the conditions provided for in the contract;

participate together with the independent-work contractor in agreeing the finished technical documentation with the respective State agencies and agencies of local self-government;

compensate the independent-work contractor for additional expenses caused by a change of base data for the fulfilment of design and survey work as a consequence of the circumstances beyond the control of the independent-work contractor;

involve the independent-work contractor to participate in the case with regard to a suit brought against the customer by a third person in connection with shortcomings in the technical documentation drawn up or survey work fulfilled.

§5. Independent-Work Contract Work for State Needs

Article 763. State Contract for Fulfilment of Independent-Work Contract Work for State Needs

1. Independent-work construction work (Article 740), and design and survey work (Article 758) intended for the satisfaction of requirements of the Russian Federation or subject of the Russian Federation and financed at the expense of the assets of the respective budgets and extrabudgetary sources shall be effectuated on the basis of a State contract for the fulfilment of independent-work contract work.

2. Under a State contract for the fulfilment of independent-work contract work for State needs (hereinafter: State contract), the independent-work contractor shall be obliged to fulfil construction, design, and other work of a production and nonproduction character connected with the construction and repair of objects and transfer them to the State customer, and the State customer shall be obliged to accept the work fulfilled and to pay for it or ensure its payment.

Article 764. Parties to State Contract

Under a State contract, the State agency possessing the necessary investment resources or organisation endowed by the respective State agency with the right to dispose of such resources shall act as the customer, and a juridical person or citizen shall be the independent-work contractor.

Article 765. Grounds and Procedure for Conclusion of State Contract

The grounds and procedure for the conclusion of a State contract shall be determined in accordance with the provisions of Articles 527 and 528 of the present Code.

Article 766. Content of State Contract

1. A State contract must contain conditions concerning the amount and cost of work subject to fulfilment, the periods for the commencement and end thereof, the amount and procedure of financing and payment for work, and the means of securing performance of the obligations of the parties.

2. In an instance when a State contract is concluded according to the results of a competition for the placement of the order for independent-work contract work for State needs, the conditions of the State contract shall be determined in accordance with the announced conditions of the competition and the proposal submitted to the competition of the independent-work contractor who is deemed to be the winner of the competition.

Article 767. Change of State Contract

1. In the event of a reduction by the respective State agencies in the established procedure of assets of the respective budget allocated to finance independent-work contract work, the parties must agree the new periods, and, if necessary, also other conditions for the fulfilment of the work. The independent-work contractor shall have the right to demand compensation of losses from the State customer caused by a change of the periods for the fulfilment of work.

2. Unless provided otherwise by a law, the changes of a State contract not connected with the circumstances specified in point 1 of the present Article shall be effectuated by agreement of the parties.

Article 768. Legal Regulation of State Contract

A law on independent-work contract work for State needs shall apply to relations relating to State contracts for the fulfilment of independent-work contract work for State needs in the part not regulated by the present Code.

Chapter 38. Fulfilment of Scientific-Research, Experimental-Construction Design,
and Technological Work

Article 769. Contracts for Fulfilment of Scientific-Research Work and Experimental-Construction Design and Technological Work

1. Under a contract for the fulfilment of scientific-research work the executor shall be obliged to carry on the scientific research stipulated by the technical planning task of the customer, and under a contract for the fulfilment of experimental construction design and technological work—to work out the model of a new manufacture, construction design documentation for it, or new technology, and the customer shall be obliged to accept the work and to pay for it.

2. A contract with the executor may encompass both the entire cycle of conducting the research, development, and manufacture of the models, or individual stages (or elements) thereof.

3. Unless provided otherwise by a law or contract, the risk of accidental impossibility of performance of contracts for the fulfilment of scientific-research work, experimental-construction design, and technological work shall be borne by the customer.

4. The conditions of contracts for the fulfilment of scientific-research work, experimental-construction design and technological work must correspond to the laws and other legal acts concerning exclusive rights (intellectual property).

Article 770. Fulfilment of Work

1. The executor shall be obliged to conduct scientific research personally. He shall have the right to involve third persons to perform the contract for the fulfilment of scientific-research work only with the consent of the customer.

2. In the event of the fulfilment of experimental-construction design or technological work the executor shall have the right, unless provided otherwise by the contract, to involve third persons in the performance thereof. The rules concerning the general independent-work contractor and subindependent-work contractor (Article 706) shall apply to relations of the executor with third persons.

Article 771. Confidentiality of Information Comprising Subject of Contract

1. Unless provided otherwise by contracts for the fulfilment of scientific-research work, experimental-construction design, and technological work, the parties shall be obliged to ensure the confidentiality of information affecting the subject of the contract, the course of its performance, and the results received. The amount of information deemed to be confidential shall be determined in the contract.

2. Each of the parties shall be obliged to publish the information received when fulfilling the work deemed to be confidential only with the consent of the other party.

Article 772. Rights of Parties to Results of Work

1. The parties to contracts for the fulfilment of scientific-research work, experimental-construction design, and technological work shall have the right to use the results of the work, including that capable of legal protection, within the limits and on the conditions provided for by the contract.

2. Unless provided otherwise by the contract, the customer shall have the right to use the results of work transferred to him by the executor, including that capable of legal protection, and the executor shall have the right to use the results of the work received by him for own needs.

Article 773. Duties of Executor

The executor in contracts for the fulfilment of scientific-research work, experimental-construction design, and technological work shall be obliged to:

fulfil work in accordance with the technical planning task agreed with the customer and transfer to the customer the results thereof within the period provided for by the contract;

agree with the customer the necessity of using the protected results of intellectual activity which belong to third persons and the acquisition of the rights for the use thereof;

eliminate with his own forces and at his expense defects permitted through his fault in the work fulfilled which might entail deviations from the technical-economic parameters provided for in the technical planning task or in the contract;

immediately inform the customer about the impossibility discovered of receiving the anticipated results or the inadvisability of continuing the work;

guarantee to the customer the transfer of results received under the contract which do not violate the exclusive rights of other persons.

Article 774. Duties of Customer

1. The customer in contracts for the fulfilment of scientific-research work, experimental-construction design, and technological work shall be obliged to:

transfer information to the executor necessary for the fulfilment of work;

accept the results of the work fulfilled and pay for them.

2. The duty of the customer to issue a technical planning task to the executor and to agree with him a programme (technical-economic parameters) or theme of the work also may be provided for by the contract.

Article 775. Consequences of Impossibility of Achieving Results of Scientific Research Work

If in the course of scientific research work the impossibility is discovered of achieving the results as a consequence of circumstances beyond the control of the executor, the customer shall be obliged to pay for the cost of the work conducted until the impossibility was elicited of receiving the results provided for by the contract for the fulfilment of the scientific-research work, but not higher than the respective part of the price of the work specified in the contract.

Article 776. Consequences of Impossibility to Continue Experimental-Construction Design and Technological Work

If in the course of the fulfilment of experimental-construction design and technological work the impossibility or inadvisability of continuing the work is discovered which arose not through the fault of the executor, the customer shall be obliged to pay for the expenditures incurred by the executor.

Article 777. Responsibility of Executor for Violation of Contract

1. The executor shall bear responsibility to the customer for a violation of contracts for the fulfilment of scientific-research work, experimental-construction design, and technological work unless it is proved that such violation occurred not through the fault of the executor (Article 401[1]).

2. The executor shall be obliged to compensate losses caused to it by the customer within the limits of the cost of the work in which the defects were elicited if it is provided by the contract that they are subject to compensation within the lim-

its of the total cost of the work under the contract. Lost advantage shall be subject to compensation in the instances provided for by the contract.

Article 778. Legal Regulation of Contracts for Fulfilment of Scientific-Research Work, Experimental-Construction Design, and Technological Work

The rules of Articles 708, 709, and 738 of the present Code shall apply to the periods for the fulfilment of and to the price of work, and also to the consequences of the failure of the customer to appear in order to receive the results of the work.

The rules of Articles 763–768 of the present Code shall apply to State contracts for the fulfilment of scientific-research work, experimental-construction design and technological work for State needs.

Chapter 39. *Compensated Rendering of Services*

Article 779. Contract of Compensated Rendering of Services

1. Under a contract of compensated rendering of services the executor shall be obliged under the planning task of the customer to render services (or perform determined actions or effectuated a determined activity), and the customer shall be obliged to pay for these services.

2. The rules of the present Chapter shall apply to contracts for the rendering of communications, medical, veterinary, auditor, consulting, and informational services, services relating to teaching, tourist servicing, and others, except for services rendered under contracts provided for by Chapters 37, 38, 40, 41, 44, 45, 46, 47, 49, 51, and 53 of the present Code.

Article 780. Performance of Contract of Compensated Rendering of Services

Unless provided otherwise by the contract of compensated rendering of services, the executor shall be obliged to render the services personally.

Article 781. Payment for Services

1. The customer shall be obliged to pay for services rendered to him within the periods and in the procedure which has been specified in the contract of compensated rendering of services.

2. In the event of the impossibility of performance which arose through the fault of the customer, the services shall be subject to payment in full unless provided otherwise by a law or by the contract of compensated rendering of services.

3. In an instance when impossibility of performance arose through circumstances for which neither of the parties is liable, the customer shall compensate the executor for the expenses actually incurred by him unless provided otherwise by a law or by the contract of compensated rendering of services.

Article 782. Unilateral Refusal to Perform Contract of Compensated Rendering of Services

1. The customer shall have the right to refuse performance of a contract of compensated rendering of services on condition of payment to the executor of expenses actually incurred by him.

2. The executor shall have the right to refuse performance of obligations under a contract of compensated rendering of services only on condition of the full compensation of losses to the customer.

Article 783. Legal Regulation of Contract of Compensated Rendering of Services

The general provisions on independent-work contract (Articles 702–729) and the provisions on domestic independent-work contract (Articles 730–739) shall apply to the contract of compensated rendering of services unless this is contrary to Articles 779–782 of the present Code, and also to the peculiarities of the subject of the contract of compensated rendering of services.

Chapter 40. Carriage

Article 784. General Provisions on Carriage

1. The carriage of goods, passengers, and baggage shall be effectuated on the basis of the contract of carriage.

2. The general conditions of carriage shall be determined by the transport charters and codes, other laws, and rules issued in accordance with them.

The conditions for the carriage of goods, passengers, and baggage by individual types of transport, and also responsibility of the parties for these carriages, shall be determined by agreement of the parties unless established otherwise by the present Code, transport charters and codes, other laws, and rules issued in accordance with them.

Article 785. Contract of Carriage of Goods

1. Under a contract for the carriage of goods the carrier shall be obliged to deliver the goods entrusted to him by the consignor to the point of destination and to issue them to the person (or recipient) empowered to receive the goods, and the consignor shall be obliged to pay the established payment for the carriage of the goods.

2. The conclusion of the contract of carriage of goods shall be confirmed by drawing up and issuing to the consignor of the goods a transport waybill (bill of lading or other document for goods provided for by the respective transport charter or code).

Article 786. Contract of Carriage of Passenger

1. Under the contract of carriage of a passenger, the carrier shall be obliged to carry the passenger to the point of destination, and in the event of the passenger handing over baggage, also to deliver the baggage to the point of destination and to issue it to the person empowered to receive the baggage; the passenger shall be obliged to pay the established payment for the travel, and in the event of handing over baggage, also for the carriage of the baggage.

2. The conclusion of a contract of carriage of a passenger shall be certified by a ticket, and the handing over of baggage by the passenger, by a baggage receipt.

The forms of ticket and baggage receipt shall be established in the procedure provided for by transport charters and codes.

3. The passenger shall have the right in the procedure provided for by the respective transport charter or codes to:

carry children with him free of charge or on privileged conditions;

travel free of charge with his hand luggage within the limits of established norms;

hand over baggage for carriage for payment according to the tariff.

Article 787. Contract of Charter

Under a contract of charter one party (owner) shall be obliged to grant to the other party (charterer) for payment all or part of the capacity of one or several means of transport for one or several voyages for the carriage of goods, passengers, and baggage.

The procedure for the conclusion of the contract of charter, and also the form of the said contract, shall be established by transport charters and codes.

Article 788. Direct Mixed Transport

The mutual relations of transport organisations in the event of the carriage of goods, passengers, and baggage by various types of transport under a single transport document (direct mixed transport), and also the procedure for the organisations of these carriages, shall be determined by agreements between the organisations of the respective types of transport concluded in accordance with a law on direct mixed (or combined) carriages.

Article 789. Carriage by Transport of Common Use

1. Carriage being effectuated by a commercial organisation shall be deemed to be carriage by transport of common use if from a law or other legal acts, it arises that this organisation is obliged to effectuate the carriage of goods, passengers, and baggage upon the recourse of any citizen or juridical person [as amended by Federal Law No. 15-ФЗ, 10 January 2003].

A list of organisations obliged to effectuate carriage deemed to be carriers by transport of common use shall be published in the established procedure.

2. The contract of carriage by transport of common use shall be a public contract (Article 426).

Article 790. Fare

1. The fare established by agreement of the parties, unless provided otherwise by a law or other legal acts, shall be recovered for the carriage of goods, passengers, and baggage.

2. Payment for the carriage of goods, passengers, and baggage by transport of common use shall be determined on the basis of tariffs confirmed in the procedure established by transport charters and codes.

3. Work and services being fulfilled by the carrier at the demand of the goods possessor and not provided for by the tariffs shall be paid by agreement of the parties.

4. The carrier shall have the right to withhold goods and baggage transferred to it for carriage to secure the fare due to him and other payments relating to carriage (Articles 359, 360) unless established otherwise by a law, other legal acts, the contract of carriage, or does not arise from the essence of the obligation.

5. In instances when in accordance with a law or other legal acts privileges or preferences have been established with regard to the fare for the carriage of goods, passengers, and baggage, the expenses incurred in connection therewith shall be compensated by [to—*WEB*] the transport organisation at the expense of assets of the respective budget.

Article 791. Supply of Means of Transport, Loading, and Unloading of Goods

1. The carrier shall be obliged to supply to the consignor of goods for loading within the period established by the application (or order) accepted by it, by the contract of carriage, or by the contract concerning the organisation of carriages, means of transport in good repair in a state fit for carriage of the respective goods.
The consignor of goods shall have the right to refuse the means of transport supplied which are not fit for the carriage of the respective goods.

2. The loading (or unloading) of the goods shall be effectuated by the transport organisation or consignor (or recipient) in the procedure provided for by the contract in compliance with the provisions established by the transport charters and codes and by the rules issued in accordance with them.

3. The loading (or unloading) of goods effectuated by the forces and means of the consignor (or recipient) of the goods must be done within the periods pro-

vided for by the contract unless such periods have been established by the transport charters and codes and with the rules issued in accordance therewith.

Article 792. Periods of Delivery of Goods, Passenger, and Baggage

The carrier shall be obliged to deliver goods, passenger, or baggage to the point of destination within the periods determined in the procedure provided for by the transport charters and codes, and in the absence of such periods, within a reasonable period.

Article 793. Responsibility for Violation of Obligations Relating to Carriage

1. In the event of the failure to perform or the improper performance of obligations relating to carriage, the parties shall bear responsibility established by the present Code, transport charters and codes, and also by agreement of the parties.

2. The agreements of transport organisations with passengers and goods possessors concerning the limitation or elimination of the responsibility of the carrier established by a law shall be invalid except for instances when the possibility of such agreements in carriages of goods has been provided for by transport charters and codes.

Article 794. Responsibility of Carrier for Failure to Provide Means of Transport and of Consignor for Failure to Use Means of Transport Provided

1. The carrier shall bear the responsibility established by transport charters and codes, and also by agreement of the parties, for the failure to supply means of transport for the carriage of goods in accordance with an accepted application (or order) or other contract, and the consignor for the failure to present the goods or failure to use the means of transport supplied for other reasons.

2. The carrier and the consignor of goods shall be relieved of responsibility in the event of the failure to supply means of transport or the failure to use the means of transport supplied if this occurred as a consequence of:

insuperable force, and also other phenomena of a natural character (fires, avalanches, floods) and military actions;

termination or limitation of the carriage of goods in determined directions established in the procedure provided for by the respective transport charter or code;

in other instances provided for by transport charters and codes.

Article 795. Responsibility of Carrier for Delay of Departure of Passenger

1. For a delay of the departure of the means of transport carrying a passenger, or for the late arrival of such means of transport at the point of destination (except for carriages on city and suburban transport) the carrier shall pay to the passenger

a fine in the amount established by the respective transport charter or code unless it is proved that the delay or lateness occurred as a consequence of insuperable force, elimination of the disrepair of the means of transport threatening the life and health of passengers, or other circumstances beyond the control of the carrier.

2. In the event of the refusal of the passenger of carriage by reason of delay of the departure of the means of transport, the carrier shall be obliged to return the fare to the passenger.

Article 796. Responsibility of Carrier for Loss, Shortage, and Damage (or Spoilage) of Goods or Baggage

1. The carrier shall bear responsibility for the failure to preserve goods or baggage which occurred after the acceptance thereof for carriage and before issuance to the consignee, person empowered by him, or person empowered to receive the baggage unless it is proved that the loss, shortage, or damage (or spoilage) of the goods or baggage occurred as a consequence of circumstances which the carrier could not prevent and the elimination of which was beyond its control.

2. Damage caused during the carriage of goods or baggage shall be compensated by the carrier:

in the event of the loss or shortage of goods or baggage—in the amount of the value of the lost or short goods or baggage;

in the event of damage (or spoilage) of the goods or baggage—in the amount by which its value was reduced, and if it is impossible to restore the damaged good or baggage—in the amount of its value;

in the event of the loss of the goods or baggage handed over to the carrier with a declaration of the value thereof—in the amount of the declared value of the goods or baggage.

The value of the goods or baggage shall be determined by proceeding from the price thereof specified on the account of the seller or provided for by the contract, and in the absence of the account or specification of the price in the contract by proceeding from the price which under comparable circumstances usually is recovered for analogous goods.

3. The carrier shall together with compensation of the established damage caused by the loss, shortage, or damage (or spoilage) of the goods or baggage return to the consignor (or recipient) the fare recovered for carriage of the lost, short delivered, spoiled, or damaged goods or baggage unless this payment is part of the value of the goods.

4. The documents concerning the reasons for the failure to preserve the goods or baggage (commercial act, act of general form, and so forth) drawn up by the carrier unilaterally shall be subject in the event of a dispute to valuation by the court together with other documents certifying to the circumstances which might

serve as grounds for the responsibility of the carrier, consignor, or recipient of the goods or baggage.

Article 797. Claims and Suits With Regard to Carriages of Goods

1. Before bringing suit against a carrier arising from the carriage of goods the presentation to him of a claim in the procedure provided for by the respective transport charter or code shall be obligatory.

2. A suit against the carrier may be brought by the consignor or consignee in the event of the full or partial refusal of the carrier to satisfy the claim or the failure to receive a reply from the carrier within a thirty-day period.

3. The period of limitations with regard to demands arising from the carriage of goods shall be established at one year from the moment determined in accordance with the transport charters or codes.

Article 798. Contracts on Organisation of Carriages

The carrier and goods possessor may, in the event of the necessity to effectuate the systematic carriages of goods, conclude long-term contracts concerning the organisation of carriages.

Under a contract of the organisation of carriage of goods the carrier shall be obliged within the established periods to accept, and the goods possessor to present, goods for carriage in the stipulated amount. The amounts, periods, and other conditions for granting means of transport and presenting goods for carriage, the procedure for the settlement of accounts, and also other conditions for the organisation of carriage, shall be determined in the contract concerning the organisation of the carriage of goods.

Article 799. Contracts Between Transport Organisations

Contracts concerning the organisation of work relating to ensuring the carriage of goods (junction agreements, contracts for centralised delivery (or export) of goods and others) may be concluded between organisations of various types of transport.

The procedure for the conclusion of such contracts shall be determined by transport charters and codes, other laws, and other legal acts.

Article 800. Responsibility of Carrier for Harm Caused to Life or Health of Passenger

The responsibility of a carrier for harm caused to the life or health of a passenger shall be determined according to the rules of Chapter 59 of the present Code unless increased responsibility of the carrier has been provided for by a law or by the contract of carriage.

Chapter 41. Transport Expediting

Article 801. Contract of Transport Expediting

1. Under a contract of transport expediting one party (expeditor) shall be obliged for remuneration and at the expense of the other party (client—consignor or consignee) to fulfil or organise the fulfilment of the services connected with the carriage of goods determined by the contract of expediting.

The duties of the expeditor to organise the carriage of goods by transport and according to the route selected by the expeditor or the client, the duty of the expeditor to conclude a contract(s) for the carriage of goods in the name of the client or in his own name, to ensure the sending and receipt of the goods, and also other duties connected with carriage may be provided for by the contract of transport expediting.

The effectuation of such operations necessary for the delivery of goods as the receipt of documents required for export or import, the fulfilment of customs and other formalities, verification of the quantity and state of goods, the loading and unloading thereof, the payment of duties, charges, and other expenses placed on the client, the keeping of goods, the receipt thereof at the point of destination, and also the fulfilment of other operations and services provided for by the contract, may be provided for by the contract of transport expediting as additional services.

2. The rules of the present Chapter shall extend also to instances when in accordance with the contract the duties of expeditor are performed by the carrier.

3. The conditions of the fulfilment of the contract of transport expediting shall be determined by agreement of the parties unless established otherwise by a law on transport-expediting activity, other laws, or other legal acts.

Article 802. Form of Contract of Transport Expediting

1. The contract of transport expediting shall be concluded in written form.

2. The client must issue a power of attorney to the expeditor if it is necessary for the fulfilment of his duties.

Article 803. Responsibility of Expeditor Under Contract of Transport Expediting

For the failure to perform or the improper performance of duties under the contract of expediting the expeditor shall bear responsibility on the grounds and in the amount which is determined in accordance with the rules of Chapter 25 of the present Code.

If the expeditor proves that a violation of the obligation has been caused by the improper performance of the contracts of carriage, the responsibility of the expeditor to the client shall be determined according to the same rules under which the respective carrier is liable to the expeditor.

Article 804. Documents and Other Information Granted to Expeditor

1. The client shall be obliged to grant to the expeditor documents and other information concerning the properties of the goods, conditions of carriage thereof, and also other information necessary for the performance of the duty by the expeditor provided for by the contract of transport expediting.

2. The expeditor shall be obliged to communicate to the client information received about defects discovered, and in the event of the incompleteness of the information to request necessary additional data from the client.

3. In the event of the failure of the client to grant the necessary information, the expeditor shall have the right not to embark upon performance of the respective duties until such information is granted.

4. The client shall bear responsibility for losses caused to the expeditor in connection with the violation of the duty relating to the granting of the information specified in point 1 of the present Article.

Article 805. Performance of Duties of Expeditor by Third Person

Unless it follows from the contract of transport expediting that the expeditor must perform his duties personally, the expeditor shall have the right to involve other persons in the performance of his duties.

The imposition of the performance of an obligation on a third person shall not relieve the expeditor from responsibility to the client for the performance of the contract.

Article 806. Unilateral Refusal of Performance of Contract of Transport Expediting

Any of the parties shall have the right to refuse to perform a contract of transport expediting, having warned the other party thereof within a reasonable period.

In the event of the unilateral refusal to perform the contract, the party which has declared the refusal shall compensate the other party for losses caused by the dissolution of the contract.

Chapter 42. Loan and Credit

§1. Loan

Article 807. Contract of Loan

1. Under a contract of loan one party (lender) shall transfer to the ownership of the other party (borrower) money or other things determined by generic

indicia, and the borrower shall be obliged to return to the lender the same amount of money (amount of loan) or equal quantity of other things received by him of the same kind and quality.

The contract of loan shall be considered to be concluded from the moment of transfer of the money or other things.

2. Foreign currency and currency valuables may be the subject of a contract of loan on the territory of the Russian Federation in compliance with the rules of Articles 140, 141, and 317 of the present Code.

Article 808. Form of Contract of Loan

1. A contract of loan between citizens must be concluded in written form if the amount thereof is not less than ten times the minimum amount of payment for labour established by a law, and in an instance when the lender is a juridical person, irrespective of the amount.

2. In confirmation of a contract of loan and its conditions a receipt of the borrower or other document certifying the transfer to him by the lender of the determined monetary amount or determined quantity of things may be presented.

Article 809. Interest Under Contract of Loan

1. Unless provided otherwise by a law or by the contract of loan, the lender shall have the right to receive interest from the borrower on the amount of the loan in the amounts and in the procedure determined by the contract. In the absence in the contract of a condition concerning the amount of interest the amount thereof shall be determined as the rate of bank interest (or rate of refinancing) existing at the place of residence of the lender, and if the lender is a juridical person, at the location thereof, on the date of payment by the borrower of the amount of the debt or corresponding part thereof.

2. In the absence of another agreement, interest shall be paid monthly until the date of return of the amount of the loan.

3. A contract of loan shall be presupposed to be without interest unless expressly provided otherwise therein in instances when:

the contract is concluded between citizens for an amount not exceeding fifty times the minimum amount of payment for labour established by a law and is not connected with the effectuation of entrepreneurial activity by either of the parties;

under the contract not money, but other things determined by generic indicia are transferred to the borrower.

Article 810. Duty of Borrower to Return Amount of Loan

1. The borrower shall be obliged to return to the lender the amount of the loan received within the period and in the procedure provided for by the contract of loan.

In instances when the period of return is not established by the contract or is determined by the moment of demand, the amount of the loan must be returned by the borrower within thirty days from the date of presentation by the lender of the demand therefor unless provided otherwise by the contract.

2. Unless provided otherwise by the contract of loan, the amount of an interest-free loan may be returned by the borrower before time.

The amount of the loan granted under interest may be returned before time with the consent of the lender or if this is provided for by the contract.

3. Unless provided otherwise by a contract of loan, the amount of the loan shall be considered to be returned at the moment of the transfer thereof to the lender or crediting of the respective monetary means in his bank account.

Article 811. Consequences of Violation by Borrower of Contract of Loan

1. Unless provided otherwise by a law or by the contract of loan, in instances when the borrower does not return the amount of the loan within the period interest shall be subject to payment on this amount in the amount provided for by Article 395(1) of the present Code from the day when it should have been returned until the day of its return to the lender irrespective of the payment of interest provided for by Article 809(1) of the present Code.

2. If return of the loan has been provided for by the contract of loan in parts (or instalment), then in the event of a violation of the period established for return of the regular part of the loan by the borrower, the lender shall have the right to require the return of the entire amount remaining before time together with the interest due.

Article 812. Contesting Contract of Loan

1. The borrower shall have the right to contest a contract of loan for its impecuniousness, proving that the money or other things in reality were not received by him from the lender or were received in a lesser quantity than specified in the contract.

2. If the contract of loan was concluded in written form (Article 808), the contesting thereof for impecuniousness by means of witness testimony shall not be permitted, except for instances when the contract was concluded under the influence of fraud, coercion, threat, or ill-intentioned agreement of a representative of the borrower with the lender or the confluence of grave circumstances.

3. If in the process of the contesting of a contract of loan by the borrower for impecuniousness it is established that the money or other things in reality were not received from the lender, the contract of loan shall be considered to be not concluded. When money or things in reality were received by the borrower from

the lender in a lesser quantity than is specified in the contract, the contract shall be considered to be concluded for this quantity of money or things.

Article 813. Consequences of Loss of Security of Obligations of Borrower

In the event of the failure of the borrower to fulfil the duties provided for by the contract of loan with regard to securing the return of the amount of the loan, and also in the event of the loss of security or a worsening of its conditions under circumstances for which the lender is not liable, the lender shall have the right to demand from the borrower the repayment of the amount of the loan before time and the payment of interest due unless provided otherwise by the contract.

Article 814. Special-Purpose Loan

1. If a contract of loan was concluded with a condition of the use by the borrower of means received for determined purposes (special-purpose loan), the borrower shall be obliged to ensure the possibility of the effectuation by the lender of control over the special-purpose use of the amount of the loan.

2. In the event of the failure to fulfil the conditions of the contract of loan by the borrower concerning the special-purpose use of the amount of the loan, and also in the event of a violation of the duties provided for by point 1 of the present Article, the lender shall have the right to demand from the borrower repayment of the amount of the loan before time and the payment of interest due unless provided otherwise by the contract.

Article 815. Bill of Exchange

In instances when in accordance with an agreement of the parties a bill of exchange is issued by the borrower certifying the unconditional obligation of the issuer (promissory note) or other payer specified in the bill of exchange (bill of exchange) to pay the monetary amounts upon the ensuing of the period provided for by the bill of exchange, the relations of the parties with regard to the bill of exchange shall be regulated by a law on the bill of exchange and promissory note.

From the moment of the issuance of the bill of exchange the rules of the present paragraph may apply to those relations insofar as they are not contrary to the law on the bill of exchange and promissory note.

Article 816. Bond

In instances provided for by a law or other legal acts a contract of loan may be concluded by means of the issuance and sale of bonds.

A bond shall be deemed to be a security certifying the right of its holder to receive from the person who issued the bond the par value of the bond within the period provided for therein or other property equivalent. A bond shall grant to its holder also the right to receive the interest fixed therein on the par value of the bond or other property rights.

The rules of the present paragraph shall apply to the relations between the person who issued the bond and its holder insofar as not provided otherwise by a law or in the procedure established by it.

Article 817. Contract of State Loan

1. Under a contract of State loan the borrower is the Russian Federation or a subject of the Russian Federation, and the lender is a citizen or juridical person.

2. State loans shall be voluntary.

3. The contract of State loan shall be concluded by means of the acquisition by the lender of issued State bonds or other State securities certifying the right of the lender to receive from the borrower the monetary assets granted to it on loan or, depending upon the conditions of the loan, other property, established interest, or other property rights within the periods provided for by the conditions of the issuance of the loan into circulation.

4. A change of the conditions of a loan issued into circulation shall not be permitted.

5. The rules concerning a contract of State loan shall apply respectively to loans issued by a municipal formation.

Article 818. Novation of Debt into Loan Obligation

1. By agreement of the parties a debt which has arisen from purchase-sale, lease of property, or other grounds may be replaced by a loan obligation.

2. The replacement of a debt by a loan obligation shall be effectuated in compliance with the requirements concerning novation (Article 414) and concluded in the form provided for the conclusion of a contract of loan (Article 808).

§2. Credit

Article 819. Credit Contract

1. Under a credit contract a bank or other credit organisation (creditor) shall be obliged to grant monetary means (credit) to the borrower in the amount and on the conditions provided for by the contract, and the borrower shall be obliged to return the monetary amount received and to pay interest on it.

2. The rules provided for by paragraph 1 of the present Chapter shall apply to relations under the credit contract unless provided otherwise by the rules of the present paragraph and does not arise from the essence of the credit contract.

Article 820. Form of Credit Contract

A credit contract must be concluded in written form.
The failure to comply with the written form shall entail the invalidity of the credit contract. Such contract shall be considered void.

Article 821. Refusal to Grant or Receive Credit

1. A creditor shall have the right to refuse to grant a credit to the borrower provided for by a credit contract wholly or partially when there are circumstances obviously testifying to the fact that the amount granted to the borrower will not be returned within the period.

2. The borrower shall have the right to refuse to receive a credit wholly or partially, having informed the creditor thereof before the period established by the contract for granting it unless provided otherwise by a law, other legal acts, or the credit contract.

3. In the event of a violation by the borrower of the duties provided for by the credit contract for the special-purpose use of the credit (Article 814), the creditor shall have the right to refuse further credits to the borrower under the contract.

§3. Goods and Commercial Credit

Article 822. Goods Credit

A contract may be concluded by the parties providing for the duty of one party to grant to the other party things determined by generic indicia (contract of goods credit). The rules of paragraph 2 of the present Chapter shall apply to such a contract unless provided otherwise by such contract or does not arise from the essence of the obligation.

The conditions concerning the quantity, assortment, sets, quality, packaging and/or packing of the things granted must be performed in accordance with the rules concerning the contract of purchase-sale of goods (Articles 465–485) unless provided otherwise by the contract of goods credit.

Article 823. Commercial Credit

1. The granting of a credit, including in the form of an advance, preliminary payment, deferral, and instalment payment for goods; work, or services (commercial credit) may be provided for by contracts whose performance is connected with the transfer to the ownership of the other party of monetary amounts or other things determined by generic indicia, unless established otherwise by a law.

2. The rules of the present Chapter shall apply to the commercial credit respectively unless provided otherwise by the rules on the contract from which the respective obligation arose and is not contrary to the essence of such obligation.

Chapter 43. Financing Under Assignment of Monetary Demand

Article 824. Contract of Financing Under Assignment of Monetary Demand

1. Under a contract of financing under assignment of a monetary demand one party (financial agent) shall transfer or shall be obliged to transfer to the other party (client) monetary means at the expense of a monetary demand of the client (creditor) to a third person (debtor) arising from goods granted by the client, the fulfilment of work by him, or the rendering of services to a third person, and the client shall assign or shall be obliged to assign this monetary demand to the financial agent.

A monetary demand against a debtor may be assigned by the client to a financial agent also for the purpose of securing the performance of an obligation of the client to the financial agent.

2. The obligations of the financial agent under a contract of financing under assignment of monetary demand may include the keeping of bookkeeping records for the client, and also the granting to the client of other financial services connected with monetary demands which are the subject of the assignment.

Article 825. Financial Agent

Banks and other credit organisations, and also other commercial organisations having an authorisation (or license) for the effectuation of activity of this type, may in the capacity of a financial agent conclude contracts of financing under assignment of monetary demand.

Article 826. Monetary Demand Assigned for Purposes of Receiving Financing

1. Both a monetary demand, the period for the payment of which has already ensued (existing demand) and the right to receive monetary means which arises in the future (future demand) may be the subject of an assignment under which financing is granted.

A monetary demand which is the subject of an assignment must be determined in the contract of the client with the financial agent such that it enables the existing demand to be identified at the moment of conclusion of the contract, and a future demand, not later than at the moment it arises.

2. When assigning a future monetary demand, it shall be considered to have passed to the financial agent after the right itself arose to receive the monetary means from the debtor which are the subject of assignment of the demand provided for by the contract. If the assignment of a monetary demand is conditioned by a determined event, it shall enter into force after the ensuing of this event.

In these instances additional formulation of the assignment of the monetary demand shall not be required.

Article 827. Responsibility of Client to Financial Agent

1. Unless provided otherwise by the contract of financing under assignment of monetary demand, the client shall bear responsibility to the financial agent for the validity of the monetary demand which is the subject of the assignment.

2. A monetary demand which is the subject of assignment shall be valid if the client possesses the right to transfer the monetary demand and at the moment of assignment of this demand circumstances were not known to him as a consequence of which the debtor has the right not to perform it.

3. A client shall not be liable for the failure to perform or the improper performance by the debtor of a demand which is the subject of an assignment in the event of the presentation thereof by the financial agent for performance unless provided otherwise by the contract between the client and the financial agent.

Article 828. Invalidity of Prohibition of Assignment of Monetary Demand

1. The assignment of a monetary demand to a financial agent shall be valid even if an agreement exists between the client and the debtor thereof concerning the prohibition or limitation thereof.

2. The provision established by point 1 of the present Article shall not relieve the client from obligations or responsibility to the debtor in connection with the assignment of the demand in violation of an agreement existing between them concerning the prohibition or limitation thereof.

Article 829. Subsequent Assignment of Monetary Demand

Unless the contract of financing under assignment of monetary demand provides otherwise, subsequent assignment of the monetary demand by the financial agent shall not be permitted.

In an instance when subsequent assignment of a monetary demand is permitted by the contract, the provisions of the present Chapter shall apply to it respectively.

Article 830. Performance of Monetary Demand by Debtor to Financial Agent

1. The debtor shall be obliged to make payment to a financial agent on condition that he received from the client or from the financial agent written notification concerning the assignment of the monetary demand to the particular financial agent and the monetary demand subject to performance has been determined in the notification, and also the financial agent to whom payment must be made has been specified.

2. At the request of the debtor, the financial agent shall be obliged within a reasonable period to submit evidence to the debtor that the assignment of the monetary demand to the financial agent has actually occurred. If the financial agent does not fulfil this duty, the debtor shall have the right to make payment with regard to the particular demand to the client in performance of his obligation to the last.

3. The performance of a monetary demand by the debtor to the financial agent in accordance with the rules of the present Article shall relieve the debtor from the respective obligation to the client.

Article 831. Rights of Financial Agent to Amounts Received from Debtor

1. If under the conditions of a contract of financing under assignment of a monetary demand the financing of a client is effectuated by means of the purchase from him of this demand by the financial agent, the last shall acquire the right to all the amounts which he receives from the debtor in performance of the demand, and the client shall not bear responsibility to the financial agent for the amounts received by him which prove to be less than the price for which the agent acquired the demand.

2. If the assignment of a monetary demand to a financial agent is effectuated for the purpose of securing the performance to him of an obligation of the client and it is not provided otherwise by the contract of financing under assignment of monetary demand, the financial agent shall be obliged to submit a report to the client and to transfer to him the amount exceeding the amount of the debt of the client securing the assignment of the demand. If the monetary means received by the financial agent from the debtor proved to be less than the amount of the debt of the client to the financial agent secured by the assignment of the demand, the client shall remain responsible to the financial agent for the balance of the debt.

Article 832. Counter Demands of Debtor

1. In the event of the financial agent applying to the debtor with a demand to make payment, the debtor shall have the right in accordance with Articles 410–412 of the present Code to present his own monetary demands as set-off which are based on the contract with the client that the debtor already had at the time when notification was received by him concerning the assignment of the demand to the financial agent.

2. Demands which the debtor might present to the client in connection with a violation by the last of the agreement concerning the prohibition or limitation of the assignment of a demand shall not have force with respect to the financial agent.

Article 833. Return to Debtor of Amounts Received by Financial Agent

1. In the event of a violation by the client of his obligations under the contract concluded with the debtor, the last shall not have the right to demand from the financial agent the return of amounts already paid to him under the demand transferred to the financial agent if the debtor has the right to receive such amounts directly from the client.

2. A debtor who has the right to receive amounts directly from the client paid to the financial agent as a result of the assignment of a demand nonetheless shall have the right to demand the return of these amounts by the financial agent if it is proved that the last did not perform his obligation to effectuate the promised payment to the client connected with the assignment or the demand or made such payment knowing about the violation by the client of the obligation to the debtor to which the payment relates connected with the assignment of the demand.

Chapter 44. Bank Deposit

Article 834. Contract of Bank Deposit

1. Under a contract of bank deposit one party (bank), having accepted a monetary amount (deposit) from the other party (depositor) or received for him, shall be obliged to return the amount of the deposit and to pay interest on it on the conditions and in the procedure provided for by the contract.

2. The contract of bank deposit in which the depositor is a citizen shall be deemed to be a public contract (Article 426).

3. To relations of the bank and depositor with regard to an account to which a deposit has been made the rules concerning the contract of bank account (Chapter 45) shall apply unless provided otherwise by the rules of the present Chapter or does not arise from the essence of the contract of bank deposit.

Juridical persons shall not have the right to credit monetary means in deposits to other persons.

4. The rules of the present Chapter relating to banks also shall apply to other credit organisations accepting deposits from juridical persons in accordance with a law.

Article 835. Right to Attract Monetary Means as Deposits

1. Banks shall have the right to attract monetary means as deposits to whom such right has been granted in accordance with an authorisation (or license) issued in the procedure established in accordance with a law.

2. In the event of the acceptance of a deposit from a citizen by a person not having the right to do so, or with a violation of the procedure established by a law

or banking rules adopted in accordance therewith, the depositor may require the immediate return of the amount of the deposit, and also payment of the interest on it provided for by Article 395 of the present Code and compensation of all losses caused to the depositor above the amount of the interest.

If the monetary means of a juridical person have been accepted by such person on the conditions of a contract of bank deposit, such contract shall be invalid (Article 168).

3. Unless established otherwise by a law, the consequences provided for by point 2 of the present Article shall also apply in instances of:

attracting monetary means of citizens and juridical persons by means of the sale of stocks and other securities to them whose issuance is deemed to be illegal;

attracting monetary means of citizens for deposit under a bill of exchange or other securities excluding the receipt by their holders of the deposit at first demand and the effectuation by the depositor of other rights provided for by the rules of the present Chapter.

Article 836. Form of Contract of Bank Deposit

1. A contract of bank deposit must be concluded in written form.

The written form of a contract of bank deposit shall be considered to be complied with if the making of the deposit is certified by a savings book, savings or deposit certificate, or other document issued by the bank to the depositor which meets the requirements provided for such documents by a law, by banking rules established in accordance therewith, and customs of business turnover used in banking practice.

2. The failure to comply with the written form of the contract of bank deposit shall entail the invalidity of this contract. Such contract shall be void.

Article 837. Types of Deposits

1. A contract of bank deposit shall be concluded on the conditions of the issuance of the deposit at first demand (demand deposit) or on conditions of the return of the deposit upon the expiry of a period determined by the contract (time deposit).

The making of deposits on other conditions of their return which are not contrary to a law may be provided for by the contract.

2. Under a contract of bank deposit of any type, the bank shall be obliged to issue the amount of the deposit or part thereof at first demand of the depositor except for deposits made by juridical persons on other conditions of return provided for by the contract.

The condition of a contract concerning the waiver by the citizen of the right to receive the deposit at first demand shall be void.

3. In instances when a time or other deposit other than a demand deposit is returned to the depositor at his demand before the expiry of the period or before the ensuing of other circumstances specified in the contract of bank deposit, interest on the deposit shall be paid in the amount corresponding to the amount of interest to be paid by the bank on demand deposits unless another amount of interest is provided for by the contract.

4. In instances when the depositor does not demand the return of the amount of a time deposit upon the expiry of the period or the amount of the deposit made on other conditions of return, upon the ensuing of the circumstances provided for by the contract, the contract shall be considered to be extended on the conditions of a demand deposit unless provided otherwise by the contract.

Article 838. Interest on Deposit

1. The bank shall pay interest to the depositor on the amount of the deposit determined by the contract of bank deposit.

In the absence in the contract of a condition concerning the amount of interest to be paid, the bank shall be obliged to pay interest in the amount determined in accordance with Article 809(1) of the present Code.

2. Unless provided otherwise by the contract of bank deposit, the bank shall have the right to change the amount of interest to be paid on demand deposits.

In the event of a reduction by the bank of the amount of interest, the new amount of interest shall apply to deposits made before depositors are notified about the reduction of interest upon the expiry of a month from the moment of the respective notice unless provided otherwise by the contract.

3. The amount of interest determined by the contract of bank deposit on a deposit made by a citizen on conditions of the issuance thereof upon the expiry of a determined period or upon the ensuing of the circumstances provided for by the contract may not be unilaterally reduced by the bank unless provided otherwise by a law. Under a contract of such bank deposit concluded by the bank with a juridical person, the amount of interest may not be unilaterally changed unless provided otherwise by a law or by the contract.

Article 839. Procedure for Calculating Interest on Deposit and Payment Thereof

1. Interest on the amount of a bank deposit shall be calculated from the day following the day of its receipt in the bank up to the day preceding the return thereof to the depositor or withdrawn from the account of the depositor on other grounds.

2. Unless provided otherwise by the contract of bank deposit, interest on the amount of a bank deposit shall be paid to the depositor at his demand upon the

expiry of each quarter separately from the amount of the deposit, and interest not demanded within this period shall increase the amount of the deposit on which the interest is calculated.

In the event of the return of the deposit all interest calculated to this moment shall be paid.

Article 840. Securing Return of Deposit

1. Banks shall be obliged to secure the return of deposits of citizens by means of obligatory insurance and in the instances provided for by a law also by other means.

The return of deposits of citizens by a bank in whose charter capital the Russian Federation and/or subjects of the Russian Federation, and also municipal formations, have more than 50% of the stocks or participatory shares of participation shall in addition be guaranteed by their subsidiary responsibility with regard to the demands of the depositor against the bank in the procedure provided for by Article 399 of the present Code.

2. The means of securing the return by the bank of deposits of juridical persons shall be determined by the contract of bank deposit.

3. When concluding a contract of bank deposit the bank shall be obliged to grant information to the depositor concerning the return of the deposit being secured.

4. In the event of the failure of the bank to fulfil the duties provided for by a law or by the contract of bank deposit with regard to securing the return of the deposit, and also in the event of the loss of security or worsening of its conditions, the depositor shall have the right to require the immediate return of the amount of the deposit from the bank, the payment of interest on it in the amount determined in accordance with Article 809(1) of the present Code, and compensation of losses caused.

Article 841. Deposit of Monetary Means by Third Persons to Account of Depositor

Unless provided otherwise by the contract of bank deposit, monetary means received at a bank in the name of the depositor from third persons shall be credited to the account on deposit with an indication of the necessary data concerning his account on deposit. In so doing it shall be presupposed that the depositor has expressed consent to receipt of the monetary means from such persons, having granted them the necessary data concerning the deposit account.

Article 842. Deposits to Benefit of Third Persons

1. A deposit may be made to a bank in the name of a determined third person. Unless provided otherwise by the contract of bank deposit, such person shall

acquire the rights of a depositor from the moment of the presentation by him of first demand to the bank based on those rights or the expression by him to the bank by other means of the intention to take advantage of such rights.

An indication of the name of the citizen (Article 19) or name of the juridical person (Article 54) to whose benefit the deposit is made shall be a material condition of the respective contract of bank deposit.

The contract of bank deposit to the benefit of a citizen deceased at the moment of conclusion of the contract or to a juridical person which does not exist at this moment shall be void.

2. Until the expression by the third person of the intention to take advantage of the rights of depositor, the person who has concluded the contract of bank deposit may take advantage of the rights of depositor with respect to the monetary means deposited by him in the deposit account.

3. The rules concerning the contract to the benefit of a third person (Article 430) shall apply to the contract of bank deposit to the benefit of a third person if this is not contrary to the rules of the present Article and the essence of the bank deposit.

Article 843. Savings Book

1. Unless provided otherwise by agreement of the parties, the conclusion of a contract of bank deposit with a citizen and the deposit of monetary means in his account on deposit shall be certified by a savings book. The issuance of an inscribed savings book or bearer savings book may be provided for by the contract of bank deposit. A bearer savings book shall be a security.

In the savings book must be specified and certified by the bank the name and location of the bank (Article 54), and if the deposit is made in a branch, also of its respective branch, the number of the account on deposit, and also all the amounts of monetary means withdrawn from the account and the balance of monetary means in the account at the moment of presentation of the savings book to the bank.

Unless a different state of the deposit is proved, the data concerning the deposit specified in the savings book shall be the grounds for the settlement of accounts regarding the deposit between the bank and the depositor.

2. The issuance of the deposit, payment of interest on it, and performance of the instructions of the depositor concerning the transfer of monetary means from the deposit account to other persons shall be effectuated by the bank upon the presentation of the savings book.

If an inscribed savings book is lost or is in a state unfit for presentation, the bank upon the application of the depositor shall issue him a new savings book.

The restoration of the rights with regard to a lost bearer savings book shall be effectuated in the procedure provided for bearer securities (Article 148).

Article 844. Savings (or Deposit) Certificate

1. A savings (or deposit) certificate shall be a security certifying the amount of the deposit made in a bank and the rights of the depositor (or holder of the certificate) to receive the amount of the deposit upon the expiry of the established period and the interest stipulated on the certificate at the bank which issued the certificate or any branch of this bank.

2. Savings (or deposit) certificates may be bearer or inscribed.

3. In the event of the presentation of a savings (or deposit) certificate for payment before time, the amount of the deposit and the interest to be paid on the deposit on demand shall be paid by the bank unless another amount of interest is established by the conditions of the certificate.

Chapter 45. Bank Account

Article 845. Contract of Bank Account

1. Under a contract of bank account the bank shall be obliged to accept and credit monetary means entering the account opened for the client (possessor of the account), fulfil instructions of the client concerning the transfer and issuance of respective amounts from the account and the conducting of other operations with regard to the account.

2. The bank may use monetary means available in the account, guaranteeing the right of the client to dispose of these assets without obstruction.

3. The bank shall not have the right to determine and control the orientations of use of the monetary means of the client nor establish other limitations of his right to dispose of monetary means at his discretion which are not provided for by a law or by the contract of bank account.

4. The rules of the present Chapter relating to banks also shall apply to other credit organisations when they conclude and perform a contract of bank account in accordance with an authorisation (or license) issued.

Article 846. Conclusion of Contract of Bank Account

1. When concluding a contract of bank account, an account shall be opened for the client or person specified by him in the bank on the conditions agreed by the parties.

2. The bank shall be obliged to conclude a contract of bank account with the client who has applied with a proposal to open the account on conditions announced by the bank for opening accounts of the particular type which correspond to the requirements provided for by a law and banking rules established in accordance therewith.

The bank shall not have the right to refuse to open the account and to perform respective operations which have been provided for by a law, the constitutive documents of the bank, or authorisation (or license) issued to it, except for instances when such refusal is caused by the lack of the possibility for the bank to accept for bank servicing or is permitted by a law or other legal acts.

In the event of the unsubstantiated avoidance of a bank to conclude a contract of bank account the client shall have the right to present a demand to it provided for by Article 445(4) of the present Code.

Article 847. Certification of Right of Disposition of Monetary Means in Account

1. The rights of persons effectuating in the name of the client instructions concerning the debiting and issuance of means from the account shall be certified by the client by means of submitting to the bank documents provided for by a law, banking rules established in accordance therewith, and by the contract of bank account.

2. The client may give an instruction to the bank concerning the withdrawal of monetary means from the account at the demand of third persons, including connected with the performance by the client of his obligations to these persons. The bank shall accept this instruction on condition of specifying therein in written form the necessary data enabling it when the respective demand is presented to identify the person having the right to present it.

3. The certification of the rights of disposition of monetary assets in the account by electronic means of payment and other documents with the use of analogues of handwritten signatures thereon (Article 160[2]), codes, passwords, and other means confirming that the instruction is given by the person empowered to do so may be provided for by the contract.

Article 848. Operations With Regard to Account to be Fulfilled by Bank

The bank shall be obliged to perform operations for the client provided for accounts of the particular type by a law, banking rules established in accordance therewith, and customs of business turnover applied in banking practice unless provided otherwise by the contract of bank account.

Article 849. Periods for Operations With Regard to Account

The bank shall be obliged to credit monetary means coming to the account of the client not later than the day following the day of receipt at the bank of the respective payment document unless a shorter period has been provided for by the contract of bank account.

The bank shall be obliged at the instruction of the client to issue or debit from the account monetary means of the client not later than the day following the day of receipt at the bank of the respective payment document unless other periods have been provided for by a law, banking rules issued in accordance therewith, or by the contract of bank account.

Article 850. Overdrawing of Account

1. In instances when in accordance with the contract of bank account the bank effectuates payments from the account notwithstanding the absence of monetary means (overdrawing of account), the bank shall be considered to have granted a credit to the client for the respective amount from the date of effectuation of such payment.

2. The rights and duties of the parties connected with overdrawing the account shall be determined by the rules concerning loan and credit (Chapter 42) unless provided otherwise by the contract of bank account.

Article 851. Payment of Expenses of Bank for Performance of Operations With Regard to Account

1. In the instances provided for by the contract of bank account the client shall pay for services of the bank with regard to the performance of operations with the monetary means in the account.

2. Payment for services of the bank provided for by point 1 of the present Article may be recovered by the bank upon the expiry of each quarter from the monetary means of the client in the account unless provided otherwise by the contract of bank account.

Article 852. Interest for Use by Bank of Monetary Means in Account

1. Unless provided otherwise by the contract of bank account, the bank shall pay interest for the use of monetary means in the account of the client, the amount of which shall be credited to the account.

The amount of interest shall be credited to the account within the periods provided for by the contract, and in an instance when such periods have not been provided for by the contract, upon the expiry of each quarter.

2. Interest specified in point 1 of the present Article shall be paid by the bank in the amount determined by the contract of bank account, and in the absence in the contract of a respective condition, in the amount usually paid by the bank for demand deposits (Article 838).

Article 853. Set-off of Counter Demands of Bank and Client With Regard to Account

Monetary demands of the bank against the client connected with overdrawing the account (Article 850) and the payment for services of the bank (Article 851), and also demands of the client against the bank concerning the payment of interest for the use of monetary means (Article 852), shall be terminated by set-off (Article 410) unless provided otherwise by the contract of bank account.

Set-off of the said demands shall be effectuated by the bank. The bank shall be obliged to inform the client about the set-off made in the procedure and within the periods which are provided for by the contract, and if the respective conditions have not been agreed by the parties, in the procedure and within the periods which are usual for banking practice of granting information to clients concerning the state of monetary means in the respective account.

Article 854. Grounds for Withdrawal of Monetary Means from Account

1. The withdrawal of monetary means from the account shall be effectuated by the bank on the basis of the instruction of the client.

2. Without the instruction of the client the withdrawal of monetary means in the account shall be permitted by decision of a court, and also in the instances established by a law or provided for by the contract between the bank and the client.

Article 855. Priority of Withdrawal of Monetary Means from Account

1. When there are monetary means in the account, the amount of which is sufficient in order to satisfy all demands presented to the account, the withdrawal of these assets from the account shall be effectuated in the sequence of receipt of instructions of the client and other documents for withdrawal (calendar priority) unless provided otherwise by a law.

2. In the event the monetary means in the account are insufficient for the satisfaction of all demands presented to it, the withdrawal of monetary means shall be effectuated in the following priority:

in first priority shall be effectuated the withdrawal with regard to documents of execution which provide for the crediting or issuance of monetary means from the account in order to satisfy demands concerning compensation of harm caused to life or health, and also demands concerning the recovery of alimony;

in second priority shall be made the withdrawal with regard to documents of execution which provide for the crediting or issuance of monetary means for the settlement of accounts regarding the payment of severance benefits and payment of labour to persons who work under a labour contract, including under a contract [контракт], and with regard to the payment of remuneration under an author's contract;

in third priority shall be made the withdrawal with regard to payment documents providing for the crediting or issuance of monetary assets for the settlement of accounts with regard to payment of labour with persons working under a labour contract, and also with regard to deductions to the Pension Fund of the Russian Federation, the Social Insurance Fund of the Russian Federation, and obligatory medical insurance funds [added by Federal Law of 12 August 1996, added to by Federal Law of 24 September 1997, as amended by Federal Law No 8–ФЗ, 10 January 2003. СЗ РФ (1994), no. 32, item 3301; (1996), no. 5, item 410; no. 34, item 4025; (1997), no. 43, item 4903; (2003), no. 2, item 160.

in fourth priority shall be made the withdrawal with regard to payment documents which provide for payments to the budget and extrabudgetary funds, deductions to which have not been provided for in the third priority [as changed by Federal Law of 12 August 1996. СЗ РФ (1996), no. 34, item 4025];

in fifth priority shall be made the withdrawal with regard to documents of execution which provide for the satisfaction of other monetary demands [as changed by Federal Law of 12 August 1996. СЗ РФ (1996), no. 34, item 4025];

in sixth priority shall be made the withdrawal with regard to other payment documents by way of calendar priority [as changed by Federal Law of 12 August 1996. СЗ РФ (1996), no. 34, item 4025].

The withdrawal of means from the account with regard to demands relegated to one priority shall be made by way of calendar priority of receipt of the documents.

Article 856. Responsibility of Bank for Improper Performance of Operations With Regard to Account

In instances of the untimely crediting to an account of monetary means received by a client or the unsubstantiated withdrawal thereof by the bank from the account, and also the failure to fulfil the instructions of the client concerning the crediting of monetary means from the account or the issuance thereof from the account, the bank shall be obliged to pay interest on this amount in the procedure and in the amount provided for by Article 395 of the present Code.

Article 857. Banking Secret

1. A bank shall guarantee the secrecy of a bank account and bank deposit, operations relating to the account, and information concerning the client.

2. Information constituting a banking secret may be granted only to the clients themselves or their representatives. Such information may be granted to State agencies and their officials exclusively in the instances and in the procedure provided for by a law.

3. In the event of the divulgence by a bank of information constituting a banking secret, the client whose rights were violated shall have the right to demand compensation of losses caused from the bank.

Article 858. Limitation of Disposition of Account

Limitation of the rights of the client to dispose of monetary means in his account shall not be permitted except for the imposition of arrest on the monetary means in the account or the suspension of operations with regard to the account in the instances provided for by a law.

Article 859. Dissolution of Contract of Bank Account

1. A contract of bank account may be dissolved upon the application of the client at any time.

2. At the demand of the bank the contract of bank account may be dissolved by a court in the following instances:

when the amount of monetary means kept in the account of the client proves to be lower than the minimum amount provided for by banking rules or the contract if such amount is not reinstated within a month from the date of the warning of the bank thereof;

in the absence of operations with regard to this account during a year unless provided otherwise by the contract.

3. The balance of monetary means in the account shall be issued to the client or upon his instruction credited to another account not later than seven days after receipt of the respective written application of the client.

4. Dissolution of a contract of bank account shall be grounds for closing the account of the client.

Article 860. Accounts of Banks

The rules of the present Chapter shall extend to correspondent accounts, correspondent subaccounts, and other accounts of banks unless provided otherwise by a law, other legal acts, or banking rules established in accordance therewith.

Chapter 46. Settlement of Accounts

§1. General Provisions on Settlement of Accounts

Article 861. Cash and Noncash Settlement Accounts

1. Settlements of accounts with the participation of citizens not connected with the effectuation by them of entrepreneurial activity may be performed with cash money (Article 140) without limitation of the amount or in a noncash procedure.

2. Settlement of accounts by juridical persons, and also settlement of accounts with the participation of citizens connected with the effectuation by them of entrepreneurial activity, shall be performed in a noncash procedure. Settlement of accounts between these persons may be performed also by cash money unless established otherwise by a law.

3. Noncash settlements of accounts shall be performed through banks and other credit organisations (hereinafter: banks) in which respective accounts have

been opened unless it arises otherwise from a law or is stipulated by the form of settlement of accounts being used.

Article 862. Forms of Noncash Settlement of Accounts

1. When effectuating noncash settlements of accounts, the settlement of accounts by payment commissions, letters of credit, cheques, and encashment shall be permitted, and also the settlement of accounts in other forms provided for by a law and by banking rules established in accordance with it, and customs of business turnover applicable in banking practice.

2. The parties under a contract shall have the right to choose and establish in the contract any of the forms of settlement of accounts specified in point 1 of the present Article.

§2. Settlements of Accounts by Payment Orders

Article 863. General Provisions on Settlements of Accounts by Payment Orders

1. When setting accounts by a payment order the bank shall be obliged on behalf of the payer at the expense of means in his account to transfer the determined monetary amount to the account of the person specified by the payer in that or in another bank within the period provided for by a law or established in accordance therewith unless a shorter period has been provided by the contract of bank account or is not determined by the customs of business turnover applicable in banking practice.

2. The rules of the present paragraph shall apply to relations connected with debiting monetary means through the bank by a person not having an account in that bank unless otherwise provided for by a law or banking rules established in accordance therewith or does not arise from the essence of these relations.

3. The procedure for the effectuation of the settlement of accounts by payment orders shall be regulated by a law, and also by banking rules established in accordance therewith and the customs of business turnover applicable in banking practice.

Article 864. Conditions of Performance by Bank of Payment Order

1. The content of a payment order and settlement of account documents submitted together with it and the form thereof must correspond to the requirements provided for by a law and the banking rules established in accordance therewith.

2. In the event the payment order fails to conform to the requirements specified in point 1 of the present Article the bank may clarify the content of the order. Such query must be made to the payer immediately upon receipt of the order. In the event of the failure to receive a reply within the period provided for by a law or

banking rules established in accordance therewith, and if there are none, within a reasonable period, the bank may leave the order without performance and return it to the payer unless provided otherwise by a law, banking rules established in accordance therewith, or the contract between the bank and the payer.

3. The order of the payer shall be performed by the bank when there are means in the account of the payer unless provided otherwise by the contract between the payer and the bank. Orders shall be performed by the bank in compliance with the priority of withdrawal of monetary means from the account (Article 855).

Article 865. Performance of Order

1. A bank which has accepted a payment order of the payer shall be obliged to transfer the respective monetary amount to the bank of the recipient of the means in order to credit it to the account of the person specified in the order within the period established by Article 863(1) of the present Code.

2. The bank shall have the right to involve other banks in order to fulfil operations with regard to transferring the monetary means to the account specified in the order of the client.

3. The bank shall be obliged immediately to inform the payer at his demand about the performance of the order. The procedure for the formalisation and requirements for the content of the notice concerning performance of the order shall be provided for by a law, banking rules established in accordance therewith, or agreement of the parties.

Article 866. Responsibility for Failure to Perform or Improper Performance of Order

1. In the event of the failure to perform or the improper performance of the order of the client the bank shall bear responsibility on the grounds and in the amounts which have been provided for by Chapter 25 of the present Code.

2. In instances when the failure to perform or the improper performance of the order occurred in connection with a violation of the rules for the performance of settlement of account operations by the bank involved in order to perform the order of the payer, the responsibility provided for by point 1 of the present Article may be placed by a court on this bank.

3. If a violation of the rules for the performance of settlement of account operations by the bank entailed the unlawful withholding of monetary means, the bank shall be obliged to pay interest in the procedure and in the amount provided for by Article 395 of the present Code.

§3. Settlements of Accounts by Letter of Credit

Article 867. General Provisions on Settlements of Accounts by Letter of Credit

1. In the event of the settlements of accounts under a letter of credit, the bank acting on behalf of the payer concerning the opening of the letter of credit and in accordance with his instruction (emitent-bank) shall be obliged to make payments of means to the recipient or to pay, accept, or discount a bill of exchange or to give a power to another bank (performing bank) to make payments of means to the recipient or to pay, accept, or discount the bill of exchange.

The rules concerning the performing bank shall apply to the emitent-bank making the payment of means to the recipient or paying, accepting, or discounting a bill of exchange.

2. In the event of the opening of a covered (or deposited) letter of credit the emitent-bank shall be obliged when opening it to transfer the amount of the letter of credit (or cover) at the expense of the payer or the credit granted to him to the disposition of the performing bank for the entire period of operation of the obligation of the emitent-bank.

In the event of the opening of an uncovered (or guaranteed) letter of credit the right shall be granted to the performing bank to withdraw the entire amount of the letter of credit from the account of the emitent-bank kept with it.

3. The procedure for the effectuation of the settlements of accounts under a letter of credit shall be regulated by a law, and also by banking rules established in accordance therewith, and by customs of business turnover applicable in banking practice.

Article 868. Revocable Letter of Credit

1. A letter of credit which may be changed or revoked by the emitent-bank without prior notice of the recipient of the assets shall be deemed to be revocable. Revocation of the letter of credit shall not create any obligations whatever of the emitent-bank to the recipient of the means.

2. The performing bank shall be obliged to effectuate payment or other operations with regard to a revocable letter of credit if at the moment of the performance thereof notice has not been received concerning the change of conditions or revocation of the letter of credit.

3. The letter of credit shall be revocable unless expressly established otherwise in the text thereof.

Article 869. Irrevocable Letter of Credit

1. A letter of credit which may not be revoked without the consent of the recipient of the means shall be deemed to be irrevocable.

2. At the request of the emitent-bank the performing bank participating in the conducting of letter of credit operations may confirm an irrevocable letter of credit (confirmed letter of credit). Such confirmation shall mean the acceptance by the performing bank of an obligation additional to the obligation of the emitent-bank to make payment in accordance with the conditions of the letter of credit.

An irrevocable letter of credit confirmed by the performing bank may not be changed or revoked without the consent of the performing bank.

Article 870. Performance of Letter of Credit

1. In order to perform a letter of credit the recipient of assets shall submit to the performing bank documents which confirm the fulfilment of all conditions of the letter of credit. In the event of a violation of even one of these conditions, performance of the letter of credit shall not be made.

2. If the performing bank has made payment or effectuated another operation in accordance with the conditions of the letter of credit, the emitent-bank shall be obliged to compensate it for expenses incurred. The said expenses, and also all other expenses of the emitent-bank connected with the performance of the letter of credit, shall be compensated by the payer.

Article 871. Refusal to Accept Documents

1. If the performing bank refused to accept documents which by external indicia do not correspond to the conditions of the letter of credit, it shall be obliged immediately to inform the recipient of the assets and the emitent-bank thereof, specifying the reasons for the refusal.

2. If the emitent-bank, having received the documents accepted by the performing bank, considers that they do not correspond by external indicia to the conditions of the letter of credit, it shall have the right to refuse to accept them and to require from the performing bank the amount of means paid to the recipient in violation of the conditions of the letter of credit, and with regard to an uncovered letter of credit, to refuse to compensate the amounts paid.

Article 872. Responsibility of Bank for Violation of Conditions of Letter of Credit

1. Responsibility for a violation of the conditions of a letter of credit to the payer shall be borne by the emitent-bank, and to the emitent-bank by the performing bank except for instances provided for by the present Article.

2. In the event of the unsubstantiated refusal of the performing bank to pay the monetary means under a covered or confirmed letter of credit responsibility to the recipient of the means may be placed on the performing bank.

3. In the event of the incorrect payment by the performing bank of monetary means under a covered or confirmed letter of credit as a consequence of a violation of the conditions of the letter of credit responsibility to the payer may be placed on the performing bank.

Article 873. Closure of Letter of Credit

1. A letter of credit in the performing bank shall be closed upon:
the expiry of the period of the letter of credit;
the statement of the recipient of the means concerning a refusal to use the letter of credit before the expiry of the period of its operation if the possibility of such refusal has been provided for by the conditions of the letter of credit;
the demand of the payer concerning the full or partial revocation of the letter of credit if such revocation is possible under the conditions of the letter of credit.

The performing bank must bring the closure of the letter of credit to the notice of the emitent-bank.

2. The unused amount of a covered letter of credit shall be subject to return to the emitent-bank immediately and simultaneously with the closure of the letter of credit. The emitent-bank shall be obliged to credit the returned amounts to the account of the payer with which the means were deposited.

§4. Settlements of Accounts by Encashment

Article 874. General Provisions on Settlements of Accounts by Encashment

1. In the event of the settlements of accounts by encashment the bank (or emitent-bank) shall be obliged on behalf of the client to effectuate at the expense of the client actions with regard to the receipt of payment and/or acceptance of payment from the payer.

2. The emitent-bank which has received a commission of the client shall have the right to involve another bank (performing bank) in order to fulfil it.

The procedure for the effectuation of the settlement of accounts by encashment shall be regulated by a law, banking rules established in accordance therewith, and customs of business turnover applicable in banking practice.

3. In the event of the failure to perform or improper performance of the commission of the client the emitent-bank shall bear responsibility to him on the grounds and in the amount which has been provided for by Chapter 25 of the present Code.

If the failure to perform or the improper performance of the commission of the client occurred in connection with a violation of the rules for performing settlement of account operations by the performing bank, responsibility to the client may be placed on this bank.

Article 875. Performance of Encashment Commission

1. In the absence of any document or of the nonconformity of documents by external indicia to an encashment commission the performing bank shall be obliged immediately to notify the person thereof from whom the encashment commission was received. In the event of the failure to eliminate the said defects the bank shall have the right to return the document without performance.

2. Documents shall be submitted to the payer in that form in which they were received except for notations and inscriptions of banks necessary in order to formalise the encashment operation.

3. If documents are subject to payment upon presentation, the performing bank must make the submission for payment immediately upon receipt of the encashment commission.

If the documents are subject to payment within a different period, the performing bank must in order to receive acceptance of the payer submit the documents for acceptance immediately upon receipt of the encashment commission, and the demand for payment must be made not later than the day of the ensuing of the period of payment specified in the document.

4. Partial payments may be accepted in instances when this has been established by banking rules or when there is special authorisation in the encashment commission.

5. The amounts received (encashed) must be immediately transferred by the performing bank to the disposition of the emitent-bank, which shall be obliged to credit these amounts to the account of the client. The performing bank shall have the right to withhold from the encashed amounts the remuneration due to it and compensation of expenses.

Article 876. Notice of Operations Conducted

1. If the payment and/or acceptance was not received, the performing bank shall be obliged to immediately notify the emitent-bank about the reasons for nonpayment or refusal of the acceptance.

The emitent-bank shall be obliged immediately to inform the client thereof, having asked for instructions from him relative to further actions.

2. In the event of the failure to receive instructions concerning further actions within the period established by banking rules, and in the absence thereof within a reasonable period, the performing bank shall have the right to return the documents to the emitent-bank.

§5. Settlements of Accounts by Cheques

Article 877. General Provisions on Settlements of Accounts by Cheques

1. A security containing an unconditional instruction of the drawer to the bank to make the payment of the amount specified therein to the payee shall be deemed to be a cheque.

2. Only the bank where the drawer has means which he has the right to dispose of by means of putting forward cheques may be specified as the payer under the cheque.

3. The revocation of a cheque before the expiry of the period for the presentation thereof shall not be permitted.

4. The issuance of a cheque shall not cancel a monetary obligation for the performance of which it is issued.

5. The procedure and conditions of the use of cheques in payment turnover shall be regulated by the present Code, and in the part not regulated by it, by other laws and by banking rules established in accordance therewith.

Article 878. Requisites of Cheque

1. A cheque must contain:
(1) the name 'cheque' incorporated into the text of the document;
(2) the order to the payer to pay the determined monetary amount;
(3) the name of the payer and indication of the account from which the payment must be made;
(4) the indication of the currency of payment;
(5) the indication of the date and place of drawing up the cheque;
(6) the signature of the person who has signed the cheque—the drawer.
The absence in the document of any of the said requisites shall deprive it of the force of a cheque.
A cheque which does not contain an indication of the place of drawing it up shall be considered as signed at the location of the drawer.
An indication concerning interest shall be considered to be not written.

2. The form of the cheque and the procedure for filling it in shall be determined by a law and by the banking rules established in accordance therewith.

Article 879. Payment of Cheque

1. A cheque shall be paid at the expense of the means of the drawer.
In the event of the depositing of means, the procedure and conditions of deposit of the assets in order to cover the cheque shall be established by banking rules.

2. A cheque shall be subject to payment by the payer on condition of the presentation thereof for payment within the period established by a law.

3. The payer under a cheque shall be obliged to certify by all means accessible to him the genuineness of the cheque, and also that the bearer of the cheque is the person empowered with regard to it.

In the event of the payment of an endorsed cheque, the payer shall be obliged to verify the correctness of the endorsements, but not the signature of the endorsers.

4. Losses which arose as a consequence of the payment by the payer of a forged, stolen, or lost cheque shall be placed on the payer or the drawer depending upon by whose fault they were caused.

5. The person who has paid the cheque shall have the right to require the transfer to him of the cheque with a notation of payment being received.

Article 880. Transfer of Rights Under Cheque

1. The transfer of rights under a cheque shall be made in the procedure established by Article 146 of the present Code in compliance with the rules provided for by the present Article.

2. An inscribed cheque shall not be subject to transfer.

3. On a transferable cheque the endorsement to the payer shall have the force of a receipt for receiving payment.

An endorsement performed by the payer shall be invalid.

The person who possesses a transferable cheque received under endorsement shall be considered to be the legal possessor thereof if he bases his right on the uninterrupted sequence of endorsements.

Article 881. Guarantee of Payment

1. Payment under a cheque may be guaranteed in full or partially by means of an aval.

Guarantee of a payment under a cheque (aval) may be given by any person except the payer.

2. An aval shall be filled in on the face side of the cheque or on an additional sheet by means of the inscription 'to consider an aval' and an indication by whom and for whom it was given. If not specified for whom it was given, it shall be considered that the aval was given for the drawer.

An aval shall be signed by the avalist with an indication of the place of his residence and date of performing the inscription, and if the avalist is a juridical person, its location and date of performing the inscription.

3. An avalist shall be liable just as he for whom the aval was given.

His obligation shall be valid even if the obligation which he guaranteed proves to be invalid on any grounds whatever other than the failure to comply with the form.

4. An avalist who has paid the cheque shall acquire the rights arising from the cheque against he for whom he gave the guarantee and against those who are obliged to the last.

Article 882. Encashment of Cheque

1. The submission of a cheque to the bank servicing the payee for encashment in order to receive payment shall be considered to be the presentation of the cheque for payment.

The payment of a cheque shall be made in the procedure established by Article 875 of the present Code.

2. The crediting of assets with regard to the encashed cheque to the account of the payee shall be made after receipt of the payment from the payer unless provided otherwise by a contract between the payee and the bank.

Article 883. Certification of Refusal to Pay Cheque

1. The refusal to pay a cheque must be certified by one of the following means:
(1) performance of a protest by a notary or drawing up of an equivalent act in the procedure established by a law;
(2) notation of the payer on the cheque concerning the refusal to pay it, indicating the date of presentation of the cheque for payment;
(3) notation of the encashing bank indicating the date that the cheque was presented in good time and not paid.

2. The protest or equivalent act must be performed before the expiry of the period for presentation of the cheque.

If the presentation of the cheque occurred on the last day of the period, the protest or equivalent act may be performed on the following work day.

Article 884. Notice of Nonpayment of Cheque

The payee shall be obliged to notify his endorsee and the drawer about the nonpayment within two work days following the day of performance of the protest or equivalent act.

Each endorsee must within two work days following the day of receipt by them of the notice bring to the information of their endorsee the notice received by him. Within the same period notice shall be sent to he who gave an aval for this person.

A person who has not been sent notice within the said period shall not lose his rights. He shall compensate losses which may occur as a consequence of the failure to notify about nonpayment of the cheque. The amount of losses to be compensated may not exceed the amount of the cheque.

319

Article 885. Consequences of Failure to Pay Cheque

1. In the event of the refusal of the payer to pay the cheque, the payee shall have the right at his choice to present the cheque to one, several, or all persons obliged under the cheque (drawer, avalist, endorsee), who shall bear joint and several responsibility to him.

2. The drawee shall have the right to demand from the said persons payment of the amount of the cheque, his costs to receive payment, and also interest in accordance with Article 395(1) of the present Code.

The same right shall belong to the person obliged under the cheque after he has paid the cheque.

3. A suit of the payee against the persons specified in point 1 of the present Article may be presented within six months from the date of the end of the period of presentation of the cheque for payment. Regress demands with regard to suits of the obliged persons against one another shall be cancelled upon the expiry of six months from the date when the respective obliged person satisfied the demand or from the date of bringing suit against him.

Chapter 47. Keeping

§1. General Provisions on Keeping

Article 886. Contract of Keeping

1. Under a contract of keeping one party (keeper) shall be obliged to keep a thing transferred to it by the other party (transferor) and to return this thing intact.

2. In a contract of keeping in which the keeper is a commercial organisation or noncommercial organisation effectuating keeping as one of the purposes of its professional activity (professional keeper) the duty of the keeper to accept a thing for keeping from the transferor within the period provided for by the contract may be provided for.

Article 887. Form of Contract of Keeping

1. A contract of keeping must be concluded in written form in the instances specified in Article 161 of the present Code. In so doing compliance with the written form shall be required for a contract of keeping between citizens (Article 161(1)[2]) if the value of the thing being transferred for keeping exceeds not less than ten times the minimum amount of payment for labour established by a law.

A contract of keeping providing for the duty of the keeper to accept a thing for keeping must be concluded in written form irrespective of the composition of the participants of this contract and the value of the thing being transferred for keeping.

The transfer of a thing for keeping under extraordinary circumstances (fire, natural disaster, sudden illness, threat of attack, and so forth) may be proved by witness testimony.

2. The simple written form of a contract of keeping shall be considered to be complied with if the acceptance of the thing for keeping is certified by the keeper by the issuance to the transferor of:

personal receipt, receipt, certificate, or other document signed by the keeper;

numbered token (or number), or other mark certifying the acceptance of the thing for keeping, if such form of confirmation of the acceptance of the thing for keeping has been provided for by a law or other legal act or is customary for the particular type of keeping.

3. The failure to comply with the simple written form of the contract of keeping shall not deprive the party of the right to refer to witness testimony in the event of a dispute concerning the identity of the thing accepted for keeping with the thing returned by the keeper.

Article 888. Performance of Duty to Accept Thing for Keeping

1. A keeper who has assumed under a contract of keeping the duty to accept a thing for keeping (Article 886[2]) shall not have the right to demand the transfer of this thing to him for keeping.

However, the transferor who has not transferred a thing for keeping within the period provided for by the contract shall bear responsibility to the keeper for losses caused in connection with the unconstituted keeping unless provided otherwise by a law or the contract of keeping. The transferor shall be relieved from this responsibility if he declares to the keeper the refusal of his services within a reasonable period.

2. Unless provided otherwise by the contract of keeping, the keeper shall be relieved from the duty to accept a thing for keeping in the event the thing is not transferred to him within the period stipulated by the contract.

Article 889. Period of Keeping

1. The keeper shall be obliged to keep a thing during the period stipulated by the contract of keeping.

2. If the period of keeping is not provided for by the contract and cannot be determined by proceeding from its conditions, the keeper shall be obliged to keep the thing until the demand thereof by the transferor.

3. If the period of keeping is determined by the moment of demand of the thing by the transferor, the keeper shall have the right upon the expiry of the period of keeping of the thing usual under the particular circumstances to require that the transferor take the thing back, having granted him a reasonable period for

this. The failure of the transferor to perform this duty shall entail the consequences provided for by Article 899 of the present Code.

Article 890. Keeping of Things with Pooling

In instances expressly provided for by the contract of keeping, things accepted for keeping from one transferor may be mixed with the things of the same kind and quality of other transferors (keeping with pooling). The quantity of things of the same kind and quality shall be returned to the transferor equally or as stipulated by the parties.

Article 891. Duty of Keeper to Ensure Preservation of Thing

1. The keeper shall be obliged to take all measures provided for by the contract of keeping in order to ensure the preservation of the thing transferred to him for keeping.

In the absence in the contract of conditions concerning such measures or of the incompleteness of these conditions, the keeper also must take for the preservation of the thing measures which correspond to the customs of business turnover and to the essence of the obligation, including the properties of the thing transferred for keeping, unless the necessity for taking such measures is excluded by the contract.

2. The keeper in any event must take measures in order to preserve a thing transferred to him, the obligatoriness of which has been provided for by a law, other legal acts, or in the procedure established by them (fire prevention, sanitary, protective, and the like).

3. If the keeping is effectuated without compensation, the keeper shall be obliged to be concerned for the things accepted for keeping no less than as for his own things.

Article 892. Use of Thing Transferred for Keeping

The keeper shall not have the right without the consent of the transferor to use a thing transferred for keeping, nor likewise to grant the possibility of use thereof to third persons, except for an instance when the use of the thing being kept is necessary in order to ensure the preservation thereof and is not contrary to the contract of keeping.

Article 893. Change of Conditions of Keeping

1. In the event of the necessity of a change of conditions of keeping things provided for by the contract of keeping, the keeper shall be obliged immediately to inform the transferor thereof and to await his answer.

If a change of conditions for keeping is necessary in order to eliminate the danger of loss, shortage, or damage of the thing, the keeper shall have the right to change the means, place, and other conditions of keeping without awaiting the answer of the transferor.

2. If during the keeping a real threat arose of the spoilage of the thing, or the thing already has been subjected to spoilage, or circumstances arose which do not enable the preservation thereof to be ensured, and the timely taking of measures by the transferor cannot be expected, the keeper shall have the right autonomously to sell the thing or part thereof at the price which has been formed at the place of keeping. If the said circumstances arose for reasons for which the keeper is not liable, he shall have the right to compensation of his expenses for the sale at the expense of the purchase price.

Article 894. Keeping of Things with Dangerous Properties

1. Things which are inflammable, explosive, or in general inherently danger-ous may, unless the transferor when handing them over for keeping warned the keeper about the properties thereof, be at any time rendered harmless or destroyed by the keeper without compensation of losses to the transferor. The transferor shall be liable for losses caused in connection with the keeping of such things to the keeper and to third persons.

In the event of the transfer of things with dangerous properties for keeping to a professional keeper, the rules provided for by paragraph one of the present point shall apply in an instance when such things were handed over for keeping under an incorrect name and the keeper when accepting them could not by means of external inspection certify their dangerous properties.

In the event of keeping for compensation in the instances provided for by the present point, the remuneration paid for keeping the things shall not be returned, and if it was not paid, the keeper may recover it in full.

2. If the things specified in point 1, paragraph one, of the present Article and accepted for keeping with the knowledge and consent of the keeper become, despite compliance with the conditions for their keeping, dangerous for sur-rounding persons or for the property of the keeper or third persons and circum-stances do not permit the keeper to require they be immediately taken back by the transferor or he does not fulfil this demand, these things may be rendered harm-less or destroyed by the keeper without compensation of losses to the transferor. The transferor shall not bear in this event responsibility to the keeper and third persons for losses caused in connection with the keeping of these things.

Article 895. Transfer of Thing for Keeping to Third Person

Unless provided otherwise by a contract of keeping, the keeper shall not have the right without the consent of the transferor to transfer a thing for keeping to a third person, except for instances when he was compelled to do so by force of cir-cumstances in the interests of the transferor and had no possibility to receive his consent.

The keeper shall be obliged immediately to inform the transferor about the transfer of the thing for keeping to a third person.

In the event of the transfer of a thing for keeping to a third person, the conditions of the contract between the transferor and the initial keeper shall retain force and the last shall be liable for the actions of the third person to which he transferred the thing for keeping as though for his own.

Article 896. Remuneration for Keeping

1. Remuneration for keeping must be paid to the keeper at the end of the keeping, and if the payment for keeping has been provided for by periods, it must be paid for by respective parts upon the expiry of each period.

2. In the event of delay of payment of remuneration for keeping of more than half of a period for which it should have been paid, the keeper shall have the right to refuse to perform the contract and require that the transferor immediately take back the thing handed over for keeping.

3. If the keeping is terminated before the expiry of the stipulated period for circumstances for which the keeper is not liable, he shall have the right to a commensurate part of the remuneration, and in the event provided for by Article 894(1) of the present Code, to the entire amount of remuneration.

If the keeping is terminated before time under circumstances for which the keeper is liable, he shall not have the right to demand remuneration for keeping, but must return to the transferor the amounts received on the account of this remuneration.

4. If upon the expiry of the period of keeping a thing in keeping is not taken back by the transferor, he shall be obliged to pay the keeper commensurate remuneration for the further keeping of the thing. This rule also shall apply in the event that the transferor is obliged to take back the thing before the expiry of the period of keeping.

5. The rules of the present Article shall apply unless provided otherwise by the contract of keeping.

Article 897. Compensation of Expenses for Keeping

1. Unless provided otherwise by the contract of keeping, the expenses of the keeper for the keeping of a thing shall be included in the remuneration for keeping.

2. In the event of keeping without compensation, the transferor shall be obliged to compensate the keeper for necessary expenses made by him for the keeping of the thing, unless provided otherwise by a law or by the contract of keeping.

Article 898. Extraordinary Expenses for Keeping

1. Expenses for the keeping of a thing which exceed ordinary expenses of this nature and which the parties could not foresee in the event of the conclusion of a contract of keeping (extraordinary expenses) shall be compensated to the keeper if the transferor gave consent to these expenses or approved them subsequently, and also in other instances provided for by a law, other legal acts, or a contract.

2. When it is necessary to make extraordinary expenses, the keeper shall be obliged to request the transferor for consent to these expenses. If the transferor does not communicate his nonconsent within the period established by the keeper, or within the time normally necessary for a reply, it shall be considered that he agreed to the extraordinary expenses.

In an instance when the keeper has made extraordinary expenses for keeping, not having received the prior consent from the transferor to these expenses, although under the circumstances of the case this was possible, and the transferor subsequently did not approve them, the keeper may demand compensation of extraordinary expenses only within the limits of the damage which might be caused to the thing if these expenses were not made.

3. Unless provided otherwise by the contract of keeping, extraordinary expenses shall be compensated in addition to the remuneration for keeping.

Article 899. Duty of Transferor to Take Thing Back

1. Upon the expiry of the stipulated period of keeping or the period granted to the keeper for receiving the thing back on the basis of Article 889(3) of the present Code, the transferor shall be obliged immediately to remove the thing transferred for keeping.

2. In the event of the failure of the transferor to perform his duty to take the thing back which was transferred for keeping, including in the event of his evading to receive the thing, the keeper shall have the right, unless provided otherwise by the contract of keeping, after warning the transferor in writing to autonomously sell the thing at the price which has formed in the place of keeping, and if the value of the thing according to a valuation exceeds one hundred minimum amounts for the payment of labour established by a law, to sell it by auction in the procedure provided for by Articles 447–449 of the present Code.

The amount received from the sale of the thing shall be transferred to the transferor, less the amounts due to the keeper, including his expenses for the sale of the thing.

Article 900. Duty of Keeper to Return Thing

1. The keeper shall be obliged to return to the transferor or to the person specified by him as the recipient that very thing which was transferred for keeping,

unless keeping with pooling has been provided for by the contract (Article 890).

2. A thing must be returned by the keeper in that state in which it was accepted for keeping, taking into account its natural deterioration, natural decrease, or other change as a consequence of its natural properties.

3. Simultaneously with the return of the thing the keeper is obliged to transfer the fruits and revenues received for the time of the keeping thereof unless provided otherwise by the contract of keeping.

Article 901. Grounds for Responsibility of Keeper

1. The keeper shall be liable for loss, shortage, or damage to things accepted for keeping on the grounds provided for by Article 401 of the present Code.

A professional keeper shall be liable for loss, shortage, or damage to things unless it is proved that the loss, shortage, or damage occurred as a consequence of insuperable force, or because of the properties of the thing, of which the keeper, in accepting it for keeping, did not know and should not have known, or as a result of intent or gross negligence of the transferor.

2. The keeper shall be liable for loss, shortage, or damage to things accepted for keeping after the duty of the transferor to take these things back ensued (Article 899[1]) only when intent or gross negligence is present on his part.

Article 902. Amount of Responsibility of Keeper

1. Losses caused to the transferor by the loss, shortage, or damaging of a thing shall be compensated by the keeper in accordance with Article 393 of the present Code unless provided otherwise by a law or the contract of keeping.

2. In the event of keeping without payment, the losses caused to the transferor by the loss, shortage, or damaging of the thing shall be compensated:

(1) for loss or shortage of the thing: in the amount of the value of the lost or short-delivered things;

(2) for the damaging of things: in the amount by which their value was reduced.

3. In an instance when as a result of the damage for which the keeper is liable the quality of the thing has been so changed that it cannot be used for its initial designation, the transferor shall have the right to refuse it and to require compensation from the keeper for the value of this thing, and also of other losses, unless provided otherwise by a law or by the contract of keeping.

Article 903. Compensation of Losses Caused to Keeper

The transferor shall be obliged to compensate the keeper for losses caused by the properties of the thing handed over for keeping if the keeper, in accepting the thing for keeping, did not know and should not have known about these properties.

Article 904. Termination of Keeping Upon Demand of Transferor

The keeper shall be obliged at the first demand of the transferor to return the thing accepted for keeping even though the period for the keeping thereof provided for by the contract has not yet ended.

Article 905. Application of General Provisions on Keeping to Individual Types Thereof

The general provisions on keeping (Articles 886–904) shall apply to individual types thereof unless the rules on individual types of keeping contained in Articles 907–926 of the present Code and in other laws have established otherwise.

Article 906. Keeping by Virtue of a Law

The rules of the present Chapter shall apply to obligations of keeping arising by virtue of a law unless other rules have been established by a law.

§2. Keeping in Goods Warehouse

Article 907. Contract of Warehouse Keeping

1. Under a contract of warehouse keeping, a goods warehouse (keeper) shall be obliged for remuneration to keep the goods transferred to it by the goods-possessor (transferor) and to return these goods intact.

A goods warehouse shall be deemed to be an organisation effectuating as entrepreneurial activity the keeping of goods and rendering services connected with keeping.

2. The written form of the contract of warehouse keeping shall be considered to be complied with if the conclusion thereof and the acceptance of the goods at the warehouse are certified by a warehouse document (Article 912).

Article 908. Keeping of Goods by Warehouse of Common Use

1. The goods warehouse shall be deemed to be a warehouse of common use if it arises from a law and other legal acts that it is obliged to accept goods for keeping from any goods-possessor [as amended by Federal Law No. 15-ФЗ, 10 January 2003].

2. A contract of warehouse keeping concluded by a goods warehouse of common use shall be deemed to be a public contract (Article 426).

Article 909. Verification of Goods During Their Acceptance by Goods Warehouse and During Keeping

1. Unless provided otherwise by the contract of warehouse keeping, a goods warehouse shall, when accepting goods for keeping, be obliged at its own expense

to inspect the goods and determine the quantity thereof (number of units or goods pieces, or measure—weight, volume) and external state.

2. A goods warehouse shall be obliged to grant to the goods-possessor the possibility during the keeping to examine the goods or examples thereof, if the keeping is effectuated by pooling, to take samples and measures necessary in order to ensure the preservation of the goods.

Article 910. Changes of Conditions of Keeping and State of Goods

1. In an instance when in order to ensure the preservation of the goods a change of the conditions of their keeping is required, a goods warehouse shall have the right to take the required measures autonomously. However, it shall be obliged to inform the goods-possessor about the measures taken if it was required to materially change the conditions of the keeping of the goods provided for by the contract of warehouse keeping.

2. In the event of discovering during keeping the damaging of a good which exceeds the limits agreed in the contract of warehouse keeping or ordinary norms of natural spoilage, the goods warehouse shall be obliged immediately to draw up an act concerning this and on the same day to notify the goods-possessor.

Article 911. Verification of Quantity and State of Good in Event of Return Thereof to Goods-Possessor

1. The goods possessor and goods warehouse shall have the right each to demand in the event of returning a good to inspect it and verify the quantity thereof. The expenses caused by this shall be borne by he who demanded inspection of the good or verification of the quantity thereof.

2. If when returning a good by the warehouse to the goods-possessor the good was not jointly inspected or verified by them, a statement concerning the shortage or damaging of the good as a consequence of the improper keeping thereof must be made to the warehouse in writing when receiving the good, and with respect to shortage or damage which could not be discovered with the ordinary means of accepting the good, within three days upon the receipt thereof.

In the absence of the statement specified in paragraph one of the present point it shall be considered, unless it is proved otherwise, that the good was returned by the warehouse in accordance with the conditions of the contract of warehouse keeping.

Article 912. Warehouse Documents

1. A goods warehouse shall issue in confirmation of the acceptance of a good for keeping one of the following warehouse documents:

dual warehouse certificate;

simple warehouse certificate;

warehouse receipt.

2. A dual warehouse certificate shall consist of two parts—the warehouse certificate and the pledge certificate (or warrant), which may be separated one from the other.

3. A dual warehouse certificate, each of the two parts thereof, and a simple warehouse certificate shall be securities.

4. A good accepted for keeping under a dual or simple warehouse certificate may be during the keeping thereof the subject of pledge by means of the pledge of the respective certificate.

Article 913. Dual Warehouse Certificate

1. On each part of a dual warehouse certificate there must be identically indicated:

(1) the name and location of the goods warehouse which accepted the good for keeping;

(2) the current number of the warehouse certificate according to the register of the warehouse;

(3) the name of the juridical person or the name of the citizen from which the good was accepted for keeping, and also the location (or place of residence) of the goods-possessor;

(4) the name and quantity of the good accepted for keeping—number of units and/or pieces of the good and/or the measure (weight, volume) of the good;

(5) the period for which the good is accepted for keeping, if such period is established, or an indication that the good is accepted for keeping on demand;

(6) the amount of remuneration for keeping or the tariffs on the basis of which it is calculated, and the procedure of payment for keeping;

(7) the date of issuance of the warehouse certificate.

Both parts of the dual warehouse certificate must have identical signatures of the empowered person and the seal of the goods warehouse.

2. A document which does not correspond to the requirements of the present Article shall not be a dual warehouse certificate.

Article 914. Rights of Holders of Warehouse and Pledge Certificates

1. The holder of a warehouse and pledge certificates shall have the right to dispose of the good being kept in the warehouse in full.

2. The holder of the warehouse certificate separated from the pledge certificate shall have the right to dispose of the good, but may not take it from the warehouse until payment of the credit issued under the pledge certificate.

3. The holder of a pledge certificate other than the holder of the warehouse certificate shall have the right of pledge to the good in the amount of the credit issued under the pledge certificate and interest thereon. In the event of the pledge of a good, a notation thereof shall be made on the warehouse certificate.

Article 915. Transfer of Warehouse and Pledge Certificates

The warehouse certificate and the pledge certificate may be transferred together or separately under inscriptions of endorsement.

Article 916. Issuance of Good Under Dual Warehouse Certificate

1. A goods warehouse shall issue the good to the holder of the warehouse and pledge certificates (dual warehouse certificate) not other than in exchange for both these certificates together.

2. A good shall be issued by the warehouse to the holder of a warehouse certificate who does not have the pledge certificate but has contributed the amount of the debt with regard to it not other than in exchange for the warehouse certificate and on condition of the presentation together with it of a receipt concerning payment of the entire amount of the debt under the pledge certificate.

3. A goods warehouse, notwithstanding the requirements of the present Article, which has issued a good to the holder of a warehouse certificate who does not have the pledge certificate and has not contributed the amount of the debt with regard to it shall bear responsibility to the holder of the pledge certificate for payment of the entire amount secured under it.

4. The holder of warehouse and pledge certificates shall have the right to demand the issuance of the good in parts. In so doing new certificates for the good remaining in the warehouse shall be issued to him in exchange for the original certificates.

Article 917. Simple Warehouse Certificate

1. A simple warehouse certificate shall be issued to the bearer.

2. A simple warehouse certificate must contain the information provided by Article 913(1), subpoints (1)(2)(4)-(7), and the last paragraph, of the present Code, and also an indication that it was issued to the bearer.

3. A document which does not correspond to the requirements of the present Article shall not be a simple warehouse certificate.

Article 918. Keeping of Things with Right to Dispose of Them

If it follows from a law, other legal acts, or the contract that a goods warehouse may dispose of goods handed over to it for keeping, the rules of Chapter 42 of the present Code shall apply to the relations of the parties concerning loan; however, the time and place of return of the goods shall be determined by the rules of the present Chapter.

§3. Specialised Types of Keeping

Article 919. Keeping in Pawnshop

1. A contract of keeping in a pawnshop of things which belong to a citizen shall be a public contract (Article 426).

2. The conclusion of a contract of keeping in a pawnshop shall be certified by the issuance by the pawnshop to the transferor of an inscribed deposit receipt.

3. A thing handed over for keeping to a pawnshop shall be subject to valuation by agreement of the parties in accordance with the prices for a thing of that type and quality usually established in trade at the moment and at the place of its acceptance for keeping.

4. A pawnshop shall be obliged to insure to the benefit of the transferor at its expense things accepted for keeping in the full amount of their valuation made in accordance with point 3 of the present Article.

Article 920. Things Not Demanded from Pawnshop

1. If a thing handed over for keeping to a pawnshop is not demanded by the transferor within the period stipulated by the agreement with the pawnshop, the pawnshop shall be obliged to keep it for two months with recovery of payment for this as provided for by the contract of keeping. Upon the expiry of this period, the undemanded thing may be sold by the pawnshop in the procedure established by Article 358(5) of the present Code.

2. From the amount received from the sale of the undemanded thing, payment for the keeping thereof shall be paid and other payments due to the pawnshop. The balance of the amount shall be returned by the pawnshop to the transferor.

Article 921. Keeping of Valuables in Bank

1. A bank may accept securities, precious metals and stones, other precious things, and other valuables, including documents, for keeping.

2. The conclusion of a contract of keeping for valuables in a bank shall be certified by the issuance by the bank to the transferor of an inscribed deposit document, the presentation of which shall be grounds for the issuance of the valuables being kept to the transferor.

Article 922. Keeping of Valuables in Individual Bank Safe

1. A contract for the keeping of valuables in a bank may provide for the keeping thereof with the use by the transferor (or client) or with the granting to him of

an individual bank safe (safe box, isolated premise in a bank) which is protected by the bank.

Under a contract for keeping of valuables in an individual bank safe the right shall be granted to the client to himself place the valuables in the safe and to remove them from the safe, for which a key to the safe must be issued to him, a card enabling the client to be identified, or other mark or document certifying the right of the client to access to the safe and that contained therein.

The right of the client to work in the bank with the valuables kept in an individual safe may be provided for by the conditions of the contract.

2. Under a contract of keeping of valuables in a bank with the use by the client of an individual bank safe, the bank shall accept valuables from the client which must be kept in the safe, effectuate control over their placement by the client in the safe and removal from the safe, and after removal shall return them to the client.

3. Under a contract of keeping of valuables in a bank with the granting of an individual bank safe to the client, the bank shall ensure to the client the possibility of placing valuables in a safe and the removal thereof from the safe without any control whatever, including control on the part of the bank.

The bank shall be obliged to effectuate control over access to the premise where the safe granted to the client is situated.

Unless provided otherwise by a contract of keeping of valuables in a bank with the granting of an individual bank safe to the client, the bank shall be relieved of responsibility for the failure to preserve that contained in the safe if it is proved that under the conditions of keeping the access of anyone to the safe without the knowledge of the client was impossible or it became possible as a consequence of insuperable force.

4. The rules of the present Code on the contract of lease shall apply to the contract on the granting of a bank safe for use to another person without responsibility of the bank for that contained in the safe.

Article 923. Keeping in Left Luggage Rooms of Transport Organisations

1. Left luggage rooms within the jurisdiction of transport organisations of common use shall be obliged to accept for keeping the things of passengers and other citizens irrespective of whether they have travel documents. A contract for the keeping of things in left luggage rooms of transport organisations shall be deemed to be a public contract (Article 426).

2. In confirmation of the acceptance of a thing for keeping in a left luggage room (except for automatic rooms) a receipt or numbered token shall be issued to the transferor. In the event of the loss of the receipt or token the thing handed over to the left luggage room shall be issued to the transferor upon presentation of evidence of the affiliation of this thing to him.

3. The period during which the left luggage room is obliged to keep things determined by the rules established in accordance with Article 784(2), paragraph two, of the present Code unless a longer period is established by agreement of the parties. Things not demanded within the specified periods the left luggage room shall be obliged to keep for another thirty days. Upon the expiry of this period the things not demanded may be sold in the procedure provided for by Article 899(2) of the present Code.

4. The losses of the transferor as a consequence of the loss, shortage, or damaging of things handed over for the left luggage room shall, within the limits of the amount of their valuation by the transferor when handing them over for keeping, be subject to compensation by the keeper within twenty-four hours from the moment of presentation of the demand concerning their compensation.

Article 924. Keeping in Check Rooms of Organisations

1. Keeping in check rooms of organisations shall be presupposed to be without compensation if remuneration for keeping is not stipulated or by other obvious means is not stipulated when handing over a thing for keeping.

The keeper of a thing handed over to a check room, irrespective of whether the keeping is effectuated for or without compensation, shall be obliged to take all measures provided for by Article 891(1) and (2) of the present Code in order to ensure the preservation of the thing.

2. The rules of the present Article also shall apply to the keeping of outer clothing, headwear, and other similar things left without handing them over for keeping by citizens at places allotted for these purposes in organisations and means of transport.

Article 925. Keeping in Hotel

1. A hotel shall be liable as keeper and without a special agreement concerning this with a person residing therein (lodger) for loss, shortage, or damaging of his things deposited at the hotel, except for money, other currency valuables, securities, and other precious things.

A thing entrusted to workers of the hotel, or a thing placed in the hotel room or other place designated for this shall be considered to be deposited in the hotel.

2. A hotel shall be liable for the loss of money, other currency valuables, securities, and other precious things of a lodger on condition that they were accepted by the hotel for keeping or were placed by the lodger in an individual safe provided to him by the hotel irrespective of whether this safe is situated in his room or in another premise of the hotel. A hotel shall be relieved from responsibility for the failure to preserve that contained in such safe if it is proved that under the conditions of keeping access of anyone to the safe without the knowledge of the lodger was impossible or became possible as a consequence of insuperable force.

3. The lodger who has discovered a loss, shortage, or damaging of his things shall be obliged without delay to declare this to the administration of the hotel. Otherwise, the hotel shall be relieved from responsibility for the failure to preserve the things.

4. An announcement made by the hotel that it does not assume responsibility for the failure to preserve things of lodgers shall not relieve it from responsibility.

5. The rules of the present Article respectively shall apply with respect to the keeping of things of citizens in motels, rest homes, boarding houses, sanatoriums, public baths, and other similar organisations.

Article 926. Keeping of Things Which Are Subject of Dispute (Sequestration)

1. Under a contract on sequestration two or several persons between whom a dispute has arisen concerning the right to a thing shall transfer this thing to a third person who assumed the duty upon the settlement of the dispute to return the thing to that person to whom it was awarded by decision of a court or by agreement of all the disputing persons (contractual sequestration).

2. A thing which is the subject of a dispute between two or several persons may be transferred for keeping by way of sequestration by decision of a court (judicial sequestration).

Either a person designated by a court or a person determined by mutual consent of the disputing parties may be the keeper under judicial sequestration. In both instances the consent of the keeper shall be required unless established otherwise by a law.

3. Both moveable and immovable things may be transferred for keeping by way of sequestration.

4. The keeper effectuating the keeping of a thing by way of sequestration shall have the right to remuneration at the expense of the disputing parties unless provided otherwise by contract or decision of the court by which the sequestration was established.

Chapter 48. Insurance

Article 927. Voluntary and Obligatory Insurance

1. Insurance shall be effectuated on the basis of contracts of property or personal insurance concluded by a citizen or juridical person (insurant) with an insurance organisation (insurer).

A contract of personal insurance shall be a public contract (Article 426).

2. In instances when the duty to insure as insurers of life, health, or property of other persons or its own civil responsibility to other persons at its own expense or

at the expense of the interested persons (obligatory insurance) has been placed by a law on the persons specified therein, the insurance shall be effectuated by means of the conclusion of contracts in accordance with the rules of the present Chapter. For insurers the conclusion of contracts of insurance on the conditions proposed by the insurant shall not be obligatory.

3. Instances of the obligatory insurance of life, health, and property of citizens at the expense of assets granted from the respective budget (obligatory State insurance) may be provided for by a law.

Article 928. Interests, the Insurance of Which is Not Permitted

1. The insurance of unlawful interests shall not be permitted.

2. The insurance of losses from participation in games, lotteries, and betting shall not be permitted.

3. The insurance of expenses which a person may be forced to make for the purpose of liberating hostages shall not be permitted.

4. The conditions of contracts of insurance which are contrary to points 1–3 of the present Article shall be void.

Article 929. Contract of Property Insurance

1. Under a contract of property insurance one party (insurer) shall be obliged for the payment stipulated by the contract (insurance premium) if the event ensues provided for in the contract (insured event) to compensate the other party (insurant) or other person to whose benefit the contract was concluded (beneficiary) for losses caused as a consequence of this event to the insured property or losses in connection with other property interests of the insurant (to pay insurance compensation) within the limits of the amount (insured amount) determined by the contract.

2. Under a contract of property insurance the following property interests in particular may be insured:
   (1) risk of loss (or perishing), shortage, or damage to determined property (Article 930);
   (2) risk of responsibility for obligations arising as a consequence of causing harm to life, health, or property of other persons, and in the instances provided for by a law, also responsibility under contracts—the risk of civil responsibility (Articles 931 and 932);
   (3) risk of losses from entrepreneurial activity as a consequence of a violation of their obligations by the contracting parties of the entrepreneur or changes of conditions of this activity under circumstances beyond the control of the entrepreneur, including the risk of the failure to receive anticipated revenues—entrepreneurial risk (Article 933).

Article 930. Insurance of Property

1. Property may be insured under a contract of insurance to the benefit of the person (insurant or beneficiary) having an interest based in a law, another legal act, or the contract in the preservation of this property.

2. A contract of insurance of property concluded in the absence of an interest of the insurant or beneficiary in the preservation of the insured property shall be invalid.

3. A contract of insurance of property to the benefit of a beneficiary may be concluded without specifying the name of the beneficiary (insurance 'on the account of he who follows').

In the event of concluding such a contract, a bearer insurance policy shall be issued to the insurant. In the event of the effectuation by the insurant or beneficiary of the rights under such contract, it shall be necessary to submit this policy to the insurer.

Article 931. Insurance of Responsibility for Causing of Harm

1. Under a contract of insurance of the risk of responsibility for obligations which arise as a consequence of the causing of harm to the life, health, or property of other persons, the risk of responsibility of the insurant himself, or another person on whom such responsibility may be placed, may be insured.

2. The person whose risk of responsibility for causing harm has been insured must be named in the contract of insurance. If this person is not named in the contract, the responsibility of the insurant himself shall be considered to be the insured risk.

3. The contract of insurance of the risk of responsibility for causing harm shall be considered to be concluded to the benefit of the persons to whom harm may be caused (beneficiary) even if the contract was concluded to the benefit of the insurant or other person responsible for causing harm or it is not said in the contract to whose benefit it was concluded.

4. In an instance when responsibility for the causing of harm was insured by virtue of the fact that the insuring thereof is obligatory, and also in other instances provided for by a law or contract of insurance of such responsibility, the person to whose benefit the contract of insurance is considered to be concluded shall have the right to present directly to the insurer a demand concerning compensation of harm within the limits of the insured amount.

Article 932. Insurance of Responsibility Under Contract

1. Insurance of the risk of responsibility for a violation of a contract shall be permitted in the instances provided for by a law.

2. Under a contract of insurance of the risk of responsibility for a violation of a contract, only the risk of responsibility of the insurant itself may be insured. A contract of insurance not corresponding to this requirement shall be void.

3. The risk of responsibility for a violation of a contract shall be considered to be insured to the benefit of the party to which under the conditions of this contract the insurant must bear respective responsibility—the beneficiary—even if the contract of insurance is concluded to the benefit of another person or it is not said therein to whose benefit it was concluded.

Article 933. Insurance of Entrepreneurial Risk

Under a contract of insurance of entrepreneurial risk only the entrepreneurial risk of the insurant itself and only to its benefit may be insured.

The contract of insurance of entrepreneurial risk of a person which is not the insurant shall be void.

A contract of insurance of entrepreneurial risk to the benefit of a person which is not the insurant shall be considered to be concluded to the benefit of the insurant.

Article 934. Contract of Personal Insurance

1. Under a contract of personal insurance one party (insurer) shall be obliged for the payment stipulated by the contract (insurance premium) to be paid by the other party (insurant) to pay one time or to pay periodically the amount stipulated by the contract (insured amount) in the event of the causing of harm to the life or health of the insurant himself or other citizen named in the contract (insured person), the attaining by him of a determined age or the ensuing in his life of another event provided for by the contract (insured event).

The right to receive the insured amount shall belong to the person to whose benefit the contract was concluded.

2. A contract of personal insurance shall be considered to be concluded to the benefit of the insured person if another person is not named in the contract as the beneficiary. In the event of the death of the person insured under the contract in which another beneficiary is not named, the heirs of the insured person shall be deemed to be the beneficiaries.

A contract of personal insurance to the benefit of a person who is not the insured person, including to the benefit of an insurant who is not the insured person, may be concluded only with the written consent of the insured person. In the absence of such consent, the contract may be deemed to be invalid upon the suit of the insured person, and in the event of the death of this person upon the suit of his heirs.

Article 935. Obligatory Insurance

1. The duty to insure may be placed by a law on the persons specified therein:

the life, health, or property of other persons determined in a law in the event of the causing of harm to their life, health, or property;

the risk of civil responsibility which may ensue as a consequence of the causing of harm to the life, health, or property of other persons or violations of contracts with other persons.

2. The duty to insure his own life or health may not be placed on a citizen according to a law.

3. In the instances provided for by a law or in the procedure established by it, the duty to insure property may be placed on juridical persons having property in economic jurisdiction or operative management which is State or municipal ownership.

4. In instances when the duty of insuring does not arise from a law, but is based on a contract, including the duty of insuring property, on a contract with the possessor of the property or constitutive documents of a juridical person which is the owner of the property, such insurance shall not be obligatory in the sense of the present Article and shall not entail the consequences provided for by Article 937 of the present Code.

Article 936. Effectuation of Obligatory Insurance

1. Obligatory insurance shall be effectuated by means of the conclusion of a contract of insurance by the person on which the duty of such insurance (as insurant) with the insurer was placed.

2. Obligatory insurance shall be effectuated at the expense of the insurant, except for obligatory insurance of passengers which in the instances provided for by a law may be effectuated at their expense.

3. Objects subject to obligatory insurance, the risks against which they must be insured, and the minimum insured amounts shall be determined by a law, and in the instance provided for by Article 935(3) of the present Code, by a law or in the procedure established by it.

Article 937. Consequences of Violation of Rules on Obligatory Insurance

1. The person to whose benefit under a law obligatory insurance must be effectuated shall have the right if it is known to it that the insurance is not effectuated to require in a judicial proceeding the effectuation thereof by the person on whom the duty of insurance has been placed.

2. If the person on which the duty of insurance has been placed does not effectuate it or has concluded a contract of insurance on conditions which worsen the

status of the beneficiary in comparison with the conditions determined by a law, it shall, in the event of the ensuing of the insured event, bear responsibility to the beneficiary on the same conditions on which the insurance compensation should have been paid under the proper insurance.

3. The amounts unfoundedly saved by a person on which the duty of insurance has been placed, thanks to the fact that it did not fulfil this duty or fulfilled it improperly, shall be recovered upon the suit of State insurance supervision agencies to the revenue of the Russian Federation, calculating interest on these amounts in accordance with Article 395 of the present Code.

Article 938. Insurer

Juridical persons having an authorisation (or license) for the effectuation of insurance of the respective type may conclude contracts of insurance as insurers.

The requirements which insurance organisations must meet, the procedure for the licensing of their activity, and the effectuation of State supervision over this activity shall be determined by laws on insurance.

Article 939. Fulfilment of Duties Under Contract of Insurance by Insurant and Beneficiary

1. The conclusion of a contract of insurance to the benefit of a beneficiary, including when he is the insured person, shall not relieve the insurant from the fulfilment of the duties under this contract, unless provided otherwise by the contract or the duties of the insurant have been fulfilled by the person to whose benefit the contract was concluded.

2. The insurer shall have the right to demand from the beneficiary, including when the insured person is the beneficiary, the fulfilment of the duties under the contract of insurance, including the duties lying on the insurant but not fulfilled by him, in the event of the presentation by the beneficiary of a demand concerning payment of the insurance compensation under a contract of property insurance or insured amount under a contract of personal insurance. The risk of the consequences of the failure to fulfil or the untimely fulfilment of the duties which should have been fulfilled earlier shall be borne by the beneficiary.

Article 940. Form of Contract of Insurance

1. The contract of insurance must be concluded in written form.

The failure to comply with the written form shall entail the invalidity of the contract of insurance, except for a contract of obligatory State insurance (Article 969).

2. A contract of insurance may be concluded by means of drawing up one document (Article 434[2]) or handing over by the insurer to the insurant on the basis of his written or oral application the insurance policy (or certificate, receipt) signed by the insurer.

In the last instance the consent of the insurant to conclude a contract on the conditions proposed by the insurer shall be confirmed by the acceptance from the insurer of the documents specified in the first paragraph of the present point.

3. The insurer when concluding a contract of insurance shall have the right to use the standard forms of the contract (or insurance policy) for individual types of insurance worked out by it or by an association of insurers.

Article 941. Insurance Under General Policy

1. The systematic insuring of various lots of property of the same kind (goods, cargoes, and the like) on the same conditions during a determined period may by agreement of the insurant with the insurer be effectuated on the basis of one contract of insurance: a general policy.

2. The insurant shall be obliged with respect to each lot of property falling under the operation of the general policy to communicate to the insurer the information stipulated by such policy within the period provided for by it, and if it is not provided for, immediately upon the receipt thereof. The insurant shall not be relieved of this duty even if at the moment of receiving such information the possibility of losses subject to compensation by the insurer already has disappeared.

3. At the demand of the insurant, the insurer shall be obliged to issue insurance policies for individual lots of property falling under the operation of the general policy.

In the event of the failure of the content of the insurance policy to conform to the general policy, preference shall be given to the insurance policy.

Article 942. Material Conditions of Contract of Insurance

1. When concluding a contract of property insurance, agreement must be reached between the insurant and the insurer concerning:

(1) the determined property or other property interest which is the object of insurance;

(2) the character of the event, in the instance of the ensuing of which the insurance is effectuated (insured event);

(3) the amount of the insured amount;

(4) the period of operation of the contract.

2. When concluding a contract of personal insurance, agreement must be reached between the insurant and the insurer concerning:

(1) the insured person;

(2) the character of the event, in the instance of the ensuing of which in the life of the insured person the insurance is effectuated (insured event);

(3) the amount of the insured amount;

(4) the period of operation of the contract.

Article 943. Determination of Conditions of Contract of Insurance in Rules of Insurance

1. The conditions on which a contract of insurance is concluded may be determined in standard rules of insurance of the respective type adopted, approved, or confirmed by the insurer or by an association of insurers (rules of insurance).

2. The conditions contained in the rules of insurance and not included in the text of the contract of insurance (or insurance policy) shall be obligatory for the insurant (or beneficiary) if the application of these rules is expressly specified in the contract (or insurance policy) and the rules themselves have been set out in a single document with the contract (or insurance policy) or on the reverse side thereof or annexed thereto. In the last event the handing over to the insurant when concluding the contract of the rules of insurance must be certified by an entry on the contract.

3. When concluding a contract of insurance, the insurant and the insurer may agree concerning the changing or excluding of individual provisions of the rules of insurance and on adding to the rules.

4. The insurant (or beneficiary) shall have the right to refer in defence of its interests to the rules of insurance of the respective type to which there is a reference in a contract of insurance (or insurance policy) even if these rules by virtue of the present Article are not obligatory for it.

Article 944. Information Granted by Insurant When Concluding Contract of Insurance

1. When concluding a contract of insurance, the insurant shall be obliged to communicate to the insurer the circumstances known to the insurant having material significance for determining the probability of the ensuing of the insured event and the amount of possible losses from the ensuing thereof (insured risk), if these circumstances are not known and should not be known to the insurer.

In any event circumstances specifically stipulated by the insurer in the standard form of a contract of insurance (or insurance policy) or in its written questionnaire shall be deemed to be material.

2. If a contract of insurance was concluded in the absence of the answers of the insurant to any questions of the insurer, the insurer may not subsequently demand dissolution of the contract or the deeming thereof to be invalid on the grounds that the respective circumstances were not communicated by the insurant.

3. If after the conclusion of the contract of insurance it is established that the insurant communicated knowingly false information to the insurer concerning

the circumstances specified in point 1 of the present Article, the insurer shall have the right to require the contract to be deemed invalid and the application of the consequences provided for by Article 179(2) of the present Code.

The insurer may not demand a contract of insurance to be deemed invalid if the circumstances concerning which the insurant was silent already have disappeared.

Article 945. Right of Insurer to Valuation of Insured Risk

1. When concluding a contract of insurance of property, the insurer shall have the right to make an inspection of the property to be insured and, when necessary, shall appoint an expert examination for the purpose of establishing its true value.

2. When concluding a contract of personal insurance, the insurer shall have the right to investigate the person to be insured in order to evaluate the actual state of his health.

3. The valuation of insured risk by the insurer on the basis of the present Article shall be not obligatory for the insurant, who shall have the right to prove otherwise.

Article 946. Secrecy of Insurance

An insurer shall not have the right to divulge information concerning the insurant, insured person, and beneficiary, state of their health, and also the property status of these persons received by it as a result of its professional activity. The insurer, for a violation of the secrecy of insurance, depending upon the nature of the violated rights and the character of the violation, shall bear responsibility in accordance with the rules provided for by Article 139 or Article 150 of the present Code.

Article 947. Insured Amount

1. The amount within whose limits the insurer is obliged to pay insurance compensation under a contract of property insurance or which it is obliged to pay under a contract of personal insurance (insured amount) shall be determined by agreement of the insurant with the insurer in accordance with the rules provided for by the present Article.

2. When insuring property or entrepreneurial risk, unless provided otherwise by the contract of insurance, the insured amount must not exceed the true value thereof (insured value). There shall be considered to be such value:

for property: its true value at the location thereof on the day of concluding the contract of insurance;

for entrepreneurial risk: the losses from entrepreneurial activity which the insurant could anticipate had the ensuing of the insured event not occurred.

3. In contracts of personal insurance and contracts of insurance of civil responsibility, the insured amount shall be determined by the parties at their discretion.

Article 948. Contesting Insured Value of Property

The insured value of property specified in a contract of insurance may not subsequently be contested, except for an instance when the insurer, not taking advantage of his right before concluding the contract to value the insured risk (Article 945[1]), was intentionally deceived with respect to this value.

Article 949. Partial Property Insurance

If in a contract of insurance of property or entrepreneurial risk the insured amount has been established lower than the insured value, the insurer shall, in the instance of the ensuing of the insured event, be obliged to compensate the insurant (or beneficiary) for part of the losses incurred by the last proportionally to the relation of the insured amount to the insured value.

A higher amount of insurance compensation, but not higher than the insured value, may be provided for by a contract.

Article 950. Additional Property Insurance

1. In the event when property or entrepreneurial risk has been insured only for part of the insured value, the insurant (or beneficiary) shall have the right to effectuate additional insurance, including from a different insurer, but so that the total insured amount under all the contracts of insurance does not exceed the insured value.

2. The failure to comply with the provisions of point 1 of the present Article shall entail the consequences provided for by Article 951(4) of the present Code.

Article 951. Consequences of Insurance in Excess of Insured Value

1. If the insured amount specified in the contract of insurance of property or entrepreneurial risk exceeds the insured value, the contract shall be void in that part of the insured amount which exceeds the insured value.

The part of the insurance premium paid in excess shall not be subject to return in this event.

2. If in accordance with the contract of insurance the insurance premium is contributed by instalments and at the moment of establishing the circumstances specified in point 1 of the present Article it was not contributed in full, the balance of the insurance contributions must be paid in the amount reduced proportionally by the reduction of the amount of the insured amount.

3. If an excess of the insured amount in a contract of insurance was a consequence of fraud on the part of the insurant, the insurer shall have the right to demand the contract be deemed to be invalid and compensation of losses caused to it by this in the amount exceeding the amount of insurance premium received by it from the insurant.

4. The rules provided for in points 1–3 of the present Article respectively shall apply also in an instance when the insured amount exceeded the insured value as a result of the insurance of one and the same object by two or several insurers (dual insurance).

The amount of insurance compensation subject to payment in this event by each of the insurers shall be reduced proportionally by the reduction of the initial insured amount under the respective contract of insurance.

Article 952. Property Insurance Against Various Insured Risks

1. Property and entrepreneurial risk may be insured against various insured risks either under one or under individual contracts of insurance, including under contracts with different insurers.

In these instances exceeding the amount of the total insured amount under all contracts over the insured value shall be permitted.

2. If the duty of insurers arises from two or several contracts concluded in accordance with point 1 of the present Article to pay insurance compensation for one and the same consequences of the ensuing of one and the same insured event, the rules provided for by Article 951(4) of the present Code shall apply to such contract in respective part.

Article 953. Co-Insurance

An object of insurance may be insured under one contract of insurance jointly by several insurers (co-insurance). If the rights and duties of each of the insurers have not been determined in such contract, they jointly and severally shall be liable to the insurant (or beneficiary) for payment of the insurance compensation under a contract of property insurance or the insured amount under a contract of personal insurance.

Article 954. Insurance Premium and Insurance Contributions

1. By insurance premium is understood payment for insurance which the insurant (or beneficiary) is obliged to pay to the insurer in the procedure and within the periods which have been established by the contract of insurance.

2. The insurer when determining the amount of insurance premium subject to payment under the contract of insurance shall have the right to apply insurance tariffs worked out by it when determining the premium to be recovered per unit of insured amount, taking into account the object of insurance and the character of the insured risk.

In the instances provided for by a law the amount of insurance premium shall be determined in accordance with insurance tariffs established or regulated by State insurance supervision agencies.

3. If the contribution of the insurance premium by instalments has been provided for by a contract of insurance, the consequences of the failure to pay the regular insurance contributions within the established periods may be determined by the contract.

4. If the insured event ensued before payment of a regular insurance contribution, the contribution of which has been delayed, the insurer shall have the right when determining the amount of insurance compensation subject to payment under a contract of property insurance or insured amount under a contract of personal insurance, to set off the amount of delayed insurance contribution.

Article 955. Replacement of Insured Person

1. In an instance when under a contract of insurance of the risk of responsibility for causing harm (Article 931) the responsibility of a person other than the insurant has been insured, the last shall have the right, unless provided otherwise by the contract, at any time before the ensuing of the insured event to replace this person by another, having informed the insurer thereof in writing.

2. The insured person named in the contract of personal insurance may be replaced by the insurant by another person only with the consent of the insured person himself and the insurer.

Article 956. Replacement of Beneficiary

An insurant shall have the right to replace the beneficiary named in the contract of insurance by another person, having informed the insurer thereof in writing. The replacement of a beneficiary under a contract of personal insurance designated with the consent of the insured person (Article 934[2]) shall be permitted only with the consent of this person.

The beneficiary may not be replaced by another person after he has fulfilled any of the duties under the contract of insurance or presented a demand to the insurer concerning payment of insurance compensation or the insured amount.

Article 957. Commencement of Operation of Contract of Insurance

1. A contract of insurance, unless provided otherwise therein, shall enter into force from the moment of payment of the insurance premium or first contribution thereof.

2. Insurance stipulated by a contract of insurance shall extend to insured events which occurred after the entry of the contract of insurance into force, unless another period for the commencement of the operation of the insurance has been provided for in the contract.

Article 958. Termination Before Time of Contract of Insurance

1. A contract of insurance shall be terminated before the ensuing of the period for which it was concluded if after its entry into force the possibility of the

ensuing of the insured event disappeared and the existence of the insured risk was terminated under circumstances other than the insured event. There shall be relegated to such circumstances, in particular:

the perishing of the insured property for reasons other than the ensuing of the insured event;

termination in the established procedure of entrepreneurial activity by the person who insured the entrepreneurial risk or risk of civil responsibility connected with such activity.

2. The insurant (or beneficiary) shall have the right to repudiate the contract of insurance at any time if at the moment of repudiation the possibility of the ensuing of the insured event had not disappeared under the circumstances specified in point 1 of the present Article.

3. In the event of the termination before time of a contract of insurance under the circumstances specified in point 1 of the present Article, the insurer shall have the right to part of the insurance premium in proportion to the time during which the insurance was in force.

In the event of the repudiation before time by the insurant (or beneficiary) of the contract of insurance, the insurance premium paid to the insurer shall not be subject to return unless provided otherwise by the contract.

Article 959. Consequences of Increase of Insured Risk in Period of Operation of Contract of Insurance

1. In the period of operation of a contract of property insurance the insurant (or beneficiary) shall be obliged immediately to communicate to the insurer significant changes in the circumstances communicated to the insurer when concluding the contract which have become known to him, if such changes may materially influence the increase of the insured risk.

Changes shall be deemed to be significant in any event which are stipulated in the contract of insurance (or insurance policy) and in the rules of insurance transferred to the insurant.

2. An insurer who has been notified about circumstances entailing an increase of the insured risk shall have the right to demand a change of the conditions of the contract of insurance or payment of an additional insurance premium commensurate to the increase of risk.

If the insurant (or beneficiary) objects to changes of the conditions of the contract of insurance or additional payment of the insurance premium, the insurer shall have the right to demand dissolution of the contract in accordance with the rules provided for by Chapter 29 of the present Code.

3. In the event of the failure of the insurant or beneficiary to perform the duty provided for in point 1 of the present Article, the insurer shall have the right to

demand dissolution of the contract of insurance and compensation of losses caused by dissolution of the contract (Article 453[5]).

4. The insurer shall not have the right to demand dissolution of the contract of insurance if the circumstances entailing an increase of the insured risk have disappeared.

5. In the event of personal insurance, the consequences of a change of insured risk in the period of operation of the contract of insurance specified in points 2 and 3 of the present Article may ensue only if they are expressly provided for in the contract.

Article 960. Transfer of Rights to Insured Property to Another Person

In the event of the transfer of rights to insured property from the person in whose interests the contract of insurance was concluded to another person the rights and duties under this contract shall pass to the person to whom the rights to the property passed, except for instances of compulsory seizure of the property on the grounds specified in Article 235(2) of the present Code and renunciation of the right of ownership (Article 236).

The person to whom the rights to insured property have passed must immediately inform the insurer thereof in writing.

Article 961. Informing Insurer About Ensuing of Insured Event

1. An insurant under a contract of property insurance shall, after the ensuing of the insured event became known to him, be obliged immediately to inform the insurer or his representative about the ensuing thereof. If a period and/or means of informing has been provided for by the contract, it must be made within the stipulated period and by the means specified in the contract.

The same duty shall lie on the beneficiary to whom the conclusion of the contract of insurance to his benefit is known if he intends to take advantage of the right to insurance compensation.

2. The failure to fulfil the duty provided for by point 1 of the present Article shall give to the insurer the right to refuse to pay the insurance compensation unless it is proved that the insurer knew in good time about the ensuing of the insured event or that the absence of information about this with the insurer could not affect his duty to pay the insurance compensation.

3. The rules provided for by points 1 and 2 of the present Article respectively shall apply to the contract of personal insurance if the insured event is the death of the insured person or the causing of harm to his health. In so doing the period of notice of the insurer established by the contract may not be less than thirty days.

Article 962. Reducing Losses from Insured Event

1. In the event the insured event ensues which is provided for by the contract of property insurance, the insurant shall be obliged to take reasonable and available measures under the circumstances which formed in order to reduce possible losses.

In taking such measures, the insurant must follow the instructions of the insurer if they have been communicated to the insurant.

2. Expenses for the purpose of reducing losses which are subject to compensation by the insurer, if such expenses were necessary or were made in order to fulfil instructions of the insurer, must be compensated by the insurer even if the respective measures proved to be unsuccessful.

Such expenses shall be compensated proportionally in the relation of the insured amount to the insured value, irrespective of whether they may exceed the insured amount together with the compensation of the other losses.

3. The insurer shall be relieved from compensation of losses which arose as a consequence of the fact that the insurant intentionally did not take reasonable and available measures in order to reduce possible losses.

Article 963. Consequences of Ensuing of Insured Event Through Fault of Insurant, Beneficiary, or Insured Person

1. An insurer shall be relieved from the payment of insurance compensation or insured amount if the insured event ensued as a consequence of the intent of the insurant, beneficiary, or insured person, except for instances provided for by points 2 and 3 of the present Article.

Instances of the relief of the insurer from the payment of insurance compensation under contracts of property insurance in the event of the ensuing of the insured event as a consequence of gross negligence of the insurant or beneficiary may be provided for by a law.

2. An insurer shall not be relieved from the payment of insurance compensation under a contract of insurance of civil responsibility for causing harm to life or health if the harm was caused through the fault of the person responsible therefor.

3. An insurer shall not be relieved from the payment of the insured amount which under a contract of personal insurance is subject to payment in the event of the death of the insured person if his death ensued as a consequence of suicide and the contract of insurance has operated at this time for not less than two years.

Article 964. Grounds for Relief of Insurer from Payment of Insurance Compensation and Insured Amount

1. Unless provided otherwise by a law or contract of insurance, the insurer shall be relieved from the payment of insurance compensation and the insured amount when the insured event has ensued as a consequence of:

the influence of a nuclear explosion, radiation, or other radioactive poisoning;
military actions, and also maneuvers or other military measures;
civil war, popular uprisings of any nature, or strikes.

2. Unless provided otherwise by a contract of property insurance, the insurer shall be relieved from the payment of insurance compensation for losses which arose as a consequence of the seizure, confiscation, requisition, arrest, or destruction of insured property at the instruction of State agencies.

Article 965. Transfer to Insurer of Rights of Insurant to Compensation of Damage (Subrogation)

1. Unless provided otherwise by the contract of property insurance, to the insurer who has paid the insurance compensation shall pass within the limits of the amount paid the right of demand which the insurant (or beneficiary) has against the person responsible for the losses compensated as a result of the insurance. However, a condition of the contract excluding the transfer to the insurer of the right of demand against a person who intentionally caused losses shall be void.

2. The right of demand which has passed to the insurer shall be effectuated by him in compliance with the rules regulating relations between the insurant (or beneficiary) and the person responsible for the losses.

3. An insurant (or beneficiary) shall be obliged to transfer to the insurer all documents and evidence and to communicate to him all information necessary for the effectuation by the insurer of the right of demand which has passed to him.

4. If the insurant (or beneficiary) has waived his right of demand against the person responsible for losses compensated by the insurer, or the effectuation of this right has become impossible through the fault of the insurant (or beneficiary), the insurer shall be relieved from the payment of insurance compensation in full or in respective part and shall have the right to demand the return of the amount of compensation paid in excess.

Article 966. Limitations with Regard to Demands Connected with Property Insurance

A suit with regard to demands arising from a contract of property insurance may be presented within two years.

Article 967. Reinsurance

1. The risk of payment of insurance compensation or the insured amount assumed by an insurer under a contract of insurance may be insured by him in full or partially with another insurer(s) under a contract of reinsurance concluded with the last.

2. The rules provided for by the present Chapter which are subject to application with respect to the insurance of entrepreneurial risk shall apply to the contract of reinsurance unless provided otherwise by the contract of reinsurance. In so doing the insurer under a contract of insurance (principal contract) which has concluded a contract of reinsurance shall be considered to be the insurant in this last contract.

3. Under reinsurance the insured under this contract shall remain responsible to the insurant under the principal contract of insurance for the payment of insurance compensation or the insured amount.

4. The subsequent conclusion of two or several contracts of reinsurance shall be permitted.

Article 968. Mutual Insurance

1. Citizens and juridical persons may insure their property and other property interests specified in Article 929(2) of the present Code on a mutual basis by means of combining in mutual insurance societies the means necessary for this.

2. The mutual insurance societies shall effectuate the insurance of property and other property interests of their members and shall be noncommercial organisations.

The peculiarities of the legal status of mutual insurance societies and the conditions of their activity shall be determined in accordance with the present Code by a law on mutual insurance.

3. Insurance of property and property interests of their members by mutual insurance societies shall be effectuated directly on the basis of membership unless the constitutive documents of the society provide for the conclusion of contracts of insurance in these instances.

The rules provided for by the present Chapter shall apply to relations relating to insurance between a mutual insurance society and its members unless provided otherwise by the law on mutual insurance, constitutive documents of the respective society, or rules of insurance established by it.

4. The effectuation of obligatory insurance by means of mutual insurance shall be permitted in the instances provided for by the law on mutual insurance.

5. A mutual insurance society may effectuate as insurer the insurance of the interests of persons who are not members of the society if such insurance activity is provided for by its constitutive documents, the society is formed in the form of a commercial organisation, has an authorisation (or license) for the effectuation of insurance of the respective type, and is liable for other demands established by a law on the organisation of insuring.

The insurance of the interests of persons who are not members of a mutual insurance society shall be effectuated by the society under contracts of insurance in accordance with the rules provided for by the present Chapter.

Article 969. Obligatory State Insurance

1. For the purpose of ensuring the social interests of citizens and the interests of the State obligatory State insurance of the life, health, and property of State employees of determined categories may be established by a law.

Obligatory State insurance shall be effectuated at the expense of means allotted for these purposes from the respective budget to ministries and other federal agencies of executive power (insurants).

2. Obligatory State insurance shall be effectuated directly on the basis of laws and other legal acts concerning such insurance by State insurance and other State organisations specified in these acts (insurers) or on the basis of contracts of insurance concluded in accordance with these acts by the insurers and insurants.

3. Obligatory State insurance shall be paid for by the insurers in the amount determined by laws and other legal acts concerning such insurance.

4. The rules provided for by the present Chapter shall apply to obligatory State insurance unless provided otherwise by laws and other legal acts concerning such insurance and does not arise from the essence of the respective relations relating to insurance.

Article 970. Application of General Rules on Insurance to Special Types of Insurance

The rules provided for by the present Chapter shall apply to relations with regard to the insurance of foreign investments against noncommercial risks, marine insurance, medical insurance, insurance of bank accounts, and insurance of pensions insofar as not established otherwise by laws on these types of insurance.

Chapter 49. Commission

Article 971. Contract of Commission

1. Under a contract of commission one party (the attorney) shall be obliged to perform in the name of and at the expense of another person (the principal) determined legal actions. The rights and duties with regard to the transaction concluded by the attorney shall arise directly with the principal.

2. A contract of commission may be concluded with or without a specification of the period during which the attorney has the right to act in the name of the principal.

Article 972. Remuneration of Attorney

1. The principal shall be obliged to pay remuneration to the attorney if this is provided for by a law, other legal acts, or the contract of commission.

In instances when a contract of commission is connected with the effectuation of entrepreneurial activity by both parties or one of them, the principal shall be obliged to pay remuneration to the attorney unless provided otherwise by the contract.

2. In the absence in a contract of commission for compensation of conditions concerning the amount of remuneration or the procedure for the payment thereof, the remuneration shall be paid after the performance of the commission in the amount determined in accordance with Article 424(3) of the present Code.

3. The attorney acting as a commercial representative (Article 184[1]) shall have the right in accordance with Article 359 of the present Code to withhold things situated with him which are subject to transfer to the principal as security for his demands under the contract of commission.

Article 973. Performance of Commission in Accordance with Instructions of Principal

1. An attorney shall be obliged to perform the commission given to him in accordance with the instructions of the principal. The instructions of the principal must be lawful, practicable, and specific.

2. An attorney shall have the right to deviate from instructions of the principal if under the circumstances of the case this is necessary in the interests of the principal and the attorney could not inquire in advance of the principal or did not receive a reply to his enquiry within a reasonable period. An attorney shall be obliged to inform the principal about deviations permitted as soon as notice became possible.

3. The right to deviate in the interests of the principal from his instructions without a preliminary enquiry about this may be granted to an attorney acting as a commercial representative (Article 184[1]). In this event the commercial representative shall be obliged within a reasonable period to inform the principal about deviations permitted unless provided otherwise by the contract of commission.

Article 974. Duties of Attorney

An attorney shall be obliged to:

personally perform the commission given to him, except for instances specified in Article 976 of the present Code;

communicate to the principal at his demand all information concerning the course of the performance of the commission;

transfer to the principal without delay everything received under the transactions concluded in performance of the commission;

upon the performance of the commission or in the event of the termination of the contract of commission before the performance thereof, return without delay the power of attorney to the attorney, the period of operation of which has not lapsed, and submit a report with documents of justification appended, if this is required according to the conditions of the contract or the character of the commission.

Article 975. Duties of Principal

1. The principal shall be obliged to issue a power(s) of attorney to the attorney for the performance of legal actions provided for by the contract of commission, except for instances provided for by Article 182(1), paragraph two, of the present Code.

2. The principal shall be obliged, unless provided otherwise by the contract, to:

compensate the attorney for costs incurred;

provide the attorney with the means necessary for the performance of the commission.

3. The principal shall be obliged without delay to accept from the attorney everything performed by him in accordance with the contract of commission.

4. The principal shall be obliged to pay remuneration to the attorney if in accordance with Article 972 of the present Code the contract of commission is to be compensated.

Article 976. Transfer of Performance of Commission

1. An attorney shall have the right to transfer the performance of a commission to another person (deputy) only in the instances and on the conditions provided for by Article 187 of the present Code.

2. The principal shall have the right to remove the deputy selected by the attorney.

3. If a possible deputy of the attorney is named in the contract of commission, the attorney shall not be liable either for his selection or for the conducting of affairs by him.

If the right of the attorney to transfer performance of a commission to another person has not been provided for in the contract, or has been provided for but the deputy is not named therein, the attorney shall be liable for the choice of the deputy.

Article 977. Termination of Contract of Commission

1. A contract of commission shall be terminated as a consequence of:
revocation of the commission by the principal;
refusal of the attorney;
death of the principal or attorney, deeming either of them to lack dispositive legal capacity, have limited dispositive legal capacity, or to be missing.

2. The principal shall have the right to revoke a commission, and the attorney to renounce it, at any time. An agreement concerning the waiver of this right shall be void.

3. A party who has repudiated a contract of commission providing for the actions of an attorney as a commercial representative must inform the other party about the termination of the contract not later than thirty days in advance unless a longer period has been provided for by the contract.

In the event of the reorganisation of a juridical person which is a commercial representative, the principal shall have the right to revoke the commission without such prior notice.

Article 978. Consequences of Termination of Contract of Commission

1. If a contract of commission is terminated before the commission is performed by the attorney in full, the principal shall be obliged to compensate the attorney for the costs incurred by him when performing the commission, and when remuneration was due to the attorney, also to pay remuneration to him commensurate with the work fulfilled by him. This rule shall not apply to the performance by the attorney of the commission after he knew or should have known about the termination of the commission.

2. The revocation of the commission by the principal shall not be grounds for compensation of losses due to the attorney by the termination of the contract of commission, except for instances of termination of the contract providing for the actions of an attorney as a commercial representative.

3. The refusal of the attorney to perform the commission of the principal shall not be grounds for the compensation of losses caused to the principal by the termination of the contract of commission, except for instances of the refusal of an attorney in conditions when the principal is deprived of the possibility to otherwise ensure his interests, and also of the refusal of the performance of the contract providing for actions of an attorney as a commercial representative.

Article 979. Duties of Heirs of Attorney and Liquidator of Juridical Person Which is Attorney

In the event of the death of an attorney, his heirs shall be obliged to notify the principal about the termination of the contract of commission and take measures necessary in order to protect the property of the principal, in particular, to preserve his things and documents, and then to transfer this property to the principal.

The same duty shall lie on the liquidator of a juridical person which is an attorney.

Chapter 50. Actions in Another's Interest Without Commission

Article 980. Conditions of Actions in Another's Interest

1. Actions without a commission, other instruction, or consent of the interested person promised in advance for the purpose of preventing harm to his person or property, performance of his obligation or in his other not unlawful interests (actions in another's interest) must be performed by proceeding from the obvious advantage or benefit and actual or probable intentions of the interested person and with the necessary concern and circumspection under the circumstances of the case.

2. The rules provided for by the present Chapter shall not apply to actions in the interest of other persons to be performed by State and municipal agencies for which such actions are one of the purposes of their activity.

Article 981. Notification of Interested Person Concerning Actions in His Interest

1. The person acting in another's interest shall be obliged at the first opportunity to notify the interested person thereof and await his decision within a reasonable period concerning approval or disapproval of the actions undertaken if only such waiting does not entail serious damage for the interested person.

2. It shall not be required to specially notify the interested citizen about actions in his interest if such actions are undertaken in his presence.

Article 982. Consequences of Approval by Interested Person of Actions in His Interest

If a person in whose interest actions are undertaken without his commission approves these actions, the rules concerning the contract of commission or other contract corresponding to the character of the actions undertaken, even if approval was oral, shall apply to relations of the parties thereafter.

Article 983. Consequences of Failure of Interested Person to Approve Actions in His Interest

1. Actions in another's interest performed after it became known to he who performed them that they are not approved by the interested person shall not entail for the last duties either with respect to those actions performed or with respect to third persons.

2. Actions for the purpose of preventing danger for the life of a person who has turned out to be in danger shall be permitted also against the will of this person, and performance of a duty with respect to someone being maintained, against the will of he on whom this duty lies.

Article 984. Compensation of Losses to Person Who Acted in Another's Interest

1. Necessary expenses and other real damage incurred by the person who acted in another's interest in accordance with the rules provided for by the present Chapter shall be subject to compensation by the interested person, except for expenses which were caused by the actions specified in Article 983(1) of the present Code.

The right to compensation of necessary expenses and other real damage shall also be preserved when the actions in another's interest did not lead to the result presupposed. However, in the event of the prevention of damage to the property of another person the amount of compensation must not exceed the value of the property.

2. Expenses and other losses of the person who acted in another's interest incurred by him in connection with actions which were undertaken after receiving the approval of the interested person (Article 982) shall be compensated according to the rules on the contract of the corresponding type.

Article 985. Remuneration for Actions in Another's Interest

The person whose actions in another's interest led to a positive result for the interested person shall have the right to receive remuneration if such right is provided for by a law, agreement with the interested person, or the customs of business turnover.

Article 986. Consequences of Transaction in Another's Interest

Duties with regard to a transaction concluded in another's interest shall pass to the person in whose interest it was concluded on condition of the approval by him of this transaction and if the other party does not object to such transfer or when concluding the transaction knew or should have known that the transaction was concluded in another's interest.

In the event of the transfer of duties under this transaction to the person in whose interests it was concluded, to the last must be transferred also the rights with regard to this transaction.

Article 987. Unfounded Enrichment as Consequence of Actions in Another's Interest

If actions directly not directed towards ensuring the interests of another person, including in an instance when the person who performed them mistakenly presupposed that he acts in his own interest, led to the unfounded enrichment of the other person, the rules provided for by Chapter 60 of the present Code shall apply.

Article 988. Compensation of Harm Caused by Actions in Another's Interest

Relations with regard to compensation of harm caused by actions in another's interest to the interested person or to third persons shall be regulated by the rules provided for by Chapter 59 of the present Code.

Article 989. Report of Person Who Acted in Another's Interest

The person who acted in another's interest shall be obliged to submit to the person in whose interests such actions were effectuated a report specifying the revenues received and expenses incurred and other losses.

Chapter 51. Commission Agency

Article 990. Contract of Commission Agency

1. Under a contract of commission agency, one party (commission agent) shall be obliged on behalf of another party (committent) to conclude for remuneration one or several transactions in his own name, but at the expense of the committent.

The commission agent shall become obliged with regard to a transaction concluded by the commission agent with a third person and shall acquire the rights, even though the committent also was named in the transaction or entered into direct relations with the third person in regard to performance of the transaction.

2. A contract of commission agency may be concluded for a determined period or without specifying the period of its operation, with or without specifying the territory of the performance thereof, with or without the obligation of the committent not to grant to third persons the right to conclude a transaction in his interests and at his expense, the conclusion of which has been entrusted to the commission agent, and with or without conditions relative to the assortment of goods which are the subject of the commission agency.

3. The peculiarities of individual types of commission agency contract may be provided for by a law and other legal acts.

Article 991. Commission Agency Remuneration

1. The committent shall be obliged to pay remuneration to the commission agent, and in the event the commission agent has assumed the guaranty for performance of the transaction by a third person (del credere), also additional remuneration in the amount and in the procedure established in the contract of commission agency.

If the amount of remuneration or the procedure for the payment thereof has not been provided for by the contract and the amount of remuneration cannot be determined by proceeding from the conditions of the contract, the remuneration shall be paid after the performance of the contract of commission agency in the amount determined in accordance with Article 424(3) of the present Code.

2. If the contract of commission agency was not performed for reasons dependent upon the committent, the commission agent shall retain the right to the commission agency remuneration, and also to compensation for expenses incurred.

Article 992. Performance of Commission Agency Commission

A commission agent shall be obliged to perform the commission assumed on the conditions most advantageous for the committent in accordance with the instructions of the committent, and in the absence of such instructions in the contract of commission agency, in accordance with the customs of business turnover or other requirements ordinarily presented.

In an instance when the commission agent has concluded a transaction on conditions more advantageous than those which were specified by the committent, the additional advantage shall be divided between the committent and the commission agent equally unless provided otherwise by agreement of the parties.

Article 993. Responsibility for Failure to Perform Transaction Concluded for Committent

1. The commission agent shall not be liable to the committent for the failure of a third person to perform a transaction concluded with him at the expense of the committent, except for instances when the commission agent did not display the necessary caution in the selection of this person or assumed a commission to perform the transaction (del credere).

2. In the event of the failure of a third person to perform a transaction concluded with him by the commission agent, the commission agent shall be obliged immediately to inform the committent thereof, to collect the necessary evidence, and also at the demand of the committent to transfer rights to him regarding such transaction in compliance with the rules concerning the assignment of a demand (Articles 382–386, 388, 389).

3. The assignment of rights to the committent with regard to a transaction on the basis of point 2 of the present Article shall be permitted irrespective of the agreement of the commission agent with the third person prohibiting or limiting such assignment. This shall not relieve the commission agent from responsibility to the third person in connection with the assignment of the right in violation of the agreement concerning the prohibition or limitation thereof.

Article 994. Sub-Commission Agency

1. Unless provided otherwise by the contract of commission agency, the commission agent shall have the right for the purpose of performance of this contract

to conclude a contract of sub-commission agency with another person while remaining responsible for the actions of the sub-commission agent to the committent.

Under a contract of sub-commission agency, the commission agent shall acquire the rights and duties of the committent with respect to the sub-commission agent.

2. Until the termination of the contract of commission agency, the committent shall not have the right without the consent of the commission agent to enter into direct relations with the sub-commission agent, unless provided otherwise by the contract of commission agency.

Article 995. Deviation from Instructions of Committent

1. The commission agent shall have the right to deviate from the instructions of the committent if under the circumstances of the case this is necessary in the interests of the committent and the commission agent could not ask the committent in advance nor receive a reply within a reasonable period to his enquiry. The commission agent shall be obliged to inform the committent about the deviations permitted as soon as notification becomes possible.

The right to deviate from his instructions without prior enquiry may be granted to the commission agent acting as an entrepreneur. In this event the commission agent shall be obliged within a reasonable period to inform the committent about the deviations permitted, unless provided otherwise by the contract of commission agency.

2. A commission agent who has sold property at a price lower than agreed with the committent shall be obliged to compensate the last for the difference unless it is proved that he had no possibility to sell the property at the agreed price and the sale for a lower price prevented more losses. In an instance when the commission agent was obliged to ask the committent in advance, the commission agent also must prove that he had no possibility to receive the consent in advance of the committent for the deviation from his instructions.

3. If the commission agent purchased property at a price higher than agreed with the committent, the committent, not wishing to accept such purchase, shall be obliged to declare this to the commission agent within a reasonable period upon receipt of notice thereof concerning the conclusion of a transaction with a third person. Otherwise, the purchase shall be deemed to be accepted by the committent.

If the commission agent communicated that he accepts the difference in price at his own expense, the committent shall not have the right to repudiate the transaction concluded for him.

Article 996. Rights to Things Which Are Subject of Commission Agency

1. Things received by the commission agent from the committent or acquired by the commission agent at the expense of the committent shall be the ownership of the last.

2. The commission agent shall have the right in accordance with Article 359 of the present Code to withhold things which he has and which are subject to transfer to the committent or to the person specified by the committent to secure his demands under the contract of commission agency.

In the event the committent is declared to be insolvent (or bankrupt), the said right of the commission agent shall terminate, and his demands against the committent within the limits of the value of things which he withheld shall be satisfied in accordance with Article 360 of the present Code equally with the demands secured by a pledge.

Article 997. Satisfaction of Demands of Commission Agent from Amounts Due to Committent

A commission agent shall have the right in accordance with Article 410 of the present Code to withhold amounts due to him under the contract of commission agency from all the amounts received by him on the account of the committent. However, the creditors of the committent who enjoy preference with respect to priority of satisfaction of their demands to the pledgeholder shall not be deprived of the right to satisfaction of these demands from the amounts withheld by the commission agent.

Article 998. Responsibility of Commission Agent for Loss, Shortage, or Damage to Property of Committent

1. A commission agent shall be liable to the committent for loss, shortage, or damage to the property of the committent situated with him.

2. If the commission agent accepts the property sent by the committent or which comes to the commission agent for the committent, and this property proves to be damaged or deficit, which may be noted from an external inspection, and also damage has been caused by someone to the property of the committent situated with the commission agent, the commission agent shall be obliged to take measures relating to the protection of the rights of the committent, to gather the necessary evidence, and immediately to notify the committent about everything.

3. A commission agent who has not insured property of the committent situated with him shall be liable to this person in instances when the committent instructed him to insure the property at the expense of the committent or the insuring of this property by the commission agency has been provided for by the contract of commission agency or the customs of business turnover.

Article 999. Report of Commission Agent

The commission agent shall be obliged to submit a report to the committent with regard to the performance of a commission and to transfer to him everything received under the contract of commission agency. The committent who has objections with regard to the report must communicate these to the commission agent within thirty days from the date of receipt of the report, unless a different period has been established by agreement of the parties. Otherwise, the report shall be considered to be accepted in the absence of another agreement.

Article 1000. Acceptance by Committent of That Performed Under the Contract of Commission Agency

The committent shall be obliged to:

accept from the commission agent everything received under the contract of commission agency;

inspect the property acquired for him by the commission agent and notify the last without delay about the defects discovered in this property;

relieve the commission agent from obligations assumed by him to a third person with regard to the performance of the commission agency commission.

Article 1001. Compensation of Expenses for Performance of Commission Agency Commission

The committent shall be obliged, in addition to payment of the commission agency remuneration, and in respective instances also additional remuneration for del credere, to compensate the commission agent for amounts expended by him for performance of the commission agency commission.

The commission agent shall not have the right to compensation of expenses for the keeping of property of the committent situated with him, unless established otherwise in a law or by the contract of commission agency.

Article 1002. Termination of Contract of Commission Agency

A contract of commission agency shall terminate as a consequence of:

refusal of the committent to perform the contract;

refusal of the commission agent to perform the contract in instances provided for by a law or the contract;

death of the commission agent, deeming him to lack dispositive legal capacity, limited dispositive legal capacity, or to be missing;

deeming of an individual entrepreneur who is a commission agent to be insolvent (or bankrupt).

In the event a commission agent is declared to be insolvent (or bankrupt), his rights and duties with regard to transactions concluded by him for the committent in performance of the instructions of the last shall pass to the committent.

Article 1003. Revocation of Commission Agency Commission by Committent

1. The committent shall have the right at any time to refuse performance of the contract of commission agency, having revoked the commission given to the commission agent. The commission agent shall have the right to demand compensation of losses caused by the revocation of the commission.

2. In an instance when the contract of commission agency was concluded without an indication of the period of its operation, the committent must inform the commission agent about the termination of the contract not later than thirty days in advance unless a more extended period of notification has been provided for by the contract.

In this event the committent shall be obliged to pay remuneration to the commission agent for transactions concluded by him before the termination of the contract, and also to compensate the commission agent for expenses incurred by him before the termination of the contract.

3. In the event of the revocation of the commission, the committent shall be obliged within the period established by the contract of commission agency, and if such period has not been established, immediately to dispose of his property located within the jurisdiction of the commission agent. If the committent does not fulfil this duty, the commission agent shall have the right to hand over the property for keeping at the expense of the committent or to sell it at the price as advantageous as possible for the committent.

Article 1004. Refusal of Commission Agent to Perform Contract of Commission Agency

1. A commission agent shall not have the right, unless provided otherwise by the contract of commission agency, to refuse the performance thereof, except for an instance when the contract was concluded without an indication of the period of its operation. In this event the commission agent must inform the committent about the termination of the contract not later than thirty days in advance, unless a more extended period of notification has been provided for by the contract.

A commission agent shall be obliged to take measures necessary in order to ensure the preservation of the property of the committent.

2. The committent must dispose of property within the jurisdiction of the commission agent within fifteen days from the date of receipt of the notification about the refusal of the commission agent to perform the commission, unless another period has been established by the contract of commission agency. If he does not fulfil this duty, the commission agent shall have the right to hand over the property for keeping at the expense of the committent or to sell it at the price as advantageous as possible for the committent.

3. Unless provided otherwise by the contract of commission agency, the commission agent who has refused to perform a commission shall retain the right to commission agency remuneration for the transactions concluded by him before termination of the contract, and also to compensation for expenses incurred by him up to that moment.

Chapter 52. Agency

Article 1005. Contract of Agency

1. Under a contract of agency one party (agent) shall be obliged for remuneration to perform on behalf of another party (principal) legal and other actions in his own name but at the expense of the principal, or in the name of and at the expense of the principal.

With regard to a transaction concluded by the agent with a third person in his own name and at the expense of the principal the agent shall acquire the rights and become obliged even though the principal also was named in the transaction or entered with the third person into direct relations with regard to performance of the transaction.

With regard to a transaction concluded by an agent with a third person in the name of and at the expense of the principal, the rights and duties shall arise directly with the principal.

2. In instances when in a contract of agency concluded in written form the general powers of an agent to conclude transactions in the name of the principal have been provided for, the last shall not have the right in relations with third persons to refer to the absence of the agent's proper powers unless it is proved that the third person knew or should have known about the limitation of the powers of the agent.

3. The contract of agency may be concluded for a determined period or without specifying the period of operation thereof.

4. The peculiarities of individual types of contract of agency may be provided for by a law.

Article 1006. Agent Remuneration

The principal shall be obliged to pay remuneration to the agent in an amount and in the procedure established in the contract of agency.

If the amount of agent remuneration has not been provided for in the contract of agency and it cannot be determined by proceeding from the conditions of the contract, the remuneration shall be subject to payment in the amount determined in accordance with Article 424(3) of the present Code.

In the absence in the contract of conditions concerning the payment of agent remuneration, the principal shall be obliged to pay remuneration within a week from the moment of the submission of the report to him by the agent for the preceding period if from the essence of the contract or customs of business turnover a different procedure for the payment of remuneration does not arise.

Article 1007. Limitations by Contract of Agency of Rights of Principal and Agent

1. The obligation of the principal not to conclude analogous contracts of agency with other agents operating on a territory determined in the contract, or to refrain from the effectuation of autonomous activity on this territory or analogous activity comprising the subject of the contract of agency, may be provided for by the contract of agency.

2. The obligation of the agent not to conclude analogous contracts of agency with other principals which must be performed on the territory wholly or partially coinciding with the territory specified in the contract may be provided for by the contract of agency.

3. The conditions of the contract of agency by virtue of which the agent has the right to sell goods, fulfil work, or render services exclusively to a determined category of purchasers (or customers) or exclusively to purchasers (or customers) having a location or place of residence on the territory determined in the contract shall be void.

Article 1008. Reports of Agent

1. In the course of the performance of the contract of agency the agent shall be obliged to submit reports to the principal in the procedure and within the periods which have been provided for by the contract. In the absence in the contract of respective conditions the reports shall be submitted by the agent as the contract is performed by him or at the end of the operation of the contract.

2. Unless provided otherwise by the contract of agency, necessary evidence of expenses made by the agent at the expense of the principal must be appended to the report of the agent.

3. The principal having objections to the report of the agent must communicate to the agent about them within thirty days from the date of receipt of the report unless a different period has been established by agreement of the parties. Otherwise the report shall be considered to be accepted by the principal.

Article 1009. Contract of Subagency

1. Unless provided otherwise by the contract of agency, an agent shall have the right for the purpose of performance of the contract to conclude a contract of subagency with another person while remaining responsible for the actions of the subagent to the principal. The duty of the agent to conclude a contract of

subagency, with or without specification of the specific conditions of such contract, may be provided for in the contract of agency.

2. A subagent shall not have the right to conclude transactions with third persons in the name of the person who is the principal under the contract of agency, except for instances when in accordance with Article 187(1) of the present Code the subagent may act on the basis of a transfer of power of attorney. The procedure and consequences of such transfer of power of attorney shall be determined according to the rules provided for by Article 976 of the present Code.

Article 1010. Termination of Contract of Agency

The contract of agency shall be terminated as a consequence of:
the refusal of one of the parties to perform the contract concluded without determination of the period of the ending of its operation;
the death of the agent, his being deemed to lack dispositive legal capacity, to have limited dispositive legal capacity, or to be missing;
the deeming of an individual entrepreneur who is an agent to be insolvent (or bankrupt).

Article 1011. Application to Agent Relations of Rules Concerning Contracts of Commission and Commission Agency

The rules provided for by Chapter 49 or Chapter 51 of the present Code shall apply to relations arising from the Contract of Agency respectively, depending upon whether the agent acts under the conditions of this contract in the name of the principal or in his own name, unless these rules are contrary to the provisions of the present Chapter or the essence of the contract of agency.

Chapter 53. Trust Management of Property

Article 1012. Contract of Trust Management of Property

1. Under a contract of trust management of property one party (founder of management) shall transfer to the other party (trustee manager) property in trust management for a determined period, and the other party shall be obliged for remuneration to effectuate the management of this property in the interests of the founder of the management or person specified therein (beneficiary).
The transfer of property to trust management shall not entail the transfer of the right of ownership thereto to the trustee manager.

2. In effectuating trust management of property, the trustee manager shall have the right to perform with respect to this property in accordance with the contract of trust management any legal and factual actions in the interests of the beneficiary.
Limitations with respect to individual actions relating to the trust management of property may be provided for by a law or contract.

3. Transactions with property transferred to trust management the trustee manager shall conclude in his own name, indicating in so doing that he acts as such manager. This condition shall be considered to be complied with if when performing actions not requiring written formalisation the other party has been informed about the performance thereof by the trustee manager in this capacity, and in written documents by making the notation 'D.U.' after the name of the trustee manager.

In the absence of an indication concerning the action of the trustee manager in this capacity, the trustee manager shall be obliged to third persons personally and shall be liable to them only with the property belonging to him.

Article 1013. Object of Trust Management

1. Enterprises and other property complexes, individual objects relegated to immovable property, securities, rights certified by paperless securities, exclusive rights, and other property may be an object of trust management.

2. Money may not be an autonomous object of trust management, except for instances provided for by a law.

3. Property in economic jurisdiction or operative management may not be transferred to trust management. The transfer to trust management of property in economic jurisdiction or operative management shall be possible only after the liquidation of the juridical person in whose economic jurisdiction or operative management the property is situated, or the termination of the right of economic jurisdiction or operative management of property and the entry thereof into the possession of the owner on other grounds provided for by a law.

Article 1014. Founder of Management

The owner of property, and in instances provided for by Article 1026 of the present Code, another person, shall be the founder of trust management.

Article 1015. Trustee Manager

1. An individual entrepreneur or commercial organisation, except for a unitary enterprise, may be a trustee manager.

In instances when the trust management of property is effectuated on the grounds provided for by a law, the trustee manager may be a citizen who is not an entrepreneur or a noncommercial organisation, except an institution.

2. Property shall not be subject to transfer in trust management to a State agency or agency of local self-government.

3. The trustee manager may not be a beneficiary under the contract of trust management of property.

Article 1016. Material Conditions of Contract of Trust Management of Property

1. There must be specified in the contract of trust management of property:

the composition of the property being transferred on trust management;

the name of the juridical person or the name of the citizen in whose interests the management of the property shall be effectuated (the founder of the management or the beneficiary);

the amount and form of remuneration to the manager, if payment of remuneration has been provided for by the contract;

the period of operation of the contract.

2. The contract of trust management of property shall be concluded for a period not exceeding five years. For individual types of property transferred in trust management other maximum periods for which the contract may be concluded may be established by a law.

In the absence of the statement by one of the parties concerning the termination of the contract regarding the end of the period of its operation, it shall be considered to be extended for the same period and on the same conditions as were provided for by the contract.

Article 1017. Form of Contract of Trust Management of Property

1. A contract of trust management of property must be concluded in written form.

2. A contract of trust management of immovable property must be concluded in the form provided for the contract of sale of immovable property. The transfer of immovable property in trust management shall be subject to State registration in the same procedure as the transfer of the right of ownership to this property.

3. The failure to comply with the form of the contract of trust management of property or the requirement concerning registration of the transfer of immovable property in trust management shall entail the invalidity of the contract.

Article 1018. Solitary Property in Trust Management

1. Property transferred in trust management shall be solitary from other property of the founder of the management, and also from property of the trustee manager. This property shall be reflected on a separate balance sheet of the trustee manager and an autonomous account shall be kept with regard to it. A separate bank account shall be opened in order to settle accounts with regard to activity connected with trust management.

2. Levy of execution for debts of the founder of the management on property transferred by him in trust management shall not be permitted, except for the

insolvency (or bankruptcy) of this person. In the event of bankruptcy of the founder of the management, the trust management of this property shall terminate and it shall be included in the bankruptcy mass.

Article 1019. Transfer in Trust Management of Property Encumbered by Pledge

1. The transfer of pledged property to trust management shall not deprive the pledgeholder of the right to levy execution on this property.

2. The trustee manager must be warned that the property transferred to him in trust management is encumbered by a pledge. If the trustee manager did not know and should not have known about the encumberment of the property by a pledge which is transferred to him in trust management, he shall have the right to demand in court the dissolution of the contract of trust management of property and payment of the remuneration due to him under the contract for one year.

Article 1020. Rights and Duties of Trustee Manager

1. The trustee manager shall effectuate within the limits provided for by a law and the contract of trust management of property the powers of the owner with respect to property transferred in trust management. The disposition of immovable property the trustee manager shall effectuate in the instances provided for by the contract of trust management.

2. The rights acquired by the trustee manager as a result of actions relating to the trust management of property shall be included in the property transferred in trust management. The duties which arise as a result of such actions of the trustee manager shall be performed at the expense of this property.

3. In order to defend the rights to property in trust management the trustee manager shall have the right to demand any elimination of a violation of his rights (Articles 301, 302, 304, and 305).

4. The trustee manager shall submit to the founder of the management and beneficiary a report concerning his activity within the periods and in the procedure established by the contract of trust management of property.

Article 1021. Transfer of Trust Management of Property

1. The trustee manager shall effectuate trust management of property personally, except for instances provided for by point 2 of the present Article.

2. The trustee manager may charge another person to perform actions in the name of the trustee manager which are necessary for the management of the property if he is empowered to do so by the contract of trust management of the property, or has received the consent of the founder to do so in written form, or is forced to do so by virtue of circumstances in order to ensure the interests of the

founder of the management or beneficiary and does not have in so doing the possibility to receive instructions of the founder of the management within a reasonable period.

The trustee manager shall be liable for actions of the entrusted person he selected as though for his own.

Article 1022. Responsibility of Trustee Manager

1. The trustee manager who has not displayed due concern in the trust management of property for the interests of the beneficiary or founder of the management shall compensate the beneficiary for lost advantage for the period of trust management of the property and the founder of the management for losses caused by the loss of or damage to the property, taking into account its natural wear and tear, and also lost advantage.

The trustee manager shall bear responsibility for losses caused unless it is proved that these losses occurred as a consequence of insuperable force or the actions of the beneficiary or the founder of the management.

2. Obligations relating to a transaction concluded by the trustee manager in excess of the powers granted to him or in violation of the limitations established for him shall be borne by the trustee manager personally. If third persons participating in the transaction did not know and should not have known about the exceeding of powers or the limitations established, the obligations which arose shall be subject to performance in the procedure established by point 3 of the present Article. The founder of the management may in this event require of the trustee manager compensation for losses incurred by him.

3. Debts relating to obligations which arose in connection with trust management of property shall be paid at the expense of this property. In the event this property is insufficient, recovery may be levied on the property of the trustee manager, and if his property also is insufficient, on the property of the founder of the management not transferred to trust management.

4. The contract of trust management of property may provide for the granting of a pledge by the trustee manager to secure compensation of losses which may be caused to the founder of the management or to the beneficiary by the improper performance of the contract of trust management.

Article 1023. Remuneration of Trustee Manager

The trustee manager shall have the right to remuneration provided for by the contract of trust management of property, and also to compensation of necessary expenses made by him in the trust management of property at the expense of revenues from the use of this property.

Article 1024. Termination of Contract of Trust Management of Property

1. A contract of trust management of property shall be terminated as a consequence of:

the death of a citizen who is a beneficiary or the liquidation of the juridical person-beneficiary unless provided otherwise by the contract;

the refusal of the beneficiary to receive advantages under the contract unless provided otherwise by the contract;

the death of the citizen who is the trustee manager, his being deemed to lack dispositive legal capacity, to have limited dispositive legal capacity, or to be missing, and also his being deemed to be insolvent (or bankrupt);

the refusal of the trustee manager or founder of the management to effectuate the trust management in connection with the impossibility of the trustee manager effectuating personally the trust management of the property;

the repudiation by the founder of the management of the contract for other reasons than those specified in point five of the present point, on condition of the payment of the remuneration stipulated by the contract to the trustee manager;

the deeming of the citizen to be insolvent (or bankrupt) who is the founder of the management.

2. In the event of the repudiation by one party of the contract of trust management of property, the other party must be notified about this three months before the termination of the contract unless another period of notification is provided by the contract.

3. In the event of the termination of a contract of trust management the property in trust management shall be transferred to the founder of the management unless provided otherwise by the contract.

Article 1025. Transfer of Securities to Trust Management

In the event of the transfer of securities to trust management, the combining of securities being transferred to trust management by different persons may be provided for.

The powers of the trust manager with regard to the disposition of securities shall be determined in the contract of trust management.

The peculiarities of the trust management of securities shall be determined by a law.

The rules of the present Article shall apply respectively to the rights certified by paperless securities (Article 149).

Article 1026. Trust Management of Property on Grounds Provided for by a Law

1. Trust management of property also may be founded:

as a consequence of the necessity for permanent management of the property of a ward in the instances provided for by Article 38 of the present Code;

on the basis of a will in which the executor of the will has been appointed;

on other grounds provided for by a law.

2. The rules provided for by the present Chapter shall apply respectively to relations relating to the trust management of property founded on the grounds specified in point 1 of the present Article unless provided otherwise by a law or does not arise from the essence of such relations.

In instances when the trust management of property is founded on the grounds specified in point 1 of the present Article, the rights of the founder of the management provided for by the rules of the present Chapter shall belong respectively to the trusteeship and guardianship agency, the executor of the will, or another person specified in a law.

Chapter 54. Commercial Concession

Article 1027. Contract of Commercial Concession

1. Under a contract of commercial concession one party (right-possessor) shall be obliged to grant to the other party (user) for remuneration for a period or without specifying a period the right to use in the entrepreneurial activity of the user the complex of exclusive rights which belong to the right-possessor, including the right to the firm name and/or commercial designation of the right-possessor, to protected commercial information, and also to other objects of exclusive rights provided for by the contract—trademark, service mark, and so forth.

2. A contract of commercial concession shall provide for the use of the complex of exclusive rights, business reputation, and commercial experience of the right-possessor within the determined amount (in particular, the establishment of the minimum and/or maximum amount of use), specifying or not the territory of use applicable to the determined sphere of entrepreneurial activity (sale of goods received from the right-possessor or to be produced by the user, effectuation of other trade activity, fulfilment of work, rendering of services).

3. Commercial organisations and citizens registered as individual entrepreneurs may be parties to a contract of commercial concession.

Article 1028. Form and Registration of Contract of Commercial Concession

1. A contract of commercial concession must be concluded in written form.

The failure to comply with the written form of a contract shall entail its invalidity. Such a contract shall be considered to be void.

2. A contract of commercial concession shall be registered by the agency which effectuated the registration of the juridical person or individual entrepreneur acting under the contract as the right-possessor.

If the right-possessor has been registered as a juridical person or individual entrepreneur in a foreign State, the registration of the contract of commercial concession shall be effectuated by the agency which effectuated the registration of the juridical person or individual entrepreneur which is the user.

In relations with third persons the parties to the contract of commercial concession shall have the right to refer to the contract only from the moment of its registration.

A contract of commercial concession for the use of an object protected in accordance with patent legislation shall be subject to registration also in a federal agency of executive power in the domain of patents and trademarks. In the event of the failure to comply with this requirement, the contract shall be considered to be void.

Article 1029. Commercial Subconcession

1. The right of the user to authorize other persons to use the complex of exclusive rights granted to him or part of this complex on conditions of a subconcession agreed by him with the right-possessor or determined in the contract of commercial concession may be provided for by the contract of commercial concession. The duty of the user to grant within a determined period to a determined number of persons the right to use the said rights on conditions of subconcession may be provided for in the contract.

A contract of commercial concession may not be concluded for a longer period than the contract of commercial concession on the basis of which it was concluded.

2. If a contract of commercial concession is invalid, the contracts of commercial subconcession concluded on the basis of it also shall be invalid.

3. Unless provided otherwise by the contract of commercial concession concluded for a period, in the event of its termination before time the rights and duties of the secondary right-possessor under a contract of commercial subconcession (user under a contract of commercial concession) shall pass to the right-possessor unless he refuses to accept the rights and duties under this contract. This rule respectively shall apply in the event of the dissolution of the contract of commercial concession concluded with specifying a period.

4. The user shall bear subsidiary responsibility for harm caused to the right-possessor by the actions of secondary users unless provided otherwise by the contract of commercial concession.

5. The rules concerning the contract of commercial concession provided for by the present Chapter shall apply to the contract of commercial subconcession unless it arises otherwise from the peculiarities of the subconcession.

Article 1030. Remuneration Under Contract of Commercial Concession

Remuneration under a contract of commercial concession may be paid by the user to the right-possessor in the form of fixed lump-sum or periodic payments, deductions from receipts, additions to the wholesale price of goods being transferred by the right-possessor for resale, or in other form provided for by the contract.

Article 1031. Duties of Right-Possessor

1. The right-possessor shall be obliged to:

transfer to the user technical and commercial documentation and to grant other information necessary to the user in order to effectuate the rights granted to him under the contract of commercial concession, and also train the user and his workers with regard to questions connected with the effectuation of these rights;

issue to the user licenses provided for by the contract, ensuring the formalisation thereof in the established procedure.

2. Unless provided otherwise by the contract of commercial concession, the right-possessor shall be obliged to:

ensure the registration of the contract of commercial concession (Article 1028[2]);

render to the user permanent technical and consultative assistance, including assistance in training and raising the qualifications of workers;

control the quality of goods (or work, services) being produced (or fulfilled, rendered) by the user on the basis of the contract of commercial concession.

Article 1032. Duties of User

Taking into account the character and peculiarities of the activity being effectuated by the user under the contract of commercial concession, the user shall be obliged to:

use when effectuating the activity provided for by the contract the firm name and/or commercial designation of the right-possessor by the means specified in the contract;

ensure the conformity of the quality of the goods produced, work fulfilled, or services rendered on the basis of the contract to the quality of analogous goods, work, or services being produced, fulfilled, or rendered directly by the right-possessor;

comply with the instructions and directions of the right-possessor aimed at ensuring the conformity of the character, means, and conditions of use of the complex of exclusive rights so that it is used by the right-possessor, including the instructions affecting the external and internal formalisation of the commercial premises to be used by the user when effectuating the rights granted to him under the contract;

render to purchasers (or customers) all additional services which they might count upon in acquiring (or ordering) the good (or work, service) directly from the right-possessor;

not divulge the secrets of production of the right-possessor and other confidential commercial information received from him;

grant the stipulated quantity of subconcessions if such duty is provided for by the contract;

inform the purchasers (or customers) by the most obvious means for them that he is using the firm name, commercial designation, trademark, service mark, or other means of individualisation by virtue of the contract of commercial concession.

Article 1033. Limitation of Rights of Parties Under Contract of Commercial Concession

1. Limitations of the right of the parties under this contract may be provided for by the contract of commercial concession, in particular there may be provided:

the obligation of the right-possessor not to grant to other persons analogous complexes of exclusive rights for the use thereof on the territory consolidated for the user or refrain from own analogous activity on this territory;

the obligation of the user not to compete with the right-possessor on the territory to which the operation of the contract of commercial concession extends with respect to entrepreneurial activity effectuated by the user with the use of the exclusive rights belonging to the right-possessor;

the refusal of the user to receive under the contracts of commercial concession analogous rights from competitors (or potential competitors) of the right-possessor;

the obligation of the user to agree with the right-possessor the location of the commercial premises to be used when effectuating the exclusive rights granted under the contract, as well as the external and internal formalisation thereof.

Limiting conditions may be deemed to be invalid upon the suit of the antimonopoly agency or other interested person if these conditions, taking into account the state of the respective market and the economic status of the parties, is contrary to antimonopoly legislation.

2. Those conditions which limit the rights of the parties under a contract of commercial concession, by virtue of which:

the right-possessor shall have the right to determine the price of the sale of the good by the user or the price of work (or services) being fulfilled (or rendered) by the user, or establish the upper or lower limit of these prices;

the user shall have the right to sell goods, fulfil work, or render services exclusively to a determined category of purchasers (or customers) or exclusively to purchasers (or customers) having a location (or place of residence) on the territory determined in the contract,

shall be void.

Article 1034. Responsibility of Right-Possessor Relating to Demands Presented to User

The right-possessor shall bear subsidiary responsibility for demands presented to the user concerning the failure of the goods (or work, services) being sold (or fulfilled, rendered) by the user under the contract of commercial concession to conform to quality.

The right-possessor shall be liable jointly and severally with the user with regard to demands presented to the user as the manufacturer of the product (or goods) of the right-possessor.

Article 1035. Right of User to Conclude Contract of Commercial Concession for New Period

1. The user who has duly performed his duties shall have the right upon the expiry of the period of the contract of commercial concession to conclude the contract for a new period on the same conditions.

2. The right-possessor shall have the right to refuse to conclude a contract of commercial concession for a new period on condition that within three years from the date of expiry of the period of the said contract he will not conclude analogous contracts of commercial concession with other persons and agrees to the conclusion of analogous contracts of commercial subconcession whose operation will extend to the same territory on which the contract which has terminated operated. If before the expiry of the three-year period the right-possessor wishes to grant someone the same rights which were granted to the user under the contract which terminated, he shall be obliged to propose to the user to conclude a new contract or to compensate losses incurred by him. When concluding a new contract, the conditions thereof must be no less favourable for the user than the conditions of the contract which terminated.

Article 1036. Change of Contract of Commercial Concession

A contract of commercial concession may be changed in accordance with the rules provided for by Chapter 29 of the present Code.

In relations with third persons the parties to a contract of commercial concession shall have the right to refer to a change of the contract only from the moment of registration of this change in the procedure established by Article 1027(2) of the present Code unless it is proved that the third person knew or should have known about the change of the contract previously.

Article 1037. Termination of Contract of Commercial Concession

1. Each of the parties to the contract of commercial concession concluded without specifying the period shall have the right at any time to repudiate the

contract, having notified the other party thereof six months in advance unless a longer period has been provided for by the contract.

2. The dissolution before time of the contract of commercial concession concluded with a specification of the period, and also the dissolution of the contract concluded without specifying the period, shall be subject to registration in the procedure established by Article 1028(2) of the present Code.

3. In the event of the termination of the rights to the firm name or commercial designation which belong to the right-possessor without a replacement thereof by new analogous rights, the contract of commercial concession shall terminate.

4. In the event the right-holder or user is declared to be insolvent (or bankrupt), the contract of commercial concession shall terminate.

Article 1038. Preservation of Contract of Commercial Concession in Force in Event of Change of Parties

1. The transfer to another person of any exclusive right whatever within the complex of exclusive rights granted to the user shall not be grounds for a change or dissolution of the contract of commercial concession. The new right-possessor shall become a party to this contract in that part of the rights and duties relating to the exclusive right transferred.

2. In the event of the death of the right-possessor his rights and duties under the contract of commercial concession shall pass to the heir on condition that it was registered or within six months from the date of opening the inheritance is registered as an individual entrepreneur. Otherwise the contract shall terminate.

The effectuation of the rights and performance of the duties of the deceased right-possessor before the acceptance of these rights and duties by the heir or before registration of the heir as an individual entrepreneur shall be effectuated by the manager appointed by the notary.

Article 1039. Consequences of Change of Firm Name or Commercial Designation of Right-Possessor

In the event of a change by the right-possessor of its firm name or commercial designation, the rights to use which is within the complex of exclusive rights, the contract of commercial concession shall operate with respect to the new firm name or commercial designation of the right-possessor unless the user requires the dissolution of the contract and compensation of losses. In the event of the continuation of the operation of the contract the user shall have the right to require commensurate reduction of the remuneration due to the right-possessor.

Article 1040. Consequences of Termination of Exclusive Right Whose Use Has Been Granted Under Contract of Commercial Concession

If in the period of operation of a contract of commercial concession the period of operation of the exclusive right has expired whose use was granted under this contract, or such right terminated on other grounds, the contract of commercial concession shall continue to operate, except for provisions relating to the terminated right, and the user, unless provided otherwise by the contract, shall have the right to require commensurate reduction of the remuneration due to the right-possessor.

In the event of the termination of the rights to the firm name or commercial designation which belong to the right-possessor, the consequences shall ensue which are provided for by Article 1037(2) and Article 1039 of the present Code.

Chapter 55. Simple Partnership

Article 1041. Contract of Simple Partnership

1. Under the contract of simple partnership (or contract concerning joint activity), two or several persons (partners) shall be obliged to combine their contributions and jointly operate without the formation of a juridical person in order to derive profit or achieve another purpose which is not contrary to law.

2. Only individual entrepreneurs and/or commercial organisations may be the parties to a contract of simple partnership concluded in order to effectuate entrepreneurial activity.

Article 1042. Contributions of Partners

1. Everything that he contributed to the common cause, including money, other property, professional and other knowledge, skills, and ability, and also business reputation and business connections, shall be deemed to be the contribution of the partner.

2. The contributions of partners shall be presupposed to be equal in value unless it follows otherwise from the contract of simple partnership or factual circumstances. The monetary valuation of the contribution of a partner shall be made by agreement between the partners.

Article 1043. Common Property of Partners

1. The property contributed by partners which they possessed by right of ownership, and also the product produced as a result of joint activity and the fruits and revenues received from such activity shall be deemed their common participatory share ownership insofar as not established otherwise by a law or the contract of simple partnership or does not arise from the essence of the obligation.

377

Property contributed by partners which they possessed on the grounds which are distinct from the right of ownership shall be used in the interests of all partners and shall comprise, together with the property in their common ownership, the common property of the partners.

2. The conducting of bookkeeping of the common property of the partners may be entrusted by them to one of the juridical persons participating in the contract of simple partnership.

3. The use of common property by the partners shall be effectuated by their common consent, and in the event of the failure to achieve consent, in the procedure established by a court.

4. The duties of the partners with regard to the maintenance of common property and the procedure for the compensation of expenses connected with the fulfilment of these duties shall be determined by the contract of simple partnership.

Article 1044. Conducting Common Affairs of Partners

1. When conducting common affairs, each partner shall have the right to act in the name of all the partners unless it is established by the contract of simple partnership that the conducting of affairs shall be effectuated by individual participants or jointly by all the participants of the contract of simple partnership.
When conducting affairs jointly the consent of all the partners shall be required for the conclusion of each transaction.

2. In relations with third persons the power of a partner to conclude a transaction in the name of all the partners shall be certified by a power of attorney issued to him by the remaining partners or by the contract of simple partnership concluded in written form.

3. In relations with third persons the partners may not refer to limitation of the rights of the partner who has concluded the transaction with regard to conducting the common affairs of the partners except for instances when they prove that at the moment of concluding the transaction the third person knew or should have known about the existence of such limitations.

4. The partner who has concluded a transaction in the name of all the partners with respect to which his right to conduct the common affairs of the partners was limited, or who has concluded in his own name a transaction in the interests of all the partners, may demand compensation for expenses made by him at his own expense if there are sufficient grounds to suppose that these transactions were necessary in the interests of all the partners. The partners who incurred losses as a consequence of such transactions shall have the right to demand compensation thereof.

5. Decisions affecting the common affairs of partners shall be adopted by the partners by their common consent unless provided otherwise by the contract of simple partnership.

Article 1045. Right of Partner to Information

Each partner shall have the right, irrespective of whether he is empowered to conduct the common affairs of the partners, to familiarise himself with all of the documentation relating to the conducting of affairs. The waiver of this right or limitation thereof, including by agreement of the partners, shall be void.

Article 1046. Common Expenses and Losses of Partners

The procedure for covering expenses and losses connected with the joint activity of the partners shall be determined by their agreement. In the absence of such agreement, each partner shall bear expenses and losses in proportion to the value of his contribution to the common cause.

An agreement wholly exempting any of the partners from participation in covering the common expenses or losses shall be void.

Article 1047. Responsibility of Partners for Common Obligations

1. Unless the contract of simple partnership is linked with the effectuation of entrepreneurial activity by its participants, each partner shall be liable for common contractual obligations with all of its property in proportion to the value of its contribution to the common cause.

Partners shall be liable jointly and severally for common obligations which arose not from the contract.

2. If the contract of simple partnership is linked with the effectuation of entrepreneurial activity by its participants, the partners shall be liable jointly and severally for all common obligations irrespective of the grounds for the arising thereof.

Article 1048. Distribution of Profit

The profit received by the partners as a result of their joint activity shall be distributed in proportion to the value of the contributions of the partners to the common cause unless provided otherwise by the contract of simple partnership or other agreement of the partners. An agreement concerning the elimination of any of the partners from participation in the profit shall be void.

Article 1049. Partition of Participatory Share of Partner at Demand of Creditors Thereof

The creditor of a participant of the contract of simple partnership shall have the right to present a demand concerning the partition of its participatory share in the common property in accordance with Article 255 of the present Code.

Article 1050. Termination of Contract of Simple Partnership

1. A contract of simple partnership shall be terminated as a consequence of:

the declaration of any of the partners to lack dispositive legal capacity, to have limited dispositive legal capacity, or to be missing unless the contract of simple partnership or subsequent agreement provides for the preservation of the contract in relations between the remaining partners;

the declaration of any of the partners to be insolvent (or bankrupt), with the exception specified in paragraph two of the present point;

the death of a partner or liquidation or reorganisation of a juridical person participating in the contract of simple partnership unless the contract or a subsequent agreement provides for the preservation of the contract in relations between the remaining partners or replacement of the deceased partner (or liquidated or reorganised juridical person) by his heirs (or legal successors);

the refusal by any of the partners of further participation in a contract of simple partnership without period, with the exception specified in paragraph two of the present point;

the dissolution of the contract of simple partnership concluded with a specification of the period, at the demand of one of the partners in relations between it and the remaining partners, with the exception specified in paragraph two of the present point;

the expiry of the period of the contract of simple partnership;

the partition of the participatory share of the partner at the demand of its creditor, with the exception specified in paragraph two of the present point.

2. In the event of the termination of a contract of simple partnership, the things transferred to the common possession and/or use of the partners shall be returned to the partner who granted them without remuneration unless provided otherwise by the agreement of the parties.

From the moment of termination of a contract of simple partnership the participants thereof shall bear joint and several responsibility for the failure to perform common obligations with respect to third persons.

The division of property in the common ownership of the partners and the common rights of demand which have arisen shall be effectuated in the procedure established by Article 252 of the present Code.

A partner who has contributed an individually specified thing to the common ownership shall have the right in the event of the termination of the contract of simple partnership to demand the return of this thing to it in a judicial proceeding on condition of compliance with the interests of the remaining partners and creditors.

Article 1051. Repudiation of Contract of Simple Partnership Without Period

A statement of repudiation by a partner of a contract of simple partnership without period must be made by him not later than three months before the proposed withdrawal from the contract.

An agreement concerning limitation of the right to repudiate a contract of simple partnership without period shall be void.

Article 1052. Dissolution of Contract of Simple Partnership at Demand of Party

Together with the grounds specified in Article 450(2) of the present Code, the party to a contract of simple partnership concluded with the specification of a period or with the specification of a purpose as a condition subsequent shall have the right to demand dissolution of the contract in relations between himself and the remaining partners for a justifiable reason with compensation to the remaining partners of the real damage caused by dissolution of the contract.

Article 1053. Responsibility of Partner with Respect to Which Contract of Simple Partnership is Dissolved

In an instance when a contract of simple partnership was not terminated as a result of the statement of one of the participants concerning repudiation of further participation therein or dissolution of the contract at the demand of one of the partners, the person whose participation in the contract has terminated shall be liable to third persons for common obligations which arose in the period of its participation in the contract as if it has remained a participant of the contract of simple partnership.

Article 1054. Nontransparent Partnership

1. It may be provided by a contract of simple partnership that its existence shall not be divulged to third persons (nontransparent partnership). The rules concerning the contract of simple partnership provided for by the present Chapter shall apply to such contract unless provided otherwise by the present Article or does not arise from the essence of the nontransparent partnership.

2. In relations with third persons each of the participants of a nontransparent partnership shall be liable with all of its property for transactions which it concluded in its name in the common interests of the partners.

3. In relations between partners, the obligations which arose in the process of their joint activity shall be considered to be common.

Chapter 56. Public Promise of Reward

Article 1055. Duty to Pay Reward

1. The person who has announced publicly the payment of monetary remuneration or the issuance of another reward (concerning the payment of a reward) to whomever performs the lawful action specified in the announcement within the period specified therein shall be obliged to pay the promised reward to anyone who performed the respective action, in particular who found a lost thing or communicated the necessary information to the person who announced the reward.

2. The duty to pay the reward shall arise on condition that the promise of the reward enables it to be established to whom it is promised. A person who has responded to the promise shall have the right to require written confirmation of the promise and shall bear the risk of consequences of the failure to present this demand if it turns out in reality that the announcement concerning the reward was not made by the person specified therein.

3. If the amount was not specified in the public promise of the reward, it shall be determined by agreement with the person who promised the reward, and in the event of a dispute, by a court.

4. The duty to pay a reward shall arise irrespective of whether the respective action was performed in connection with the announcement made or irrespective thereof.

5. In instances when an action specified in an announcement was performed by several persons, the right to receive the reward shall be acquired by he who performed the respective action first.

If the action specified in the announcement was performed by two or more persons and it is impossible to determine which of them performed the respective action first, and also if the action was performed by two or more persons simultaneously, the reward shall be divided between them equally or in another amount provided for by an agreement between them.

6. Unless provided otherwise in the announcement concerning the reward and does not arise from the character of the action specified therein, the conformity of the action fulfilled to the requirements contained in the announcement shall be determined by the person who publicly promised the reward, and in the event of a dispute, by a court.

Article 1056. Revocation of Public Promise of Reward

1. A person who has announced publicly the payment of a reward shall have the right in the same form to renounce this promise except for instances when in

the announcement itself the inadmissability of a renunciation has been provided for or arises therefrom or a determined period has been given for the performance of the action for which the reward is promised, or at the moment of the announcement concerning the renunciation one or several persons who responded already had fulfilled the action specified in the announcement.

2. The revocation of a public promise of a reward shall not relieve he who announced the reward from compensation of the expenses of persons who responded incurred by them in connection with the performance of the action specified in the announcement within the limits of the reward specified in the announcement.

Chapter 57. Public Competition

Article 1057. Organisation of Public Competition

1. A person who has announced publicly the payment of monetary remuneration or issuance of other reward (or concerning the payment of a reward) for the best fulfilment of work or achievement of other results (public competition) must pay (or issue) the stipulated reward to whomever in accordance with the conditions for conducting the competition is deemed to be the winner thereof.

2. The public competition must be directed towards the achievement of some socially useful purposes.

3. A public competition may be open when the proposal of the organiser of the competition to take part therein is addressed to all who wish to do so by means of an announcement in the press or other mass media, or be closed when the proposal to take part in the competition is sent to a specified group of persons at the choice of the organiser of the competition.

An open competition may be conditioned by the preliminary qualification of the participants thereof when a preliminary selection of persons who wish to take part therein is conducted by the organiser of the competition.

4. The announcement concerning a public competition must contain at least the conditions providing for the essence of the task, the criteria and procedure for evaluating the results of the work or other achievements, the place, period, and procedure for submitting them, the amount and form of the reward, and also the procedure and periods for the announcement of the results of the competition.

5. To a public competition containing an obligation to conclude a contract with the winner of the competition shall apply the rules provided for by the present Chapter insofar as not provided otherwise by Articles 447–449 of the present Code.

Article 1058. Change of Conditions and Cancellation of Public Competition

1. A person who has announced a public competition shall have the right to change its conditions or to cancel a competition only during the first half of the period established for the submission of the work.

2. A notice concerning a change of the conditions or cancellation of the competition must be made by the same means as the competition was announced.

3. In the event of a change of the conditions of a competition or the cancellation thereof, the person who has announced the competition must compensate the expenses incurred by any person who fulfilled the work provided for in the announcement until it became known or should have become known to him that the conditions of the competition changed or of the cancellation thereof.

The person who has announced a competition shall be relieved of the duty to compensate expenses if it is proved that the said work was fulfilled not in connection with the competition, in particular, before the announcement of the competition, or knowingly did not conform to the conditions of the competition.

4. If in the event of the change of the conditions of a competition or in the event of the cancellation thereof the requirements specified in points 1 or 2 of the present Article were violated, the person who announced the competition must pay a reward to those who fulfilled the work which satisfies the conditions specified in the announcement.

Article 1059. Decision Concerning Payment of Reward

1. A decision concerning the payment of a reward must be rendered and communicated to the participants of a public competition in the procedure and within the periods established in the announcement concerning the competition.

2. If the results specified in the announcement were achieved in a work fulfilled jointly by two or more persons, the reward shall be distributed in accordance with the agreement reached between them. If such agreement is not reached, the procedure for the distribution of the reward shall be determined by a court.

Article 1060. Use of Works of Science, Literature, and Art Conferred with Award

If the subject of a public competition consists of the creation of a work of science, literature, or art and the conditions of the competition do not provide otherwise, the person who has announced the public competition shall acquire a preferential right to conclude with the author of the work on which an award is conferred a contract concerning the use of the work with the payment to him a respective remuneration for this.

Article 1061. Return to Participants of Public Competition of Works Submitted

The person who has announced a public competition shall be obliged to return to the participants of the competition works not conferred with an award unless provided otherwise by the announcement concerning the competition and does not arise from the character of the work fulfilled.

Chapter 58. Conducting Games and Betting

Article 1062. Demands Connected with the Organisation of Games and Betting and Participation Therein

The demands of citizens and juridical persons connected with the organisation of games and betting or participation therein shall not be subject to judicial defence, except for the demands of persons who took part in games or betting under the influence of deceit, coercion, threat, or ill-intentioned agreement of their representative with the organiser of the games or betting, and also the demands specified in Article 1063(5) of the present Code.

Article 1063. Conducting Lotteries, Totalisers, and Other Games by State and Municipal Formations or Upon Authorisation Thereof

1. Relations between the organisers of lotteries, totalisers (or mutual betting), and other games based on risk—the Russian Federation, subjects of the Russian Federation, municipal formations, persons who have received an authorisation from an empowered State or municipal agency—and the participants of the games are based on a contract [as amended by Federal Law No. 15-ФЗ, 10 January 2003].

2. In the instances provided for by the rules of the organisation of the games the contract between the organiser and the participant of the games shall be formalised by the issuance of a lottery ticket, receipt, or other document.

3. The proposal concerning the conclusion of a contract provided for by point 1 of the present Article must include the conditions concerning the period of conducting the games and the procedure for the determination of the winnings and the amount thereof.

In the event of a refusal of the organiser of the games to conduct them within the established period, the participants of the games shall have the right to demand compensation from the organiser thereof for real damage sustained because of the cancellation or postponement of the period thereof.

4. Persons who in accordance with the conditions of conducting a lottery, totaliser, or other games are deemed to be winners must be paid the winnings by the organiser of the games in the amount, form (monetary or in kind), and period provided for by the conditions of conducting the games, and if the period is not

specified in these conditions, not later than ten days from the moment of determining the results of the games.

5. In the event of the failure of the organiser of the games to perform the duties specified in point 4 of the present Article, the participant who won the lottery, totaliser, or other games shall have the right to demand the payment of the winnings from the organiser of the games, and also the compensation of losses caused by a violation of the contract on the part of the organiser.

Chapter 59. Obligations as Consequence of Causing Harm

§1. General Provisions on Compensation of Harm

Article 1064. General Grounds of Responsibility for Causing of Harm

1. Harm caused to the person or property of a citizen, and also harm caused to the property of a juridical person, shall be subject to compensation in full by the person who caused the harm.

The duty of compensation of harm may be placed by a law on a person who is not the causer of the harm.

The duty of the causer of harm to pay contributory compensation to a victim above the compensation of harm may be established by a law or by a contract.

2. A person who has caused harm shall be relieved from compensation of harm if it is proved that the harm was caused not through his fault. Compensation of harm in the absence of the fault of the causer of harm also may be provided for by a law.

3. Harm caused by lawful actions shall be subject to compensation in the instances provided for by a law.

Compensation of harm may be refused if the harm was caused at the request or with the consent of the victim, and the actions of the causer of the harm do not violate moral principles of society.

Article 1065. Warning of Causing of Harm

1. The danger of causing harm in the future may be grounds for a suit concerning the prohibition of the activity creating such a danger.

2. If the harm caused is the consequence of the operation of an enterprise, installation, or other production activity which continues to cause harm or threatens new harm, a court shall have the right to oblige the defendant, in addition to compensation of harm, to suspend or terminate the respective activity.

A court may reject a suit concerning the suspension or termination of the respective activity only if the suspension or termination thereof is contrary to social interests. The refusal to suspend or terminate such activity shall not deprive the victim of the right to compensation for the harm caused by this activity.

Article 1066. Causing Harm in State of Necessary Defence

Harm caused in a state of necessary defence, unless the limits thereof were exceeded, shall not be subject to compensation.

Article 1067. Causing Harm in State of Extreme Necessity

Harm caused in a state of extreme necessity, that is, in order to eliminate a danger threatening the causer of the harm himself or other persons, if this danger under the given circumstances could not be eliminated by other means, must be compensated by the person who caused the harm.

Taking into account the circumstances under which such harm was caused, a court may place the duty of compensating it on a third person in whose interests the causer of the harm acted or relieve both this third person and the causer of the harm fully or partially from compensation of the harm.

Article 1068. Responsibility of Juridical Person or Citizen for Harm Caused by Worker Thereof

1. A juridical person or citizen shall compensate harm caused by a worker thereof when performing labour (or employment, official) duties.

With regard to the rules provided for by the present Chapter, citizens fulfilling work on the basis of a labour contract (or *kontrakt*), and also citizens fulfilling work under a civil-law contract, if in so doing they acted or should have acted under a planning task of the respective juridical person or citizen and under its control for the safe conducting of work, shall be deemed to be workers.

2. Economic partnerships and production cooperatives shall compensate harm caused by their participants (or members) when the last effectuate entrepreneurial, production, or other activity of the partnership or cooperative.

Article 1069. Responsibility for Harm Caused by State Agencies, Agencies of Local Self-Government, and Also the Officials Thereof

Harm caused to a citizen or juridical person as a result of the illegal actions (or failure to act) of State agencies, agencies of local self-government, or officials of these agencies, including as a result of the issuance of an act of a State agency or agency of local self-government which does not correspond to a law or other legal act, shall be subject to compensation. Harm shall be compensated at the expense, respectively, of the treasury of the Russian Federation, treasury of the subject of the Russian Federation, or treasury of the municipal formation.

Article 1070. Responsibility for Harm Caused by Illegal Actions of Agencies of Inquiry or Preliminary Investigation, Procuracy, and Court

1. Harm caused to a citizen as a result of the illegal conviction, illegal bringing to criminal responsibility, illegal application of confinement under guard or

written undertaking not to leave as a measure of restraint, or the illegal imposition of an administrative sanction in the form of arrest or correctional tasks shall be compensated at the expense of the treasury of the Russian Federation, and in the instances provided for by a law, at the expense of the treasury of the subject of the Russian Federation or treasury of municipal formation, in full irrespective of the fault of the officials of the agencies of inquiry or preliminary investigation, procuracy, and court in the procedure provided for by a law.

2. Harm caused to a citizen or juridical person as a result of the illegal activity of agencies of inquiry or preliminary investigation, and procuracy which does not entail the consequences provided for by point 1 of the present Article shall be compensated on the grounds and in the procedure which has been provided for by Article 1069 of the present Code. The harm caused when effectuating justice shall be compensated if the fault of the judge has been established by the judgment of a court which has entered into legal force.

Article 1071. Agencies and Persons Acting in Name of Treasury When Compensating Harm at its Expense

In instances when in accordance with the present Code or other laws the harm caused is subject to compensation at the expense of the treasury of the Russian Federation, treasury of a subject of the Russian Federation, or treasury of a municipal formation, the respective financial agencies shall act in the name of the treasury unless in accordance with Article 125(3) of the present Code this duty is placed on another agency, juridical person, or citizen.

Article 1072. Compensation of Harm by Person Who Has Insured His Responsibility

A juridical person or citizen who has insured his responsibility by way of voluntary or obligatory insurance to the benefit of the victim (Article 931, Article 935[1]), when insurance compensation is insufficient in order to compensate the harm caused in full, shall compensate the difference between the insurance compensation and the actual amount of damage.

Article 1073. Responsibility for Harm Caused by Minors Up to Fourteen Years of Age

1. For harm caused by a minor who has not attained fourteen years of age (youth) his parents (or adoptive parents) or trustees shall be liable unless it is proved that the harm arose not through their fault.

2. If a youth who needs trusteeship is situated in a respective nurturing or medical institution, institution for the social defence of the populace, or other analogous institution which by virtue of a law is his trustee (Article 35), this institution shall be obliged to compensate the harm caused by the youth unless it is proved that the harm arose not through the fault of the institution.

3. If a youth has caused harm at the time when he was under the supervision of an educational, nurturing, medical, or other institution obliged to effectuate supervision over him, or a person effectuating supervision on the basis of a contract, this institution or person shall be liable for the harm unless it is proved that the harm arose not through his fault in effectuating the supervision.

4. The duty of the parents (or adoptive parents), trustees, educational, nurturing, treatment, and other institutions relating to compensation of harm caused to a youth shall not terminate with the attainment by the youth of majority or receipt by him of property sufficient to compensate the harm.

If the parents (or adoptive parents), trustees, or other citizens specified in point 3 of the present Article have died or do not have sufficient means in order to compensate the harm caused to the life or health of the victim, and the causer of the harm himself having full dispositive legal capacity possesses such means, the court shall have the right, taking into account the property status of the victim and the causer of the harm, and also other circumstances, to adopt a decision concerning compensation of harm fully or partially at the expense of the causer of the harm himself.

Article 1074. Responsibility for Harm Caused by Minors in Age from Fourteen to Eighteen Years

1. A minor from fourteen to eighteen years of age shall bear responsibility on the general grounds autonomously for harm caused.

2. In an instance when a minor from fourteen to eighteen years of age has no revenues or other property sufficient to compensate harm, the harm must be compensated fully or in the insufficient part by his parents (or adoptive parents) or guardian unless they prove that the harm arose not through their fault.

If a minor from fourteen to eighteen years of age who needs guardianship is situated in a respective nurturing or treatment institution, institution for the social defence of the populace, or other analogous institution which by virtue of a law is his guardian (Article 35), this institution shall be obliged to compensate the harm fully or in the insufficient part unless it is proved that the harm arose not through its fault.

3. The duty of parents (or adoptive parents), guardian, and respective institution with regard to the compensation of harm caused by a minor from fourteen to eighteen years of age shall terminate upon the attainment of majority by the causer of the harm or in instances when upon attaining majority he has revenues or other property sufficient to compensate the harm or when he before attaining majority has acquired dispositive legal capacity.

Article 1075. Responsibility of Parents Deprived of Parental Rights for Harm Caused by Minors

A court may place on a parent deprived of parental rights responsibility for harm caused by the minor children thereof for three years after the deprivation of the parent of parental rights if the behaviour of the child which entailed the causing of the harm was a consequence of the improper effectuation of parental duties.

Article 1076. Responsibility for Harm Caused by Citizen Deemed to Lack Dispositive Legal Capacity

1. Harm caused by a citizen deemed to lack dispositive legal capacity shall be compensated by his trustee or the organisation obliged to effectuate supervision over him unless they prove that the harm arose not through their fault.

2. The duty of a trustee or the organisation obliged to effectuate supervision with regard to compensation of harm caused by a citizen deemed to lack dispositive legal capacity shall not terminate in the event of the last being deemed to have dispositive legal capacity.

3. If the trustee is deceased or does not have sufficient means for the compensation of harm caused to the life or health of a victim, and the causer of harm himself possesses such means, the court shall have the right, taking into account the property status of the victim and the causer of the harm, and also other circumstances, to adopt a decision concerning compensation of harm fully or partially at the expense of the causer of the harm himself.

Article 1077. Responsibility for Harm Caused by Citizen Deemed to Have Limited Dispositive Legal Capacity

Harm caused by a citizen limited in dispositive legal capacity as a consequence of the abuse of alcoholic beverages or narcotic means shall be compensated by the causer of the harm himself.

Article 1078. Responsibility for Harm Caused by Citizen Not Capable of Understanding Significance of His Actions

1. A citizen who has dispositive legal capacity, and also a minor from fourteen to eighteen years of age, who has caused harm in such a state that he could not understand the significance of his actions or direct them shall not be liable for the harm caused by him.

If harm is caused to the life or health of the victim, the court may, taking into account the property status of the victim and the causer of the harm, and also other circumstances, place the duty with regard to compensation of harm fully or partially on the causer of the harm.

2. The causer of harm shall not be relieved from responsibility if he has brought himself into a state in which he could not understand the significance of his actions or direct them by the use of alcoholic beverages, narcotic means, or other means.

3. If the harm was caused by a person who could not understand the significance of his actions or direct them as a consequence of mental disturbance, the duty to compensate the harm may be placed by a court on a spouse, parents, or children who have reached majority and who have labour capacity and reside jointly with this person, and who knew about the mental disturbance of the causer of the harm but did not raise the question of deeming him to lack dispositive legal capacity.

Article 1079. Responsibility for Harm Caused by Activity Creating Increased Danger for Surrounding Persons

1. Juridical persons and citizens whose activity is connected with an increased danger for surrounding persons (use of means of transport, mechanisms, high tension electric power, atomic power, explosive substances, virulent poisons, etc; effectuation of construction and other activity connected therewith, and others) shall be obliged to compensate the harm caused by a source of increased danger unless it is proved that the harm arose as a consequence of insuperable force or the intent of the victim. The possessor of a source of increased danger may be relieved by a court from responsibility fully or partially also on the grounds provided for by Article 1083(2) and (3) of the present Code.

The duty of compensation of harm shall be placed on the juridical person or citizen who possesses the source of increased danger by right of ownership, right of economic jurisdiction, or right of operative management, or other legal basis (right of lease, power of attorney for the right to drive means of transport, by virtue of the regulation of a respective agency concerning the transfer of the source of increased danger to it, and the like).

2. The possessor of a source of increased danger shall not be liable for harm caused by this source if it is proved that the source left his possession as a result of the unlawful actions of other persons. Responsibility for harm caused by the source of increased danger in such instances shall be borne by the person who unlawfully possessed the source. If there is fault of the possessor of a source of increased danger in the unlawful seizure of his source from his possession, responsibility may be placed both on the possessor and on the person who unlawfully took possession of the source of increased danger.

3. The possessors of sources of increased danger shall bear responsibility jointly and severally for harm caused as a result of the interaction of these sources (collisions of means of transport, and others) to third persons on the grounds provided for by point 1 of the present Article.

The harm caused as a result of the interaction of the sources of increased danger to their possessors shall be compensated on the general grounds (Article 1064).

Article 1080. Responsibility for Jointly Caused Harm

Persons who have caused harm jointly shall be liable to the victim jointly and severally.

Upon the application of the victim and in his interests a court shall have the right to place on the persons who have caused harm jointly responsibility in participatory shares determined according to the rules provided for by Article 1081(2) of the present Code.

Article 1081. Right of Regression Against Person Who Caused Harm

1. The person who has compensated harm caused by another person (or by a worker when performing his employment, official, or other labour duties, by a person driving means of transport, and so forth) shall have the right of counter demand (regression) against this person in the amount of the compensation paid unless another amount has been established by a law.

2. The causer of harm who has compensated harm jointly caused shall have the right to demand from each of the other causers of harm a participatory share of the compensation paid to the victim in the amount corresponding to the degree of fault of this causer of harm. If it is impossible to determine the degree of fault, the participatory shares shall be deemed to be equal.

3. The Russian Federation, subject of the Russian Federation, or municipal formation in the event of compensation by them of harm caused by an official of agencies of inquiry or preliminary investigation, the procuracy, or a court (Article 1070[1]) shall have the right of regression against this person if his fault has been established by the judgment of a court which has entered into legal force.

4. The persons who have compensated harm on the grounds specified in Articles 1073–1076 of the present Code shall not have the right of regression against the person who caused the harm.

Article 1082. Means of Compensation of Harm

In satisfying the demand concerning compensation of harm, the court in accordance with the circumstances of the case shall oblige the person responsible for causing the harm to compensate the harm in kind (to provide a thing of the same kind and quality, to rectify a damaged thing, and so forth) or to compensate the losses caused (Article 15[2]).

Article 1083. Taking Account of Fault of Victim and Property Status of Person Who Caused Harm

1. Harm which arose as a consequence of the intent of the victim shall not be subject to compensation.

2. If the gross negligence of the victim himself furthered the arising or the increasing of the harm, depending on the degree of fault of the victim and the causer of the harm the amount of compensation must be reduced.

In the event of gross negligence of the victim and the absence of fault of the causer of the harm in instances when his responsibility ensues irrespective of fault, the amount of compensation must be reduced or compensation of harm may be refused unless provided otherwise by a law. In the event of causing harm to the life or health of a citizen, refusal of compensation of harm shall not be permitted.

The fault of the victim shall not be taken into account when compensating additional expenses (Article 1085[1]), when compensating harm in connection the death of a breadwinner (Article 1089), and also when compensating expenses for burial (Article 1094).

3. A court may reduce the amount of compensation of harm caused by a citizen by taking into account his property status, except for instances when the harm was caused by actions committed intentionally.

§2. Compensation of Harm Caused to Life or Health of Citizen

Article 1084. Compensation of Harm Caused to Life or Health of Citizen When Performing Contractual or Other Obligations

Harm caused to the life or health of a citizen when performing contractual obligations, and also when performing duties of military service, service in the police, and other respective duties shall be compensated according to the rules provided for by the present Chapter unless a higher amount of responsibility has been provided by a law or by a contract.

Article 1085. Amount and Character of Compensation of Harm Which Caused Impairment of Health

1. When mutilation or other impairment of health is caused to a citizen, the earnings (or revenue) lost by the victim which he had or could be determined to have, and also additional expenses incurred caused by the impairment of health, including expenses for treatment, additional nourishment, acquisition of medicines, prosthetics, outside care, sanitorium-resort treatment, the acquisition of special means of transport, and training for another profession shall be subject to compensation if it is established that the victim needs these types of assistance and care and does not have the right to receive them free of charge.

2. When determining lost earnings (or revenue), the disability pension assigned to the victim in connection with the mutilation or other impairment of health, and likewise other pensions, benefits, and other similar payments assigned both before and after the causing of harm to health shall not be taken into account and shall not entail a reduction of the amount of compensation of harm (or shall not be set-off at the expense of compensation of harm). Earnings (or revenue) received by the victim after the impairment of health also shall not be set-off at the expense of compensation of harm.

3. The extent and amount of compensation of harm due to the victim in accordance with the present Article may be increased by a law or by a contract.

Article 1086. Determination of Earnings (or Revenue) Lost as Result of Impairment of Health

1. The amount of earnings (or revenue) lost by a victim which is subject to compensation shall be determined in percentages of his average monthly earnings (or revenue) before the mutilation or other impairment of health or before this loss of his capacity to labour, of the respective degree of loss by the victim of professional capacity to labour, and in the absence thereof, the degree of loss of general capacity to labour.

2. Within lost earnings (or revenue) of the victim shall be included all types of payment for his labour under labour and civil-law contracts both at the principal place of work and second jobs levied with income tax. Payments of an extraordinary character, in particular, contributory compensation for unused leave and severance benefit in the event of dismissal shall not be taken into account. The benefit paid for the period of temporary lack of labour capacity or pregnancy and birth leave shall be taken into account. Revenues from entrepreneurial activity, and also author's royalty, shall be included in the lost earnings, the revenues from entrepreneurial activity being included on the basis of tax inspectorate data.

All types of earnings (or revenue) shall be taken into account in the amounts credited before the withholding of taxes.

3. The average monthly earnings (or revenue) of the victim shall be calculated by means of dividing the total amount of his earnings (or revenue) for the twelve months of work which preceded the impairment of health by twelve. In an instance when the victim at the time of causing the harm worked less than twelve months, the average monthly earnings (or revenue) shall be calculated by means of dividing the total amount of earnings (or revenue) for the number of months actually worked which preceded the impairment of health by the number of these months.

Months not worked in full by the victim shall at his wish be replaced by the preceding months fully worked or by excluding them from the calculation when it is impossible to substitute them.

4. In an instance when the victim at the moment of causing of the harm was not working, the earnings before dismissal or the usual amount of remuneration of a worker of his qualifications in the particular locality shall, at his wish, be taken into account, but not less than the amount of living minimum of the able-bodied population as a whole established in accordance with a law with regard to the Russian Federation [as amended by Law of 26 November 2002. СЗ РФ (2002), no. 48, item 4737].

5. If stable changes improving his financial status (increase of earnings in the post held, he was transferred to higher-paid work, took up the job after completing day education, and in other instances when the stability of the change or possibility of change of payment of labour of the victim is proved) occurred in the earnings (or revenue) of the victim before the mutilation or other impairment of health was caused to him, when determining his average monthly earnings (or revenue) only the earnings (or revenue) which he received or should have received after the respective changes shall be taken into account.

Article 1087. Compensation of Harm in Event of Impairment of Health of Person Who Has Not Attained Majority

1. In the event of the mutilation or other impairment of health of a minor who has not attained fourteen years (youth) and does not have earnings (or revenue), the person responsible for the harm caused shall be obliged to compensate the expenses caused by the impairment of health.

2. Upon the attainment of fourteen years by a youth-victim, and also in the event of the causing of harm to a minor of from fourteen to eighteen years in age who does not have earnings (or revenue), the person responsible for the harm caused shall be obliged to compensate the victim for, in addition to the expenses caused by the impairment of health, also the harm connected with the loss of or reduction of his labour capacity, proceeding from the amount of living minimum of the able-bodied population as a whole established in accordance with a law with regard to the Russian Federation [as amended by Law of 26 November 2002. СЗ РФ (2002), no. 48, item 4737].

3. If at the time of the impairment to his health a minor had earnings, the harm shall be compensated by proceeding from the amount of these earnings, but not lower than the amount of living minimum of the able-bodied population as a whole established in accordance with a law with regard to the Russian Federation [as amended by Law of 26 November 2002. СЗ РФ (2002), no. 48, item 4737].

4. After the commencement of labour activity a minor whose health was previously caused harm shall have the right to demand an increase of the amount of compensation of harm by proceeding from the earnings received by him, but not less than the amount of remuneration established for the post occupied by him or the earnings of a worker of the same qualifications at his place of work.

Article 1088. Compensation of Harm to Persons Who Have Incurred Damage as Result of Death of Breadwinner

1. In the event of the death of the victim (breadwinner), the right to compensation shall be had by:

persons lacking dispositive labour capacity who were dependent on the deceased or had on the day of his death the right to receive maintenance from him;

a child of the deceased born after his death;

one of the parents, spouse, or other member of the family, irrespective of his labour capacity, who did not work and cared for the children, grandchildren, brothers and sisters who had not attained fourteen years who were dependent on him or although they had attained the said age but in the opinion of medical agencies need outside care by reason of their state of health;

persons who were dependent on the deceased and came to lack labour capacity within five years after his death.

One of the parents, spouse, or other member of the family who does not work and cares for children, grandchildren, brothers, and sisters of the deceased and came to lack labour capacity during the period of effectuating care shall retain the right to compensation of harm after ending the care of these persons.

2. The harm shall be compensated to:

a minor: until attaining eighteen years of age;

pupils older than eighteen years: until completion of study in day educational institutions, but not more than up to twenty-three years of age;

women older than fifty-five years and men older than sixty years: for life;

disabled persons: for the period of disability;

one of the parents, spouse, or other member of the family engaged in caring for children, grandchildren, brothers, and sisters dependent on the deceased: until they attain fourteen years of age or the state of health changes.

Article 1089. Amount of Compensation of Harm Incurred in Event of Death of Breadwinner

1. Harm shall be compensated to persons who have the right to compensation for harm in connection with the death of a breadwinner in the amount of that participatory share of earnings (or revenue) of the deceased determined according to the rules of Article 1086 of the present Code which they have received or had the right to receive for their maintenance while alive. When determining the compensation for harm to these persons within the revenues of the deceased, together with earnings (or revenue) shall be included a pension, maintenance for life, and other similar payments being received by them while alive.

2. When determining the amount of compensation for harm, pensions assigned to persons in connection with the death of a breadwinner, and likewise other types of pensions assigned both before and after the death of a breadwinner,

and also earnings (or revenue) and stipend received by these persons, shall not be taken into consideration at the expense of the compensation for harm.

3. The amount of compensation established for each of those having the right to compensation for harm in connection with the death of a breadwinner shall not be subject to further recalculation except for instances of:

the birth of a child after the death of a breadwinner;

the assignment or termination of the payment of compensation to persons caring for children, grandchildren, brothers, and sisters of a deceased breadwinner.

The amount of compensation may be increased by a law or by a contract.

Article 1090. Subsequent Change of Amount of Compensation of Harm

1. A victim who has partially lost labour capacity shall have the right at any time to require from the person on whom the duty of compensation of harm has been placed a respective increase of the amount of his compensation if the labour capacity of the victim has decreased thereafter in connection with the causing of impairment of health in comparison with that which remained at the moment of awarding compensation for harm to him.

2. A person on whom the duty to compensate harm has been placed which was caused to the health of the victim shall have the right to demand a respective reduction of the amount of compensation if the labour capacity of the victim increased in comparison with that which he had at the moment compensation for harm was awarded.

3. The victim shall have the right to demand an increase of the amount of compensation of harm if the property status of the citizen on whom the duty to compensate harm was placed has improved, and the amount of compensation was reduced in accordance with Article 1083(3) of the present Code.

4. A court may at the demand of a citizen who caused harm reduce the amount of compensation for harm if his property status in connection with disability or attainment of pension age was worsened in comparison with the status at the moment of awarding compensation for harm, except for instances when the harm was caused by actions committed intentionally.

Article 1091. Increase of Amount of Compensation of Harm in Connection with Increased Cost of Living [as amended by Law of 26 November 2002, СЗ РФ (2002), no. 48, item 4737].

Amounts to be paid to citizens for compensation of harm caused to the life or health of a victim shall be subject, in the event of an increase in the cost of living, to indexation in the procedure established by a law (Article 318) [as amended by Law of 26 November 2002 СЗ РФ (2002), no. 48, item 4737].

Article 1092. Payments for Compensation of Harm

1. Compensation of harm caused by a reduction of labour capacity or the death of a victim shall be made by monthly payments.

When there are justifiable reasons, the court, taking into account the possibilities of the causer of harm, may at the demand of the citizen who has the right to compensation of harm award him the payments due as a single payment, but for not more than three years.

2. Amounts in compensation of additional expenses (Article 1085[1]) may be awarded for a future period within the limits of the periods determined on the basis of the opinion of medical expert examination, and also, when necessary, advance payment of the cost of the respective services and property, including the acquisition of passes, payment for travel, and payment for special means of transport.

Article 1093. Compensation of Harm in Event of Termination of Juridical Person

1. In the event of the reorganisation of a juridical person deemed in the established procedure to be responsible for harm caused to life or health, the duty relating to the payment of the respective payments shall be borne by its legal successor. Demands concerning compensation of harm shall also be presented to it.

2. In the event of the liquidation of a juridical person deemed in the established procedure to be responsible for harm caused to life or health, the respective payments must be capitalised for the payment thereof to the victim according to the rules established by a law or other legal acts.

Other instances under which the capitalisation of payments may be made also may be established by a law or by other legal acts.

Article 1094. Compensation of Expenses for Burial

Persons responsible for harm which caused the death of the victim shall be obliged to compensate necessary expenses for the burial to the person who incurred those expenses.

The benefit for burial received by citizens who incurred these expenses shall not be set-off at the expense of compensation of harm.

§3. Compensation of Harm Caused as Consequence of Defects of Good,
Work, or Service

**Article 1095. Grounds for Compensation of Harm Caused as
Consequence of Defects of Good, Work, or Service**

Harm caused to the life, health, or property of a citizen or the property of a
juridical person as a consequence of design, prescription, or other defects of a
good, work, or service, and also as a consequence of unreliable or insufficient
information concerning the good (or work, service), shall be subject to compensation by the seller or the manufacturer of the good or by the person who fulfilled
the work or rendered the service (executor) irrespective of their fault and whether
the victim was in contractual relations with them or not.

The rules provided for by the present Article shall be applied only in the
instances of the acquisition of the good (or fulfilment of work, rendering of service) for consumption purposes, and not for use in entrepreneurial activity.

**Article 1096. Persons Responsible for Harm Caused as Consequence of
Defects of Good, Work, or Service**

1. Harm caused as a consequence of the defects of a good shall be subject to
compensation at the choice of the victim by the seller or the manufacturer of the
good.

2. Harm caused as a consequence of the defects of work or service shall be subject to compensation by the person who fulfilled the work or rendered the service
(executor).

3. Harm caused as a consequence of the failure to grant full or reliable
information concerning the good (or work, service) shall be subject to compensation by the persons specified in points 1 and 2 of the present Article.

**Article 1097. Periods for Compensation of Harm Caused as Result of
Defects of Good, Work, or Service**

1. Harm caused as a consequence of defects of a good, work, or service shall be
subject to compensation if it arose during the established period of fitness or service period of a good (or work, service), and if the period of fitness or service
period is not established, within ten years from the date of production of the good
(or work, service) [as amended 17 December 1999. СЗ РФ (1999), no. 51, item
6288].

2. Irrespective of the time of causing, harm shall be subject to compensation if:
in violation of the requirements of a law the period of fitness or service period
has not been established;
the person to whom the good was sold, for whom work was fulfilled, or to
whom a service was rendered was not warned about necessary actions upon the

expiry of the period of fitness or service period and possible consequences in the event of the failure to fulfil the said actions, or he was not granted full and reliable information about the good (or work, service) [as amended 17 December 1999. СЗ РФ (1999), no. 51, item 6288].

Article 1098. Grounds for Relief from Responsibility for Harm Caused as Consequence of Defects of Good, Work, or Service

The seller or manufacturer of a good or executor of work or service shall be relieved from responsibility if it is proved that the harm arose as a consequence of insuperable force or a violation by the consumer of the established rules for the use of the good or the results of the work, service, or keeping thereof.

§4. Contributory Compensation of Moral Harm

Article 1099. General Provisions

1. The grounds and amount of contributory compensation to a citizen of moral harm shall be determined by the rules provided for by the present Chapter and Article 151 of the present Code.

2. Moral harm caused by actions (or failure to act) which violate the property rights of a citizen shall be subject to contributory compensation in the instances provided for by a law.

3. Contributory compensation of moral harm shall be effectuated irrespective of the property harm subject to compensation.

Article 1100. Grounds for Contributory Compensation of Moral Harm

Contributory compensation of moral harm shall be effectuated irrespective of the fault of the causer of the harm in instances when:

the harm is caused to the life or health of a citizen by a source of increased danger;

the harm is caused to a citizen as a result of his illegal conviction, illegal bringing to criminal responsibility, illegal application of confinement under guard or undertaking not to leave as a measure of restraint, or illegal imposition of an administrative sanction in the form of arrest or correctional tasks;

the harm was caused by the dissemination of information defaming honour, dignity, and business reputation;

in other instances provided for by a law.

Article 1101. Means and Amount of Contributory Compensation of Moral Harm

1. Contributory compensation of moral harm shall be effectuated in monetary form.

2. The amount of contributory compensation of moral harm shall be determined by a court depending upon the character of physical and moral suffering

caused to the victim, and also the degree of fault of the causer of the harm in the instances when the fault is the grounds for the compensation of harm. When determining the amount of contributory compensation of harm, the requirements of reasonableness and justness must be taken into account.

The character of physical and moral sufferings shall be valued by the court by taking into account the factual circumstances under which the moral harm was caused and the individual peculiarities of the victim.

Chapter 60. Obligations as Consequence of Unfounded Enrichment

Article 1102. Duty to Return Unfounded Enrichment

1. A person who without grounds established by a law, other legal acts, or transaction acquired or saved property (acquirer) at the expense of another person (victim) shall be obliged to return to the last property unfoundedly acquired or saved (unfounded enrichment), except for instances provided for by Article 1109 of the present Code.

2. The rules provided for by the present Chapter shall be applied irrespective of whether the unfounded enrichment is a result of the behaviour of the acquirer of the property, the victim himself, third persons, or occurred outside their will.

Article 1103. Correlation of Demands Concerning Return of Unfounded Enrichment with Other Demands Concerning Defence of Civil Rights

Insofar as not established otherwise by the present Code, by other laws, or by other legal acts and does not arise from the essence of the respective relations, the rules provided for by the present Chapter shall be subject to application also to demands:

(1) concerning the return of that performed under an invalid transaction;

(2) concerning the demanding and obtaining of property by the owner from another's illegal possession;

(3) by one party in an obligation to another concerning the return of that performed in connection with this obligation;

(4) concerning compensation of harm, including that caused by the behaviour not in good faith of the person enriched.

Article 1104. Return of Unfounded Enrichment in Kind

1. Property comprising unfounded enrichment of the acquirer must be returned to the victim in kind.

2. The acquirer shall be liable to the victim for any, including also for any accidental, shortage or deterioration of the unfoundedly acquired or saved property which occurred after he knew or should have known about the unfoundedness of the enrichment. Before that moment he shall be liable only for intent and gross negligence.

Article 1105. Compensation of Value of Unfounded Enrichment

1. In the event it is impossible to return in kind property unfoundedly received or saved, the acquirer must compensate the victim for the real value of this property at the moment of its acquisition, and also the losses caused by the subsequent change of the value of this property if the acquirer did not compensate its value immediately after he knew about the unfounded enrichment.

2. A person who unfoundedly and temporarily used another's property without the intention to acquire it or another's services must compensate the victim for that which he saved as a consequence of such use at the price which existed at the time when the use ended and in that place where it occurred.

Article 1106. Consequences of Unfounded Transfer of Right to Another Person

A person who has transferred by means of assignment of a demand or otherwise a right belonging to him to another person on the basis of a nonexistent or invalid obligation shall have the right to demand the restoration of his previous position, including the return to him of documents certifying the transferred right.

Article 1107. Compensation to Victim of Revenues Not Received

1. The person who unfoundedly received or saved property shall be obliged to return or to compensate the victim for all revenues which he derived or should have derived from this property from the time when he knew or should have known about the unfounded enrichment.

2. Interest shall be subject to being calculated on the amount of unfounded monetary enrichment for the use of another's means (Article 395) from the time when the acquirer knew or should have known about the unfoundedness of the receipt or savings of monetary means.

Article 1108. Compensation of Expenditures on Property Subject to Return

In the event of the return of property unfoundedly received or saved (Article 1104) or compensation of the value thereof (Article 1105), the acquirer shall have the right to demand compensation from the victim of necessary expenditures incurred for the maintenance and preservation of property from the time from which he is obliged to return revenues (Article 1106) [Article 1107– *WEB*], setting off the advantages received by him. The right to compensation of expenditures shall be lost in an instance when the acquirer intentionally withheld property which is subject to return.

Article 1109. Unfounded Enrichment Not Subject to Return

There shall not be subject to return as unfounded enrichment:

(1) property transferred in performance of an obligation before the ensuing of the period of performance unless provided otherwise by the obligation;

(2) property transferred in performance of an obligation upon the expiry of the period of limitations;

(3) earnings and payments equated thereto, pensions, benefits, stipends, compensation of harm caused to life or health, alimony and other monetary amounts granted to a citizen as means for existence, in the absence of lack of good faith on his part and mathematical error;

(4) monetary amounts and other property granted in performance of a nonexistent obligation if the acquirer proves that the person requiring the return of the property knew about the absence of the obligation or granted the property for the purposes of philanthropy.

PART THREE

SECTION V. INHERITANCE LAW

Chapter 61. *General Provisions on Inheritance*

Article 1110. Inheriting

1. In the event of inheriting, the property of a deceased (inheritance, inheritance property) shall pass to other persons by way of universal legal succession, that is, in unchanged form as a single whole and at one and the same moment, unless it follows otherwise from the rules of the present Code.

2. Inheriting shall be regulated by the present Code and by other laws, and in the instances provided for by a law, by other legal acts.

Article 1111. Grounds of Inheritance

Inheriting shall be effectuated by will and by operation of law.

Inheriting by operation of law shall occur when and insofar as it is not changed by a will, and also in other instances established by the present Code.

Article 1112. Inheritance

Within the composition of an inheritance shall be the things and other property, including property rights and duties, which belonged to the decedent on the day of opening the inheritance.

Rights and duties shall not be within the compsition of an inheritance which are inextricably connected with the person of the decedent, in particular, the right to alimony, the right to compensation of harm caused to the life or health of a citizen, and also rights and duties whose transfer by way of inheriting is not permitted by the present Code or by other laws.

Article 1113. Opening of Inheritance

An inheritance shall open with the death of a citizen. The declaration by a court of a citizen to be deceased shall entail the same legal consequences as the death of the citizen.

Article 1114. Time of Opening of Inheritance

1. The day of opening an inheritance shall be the day of death of the citizen. When declaring a citizen to be deceased, the day of opening an inheritance shall be the day of entry into legal force of the decision of the court concerning the

declaring of the citizen to be deceased, and when in accordance with Article 45(3) of the present Code the day of death of a citizen is deemed to be the day of his supposed perishing,—the day of death specified in the decision of the court.

2. Citizens deceased on the same day shall be considered for the purposes of inheritance legal succession to be deceased simultaneously and shall not inherit one after the other. The heirs of each of them shall be called to inherit in so doing.

Article 1115. Place of Opening of Inheritance

The place of opening an inheritance shall be the last place of residence of the decedent (Article 20).

If the last place of residence of a decedent who possessed property on the territory of the Russian Federation is unknown or situated beyond the limits thereof, the place of opening an inheritance in the Russian Federation shall be deemed to be the location of such inheritance property. If such inheritance property is situated in various places, the place of opening the inheritance shall be the location of the immoveable property within its composition or of the most valuable part of the immoveable property, and in the absence of immoveable property—the location of moveable property or most valuable part thereof. The value of the property shall be determined by proceeding from its market value.

Article 1116. Persons Who May Be Called to Inherit

1. Citizens alive on the day of opening an inheritance, as well as conceived during the life of the decedent and born alive after the opening of the inheritance, may be called to inherit.

Juridical persons specified therein and existing on the day of opening an inheritance also may be called to inherit by will.

2. The Russian Federation, subjects of the Russian Federation, municipal formations, foreign States, and international organisations may be caued to inherit by will, and to inherit by operation of law—the Russian Federation in accordance with Article 1151 of the present Code.

Article 1117. Unworthy Heirs

1. Citizens who by their intentional unlawful actions directed against the decedent, any of the heirs thereof, or against the effectuation of the last will of the decedent expressed in the will, facilitated or attempted to faciliate the calling of themselves or other persons to inherit, or facilitated or attempted to facilitate an increase of the participatory share of the inheritance due to them or to other persons, if these circumstances have been confirmed in a judicial proceeding, shall not inherit either by operation of law or by will. However, citizens to whom the decedent after their loss of the right to inherit bequeathed property shall have the right to inherit this property.

Parents after children with respect to whom the parents were deprived of parental rights in a judicial proceeding and not restored in such rights on the day of opening the inheritance shall not inherit by operation of law.

2. Upon the demand of an interested person, a court shall remove from inheriting by law citizens who have maliciously avoided the fulfilment of duties placed on them by virtue of a law with regard to maintenance of the decedent.

3. A person who does not have the right to inherit or removed from inheriting on the basis of the present Article (unworthy heir) shall be obliged to return in accordance with the rules of Chapter 60 of the present Code all property unfoundedly received by him from the composition of the inheritance.

4. The rules of the present Article shall extend to heirs having the right to an obligatory participatory share in the inheritance.

5. The rules of the present Article respectively shall apply to a testamentary legacy (Article 1137). When the subject of a testamentary legacy was the fulfilment of determined work for an unworthy recipient of a testamentary legacy or rendering to him of a determined service, the last shall be obliged to compensate the heir who has performed the testamentary legacy for the value of work fulfilled for the unworthy recipient of the testamentary legacy or service rendered to him.

Chapter 62. Inheriting by Will

Article 1118. General Provisions

1. One may dispose of property in the event of death only by means of concluding a will.

2. A will may be concluded by a citizen possessing at the moment of conclusion thereof dispositive legal capacity in full.

3. A will must be concluded personally. The conclusion of a will through a representative shall not be permitted.

4. The dispositions of only one citizen may be contained in a will. The conclusion of a will by two or more citizens shall not be permitted.

5. A will shall be a unilateral transaction which creates rights and duties after the opening of the inheritance.

Article 1119. Freedom of Will

1. A testator shall have the right at his discretion to bequeath property to any persons, by any means to determine the participatory shares of heirs in the inheritance, to deprive of an inheritance one, several, or all heirs by operation of law, without specifying the reasons for such deprivation, and also to include in the will

any dispositions provided for by the rules of the present Code concerning inheriting and to revoke or to change a will concluded.

The freedom of will shall be limited by the rules concerning an obligatory participatory share in an inheritance (Article 1149).

2. A testator shall not be obliged to communicate to anyone the content, conclusion, change, or revocation of a will.

Article 1120. Right to Bequeath Any Property

A testator shall have the right to conclude a will containing a disposition concerning any property, including that which he may acquire in future.

A testator may dispose of his property or any part thereof, having drawn up one or several wills.

Article 1121. Designation and Subdesignation of Heir in Will

1. A testator may conclude a will to the benefit of one or several persons (Article 1116) both within and not within the group of heirs by operation of law.

2. A testator may specify in a will another heir (subdesignation of heir) if the heir designated by him in the will or an heir of the testator by operation of law dies before the opening of the inheritance, or simultaneously with the testator, or after the opening of the inheritance without succeeding to accept it, or does not accept the inheritance for other reasons or renounces it, or will not have the right to inherit or will be removed from inheriting as unworthy.

Article 1122. Participatory Share of Heirs in Bequeathed Property

1. Property bequeathed to two or several heirs without specifying their participatory shares in the inheritance and without specifying which things or rights within the composition of the inheritance are intended for which of the heirs shall be considered to be bequeathed to the heirs in equal participatory shares.

2. A specification in the will of parts of an indivisible thing (Article 133) intended for each of the heirs in kind shall not entail the invalidity of the will. Such thing shall be considered to be bequeathed in equal participatory shares corresponding to the value of these parts. The procedure for the use by heirs of this indivisible thing shall be established in accordance with the parts of that thing intended for them in the will.

The participatory shares of the heirs and the procedure for use of such thing shall be specified with the consent of the heirs, in accordance with the present Article, in the certificate on the right to an inheritance with respect to an indivisible thing bequeathed by parts in kind. In the event of a dispute between the heirs, their participatory shares and the procedure for the use of an indivisible thing shall be determined by a court.

Article 1123. Secrecy of Will

A notary, other person certifying a will, interpreter, executor of a will, witnesses, and also citizen signing a will in place of the testator, shall not have the right before the opening of an inheritance to divulge information concerning the content of the will, conclusion thereof, change, or revocation.

In the event of a violation of the secrecy of a will, the testator shall have the right to demand contributory compensation of moral harm, and also to take advantage of other means of defence of civil rights provided for by the present Code.

Article 1124. General Rules Affecting Form and Procedure for Concluding Will

1. A will must be drawn up in written form and certified by a notary. Certification of a will by other persons shall be permitted in the instances provided for by Article 1125(7), Article 1127, and Article 1128(2) of the present Code.

The failure to comply with the rules established by the present Code concerning written form of a testament and certification thereof shall entail the invalidity of the will.

Drawing up a will in simple written form shall be permitted only by way of exception in the instances provided for by Article 1129 of the present Code.

2. When in accordance with the rules of the present Code witnesses are present during the drawing up, signature, and certification of a will or when the will is transferred to a notary, there may not be such witnesses and there may not sign the will in place of the testator:

the notary or other person certifying the will;

the person to whose benefit the will has been drawn up or a testamentary legacy has been made, the spouse of such person, his children, and parents;

citizens not possessing dispositive legal capacity in full;

illiterate persons;

citizens with physical defects which clearly do not enable them to be aware in full of the essence of what is happening;

persons not possessing in sufficient degree the language in which the will has been drawn up, except for instances when a closed will is drawn up.

3. When in accordance with the rules of the present Code the presence of a witness is obligatory during the drawing up, signature, and certification of a will or when transferring it to a notary, the absence of a witness when performing the said actions shall entail the invalidity of the will, and the failure of the witness to conform to the requirements established by point 2 of the present Article may be grounds for deeming a will to be invalid.

4. On a will must be specified the place and date of certification thereof, except for the instance provided for by Article 1126 of the present Code.

Article 1125. Notarially-Certified Will

1. A notarially-certified will must be written out by the testator or written down from his words by the notary. When writing out or writing down the will technical means (computer, typewriter, and others) may be used.

2. A will written down by a notary from the words of the testator must before signature thereof be read out in full by the testator in the presence of the notary. If the testator is not in a state personally to read out the will, the text thereof shall be read aloud for him by the notary, concerning which a respective notation shall be made on the will specifying the reasons for which the testator could not personally read out the will.

3. A will must be signed by the testator in his own hand.

If the testator by virtue of physical defects, grave illness, or illiteracy cannot sign the will in his own hand, at his request it may be signed by another citizen in the presence of the notary. The reasons for which the testator could not sign the will in his own hand must be specified on the will, as well as the surname, forename, and patronymic and place of residence of the citizen who signed the will at the request of the testator in accordance with a document certifying the identity of this citizen.

4. When drawing up and notarially-certifying a will, a witness may be present at the wish of the testator.

If a will is drawn up and certified in the present of a witness, it must be signed by him and on the will must be specified the surname, forename, patronymic, and place of residence of the witness in accordance with a document certifying his identity.

5. A notary shall be obliged to warn the witness, and also a citizen signing the will in place of the testator, about the need to comply with secrecy of the will (Article 1123).

6. When certifying a will, a notary shall be obliged to explain to the testator the content of Article 1149 of the present Code and to make a respective inscription about this on the will.

7. When the right to perform notarial actions has been granted by a law to officials of agencies of local self-government and to officials of consular institutions of the Russian Federation, a will may be certified in place of the notary by the respective official in compliance with the rules of the present Code concerning the form of the will, procedure for notarial certification thereof, and secrecy of the will.

Article 1126. Closed Will

1. A testator shall have the right to conclude a will without providing in so doing to other persons, including a notary, the possibility to familiarise themselves with the content thereof (closed will).

2. A closed will must be written out in the own hand and signed by the testator. The failure to comply with these rules shall entail the invalidity of the will.

3. A closed will in a sealed envelope shall be transferred to a notary in the presence of two witnesses, who shall place on the envelope their signatures. The envelope signed by the witnesses shall be sealed in their presence by the notary in another envelope, on which the notary shall make an inscription containing information concerning the testator from whom the notary has accepted the closed will, place and date of acceptance thereof, surname, forename, and patronymic, and place of residence of each witness in accordance with a document certifying identity.

In accepting from the testator an envelope with a closed will the notary shall be obliged to explain to the testator the content of point 2 of the present Article and Article 1149 of the present Code and to make a respective inscription thereof on the second envelope, and also to issue to the testator a document confirming acceptance of the closed will.

4. Upon the presentation of a certificate concerning the death of the person who concluded a closed will, the notary not later than fifteen days from the day of submission of the certificate shall open the envelope with the will in the presence of not less than two witnesses and interested person who wished to be present from among the heirs by operation of law. After opening the envelope, the text of the will contained therein shall be read out at once by the notary, after which the notary shall draw up and together with the witnesses sign a protocol certifying the opening of the envelope with the will and containing the full text of the will. The original of the will shall be kept by the notary. A notarially-certified copy of the protocol shall be issued to the heirs.

Article 1127. Wills Equated to Notarially-Certified Wills

1. There shall be equated to notarially-certified wills:

(1) wills of citizens situated for care in hospitals, military hospitals, and other inpatient treatment institutions or residing in homes for the aged and disabled certified by chief doctors, their deputies for the medical section or duty doctors of those hospitals, military hospitals, and other inpatient treatment institutions, and also heads of military hospitals, directors, or chief doctors of homes for the aged and disabled;

(2) wills of citizens situated during voyages on vessels sailing under the State Flag of the Russian Federation certified by the masters of those vessels;

(3) wills of citizens on prospecting, arctic, or other similar expeditions certified by the heads of those expeditions;

(4) wills of military servicemen, and in stationing centres for military units where there are no notaries, also the wills of civilian persons working in those units, members of their families, and members of the families of military servicemen, certified by the commanders of the military units;

(5) wills of citizens in places of deprivation of freedom certified by the heads of places of deprivation of freedom.

2. A will equated to a notarially-certified will must be signed by the testator in the presence of the person certifying the will and a witness also signing the will.

The remainder of the rules of Articles 1124 and 1125 of the present Code respectively shall apply to such will.

3. A will certified in accordance with the present Article must be as soon as this is possible be sent by the person certifying the will through justice agencies to a notary at the place of residence of the testator. If the place of residence of the testator is known to the person certifying the will, the will shall be sent directly to the respective notary.

4. If in any of the instances provided for by point 1 of the present Article a citizen intending to conclude a will expresses the wish to invite a notary for this and there is a reasonable possibility to fulfil this wish, the persons who have been granted the right to certify a will in accordance with the said point shall be obliged to take all measures to invite the notary to the testator.

Article 1128. Testamentary Dispositions of Rights to Monetary Means in Banks

1. The rights to monetary means deposited by a citizen as a contribution or in any other account of a citizen in a bank may at the discretion of the citizen be bequeathed either in the procedure provided for by Articles 1124–1127 of the present Code, or by means of the conclusion of a testamentary disposition in written form in that branch of the bank in which this account is situated. With respect to means in the account such testatmentary disposition shall have the force of a notarially-certified will.

2. A testamentary disposition of rights to monetary means in a bank must be signed in the testator's own hand, specifying the date of drawing up thereof and certified by an employee of the bank having the right to accept for performance dispositions of a client with respect to means in his account. The procecure for concluding testamentary dispositions of monetary means in banks shall be determined by the Government of the Russian Federation.

3. Rights to monetary means with respect to which a testamentary disposition has been concluded in a bank shall be within the composition of the inheritance

and shall be inherited on the general grounds in accordance with the rules of the present Code. These means shall be issued to heirs on the basis of a certificate on the right to an inheritance and in accordance with it, except for instances provided for by Article 1174(3) of the present Code.

4. The rules of the present Article respectively shall apply to other credit organisations which have been granted the right to attract monetary means of citizens as contributions or in other accounts.

Article 1129. Will in Extraordinary Circumstances

1. A citizen who is in a position clearly threatening his life and by virtue of extraordinary circumstances which have formed is deprived of the possibility to conclude a will in accordance with the rules of Articles 1124–1128 of the present Code may set out their last will with respect to their property in simple written form.

The setting out by a citizen of last will in simple written form shall be deemed to be his will if the testator in the presence of two witnesses in his own hand wrote out and signed a document, from the content of which it follows that it represents a will.

2. A will concluded in the circumstances specified in paragraph one of point 1 of the present Article shall lose force if the testator within a month after the termination of these circumstances does not take advantage of the possibility to conclude a will in another form provided for by Articles 1124–1128 of the present Code.

3. A will concluded in extraordinary circumstances in accordance with the present Article shall be subject to performance only on condition of confirmation by a court upon the demand of interested persons of the fact of the conclusion of the will in extraordinary circumstances. The said demand must be declared before the expiry of the period established for acceptance of the inheritance.

Article 1130. Revocation and Change of Will

1. A testator shall have the right to revoke or change a will drawn up by him at any time after the conclusion thereof, without specifying in so doing the reasons for the revocation or change thereof.

In order to revoke or change a will the consent of no one shall be required, including persons designated as heirs in the will being revoked or changed.

2. A testator shall have the right by means of a new will to vacate the previous will as a whole or to change it by means of revocation or change of individual testatmentary dispositions contained therein.

A subsequent will not containing direct instructions concerning revocation of a previous will or individual testatmentary dispositions contained therein shall vacate this previous will wholly or in the part in which it is contrary to the subsequent will.

A will vacated wholly or partially by a subsequent will shall not be reinstated if the subsequent will is revoked by the testator wholly or in respective part.

3. In the event of the invalidity of a subsequent will, inheriting shall be effectuated in accordance with the previous will.

4. A will may be revoked also by means of an instruction concerning the revocation thereof. An instruction concerning the revocation of the will must be concluded in the work established by the present Code for the conclusion of the will. The rules of point 3 of the present Article shall apply respectively to an instruction concerning the revocation of a will.

5. A will concluded in extraordinary circumstances (Article 1129) may be revoked or changed only by a will.

6. A testamentary disposition in a bank (Article 1128) may be revoked or changed only by a testamentary disposition of the rights to monetary means in the respective bank.

Article 1131. Invalidity of Will

1. In the event of a violation of the provisions of the present Code entailing the invalidity of a will, depending upon the grounds of the invalidity, a will shall be invalid by virtue of being deemed to be such by a court (contestable will) or irrespective of deeming such (null will).

2. A will may be deemed by a court to be invalid upon the suit of the person whose rights or legal interests have been violated by this will.

The contesting of a will before the opening of an inheritance shall not be permitted.

3. Slips of the pen and other insignificant violations of the procedure for drawing up thereof, signature, or certification, if it is established by a court that they do not influence the understanding of the expression of will of the testator, may not serve as grounds for invalidity of a will.

4. Both a will as a whole and individual testamentary dispositions contained therein may be invalid. The invalidity of individual dispositions contained in a will shall not affect the remaining part of the will if it can be supposed that it would have been included in the will also in the absence of the dispositions which are invalid.

5. The invalidity of a will shall not deprive persons specified therein as heirs or recipients of a legacy of the right to inherit by law or on the basis of another valid will.

Article 1132. Interpretation of Will

When a will is interpreted by a notary, executor of the will, or by a court the literal meaning of the words and expressions contained in the will shall be taken into account.

In the event of an ambiguity in the literal meaning of any provision of a will, it shall be established by means of contrasting this provision with other provisions and the meaning of the will as a whole. In so doing the fullest effectuation of the supposed will of the testator must be ensured.

Article 1133. Performance of Will

The performance of a will shall be effectuated by the heirs by will, except for instances when the performance thereof wholly or in determined part shall be effectuated by the executor of the will (Article 1134).

Article 1134. Executor of Will

1. A testator may charge performance of a will to a citizen-executor specified by him in the will, irrespective of whether this citizen is an heir.

The consent of the citizen to be an executor of the will shall be expressed by this citizen in an inscription in his own hand on the will itself, or in a statement appended to the will, or in a statement given to a notary witin a month from the day of opening the inheritance.

A citizen shall be deemed also to have given consent to be the executor of a will if he within a month from the day of opening an inheritance actually commenced to execute the will.

2. After the opening of an inheritance a court may free the executor of a will from his duties either at the request of the executor himself or at the request of the heirs when there are circumstances obstructing the performance of these duties by the citizen.

Article 1135. Powers of Executor of Will

1. The powers of the executor of a will shall be based on the will by which he was designated executor and shall be certified by a certificate issued by a notary.

2. Unless provided otherwise in the will, the executor of a will must take measures necessary for performance of the will, including to:

(1) ensure the transfer to the heirs of inheritance property due to them in accordance with the will of the testator expressed in the will and a law;

(2) take autonomously or through a notary measures for the protection of the inheritance and management thereof in the interests of the heirs;

(3) receive monetary means due to the testator and other property for transfer thereof to heirs, if this property is not subject to transfer to other persons (Article 1183[1]);

(4) perform a testamentary imposition or demand from heirs the performance of a testamentary legacy (Article 1137) or testamentary imposition (Article 1139).

3. The executor of a will shall have the right in his own name to conduct affairs connected with the performance of a will, including in a court and other State agencies and State institutions.

Article 1136. Compensation of Expenses Connected with Performance of Will

The executor of a will shall have the right to compensation at the expense of the inheritance of necessary expenses connected with performance of the will, and also to receipt remuneration above the expenses at the expense of the inheritance if this has been provided for by the will.

Article 1137. Testamentary Legacy

1. A testator shall have the right to place on one or several heirs by will or by operation of law the performance at the expense of the inheritance of any duty of a property character to the benefit of one or several persons (recipient of legacy), who shall acquire the right to demand performance of the duty (testamentary legacy).

A testamentary legacy must be established in the will.

The content of a will may be exhausted by the testamentary legacy.

2. The subject of a testamentary legacy may be the transfer to the recipient of a legacy in ownership, in possession in another right to a thing, or for use things within the position of an inheritance, transfer to the recipient of the legacy of a property right within the composition of the inheritance, acquisition for the recipient of the legacy and transfer to him of other property, the fulfilment for him of determined work or the rendering to him of a determined service, or the effectuation to the benefit of the recipient of the legacy of period payments, and so on.

In particular, on a heir to whom a dwelling house, apartment, or other dwelling premise passes the testator may place the duty to provide to another person for the period of life of that person or for another period the right to use this premise or a determined part thereof.

In the event of subsequent transfer of the right of ownership to property within the composition of an inheritance to another person, the right of use of this property granted under testamentary legacy shall retain force.

3. To relations between the recipient of a legacy (creditor) and heir on whom the testamentary legacy has been placed (debtor) shall apply the provisions of the present Code on obligations, unless follows otherwise from the rules of the present Section and essence of the testamentary legacy.

4. The right to receive a testamentary legacy shall operate for three years from the day of opening the inheritance and shall not pass to other persons. However, the recipient of a legacy in a will may subdesignate another recipient of the legacy if the recipient of the legacy designated in the will dies before the opening of inheritance or simultaneously with the decedent, or refuses to accept the testamentary legacy, or does not take advantage of his right to receive the testamentary legacy, or is deprived of the right to receive the testamentary legacy in accordance with the rules of Article 1117(5) of the present Code.

Article 1138. Performance of Testamentary Legacy

1. An heir on whom a testamentary legacy has been placed by a testator must perform it within the limits of the value of the inheritance which has passed to him, deducting the debts of the testator with regard to it.

If an heir on whom a testamentary legacy has been placed has the right to an obligatory participatory share in an inheritance, his duty to perform the legacy shall be limited to the value of the inheritance which has passed to him which exceeds the amount of his obligatory participatory share.

2. If a testamentary legacy has been placed on several heirs, such legacy shall encumber the right of each of them to the inheritance commensurately to the participatory share thereof in the inheritance insofar as not provided otherwise by the will.

3. If the recipient of a legacy is deceased before the opening of the inheritance or simultaneously with the testator, or refused to accept the testamentary legacy (Article 1160), or did not take advantage of his right to receive the testamentary legacy within three years from the day of opening of the inheritance, or was deprived of the rgith to receive the testamentary legacy in accordance with the rules of Article 1117 of the present Code, the heir obliged to perform the testamentary legacy shall be relieved from this duty, except for the instance when the recipient of the legacy has subdesignated another recipient of the legacy.

Article 1139. Testamentary Imposition

1. A testator may in a will place on one or several heirs by will or by operation of will the duty to perform any action of a property or nonproperty character directed toward the effectuation of a generally-useful purpose (testamentary imposition). Such duty may be placed on the executor of the will on condition of singling out in the will part of the inheritance property for performance of the testamentary imposition.

A testator shall have the right also to place on one or several heirs the duty to maintain household animals belonging to the testator, and also to effectuate necessary supervision and care over them.

2. To a testamentary imposition, the subject of which are actions of a property character, shall apply respectively the rules of Article 1138 of the present Code.

3. Interested persons, the executor of the will, and any of the heirs shall have the right to demand performance of a testamentary imposition in a judicial proceeding unless provided otherwise by the will.

Article 1140. Transfer to Other Heirs of Duty to Perform Testamentary Legacy or Testamentary Imposition

If as a consequence of circumstances provided for by the present Code the participatory share of an inheritance due to the heir on whom the duty to perform a testamentary legacy or testamentary imposition was placed passes to other heirs, the last, insofar as does not follow otherwise from the will or a law shall be obliged to perform such legacy or such imposition.

Chapter 63. Inheriting by Operation of Law

Article 1141. General Provisions

1. Heirs by operation of law shall be called to inherit by way of the priority provided for by Articles 1142–1145 and 1148 of the present Code.

The heirs of each subsequent priority shall inherit if there are no heirs of preceding priorities, that is, if the heirs of preceding priorities are lacking, or no one of them has the right to inherit, or they all have been removed from inheriting (Article 1117), or deprived of inheritance (Article 1119[1]), or none of them has accepted the inheritance, or they all have refused the inheritance.

2. The heirs of another priority shall inherit in equal participatory shares, except for heirs inheriting by right of representation (Article 1146).

Article 1142. Heirs of First Priority

1. Children, the spouse, and parents of the decedent shall be heirs of the first priority by operation of law.

2. Grandchildren of the decedent and descendants thereof shall inherit by right of representation.

Article 1143. Heirs of Second Priority

1. If there are no heirs of the first priority, the heirs of the second priority by operation of law shall be full or half brothers and sisters of the decedent, grandfather and grandmother thereof both on the side of the father and on the side of the mother.

2. Children of full and half brothers and sisters of the decedent (nephews and nieces of the decedent) shall inherit by right of representation.

Article 1144. Heirs of Third Priority

1. If there are no heirs of the first and second priorities, the heirs of the third priority by operation of law shall be full or half brothers and sisters of parents of the decedent (uncles and aunts of the decedent).

2. First cousins of the decedent shall inherit by right of representation.

Article 1145. Heirs of Subsequent Priorities

1. If there are no heirs of the first, second, or third priorities (Articles 1142–1144), relatives of the decedent of the third, fourth, and fifth degrees of kinship who are not relegated to heirs of the preceding priorities shall receive the right to inherit by operation of law.

The degree of kinship shall be determined by the date of births separating the relatives one from the other. The birth of the decedent himself shall not be within this number.

2. In accordance with point 1 of the present Article there shall be called to inherit:

as heirs of the fourth priority, relatives of the third degree of kinship—great-grandfather and great-grandmother of the decedent;

as heirs of the fifth priority, relatives of the fourth degree of kinship—children of full cousins of the decedent (grandchildren once removed) and own brothers and sisters of his grandfather and grandmother (grandfathers and grandmothers once removed);

as heirs of the sixth priority, relatives of the fifth degree of kinship—children of grandchildren once removed of the decedent (great-grandchildren once removed), children of his brothers and sisters once removed (nephews and nieces once removed), and children of his grandfathers and grandmothers once removed (uncles and aunts once removed).

3. If there are no heirs of the preceding priorities, stepchildren, stepdaughters, step-father, and step-mother of the decedent shall be called to inherit as heirs of the seventh priority by operation of law.

Article 1146. Inheriting by Right of Representation

1. The participatory share of an heir by operation of law who has died before the opening of an inheritance or simultaneously with the decedent shall pass by right of representation to his respective descendants in the instances provided for by Article 1142(2), Article 1143(2), and Article 1144(2) of the present Code, and shall be divided between them equally.

2. Descendants of an heir by operation of law deprived by the decedent of inheritance (Article 1119[1]) shall not inherit by right of representation.

3. Descendants of an heir who dies before the opening of the inheritance or simultaneously with the decedent and who did not have the right to inherit in accordance with Article 1117(1) of the present Code shall not inherit by right of representation.

Article 1147. Inheriting by Adopted and Adoptive Persons

1. When inheriting by operation of law, an adopted person and his descendant, on one side, and the adoptive person and his relatives, on the other, shall be equated to relatives by origin (blood relatives).

2. An adopted person and his descendant shall not inherit by operation of law after the death of the parents of the adopted person and others of his relatives by origin, and the parents of the adopted person and others of his relatives by origin shall not inherit by operation of law after the death of the adopted person and his descendant, except for instances specified in point 3 of the present Article.

3. When in accordance with the Family Code of the Russian Federation an adopted person retains by decision of a court relations with one of the parents or other relatives by origin, an adopted person and his descendant shall inherit by operation of law after the death of these relatives, and the last shall inherit by operation of law after the death of the adopted person and his descendant.

Inheriting in accordance with the present point shall not exclude inheriting in accordance with point 1 of the present Article.

Article 1148. Inheriting by Dependents Not Having Labour Capacity of Decedent

1. Citizens relegated to heirs by operation of law specified in Articles 1143–1145 of the present Code who lack labour capacity on the day of opening the inheritance but are not within the group of heirs of that priority which is called to inherit shall inherit by operation of law together and equally with the heirs of that priority if not less than a year before the death of the decedent they were dependent on him, irrespective of whether they resided jointly with the decedent or not.

2. To heirs by operation of law shall be relegated citizens who are not within the group of heirs specified in Articles 1142–1145 of the present Code but on the day of opening the inheritance did not have labour capacity and not less than a year before the death of the decedent were dependent on him and resided jointly with him. When there are other heirs by operation of law they shall inherit together and equally with heirs of that priority which is called upon to inherit.

3. In the absence of other heirs by operation of law dependents not having labour capacity of the decedent specified in point 2 of the present Article shall inherit autonomously as heirs of the eighth priority.

Article 1149. Right to Obligatory Participatory Share in Inheritance

1. Minors and children not having labour capacity of the decedent, the spouse and parents thereof not having labour capacity, and also dependents not having labour capacity of the decedent shall be subject to being called to inherit on the basic of Article 1148(1) and (2) of the present Code, and shall inherit irrespective of the content of the will not less than half of the participatory share which would have been due to each of them when inheriting by operation of law (obligatory participatory share).

2. The right to an obligatory participatory share in an inheritance shall be satisfied from the remaining unbequeathed part of inheritance property, even if this leads to a reduction of the rights of other heirs by operation of law to this part of the property, and in the event the unbequeathed part of the property is not sufficient to effectuate the right to an obligatory participatory share—from that part of the property which has been bequeathed.

3. To an obligatory participatory share shall be credited all that an heir having the right to such participatory share receives from the inheritance on any ground whatsoever, including the value of a testamentary legacy established to the benefit of such heir.

4. If effectuation of the right to an obligatory participatory share to an inheritance entails the impossibility to transfer to the heir by will property to which the heir having the right to an obligatory participatory share was not used during the life of the decedent and the heir by will used for habitation (dwelling house, apartment, other dwelling premise, dacha, and the like) or used as the principal source of receiving means for existence (implements of labour, creative studio, and the like), a court may, taking into account the property position of the heirs having the right to an obligatory participatory share, reduce the amount of the obligatory participatory share or refuse to award it.

Article 1150. Rights of Spouse When Inheriting

The right to inherit belonging to a surviving spouse of the decedent by virtue of a will or law shall not diminish the right thereof to part of the property acquired during marriage with the decedent and being their joint ownership. The participatory share of the deceased spouse in this property determined in accordance with Article 256 of the present Code shall be within the composition of the inheritance and pass to heirs in accordance with the rules established by the present Code.

Article 1151. Inheriting of Escheat Property

1. If heirs are lacking both by operation of law and by will, or none of the heirs has the right to inherit, or all heirs have been removed from inheriting (Article

1117), or none of the heirs had accepted the inheritance, or all the heirs have refused the inheritance and in so doing none of them has specified that he is refusing in favour of another heir (Article 1158), the property of the deceased shall be considered to be escheat.

2. Escheat property shall pass by way of inheriting by operation of law to the ownership of the Russian Federation.

3. The procedure for inheriting and recording of escheat property, and also the procedure for the transfer thereof to the ownership of subjects of the Russian Federation or ownership of municipal formations, shall be determined by a law.

Chapter 64. Acquisition of Inheritance

Article 1152. Acceptance of Inheritance

1. In order to acquire an inheritance an heir must accept it.
In order to acquire escheat property (Article 1151) acceptance of an inheritance is not required.

2. Acceptance by an heir of part of the inheritance shall mean acceptance of the entire inheritance due to him, irrespective of what it consists of or where it was situated.
When calling an heir to inherit simultaneously on several grounds (by will and by operation of law, or by way of inheritance transmission and as a result of opening an inheritance, and the like), an heir may accept the inheritance due to him with regard to one of these grounds, several of them, or all grounds.
Acceptance of an inheritance on condition or with reservations shall not be permitted.

3. Acceptance of an inheritance by one or several heirs shall not mean acceptance of the inheritance by remaining heirs.

4. An accepted inheritance shall be deemed to belong to the heir from the day of opening the inheritance irrespective of the time of actual acceptance, and also irrespective of the moment of State registration of the right of the heir to inherited property when such right is subject to State registration.

Article 1153. Means of Acceptance of Inheritance

1. The acceptance of an inheritance shall be effectuated by the filing at the place of opening the inheritance with a notary or official empowered in accordance with a law to issue a certificate on the right to an inheritance an application of the heir to accept the inheritance or application of an heir concerning the issuance of a certificate on the right to an inheritance.
If the application of the heir is transferred to the notary by another person or is sent by post, the signature of the heir on the application must be certified by a

notary, official empowered to perform notarial actions (Article 1125[7]), or person empowered to certify a power of attorney in accordance with Article 185(3) of the present Code.

The acceptance of an inheritance through a representative shall be possible if the power to accept an inheritance is specially provided for in the power of attorney. A power of attorney shall not be required for the acceptance of an inheritance by a legal representative.

2. It shall be deemed, so long as not proved otherwise, that an heir has accepted an inheritance if he has performed actions testifying to the actual acceptance of the inheritance, in particular if the heir:

entered into possession or management of inheritance property;

took measures with regard to preservation of inheritance property and defence thereof against infringements or claims of third persons;

made expenses at his own account for maintenance of inheritance property;

paid debts at his own account of the decedent or received monetary means from third persons due to the decedent.

Article 1154. Period for Acceptance of Inheritance

1. An inheritance may be accepted within six months from the day of opening the inheritance.

In the event of the opening of an inheritance on the day of the supposed perishing of a citizen (Article 1114[1]), the inheritance may be accepted within six months from the day of entry into legal force of the decision of the court declaring him to be deceased.

2. If the right to inherit arises for other persons as a consequence of a refusal by an heir of an inheritance or elimination of an heir on the grounds established by Article 1117 of the present Code, such persons may accept the inheritance within six months from the day the right to inherit arose with them.

3. Persons for whom the right to inherit arises only as a consequence of the failure of another heir to accept an inheritance may accept the inheritance within three months from the day of the end of the period specified in point 1 of the present Article.

Article 1155. Acceptance of Inheritance Upon Expiry of Established Period

1. Upon the application of the heir who has allowed the period to lapse established for acceptance of an inheritance (Article 1154), a court may restore this period and deem the heir to have accepted the inheritance if the heir did not know and should not have known about the opening of the inheritance or allowed this period to lapse for other justifiable reasons and on condition that the heir who allowed the period to lapse established for acceptance of an inheritance had recourse to a court within six months after the reasons for the lapse of his period disappeared.

Upon deeming an heir to have acceptance an inheritance, the court shall determine the participatory share of all heirs in the inheritance property and, when necessary, determine the measures with regard to defence of the rights of the new heir to receive the participatory share of the inheritance due to him (point 3 of the present Article). Previously issued certificates on the right to an inheritance shall be deemed by the court to be invalid.

2. An inheritance may be accepted by an heir upon the expiry of the period established for the acceptance thereof without recourse to a court on condition of the consent in written form to this of all the remaining heirs who have accepted the inheritance. If such consent in written form is given by the heirs not in the presence of a notary, their signatures on documents concerning consent must be certified in the procedure specified in Article 1153(1), paragraph two, of the present Code. The consent of heirs shall be grounds for annulment by the notary of a previously issued certificate on the right to inheritance and the basis for the issuance of a new certificate.

If on the basis of a previously issued certificate State registration was effectuated of the rights to immoveable property, the decree of the notary concerning the annulment of the previously issued certificate and the new certificate shall be the basis for making respective changes in the entry on State registration.

3. An heir who has accepted an inheritance after the expiry of the established period in compliance with the rules of the present Article shall have the right to receive the inheritance due to him in accordance with the rules of Articles 1104, 1105, 1107, and 1108 of the present Code which, in the event specified in point 2 of the present Article, shall apply insofar as not provided otherwise in an agreement among the heirs concluded in written form.

Article 1156. Transfer of Right to Acceptance of Inheritance (Inheritance Transmission)

1. If an heir called to inherit by will or by operation of law dies after the opening of the inheritance without succeeding to accept it in the estalbished period, the right to accept the inheritance due to him shall pass to his heirs by operation of law, and if all inheritance property was bequeathed—to his heirs by will (inheritance transmission). The right to accept an inheritance by way of inheritance transmission shall not be within the composition of the inheritance opened after the death of such heir.

2. The right to accept an inheritance belonging to a deceased heir may be effectuated by his heirs on the general grounds.

If the part of the period established for acceptance of an inheritance comprises less than three months after the death of an heir, it shall be extended up to three months.

Upon the expiry of the period established for acceptance of an inheritance the heirs of the deceased heir may be deemed by a court to have accepted the inheritance in accordance with Article 1155 of the present Code if the court finds the reasons for the lapse of this period by them to be justifiable.

3. The right of an heir to accept part of an inheritance as an obligatory participatory share (Article 1149) shall not pass to his heirs.

Article 1157. Right to Refuse Inheritance

1. An heir shall have the right to refuse an inheritance to the benefit of other persons (Article 1158) or without specifying the persons to whose benefit he is refusing the inheritance.

In the event of inheriting escheat property a refusal of the inheritance shall not be permitted.

2. An heir shall have the right to refuse an inheritance within the period established for acceptance of the inheritance (Article 1154), including when he already has accepted the inheritance.

If an heir has committed actions testifying to the actual acceptance of the inheritance (Article 1153[2]), a court may upon the application of this heir deem him to have refused the inheritance also upon the expiry of the established period if it finds the reasons for the lapse of such period to be justified.

3. A refusal of an inheritance may not be subsequently changed or taken back.

4. A refusal of an inheritance when the heir is a minor or is a citizen who does not have dispositive legal capacity or has limited dispositive legal capacity shall be permitted with the prior authorisation of a trusteeship and guardianship agency.

Article 1158. Refusal of Inheritance to Benefit of Other Persons and Refusal of Part of Inheritance

1. An heir shall have the right to refuse an inheritance to the benefit of other persons from among the heirs by will or heirs by operation of law of any priority who are not deprived of inheritance (Article 1119[1]), including to the benefit of those who have been called to inherit by right of representation or by way of inheritance transmission (Article 1156).

Refusal to the benefit of any of the said persons shall not be permitted:

from property inheritable by will, if all the property of the decedent was bequeathed to the heirs designated by him;

from an obligatory participatory share in the inheritance (Article 1149);

if the heir has subdesignated an heir (Article 1121).

2. A refusal of inheritance to the benefit of persons not specified in point 1 of the present Article shall not be permitted.

A refusal of an inheritance with reservations or on condition also shall not be permitted.

3. A refusal of part of an inheritance due to an heir shall not be permitted. However, if an heir is called to inherit simultaneously on several grounds (by will and by operation of law, or by way of inheritance transmission and as a result of the opening of an inheritance, and so on), he shall have the right to refuse the inheritance due to him under one of these grounds, several of them, or all the grounds.

Article 1159. Means of Refusal of Inheritance

1. Refusal of an inheritance shall be performed by filing at the place of opening the inheritance with the notary or official empowered in accordance with a law to issue a certificate on the right to inherit a statement of the heir concerning the refusal of the inheritance.

2. When a statement concerning refusal of an inheritance is filed with the notary not by the heir himself, but by another person or is sent by post, the signature of the heir on such statement must be certified in the procedure established by Article 1153(1), paragraph two, of the present Code.

3. Refusal of an inheritance through a representative shall be possible if the power for such refusal has been specially provided for in the power of attorney. A power of attorney shall not be required for a legal representative to refuse an inheritance.

Article 1160. Right to Refuse Receipt of Testamentary Legacy

1. The recipient of a legacy shall have the right to refuse receipt of a testamentary legacy (Article 1137). In so doing the refusal to the benefit of another person, refusal with reservations, or on condition shall not be permitted.

2. If the recipient of a legacy is simultaneously an heir, his right provided for by the present Article shall not depend upon his right to accept or to refuse the inheritance.

Article 1161. Increment of Inheritance Participatory Shares

1. If an heir does not accept an inheritance, refuses an inheritance without having specified in so doing that he is refusing to the benefit of another heir (Article 1158), has no righ to inherit, or will be eliminated from inheriting on the grounds established by Article 1117 of the present Code, or as a consequence of the invalidity of a will, the part of the inheritance which would have been due to such heir who has fallen away shall pass to the heirs by operation of law called to inherit in proportion to their inheritance participatory shares.

However, when the decedent has bequeathed all property to heirs designated by him, the part of the inheritance due to an heir who has refused the inheritance or who has fallen away for the other specified grounds shall pass to the remaining heirs by will in proportion to their inheritance participatory shares unless the distribution of this part of the inheritance has been provided otherwise by the will.

2. The rules contained in point 1 of the present Article shall not apply if the heir who has refused the inheritance or fallen away on other grounds has sub-designated an heir (Article 1121[2]).

Article 1162. Certificate on Right to Inheritance

1. A certificate on the right to an inheritance shall be issued at the place of opening the inheritance by a notary or official empowered in accordance with a law to perform such notarial action.

The certificate shall be issued upon the application of an heir. At the wish of heirs the certificate may be issued to all heirs together or to each heir individually for all the inheritance property as a whole or for individual parts thereof.

A certificate shall be issued in the same procedure also in the event of the transfer of escheat property by way of inheriting to the Russian Federation (Article 1151).

2. In the event inheritance property is elicited after the issuance of a certificate on the right to an inheritance for which such certificate was not issued, an additonal certificate on the right to an inheritance shall be issued.

Article 1163. Periods for Issue of Certificate on Right to Inheritance

1. A certificate on the right to an inheritance shall be issued to heirs at any time upon the expiry of six months from the day of opening an inheritance, except for instances provided for by the present Code.

2. When inheriting either by operation of law or by will a certificate on the right to inheritance may be issued before the expiry of six months from the day of opening the inheritance if there is reliable data that, besides the persons who have had recourse for the issuance of the certificate, there are no other heirs having the right to an inheritance of respective part thereof.

3. The issuance of a certificate on the right to an inheritance shall be suspended by decision of a court, and also when there is a conceived but as yet unborn heir.

Article 1164. Common Ownership of Heirs

When inheriting by operation of law, if the inheritance property passes to two or several heirs, and when inheriting by will, if it was bequeathed to two or seveal heirs without specifying the specific property to be inherited by each of them, the inheritance property shall ensue from the day of opening the inheritance to the common participatory share ownership of the heirs.

The provisions of Chapter 16 of the present Code on common participatory share ownership, taking into account the rules of Articles 1165–1170 of the present Code, shall apply to the common ownership of heirs in the inheritance property. However, in the event of the division of inheritance property the rules of Articles 1168–1170 of the present Code shall apply for three years from the day of opening of the inheritance.

Article 1165. Division of Inheritance by Agreement Between Heirs

1. Inheritance property which is in common participatory share ownership of two or several heirs may be divided by agreement between them.

The rules of the present Code concerning the form of transactions and the form of contracts shall apply to an agreement concerning the division of an inheritance.

2. An agreement concerning the division of an inheritance in the composition of which there is immoveable property, including an agreement concerning the apportionment from the inheritance of the participatory share of one or several heirs, may be concluded by the heirs after the issuance to them of the certificate on the right to an inheritance.

State registration of the rights of heirs to immoveable property with respect to which an agreement has been concluded concerning the division of an inheritance shall be effectuated on the basis of an agreement concerning the division of an inheritance and previously issued certificate concerning the right to an inheritance, and in the event State registration of the rights of heirs to immoveable property was effectuated before the conclusion by them of an agreement concerning the division of an inheritance, on the basis of an agreement concerning the division of the inheritance.

3. The failure of the division of an inheritance effectuated by heirs in an agreement concluded by them to conform to the participatory shares due to the heirs specified in the certificate on the right to an inheritance may not entail a refusal of State registration of their rights to immoveable property received as a result of the division of the inheritance.

Article 1166. Protection of Interests of Child When Dividing Inheritance

When there is a conceived but as yet unborn heir, the division of an inheritance may be effectuated only after the birth of that heir.

Article 1167. Protection of Legal Interests of Minors, Citizens Lacking Dispositive Legal Capacity and Limited in Dispositive Legal Capacity When Dividing Inheritance

When there are among the heirs minors and citizens lacking dispositive legal capacity or having limited dispositive legal capacity, the division of an inheritance shall be effectuated in compliance with the rules of Article 37 of the present Code.

For the purposes of protection of the legal interests of the said heirs, the trustee-ship and guardianship agency must be notified about drawing up an agreement concerning division of the inheritance (Article 1165) and consideration in a court of a case concerning division of the inheritance.

Article 1168. Preferential Right to Indivisible Thing When Dividing Inheritance

1. An heir who possessed jointly with the decedent the right of common ownership to an indivisible thing (Article 133), the participatory share in the right to which is within the composition of the inheritance, shall have when dividing the property a preferential right to receive at the expense of his inheritance participatory share the thing in common ownership before heirs who previously were not participants of common ownership irrespective of whether they used this thing or not.

2. An heir who constantly used an indivisible thing (Article 133) within the composition of the inheritance shall have when dividing the property a preferential right to receive at the expense of his inheritance participatory share this thing before heirs who have not used this thing and were not previously participants of common ownership therein.

3. If within the composition of an inheritance there is a dwelling premise (dwelling house, apartment, and the like), the division of which in kind is impossible, when dividing the inheritance the heirs who resided in this dwelling premise on the day of opening the inheritance and who do not have another dwelling premise shall have a preferential right before other heirs who are not owners of the dwelling premise within the composition of the inheritance to receive this dwelling premise at the expense of their inheritance participatory shares.

Article 1169. Preferential Right to Ordinary Household Articles When Dividing Inheritance

An heir who lived on the day of opening the inheritance jointly with the decedent shall have when dividing the inheritance a preferential right to receive at the expense of his inheritance participatory share the ordinary domestic household articles.

Article 1170. Contributory Compensation for Incommensurateness of Inheritance Property from Inheritance Participatory Shares

1. The incommensurateness of inheritance property, the preferential right to receive which is declared by an heir on the basis of Article 1168 or 1169 of the present Code, with the inheritance participatory share of this heir, shall be eliminated by the transfer of this heir to the remaining heirs of other property from the composition of the inheritance or granting of other contributory compensation, including payment of a respective monetary amount.

2. Unless established otherwise by agreement between all the heirs, the effectuation by any of them of a preferential right shall be possible after the provision of respective contributory compensation to the other heirs.

Article 1171. Protection and Management of Inheritance

1. In order to defend the rights of heirs, recipients of legacies, and other interested persons, measures specified in Articles 1172 and 1173 of the present Code, and other necessary measures with regard to the protection and management of the inheritance, by the executor of the will or notary at the place of opening the inheritance.

2. A notary shall take measures with regard to the protection and management of an inheritance upon the application of one or several heirs, executor of the will, agencies of local self-government, trusteeship and guardianship agency, or other persons acting in the interests of preserving inheritance property. When the executor of a will has been appointed (Article 1134), the notary shall take measures with regard to protection and management of the inheritance by agreement with the executor of the will.

The executor of a will shall take measures with regard to the protection and management of an inheritance autonomously or at the request of one or several heirs.

3. For the purposes of eliciting the composition of an inheritance and protection thereof, banks, other credit organisations, and other juridical persons shall be obliged at the request of the notary to notify him about information available with these persons concerning property belonging to the decedent. The notary may communicate information received only to the executor of the will and heirs.

4. A notary shall effectuate measures with regard to the protection and management of an inheritance during the period determined by the notary, taking into account the character and valuables of the inheritance, and also the time necessary for heirs to enter into possession of the inheritance, but not more than six months, and in instances provided for by Article 1154(2) and (3) and Article 1156(2) of the present Code, not more than nine months from the day of opening the inheritance.

The executor of a will shall effectuate measures with regard to protection and management of the inheritance during the period necessary for performance of the will.

5. When inheritance property is situated in various places, the notary at the place of opening the inheritance shall send through justice agencies to the notary at the location of the respective part of the inheritance property a commission binding for execution concerning the protection and management of this property. If it is known by the notary at the place of opening the inheritance by whom

measures with regard to protection of the property should be taken, such commission shall be sent to the respective notary or official.

6. The procedure for the protection and management of inheritance property, including the procedure for the inventory of the inheritance, shall be determined by legislation on the notariat. The maximum amounts of remuneration under a contract of keeping of inheritance property and contract of trust management of inheritance property shall be established by the Government of the Russian Federation.

7. When the right to perform notarial actions has been granted by a law to officials of agencies of local self-government and officials of consular institutions of the Russian Federation, necessary measures for the protection and management of an inheritance may be taken by the respective official.

Article 1172. Measures with Regard to Protection of Inheritance

1. In order to protect an inheritance, a notary shall make an inventory of inheritance property in the presence of two witnesses who meet the requirements established by Article 1124(2) of the present Code.

The executor of the will, heirs, and in respective instances, representatives of trusteeship and guardianship agencies, may be present when the inventory of property is made.

Upon the application of persons specified in paragraph two of the present point a valuation of the inheritance property must be made by agreement between the heirs. In the absence of agreement, the valuation of the inheritance property or that part thereof with respect to which agreement is not reached shall be made by an independent valuer at the expense of the person requiring valuation of the inheritance property, with subsequent distribution of these expenses between the heirs in proportion to the value of inheritance received by each of them.

2. Cash money within the composition of the inheritance shall be placed on deposit with the notary, and hard currency valuables, precious metals and stones, manufactures thereof, and securities not requiring management shall be transferred to a bank for keeping under a contract in accordance with Article 921 of the present Code.

3. If it has become known to a notary that a weapon is within the composition of the inheritance, he shall inform internal affairs agencies thereof.

4. Property within the composition of an inheritance and not specified in points 2 and 3 of the present Article, unless it required management, shall be transferred by the notary under a contract for keeping to someone of the heirs, and, when it is impossible to transfer it to heirs—to another person at the discretion of the notary.

When an inheritance is effectuated by a will in which the executor of the will is designated, the keeping of the said property shall be ensured by the executor of the will autonomously or by means of the conclusion of a contract with any of the heirs or other person at the discretion of the executor of the will.

Article 1173. Trust Management of Inheritance Property

If within the composition of an inheritance there is property requiring not only protection, but also management (enterprise, participatory share in charter (or contributed) capital of an economic partnership or society, securities, exclusive rights, and the like), a notary shall in accordance with Article 126 of the present Code as the founder of trust management conclude a contract of trust management for this property.

When inheriting is effectuated by will in which an executor of the will has been designated the rights of founder of trust management shall belong to the executor of the will.

Article 1174. Compensation of Expenses Caused by Death of Decedent and Expenses for Protection and Management of Inheritance

1. Necessary expenses caused by the fatal illness of the decedent, expenses for his dignified burial, including necessary expenses for payment of the site of interment of the decedent, expenses for the protection and management of the inheritance, and also expenses connected with execution of the will, shall be compensated at the expense of the inheritance within the limits of the value thereof.

2. Demands concerning compensation of expenses specified in point 1 of the present Article may be presented to heirs who have accepted the inheritance, and until acceptance of the inheritance—to the executor of the will or against the inheritance property.

Such expenses shall be compensated before payment of debts to creditors of the decedent and within the limits of the value of the inheritance property which passed to each of the heirs. In so doing expenses shall be compensated in first priority caused by the illness and burial of the decedent, in second—expenses for the protection and management of the inheritance, and in third—expenses connected with execution of the will.

3. In order to effectuate expenses for a dignified burial of the decedent any monetary means belonging to him may be used, including on deposit or in accounts in banks.

Banks in whose deposits or accounts monetary means of the decedent are situated shall be obliged upon decree of a notary to provide them to the person specified in the decree of the notary for payment of the said expenses.

An heir to whom monetary means have been bequeathed which are on deposit or in any other accounts of the decedent in banks, including when they have been bequeathed by means of testamentary disposition in a bank (Article 1128), shall have the right at any time before the expiry of six months from the day of opening the inheritance to receive from the contribution or from the account of the decedent monetary means necessary for his burial.

The amount of means to be issued on the basis of the present point by a bank for burial to an heir or person specified in the decree of a notary may not exceed one hundred minimum amounts of payment for labour established by a law on the day of recourse for receipt of these means.

The rules of the present point respectively shall apply to other credit institutions which have been granted the right to attract monetary means of citizens in deposits or other accounts.

Article 1175. Responsibility of Heirs for Debts of Decedent

1. Heirs who have accepted an inheritance shall be liable for debts of the decedent jointly and severally (Article 323).

Each of the heirs shall be liable for debts of the decedent within the limits of the value of inherited property which has passed to him.

2. An heir who has accepted an inheritance by way of inheritance transmission (Article 1156) shall be liable within the limits of the value of this inherited property for debts of the decedent to whom this property belonged and shall not be liable with this property for debts of the heir from whom the right to accept the inheritance passed to him.

3. Creditors of the decedent shall have the right to present their demands to heirs who accepted the inheritance within the limits of the periods of limitation established for respective demands. Until acceptance of the inheritance demands of creditors may be presented to the executor of the will or against the inheritance property. In the last event the court shall suspend consideration of the case until acceptance of the inheritance by heirs or transfer of escheat property by way of inheriting to the Russian Federation.

When demands are presented by creditors of the decedent the period of limitations established for the respective demands shall not be subject to interruption, suspension, or restoration.

Chapter 65. *Inheriting of Individual Types of Property*

Article 1176. Inheriting of Rights Connected with Participation in Economic Partnerships and Societies and Production Cooperatives

1. Within the composition of an inheritance of a participant of a full partnership or full partner in a limited partnership, participant of a society with limited or

additional responsibility, and member of a production cooperative shall be the participatory share (or share) of this participant (or member) in the contributed (or charter) capital (or property) of the respective partnership, society, or cooperative.

If in accordance with the present Code, other laws, or constitutive documents of an economic partnership or society or production cooperative the consent of the other members of the partnership or society or members of a cooperative is required for entry of an heir into the economic partnership or production cooperative or transfer of a participatory share to an heir in the charter capital of an economic society and such consent is refused to the heir, he shall have the right to receive from the economic partnership or society or production cooperative the actual value of the inherited participatory share (or share) or part of the property corresponding to it in the procedure provided for by the rules of the present Code applicable to the said instance, other laws, or constitutive documents of the respective juridical person.

2. Within the composition of an inheritance of a contributor to a limited partnership shall be his participatory share in the contributed capital of that partnership. The heir to whom this participatory share passed shall become a contributor of the limited partnership.

3. Within the composition of an inheritance of a participant of a joint-stock society shall be the stocks which belonged to him. The heirs to whom these stocks have passed shall become participants of the joint-stock society.

Article 1177. Inheriting of Rights Connected with Participation in Consumer Cooperative

1. Within the composition of an inheritance of a member of a consumer cooperative shall be his share.

The heir of a member of a housing, dacha, or other consumer cooperative shall have the right to be admitted to membership in the respective cooperative. Such heir may not be refused admission to membership of the cooperative.

2. Deciding the question of who of the heirs may be admitted to membership of a consumer cooperative when the share of the decedent has passed to several heirs, and also the procedure, means, and periods for payment to heirs who do not become members of the cooperative the amounts due to them or issuance instead of property in kind shall be determined by legislation on consumer cooperatives and constitutive documents of the respective cooperative.

Article 1178. Inheriting of Enterprise

An heir who on the day of opening an inheritance is registered as an individual entrepreneur, or a commercial organisation which is an heir by will, shall have when dividing the inheritance a preferential right to receive at the expense of his inheritance participatory share the enterprise (Article 132) within the composition of the inheritance in compliance with the rules of Article 1170 of the present Code.

When none of the heirs has the said preferential right or does not take advantage of it, the enterprise within the composition of the inheritance shall not be subject to division and shall ensue to the common participatory share ownership of the heirs in accordance with the inheritance participatory shares due to them, unless provided otherwise by agreement of the heirs who have accepted the inheritance within the composition of which the enterprise is.

Article 1179. Inheriting of Property of Member of Peasant (or Farmer) Economy

1. After the death of any member of a peasant (or farmer) economy, an inheritance shall be opened and inheriting shall be effectuated on the general grounds in compliance in so doing with the rules of Articles 253–255 and 257–259 of the present Code.

2. If an heir of the deceased member of a peasant (or farmer) economy himself is not a member of this economy, he shall have the right to receive contributory compensation commensurate to the participatory share inherited by him in the property in the common joint ownership of members of the economy. The period for payment of contributory compensation shall be determined by agreement of the heir with members of the economy, and in the absence of agreement, by a court, but may not exceed one year from the day of opening the inheritance. In the absence of agreement between the members of the economy and the said heir about otherwise, the participatory share of the decedent in this property shall be considered to be equal to the participatory shares of the other members of the economy. In the event of the admission of the heir to membership in the economy, the said contributory compensation shall not be paid to him.

3. When after the death of a member of a peasant (or farmer) economy this economy is terminated (Article 258[1]), including in connection with the fact that the decedent was the sole member of the economy and among his heirs there are no persons wishing to continue to conduct the peasant (or farmer) economy, the property of the peasant (or farmer) economy shall be subject to division between the heirs according to the rules of Article 258 and 1182 of the present Code.

Article 1180. Inheriting of Things Limited in Circulability

1. A weapon, virulent and poisonous substances, narcotic and psychotropic means, and other things limited in circulability (Article 129(2), paragraph two) which belonged to the decedent shall be within the composition of the inheritance and shall be inherited on the general grounds established by the present Code. A special authorisation shall not be required to accept an inheritance within those composition there are such things.

2. Measures with regard to the protection of things of limited circulability within the composition of an inheritance until an heir receives the special authorisation for these things shall be effectuated in compliance with the procedure established by a law for the respective property.

In the event of a refusal to issue the said authorisation to an heir, his right of ownerhsip to such property shall be subject to termination in accordance with Article 238 of the present Code, and the amounts received from the realisation of the property transferred to the heir, deducting expenses for the realisation thereof.

Article 1181. Inheriting of Land Plots

A land plot belonging to the decedent by right of ownership or right of inheritable possession of the land plot for life shall be within the composition of the inheritance and shall be inherited on the general grounds established by the present Code. Special authorisation shall not be required for acceptance of an inheritance within whose composition the said property is.

In the event of inheriting a land plot or right of inheritable possession of a land plot for life, the surface (earth) layer, enclosed waters, and forest and flora situated thereon within the boundaries of this land plot also shall pass by inheritance.

Article 1182. Peculiarities of Division of Land Plot

1. The division of a land plot belonging to heirs by right of common ownership shall be effectuated by taking into account the minimum size of a land plot established for plots of the respective special-purpose designation.

2. In the event it is impossible to divide a land plot in the procedure established by point 1 of the present Article, the land plot shall pass to the heir having a preferential to receive this land plot at the expense of his inheritance participatory share. Contributory compensation to the remaining heirs shall be granted in the procedure established by Article 1170 of the present Code.

In the event when no one of the heirs has a preferential right to receive the land plot or does not take advantage of this right, the possession, use, and disposition of the land plot shall be effectuated by the heirs on the conditions of common participatory share ownership.

Article 1183. Inheriting of Unpaid Amounts Granted to Citizen as Means for Existence

1. The right to receive amounts of earnings and payments equated thereto, pensions, stipends, benefits for social insurance, compensation of harm caused to life or health, alimony, and other monetary amounts granted to a citizen as means for existence which are subject to payment to the decedent but not received by him while alive for any reason whatsoever shall belong to members of his family who resided jointly with the deceased, and also to his dependents without labour capacity irrespective of whether they resided or did not reside jointly with the deceased.

2. Demands concerning the payment of amounts on the basis of point 1 of the present Article must be presented to obliged persons within four months from the day of opening the inheritance.

3. In the absence of persons having the right on the basis of point 1 of the present Article to receive amounts not paid to the decedent, or in the event of the failure of these persons to present demands concerning payment of the said amounts within the established period, the respective amounts shall be included in the composition of the inheritance and shall be inherited on the general grounds established by the present Code.

Article 1184. Inheriting of Property Granted to Decedent by State or Municipal Formation on Privileged Conditions

Means of transport and other property granted by the State or municipal formation on privileged conditions to the decedent in connection with his disability or other similar circumstances shall be within the composition of the inheritance and shall be inherited on the general grounds established by the present Code.

Article 1185. Inheriting of State Awards, Marks of Honour, and Commemorative Marks

1. State awards which have been conferred on the decedent and to which legislation on State awards of the Russian Federation extends shall not be within the composition of an inheritance. The transfer of the said awards after the death of the awardee to other persons shall be effectuated in the procedure established by legislation on State awards of the Russian Federation.

2. State awards belonging to a decedent to which legislation on State awards of the Russian Federation does not extend, honorary, commemorative, and other marks, including awards and marks within collections, shall be within the composition of an inheritance and shall be inherited on the general grounds established by the present Code.

SECTION VI. INTERNATIONAL PRIVATE LAW

Chapter 66. General Provisions

Article 1186. Determination of Law Subject to Application to Civil-Law Relations with Participation of Foreign Persons or Civil-Law Relations Complicated by Another Foreign Element

1. The law subject to application to civil-law relations with the participation of foreign citizens or foreign juridical persons or civil-law relations complicated by another foreign element, including in instances when the object of civil rights is

situated abroad, shall be determined on the basis of international treaties of the Russian Federation, the present Code, other laws (Article 3[2]), and customs recognised in the Russian Federation.

The peculiarities of determining the law subject to application by an international commercial arbitration tribunal shall bestablished by the Law on International Commercial Arbitration.

2. If in accordance with point 1 of the present Article it is impossible to determine the law subject to application, the law of the country from which the civil-law relation complicated by a foreign element is most closely connected shall apply.

3. If an international treaty of the Russian Federation contains material-law norms subject to application to the respective relation, determination on the basis of conflicts norms of the law subject to application to questions completely regulated by such material-law norms shall be excluded.

Article 1187. Classification of Legal Concepts When Determining Law Subject to Application

1. When determining the law subject to application, the interpretation of legal concepts shall be effectuated in accordance with Russian Law unless provided otherwise by a law.

2. If when determining the law subject to application, legal concepts requiring classification are unknown to Russian Law or known in another verbal symbol or with other content and cannot be determined by means of interpretation in accordance with Russian Law, foreign law may be applied when classifying them.

Article 1188. Application of Law of Country with Multiplicity of Legal Systems

When the law of a country in which several legal systems operate is subject to application, the legal system determined in accordance with the law of that country shall be applied. If it is impossible to determine in accordance with the law of that country which of the legal systems is subject to application, the legal system with which the relation is most closely connected shall apply.

Article 1189. Reciprocity

1. Foreign law shall be subject to application in the Russian Federation irrespective of whether Russian Law is applied in the respective foreign State to relations of such nature, except for instances when the application of foreign law on the principle of reciprocity is provided for by a law.

2. When the application of foreign law depends upon reciprocity, it shall be presupposed that it exists unless proved otherwise.

Article 1190. Renvoi

1. Any reference to foreign law in accordance with the rules of the present Section must be considered as a reference to material, and not to conflicts, law of the respective country, except for instances provided for by point 2 of the present Article.

2. Renvoi of foreign law may be applied in instances of reference to Russian Law determining the legal status of a natural person (Articles 1195–1200).

Article 1191. Establishment of Content of Norms of Foreign Law

1. When applying foreign law, a court shall establish the content of its norms in accordance with their official interpretation, practice of application, and doctrine in the respective foreign State.

2. For the purposes of establishing the content of norms of foreign law a court may apply in the established procedure for assistance and explanation to the Ministry of Justice of the Russian Federation and other competent agencies or organisations in the Russian Federation and abroad or enlist experts.

Persons participating in a case may submit documents confirming the content of norms of foreign law to which they refer in substantiation of their demands or objections and otherwise facilitate the court in establishing the content of these norms.

With regard to demands connected with the effectuation by the parties of entrepreneurial activity the burden of proof of the content of norms of foreign law may be placed by the court on the parties.

3. If the content of norms of foreign law, notwithstanding the measures undertaken in accordance with the present Article, is not established within reasonable periods, Russian Law shall be applied.

Article 1192. Application of Imperative Norms

1. The rules of the present Section shall not affect the operation of those imperative norms of legislation of the Russian Federation which as a consequence of the indications in the imperative norms themselves or in view of their special significance, including in order to ensure rights and interests protected by a law of participants of civil turnover, regulate respective relations irrespective of the law subject to application.

2. When applying the law of any country according to the rules of the present Section, a court may take into account imperative norms of law of the other country having a close connection with the relation if according to the law of that country such norms must regulate the respective relations irrespective of the law subject to application. In so doing the court must take into account the

designation and character of such norms, as well as the consequences of their application or nonapplication.

Article 1193. Clause Concerning Public Policy

A norm of foreign law subject to application in accordance with the rules of the present Section shall not be applied in exceptional instances when the consequences of its application would clearly be contrary the foundations of the legal order (public policy) of the Russian Federation. In this event the respective norm of Russian Law shall be applied when necessary.

A refusal to apply a norm of foreign law may not be based only on the distinction of the legal, political, or economic system of the respective foreign State from the legal, political, or economic system of the Russian Federation.

Article 1194. Retorsion

Retaliatory limitations (retorsion) with respect to property and personal nonproperty rights of citizens and juridical persons of those States in which there are special limitations of property and personal nonproperty rights of Russian citizens and juridical persons may be established by the Government of the Russian Federation.

Chapter 67. Law Subject to Application When Determining Legal Status of Persons

Article 1195. Personal Law of Natural Person

1. The law of the country of which this person is a citizen shall be considered to be the personal law of a natural person.

2. If a person together with Russian citizenship also has foreign citizenship, his personal law shall be Russian Law.

3. If a foreign citizen has a place of residence in the Russian Federation, his personal law shall be Russian Law.

4. When a person has several foreign citizenships, the law of the country in which this person has a place of residence shall be considered to be the personal law.

5. The personal law of a stateless person shall be considered to be the law of the country in which this person has a place of residence.

6. The personal law of a refugee shall be considered to be the law of the country who granted asylum to him.

Article 1196. Law Subject to Application When Determining Civil Legal Capacity of Natural Person

The civil legal capacity of a natural person shall be determined by his personal law. In so doing, foreign citizens and stateless persons shall enjoy in the Russian

Federation civil legal capacity equally with Russian citizens, except for instances established by a law.

Article 1197. Law Subject to Application When Determining Civil Dispositive Legal Capacity of Natural Person

1. The civil dispositive legal capacity of a natural person shall be determined by his personal law.

2. A natural person who does not possess civil dispositive legal capacity according to his own personal law shall not have the right to refer to the absence of his dispositive legal capacity if he has dispositive legal capacity according to the law of the place of conclusion of a transaction, except for instances when it is proved that the other party knew or knowingly should have known about the absence of dispositive legal capacity.

3. The deeming in the Russian Federation of a natural person to lack dispositive legal capacity or to have limited dispositive legal capacity shall be subordinate to Russian Law.

Article 1198. Law Subject to Application When Determining Rights of Natural Person to Name

The rights of a natural person to a name, the use and defence thereof shall be determined by his personal law, unless provided otherwise by the present Code or other laws.

Article 1199. Law Subject to Application to Trusteeship and Guardianship

1. Trusteeship or guardianship over minors, persons who have reached majority and are lacking dispositive legal capacity, or limited in dispositive legal capacity shall be established and revoked according to the personal law of the person with respect to whom the trusteeship or guardianship is established or revoked.

2. The duty of a trustee (or guardian) to accept a trusteeship (or guardianship) shall be determined according to the personal law of the person appointed as trustee (or guardian).

3. Relations between the trustee (or guardian) and the person under a trusteeship (or guardianship) shall be determined according to the law of the country, an institution of which appointed the trustee (or guardian). However, when a person under trusteeship (or guardianship) has a place of residence in the Russian Federation, Russian Law shall apply if it is more favourable for this person.

Article 1200. Law Subject to Application When Deeming Natural Person to be Missing and When Declaring Natural Person to be Deceased

The deeming in the Russian Federation of a natural person to be missing and the declaration of a natural person to be deceased shall be subordinate to Russian Law.

Article 1201. Law Subject to Application When Determining Possibility of Natural Person to Engage in Entrepreneurial Activity

The right of a natural person to engage in entrepreneurial activity without the formation of a juridical person as an individual entrepreneur shall be determined according to the law of the country where such natural person has registered as an individual entrepreneur. If this rule cannot be applied in view of the absence of obligatory registration, the law of the country of the principal place of effectuating entrepreneurial activity shall apply.

Article 1202. Personal Law of Juridical Person

1. The personal law of a juridical persons shall be considered to be the law of the country where the juridical person is founded.

2. There shall be determined, in particular, on the basis of the personal law of a juridical person:
    (1) the status of the organisation as a juridical person;
    (2) the organisational-legal form of the juridical person;
    (3) the requirements for the name of the juridical person;
    (4) the questions of the creation, reorganisation, and liquidation of the juridical person, including questions of legal succession;
    (5) the content of the legal capacity of the juridical person;
    (6) the procedure for the acquisition by the juridical person of civil rights and the assuming of civil duties;
    (7) the internal relations, including relations of a juridical person with the participants thereof;
    (8) the capacity of a juridical person to be liable for its obligations.

3. A juridical person may not refer to a limitation of powers of its organ or representative to conclude transactions unknown to the law of the country in which the organ or representative of the juridical person has concluded a transaction, except for instances when it is proved that the other party to the transaction knew or knowingly should have known about the said limitation.

Article 1203. Personal Law of Foreign Organisation Which is Not Juridical Person Under Foreign Law

The personal law of a foreign organisation which is not a juridical person according to foreign law shall be considered to be the law of the country where this organisation was founded.

To the activity of such organisation, if Russian Law is applicable, respectively shall apply the rules of the present Code which regulate the activity of juridical persons, unless arises otherwise from a law, other legal acts, or the essence of the relation.

Article 1204. Participation of State in Civil-Law Relations Complicated by Foreign Element

The rules of the present Section shall apply on the general grounds, unless established otherwise by a law, to civil-law relations complicated by a foreign element with the participation of a State.

Chapter 68. Law Subject to Application to Property and
Personal Nonproperty Relations

Article 1205. General Provisions on Law Subject to Application to Rights to Thing

1. The content of the right of ownership and other rights to things to immoveable and moveable property and the effectuation and defence thereof shall be effectuated according to the law of the country where this property is situated.

2. The affiliation of property to immoveable or moveable things shall be determined according to the law where this property is situated.

Article 1206. Law Subject to Application to Arising and Termination of Rights to Thing

1. The arising and termination of the right of ownership and other rights to things in property shall be determined according to the law of the country where this property is situated at the moment when the action or other circumstance occurred which served as grounds for the arising or termination of the right of ownership and other rights to a thing, unless provided otherwise by a law.

2. The arising and termination of the right of ownership and other rights to things with regard to a transaction concluded with respect to moveable property en route shall be determined according to the law of the country form which this property was sent, unless provided otherwise by a law.

3. The arising of the right of ownership and other rights to things in property by virtue of acquisitive prescription shall be determined according to the law of the country where the property was situated at the moment of the ending of the period of acquisitive prescription.

Article 1207. Law Subject to Application to Rights to Thing in Vessels and Outer Space Objects

The law of the country where these vessels and objects have been registered shall apply to the right of ownership and other rights to things in aircraft and sea-going vessels, vessels of internal navigation, and outer space objects subject to State registration and to the effectuation and defence thereof.

Article 1208. Law Subject to Application to Limitations

Limitations shall be determined according to the law of the country subject to application to the respective relation.

Article 1209. Law Subject to Application to Form of Transaction

1. The form of a transaction shall be subordinated to the law of the place of conclusion thereof. However, a transaction concluded abroad may not be deemed to be invalid as a consequence of the failure to comply with the form if the requirements of Russian Law have been complied with.

The rules provided for by paragraph one of the present point shall apply to the form of a power of attorney.

2. The form of a foreign economic transaction, if one of the parties to which is a Russian juridical person, shall be subordinated, irrespective of the place of conclusion of tis transaction, to Russian Law. This rule shall apply also in instances when one of the parties to such a transaction is a natural person effectuating entrepreneurial activity whose personal law in accordance with Article 1195 of the present Code is Russian Law.

3. The form of a transaction with respect to immoveable property shall be subordinated to the law of the country where this property is situated, and with respect to immoveable property which has been entered in the State register in the Russian Federation, to Russian Law.

Article 1210. Choice of Law by Parties of Contract

1. The parties of a contract may when concluding a contract or subsequently choose by agreement between them the law which shall be subject to application to their rights and duties under this contract. The law chosen by the parties shall apply to the arising and termination of the right of ownership and other rights to things to moveable property without prejudice to the rights of third persons.

2. An agreement of the parties concerning the choice of law subject to application must be directly expressed or should definitely arise from the conditions of the contract or the aggregate of circumstances of the case.

3. The choice by the parties of the law subject to application made after conclusion of the contract shall have retroactive force and shall be considered to be valid without prejudice to the rights of third persons from the moment of conclusion of the contract.

4. The parties to a contract may choose the law subject to application either for the contract as a whole or for individual parts thereof.

5. If from the aggregate of circumstances of the case which existed at the moment of choice of the law subject to application it follows that the contract is actually connected only with one country, the choice by the parties of the law of another country may not affect the operation of imperative norms of the country with which the contract is actually connected.

Article 1211. Law Subject to Application to Contract in Absence of Agreement of Parties on Choice of Law

1. In the absence of an agreement of the parties concerning the law subject to application to a contract, the law of the country with which the contract is most closely connected shall be applied.

2. The law of the country where the place of residence or principal place of activity of the party is situated which effectuates the performance having decisive significance for the content of the contract shall be considered the law of the country with which the contract is most closely connected unless arises otherwise from a law, the conditions or essence of the contract, or the aggregate of circumstances of the matter.

3. The party which effectuates performance having decisive significance for the content of a contract shall be deemed, unless arises otherwise from a law, the conditions or essence of the contract, or the aggregate of circumstances of the matter, to be the party which is, in particular:
    (1) the seller—in a contract of purchase-sale;
    (2) the donor—in a contract of gift;
    (3) the lessor—in a contract of lease;
    (4) the lender—in a contract of uncompensated use;
    (5) the independent-work contractor—in a contract of independent-work;
    (6) the carrier—in a contract of carriage;
    (7) the expeditor—in a contract of transport expedition;
    (8) the lender (or creditor)—in a contract of loan (or credit contract);
    (9) the financial agent—in a contract of financing under assignment of monetary demand;
    (10) the bank—in a contract of bank deposit and contract of bank account;
    (11) the keeper—in a contract of keeping;
    (12) the insurer—in a contract of insurance;
    (13) the attorney—in a contract of commission;
    (14) the commission agent—in a contract of commission agency;
    (15) the agent—in an agency contract;
    (16) the right-possessor—in a contract of commercial concession;
    (17) the pledgor—in a contract on pledge;
    (18) the surety—in a contract of suretyship;
    (19) the licensor—in a license contract.

4. The law of the country with which a contract is most closely connected shall be considered to be, unless arises otherwise from a law, the conditions or essence of the contract, or aggregate of circumstances of the matter, in particular:

(1) with respect to a contract of construction independent-work and contract of independent-work for the fulfilment of design and survey work—the law of the country where the results provided for by the respective contract are basically created;

(2) with respect to a contract of simple partnership—the law of the country where the activity of such partnership is basically effectuated;

(3) with respect to a contract concluded at an auction, competition, or exchange—the law of the country where the auction or competition is conducted or the exchange is situated.

5. To a contract containing elements of various contracts shall apply, unless arises otherwise from a law, the conditions or essence of the contract, or the aggregate of circumstances of the matter, the law of the country with which this contract, considered as a whole, is most closely connected.

6. If trade terms have been used in a contract accepted in international turnover, in the absence in the contract of other indications, it shall be considered that the application to their relations of customs of business turnover designated by the respective trade terms has been agreed by the parties.

Article 1212. Law Subject to Application to Contract with Participation of Consumer

1. The choice of law subject to application to a contract, a party to which is a natural person using, acquiring, or ordering, or having the intention to use, acquire, or order moveable things (or work, services) for personal, family, household, and other needs not connected with the effectuation of entrepreneurial activity may not entail the deprivation of such natural person (or consumer) of the defence of his rights granted by imperative norms of law of the country of the place of residence of the consumer if one of the following circumstances occurred:

(1) the conclusion of the contract was preceded in that country by an offer addressed to the consumer, or an advertisement and the consumer committed actions in that country necessary in order to conclude a contract;

(2) the contracting party of the consumer or representative of the contracting party received the order from the consumer in that country;

(3) the order for the acquisition of moveable things, fulfilment of work, or rendering of services was made by the consumer in another country, the visiting of which was initiated by the contracting party of the consumer for the purposes of persuading the consumer to conclude the contract.

2. In the absence of an agreement of the parties concerning the law subject to application and in the presence of circumstances specified in point 1 of the

present Article, the law of the country of the place of residence of the consumer shall apply to the contract with the participation of the consumer.

3. The rules established by points 1 and 2 of the present Article shall not apply to:
(1) a contract of carriage;
(2) a contract concerning the fulfilment of work or rendering of services, if the work should be fulfilled or the services should be rendered exclusively in another country than the country of the place of residence of the consumer.

The exceptions provided for by the present Article shall not extend to contracts concerning the rendering of services for a total price with regard to carriage and accommodation (irrespective of including the cost of other services in the total price), in particular, contracts in the sphere of tourist servicing.

Article 1213. Law Subject to Application to Contract with Respect to Immovable Property

1. In the absence of agreement of the parties concerning the law subject to application to a contract with respect to immovable property, the law of the country with which the contract is most closely connected shall apply. The law of the country with which such contract is most closely connected shall be considered, unless arises otherwise from a law, conditions of essence of the contract, or aggregate of circumstances of the matter, to be the law of the country where the immovable property is situated.

2. Russian Law shall apply to contracts with respect to land plots, subsoil plots, solitary water objects, and other immovable property situated on the territory of the Russian Federation.

Article 1214. Law Subject to Application to Contract on Creation of Juridical Person with Foreign Participation

The law of the country in which according to the contract a juridical person is subject to founding shall apply to the contract on the creation of a juridical person with foreign participation.

Article 1215. Sphere of Operation of Law Subject to Application to Contract

The law subject to application to a contract in accordance with the rules of Articles 1210–1215 and 1216 of the present Code shall be determined, in particular, by:
(1) interpretation of the contract;
(2) rights and duties of the parties of the contract;
(3) performance of the contract;
(4) consequences of failure to perform or improper performance of the contract;

(5) termination of the contract;

(6) consequences of invalidity of the contract.

Article 1216. Law Subject to Application to Assignment of Demand

1. The law subject to application to an agreement etween the initial and new creditors concerning the assignment of a demand shall be determined in accordance with Article 1211(1) and (2) of the present Code.

2. The admissability of the assignment of a demand, relations between the new creditor and the debtor, conditions under which this demand may be presented to the debtor by the new creditor, and also the question concerning proper performance of the obligation by the debtor, shall be determined according to the law subject to application to the demand which is the subject of the assignment.

Article 1217. Law Subject to Application to Obligations Arising from Unilateral Transactions

To obligations arising from unilateral transactions, unless arises otherwise from a law, the conditions or essence of the transaction, or aggregate of circumstances of the matter, the law of the country shall apply where the place of residence or principal place of activity of the party assuming the obligation under the unilateral transaction is situated.

The period of operation of a power of attorney and the grounds for termination thereof shall be determined according to the law of the country where the power of attorney was issued.

Article 1218. Law Subject to Application to Relations with Regard to Payment of Interest

The grounds for the recovery, procedure for calculating, and amount of interest with regard to monetary obligations shall be determined according to the law of the country subject to application to the respective obligation.

Article 1219. Law Subject to Application to Obligations Arising as Consequence of Causing of Harm

1. To obligations arising as a consequence of the causing of harm shall apply the law of the country where the action or other circumstance occurred serving as the grounds for the demand concerning compensation of harm. When as a result of such action or other circumstance harm has ensued in another country, the law of that country may be applied if the causer of harm foresaw or should have foreseen the ensuing of harm in that country.

2. To obligations arising as a consequence of the causing of harm abroad, if the parties are citizens or juridical persons of the same country, the law of that country shall apply. If the parties of such obligation are not citizens of the same country but have a place of residence in the same country, the law of that country shall apply.

3. After the commission of an action or ensuing of other circumstance which entailed the causing of harm the parties may agree concerning the application to the obligation which arose as a consequence of causing harm the law of the country of the court.

Article 1220. Sphere of Operation of Law Subject to Application to Obligations Arising as Consequence of Causing of Harm

On the basis of the law subject to application to obligations arising as a consequence of the causing of harm there shall be determined, in particular:

(1) the capacity of the person to bear responsibility for harm caused;

(2) the placing of responsibility for harm on a person who is not the causer of harm;

(3) the grounds of responsibility;

(4) the ground of limitation of responsibility and relief therefrom;

(5) the means of compensation of harm;

(6) the extent and amount of compensation of harm.

Article 1221. Law Subject to Application to Responsibility for Harm Caused as Consequence of Defects of Good, Work, or Service

1. To demands concerning compensation of harm caused as a consequence of defects of a good, work, or service there shall apply at the choice of the victim:

(1) the law of the country where the seller or manufacturer of the good or other causer of harm has a place of residence or principal place of activity;

(2) the law of the country where the victim has a place of residence or principal place of activity;

(3) the law of the country where work was fulfilled, service rendered, or law of the country where the good was acquired.

The choice by the victim of the law provided for by subpoint 2 or 3 of the present point may be recognised only if the causer of harm does not prove that the good came to the respective country without its consent.

2. If the victim has not taken advantage of the right of choice granted to him by the present Article, the law subject to application shall be determined in accordance with Article 1219 of the present Code.

3. The rules of the present Article respectively shall apply to demands concerning compensation of harm caused as a consequence of unreliable or insufficient information concerning the good, work or service.

Article 1222. Law Subject to Application to Obligations Arising as Consequence of Unfair Competition

To obligations arising as a consequence of unfair competition shall apply the law of the country whose market is affected by such competition, unless arises otherwise from a law or the essence of the obligation.

Article 1223. Law Subject to Application to Obligations Arising as Consequence of Unfounded Enrichment

1. The law of the country where enrichment occurred shall apply to obligations arising as a consequence of unfounded enrichment.

The parties may agree about the application to such obligations of the law of the country of the court.

2. If unfounded enrichment arose in connection with an existing or proposed legal relation under which property was acquired or saved, to obligations arising as a consequence of such unfounded enrichment shall apply the law of the country to which this legal relation was or could be subordinated.

Article 1224. Law Subject to Application to Relations with Regard to Inheriting

1. Relations with regard to inheriting shall be determined according to the law of the country where the decedent had his last place of residence, unless provided otherwise by the present Article.

The inheriting of immoveable property shall be determined according to the law of the country where this property is situated, and the inheriting of immoveable property which has been entered in a State register in the Russian Federation,—according to Russian Law.

2. The capacity of a person to draw up and revoke a will, including with respect to immoveable property, and also the form of such will or act of revocation thereof, shall be determined according to the law of the country where the testator had a place of residence at the moment of drawing up such will or act. However, the will or revocation thereof may not be deemed invalid as a consequence to comply with the form if it satisfies the requirements of the place of drawing up the will or act of revocation thereof or the requirements of Russian Law.

FEDERAL LAW ON THE INTRODUCTION INTO OPERATION OF PART ONE OF THE CIVIL CODE OF THE RUSSIAN FEDERATION

[Federal Law of 30 November 1994,
No. 52-ФЗ, as amended by Federal Law
of 16 April 2001, No. 45-ФЗ; and Federal Law
of 26 November 2001, No. 147-ФЗ.
СЗ РФ (1994), no. 32, item 3302; (2001),
no. 17, item 1644; no. 49, item 4553]

Article 1. To introduce Part One of the Civil Code of the Russian Federation (hereinafter: Part One of the Code) into operation from 1 January 1995, except for the provisions for which the present Federal Law has established other periods of introduction into operation.

Article 2. To deem to have lost force from 1 January 1995:

the preamble, Section I 'General Provisions', Section II 'Law of Ownership' and Section III 'Law of Obligations', Subsection I 'General Provisions on Obligations' confirmed by the Law of the RSFSR of 11 June 1964 'On Confirmation of the Civil Code of the RSFSR' (*Ведомости Верховного Совета РСФСР* (1964), no. 24, item 406; (1966), no. 32, item 771; (1972), no. 33, item 825; (1973), no. 51, item 1114; (1974), no. 51, item 1346; (1977), no. 6, item 129; (1987), no. 9, item 250; (1988), no. 1, item 1; no. 16, item 476; (1990), no. 3, item 78; *Ведомости Съезда народных депутатов РСФСР и Верховного Совета РСФСР* (1991), no. 15, item 494; *Ведомости Съезда народных депутатов Российской Федерации и Верховного Совета Российской Федерации* (1992), no. 29, item 1689; no. 34, item 1966);

Articles 4 and 5, Article 6 (in that part of the rules established by Article 79 of the Civil Code of the RSFSR), Articles 7–13 of the Edict of the Presidium of the Supreme Soviet of the RSFSR of 12 June 1964 'On the Procedure for the Introduction into Operation of the Civil and Civil Procedure Codes of the RSFSR' (*Ведомости Верховного Совета РСФСР* (1964), no. 24, item 416; (1987), no. 9, item 250);

the Law of the RSFSR of 24 December 1990 'On Ownership in the RSFSR' (*Ведомости Съезда народных депутатов РСФСР и Верховного Совета РСФСР* (1990), no. 30, item 416; *Ведомости Съезда народных депутатов Российской Федерации и Верховного Совета Российской Федерации* (1992), no. 34, item 1966);

the Decree of the Supreme Soviet of the RSFSR of 24 December 1990 'On the Introduction into Operation of the Law of the RSFSR "On Ownership in the RSFSR" ' (*Ведомости Съезда народных депутатов РСФСР и Верховного Совета РСФСР* (1990), no. 30, item 417);

the Law of the RSFSR of 25 December 1990 'On Enterprises and Entrepreneurial Activity' (*Ведомости Съезда народных депутатов РСФСР и Верховного Совета РСФСР* (1990), no. 30, item 418; *Ведомости Съезда народных депутатов Российской Федерации и Верховного Совета Российской Федерации* (1992), no. 34, item 1966; (1993), no. 32, item 1231, 1256), except for Articles 34 and 35.

Article 3. There shall not apply from 1 January 1995 on the territory of the Russian Federation:

Section I 'General Provisions', Section II 'Right of Ownership. Other Rights to Thing', and Chapter 8 'General Provisions on Obligations' of Section III 'Law of Obligations' of the Fundamental Principles of Civil Legislation of the USSR and Republics (*Ведомости Съезда народных депутатов СССР и Верховного Совета СССР* (1991), no. 26, item 733);

point 4, paragraph three, and point 4 of the Decree of the Supreme Soviet of the Russian Federation of 3 March 1993 'On Certain Questions of the Application of Legislation of the USSR on the Territory of the Russian Federation' (*Ведомости Съезда народных депутатов Российской Федерации и Верховного Совета Российской Федерации* (1993), no. 11, item 393).

Article 4. Until laws and other legal acts prevailing on the territory of the Russian Federation are brought into conformity with Part One of the Code, the laws and other legal acts of the Russian Federation, and also the Fundamental Principles of Civil Legislation of the USSR and Republics and other acts of legislation of the USSR prevailing on the territory of the Russian Federation shall, within the limits and in the procedure provided for by the Constitution of the Russian Federation, the Decree of the Supreme Soviet of the RSFSR of 12 December 1991 'On the Ratification of the Agreement on the Creation of the Commonwealth of Independent States', the decrees of the Supreme Soviet of the Russian Federation of 14 July 1992 'On the Regulation of Civil Law Relations in the Period of Implementing Economic Reform' and of 3 March 1993 'On Certain Questions of the Application of Legislation of the USSR on the Territory of the Russian Federation', apply insofar as they are not contrary to Part One of the Code.

Normative acts, issued before the introduction into operation of Part One of the Code, of the Supreme Soviet of the RSFSR and the Supreme Soviet of the Russian Federation which are not laws, and normative acts of the Presidium of the Supreme Soviet of the RSFSR, President of the Russian Federation, and Government of the Russian Federation, and also normative acts of the Supreme Soviet of the USSR applicable on the territory of the Russian Federation which are not laws, and normative acts of the Presidium of the Supreme Soviet of the USSR, President of the USSR, and Government of the USSR with regard to questions which according to Part One of the Code may be regulated only by federal laws shall operate until the introduction into operation of the respective laws [as amended by Federal Law of 26 November 2001. *C3 РФ* (2001), no. 49, item 4553].

Article 5. Part One of the Code shall apply to civil law relations which arose after the introduction thereof into operation.

With regard to civil law relations which arose before the introduction thereof into operation, Part One of the Code shall apply to those rights and duties which arise after the introduction thereof into operation.

Article 6.

1. Chapter 4 of the Code shall be introduced into operation from the day of official publication of Part One of the Code. From that day commercial organisations may be created exclusively in those organisational-legal forms which have been provided for them by Chapter 4 of the Code.

The creation of juridical persons after the official publication of Part One of the Code shall be effectuated in the procedure provided for by Chapter 4 of the Code unless it arises otherwise from Article 8 of the present Federal Law.

2. The norms of Chapter 4 of the Code on the full partnership (Articles 69–81), limited partnership (Articles 82–86), limited responsibility society (Articles 87–94), and joint-stock society (Articles 96–104) shall apply respectively to full partnerships, mixed partnerships, limited responsibility partnerships, joint-stock societies of the closed-type, and joint-stock societies of the open-type created before the official publication of Part One of the Code.

The constitutive documents of these economic partnerships and societies shall operate, until they are brought into conformity with the norms of Chapter 4 of the Code, in that part which is not contrary to the said norms.

3. The constitutive documents of full partnerships and mixed partnerships created before the official publication of Part One of the Coe shall be subject to being brought into conformity with the norms of Chapter 4 of the Code not later than 1 July 1995.

4. The constitutive documents of limited responsibility partnerships, joint-stock societies, and production cooperatives created before the official publication

of Part One of the Code shall be subject to being brought into conformity with the norms of Chapter 4 of the Code on limited responsibility societies, joint-stock societies, and on production cooperatives in the procedure and within the periods which will be determined respectively when the laws on limited responsibility societies, joint-stock societies, and production coooperatives are adopted.

5. Individual (family) private enterprises, and also enterprises created by economic partnerships and societies, social and religious organisations, associations, philanthropic foundations, and other enterprises which are not in State or municipal ownership based on the right of full economic jurisdiction shall be subject before 1 July 1999 to transformation into economic partnerships, societies, or cooperatives or to liquidation. Upon the expiry of this period the enterprises shall be subject to liquidation in a judicial proceeding at the demand of the agency effectuating State registration of the respective juridical persons, tax agency, or procurator.

The norms of the Code on unitary enterprises based on the right of operative management (Articles 113, 115, 296, 297) shall apply to the said enterprises until their transformation or liquidation, taking into account the fact that the founders thereof shall be the owners of their property.

6. The norms of the Code on unitary enterprises based on the right of economic jurisdiction (Articles 113, 114, 294, 295, 299, 300) and unitary enterprises based on the right of operative management (Articles 113, 115, 296, 297, 299, 300) shall apply respectively to State and municipal enterprises based on the right of full economic jurisdiction, and also to federal treasury enterprises, created before the official publication of Part One of the Code.

The constitutive documents of these enterprises shall be subject to being brought into conformity with the norms of Part One of the Code in the procedure and within the periods which will be determined when the law on State and municipal unitary enterprises is adopted.

7. Associations of commercial organisations not effectuating entrepreneurial activity and created before the official publication of Part One of the Code in the form of partnerships or joint-stock societies shall have the right to retain the respective form or may be transformed into associations or unions of commercial organisations (Article 121).

Article 7. Juridical persons specified in Article 6(2)-(7) of the present Federal Law, and also peasant (or farmers's) economies, shall be relieved from the payment of the registration fee when registering changes of their legal status in connection with bringing it into conformity with the norms of Part One of the Code.

Article 8. Until the introduction into operation of a law on the registration of juridical persons and the law on the registration of rights to immovable property and transactions with them, the prevailing procedure for the registration of juridical persons and the registration of immovable property and transactions with them shall apply.

Article 9. The norms of the Code on the grounds and consequences of the invalidity of transactions (Articles 162, 165–180) shall apply to transactions, the demands concerning the recognition of the invalidity and consequences of the invalidity of which are considered by a court, arbitrazh court, or arbitration court after 1 January 1995 irrespective of the time the respective transactions were concluded.

Article 10. The periods of limitations established by Part One of the Code and the rules for calculating them shall apply to those demands, the periods for presenting which provided for by legislation which was previously in force have not expired before 1 January 1995 [as amended by Federal Law of 26 November 2001. СЗ РФ (2001), no. 49, item 4553].

The period of limitations established for the respective suits of legislation previously in force shall apply to the suit provided for by Article 181(2) of the Code concerning the recognition of a contested transaction to be invalid and concerning the consequences of the invalidity thereof, the right to present which arose before 1 January 1995.

Article 11. The operation of Article 234 of the Code (acquisitive prescription) shall also extend to instances when the possession of property commenced before 1 January 1995 and continues to the moment of the introduction of Part One of the Code into operation.

Article 12. The procedure for the conclusion of contracts established by Chapter 28 of the Code shall apply to contracts, the proposals to conclude which were sent after 1 January 1995.

Article 13. The norms of Chapter 17 of Part One of the Code in the part affecting transactions with land plots of agricultural lands shall be introduced into operation from the day of the introduction into operation of the Land Code of the Russian Federation and the Law on Turnover of Lands of Agricultural Designation [as amended by Federal Law of 16 April 2001. СЗ РФ (2001), no. 17, item 1644].

Article 14. The peculiarities of the creation and activity of agricultural cooperatives (production, processing, servicing agricultural producers) shall be determined by a law on agricultural cooperative societies.

FEDERAL LAW ON THE INTRODUCTION INTO OPERATION OF PART TWO OF THE CIVIL CODE OF THE RUSSIAN FEDERATION

[Federal Law of 26 January 1996, No. 15-ФЗ; as amended
by Federal Law of 26 November 2001, No. 147-ФЗ.
СЗ РФ (1996), no. 5, item 411; (2001),
no. 49, item 4553]

Article 1. Part Two of the Civil Code of the Russian Federation (hereinafter: Part Two of the Code) shall be introduced into operation from 1 March 1996.

Article 2. To deem to have lost force from 1 March 1996:

Section III 'Law of Obligations' of the Civil Code of the RSFSR, confirmed by Law of the RSFSR of 11 June 1964 'On the Confirmation of the Civil Code of the RSFSR' (*Ведомости Верховного Совета РСФСР* (1964), no. 24, item 406; (1969), no. 23, item 783; (1970), no. 26, item 511; (1972), no. 33, item 825; (1973), no. 51, item 1114; (1977), no. 6, item 129; (1985), no. 9, item 305; (1986), no. 23, item 638; (1987), no. 9, item 250; (1988), no. 1, item 1; *Ведомости Съезда народных депутатов Российской Федерации и Верховного Совета Российской Федерации* (1992), no. 15, item 768; no. 34, item 1966; (1993), no. 4, item 119; *Собрание законодательства Российской Федерации* (1994), no. 3, item 3302);

points 3, 6, and 15 of the Edict of the Presidium of the Supreme Soviet of the RSFSR of 12 June 1964 'On the Procedure for the Introduction into Operation of the Civil and Civil Procedure Codes of the RSFSR' (*Ведомости Верховного Совета РСФСР* (1964), no. 24, item 416);

Decree of the Supreme Soviet of the Russian Federation of 13 February 1992 'On the Introduction into Operation of the Statute on Cheques' (*Ведомости Съезда народных депутатов Российской Федерации и Верховного Совета Российской Федерации* (1992), no. 24, item 1283);

points 3–7, and also point 8 (with respect to the application of the provisions of Chapter 13 of the Fundamental Principles of Civil Legislation of the USSR and Republics) of the Decree of the Supreme Soviet of the Russian Federation of

3 March 1993 'On Certain Questions of the Application of Legislation of the USSR on the Territory of the Russian Federation' (*Ведомости Съезда народных депутатов Российской Федерации и Верховного Совета Российской Федерации* (1993), no. 11, item 393; *Собрание законодательства Российской Федерации* (1994), no. 32, item 3302).

Article 3. From 1 March 1996 there shall not apply on the territory of the Russian Federation:

Section III 'Law of Obligations' of the Fundamental Principles of Civil Legislation of the USSR and Republics' (*Ведомости Съезда народных депутатов СССР и Верхогвного Совета СССР* (1991), no. 26, item 733);

Fundamental Principles of Legislation of the USSR and Union Republics on Lease of 23 November 1989 (*Ведомости Съезда народных депутатов СССР и Верховного Совета СССР* (1989), no. 25, item 481; (1991), no. 12, item 325);

Decree of the Supreme Soviet of the USSR of 23 November 1989 'On the Procedure for the Introduction into Operation of the Fundamental Principles of the USSR and Union Republics on Lease' (*Ведомости Съезда народных депутатов СССР и Верховного Совета СССР* (1989), no. 25, item 482).

Article 4. Until laws and other legal acts operating on the territory of the Russian Federation are brought into conformity with Part Two of the Code, the laws and other legal acts of the Russian Federation, and also acts of legislation of the USSR operating on the territory of the Russian Federation within the limits and in the procedure provided for by legislation of the Russian Federation shall be applied insofar as they are not contrary to Part Two of the Code.

Normative acts of the Supreme Soviet of the RSFSR and Supreme Soviet of the Russian Federation which are not laws, and normative acts of the Presidium of the Supreme Soviet of the RSFSR, President of the Russian Federation, and Government of the Russian Federation, and also normative acts of the Supreme Soviet of the USSR applicable on the territory of the Russian Federation which are not laws, and normative acts of the Presidium of the Supreme Soviet of the USSR, President of the USSR, and Government of the USSR with regard to questions which according to Part Two of the Code may be regulated only by federal laws, shall operate until the introduction into operation of the respective laws [as amended by Federal Law of 26 November 2001. СЗ РФ (2001), no. 49, item 4553].

Article 5. Part Two of the Code shall apply to relations of obligations which arose after the introduction thereof into operation. With regard to relations of obligations which arose before 1 March 1996, Part Two of the Code shall apply to those rights and duties which arise after the introduction thereof into operation.

The periods of limitations established by Part Two of the Code and the rules for calculating them shall apply to those demands, the periods for presenting which provided for by legislation which was previously in force have not expired before 1 March 1996 [added by Federal Law of 26 November 2001. СЗ РФ (2001), no. 49, item 4553].

Article 6. The norms of Part Two of the Code concerning the procedure for the conclusion and the form of contracts of individual types, and also concerning the State regulation thereof, shall apply to contracts, the proposals to conclude which were sent after the introduction into operation of Part Two of the Code. The norms of Part Two of the Code concerning the form of contracts of individual types, and also concerning the State registration thereof, shall apply to contracts, proposals to conclude which were sent before 1 March 1996 and which were concluded after 31 March 1996.

Until the introduction into operation of the law on the registration of rights to immoveable property and transactions therewith, the prevailing procedure for the registration of transactions with immoveable property shall apply.

The norms of Part Two of the Code determining the content of contracts of individual types shall apply to contracts concluded after the introduction thereof into operation.

Article 7. Until the introduction into operation of a federal law on the registration of rights to immoveable property and transactions therewith, the rules concerning the obligatory notarial certification of such contracts established by legislation before the introduction into operation of Part Two of the Code shall retain force for contracts provided for by Articles 550, 560, and 574 of the Civil Code of the Russian Federation.

Article 8. The norms of Part Two of the Code which are binding on the parties to a contract concerning the grounds, consequences, and procedure for the dissolution of contracts of individual types shall also apply to contracts which continue to operate after the introduction into operation of Part Two of the Code, irrespective of the date of conclusion thereof.

The norms of Part Two of the Code which are binding upon the parties to a contract concerning responsibility for a violation of contractual obligations shall apply if the respective violations were permitted after the introduction into operation of Part Two of the Code, except for instances when in contracts concluded before 1 March 1996 other responsibility for such violations was provided for.

Article 9. In instances when one of the parties to an obligation is a citizen using, acquiring, ordering, or having the intention to acquire or order goods (or work services) for personal domestic needs, such citizen shall enjoy the rights of a party to an obligation in accordance with the Civil Code of the Russian Federation, and also the rights granted to a consumer by the Law of the Russian Federation 'On the Defence of the Rights of Consumers' and other legal acts issued in accordance therewith.

Article 10. Until the establishment of the conditions for licensing activity of financial agents (Article 825 of the Civil Code of the Russian Federation), the existing procedure for the effectuation of their activity shall be retained.

Article 11. The operation of Article 835(2) and (3) of the Civil Code of the Russian Federation shall also extend to instances when relations connected with the attraction of monetary means for contributions which arose before the introduction of Part Two of the Civil Code into operation and are preserved at the moment of the introduction of Part Two of the Code into force.

Article 12. The operation of Articles 1069 and 1070 of the Civil Code of the Russian Federation shall extend also to instances when the causing of harm to the victim occurred before 1 March 1996 but not earlier than 1 March 1993 and the harm caused remained uncompensated.

The operation of Articles 1085–1094 of the said Code shall extend also to instances when the causing of harm to the life and health of a citizen occurred before 1 March 1996 but not earlier than 1 March 1993 and the harm caused remained uncompensated.

Article 13. The norms of Part Two of the Code in the part affecting transactions with land plots shall apply to the extent that their turnover is permitted by land legislation.

FEDERAL LAW ON THE INTRODUCTION INTO OPERATION OF PART THREE OF THE CIVIL CODE OF THE RUSSIAN FEDERATION

[Federal Law of 26 November 2001, No. 147-ФЗ.
СЗ РФ (2001), no. 49, item 4553]

Article 1. To introduce Part Three of the Civil Code of the Russian Federation (hereinafter: Part Three of the Code) into operation from 1 March 2002.

Article 2. To deem to have lost force from 1 March 2002:

Section VII 'Inheritance Law' and Section VIII 'Legal Capacity of Foreign Citizens and Stateless Persons. Application of Civil Laws of Foreign States and International Treaties' of the Civil Code of the RSFSR (*Ведомости Верховного Совета РСФСР* (1964), no. 24, item 406);

point 16 of the Edict of the Presidium of the Supreme Soviet of the RSFSR of 12 June 1964 'On the Procedure for the Introduction into Operation of the Civil and Civil Procedure Codes of the RSFSR' (*Ведомости Верховного Совета РСФСР* (1964), no. 24, item 416);

point 1 and paragraphs nine and ten of point 2 of Section I of the Edict of the Presidium of the Supreme Soviet of the RSFSR of 18 December 1974 'On Changes and Deeming to Have Lost Force Certain Legislative Acts of the RSFSR in Connection with the Introduction into Operation of the Law of the RSFSR on the State Notariat' (*Ведомости Верховного Совета РСФСР* (1974), no. 51, item 1346);

Section I of the Edict of the Presidium of the Supreme Soviet of the RSFSR of 14 June 1977 'On Making Changes in and Additions to the Civil and Civil Procedure Codes of the RSFSR' (*Ведомости Верховного Совета РСФСР* (1977), no. 24, item 586);

points 74–76 of Section I of the Edict of the Presidium of the Supreme Soviet of the RSFSR of 24 February 1987 'On Making Changes in and Additions to the Civil Code of the RSFSR and Certain Other Legislative Acts of the RSFSR' (*Ведомости Верховного Совета РСФСР* (1987), no. 9, item 250);

point 9 of the Decree of the Supreme Soviet of the Russian Federation of 3 March 1993, No. 4604-I 'On Certain Questions of the Application of Legislation of the USSR on the Territory of the Russian Federation' (*Ведомости Съезда народных депутатов Российской Федерации и Верховного Совета Российской Федерации* (1993), no. 11, item 393);

Federal Law of 14 May 2001, No. 51-ФЗ 'On Making Changes in and Additions to Article 532 of the Civil Code of the RSFSR' (*Собрание законодательства Российской Федерации* (2001), no. 21, item 2060).

Article 3. From 1 March 2002 Section VI 'Inheritance Law' and Section VII 'Legal Capacity of Foreign Citizens and Juridical Persons. Application of Civil Laws of Foreign States and International Treaties' of the Fundamental Principles of Civil Legislation of the USSR and Republics (*Ведомости Съезда народных депутатов СССР и Верховного Совета СССР* (1991), no. 26, item 733) shall not apply on the territory of the Russian Federation.

Article 4. Until the bringing of laws and other legal acts operating on the territory of the Russian Federation into conformity with Part Three of the Code, laws and other legal acts of the Russian Federation, and also acts of legislation of the USSR operating on the territory of the Russian Federation within the limits and in the procedure which have been provided for by legislation of the Russian Federation, shall apply insofar as they are not contrary to Part Three of the Code.

Normative acts, issued before the introduction into operation of Part Three of the Code, of the Supreme Soviet of the RSFSR and Supreme Soviet of the Russian Federation which are not laws, and normative acts of the Presidium of the Supreme Soviet of the RSFSR, President of the Russian Federation, and Government of the Russian Federation, and also normative acts of the Supreme Soviet of the USSR applicable on the territory of the Russian Federation which are not laws, and normative acts of the Presidium of the Supreme Soviet of the USSR, President of the USSR, and Government of the USSR with regard to questions which according to Part Three of the Code may be regulated only by federal laws, shall operate until the introduction into operation of the respective laws.

Article 5. Part Three of the Code shall apply to civil-law relations which arose after the introduction thereof into operation.

With regard to civil-law relations which arose before the introduction into operation of Part Three of the Code, Section V 'Inheritance Law' shall apply to those rights and duties which arise after the introduction thereof into operation.

Article 6. With regard to an inheritance opened before the introduction into operation of Part Three of the Code, the group of heirs by operation of law shall be determined in accordance with the rules of Part Three of the Code if the period for acceptance of the inheritance has not expired on the day of introduction into operation of Part Three of the Code or if the said period has expired, but on the day of introduction into operation of Part Three of the Code the inheritance was not accepted by anyone of the heirs specified in Articles 532 and 548 of the Civil Code of the RSFSR, a certificate on the right to an inheritance was not issued by the Russian Federation, subject of the Russian Federation, or municipal formation or the inheritance property had not passed to their ownership on other grounds established by a law. In these instances the persons who could not be heirs by operation of law in accordance with the rules of the Civil Code of the RSFSR but are such according to the rules of Part Three of the Code (Articles 1142–1148) may accept the inheritance within six months from the day of introduction into operation of Part Three of the Code.

In the absence of heirs specified in Articles 1142–1148 of the Civil Code of the Russian Federation, or if none of the heirs has the right to inherit, or all the heirs have been removed from inheriting (Article 1117 of the Civil Code of the Russian Federation), or none of the heirs has accepted the inheritance, or all the heirs have refused the inheritance and in so doing none of them has specified that he is refusing to the benefit of another heir, the rules on inheriting of escheat property shall apply established by Article 1151 of the Civil Code of the Russian Federation.

Article 7. The rules concerning the grounds of invalidity of a will which were in force on the day of the conclusion of the will shall apply to wills concluded before the introduction into operation of Part Three of the Code.

Article 8. The rules concerning the obligatory participatory share in an inheritance established by Part Three of the Code shall apply to wills concluded after 1 March 2002.

Article 9. To make the following changes in Federal Law of 30 November 1994, No. 52-ФЗ 'On the Introduction into Operation of Part One of the Civil Code of the Russian Federation' (*Собрание законодательства Российской Федерации* (1994), no. 32, item 3302):

Article 4, paragraph two, to set out in the following version:

'Normative acts, issued before the introduction into operation of Part One of the Code, of the Supreme Soviet of the RSFSR and the Supreme Soviet of the Russian Federation which are not laws, and normative acts of the Presidium of the Supreme Soviet of the RSFSR, President of the Russian Federation, and Government of the Russian Federation, and also normative acts of the Supreme Soviet of the USSR applicable on the territory of the Russian Federation which are not laws, and normative acts of the Presidium of the Supreme Soviet of the USSR, President of the USSR, and Government of the USSR with regard to questions which according to Part One of the Code may be regulated only by federal laws shall operate until the introduction into operation of the respective laws.';

Article 10, paragraph one, to set out in the following version:

'The periods of limitations established by Part One of the Code and the rules for calculating them shall apply to those demands, the periods for presenting which provided for by legislation which was previously in force have not expired before 1 January 1995.'.

Article 10. To make the following change and addition to Federal Law of 26 January 1996, No. 15-ФЗ 'On the Introduction into Operation of Part Two of the Civil Code of the Russian Federation' (*Собрание законодательства Российской Федерации* (1996), no. 5, item 411):

Article 4, paragraph two, to set out in the following version:

'Normative acts, issued before the introduction into operation of Part Two of the Code, of the Supreme Soviet of the RSFSR and Supreme Soviet of the Russian Federation which are not laws, and normative acts of the Presidium of the Supreme Soviet of the RSFSR, President of the Russian Federation, and Government of the Russian Federation, and also normative acts of the Supreme Soviet of the USSR applicable on the territory of the Russian Federation which are not laws, and normative acts of the Presidium of the Supreme Soviet of the USSR, President of the USSR, and Government of the USSR with regard to questions which according to Part Two of the Code may be regulated only by federal laws, shall operate until the introduction into operation of the respective laws.';

to add to Article 5 a paragraph of the following content:

'The periods of limitations established by Part Two of the Code and the rules for calculating them shall apply to those demands, the periods for presenting which provided for by legislation which was previously in force have not expired before 1 March 1996.'

GLOSSARY OF RUSSIAN LEGAL TERMS

аванс	advance payment
автор	author
авторство	authorship
агент	agent
аккредитив	letter of credit
~ документарный	documentary letter of credit
~ товарный	goods letter of credit
актив	asset
акцепт	acceptance
~ по умолуанию	tacit acceptance
акцессорный	accessory
акция	stock
аналогия	analogy
~ закона	analogy of law [lex]
~ права	analogy of law [jus]
аннулирование	annulment
аренда	lease
арендатор	lessee
арендодателв	lessor
база данных	data base
банкротство	bankruptcy
безвестно отсутствуюшим	missing
бездействие	failure to act
бенефициар	beneficiary
бессроуно	in perpetuity
бесхозный	masterless
благоразумиз	prudence
бонитет	estimated yield power
бремя	burden
буква	letter

варрант	warrant
вексель	bill of exchange
~ простой	promissory note
вешание	broadcast
~ кабельное	broadcast by cable
~ эфирное	broadcast on air
вешь	thing
~ неделимый	indivisible
~ сложный	complex
взаимность	reciprocity
взнос	1. contribution; 2. premium
~ страховой	insurance premium
вина	1. fault; 2. guilt
вклад	deposit
владелец	possessor
владение	possession
вмешательство	interference
внедоговорный	extracontractual
внук	grandchild
возведение	erection
возврат	1. return; 2. refund
возмешение	compensation
вознаграждение	remuneration
возникновение	arising
возражение	objection
~ должника	objection of debtor
возраст	age
волеизъявление	expression of will
воля	will
воспроизведение	reproduction
восстановление	1. restoration; 2. reinstatement
вред	harm
вступление	entry
~ в силу	entry into force
выбор	choice
выгода	advantage
~ упушенная	lost advantage
выдача	1. issuance; 2. extradition
выделение	apportionment
выкуп	redemption
вымышление	fictitious
выполнение	fulfilment

выручка	receipt
выход	withdrawal
вычет	deduction
гарант	guarantor
гарантия	guarantee
~ банковская	bank guarantee
глава	1. head; 2. chapter
год	year
годность	fitness
годовой	yearly
гонорар	royalty
государство	State
гражданин	citizen
гражданский	civil
гражданство	citizenship
граница	boundary
~ государственная	State boundary
графика	graphics
грубый	1. flagrant; 2. gross
груз	1. cargo; 2. freight
давность	limitation
~ пргрузобретательская	acquisitive prescription
данные	data
дарение	gift
дата	date
двусторонний	bilateral
дееспособность	dispositive legal capacity
действие	1. operation; 2. action
деликт	delict
день	day
деньги	money
~ наличные	cash
депонирование	deposit
держатель	holder
дестинатор	designee
деяние	act
деятельность	activity
~ предпринимательская	entrepreneurial activity
~ совместная	joint activity

добросовестность	good faith
доверенность	power of attorney
доверитель	principal
доверительный	trust
договор	1. contract; 2. treaty
~ международный	international treaty
доказательство	evidence
доказывание	proof
документ	document
долг	1. debt; 2. duty
долгосрочный	long-term
должник	debtor
~ основной	principal debtor
доля	participatory share
досрочный	before time
достоверность	reliability
доход	revenue
дубликат	duplicate
душевнобольной	mentally-ill
заблуждение	delusion
завешание	1. will; 2. testament
завешатель	testator
задаток	deposit
заем	loan
заемшик	borrower
заказчик	customer
заключение	1. conclusion; 2. opinion
~ выгодное	advantageous conclusion
закон	law [lex]
~ по закону	by operation of law
збконодательство	legislation
залог	pledge
залогодатель	pledgor
залогодержатель	pledgeholder
замена	1. replacement; 2. substitution
замирение	reconciliation
записсь	1. entry; 2. recording; 3. writing down
запоздание	lateness
застройка	build
~ право	right to build
зачатый	conceived

зачет	set-off
зашита	defence
заявление	1. statement; 2. application
земля	land
злоупотребление	abuse
значение	1. meaning; 2. significance
иждивенец	dependent
изготовитель	manufacturer
изготовление	manufacture
издержка	cost
~ судебная	court cost
износ	wear and tear
изображение	1. image; 2. depiction
изъятие	1. seizure; 2. withdrawal; 3. exception
имушество	property
имя	name
инвалид	disabled person
индоссамент	endorsement
~ бланкетный	endorsement in blank
индоссатор	endorsee
инкассо	encashment
ипотека	mortgage
~ обшая	common mortgage
ипотекарь	mortgageholder
иск	suit
исключение	exception
исключительный	1. exclusive; 2. exceptional
исполнение	performance
~ служебное	employee performance
исполнитель	1. executor; 2. performer
использование	use
исправление	correction
истечение	expiry
истребование	demanding and obtaining
исчисление	calculation
казна	treasury
качество	quality
квартал	quarter
клал	treasure

клиент	client
книжка	book
~ сберегательная	savings book
кодекс	code
количество	quantity
комиссионер	commission agent
комиссия	commission
комитент	committent
коммерсант	merchant
компенсация	contributory compensation
компетенция	competence
конверсия	conversion
~ сделки	conversion of transaction
конкурс	competition
контокоррент	account current
контроль	control
конкуренция	competition
конфиденциальность	confidentiality
кормилец	1. breadwinner; 2. provider
кредит	credit
~ потребительский	consumer
кредитор	creditor
легат	legacy
лечение	treatment
лизинг	finance lease
лизингдатель	finance lessor
лизингополучатель	finance lessee
ликвидация	liquidation
ликвидность	liquidity
литература	literature
литография	lithography
лицензия	license
лицо	person
~ дееспособное	person with dispositive legal capacity
~ физическое	natural person
~ юридическое	juridical person
личность	1. person; 2. individual; 3. identity
личный	personal
лишение	deprivation
~ свободы	deprivation of freedom
лотерея	lottery

маклер	broker
~ жилишный	housing broker
~ по займу	loan broker
~ торговый	trade broker
малолетние	youth
мена	barter
мера	measure
местный	local
местонахождение	1. location; 2. whereabouts
месяц	month
мнимый	fictitious
момент	moment
наблюдательный	supervisory
~ орган	supervisory organ
надзор	supervision
назначение	1. appointment; 2. designation
наименование	name
~ фирменное	firm name
накладная	waybill
налог	tax
намерение	intention
нарушение	1. violation; 2. breach
насилие	force
наследник	heir
наследование	inheriting
наследство	inheritance
настояший	present
наступление	ensuing
наука	science
находка	find
небрежность	negligence
невозможность	impossibility
недействительность	invalidity
неделимость	indivisibility
недопустимость	inadmissability
незаконный	illegal
неимушественный	nonproperty
неисполнение	1. failure to perform; 2. nonperformance
некачественный	poor-quality
нематериальный	1. nonmaterial; 2. intangible
необратимость	irreversibility

непреодолимый	insuperable
неосторожность	carelessness
непередаваемость	nontransferability
неплатежеспособность	insolvency
непредпринимательское	nonentrepreneurial
непрерывный	uninterrupted
неравноценность	nonequivalence of value
несовершеннолетний	minor
несоразмерность	incommensurateness
неустойка	penalty
неявка	failure to appear
ничтожность	nullity
норма	norm
нотариальный	notarial
нотариус	notary
нравственность	morality
обеспечение	1. security; 2. ensuring; 3. provision
обешание	promise
~ публичное	public promise
обжалование	appeal
обзор	survey
облигация	bond
обман	fraud
обогашение	enrichment
~ неосновательное	unsubstantiated enrichment
обособленный	solitary
овременение	encumberment
обслуживание	servicing
~ туристическое	tourist servicing
обстоятельство	circumstance
обшественный	1. social; 2. public
обшество	society
~ акционерное	joint-stock society
~ хозяйственное	economic society
обший	1. common; 2. general
~ суд	common court
~ ее имушество	common property
объект	object
~ авторского права	object of author's right
обычай	custom
~ делового оборота	custom of business turnover

~ торговый	trade custom
обязанность	duty
обязательный	obligatory
обюязательство	obligation
~ альтернативное	alternative obligation
~ денежное	monetary obligation
~ долговое	debt obligation
оговорка	1. reservation; 2. stipulation
ограничение	limitation
опасность	danger
опека	trusteeship
опекун	trustee
определение	1. determination; 2. definition; 3. ruling
опровержение	refutation
опубликование	publication
опцион	option
орган	1. agency; 2. organ
~ государственный	State agency
~ управленческий	management organ
организация	organisation
основание	ground(s)
особенность	1. peculiarity; 2. distinctiveness
оспаривание	contesting
ответственность	responsibility
~ имушественная	property responsibility
~ солидарная	joint and several responsibility
отказ	1. refusal; 2. legacy
~ завешательный	testamentary legacy
открытие	1. opening; 2. discovery
~ наследства	opening of inheritance
отмена	1. abolition; 2. revocation; 3. repeal
отметка	notation
отношение	relation
отправитель	consignor
отстранение	removal
~ от должности	removal from office
отсутствие	absence
отцовство	1. paternity; 2. fatherhood
отчет	report
отчуждатель	alienator
отчуждение	alienation
оферент	offeror

оферта	offer
охрана	protection
ошибка	1. mistake; 2. error
пай	share
пассив	liability
патронаж	home visiting
перевод	1. translation; 2. transfer
~ долга	transfer of debt
перевозка	carriage
перевозчик	carrier
передача	1. transfer; 2. transmission
передоверие	transfer of power of attorney
перемена	change
~ имени	change of name
плата	payment
плод	fruit
поверенный	attorney
повестка дня	agenda
повреждение	damage
подзаконный	subordinate
подопечный	ward
подпись	signature
подразделение	subdivision
подряд	independent-work
подрядчик	independent-work contractor
пожар	fire
пожертвование	donation
покладжедатель	transferor (for keeping)
полис	policy
полмесяца	half-month
положение	1. provision; 2. statute; 3. status; 4. situation
полномочие	power
полугодие	half-year
получатель	recipient
пользование	use
помошник	1. assistant; 2. aide
понятие	concept
попечитель	guardian
попечительство	guardianship
пороки	defects

поручение	commission
поручительство	suretyship
порядок	1. procedure; 2. order; 3. proceeding
последствие	consequence
посредник	intermediary
посредничество	mediation
постановление	decree
потерпевший	victim
правило	rule
правительство	government
правление	1. board; 2. rule
право	1. law [jus]; 2. right
~ авторское	author's right
~ соседское	right of neighbour
~ обязательственное	law of obligations
прбвомерный	lawful
правомочие	competence
правомочность	competence
правопорядок	legal order
правопреемство	legal succession
правоспособность	legal capacity
правосудие	justice
предварительный	1. preliminary; 2. prior
предел	limit
~ за пределмаи	beyond the limits of
предложение	1. proposal; 2. sentence
предмет	1. article; 2. subject
~ залога	subject of pledge
преднамеренно	deliberately
предоставление	1. granting; 2. provision
предписание	prescription
предпринимательство	entrepreneurship
предприятие	enterprise
представитель	representative
~ торговый	trade representative
представительство	representation
предупреждение	1. warning; 2. caution
предусмотренный	provided for
предусмотрительный	prudent
предъявитель	bearer
презумпция	presumption
~ авторства	presumption of authorship

преимушественный	preferential
~ право	preferential right
прекрашение	termination
~ договора	termination of contract
пресечение	restraint
претензия	claim
прибыл	profit
приговор	judgment
пригодность	fitness
признак	indicator
~ родовой	generic indicator
признание	1. recognition; 2. deeming
~ права	recognition of right
применение	application
принадлежность	1. affiliation; 2. appurtenance
принудительный	compulsory
принципал	principal
принятие	1. adoption; 2. acceptance
приобретатель	acquirer
приобретение	acquisition
приоритет	priority
приостбновление	suspension
прирост	increment
присвоение	appropriation
притворный	sham
причина	1. reason; 2. cause
проверка	verification
~ поштучная	verification by piece
продажа	sale
продление	extension
продукт	product
произведение	work
~ искусства	work of art
производный	derivative
промзводство	production
пропорциональный	proportional
просрочка	delay
прошение	forgiveness
~ долга	forgiveness of debt
проценты	interest
пункт	point
~ населённый	population centre

работа	work
разбирательство	examination
~ судебное	judicial examination
развод	divorce
раздел	1. division; 2. separation; 3. section
разрешение	1. authorisation; 2. settlement
~ споров	settlement of disputes
разумный	reasonable
рбспоряжение	1. regulation; 2. disposition
распределение	distribution
распространение	1. dissemination; 2. spreading
рассмотрение	consideration
расстройство	distress, derangement
расторжение	dissolution
расход	expense
регистрация	registration
~ актов гражданского состояния	registration of acts of civil status
регресс	regression
реестр	register
~ публичный	public register
реквизит	requisite
рекламация	claim for replacement
ремонт	repair
репутация	reputation
~ деловая	business reputation
решение	decision
риск	risk
родитель	parent
родственник	relative
рождение	birth
сальдо	balance
самостоятельный	autonomous
самоуправный	arbitrary
сбор	charge
сведение	information
свидетельство	certificate
~ склдское	warehouse certificate
свобода	freedom
~ договора	freedom of contract
свободный	free
сделка	transaction

семья	family
сервитут	servitude
сила	1. force; 2. power
~ непреодолимая	insuperable force
склад	warehouse
слабоумный	feeble-minded
смежный	neighbouring
смерть	death
смета	estimate
соавтор	co-author
соавторство	co-authorship
соблюдение	compliance
собрание	1. meeting; 2. assembly
~ обшее	general meeting
собственник	owner
собственность	ownership
~ индивидуальная	individual ownership
~ интеллектуальная	intellectual property
~ обшая	common ownership
~ частная	private ownership
совокупность	aggregate
согласие	consent
соглашение	agreement
содержание	1. maintenance; 2. content
~ пожизненное	maintenance for life
создание	creation
солидарный	joint and several
сонаследник	co-heir
сообшение	1. communication; 2. notification
состав	composition, membership
~ ы	constituent elements
союз	union
~ профессиональный	trade union
списание	1. withdrawal; 2. writing off
спор	dispute
спорность	disputation
средство	means
~ массово информации	mass media
срок	1. period; 2. term
~ давности	period of limitation
ссудодатель	lender
ссудополучатель	borrower

ставка	rate
~ процентная	interest rate
статья	1. article; 2. item
стоимость	value, cost
сторона	1. party; 2. side
страхование	insurance
~ жизни	life insurance
~ обязатеяьное	obligatory insurance
страхователь	insurant
страховшик	insurer
структура	structure
субаренда	sublease
субарендатор	sublessee
суд	court
судебный	judicial
сумма	amount
сушество	essence
сушность	essence
сфера	sphere
счёт	account
~ расчетный	settlement account
~ текуший	current account
тайна	secrecy, secret
текст	text
территория	territory
течение	running
~ срока	running of period
товар	good
товаришество	partnership
толкование	interpretation
торги	sale(s)
~ публичные	public sale
традиция	tradition
трансмиссия	transmission
~ наследственная	inheritance transmission
требование	1. demand; 2. requirement
труд	labour
убыток	loss
увеwomление	notification

удовлетворение	satisfaction
удочерение	adoption [of female]
узуфрукт	usufruct
узуфруктарий	usufructuary
указ	edict
уменьшение	reduction
уничтжение	destruction
управление	1. administration; 2. management
~ доверительное	trust management
уравнение	equalisation
условие	condition
~ негативное	negative condition
~ отлагательное	condition resolutory
~ отменительное	condition suspensive
~ позитивное	positive condition
~ стандартное	standard condition
устав	charter
установление	establishment
устный	oral
устранение	elimination
уступка	assignment
усыновление	adoption [of male]
утрата	loss
~ груза	loss of cargo
~ прав	loss of rights
уход	1. care; 2. departure
участие	participation
участник	participant
участок	plot
учредительный	1. founding; 2. constitutive
~ документ	constitutive document
учреждение	institution
~ кредитное	credit institution
филиал	branch
фонд	1. fund; 2. foundation
фонограмма	phonogram
форма	form
~ письменная	written form
фотография	photograph
франчайзинг	franchising
франчайзодатель	franchiser

франчайзополучатель	franchisee
функция	function
хозяйственник	master
хореография	choreography
хранение	keeping
~ ответственное	custody
хранитель	keeper
часть	1. part; 2. paragraph
~ по частям	in parts
~ составная	integral part
чек	cheque
человек	1. person; 2. man
~ свободы человека и гражданина	freedoms of man and citizen
членство	membership
цель	1. purpose; 2. aim
цена	price
ценные бумаги	securities
экземпляр	example
эксклюзивный	exclusive
экспедитор	expeditor
экспедиция	expediting
~ транспортная	transport expediting
эмансипация	emancipation
эскмз	sketch
юстиция	justice
~ ии министерство	Ministry of Justice

SUBJECT INDEX

Numbers refer to respective Articles of the Civil Code.

confiscation (конфискация), 235, 243
conflicts norms (коллизионные нормы), 1186,
1190
confluence of grave circumstances (стечение
тяжелых обстоятельств), 179, 812
consent (согласие), of causer of harm, 1221; of citizen, 41, 1134; of commission agent, 994; of committent, 995; of cooperative, 111; of creditor,
391, 562, 657; of customer, 709, 711, 713, 735,
743, 749, 754, 759, 760, 762, 770; of debtor,
313, 382, 388; of depositor, 841; of donor, 582;
of electric power organisation, 545; of founder of
trust management, 1021; of full partners, 71–73,
78, 79; of general independent-work contractor,
706; of guardian, 30, 33, 35, 37, 176; of heirs,
1122, 1155; of insurant, 940; of insured person,
934, 955, 956; of interested person, 980; of
keeper, 894; of landlord, 678, 679, 685, 686; of
legal representative, 28; of lender, 696, 698, 810;
of lessee, 638, 647; of lessor, 615, 623, 638, 660,
670; of members of association, 123; of minor,
26, 175; of owner, 233, 270, 279, 286, 295, 297,
335, 552, 576, 652; of parents, 27; of participants
of limited responsibility society, 93, 94; of participants of participatory share ownership, 250, 252;
of parties to dispute, 926; of parties to transaction,
184, 443, 457, 460, 508, 512, 604, 727, 771; of
partners, 1043, 1044, 1176; of pledgeholder, 345,
346; of pledgor, 356; of recipient of letter of
credit, 869; of rights-possessor, 138; of State agencies, 57; of State customer, 530; of stockholders,
97; of surety, 367; of tenant, 680, 681, 687; of
third persons, 349, 430; of transferor, 892, 895,
898; of trustee, 35; of trusteeship and guardianship agency, 293; of victim, 1064; of ward, 41
consignor (отправитель), of good, 785, 791, 794,
796, 797, 801
constitutional system, 1
constitutive documents (учредительные
документы), content of, 49, 52–57, 61, 63,
173; inheritance and, 1176, 1177; insurance and,
935; of additional responsibility society, 95; of
association, 121–123; of bank, 846; of consumer
cooperative, 116; of economic partnership or society, 67; of foundation, 119; of full partnership,
69–74, 76, 78; of institution, 298; of joint-stock
society, 98; of limited partnership, 83, 85, 86; of
limited responsibility society, 87, 89, 93, 94; of
mutual insurance society, 968; of production
cooperative, 107, 108; of social and religious orga-
nizations, 213; of unitary enterprise, 113–115;
ultra vires, 174
contributory compensation (компенсация), independent-work contractor and, 709, 729; inheritance property and, 1170, 1178, 1179, 1182;
leave and, 1086; moral harm and, 12, 151, 1064,

1099–1101; participatory share ownership and,
247, 252, 258; rent and, 605
construction (строительство), object of,
741–743, 752, 754, 755
construction independent-work (строительный
подряд), acceptance of work under, 753; additional duties under, 747; contract of, 740, 742,
743, 745–747, 749, 750, 753, 757, 1211; control
over, 748, 749; cooperation of parties under, 750;
general provisions of, 702; guarantees under, 755;
State needs and, 763; suspension of work under,
752
construction norms and rules (строительные
нормы и правила), 754
consumer cooperative (потребительский
кооператив), bankruptcy and, 65; charter of,
116; inheritance and, 1177; liquidation of, 61,
116; ownership of, 218; noncommercial juridical
person, 50
consumer society (потребительское
общество), 116
consumer union (потребительский союз), 116
contract (договор), applicable law and,
1210–1215; arising of civil rights and duties
under, 8, 223, 307; change of, 428, 430,
450–453, 523, 534, 540, 546, 565, 566, 617,
675, 744, 1036, 1038; compensated and uncompensated, 423; concept of, 420; conclusion of,
421, 425, 426, 429, 432–437, 440–448,
1210–1212; place of conclusion of, 732; customs
of business turnover contrary to, 5; freedom of, 1,
421; inheritance and, 1165; interpretation of,
431; law and, 422; operation of, 425; preliminary,
429; principal, 429, 967; public, 426; refusal to
conclude, 426, 507, 530, 621, 1035; refusal to
extend, 684; refusal to perform, 396, 405, 406,
450, 463, 467, 468, 475, 480, 483, 484, 486,
490, 495–497, 500, 503, 509, 511, 515, 523,
533, 546, 556, 573, 577, 579, 592, 610, 627,
655, 699, 709, 715–717, 719, 723, 731, 737,
745, 782, 795, 806, 821, 888, 896, 958, 977,
978, 1002,–1004, 1010, 1024, 1037, 1050, 1051,
1053; transactions and, 421; violation of, 929,
932, 935
contribution (вклад), inadmissability of contribut-
ing lease rights under contract of rental to charter
capital, 631; of lease rights to charter capital, 615;
of partner to limited partnership, 66, 82–86; of
partner under contract of simple partnership,
1041, 1042, 1046–1048; preferential right to
receive, 86; prohibition against division of in unitary enterprise, 113; responsibility for violation of
duty to make, 70, 83, 87, 89; right of ownership
to, 213; to charter capital, 66, 67, 70, 73, 83, 85,
87, 89, 90, 95, 295; to farmer economy, 259; valuation of, 66